Lecture Notes in Computer Science   8977

Commenced Publication in 1973
Founding and Former Series Editors:
Gerhard Goos, Juris Hartmanis, and Jan van Leeuwen

More information about this series at http://www.springer.com/series/7407

Adrian-Horia Dediu · Enrico Formenti
Carlos Martín-Vide · Bianca Truthe
(Eds.)

Language and Automata Theory and Applications

9th International Conference, LATA 2015
Nice, France, March 2–6, 2015
Proceedings

 Springer

Editors
Adrian-Horia Dediu
Rovira i Virgili University
Tarragona
Spain

Enrico Formenti
Nice Sophia Antipolis University
Nice
France

Carlos Martín-Vide
Rovira i Virgili University
Tarragona
Spain

Bianca Truthe
Justus-Liebig-Universität
Gießen
Germany

ISSN 0302-9743
Lecture Notes in Computer Science
ISBN 978-3-319-15578-4
DOI 10.1007/978-3-319-15579-1

ISSN 1611-3349   (electronic)

ISBN 978-3-319-15579-1   (eBook)

Library of Congress Control Number: 2015932508

LNCS Sublibrary: SL 1 – Theoretical Computer Science and General Issues

Springer Cham Heidelberg New York Dordrecht London

Printed on acid-free paper

Springer International Publishing AG Switzerland is part of Springer Science+Business Media
(www.springer.com)

Preface

These proceedings contain the papers that were presented at the 9th International Conference on Language and Automata Theory and Applications (LATA 2015), held in Nice, France, during March 2–6, 2015.

The scope of LATA is rather broad, including: algebraic language theory; algorithms for semi-structured data mining; algorithms on automata and words; automata and logic; automata for system analysis and program verification; automata networks; automata, concurrency, and Petri nets; automatic structures; cellular automata; codes; combinatorics on words; computational complexity; data and image compression; descriptional complexity; digital libraries and document engineering; foundations of finite state technology; foundations of XML; fuzzy and rough languages; grammars (Chomsky hierarchy, contextual, unification, categorial, etc.); grammatical inference and algorithmic learning; graphs and graph transformation; language varieties and semigroups; language-based cryptography; parallel and regulated rewriting; parsing; patterns; power series; string and combinatorial issues in bioinformatics; string processing algorithms; symbolic dynamics; term rewriting; transducers; trees, tree languages, and tree automata; unconventional models of computation; weighted automata.

LATA 2015 received 115 submissions. Most of the papers were given at least three reviews by Program Committee members or by external referees. After a thorough and vivid discussion phase, the committee decided to accept 53 papers (which represents an acceptance rate of 46.09%). The conference program also included five invited talks. Part of the success in the management of such a large number of submissions was due to the excellent facilities provided by the EasyChair conference management system.

We would like to thank all invited speakers and authors for their contributions, the Program Committee and the reviewers for their cooperation, and Springer for its very professional publishing work.

December 2014

Adrian-Horia Dediu
Enrico Formenti
Carlos Martín-Vide
Bianca Truthe

Organization

LATA 2015 was organized by the National Scientific Research Center – CNRS, I3S, UMR 7271 Nice Sophia Antipolis University and the Research Group on Mathematical Linguistics – GRLMC, from Rovira i Virgili University, Tarragona.

Program Committee

Andrew Adamatzky	University of the West of England, Bristol, UK
Andris Ambainis	University of Latvia, Riga, Latvia
Franz Baader	Dresden University of Technology, Germany
Rajesh Bhatt	University of Massachusetts Amherst, USA
José-Manuel Colom	University of Zaragoza, Spain
Bruno Courcelle	University of Bordeaux, France
Erzsébet Csuhaj-Varjú	Eötvös Loránd University, Budapest, Hungary
Aldo de Luca	University of Naples Federico II, Italy
Susanna Donatelli	University of Turin, Italy
Paola Flocchini	University of Ottawa, Canada
Enrico Formenti	University of Nice Sophia Antipolis, France
Tero Harju	University of Turku, Finland
Monika Heiner	Brandenburg University of Technology, Cottbus, Germany
Yiguang Hong	Chinese Academy of Sciences, Beijing, China
Kazuo Iwama	Kyoto University, Japan
Sanjay Jain	National University of Singapore, Singapore
Maciej Koutny	Newcastle University, UK
Antonín Kučera	Masaryk University, Brno, Czech Republic
Thierry Lecroq	University of Rouen, France
Salvador Lucas	Polytechnic University of Valencia, Spain
Veli Mäkinen	University of Helsinki, Finland
Carlos Martín-Vide (Chair)	Rovira i Virgili University, Tarragona, Spain
Filippo Mignosi	University of L'Aquila, Italy
Victor Mitrana	Polytechnic University of Madrid, Spain
Ilan Newman	University of Haifa, Israel
Joachim Niehren	INRIA Lille – Nord Europe, France
Enno Ohlebusch	University of Ulm, Germany
Arlindo Oliveira	University of Lisbon, Portugal
Joël Ouaknine	University of Oxford, UK
Wojciech Penczek	Polish Academy of Sciences, Warsaw, Poland

Dominique Perrin   University Paris-Est Marne-la-Vallée, France
Alberto Policriti   University of Udine, Italy
Sanguthevar Rajasekaran   University of Connecticut, Storrs, USA
Jörg Rothe   University of Düsseldorf, Germany
Frank Ruskey   University of Victoria, Canada
Helmut Seidl   Technical University of Munich, Germany
Ayumi Shinohara   Tohoku University, Sendai, Japan
Bernhard Steffen   Technical University of Dortmund, Germany
Frank Stephan   National University of Singapore, Singapore
Paul Tarau   University of North Texas, Denton, USA
Andrzej Tarlecki   University of Warsaw, Poland
Jacobo Torán   University of Ulm, Germany
Frits Vaandrager   Radboud University Nijmegen, The Netherlands
Jaco van de Pol   University of Twente, Enschede, The Netherlands
Pierre Wolper   University of Liège, Belgium
Zhilin Wu   Chinese Academy of Sciences, Beijing, China
Slawomir Zadrozny   Polish Academy of Sciences, Warsaw, Poland
Hans Zantema   Eindhoven University of Technology,
                            The Netherlands

Organizing Committee

Sébastien Autran   University of Nice Sophia Antipolis, France
Adrian-Horia Dediu   Rovira i Virgili University, Tarragona, Spain
Enrico Formenti   University of Nice Sophia Antipolis,
                            France (Co-chair)
Sandrine Julia   University of Nice Sophia Antipolis, France
Carlos Martín-Vide   Rovira i Virgili University, Tarragona, Spain (Co-
                            chair)
Christophe Papazian   University of Nice Sophia Antipolis, France
Julien Provillard   University of Nice Sophia Antipolis, France
Pierre-Alain Scribot   University of Nice Sophia Antipolis, France
Bianca Truthe   University of Giessen, Germany
Lilica Voicu   Rovira i Virgili University, Tarragona, Spain

External Reviewers

Amparore, Elvio Gilberto   Cain, Alan
Babu, Jasine   Caron, Pascal
Belazzougui, Djamal   Carpi, Arturo
Blahoudek, František   Casagrande, Alberto
Bollig, Benedikt   Costa Florencio, Christophe
Bonacina, Ilario   Cunial, Fabio
Bredereck, Robert   De Luca, Alessandro
Bresolin, Davide   Delecroix, Vincent

Dennunzio, Alberto
Didier, Gilles
Droste, Manfred
Durand, Irène Anne
Dömösi, Pál
Escobar, Santiago
Fici, Gabriele
Fox, Nathan
Gagie, Travis
Gazdag, Zsolt
Giaquinta, Emanuele
Gog, Simon
Gurski, Frank
Götze, Doreen
Halava, Vesa
Hamid, Fahmida
Hertrampf, Ulrich
Hoffmann, Stefan
Hoogeboom, Hendrik Jan
Howar, Falk
Hugot, Vincent
Irvine, Veronika
Isberner, Malte
Janicki, Ryszard
Karhumäki, Juhani
Kawano, Yu
Kiefer, Stefan
Knapik, Michał
Kociumaka, Tomasz
Lazar, Katalin A.
Lempp, Steffen
Leporati, Alberto
Li, Rui
Liu, Fei
Luttik, Bas
Madelaine, Guillaume
Manea, Florin
Manuel, Amaldev
Marzi, Francesca
Mathew, Rogers
Mazzei, Alessandro
Mercaş, Robert
Męski, Artur
Mignot, Ludovic
Mikulski, Łukasz
Miller, Joseph S.
Miltersen, Peter Bro

Moreira, Nelma
Nicolae, Marius
Niewiadomski, Artur
Novotný, Petr
Oliveira, Igor Carboni
Otto, Friedrich
Perevoshchikov, Vitaly
Pommereau, Franck
Porreca, Antonio E.
Prezza, Nicola
Prieur-Gaston, Élise
Puglisi, Simon
Quaas, Karin
Ramsay, Steven
Rehak, Vojtech
Restivo, Antonio
Rey, Anja
Reynier, Pierre-Alain
Rohr, Christian
Romashchenko, Andrei
Saha, Subrata
Sakarovitch, Jacques
Salzer, Gernot
Sattler, Uli
Scheftelowitsch, Dimitri
Schulz, Marcel
Schwarick, Martin
Skaruz, Jarosław
Sobocinski, Pawel
Song, Siang Wun
Souto, André
Sproston, Jeremy
Steggles, Jason
Szreter, Maciej
Tamaki, Suguru
Tichler, Krisztián
Tivoli, Massimo
Tomescu, Alexandru I.
Ulidowski, Irek
Vanier, Pascal
Vicedomini, Riccardo
Villanueva, Alicia
Vitacolonna, Nicola
Wang, Weifeng
Willemse, Tim
Xia, Mingji
Zhang, Kuize

Contents

Combinatorics on Words

Complexity and Recursive Functions

Compression, Inference, Pattern Matching, and Model Checking

Graphs, Term Rewriting, and Networks

Transducers, Tree Automata, and Weighted Automata

Invited Talks

Automated Synthesis of Application-Layer Connectors from Automata-Based Specifications

Marco Autili[1], Paola Inverardi[1], Filippo Mignosi[1],
Romina Spalazzese[2], and Massimo Tivoli[1](✉)

[1] Dipartimento di Ingegneria e Scienze dell'Informazione e Matematica,
Università dell'Aquila, L'Aquila, Italy
{marco.autili,paola.inverardi,filippo.mignosi,massimo.tivoli}@univaq.it
[2] Department of Computer Science, Malmö University, Malmö, Sweden
romina.spalazzese@mah.se

Abstract. The heterogeneity characterizing the systems populating the Ubiquitous Computing environment prevents their seamless interoperability. Heterogeneous protocols may be willing to cooperate in order to reach some common goal even though they meet dynamically and do not have a priori knowledge of each other. Despite numerous efforts have been done in the literature, the automated and run-time interoperability is still an open challenge for such environment. We consider interoperability as the ability for two Networked Systems (NSs) to communicate and correctly coordinate to achieve their goal(s).

In this paper, we report the main outcomes of our past and recent research on automatically achieving protocol interoperability via connector synthesis. We consider application-layer connectors by referring to two conceptually distinct notions of connector: *coordinator* and *mediator*. The former is used when the NSs to be connected are already able to communicate but they need to be specifically coordinated in order to reach their goal(s). The latter goes a step forward representing a solution for both achieving correct coordination and enabling communication between highly heterogeneous NSs.

In the past, most of the works in the literature described efforts to the automatic synthesis of coordinators while, in recent years the focus moved also to the automatic synthesis of mediators. By considering our past experience on the automatic synthesis of coordinators and mediators as a baseline, we conclude by overviewing a formal method for the automated synthesis of mediators that allows to relax some assumptions state-of-the-art approaches rely on, and characterize the necessary and sufficient interoperability conditions that ensure the mediator existence and correctness.

1 Introduction

The heterogeneity characterizing the systems that populate the ubiquitous computing environment prevents their seamless interoperability. Systems with heterogeneous protocols may be willing to cooperate in order to reach some common goal even though they do not have a priori knowledge of each other.

© Springer International Publishing Switzerland 2015
A.-H. Dediu et al. (Eds.): LATA 2015, LNCS 8977, pp. 3–24, 2015.
DOI: 10.1007/978-3-319-15579-1_1

The term *protocol* refers to *interaction protocols* or *observable protocols*. A protocol is the sequences of messages visible at the interface level which a system exchanges with other systems. In this paper we consider *application-layer* protocols as opposed to *midlleware-layer* protocols [31].

By referring to the notion of *interoperability* introduced in [28], the problem we address in this paper, is related *to automatically achieve the interoperability between heterogeneous protocols in the ubiquitous computing environment.*

With interoperability, we mean the ability of heterogeneous protocols to communicate and correctly coordinate to achieve their goal(s). The communication is expressed as synchronization, i.e., two systems communicate if they are able to synchronize on "common actions". Coordination is expressed by the achievement of a specified goal, i.e., two systems succeed in coordinating if they interact through synchronization according to the achievement of their goal(s). Communication that requires a complex protocol interaction can be regarded as a simple form of coordination. Indeed, application level protocols introduce a notion of communication that goes beyond single basic synchronization and may require a well defined sequence of synchronizations to be achieved.

In order to make communication and correct coordination between heterogeneous protocols possible, we focus on methods, and related tools, for the *automatic synthesis of application-layer connectors*. In this paper, we report our past and recent work on devising automatic connector synthesis techniques in the domains of *Component Based Software Engineering* (CBSE) and *Ubiquitous Computing* (UbiComp), respectively. The work carried on within the CBSE domain can be considered as a baseline for the work done in the UbiComp domain. However, it is worth mentioning that these two research contributions address two distinct sub-problems of the automatic connector synthesis problem.

In particular, in the CBSE domain, we used automatic connector synthesis in order to face the so-called *component assembly* problem. This problem can be considered as an instance of the general interoperability problem where the issue of enabling communication is assumed to be already solved. The focus, in the component assembly problem, is on how to coordinate the interactions of already communicating black-box components so that the resulting system is free from possible deadlocks and it satisfies a goal specified in terms of *coordination policies*. Since components are black-box, this is done by introducing in the system a software *coordinator*. It is an additional component and it is synthesized to intercept all other components interactions in order to prevent deadlocks and those interactions that violate the coordination policies. Coordination policies are routing policies usually specified in some automata-based or temporal logic formalism. Thus, a *coordinator* can be considered as a specific notion of connector, i.e., a *coordination connector*.

In the UbiComp domain, the granularity of a system shifts from the granularity of a system of components (as in the CBSE domain) to the one of a *System-of-Systems* (SoS) [27]. A SoS is characterized by an assembly of a wide variety of building blocks. Thus, in the UbiComp domain, enabling communication between heterogeneous NSs regardless, at a first stage, possible coordination mismatches, becomes a primary concern. This introduces another specific notion of connector, i.e., the notion of *mediator* seen as a *communication connector*.

Achieving correct communication and coordination among heterogeneous NSs means achieving interoperability among them. The interoperability problem and the specific notions of connector (e.g., coordinator or mediator) that can be used to solve it, or part of it, have been the focus of extensive studies within different research communities. Protocol interoperability come from the early days of networking and different efforts, both theoretical and practical, have been done to address it in several areas including: protocol conversion, component adaptors, Web services mediation, theories of connectors, wrappers, bridges, and interoperability platforms.

Despite the existence of numerous solutions in the literature, most of them are focused on coordinator synthesis and less effort has been devoted to the automatic synthesis of mediators. In particular, most of these approaches: (i) assume the communication problem solved by considering protocols already able to communicate; (ii) are informal making automatic reasoning impossible; (iii) follow a semi-automatic process for the mediator synthesis requiring human intervention; (iv) consider only few possible mismatches.

Our recent work on mediator synthesis has been devoted to the definition of a theory of mediators with related supporting methods and tools. In particular, our recent work has led us *to design automated model-based techniques and tools to support the mediator synthesis process*, from protocol abstraction to protocol matching and protocol mapping.

The remainder of the paper is organized as follows. Section 2 sets the context of the work reported in this paper. In particular, by means of two examples, this section clarifies the distinction between the notions of coordinator and mediator. Sections 3 and 4 describe different approaches for the automatic synthesis of coordinators and of mediators, respectively. The coordinator synthesis approaches that are discussed in Section 3 represents the *baseline* chosen from the state-of-the-art for the approaches described in Section 4. We conclude Section 4 by proposing a formal method for the automated synthesis of mediators that permits to relax some assumptions state-of-the-art approaches rely on, and characterize the necessary and sufficient interoperability conditions that ensure the mediator existence and correctness. The method relies on both *transducer theory* [11] and *Mazurkiewicz trace theory* [24]. Section 5 discusses related works in the areas of both coordinator and mediator synthesis. Section 6 concludes the chapter and outlines our future research perspectives.

2   Setting the Context

A common assumption of the connector synthesis techniques discussed in this paper is related to the possibility to characterize the interaction protocol of a system by means of an automata-based specification, e.g., a *Labeled Transition System* (LTS) [33]. Note that this assumption is supported by the increasing proliferation of techniques for software model elicitation (see [14, 21, 26, 36, 37, 43] just to cite a few).

The interaction protocol of a NS expresses the order in which input and output actions are performed while the NS interacts with its environment. Input actions model methods that can be called as well as the return values from such calls, or the reception of messages from communication channels. Output actions model method calls, message transmission via communication channels, or exceptions that occur during methods execution.

As introduced in Section 1, our focus is on the automatic synthesis of *application-layer connectors*. Our notion of protocol abstracts from the content of the exchanged data, i.e., values of method/operation parameters, return values, or content of messages. This means that we are interested in harmonizing the behavior protocol (e.g., scheduling of operation calls) of heterogeneous NSs rather than performing mediation of communication primitives or of data encoding/decoding that are issues related to the synthesis of middleware-layer connectors [31].

The interoperability problem concerns the problem of both enabling *communication* and achieving correct *coordination*. In our past research we addressed correct coordination by assuming communication already solved. This is done via automatic coordinator synthesis (Section 3). Instead, in our more recent and current research we focus on the whole interoperability problem by devising methods and tools for the automatic mediator synthesis (Section 4).

In order to better clarify the distinction between the notions of coordinator and mediator, in the following two sub-sections, we describe two simple yet significant examples of the kinds of interoperability problems that can be solved by using coordinators (Section 2.1) and mediators (Section 2.2), respectively.

2.1   The Need for Coordinators: The Shared Resource Scenario

To better illustrate protocol coordination and the related underlying problems, we introduce the Shared Resource scenario. This explanatory example is concerned with the automatic assembly of a client-server component-based system. This system is formed by three components: two clients, respectively denoted as C1 and C2, and one server denoted as C3 (the component controlling the Shared Resource). This example, although very simple, exhibits coordination problems that exemplify the kind of problems that coordinator synthesis can solve. For instance, here, the problem is due to the presence of *race conditions* in accessing a shared resource.

Let us assume that we want to assemble a system formed by C1, C2, and C3. In doing so, we want to automatically prevent possible deadlocks and guarantee a specified coordination policy, hence, guaranteeing that the system's goal is reached.

Figure 1 represents the behavior of each component in terms of an LTS. Each LTS models the component observable behavior in an intuitive way. Each state of an LTS represents a state of the component and the state S0 represents its initial state. Each action or complementary action performed by interacting with the environment of the component (i.e., all other components in parallel) is represented as a label of a transition into a new state. Actions are input or

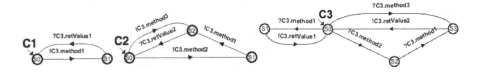

Fig. 1. Components' behavior for the Shared Resource scenario

output. Within an LTS of a component, the label of an input action is prefixed by the question mark "?" (e.g., `?C3.retValue1` of C1). The label of an output action is prefixed by the exclamation mark "!" (e.g., `!C3.method2` of C2).

The interface of server C3 exports three methods denoted as `C3.method1`, `C3.method2`, and `C3.method3`, respectively. While `C3.method2` has no return value, `C3.method1` and `C3.method3` can return some value. `C3.method1` returns two possible return values denoted as `C3.retValue1`, and `C3.retValue2`. The former is returned when a call of `C3.method1` has not preceded by a call of `C3.method2`. Otherwise, the latter is returned. `C3.method3` returns only one value, i.e., `C3.retValue2`. The two clients perform method calls according to the server interface.

It is worthwhile noticing that the described component interfaces syntactically match since either they already match or suitable component wrappers have been previously developed by the system assembler. As stated above, the problem of enabling communication is here considered as already solved. We recall that, in coordinator synthesis, the focus is on automatically preventing interaction protocol mismatches rather than enabling communication.

In this example deadlocks can occur because of a race condition among C1 and C2. In fact, one client (i.e., C2) performs a call of `C3.method2`, hence leading the server C3 in a state in which it expects a call of `C3.method1`. While C2 is attempting to perform the call of `C3.method1`, the other client (i.e., C1) performs such a call. In this scenario C1, C2, and C3 are in the state S1, S1, and S3 of their LTSs. Now, C3 expects to return `C3.retValue2` as return value of `C3.method1` but C2 is still waiting to perform a call of `C3.method1` and C1 expects a different return value. Thus, a coordination mismatch occurs and it results in an deadlock in the interaction between C1, C2, and C3.

This mismatch can be solved by synthesizing a software coordinator that supervises components' interactions by preventing the deadlock [7,50,51]. At the level of the coordinator's actual code, the coordinator is synthesized as a multi-threaded component that creates a thread for each request and for each caller performing such a request. Preventing, or solving if possible, deadlocks corresponds to put in a *waiting* state the thread that handles the request leading to the deadlock state and performed by the identified caller. Thus the coordinator will return the control to the caller only when it reaches a state in which the blocked request is possible[1]. Such multi-threaded servers are supported by existing component technologies such as COM/DCOM or CORBA.

Another coordination issue that one can note is that, e.g., C1 can always obtain the access to the shared resource, while C2 never obtains it since C2 can always require the access whenever the resource is already "lock" by C1. In other words, C3 cannot be fair in providing the access to the shared resource it supervises. To solve this issue, a software coordinator can be automatically synthesized so to enforce an *alternating protocol* policy [51] on the components' interaction. The coordinator allows only the alternating access of C1 and C2 to the shared resource.

2.2   The Need for Mediators: the Purchase Order Scenario

To better illustrate protocol mediation and the related underlying problems that we characterized in [47], in the following we introduce the Purchase Order scenario from the Semantic Web Service (SWS) Challenge[2] [38]. It represents a typical real-world problem that is both close to industrial reality and practical. This scenario highlights the various mismatches that can be encountered when making heterogeneous systems interoperable. The scenario considers two NSs implemented as *prosumer*[3] Web Services (WSs) by using different protocols: the *Moon Client* (*MC*) and the *Blue Service* (*BS*).

MC and *BS*, as prosumers, require to deal with all the possible type of WSDL[4] operations that WSs can support: (i) **one-way** - the WS receives a message (input only); (ii) **request-response** - the WS receives a message (input), and sends a correlated message (output); (iii) **solicit-response** - the WS sends a message (output), and receives a correlated message (input); (iv) **notification** - the WS sends a message (output only).

Figures 2 and 3 represent the WSDL interfaces of *MC* and *BS*, respectively, plus a description of each interface operation. As prosumers, the WSDL interfaces of *MC* and *BS* specify both their *required* and *provided* interfaces. A required interface defines the set of operations (solicit-response and notification) that a WS expects to invoke on an ideal service provider. A provided interface defines the set of operations (request-response and one-way) that can be invoked on the WS by its clients. As far as the interaction protocol of *MC* and *BS* is concerned, the operations in the figures are listed in the order they are performed while the two WSs interact with their environment.

MC cannot communicate with *BS* due to the following protocol mismatches, of two different types.

---

[1] Meaning that, this time, that request performed from that caller does not lead to a deadlock.

[2] http://sws-challenge.org/.

[3] A prosumer is both a *consumer* and a *provider* of service operations.

[4] Web Services Description Language (WSDL) 1.1 - W3C Note 15 March 2001 - http://www.w3.org/TR/wsdl.

Operation	WSDL Op. Type	Input Message	Output Message	Description
Login	solicit-response	LoginResp	LoginReq	Perform authentication to prove that it is an authorized customer
CreateOrder	solicit-response	CreateOrdResp	CreateOrdReq	Create an order by starting with an empty cart
SelectItem	solicit-response	SelectItemResp	SelectItemReq	Perform the seleciton of the items to be added to the cart
SetQuantity	notification		SetQuantReq	Set the required quantity for each item to be added to the cart
CloseOrder	request-response	CloseOrderReq	CloseOrderResp	MC is asked to close the order (from the Moon Service)
ConfirmOrder	one-way	ConfOrderReq		MC is informed (by the Moon Service) about the confirmation of the order
Payment	solicit-response	PayResp	PayReq	Perform the payment of the confirmed order

Fig. 2. MC WSDL interface

Communication mismatches concern the semantics and granularity of the protocol actions. For instance, a client of BS provides its identifier while placing the order, whereas MC has to authenticate before performing any operation. Furthermore, BS provides a single operation to add an item, with the needed quantity, to the order, whereas MC expects to use two different operations, one for the addition and one for the quantity specification. To solve these kind of mismatches it is necessary to *align the two protocols to the same language*, e.g., by using a suitable *wrapper component*.

Operation	WSDL Op. Type	Input Message	Output Message	Description
Order	request-response	OrderReq	OrderResp	Perform authentication and start the order by creating an empty cart
AddItem	request-response	AddItemReq	AddItemResp	Add a single item to the cart by specifying also the related quantity
Confirmation	notification		ConfOrderReq	Confirm the order
CloseOrder	solicit-response	CloseOrderResp	CloseOrderReq	Ask for closing the order
DoBill	request-response	DoBillReq	DoBillResp	Perform the payment of the confirmed order

Fig. 3. BS WSDL interface

Coordination mismatches concern the control structure of the (aligned) protocols and can be solved by means of a *mediator that can mediate the conversation between the two protocols so that they can actually interact*. For instance, BS requires its clients to confirm the ordered items and then place the order, whereas MC expects to confirm the ordered items only once the order is placed. Finally, BS allows the addition of several kinds of items in the same order, whereas MC performs the addition of only one kind of item per order.

The mediation logic should account for the data that define the structure of the messages exchanged by the respective input/output actions. For example, the message associated to the request of AddItem is an *aggregate* of the messages associated to the requests of SelectItem and SetQuantity. The methods described in Section 4 have been conceived in order to automatically infer, among the others, this aggregation relation by, e.g., assuming ontology knowledge or weaker semantic information on input/output messages correlation.

3   Automatic Synthesis of Application-layer Coordinators

This section provides an overview of two different approaches to the automatic synthesis of application-layer coordinators. We first introduce each approach by outlining their commonalities and differences. Then, in Sections 3.1 and 3.2, we give an overview of each approach.

Section 3.1 describes a method for the correct (with respect to coordination mismatches) and automatic assembly of component-based systems via centralized coordinator synthesis [51]. In this context, by considering communication issues already solved, the interoperability problem introduced in Section 1 can be rephrased as follows: *given a set of communicating components, C, and a set of behavioral properties, P, automatically derive a deadlock-free assembly, A, of these components which guarantees every property in P, if possible.* The assembly A is a composition of the components in C plus a synthesized coordinator. The coordinator is synthesized as an additional component which intercepts all the component interactions so as to control the exchange of messages with the aim of preventing possible deadlocks and those interactions that violate the properties in P. In [51] this problem is addressed by showing how to automatically synthesize the implementation of a centralized coordinator.

Unfortunately, in a distributed environment it is not always possible or convenient to introduce a centralized coordinator. For example, existing distributed systems might not allow the introduction of an additional component (i.e., the coordinator) which coordinates the information flow in a centralized way. Moreover, the coordination of several components might cause loss of information and bottlenecks hence slowing down the response time of the centralized coordinator. Conversely, building a distributed coordinator might extend the applicability of the approach to large-scale contexts.

To overcome the above limitations, in [7], an extension of the previous method is proposed. This extension is discussed in Section 3.2. The aim of the proposed extension is to automatically synthesize a distributed coordinator into a set of wrappers (local coordinators), one for each component whose interaction has to be controlled. The distributed coordinator synthesis approach has various advantages with respect to the synthesis of centralized coordinators. The most relevant ones are: (i) no centralized point of information flow exists; (ii) the degree of parallelism of the system without the coordinator is maintained; and (iii) all the domain-specific deployment constraints imposed on the centralized coordinator can be removed.

Beyond the above two methods that we discuss in this paper, in the past, we also defined automated coordinator synthesis techniques for both evolvable systems [41] and real-time systems [50].

Our coordinator synthesis methods have been all applied to real case studies in the domains of COM/DCOM and J2EE applications. This experimentation has been carried on through the SYNTHESIS tool [6] that implements all the above mentioned methods.

3.1   Automatic Synthesis of Centalized Coordinators

SYNTHESIS is a technique equipped with a tool [6] that permits to assemble a component-based application in a deadlock-free way [7,51]. Starting from a set of black-box components, SYNTHESIS assembles them together according to a so called coordinator-based architecture by synthesizing a coordinator that guarantees deadlock-free interactions among components. The code that implements the coordinator is automatically derived directly from the components' interfaces. Synthesis assumes a partial knowledge of the components' interaction behavior described as finite state automata plus the knowledge of a specification of the system to be assembled given in terms of Message Sequence Charts (MSCs) [1]. Under these hypotheses, SYNTHESIS automatically derives the assembling code of the coordinator for a set of components. The coordinator is derived in such a way to obtain a failure-free system. It is shown that the coordinator-based system is equivalent according to a suitable equivalence relation to the initial one once depurate of all the failure behaviors. The initial coordinator is a *no-op* coordinator that serves to model all the possible component interactions (i.e., the failure-free and the failing ones). Acting on the initial coordinator is enough to automatically prevent both deadlocks and other kinds of failure hence obtaining the failure-free coordinator.

As illustrated in Figure 4, the SYNTHESIS framework realizes a form of system adaptation. The initial software system is changed by inserting a new component, the coordinator, in order to prevent interactions failures.

The framework makes use of the following models and formalisms. An architectural model, namely coordinator-based architecture, which constrains the way components can interact, by forcing interaction to go through the coordinator. A set of behavioral models for the components. Each model describes a specific view of the component interaction protocol in the form of an LTS. A behavioral equivalence on LTS to establish the equivalence among the original system and the adapted/coordinated one. MSCs are used to specify the integration failure to be avoided. Then LTSs and *LTS synchronous product* [5,33] plus a notion of *behavioral refinement* [39] are used to synthesize the failure-free coordinator specification, as it is described in detail in [51]. As already mentioned, from the coordinator specification the actual code can be automatically derived as a centralized component [51].

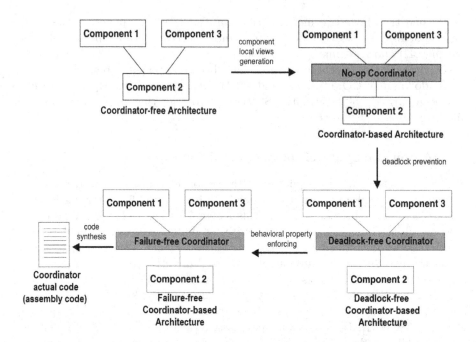

Fig. 4. Automatic synthesis of centralized failure-free coordinators

3.2   Automatic Synthesis of Distributed Coordinators

As an extension of the method described in Section 3.1, the method that we discuss in this section assumes as input (see Figure 5): (i) a behavioral specifi cation of the coordinator-free system formed by communicating components. It is given as a set $\{C_1, \ldots, C_n\}$ of LTSs (one for each component). The behavior of the system is modeled by composing in parallel all the LTSs and by forcing synchronization on common events; (ii) the specification of the desired behavior that the system must exhibit. This is given in terms of an extended LTS, from now on denoted by P_{LTS}.

These two inputs are then processed in two main steps:

1. by taking into account all component LTSs, we automatically derive the LTS that models the behavior of a centralized deadlock-free coordinator. This first step is inherited from the approach described in Section 3.1. Whenever P_{LTS} ensures itself deadlock-freeness and its traces are all traces of the centralized coordinator LTS, such a step is not required and, hence, the centralized coordinator cannot be generated. By avoiding the generation of the centralized coordinator, the method's complexity is polynomial in the number of states of P_{LTS}. The first step terminates by checking whether enforcing P_{LTS} is possible or not. This check is implemented by a specific notion of refinement. In our method, we use a suitable notion of *strong simulation* [39] to check the refinement relation between two LTSs.

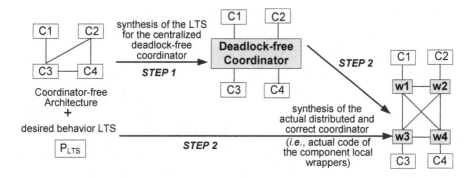

Fig. 5. Automatic synthesis of distributed failure-free coordinators

2. In the second step, let K be the LTS of the centralized coordinator. If K has been generated and it has been checked that P_{LTS} can be enforced on it, our method explores K looking for those states representing the *last chance* before entering an execution trace that leads to a deadlock. For instance, in Figure 6, the state S4 represents the last chance state before incurring in the deadlock state S7. This information is crucial for deadlock prevention purposes. The search of the last chance states aims at storing into the local wrappers the states of the components that could lead the system to a deadlock by means of a so called *critical action*. The idea is therefore not to allow a component to perform a critical action before being sure that the system will not reach a deadlock state.

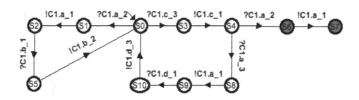

Fig. 6. An example of a centralized coordinator LTS in SYNTHESIS

By interacting with the SYNTHESIS tool, the user can tag component actions as either *controllable* or *uncontrollable* by the external environment. If such a critical action is controllable then it can be discarded. Otherwise, if it is uncontrollable, SYNTHESIS performs a *controller synthesis step* [15,44] that "backtracks" by looking for the first controllable action that can be discarded to prevent the execution of the critical action. After the execution of this search, the set of last chance states and associated critical actions are stored in a table, one for each component wrapper.

The second step also explores P_{LTS} to retrieve information crucial for undesired behavior prevention. The aim here is to split and distribute P_{LTS} in a way that each local wrapper knows which actions the wrapped component is *allowed* to execute.

Referring to Figures 6 and 7 for instance, the wrapper of component $C3$ must not allow the component to send the request C1.a, if the current global state of the system matches the state S0 in P_{LTS}, hence enforcing the desired behavior modeled by P_{LTS}. In particular, the label $\{!-C1.a\_2,!-C1.a\_3\}$ of the loop on S0 denotes two loops, one labeled with !−C1.a_2 and one labeled with !−C1.a_3. The action !C1.a_3 denotes an output action C1.a by C3; !−C1.a_3 represents *its neagation*, i.e., all possible actions different from it.

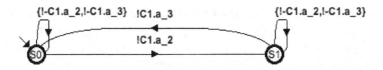

Fig. 7. An example of a desired behavior LTS in SYNTHESIS

The sets of *last chance states* and *allowed actions* are stored and, subsequently, used by the local wrappers as basis for correctly synchronizing with each other by exchanging additional communication. In other words, the local wrappers interact with each other to restrict the components' standard communication (modeled by K) by allowing only the part of the communication that is correct with respect to deadlock-freeness and P_{LTS}. By decentralizing K, the local wrappers preserve parallelism of the components forming the system.

4   Automatic Synthesis of Application-Layer Mediators

This section describes our recent works on the automatic synthesis of application-layer mediators. The approach presented in Section 4.1 discusses a theory of mediating connectors and an automated synthesis method for achieving on-the-fly interoperability (both communication and correct coordination) between heterogeneous protocols. The method discussed in Section 4.2 allows the automatic synthesis of a connector (serving as both coordinator and mediator) which is described in a modular form, hence enabling connector evolution and maintenance. Finally, Section 4.3 discusses a theoretical approach for the automated synthesis of mediators which permits to relax some of the assumptions state-of-the-art approaches rely on. The discussed formalization relies on *transducer theory* [11] and *Mazurkiewicz trace theory* [23].

4.1   Automated Synthesis of Mediating Connectors for On-the-fly Interoperability

In this section we overview our methodology and theory for the automated synthesis of **mediating connectors** (also called **mediators** or **connectors**) for on-the-fly interoperability [29,45,48]. Our *emergent* mediator is automatically elicited and synthesized. It makes the communication and correct coordination between heterogeneous protocols possible despite a set of mismatches that we characterized in [46,47] together with their related mediating connector patterns.

We focus on *compatible* or *functionally matching* protocols, i.e., heterogeneous protocols that realize complementary functionalities (see [10] for the inference of such high level compatibility). These protocols can *potentially* communicate and correctly coordinate by performing complementary sequences of messages and data (or complementary conversations). However, communication and correct coordination might not be achieved because of mismatches [47] (heterogeneity). For example, protocol languages (both messages and data) can have: (i) different granularity, or (ii) different alphabets. Protocols' behavior may have different sequences of messages and data because of (a.1) the order in which they are performed by a protocol is different from the order in which they are performed by the other protocol; (a.2) interleaved messages and data related to third parties communications i.e., with other systems or with the environment. In some cases, as for example (i), (ii) and (a.1), it is necessary to properly perform a manipulation of the two languages. In case (a.2) it is necessary to provide an abstraction of the two sequences of messages and data that results in sequences containing only those messages and data that are relevant to the communication. The scenario in Section 2.2 illustrates some of the mismatches that we identified in [47]. A mediator is then a protocol that allows the communication and correct coordination among compatible protocols by reconciling/mediating their differences.

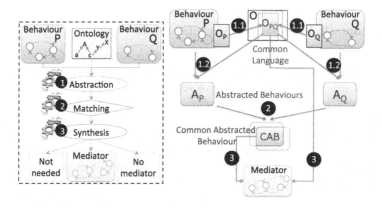

Fig. 8. Overview of our methodology

Figure 8 shows an overview of the approach. Inside the big dashed box on the left, there are the main elements of the methodology that takes as input (i) LTS-based specifications of interaction protocols modeling both messages and input/output data, and (ii) a semantic characterization of their messages and data through ontologies (in the figure they are called Behaviour P, Q and Ontology, respectively). The approach includes three steps called ❶ Abstraction, ❷ Matching and ❸ Synthesis after which it returns as output either the automatically synthesised *mediator* or that the mediator is *not needed* because the protocols can already communicate and coordinate, or that there is *no mediator* because the protocols are not compatible. Outside the dashed box, right part of the figure, there are some details about how the overall process and theory work. During the Abstraction step (❶), two substeps are performed: **(1.1)** identification of the Common Language (O_{PQ}) through *reasoning and inference on the ontology* and, if this exists, **(1.2)** abstraction of the behaviours of protocols (P, Q) into Abstract Behaviours (A_P, A_Q) driven by the Common Language. The Abstraction step finds how to align the languages by reasoning and solving both communication and coordination mismatches[47]. During the Matching step (❷), the approach looks for a Common Abstracted Behaviour (CAB) by reasoning on the mismatches once again. If CAB exists, it means that the protocols (P, Q) can communicate, that a Mediator exists and it is automatically synthesised by the Synthesis step (❸), thanks to CAB and O_{PQ}. The emergent synthesised mediator is correct-by-construction.

An implementation of this theory has been done and applied in [9]. We have also conducted several further investigation by considering both *functional interoperability* and *non functional interoperability* during the synthesis process, i.e., modeling and taking non functional concerns into account. Results are presented in [12,13] with a focus on dependability and performance arising from the execution environment, while in [22] with a focus on user performance requirements. In [22], the synthesised connector is self-adaptive with respect to run-time changes in the performance requirements.

4.2   Automatic Synthesis of Modular Connectors

In this section we overview a method for the automatic synthesis of *modular connectors* described in [30]. A modular connector is a composition of independent mediators. Each mediator is a primitive sub-connector that realizes a mediation pattern, which corresponds to the solution of a recurring protocol mismatch. The advantage of our connector decomposition is twofold: (i) it is *correct*, i.e., as for its monolithic version, a modular connector performs a mediation that is free from possible mismatches; and (ii) it promotes connector *evolution*, hence also easing code synthesis and maintenance. As described in [30], to show (i), we defined the semantics of protocols (as well as of mediators and connectors) by using a revised version of the *Interface Automata* (IA) theory [4]. Then, we proved that a modular connector for two protocols P and R enjoys the same correctness properties of the monolithic connector obtained by expressing the synthesis problem as a *quotient* problem between P and R [20]. Concerning the set of considered mediation patterns and, hence, connector modularization, our synthesis method relies on a revised version, namely $\mathcal{AP}(\mathcal{A})$, of the connector algebra described in [8]. It is an algebra for reasoning about protocol mismatches where basic mismatches can be solved by suitably defined primitives, while complex mismatches can be settled by composition operators that build connectors out of simpler ones. We revise the original algebra by adding an iterator operator and by giving its semantics in terms of our revised IA theory. For (ii), we used the above introduced Purchase Order scenario to illustrate that relevant changes can be applied on a modular connector by acting on its constituent mediators, without entirely re-synthesizing its protocol.

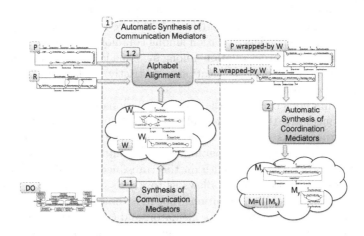

Fig. 9. Overview of the method

Figure 9 pictorially shows the synthesis phases (as rounded-corner rectangles) with their related input/output artefacts. The numbers denote the order in which the phases are carried out. The first phase splits into two sub-phases (1.1 and 1.2); it takes as input a domain ontology DO, for IPSs P and R, and automatically synthesizes a set, W, of *Communication Mediators* (CMs). CMs are represented as terms in $\mathcal{AP}(\mathcal{A})$ and they are responsible for solving communication mismatches. In particular, the CMs in W are used as wrappers for P and R so to "align" their different alphabets to the same alphabet. Roughly speaking, the goal of this phase is to make two heterogeneous protocols "speak" the same language. To this aim, the synthesized CMs translate an action from an alphabet into a certain sequence of actions from another alphabet. However, despite the achieved alphabet alignment, coordination mismatches are still possible; the second phase is for solving such mismatches. The synthesis of *COordination Mediators* (COMs) is carried out by reasoning on the traces of the "wrapped" P and R. As detailed in [30], for all pairs of traces, if possible, a COM that makes the two traces interoperable is synthesized as a term in $\mathcal{AP}(\mathcal{A})$. The parallel composition of the synthesized COMs represents, modulo alphabet alignment, the correct modular connector for P and R.

4.3   Synthesis of Protocol Mediators as Composition of Transducers

In this section we briefly report on a synthesis approach that we are recently investigating in order to overcome some limitations of the above methods. In particular, a strong common assumption of the above approaches is related to the availability of a domain ontology mapping one protocol ontology into the other. The domain ontology defines the relations that hold between the various concepts used by systems belonging to the same application domain. Thus, assuming the existence of such ontology means to consider as already solved a crucial, yet mostly difficult, part of the mediator synthesis problem. Instead, in this research, we formalized a method for the automated synthesis of protocol mediators that automatically infers the semantic relations that, in the state-of-the-art works, are explicit in the domain ontology. In order to do this we require the user to specify the Mazurkiewicz independence relation among common symbols in the protocol signatures. That is, the order relation among symbols joins commutativity. Note that providing the independence relation is a strictly weaker assumption than having a domain ontology.

Then, our methods combines the inferred relations in order to synthesize a model of the mediator that is amenable to be automatically treated for code generation purposes. As part of the method formalization, we provided a rigorous characterization of the interoperability notion and, hence, of the synthesis problem. This characterization relies on the transducer formalism [11] and on Mazurkiewicz trace theory [24]. It allows us to: (i) clearly express the necessary and sufficient interoperability conditions that must hold, for the protocols to be mediated, to ensure the existence of a mediator, and (ii) show that our synthesis method is sound. In this direction, we defined a novel notion of composition of transducers, which allows us to automatize a mediator existence check. Although different, this notion is related to other notions of automata compositions, such as Interface Automata (IA) [4], and to the best of our knowledge, it has never been considered in the transducer theory.

Within our approach, by means of a suitable behavioral mapping, transducers are used to model interaction protocols in terms of sequences of input/output actions with the external environment. The newly introduced notion of parallel composition between transducers is then leveraged to automatically check the existence of a mediator and to synthesize it accounting for solvable communication and coordination mismatches.

Informally, a transducer can be defined as an automaton that reads input words on one tape, and prints output words on a second tape. A word goes either from the initial state to a final state or in between final states.

In Figure 10, we show a simple composition of two transducers. Transducers have a graphical representation very similar to the usual representation of finite automata. Each state q is represented by a circle, labeled with q, and possibly marked with the name of the associated task. Each transition $t=(q,i,o,q')$ is represented as an arrow directed from q to q' and labeled with i/o. ϵ denotes the *empty word*. The initial state has an incoming arrow and it is always marked as *start*. Final states are doubly circled.

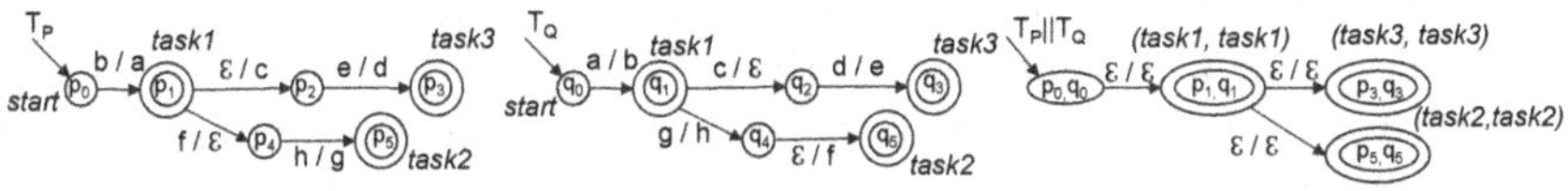

Fig. 10. Two transducers, T_P and T_Q, and their parallel composition $T_P \| T_Q$

Intuitively, the parallel composition of two transducers, T_1 and T_2, contains a transition from a state to a final state whenever T_1 recognizes an input (resp., output) word that, modulo reordering of some common symbols and consumption of some unshared ones, is the same as the output (resp., input) word recognized by T_2, and vice versa. Our notion of parallel composition ensures that if two protocols do not match on at least a common word, then they cannot be mediated.

For space reasons, we omit the formal definition but it is worth noticing that this notion of composition makes use of the intersection of trace languages, which is in general undecidable [2]. To cope with this undecidability result, several attempts are under investigation, e.g., by:

- allowing transducers to visit cycles only a bounded number of times. In this case, it is enough to compute the *Foata normal form* for Mazurkiewicz traces [23,42] of all the words in the paths and check for equality.
- considering a transitive independence relation. In this case, there are algorithms [3] that allows our synthesis method to build the composition.
- considering an independence relation that is a transitive forest [2]. In this case, there are algorithms that allows to decide the existence of a mediator by showing the non-emptiness of the composition.
- considering transducers where all the paths (i.e., sequences of states) of the form $q_0^1 \ldots q_i \ldots q_j \ldots$, with $q_i = q_j$ and q_0^1 initial, are such that q_h is final and $i \leq h \leq j$. This restriction is motivated by the fact that, when there is a cycle, a task must be accomplished within the cycle. Again, this permits to compute the *Foata normal form* of all the words in the paths and check for equality.

5   Related Works

Interoperability, considered as the ability to correctly coordinate and mediate components' interaction, have been investigated in several contexts, among which integration of heterogeneous data sources [54], architectural patterns [17], patterns of connectors [49], Web services [32,35], and algebra to solve mismatches [25]. For the sake of brevity, we discuss only the works, from the different contexts, closest to our methods.

The interoperability of protocols have received attention since the early days of networking. Indeed many efforts have been done in several directions including for example formal approaches to protocol conversion, like in [18,34].

The seminal work in [55] is strictly related to the notions of mediator presented in this paper. Compared to our connector synthesis, this work does not allow to deal with ordering mismatches and different granularity of the languages (solvable by the split and merge primitives).

Recently, with the emergence of web services and advocated interoperability, the research community has been studying solutions to the automatic mediation of business processes [52,53]. However, most solutions are discussed informally, making it difficult to assess their respective advantages and drawbacks.

In [49] the authors present an approach for formally specifying connector wrappers as protocol transformations, modularizing them, and reasoning about their properties, with the aim to resolve component mismatches. In [16] the authors present an algebra for five basic stateless connectors that are symmetry, synchronization, mutual exclusion, hiding and inaction. They also give the operational, observational and denotational semantics and a complete normal-form axiomatization. The presented connectors can be composed in series and in parallel. Although these formalizations supports connector modularization, automated synthesis is not treated at all hence keeping the focus only on connector design and specification.

In [40], the authors use game theory for checking whether incompatible component interfaces can be made compatible by inserting a converter between them which satisfies specified requirements. This approach is able to automatically synthesize the converter. In contrast to our methods, their method needs as input a deadlock-free specification of the requirements that should be satisfied by the adaptor, by delegating to the user the non-trivial task of specifying that.

In other work in the area of component adaptation [19], it is shown how to automatically generate a concrete adaptor from: (i) a specification of component interfaces, (ii) a partial specification of the components interaction behavior, (iii) a specification of the adaptation in terms of a set of correspondences between actions of different components and (iv) a partial specification of the adaptor. The key result is the setting of a formal foundation for the adaptation of heterogeneous components that may present mismatching interaction behavior. Assuming a specification of the adaptation in terms of a set of correspondences between methods (and their parameters) of two components requires to know many implementation details (about the adaptation) that we do not want to consider in order to synthesize a connector.

6   Conclusion and Future Perspectives

Interoperability is a key requirement for heterogeneous protocols within ubiquitous computing environments where networked systems *meet dynamically* and need to *interoperate without a priori knowledge* of each other. Although numerous efforts has been done in many different research areas, protocol interoperability is still an open challenge.

In our recent research, we concentrated on the automatic synthesis of mediators between compatible protocols which enables them to communicate.

We proposed *rigorous techniques* to automatically reason about and compose the behavior of networked systems that aim at fulfilling some common goal.

The reasoning permits to find a way to achieve communication and to build the related mediation solution. Our current work puts the emphasis on "the elicitation of a way to achieve communication". In particular, we contributed with: (i) a rigorous characterization of the interoperability notion as the necessary and sufficient conditions that must hold for the existence of a connector between two heterogeneous protocols; and (ii) the provision of automated methods and tools to solve communication and coordination mismatches, even without assuming the existence of a domain ontology that represents a common strong assumption in the state of the art. We have started to show, through their application to real world case studies that our methods are viable and sound. All the synthesized connector models are suitable for the automatic generation of the connector actual code.

As future work, we intend to establish what are the applicability boundaries of our synthesis methods in terms of both their complexity and expressive power. More in general, with our formalizations we would like to contribute to determine which class of interoperability mismatches can be automatically solved. In the future, we also plan to: (i) study run-time techniques towards efficient synthesis; (ii) scale the synthesis process up to an arbitrary number of protocols; and (iii) ensure dependability. Towards this direction we have already done some investigations by considering both functional interoperability and non-functional interoperability during the synthesis process, i.e., modeling and taking non functional concerns into account. Preliminary results are presented in [12,13] with a focus on dependability and performance arising from the execution environment, while in [22] with a focus on user performance requirements. In [22], the synthesised mediator is self-adaptive with respect to run-time changes in the performance requirements.

References

1. ITU Telecommunication Standardisation sector, ITU-T reccomendation Z.120. Message Sequence Charts (MSC 1996). Geneva
2. Aalbersberg, I.J., Hoogeboom, H.J.: Characterizations of the decidability of some problems for regular trace languages. Mathematical Systems Theory **22**(1), 1–19 (1989)
3. Aalbersberg, I.J., Welzl, E.: Trace languages defined by regular string languages. ITA **20**(2), 103–119 (1986)
4. de Alfaro, L., Henzinger, T.A.: Interface automata. In: ESEC/FSE (2001)
5. Arnold, A.: Finite Transition Systems. International Series in Computer Science. Prentice Hall International (UK) (1989)
6. Autili, M., Inverardi, P., Navarra, A., Tivoli, M.: Synthesis: A tool for automatically assembling correct and distributed component-based systems. In: 29th International Conference on Software Engineering (ICSE 2007), Minneapolis, MN, USA, pp. 784–787. IEEE Computer Society (2007). http://doi.ieeecomputersociety.org/10.1109/ICSE.2007.84
7. Autili, M., Mostarda, L., Navarra, A., Tivoli, M.: Synthesis of decentralized and concurrent adaptors for correctly assembling distributed component-based systems. Journal of Systems and Software **81**(12), 2210–2236 (2008)

8. Autili, M., Chilton, C., Inverardi, P., Kwiatkowska, M., Tivoli, M.: Towards a connector algebra. In: Margaria, T., Steffen, B. (eds.) ISoLA 2010, Part II. LNCS, vol. 6416, pp. 278–292. Springer, Heidelberg (2010)
9. Bennaceur, A., Issarny, V., Spalazzese, R., Tyagi, S.: Achieving interoperability through semantics-based technologies: the instant messaging case. In: Cudré-Mauroux, P., et al. (eds.) ISWC 2012, Part II. LNCS, vol. 7650, pp. 17–33. Springer, Heidelberg (2012). http://dx.doi.org/10.1007/978-3-642-35173-0_2
10. Bennaceur, A., Johansson, R., Moschitti, A., Spalazzese, R., Sykes, D., Saadi, R., Issarny, V.: Inferring affordances using learning techniques. In: Moschitti, A., Scandariato, R. (eds.) EternalS 2011. CCIS, vol. 255, pp. 79–87. Springer, Heidelberg (2012). http://dx.doi.org/10.1007/978-3-642-28033-7_7
11. Berstel, J., Boasson, L.: Transductions and context-free languages, pp. 1–278. Teubner (1979)
12. Bertolino, A., Calabrò, A., Di Giandomenico, F., Nostro, N., Inverardi, P., Spalazzese, R.: On-the-fly dependable mediation between heterogeneous networked systems. In: Escalona, M.J., Cordeiro, J., Shishkov, B. (eds.) ICSOFT 2011. CCIS, vol. 303, pp. 20–37. Springer, Heidelberg (2013). http://dx.doi.org/10.1007/978-3-642-36177-7_2
13. Bertolino, A., Inverardi, P., Issarny, V., Sabetta, A., Spalazzese, R.: On-the-fly interoperability through automated mediator synthesis and monitoring. In: Margaria, T., Steffen, B. (eds.) ISoLA 2010, Part II. LNCS, vol. 6416, pp. 251–262. Springer, Heidelberg (2010)
14. Bertolino, A., Inverardi, P., Pelliccione, P., Tivoli, M.: Automatic synthesis of behavior protocols for composable web-services. In: Proc. of ESEC/FSE (2009)
15. Brandin, B., Wonham, W.: Supervisory control of timed discrete-event systems. IEEE Transactions on Automatic Control $39(2)$ (1994)
16. Bruni, R., Lanese, I., Montanari, U.: A basic algebra of stateless connectors. Theor. Comput. Sci. $366(1)$, 98–120 (2006)
17. Buschmann, F., Meunier, R., Rohnert, H., Sommerlad, P., Stal, M.: Pattern-Oriented Software Architecture, Volume 1: A System of Patterns. Wiley, Chichester (1996)
18. Calvert, K.L., Lam, S.S.: Formal methods for protocol conversion. IEEE Journal on Selected Areas in Communications $8(1)$, 127–142 (1990)
19. Canal, C., Poizat, P., Salaün, G.: Model-based adaptation of behavioral mismatching components. IEEE Trans. Software Eng. $34(4)$, 546–563 (2008)
20. Chen, T., Chilton, C., Jonsson, B., Kwiatkowska, M.: A compositional specification theory for component behaviours. In: Seidl, H. (ed.) ESOP 2012. LNCS, vol. 7211, pp. 148–168. Springer, Heidelberg (2012)
21. Dallmeier, V., Knopp, N., Mallon, C., Fraser, G., Hack, S., Zeller, A.: Automatically generating test cases for specification mining. IEEE TSE $38(2)$ (2012)
22. Di Marco, A., Inverardi, P., Spalazzese, R.: Synthesizing self-adaptive connectors meeting functional and performance concerns. In: Proceedings of the 8th International Symposium on Software Engineering for Adaptive and Self-Managing Systems, SEAMS 2013, pp. 133–142. IEEE Press, Piscataway (2013). http://dl.acm.org/citation.cfm?id=2487336.2487358
23. Diekert, V., Muscholl, A.: Trace theory. In: Encyclopedia of Parallel Computing, pp. 2071–2079 (2011)
24. Diekert, V., Rozenberg, G.: The Book of Traces. World Scientific (1995)
25. Dumas, M., Spork, M., Wang, K.: Adapt or perish: algebra and visual notation for service interface adaptation. In: Dustdar, S., Fiadeiro, J.L., Sheth, A.P. (eds.) BPM 2006. LNCS, vol. 4102, pp. 65–80. Springer, Heidelberg (2006)

26. Ernst, M.D., Cockrell, J., Griswold, W.G., Notkin, D.: Dynamically discovering likely program invariants to support program evolution. IEEE Trans. Software Eng. **27**(2) (2001)

27. Feiler, P., Gabriel, R.P., Goodenough, J., Lingerand, R., Longstaff, T., Kazman, R., Klein, M., Northrop, L., Schmidt, D., Sullivan, K., Wallnau, K.: Ultra-Large-Scale Systems: The Software Challenge of the Future (2006)

28. Blair, G.S., Paolucci, M., Grace, P., Georgantas, N.: Interoperability in complex distributed systems. In: Bernardo, M., Issarny, V. (eds.) SFM 2011. LNCS, vol. 6659, pp. 1–26. Springer, Heidelberg (2011). http://dx.doi.org/10.1007/978-3-642-21455-4

29. Inverardi, P., Issarny, V., Spalazzese, R.: A theory of mediators for eternal connectors. In: Margaria, T., Steffen, B. (eds.) ISoLA 2010, Part II. LNCS, vol. 6416, pp. 236–250. Springer, Heidelberg (2010)

30. Inverardi, P., Tivoli, M.: Automatic synthesis of modular connectors via composition of protocol mediation patterns. In: Proceedings of ICSE 2013 (2013)

31. Issarny, V., Bennaceur, A., Bromberg, Y.-D.: Middleware-layer connector synthesis: beyond state of the art in middleware interoperability. In: Bernardo, M., Issarny, V. (eds.) SFM 2011. LNCS, vol. 6659, pp. 217–255. Springer, Heidelberg (2011)

32. Jiang, F., Fan, Y., Zhang, X.: Rule-based automatic generation of mediator patterns for service composition mismatches. In: Proceedings of the 2008 The 3rd International Conference on Grid and Pervasive Computing - Workshops, pp. 3–8. IEEE Computer Society, Washington, DC (2008). http://portal.acm.org/citation.cfm?id=1381299.1381352

33. Keller, R.M.: Formal verification of parallel programs. Commun. ACM **19**(7), 371–384 (1976)

34. Lam, S.S.: Correction to "protocol conversion". IEEE Trans. Software Eng. **14**(9), 1376 (1988)

35. Li, X., Fan, Y., Wang, J., Wang, L., Jiang, F.: A pattern-based approach to development of service mediators for protocol mediation. In: Proceedings of WICSA 2008, pp. 137–146. IEEE Computer Society (2008)

36. Lo, D., Mariani, L., Santoro, M.: Learning extended fsa from software: An empirical assessment. J. Syst. Softw. **85**(9) (2012)

37. Lorenzoli, D., Mariani, L., Pezzè, M.: Automatic generation of software behavioral models. In: Proc. of ICSE 2008 (2008)

38. Margaria, T.: The semantic web services challenge: Tackling complexity at the orchestration level. In: ICECCS 2008 (2008)

39. Milner, R.: Communication and Concurrency. Prentice Hall, New York (1989)

40. Passerone, R., de Alfaro, L., Henzinger, T.A., Sangiovanni-Vincentelli, A.L.: Convertibility verification and converter synthesis: two faces of the same coin. In: Proceedings of the 2002 IEEE/ACM International Conference on Computer-Aided Design, ICCAD 2002, pp. 132–139 (2002)

41. Pelliccione, P., Tivoli, M., Bucchiarone, A., Polini, A.: An architectural approach to the correct and automatic assembly of evolving component-based systems. Journal of Systems and Software **81**(12), 2237–2251 (2008)

42. Perrin, D.: Partial commutations. In: Proceedings of 16th International Colloquium on Automata, Languages and Programming, ICALP 1989, Stresa, Italy, July 11–15, pp. 637–651 (1989)

43. Raffelt, H., Steffen, B., Berg, T., Margaria, T.: Learnlib: a framework for extrapolating behavioral models. Int. J. Softw. Tools Technol. Transf. **11**(5) (2009)

44. Ramadge, P., Wonham, W.: Supervisory control of a class of discrete event processes. Siam J. Control and Optimization **25**(1) (1987)
45. Spalazzese, R.: A Theory of Mediating Connectors to achieve Interoperability. Ph.D. thesis, University of L'Aquila, April 2011
46. Spalazzese, R., Inverardi, P.: Components interoperability through mediating connector pattern. In: WCSI 2010, arxiv.org/abs/1010.2337; EPTCS 37, pp. 27–41 (2010)
47. Spalazzese, R., Inverardi, P.: Mediating connector patterns for components interoperability. In: Babar, M.A., Gorton, I. (eds.) ECSA 2010. LNCS, vol. 6285, pp. 335–343. Springer, Heidelberg (2010)
48. Spalazzese, R., Inverardi, P., Issarny, V.: Towards a formalization of mediating connectors for on the fly interoperability. In: Proceedings of the Joint Working IEEE/IFIP Conference on Software Architecture and European Conference on Software Architecture (WICSA/ECSA 2009), pp. 345–348 (2009)
49. Spitznagel, B., Garlan, D.: A compositional formalization of connector wrappers. In: ICSE, pp. 374–384 (2003)
50. Tivoli, M., Fradet, P., Girault, A., Goessler, G.: Adaptor synthesis for real-time components. In: Grumberg, O., Huth, M. (eds.) TACAS 2007. LNCS, vol. 4424, pp. 185–200. Springer, Heidelberg (2007)
51. Tivoli, M., Inverardi, P.: Failure-free coordinators synthesis for component-based architectures. Science of Computer Programming **71**(3), 181–212 (2008)
52. Vaculín, R., Neruda, R., Sycara, K.: An agent for asymmetric process mediation in open environments. In: Kowalczyk, R., Huhns, M.N., Klusch, M., Maamar, Z., Vo, Q.B. (eds.) SOCASE 2008. LNCS, vol. 5006, pp. 104–117. Springer, Heidelberg (2008)
53. Vaculín, R., Sycara, K.: Towards automatic mediation of OWL-S process models. In: IEEE International Conference on Web Services, pp. 1032–1039 (2007)
54. Wiederhold, G., Genesereth, M.: The conceptual basis for mediation services. IEEE Expert: Intelligent Systems and Their Applications **12**(5), 38–47 (1997)
55. Yellin, D.M., Strom, R.E.: Protocol specifications and component adaptors. ACM Trans. Program. Lang. Syst. **19** (1997)

Automated Program Verification

Azadeh Farzan[2], Matthias Heizmann[1]([⊠]), Jochen Hoenicke[1],
Zachary Kincaid[2], and Andreas Podelski[1]

[1] University of Freiburg, Freiburg im Breisgau, Germany
`heizmann@informatik.uni-freiburg.de`
[2] University of Toronto, Toronto, Canada

Abstract. A new approach to program verification is based on automata. The notion of automaton depends on the verification problem at hand (nested word automata for recursion, Büchi automata for termination, a form of data automata for parametrized programs, etc.). The approach is to first construct an automaton for the candidate proof and then check its validity via automata inclusion. The originality of the approach lies in the construction of an automaton from a correctness proof of a given sequence of statements. A sequence of statements is at the same time a word over a finite alphabet and it is (a very simple case of) a program. Just as we ask whether a word has an accepting run, we can ask whether a sequence of statements has a correctness proof (of a certain form). The automaton accepts exactly the sequences that do.

1 Introduction

The verification of a program can often be divided into two steps: 1) the construction of a candidate proof and 2) the check of the validity of the candidate proof for the given program. An example is the construction of a Floyd-Hoare style annotation and the check of its inductiveness. In a new approach to program verification, the candidate proof in Step 1 comes in the form of an automaton and Step 2 is reduced to an automata inclusion test. If the inclusion test succeeds, the program is proven correct. The approach lends itself to a verification algorithm in the form of a loop: the automaton for the candidate proof is constructed incrementally until the inclusion holds (see also Figure 1 in Section 2).

The approach introduces a novel separation between

- the symbolic reasoning about data and
- the automata-theoretic reasoning about control.

By data we mean the values of program variables (e.g., integers) which are read and written by program statements. Examples of statements are tests of conditions and updates. We apply symbolic reasoning to mechanize the analysis of the data and produce a correctness proof for the sequence of statements. For example, we can first translate the sequence of statements into a logical formula (in the logical theory corresponding to the data domain) and then apply a dedicated decision procedure (as implemented by an SMT solver).

A.-H. Dediu et al. (Eds.): LATA 2015, LNCS 8977, pp. 25–46, 2015.
DOI: 10.1007/978-3-319-15579-1_2

The originality of the approach lies in the construction of the automaton from a correctness proof of a given sequence of statements. The construction relies on the following observation: one can first decompose the correctness proof into its base components and then *rearrange* the base components to obtain a correctness proof for a new sequence built up from the same set of statements. In the automaton that we construct, the non-determinism reflects the combinatorial choice of ways to rearrange the base components. A sequence of statements is at the same time a word over a finite alphabet and it is (a very simple case of) a program. Just as we ask whether a word has an accepting run, we can ask whether a sequence of statements has a correctness proof (one which can be obtained by rearranging the base components). The automaton accepts exactly the sequences that do.

The control of the program can be expressed through a graph, the so-called control flow graph of the program. The paths in the graph define the set of sequences of statements that are possible according to the control flow alone (i.e., ignoring the data and ignoring in particular the outcome of tests of conditions). It is this set of sequences which is the language recognized by the *program automaton* (we here use the finite set of the statements in the program as the alphabet and sequences of statements as words).

The control of the program comes in only in Step 2. We test the inclusion between the program automaton and the automaton for the candidate proof. The inclusion means that each sequence of statements that is possible according to the control flow of the program has a correctness proof.

Roadmap. In the remainder of this paper, we will instantiate the approach for six different verification problems. Each verification problem requires a specific class of automata. In the table below, *unbounded parallelism* refers to programs with an unbounded number of threads, and *predicate automata* are a new version of data automata that we introduce. The term *proofs that count* refers to programs whose verification involves the task to synthesize ghost variables that count. Each verification problem poses a new challenge in finding an appropriate notion of automata for the program and for the candidate proof, a way of representing and constructing the automata, and finally an algorithm for checking automata inclusion. We will explain each challenge and our approach to the solution informally, by way of examples. For technical details, we refer to the corresponding paper.

	verification problem	inclusion problem	reference
Section 2	sequential programs	nondeterministic finite automata	[10, 12]
Section 3	termination	Büchi automata	[13]
Section 4	recursion	nested word automata	[11]
Section 5	concurrency	alternating finite automata	[4]
Section 6	unbounded parallelism	predicate automata	[6]
Section 7	proofs that count	Petri net $\subseteq$ counting automaton	[5]

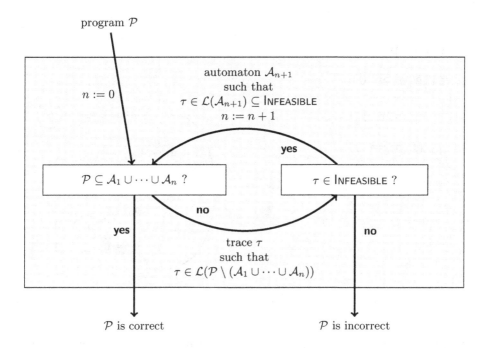

Fig. 1. Automated program verification

2   Sequential Programs: Nondeterministic Finite Automata

In this section we instantiate the approach for verifying sequential programs. We present two ways to construct an automaton from the correctness proof of a sequence of statements. We often refer to a sequence of statements as a *trace*.

Automata from unsatisfiable cores. The program $\mathcal{P}_{ex1}$ in Figure 2 is the adaptation of an example in [14]. The original program in [14] allocates a pointer p and then enters a while loop which *uses* p and conditionally *frees* p. The original correctness property in [14] is "the pointer p is not *used* after it has been *freed*."

In our setting we use **assert** statements to define the correctness of the program executions. In the example of $\mathcal{P}_{ex1}$, an *incorrect* execution would start with a non-zero value for the variable p and, at some point, enter the body of the while loop when the value of p is 0 (and the execution of the **assert** statement *fails*).

Informally, we can argue the correctness of $\mathcal{P}_{ex1}$ rather directly if we split the executions into two cases, namely according to whether the **then** branch of the conditional gets executed at least once during the execution or it does not. If not, then the value of p is never changed and remains non-zero (and the assert statement cannot fail). If the **then** branch of the conditional is executed, then the value of n is 0, the statement **n--** decrements the value of n from 0 to -1,

ℓ_0: assume p != 0;

ℓ_1: while(n >= 0)

   {

ℓ_2:   assert p != 0;

     if(n == 0)

     {

ℓ_3:   p := 0;

     }

ℓ_4:   n--;

  }

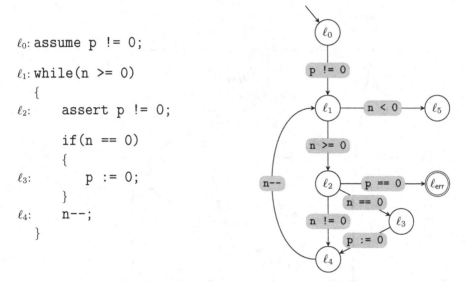

Fig. 2. Example program $\mathcal{P}_{\text{ex1}}$

and the while loop will exit directly, without executing the assert statement. It turns out that our verification algorithm will automatically reproduce this case split.

The algorithm starts with the sequence of statements on some path from ℓ_0 to ℓ_{err} in the *control flow graph* of $\mathcal{P}_{\text{ex1}}$ (see Figure 2). We take the shortest path which goes from ℓ_0 to ℓ_{err} via ℓ_1 and ℓ_2. The sequence of statements on this path is *infeasible* because it is not possible to execute the assume statements p!=0 and p==0 without an update of p in between. Formally, the formula obtained by translating the sequence of statements is unsatisfiable, and the conjuncts $p \neq 0$ and $p = 0$ form an *unsatisfiable core* of the formula.

We construct the automaton $\mathcal{A}_1$ in Figure 3 by first constructing an automaton that accepts only the sequence of the assume statements p!=0 and p==0 and then adding a number of self-loops. The idea behind the construction is that the sequence of statements remains infeasible if we add any statement before or after and any statement other than an update of p in-between.

The automaton $\mathcal{A}_1$ does not accept a sequence of statements *with* an update of p in between the statements p!=0 and p==0. The shortest path from ℓ_0 to ℓ_{err} with such a sequence of statements goes from ℓ_2 to ℓ_{err} after it has gone from ℓ_2 to ℓ_3 once before. The sequence of statements on this path is again infeasible: it is not possible to execute the assume statement n==0, the update statement n--, and then the assume statement n>=0 (without an update of n between n==0 and n-- and between n-- and n>=0).

We construct the automaton $\mathcal{A}_2$ depicted in Figure 3 in the analogous way. Now, the unsatisfiable core corresponds to the sequence of the statements n==0, n--, and n>=0. Thus, we first construct an automaton that accepts only this

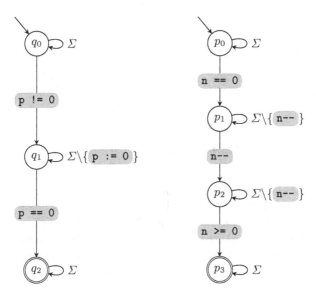

Fig. 3. Automata $\mathcal{A}_1$ and $\mathcal{A}_2$ whose union forms a proof of correctness for $\mathcal{P}_{\text{ex1}}$ (an edge labeled with Σ means a transition reading any letter, an edge labeled with $\Sigma \backslash \{ \text{p := 0} \}$) means a transition reading any letter except for p := 0)

sequence and then add a number of self-loops. Now we are careful to not add a self-loop with an update of n.

To summarize, we have twice taken a path from ℓ_0 to ℓ_{err} and constructed an automaton from the unsatisfiable core of the proof of the infeasibility of the sequence of statements on the path.

The control flow graph $\mathcal{P}_{\text{ex1}}$ defines an automaton that recognizes the set of all sequences of statements on paths from ℓ_0 to ℓ_{err}. We can thus check that all such sequences are accepted by one of the two automata by testing the inclusion

$$\mathcal{P}_{\text{ex1}} \subseteq \mathcal{A}_1 \cup \mathcal{A}_2.$$

Automata from sets of Hoare triples. It is "easy" to justify the construction of the automata $\mathcal{A}_1$ and $\mathcal{A}_2$ in the example above: the infeasibility of a sequence of statements (such as the sequence p!=0 p==0) is preserved if one adds statements that do not modify any of the variables of the statements in the sequence (here, the variable p).

The example of the program $\mathcal{P}_{\text{ex2}}$ in Figure 4 shows that sometimes a more involved justification is required. The sequence of the two statements x:=0 and x==-1 (which labels a path from ℓ_0 to ℓ_{err}) is infeasible. However, the statement x++ does modify the variable that appears in the two statements. So how can we account for the paths that loop in ℓ_2 taking the edge labeled x++ one or more times? We need to construct an automaton that covers the case of those paths, but we can no longer base the construction solely on unsatisfiable cores.

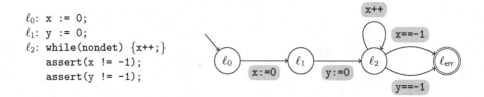

```
ℓ₀: x := 0;
ℓ₁: y := 0;
ℓ₂: while(nondet) {x++;}
    assert(x != -1);
    assert(y != -1);
```

Fig. 4. Example program $\mathcal{P}_{ex2}$

We must base the construction of the automaton on a more powerful form of correctness argument: Hoare triples. The four Hoare triples below are sufficient to prove the infeasibility of all those paths. They express that the assertion $x \geq 0$ holds after the update `x:=0`, that it is *invariant* under the updates `y:=0` and `x++`, and that is blocks the execution of the assume statement `x==-1`.

$$\{ \text{true} \} \text{ x:=0 } \{x \geq 0\}$$
$$\{x \geq 0\} \text{ y:=0 } \{x \geq 0\}$$
$$\{x \geq 0\} \text{ x++ } \{x \geq 0\}$$
$$\{x \geq 0\} \text{ x==-1 } \{ \text{false} \}$$

The automaton $\mathcal{A}_1$ in Figure 5 has four transitions, one for each Hoare triple. It has three states, one for each assertion: the initial state q_0 for *true*, the state q_1 for $x \geq 0$, the (only) final state q_2 for *false*. The construction of such a *Floyd-Hoare automaton* generalizes to any set of Hoare triples. The resulting automaton can have arbitrary loops. In contrast, an automaton constructed as in the preceding example can only have self-loops.

In our implementation [8], the set of Hoare triples comes from an interpolating SMT solver such as [2] which generates the assertion $x \geq 0$ from the infeasibility proof.

The four Hoare triples below are sufficient to prove the infeasibility of all paths that reach the error location via the edge labeled with `y==-1`.

$$\{ \text{true} \} \text{ x:=0 } \{ \text{true} \}$$
$$\{ \text{true} \} \text{ y:=0 } \{y \geq 0\}$$
$$\{y \geq 0\} \text{ x++ } \{y \geq 0\}$$
$$\{y \geq 0\} \text{ y==-1 } \{ \text{false} \}$$

We use them in the same way as above in order to construct the automaton $\mathcal{A}_2$ in Figure 5. The two automata are sufficient to prove the correctness of the program; i.e., $\mathcal{P}_{ex2} \subseteq \mathcal{A}_1 \cup \mathcal{A}_2$.

We could have based the construction of the automaton $\mathcal{A}_2$ in Figure 5 on the unsatisfiable core of the infeasibility proof, as in the example of $\mathcal{P}_{ex2}$. Intuitively, we do not need to know the precise form of the assertion $y \geq 0$ in order to know that it is invariant under `x++`. It is sufficient to know that the variabe x does not occur in the assertion (which we can assume because x does not appear in the unsatisfiable core).

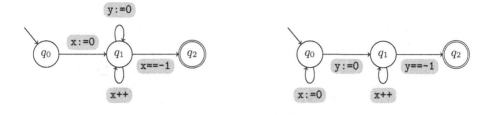

Fig. 5. Automata $\mathcal{A}_1$ and $\mathcal{A}_2$ for $\mathcal{P}_{ex2}$

To summarize, we have presented two ways to construct an automaton from the correctness proof of a sequence of statements. The first gets away without the synthesis of assertions, but the second is more general and leads to a complete verification method [10].

In the verification algorithm depicted in Figure 1, the union of the automata constructed from the correctness proofs for sequences of statements is constructed incrementally until the inclusion holds. In our implementation [8], we need not construct the union explicitly. Instead, we can incrementally construct the difference automaton $\mathcal{P} \setminus (\mathcal{A}_1 \cup \cdots \cup \mathcal{A}_n)$.

3   Termination: Büchi Automata

In this section we present how we use Büchi automata to construct a termination proof of a program.

In the presence of loops with branching or nesting, the termination proof has to account for all possible interleavings between the different paths through the loop. If the program is *lasso-shaped* (a stem followed by a single loop without branching), the control flow is trivial: there is only one path. Consequently, the termination proof can be very simple. Many procedures are specialized to lasso-shaped programs and derive a simple termination proof rather efficiently [9,16,18]. The relevance of lasso-shaped programs stems from their use as the representation of an *ultimately periodic* infinite trace through the control flow graph of a program with arbitrary nesting (the *period*, i.e., the cycle of the lasso, may itself go through a sequence of loops in the program).

We can explain our algorithm informally using the program $\mathcal{P}^{\text{sort}}$ depicted in Figure 6 which is an implementation of bubblesort. We begin by picking some ω-trace of $\mathcal{P}^{\text{sort}}$. We take the trace that first enters the outer while loop and then takes the inner while loop infinitely often. We denote this trace using the ω-regular expression OUTER.INNER$^\omega$. We see that this trace is terminating: its termination can be shown using the linear ranking function $f(\mathtt{i},\mathtt{j}) = \mathtt{i} - \mathtt{j}$. Moreover, we see that this ranking function is applicable not only to this trace, but to all traces that eventually always take the inner loop. Such traces can be represented by the ω-regular expression

$$(\text{INNER} + \text{OUTER})^*.\text{INNER}^\omega \ . \tag{1}$$

```
program sort(int i)
ℓ₁: while (i>0)
ℓ₂:    int j:=1
ℓ₃:    while(j<i)
       //   if (a[j]>a[i])
       //      swap(a[j],a[i])
ℓ₄:       j++
ℓ₅:    i--
```

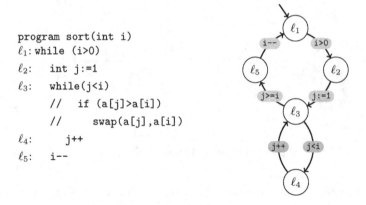

Fig. 6. Program $\mathcal{P}^{\texttt{sort}}$ which is an implementation of bubblesort

Now, let us pick another ω-trace from $\mathcal{P}^{\texttt{sort}}$. This time we take the trace that always takes the outer while loop. We see that this trace is terminating. Its termination can be shown using the linear ranking function $f(\texttt{i}, \texttt{j}) = \texttt{i}$. Moreover, we see that this ranking function is applicable not only to this trace, but to all traces that take the outer while loop infinitely often, as represented by the ω-regular expression

$$(\text{INNER}^*.\text{OUTER})^\omega \ . \tag{2}$$

Finally, we consider the set of all ω-trace of the program $\mathcal{P}^{\texttt{sort}}$

$$(\text{OUTER} + \text{INNER})^\omega,$$

check that each trace has the form (1) or has the form (2), and conclude that $\mathcal{P}^{\texttt{sort}}$ is terminating.

The approach is based on the notion of an ω-*trace*, which is an infinite sequence of program statements $\pi = s_1 s_2 \ldots$. Like in the section before, we assume that the statements are taken from a given finite set of program statements Σ. If we consider Σ as an alphabet and each statement as a letter, then an ω-*trace* is an infinite word over this alphabet. For example, we can write the alphabet of our running example $\mathcal{P}^{\texttt{sort}}$ as $\Sigma_{sort} = \{\, \texttt{i>0}\,, \texttt{j:=1}\,, \texttt{j<i}\,, \texttt{j++}\,, \texttt{j>=i}\,, \texttt{i--}\,\}$ and $\pi = \texttt{j<i}\,\texttt{j:=1}\,.(\,\texttt{j:=1}\,\texttt{j++}\,\texttt{j:=1}\,)^\omega$ is an ω-trace.

We call an ω-trace *terminating* if it does not correspond to any possible execution (i.e., if there is no starting state such that all statements in the trace can be executed). The ω-traces $(\,\texttt{x<0}\,\texttt{x:=1}\,)^\omega$ and $(\,\texttt{x>=0}\,\texttt{x--}\,)^\omega$ are terminating. In the first one, already the finite prefix $\texttt{x<0}\,\texttt{x:=1}\,\texttt{x<0}$ does not correspond to any possible execution. In the second, every finite prefix has a possible execution (for a prefix of length $2n$, take a starting state where x is greater than $n-1$).

As before, a program is represented as a control flow graph whose edges are labeled with statements (one node is singled out as the initial nodes; here, there are no error nodes). We may view a program $\mathcal{P} = \langle \mathsf{Loc}, \delta, \ell_{\mathsf{init}} \rangle$ as a *Büchi automaton* where every state is a final state. We call the program $\mathcal{P}$ *terminating* if each of its ω-traces is terminating.

We define a *module* to be a restricted form of Büchi automaton which has exactly one final state. A Büchi automaton of this form recognizes an ω-regular language of the form $U.V^\omega$, where U and V are regular languages over the alphabet of statements $U, V \subseteq \Sigma^*$.

A *fair ω-trace* of a module $\mathcal{P}$ is an ω-trace that labels a fair path in the graph of $\mathcal{P}$, i.e., a path that visits the distinguished location ℓ_{fin} infinitely often. We call the module $\mathcal{P}$ *terminating* if each of its fair ω-traces is terminating. A *non-fair* ω-trace of a terminating module (i.e., an ω-trace that labels a path in its control flow graph without satisfying the fairness constraint) can be non-terminating.

We define a *certified module* to be a module that is equipped with a termination argument. The termination argument consists of two parts: a ranking function and an annotation of the module's location with assertions that certify that the ranking function decreases every time the final location ℓ_{fin} is visited. The certificate ensures that the module is terminating (every fair ω-trace of the module terminates).

The figure on the right depicts a certified module ($\mathcal{P}_1^{\mathsf{sort}}$, $f, \mathcal{I}$) where f is the ranking function $f(i,j) = i - j$ and $\mathcal{I}$ is the mapping of locations to predicates indicated by writing the predicate beneath the location.

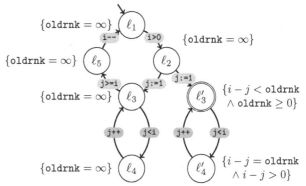

The variables oldrnk is an auxiliary variable whose value is the value of the ranking function at the previous visit of the final location. We note that for each transition $(\ell, \mathit{st}, \ell')$ the corresponding triple $\{\mathcal{I}(\ell)\}$ st $\{\mathcal{I}(\ell')\}$ is a valid Hoare triple (with the understanding that outgoing transitions of final states implicitly assign $\mathsf{oldrnk} := f(i, j)$).

4 Recursion: Nested Word Automata

A new verification method for recursive programs is based on the theory of nested words [1]. The verification method constructs a nested word automaton from an inductive sequence of "nested interpolants", i.e., an inductive annotation for the "nested trace" of the recursive program with assertions. Such an annotation may come from an interpolating SMT solver such as [2].

The theory of *nested word automata* offers an interesting potential as an alternative to the low-level view of a recursive program as a stack-based device that defines a set of traces. A nested word expresses not only the linear order of a trace but also the nesting of calls and returns. Regular languages of nested words enjoy the standard properties of regular language theory, of which we will use the closure under intersection and complement, and the decidability of emptiness [1].

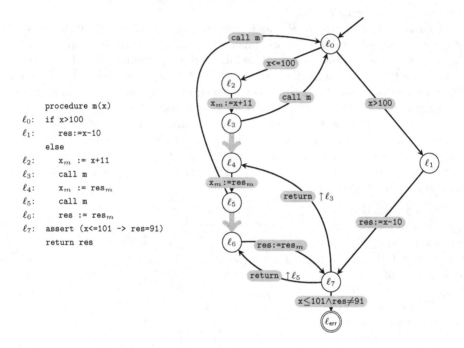

```
        procedure m(x)
ℓ₀:  if x>100
ℓ₁:     res:=x-10
        else
ℓ₂:     xₘ := x+11
ℓ₃:     call m
ℓ₄:     xₘ := resₘ
ℓ₅:     call m
ℓ₆:     res := resₘ
ℓ₇:  assert (x<=101 -> res=91)
        return res
```

Fig. 7. McCarthy's 91 function with correctness specification given as pseudocode and recursive control flow graph $\mathcal{P}^{91}$. The program is correct if the assert statement never fails resp. if there is no feasible trace from the initial location ℓ_0 to the error location ℓ_{err}. In a different reading, the graph presents a nested word automaton, the *control automaton* $\mathcal{P}^{91}$.

Figure 7 shows an implementation of McCarthy's 91 function,

$$m(x) = \begin{cases} x - 10 & \text{if } x > 100 \\ m(m(x + 11)) & \text{if } x \le 100 \end{cases}$$

together with the correctness specification (if the argument x is not greater than 101, the function returns 91).

Following [19], we present a recursive program formally as a *recursive control flow graph*; see Figure 7 for an example. Each node is a program location ℓ. Each edge is labeled with a statement s, which is either an assignment $\boxed{\texttt{y:=t}}$, an assume $\boxed{\varphi}$, a call $\boxed{\texttt{call p}}$, or a return $\boxed{\texttt{return p}}$. We note that transitions labeled with `return p` have two predecessors. First the exit location of the called procedure, second the location of the corresponding procedure call. In Figure 7 we label edges additionally with $\uparrow\ell$ (reminiscent to pop transitions in a pushdown automaton) to denote the location of the corresponding call transition.

Following [1], a *nested word* over an alphabet Σ is a pair $(w, \leadsto)$ consisting of a word $w = a_0 \ldots a_{n-1}$ over the alphabet Σ and the *nesting relation* $\leadsto$ (a binary relation between the n positions of w). We can use the nesting relation $i \leadsto j$ to express that i is the position of a call and j the position of the matching return.

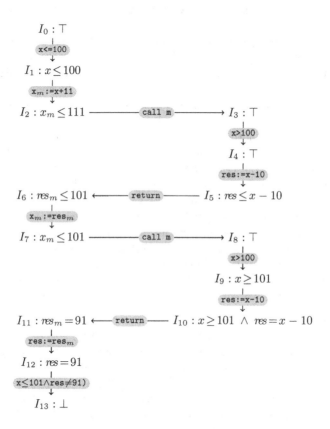

Fig. 8. Error trace of $\mathcal{P}^{91}$ in Figure 7 annotated with an inductive sequence of state assertions that prove infeasibility of this trace

In Figure 8, we present an error trace (a nested word accepted by $\mathcal{P}^{91}$) that is infeasible. The trace is interleaved with an inductive sequence of state assertions that proves infeasibility of this trace. The sequence of state assertions is modular in the sense that each state assertion describes only local states of the current calling context.

The global state of the program can be obtained using the nesting relation of the nested word. We call such a sequence of state assertions a sequence of nested interpolants. A method that uses Craig interpolation to compute a sequence of nested interpolants is presented in [11]. We note that we cannot apply Craig interpolation directly to a nested trace, because then variables of parent contexts may occur in the current context. Using nested interpolants and nested word automata we can use the scheme presented in Section 2 to analyze programs with procedures in a modular way.

5 Concurrent Programs: Alternating Finite Automata

In principle, one could apply the method developed in Section 2 to verify concurrent programs with shared memory. If each thread of a program is represented as an NFA, then their Cartesian product gives an NFA which recognizes the set of interleaved traces of the program. The challenge posed by concurrency is that the size of this Cartesian product is exponential in the number of threads, and the number of interleaved traces is greater still.

In [4], we propose a method for overcoming the exponential explosion problem using a novel proof system which is based on the notion of an *inductive data flow graph (iDFG)*. An iDFG is a data flow graph with incorporated inductive assertions. It accounts for a set of dependencies between data operations in interleaved traces. It stands as a representation for the set of traces which give rise to these dependencies, and acts as certificate that each of these traces is infeasible. This set of traces can be recognized by an *alternating finite automaton (AFA)*, enabling the reduction of the iDFG proof checking problem to a language inclusion problem for AFA. This problem suffers from high worst-case complexity (PSPACE-complete), but this is vastly superior to the exponential space complexity (not just in the worst case) of constructing the Cartesian product.

We will use the Ticket mutual exclusion protocol as an example to illustrate iDFGs. The program has two global variables, t and s, representing a ticket counter and service counter, respectively. We suppose that the protocol is executed by three threads (Threads 1, 2, and

> Thread i
> $\ell_{i,1}:$ $m_i := t++$
> $\ell_{i,2}:$ $[m_i \leq s]$
>   // critical section
> $\ell_{i,3}:$ $s := s + 1$

3), where each Thread i is executing the sequence of three instructions shown to the right. The program begins in a state where s and t are both zero. To execute the protocol, a thread first acquires its (unique ticket) and stores it in the local variable m_i ($\ell_{i,1}$), then waits until the service counter reaches its ticket to enter its critical section ($\ell_{i,2}$), and then finally leaves its critical section by incrementing the service counter ($\ell_{i,3}$). The property we wish to prove is mutual exclusion: no two threads may be in the critical section at the same time. We accomplish this by proving that every trace which *violates* mutual exclusion is infeasible.

One trace of the program which violates mutual exclusion (Thread 2 and Thread 3 both end in their critical sections) is pictured (on the left hand side) in Figure 9, along with a Hoare-style proof of its infeasibility. To its right is an iDFG, which represents the *essence* of this proof: the trace is infeasible because Thread 3 enters its critical section when it is Thread 2's turn to do so. The iDFG abstracts away the details of the Hoare-style proof which are irrelevant to this essential argument, such as the relative order between the events s++ and m₃ := t++, or whether the events $[m_1 \leq s]$ or $[m_2 \leq s]$ occur at all. The graph is labeled with program assertions on each edge, where each incoming edge represents a pre-condition, and each outgoing edge a post-condition. Bifurcation in the graph represents pre-conditions which can potentially be achieved in parallel. For example, consider the two incoming edges to $[m_3 \leq s]$: it does not matter in which order the pre-conditions $\{s = 1\}$ and $\{m_3 > 1\}$ are achieved; as long as both hold when $[m_3 \leq s]$ is executed, then the resulting state will satisfy the post-condition $\{false\}$. The assertions are *inductive* in the sense that each node corresponds to a valid Hoare triple, where the pre-condition is the conjunction of the labels of all incoming edges, and the post-condition is the conjunction of the labels of all outgoing edges.

Each edge in the iDFG represents a constraint on the traces which are recognized by the iDFG. For example, the edge $\text{s++} \xrightarrow{\{s=1\}} [m_3 \leq s]$ indicates that s++ must appear before $[m_3 \leq s]$ in the trace, and every instruction which appears in between must leave the assertion $\{s = 1\}$ invariant. A trace is recognized by the iDFG when it satisfies all of these constraints. The inductiveness condition for the assertion labels ensures that every trace which is recognized by the iDFG is infeasible.

A more operational view of the language of traces recognized by an iDFG can be given by translation into an AFA. AFAs may be understood as a generalization of nondeterministic finite automata. We may think of NFAs as having a transition function which maps each state and letter to a disjunction of states, with the interpretation that at least one of them must lead to an accepting state for the input word to be accepted. AFAs generalize this by also allowing *conjunctions* of states, with the interpretation that *all* states must lead to an accepting state for the input word to be accepted (i.e., the transition function maps each state and letter to a (positive) propositional formula where the propositions are states).

For any iDFG we may construct an AFA that recognizes the set of all traces τ such that the *reversal* of τ is recognized by the iDFG. Each assertion in the iDFG corresponds to a state of the AFA and each iDFG node corresponds to an AFA transition. Since the AFA accepts the reversed language, each node in the iDFG should be read as a *backwards* transition. This allows the bifurcation in the iDFG to be interpreted using conjunction. For example, iDFG node labeled $[m_3 \leq s]$ indicates that starting in the state $\{false\}$, we may read the letter $[m_3 \leq s]$ and transition to *both* $\{s = 1\}$ and $\{m_3 > 1\}$, and must accept along each path. More explicitly, the transition rule corresponding to this vertex is as follows:

$$\delta(\{false\}, [m_3 \leq s]) = \{s = 1\} \wedge \{m_3 > 1\} \ .$$

A complete iDFG proof for the 3-thread Ticket protocol is given in Figure 10. This iDFG illustrates the need for disjunction as well as conjunction. Consider that there are two nodes of the iDFG which are labeled $[m_3 \leq s]$ and which

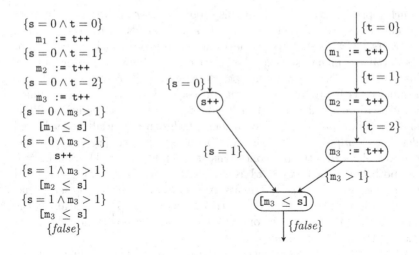

Fig. 9. Example trace with a Hoare-style proof and iDFG proof

have {*false*} as a post-condition. This means that there are *two* transition rules corresponding to reading the letter [$m_3 \leq s$] at the state {*false*}. The transition rules can be combined by disjunction, yielding the following transition function:

$$\delta(\{false\}, [m_3 \leq s]) = (\{s = 0\} \wedge \{m_3 > 0\}) \vee (\{s = 1\} \wedge \{m_3 > 1\}) .$$

The main appeal of iDFGs is that they are succinct proof objects for concurrent programs. Generalizing the Ticket example to N threads, the iDFG proof has $O(N^2)$ vertices, while the product control flow graph has $O(3^N)$. In [4], we make the claim of succinctness more general and formal by defining a measure of *data complexity* and showing that iDFG proofs are polynomial in this measure. Intuitively, this succinctness is possible because iDFGs represent only the *data flow* of the program, and abstract away control features that are irrelevant to the proof. This approach shifts the burden of the exponential explosion incurred by concurrency towards the check whether all program traces are represented, which is an automata-theoretic problem.

6   Unbounded Parallelism: Predicate Automata

The preceding section discusses a method for attacking the problem that the size of the automaton for a concurrent program is exponential in the number of threads. For many programs (filesystems, device drivers, web servers, ...), the number of threads is not statically known, or may increase without bound during the course of the program's execution. For such a program, the Cartesian product is *infinite* (as is the alphabet of program instructions), and the set of program traces is not a regular language. Thus, the problem of unbounded parallelism

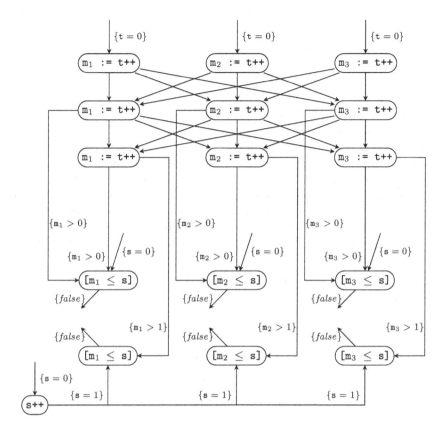

Fig. 10. Complete iDFG proof for the 3-thread Ticket protocol

is not merely one of high complexity, and we must develop new technology to address it.

In [6], we present *proof spaces*, a proof system which generalizes iDFGs to allow unboundedly many threads. The proof checking problem for proof spaces is carried out using *predicate automata*, which are an infinite-state (and infinite-alphabet) generalization of alternating finite automata.

We will start by demonstrating proof spaces on a simple example. Consider a program in which an arbitrary number of threads concurrently execute the code below. The goal is to verify that, if $g \geq 1$ holds initially, then it will always hold (regardless of how many threads are executing).

```
global int g
local  int x
1: x := g;
2: g := g+x;
```

Consider the set of the Hoare triples (A) - (D) given below.

$$(\text{A}) \qquad\qquad \{g \geq 1\}\ \langle x := g : 1\rangle \qquad \{x(1) \geq 1\}$$
$$(\text{B}) \quad \{g \geq 1 \wedge x(1) \geq 1\}\ \langle g := g + x : 1\rangle\ \{g \geq 1\}$$
$$(\text{C}) \qquad\qquad \{g \geq 1\}\ \langle x := g : 1\rangle \qquad \{g \geq 1\}$$
$$(\text{D}) \qquad\qquad \{x(1) \geq 1\}\ \langle x := g : 2\rangle \qquad \{x(1) \geq 1\}$$

Here we use $x(1)$ to refer to Thread 1's copy of the local variable x, and $\langle x := g : 1\rangle$ to indicate the instruction x := g executed by Thread 1.

The question of how such Hoare triples can be generated automatically is discussed in more detail in [6]; for our present purposes, we suppose that they are received from an oracle. We pose the question: given a set of ordinary Hoare triples (of the type one might expect to generate using sequential verification techniques), *what can we do with them?* We consider a deductive system in which these triples are taken as axioms, and the only rules of inference are *sequencing, symmetry,* and *conjunction.* These rules are easily illustrated with concrete examples:

- *Sequencing* composes two Hoare triples sequentially. For example, sequencing (A) and (D) yields

$$(\text{A} \circ \text{D}) \quad \{g \geq 1\}\ \langle x := g : 1\rangle\langle x := g : 2\rangle\ \{x(1) \geq 1\}$$

- *Symmetry* permutes thread identifiers. For example, renaming (A) and (C) (mapping $1 \mapsto 2$) yields

$$(\text{A'})\ \{g \geq 1\}\ \langle x := g : 2\rangle\ \{x(2) \geq 1\}$$
$$(\text{C'})\ \{g \geq 1\}\ \langle x := g : 2\rangle\ \{g \geq 1\}$$

and, renaming (D) (mapping $1 \mapsto 2$ and $2 \mapsto 1$) yields

$$(\text{D'})\ \{x(2) \geq 1\}\ \langle x := g : 1\rangle\ \{x(2) \geq 1\}$$

- *Conjunction* composes two Hoare triples by conjoining pre- and postconditions. For example, conjoining (A') and (C') yields

$$(\text{A'} \wedge \text{C'})\ \{g \geq 1\}\ \langle x := g : 2\rangle\ \{g \geq 1 \wedge x(2) \geq 1\}$$

and conjoining (A) and (D') yields (A $\wedge$ D')

$$\{g \geq 1 \wedge x(2) \geq 1\}\ \langle x := g : 1\rangle\ \{x(1) \geq 1 \wedge x(2) \geq 1\}$$

Naturally, the deductive system may apply inference rules to deduced Hoare triples as well: for example, by sequencing (A' $\wedge$ C') and (A $\wedge$ D'), we get the Hoare triple

$$\{g \geq 1\}\ \langle x := g : 2\rangle\langle x := g : 1\rangle\ \{x(1) \geq 1 \wedge x(2) \geq 1\}$$

A *proof space* is a set of valid Hoare triples which is closed under sequencing, symmetry, and conjunction (that is, it is a *theory* of this deductive system). Any finite set of valid Hoare triples generates an infinite proof space by considering those triples to be axioms and taking their closure under deduction; we call such a finite set of Hoare triples a *basis* for the generated proof space. Fixing a pre-condition φ_{pre} and a post-condition φ_{post} (for instance, taking both to be $\mathsf{g} \geq 1$ for our example), a proof space can be said to recognize all of those traces τ such that $\{ \varphi_{\mathsf{pre}} \} \tau \{ \varphi_{\mathsf{post}} \}$ belongs to the space.

As with iDFGs, we can give a more operational view of the traces recognized by a proof space using automata. For this purpose, we developed the notion of *predicate automata* (PA), an infinite-state, infinite-alphabet generalization of alternating finite automata (closely related to *alternating register automata* [3,7,15,17]). If one conceives of alternating finite automata as the automata of propositional logic, then predicate automata may be thought of as the automata for first-order logic. A PA A is equipped with a finite vocabulary of *predicates*, and its states are propositions over this vocabulary (i.e., if p is a binary predicate symbol of A, then $p(1, 2)$ is a state of A). The transition function of a PA maps each predicate symbol and letter to a positive Boolean formula over its vocabulary. For example, the transition

$$\delta(p(i,j), a : k) = (p(i,j) \wedge i \neq k) \vee (q(i) \wedge q(j) \wedge i = k)$$

indicates that, if the PA is at state $p(1, 2)$ and reads $a : 2$, then it transitions to $p(1, 2)$; if it then reads $a : 1$, then it transitions to *both* the state $q(1)$ and $q(2)$..

From a finite basis B of Hoare triples, we may construct a predicate automaton which recognizes the same traces as the proof space generated by B. Each n-thread assertion which appears in the basis corresponds to an n-ary predicate, and each Hoare triple in the basis corresponds to a transition. For example, the Hoare triple (B) corresponds to the PA transition

$$\delta(\{\mathsf{g} \geq 1\}, \mathsf{g} := \mathsf{g} + \mathsf{x} : k) = \{\mathsf{g} \geq 1\} \wedge \{\mathsf{x}(k) \geq 1\}$$

(where $\{\mathsf{g} \geq 1\}$ is a nullary predicate and $\{\mathsf{x}(k) \geq 1\}$ is a unary predicate).

The proof checking problem for proof spaces reduces to the inclusion problem for PA. Although this problem is undecidable in general, [6] gives a semi-algorithm which is a decision procedure for the special case of PAs where each predicate symbol in its vocabulary has arity at most one.

7 Proofs that Count: Petri Nets

Consider the program that consists of an arbitrary number of threads whose control flow graph is pictured below. The (global) integer variables s and t are initially 0. The task is to automatically construct a proof that the error location ℓ_{error} is unreachable (i.e., the program satisfies the specification $s = t = 0/false$). This deceptively simple property is surprisingly difficult to prove correct using automated techniques.

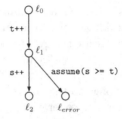

We illustrate the difficulty of proving this example by informally applying the technique from Section 2. We begin by sampling an error trace from the program, say (a trace that involves two threads)

$$\tau = \texttt{t++; t++; s++; assume(s >= t)}$$

A correctness proof for τ is a sequence of intermediate assertions, shown below in Figure 11(a). We may generalize the proof to apply to a language of traces, as shown in the NFA in Figure 11(b).

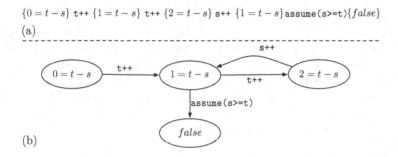

Fig. 11. Proof for the sample trace τ

This automaton does not yet accept every trace of the program. We could continue by sampling a new trace, for instance

$$\tau' = \texttt{t++; t++; t++; s++; assume(s >= t)},$$

but it is already clear that this strategy is doomed to fail. There is no regular language which contains all the program traces and which does not contain incorrect traces. Similarly, there is no finitely-generated proof space which proves the correctness of every trace.

A *counting argument* (in the context of formal methods) is a program proof that makes use of one or more *counters*, which are not part of the program itself, but which are useful for abstracting program behavior. One informal argument for correctness is as follows: a global, inductive invariant for this program is that the number of threads at line ℓ_1 (i.e., after executing $\texttt{t++}$ but before executing s++), let us call this k, is equal to the difference $t - s$. Since the number of threads at line ℓ_1 is non-negative, we must always have $s \leq t$, and ℓ_{error} must be unreachable. This counting argument is clear and simple to our human intuition, but *how can we take this intuition and formalize it into a mechanically constructed proof?* This question was investigated and answered in [5].

Our solution to this problem is pictured to the right. This *counting proof* consists of a counting automaton A (a kind of restricted counter machine) paired with an annotation φ mapping the states of A to assertions. The counting automaton A is a finite automaton equipped with a $\mathbb{N}$-valued counter denoted k (initially 0). Each transition of the automaton is equipped with an action for k, which may be inc (increment the counter), dec (decrement, but block unless the

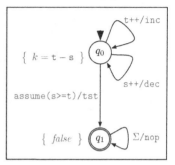

counter is ≥ 1), tst (block unless the counter is ≥ 1), or nop (do nothing). The annotation φ associates with each state of this automaton a formula over the program variables and the counter variable k. This annotation is *inductive* in the sense that each transition is associated with a valid Hoare triple: for example,

$$\{k = t - s\} \qquad \texttt{t++; k++} \qquad \{k = t - s\}$$
$$\{k = t - s\} \qquad \texttt{s++; k--} \qquad \{k = t - s\}$$
$$\{k = t - s\} \; \texttt{assume(s>=t); assume(k>=1)} \; \{false\}$$

A trace is accepted by A if it labels a path from q_0 (the initial state) to q_1 (the final state), and none of the counter actions block. Every trace which is accepted by A is associated with a sequence of assertions (thus proving its correctness). This sequence is obtained from the accepting run of A by taking, for each position in the run, the assertion at the current state with k replaced by its current value. For example, the proof for the trace τ above is as follows:

$$\{0 = t - s\} \quad \overset{\texttt{t++}}{\underset{\texttt{inc}}{\longrightarrow}} \quad \{1 = t - s\} \quad \overset{\texttt{t++}}{\underset{\texttt{inc}}{\longrightarrow}} \quad \{2 = t - s\} \quad \overset{\texttt{s++}}{\underset{\texttt{dec}}{\longrightarrow}} \quad \{1 = t - s\} \quad \overset{\texttt{assume(s>=t)}}{\underset{\texttt{tst}}{\longrightarrow}} \quad \{false\}$$
$$q_0, k = 0 \qquad\qquad q_0, k = 1 \qquad\qquad q_0, k = 2 \qquad\qquad q_0, k = 1 \qquad\qquad q_1, k = 1$$

This counting proof works not only for the trace τ, but for *every* trace of the program (that is, the proof is enough to show that ℓ_{error} is unreachable). The key to this proof is the use of the counter variable k, which counts the number of t++ statements in excess of s++ statements along a trace. Using this auxiliary counter allows us to make a simple, succinct argument for the correctness of this program.

The essential idea for constructing counting proofs is to encode the problem as an SMT query. Our encoding requires us to specify the "size" of the candidate proof to find (e.g., the number of states that may be used), and will always succeed if a proof of that size exists. The main insight behind our proof construction procedure is that by looking for *small* proofs, we can force an SMT solver to synthesize nontrivial counting arguments. For example, we can force an SMT

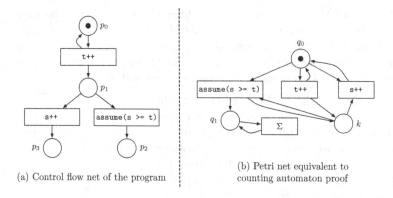

(a) Control flow net of the program

(b) Petri net equivalent to
counting automaton proof

Fig. 12. The language of Petri net (a) is included in the language of the deterministic Petri net (b)

solver to "discover" the need to count the number of t++ statements in excess of s++ statements in the proof above completely automatically, simply by asking for a proof with 2 states.

The idea behind proof checking is based on the observation that counting automata can be converted into deterministic labeled Petri nets (Figure 12(b)). Similarly, the language of program traces can be represented by a Petri net (Figure 12(a)). The final step of the correctness argument is performed by showing that the language of the Petri net for the program is included in the language of the deterministic Petri net for the counting automaton, a problem which is known to be decidable.

8   Conclusion

We have described several instances of a new approach to program verification which constructs and checks automata. We have shown that, in order to instantiate the approach for a specific verification problem, one has to come up with the appropriate notion of automaton, one has to define the construction of an automaton from the proof of a sequence of statements (i.e., a trace), one has to define the *program automaton* which recognizes the set of *error traces*, and one has to present an algorithm for solving the corresponding automata inclusion problem.

There are several interesting verification problems (timed systems, hybrid systems, game-theoretic properties, termination for unbounded parallelism, ...) where the question whether an appropriate notion of automaton exists, is still open and we do not know whether the approach can be instantiated.

Conversely, given a notion of automaton, one may ask whether there exists a verification problem for which this notion may be useful. For example, in some restricted cases in Section 5 it may be useful to define the denotation of an iDFG as a set of trees and replace alternating finite automata by tree automata.

An obvious topic for future research is the scalability of the automata operations used in the approach: the check of automata inclusion, minimization for the incremental construction of the difference automaton, etc.

Finally, the construction of an automaton from the proof of a sequence of statements may be interesting in settings other than verification. For example, given a failed test for a program with a bug, we can again construct an automaton and use the automaton for the *diagnosis* of the bug [20].

Acknowledgements. We would like to thank Jürgen Christ for comments and discussions.

References

1. Alur, R., Madhusudan, P.: Adding nesting structure to words. JACM **56**(3) (2009)
2. Christ, J., Hoenicke, J., Nutz, A.: Proof tree preserving interpolation. In: Piterman, N., Smolka, S.A. (eds.) TACAS 2013. LNCS, vol. 7795, pp. 124–138. Springer, Heidelberg (2013)
3. Demri, S., Lazić, R.: LTL with the freeze quantifier and register automata. ACM Trans. Comput. Logic **10**(3), 16:1–16:30 (2009)
4. Farzan, A., Kincaid, Z., Podelski, A.: Inductive data flow graphs. In: POPL, pp. 129–142. ACM (2013)
5. Farzan, A., Kincaid, Z., Podelski, A.: Proofs that count. In: POPL, pp. 151–164. ACM (2014)
6. Farzan, A., Kincaid, Z., Podelski, A.: Proof spaces for unbounded parallelism. In: POPL. ACM (2015)
7. Figueira, D.: Alternating register automata on finite words and trees. Logical Methods in Computer Science **8**(1) (2012)
8. Heizmann, M., et al.: Ultimate automizer with unsatisfiable cores (competition contribution). In: Ábrahám, E., Havelund, K. (eds.) TACAS 2014. LNCS, vol. 8413, pp. 418–420. Springer, Heidelberg (2014)
9. Heizmann, M., Hoenicke, J., Leike, J., Podelski, A.: Linear ranking for linear lasso programs. In: Van Hung, D., Ogawa, M. (eds.) ATVA 2013. LNCS, vol. 8172, pp. 365–380. Springer, Heidelberg (2013)
10. Heizmann, M., Hoenicke, J., Podelski, A.: Refinement of trace abstraction. In: Palsberg, J., Su, Z. (eds.) SAS 2009. LNCS, vol. 5673, pp. 69–85. Springer, Heidelberg (2009)
11. Heizmann, M., Hoenicke, J., Podelski, A.: Nested interpolants. In: POPL, pp. 471–482. ACM (2010)
12. Heizmann, M., Hoenicke, J., Podelski, A.: Software model checking for people who love automata. In: Sharygina, N., Veith, H. (eds.) CAV 2013. LNCS, vol. 8044, pp. 36–52. Springer, Heidelberg (2013)
13. Heizmann, M., Hoenicke, J., Podelski, A.: Termination analysis by learning terminating programs. In: Biere, A., Bloem, R. (eds.) CAV 2014. LNCS, vol. 8559, pp. 797–813. Springer, Heidelberg (2014)
14. Junker, M., Huuck, R., Fehnker, A., Knapp, A.: SMT-based false positive elimination in static program analysis. In: Aoki, T., Taguchi, K. (eds.) ICFEM 2012. LNCS, vol. 7635, pp. 316–331. Springer, Heidelberg (2012)

15. Kaminski, M., Francez, N.: Finite-memory automata. Theor. Comput. Sci. **134**(2), 329–363 (1994)
16. Leike, J., Heizmann, M.: Ranking templates for linear loops. In: Ábrahám, E., Havelund, K. (eds.) TACAS 2014. LNCS, vol. 8413, pp. 172–186. Springer, Heidelberg (2014)
17. Neven, F., Schwentick, T., Vianu, V.: Finite state machines for strings over infinite alphabets. ACM Trans. Comput. Logic **5**(3), 403–435 (2004)
18. Podelski, A., Rybalchenko, A.: A complete method for the synthesis of linear ranking functions. In: Steffen, B., Levi, G. (eds.) VMCAI 2004. LNCS, vol. 2937, pp. 239–251. Springer, Heidelberg (2004)
19. Reps, T.W., Horwitz, S., Sagiv, S.: Precise interprocedural dataflow analysis via graph reachability. In: POPL, pp. 49–61. ACM (1995)
20. Schäf, M., Schwartz-Narbonne, D., Wies, T.: Explaining inconsistent code. In: ESEC/FSE, pp. 521–531. ACM (2013)

Hankel Matrices: From Words to Graphs
(Extended Abstract)

Johann A. Makowsky[✉] and Nadia Labai

Department of Computer Science, Technion - Israel Institute
of Technology, Haifa, Israel
{janos,nadia}@cs.technion.ac.il

Abstract. We survey recent work on the use of Hankel matrices $H(f, \square)$ for real-valued graph parameters f and a binary sum-like operation $\square$ on labeled graphs such as the disjoint union and various gluing operations of pairs of laeled graphs. Special cases deal with real-valued word functions. We start with graph parameters definable in Monadic Second Order Logic MSOL and show how MSOL-definability can be replaced by the assumption that $H(f, \square)$ has finite rank. In contrast to MSOL-definable graph parameters, there are uncountably many graph parameters f with Hankel matrices of finite rank. We also discuss how real-valued graph parameters can be replaced by graph parameters with values in commutative semirings.

In this talk we survey recent work done together with the first author's former and current graduate students B. Godlin, E. Katz, T. Kotek, E.V. Ravve, and the second author on the definability of word functions and graph parameters and their Hankel matrix. There are three pervasive themes.

- Definability of word functions and graph parameters f in some logical formalism $\mathcal{L}$ which is a fragment of Second Order Logic SOL, preferably Monadic Second Order Logic MSOL, or CMSOL, i.e., $MSOL$ possibly augmented with modular counting quantifiers;
- Replacing the definability of f by the assumption that certain Hankel matrices have finite rank; and
- Replacing the field of real numbers $\mathbb{R}$ by arbitrary commutative rings or semirings $\mathcal{S}$.

1 Hankel Matrices

In linear algebra, a *Hankel matrix*, named after Hermann Hankel, is a square matrix with constant skew-diagonals. In automata theory, a *Hankel matrix* $H(f, \circ)$ is an infinite matrix where the rows and columns are labeled with words w over a fixed

---

J.A. Makowsky: Partially supported by a grant of Technion Research Authority.

N. Labai: Partially supported by a grant of the Graduate School of the Technion.

A.-H. Dediu et al. (Eds.): LATA 2015, LNCS 8977, pp. 47–55, 2015.
DOI: 10.1007/978-3-319-15579-1_3

alphabet Σ, and the entry $H(f, \circ)_{u,v}$ is given by $f(u \circ v)$. Here $f : \Sigma^* \to \mathbb{R}$ is a real-valued word function and $\circ$ denotes concatenation. A classical result of G.W. Carlyle and A. Paz [3] in automata theory characterizes real-valued word functions f recognizable by weighted (aka multiplicity) automata (WA-recognizable) in algebraic terms:

Theorem 1 (G.W. Carlyle and A. Paz, 1971).
A word function is WA-recognizable iff its Hankel matrix has finite rank.

Hankel matrices for graph parameters (aka connection matrices) were introduced by L. Lovász [33] and used in [14,34] to study real-valued partition functions of graphs. In [14,34] the role of concatenation is played by k-connections of k-graphs, i.e., graphs with $v_1, \ldots, v_k$ distinguished vertices. Given two k-graphs G, G' the k-connections $G \sqcup_k G'$ is defined by taking first the disjoint union of G and G' and then identifying corresponding labeled vertices. The Hankel matrix $H(f, \sqcup_k)$ is the infinite matrix where rows and columns are labeled by k-graphs and the entry $H(f, \sqcup_k)_{G,G'}$ is given by $f(G \sqcup_k G')$. We say that f has *finite connection rank* if all the matrices $H(f, \sqcup_k)$ have finite rank.

Partition functions are defined by counting weighted homomorphisms, which in some way generalize weighted automata. Let $H = (V(H), E(H))$ be a fixed graph, and let $\alpha : V(H) \to \mathbb{R}$ and $\beta : E(H) \to \mathbb{R}$ be real-valued functions (weights). For a graph $G = (V(G), E(G))$ we define

$$Z_{H,\alpha,\beta}(G) = \sum_{h:G \to H} \prod_{(v) \in V(G)} \alpha(h(v)) \cdot \prod_{(u,v) \in E(G)} \beta(h(u), h(v))$$

M. Freedman, L. Lovász and A. Schrijver [14] give the following characterization of partition functions:

Theorem 2 (M. Freedman, L. Lovász and A. Schrijver, 2007).
A real-valued graph parameter f can be presented as a partition function

$$f(G) = Z_{H,\alpha,\beta}(G)$$

for some H, α, β, iff all its connection matrices $H(f, \sqcup_k)$ have finite rank and are positive definite.

In [34] many variations of this theorem are discussed using different notions of connections of labeled graphs.

2   Definability in MSOL via Guiding Examples

Second Order Logic SOL allows quantification over vertices, edges and relations thereof. Monadic Second Order Logic MSOL allows quantification only over unary relations over the universe. In the case of graphs we have to distinguish between graphs G as structures where the universe $V(G)$ consists of vertices only and edges are given by the edge relation $E(G)$, and hypergraphs, where the universe consists of vertices and edges, and the hyperedges are given by an incidence relation $R(G)$. The notion of *definability of graph parameters and graph polynomials in* SOL *and* MSOL was first introduced in [7] and extensively studied in [12,17,18,22,24–26,36]. Later these studies also included MSOL augmented by modular counting quantifiers $D_{m,k} x\ \Phi(x)$ which assert that there are, modulo m, exactly k elements satisfying ϕ. This logic is denoted by CMSOL.

The set of real-valued graph parameters definable in SOL (MSOL, CMSOL) is denoted by SOLEVAL (MSOLEVAL, CMSOLEVAL), as they are evaluations of SOL-definable (MSOL, CMSOL-definable) graph polynomials. Words are treated here as special cases of labeled graphs.

Our first examples use *small*, i.e., polynomial sized sums and products:

(i) The cardinality $|V|$ of V is FOL-definable by

$$|V| = \sum_{v \in V} 1$$

(ii) The number of connected components of a graph G, $k(G)$ is MSOL-definable by

$$k(G) = \sum_{C \subseteq V:\text{component}(C)} 1$$

where component(C) says that C is a connected component.

(iii) The graph polynomial $X^{k(G)}$ is MSOL-definable by

$$X^{k(G)} = \prod_{c \in V:\text{first}-\text{in}-\text{comp}(c)} X$$

if, in addition, we have a linear order on the vertices and $first - in - comp(c)$ says that c is a first element in a connected component.

Our next examples use possibly *large*, i.e., exponential sized sums:

(iv) The number of cliques $\sharp\text{Clique}(G)$ in a graph is MSOL-definable by

$$\sharp\text{Clique}(G) = \sum_{C \subseteq V:\text{clique}(C)} 1$$

where clique(C) says that C induces a complete graph.

(v) Similarly "the number of maximal cliques" $\sharp\text{MClique}(G)$ is MSOL-definable by

$$\sharp\text{MClique}(G) = \sum_{C \subseteq V:\text{maxclique}(C)} 1$$

where maxclique(C) says that C induces a maximal complete graph.

(vi) The clique number of G, $\omega(G)$ is is SOL-definable by

$$\omega(G) = \sum_{C \subseteq V:\text{largest}-\text{clique}(C)} 1$$

where $largest - clique(C)$ says that C induces a maximal complete graph of largest size.

An inductive definition of a fragment $\mathcal{L}$ of SOLEVAL can be sketched as follows:

Definition 3. *Let $\mathcal{R}$ be a (polynomial) ring. A numeric graph parameter p : Graphs $\rightarrow \mathcal{R}$ is $\mathcal{L}$-definable if it can be defined inductively as follows:*

- *Monomials are of the form $\prod_{\bar{v}:\phi(\bar{v})} t$ where t is an element of the ring $\mathcal{R}$ and ϕ is a formula in $\mathcal{L}$ with first order variables $\bar{v}$.*
- *Polynomials are obtained by closing under small products, small sums, and large sums.*

Usually, *summation* is allowed over *second order variables*, whereas *products* are over *first order variables* only.

Our definition of SOLEVAL is somewhat reminiscent to the definition of *Skolem's Lower Elementary Functions*, [37–39].

3   The Finite Rank Theorem

In [18] the following Finite Rank Theorem is proved:

Theorem 4 (Finite Rank Theorem).
Let f be a real-valued graph parameter definable in CMSOL. Then f has finite connection rank.

The same holds for a wider class of Hankel matrices arising from sum-like binary operations on labeled graphs. A binary operation on labeled graphs is *sum-like* if it can be obtained from the disjoint union of two graphs by applying a quantifier-free scalar transduction, see e.g. [4,35].

If we consider words instead of graphs, also the converse holds for the Hankel matrix of concatenation, [28,29]:

Theorem 5 (NL and JAM, 2013). *A real-valued word function f is definable in MSOL iff its Hankel matrix for concatenation has finite rank.*

These results are reminiscent to results by [10], but their logical formalism differs from ours and was introduced later than MSOL-definability of graph parameters, [7].

The Finite Rank Theorem can also be used to show non-definability, [23,24], which gives a more convenient and versatile tool than the usual methods involving Ehrenfeucht-Fraïssé games.

4   Meta-Theorems Using Logic

The notions of path-width, tree-width and clique-width are the most used notions of width of graphs, cf. [20]. Widths are graph parameters with non-negative integer values. The exact definition is not needed here. What matters is that graphs can have unbounded width of either kind. Classes of bounded path-width have bounded tree-width, which in turn have bounded clique-width, but not conversely.

B. Courcelle's celebrated theorem for graph properties and graph classes of bounded tree-width [4,9] says that on graph classes of bounded tree-width, MSOL-definable graph properties can be decided in linear time.

In [5], [6, Theorem 4], [7, Theorem 31], this is extended to graph parameters and bounded clique-width:

Theorem 6 (B. Courcelle, JAM, and U. Rotics, 1998). *Let f be a CMSOL-definable graph parameter with values in a ring $\mathcal{R}$. Then f can be computed in polynomial time[1] on graph classes of bounded clique-width.*

As a generalization of graph classes of given tree-width or clique-width, the notion of CMSOL-inductive classes of graphs was introduced in [35]. Special cases of CMSOL-inductive classes are the *sum-like inductive classes*.

Definition 7 (Sum-like Inductive). *$\mathcal{C}$ is sum-like inductive if it is inductively defined using a finite set of basic labeled graphs $G_j, j \leq J$ and a finite set of sum-like binary operations $\square_i, i \leq I$. In other words, each $G_j, j \leq J$ is in $\mathcal{C}$, and whenever $H_1, H_2 \in \mathcal{C}$ then also $\square_i(H_1, H_2) \in \mathcal{C}$ for all $i \leq I$.*

The classes of graphs of fixed tree-width (path-width, clique-width) are all sum-like inductive, cf. [35]. Other examples of sum-like inductive classes of labeled graphs can be found using various graph grammars, cf. [15,16,35]. In the framework of sum-like inductive classes, Theorem 6 can be stated in model theoretic terms, [35, Theorem 6.6].

Theorem 8 (JAM, 2004/14).
Let $\mathcal{C}$ be sum-like inductive[2], and f be a graph parameter in CMSOLEVAL. Then the computation of $f(G)$ is Fixed Parameter Tractable[3] in the size of the parse tree witnessing that $G \in \mathcal{C}$.

5 Eliminating Logic

L. Lovász, in [34], also noted that Hankel matrices can be used to make Courcelle's Theorem *logic-free* for the case of bounded tree-width by replacing MSOL-definability by a finiteness condition on the rank of its connection matrices. In addition, graph parameters are allowed to take values in an arbitrary field $\mathcal{K}$.

---

[1] For real-valued graph parameters we have to be careful abut the model of computation. Either we work in a Turing computable subfield of $\mathbb{R}$, or we use the computational model of Blum-Shub-Smale BSS, cf. [1].

[2] Originally the theorem was stated for MSOL-smooth operations. The proof I had in mind in [35] only works for sum-like operations. However, it is not known whether there are MSOL-smooth operations which are not sum-like.

[3] A graph parameter is Fixed Parameter Tractable (FPT), if it can be computed in time $O(c(k) \cdot n^{d(k)})$ where n is the size of the graph, and $c(k), d(k)$ are functions depending on the parameter k, but independent of the size of the graph, cf. [9,13]. Here the parameter is hidden in the fact that $\mathcal{C}$ is CMSOL-inductive.

Theorem 9 (L. Lovász, 2007).
Let $\mathcal{K}$ be a field and let f be a $\mathcal{K}$-valued graph parameter with finite connection rank. Then f can be computed in linear time on graph classes of bounded tree-width.

In [30,31] this is extended to make Theorem 6 logic-free for the case of bounded clique-width. To do this one defines a suitable sum-like *binary* operation $\eta_{P,Q}$ on graphs with additional unary predicates $P(G), Q(G)$ on the vertices $V(G)$. $\eta_{P,Q}(G_1, G_2)$ is the disjoint union of G_1 and G_2 augmented with all the edges from

$$E_{P,Q} = \{(u,v) \in (V(G_1) \sqcup V(G_2))^2 : u \in P(G_1) \sqcup P(G_2) \text{ and } v \in Q(G_1) \sqcup Q(G_2)\}$$

In words $\eta_{P,Q}(G_1, G_2)$ is the disjoint union of G_1 and G_2 augmented with all the edges with one vertex in $P(G_1 \cup G_2)$ and one vertex in $Q(G_1 \cup G_2)$.

Theorem 10 (NL and JAM, 2014). *Let f be a real-valued graph parameter with $H(f, \eta_{P,Q})$ of finite rank. Then f can be computed in polynomial time on graph classes of bounded clique-width.*

In [27] this is further extended to make Theorem 8 also logic-free. A detailed discussion will appear in [32]. For this extension we introduce the notion of *linearly linked Hankel matrices*.

Definition 11 (Linearly Linked Hankel Matrices). *Let $\square_i, i \leq I$ be finitely many binary operations on labeled graphs, and let $G_j, j \leq J$ be a finite set of basic graphs. $p_k, k \leq K$ be finitely many real-valued graph parameters. For a labeled graph H let $\bar{p}(H)$ denote the vector $(p_1(H), \ldots, p_K(H))$.*

1. *$\mathcal{C}$ is inductively defined using $G_j, j \in J$ and $\square_i, i \leq I$ if each $G_j, j \in J$ is in $\mathcal{C}$, and whenever $H_1, H_2 \in \mathcal{C}$ then also $\square_i(H_1, H_2) \in \mathcal{C}$. Here $\square_i$ does not have to be sum-like.*
2. *The Hankel matrices $H(p_k, \square_i, i \leq I, j \leq J$ are linearly linked if the following hold:*
   (a) *For each $p_k, k \leq K$ and $\square_i, i \leq I$ the Hankel matrices $H(p_k, \square_i)$ are of finite rank.*
   (b) *For each $i \leq I$ there is a matrix P_i such that for all graphs H_1, H_2*

$$\bar{p}(\square_i(H_1, H_2)) = P_i \cdot \bar{p}(\square_1(H_1, H_2))$$

Theorem 12 (NL and JAM, 2014).
Let $\mathcal{C}$ be inductively defined using $G_j, j \in J$ and $\square_i, i \leq I$, and let $p_k, k \leq K$ be finitely many graph parameters, such that the Hankel matrices $H(p_k, \square_i), i \leq I, j \leq J$ are linearly linked. Then for graphs $H \in \mathcal{C}$ with parse-tree $pt(H)$, all the graph parameters $p_k, k \leq K$ can be computed in polynomial time in the size of $pt(H)$.

Theorem 12 is a proper generalization of Theorem 8:

Proposition 13. *If C is sum-like inductive using $G_j, j \in J$ and $\Box_i, i \leq I$, and $f \in$ CMSOLEVAL, there are finitely many graph parameters $p_1, \ldots, p_K$ such that all the Hankel matrices $H(p_k, \Box_i), i \leq I, j \leq J$ are linearly linked.*

To prove Proposition 13 one uses the Bilinear Reduction Theorem from [35], which is proven in full detail as [24, Theorem 8.7].

In the logical versions of these theorems there are only countably many CMSOL-definable graph parameters. However, there are uncountably many graph parameters with finite rank Hankel matrices even for the disjoint union of graphs. Hence, in contrast to the case of word functions, the finiteness assumption on the rank does not imply MSOL-definability. Furthermore, eliminating logic from these theorems allows us to separate the algebraic character of the proof from its logical part given by the Finite Rank Theorem for sum-like operations.

6 From Fields to Semirings

Finally, we discuss how to formulate Theorem 8 both logic-free and for graph parameters with values in a commutative semiring. A motivating example for this shift of perspective is the clique number $\omega(G)$ of a graph G, which has infinite connection rank over the reals, but finite row-rank in the tropical semiring $\mathcal{T}_{\max}$, the max-plus algebra defined over the reals. There are several notions of rank for matrices over commutative semirings. All of them coincide in the case of a field, and some of them coincide in the tropical case, [2, 8, 19].

In [30, 31] Lovász's Theorem is generalized to graph parameters with values in the tropical semirings rather than a field, and graph classes of bounded clique-width. There we work with two specific notions: row-rank in the tropical case, and a finiteness condition introduced by G. Jacob [21], which we call J-finiteness, in the case of arbitrary commutative semirings.

Theorem 14 (NL and JAM, 2014). *Let f be a graph parameter with values in $\mathcal{T}_{\max}$ with $H(f, \eta_{P,Q})$ of finite row rank. Then f can be computed in polynomial time on graph classes of bounded clique-width.*

In the case of graph parameters with values in arbitrary commutative semirings, this remains true for graph classes of bounded linear clique-width, cf. [11]. Linear clique-width relates to clique-width like path-width relates to tree-width.

References

1. Blum, L., Cucker, F., Shub, M., Smale, S.: Complexity and Real Computation. Springer (1998)
2. Butkovič, P.: Max-linear Systems: Theory and Algorithms. Springer Monographs in Mathematics. Springer (2010)
3. Carlyle, J.W., Paz, A.: Realizations by stochastic finite automata. J. Comp. Syst. Sc. **5**, 26–40 (1971)
4. Courcelle, B., Engelfriet, J.: Graph Structure and Monadic Second-order Logic, a Language Theoretic Approach. Cambridge University Press (2012)

5. Courcelle, B., Makowsky, J.A., Rotics, U.: Linear Time Solvable Optimization Problems on Graphs of Bounded Clique Width. In: Hromkovič, J., Sýkora, O. (eds.) WG 1998. LNCS, vol. 1517, pp. 1–16. Springer, Heidelberg (1998)
6. Courcelle, B., Makowsky, J.A., Rotics, U.: Linear time solvable optimization problems on graphs of bounded clique-width. Theory of Computing Systems **33**(2), 125–150 (2000)
7. Courcelle, B., Makowsky, J.A., Rotics, U.: On the fixed parameter complexity of graph enumeration problems definable in monadic second order logic. Discrete Applied Mathematics **108**(1–2), 23–52 (2001)
8. Cuninghame-Green, R.A., Butkovič, P.: Bases in max-algebra. Linear Algebra and its Applications **389, 107–120 (2004)**
9. Downey, R.G., Fellows, M.F., Parametrized Complexity. Springer (1999)
10. Droste, M., Gastin, P.: Weighted automata and weighted logics. Theor. Comput. Sci. **380**(1–2), 69–86 (2007)
11. Fellows, M.R., Rosamond, F.A., Rotics, U., Szeider, S.: Proving NP-hardness for clique width i: Non-approximability of linear clique-width. Electronic Colloquium on Computational Complexity (2005)
12. Fischer, E., Kotek, T., Makowsky, J.A.: Application of logic to combinatorial sequences and their recurrence relations. In: Grohe, M., Makowsky, J.A. (eds), Model Theoretic Methods in Finite Combinatorics, vol. 558. Contemporary Mathematics, pp. 1–42. American Mathematical Society (2011)
13. Flum, J., Grohe, M.: Parameterized complexity theory. Springer (2006)
14. Freedman, M., Lovász, L., Schrijver, A.: Reflection positivity, rank connectivity, and homomorphisms of graphs. Journal of AMS 20, 37–51 (2007)
15. Glikson, A., Verification of generally intractable graph properties on graphs generated by graph grammars. Master's thesis, Technion - Israel Institute of Technology, Haifa, Israel (2004)
16. Glikson, A., Makowsky, J.A.: *NCE* Graph Grammars and Clique-Width. In: Bodlaender, H.L. (ed.) WG 2003. LNCS, vol. 2880, pp. 237–248. Springer, Heidelberg (2003)
17. Godlin, B., Katz, E., Makowsky, J.A.: Graph polynomials: From recursive definitions to subset expansion formulas. Journal of Logic and Computation **22**(2), 237–265 (2012)
18. Godlin, B., Kotek, T., Makowsky, J.A.: Evaluations of Graph Polynomials. In: Broersma, H., Erlebach, T., Friedetzky, T., Paulusma, D. (eds.) WG 2008. LNCS, vol. 5344, pp. 183–194. Springer, Heidelberg (2008)
19. Guterman, A.E.: Matrix invariants over semirings. In: Hazewinkel, M. (ed.) Handbook of Algebra, vol. 6. Handbook of Algebra, pp. 3–33. North-Holland (2009)
20. Hlinený, P., Oum, S., Seese, D., Gottlob, G.: Width parameters beyond tree-width and their applications. Comput. J. **51**(3), 326–362 (2008)
21. Jacob, G.: Représentations et substitutions matricielles dans la théorie algébrique des transductions. PhD thesis, Université de Paris, VII (1975)
22. Kotek, T.: Definability of combinatorial functions. PhD thesis, Technion - Israel Institute of Technology, Haifa, Israel (March 2012)
23. Kotek, T., Makowsky, J.A.: Connection matrices and the definability of graph parameters. CSL 2012, pp. 411–425 (2012)
24. Kotek, T., Makowsky, J.A.: Connection matrices and the definability of graph parameters. Logical Methods in Computer Science 10(4) (2014)

25. Kotek, T., Makowsky, J.A., Ravve, E.V.: A computational framework for the study of partition functions and graph polynomials (abstract). In: Negru, V., et al. (eds) SYNASC 2012, Proceedings of the International Symposium on Symbolic and Numeric Algorithms for Scientific Computing, SYNASC, Timisoara, Romania, page in press. IEEE Computer Society (2013)

26. Kotek, T., Makowsky, J.A., Zilber, B.: On counting generalized colorings. In: Grohe, M., Makowsky, J.A. (eds) Model Theoretic Methods in Finite Combinatorics, vol. 558. Contemporary Mathematics, pp. 207–242. American Mathematical Society (2011)

27. Labai, N.: Hankel matrices and definability of graph parameters. Master's thesis, Technion - Israel Institute of Technology, Haifa, Israel (2015)

28. Labai, N., Makowsky, J.A.: Weighted automata and monadic second order logic. In: Proceedings Fourth International Symposium on Games, Automata, Logics and Formal Verification, GandALF 2013, Borca di Cadore, Dolomites, Italy, August 29-31, pp. 122–135 (2013)

29. Labai, N., Makowsky, J.A.: Weighted automata and monadic second order logic. arXiv preprint arXiv:1307.4472 (2013)

30. Labai, N., Makowsky, J.A.: Finiteness conditions for graph algebras over tropical semirings. arXiv preprint arXiv:1405.2547 (2014)

31. Labai, N., Makowsky, J.A.: Tropical graph parameters. In: 26th International Conference on Formal Power Series and Algebraic Combinatorics (FPSAC 2014), DMTCS Proceedings, pp. 357–368 (2014)

32. Labai, N., Makowsky, J.A.: to be determined. (in preparation, 2015)

33. Lovász, L.: Connection matrics. In: Grimmet, G., McDiarmid, C. (eds) Combinatorics, Complexity and Chance, A Tribute to Dominic Welsh, pp. 179–190. Oxford University Press (2007)

34. Lovász, L.: Large Networks and Graph Limits, vol. 60. Colloquium Publications. AMS (2012)

35. Makowsky, J.A.: Algorithmic uses of the Feferman-Vaught theorem. Annals of Pure and Applied Logic **126**(1-3), 159–213 (2004)

36. Makowsky, J.A., Kotek, T., Ravve, E.V.: A computational framework for the study of partition functions and graph polynomials. In: Proceedings of the 12th Asian Logic Conference 2011, pp. 210–230. World Scientific (2013)

37. Péter, R., Földes, I.: Recursive functions. Academic Press, New York (1967)

38. Rose, H.E.: Subrecursion: functions and hierarchies. Clarendon Press, Oxford (1984)

39. Skolem, T.: Proof of some theorems on recursively enumerable sets. Notre Dame J. Formal Logic **3**, 65–74 (1963)

Complexity Classes for Membrane Systems: A Survey

Giancarlo Mauri[✉], Alberto Leporati, Luca Manzoni,
Antonio E. Porreca, and Claudio Zandron

Dipartimento di Informatica, Sistemistica e Comunicazione,
Università degli Studi di Milano-Bicocca, Viale Sarca 336/14, 20126 Milano, Italy
{mauri,leporati,luca.manzoni,porreca,zandron}@disco.unimib.it

Abstract. The computational power of membrane systems, in their different variants, can be studied by defining classes of problems that can be solved within given bounds on computation time or space, and comparing them with usual computational complexity classes related to the Turing Machine model. Here we will consider in particular membrane systems with active membranes (where new membranes can be created by division of existing membranes). The problems related to the definition of time/space complexity classes for membrane systems will be discussed, and the resulting hierarchy will be compared with the usual hierarchy of complexity classes, mainly through simulations of Turing Machines by (uniform families of) membrane systems with active membranes.

1 Introduction

Membrane systems (or P systems) were introduced in [12] as a class of distributed parallel computing devices inspired by the functioning of living cells. The basic model consists of several *membranes*, hierarchically embedded in a main membrane called *skin membrane*. The membranes delimit *regions* and can contain *objects*, which evolve according to given *evolution rules* associated with the regions. Such rules are applied in a nondeterministic and maximally parallel manner: at each step, all the objects which can evolve should evolve. If we let the system evolve starting from a given initial configuration, we obtain a computation. The computation halts if no further rule can be applied, and the objects expelled through the skin membrane (or collected inside a specified output membrane) are the result of the computation.

An obvious question that arises concerns the definition of classes of (decision) problems that can be solved by P systems within given bounds of time/space, adapting to the framework of P systems notions from classical structural complexity theory, and the possibility of attacking computationally hard problems (e.g., **NP**-complete problems) using a polynomial amount of resources.

Many variants of P systems have been defined, adding further features or capabilities to the basic model (see, e.g., [14]). Here, we will focus on P systems with active membranes, introduced in [13] as a variant of P systems where the membranes play an active role in the computations: an electrical charge, that

© Springer International Publishing Switzerland 2015
A.-H. Dediu et al. (Eds.): LATA 2015, LNCS 8977, pp. 56–69, 2015.
DOI: 10.1007/978-3-319-15579-1_4

can be positive $(+)$, neutral (0), or negative $(-)$, is associated with each membrane, and the application of the evolution rules can be controlled by means of these electrical charges. Moreover, new membranes can be created during the computation by division of existing ones. This latter feature makes them extremely efficient from a computational complexity standpoint: using an exponentially increasing number of membranes that evolve in parallel, they can solve **NP**-complete and even **PSPACE**-complete problems [3,21] in polynomial time. On the other hand, if division of membranes is not allowed the efficiency apparently decreases: the so-called Milano theorem [23] tells us that no **NP**-complete problem can be solved in polynomial time without using division rules, unless **P** = **NP** holds. When division rules operate only on *elementary* membranes (i.e. membranes not containing other membranes), such systems are still able to efficiently solve **NP**-complete problems [15,23]. More recently, in [19] it was proved that all problems in $\mathbf{P}^{\mathbf{PP}}$ (a possibly larger class including the polynomial hierarchy) can also be solved in polynomial time using P systems with elementary membrane division.

A survey of results concerning time complexity classes for P systems can be found in [14], chapter 12.

Here we will instead focus mainly on *space* complexity. A measure of space complexity for P systems has been introduced in [16]: the space required by a P system is the maximal size it can reach during any legal computation, defined as the sum of the number of membranes and the number of objects. Under this notion of space complexity, in [18] it has been proved, by mutual simulation of P systems and Turing Machines, that the class of problems solvable in polynomial space by P systems with active membranes, denoted by **PMCSPACE$_{\mathbf{AM}}$**, coincides with **PSPACE**. This equivalence has been subsequently extended to exponential space [1], by proving that the class **EXPMCSPACE$_{\mathbf{AM}}$** of problems solvable in exponential space by P systems with active membranes coincides with **EXPSPACE**.

A further improvement has been given in [2], showing with different simulation techniques that arbitrary single-tape Turing Machines can be simulated by uniform families of P systems with a cubic slowdown and a quadratic space overhead; hence, the classes of problems solvable by P systems with active membranes and by Turing Machines coincide up to a polynomial with respect to space complexity. This equivalence allows us to adapt to P systems existing theorems about the space complexity of Turing machines, such as Savitch's theorem and the space hierarchy theorem.

The results recalled above show that P systems and Turing Machines are equivalent up to a polynomial from the space complexity point of view, assuming the available space is at least polynomial. In [5,8,20] P systems with stronger space restrictions, i.e. working in sublinear or even constant space, have been considered. Obviously, the definition of space must be revised, making the input read-only, and removing its size from the calculation of space, as usually done for Turing Machines working in sub linear space. A first result shows that **DLOGTIME**–uniform P systems with active membranes, using a logarithmic amount of space, are able to simulate logarithmic-space deterministic Turing machines, and thus to solve all problems in the class **L**. Then, in [8] it is pointed out that, while logarithmic-space Turing machines can only generate a polynomial number of distinct configurations, P systems working in logarithmic space have *exponentially* many potential ones, and thus they can be exploited to solve computational problems that are harder than those in **L**. In particular, polynomial-space Turing machines can be simulated by means of P systems with active membranes using only logarithmic auxiliary space, thus obtaining a characterization of **PSPACE**. Finally, in [5] it is proved that, quite surprisingly, a constant amount of space is sufficient (and trivially necessary) to solve all problems in **PSPACE**.

Another possibility is to consider P systems with restrictions on the available set of rules. An interesting case consists in limiting membrane division to elementary membranes (not containing further membranes). We can give a complete charac-terisation of the class of problems that can be solved by P systems with elementary division working in polynomial time, showing that it coincides with the class $\mathbf{P^{\#P}}$.

The paper is structured as follows. In Section 2 the basic definitions concerning P systems and the relevant complexity classes are briefly recalled. Section 3 contains the main results concerning the relationships between space complexity classes for P systems and Turing Machines, and corollaries obtained by re-interpreting some classic statements about Turing Machines in the membrane computing framework. The main results about systems working in sublinear space are then presented in Section 4 while Section 5 discusses the case of systems without non elementary divi-sion rules.

2   P Systems with Active Membranes

We recall here some basic definitions on P systems with active membranes; more details can be found in ([14], chapters 11–12).

Definition 1. *A P system with active membranes of initial degree $d \geq 1$ is a tuple* $\Pi = (\Gamma, \Lambda, \mu, w_1, \ldots, w_d, R)$, *where:*

- *Γ is an alphabet, i.e., a finite non-empty set of symbols, usually called* objects;
- *Λ is a finite set of* labels;
- *μ is a membrane structure (i.e., a rooted* unordered *tree, usually represented by nested brackets) consisting of d membranes enumerated by $1, \ldots, d$; furthermore, each membrane is labeled by an element of Λ, not necessarily in a one-to-one way, and possesses an* electrical charge *(or polarization), which can be either neutral (0), positive (+) or negative (−)*
- *$w_1, \ldots, w_d$ are strings over Γ, describing the initial multisets of objects placed in the d regions of μ;*
- *R is a finite set of rules of the following kinds:*
    - Object evolution rules, *of the form $[a \rightarrow w]_h^\alpha$ They can be applied inside a membrane labeled by h, having charge α and containing an occurrence of the object a; the object a is rewritten into the multiset w (i.e., a is removed from the multiset in h and replaced by the objects in w).*

- Send-in communication rules, *of the form* $a \left[\; \right]_h^\alpha \rightarrow [b]_h^\beta$
  They can be applied to a membrane labeled by h, having charge α and such that the external region contains an occurrence of the object a; the object a is sent into h becoming b and, simultaneously, the charge of h is changed to β.
- Send-out communication rules, *of the form* $[a]_h^\alpha \rightarrow \left[\; \right]_h^\beta b$
  They can be applied to a membrane labeled by h, having charge α and containing an occurrence of the object a; the object a is sent out from h to the outside region becoming b and, simultaneously, the charge of h is changed to β.
- Dissolution rules, *of the form* $[a]_h^\alpha \rightarrow b$
  They can be applied to a membrane labeled by h, having charge α and containing an occurrence of the object a; the membrane h is dissolved and its contents are left in the surrounding region unaltered, except that an occurrence of a becomes b.
- Elementary division rules, *of the form* $[a]_h^\alpha \rightarrow [b]_h^\beta [c]_h^\gamma$
  They can be applied to a membrane labeled by h, having charge α, containing an occurrence of the object a but having no other membrane inside (an elementary membrane); the membrane is divided into two membranes having label h and charge β and γ; the object a is replaced, respectively, by b and c while the other objects in the initial multiset are copied to both membranes.
- Nonelementary division rules, *of the form*

$$\left[\, \left[\;\right]_{h_1}^+ \cdots \left[\;\right]_{h_k}^+ \left[\;\right]_{h_{k+1}}^- \cdots \left[\;\right]_{h_n}^- \,\right]_h^\alpha \rightarrow \left[\, \left[\;\right]_{h_1}^\delta \cdots \left[\;\right]_{h_k}^\delta \,\right]_h^\beta \left[\, \left[\;\right]_{h_{k+1}}^\epsilon \cdots \left[\;\right]_{h_n}^\epsilon \,\right]_h^\gamma$$

They can be applied to a membrane labeled by h, having charge α, containing the positively charged membranes $h_1, \ldots, h_k$, the negatively charged membranes $h_{k+1}, \ldots, h_n$, and possibly some neutral membranes. The membrane h is divided into two copies having charge β and γ, respectively; the positive children are placed inside the former membrane, their charge changed to δ, while the negative ones are placed inside the latter membrane, their charges changed to ϵ. Any neutral membrane inside h is duplicated and placed inside both copies.

The class of P systems with active membranes will be denoted by $\mathcal{AM}$. We will also consider the subclass $\mathcal{AM}(-d, -n)$, whose elements do not use dissolution or nonelementary division rules. These P systems compute by applying the rules to iteratively modify their membrane structure and the multisets of objects contained in membranes, as defined below.

Definition 2. *Let Π be a P system with active membranes. Then:*

- *A* configuration *of Π is given by its current membrane structure, including the electrical charges, together with the multisets located in the corresponding regions. In particular, the* initial configuration *is given by the membrane structure μ, where all the membranes are neutral, and the initial contents $w_1, \ldots, w_d$, of its membranes.*
- *A* computation step *changes the current configuration according to the following principles:*

- *Each object and membrane can be subject to at most one rule per step, except for object evolution rules (inside each membrane any number of evolution rules can be applied simultaneously).*
- *The application of rules is* maximally parallel*: each object appearing on the left-hand side of evolution, communication, dissolution or elementary division rules must be subject to exactly one of them (unless the current charge of the membrane prohibits it). The same principle applies to each membrane that can be involved in communication, dissolution, elementary or nonelementary division rules. In other words, the only objects and membranes that do not evolve are those associated with no rule, or only to rules that are not applicable due to the electrical charges.*
- *When several conflicting rules can be applied in the same step, a single rule is chosen nondeterministically from the set of applicable rules. This nondeterministic choice occurs independently for each copy of the object in the membrane. This implies that multiple possible configurations can be reached as the result of a computation step.*
- *While all the chosen rules are considered to be applied simultaneously during each computation step, they are logically applied in a bottom-up fashion: first, all evolution rules are applied to the elementary membranes, then all communication, dissolution and division rules; then the application proceeds towards the root of the membrane structure. In other words, each membrane evolves only after its internal configuration has been updated.*
- *The outermost membrane cannot be divided or dissolved, and any object sent out from it cannot re-enter the system again.*
- A halting computation *of the P system Π is a finite sequence of configurations $\mathcal{C} = (\mathcal{C}_0, \ldots, \mathcal{C}_k)$, where $\mathcal{C}_0$ is the initial configuration, every $\mathcal{C}_{i+1}$ is reachable by $\mathcal{C}_i$ via a single computation step, and no rules can be applied anymore in $\mathcal{C}_k$. A non-halting* computation $\mathcal{C} = (\mathcal{C}_i : i \in \mathbb{N})$ *consists of infinitely many configurations, again starting from the initial one and generated by successive computation steps, where the applicable rules are never exhausted.*

P systems can be used as language *recognisers* by employing two distinguished objects **yes** and **no**; exactly one of these must be sent out from the outermost membrane in the last step of each computation, in order to signal acceptance or rejection, respectively; we also assume that all computations are halting. If all computations starting from the same initial configuration are accepting, or all are rejecting, the P system is said to be *confluent*. If this is not necessarily the case, then we have a *non-confluent* P system, and the overall result is established as for nondeterministic Turing Machines: it is acceptance iff at least an accepting computation exists.

In order to solve decision problems (i.e., decide languages), we use *families* of recogniser P systems $\Pi = \{\Pi_x : x \in \Sigma^\star\}$. Each input x is associated with a P system Π_x that decides the membership of x in the language $L \subseteq \Sigma^\star$ by accepting or rejecting. The mapping $x \mapsto \Pi_x$ must be efficiently computable for each input length, as discussed in detail in [10].

Definition 3. *Let $\mathcal{E}$, $\mathcal{F}$ be classes of functions over strings. A family of P systems $\boldsymbol{\Pi} = \{\Pi_x : x \in \Sigma^\star\}$ is said to be $(\mathcal{E}, \mathcal{F})$-uniform if the mapping $x \mapsto \Pi_x$ can be described by two functions $F \in \mathcal{F}$ (for "family") and $E \in \mathcal{E}$ (for "encoding") as follows:*

- *$F(1^n) = \Pi_n$, where n is the length of the input x and Π_n is a common P system for all inputs of length n, with a distinguished input membrane.*
- *$E(x) = w_x$, where w_x is a multiset encoding the specific input x.*
- *Finally, Π_x is simply Π_n with w_x added to the multiset placed inside its input membrane.*

In particular, a family $\boldsymbol{\Pi}$ is said to be $(\mathbf{L}, \mathbf{L})$-uniform if the functions E and F can be computed by a deterministic Turing machine in logarithmic space.

Any explicit encoding of Π_x is allowed as output of the construction, as long as the number of membranes and objects represented by it does not exceed the length of the whole description, and the rules are listed one by one. This restriction is enforced in order to mimic a (hypothetical) realistic process of construction of the P systems, where membranes and objects are presumably placed in a constant amount during each construction step, and require actual physical space proportional to their number; see also [10] for further details on the encoding of P systems.

Finally, we describe how space complexity for families of recogniser P systems is measured, and define the related complexity classes [16].

Definition 4. *Let $\mathcal{C}$ be a configuration of a P system Π. The size $|\mathcal{C}|$ of $\mathcal{C}$ is defined as the sum of the number of membranes in the current membrane structure and the total number of objects they contain. If $\boldsymbol{\mathcal{C}} = (\mathcal{C}_0, \ldots, \mathcal{C}_k)$ is a halting computation of Π, then the space required by $\boldsymbol{\mathcal{C}}$ is defined as*

$$|\boldsymbol{\mathcal{C}}| = \max\{|\mathcal{C}_0|, \ldots, |\mathcal{C}_k|\}$$

or, in the case of a non-halting computation $\boldsymbol{\mathcal{C}} = (\mathcal{C}_i : i \in \mathbb{N})$,

$$|\boldsymbol{\mathcal{C}}| = \sup\{|\mathcal{C}_i| : i \in \mathbb{N}\}.$$

Non-halting computations might require an infinite amount of space (in symbols $|\boldsymbol{\mathcal{C}}| = \infty$): for example, if the number of objects strictly increases at each computation step. The space required by Π itself is then

$$|\Pi| = \sup\{|\boldsymbol{\mathcal{C}}| : \boldsymbol{\mathcal{C}} \text{ is a computation of } \Pi\}.$$

Notice that $|\Pi| = \infty$ might occur if either Π has a non-halting computation requiring infinite space (as described above), or Π has an infinite set of halting computations, such that for each bound $b \in \mathbb{N}$ there exists a computation requiring space larger than b.

Finally, let $\boldsymbol{\Pi} = \{\Pi_x : x \in \Sigma^\star\}$ be a family of recogniser P systems, and let $s \colon \mathbb{N} \to \mathbb{N}$. We say that $\boldsymbol{\Pi}$ operates within space bound s iff $|\Pi_x| \leq s(|x|)$ for each $x \in \Sigma^\star$.

Definition 5. *Let* $f \colon \mathbb{N} \rightarrow \mathbb{N}$ *be a function. By* $(\mathbf{L}, \mathbf{L})$-$\mathbf{MC}_{\mathcal{D}}(f(n))$ *(respectively,* $(\mathbf{L}, \mathbf{L})$-$\mathbf{MCSPACE}_{\mathcal{D}}(f(n))$*) we denote the class of languages which can be decided by* $(\mathbf{L}, \mathbf{L})$-*uniform families of confluent P systems of type* $\mathcal{D}$ *(with* $\mathcal{D} = \mathcal{AM}$ *or* $\mathcal{D} = \mathcal{AM}(-d, -ne)$*) within time (resp., space) bound* f*. The corresponding classes for non-confluent P systems are* $(\mathbf{L}, \mathbf{L})$-$\mathbf{NMC}_{\mathcal{D}}(f(n))$ *and* $(\mathbf{L}, \mathbf{L})$-$\mathbf{NMCSPACE}_{\mathcal{D}}(f(n))$*.*

As usual, the above definitions can be generalized to sets of functions by union, defining $(\mathbf{L}, \mathbf{L})$-$\mathbf{PMC}_{\mathcal{D}}$ (resp., $(\mathbf{L}, \mathbf{L})$-$\mathbf{PMCSPACE}_{\mathcal{D}}$) as the union of the classes $(\mathbf{L}, \mathbf{L})$-$\mathbf{MC}_{\mathcal{D}}(f(n))$ (resp., $(\mathbf{L}, \mathbf{L})$-$\mathbf{MCSPACE}_{\mathcal{D}}(f(n))$) with $f(n)$ polynomial, and in analogous way $(\mathbf{L}, \mathbf{L})$-$\mathbf{EXPMC}_{\mathcal{D}}$ ($(\mathbf{L}, \mathbf{L})$-$\mathbf{EXPMCSPACE}_{\mathcal{D}}$) for exponential $f(n)$, $(\mathbf{L}, \mathbf{L})$-$\mathbf{LMC}_{\mathcal{D}}$ ($(\mathbf{L}, \mathbf{L})$-$\mathbf{LMCSPACE}_{\mathcal{D}}$) for $f(n)$ logarithmic, and finally $(\mathbf{L}, \mathbf{L})$-$\mathbf{kEXPMC}_{\mathcal{D}}$ ($(\mathbf{L}, \mathbf{L})$-$\mathbf{NEXPMCSPACE}_{\mathcal{D}}$) for $f(n)$ superexponential of level k.

3   Space Complexity: $\mathcal{AM}$ vs $\mathcal{TM}$

We will compare the above classes with the well known space complexity classes for Turing Machines ($\mathcal{TM}$) **PSPACE** and **EXPSPACE**; for the precise definitions and properties of Turing Machines and related complexity classes we refer the reader to [11]. The results reported below essentially establish the equivalence (up to a polynomial) of many space complexity classes defined in terms of P systems with active membranes and of Turing Machines, and are based on mutual simulation of the two computation models. The detailed description of simulations cannot be given here, and can be found in [2].

Theorem 6. *Let* M *be a single-tape deterministic Turing machine working in time* $t(n)$ *and space* $s(n)$*, including the space required for its input. Then there exists an* $(\mathbf{L}, \mathbf{L})$-*uniform family* Π *of confluent P systems with restricted elementary active membranes working in time* $O(t(n)s(n)^2) \subseteq O(t(n)^3)$ *and space* $O(s(n)^2)$ *such that* $L(M) = L(\Pi)$*.*

Notice that the uniformity condition for Π ensures that the P systems themselves carry out the simulation of the Turing machine M, as opposed to the machines constructing them, whenever the problem they solve is outside $\mathbf{L}$, e.g., in the case of **PSPACE**-hard problems. Theorem 6 can be also compared to the previously known simulations of polynomial-space P systems [19, 22], requiring only $O(t(n))$ time and $O(s(n))$ space, and exponential-space P systems [1], requiring $O(t(n)^2 \log t(n))$ time and $O(s(n) \log s(n))$ space, which are more efficient in those restricted cases.

The simulation of Theorem 6 can be made faster, and also generalised to nondeterministic Turing machines, by using non-confluent P systems.

Theorem 7. *Let* M *be a single-tape, possibly nondeterministic Turing machine working in time* $t(n)$ *and space* $s(n)$*, including the space required for its input. Then there exists an* $(\mathbf{L}, \mathbf{L})$-*uniform family* Π *of non-confluent P systems with restricted elementary active membranes working in time* $O(t(n)s(n)) \subseteq O(t(n)^2)$ *and space* $O(s(n)^2)$ *such that* $L(M) = L(\Pi)$*.*

From Theorem 6 we obtain inclusions of complexity classes for Turing machines and P systems when the space bound is at least linear (since we are dealing with single-tape Turing machines).

Theorem 8. *For every function $f(n) \in \Omega(n)$ the following inclusions hold:*

$$
\begin{aligned}
\mathbf{TIME}(f(n)) \quad &\subseteq (\mathbf{L}, \mathbf{L})\text{-}\mathbf{MC}_{\mathcal{AM}(-d,-n)}(O(f(n)^3)) \\
&\subseteq (\mathbf{L}, \mathbf{L})\text{-}\mathbf{MC}_{\mathcal{AM}}(O(f(n)^3)) \\
\mathbf{SPACE}(f(n)) \quad &\subseteq (\mathbf{L}, \mathbf{L})\text{-}\mathbf{MCSPACE}_{\mathcal{AM}(-d,-n)}(O(f(n)^2)) \\
&\subseteq (\mathbf{L}, \mathbf{L})\text{-}\mathbf{MCSPACE}_{\mathcal{AM}}(O(f(n)^2)).
\end{aligned}
$$

Let us now recall how P systems may be simulated by Turing Machines with a polynomial space overhead [18].

Theorem 9. *Let Π be a $(\mathbf{L}, \mathbf{L})$-uniform confluent (resp., non-confluent) family of recogniser P systems with active membranes working in space $s(n)$; let $t(n) \in$ poly be the time complexity of the Turing machine F computing the mapping $1^n \mapsto \Pi_n$. Then, Π can be simulated by a deterministic (resp., nondeterministic) Turing machine working in space $O(s(n)t(n)\log s(n))$.*

By combining Theorem 6 and Theorem 9, we can prove *equality* between space complexity classes for P systems and Turing machines under some (not very restrictive) assumptions on the set of space bounds we are interested in.

Theorem 10. *Let $\mathcal{F}$ be a class of functions $\mathbb{N} \to \mathbb{N}$ such that*

- *$\mathcal{F}$ contains the identity function $n \mapsto n$;*
- *If $s(n) \in \mathcal{F}$ and $p(n)$ is a polynomial, then there exists some $f(n) \in \mathcal{F}$ with $f(n) \in \Omega(p(s(n)))$.*

Then $\mathbf{SPACE}(\mathcal{F}) = (\mathbf{L}, \mathbf{L})\text{-}\mathbf{MCSPACE}_{\mathcal{AM}}(\mathcal{F})$. In particular, we have the following equalities:

$$
\begin{aligned}
\mathbf{PSPACE} &= (\mathbf{L}, \mathbf{L})\text{-}\mathbf{PMCSPACE}_{\mathcal{AM}} \\
\mathbf{EXPSPACE} &= (\mathbf{L}, \mathbf{L})\text{-}\mathbf{EXPMCSPACE}_{\mathcal{AM}} \\
\mathbf{2EXPSPACE} &= (\mathbf{L}, \mathbf{L})\text{-}\mathbf{2EXPMCSPACE}_{\mathcal{AM}} \\
k\mathbf{EXPSPACE} &= (\mathbf{L}, \mathbf{L})\text{-}k\mathbf{EXPMCSPACE}_{\mathcal{AM}}.
\end{aligned}
$$

Another consequence of the possibility of P systems to simulate Turing machines with a polynomial overhead and vice versa is that we are now able to translate theorems about the space complexity of Turing machines into theorems about P systems. As an example, the following two corollaries can be proved almost immediately for large enough space complexity bounds[1].

---

[1] Corollaries 11 and 12 can be proved, in the restricted polynomial and exponential space cases, with tighter space bounds by using the simulations in [19] and [1] respectively (see also [22]).

Corollary 11 (Savitch's Theorem for P Systems). *For each function $s(n)$ growing faster than every polynomial we have* $(\mathbf{L}, \mathbf{L})\text{-}\mathbf{NMCSPACE}_{\mathcal{AM}}(s(n)) \subseteq (\mathbf{L}, \mathbf{L})\text{-}\mathbf{MCSPACE}_{\mathcal{AM}}(O(s(n)^8 \log^4 s(n)))$.

As a consequence, for all classes of functions $\mathcal{F}$ satisfying the hypotheses of Theorem 10 we have

$$(\mathbf{L}, \mathbf{L})\text{-}\mathbf{MCSPACE}_{\mathcal{AM}}(\mathcal{F}) = (\mathbf{L}, \mathbf{L})\text{-}\mathbf{NMCSPACE}_{\mathcal{AM}}(\mathcal{F}).$$

Corollary 12 (Space Hierarchy Theorem for P Systems). *Let $s(n)$ be a function growing faster than every polynomial, and let $f(n)$ be a space-constructible function such that $O(s(n)^2 \log s(n)) \subseteq o(f(n))$. Then*

$$(\mathbf{L}, \mathbf{L})\text{-}\mathbf{MCSPACE}_{\mathcal{AM}}(s(n)) \subsetneq (\mathbf{L}, \mathbf{L})\text{-}\mathbf{MCSPACE}_{\mathcal{AM}}(O(f(n)^2)).$$

4   Sublinear Space Complexity

Having proved the equivalence (up to a polynomial) of the space complexity of Turing machines and P systems working in space $\Omega(n)$, we are also interested in finding out the behaviour of P systems with stronger restrictions. However, the measure of space complexity employed up to now, given by the largest amount of membranes and objects reached during the computation, is not suitable for this task. The multiset w_x encoding the input string x is, in general, already of polynomial size with respect to $|x|$; requiring it to be smaller would make the mapping $x \mapsto w_x$, and consequently the mapping $x \mapsto \Pi_x$, non-injective.

Hence, to analyse the complexity of sublinear-space P systems we use an expedient similar to one usually employed with Turing machines: making the input read-only, and removing its size from the calculation of space [20].

More formally, we partition the alphabet of P systems into two disjoint sets $\Delta \cup \Gamma$, where Δ is the *input alphabet* and Γ the *working alphabet*, and require the multiset w_x encoding $x \in \Sigma^*$ to be defined over Δ. Furthermore, we require that input objects are either rewritten into themselves by the rules, or into objects of Γ: for evolution rules $[a \to w]_h^\alpha$, at most one object $b \in \Delta$ may appear in w, and only if $b = a$; for rules $a \ [\ ]_h^\alpha \to [b]_h^\beta$, $[a]_h^\alpha \to [\ ]_h^\beta b$, and $[a]_h^\alpha \to b$, if $b \in \Delta$ then $b = a$; for division rules $[a]_h^\alpha \to [b]_h^\beta [c]_h^\gamma$, if $b \in \Delta$ (resp., $c \in \Delta$) then $b = a$ and $c \notin \Delta$ (resp., $c = a$ and $b \notin \Delta$). This ensures that no new input objects are created during the computation, and the existing ones are either just moved around the membrane structure, or rewritten into working alphabet objects. Finally, we define the space required by a configuration of a P systems to be the sum of the number of membranes and the number of *working alphabet objects only*.

A preliminary consequence of this definition of space complexity is that P systems with active membranes working in $O(\log n)$ space are able to simulate Turing machines working in $O(\log n)$ space [20]. The input objects of the P systems correspond to symbols of the input tape of the Turing machine, subscripted with indices denoting their position in the string in order not to lose information about ordering. An object of the P system encodes both the state q of the Turing machine and the position i on the input tape; the input multiset (i.e., the Turing machine input tape) is queried by writing the number i in binary on the charges of $O(\log n)$ nested membranes; this allows the movement of the object denoting the i-th input symbol (and only that object) to a specified region, when it can be finally read by the state-object. Further $O(\log n)$ membranes and objects directly encode the configuration of the working tape of the Turing machine.

The existence of this simulation proves the following inclusion [20]:

Theorem 13. **L** $\subseteq$ **(DLOGTIME, DLOGTIME)-LMCSPACE**$_{\mathcal{AM}}$.

Here the uniformity condition is **DLOGTIME**, a very weak uniformity condition usually employed for Boolean circuits [9]. This is necessary in order to disallow the machine providing the uniformity condition to perform the simulation itself, which would be possible when using the usual **P** (or even **L**) uniformity condition.

The simulation we just described was relatively straightforward, since the $O(\log n)$ read-write space of the simulated Turing machine can be easily represented by a similar amount of space in the simulating P system. However, a more sophisticated solution allows a polynomial amount of read-only input objects to simulate a polynomial amount of read-write tape cells [8].

Consider an object a_i, encoding the fact that the i-th symbol of the tape of a Turing machine M is a; this object is located within a certain membrane h of the P system simulating M. When M rewrites the symbol in position i on the tape as b, rather than rewriting a_i (which would need to be rewritten into a working alphabet object, thus increasing the space consumption), we can *move* it to a different region k of the P system. Essentially, we establish an injective mapping between regions and objects (in the example, h corresponds to a and k to b), and the symbol on the i-th tape cell of M is encoded by the *location* of a_i. Only the subscript i of a_i is meaningful (except during the initialisation of the system, to choose whether to move a_i to h rather than to k). The regions associated with tape symbols, such as h and k, are once again surrounded by $O(\log n)$ membranes, in whose charges a state-object writes the index i of the tape cell currently scanned by M. This allows us to isolate the corresponding object, as in the previous simulation.

As an immediate consequence, we obtain the identification of logarithmic space for P systems with polynomial space for Turing machines and, perhaps counter-intuitively, with polynomial space for P systems: the latter turns out to be always exponentially wasteful [8].

Theorem 14. **(L, L)-LMCSPACE**$_{\mathcal{AM}}$ = **(L, L)-PMCSPACE**$_{\mathcal{AM}}$ = **PSPACE**.

Here the logarithmic-space uniformity condition suffices to obtain a meaningful result, since **L** provably differs from **PSPACE**.

The last simulation we described requires only a constant number of non-input objects: the logarithmic space complexity is entirely due to the membranes encoding the tape head position. However, a simple combinatorial argument shows that, even with a constant number of membranes, moving around a polynomial number of read-only input objects may create exponentially many configurations, the same as polynomial-space Turing machines. The question is whether these configurations can be generated in a controlled (i.e., deterministic, or at least confluent) way and, more specifically, if we can isolate the object encoding the i-th tape symbol for any i and for any input string. If so, we would be able to simulate **PSPACE** in *constant* space with P systems.

Surprisingly, this turns out to be the case [5]. The $O(\log n)$-depth machinery previously used to isolate the object with subscript i can be replaced by confluent nondeterminism and timers. The idea is to pick nondeterministically one input object x_j at a time, rewrite it into itself *and* a timer-object counting from j down to 0. Simultaneously, a state-object similar to that of the previous simulations counts from i (the index of the object we are trying to select) down to 0. We can check whether $i = j$ by letting the first object to reach 0 signal to the other by changing the charge of a membrane. If $i \neq j$, the object x_j is moved to a temporary membrane, and another object x_k is selected. Eventually, the object x_j will be selected (thus ensuring confluence) and, by keeping track of the region it came from, it will be possible to deduce the identity of the tape symbol it represents.

This simulation only requires a constant number of membranes and non-input objects, proving the following result [5].

Theorem 15. $(\mathbf{L}, \mathbf{L})\text{-}\mathbf{MCSPACE}_{\mathcal{AM}}(O(1)) = \mathbf{PSPACE}$.

Hence, from the point of view of space complexity, non-wasteful P systems either use constant space, or super-polynomial space.

This latter result may cause some doubts regarding the definition of space complexity: does it really capture our intuitive idea of size of a P system as a biologically inspired device? After all, the distinctiveness of $m = |\Gamma|$ object types (chemical species) and of $k = |\Lambda|$ labels (membrane types) in the P system model must correspond to a physical distinctiveness of the objects they model, which we may assume to require a non-constant amount of space. On the other hand, the definitions of space employed up to now in this paper assign unitary volume to each object and each membrane. A more sophisticated (but still mathematically simple) way to measure space would be to assign to each non-input object and each membrane a volume proportional to the amount of information they encode: $\log m$ and $\log k$, respectively. Adopting this more detailed measure of space complexity implies that the last simulation we described would run in logarithmic space, rather than constant space.

5   Elementary Division and Counting Problems

An interesting restriction of P systems with active membranes is obtained by limiting membrane division to elementary membranes (not containing further membranes). The fact that this kind of division suffices to solve **NP**-complete problems in polynomial time is one of the oldest results in membrane computing [23], in symbols:

Theorem 16. NP $\subseteq$ PMC$_{\mathcal{AM}(-n)}$.

The main idea is to evaluate Boolean formulae φ in conjunctive normal form by having objects $x_1, \ldots, x_m$, corresponding to the variables of φ, in an elementary membrane h; the membrane is then divided repeatedly by using rules $[x_i]_h^0 \to [t_i]_h^0 \, [f_i]_h^0$ for $1 \leq i \leq m$. Interpreting t_i (resp., f_i) as a true (resp., false) truth assignment to x_i, we obtain 2^m copies of membrane h after m steps, each one of them containing a distinct truth assignment for φ. The truth assignments are then evaluated in parallel, also in polynomial time, by using evolution and communication rules, and an object t is sent out from each copy of membrane h containing a satisfying assignment. Clearly, a copy of t exists if and only if φ is satisfiable.

This algorithm can be extended [17] to *count* the truth assignments, or more specifically to check whether the number of assignments exceeds a given threshold k. This can be accomplished by having another set of membranes perform binary division until there are exactly k of them, and have each of these membranes absorb a copy of t as soon as it is sent out from a membrane h. If there is any instance of t left after this "deletion" phase, then the number of truth assignments satisfying the formula exceeds k. Since the satisfiability threshold problem is **PP**-complete, we obtain the following result:

Theorem 17. PP $\subseteq$ PMC$_{\mathcal{AM}(-n)}$.

This result can be further improved by using the counting P systems as "modules" in a larger P system simulating a polynomial-time Turing machine [19]. These P system modules are literally used as sub-routines by sending them an input (computed by the Turing machine being simulated) by means of communication rules. Since the modules operate in polynomial time, this allows us to simulate Turing machines with oracles for **PP** (or, equivalently, **#P**) problems with a polynomial slowdown:

Theorem 18. P$^{\#P}$ $\subseteq$ PMC$_{\mathcal{AM}(-n)}$.

The **P$^{\#P}$** lower bound turns out to be optimal [7]. Indeed, we can simulate a P system with elementary membrane division working in polynomial time as follows. It is a well-known result that non-dividing membranes can be simulated directly by a polynomial-time Turing machine [23]; the key observation here is that we do *not* need to simulate the dividing elementary membranes directly, but we only need to know what objects they send out or absorb during each computation step. This would allow us to update the configuration of the outermost membrane (which cannot divide, by definition) and obtain the result of the computation of the P system. The multisets entering and exiting the elementary membranes can be computed by having oracles for queries substantially equivalent to "How many instances of a are output (resp., absorbed) by membranes having label h at computation step t?". Since this can be answered by a **#P** oracle, we obtain a complete characterisation of the computing power of P systems with elementary division working in polynomial time:

Theorem 19. $\mathrm{PMC}_{\mathcal{AM}(-\mathrm{n})} = \mathbf{P}^{\#\mathbf{P}}$.

The conjecture here is that this equivalence can be generalised to P systems with non-elementary membrane division of constant depth by increasing the power of the oracle. Let the counting hierarchy [4] be defined by $\mathbf{C}_0\mathbf{P} = \mathbf{P}$ and $\mathbf{C}_{d+1}\mathbf{P} = \mathbf{PP}^{\mathbf{C}_d\mathbf{P}}$. Then, a preliminary result [6] shows that P systems of depth d, using non-elementary membrane division and working in polynomial time are at least as powerful as $\mathbf{P}^{\mathbf{C}_d\mathbf{P}}$. The outermost membrane simulates a polynomial-time deterministic Turing machine, and the membrane substructure of depth $d-1$ it contains simulates a $\mathbf{C}_d\mathbf{P}$ Turing machine. This, in turn, contains a membrane substructure of depth $d-2$ simulating a $\mathbf{C}_{d-1}\mathbf{P}$ Turing machine, and so on recursively. As a consequence, P systems of depth $O(1)$ are at least as powerful as $\mathbf{CH} = \bigcup_{d=0}^{\infty} \mathbf{C}_d\mathbf{P}$.

References

1. Alhazov, A., Leporati, A., Mauri, G., Porreca, A.E., Zandron, C.: The computational power of exponential-space P systems with active membranes. In: Martínez-del-Amor, M.A., Păun, Gh., Pérez-Hurtado, I., Romero-Campero, F.J. (eds.) Proceedings of the Tenth Brainstorming Week on Membrane Computing, vol. I, pp. 35–60. Fénix Editora (2012)
2. Alhazov, A., Leporati, A., Mauri, G., Porreca, A.E., Zandron, C.: Space complexity equivalence of P systems with active membranes and Turing machines. Theoretical Computer Science **529**, 69–81 (2014)
3. Alhazov, A., Martín-Vide, C., Pan, L.: Solving a PSPACE-complete problem by recognizing P systems with restricted active membranes. Fundamenta Informaticae **58**(2), 67–77 (2003)
4. Hemaspaandra, L.A., Ogihara, M.: The Complexity Theory Companion. Texts in Theoretical Computer Science. Springer (2002)
5. Leporati, A., Manzoni, L., Mauri, G., Porreca, A.E., Zandron, C.: Constant-space P systems with active membranes. Fundamenta Informaticae **134**(1–2), 111–128 (2014)
6. Leporati, A., Manzoni, L., Mauri, G., Porreca, A.E., Zandron, C.: Membrane division, oracles, and the counting hierarchy. Fundamenta Informaticae **137**, 1–15 (2015)
7. Leporati, A., Manzoni, L., Mauri, G., Porreca, A.E., Zandron, C.: Simulating elementary active membranes, with an application to the P conjecture. In: Gheorghe, M., Rozenberg, G., Salomaa, A., Sosík, P., Zandron, C. (eds.) CMC 2014. LNCS, vol. 8961, pp. 284–299. Springer, Heidelberg (2014)
8. Leporati, A., Mauri, G., Porreca, A.E., Zandron, C.: A gap in the space hierarchy of P systems with active membranes. Journal of Automata, Languages and Combinatorics **19**(1–4), 173–184 (2014)
9. Mix Barrington, D.A., Immerman, N., Straubing, H.: On uniformity within NC^1. Journal of Computer and System Sciences **41**(3), 274–306 (1990)
10. Murphy, N., Woods, D.: The computational power of membrane systems under tight uniformity conditions. Natural Computing **10**(1), 613–632 (2011)
11. Papadimitriou, C.H.: Computational Complexity. Addison-Wesley (1993)
12. Păun, Gh.: Computing with membranes. Journal of Computer and System Sciences **61**(1), 108–143 (2000)
13. Păun, Gh.: P systems with active membranes: Attacking NP-complete problems. Journal of Automata, Languages and Combinatorics **6**(1), 75–90 (2001)

14. Păun, Gh., Rozenberg, G., Salomaa, A. (eds.): The Oxford Handbook of Membrane Computing. Oxford University Press (2010)
15. Pérez-Jiménez, M.J., Romero-Jiménez, A., Sancho-Caparrini, F.: Complexity classes in models of cellular computing with membranes. Natural Computing **2**(3), 265–284 (2003)
16. Porreca, A.E., Leporati, A., Mauri, G., Zandron, C.: Introducing a space complexity measure for P systems. International Journal of Computers, Communications & Control **4**(3), 301–310 (2009)
17. Porreca, A.E., Leporati, A., Mauri, G., Zandron, C.: Elementary active membranes have the power of counting. International Journal of Natural Computing Research **2**(3), 329–342 (2011)
18. Porreca, A.E., Leporati, A., Mauri, G., Zandron, C.: P systems with active membranes working in polynomial space. International Journal of Foundations of Computer Science **22**(1), 65–73 (2011)
19. Porreca, A.E., Leporati, A., Mauri, G., Zandron, C.: P systems simulating oracle computations. In: Gheorghe, M., Păun, G., Rozenberg, G., Salomaa, A., Verlan, S. (eds.) CMC 2011. LNCS, vol. 7184, pp. 346–358. Springer, Heidelberg (2012)
20. Porreca, A.E., Leporati, A., Mauri, G., Zandron, C.: Sublinear-space P systems with active membranes. In: Csuhaj-Varjú, E., Gheorghe, M., Rozenberg, G., Salomaa, A., Vaszil, G. (eds.) CMC 2012. LNCS, vol. 7762, pp. 342–357. Springer, Heidelberg (2013)
21. Sosík, P.: The computational power of cell division in P systems: Beating down parallel computers? Natural Computing **2**(3), 287–298 (2003)
22. Valsecchi, A., Porreca, A.E., Leporati, A., Mauri, G., Zandron, C.: An efficient simulation of polynomial-space turing machines by P systems with active membranes. In: Păun, G., Pérez-Jiménez, M.J., Riscos-Núñez, A., Rozenberg, G., Salomaa, A. (eds.) WMC 2009. LNCS, vol. 5957, pp. 461–478. Springer, Heidelberg (2010)
23. Zandron, C., Ferretti, C., Mauri, G.: Solving NP-complete problems using P systems with active membranes. In: Antoniou, I., Calude, C.S., Dinneen, M.J. (eds.) Unconventional Models of Computation, UMC 2000, Proceedings of the Second International Conference, pp. 289–301. Springer (2001)

The Shuffle Product: New Research Directions

Antonio Restivo[✉]

Dipartimento di Matematica e Informatica, Università di Palermo,
Palermo, Italy
restivo@math.unipa.it

Abstract. In this paper we survey some recent researches concerning the shuffle operation that arise both in Formal Languages and in Combinatorics on Words.

Keywords: Shuffle · Intermixed languages · Star-free languages · Shuffle squares

1 Introduction

Parallel composition of words appears to be an important issue both in the theory of concurrency and in formal languages. Usually it is modeled by the shuffle operation. The shuffle $u \sqcup\!\sqcup v$ of words u and v is the finite set of words obtained by merging the words u and v from left to right, but choosing the next symbol arbitrarily from u or v. In other words, the shuffle operator returns the set of all possible interleaving of symbols in u and v. For example, the shuffle of ab and cd is the set $\{abcd, acbd, acdb, cabd, cadb, cdab\}$. This definition naturally extends to languages. In a more formal way, the *shuffle product* (or simply *shuffle*) of two languages L_1, L_2 over the alphabet Σ is the language

$$L_1 \sqcup\!\sqcup L_2 = \{u_1 v_1 ... u_n v_n \in \Sigma^* \,|\, n \geq 1, u_1 ... u_n \in L_1, v_1 ... v_n \in L_2\}.$$

It is known that the shuffle product is an associative and commutative operation, which is also distributive over the union. Other variants of the shuffle product, related to various forms of synchronization, are known in literature (cf [2]), but are not considered in this paper.

The initial work on shuffle arose out of formal language theory. To the best of our knowledge, the shuffle product was first used in this context by Ginsburg and Spanier [11]. Early research with applications to concurrent processes can be found in [34] and [22]. Various problems concerning the complexity of the shuffle operation were investigated in the 1980's in [23,26,27,37]. Interestingly, it was observed in [21] that some aspects of the shuffle product bear strong similarities with genetic recombination.

In recent years new problems on the shuffle operation have been proposed, partly motivated by their applications to concurrent systems (cf. [3,4,6,16,17,33]).

Antonio Restivo—Partially supported by Italian MIUR Project PRIN 2010LYA9RH, "Automi e Linguaggi Formali: Aspetti Matematici e Applicativi".

A.-H. Dediu et al. (Eds.): LATA 2015, LNCS 8977, pp. 70–81, 2015.
DOI: 10.1007/978-3-319-15579-1_5

In this paper we survey some recent researches on the shuffle that arise both in formal language theory and in combinatorics on words. In the first section we report recent investigations into the language generation power of the shuffle operation, when used in combination with the more conventional operations of union, intersection, complement, concatenation and Kleene star. In the second section we consider very recent combinatorial and algorithmic problems related to the shuffle. Some of them concern the complexity and the avoidability of shuffle squares.

2 Formal Languages

A very general problem in the theory of formal languages is, given a "basis" of languages and a set of operations, to characterize the family of languages expressible from the "basis" by using the operations. In practice, a basis of languages will consists of a set of very simple languages, such as the languages of the form $\{a\}$, where a is a letter of the alphabet. In the theory of *regular* languages, the operations taken into account are usually the Boolean operations, the concatenation and the (Kleene) star operation.

In this setting, two families of languages play a fundamental role: the family *REG* of regular languages, and the family *SF* of star-free languages. *REG* is defined as the smallest family of languages containing the languages of the form $\{a\}$, where a is a letter, and $\{\epsilon\}$, where ϵ is the empty word, and closed under union, concatenation and star. It is well known that the family *REG* is closed also under all Boolean operations. The family *SF* of star-free languages is the smallest family of languages containing the languages of the form $\{a\}$ and $\{\epsilon\}$, and closed under Boolean operations and concatenation.

Another operation that plays an important role in the theory of formal languages is the *shuffle* operation. It is well known (cf [10]) that the family *REG* of regular languages is closed under shuffle. The study of subfamilies of regular languages closed under shuffle is a difficult problem, partly motivated by its applications to the modeling of process algebras [1] and to program verification.

In particular, we here consider the smallest family of languages containing the languages of the form $\{a\}$ and $\{\epsilon\}$, and closed under Boolean operations, concatenation and shuffle. Let us call *intermixed* the languages in this family, which is denoted by *INT*. It is perhaps surprising that the following important problem in the theory of regular languages is still open, and to a large extent unexplored.

Problem 1. *Give a (decidable) characterization of the family INT.*

In this section we discuss this problem: we present some partial results and we introduce new special problems as possible steps in the characterization of the family *INT*. Such partial results and special problems show the deep connections of Problem 1 with other relevant aspects of formal languages theory. The results presented in this section are essentially based on the papers [3], [6] and [15].

2.1   Star-Free and Intermixed Languages

In [3] it is proved the following theorem showing that the family INT of inter-mixed languages is strictly included in the family REG of regular languages and strictly contains the family SF of star-free languages.

Theorem 1. $SF \subsetneq INT \subsetneq REG$

Moreover, in [3] it is shown that the family INT is closed under quotients, but it is not closed under inverse morphism. Therefore, the family INT is not a variety of languages (cf [30] and [31]), and so it cannot be characterized in terms of syntactic monoids.

Let us recall (cf [28]) that a language $L \subseteq \Sigma^*$ is said to be *aperiodic*, or *non-counting*, if there exists an integer $n > 0$ such that for all $x, y, z \in \Sigma^*$ one has

$$xy^n z \in L \Leftrightarrow xy^{n+1}z \in L.$$

A fundamental theorem of Schutzenberger states that *a regular language is star-free if and only if it is aperiodic*.

The strict inclusion between the families SF and INT implies that the shuffle of two star-free languages in general is not star-free. This means, roughly speaking, that *the shuffle creates periodicities*.

In order to enlighten on the difficult Problem 1, we consider the following

Problem 2. *Determine conditions under which the shuffle of two star-free languages is star-free too.*

A first simple condition is obtained in [6] by introducing a weaker version of the shuffle product, called *bounded shuffle*.

Let k be a positive integer. The *k-shuffle* of two languages $L_1, L_2 \subseteq \Sigma^*$ is defined as follows:

$$L_1 \sqcup\!\sqcup_k L_2 = \{u_1 v_1 ... u_m v_m | m \le k, u_1 ... u_m \in L_1, v_1 ... v_m \in L_2\}.$$

Any k-shuffle is called *bounded shuffle*. It is not difficult to show that the family REG of regular languages is closed under bounded shuffle. In [6] it is proved the following theorem.

Theorem 2. *SF is closed under bounded shuffle, i.e. if $L_1, L_2 \in SF$ then $L_1 \sqcup\!\sqcup_k L_2 \in SF$, for any $k \ge 1$.*

One can derive the following corollary.

Corollary 3. *The shuffle of a star-free language and a finite language is star-free.*

In the following subsections we report some partial results to Problem 2, that highlight the role of the notions of *commutativity* and *ambiguity* in the problem.

2.2 Partial Commutations

The family *SF* is closed under concatenation and it is not closed under shuffle. What is the difference between concatenation and shuffle? Actually, the shuffle $u \sqcup\!\sqcup v$ of words u and v can be obtained from their product uv by allowing the commutation of the letters occurring in u with the letters occurring in v (*external* commutation), keeping the letters in each word in order, i.e. preventing their (*internal*) commutation. In the following results we show that such a discrepancy between *internal* and *external* commutations is at the origin of the fact that the shuffle *"creates periodicity"*. Next results illustrates this phenomenon.

Let us start with an interesting result known in the commutative case. We denote by $[u]$ the *commutative closure* of a word u, which is the set of words commutatively equivalent to u. A language L is *commutative* if, for every word $u \in L$, $[u]$ is contained in L. The following result was proved by J.F. Perrot in [29].

Theorem 4. *The shuffle of two commutative star-free languages is star-free.*

This result settles a special case in which there is, roughly speaking, an agreement between internal and external commutations.

In order to examine the phenomenon in its generality we introduce an operation between languages, that generalizes at the same time concatenation and shuffle, and we investigate the closure of *SF* with respect to this operation. The new operation is defined by introducing a partial commutation between the letters of the alphabet, and its appropriate setting is the theory of *traces* (cf [9]).

Let Γ be a finite alphabet and let $\theta \subseteq \Gamma \times \Gamma$ be a symmetric and irreflexive relation called the *(partial) commutation relation*. We consider the congruence $\sim_\theta$ of Γ^* generated by the set of pairs (ab, ba) with $(a, b) \in \theta$. If $L \subseteq \Gamma^*$ is a language, $[L]_\theta$ denoted the closure of L by $\sim_\theta$, and L is *closed by* $\sim_\theta$ if $L = [L]_\theta$. The closed subsets of Γ^* are called *trace languages*.

Let now L_1 and L_2 be two languages over the alphabet Σ

Let us consider two disjoint copies Σ_1 and Σ_2 of the alphabet Σ, i.e. such that $\Sigma_1 \cap \Sigma_2 = \emptyset$, and the isomorphism σ_1 from Σ_1^* to Σ^* and σ_2 from Σ_2^* to Σ^*.

Let L_1' (L_2' resp.) be the subset of Σ_1^* (Σ_2^* resp.) corresponding to L_1 (L_2 resp.) under the isomorphism σ_1 (σ_2 resp.). Let us consider the morphism $\sigma :$ $(\Sigma_1 \cup \Sigma_2)^* \to \Sigma^*$ defined as follows:

$$\sigma(a) = \begin{cases} \sigma_1(a), \text{ if } a \in \Sigma_1^*; \\ \sigma_2(a), \text{ if } a \in \Sigma_2^*. \end{cases}$$

Let θ be of the form $\theta \subseteq \Sigma_1 \times \Sigma_2$. The θ-product (denoted by $\sqcup\!\sqcup_\theta$) of the languages $L_1, L_2 \subseteq \Sigma^*$ is defined as follows:

$$L_1 \sqcup\!\sqcup_\theta L_2 = \sigma([L_1' L_2']_\theta).$$

Remark that the product (concatenation) and the shuffle correspond to two special (extremal) cases of the θ-product. Indeed, if $\theta = \emptyset$ then $L_1 \sqcup\!\sqcup_\theta L_2 = L_1 L_2$, and, if $\theta = \Sigma_1 \times \Sigma_2$, then $L_1 \sqcup\!\sqcup_\theta L_2 = L_1 \sqcup\!\sqcup L_2$.

The partial commutation $\theta \subseteq \Sigma_1 \times \Sigma_2$ induces a partial commutation θ' on Σ defined as follows: if $(a, b) \in \theta$ the $(\sigma_1(a), \sigma_2(b)) \in \theta'$.

In [15] it is proved the following theorem.

Theorem 5. *Let L_1 and L_2 be two languages closed under θ', i.e., $[L_1]_{\theta'} = L_1$ and $[L_2]_{\theta'} = L_2$. If L_1 and L_2 are star-free, then $L_1 \sqcup\!\sqcup_\theta L_2$ is star-free.*

The theorem states, roughly speaking, that, if *internal* commutation θ' (i.e., the commutations allowed inside words in each of the languages L_1 and L_2) "coincides" with the *external* commutation θ (i.e., the commutations between the letters in words of L_1 and the letters in words of L_2), then the θ-product preserves the star-freeness.

Special cases of the previous theorem are the well known result that the concatenation of two star-free languages is star-free (corresponding to the case $\theta = \emptyset$), and the result of J.F. Perrot (cf Theorem 4) that *the shuffle of two* commutative *star-free languages is star-free* (corresponding to the case $\theta = \Sigma_1 \times \Sigma_2$).

A recent paper of Cano, Guaiana and Pin [5] contains some other interesting results on regular languages closed under partial commutations, which are related to the problems raised in this section.

2.3   Unambiguous Star-Free Languages

In this section we investigate some conditions for Problem 2, related to the unambiguity of the product of languages.

A language $L \subseteq \Sigma^*$ is a *marked product* of the languages $L_0, L_1, ..., L_n$ if

$$L = L_0 a_1 L_1 a_2 L_2 ... a_n L_n,$$

for some letters $a_1, a_2, ..., a_n$ of Σ.

It is known (cf [32]) that the family *SF* of star-free languages is the smallest Boolean algebra of languages of Σ^* which is closed under marked product.

A marked product $L = L_0 a_1 L_1 a_2 L_2 ... a_n L_n$ is said to be *unambiguous* if every word u of L admits a *unique* decomposition

$$u = u_0 a_1 u_1 ... a_n u_n,$$

with $u_0 \in L_0, u_1 \in L_1, ..., u_n \in L_n$.

For instance, the marked product $\{a, c\}^* a \{\epsilon\} b \{b, c\}^*$ is unambiguous.

Let us define the family *USF* of *unambiguous star-free* languages as the smallest Boolean algebra of languages of Σ^* containing the languages of the form A^*, for $A \subseteq \Sigma$, which is closed under unambiguous marked product (cf [32]).

The family *USF* is a very robust class of languages: it is a *variety* of languages and it admits several other nice characterizations (see [36] for a survey).

Moreover, it can be shown that *USF* is strictly included in *SF*, and so we have the following chain of inclusions:

$$USF \subsetneq SF \subsetneq INT \subsetneq REG.$$

The following theorem, proved in [6], shows the role of unambiguity in Problem 2.

Theorem 6. *If L_1 and L_2 are unambiguous star-free languages, then $L_1 \sqcup\!\sqcup L_2$ is star-free.*

Other interesting results on varieties of languages closed under the shuffle product are in [12,13].

2.4   Cyclic Submonoids

The languages in the family *USF* can be described by regular expressions in which the star operation is restricted to subsets of the alphabet. Furthermore, Theorem 6 states that the shuffle of languages in this family is star-free. Hence, the critical situations, with respect to Problem 2, occur with languages corresponding to regular expressions in which the star operation is applied to concatenation of letters. So, in this section, we consider the shuffle of languages of the form u^*, where u is a word of Σ^*. Actually, such languages correspond to *cyclic submonoids* of Σ^*.

The special interest of such languages in our context is shown by the following theorem, proved in [3].

Theorem 7. *If the word u contains more than one letter, then the language u^* is intermixed.*

Remark that in [3] in order to prove the strict inclusion $INT \subsetneq REG$, it is shown that the regular language $(a^2)^*$ is not intermixed, and this is again a language of the form u^*. Moreover, next theorem, firstly proved by McNaughton and Papert in [28], shows that the combinatorial properties of the word u play a role in Problem 2. Let us first introduce a definition. A word $u \in \Sigma^*$ is *primitive* if it is not a proper power of another word of Σ^*, i.e., if the condition $u = v^n$, for some word v and integer n, implies that $u = v$ and $n = 1$.

Theorem 8. *The language u^* is star-free if and only if u is a primitive word.*

We now consider the shuffle $u^* \sqcup\!\sqcup v^*$ of two cyclic submonoids generated by the words u and v, respectively. If u and v are primitive words then, by the previous theorem, u^* and v^* are star-free languages. Remark that the languages u^* and v^* do not belong to *USF*, and their shuffle, in general, is not star-free. Here we are interested to the conditions under which the language $u^* \sqcup\!\sqcup v^*$ is star-free.

Let us consider some examples. If $u = b$ and $v = ab$, the language $b^* \sqcup\!\sqcup (ab)^* = (b + ab)^*$ is star-free. Let us consider now $u = aab$ and $v = bba$, the language $(aab)^* \sqcup\!\sqcup (bba)^*$ is not star-free. Indeed the language

$$((aab)^* \sqcup\!\sqcup (bba)^*) \cap (ab)^* = ((ab)^3)^*$$

is not star-free, by the Theorem 8

Problem 3. *Characterize the pairs of primitive words $u, v \in \Sigma^*$ such that $u^* \sqcup\!\sqcup v^*$ is a star-free language.*

3   Combinatorics on Words

This last problem in the previous section is closely related to some relevant questions in combinatorics on words. Recall that combinatorics on words is a fundamental part of the theory of words and languages. It is deeply connected to numerous different fields of mathematic and its applications, and it emphasizes the algorithmic nature of many problems on words (cf [24]).

3.1   Lyndon-Schutzenberger Problems for the Shuffle

Some important problems in combinatorics on words pertain to the non primitive words that appear in the set u^+v^+, where u and v are primitive words.

A remarkable result in this direction is the famous Lyndon-Schutzenberger theorem (cf [25]), originally formulated for the free groups.

Theorem 9. *If u and v are distinct primitive words, then the word $u^n v^m$ is primitive for all $n, m \geq 2$.*

The next theorem, proved by Shyr and Yu ([35]), can be considered as a light improvement of the previous result.

Theorem 10. *If u and v are distinct primitive words, then there is at most one non-primitive word in the language u^+v^+.*

Problem 3 is, in a certain sense, related to those considered in the above theorems, with the difference that we here take into account the shuffle of the two languages u^+ and v^+, instead of their concatenation. Actually, Problem 3 leads to investigate the non-primitive words that appear in the language $u^+ \shuffle v^+$, where u and v are primitive words. In particular, we are interested to investigate the exponents of the powers that appear in $u^+ \shuffle v^+$.

Let us introduce further notation. Let us denote by Q the set of primitive words. Consider words $u, v, w \in Q$ such that $(u^+ \shuffle v^+) \cap w^+ \neq \phi$, and let $p(u, v, w)$ be the positive integer k such that

$$(u^+ \shuffle v^+) \cap w^+ = (w^k)^+.$$

For $u, v \in Q$, let us define the set of integers

$$P(u, v) = \{p(u, v, w) \mid w \in Q\}.$$

For instance, if we consider the words $u = a^{10}b$, $v = b$, then $P(u, v) = \{1, 2, 5, 10\}$.

The following problem is closely related to Problem 3.

Problem 4. *Given two primitive words u, v, characterize the set $P(u, v)$ in terms of the combinatorial properties of u and v.*

Actually, with the notation here introduced, Problem 3 can be restated as follows: *Characterize the pairs of primitive words $u, v \in \Sigma^*$ such that $P(u,v) = \{1\}$.*

A more elementary problem is to investigate the powers that appear in the language $u \sqcup\!\sqcup v$, where u and v are primitive words. For instance, if $u = aab$ and $v = bba$, then $(ab)^3 \in u \sqcup\!\sqcup v$; if $u = b$ and $v = aaaab$, then $(aab)^2 \in u \sqcup\!\sqcup v$; if $u = b$ and $v = aaaaab$, then $u \sqcup\!\sqcup v$ does not contain proper powers.

Problem 5. *Characterize the pairs of primitive words (u, v) such that $u \sqcup\!\sqcup v$ does not contain proper powers, i.e $u \sqcup\!\sqcup v \subset Q$.*

3.2 Shuffle Squares

Closely related to the last problem in the previous section is the notion of *shuffle square*. Recall that a word u is a *square* if there exists a word v such that $u = v^2$. A word u is a *shuffle square* if there exists a word v such that $u \in v \sqcup\!\sqcup v$. For instance, the word $u = bbabbabb$ is a shuffle square, because $u \in v \sqcup\!\sqcup v$, where $v = babb$:

$$u = \underline{bb}a\underline{bb}abb,$$

where the letters of one occurrence of v are underlined for ease of reading. The word v is called a *shuffle root* of u. Remark that the shuffle root is, in general, not unique. For instance, the word $bbab$ is another shuffle root of $bbabbabb$.

Problem 6. *Characterize the shuffle squares having a unique shuffle root*

Another remark is that a shuffle square, i.e. a word u such that $u \in v \sqcup\!\sqcup v$ for some word v, can occur *twice* in $v \sqcup\!\sqcup v$, as shown by the following example. The word $abcbacabacbabcbacabcacbc$ occurs twice in $v \sqcup\!\sqcup v$, where $v = abcbacabacbc$:

$$\underline{abcbacabac}b\underline{ab}cbacab\underline{c}ac\underline{bc}$$

$$abcba\underline{cabac}ba\underline{bc}bacab\underline{cacbc}.$$

Recently several papers have given special attention to shuffle squares. Harju [17] studied shuffle of *square-free* words, i.e. words that have no factors of the form vv for some non-empty word v. More results on square-free shuffles were obtained independently by Harju and Müller [18], and Currie and Saari [8], who proved, in particular, the following result about the existence of arbitrarily long square-free words that are shuffle squares.

Theorem 11. *For every integer $n \geqslant 3$ there exists a square-free ternary word v of length n such that $v \sqcup\!\sqcup v$ contains a square-free word.*

The most intriguing questions perhaps concern the *complexity* of shuffle squares. We consider the following two problems:

(1) Given words u and v, is $u \in v \sqcup\!\sqcup v$?

(2) Given a word u, does there exists a word v such that $u \in v \sqcup\!\sqcup v$?

Actually, as we shall see, these two problems dramatically differ in complexity. The complexity of the first problem was proved by van Leeuwen and Nivat (cf [23]):

Theorem 12. *Given words u, v_1 and v_2, it can be tested $O(|u|^2/log(|u|))$ time whether or not $u \in v_1 \sqcup\!\sqcup v_2$.*

As to concern the second problem, recently the following result has been independently proved by Buss and Soltys [4] and by Rizzi and Vialette [33].

Theorem 13. *Given a word u, it is NP-complete to decide whether u is a shuffle square, i.e. whether there exists a word v such that $u \in v \sqcup\!\sqcup v$.*

Actually in [4] it is proved that the problem to decide whether a word is a shuffle square is NP-complete for an alphabet with 9 letters, and it is claimed that this can be improved to 7. It remains open the problem to determine the smallest cardinality of the alphabet for which the problem is NP-complete. In particular, it remains open the following problem.

Problem 7. *How hard is the problem to check whether a binary word is a shuffle square?*

Notice that in [20] it is claimed without proof that checking whether a binary word is a shuffle square is NP-complete.

A closely related problem, concerning the shuffle of a word with its reverse, has been approached by Henshall , Rampersad and Shallit in [19], where they prove the following theorem. Recall that a word u is an *abelian square* if $u = vv'$, where v' is a permutation of v. For any word $v = a_1 a_2 \cdots a_n$, where the a_i's are letters, denote by v^R the *reverse* of v, i.e $v = a_n \cdots a_2 a_1$.

Theorem 14. *Given a word u, if there exists a word v such that $u \in v \sqcup\!\sqcup v^R$, then u is an abelian square. Conversely, if u is a binary abelian square, then there exists a word v such that $u \in v \sqcup\!\sqcup v^R$.*

Remark that the converse does not hold if u is not a binary word, as shown by the following example: the word $u = abcabc$ is an abelian square, but does not exist any v such that $u \in v \sqcup\!\sqcup v^R$.

A consequence of previous theorem is that it is polynomial-time solvable to decide whether a *binary* word is the shuffle of another word with its reverse. Rizzi and Vialette completed in [33] the result of [19], and showed the following theorem for unbounded alphabet words.

Theorem 15. *Given a word u over an arbitrary alphabet, there is a polynomial time algorithm to decide whether there exists a word v such that $u \in v \sqcup\!\sqcup v^R$.*

The following problems concern the *avoidability of shuffle squares*. In the Dagstuhl Seminar *Combinatorics and Algorithmics on Strings*, March 9-14, 2014, Karhumaki asked the question whether shuffle squares are avoidable, i.e. whether there exist words of arbitrary length, over a finite alphabet, that do not contain as factor a shuffle square. In the course of the same seminar Currie [7] gave a positive answer to the question: he showed, by using the Lovasz local lemma, that shuffle squares are avoidable over an alphabet of size $\lceil e^{115} \rceil$. This result has been recently improved by Guegan and Ochem in [16], where they prove the following result.

Theorem 16. *Shuffle squares are avoidable over an alphabet of size 7.*

The proof of Guegan and Ochem, as that of Currie, is non constructive: it uses the *entropy compression* method in the general framework developed by Gonçalves, Montassier and Pinlou in [14]. The following questions naturally arise:

Problem 8. *What is the minimal alphabet size that allows to avoid shuffle squares?*

Problem 9. *Give an explicit construction of an infinite word that avoids shuffle squares*

One can also consider powers larger than squares. A word u is a *shuffle cube* if there exists a word v such that $u \in v \sqcup\!\sqcup v \sqcup\!\sqcup v$. For instance, the word $u = babaabababab$ is a shuffle cube, because $u \in v \sqcup\!\sqcup v \sqcup\!\sqcup v$, where $v = baab$:

$$u = \underline{ba}\overline{b}\underline{aa}\overline{ab}\overline{a}\underline{ba}\overline{b}ab.$$

In the same way, a word v is a *shuffle k-power*, for some positive integer k, if there exist a word v such that $u \in v \sqcup\!\sqcup v \sqcup\!\sqcup v \sqcup\!\sqcup \ldots \sqcup\!\sqcup v$, k times. One can then ask the question whether shuffle cubes (or, more generally, shuffle k-powers) are avoidable over an alphabet of size m. In other terms, the question is whether there exists a word of an arbitrary length, over an alphabet of size m, that does not contain as factor a shuffle cube.

Remark that, contrary to the case of the usual product, in general a shuffle cube does not contain as factor a shuffle square, as shown by the following example found by Romeo Rizzi (private communication). Consider the word

$$u = abcabadcdbcd.$$

It easy to see u is a shuffle cube: $u \in v \sqcup\!\sqcup v \sqcup\!\sqcup v$, where $v = abcd$. If a factor of u is a shuffle square, then it must contain an even number of occurrences of each letter. There are only three factors of u having this property, and one can easily check that they are not shuffle squares. Hence, the avoidability of shuffle cubes over a finite alphabet cannot be obtained as a consequence of Theorem 16. This leads to the following

Problem 10. *Prove that shuffle cubes are avoidable over a finite alphabet*

References

1. Baeten, J., Weijland, W.: Process Algebra. Cambridge University Press (1990)
2. ter Beek, M.H., Martin-Vide, C., Mitrana, V.: Synchronized shuffles. Theoretical Computer Science **341**(1-3), 263–275 (2005)
3. Berstel, J., Boasson, L., Carton, O., Pin, J.É., Restivo, A.: The expressive power of the shuffle product. Inf. Comput. **208**(11), 1258–1272 (2010)
4. Buss, S., Soltys, M.: Unshuffling a square is np-hard. J. Comput. Syst. Sci. **80**(4), 766–776 (2014)

5. Gomez, A.C., Guaiana, G., Pin, J.É.: Regular languages and partial commutations. Information and Computation **230**, 76–96 (2013)
6. Castiglione, G., Restivo, A.: On the shuffle of star-free languages. Fundam. Inform. **116**(1–4), 35–44 (2012)
7. Currie, J.: Shuffle squares are avoidable. Manuscript
8. Currie, J.D., Saari, K.: Square-free words with square-free self-shuffles. Electr. J. Comb., 1–9 (2014)
9. Diekert, V., Rozenberg, G.: The Book of Traces. World Scientific, River Edge (1995)
10. Eilenberg, S.: Automata, Languages, and Machines. Academic Press Inc., Orlando (1976)
11. Ginsburg, S., Spanier, E.H.: Mappings of languages by two-tape devices. J. ACM **12**(3), 423–434 (1965)
12. Gómez, A.C., Pin, J.É.: Shuffle on positive varieties of languages. Theoretical Computer Science **312**(23), 433–461 (2004)
13. Gómez, A.C., Pin, J.É.: A robust class of regular languages. In: Ochmański, E., Tyszkiewicz, J. (eds.) MFCS 2008. LNCS, vol. 5162, pp. 36–51. Springer, Heidelberg (2008)
14. Gonçalves, D., Montassier, M., Pinlou, A.: Entropy compression method applied to graph colorings. CoRR abs/1406.4380 (2014). http://arxiv.org/abs/1406.4380
15. Guaiana, G., Restivo, A., Salemi, S.: Star-free trace languages. Theor. Comput. Sci. **97**(2), 301–311 (1992)
16. Guegan, G., Ochem, P.: Avoiding shuffle squares using entropy compression. 15e Journées Montoises d'Informatique Théorique (2014)
17. Harju, T.: A note on square-free shuffles of words. In: Karhumäki, J., Lepistö, A., Zamboni, L. (eds.) WORDS 2013. LNCS, vol. 8079, pp. 154–160. Springer, Heidelberg (2013)
18. Harju, T., Müller, M.: Square-free shuffles of words. CoRR abs/1309.2137 (2013). http://arxiv.org/abs/1309.2137
19. Henshall, D., Rampersad, N., Shallit, J.: Shuffling and unshuffling. Bulletin of the EATCS **107**, 131–142 (2012)
20. Aoki, H., Uehara, R., Yamazaki, K.: Expected length of longest common subsequences of two biased random strings and its application. In: LA Symposium (2001)
21. Kececioglu, J., Gusfield, D.: Reconstructing a history of recombinations from a set of sequences. Discrete Applied Mathematics **88**(13), 239–260 (1998). computational Molecular Biology DAM - CMB Series
22. Kimura, T.: An algebraic system for process structuring and interprocess communication. In: STOC 1976, pp. 92–100 (1976)
23. van Leeuwen, J., Nivat, M.: Efficient recognition of rational relations. Information Processing Letters **14**(1), 34–38 (1982)
24. Lothaire, M.: Algebraic Combinatorics on Words. Cambridge University Press (2002)
25. Lyndon, R.C., Schützenberger, M.P.: The equation $a^m = b^n c^p$ in a free group. Michigan Math. J. **9**(4), 289–298 (1962)
26. Mansfield, A.: An algorithm for a merge recognition problem. Discrete Applied Mathematics **4**(3), 193–197 (1982)
27. Mansfield, A.: On the computational complexity of a merge recognition problem. Discrete Applied Mathematics **5**(1), 119–122 (1983)
28. McNaughton, R., Papert, S.: Counter-Free Automata. MIT Press, Cambridge (1971)

29. Perrot, J.F.: Varietes de langages et operations. Theor. Comput. Sci. **7**, 197–210 (1978)
30. Pin, J.É.: Varieties of formal languages. North Oxford, London and Plenum (1986). (Traduction de Variétés de langages formels)
31. Pin, J.-É.: Syntactic semigroups. In: Rozenberg, G., Salomaa, A. (eds.) Handbook of Formal Languages, chap. 10, vol. 1, pp. 679–746. Springer (1997)
32. Pin, J.É.: Theme and variations on the concatenation product. In: Winkler, F. (ed.) CAI 2011. LNCS, vol. 6742, pp. 44–64. Springer, Heidelberg (2011)
33. Rizzi, R., Vialette, S.: On recognizing words that are squares for the shuffle product. In: Bulatov, A.A., Shur, A.M. (eds.) CSR 2013. LNCS, vol. 7913, pp. 235–245. Springer, Heidelberg (2013)
34. Shaw, A.: Software descriptions with flow expressions. IEEE Transactions on Software Engineering **SE–4**(3), 242–254 (1978)
35. Shyr, H.J., Yu, S.S.: Non-primitive words in the language p^+q^+. Soochow J. Math. **20**(4), 535–546 (1994)
36. Tesson, P., Therien, D.: Diamonds are forever: The variety da. In: Semigroups, Algorithms, Automata and Languages, Coimbra (Portugal) 2001, pp. 475–500. World Scientific (2002)
37. Warmuth, M.K., Haussler, D.: On the complexity of iterated shuffle. Journal of Computer and System Sciences **28**(3), 345–358 (1984)

Algorithms

Average-Case Optimal Approximate Circular String Matching

Carl Barton[1][(✉)], Costas S. Iliopoulos[1,2], and Solon P. Pissis[1]

[1] Department of Informatics, King's College London, The Strand, London, UK
{carl.barton,costas.iliopoulos,solon.pissis}@kcl.ac.uk
[2] Department of Mathematics and Statistics, University of Western Australia,
35 Stirling Highway, Perth, Australia

Abstract. Approximate string matching is the problem of finding all factors of a text t of length n that are at a distance at most k from a pattern x of length m. Approximate circular string matching is the problem of finding all factors of t that are at a distance at most k from x *or* from any of its rotations. In this article, we present a new algorithm for approximate circular string matching under the edit distance model with optimal average-case search time $\mathcal{O}(n(k + \log m)/m)$. Optimal average-case search time can also be achieved by the algorithms for multiple approximate string matching (Fredriksson and Navarro, 2004) using x and its rotations as the set of multiple patterns. Here we reduce the preprocessing time and space requirements compared to that approach.

Keywords: Algorithms on automata and words · Average-case complexity · Average-case optimal · Approximate string matching

1 Introduction

In order to provide an overview of our results and algorithms, we begin with a few definitions, generally following [4]. We think of a *string* x of *length* n as an array $x[0 \mathinner{.\,.} n-1]$, where every $x[i]$, $0 \le i < n$, is a *letter* drawn from some fixed *alphabet* Σ of size $\sigma = \mathcal{O}(1)$. By a *q-gram* we refer to any string $x \in \Sigma^q$. The *empty string* of length 0 is denoted by ε. A string x is a *factor* of a string y if there exist two strings u and v, such that $y = uxv$. Consider the strings $x, y, u,$ and v, such that $y = uxv$. If $u = \varepsilon$, then x is a *prefix* of y. If $v = \varepsilon$, then x is a *suffix* of y. Let x be a non-empty string of length n and y be a string. We say that there exists an *occurrence* of x in y, or, more simply, that x *occurs in* y, when x is a factor of y. Every occurrence of x can be characterised by a position in y. Thus we say that x occurs at the *starting position* i in y when $y[i \mathinner{.\,.} i+n-1] = x$. Given a string x of length m and a string y of length $n \ge m$, the *edit distance*, denoted by $\delta_E(x, y)$, is defined as the minimum total cost of operations required to transform one string into the other. For simplicity, we only count the number of edit operations, considering the cost of each to be 1 [15]. The allowed edit operations are as follows:

Solon P. Pissis - Supported by a London Mathematical Society grant (no. 51303).

A.-H. Dediu et al. (Eds.): LATA 2015, LNCS 8977, pp. 85–96, 2015.
DOI: 10.1007/978-3-319-15579-1_6

- *Insertion*: insert a letter in y, not present in x; (ε, b), $b \neq \varepsilon$
- *Deletion*: delete a letter in y, present in x; (a, ε), $a \neq \varepsilon$
- *Substitution*: replace a letter in y with a letter in x; (a, b), $a \neq b$, and $a, b \neq \varepsilon$.

We write $x \equiv_k^E y$ if the edit distance between x and y is at most k. Equivalently, if $x \equiv_k^E y$, we say that x and y have at most k *differences*. We refer to the *standard dynamic programming matrix* of x and y as the matrix defined by $D[i, 0] = i$, $0 \leq i \leq m$, $D[0, j] = j$, $0 \leq j \leq n$

$$D[i, j] = \min \begin{cases} D[i-1, j-1] + (1 \text{ if } x[i-1] \neq y[j-1]) \\ D[i-1, j] + 1 \\ D[i, j-1] + 1 \end{cases}, 1 \leq i \leq m, 1 \leq j \leq n.$$

Similarly we refer to the *standard dynamic programming algorithm* as the algorithm to compute the edit distance between x and y through the above recurrence in time $\mathcal{O}(mn)$. Given a non-negative integer threshold k for the edit distance, this can be computed in time $\mathcal{O}(mk)$ [17]. We say that there exists an *occurrence* of x in y with at most k differences, or, more simply, that x *occurs in* y with at most k differences, when $u \equiv_k^E x$ and u is a factor of y.

A circular string of length n can be viewed as a traditional linear string which has the left- and right-most symbols wrapped around and stuck together in some way. Under this notion, the same circular string can be seen as n different linear strings, which would all be considered equivalent. Given a string x of length n, we denote by $x^i = x[i..n-1]x[0..i-1]$, $0 < i < n$, the i-th *rotation* of x and $x^0 = x$. Consider, for instance, the string $x = x^0 = \text{abababbc}$; this string has the following rotations: $x^1 = \text{bababbca}$, $x^2 = \text{ababbcab}$, $x^3 = \text{babbcaba}$, $x^4 = \text{abbcabab}$, $x^5 = \text{bbcababa}$, $x^6 = \text{bcababab}$, $x^7 = \text{cabababb}$.

This type of structure occurs in the DNA of viruses, bacteria, eukaryotic cells, and archaea. In [9], it was noted that, due to this, algorithms on circular strings may be important in the analysis of organisms with such structure. For instance, circular strings have been studied before in the context of sequence alignment. In [5,14], algorithms for multiple circular sequence alignment were presented. Here we consider the problem of finding occurrences of a pattern x of length m with circular structure in a text t of length n with linear structure. This is the problem of *circular string matching*.

The problem of exact circular string matching has been considered in [16], where an $\mathcal{O}(n)$-time algorithm was presented. The approach presented in [16] consists of preprocessing x by constructing a *suffix automaton* of the string xx, by noting that every rotation of x is a factor of xx. Then, by feeding t into the automaton, the lengths of the longest factors of xx occurring in t can be found by the links followed in the automaton in time $\mathcal{O}(n)$. In [6], an average-case optimal algorithm for exact circular string matching was presented and it was also shown that the average-case lower bound for single string matching of $\Omega(n \log_\sigma m/m)$ also holds for circular string matching. Very recently, in [3], the authors presented two fast average-case algorithms based on word-level parallelism. The first algorithm requires average-case time $\mathcal{O}(n \log_\sigma m/w)$, where w is the number of bits in the computer word. The second one is based on a mixture of word-level parallelism and q-grams. The authors showed that with the addition of q-grams, and by setting $q = \Theta(\log_\sigma m)$, an average-case optimal time of $\mathcal{O}(n \log_\sigma m/m)$ is achieved. Indexing circular patterns [12] based on the construction of *suffix tree*—have also been considered.

The aforementioned algorithms for the exact case have the disadvantage that they cannot be applied in a biological context since single nucleotide polymorphisms and errors introduced by wet-lab sequencing platforms might have occurred in the sequences; also it is not clear whether they could easily be adapted to deal with the approximate case. For the rest of the article, we assume that each position in the text t is uniformly randomly drawn from Σ, and consider the following problem.

APPROXIMATECIRCULARSTRINGMATCHING
Input: a pattern x of length m, a text t of length $n > m$, and an integer threshold $k < m$
Output: all factors u of t such that $u \equiv_k^E x^i$, $0 \leq i < m$

Similar to the exact case [6], it can be shown that the average-case lower bound for single approximate string matching of $\Omega(n(k + \log_\sigma m)/m)$ [2] also holds for approximate circular string matching under the edit distance model. Recently, we have presented average-case $\mathcal{O}(n)$-time algorithms for approximate circular string matching which are also very efficient in practice [1]. In [10], an algorithm with $\mathcal{O}(\frac{nk \log m}{m})$ average-case search time was presented. To achieve average-case optimality, one could use the algorithms for multiple approximate string matching, presented in [8], for matching the $r = m$ rotations of x with $\mathcal{O}(n(k + \log_\sigma rm)/m)$ average-case search time, only if $k/m < 1/2 - \mathcal{O}(1/\sqrt{\sigma})$ and $r = \mathcal{O}(\min(n^{1/3}/m^2, \sigma^{o(m)}))$. Therefore the focus of this article is on a more *direct* algorithm which also improves on the preprocessing time and space complexity.

Our Contribution. In this article, we present a new average-case optimal algorithm for approximate circular string matching, under the edit distance model, that reduces the preprocessing time and space requirements compared to previous algorithms with optimal average-case search time. These savings are around $\mathcal{O}(m^2)$ or more in all cases.

2   Algorithm

In this section, we present our algorithm for approximate circular string matching under the edit distance model. The presented algorithm consists of two distinct schemes: the *searching* scheme, which determines if the currently considered text window potentially has a valid occurrence; in case the window *may* contain a valid occurrence, we are required to check the window for valid occurrences of the pattern or any of its rotations; this is done through the *verification* scheme.

Intuitively, the algorithm considers a *sliding window* of length m-k of the text, and reads q-grams backwards from the end of the window until it is likely to have found enough *differences* to skip the entire window. That is, we wish to make the probability of a verification being triggered sufficiently unlikely whilst also ensuring we can shift the window a reasonable amount.

The rest of this section is structured as follows. We first present an efficient incremental string comparison technique which forms the basis of the verification scheme. We then present the searching scheme of our algorithm which requires a preprocessing step. In fact, this preprocessing step is similar to the verification scheme. Finally, we show how plugging these schemes together results in a new average-case optimal algorithm for approximate circular string matching.

2.1   Verification Scheme

The verification scheme of our algorithm is based on incremental string comparison techniques. First we give an introduction to these techniques; and then explain how we use them in the verification scheme. The incremental string comparison problem was introduced in the pioneering work of Landau *et al* [13]. The authors considered the following problem: given the edit distance between two strings A and B, how can the edit distance between A and bB or Bb be efficiently derived, where b is an additional letter. Given a threshold on the number of differences k, they solve this problem and allow prepending and appending of letters in time $\mathcal{O}(k)$ per operation. Later the authors of [11] considered a generalisation of this problem with the aim of computing all maximal gapped palindromes in a string. The problem considered is a generalisation of the incremental string comparison problem considered in [13] as it considers how to efficiently derive the edit distance when prefixes are deleted and letters are prepended to A or B. The solution proposed in [11] also has a time complexity of $\mathcal{O}(k)$ per operation. The solution for the generalised incremental string comparison problem forms the basis of our verification step. The technique lends itself more naturally to circular string matching due to the increased flexibility it provides. We begin by recalling some of the main results from [11] required for our algorithm.

The main idea in both [13] and [11] is the efficient computation of the so-called h-waves. In the standard dynamic programming matrix for two strings x and y, we say that a cell $D[i, j]$ is on the diagonal d *iff* $j - i = d$. For each diagonal, we may have a lowest cell with value h; if $D[i, j] = h$ and $D[i+1, j+1] = h+1$ then $D[i, j]$ is this cell for diagonal $j - i$. The h-wave, for all $0 \leq h \leq k$, is the position of all these cells across all diagonals, that is, a list H_h of length $\mathcal{O}(k)$, where each entry is a pair (i, j) such that $D[i, j] = h$ and $D[i + 1, j + 1] = h + 1$. Note that the i-th wave can only contain entries on diagonal zero and the i diagonals either side of it, so for $0 \leq i \leq k$ every wave has size $\mathcal{O}(k)$. Both incremental string comparison techniques show some bounds on the possible values of the cells on h-waves and how to efficiently compute them. These h-waves define the entire dynamic programming matrix due to the monotonicity properties of the matrix. For any diagonal d, if we know the position of the lowest cell on d with value h and $h + 1$, then we also know the value of every cell between these two cells: it must be $h + 1$. So given the h-waves of the matrix, for all $0 \leq h \leq k$, we have all the information that is in the standard dynamic programming matrix. The key result from our perspective is the following.

Let $\mathsf{cat}(u', u)$ denote the string obtained by concatenating u' and u, where $u, u' \in \Sigma^+$. Let $\mathsf{del}(\alpha, u)$ denote the string obtained by deleting the prefix of length α of u. Let D' denote the standard dynamic programming matrix for strings $\mathsf{cat}(\mathsf{A}', \mathsf{A})$ and $\mathsf{del}(t_2, \mathsf{B})$, where $|\mathsf{A}'| = t_1$.

Theorem 1 ([11]). *The 0-wave, 1-wave, ... , and k-wave of matrix D' can be computed in time $\mathcal{O}((t_1 + t_2)k)$.*

If a window of the text triggers a verification then we have a window of length $m - k$ such that there exist some q-grams of the window that occur in x or its rotations with at most k differences in total. When we verify a window, we check for occurrences of pattern x starting at every position in the window. For each position, we may have a factor of length at most $m + k$ representing an occurrence, meaning we must consider a factor w of the text of length $2m$ which we refer to as a *block*. This ensures we avoid missing any occurrences at the $m - k$ starting positions as $(m - k) + (m + k) = 2m$.

For each possible starting position i, $0 \leq i < m - k$, we compute the 0-wave, 1-wave, ... , and k-wave for x and $w' = w[i\,..\,2m - 1]$, the suffix of w starting at position i. To check if we have an occurrence, we must check the k-wave H_k. We iterate through each entry in the k-wave H_k; and if H_k has missing entries or contains entries on the last row of the matrix, then x occurs in w with at most k differences.

Similarly we can check for the occurrences of the rotations of x using the incremental string comparison techniques. We are now ready to outline the verification scheme, denoted by function VER. Given the pattern x of length m, an integer threshold $k < m$, and a block w of length $2m$ of the text t, function VER finds all factors u of w such that $u \equiv_k^E x^i$, $0 \leq i < m$. If any diagonal has no entry on the k-wave then that diagonal reached the last row of the matrix with less than k differences; this means x occurs in w with less than k differences.

Function *VER*$(x, m, k, w, 2m)$

> Compute the edit distance between x and $w' = w[0\,..\,2m - 1]$ with at most k differences using the standard dynamic programming algorithm;
>
> Check for any occurrences using D, and if found, **report** an occurrence at position 0;
>
> **foreach** $i \in \{1, m - k - 1\}$ **do**
>
>> **foreach** $j \in \{1, m\}$ **do**
>>
>>> Construct rotation x^j of x by removing the first letter of x^{j-1} and appending it to the end of x^{j-1};
>>>
>>> Compute the edit distance between x^j and $w' = w[i\,..\,2m - 1]$ using the incremental string comparison techniques;
>>>
>>> Check for any occurrences using H_k, and if found, **report** an occurrence at the current position i being checked;

Lemma 2. *Given the pattern x of length m, an integer threshold $k < m$, and string w of length $2m$, function* VER *requires time $\mathcal{O}(m^2 k)$.*

Proof. Computing the edit distance between x and $w[0 .. 2m - 1]$ with at most k differences takes time $\mathcal{O}(mk)$ using the standard dynamic programming algorithm. By Theorem 1, computing the edit distance between all the rotations of the pattern and $w[i .. 2m-1]$ for a single position in w requires $\mathcal{O}(mk)$; and there are $\mathcal{O}(m)$ positions in w. In total, the time is $\mathcal{O}(mk + m^2 k)$, that is $\mathcal{O}(m^2 k)$. □

2.2   Searching Scheme

The searching scheme of the presented algorithm requires the preprocessing and indexing of the pattern x. We first present the preprocessing required and then present the searching technique itself.

Preprocessing. We build a q-gram index in a similar way as that proposed by Chang and Marr in [2]. Intuitively, we wish to determine the minimum possible edit distance between every q-gram and any factor of x or its rotations. Equivalently we find the minimum possible edit distance between every q-gram and any *prefix* of a factor of length $2q$ of x and the suffixes of length 1 to $2q$ of x or its rotations. An index like this allows us to lower bound the edit distance between a window of the text and x or its rotations without computing the edit distance between them. To build this index, we generate every string of length q on Σ, and find the minimum edit distance between it and all prefixes of factors of length $2q$ of x or its rotations. This information can easily be stored by generating a numerical representation of the q-gram and storing the minimum edit distance in an array at this location. If we know the numerical representation, we can then look up any entry in constant time.

We determine the edit distance using the preprocessing scheme, denoted by function PRE, which is similar to the verification scheme (function VER).

Given the string $x' = x[0 .. m - 1]x[0 .. m - 2]$ of length $2m - 1$, function PRE finds the minimum edit distance between every q-gram on Σ, generated in increasing order, and any factor u of length $2q$ of x' and its suffixes of length 1 to $2q$.

Lemma 3. *Given the string $x' = x[0 .. m - 1]x[0 .. m - 2]$ of length $2m - 1$ on Σ, $\sigma = |\Sigma|$, and $q < m$, function* PRE *requires time $\mathcal{O}(\sigma^q mq)$ and space $\mathcal{O}(\sigma^q)$.*

Proof. The time required for initialising array M is $\mathcal{O}(\sigma^q)$. The time required for computing the edit distance between $x'[0 .. 2q - 1]$ and s is $\mathcal{O}(q^2)$ using the standard dynamic programming algorithm. By Theorem 1, computing the edit distance between all $2q$-grams of x' and s requires time $\mathcal{O}(mq)$. There exist $\mathcal{O}(\sigma^q)$ possible q-grams on Σ and so, in total, the time complexity is $\mathcal{O}(\sigma^q mq)$. Keeping array M in memory requires space $\mathcal{O}(\sigma^q)$. □

Function $PRE(x', 2m - 1, q, \sigma)$

$\quad$ $\mathsf{M}[0 .. \sigma^q - 1] \leftarrow 0$;

$\quad$ $j \leftarrow 0$;

$\quad$ **foreach** $s \in \Sigma^q$ **do**

$\quad\quad$ Compute the edit distance between $u = x'[0 .. 2q - 1]$ and s using the standard dynamic programming algorithm. Set $E_{\min}$ equal to the minimum edit distance between s and any prefix of u using D;

$\quad\quad$ **foreach** $i \in \{1, 2m - q - 1\}$ **do**

$\quad\quad\quad$ $u \leftarrow x'[i .. \min\{i + 2q - 1, 2m - 2\}]$;

$\quad\quad\quad$ Compute the edit distance E' between u and s using the incremental string comparison techniques. Set E' equal to the minimum edit distance between s and any prefix of u using H_q;

$\quad\quad\quad$ **if** $E' < E_{\min}$ **then** $E_{\min} \leftarrow E'$;

$\quad\quad$ $\mathsf{M}[j] \leftarrow E_{\min}$;

$\quad\quad$ $j \leftarrow j + 1$;

$\quad$ **return** M;

Searching. In the search phase we wish to read backwards enough q-grams from a window of size $m - k$ that the probability we must verify the window is small and the amount we can shift the window by is sufficiently large. We now recall some important lemmas from [2] that we will use in the analysis of our algorithm.

Lemma 4 ([2]). *The probability that two q-grams on Σ, one being uniformly random, have a common subsequence of length $(1 - c)q$ is at most $\frac{a\sigma^{-dq}}{q}$, where $a = (1 + o(1))/(2\pi c(1 - c))$ and $d = 1 - c + 2c\log_\sigma c + 2(1 - c)\log_\sigma(1 - c)$. The probability decreases exponentially for $d > 1$, which holds if $c < 1 - \frac{e}{\sqrt{\sigma}}$.*

Lemma 5 ([2]). *If s is a q-gram occurring with less than cq differences in a given string u, $|u| \geq q$, s has a common subsequence of length $q - cq$ with some q-gram of u.*

By Lemmas 4 and 5, we know that the probability of a random q-gram occurring in a string of length m with less than cq differences is no more than $ma\sigma^{-dq}/q$ as we have $m - q + 1$ q-grams in the string. For circular string matching this is not sufficient. To ensure that we have the q-grams of all possible rotations of pattern x, we instead consider the string $x' = x[0 .. m - 1]x[0 .. m - 2]$ and extract the q-grams from x'. We may have up to $2m - q$ q-grams, but to simplify the analysis we assume we have $2m$ and so the probability becomes $2ma\sigma^{-dq}/q$.

In the case when we read $k/(cq)$ q-grams, we know that with probability at most $(k/(cq))2ma\sigma^{-dq}/q$ we have found less than k differences. This does not permit us to discard the window if all q-grams occur with at most cq differences. To fix this, we instead read $1 + k/(cq)$ q-grams. If any q-gram occurs with less than cq differences, we will need to verify the window; but if they all occur with at least cq differences, we must exceed the threshold k and can shift the window. When shifting the window we have the case that we shift after verifying the window and the case that the differences exceed k so we do not verify the window. If we have verified the window, we can shift past the last position we checked for an occurrence: we can shift by $m-k$ positions. If we have not verified the window, as we read a fixed number of q-grams, we know the minimum-length shift we can make is one position past this point. The length of this shift is at least $m-k-(q+k/c)$ positions. This means we will have at most $\frac{n}{m-k-(q+k/c)} = \mathcal{O}(\frac{n}{m})$ windows. The previous statement is only true assuming $m-q > k+k/c$, as then the denominator is positive. From there we see that we also have the condition that $q+k+k/c$ can be at most ϵm, where $\epsilon < 1$, so the denominator will be $\mathcal{O}(m)$. This puts a slightly stricter condition on c, that is, $c > \frac{k}{\epsilon m-q-k}$.

We can see that, for each window, we verify with probability at most $(1 + k/(cq))2ma\sigma^{-dq}/q$, where $a = (1+o(1))/(2\pi c(1-c))$ and $d = 1-c+2c\log_\sigma c + 2(1-c)\log_\sigma(1-c)$. So the probability that a verification is triggered is

$$\frac{(1+k/(cq))2ma\sigma^{-dq}}{q}.$$

Because by Lemma 2, verification takes time $\mathcal{O}(m^2k)$, then per window, the expected cost is

$$\frac{(1+k/(cq))2ma\sigma^{-dq}\mathcal{O}(m^2k)}{q} = \mathcal{O}(\frac{(q+k)m^3ka\sigma^{-dq}}{q^2}).$$

We wish to ensure that the probability of verifying a window is small enough that the average work done is no more than the work we must do if we skip a window without verification. When we do not verify a window, we read $1 + k/(cq)$ q-grams and shift the window. This means that we read $q+k/c = \mathcal{O}(q+k)$ letters. So a sufficient *condition* is the following:

$$\frac{(q+k)m^3ka\sigma^{-dq}}{q^2} = \mathcal{O}(q+k).$$

Or equivalently the below expression, where f is the constant of proportionality:

$$\frac{(q+k)m^3ka\sigma^{-dq}}{q^2} \le f(q+k).$$

By rearranging and setting $f = \sigma$ we get the condition on the value of q below:

$$q \ge \frac{3\log_\sigma m + \log_\sigma k + \log_\sigma a - 2\log_\sigma q}{d}.$$

From the condition on q we can see that it is sufficient to pick $q = \Theta(\log_\sigma km)$, so asymptotically on m we get the following:

$$q \ge \frac{3\log_\sigma m + \log_\sigma k - \mathcal{O}(\log_\sigma \log_\sigma km)}{d}.$$

Therefore, for sufficiently large m, the below condition is sufficient for optimality, where $d = 1 - c + 2c \log_\sigma c + 2(1 - c) \log_\sigma (1 - c)$:

$$q = \frac{3 \log_\sigma m + \log_\sigma k}{d}.$$

For this analysis to hold we must be able to read the required number of q-grams to ensure the probability of verifying a window is small enough to negate the work of doing it. Note that the above probability is the probability that at least one of q-grams match with less than cq differences. To ensure we have enough unread random q-grams in the window for Lemma 5 to hold in the above analysis the window must be of size $m - k \geq 2q + 2k/c$. Now we consider the case where $2q + 2k/c > m - k \geq 2q + k/c$. If we have just verified a window then we have enough new random q-grams and our analysis holds. If we have just shifted then we know that all the q-grams we previously read matched with at least cq differences and we have between 1 and k/qc q-grams and the probability that one of these matches with less than cq difference is less than in the analysis above so it holds.

The condition $m - k \geq 2q + k/c$ implies a condition on c, it must be the case that $c \geq \frac{k}{m-k-2q}$. This condition on c is weaker than our previous condition on c, so to determine the *error ratio* $\frac{k}{m}$, we use the stronger condition. Additionally, from Lemma 4, we know that $c < 1 - \frac{e}{\sqrt{\sigma}}$. So we must pick a value for c subject to $\frac{k}{em-k-q} \leq c < 1 - \frac{e}{\sqrt{\sigma}}$. This inequality implies a limit on the error ratio for which our algorithm is optimal. Clearly it must be the case that $\frac{k}{em-k-q} < 1 - \frac{e}{\sqrt{\sigma}}$ for $\epsilon < 1$. Rearranging the inequality implies the following sufficient condition on our error ratio:

$$\frac{2k}{m} < \epsilon - \frac{q}{m} - \frac{\epsilon e}{\sqrt{\sigma}} + \frac{qe}{m\sqrt{\sigma}} + \frac{ke}{m\sqrt{\sigma}}.$$

From here we can factorise and divide everything by 2 to get the following:

$$\frac{k}{m} < \frac{\epsilon}{2} - \frac{q}{2m} - \frac{e}{2\sqrt{\sigma}}(\epsilon - \frac{q}{m} - \frac{k}{m}).$$

So asymptotically on m we have:

$$\frac{k}{m} < \frac{\epsilon}{2} - \mathcal{O}(\frac{1}{\sqrt{\sigma}}).$$

Note that this technique can work for any ratio which satisfies $\frac{k}{m} < \frac{1}{2} - \mathcal{O}(\frac{1}{\sqrt{\sigma}})$. For any ratio below this, pick a large enough value for ϵ such that asymptotically on m the algorithm will work in the claimed search time. By choosing a suitable value for c and $q \geq \frac{3 \log_\sigma m + \log_\sigma k}{d}$ we obtain the following result.

Theorem 6. *The problem* APPROXIMATECIRCULARSTRINGMATCHING *can be solved in optimal average-case search time* $\mathcal{O}(n(k + \log_\sigma m)/m)$.

3   Comparison with Existing Algorithms

To the best of our knowledge, the only other algorithms to achieve optimal average-case search time for approximate circular string matching are the algorithms presented in [8] for multiple approximate string matching. In the analysis of the algorithms in [8] it is assumed that all patterns are random. In [7] the authors re-analyse their algorithms for the problem of circular string matching with the same preprocessing and space costs. In this section, we analyse these results and compare them with our own. We refer to the algorithm presented in Section 2 as BIP. Due to the constant c in the value of q from Lemma 4, the exact preprocessing and space costs for these algorithms depend on the chosen value for c. It is however possible to determine the minimum savings we make based on the value of q used in all algorithms.

Applying the algorithms in [8] to approximate circular string matching requires a reduction to multiple approximate string matching for matching the m rotations of x. The first algorithm in [8] has the following time complexity:

$$\mathcal{O}(n(k + \log_\sigma rm)/m).$$

By setting $r = m$ this matches our search time and the result is valid when $k/m < 1/2 - \mathcal{O}(1/\sqrt{\sigma})$, $r = \mathcal{O}(\min(n^{1/3}/m^2, \sigma^{o(m)}))$, and we have $\mathcal{O}(\sigma^q)$ space available, where q is subject to the constraint:

$$q \geq \frac{4\log_\sigma m + 2\log_\sigma r}{d}.$$

Again by setting $r = m$ this becomes $q \geq \frac{6\log_\sigma m}{d}$ and the preprocessing time is $\mathcal{O}(\sigma^q m^2)$. We will refer to this algorithm as FN1. The second algorithm, presented in [8], has the same preprocessing cost and requires space $\mathcal{O}(\sigma^q m)$. We will refer to this algorithm as FN2. The important difference between FN1 and FN2 comes in the condition on q which is slightly lower for FN2:

$$q \geq \frac{3\log_\sigma m + \log_\sigma r + \log_\sigma(m + \log_2 r)}{d}.$$

Again, setting $r = m$ this becomes:

$$q \geq \frac{4\log_\sigma m + \log_\sigma(m + \log_2 m)}{d}.$$

To simplify the comparison between these approaches, we will ignore the factor of $\log_2 m$, and simply say that the value of q for algorithm FN2 is greater than or equal to $\frac{5\log_\sigma m}{d}$. This is lower than the sufficient requirement, so any saving we make using this value must be at least as good or better in reality.

First let us consider FN1. The preprocessing requirement of BIP is $\mathcal{O}(\sigma^q mq)$, so before any savings made due to the value of q for BIP, we have reduced the preprocessing cost by a factor of $\mathcal{O}(\frac{m}{q})$. Given the condition on q for BIP, it is clear that even in the worst case, when $k = \mathcal{O}(m)$, BIP will make a saving of at least $2\log_\sigma m$ on the value of q. This corresponds to an additional saving of

$\mathcal{O}(m^2)$ in preprocessing time bringing the total to $\mathcal{O}(\frac{m^3}{q})$ and $\mathcal{O}(m^2)$ in space. In the case of FN2, we make a saving of at least $\log_\sigma m$ on the value of q. This corresponds to a total saving of $\mathcal{O}(\frac{m^2}{q})$ in preprocessing time and $\mathcal{O}(m^2)$ in space. It should be noted that this is a pessimistic analysis of the savings as we have assumed $k = \mathcal{O}(m)$ and $d = 1$, although it must hold that $d < 1$. Note that the standard dynamic programming algorithm can be used with runtime $\mathcal{O}(m^3)$ for verification and $\mathcal{O}(\sigma^q m q^2)$ for preprocessing. The speed-ups mentioned in the previous section remain significant as we assumed that $k = \mathcal{O}(m)$. We still achieve a preprocessing speed up of at least $\mathcal{O}(m^2)$ and $\mathcal{O}(m)$ against FN1 and FN2, respectively. Table 1 corresponds to this analysis.

Table 1. Comparison of average-case optimal approximate circular string matching algorithms

Algorithm	Error Ratio (k/m)	Space	Preprocessing Time	Condition on q
FN1	$\frac{1}{2} - \mathcal{O}(\frac{1}{\sqrt{\sigma}})$	$\mathcal{O}(\sigma^q)$	$\mathcal{O}(\sigma^q m^2)$	$\frac{6\log_\sigma m}{d}$
FN2	$\frac{1}{2} - \mathcal{O}(\frac{1}{\sqrt{\sigma}})$	$\mathcal{O}(\sigma^q m)$	$\mathcal{O}(\sigma^q m^2)$	$\frac{4\log_\sigma m + \log_\sigma(m + \log_2 m)}{d}$
BIP	$\frac{1}{2} - \mathcal{O}(\frac{1}{\sqrt{\sigma}})$	$\mathcal{O}(\sigma^q)$	$\mathcal{O}(\sigma^q m q)$	$\frac{3\log_\sigma m + \log_\sigma k}{d}$

4   Final Remarks

In this article, we presented a new average-case optimal algorithm for approximate circular string matching. To the best of our knowledge, this algorithm is the first average-case optimal algorithm specifically designed for this problem. Other average-case optimal algorithms exist but with higher preprocessing and space requirements than the presented algorithm. Additionally the considered problem is solved in a more direct fashion, that is, with no reduction to multiple approximate string matching by taking greater advantage of the similarity of the rotations of the pattern.

Our immediate target is twofold:

- first, we plan on tackling the problem of multiple approximate circular string matching. We will try to generalise the approach we have taken here to see if it leads to an average-case optimal algorithm in this case.
- second, we plan on implementing the presented algorithm. We will then compare the respective implementation to other average- and worst-case approaches.

References

1. Barton, C., Iliopoulos, C.S., Pissis, S.P.: Fast algorithms for approximate circular string matching. Algorithms for Molecular Biology **9**(1), 9 (2014). http://www.almob.org/content/9/1/9

2. Chang, W.I., Marr, T.G.: Approximate string matching and local similarity. In: Crochemore, M., Gusfield, D. (eds.) CPM 1994. LNCS, vol. 807, pp. 259–273. Springer, Heidelberg (1994)
3. Chen, K.H., Huang, G.S., Lee, R.C.T.: Bit-Parallel Algorithms for Exact Circular String Matching. The Computer Journal (2013)
4. Crochemore, M., Hancart, C., Lecroq, T.: Algorithms on Strings. Cambridge University Press, New York (2007)
5. Fernandes, F., Pereira, L., Freitas, A.T.: CSA: An efficient algorithm to improve circular DNA multiple alignment. BMC Bioinformatics 10(1), 1–13 (2009)
6. Fredriksson, K., Grabowski, S.: Average-optimal string matching. Journal of Discrete Algorithms 7(4), 579–594 (2009)
7. Fredriksson, K., Mäkinen, V., Navarro, G.: Flexible music retrieval in sublinear time. International Journal of Foundations of Computer Science 17(06), 1345–1364 (2006). http://www.worldscientific.com/doi/abs/10.1142/S0129054106004455
8. Fredriksson, K., Navarro, G.: Average-optimal single and multiple approximate string matching. Journal of Experimental Algorithmics 9, December 2004. http://doi.acm.org/10.1145/1005813.1041513
9. Gusfield, D.: Algorithms on Strings. Cambridge University Press, Trees and Sequences (1997)
10. Hirvola, T., Tarhio, J.: Approximate online matching of circular strings. In: Gudmundsson, J., Katajainen, J. (eds.) SEA 2014. LNCS, vol. 8504, pp. 315–325. Springer, Heidelberg (2014)
11. Hsu, P.-H., Chen, K.-Y., Chao, K.-M.: Finding all approximate gapped palindromes. In: Dong, Y., Du, D.-Z., Ibarra, O. (eds.) ISAAC 2009. LNCS, vol. 5878, pp. 1084–1093. Springer, Heidelberg (2009)
12. Iliopoulos, C.S., Rahman, M.S.: Indexing circular patterns. In: Nakano, S., Rahman, M.S. (eds.) WALCOM 2008. LNCS, vol. 4921, pp. 46–57. Springer, Heidelberg (2008)
13. Landau, G.M., Myers, E.W., Schmidt, J.P.: Incremental string comparison. SIAM Journal of Computing 27(2), 557–582 (1998)
14. Lee, T., Na, J.C., Park, H., Park, K., Sim, J.S.: Finding optimal alignment and consensus of circular strings. In: Amir, A., Parida, L. (eds.) CPM 2010. LNCS, vol. 6129, pp. 310–322. Springer, Heidelberg (2010)
15. Levenshtein, V.I.: Binary codes capable of correcting deletions, insertions, and reversals. Tech. Rep. 8 (1966)
16. Lothaire, M.: Applied Combinatorics on Words. Cambridge University Press (2005)
17. Ukkonen, E.: On approximate string matching. In: Karpinski, M. (ed.) FCT 1983. LNCS, vol. 158, pp. 487–495. Springer, Heidelberg (1983)

An Efficient Best-Trees Algorithm for Weighted Tree Automata over the Tropical Semiring

Johanna Björklund, Frank Drewes$^{(\boxtimes)}$, and Niklas Zechner

Department of Computing Science, Umeå University, 901 87 Umeå, Sweden
{johanna,drewes,zechner}@cs.umu.se

Abstract. We generalise a search algorithm by Mohri and Riley from strings to trees. The original algorithm takes as input a weighted automaton M over the tropical semiring, together with an integer N, and outputs N strings of minimal weight with respect to M. In our setting, M defines a weighted tree language, again over the tropical semiring, and the output is a set of N trees with minimal weight. We prove that the algorithm is correct, and that its time complexity is a low polynomial in N and the relevant size parameters of M.

1 Introduction

Tree automata are useful in natural language processing (NLP), not least to describe the derivation trees of context-free grammars in an automata-theoretic way. Since data-driven approaches were made feasible through the availability of large-scale corpora, weighted grammars have increased in popularity, and with them, so have weighted tree automata [4]. The weights assigned to transitions in these devices allow analyses to be computed together with, for example, an associated confidence level or a probability. This is helpful when we want to assess the quality of an analysis, or when there are several competing analyses to choose between.

At a higher level of abstraction, selecting the right analysis consists in optimising some objective function f over the set A of all possible analyses. Huang and Chiang [6] observe that it may not be tractable to compute $f(a)$ for every single $a \in A$, but that we may obtain a satisfactory approximation by first ranking the elements of A according to a simpler function, computing an N-best list $a_1, \ldots, a_N$ according to this ranking, and finally optimising f over $\{a_1, \ldots, a_N\}$. Examples include reranking the hypotheses produced by parsers or translation systems, where the reranking is based on auxiliary language models or evaluation scores orthogonal to the first round of analysis; see, e.g. [3,9].

There are other situations in which an N-best analysis can be used for approximation. Suppose for instance that the analysis is computed by a cascade of computational modules, a common architecture for NLP systems [6]. Each module typically comes with its own objective function, and the goal is to optimise these jointly. Although it might not be possible to compute the full set of outputs from each module, we may again settle for the N best outputs from each module, and

© Springer International Publishing Switzerland 2015
A.-H. Dediu et al. (Eds.): LATA 2015, LNCS 8977, pp. 97–108, 2015.
DOI: 10.1007/978-3-319-15579-1_7

propagate them downstream. In their paper Huang and Chiang provide several examples of this technique, including joint parsing and semantic role labeling, and combined information extraction and coreference resolution.

In the majority of the above-mentioned applications, the weights represent probabilities and are as such taken from the interval of real values between zero and one. However, for the sake of numerical precision, negative log likelihoods are used in the actual computations, and the min operation is used to find the most likely analysis. This makes the min-plus semiring (or *tropical semiring*) ($\mathbb{R}_+ \cup \{\infty\}, \min, +, 0, \infty$) an appropriate structure for transition weights. Alternatively, the max-plus semiring ($\mathbb{R}_+ \cup \{-\infty\}, \max, +, 0, -\infty$) may be used.

In this paper, we focus on the case where trees are associated with weights by means of a weighted tree automaton (wta) over the tropical semiring. Thus, the weight of a computation, called a run, is the sum of the weights of the rules applied, and the weight of a tree is the minimum of the set of all runs on that tree. Note that the latter is only relevant if the automaton is nondeterministic. In [6] Huang and Chiang give an $O(m + D \cdot N \log N)$ algorithm for (essentially) finding a set S of N best runs in an acyclic wta, where m is the number of transitions and D is the size of the largest run in S. However, as pointed out by Mohri and Riley [8], one would usually rather determine the N best *trees*, because the trees correspond to the analyses and it is not very useful to obtain the same analysis twice in an N-best list just because it corresponds to several distinct runs of the nondeterministic automaton that implements the weight assignment. Unfortunately, determining the N best trees is a harder problem. Part of the difficulty lies in the fact that weighted automata are not closed under determinisation. In fact, both in the string and in the tree case the set of weighted languages recognisable by deterministic weighted automata is a proper subset of those recognisable by nondeterministic weighted automata. When the standard determinisation algorithm is applied to an automaton of this kind, the algorithm will not terminate but continue forever to build up an ever-increasing state space.

Mohri and Riley [8] solve the problem of finding the N best strings, where the input is a weighted string automaton (wsa) over the tropical semiring (and the number N). To avoid computing redundant paths, they apply Dijkstra's N-shortest paths algorithm to a determinised version of the input automaton. Their algorithm applies the determinisation algorithm under a lazy evaluation scheme to guarantee termination and keep the running time polynomial. We generalise this algorithm to weighted tree languages, while simplifying the technique by working directly with the input automaton rather than an on-the-fly determinisation. The frontier is no longer a set of paths, but rather a set of trees that are combined and recombined into new trees to drive the search. This increased dimensionality creates an efficiency problem which we solve by a pruning technique.

Owing to space limitations, some of the proofs had to be abridged or left out. A detailed treatment can be found in [1], which can be downloaded from http://www8.cs.umu.se/research/uminf/index.cgi?year=2014&number=22.

2   Preliminaries

We write $\mathbb{N}$ for the set of nonnegative integers and $\mathbb{R}_+$ for the set of non-negative reals; $\mathbb{R}_+^\infty$ denotes $\mathbb{R}_+ \cup \{\infty\}$. For $n \in \mathbb{N}$, $[n] = \{1, \dots, n-1\}$. In particular, $[0] = \emptyset$. The number of elements of a (possibly infinite) set S is written $|S|$, and the powerset of S is denoted by $pow(S)$. The empty string is denoted by λ.

The estimation of the running time of our algorithm contains the factor $\log r$, where r is the maximum rank of symbols in the ranked alphabet considered (see below for the definitions). To avoid the technical problem that $\log 1 = 0$ we use the convention that, throughout this paper, $\log r$ abbreviates $\max(1, \log r)$.

For a set A, an A-labelled *tree* is a function $t\colon D \to A$ where $D \subseteq \mathbb{N}^*$ is such that, for every $v \in D$, there exists a $k \in \mathbb{N}$ with $\{i \in \mathbb{N} \mid vi \in D\} = [k]$. We call D the *domain* of t and denote it by $dom(t)$. An element v of $dom(t)$ is called a *node of t*, and k is the *rank of v*. The *subtree* of $t \in T_\Sigma$ rooted at v is the tree t/v defined by $dom(t/v) = \{u \in \mathbb{N}^* \mid vu \in dom(t)\}$ and $t/v(u) = t(vu)$ for every $u \in \mathbb{N}^*$. If $t(\lambda) = f$ and $t/i = t_i$ for all $i \in [k]$, where k is the rank of λ in t, then we denote t by $f[t_1, \dots, t_k]$. If $k = 0$, then $f[]$ is usually abbreviated as f. In other words, a tree t with domain $\{\lambda\}$ is identified with $t(\lambda)$.

A *ranked alphabet* is a finite set of symbols $\Sigma = \bigcup_{k \in \mathbb{N}} \Sigma_{(k)}$ which is partitioned into pairwise disjoint subsets $\Sigma_{(k)}$. For every $k \in \mathbb{N}$ and $f \in \Sigma_{(k)}$, the *rank* of f is $rank(f) = k$. The set T_Σ of all trees over Σ consists of all Σ-labelled trees t such that the rank of every node $v \in dom(t)$ coincides with the rank of $t(v)$. For a set T of trees we denote by $\Sigma(T)$ the set of all trees $f[t_1, \dots, t_k]$ such that $f \in \Sigma_{(k)}$ and $t_1, \dots, t_k \in T$.

Let Σ be a ranked alphabet and let $\square \notin \Sigma$ be a special symbol of rank 0. The set of *contexts over Σ* is the set C_Σ consisting of all $c \in T_{\Sigma \cup \{\square\}}$ such that there is exactly one $v \in dom(c)$ with $c(v) = \square$. The substitution of a tree t for $\square$ in c is defined as usual, and is denoted by $c[t]$.

A *weighted tree language* over the tropical semiring is a mapping $L\colon T_\Sigma \to \mathbb{R}_+^\infty$, where Σ is a ranked alphabet. Such languages can be specified by the use of so-called weighted tree automata (wta), of which there exist variants with *final weights* and with *final states*. As shown by Borchardt [2] these two variants are equivalent, and going from final weights to final states only requires a single additional state (which becomes the unique final state) and, in the worst case, twice as many transitions. This means that all results shown in this paper, including the running time estimations, hold for both types of wta.

Formally, a *weighted tree automaton* is a system $M = (Q, \Sigma, \delta, Q_f)$ where

- Q is a finite set of *states* which are considered as symbols of rank 0;
- Σ is a ranked alphabet of *input symbols* disjoint with Q;
- $\delta\colon \Sigma(Q) \times Q \to \mathbb{R}_+^\infty$ is the *transition function*; and
- $Q_f \subseteq Q$ is the set of *final states*.

Note that the transition function δ can be specified as a set of all transition rules $f[q_1, \dots, q_k] \xrightarrow{w} q$ such that $\delta(f[q_1, \dots, q_k], q) = w \neq \infty$. In particular, transition rules whose weight is ∞ are not mentioned explicitly. In the following, we let $|\delta|$ denote the number of transition rules describing δ.

For convenience, we define the behaviour of M on trees in $T_{\Sigma \cup Q}$ as opposed to just T_Σ, where states are considered to be symbols of rank 0: The set of *runs* of M on $t \in T_{\Sigma \cup Q}$ is the set of all Q-labelled trees $\pi \colon dom(t) \to Q$ such that $\pi(v) = t(v)$ for all $v \in dom(t)$ with $t(v) \in Q$. A run π is *accepting* if $\pi(\lambda) \in Q_f$. The *weight* of a run π on a tree $t = f[t_1, \ldots, t_k]$ is defined as

$$w(\pi) = \sum_{v \in dom(t),\, t(v) \in \Sigma_{(k)}} \delta(t(v)[\pi(v1) \cdots \pi(vk)], \pi(v)) \ .$$

Now, let $M(t) = \min \{w(\pi) \mid \pi \text{ is an accepting run of } M \text{ on } t\}$ for every tree $t \in T_{\Sigma \cup Q}$. This defines the weighted tree language $\mathcal{W}_M \colon T_\Sigma \to \mathbb{R}_+^\infty$ *recognised* by M, namely $\mathcal{W}_M(t) = M(t)$ for all $t \in T_\Sigma$.

The problem we are concerned with in this paper is to compute N trees of minimal weight according to M. For $N \in \mathbb{N}$, an acceptable solution is a set $T = \{t_1, \ldots, t_N\} \subseteq T_\Sigma$ such that $M(t_i) \leq M(t)$ for all $i \in [N]$ and $t \in T_\Sigma \setminus T$. Similarly, for $N = \infty$, we seek an infinite set $T = \{t_1, t_2, \ldots\} \subseteq T_\Sigma$ with $M(t_i) \leq M(t)$ for all $i \geq 1$ and $t \in T_\Sigma \setminus T$.

3   The Algorithm

We now develop our algorithm for computing N minimal trees with respect to a given wta. This will be done in two steps: First a basic version is developed, and second it is turned into a more efficient one by means of a pruning strategy. Correctness and efficiency will be studied in Section 5. Throughout the paper, let $M = (Q, \Sigma, \delta, Q_f)$ be the wta given as input to the search algorithm. We will use the letters m, n, and r to denote the number $|\delta|$ of transition rules, the number $|Q|$ of states, and the maximum rank of symbols in Σ.

Our algorithm explores its search space recursively. The frontier of the explored part is organised as a priority queue. The algorithm iteratively selects a promising tree t from the queue, considers t for output, puts it into a set T of explored trees, and finally expands the frontier by all trees in $\Sigma(T)$ which have at least one occurrence of t as a direct subtree. For $t \in T \subseteq T_\Sigma$ this expansion is defined as

$$expand(T, t) = \{f[t_1, \ldots, t_k] \in \Sigma(T) \mid t_i = t \text{ for at least one } i \in [k]\} \ .$$

To define our algorithm, it is convenient to consider two wta M^q and M_q, for every $q \in Q$. The wta M^q is simply given by $M^q = (Q, \Sigma, \delta, \{q\})$, i.e. q becomes the unique final state. The wta M_q is given by $M_q = (Q, \Sigma \cup \{\Box\}, \delta \cup \{\Box \xrightarrow{0} q\}, Q_f)$. Note that $M_q(c) = M(c[\![q]\!])$ for all $c \in C_\Sigma$ and $q \in Q$.

The priority of a tree t in our queue is primarily determined by the minimal value of $M(c[\![t]\!])$, where c ranges over all possible contexts. To determine this, we compute for every $q \in Q$ the minimal value of $M_q(c) + M^q(t)$. Since M^q denotes the wta obtained from M by taking q as the unique final state, $M^q(t)$ is the minimal weight of all runs on t whose root state is q. Since $M_q(c)$ is independent of t, a c that minimises it can be calculated in advance using, e.g., Knuth's extension of Dijkstra's algorithm [7]. This yields the following lemma.

Lemma 1. *A family of contexts* $(c_q)_{q \in Q}$ *such that* $M_q(c_q) = \min \{M_q(c) \mid c \in C_\Sigma\}$ *for each* $q \in Q$, *can be computed in time* $O(mr \cdot (\log n + r))$.

In the rest of the paper, we frequently make use of the contexts c_q, assuming that they have been computed for all $q \in Q$. For a tree t in the frontier of our search space we are, intuitively, interested in the tree $c[\![t]\!]$ that has the least possible weight. Clearly, c can be assumed to be one of the contexts c_q. Thus, our aim has to be to determine the state q that minimises the weight of $c_q[\![t]\!]$.

Definition 2 (Optimal state). *The mapping* $optset \colon T_\Sigma \to pow(Q)$ *is defined by*

$$optset(t) = \{q \in Q \mid M_q(c_q) + M^q(t) = \min_{c \in C_\Sigma} M(c[\![t]\!])\} \ .$$

In addition, let $opt(t)$ *denote an arbitrary but fixed element of* $optset(t)$, *for every* $t \in T_\Sigma$.

We can now give our basic algorithm. Rather than formulating the algorithm for arbitrary wta, we formulate it only for wta computing *monotone* weighted tree languages. Here, a weighted tree language L is called *monotone* if, for all trees $t \in T_\Sigma$ and all $c \in C_\Sigma \setminus \{\square\}$, $L(t) \neq \infty$ implies $L(c[\![t]\!]) \geq L(t)$. To see that this does not diminish the usefulness of the algorithm, notice that an arbitrary input wta M can be made monotone as follows. We introduce a new symbol *out* of rank 1 and turn M into M' such that $M'(t) = \infty$ and $M'(out[t]) = M(t)$ for all $t \in T_\Sigma$. This can easily be achieved by adding a new state q_f, which becomes the unique final state, and transitions $out[q] \xrightarrow{0} q_f$ for $q \in Q_f$. Then M' is monotone and if $out[t_1], \ldots, out[t_N]$ are N trees of minimal weight with respect to M', then $t_1, \ldots, t_N$ are minimal with respect to M.

Algorithm 1. Enumerate N trees of minimal weight in ascending order for a wta M such that $\mathcal{W}_M$ is monotone

---

```
 1: procedure BestTreesBasic(M, N)
 2:     T ← ∅; K ← ∅
 3:     enqueue(K, Σ₀)
 4:     i ← 0
 5:     while i < N ∧ K nonempty do
 6:         t ← dequeue(K)
 7:         T ← T ∪ {t}
 8:         if M(t) = Δ(t) then
 9:             output(t)
10:             i ← i + 1
11:         end if
12:         enqueue(K, expand(T, t))
13:     end while
14: end procedure
```

Our basic algorithm is presented in Algorithm 1. It maintains three data structures: T is a set of trees that represents the explored search space, K is a priority queue of trees in $\Sigma(T)$, and C is a table containing the value $M^q(t)$, for all $q \in Q$ and $t \in T \cup K$. The table C can easily be updated whenever new trees are added to K. The priority order $\leq_K$ of K is given by

$$t <_K t' \Rightarrow \Delta(t) < \Delta(t') \text{ or } \Delta(t) = \Delta(t') \text{ and } t <_{lex} t'$$
$$\text{where } \Delta(s) = M(c_{opt(s)}[\![s]\!]) \text{ for all } s \in T_\Sigma.$$

Here, $<_{lex}$ is any lexical order that orders trees first by size and then lexically. Note that the output condition in Line 8 cannot be replaced by the more intuitive $M(t) < \infty$ because it has to cover the case where $\Delta(t) = \infty$ (which happens if there are fewer than N trees of finite weight).

Unfortunately, Algorithm 1 builds a large number of trees and is thus not very efficient. Therefore, we now give a more efficient version that works by repeatedly pruning the priority queue.

The idea of the pruning step is that a tree s can be discarded from the queue if we already have, for every state $q \in opt set(s)$, at least N other trees $t <_K s$ such that $q \in opt set(t)$. Intuitively, in this case we have sufficiently many good alternatives to s in the formation of a set of minimal trees, so that s will not be needed. A polynomial runtime is thus obtained by applying the new procedure *Prune* (see Algorithm 2) in Lines 3 and 12 of Algorithm 1. This leads to Algorithm 3.

Algorithm 2. Prune the priority queue

1: **procedure** *Prune*(T, K)
2:    **for** $s \in K$ **do**
3:        **if** $|\{t \in T \cup K \mid q \in opt set(t) \text{ and } t <_K s\}| \geq N$ for all $q \in opt set(s)$ **then**
4:            discard(K, s)
5:        **end if**
6:    **end for**
7: **end procedure**

---

4   Example

Let us have a look at an example. We consider the input automaton M in Figure 1, where $\Sigma_{(0)} = \{a, b\}$ and $\Sigma_{(2)} = \{\circ\}$. Assume that the lexical ordering places a before b before $\circ$, let $\|t\|_\sigma$ ($\sigma \in \Sigma_{(0)}$) denote the number of occurrences of σ in a tree $t \in T_\Sigma$, and let $\|t\|$ denote the total number of leaves of t. Then

$$M(t) = \begin{cases} \|t\| + \min(\|t\|_a, \|t\|_b) & \text{if } \|t\| \text{ is even} \\ \infty & \text{otherwise .} \end{cases}$$

Lemma 1 gives $c_{p_a} = \circ[\Box, b]$, $c_{p_b} = \circ[\Box, a]$, and $c_{q_a} = c_{q_b} = \Box$. We note that

- $M_{p_\sigma}(c_{p_\sigma}) = 1$ and $M_{q_\sigma}(c_{q_\sigma}) = 0$ for $\sigma \in \Sigma_{(0)}$, and

Algorithm 3. Compute N trees of minimal weight for a wta M s.t. $\mathcal{W}_M$ is monotone

```
1: procedure BestTrees(M, N)
2:     T ← ∅; K ← ∅
3:     Prune(T, enqueue(K, Σ₀))
4:     i ← 0
5:     while i < N ∧ K nonempty do
6:         t ← dequeue(K)
7:         T ← T ∪ {t}
8:         if M(t) = Δ(t) then
9:             output(t)
10:            i ← i + 1
11:        end if
12:        Prune(T, enqueue(K, expand(T, t)))
13:    end while
14: end procedure
```

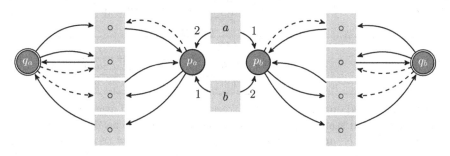

Fig. 1. The input wta considered as an example. The input alphabet is $\Sigma_{(0)} \cup \Sigma_{(2)}$, where $\Sigma_{(0)} = \{a, b\}$ and $\Sigma_{(2)} = \{\circ\}$. Round nodes (with double circles if final) represent states, and squares represent transitions. The consumed input symbols are shown inside the squares. Solid arcs point to the right-hand side of the transition in question and are labelled with the weight of the transition unless it is zero. In the case of input symbol $\circ$ the two states in the left-hand side of a transition are indicated by incoming solid and dashed arcs. Since the wta is symmetric, the latter distinction is, in fact, irrelevant.

- letting $\bar{a} = b$ and $\bar{b} = a$ we have, for all $t \in T_\Sigma$,

$$optset(t) = \begin{cases} \{p_\sigma \mid \sigma \in \Sigma_{(0)} \text{ and } \|t\|_\sigma \leq \|t\|_{\bar\sigma}\} & \text{if } \|t\| \text{ is odd} \\ \{q_\sigma \mid \sigma \in \Sigma_{(0)} \text{ and } \|t\|_\sigma \leq \|t\|_{\bar\sigma}\} & \text{if } \|t\| \text{ is even} \end{cases} .$$

To increase readability, let us denote trees in T_Σ without the binary symbol $\circ$. For example, $\circ[\circ[a, b], b]$ will be denoted by $[[a, b], b]$. To find $N = 3$ minimal trees with respect to M, Algorithm 3 proceeds as follows:

After initialisation, $T = \emptyset$ and K contains a and b, where $a <_K b$ as $\Delta(a) = 2 = \Delta(b)$ and $a <_{lex} b$. The iterations of the 'while' loop proceed as follows:

Step 1: dequeue a	
Output: none (as $\Delta(a) = 2 \neq M(a)$)	Expand with: $[a, a]$
T: $\boxed{a}$	K: $\boxed{b}$, $\boxed{[a, a]}$

Step 2: dequeue b	
Output: none (as $\Delta(b) = 2 \neq M(b)$)	Expand with: $\boxed{[b, b]}$, $\boxed{[b, a]}$, $\boxed{[a, b]}$
T: $\boxed{a}$, $\boxed{b}$	K: $\boxed{[a, a]}$, $\boxed{[b, b]}$, $\boxed{[a, b]}$, $\boxed{[b, a]}$

Step 3: dequeue $[a, a]$	
Output: $[a, a]$ (as $\Delta([a, a]) = 2 = M([a, a])$)	Expand with: $\boxed{[[a, a], [a, a]]}$, $\boxed{[[a, a], a]}$, $\boxed{[[a, a], b]}$, $\boxed{[a, [a, a]]}$, $\boxed{[b, [a, a]]}$
T: $\boxed{a}$, $\boxed{b}$, $\boxed{[a, a]}$	K: $\boxed{[b, b]}$, $\boxed{[a, b]}$, $\boxed{[b, a]}$, $\boxed{[a, [a, a]]}$, $\boxed{[[a, a], a]}$, $\boxed{[[a, a], [a, a]]}$, $\boxed{[b, [a, a]]}$, $\boxed{[[a, a], b]}$

Here, the greyed out boxes indicate trees that are pruned away. For instance, the unique optimal state of $[[a, a], [a, a]]$ is q_b, which is also an optimal state of the trees $[a, a]$, $[a, b]$, and $[b, a]$ having a higher priority. The trees $[b, [a, a]]$ and $[[a, a], b]$, whose optimal state is p_b, are superseded by a, $[a, [a, a]]$, and $[[a, a], a]$. Note that $[[a, a], [a, a]] <_K [b, [a, a]]$ despite the fact that $[[a, a], [a, a]] <_{lex} [b, [a, a]]$, because $\Delta([[a, a], [a, a]]) = 4$ whereas $\Delta([b, [a, a]]) = 5$. Pruned trees of Step 3 are not added to the next table.

Step 4: dequeue $[b, b]$	
Output: $[b, b]$ (as $\Delta([b, b]) = 2 = M([b, b])$)	Expand with: $\boxed{[[b, b], [b, b]]}$, $\boxed{[[b, b], a]}$, $\boxed{[[b, b], b]}$, $\boxed{[[b, b], [a, a]]}$, $\boxed{[a, [b, b]]}$, $\boxed{[b, [b, b]]}$, $\boxed{[[a, a], [b, b]]}$
T: $\boxed{a}$, $\boxed{b}$, $\boxed{[a, a]}$, $\boxed{[b, b]}$	K: $\boxed{[a, b]}$, $\boxed{[b, a]}$, $\boxed{[a, [a, a]]}$, $\boxed{[[a, a], a]}$, $\boxed{[b, [b, b]]}$, $\boxed{[[b, b], b]}$, $\boxed{[[b, b], [b, b]]}$, $\boxed{[a, [b, b]]}$, $\boxed{[[b, b], a]}$, $\boxed{[[a, a], [b, b]]}$, $\boxed{[[b, b], [a, a]]}$

In the next step, $[a, b]$ is dequeued and written to the output, and the algorithm terminates.

5   Correctness and Efficiency

Let us now establish the correctness of Algorithms 1 and 3, and then study the efficiency of the latter. For this, we assume that $\Sigma \neq \Sigma_{(0)}$, so that T_Σ is infinite and hence N trees of minimal weight can always be found. It is clear that Algorithm 1 is correct if $\Sigma = \Sigma_{(0)}$ and terminates within $O(m)$ steps in this case. Throughout this section we will write $BestTreesBasic(M, N) = t_1, t_2, \ldots, t_l$ or $BestTreesBasic(M, N) = t_1, t_2, \ldots$ (and similarly for $BestTrees$) if running Algorithm 1 with the inputs M and N results in the (finite or infinite) sequence $t_1, t_2, \ldots, t_l$ or $t_1, t_2, \ldots$ of output trees.

Using the following lemma, which is not difficult to show, we can prove the correctness of Algorithm 1.

Lemma 3.

(1) Algorithm 1 never dequeues the same tree twice.
(2) If Algorithm 1 dequeues a tree in $t \in T_\Sigma$, then it has previously dequeued all trees in $s \in T_\Sigma$ such that $s <_K t$. In particular, if a tree in $t \in T_\Sigma$ is dequeued, then all trees $s \in T_\Sigma$ with $\Delta(s) < \Delta(t)$ have been dequeued earlier.

Theorem 4 (Correctness of Alg. 1). *For all $N \in \mathbb{N}$, $BestTreesBasic(M, N)$ terminates and returns N trees of minimal weight according to the wta M. Moreover, $BestTreesBasic(M, \infty) = t_1, t_2, \ldots$ consists of pairwise distinct trees such that, for each $i \in \mathbb{N}$ and every tree $t \in T_\Sigma \setminus \{t_1, \ldots, t_i\}$, $M(t) \geq M(t_i)$.*

Proof. Clearly, the first statement of the theorem is a consequence of the second. By Lemma 3(1), the output trees of $BestTreesBasic(M, \infty)$ are pairwise distinct. To prove that $BestTreesBasic(M, \infty)$ outputs an infinite sequence of trees we show that, after any number of iterations, only a finite number of additional iterations can be made until a tree is written to the output (i.e., until Line 9 is reached). Suppose that a tree t is dequeued. Then the tree $t' = c_{opt(t)}[\![t]\!]$ satisfies $M(t') = \Delta(t') = \Delta(t)$. By Lemma 3(2) and the definition of $\leq_K$, no tree s with $|s| > |t'|$ will be dequeued before t' (as it would imply $s >_{lex} t'$). By Lemma 3(1) this means that t' will eventually be dequeued. Since $M(t') = \Delta(t')$, the condition in Line 8 is satisfied and t' is written to the output in Line 9.

To complete the proof, assume that $BestTreesBasic(M, \infty) = t_1, t_2, \ldots$ and consider some $i \in \mathbb{N}$ and a tree $t \in T_\Sigma \setminus \{t_1, \ldots, t_i\}$. To show that $M(t) \geq M(t_i)$, assume that $M(t) < \infty$, because otherwise the assertion is trivially true. Now, recall the assumption that $\mathcal{W}_M$ is monotone. Since $M(t) < \infty$, it implies that $M(c[\![t]\!]) \geq M(t)$ for all contexts c, which means that $M(t) = \Delta(t)$ (because $M(\Box[\![t]\!]) = M(t)$). We also have $M(t_i) = \Delta(t_i)$, because Algorithm 1 outputs t_i only if the condition in Line 8 is satisfied. However, by Lemma 3(2) we have $\Delta(t) \geq \Delta(t_i)$ since t would otherwise have been dequeued before t_i, and would thus have been written to the output at that stage (because $M(t) = \Delta(t)$). Hence, $M(t) = \Delta(t) \geq \Delta(t_i) = M(t_i)$, which finishes the proof. $\Box$

Based on the correctness of Algorithm 1 we can now go on to prove the correctness of Algorithm 3 and study its efficiency. In the following, let us say that a tree $s \in T_\Sigma$ is *discarded* in a run of *BestTrees*(M, N) if it, at some stage, is considered in Line 2 of Algorithm 2, fulfills the pruning condition in Line 3, and is consequently removed from the queue in Line 4. Further, call a tree $t \in T_\Sigma$ *inactive* (with respect to the considered run of *BestTrees*(M, N)) if it contains a discarded subtree. Naturally, a tree that is not inactive is called *active*.

Lemma 5. *Let BestTreesBasic*$(M, \infty) = t_1, t_2, \ldots$ *and consider the execution of BestTrees*(M, N) *for some $N > 0$. Let $l \in \mathbb{N} \cup \{\infty\}$ be the number of active trees among $t_1, t_2, \ldots$, and let i_j be such that t_{i_j} is the jth active tree in $t_1, t_2, \ldots$, for all $j \leq l$. Then BestTrees*$(M, N) = t_{i_1}, t_{i_2}, \ldots, t_{i_{\min(l, N)}}$.

Using this, the correctness of Algorithm 3 is established.

Theorem 6 (Correctness of Alg. 3). *For all $N \in \mathbb{N}$, BestTrees*(M, N) *terminates and returns N trees of minimal weight according to the input wta M. Moreover, BestTrees*$(M, \infty) = t_1, t_2, \ldots$ *consists of pairwise distinct trees such that, for each $i \in \mathbb{N}$ and every tree $t \in T_\Sigma \setminus \{t_1, \ldots, t_i\}$, $M(t) \geq M(t_i)$.*

Proof. The second statement is correct by Theorem 4, because the behaviours of both algorithms are obviously identical for $N = \infty$.

To prove the first statement, assume that *BestTreesBasic*$(M, \infty) = t_1, t_2, \ldots$ and, using Lemma 5, that *BestTrees*$(M, N) = t_{i_1}, \ldots, t_{i_l}$ for some $l \leq N$. We show that $\{t_{i_1}, \ldots, t_{i_l}\} = \{t_1, \ldots, t_N\}$. Let $\Theta = \{t_1, \ldots, t_N\} \setminus \{t_{i_1}, \ldots, t_{i_l}\}$. By Lemma 5 each tree in Θ is inactive. Let us assume that $\Theta \neq \emptyset$, and let k be the least index such that $t_k \in \Theta$. In other words, t_k is the first tree among $t_1, \ldots, t_N$ containing a discarded subtree. Since t_k is one of the output trees of Algorithm 1 we have $\Delta(t_k) = M(t_k)$. Let $t_k = c[\![s]\!]$, where s is one of the discarded subtrees of t_k. Thus, s is inactive but all its proper subtrees are active.

To finish the proof, let $v \in dom(c)$ be the node of c such that $c/v = \square$, and consider a minimal run π on t_k, where $q = \pi(v)$. We know that $M(c'[\![s]\!]) \geq M(t_k)$ for all $c' \sqsubset C_\Sigma$ because otherwise $c'[\![s]\!] \in \{t_1, \ldots, t_{k-1}\}$, which would contradict the choice of k since $c'[\![s]\!]$ contains the discarded subtree s. In other words,

$$q \in optset(s), \ M_q(c) = \min\{M_q(c') \mid c' \in C_\Sigma\} \text{ and } M(t_k) = M(c_q[\![s]\!]) \ . \quad (1)$$

Since s was discarded during the execution of Algorithm 3, we know further that $T \cup K$, from that point onward, always contained N pairwise distinct trees $u_1, \ldots, u_N$ such that $u_i <_K s$ and $q \in optset(u_i)$.

We distinguish two cases, deriving a contradiction in each case and thus proving that t_k cannot exist:

1. If $M(c_q[\![u_i]\!]) < M(c[\![u_i]\!])$ for some $i \in [N]$ (i.e., c is not a context that Lemma 1 could have computed instead of c_q) then it follows from the equations $M(c_q[\![u_i]\!]) = M_q(c_q) + M^q(u_i)$ and $M(c[\![u_i]\!]) \leq M_q(c) + M^q(u_i)$ that $M_q(c_q) < M_q(c)$ and thus $M(c_q[\![s]\!]) < M(t_k)$, contradicting (??).
2. If $M(c[\![u_i]\!]) = M(c_q[\![u_i]\!]) = \Delta(u_i)$ for all $i \in N$, let us consider some $i \in [N]$. If $s <_{lex} u_i$ despite the fact that $u_i <_K s$, then $\Delta(u_i) < \Delta(s)$. Consequently, $M(c[\![u_i]\!]) = M(c_q[\![u_i]\!]) < M(t_k)$, which gives us $c[\![u_i]\!] <_K t_k$. If, on the

contrary, $u_i <_{lex} s$ and thus $c[\![u_i]\!] <_{lex} c[\![s]\!]$, then $\Delta(c[\![u_i]\!]) \leq M(c[\![u_i]\!]) = \Delta(u_i) \leq \Delta(s) = M(t_k) = \Delta(t_k)$ gives us again $c[\![u_i]\!] <_K t_k$.

We have thus shown that $c[\![u_i]\!] <_K t_k$ for all $i \in [N]$. By Lemma 3(2), all of these N pairwise distinct trees occur among $t_1, \ldots, t_{k-1}$, which is impossible because $k \leq N$. $\qquad\square$

Let us now discuss the worst-case efficiency of *BestTrees*. A consequence of the pruning is that T can only grow to contain $N \cdot n$ trees, since at this point, the pruning will discard everything that is left in the queue.[1] Since each execution of the 'while' loop increases the size of T, this means that the body of the 'while' loop in *BestTrees* is executed at most $N \cdot n$ times.

Lemma 7. *Prune$(K, Expand(T, t))$ is computable in time*

$$O\big(\max(m \cdot (Nr + r \log r + N \log N), Nn^2)\big) \ .$$

Proof (Sketch). In order to implement pruning efficiently, we have to avoid the explicit computation of *Expand(T, t)*. Let us denote the subset of transition rules in δ that lead to the state q by δ_q. We first compute, for every $q \in Q$, an ordered list of (at most) N trees $s_1, \ldots, s_N$ in *Expand(T, t)* such that $M^q(s_1) \leq \cdots \leq M^q(s_N)$ and $M^q(s) \geq M^q(s_N)$ for all $s \in Expand(T, t) \setminus \{s_1, \ldots, s_N\}$. To do this, consider every rule $\rho = (f[q_1, \ldots, q_k] \xrightarrow{w} q) \in \delta_q$ in turn and build a weighted edge-labelled digraph G_ρ having nodes $u_0, \ldots, u_k$ and $v_0, \ldots, v_k$ and the following edges for every $i \in [k]$:[2]

> For every $s' \in T \setminus \{t\}$ there are edges with label s' and weight $M^{q_i}(s')$ from u_{i-1} to u_i and from v_{i-1} to v_i. In addition, there are edges with label t and weight $M^{q_i}(t)$ from both u_{i-1} and v_{i-1} to v_i.

A path from u_1 to v_{k+1} in G_ρ which is labelled $t_1 \cdots t_k$ corresponds to the tree $f[t_1, \ldots, t_k] \in Expand(T, t)$. Note that t occurs among $t_1, \ldots, t_k$ since only t-labelled edges lead from u_i to v_{i+1}. The weight of the path (plus w) is the weight of a minimal run π on $f[t_1, \ldots, t_k]$ with $\pi(\lambda) = q$ and $\pi(i) = q_i$ for all $i \in [k]$. Since G_ρ has $O(r)$ nodes and $O(Nnr)$ edges, N paths of minimal weight can be computed by Eppstein's algorithm [5] in time $O(Nnr + r \log r + N \log N)$. We can improve this to $O(Nr + r \log r + N \log N)$ by including in G_ρ, for every pair of nodes, only N edges of minimal weight between those nodes. Clearly, only these edges can be on the N paths of minimal weight. For every rule ρ, this gives rise to an ordered list L_ρ of (at most) N trees. The time required for this is $O(m \cdot (Nr + r \log r + N \log N))$ in total. Together with each tree in the computed lists L_ρ, we keep track of the corresponding weight in order to be able to implement the following steps.

In the next step, the lists obtained for rules with the same right-hand side q are merged into a single list. Finally, a similar procedure merges the n lists obtained in the previous step with the trees in K. Since the queue contains at

---

[1] See [1] for an explicit proof of this fact.

[2] The nodes v_0 and u_k are superfluous but simplify the description of G_ρ.

most Nn elements, the time required for this is $O(Nn^2)$ if we implement K as a linked list. $\square$

Using Lemmas 1 and 7 as well as the fact that the main loop of Algorithm 3 is executed at most Nn times, we obtain Theorem 8.

Theorem 8. *BestTrees(M, N) runs in time*

$$O\big(\max(Nmn \cdot (Nr + r \log r + N \log N), N^2 n^3, mr^2)\big).$$

It may be worthwhile to notice that the set T of Algorithm 3 is subtree closed, meaning that $t_1, \ldots, t_k \in T$ for every tree $f[t_1, \ldots, t_k] \in T$. Since all output trees of Algorithm 3 are in T, this means that the output of Algorithm 3 can be represented as a packed forest with $|T|$ nodes, i.e., of size $\leq N \cdot n$.

6   Conclusion and Future Work

Future work includes the implementation and integration of the algorithm into an open-source library for formal tree languages. On the theoretical side, we are interested in seeing further generalisations of the search algorithm, for example, from trees to directed acyclic graphs, or from the tropical semiring to some encompassing family of extremal semirings.

Acknowledgments. We thank Loek Cleophas and the referees for reading the manuscript carefully and making numerous useful suggestions.

References

1. Björklund, J., Drewes, F., Zechner, N.: An efficient best-trees algorithm for weighted tree automata over the tropical semiring. Report UMINF 14.22, Umeå University (2014)
2. Borchardt, B.: A pumping lemma and decidability problems for recognizable tree series. Acta Cybernetica **16**, 509–544 (2004)
3. Collins, M.: Discriminative reranking for natural language parsing. In: Computational Linguistics, pp. 175–182. Morgan Kaufmann (2000)
4. Droste, M., Kuich, W., Vogler, H. (eds.): Handbook of Weighted Automata. Springer (2009)
5. Eppstein, D.: Finding the k shortest paths. SIAM J. Computing **28**(2), 652–673 (1998)
6. Huang, L., Chiang, D.: Better k-best parsing. In: Proceedings of the Conference on Parsing Technology 2005, pp. 53–64. Association for Computational Linguistics (2005)
7. Knuth, D.E.: A generalization of Dijkstra's algorithm. Information Processing Letters **6**, 1–5 (1977)
8. Mohri, M., Riley, M.: An efficient algorithm for the n-best-strings problem. In: Proceedings of the Conference on Spoken Language Processing (2002)
9. Shen, L.: Discriminative reranking for machine translation. In: Proceedings of HLT-NAACL 2004, pp. 177–184 (2004)

Construction of a de Bruijn Graph for Assembly from a Truncated Suffix Tree

Bastien Cazaux[1], Thierry Lecroq[2], and Eric Rivals[1(✉)]

[1] L.I.R.M.M. and Institut Biologie Computationnelle, Université de Montpellier II,
CNRS U.M.R. 5506, Montpellier, France
{cazaux,rivals}@lirmm.fr
[2] LITIS EA 4108, NormaStic CNRS FR 3638, Université de Rouen, Rouen, France
thierry.lecroq@univ-rouen.fr

Abstract. In the life sciences, determining the sequence of bio-molecules is essential step towards the understanding of their functions and interactions inside an organism. Powerful technologies allows to get huge quantities of short sequencing reads that need to be assemble to infer the complete target sequence. These constraints favour the use of a version de Bruijn Graph (DBG) dedicated to assembly. The de Bruijn Graph is usually built directly from the reads, which is time and space consuming. Given a set R of input words, well-known data structures, like the generalised suffix tree, can index all the substrings of words in R. In the context of DBG assembly, only substrings of length $k + 1$ and some of length k are useful. A truncated version of the suffix tree can index those efficiently. As indexes are exploited for numerous purposes in bioinformatics, as read cleaning, filtering, or even analysis, it is important to enable the community to reuse an existing index to build the DBG directly from it. In an earlier work we provided the first algorithms when starting from a suffix tree or suffix array. Here, we exhibit an algorithm that exploits a reduced version of the truncated suffix tree and computes the DBG from it. Importantly, a variation of this algorithm is also shown to compute the contracted DBG, which offers great benefits in practice. Both algorithms are linear in time and space in the size of the output.

Keywords: Stringology · Text algorithms · Indexing data structures · de Bruijn graph · Assembly · Space complexity · Dynamic update

1 Introduction

The de Bruijn Graph (DBG) serves in bioinformatics and genomics to assemble the sequence of large molecules from a huge set of short sequencing reads. In this context, only the substrings of length, say k, of the reads form the nodes of the

---

This work is supported by ANR Colib'read (http://colibread.inria.fr) (ANR-12-BS02-0008) and by Défi MASTODONS SePhHaDe (http://www.lirmm.fr/mastodons) from CNRS.

A.-H. Dediu et al. (Eds.): LATA 2015, LNCS 8977, pp. 109–120, 2015.
DOI: 10.1007/978-3-319-15579-1_8

DBG (unlike in the original DBG). These substrings are termed k-mers in biology or k-grams in computer science. An arc links two k-mers whenever they overlap by $(k-1)$ symbols at (necessarily) successive positions in a read. The assembly DBG is then traversed searching for long paths, which will form the *contigs*, i.e. the sequence of sub-regions of the molecule. However, in non repetitive regions, the layout of the reads dictate a single path of k-mers without bifurcations. Any simple path between an in-branching node and the next out-branching node, can then be contracted into a single arc without loosing any information on the graph structure. The sequence of such simple paths are called *unitigs* (the contraction from unique and contigs). The version of the DBG where all such "non-branching" paths are condensed into an arc is termed the Contracted DBG (CDBG).

Given the extreme throughput delivered by nowadays sequencing machines, it is crucial to enable fast construction of the CDBG. It is also desirable to build the DBG or its contracted version directly from space efficient indexing data structures, since the read set has often been filtered and mined for patterns representing errors prior to assembly using such data structures. It occurs for instance in a preprocessing phase of sequencing errors removal using a *generalised suffix tree* of the reads [15]. The use of an indexing data structure (or index for short) also allows to compute and store additional information into the nodes of the DBG: the coverage of a k-mer, i.e. the number of reads in the layout covering that k-mer. Unexpected variations of the coverage permits to detect sequencing errors, to distinguish between classes of point mutations [13], but also to disentangle repetitive sequence regions. Indeed, the reads coming from the distinct but similar copies of a repeat tend to collapse into a single assembly region and increase abnormally the local coverage. Actually, the DBG is itself used as a data structure to seek graph patterns representing mutation, large insertions/deletions, or chromosomal rearrangements [14].

In a first attempt towards constructing the DBG from an index, Cazaux *et al.* gave recently two algorithms for building it from either a Generalised Suffix Tree (GST) or a generalised Suffix Array [4]. Indeed, a subset of the suffix tree nodes represent either exactly one k-mer or its shortest extension; hence, this subset is isomorphic to the set of nodes of the DBG. Moreover, following a suffix link and then going down the tree at most once allows to traverse from one node to its neighbours in the DBG. Hence, the arcs of the DBG can be simulated on the GST or computed using it. This summarises the basis of the DBG construction algorithms, which require linear time in the input length (i.e. the cumulated sum of the read lengths). Importantly, it was also shown that the contraction of the arcs in the DBG can be computed in linear time during the construction: this gave the first linear time CDBG construction algorithm [4].

However, in practice the size of the Generalised Suffix Tree remains prohibitive for large read sets. In the light of these algorithms, it is clear that many nodes of the GST among those having a string depth either larger than the order k or strictly smaller than $k-1$ are useless. Truncated Suffix Trees (TST) have been introduced to index only a subset of factors below a certain string depth

[10,16]. TST avoid storing strings longer than a limit length k, but still include all suffixes shorter than k. In practice, when the reads are numerous and short, which occurs in a majority of large sequencing applications, the memory wasted for such nodes is important. With an Illumina sequencing experiment, the typical number of reads $n := 10^8$, the read length is 100 and several values are used for k up to 64. In such a case, one stores $n \times (k - 2)$ useless nodes. We set out to first find an algorithm for a reduced version of the TST that avoids those nodes, second to show that the DBG and CDBG can both be built in time and space that are linear in the size of the DBG, rather than in the cumulated length of the reads. The paper is organised as follows. Below we list related works. In Section 2, we define a simple condition that a set of input strings must satisfy to allow building a generalised index and sketch a modification of McCreight's algorithm [9] for doing so. In Section 3, we introduce the reduced truncated suffix tree and specialise the previous algorithm for constructing it efficiently. Finally, in Section 4 we show how to construct both the de Bruijn Graph and its contracted version in optimal time from the reduced truncated suffix tree. We then conclude mentioning lines of future work.

1.1   Related Works

Suffix trees are well-known indexing data structures that enable to store and retrieve all the factors of a given string. They can be adapted to a finite set of strings and are then called generalised suffix trees (GST). They can be built in linear time and space. They have been widely studied and used in a large number of applications (see [1,8]). In some applications it is not required to consider the full set of suffixes, since one may only be interested in factors of length bounded by a given constant. These factors are actually prefixes of suffixes. In 2003, Na *et al.* [10] introduced the truncated suffix trees which only stores the factors of length at most k in a context of lossless data compression. They gave linear time algorithms for directly constructing truncated suffix trees. They present experimental results showing that on various kinds of strings and different values of k that truncated suffix trees have much less nodes than suffix trees. Truncated suffix trees have been generalised to set of strings and use for performing efficient pattern matching in biological sequences [16].

DBGs are heavily exploited for genome assembly in bioinformatics [12], where several compact data structures for storing DBGs have been developed [2,6] including probabilistic ones [5]. The emphasis is placed on the practical space needed to store the DBGs in memory. Moreover, some recent assembly algorithms put forward the advantage of using for the same input, multiple DBGs with increasing orders [11], thereby emphasising the need for dynamically updating the DBGs.

From now on, the input of our problem consists of an integer $k > 0$ and $R = \{w_1, \ldots, w_n\}$ a set of n finite words over a finite alphabet. We will consider indexing the substrings of words in R; hence, all indexes used are *generalised* indexes [8]. For simplicity, we may omit this adjective.

2 Set of Chains of Suffix-Dependant Strings and Tree

Here, we introduce the notion of *suffix dependence* between strings, and the notion of *chain of suffix-dependant strings* in order to define a unified index that generalises both the suffix tree [9] and the truncated suffix tree [10]. First, we introduce a notation on strings.

Notation on Strings. We consider finite strings (also termed words or sequences) over a finite alphabet Σ. For a string w, $|w|$ denotes the *length* of w. For any integers i and j such that $i \leq j$, $[i, j]$ denotes the interval of integers between i and j. For any $i \leq j$ in $[1, |w|]$, $w[i, j]$ is the substring of w beginning at position i and ending at position j. Then, $w[1, i]$ is called a *prefix* of w, while $w[i, |w|]$ is a *suffix* of w. For a set of strings A, the *norm* of A, denoted $||A||$, is

$$||A|| = \sum_{a \in A} |a|.$$

Now, let us define the concept of suffix-dependant strings and of chains of suffix-dependant strings.

Definition 1. *1. A string x is said to be* suffix-dependant *of another string y if $x[2..|x|]$ is prefix of y.*

2. Let w be a string and m be a positive integer smaller than $|w| - 1$. A m-tuple of m strings $(x_1, \ldots, x_m)$ is a chain of suffix-dependant strings *of w if x_1 is a prefix of w and for each $i \in [2, m]$, x_i is a prefix of $w[i, |w|]$ such that $|x_i| \geq |x_{i-1}| - 1$.*

For a set of strings $R = \{w_1, \ldots, w_n\}$, let $\mathcal{S} = \{C_1, \ldots, C_n\}$ be a set of tuples such that for each $i \in [1, n]$, C_i is a chain of suffix-dependant strings of the string w_i. For $i \in [1, n]$ and $j \in [1, |C_i|]$, $C_i[j]$ is the jth string of the tuple C_i. Let be $\widehat{\mathcal{S}} = \{\widehat{C_1}, \ldots, \widehat{C_n}\}$ the set of tuples such that for each $i \in [1, n]$ and $j \in [1, |C_i|]$, $\widehat{C_i}[j] = |C_i[j]|$, i.e. $\widehat{\mathcal{S}}$ contains tuples of lengths.

With $\widehat{\mathcal{S}}$ and R, we can easily compute $\mathcal{S}$. In the sequel, we use $\mathcal{S}$ to demonstrate our results, and $\widehat{\mathcal{S}}$ to state the complexities of algorithms. Indeed, in the case where C_i is the tuple of each suffix of w_i, the size of C_i is linear in $|w_i|^2$ but $\widehat{C_i}$ is linear in $|w_i|$.

Let w be a string; w may occur in distinct tuples of $\mathcal{S}$. Thus, we define $N(w)$ the set of (i, j) such that $w = C_i[j]$. In other words, $N(w)$ is the set of coordinates of the elements of $\mathcal{S}$ that are equal to w.

We define a contracted version of the well-known Aho-Corasick tree [8]. In fact, we apply nearly the same contraction process that turns a trie of a word into its compact Suffix Tree [8]. Consider the Aho-Corasick tree of R, in which each node represents a prefix of word in R. We contract the non-branching parts of the branches except that we keep all nodes representing a word that belongs to a tuple in $\mathcal{S}$. From now on, let $T(\mathcal{S})$ denote this contracted version of the Aho-Corasick tree of R.

$\mathcal{N}$ and $\mathcal{L}$ denote respectively the set of nodes and the set of leaves of $T(\mathcal{S})$. Furthermore, we define for each node v of $T(\mathcal{S})$ two weights:

- $s(v)$ is the number of times that an element of a tuple of $\mathcal{S}$ is equal to the word represented by v (i.e. $s(v) := |N(v)|$).
- $t(v)$ is the number of times that the first element of a tuple of $\mathcal{S}$ is equal to the word represented by v (i.e. $t(v) := |\{(i,1) \in N(v) \mid i \in [1,n]\}|$).

Let w be a string, we put $Succ(w) = \{(i,j) \mid (i,j-1) \in N(w) \text{ and } j \leq |C_i|\}$. We define $\mathcal{F}$ as the subset of $\mathcal{L}$ such that:

$$\mathcal{F} := \{u \in \mathcal{L} \mid \exists C \in S \text{ and } j < |C| \text{ such that } u = C[j]\}$$

It is equivalent to say that $\mathcal{F} = \{u \in \mathcal{L} \mid Succ(u) \text{ is not empty}\}$. A mapping m from $\mathcal{F}$ to $\mathcal{N}$ is called *possible link* if for each node v in $\mathcal{F}$, $\exists (i,j) \in Succ(v)$ such that $m(v) = C_i[j]$.

Below we present an algorithm that constructs $T(\mathcal{S})$, and computes for each node v in $\mathcal{N}$, the weights $s(v)$ and $t(v)$ and a possible link P_0.

Algorithm to Construct the $T(\mathcal{S})$. Now, we give an algorithm to construct $T(\mathcal{S})$. We use the version of McCreight's algorithm given by Na et al. [10] on our input and we build for each leaf v, $s(v)$, $t(v)$ and $P_0(v)$. For building $T(\mathcal{S})$, we start with a tree that contains only the root. Then, for each word w in every chain C, we create or update (if it exists) the node w as follows. Assume that we keep in memory the node v that has been processed just before w.

If w is the first word of C, we go down from the root by comparing w to the labels of the tree. If we create the node w, $s(w)$ and $t(w)$ are initialised to 1, and $P_0(w)$ to *nil*. If w already exists on the tree, we increment $s(w)$ and $t(w)$ by 1.

If w is not the first word of C, we start from v, and as in McCreight's algorithm, we create or arrive on the node representing w. If we need to create this node, $s(w)$ is initialised to 1, $t(w)$ to 0, and $P_0(w)$ to *nil*. Otherwise, we add 1 to $s(w)$. We set $P_0(v) = w$.

The loop continues with the next word until the end, and we obtain $T(\mathcal{S})$.

Theorem 2. *For a set of chain of suffix-dependant strings $\mathcal{S}$, we can construct $T(\mathcal{S})$ in $O(||R||)$ time and space.*

Proof. To begin with, let us to prove that $T(\mathcal{S})$ is in $O(||R||)$ space. Its number of leaves equals $\sum_{C \in \mathcal{S}} |C|$. Hence, its number of nodes is at most $2\sum_{C \in \mathcal{S}} |C| - 1 \leq 2||R||$, and its number of edges is at most $2||R||$. Thus the size of $T(\mathcal{S})$ is in $O(||R||)$.

Clearly, the construction algorithm of $T(\mathcal{S})$ computes both weights $s(.)$ and $t(.)$, and the possible link $P_0(.)$ correctly. For the complexity, for each chain of suffix-dependant C_i of $\mathcal{S}$, the length of the traverse path on the tree is equal to $|w_i|$, thanks to the use of the suffix links. Thus as in McCreight's algorithm, the complexity is in $O(||R||)$.

Now, we are equipped with an algorithm that builds $T(\mathcal{S})$ for any set of chains of suffix-dependant strings. Let us review some instances of sets S, for which $T(\mathcal{S})$ is in fact a well-known tree.

- If $\mathcal{C} := \cup_{w \in R}\{\text{tuple of suffixes of } w\}$, then $T(\mathcal{C})$ is the Generalised Suffix Tree of R (see Figure 1a). We have that the restrained mapping $sl(.)$ is an example of a possible link.
- If $B_k := \cup_{w \in R}\{\text{tuple of } k\text{-mer of } w \text{ and suffixes of length } k' < k \text{ of } w\}$, then $T(B_k)$ is the generalised k-truncated suffix tree of R, as defined in [16] (which generalizes the k-truncated suffix tree of Na et al. [10]).
- If $A_k := \cup_{w \in R}\{\text{tuple of } k + 1\text{-mer of } w \text{ and suffixes of length } k \text{ of } w\}$, then $T(A_k)$ is the truncated suffix tree that we define below in Section 3 (see Figure 1b).

3   Our Truncated Suffix Tree

First, let us introduce a notation about trees that index strings and whose edges are labelled with strings.

Definition 3. *For a node v of a tree, $f(v)$ denotes the parent node of v, and $Children(v)$ its set of children. The depth of v is the length of the unique path between the root and v in the tree. Each represent a unique word: that made up by the concatenation of the label of edges along this path. The notion of node and the word it represents are confounded (we used one for the other). For a $T(\mathcal{S})$ and a word w, $\lceil w \rceil$ is the node v with the shortest depth of the $T(\mathcal{S})$ such that w is a prefix of v. If this node does not exist, $\lceil w \rceil$ does not exists. For a node u of the $T(\mathcal{S})$, $sl(u)$ is the node $\lceil u[2, |u|] \rceil$.*

For a set of words $R = \{w_1, w_2, \ldots, w_n\}$ and an integer $k > 0$, we define the following notation.

Definition 4

1. $F_k(R)$ is the set of substrings of length k of words of R.
2. $Suff_k(R)$ is the set of suffixes of length k of words of R.
3. For all $i \in [1, |R|]$ and $j \in [1, |w_i| - k + 1]$, $A_{k,i}$ denotes the tuple such that its jth element is defined by

$$A_{k,i}[j] := \begin{cases} w_i[j, j + k] & \text{if } j \leq |w_i| - k \\ w_i[j, |w_i|] & \text{otherwise.} \end{cases}$$

4. and finally A_k is the set of these tuples: $A_k := \bigcup_{i=1}^{n} A_{k,i}$.

Proposition 5.   *1. $A_{k,i}$ is a chain of suffix-dependant strings of w_i.*
2. Moreover, $\{w \in A_{k,i} \mid A_{k,i} \in A_k\} = F_{k+1}(R) \cup Suff_k(R)$.

Proof.   1. For all $j \in [1, |A_{k,i}| - k]$, it is easy to see that $A_{k,i}[j]$ is a suffix-dependant string of $A_{k,i}[j + 1]$.

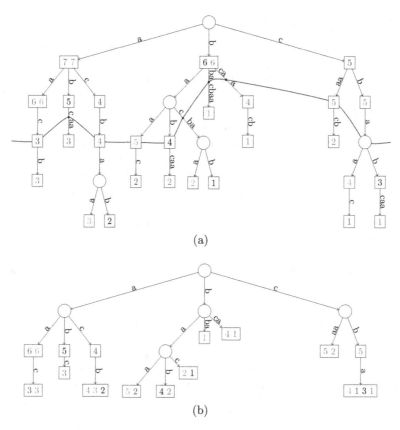

(a)

(b)

Fig. 1. (a) The generalised suffix tree for the set of words {*bacbab*, *bbacbaa*, *bcaacb*, *cbaac*, *cbabcaa*}. The part above the green line corresponds to the TST $T(A_2)$, which is shown in (b). (b) The truncated suffix tree $T(A_2)$ for the same set of words.

2. For the second point

$$\{w \in A_{k,i} \mid A_{k,i} \in A_k\} = \bigcup_{i=1}^{n} \left(\bigcup_{j=1}^{|w_i|-k+1} \{A_{k,i}[j]\} \right)$$

$$= \bigcup_{i=1}^{n} \left(\bigcup_{j=1}^{|w_i|-k} \{A_{k,i}[j]\} \bigcup \{A_{k,i}[|w_i| - k + 1]\} \right)$$

$$= \bigcup_{i=1}^{n} \left(F_{k+1}(\{w_i\}) \bigcup \text{Suff}_k(\{w_i\}) \right)$$

$$= F_{k+1}(R) \cup \text{Suff}_k(R)$$

By applying the algorithm described in Section 2 to the set A_k (Definition 4), and by using Theorem 2, we get the following result.

Corollary 6. *We can construct $T(A_k)$ in $O(\|R\|)$ time and space.*

3.1   Experimental Results

We tested the two data structures GST and TST on real biological data. We considered a set of 2249632 Illumina reads of yeast of length 101 and performed tests for subsets of size 100, 1000, 10000, 100000, 1000000 and for the whole set. We counted the number of nodes of the GST and of the TST for various values of k (5, 10, 20 and 40). We used the gsuffix[1] of [16]. It should be noted that their implementation of the TST stores all the suffixes shorter than k producing thus more nodes than our TST. Table 1 show the results. It can be seen that for small sets, TSTs do not save many nodes compare to the GST except for very small values of k but that for large sets TSTs save a lot of nodes for small values of k, they save more than two third of nodes for $k = 20$ and almost half of the nodes for $k = 40$. We also performed experiments with longer reads from Pacific Biosciences technology (not shown here). In this case, as expected, TSTs save less nodes than for Illumina reads.

Table 1. Number of nodes of the GST vs the TST for $k = 5, 10, 20, 40$ and the percentage compare to the GST for Illumina reads of length 101

♯reads	100	1000	10000
ST	14382	135558	1320811
TST ($k = 5$)	1352 (9.40%)	1365 (1.00%)	1365 (0.10%)
TST ($k = 10$)	14100 (98.03%)	120602 (88.96%)	677153 (51.26%)
TST ($k = 20$)	14347 (99.75%)	133204 (98.26%)	1263803 (95.68%)
TST ($k = 40$)	14382 (100.00%)	134316 (99.08%)	1291685 (97.79%)

♯reads	100000	1000000	2249632
ST	12354838	103555389	216725799
TST ($k = 5$)	1365 (0.01%)	1365 (0.001%)	1365 (0.0006%)
TST ($k = 10$)	1315886 (10.65%)	1396675 (1.34%)	1397752 (0.64%)
TST ($k = 20$)	10549607 (85.38%)	49389538 (47.69%)	69248532 (31.95%)
TST ($k = 40$)	11337038 (91.76%)	69375578 (66.99%)	117282522 (54.11%)

4   De Bruijn Graph via the Truncated Suffix Tree

Here, we describe an algorithm that builds the de Bruijn Graph of a set of words R starting from the generalised truncated suffix tree of R. Note that this DBG differs from the original graph as defined by de Bruijn in the field of word combinatorics [3]. The DBG studied here serves for genome assembly and for approximating the well-known Shortest Superstring problem [7].

4.1   De Bruijn Graph

Let k be a positive integer and $R := \{w_1, \ldots, w_n\}$ be a set of n words. We use the definition of a de Bruijn graph stated in [4].

---

[1] http://gsuffix.sourceforge.net/gsuffix-docs/main.html

Definition 7 (de Bruijn graph). *The* de Bruijn graph *of order k for R, denoted by dBG_k^+, is a directed graph, whose vertices are the k-mers of words of S and where an arc links u to v if and only if u and v are two successive k-mers of a word of R.*

Another slightly different definition is sometimes used in which the k-mers must overlap by $(k-1)$ symbols, but must not necessarily occur in the same read. This relaxed definition introduces "false" arcs, but is easier to build and store in memory. All our results can easily be adapted to that definition.

Proposition 8 states that there does not exist any leaf in $T(A_k)$ representing a word strictly shorter than k.

Proposition 8. *Let v be a leaf of $T(A_k)$. Then $|v| = k$ or $|v| = k + 1$.*

Proof. For all $w_i \in R$ and $j \in [1, |w_i| - k + 1]$, $|A_{k,i}[j]| = k$ or $k + 1$.

We set $Init_{R,k} = \{v \in V_{T(A_k)} \mid |v| \geq k$ and $|f(v)| < k\}$. For a possible link P_0, we define the mapping P from $\mathcal{F}$ to $\mathcal{N}$. $\mathcal{F}$, $\mathcal{N}$ and $\mathcal{L}$ have the same definition as before, but applied to the $T(A_k)$. $\mathcal{F}$ can be seen in this case as the set of leaves of length $k + 1$ of $T(A_k)$. We define the mapping P as follows:

$$P : \mathcal{F} \longrightarrow \mathcal{N}$$
$$v \mapsto \begin{cases} P_0(v) & \text{if } P_0(v) \in Init_{R,k} \\ f(P_0(v)) & \text{otherwise} \end{cases}$$

The mapping P can be constructed in linear time in $O(||R||)$. In fact, for each $v \in \mathcal{F}$, $P(v)$ can be constructed in $O(1)$ because in this case, $P_0(v) \in Init_{R,k} \Leftrightarrow |f(P_0(v))| \neq k$. As $|\mathcal{F}| \leq ||R||$, we can construct P for all elements of $\mathcal{F}$ in $O(||R||)$. Indeed, it is enough to look the length of the parent of $P_0(v)$ to decide if $P_0(v)$ is in $Init_{R,k}$.

Proposition 9. *Le be $v \in \mathcal{L}$, $P(v) \in Init_{R,k}$ and $P(v) = sl(v)$ if $sl(v)$ exists.*

Proof. Let be $v \in \mathcal{L}$. If $v \in \mathcal{F}$ and $P_0(v) \notin Init_{R,k}$, $|f(P_0(v))| = k$ and thus $P(v) = f(P_0(v)) \in Init_{R,k}$. According to the definitions of a possible link P, and of A_k, for any node v in $\mathcal{L}$, $P(v)$ is the shortest node of $T(A_k)$ such that v is a prefix of $P(v)$. Hence, $P(v) = sl(v)$.

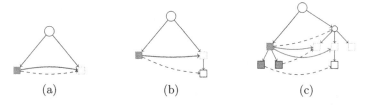

(a)   (b)   (c)

Fig. 2. The different cases of the definition of E^+. For a node v (in green), the curved solid arrows are the arcs of E^+, the dashed arrows are the suffix links of the nodes, the dashed nodes are nodes of $Init_{R,k}$. (a) and (b) correspond to the first part of the definition while (c) corresponds to the second part.

Let us define, E_k^+, a set of arcs. Two cases arise depending on whether the starting node v represents a word of length k or $k+1$.

$$E_k^+ = \left(\cup_{|v|=k+1}(v, P(v))\right) \cup \left(\cup_{|v|=k}\left(\cup_{u \in Children(v)}(v, P(u))\right)\right)$$

with $v \in Init_{R,k}$. Figure 2 illustrates the alternative cases in the definition of E_k^+, which is the union of these cases.

Proposition 10. $(Init_{R,k}, E_k^+)$ *is isomorphic to* dBG_k^+ *of* R.

Proof. The proof is identical to that of [4, Theorem 1].

Figure 3 shows an example of de Bruijn graph of order 2 built from $T(A_2)$.

Proposition 11. *The size of* $T(A_k)$ *is linear in the size of* dBG_k^+.

Proof. Let b be the application from $\mathcal{F}$ to E_k^+ such that

$$b(v) = \begin{cases} (v, P(v)) & \text{if } |f(v)| \neq k \\ (f(v), P(v)) & \text{if } |f(v)| = k. \end{cases}$$

By Proposition 10, b is a bijection. As $|\mathcal{L} \setminus \mathcal{F}| \leq |R|$, $|\mathcal{L}|$ is linear in the size of dBG_k^+.

From R, we first build the reduced TST, and then build dBG_k^+ from it. By Proposition 11, we get the following theorem.

Theorem 12. *For a set of words R, we can construct dBG_k^+ in $O(\|R\|)$ time and in $O(size(dBG_k^+))$ space.*

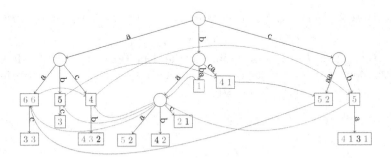

Fig. 3. The de Bruijn graph of order 2 built on $T(A_2)$. The solid curved arrows are the edges corresponding to the first part of the definition of E_k^+, while those in blue correspond to the second part.

4.2   A Contracted de Bruijn Graph

Let e an arc of dBG_k^+, $e = (x, y)$ is said to be *reducible* if the outdegree of x and the indegree of y are at most 1.

Definition 13. *Let k be a positive integer and $R := \{w_1, \ldots, w_n\}$ be a set of n words. The* contracted de Bruijn graph *of order k for R, denoted by $CdBG_k^+$, is the de Bruijn graph of order k where each reducible arc is contracted.*

Let z be a word, $Support_R(z)$ is the set of pairs (i, j), where z is the substring $w_i[j, j + |s| - 1]$. $Support_R(z)$ is called the support of z in R.

Proposition 14. *For each leaf v of $T(A_k)$, $s(v)$ is the size of the support of v in R and $t(v)$ is the size of the set $\big(Support_R(v) \cap \{(i, 1) \mid 1 \leq i \leq n\}\big)$.*

Hence, we obtain the following theorem.

Theorem 15. *For a set of words R, we can construct $CdBG_k^+$ in $O(\|R\|)$ time and in $O(size(dBG_k^+))$ space.*

Instead of using $T(A_k)$ to build dBG_k^+ or $CdBG_k^+$, we could have taken $T(B_{k+1})$. Indeed, $T(B_{k+1})$ is the tree $T(A_k)$ with additional leaves representing all suffixes shorter than $(k - 1)$ of the words in R. These leaves make $T(B_{k+1})$ linear in $\|R\|$, but not in the size of dBG_k^+.

5   Conclusion

First, we provided a unified version of the construction of a generalised index for a set of input words. It only requires that the elements of the chain are suffix-dependant. This general framework was applied to build a reduced version of a Truncated Suffix Tree (TST), which given a parameter k does not contain any node representing a string smaller than $k - 1$ compared to the TST of [10,16]. Our algorithm does not build those nodes and thus does not need to remove them afterwards, as would be the case with the TST of [10,16]. This feature is crucial since otherwise, the index data structure could not be linear in the size of the de Bruijn Graph of order k. Moreover, it is not trivial to modify the algorithm of [10] to avoid building those nodes, because it uses their suffix links during construction. A natural question arise: does our framework remain valid if one relaxes the assumption of suffix-dependency?

Second, we show that the de Bruijn Graph (DBG) that is heavily exploited in the context of genome assembly can be directly constructed from the reduced TST in a time and a space linear in the size of the output. This part builds upon our earlier work, which addressed the question of DBG construction from a Suffix Tree and Suffix Array [4]. Moreover, we provide an algorithm to build the contracted DBG, which is used in practice and directly yields as a by-product of the contraction, a class of reliable contigs, called the unitigs[2]. Starting from the reduced truncated

---

[2] Usually the result of an assembly is not the complete sequence, rather a set of assembled sequences called contigs.

tree ensures that the index encodes as much nodes as the output DBG. However, as information on nodes representing substrings of length $< k - 1$ are missing, one looses the dynamicity obtained in [4, Section6]: our reduced TST cannot support the DBG construction for both order k and $k - 1$. Nevertheless, as our algorithms remain valid for the original TST [10,16], which stores all the information for any substrings $\leq k$, we can dynamically update the DBG for any order $< k$. Both the questions of 1/ dynamically updating the de Bruijn Graph when the order changes by more than one unit, and of 2/ constructing the DBG from compressed indexes make thrilling lines of future work.

References

1. Apostolico, A.: The myriad virtues of suffix trees. In: Apostolico, A., Galil, Z. (eds.) Combinatorial Algorithms on Words. NATO Advanced Science Institutes. Series F, vol. 12, pp. 85–96. Springer (1985)
2. Bowe, A., Onodera, T., Sadakane, K., Shibuya, T.: Succinct de bruijn graphs. In: Raphael, B., Tang, J. (eds.) WABI 2012. LNCS, vol. 7534, pp. 225–235. Springer, Heidelberg (2012)
3. de Bruijn, N.: On bases for the set of integers. Publ. Math. Debr. 1, 232–242 (1950)
4. Cazaux, B., Lecroq, T., Rivals, E.: From indexing data structures to de bruijn graphs. In: Kulikov, A.S., Kuznetsov, S.O., Pevzner, P. (eds.) CPM 2014. LNCS, vol. 8486, pp. 89–99. Springer, Heidelberg (2014)
5. Chikhi, R., Rizk, G.: Space-efficient and exact de Bruijn graph representation based on a Bloom filter. Algorithms for Molecular Biology 8, 22 (2013)
6. Conway, T.C., Bromage, A.J.: Succinct data structures for assembling large genomes. Bioinformatics 27(4), 479–486 (2011)
7. Golovnev, A., Kulikov, A.S., Mihajlin, I.: Approximating shortest superstring problem using de bruijn graphs. In: Fischer, J., Sanders, P. (eds.) CPM 2013. LNCS, vol. 7922, pp. 120–129. Springer, Heidelberg (2013)
8. Gusfield, D.: Algorithms on strings, trees and sequences: computer science and computational biology. Cambridge University Press, Cambridge (1997)
9. McCreight, E.: A space-economical suffix tree construction algorithm. J. of Association for Computing Machinery 23(2), 262–272 (1976)
10. Na, J.C., Apostolico, A., Iliopoulos, C.S., Park, K.: Truncated suffix trees and their application to data compression. Theoretical Computer Science 304(1–3), 87–101 (2003)
11. Peng, Y., Leung, H.C.M., Yiu, S.M., Chin, F.Y.L.: IDBA – A practical iterative de bruijn graph de novo assembler. In: Berger, B. (ed.) RECOMB 2010. LNCS, vol. 6044, pp. 426–440. Springer, Heidelberg (2010)
12. Pevzner, P., Tang, H., Waterman, M.: An Eulerian path approach to DNA fragment assembly. Proc. Natl. Acad. Sci. USA 98(17), 9748–9753 (2001)
13. Philippe, N., Salson, M., Commes, T., Rivals, E.: CRAC: an integrated approach to the analysis of RNA-seq reads. Genome Biology 14(3), R30 (2013)
14. Rizk, G., Gouin, A., Chikhi, R., Lemaitre, C.: Mindthegap: integrated detection and assembly of short and long insertions. Bioinformatics (2014)
15. Salmela, L.: Correction of sequencing errors in a mixed set of reads. Bioinformatics 26(10), 1284–1290 (2010)
16. Schulz, M.H., Bauer, S., Robinson, P.N.: The generalised k-truncated suffix tree for time-and space-efficient searches in multiple DNA or protein sequences. International J. of Bioinformatics Research and Applications 4(1), 81–95 (2008)

Frequent Pattern Mining
with Non-overlapping Inversions

Da-Jung Cho, Yo-Sub Han$^{(\boxtimes)}$, and Hwee Kim

Department of Computer Science, Yonsei University, 50, Yonsei-Ro,
Seodaemun-Gu, Seoul 120-749, Republic of Korea
{dajung,emmous,kimhwee}@cs.yonsei.ac.kr

Abstract. Frequent pattern mining is widely used in bioinformatics
since frequent patterns in bio sequences often correspond to residues
conserved during evolution. In bio sequence analysis, non-overlapping
inversions are well-studied because of their practical properties for local
sequence comparisons. We consider the problem of finding frequent patterns in a bio sequence with respect to non-overlapping inversions, and
design efficient algorithms.

Keywords: String processing algorithms · Frequent pattern mining ·
Non-overlapping inversions

1 Introduction

Agrawal et al. [1] studied the frequent pattern mining for finding associations
among market products and increasing profit. For example, frequent patterns in
customer behavior are useful for setting affordable product price, promotion and
store layout. They investigated the problem of finding meaningful associations
over market transactions. Frequent patterns are also useful in other domains
including sequential data, bio sequences or strings [10–12,14,18,20].

In bioinformatics, frequent motifs in DNA or protein sequences often correspond to residues conserved during evolution due to an important structural or
functional role [18]. Note that traditional pattern mining algorithms are not suitable for bio sequences, since they cope with a large number of items and short
sequence lengths [11]. Wang et al. [18] first proposed an algorithm that finds
sequential patterns on bio sequences. Recently, Liao and Chen [12] designed an
algorithm for the problem with gaps—regions not conserved in evolution.

From a biological aspect, an inversion—breakage and rearrangement within
itself—is one of the most important operations since such an event produces
new gene sequences from an original gene sequence and sometimes causes a disease [13]. Schöniger and Waterman [15] introduced a simplification hypothesis
that all regions involved in the inversions do not overlap. This hypothesis—
non-overlapping inversions—is realistic for local DNA comparisons on relatively
closed sequences [17]. On the string with non-overlapping inversions,
Chen et al. [4] designed an $O(n^4)$ algorithm to solve the alignment with non-
overlapping inversions, which was improved to $O(n^3)$ by Vellozo et al. [17]. Amir

© Springer International Publishing Switzerland 2015
A.-H. Dediu et al. (Eds.): LATA 2015, LNCS 8977, pp. 121–132, 2015.
DOI: 10.1007/978-3-319-15579-1_9

and Porat [2] proposed an $O(n^2)$ approximation algorithm for the problem, and Cho et al. [6] proposed an $O(n^3)$ algorithm for the modified problem where inversions occur to both strings. For the pattern matching, Cantone et al. [3] proposed $O(nm)$ algorithm to solve the pattern matching with non-overlapping inversions. On formal language theory, researchers [5,8] have studied properties and decision problems of formal languages considering inversions. This leads us to consider frequent pattern mining problem on a string with non-overlapping inversions.

Due to the irregularity of gene evolution, rearrangements—for instance, non-overlapping inversions—may occur in conserved regions P. Suppose that a sequence S_{i+1} is obtained from a sequence S_i that has conserved regions, and non-overlapping inversions occur during the evolution from S_i to S_{i+1} (See Fig. 1 for an illustrative example of this phenomenon.). We search for a conserved region P in S_{i+1}, which now may have been modified by non-overlapping inversions from S_i. Note that often we do not have all evolution sequences of a gene. For instance, here S_{i+1} is a mere sequence that we have and, thus, we do not know where exactly non-overlapping inversions occur in S_{i+1}. This makes the problem of finding similar or same pattern occurrences in S_i challenging when we have only S_{i+1} and the fact that S_{i+1} is generated from S_i by some non-overlapping inversions.

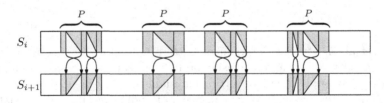

Fig. 1. Let S_i be a gene sequence of the ith generation and S_{i+1} be a gene sequence of the $i+1$th generation. During the evolution from S_i to S_{i+1}, non-overlapping inversions flip subsequences of the conserved regions P.

We formulate our problem as a frequent pattern mining problem on a string: Given a text T of length n over an alphabet Σ, a pattern length m and a pattern occurrence threshold r, our goal is to compute the set of all patterns P of length m that occur in T at least r times when we allow non-overlapping inversions on P. Note that P may not be a substring or a subsequence of T, which is different from other frequent pattern mining problems in the literature. We first compute a set of all possible substrings T_i of length m and construct digraphs G_i representing all strings that can be generated by non-overlapping inversions on T_i. Next, we overlay all such G_i's and obtain a weighted multidigraph G. Then, we find all paths in G with the bottleneck lower bound r—each path represents P and the bottleneck is the number of occurrences of P. We show that we can find all patterns in $O(nm^2 + Cm)$ time using $O(m)$ space, where C is the number of matching patterns. If we want to store all matching patterns instead of reporting them, then we can construct a DFA that recognizes the set of all such patterns in $O(nm^2)$ time using $O(m)$ space.

2 Preliminaries

Let $A[a_1][a_2]\cdots[a_n]$ be an n-dimensional array, where the size of each dimension is a_i for $1 \leq i \leq n$. Let $A[i_1][i_2]\cdots[i_n]$ be the element of A with indices $(i_1, i_2, \ldots, i_n)$. Given a finite set Σ of characters and a string w over Σ, we use $|w|$ to denote the length of w and $w[i]$ to denote the symbol of w at position i. We use $w[i:j]$ to denote the substring $w[i]\cdots w[j]$, where $0 < i \leq j$.

For a finite set Σ of characters, Σ^* denotes the set of all strings over Σ. A language over Σ is any subset of Σ^*. The symbol $\emptyset$ denotes the empty language and the symbol λ denotes the null string. A finite-state automaton (FA) A is specified by $A = (Q, \Sigma, \delta, s, F)$, where Q is a set of states, Σ is an alphabet, $\delta \subseteq Q \times \Sigma \times Q$ is a set of transitions, $s \in Q$ is the start state and $F \subseteq Q$ is a set of final states. For a transition $\delta(p, \sigma) = q$, we say that p has an *out-transition* and q has an *in-transition*. Moreover, we call q a *target state* of p. A string w is accepted by A if there is a labeled path from s to a final state in F such that the path spells out w. The language $L(A)$ of an FA A is the set of all strings accepted by A. If $|\{\delta(p, \sigma)\}| = 1$ for all $p \in Q$ and $\sigma \in \Sigma$, we say that A is a deterministic finite-state automaton (DFA); otherwise, A is a nondeterministic finite-state automaton (NFA). For more knowledge in automata theory, the reader may refer to textbooks [16,19].

We consider a biological operation *inversion* θ and denote by $\theta(w)$ the reverse of a string w. We define an inversion operation $\theta_{(i,j)}$ for a given range (i, j) to be $\theta_{(i,j)}(w) = \theta(w[i:j])$. When the context is clear, we denote $\theta_{(i,j)}$ as (i, j). We say that the length of (i, j) is $j - i + 1$. We define a sequence $\Theta = ((p_1, q_1), (p_2, q_2), \ldots, (p_k, q_k))$ of inversions for a string w to be *non-overlapping* (NOI-sequence for short) if it satisfies the following conditions: For $1 \leq i \leq k$, $p_1 \geq 1$, $q_k \leq |w|$, $p_i \leq q_i$ and $p_{i+1} \geq q_i + 1$ for $1 \leq i \leq k - 1$. For the sake of easier explanation of our algorithms, for any given index i, we assume that there always exists an inversion whose range covers i; namely, $p_1 = 1$, $q_k = |w|$ and $p_{i+1} = q_i + 1$, since a non-inversed range (i, j) can be represented by a sequence $((i, i), (i+1, i+1), \ldots, (j, j))$ of inversions. Now, in summary, given an NOI-sequence $\Theta = ((p_1, q_1), (p_2, q_2), \ldots, (p_k, q_k))$ and a string w, $\Theta(w) = \theta(w[p_1 : q_1])\theta(w[p_2 : q_2])\cdots\theta(w[p_k : q_k])$.

An undirected graph $G = (V, E)$ consists of a finite nonempty set V of nodes and a set E of unordered pairs of distinct nodes of V. Each pair $e = \{u, v\}$ of nodes in E is an edge of G and e is said to join u and v. A directed graph or digraph D consists of a finite nonempty set V of nodes and a set E of ordered pairs of nodes. For an edge $e = (u, v)$ of a digraph, we say that e is from node u to node v. A multidigraph is a digraph where more than one edge can join two nodes. The reader may refer to Harary [7] for more details in graph theory.

Given two strings X and Y of the same length, we say that X and Y have an *alignment with non-overlapping inversions* (NOI-alignment for short) if there exists an NOI sequence Θ such that $\Theta(Y) = X$. Given a text T of length n and a pattern P of length m, the *NOI-occurrence* $Occ(T, P)$ of T and P is the number of indices i where $T_i = T[i : i+m-1]$ has an NOI-alignment with P.

Definition 1 (Frequent Pattern Mining with Non-overlapping Inversions). *Given a text T of length n over Σ, a pattern length m and a minimum number r of pattern occurrences, find all pairs $(P \in \Sigma^m, Occ(T, P))$ where $Occ(T, P) \geq r$.*

3   The Algorithm

Given a text T, our algorithm starts from inspecting all substrings T_i of T. We first compute a set of all NOI-alignment strings for T_i and construct a digraph G_i that represents the set. Then we overlay all G_i's and construct a weighted digraph G, and find all frequent patterns P from G, where $Occ(T, P) \geq r$. We construct inversion fragment tables for all substrings T_i.

Definition 2. *Given a text T, an index i and a pattern length m, the* inversion fragment table *(IFT for short) is defined as follows:*

$$F_i[j][k] = \begin{cases} ((k, j), T_i[k]) & \text{if } k \leq j, \\ ((j, k), T_i[k]) & \text{otherwise} \end{cases}$$

for $1 \leq j, k \leq m$.

We call all elements in $F_i[j][k]$ *inversion fragments* (IFs for short) of T_i. For an IF $\mathbb{F} = ((p, q), \sigma)$, we say that $\mathbb{F}$ *yields* the character σ. For a sequence of IFs $\mathbb{F}_1, \ldots, \mathbb{F}_l$, where $\mathbb{F}_i$ yields σ_i, we say that the sequence yields a string $\sigma_1 \cdots \sigma_l$. Fig. 2 shows an example of an IFT.

F_1	1	2	3	4
1	$((1,1), A)$	$((1,2), A)$	$((1,3), A)$	$((1,4), A)$
2	$((1,2), G)$	$((2,2), G)$	$((2,3), G)$	$((2,4), G)$
3	$((1,3), C)$	$((2,3), C)$	$((3,3), C)$	$((3,4), C)$
4	$((1,4), T)$	$((2,4), T)$	$((3,4), T)$	$((4,4), T)$

Fig. 2. IFT F_1 for $T = AGCTA$ and $m = 4$. Shaded cells denote IFs for $\theta_{(2,4)}(T_1)$.

IFs become useful for computing a substring created by an inversion because of the following property of the inversion operation:

Observation 3. *For a text T, an index i and its IFT F_i, a sequence $F_i[j][k]$, $F_i[j+1][k-1], \ldots, F_i[k-1][j+1], F_i[k][j]$ of IFs yields $\theta_{(j,k)}(T_i)$.*

From Observation 3, we know that if we apply $\theta_{(\min(j,k),\max(j,k))}$ to T_i, then σ yielded by $F_i[j][k]$ becomes the jth character of the result string. Using Observation 3, we construct a digraph G_i for T and i by Algorithm 1, which we call an *inversion graph*. In an inversion graph, each node (j, σ) represents that there exists an NOI-alignment between P and T_i where $P[j] = \sigma$. Each edge $((j, \sigma_1), (j+1, \sigma_2), \sigma_0, f = c$ or $v)$ represents that if $P[j-1] = \sigma_0$ and $P[j] = \sigma_1$, then we can set $P[j+1] = \sigma_2$ to ensure that P and T_i have an NOI-alignment. The last element f is the flag that indicates whether $P[j]$ and $P[j+1]$ are from a single inversion on the text (v) or from two adjacent inversions (c). Fig. 3 shows an example of an inversion graph.

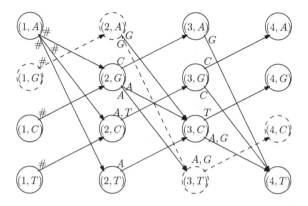

Fig. 3. An inversion graph G_1 for $T = AGCTA$ (flags of the edges are omitted). The dashed path represents $P = GATC$, which has an NOI-alignment with $T_1 = AGCT$.

We can retrieve all strings that have an NOI-alignment with T_i from G_i.

Theorem 4. *Given a text T, an index i, a pattern length m and an inversion graph G_i, a string P has an NOI-alignment with T_i if and only if*

1. $((1, P[1]), (2, P[2]), \#, f) \in E_i$ *and*
2. $((j, P[j]), (j+1, P[j+1]), P[j-1], f) \in E_i$ *for* $2 \le j \le m - 1$.

In the literature, people often assume that the alphabet size is constant, since Σ is a finite set. Then, it is straightforward to verify that Algorithm 1 runs in $O(m^3)$ time. Since the size of F_i is $O(m^2)$ and the size of E_i is $O(m)$, Algorithm 1 requires $O(m^2)$ space. We improve the runtime of Algorithm 1 by relying on the following property of IFTs.

Observation 5. *For* $1 \le i < n - m$ *and* $2 \le j, k \le m$,

1. *If* $F_i[j][k] = ((j, k), \sigma)$, *then* $F_{i+1}[j-1][k-1] = ((j-1, k-1), \sigma)$.
2. *If* $F_i[j][k] = ((k, j), \sigma)$, *then* $F_{i+1}[j-1][k-1] = ((k-1, j-1), \sigma)$.

Algorithm 1. ConstructInversionGraph

Input: Text T over Σ of size t, index i and a pattern length m
Output: Digraph $G_i = (V, E_i)$, where $V = \{1, 2, \ldots, m\} \times \Sigma$ and
$\qquad E_i \subset V \times V \times (\Sigma \cup \{\#\}) \times \{v, c\}$

1 construct F_i
2 $V \leftarrow \{1, 2, \ldots, m\} \times \Sigma$
3 **for** $j \leftarrow 1$ **to** $m - 1$ **do**

    /* check the case that an inversion ends at index j */
4   **for** $k \leftarrow 1$ **to** j **do**
5   let σ_1 be yielded by $F_i[j][k]$
        /* find all first characters of inversions starting from
           index $j + 1$ and record the edge   */
6   **for** $l \leftarrow j + 1$ **to** m **do**
7   let σ_2 be yielded by $F_i[j+1][l]$
8   **if** $j = 1$ **then**
9   add $((j, \sigma_1), (j+1, \sigma_2), \#, c)$ to E_i

10   **else**
11   **for each** σ_0 *where* $((j-1, \sigma_0), (j, \sigma_1), \sigma', f) \in E_i$ **do**
12   add $((j, \sigma_1), (j+1, \sigma_2), \sigma_0, c)$ to E_i

    /* check the case that index j is within the range of an
       inversion   */
13   **for** $k \leftarrow 2$ **to** m **do**
        /* find the next character in the inversion and record the
           edge   */
14   let σ_1 be yielded by $F_i[j][k]$ and σ_2 be yielded by $F_i[j+1][k-1]$
15   **if** $j = 1$ **then**
16   add $((j, \sigma_1), (j+1, \sigma_2), \#, v)$ to E_i

17   **else if** $k \neq m$ **then**
18   add $((j, \sigma_1), (j+1, \sigma_2), \sigma_0, v)$ to E_i, where $F_i[j-1][k+1]$ yields σ_0

19   **if** $j < k$ *and* $j \neq 1$ **then**
20   **for each** σ_0 *where* $((j-1, \sigma_0), (j, \sigma_1), \sigma', c) \in E_i$ **do**
21   add $((j, \sigma_1), (j+1, \sigma_2), \sigma_0, v)$ to E_i

22 **return** (V, E_i)

From Observation 5, we know that G_i and G_{i+1}, constructed from F_i and F_{i+1} respectively, have common edges (See Fig. 4.). We make use of these edges to reduce the construction time of G_{i+1} by adding a new label (k, k') for each edge $((j, \sigma_1), (j+1, \sigma_2), \sigma_0, f) \in E_i$, representing that the edge is from IFs $F_i[j][k]$ and $F_i[j+1][k']$. Moreover, in Algorithm 1, F_i is only used to compute characters yielded by IFs. Since $F_i[j][k]$ always yields $T_i[k]$, we do not need to construct F_i in the algorithm. By modifying Algorithm 1, we obtain a new algorithm with improved time and space complexity. Algorithm 2 preserves all edges in E_{i-1} that are made from IFs in F_i and runs Algorithm 1 only for edges that are not in E_{i-1} but in E_i. While adding new edges to E_i, Algorithm 2 does not use F_i. Thus, Algorithm 2 constructs the same G_i as Algorithm 1 except for additional labels.

Algorithm 2. ConstructFastInversionGraph

Input: Text T over Σ of size t, index i, pattern length m and inversion graph
$\quad\quad G_{i-1} = (V, E_{i-1})$
Output: Digraph $G_i = (V, E_i)$, where $V = \{1, 2, \ldots, m\} \times \Sigma$ and
$\quad\quad E_i \subset V \times V \times (\Sigma \cup \{\#\}) \times \{v, c\} \times \{1, 2, \ldots, m\}^2$

```
/* retrieve common edges from E_{i-1}                              */
```
1 **for each** $((j, \sigma_1), (j{+}1, \sigma_2), \sigma_0, f, k, k') \in E_{i-1}$ **do**
2 $\quad$ **if** $2 \le j \le m-1$ *and* $2 \le k, k'$ **then**
3 $\quad\quad$ add $((j{-}1, \sigma_1), (j, \sigma_2), \sigma_0, f, k{-}1, k'{-}1)$ to E_i

4 **for** $j \leftarrow 1$ **to** $m-1$ **do**
```
    /* add only new edges for j using a part of Algorithm 1        */
```
5 $\quad$ **for** $k \leftarrow 1$ **to** j **do**
6 $\quad\quad$ $\sigma_1 \leftarrow T_i[k], \sigma_2 \leftarrow T_i[m]$
7 $\quad\quad$ **if** $j = 1$ **then** add $((1, \sigma_1), (2, \sigma_2), \#, c, k, m)$ to E_i **else**
8 $\quad\quad\quad$ **for each** σ_0 *where* $((j{-}1, \sigma_0), (j, \sigma_1), \sigma', f, k', k) \in E_i$ **do**
9 $\quad\quad\quad\quad$ add $((j, \sigma_1), (j{+}1, \sigma_2), \sigma_0, c, k, m)$ to E_i

```
    /* add edges for j = m - 1 using a part of Algorithm 1         */
```
10 $\quad$ **if** $j = m-1$ **then**
11 $\quad\quad$ **for** $k \leftarrow 2$ **to** $m-1$ **do**
12 $\quad\quad\quad$ $\sigma_1 \leftarrow T_i[k], \sigma_2 \leftarrow T_i[k{-}1]$
13 $\quad\quad\quad$ add $((m{-}1, \sigma_1), (m, \sigma_2), T_i[k{+}1], v, k, k{-}1)$ to E_i

14 $\quad\quad$ $\sigma_1 \leftarrow T_i[m{-}1], \sigma_2 \leftarrow T_i[m]$
15 $\quad\quad$ **for each** σ_0 *where* $((m{-}2, \sigma_0), (j, \sigma_1), \sigma', c, k', k) \in E_i$ **do**
16 $\quad\quad\quad$ add $((j, \sigma_1), (j{+}1, \sigma_2), \sigma_0, v, k, k{-}1)$ to E_i

17 **return** (V, E_i)

F_1	1	2	3	4
1	$((1,1), A)$	$((1,2), A)$	$((1,3), A)$	$((1,4), A)$
2	$((1,2), G)$	$((2,2), G)$	$((2,3), G)$	$((2,4), G)$
3	$((1,3), C)$	$((2,3), C)$	$((3,3), C)$	$((3,4), C)$
4	$((1,4), T)$	$((2,4), T)$	$((3,4), T)$	$((4,4), T)$

F_2	1	2	3	4
1	$((1,1), G)$	$((1,2), G)$	$((1,3), G)$	$((1,4), G)$
2	$((1,2), C)$	$((2,2), C)$	$((2,3), C)$	$((2,4), C)$
3	$((1,3), T)$	$((2,3), T)$	$((3,3), T)$	$((3,4), T)$
4	$((1,4), A)$	$((2,4), A)$	$((3,4), A)$	$((4,4), A)$

Fig. 4. Comparison of F_1 and F_2 for $T = AGCTA$. Shaded cells denote IFs from F_1 and F_2 that satisfy the properties of Observation 5.

Since moving common edges from E_{i-1} to E_i takes $O(m)$ time and adding new edges takes $O(m^2)$ time, Algorithm 2 runs in $O(m^2)$ time using $O(m)$ space. Note that G_i represents all possible strings that have NOI-alignments with T_i. We overlay all G_i's to construct an *accumulated inversion graph* G for T by Algorithm 3 (See Fig. 5 for an example.). Note that Algorithm 3 requires $O(nm^2)$ time and $O(m)$ space.

Algorithm 3. ConstructAccumulatedInversionGraph

Input: Text T of length n over Σ of size t and a pattern length m
Output: Weighted multidigraph $G = (V, E)$, where $V = \{1, 2, \ldots, m\} \times \Sigma$ and
$\qquad E \subset V \times V \times (\Sigma \cup \{\#\}) \times \{1, 2, \ldots, n - m + 1\}$

1 $V \leftarrow \{1, 2, \ldots, m\} \times \Sigma$
2 **for each** $(i, \sigma_1), (i+1, \sigma_2) \in V$, $\sigma_0 \in \Sigma$ **do**
3 $\quad \llcorner$ add $((i, \sigma_1), (i+1, \sigma_2), \sigma_0, 0)$ to E.

4 **for** $i \leftarrow 1$ **to** $n - m + 1$ **do**
  $\qquad$ /* modify ConstructInversionGraph not to use F_1 $\qquad\qquad$ */
5 $\quad$ **if** $i = 1$ **then** ConstructInversionGraph(T, i, m) **else**
  $\qquad$ ConstructFastInversionGraph(T, i, m, G_{i-1}) **for each** $(v_1, v_2, \sigma, f, k, k') \in E_i$
  $\qquad$ **do**
6 $\quad\quad \llcorner$ change $(v_1, v_2, \sigma, g) \in E$ to $(v_1, v_2, \sigma, g + 1)$

7 **return** (V, E)

For an edge $((i, \sigma_i), (i+1, \sigma_{i+1}), \sigma_{i-1}, g_i) \in E$, we call g_i the *weight* of the edge and σ the *preceding symbol* of the edge. We also say that the edge is from index i to index $i+1$. For a path $((1, \sigma_1), (2, \sigma_2), \#, g_1), ((2, \sigma_2), (3, \sigma_3), \sigma_1, g_2), \ldots, ((m-1, \sigma_{m-1}), (m, \sigma_m), \sigma_{m-2}, g_m)$, we call $\min(g_1, g_2, \ldots, g_m)$ the *minimum weight* of the path. From the construction of G, we have the following statement:

Lemma 6. *For a text T of length n and a pattern length m, let $P \in \Sigma^m$ be a pattern such that $((1, P[1]), (2, P[2]), \#, g_1) \in E$ and $((j, P[j]), (j+1, P[j+1]), P[j-1], g_j) \in E$ for $2 \le j \le m - 1$. Then $Occ(T, P) = \min(g_j)$ for $1 \le j \le m - 1$.*

Now, our goal is to find all paths from index 1 to index m, where the minimum weight of each path is greater than or equal to r. We reduce the resulting accumulated inversion graph so that any path from index 1 to index m in the graph satisfies the condition based on a modified Kruskal's Algorithm [9]. We sort edges by ascending order with respect to weights as in Kruskal's Algorithm. Then we repeatedly remove an edge with the minimum weight until all remaining edges have weights greater than or equal to r. Once we remove an edge e, we then remove all adjacent edges of e that cannot be in a path anymore because of the removal of e. Algorithm 4 removes edges from G to return a graph G', where any path from index 1 to index m represents a pattern P such that $Occ(T, P) \ge r$. Fig. 6 is an example of G' for $T = AGCTAGCTAG$ and $r = 3$.

We prove that any path from index 1 to index m in the resulting graph represents a pattern P such that $Occ(T, P) \ge r$.

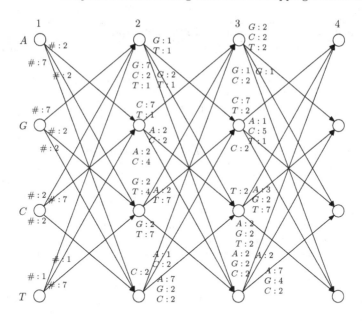

Fig. 5. An accumulated inversion graph G for $T = AGCTAGCTAG$

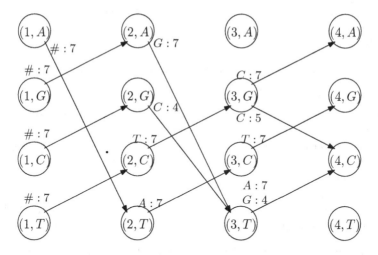

Fig. 6. A reduced accumulated inversion graph G' for $T = AGCTAGCTAG$ and $r = 3$

Theorem 7. *Suppose Algorithm 4 returns $G' = (V, E')$. Let $P \in \Sigma^m$ be a pattern such that $((1, P[1]), (2, P[2]), \#, g_1) \in E'$ and $((j, P[j]), (j+1, P[j+1]), P[j-1], g_j) \in E$ for $2 \le j \le m - 1$. Then $Occ(T, P) \ge r$.*

Next, we analyze the time and space complexity of the algorithm.

Lemma 8. *Algorithm 4 runs in $O(nm^2)$ time using $O(m)$ space.*

Algorithm 4. ReduceAccumulatedInversionGraph

Input: Text T of length n over Σ of size t, a pattern length m and a pattern
 occurrence threshold r

Output: Weighted multidigraph $G' = (V, E')$, where $V = \{1, 2, \ldots, m\} \times \Sigma$ and
 $E \subset V \times V \times (\Sigma \cup \{\#\}) \times \{1, 2, \ldots, n - m + 1\}$

1 ConstructAccumulatedInversionGraph(T, m)
2 $E' \leftarrow E$
3 sort E' by ascending order with respect to weights
4 **while** *there exists an edge in E' with weight less than r* **do**
5      $e \leftarrow ((i, \sigma_1), (i+1, \sigma_2), \sigma_0, g)$ be the edge with minimum weight in E'
6      $R \leftarrow \emptyset$ // **set of edges to remove**
7   **if** *e is the only edge from index i to index $i + 1$ in E'* **then return** $(V, \emptyset)$
     else add e to R **while** $R \neq \emptyset$ **do**
8   **for each** $e' \leftarrow ((j, \sigma_1'), (j+1, \sigma_2'), \sigma_0', g') \in R$ **do**
9   **if** *e' is the only edge from node (j, σ_1') to node $(j+1, \sigma_2')$ in E'* **then**
10   add all edges from node $(j+1, \sigma_2')$ to index $j+2$ with preceding
               symbol σ_1' in E' to R
11   **if** *there is no edge from node (j, σ_1') to index $j+1$ in E'* **then**
12   add all edges from index $j-1$ to node (j, σ_1') in E' to R

13   **if** *e' is the only edge from node (j, σ_1') to index $j+1$ with preceding
            symbol σ_0' in E'* **then**
14   add all edges from node $(j-1, \sigma_0')$ to node (j, σ_1') in E' to R

15   remove e' from E' and R

16 **return** (V, E')

It requires at least $O(Cm)$ to report all patterns with the occurrence greater
than or equal to r, where C is the number of such patterns. From G', we can
find all such patterns by simple depth-first search in $O(Cm)$. On the other hand,
if we want to identify all matching patterns instead of reporting them, then we
can convert G' to a DFA A', where $L(A')$ is the set of all such patterns. For
a weighted multidigraph $G' = (V, E')$, where $V = \{1, 2, \ldots, m\} \times \Sigma$ and $E \subset
V \times V \times (\Sigma \cup \{\#\}) \times \{1, 2, \ldots, n - m + 1\}$, we construct a DFA $A' = (Q, \Sigma, \delta, s, F)$
by the following procedure:

1. $Q = s \cup \{\#\} \times \Sigma \times \{1\} \cup \Sigma \times \Sigma \times \{2, \ldots, m\}$.
2. $F = \Sigma \times \Sigma \times \{m\}$.
3. $\delta(s, \sigma) = (\#, \sigma, 1)$ for all $((1, \sigma), (2, \sigma'), \#, g) \in E'$, where $\sigma, \sigma' \in \Sigma$ and
 $g \geq 0$.
4. $\delta((\sigma_0, \sigma_1, i), \sigma_2) = (\sigma_1, \sigma_2, i+1)$ for all $e' \leftarrow ((i, \sigma_1), (i+1, \sigma_2), \sigma_0, g) \in E'$,
 where $\sigma_0, \sigma_1, \sigma_2 \in \Sigma$, $g \geq 0$ and $1 \leq i \leq m - 1$.

From the construction of the transition function, it is straightforward that
A' is a DFA and $L(A')$ is equal to the set of all patterns with the occurrence
greater than or equal to r. The construction of A' requires $O(m)$ time using
$O(m)$ space. Then, we establish the following theorem.

Theorem 9. *Given a text T of length n over an alphabet Σ, a pattern length m and a pattern occurrence threshold r, we can solve frequent pattern mining with non-overlapping inversions in $O(nm^2 + Cm)$ time using $O(m)$ space, where C is the number of patterns to find. Moreover, we can construct a DFA that recognizes the set of all patterns to find in $O(nm^2)$ time using $O(m)$ space.*

4   Conclusions

We have considered non-overlapping inversions in frequent pattern mining. We have proposed a graph-based algorithm that finds all patterns P where $Occ(T, P) \geq r$ in $O(nm^2 + Cm)$ time using $O(m)$ space, where m is the desired pattern length, n is the size of an input text T, r is the pattern occurrence threshold and C is the number of patterns to find. Moreover, we have proposed an algorithm that constructs a DFA recognizing the set of all matching patterns in $O(nm^2)$ time using $O(m)$ space. Notice that we need to examine all possible strings that have an NOI-alignment with any T_i to find P. Since the number of inversions in all T_i's is $O(nm^2)$, it would be challenging to design an algorithm that runs faster than $O(nm^2 + Cm)$. Since the runtime to find all patterns depends on C, it is an interesting open question to establish the lower and the upper bound of C. Another future direction is to find frequent patterns from a bio sequence under other biological operations and combined operations.

Acknowledgments. This research was supported in part by the Basic Science Research Program through NRF funded by MEST (2012R1A1A2044562) and by the Yonsei University Future-leading Research Initiative of 2014. Kim was supported by NRF Grant funded by the Korean Government (NRF-2013-Global Ph.D. Fellowship Program).

References

1. Agrawal, R., Imieliński, T., Swami, A.: Mining association rules between sets of items in large databases. ACM SIGMOD Record **22**(2), 207–216 (1993)
2. Amir, A., Porat, B.: Pattern matching with non overlapping reversals - approximation and on-line algorithms. In: Cai, L., Cheng, S.-W., Lam, T.-W. (eds.) Algorithms and Computation. LNCS, vol. 8283, pp. 55–65. Springer, Heidelberg (2013)
3. Cantone, D., Cristofaro, S., Faro, S.: Efficient string-matching allowing for non-overlapping inversions. Theoretical Computer Science **483**(29), 85–95 (2013)
4. Chen, Z.Z., Gao, Y., Lin, G., Niewiadomski, R., Wang, Y., Wu, J.: A space-efficient algorithm for sequence alignment with inversions and reversals. Theoretical Computer Science **325**(3), 361–372 (2004)
5. Cho, D.-J., Han, Y.-S., Kang, S.-D., Kim, H., Ko, S.-K., Salomaa, K.: Pseudo-inversion on formal languages. In: Ibarra, O.H., Kari, L., Kopecki, S. (eds.) UCNC 2014. LNCS, vol. 8553, pp. 93–104. Springer, Heidelberg (2014)
6. Cho, D.J., Han, Y.S., Kim, H.: Alignment with non-overlapping inversions and translocations on two strings. Theoretical Computer Science (in press)
7. Harary, F.: Graph Theory. Addison-Wesley series in mathematics. Perseus Books (1994)

8. Ibarra, O.H.: On decidability and closure properties of language classes with respect to bio-operations. In: Murata, S., Kobayashi, S. (eds.) DNA 2014. LNCS, vol. 8727, pp. 148–160. Springer, Heidelberg (2014)

9. Kruskal, Jr., J.B.: On the shortest spanning subtree of a graph and the traveling salesman problem. Proceedings of the American Mathematical Society **7**(1), 48–50 (1956)

10. Kum, H.C., Pei, J., Wang, W., Duncan, D.: Approxmap: Approximate mining of consensus sequential patterns. In: Proceedings of the 2nd SIAM International Conference on Data Mining, pp. 311–315 (2003)

11. Liao, V.C.C., Chen, M.S.: DFSP: a Depth-First SPelling algorithm for sequential pattern mining of biological sequences. Knowledge and Information Systems **38**(3), 623–639 (2013)

12. Liao, V.C.C., Chen, M.S.: Efficient mining gapped sequential patterns for motifs in biological sequences. BMC Systems Biology **7**(4), 1–13 (2013)

13. Lupski, J.R.: Genomic disorders: structural features of the genome can lead to DNA rearrangements and human disease traits. Trends in Genetics **14**(10), 417–422 (1998)

14. Sagot, M.-F.: Spelling approximate repeated or common motifs using a suffix tree. In: Lucchesi, C.L., Moura, A.V. (eds.) LATIN 1998. LNCS, vol. 1380, pp. 374–390. Springer, Heidelberg (1998)

15. Schöniger, M., Waterman, M.S.: A local algorithm for DNA sequence alignment with inversions. Bulletin of Mathematical Biology **54**(4), 521–536 (1992)

16. Shallit, J.: A Second Course in Formal Languages and Automata Theory. Cambridge University Press (2008)

17. Vellozo, A.F., Alves, C.E.R., do Lago, A.P.: Alignment with non-overlapping inversions in $O(n^3)$-time. In: Bücher, P., Moret, B.M.E. (eds.) WABI 2006. LNCS (LNBI), vol. 4175, pp. 186–196. Springer, Heidelberg (2006)

18. Wang, K., Xu, Y., Yu, J.X.: Scalable sequential pattern mining for biological sequences. In: Proceedings of the 13th ACM International Conference on Information and Knowledge Management, pp. 178–187 (2004)

19. Wood, D.: Theory of Computation. Harper & Row (1986)

20. Zhu, F., Yan, X., Han, J., Yu, P.S.: Efficient discovery of frequent approximate sequential patterns. In: Proceedings of the 7th IEEE International Conference on Data Mining, pp. 751–756 (2007)

A Parallel Algorithm for Finding All Minimal Maximum Subsequences via Random Walk

H.K. Dai$^{(\boxtimes)}$ and Z. Wang

Computer Science Department, Oklahoma State University, Stillwater,
Oklahoma 74078, USA
{dai,wzhu}@cs.okstate.edu

Abstract. A maximum(-sum) contiguous subsequence of a real-valued sequence is a contiguous subsequence with the maximum cumulative sum. A minimal maximum contiguous subsequence is a minimal contiguous subsequence among all maximum ones of the sequence. We have designed and implemented a domain-decomposed parallel algorithm on cluster systems with Message Passing Interface that finds all successive minimal maximum subsequences of a random sample sequence from a normal distribution with negative mean. Our study employs the theory of random walk to derive an approximate probabilistic length bound for minimal maximum subsequences in an appropriate probabilistic setting, which is incorporated in the algorithm to facilitate the concurrent computation of all minimal maximum subsequences in hosting processors. We also present a preliminary empirical study of the speedup and efficiency achieved by the parallel algorithm with synthetic random data.

Keywords: All maximum subsequences · Theory of random walk · Message passing interface · Parallel random access machine model

1 Preliminaries

Algorithmic and optimization problems in sequences and trees arise in widely varying domains such as bioinformatics and information retrieval. Large-scale (sub)sequence comparison, alignment, and analysis are important research areas in computational biology. Time- and space-efficient algorithms for finding multiple contiguous subsequences of a real-valued sequence having large cumulative sums help identify statistically significant subsequences in biological sequence analysis with respect to an underlying scoring scheme – an effective filtering pre-process even with simplistic random-sequence models of independent residues.

For a real-valued sequence $X = (x_\eta)_{\eta=1}^n$, the cumulative sum of a non-empty contiguous subsequence $(x_\eta)_{\eta=i}^j$, where i and j are in the index range $[1, n]$ with $i \leq j$, is $\sum_{\eta=i}^j x_\eta$ (and that of the empty sequence is 0). All subsequences addressed in our study are contiguous in real-valued sequences; the terms "subsequence" and "supersequence" will hereinafter abbreviate "contiguous subsequence" and "contiguous supersequence", respectively. A maximum(-sum) subsequence of X is one with the maximum cumulative sum. A minimal maximum subsequence of X is a

© Springer International Publishing Switzerland 2015
A.-H. Dediu et al. (Eds.): LATA 2015, LNCS 8977, pp. 133–144, 2015.
DOI: 10.1007/978-3-319-15579-1_10

minimal subsequence (with respect to subsequential containment) among all maximum subsequences of X.

Very often in applications it is required to find many or all pairwise disjoint subsequences having cumulative sums above a prescribed threshold. Observe that subsequences having major overlap with a maximum subsequence tend to have good cumulative sums. Intuitively, we define the sequence of all successive minimal maximum subsequences $(S_1, S_2, \ldots)$ of X inductively as follows: (1) The sequence S_1 is a (non-empty) minimal maximum subsequence of X, and (2) Assume that the sequence $(S_1, S_2, \ldots, S_i)$ of non-empty subsequences of X, where $i \geq 1$, has been constructed, the subsequence S_{i+1} is a (non-empty) minimal subsequence (with respect to subsequential containment) among all non-empty maximum subsequences (with respect to cumulative sum) that are disjoint from each of $\{S_1, S_2, \ldots, S_i\}$.

Efficient algorithms for computing the sequence of all successive minimal maximum subsequences of a given sequence are essential for statistical inference in large-scale biological sequence analysis. In biomolecular sequences, high (sub)sequence similarity usually implies significant structural or functional similarity. When incorporating good scoring schemes, this provides a powerful statistical paradigm for identifying biologically significant functional regions in biomolecular sequences [8], such as transmembrane regions and deoxyribonucleic acid-binding domains [6] in protein analyses. The non-positivity of the expected score of a random single constituent tends to delimit unrealistic long runs of contiguous positive scores.

We design and implement a domain-decomposed parallel algorithm on cluster systems with Message Passing Interface that finds all successive minimal maximum subsequences of a random sample sequence from a normal distribution with negative mean. A brief summary of a preliminary empirical study of the speedup and efficiency achieved by the parallel algorithm is also presented. Our study is motivated by the linear-time sequential algorithm [8] and a logarithmic-time and optimal-work parallel algorithm on the parallel random access machine (PRAM) [3] for this computation problem.

For computing a single (minimal) maximum subsequence of a length-n real-valued sequence of X, a simple sequential algorithm solves this problem in $O(n)$ optimal time. A parallel algorithm [1] on the PRAM model solves the single maximum subsequence problem in $O(\log n)$ parallel time using a total of $O(n)$ operations (work-optimal). A generalization of the problem and the selection problem is the sum-selection that, for given input length-n sequence X, range-bound $[l, u]$, and rank k, finds a subsequence of X such that the rank of its cumulative sum is k among all subsequences with cumulative sum in $[l, u]$. A randomized algorithm [7] solves the sum-selection problem in expected $O(n \log(u - l))$ time.

For the problem of finding the sequence of all successive minimal maximum subsequences of a length-n real-valued sequence X, a recursive divide-and-conquer strategy can apply the linear-time sequential algorithm above to compute a minimal maximum subsequence of X whose deletion results in a prefix and a suffix for recursion. The algorithm has a (worst-case) time complexity of $\Theta(n^2)$. Empirical analyses of the algorithm [8] on synthetic data sets (sequences of independent and identically distributed uniform random terms with negative mean) and score sequences of genomic data indicate that the running time grows at $\Theta(n \log n)$.

In order to circumvent the iterative dependency in computing the sequence of all successive minimal maximum subsequences, Ruzzo and Tompa [8] prove a structural characterization of the sequence as follows. Denote by $\mathrm{Max}(X)$ the set of all successive minimal maximum subsequences or their corresponding index subranges (when the context is clear) of a real-valued sequence X.

Theorem 1. [8] *For a non-empty real-valued sequence X, a non-empty subsequence S of X is in $\mathrm{Max}(X)$ if and only if: (1) [monotonicity] the subsequence S is monotone: every proper subsequence of S has its cumulative sum less than that of S, and (2) [maximality of monotonicity] the subsequence S is maximal in X with respect to monotonicity, that is, every proper supersequence of S contained in X is not monotone.*

Hence, we also term $\mathrm{Max}(X)$ as the set of all maximal monotone subsequences of X. This gives a structural decomposition of X into $\mathrm{Max}(X)$: (1) every non-empty monotone subsequence of X is contained in a maximal monotone subsequence in $\mathrm{Max}(X)$; in particular, every positive term of X is contained in a maximal monotone subsequence in $\mathrm{Max}(X)$, and (2) the set $\mathrm{Max}(X)$ is a pairwise disjoint collection of all maximal monotone subsequences of X.

Based on the structural characterization of $\mathrm{Max}(X)$, Ruzzo and Tompa present a sequential algorithm that computes $\mathrm{Max}(X)$ in $O(n)$ optimal sequential time and $O(n)$ space (worst case). Alves, Cáceres, and Song [2] develop a parallel algorithm for computing $\mathrm{Max}(X)$ of a length-n sequence X on the bulk synchronous parallel/coarse grained multicomputer model of p processors in $O(\frac{n}{p})$ computation time and $O(1)$ communication rounds.

In the following section, we introduce other structural decompositions of a sequence X that lead to computing $\mathrm{Max}(X)$ with: (1) a parallel algorithm on the PRAM model [3] in logarithmic parallel time and optimal linear work, and (2) a domain-decomposed parallel algorithm implemented on cluster systems with Message Passing Interface. This paper presents the skeletons for the main results without lengthy derivations and proofs, which are detailed in the full version.

2   Structural Decompositions of X Leading to Max(X)

For a real-valued sequence $X = (x_\eta)_{\eta=1}^n$, denote by $s_i(X)$ the i-th prefix sum $\sum_{\eta=1}^i x_\eta$ of X for $i \in [1, n]$, and $s_0(X) = 0$. We abbreviate the prefix sums $s_i(X)$ to s_i for all $i \in [1, n]$ when the context is clear. For a subsequence Y of X, denote by $\alpha(Y; X)$, $\beta(Y; X)$, and $\gamma(Y; X)$ its starting index, ending index, and index subrange $[\alpha(Y; X), \beta(Y; X)]$ ($\gamma(Y; X) = \emptyset$ if Y is empty) in the context of X, respectively, and by $\gamma_+(Y; X)$ the set of all indices in $\gamma(Y; X)$ yielding positive terms of Y. When considering the subsequence Y as a sequence in its own context we abbreviate $\alpha(Y;Y)$, $\beta(Y;Y)$, $\gamma(Y;Y)$, and $\gamma_+(Y;Y)$ to $\alpha(Y)$, $\beta(Y)$, $\gamma(Y)$, and $\gamma_+(Y)$, respectively.

The following characterization of monotonicity [3] yields an effective computation of the index subrange of a non-trivial monotone subsequence containing a given term of X.

Lemma 1. *Let X be a non-empty real-valued sequence and Y be a non-empty subsequence of X (with index subrange $[\alpha(Y;X), \beta(Y;X)]$). The following statements are equivalent:*

1. *Y is monotone in X.*
2. *The starting prefix sum $s_{\alpha(Y;X)-1}(X)$ of Y is the unique minimum and the ending prefix sum $s_{\beta(Y;X)}(X)$ of Y is the unique maximum of all $s_i(X)$ for all $i \in [\alpha(Y;X) - 1, \beta(Y;X)]$.*
3. *All non-empty prefixes and non-empty suffixes of Y have positive cumulative sums.*

The key to the parallel implementation [3] of finding $\mathrm{Max}(X)$ for a length-n sequence $X = (x_\eta)_{\eta=1}^n$ lies in the concurrent computation of the ending index of the maximal monotone subsequence constrained with the starting index $i \in \gamma(X)$. Lemma 1 suggests to consider only positive terms x_i of X for the desired computation. Let $\epsilon : \gamma_+(X) \to \gamma(X)$ be the function that $\epsilon(i)$ denotes the ending index of the maximal monotone subsequence of X constrained with the starting index i. The concurrent computation of ϵ via the computations of all-nearest-smaller-values and range-minima, when applied to all the positive terms x_i in X, generates the statistics $\mathrm{Mon}(X) = \{[i, \epsilon(i)] \mid i \in \gamma_+(X)\}$ for the set of all index subranges of all maximal monotone subsequences of X constrained with given positive starting terms. The following theorem [3] reveals the structural decomposition of X into $\mathrm{Mon}(X)$, which refines $\mathrm{Max}(X)$ and provides a basis for a parallel computation of $\mathrm{Max}(X)$ from $\mathrm{Mon}(X)$.

Theorem 2. *For a real-valued sequence X, $\mathrm{Mon}(X)$ enjoys the following parenthesis structure:*

1. *Every positive term of X has its index as the starting index of a unique index subrange in $\mathrm{Mon}(X)$,*
2. *For every pair of index subranges in $\mathrm{Mon}(X)$, either they are disjoint or one is a subrange of another, and*
3. *For every maximal monotone subsequence of X in $\mathrm{Max}(X)$, its index subrange is in $\mathrm{Mon}(X)$.*

Our current work on Max-computation includes adapting the logarithmic-time optimal-work parallel algorithm on practical parallel systems. However, in view of the efficient linear-time sequential algorithm [8], we devise and implement a domain-decomposed parallel algorithm computing Max that employs the optimal sequential algorithm in subsequence-hosting processors.

An ideal domain decomposition of a sequence X is a partition of X into a pairwise disjoint family $\mathcal{X}$ of non-empty subsequences of X that are length-balanced and Max-independent: $\mathrm{Max}(X) = \cup_{Y \in \mathcal{X}} \mathrm{Max}(Y)$ (Y as a sequence in its own right). We first finds a sufficient condition for the Max-independence that can be computed locally in subsequence-hosting processors. The characterization of monotonicity in Lemma 1 suggests to consider the following two functions on indices of positive terms of X with index range $\gamma(X)$ $(= [1, n]))$. Let $\mathrm{rm}_X :$ $\gamma_+(X) \rightarrow [\alpha(X) + 1, \beta(X)] \cup \{\beta(X) + 1\}$ $(= [2, n + 1])$ denote the nearest-smaller-or-equal right-match of the prefix sum s_{i-1} of X:

$$\mathrm{rm}_X(i) = \begin{cases} \min\{\eta \in [i + 1, \beta(X)] \mid s_{i-1} \geq s_\eta\} & \text{if the minimum exists,} \\ \beta(X) + 1\, (= n + 1) & \text{otherwise.} \end{cases}$$

A symmetric analogue of rm_X is the nearest-smaller left-match function lm_X. Note that the families $\{[\mathrm{lm}_X(i), i] \mid i \in \gamma_+(X)\}$ and $\{[i, \mathrm{rm}_X(i)] \mid i \in \gamma_+(X)\}$ satisfy the parenthesis structure similar to that of Mon – but permitting abutting index subranges (at subrange ends) in the lm_X-family. Both lm_X and rm_X help locate the (minimum) starting and (maximum) ending indices, respectively, of a maximal monotone subsequence of X containing the positive term x_i: determine if a merge of multiple maximal monotone subsequences covering the index subrange $[\mathrm{lm}_X(i), i]$ may occur.

Lemmas 2 and 3 give a sufficient condition for the Max-independence of a partition of X based on a local computation of rm_{X_i} and its intuitive equivalence by the X_i-hosting processor for each $i \in \{1, 2, \ldots, n\}$.

Lemma 2. *Let $(X_\eta)_{\eta=1}^m$ be a sequential partition of a real-valued sequence X with X_η, for $\eta = 1, 2 \ldots, m$, represented as a sequence in its own right over its index range $\gamma(X_\eta)$. If the partition satisfies the* rm-*closure condition: for all $i \in \{1, 2, \ldots, m - 1\}$ and all $j \in \gamma_+(X_i)$, $\mathrm{rm}_{X_i}(j) \in [j + 1, \beta(X_i)]$, then the partition is* Max-*independent:* $\mathrm{Max}(X) = \cup_{\eta=1}^m \mathrm{Max}(X_\eta)$.

Lemma 3. *For a non-empty real-valued sequence Y, the right-match function $\mathrm{rm}_Y : \gamma_+(Y) \rightarrow [\alpha(Y) + 1, \beta(Y)] \cup \{\beta(Y) + 1\}$ satisfies the* rm-*closure condition stated in Lemma 2 (for all $j \in \gamma_+(Y)$, $\mathrm{rm}_Y(j) \in [j + 1, \beta(Y)]$) if and only if the sequence Y satisfies the minimum prefix-sum condition: the ending prefix sum of Y, $s_{\beta(Y)}(Y)$, is a global minimum of all $s_i(Y)$ for all $i \in [\alpha(Y) - 1, \beta(Y)]$.*

The minimum prefix-sum condition, equivalent to the rm-closure condition as shown in Lemma 3, exposes a stringent sufficiency for Max-independence of a priori sequential partition of a sequence X: for all $i \in \{1, 2, \ldots, m-1\}$, the ending prefix sum is a global minimum of all prefix sums of X_i. We incorporate the minimum prefix-sum condition into constructing a posteriori sequential partition of X that forms the basis in designing a domain-decomposed parallel algorithm in computing $\mathrm{Max}(X)$.

For two sequences X and Y, denote the concatenation of X and Y by the juxtaposition XY. Let X be a non-empty real-valued sequence with a sequential partition $\mathcal{P}(X) = (X_1, X_{1,2}, X_2, X_{2,3}, X_3, \ldots, X_{m-1}, X_{m-1,m}, X_m)$. For notational simplicity, let $X_{0,1} = \emptyset$ and $X_{m,m+1} = \emptyset$.

For every $i \in \{1, 2, \ldots, m - 1\}$, denote by β_i^* the maximum/right-most index $\eta \in \gamma_+(X_{i-1,i}X_i)$, if non-empty, such that $s_{\eta-1}(X_{i-1,i}X_i)$ is the minimum prefix sum of those of $X_{i-1,i}X_i$ over $\gamma_+(X_{i-1,i}X_i)$; that is,

$$\beta_i^* = \max \arg \min\{s_{\eta-1}(X_{i-1,i}X_i) \mid \eta \in \gamma_+(X_{i-1,i}X_i) \ (\neq \emptyset)\}.$$

The sequential partition $\mathcal{P}(X)$ satisfies the rm-locality condition if for every $i \in \{1, 2, \ldots, m-1\}$ with non-empty $\gamma_+(X_{i-1,i}X_i)$, $\mathrm{rm}_{X_{i-1,i}X_iX_{i,i+1}}(\beta_i^*) \in [\beta_i^* + 1, \beta(X_{i-1,i}X_iX_{i,i+1})]$.

The rm-localized sequential partition $\mathcal{P}(X)$ derives a Max-independent partition $\tilde{\mathcal{P}}(X) = (X_{i-1,i}''X_iX_{i,i+1}')_{i=1}^m$ where $X_{i-1,i}''$ and $X_{i,i+1}'$ are respectively the suffix of $X_{i-1,i}$ and prefix of $X_{i,i+1}$ that are determined by rm-computation as follows. Recall that $X_{0,1} = \emptyset$ and $X_{m,m+1} = \emptyset$, let $X_{0,1}'' = \emptyset$ and $X_{m,m+1}' = \emptyset$ accordingly. For every $i \in \{1, 2, \ldots, m-1\}$, define $X_{i,i+1}'$ as:

$$\begin{cases} \emptyset \text{ if } \gamma_+(X_{i-1,i}X_i) = \emptyset \ \vee \\ \quad \mathrm{rm}_{X_{i-1,i}X_iX_{i,i+1}}(\beta_i^*) \in [\beta_i^* + 1, \beta(X_{i-1,i}X_i; X_{i-1,i}X_iX_{i,i+1})], \\ \text{the prefix of } X_{i,i+1} \text{ with} \\ \quad \text{index subrange } [\alpha(X_{i,i+1}; X_{i-1,i}X_iX_{i,i+1}), \mathrm{rm}_{X_{i-1,i}X_iX_{i,i+1}}(\beta_i^*)] \\ \quad \text{if } \gamma_+(X_{i-1,i}X_i) \neq \emptyset \wedge \mathrm{rm}_{X_{i-1,i}X_iX_{i,i+1}}(\beta_i^*) \in \gamma(X_{i,i+1}; X_{i-1,i}X_iX_{i,i+1}), \end{cases}$$

and $X_{i,i+1}''$ to be the (remaining) suffix of $X_{i,i+1}$ such that $X_{i,i+1}'X_{i,i+1}'' = X_{i,i+1}$. Note that the first case in defining $X_{i,i+1}'$ may be absorbed into the second case.

Theorem 3. *Let X be a non-empty real-valued sequence with an rm-localized sequential partition $\mathcal{P}(X) = (X_1, X_{1,2}, X_2, X_{2,3}, X_3, \ldots, X_{m-1}, X_{m-1,m}, X_m)$ and its derived sequential partition $\tilde{\mathcal{P}}(X) = (X_{\eta-1,\eta}''X_\eta X_{\eta,\eta+1}')_{\eta=1}^m$. Then:*

1. *$\tilde{\mathcal{P}}(X)$ is Max-independent: $\mathrm{Max}(X) = \cup_{\eta=1}^m \mathrm{Max}(X_{\eta-1,\eta}''X_\eta X_{\eta,\eta+1}')$, and*
2. *For all $i \in \{1, 2, \ldots, m\}$, $\mathrm{Max}(X_{i-1,i}''X_iX_{i,i+1}') = \mathrm{Max}(X_{i-1,i}X_iX_{i,i+1}') - \{Y \in \mathrm{Max}(X_{i-1,i}X_iX_{i,i+1}') \mid \alpha(Y; X_{i-1,i}X_iX_{i,i+1}') \in \gamma(X_{i-1,i}'; X_{i-1,i}X_iX_{i,i+1}')\}$; so $\mathrm{Max}(X) = \cup_{\eta=1}^m (\mathrm{Max}(X_{\eta-1,\eta}X_\eta X_{\eta,\eta+1}') - \{Y \in \mathrm{Max}(X_{\eta-1,\eta}X_\eta X_{\eta,\eta+1}') \mid \alpha(Y; X_{\eta-1,\eta}X_\eta X_{\eta,\eta+1}') \in \gamma(X_{\eta-1,\eta}'; X_{\eta-1,\eta}X_\eta X_{\eta,\eta+1}')\})$.*

3   Probabilistic Analysis of the Locality Condition

The structural decomposition of a non-empty real-valued sequence X in Theorem 3 suggests a basis for an ideal decomposition of X with length-balance and Max-independence – provided the decomposition satisfies the rm-locality condition. While the rm-localized decomposition $\tilde{\mathcal{P}}(X)$ is the (derived) sequential partition $(X_{\eta-1,\eta}''X_\eta X_{\eta,\eta+1}')_{\eta=1}^m$ in m pairwise disjoint subsequences, our domain-decomposed parallel algorithm computing $\mathrm{Max}(X)$ will employ m processors with the i-th processor hosting the subsequence $X_{i-1,i}X_iX_{i,i+1}$ for $i \in \{1, 2, \ldots, m\}$. The subsequences $X_{i-1,i}X_iX_{i,i+1}$ and $X_{i,i+1}X_{i+1}X_{i+1,i+2}$ hosted in successive i-th and $(i+1)$-th processors have the common subsequence $X_{i,i+1}$ that serves as a buffer to capture the rm-locality originated from $X_{i-1,i}X_i$ and a floating separation between successive Max-sets: $\mathrm{Max}(X_{i-1,i}''X_iX_{i,i+1}')$ and $\mathrm{Max}(X_{i,i+1}''X_{i+1}X_{i+1,i+2}')$. A longer common subsequence facilitates the satisfiability of the rm-locality of the preceding subsequence while a shorter one avoids redundant computation among successive processors.

In this section we analyze the length bound of the common subsequences probabilistically for random sequences of normally-distributed terms – via the theory of random walk. Let $X_1, X_2, \ldots$ be a sequence of pairwise independent and identically distributed random variables. Denote by $(S_\eta)_{\eta=0}^{\infty}$ the sequence of prefix-sum random variables with $S_0 = 0$ and $S_i = \sum_{\eta=1}^{i} X_\eta$ for $i \geq 1$, which corresponds to a general random walk for which S_i gives the position at epoch/index i. A record value occurs at (random) epoch $i \geq 1$ corresponds to the probabilistic event "$S_i > S_\eta$ for each $\eta \in [0, i-1]$". For every positive integer j, the j-th strict ascending ladder epoch random variable is the index of the j-th occurrence of the probabilistic event above. We define analogously the notions of: (1) strict descending ladder epochs by reversing the defining inequality from ">" to "<", and (2) weak ascending and weak descending epochs by replacing the defining inequalities by "$\geq$" and "$\leq$", respectively.

The first strict ascending ladder epoch is the random index of the first entry into $(0, +\infty)$, and the continuation of the random walk beyond this epoch is a probabilistic replica of the entire random walk. Other variants of (strict/weak, ascending/descending) ladder epoch yield similar behavior.

Viewing the sequence X in the Max-computation in an appropriate probabilistic setting studied below and following the above-stated denotations and construction of the Max-independent sequential partition $\tilde{\mathcal{P}}(X)$ from an rm-localized sequential partition $\mathcal{P}(X)$, we: (1) see intuitively that the random index-difference $\mathrm{rm}_{X_{i-1,i}X_iX_{i,i+1}}(\beta_i^*) - \beta_i^* + 1$ behaves like the first weak descending ladder epoch T of the underlying random walk (yielding $\sum_{\eta=1}^{\kappa} y_{\eta+\beta_i^*-1}$ for $\kappa = 0, 1, \ldots$) conditional on the probabilistic event "the positivity of the first term $y_{\beta_i^*}$" – with finite variance (and mean), and (2) develop a probabilistic upper bound on the length of the common subsequences in $\tilde{\mathcal{P}}(X)$ via the mean and variance of a variant of the first ladder epoch.

Remark 1. Ideally in $\tilde{\mathcal{P}}(X)$, we desire that:

$$|X_{i,i+1}| \,(= |[\alpha(X_{i,i+1}; X_{i-1,i}X_iX_{i,i+1}), \mathrm{rm}_{X_{i-1,i}X_iX_{i,i+1}}(\beta_i^*)]|)$$
$$\leq |[\beta_i^*, \mathrm{rm}_{X_{i-1,i}X_iX_{i,i+1}}(\beta_i^*)]| = \mathrm{rm}_{X_{i-1,i}X_iX_{i,i+1}}(\beta_i^*) - \beta_i^* + 1.$$

Thus, if we select the common subsequence $X_{i,i+1}$ such that $|X_{i,i+1}| \geq \lceil \mathrm{E}(T) + \delta\sqrt{\mathrm{Var}(T)}\rceil$ for some positive real δ, then the following two probabilistic events satisfy the subset-containment:

$$\text{"}\mathrm{rm}_{X_{i-1,i}X_iX_{i,i+1}}(\beta_i^*) - \beta_i^* + 1 \geq |X_{i-1,i}|\text{"}$$
$$\subseteq \text{"}(\mathrm{rm}_{X_{i-1,i}X_iX_{i,i+1}}(\beta_i^*) - \beta_i^* + 1) - \mathrm{E}(T) \geq \delta\sqrt{\mathrm{Var}(T)}\text{"},$$

and, in accordance with Chebyshev's inequality,

$$\mathrm{pr}\,(\text{random index-difference } \mathrm{rm}_{X_{i-1,i}X_iX_{i,i+1}}(\beta_i^*) - \beta_i^* + 1 \geq |X_{i-1,i}|)$$
$$\leq \mathrm{pr}\,(T - \mathrm{E}(T) \geq \delta\sqrt{\mathrm{Var}(T)}) \leq \mathrm{pr}\,(|T - \mathrm{E}(T)| \geq \delta\sqrt{\mathrm{Var}(T)}) \leq \frac{1}{\delta^2}.$$

These will be applied to bound the likelihood of (non-)satisfiability of the rm-locality condition for $\mathcal{P}(X)$.

We now relate the conditional weak descending ladder epoch T to the unconditional one and then, in an appropriate probabilistic setting, the means and variances of the two random variables.

For a sequence of pairwise independent and identically distributed random variables $X_1, X_2, \ldots$ and its associated random-walk sequence $(S_\eta)_{\eta=0}^\infty$ of prefix-sum random variables, denote by T_1 its first weak descending ladder epoch. Assume hereinafter that $(X_\eta)_{\eta=1}^\infty$ follows a common random variable X_1 with $\mathrm{pr}\,(X_1 > 0) \geq 0$. For notational simplicity, denote by p and $\bar{p}\,(=1-p)$ the probabilities $\mathrm{pr}\,(X_1 > 0)$ and $\mathrm{pr}\,(X_1 \leq 0)$, respectively.

The unconditional and conditional ladder epochs T_1 and $T\,(= T_1 \mid X_1 > 0)$ have sample spaces of $\{1, 2, \ldots\}$ and $\{2, 3, \ldots\}$, respectively, and for every $t \in \{2, 3, \ldots\}$,

$$\mathrm{pr}\,(T = t) = \mathrm{pr}\,(T_1 = t \mid X_1 > 0) = \frac{\mathrm{pr}\,(T_1 = t \cap X_1 > 0)}{\mathrm{pr}\,(X_1 > 0)} = \frac{1}{p}\,\mathrm{pr}\,(T_1 = t)$$

due to the subset-containment of the probabilistic events: "$T_1 = t\,(\geq 2)$" $\subseteq$ "$X_1 > 0$".

Lemma 4. *Assume that the variance, hence the mean, of the unconditional weak descending ladder epoch T_1 exist. The means and variances of the unconditional and conditional ladder epochs T_1 and $T = T_1 \mid X_1 > 0$ are related as follows:*

$$(1)\ \mathrm{E}(T) = \frac{1}{p}\mathrm{E}(T_1) - \frac{\bar{p}}{p}\quad and\quad (2)\ \mathrm{Var}(T) = \frac{1}{p}\mathrm{Var}(T_1) - \bar{p}(\frac{1}{p}(E(T_1) - 1))^2.$$

Remark 2. Remark 1 and Lemma 4 suggest to seek lower and upper bounds on $\mathrm{E}(T_1)$ and an upper bound on $\mathrm{Var}(T_1)$ for their use with the mean- and variance-relationships – which translate to non-trivial bounds on $\mathrm{E}(T)$ and $\mathrm{Var}(T)$. Note that, by the assumption of $\mathrm{pr}\,(X_1 > 0)$, we have $\mathrm{E}(T_1) > 1$.

For our Max-computing problem, we assume hereinafter (unless explicitly stated otherwise) that the sequence $X = (x_\eta)_{\eta=1}^n$ is a random sample from a normal distribution with mean $-a$ and variance b^2 for some positive reals a and b. That is, a sequence of pairwise independent and identically distributed random variables $X_1, X_2, \ldots$ with a common normal distribution with mean $-a$ and variance b^2 gives rise to the observed values $x_1, x_2, \ldots$. In applications, the knowledge of the mean and variance of the common random variable is known (see a uniformly-distributed case studied in [8]) or can be approximated.

The negativity of the mean $(-a)$ of the underlying normal distribution is desired in order to avoid yielding unrealistically long minimal maximum subsequences for viable applications. Formally for the induced random-walk sequence $(S_\eta)_{\eta=0}^\infty$ of $(X_\eta)_{\eta=1}^\infty$, since $\mathrm{E}(X_1)$ is finite and negative, the first (weak descending) ladder epoch T_1 has a proper probability distribution with finite mean and the random walk drifts to $-\infty$. For notational simplicity, denote by λ the "mean to standard deviation" ratio $\frac{\mathrm{E}(X_1)}{\sqrt{\mathrm{Var}(X_1)}}$; $\lambda = \frac{-a}{b}$ for a common normal distribution X_1 with mean $-a$ and standard deviation b.

Theorem 4. *For a sequence of pairwise independent and identically distributed random variables $(X_\eta)_{\eta=1}^\infty$ with a negative (common) finite mean $\mathrm{E}(X_1)$ and a positive probability $p\,(=\,\mathrm{pr}\,(X_1 > 0))$, the unconditional and conditional first weak descending epochs, T_1 and $T\,(=\,T_1 \mid X_1 > 0)$ respectively, satisfy the followings:*

1. *[General Case: Means] For T_1:$\mathrm{E}(T_1) = \exp(\sum_{\eta=1}^\infty \frac{\mathrm{pr}\,(S_\eta > 0)}{\eta})$; for T: $\mathrm{E}(T) =$*
   $\frac{1}{p}\exp(\sum_{\eta=1}^\infty \frac{\mathrm{pr}\,(S_\eta > 0)}{\eta}) - \frac{\bar{p}}{p}.$
2. *[Normally-Distributed Case: Means] For a common normal distribution of $(X_\eta)_{\eta=1}^\infty$ with mean $-a$ and variance b^2 for some positive reals a and b and for every positive integer l, denote $B(\lambda, l, \eta) = 1 - \exp(-\frac{\lambda^2}{2\sin^2(\eta\pi/(2l))})$ for $\eta \in \{1, 2, \dots, l\}$, then:*

$$\text{for } T_1: 1 < (\prod_{\eta=1}^{l-1} B(\lambda, l, \eta))^{-\frac{1}{2l}} \leq \mathrm{E}(T_1) \leq (\prod_{\eta=1}^{l} B(\lambda, l, \eta))^{-\frac{1}{2l}};$$

$$\text{for } T: \frac{1}{p}(\prod_{\eta=1}^{l-1} B(\lambda, l, \eta))^{-\frac{1}{2l}} - \frac{\bar{p}}{p} \leq \mathrm{E}(T) \leq \frac{1}{p}(\prod_{\eta=1}^{l} B(\lambda, l, \eta))^{-\frac{1}{2l}} - \frac{\bar{p}}{p}.$$

For our purpose in this study, we consider $l = 6$, and denote by μ' and μ'' the lower and upper bounds on the mean $\mathrm{E}(T_1)$ obtained in Theorem 4.

Remark 3. The range-constraint on $\mathrm{E}(T_1)$: $\mathrm{E}(T_1) \in [\mu', \mu'']$ induces an upper bound on $\mathrm{Var}(T_1)$ via some stochastic relationships of the first- and second-order moments of the first weak descending ladder epoch T_1, its associate (first weak descending) ladder height S_{T_1}, and the common distribution X_1 of the underlying random walk.

The following scenario will appear in upper-bounding $\mathrm{Var}(T_1)$ and $\mathrm{Var}(T)$: a quadratic polynomial Q with negative leading coefficient and two distinct real roots r' and r'' ($r' < r''$) serves as an upper bound on a nonnegative quantity v (such as a variance): $0 \leq v \leq Q(s)$ where s is a real-valued statistics – which induces a range-constraint: $s \in [r', r'']$.

Denote by q_1 and q the two quadratic polynomial forms that represent upper bounds on $\mathrm{Var}(T_1)$ and $\mathrm{Var}(T)$, respectively, in Theorem 5 below:

1. $q_1(t) = 2(-t^2 + (1 + \frac{2}{\lambda^2})t)$ with distinct real roots r_1' and r_1'' ($r_1' < r_1''$), and
2. $q(t) = -(2 + \frac{\bar{p}}{p^2})t^2 + 2(1 + \frac{2}{\lambda^2} + \frac{\bar{p}}{p^2})t - \frac{\bar{p}}{p^2}$ with distinct real roots r' and r'' ($r' < r''$).

Theorem 5. *For a sequence of pairwise independent and identically distributed random variables $(X_\eta)_{\eta=1}^\infty$ with a negative (common) finite mean $\mathrm{E}(X_1)$, a finite (common) third-order absolute moment $\mathrm{E}(|X_1|^3)$, and a positive probability $p\,(=\,\mathrm{pr}\,(X_1 > 0))$, the unconditional and conditional first weak descending epochs T_1 and $T\,(=\,T_1 \mid X_1 > 0)$ respectively, satisfy the followings:*

1. [*General Case: Means and Variances*] *For* T_1: $r' \leq E(T_1) \leq r''$ *and*

$$\text{Var}(T_1) < q_1(E(T_1)) = 2(-E(T_1)^2 + (1 + \frac{2}{\lambda^2})E(T_1));$$

for T: $\frac{1}{p}r' - \frac{\bar{p}}{p} \leq E(T) \leq \frac{1}{p}r'' - \frac{\bar{p}}{p}$ *and*

$$\text{Var}(T) < q(E(T_1)) = -(2 + \frac{\bar{p}}{p^2})E(T_1)^2 + 2(1 + \frac{2}{\lambda^2} + \frac{\bar{p}}{p^2})E(T_1) - \frac{\bar{p}}{p^2}.$$

2. [*Normally-Distributed Case: Means and Variances*] *With a common normal distribution of* $(X_\eta)_{\eta=1}^{\infty}$ *with mean* $-a$ *and variance* b^2 *for some positive reals* a *and* b:

$$\text{for } T_1: \mu' \leq E(T_1) \leq \mu'' \text{ and } \text{Var}(T_1) < q_1(E(T_1));$$
$$\text{for } T: \frac{1}{p}\mu' - \frac{\bar{p}}{p} \leq E(T) \leq \frac{1}{p}\mu'' - \frac{\bar{p}}{p} \text{ and } \text{Var}(T) < q(E(T_1)).$$

4   Max-Algorithms, Performance, and Conclusion

We have implemented a Max-computing parallel algorithm on cluster systems in which subsequence-hosting processors employ an optimal linear-time sequential algorithm Max_Sequential (which is detailed in the full version) for local Max-computation. Improvements to the algorithms and work in progress will be addressed in the conclusion. The algorithms implemented with Message Passing Interface (MPI) are available from the authors.

The performance of the parallel algorithm Max_Parallel is assessed in a preliminary empirical study on a cluster with synthetic random data as follows: (1) $N = 100$ trial-sequences, each is a random sample/sequence of length $n = 5 \cdot 10^6$ from a normal distribution with mean   0.25 and variance 1.0, and (2) Performance measures in (absolute) speedup and efficiency of Max_Parallel are collected in two sets of mean-statistics: (2.1) the set of conditional mean-statistics on "success" scenario (satisfiability of the rm-locality condition for the first $(p-1)$ processors) from N trial-sequences and the Max-computing by (local) Max_Sequential in Max_Parallel: Steps $1 - 3$, and (2.2) the set of unconditional ones for Max_Parallel: all steps.

Based on the optimal sequential-time algorithm [8], the (mean) optimal sequential time for Max-computation of a length-n sequence, $T^*(n)$, is approximately 0.155881 sec for the synthetic random data prepared in item 1 above (when averaged over $N = 100$ sequences).

Table 1 summarizes the above-stated two sets of mean-statistics of the running time, speedup, and efficiency of Max_Parallel for $\delta = 3$ (in Remark 1 and Max_Parallel: Step 1) and m processors with $m \in \{1, 2, 4, 8, 16, 32, 64\}$: $T_m(n)$ (in seconds), $S_m(n) = \frac{T^*(n)}{T_m(n)}$, and $E_m(n) = \frac{T_1(n)}{mT_m(n)}$, respectively.

Since pr (satisfiability of rm-locality for single processor) $\geq 1 - \frac{1}{\delta^2}$ $(= \frac{8}{9})$, the expected number N_s of "successes" from N trial-sequences is bounded below:

$N_s \geq N(1 - \frac{1}{\delta^2})^{m-1}$. The empirical and statistical results tabulated in the two columns: (expected) N_s and empirical-N_s show that the constraints on $\mathrm{E}(T)$ and $\mathrm{Var}(T)$ (Theorem 5: part 2) in bounding $E(T) + \delta\sqrt{\mathrm{Var}(T)}$ (Max_Parallel: Step 1) serves as a good lower-bound predictor for N_s. For the conditional statistics on "success" scenario, the speedup and efficiency are close to their theoretical bounds of m and 1, respectively. For the unconditional ones, even for a small δ ($= 3$), the speedup and efficiency exceed $\frac{3}{4}$ of their theoretical bounds, except for $m = 64$. The speedup and efficiency performance of an improved Max_Parallel depends on the extent of resolving violations of rm-locality among neighbor processors and tradeoffs involving δ and m.

Algorithm Max_Parallel.

Require: A length-n real-valued sequence X (which is a random sample satisfying the assumptions in Theorem 5: part 2) and a prescribed probability threshold δ (Remark 1: Chebyshev's inequality).

Ensure: The sequence of all successive minimal maximum subsequences (all maximal monotone subsequences) of X.

1: Construct sequential partition $\mathcal{P}(X) = (X_1, X_{1,2}, X_2, X_{2,3}, X_3, \ldots, X_{m-1}, X_{m-1,m}, X_m)$ of X (stated in Section 3) such that: (1) for all $i \in \{1, 2, \ldots, m\}$, processor P_i hosts the subsequence $X_{i,i-1}X_iX_{i,i+1}$ in a length-balanced manner except possibly for the last processor P_m, and (2) for all $i \in \{1, 2, \ldots, m-1\}$, $|X_{i,i+1}|$ is the least upper bound of $\lceil \mathrm{E}(T) + \delta\sqrt{\mathrm{Var}(T)} \rceil$ computed via Theorem 5: part 2;

2: {Decide if $\mathcal{P}(X)$ is an rm-localized partition:}

    2.1: **for all** $i \in \{1, 2, \ldots, m\}$

        $\{1 \leq i \leq m-1$: processor P_i computes:

            $is\_rmLocalized_i := (\gamma_+(X_{i-1,i}X_i) = \emptyset) \vee$

            $(\mathrm{rm}_{X_{i-1,i}X_iX_{i,i+1}}(\beta_i^*) \in [\beta_i^* + 1, \beta(X_{i-1,i}X_iX_{i,i+1})]);$

        $i = m$:   processor P_m computes:

            $is\_rmLocalized_m := \mathrm{true};\}$

    2.2: Compute $is\_rmLocalized := \wedge_{\eta=1}^{m-1} is\_rmLocalized_\eta$ using prefix-sum function;

    2.3: **for all** $i \in \{1, 2, \ldots, m\}$ processor P_i updates:

        $is\_rmLocalized_i := is\_rmLocalized;$

3: {If $\mathcal{P}(X)$ is rm-localized, then compute $\mathrm{Max}(X)$ via Theorem 3: determine $X'_{i,i+1}$ for all $i \in \{1, 2, \ldots, m-1\}$ and compute $\mathrm{Max}(X''_{i-1,i}X_iX'_{i,i+1})$ for all $i \in \{1, 2, \ldots, m\}$:}

    for all $i \in \{1, 2, \ldots, m\}$ processor P_i updates:

    if $is\_rmLocalized_i$ **then**

        $\{1 \leq i \leq m-1$: processor P_i sends $\mathrm{rm}_{X_{i-1,i}X_iX_{i,i+1}}(\beta_i^*)$ to processor P_{i+1};

            processor P_{i+1} receives $\mathrm{rm}_{X_{i-1,i}X_iX_{i,i+1}}(\beta_i^*)$;

        $i = m$:   **null**;$\}$

    Invokes Max_Sequential to compute $\mathrm{Max}(X''_{i-1,i}X_iX'_{i,i+1})$;

    else goto Step 4;

4: Invoke a parallel algorithm adapted from the Max-computing PRAM-algorithm [3] in which two embedded problems are solved by parallel algorithms implemented with MPI: "all nearest smaller values" [4] and "range-minima" [5];

Table 1. Preliminary empirical study of speedup and efficiency of Max_Parallel

	mean-statistics over N		conditional on "success" scenario: mean-statistics over observed N_s			unconditional: mean-statistics over N		
m	N_s	observed N_s	$T_m(n)$	$S_m(n)$	$E_m(n)$	$T_m(n)$	$S_m(n)$	$E_m(n)$
1	100.00	100	0.156833	0.9939	1.0000	0.156835	0.9939	1.0000
2	88.89	98	0.078377	1.9889	1.0005	0.078712	1.9804	0.9963
4	70.23	95	0.039663	3.9301	0.9885	0.040095	3.8878	0.9779
8	43.85	81	0.020464	7.6173	0.9580	0.021470	7.2604	0.9131
16	17.09	72	0.010410	14.9742	0.9416	0.011246	13.8610	0.8716
32	2.60	43	0.005312	29.3451	0.9226	0.006318	24.6725	0.7757
64	0.06	21	0.003002	51.9257	0.8163	0.005047	30.8859	0.4855

Our work in progress includes a comparative empirical/probabilistic study based on current implementation and refining the algorithms to detect and resolve violations of rm-locality among near-neighbor processors. There are two directions for general theoretical developments. First, the length bound of the common subsequences (to capture the rm-locality) is achieved via explicit bounds on the mean/variance of the first ladder epoch in the underlying random walk with normal distribution. This leads to a deserving study for general probability distribution. Second, there are other notions of (minimal) maximality for ranking subsequences of a real-valued sequence, developing efficient parallel algorithms for their computation is interesting.

References

1. Akl, S.G., Guenther, G.R.: Applications of Broadcasting with Selective Reduction to the Maximal Sum Subsegment Problem. International Journal of High Speed Computing **3**(2), 107–119 (1991)
2. Alves, C.E.R., Cáceres, E.N., Song, S.W.: Finding All Maximal Contiguous Subsequences of a Sequence of Numbers in $O(1)$ Communication Rounds. IEEE Transactions on Parallel and Distributed Systems **24**(3), 724–733 (2013)
3. Dai, H.-K., Su, H.-C.: A parallel algorithm for finding all successive minimal maximum subsequences. In: Correa, J.R., Hevia, A., Kiwi, M. (eds.) LATIN 2006. LNCS, vol. 3887, pp. 337–348. Springer, Heidelberg (2006)
4. He, X., Huang, C.-H.: Communication Efficient BSP Algorithm for All Nearest Smaller Values Problem. Journal of Parallel and Distributed Computing **61**(10), 1425–1438 (2001)
5. JáJá, J.: An Introduction to Parallel Algorithms. Addison-Wesley (1992)
6. Karlin, S., Brendel, V.: Chance and Statistical Significance in Protein and DNA Sequence Analysis. Science **257**(5066), 39–49 (1992)
7. Lin, T.-C., Lee, D.T.: Randomized Algorithm for the Sum Selection Problem. Theoretical Computer Science **377**(1–3), 151–156 (2007)
8. Ruzzo, W.L., Tompa, M.: A linear time algorithm for finding all maximal scoring subsequences. In: The Seventh International Conference on Intelligent Systems for Molecular Biology, pp. 234–241. International Society for Computational Biology (1999)

Building Bridges Between Sets of Partial Orders

Hernán Ponce-de-León[1]([⊠]) and Andrey Mokhov[2]

[1] Helsinki Institute for Information Technology HIIT and Department of Computer Science and Engineering, School of Science, Aalto University, Espoo, Finland
hernan.poncedeleon@aalto.fi
[2] School of Electrical and Electronic Engineering, Newcastle University, Newcastle upon Tyne, UK
andrey.mokhov@ncl.ac.uk

Abstract. Partial orders are a fundamental mathematical structure capable of representing true concurrency and causality on a set of atomic events. In this paper we study two mathematical formalisms capable of the compressed representation of sets of partial orders: Labeled Event Structures (LESs) and Conditional Partial Order Graphs (CPOGs). We demonstrate their advantages and disadvantages and propose efficient algorithms for transforming a set of partial orders from a given compressed representation in one formalism into an equivalent representation in another formalism without the explicit enumeration of each scenario. These transformations reveal the superior expressive power of CPOGs as well as the cost of this expressive power. The proposed algorithms make use of an intermediate mathematical formalism, called Conditional Labeled Event Structures (CLESs), which combines the advantages of LESs and CPOGs. All three formalisms are compared on a number of benchmarks.

1 Introduction

Partial orders – the protagonists of this paper – play a fundamental role in the concurrency theory. The concept has a very simple definition: a partial order is a reflexive, antisymmetric and transitive relation $\leq$ on a set of elements S. Two distinct elements $a, b \in S$ can be either ordered ($a \leq b$ or $b \leq a$) or concurrent ($a \not\leq b$ and $b \not\leq a$). Partial orders arise in numerous application areas such as model checking, process mining, concurrent programming, and VLSI design to name but a few. In this paper we do not focus on a particular application area, however, we use partial orders coming from the VLSI design domain as real-life benchmarks (specifically we use partial orders corresponding to processor instructions and on-chip communication protocols).

A single partial order can capture a single behavioral scenario of a modeled system. However, real-life systems rarely exhibit just a single scenario; in fact, we routinely design systems exhibiting millions of scenarios, each being a partial

This research was done while the author was preparing his thesis at INRIA and LSV, École Normale Supérieure de Cachan and CNRS, France.

A.-H. Dediu et al. (Eds.): LATA 2015, LNCS 8977, pp. 145–160, 2015.
DOI: 10.1007/978-3-319-15579-1_11

order defined on a subset of events that may occur in a system. How do we represent all of those partial orders? One can, of course, simply list them explicitly but this is clearly not a scalable solution – 6.6 trillion different partial orders can be defined on just 10 events!

In this paper we study two mathematical formalisms to compactly represent sets of partial orders: Labeled Event Structures (LESs) [6] and Conditional Partial Order Graphs (CPOGs) [5]. The two formalisms are significantly different from each other, hence one cannot directly use them together: conversion from one formalism to another without an intermediate uncompression step is non-trivial. As will be demonstrated in Section 4, different formalisms may be preferable in different application domains. For example, LESs can typically be obtained from Petri Net specifications via unfolding, while CPOGs naturally come from hardware specifications and implementations, where partial orders are pre-encoded with Boolean vectors (low-level signals, instruction opcodes, etc.).

This brings us to the main contribution of this paper: we present two direct transformation algorithms (Section 5) for converting compressed sets of partial orders from LESs to CPOGs and from CPOGs to LESs without an intermediate uncompression. The presented transformations reveal the superior expressive power of CPOGs as well as the cost of this expressive power: CPOGs are often more demanding from the algorithmic complexity point of view. The proposed algorithms make use of a new mathematical formalism, called Conditional Labeled Event Structures (CLESs), which combines the advantages of LESs and CPOGs. The CLES formalism makes it possible to directly combine sets of partial orders represented in LESs and CPOGs, thereby improving their interoperability.

To the best of the authors' knowledge, no other mathematical model has been directly used for the task of compressed representation of sets of partial orders, hence we only build one (bidirectional) bridge between LESs and CPOGs. If one would like to use other models for this task (for example Petri Nets or Message Sequence Charts), it is possible to reuse existing bridges to connect to the body of our work, e.g., one can obtain a LES from a Petri Net via its unfolding [3].

2   Preliminaries

This section introduces two formalisms that compactly represent partial orders: Labeled Event Structures [6] and Conditional Partial Order Graphs [5].

2.1   Labeled Event Structures

Event Structures[1] can represent several execution scenarios of a system by means of so called *configurations*. We study their widely used extension, called *Labeled Event Structures*, whose events are labeled with actions over a fixed alphabet L.

---

[1] In this article, we restrict to prime event structures.

Definition 1. *A labeled event structure (LES) over alphabet L is a tuple $\mathcal{E} = (E, \leq, \#, \lambda)$ where E is a set of events; $\leq\ \subseteq E \times E$ is a partial order (called causality) satisfying the property of finite causes, i.e. $\forall e \in E : |\{e' \in E \mid e' \leq e\}| < \infty$; $\# \ \subseteq E \times E$ is an irreflexive symmetric relation (called conflict) satisfying the property of conflict heredity, i.e. $\forall e, e', e'' \in E : e \# e' \wedge e' \leq e'' \Rightarrow e \# e''$; and $\lambda : E \to L$ is a labeling function.*

Remark 1. Note that in most cases one only needs to consider reduced versions of relations $\leq$ and $\#$, which we will denote $\leq_r$ and $\#_r$, respectively. Formally, $\leq_r$ (which we call *direct causality*) is the transitive reduction of $\leq$, and $\#_r$ (*direct conflict*) is the smallest relation inducing $\#$ through the property of conflict heredity. In practice $|\leq_r|$ and $|\#_r|$ are often a lot smaller than $|\leq|$ and $|\#|$, however, in the worst case $|\leq_r| = \Theta(|\leq|)$ and $|\#_r| = \Theta(|\#|)$, therefore the speed up gained by using the reduced relations does not affect the worst case performance of the presented algorithms.

A *configuration* is a computation state of a LES. It is represented by a set of events that have occurred in the computation. If an event is present in a configuration, then so must all the events on which it causally depend. Moreover, a configuration does not contain conflicting events.

Definition 2. *A configuration of a LES $\mathcal{E} = (E, \leq, \#, \lambda)$ is a set $C \subseteq E$ that is causally closed, i.e. $e \in C \Rightarrow \forall e' \leq e : e' \in C$, and conflict-free, i.e. $e \in C$ and $e \# e'$ imply $e' \notin C$. The set of maximal (w.r.t. set inclusion) configurations of $\mathcal{E}$ is denoted by $\Omega(\mathcal{E})$.*

In this paper we only deal with LESs whose configurations do not contain two events with the same label. With such a restriction one can associate to every configuration C a partial order whose elements are $\lambda(C)$ (where λ is lifted to sets) and causality is inherited from $\leq$. We will denote such partial order as $\pi(C)$ and lift π to sets of configurations.

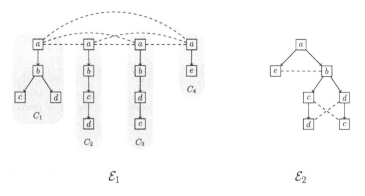

Fig. 1. Two Labeled Event Structures representing the same set of partial orders

The local configuration $[e]$ of an event e is a set of events on which it causally depends, i.e. $[e] \triangleq \{e' \in E \mid e' \leq e\}$; and its future $\lfloor e \rfloor$ is the set of events that causally depend on it, i.e. $\lfloor e \rfloor \triangleq \{e' \in E \mid e < e'\}$. Since a configuration together with the causality relation form a partial order, one can consider a LES $\mathcal{E}$ as a compressed representation of the set of partial orders induced by the maximal configurations $\Omega(\mathcal{E})$.

Fig. 1 shows an LES $\mathcal{E}_1$ defined on alphabet $L = \{a, b, c, d, e\}$ which contains four maximal configurations C_1-C_4. Note that throughout this paper we only show direct causality (by arrows) and direct conflicts (by dashed lines) on diagrams for clarity (events that belong to different configurations C_1-C_4 are all in conflict pairwise). It can be observed that not much compression is achieved by $\mathcal{E}_1$. The LES $\mathcal{E}_2$ represents the same set of partial orders, i.e. $\pi(\Omega(\mathcal{E}_1)) = \pi(\Omega(\mathcal{E}_2))$, and it is more compact.

2.2   Conditional Partial Order Graphs

A *Conditional Partial Order Graph* (CPOG) is a quintuple $H = (V, A, X, \phi, \rho)$, where V is a set of vertices, $A \subseteq V \times V$ is a set of arcs between them, and X is a set of operational variables. An opcode is an assignment $(x_1, x_2, \ldots, x_{|X|}) \in \{0, 1\}^{|X|}$ of these variables; X can be assigned only those opcodes which satisfy the restriction function ρ of the graph, i.e. $\rho(x_1, x_2, \ldots, x_{|X|}) = 1$. Function ϕ assigns a Boolean condition ϕ_z to every vertex and arc $z \in V \uplus A$ of the graph.

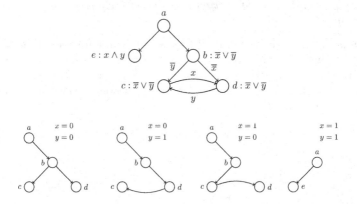

Fig. 2. Conditional Partial Order Graph and the corresponding set of partial orders

Fig. 2 (top) shows an example of a CPOG containing 5 vertices and 6 arcs; there are two operational variables x and y; the restriction function is $\rho = 1$, hence, all four opcodes $(x, y) \in \{0, 1\}^2$ are allowed. Vertices and arcs labeled by 1 are called unconditional (conditions equal to 1 are not depicted in the graph). The purpose of vertex and arc conditions is to 'switch off' some vertices and/or arcs in the graph according to the given opcode. This makes CPOGs capable of containing multiple *projections* as shown in Fig. 2 (bottom). The leftmost projection is obtained by keeping in the graph only those vertices and arcs whose conditions evaluate to Boolean 1 after substitution of the operational variables x and y with Boolean 0. Hence, vertex e disappears, because its condition evaluates to 0: $\phi_e = x \wedge y = 0$. Arcs $\{c \to d, d \to c\}$ disappear for the same reason. Note also that although the condition on arc $a \to e$ evaluates to 1 (in fact it is constant 1) the arc is still excluded from the projection because one of the vertices it connects (vertex e) is excluded and an arc cannot appear in a graph without one of its adjacent vertices.

Each projection is treated as a partial order specifying a behavioral scenario of a modeled system. Potentially, a CPOG $H = (V, A, X, \phi, \rho)$ can specify an exponential number of different partial orders on events V according to $2^{|X|}$ possible opcodes. We will use notation $H_{|\psi}$ to denote a projection of a CPOG H under an opcode $\psi = (x_1, x_2, \ldots x_{|X|})$. A projection $H_{|\psi}$ is called *valid* iff opcode ψ is allowed by the restriction function, i.e. $\rho(x_1, x_2, \ldots x_{|X|}) = 1$, and the resulting graph is acyclic. The latter requirement guarantees that the graph defines a partial order. A CPOG H is *well-formed* iff every allowed opcode produces a valid projection. The graph H in Fig. 2 is well-formed, because $H_{|x,y=0}, H_{|x=0,y=1}, H_{|x=1,y=0}$ and $H_{|x,y=1}$ are valid. A well-formed graph H therefore defines a set of partial orders $P(H)$.

Complexity. The original definition of CPOG complexity [5] is simply the total count of literals used in all the conditions: $\sum_{e \in V \uplus A} |\phi_e|$, where $|\phi|$ denotes the count of literals in condition ϕ, e.g., $|\overline{x} \wedge \overline{y}| = 2$ and $|1| = 0$. The complexity of the CPOG shown in Fig. 2 is thus equal to 10 according to this definition. We argue that this definition is not very useful in practice, because it does not take into account the fact that some of the conditions coincide. Intuitively, since $\phi_b = \phi_c = \phi_d = \overline{x} \vee \overline{y}$ we can compute condition $\overline{x} \vee \overline{y}$ only once and reuse the result three times. Furthermore, one can notice that conditions $\phi_b = \overline{x} \vee \overline{y}$ and $\phi_e = x \wedge y$ are not very different from each other; in fact $\phi_b = \neg\phi_e$, therefore having computed ϕ_e we can efficiently compute ϕ_b by a single inversion operation. In Section 4 we introduce an improved measure of complexity (based on Boolean circuits) which is free from the above shortcomings.

3   Enriched and Conditional LESs

A LES can represent several partial orders by means of its maximal configurations. CPOGs provide an additional mapping between partial orders and the corresponding opcodes, that is, given an opcode ψ satisfying the restriction function of a well-formed CPOG H, one can obtain the corresponding partial order as a projection $H_{|\psi}$. In the next subsection we show that a similar correspondence between opcodes and partial orders can be established by LESs if we enrich them with additional information on conflict resolution.

3.1   Enriched Labeled Event Structures

Partial orders are represented by maximal configurations of a LES, therefore to extract a partial order from a LES one needs to resolve event conflicts in a certain way. We enrich LESs with a total order on the conflicts and restrict the way conflicts can be resolved, leading to *Enriched Labeled Event Structures.*

Definition 3. *An Enriched Labeled Event Structure (ELES) over alphabet L is a tuple $\mathcal{E} = (E, \leq, \#, \lambda, \mathcal{L}, \mathcal{V})$ where $(E, \leq, \#, \lambda)$ is a labeled event structure, $\mathcal{L}$ is a total order on $\#$ and $\mathcal{V}$ is a set of vectors of length $|\mathcal{L}|$.*

A *conflict solver* is a vector $v \in \{0, 1\}^{|\mathcal{L}|}$ indicating which event is chosen in each conflicting pair (conflict $\mathcal{L}[i]$ is resolved by $v[i]$'s event in the conflict). Not every conflict solver is acceptable as illustrated in Fig. 3: any solver that chooses d_2 over d_1 must also choose c_1, because c_2 is in future of d_1; therefore vector 111 is disallowed. This is not the only restriction. If an event is a part of more than one conflict, whenever we choose it w.r.t. one conflict, we must also choose it w.r.t. to the others. Let $\overline{E}$ denote events that are not selected by a conflict solver v, i.e. $\overline{E} = \{e \in E \mid \exists i, j : v[i] = j \wedge \mathcal{L}[i][1 - j] = e\}$, the conflict solver is *valid* iff it generates a maximal configuration, i.e. $E \backslash \lfloor \overline{E} \rfloor \in \Omega(\mathcal{E})$. The set $\mathcal{V}$ in the definition of ELESs contains all valid conflict solvers.

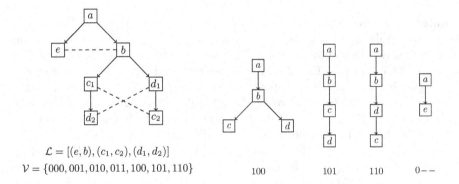

$$\mathcal{L} = [(e, b), (c_1, c_2), (d_1, d_2)]$$
$$\mathcal{V} = \{000, 001, 010, 011, 100, 101, 110\}$$

Fig. 3. An ELES and its conflict solvers

Example 1. Consider the ELES shown in Fig 3. If event e is chosen ($v[1] = 0$), the resolution of the other conflicts becomes unimportant since the configuration obtained is already maximal. However if event b is chosen ($v[1] = 1$), other conflicts need to be resolved, hence $\mathcal{V} = \{000, 001, 010, 011, 100, 101, 110\}$.

The following proposition characterizes the set of valid conflict solvers for a given labeled event structure (the proof can be found in [8]).

Proposition 1. *Let $\mathcal{E} = (E, \leq, \#, \lambda, \mathcal{L}, \mathcal{V})$ be such that for every $v \in \mathcal{V}$, if $v[i] = j$ and $\mathcal{L}[i][j] = e$, then $\forall h, k : \mathcal{L}[h][k] = e$ implies $v[h] = k$ and $\forall e' \in [e], h, k : \mathcal{L}[h][k] = e'$ implies $v[h] = k$. Then $\mathcal{V}$ is a set of valid conflict solvers.*

The above result shows how to compute a set of valid conflicts solvers for a LES and therefore each LES can be easily extended into the corresponding ELES. This means that both LESs and CPOGs can be used when one needs to store partial orders in a compressed form and access them by providing the corresponding opcodes. In the rest of the paper we will focus on LESs; however, all presented results also hold for their enriched counterparts.

3.2   Conditional Labeled Event Structures

The acyclicity of LESs often introduces redundancy in events: vertex c from the CPOG in Fig. 2 needs to be represented by two events (c_1 and c_2) in the LES of Fig. 3. In order to avoid this redundancy, we follow ideas of CPOGs and label elements of a LES (events and relations) by Boolean conditions in order to represent several LESs with one *Conditional Labeled Event Structure*. The next section shows that CLESs are of particular interest when transforming LESs into CPOGs and vice versa.

Definition 4. *A Conditional Labeled Event Structure (CLES) over alphabet L is a tuple $\mathcal{E} = (E, \leq, \#, \lambda, X, \phi, \rho)$ where E are events; $\leq$ is a set of arcs; $\#$ represents conflicts; λ labels events; X is a set of operational variables; ϕ assigns Boolean conditions to $E, \leq$ and $\#$; and ρ is the restriction function.*

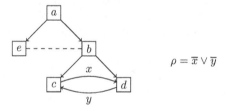

Fig. 4. Conditional Labeled Event Structure

A *well-formed* CLES is such that its projection on a valid opcode (allowed by the restriction function) generates a LES, i.e. $\leq$ becomes acyclic. CLESs generalize both CPOGs and LESs: if conflicts are dropped we get a CPOG; if the structure is acyclic and conditions are dropped, we get a LES.

The CLES in Fig. 4 represents the same partial orders as the CPOG and the LES in Figs. 2 and 3. If we compare it to the CPOG, a conflict is introduced, but the number of Boolean conditions is reduced. Comparing it to the LES, one can see that not only the number of events is reduced, but also the number of conflicts. The cardinality of the causality relation is preserved, but the information about Boolean labeling needs to be stored, i.e. $\phi_{c \leq d} = x$ and $\phi_{d \leq c} = y$. In the next section we introduce a complexity measure for such conditional, or parameterized, structures that we will use to compare (Enriched) LESs, CPOGs and CLESs to each other.

4 Parameterized Structures

The formalisms we presented in the previous sections can be used for the compressed representation of sets of partial orders. The key feature of these formalisms is the support for *conditional elements,* i.e. elements labeled with Boolean conditions.

Definition 5. *A mathematical structure over a set of elements S is called a parameterized structure if the elements are labeled with Boolean conditions ϕ : $S \rightarrow \Phi$, where Φ is a set of predicates (Boolean functions) on X, that is $\Phi \subseteq X \rightarrow \{0,1\}$.*

A CPOG is a parameterized structure whose elements are vertices and arcs. Events and causality/conflict relations are elements of both LESs and CLESs, but every LES element is labeled by 1 while CLES elements can be labeled by arbitrary conditions.

Below we define a complexity measure for parameterized structures that we will use to compare compactness of CPOGs, LESs and CLESs in our experiments.

Complexity Measure. Instead of treating each predicate in Φ separately let us construct a *Boolean circuit* [10] that computes all of them together and makes use of shared intermediate results. This is exactly what happens in practice regardless of whether a parameterized structure is used for verification purposes or in hardware synthesis. The *decoding complexity* of a predicate set Φ is the number of variables in Φ plus the number of gates in the smallest circuit[2] computing all predicates.

Definition 6. *The Complexity of a parameterized structure with predicate set Φ on a set of elements S is the decoding complexity of Φ plus the number of elements in S.*

Fig. 5 shows a circuit that computes predicates in $\Phi = \{1, \overline{x} \vee \overline{y}, x \wedge y, x, y, \overline{x}, \overline{y}\}$ required for the CPOG shown in Fig. 2. Note that trivial conditions 1, x and y require no computation at all and are therefore omitted in the diagram. We do not need a circuit to compute conditions of a LES which are always 1; only a single NAND gate is required for the CLES in Fig. 4. Therefore, the CPOG complexity is considered to be equal to 17 (2 variables + 4 gates + 5 vertices + 6 arcs); the LES complexity is 16 (7 events + 6 direct causality arcs + 3 direct conflicts); finally, the CLES complexity is 15 (2 variables + 1 gate + 5 vertices + 6 direct causality arcs + 1 direct conflict).

Comparison of Parameterized Structures. We compare LESs, CPOGs and CLESs on a number of benchmarks coming from the VLSI design domain, in

---

[2] In our experiments we restrict the number of inputs of each gate to 2. Since finding the smallest circuit is a very hard problem, we use approximation of the circuit complexity measure [1].

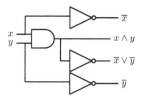

Fig. 5. Circuit computing conditions for CPOG in Fig. 2

particular, on-chip communication controllers [2] and processor microarchitectures [4]. We observed that a CPOG often has a lower complexity than a corresponding LES, however, the opposite can also be true. Since every CPOG is a CLES with $\# = \emptyset$ and every LES is a CLES with $\phi = 1$, CLESs have at most the same complexity as CPOGs and LESs.

Example 2. Phase encoders [2] are communication controllers capable of generating all permutations of n events. They are badly handled by acyclic structures as can be seen in Fig. 6 (right). The LES for a phase encoder with $n = 3$ has complexity 33 while its corresponding CPOG has complexity 15. In general, the complexity of CPOGs for phase encoders grows quadratically with n, while the complexity of LESs grows exponentially: one can see that the LES for a phase encoder of size n must have $n!$ events on its lowest level. In fact, a LES must contain at least as many events as there are partial orders in it.

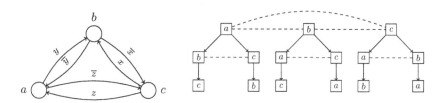

Fig. 6. Phase encoder for $n = 3$ represented by a CPOG (left) and a LES (right)

Example 3. Decision trees [9] are binary trees that can be used to model choices and their consequences. LESs for decision trees are smaller than CPOGs as the number of direct conflicts is smaller than the decoding complexity for conditions. This is illustrated in Fig. 7 where the LES on the right has complexity 16, while the complexity of the CPOG is 21. Asymptotically the complexity of both LESs and CPOGs grows linearly with the size of decision trees, so in this example LESs are better by just a constant factor. In general, as we will demonstrate in Section 5, the complexity of a CPOG never exceeds the complexity of the corresponding LES by more than just a constant factor.

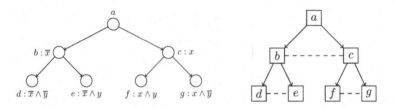

Fig. 7. A decision tree represented by a CPOG (left) and a LES (right)

Example 4. Trees of phase encoders are a combination of decision trees of height h and phase encoders with n actions: after h choices are made, all permutations of n events are possible. CLESs are strictly smaller than both CPOGs and LESs in this example, as shown in Fig. 8: the CPOG, the LES and the CLES have complexity 35, 52, and 30, respectively.

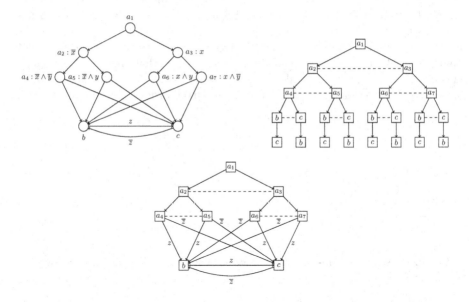

Fig. 8. A tree of phase encoders represented by a CPOG, a LES and a CLESs

Table 1 provides a summary of our experimental comparison of the complexity of CPOGs, LESs and CLESs. We compressed different sets of partial orders: phase encoders, decision trees, trees of phase encoders, as well as several sets of processor instructions (from ARM Cortex M0 and Intel 8051 processors [4]).

5   Transformations

This section presents the main contribution of this work: algorithms for transforming LESs into CPOGs and vice versa without performing an intermediate

Table 1. Experimental results

Name	Scenarios	Complexity		
		CPOG	LES	CLES
	24	24	158	24
Phase encoder	120	35	825	35
	720	48	5001	48
	8	44	39	36
Decision tree	16	97	76	76
	32	195	156	156
	16	70	108	58
Trees of phase encoders	29	43	158	41
	32	136	220	114

Name	Scenarios	Complexity		
		CPOG	LES	CLES
	5	26	28	26
	6	27	35	27
	7	26	38	26
ARM Cortex M0	8	28	43	28
	9	28	46	28
	10	29	46	29
	11	30	50	30
	5	34	48	34
	6	35	52	35
	7	36	56	36
Intel 8051	8	37	56	37
	9	46	71	46
	10	47	81	47
	11	51	90	51

uncompression step. Note that both algorithms make use of CLESs as an intermediate representation, which is not essential but convenient. The proofs of the results in this section can be found in [8].

From LESs to CPOGs. Every LES can be seen as an acyclic CLES where vertices and arcs are labeled by 1. If conflicts are removed from the CLES, an acyclic CPOG is obtained which can then be folded to remove redundant vertices. In order to preserve the information about conflicts, conflicting events need to be labeled by Boolean conditions in such a way that they cannot belong to the same projection. Proposition 1 shows that whenever an event is selected in one conflict, it must be selected in all other conflicts it participates in, along with all of its causal predecessors. This can be encoded in the restriction function of the resulting CPOG as follows[3]:

$$\rho = \left(\bigwedge_{e \# f} \neg \phi_e \vee \neg \phi_f \right) \left(\bigwedge_{e \leq f} \phi_f \Rightarrow \phi_e \right) \tag{1}$$

For the example shown in Fig. 3 this generates the following restriction function:

$$(v_b \Rightarrow v_a) \wedge (v_e \Rightarrow v_a) \wedge (v_{c_1} \Rightarrow v_b) \wedge (v_{d_1} \Rightarrow v_b) \wedge (v_{d_2} \Rightarrow v_{c_1}) \wedge (v_{c_2} \Rightarrow v_{d_1}) \wedge$$
$$(\overline{v}_e \vee \overline{v}_b) \wedge (\overline{v}_{c_1} \vee \overline{v}_{c_2}) \wedge (\overline{v}_{d_1} \vee \overline{v}_{d_2})$$

---

[3] Some optimization techniques allow to consider only direct causality and direct conflicts. We make use of this observation in our further examples.

By employing a SAT solver one can easily check that the above is satisfied by the following assignments which correspond to maximal configurations of the LES:

$$v_a = v_b = v_{c_1} = v_{d_1} = 1, v_{c_2} = v_{d_2} = v_e = 0$$
$$v_a = v_b = v_{d_1} = v_{c_2} = 1, v_{c_1} = v_{d_2} = v_e = 0$$
$$v_a = v_b = v_{c_1} = v_{d_2} = 1, v_{c_2} = v_{d_1} = v_e = 0$$
$$v_a = v_e = 1, v_b = v_{c_1} = v_{c_2} = v_{d_1} = v_{d_2} = 0$$

However not only maximal configurations satisfy the function, for example, the empty configuration clearly satisfies it as well: $v_a = v_b = v_{c_1} = v_{c_2} = v_{d_1} = v_{d_2} = v_e = 0$.

Since we do not want such non-maximal configurations to be allowed by the restriction function, we need to further elaborate it. A configuration is maximal if and only if, for every event $e \in E$ one of the following conditions holds: *(i)* event e belongs to the configuration; *(ii)* there exist an event f which belongs to the configuration and prevents e. The restriction function (1) can now be refined to allow only maximal configurations:

$$\rho = (\bigwedge_{e \# f} \neg \phi_e \vee \neg \phi_f)(\bigwedge_{e \leq f} \phi_f \Rightarrow \phi_e)(\bigwedge_{e \in E} \phi_e \vee \bigvee_{e \# f} \phi_f) \qquad (2)$$

Coming back to the example in Fig. 3, the refined restriction function (2) has only four satisfying assignments that represent the four maximal configurations of the LES.

Once conditions are assigned to events, arcs also need to be labeled before folding the result into a CPOG: we label each arc by the conjunction of the conditions of the events it connects to make sure an arc appears only if both of the events do. The resulting CLES may contain several events labeled by the same action, which is redundant for CPOGs. Such events can be merged and the resulting condition is the disjunction of conditions of the original events. This transformation method is summarized in Algorithm 1.

The scenarios represented by the CPOG obtained by merging events coincide with the maximal configurations of the LES.

Theorem 1. *Given an LES $\mathcal{E} = (E, \leq, \#, \lambda)$, Algorithm 1 constructs a CPOG $H = (V, A, X, \phi, \rho)$ such that $P(H) = \pi(\Omega(\mathcal{E}))$.*

Proof. The algorithm transforms the LES $\mathcal{E}$ into a CLES G and the later into a CPOG H. The proof shows that *i)* projections of G coincide with maximal configurations of $\mathcal{E}$; and *ii)* $G_{|\psi} = H_{|\psi}$ for any valid opcode ψ. The complete proof can be found in [8].

The complexity of the CPOG constructed by this procedure is linear with respect to the size of the original LES, as stated by the following theorem.

Theorem 2. *Given an LES $\mathcal{E} = (E, \leq, \#, \lambda)$, Algorithm 1 constructs a CPOG $H = (V, A, X, \phi, \rho)$ of complexity $\Theta(|\mathcal{E}|)$.*

Algorithm 1. Transforming a LES into a CPOG

Require: $\mathcal{E} = (E, \leq, \#, \lambda)$ and a set of Boolean variables $\{x_1, \ldots, x_{|E|}\}$
Ensure: $H = (V, A, X, \phi, \rho)$ such that $P(H) = \Omega(\mathcal{E})$
1: $V = E, A = \leq$
2: $\forall v \in V : \phi_v = x_v$
3: $\forall e = (v_1, v_2) \in A : \phi_e = v_1 \land v_2$
4: **while** $\exists v_1, v_2 \in V : \lambda(v_1) = \lambda(v_2)$ **do** (for $v \notin V$)
5:     $V = V \backslash \{v_1, v_2\} \cup \{v\}$
6:     $\forall v' \in V : v' \leq v \Leftrightarrow v' \leq v_1 \lor v' \leq v_2$
7:     $\forall v' \in V : v \leq v' \Leftrightarrow v_1 \leq v' \lor v_2 \leq v'$
8:     $\phi_v = \phi_{v_1} \lor \phi_{v_2}$
9: $\rho = (\bigwedge\limits_{e \# f} \neg\phi_e \lor \neg\phi_f)(\bigwedge\limits_{e \leq f} \phi_f \Rightarrow \phi_e)(\bigwedge\limits_{e \in E} \phi_e \lor \bigvee\limits_{e \# f} \phi_f)$
10: **return** $H = (V, A, X, \phi, \rho)$

Proof. The resulting CPOG contains at most the same number of vertices and arcs as the original LES. Moreover, $|X| = |E|$, $|\phi| = O(|E|)$, and the size of the restriction function ρ is linear with respect to $|\mathcal{E}|$ as seen from (2). The complete proof can be found in [8].

From CPOGs to LESs. In order to transform a CPOG into an LES, the graph is unfolded (in order to obtain an acyclic structure) while keeping conditions that will be replaced by conflicts in the final LES. For this, a CLES is constructed as an intermediate structure. We start from an empty CLES (that containing no events) and at each iteration, we compute the set of possible extensions. To decide if an instance of vertex $a \in V$ is a possible extension, we need to find a set of predecessor events $P \subseteq E$ such that *(i)* the vertex is active; *(ii)* instances of its predecessors and their corresponding arcs are active; *(iii)* if an event is not a predecessor, then either it is not active or its corresponding arc is not active; *(iv)* the instance of the vertex is different to any other in the prefix. This is captured by the following formula for each vertex a:

$$\phi_a \land (\bigwedge_{\substack{e_b \in P \\ b \to a \in A}} \phi_{e_b} \land \phi_{b \to a})(\bigwedge_{\substack{e_b \in E \backslash P \\ b \to a \in A}} \neg\phi_{e_b} \lor \neg\phi_{b \to a})(\bigwedge_{e_a \in E} \neg\phi_{e_a}) \qquad (3)$$

Whenever such a combination exists and the formula reduces to ϕ, we add the event to the unfolding, appropriately connecting it to P and labeling it by the ϕ. The unfolding procedure finishes when (3) is no longer satisfiable. Finally, conditions are replaced by conflicts: for every pair of mutually exclusive events, their Boolean conditions are removed and conflict $e_a \# e_b$ is added.

Algorithm 2 shows the complete transformation procedure. Function $PE(\mathcal{E}, H)$ takes the current unfolding $\mathcal{E}$ and CPOG H, and returns a set pe of possible extensions satisfying (3), and for each $e \in pe$, its set of predecessors P_e, label $\lambda^{pe}(e)$, and condition ϕ_e^{pe}.

We can use a SAT-solver to 'guess' a combination of an event a and a predecessor set P satisfying (3).

Algorithm 2. Transforming a CPOG into a LES

Require: a well-formed CPOG $H = (V, A, X, \phi, \rho)$
1: $\mathcal{E} = (E, \leq, \#, \lambda, X, \phi, \rho) := (\emptyset, \emptyset, \emptyset, \emptyset, X, \emptyset, \rho)$
2: $(pe, \{P_e\}_{e \in pe}, \lambda^{pe}, \phi^{pe}) := PE(\mathcal{E}, H)$
3: **while** $pe \neq \emptyset$ **do**
4:    add some $e \in pe$ to E and set $P_e \leq e, \lambda(e) = \lambda^{pe}(e)$ and $\phi_e = \phi_e^{pe}$
5: **while** $\exists e_a, e_b \in E : \neg \phi_{e_a} \vee \neg \phi_{e_b}$ **do**
6:    set $e_a \# e_b$
       return $(E, \leq, \#, \lambda)$

Proposition 2. *Given a CPOG and a prefix of its unfolding, deciding if an instance of a vertex is a possible extension is NP-hard.*

Proof. Consider a CPOG with a single vertex v having condition ϕ. Deciding if v is a possible extension requires checking ϕ for being a contradiction. See more details in [8]. □

The unfolding algorithm is deterministic: the resulting LES does not depend on the order in which events are added into the unfolding due to the following result proved in [8].

Proposition 3. *Let E be the current set of events of the unfolding and $e_a \neq e_b$ two possible extensions, then e_b is a possible extension of $E \cup \{e_a\}$.*

Example 5. Consider the CPOG shown in Fig. 2. The unfolding procedure starts with $E = \emptyset$ and keeps checking vertices of the CPOG for possible extensions. At start, only vertex a can be added. For example, the constraint imposed by non-predecessors in (3) will include $\neg \phi_{a \to b} = 0$ for vertex b, hence it is not a possible extension at start. We proceed by adding event e_a^0 to the unfolding with $\phi_{e_a^0} = 1$. When we recompute the possible extensions, formula (3) reduces to $\overline{x} \vee \overline{y}$ and $x \wedge y$ for vertices b and c, respectively, therefore events e_b^0 and e_e^0 are added with e_a^0 as their predecessor and with $\phi_{e_b^0} = \overline{x} \vee \overline{y}$ and $\phi_{e_e^0} = x \wedge y$. At this point $E = \{e_a^0, e_b^0, e_e^0\}$ and we find that c and d are possible extensions adding events e_c^0 and e_d^0 with event e_b^0 as the predecessor and conditions $\phi_{e_c^0} = \overline{y}$ and $\phi_{e_d^0} = \overline{x}$. Now $E = \{e_a^0, e_b^0, e_c^0, e_d^0, e_e^0\}$ and we find that c and d are possible extensions again. Two new events e_c^1 and e_d^1 are added. Finally, as E grows to $\{e_a^0, e_b^0, e_c^0, e_c^1, e_d^0, e_d^1, e_e^0\}$, formula (3) becomes unsatisfiable and the unfolding procedure is finished. Conditions of events e_b^0 and e_e^0 are mutually exclusive: $(x \wedge y) \wedge (\overline{x} \vee \overline{y}) = 0$, therefore we add conflict $e_b^0 \# e_e^0$. Due to the same reasoning, conflicts $e_c^0 \# e_c^1$ and $e_d^0 \# e_d^1$ are added. Finally, when all Boolean conditions are removed from the CLES, the resulting LES is that of Fig. 3.

The result below shows that the unfolding algorithm is correct, i.e. it preserves set of partial orders.

Theorem 3. *Let $H = (V, A, X, \phi, \rho)$ be a well-formed CPOG and $\mathcal{E} = (E, \leq, \#, \lambda)$ the LES obtained by the unfolding procedure, then $\pi(\Omega(\mathcal{E})) = P(H)$.*

Proof. The algorithm transforms the CPOG H into a CLES G and the later into a LES $\mathcal{E}$ by replacing conditions by conflicts. The proof shows that *i)* projections of H and G over a valid opcode coincide; and *ii)* projections over G and maximal configurations of $\mathcal{E}$ coincide. The complete proof can be found in [8].

As one can see, the transformation procedure from CPOGs to LESs is significantly more computationally intensive: unravelling CPOGs requires the use of a SAT solver. Fortunately, the SAT instances that need to be solved are similar to each other, therefore one can use incremental SAT solving techniques [11] to speed up the algorithm.

6   Conclusion

The paper discusses the use of two models (LESs and CPOGs) for the compressed representation of sets of partial orders. We show that LESs work well on most practical examples, however, due to their acyclic nature they cannot efficiently handle the cases where sets of partial orders contain many permutations defined on the same set of events. These cases are very well handled by CPOGs, however, the use of Boolean conditions for resolving conflicts makes them less intuitive and more demanding from the algorithmic complexity point of view. In particular, most interesting questions about CPOGs are NP-hard. The advantages of both models are combined by CLESs which are used as an intermediate formalism by the presented algorithms transforming a set of partial orders from a given compressed representation in a LES or a CPOG into an equivalent compressed representation in the other formalism without the explicit enumeration of all partial orders.

Further work includes optimization of the presented algorithms, their integration with Workcraft EDA suite [7], and validation on larger case studies coming from process mining and VLSI design domains.

Acknowledgments. This work was partially supported by EPSRC research grant UNCOVER (EP/K001698/1).

References

1. Berkeley Logic Synthesis and Verification Group: ABC: A System for Sequential Synthesis and Verification, Release 70930. http://www.eecs.berkeley.edu/alanmi/abc/
2. D'Alessandro, C., Mokhov, A., Bystrov, A.V., Yakovlev, A.: Delay/phase regeneration circuits. In: 13th IEEE International Symposium on Asynchronous Circuits and Systems (ASYNC 2007), 12–14 March 2006, Berkeley. pp. 105–116. IEEE Computer Society (2007). http://doi.ieeecomputersociety.org/10.1109/ASYNC.2007.14
3. Esparza, J., Römer, S., Vogler, W.: An improvement of McMillan's unfolding. In: Margaria, T., Steffen, B. (eds.) TACAS 1996. LNCS, vol. 1055, pp. 87–106. Springer, Heidelberg (1996)

4. Mokhov, A., Iliasov, A., Sokolov, D., Rykunov, M., Yakovlev, A., Romanovsky, A.: Synthesis of processor instruction sets from high-level ISA specifications. IEEE Trans. Computers **63**(6), 1552–1566 (2014). http://doi.ieeecomputersociety.org/10.1109/TC.2013.37

5. Mokhov, A., Yakovlev, A.: Conditional partial order graphs: Model, synthesis, and application. IEEE Trans. Computers **59**(11), 1480–1493 (2010)

6. Nielsen, M., Plotkin, G.D., Winskel, G.: Petri nets, event structures and domains, part I. Theoretical Computer Science **13**, 85–108 (1981)

7. Poliakov, I., Sokolov, D., Mokhov, A.: Workcraft: a static data flow structure editing, visualisation and analysis tool. In: Kleijn, J., Yakovlev, A. (eds.) ICATPN 2007. LNCS, vol. 4546, pp. 505–514. Springer, Heidelberg (2007)

8. Ponce de León, H., Mokhov, A.: Building bridges between sets of partial orders. (2014), technical report. http://hal.inria.fr/hal-01060449 (Visited on September 4, 2014)

9. Sung-hyuk, C.: A genetic algorithm for constructing compact binary decision trees. In: International Journal of Information Security and Privacy. pp. 32–60 (2010)

10. Wegener, I.: The Complexity of Boolean Functions. Johann Wolfgang Goethe-Universitat (1987)

11. Zhang, L., Malik, S.: The Quest for Efficient Boolean Satisfiability Solvers. In: Brinksma, E., Larsen, K.G. (eds.) CAV 2002. LNCS, vol. 2404, pp. 17–36. Springer, Heidelberg (2002)

Complexity of Road Coloring
with Prescribed Reset Words

Vojtěch Vorel[1][(✉)] and Adam Roman[2]

[1] Faculty of Mathematics and Physics, Charles University, Malostranské nám. 25,
Prague, Czech Republic
vorel@ktiml.mff.cuni.cz
[2] Institute of Computer Science, Jagiellonian University, Lojasiewicza 6,
30-348 Krakow, Poland
roman@ii.uj.edu.pl

Abstract. By the Road Coloring Theorem (Trahtman, 2008), the edges
of any given aperiodic directed multigraph with a constant out-degree can
be colored such that the resulting automaton admits a reset word. There
may also be a need for a particular reset word to be admitted. For certain
words it is NP-complete to decide whether there is a suitable coloring. For
the binary alphabet, we present a classification that separates such words
from those that make the problem solvable in polynomial time. The classification differs if we consider only strongly connected multigraphs. In this
restricted setting the classification remains incomplete.

Keywords: Algorithms on automata and words · Road coloring
theorem · Road coloring problem · Reset word · Synchronizing word

1 Introduction

Questions about synchronization of finite automata have been studied since the
early times of automata theory. The basic concept is very natural: we want to
find an input sequence that would get a given machine to a unique state, no
matter in which state the machine was before. Such sequence is called a *reset
word*. If an automaton has a reset word, we call it a *synchronizing automaton*.

In the study of *road coloring*, synchronizing automata are created from
directed multigraphs through edge coloring. A directed multigraph is said to be
admissible, if it is aperiodic and has a constant out-degree. A multigraph needs
to be admissible in order to have a synchronizing coloring. Given an alphabet I
and an admissible graph with out-degrees $|I|$, the following questions arise:

1. Is there a coloring such that the resulting automaton has a reset word?
2. Given a number $k \geq 1$, is there a coloring such that the resulting automaton
   has a reset word of length at most k?

Vojtěch Vorel—Supported by the Czech Science Foundation grant GA14-10799S.
Adam Roman—Supported in part by Polish MNiSW grant IP 2012 052272.

© Springer International Publishing Switzerland 2015
A.-H. Dediu et al. (Eds.): LATA 2015, LNCS 8977, pp. 161–172, 2015.
DOI: 10.1007/978-3-319-15579-1_12

3. Given a word $w \in I^{\star}$, is there a coloring such that w is a reset word of the resulting automaton?
4. Given a set of words $W \subseteq I^{\star}$, is there a coloring such that some $w \in W$ is a reset word of the resulting automaton?

For the first question it was conjectured in 1977 by Adler, Goodwyn, and Weiss [1] that the answer is always *yes*. The conjecture was known as the *Road Coloring Problem* until Trahtman [5] in 2008 found a proof, turning the claim into the *Road Coloring Theorem*.

The second question was initially studied in the paper [3] presented at LATA 2012, while the yet-unpublished papers [2] and [6] give closing results: The problem is NP-complete for any fixed $k \geq 4$ and any fixed $|I| \geq 2$. The instances with $k \leq 3$ or $|I| = 1$ can be solved by a polynomial-time algorithm.

The third question is the subject of the present paper. We show that the problem becomes NP-complete even if restricted to $|I| = 2$ and $w = abb$ or to $|I| = 2$ and $w = aba$, which may seem surprising. Moreover, we provide a complete classification of binary words: The NP-completeness holds for $|I| = 2$ and any $w \in \{a, b\}^{\star}$ that does not equal a^k, b^k, $a^k b$, nor $b^k a$ for any $k \geq 1$. On the other hand, for any w that matches some of these patterns, the restricted problem is solvable in polynomial time.

The fourth question was raised in [2] and it was emphasized that there are no results about the problem. Our results about the third problem provide an initial step for this direction of research.

It is an easy but important remark that the Road Coloring Theorem holds generally if and only if it holds for strongly connected graphs. It may seem that strong connectivity can be safely assumed even if dealing with other problems related to road coloring. Surprisingly, we show that this does not hold for complexity issues. If P is not equal to NP, the complexity of the third problem for strongly connected graphs differs from the basic third problem in the case of $w = abb$. However, for the strongly connected case we are not able to provide a complete characterization as described above, we give only partial results.

Due to the page limit, some proofs are omitted or shortened. The results are presented in Sections 3 and 4.

2   Preliminaries

2.1   Automata and Synchronization

For $u, w \in I^{\star}$ we say that u is a *prefix*, a *suffix*, or a *factor* of w if $w = uv$, $w = vu$, or $w = vuv'$ for some $v, v' \in I^{\star}$, respectively.

A *deterministic finite automaton* is a triple $A = (Q, I, \delta)$, where Q and I are finite sets and δ is an arbitrary mapping $Q \times I \to Q$. Elements of Q are called *states*, I is the *alphabet*. The *transition function* δ can be naturally extended to $Q \times I^{\star} \to Q$, still denoted by δ, slightly abusing the notation. We extend it also by defining

$$\delta(S, w) = \{\delta(s, w) \mid s \in S\}$$

for each $S \subseteq Q$ and $w \in I^*$. If $A = (Q, I, \delta)$ is fixed, we write $r \xrightarrow{x} s$ instead of $\delta(r, x) = s$.

For a given automaton $A = (Q, I, \delta)$, we call $w \in I^*$ a *reset word* if $|\delta(Q, w)| = 1$. If such a word exists, we call the automaton *synchronizing*. Note that each word having a reset word as a factor is also a reset word.

2.2   Road Coloring

In the rest of the paper we use the term *graph* for a directed multigraph. A graph is:

1. *aperiodic*, if 1 is the only common divisor of all the lengths of cycles,
2. *admissible*, if it is aperiodic and all its out-degrees are equal,
3. *road colorable*, if its edges can be labeled such that a synchronized deterministic finite automaton arises.

Naturally, we identify a coloring of edges with a transition function δ of the resulting automaton. It is not hard to observe that any road colorable graph is admissible. In 1977 Adler, Goodwyn, and Weiss [1] conjectured that the backward implication holds as well. Their question became known as the Road Coloring Problem and a positive answer was given in 2008 by Trahtman [5].

For any alphabet I and $w \in I^*$, by $\mathbb{G}_w^{|I|}$ we denote the set of graphs with all out-degrees equal to $|I|$ such that there exists a coloring δ with $|\delta(Q, w)| = 1$. In this paper we work with the following computational problem:

SRCW (*Synchronizing road coloring with prescribed reset words*)

Input: Alphabet I, graph $G = (Q, E)$ with out-degrees $|I|$, $W \subseteq I^*$

Output: Is there a $w \in W$ such that $G \in \mathbb{G}_w^{|I|}$?

In this paper we study the restrictions to one-element sets W, which means that we consider the complexity of the sets $\mathbb{G}_w^{|I|}$ themselves.

Restrictions are denoted by subscripts and superscripts: $\text{SRCW}_{k,X}^{\mathcal{M}}$ denotes SRCW restricted to inputs with $|I| = k$, $W = X$, and $G \in \mathcal{M}$, where $\mathcal{M}$ is a class of graphs. By $\mathcal{SC}$ we denote the class of strongly connected graphs. Having a graph $G = (Q, E)$ fixed, by $d_G(s, t)$ we denote the length of shortest directed path from $s \in Q$ to $t \in Q$ in G. For each $k \geq 0$ we denote

$$V_k(q) = \{s \in Q \mid d_G(s, q) = k\}.$$

Having $R \subseteq Q$, let $G[R]$ denote the induced subgraph of G on the vertex set R. If a graph G has constant out-degree $|I|$, a vertex $v \in Q$ is called a *sink state* if there are $|I|$ loops on v. By $\mathcal{Z}$ we denote the class of graphs having a sink state. The following lemma can be easily proved by a reduction that adds a chain of $|u|$ new states to each state of a graph:

Lemma 1. *Let $|I| \geq 1$ and $u, w \in \{a, b\}^*$. Then:*

1. *If $\text{SRCW}_{k,\{w\}}$ is NP-complete, so is $\text{SRCW}_{k,\{uw\}}$.*
2. *If $\text{SRCW}_{k,\{w\}}^{\mathcal{Z}}$ is NP-complete, so is $\text{SRCW}_{k,\{uw\}}^{\mathcal{Z}}$.*

3   A Complete Classification of Binary Words According to Complexity of SRCW$_{2,\{w\}}$

The theorem below presents one of the main results of the present paper. Assuming that P does not equal NP, it introduces an exact dichotomy concerning the words over binary alphabets. Let us fix the following partition of $\{a,b\}^{\star}$:

$$T_1 = \left\{a^k, b^k \mid k \geq 0\right\}, \qquad T_3 = \left\{a^l b^k, b^l a^k \mid k \geq 2, l \geq 1\right\},$$
$$T_2 = \left\{a^k b, b^k a \mid k \geq 1\right\}, \qquad T_4 = \{a,b\}^{\star} \setminus (T_1 \cup T_2 \cup T_3).$$

For the NP-completeness reductions throughout the present paper we use a suitable variant of the satisfiability problem. The following can be verified using the Schaefer's dichotomy theorem [4]:

Lemma 2. *It holds that W-SAT is NP-complete.*

W-SAT	
Input:	Finite set X of *variables*, finite set $\Phi \subseteq X^4$ of *clauses*.
Output:	Is there an assignment $\xi : X \to \{0,1\}$ such that for each clause $(z_1, z_2, z_3, z_4) \in \Phi$ it holds that: (1) $\xi(z_i) = 1$ for some i, (2) $\xi(z_i) = 0$ for some $i \in \{1,2\}$, (3) $\xi(z_i) = 0$ for some $i \in \{3,4\}$?

In this section we use reductions from W-SAT to prove the NP-completeness of SRCW$_{2,\{w\}}$ for each $w \in T_3$ and $w \in T_4$. In the case of $w \in T_4$ the reduction produces only graphs having sink states. This shows that for $w \in T_4$ the problem SRCW$_{2,\{w\}}^{\mathcal{Z}}$ is NP-complete as well, which turns out to be very useful in Section 4, where we deal with strongly connected graphs. For $w \in T_3$ we also prove NP-completeness, but we use automata without sink states. We show that the cases with $w \in T_1 \cup T_2$ are decidable in polynomial time.

In all the figures below we use bold solid arrows and bold dotted arrows for the letters a and b respectively.

Theorem 3. *Let $w \in \{a,b\}^{\star}$.*

1. *If $w \in T_1 \cup T_2$, the problem SRCW$_{2,\{w\}}$ is solvable in polynomial time.*
2. *If $w \in T_3 \cup T_4$, the problem SRCW$_{2,\{w\}}$ is NP-complete. Moreover, if $w \in T_4$, the problem SRCW$_{2,\{w\}}^{\mathcal{Z}}$ is NP-complete.*

Proof for $w \in T_1$. It is easy to see that $G \in \mathbb{G}_{a^k}$ if and only if there is $q_0 \in Q$ such that there is a loop on q_0 and for each $s \in Q$ we have $\mathrm{d}_G(s, q_0) \leq k$.   □

Proof for $w \in T_2$. For a fixed $q_0 \in Q$, we denote $Q_1 = \{s \in Q \mid s \longrightarrow q_0\}$ and

$$R = \{s \in Q_1 \mid H_1 \text{ has a cycle reachable from } s\},$$

where H_1 is obtained from $G[Q_1]$ by decreasing multiplicity by 1 for each edge ending in q_0. If $q_0 \notin Q_1$, we have $H_1 = G[Q_1]$. Let us prove that $G \in \mathbb{G}_{a^k b}$ if and only if there is $q_0 \in Q$ such that:

1. It holds that $\mathrm{d}_G(s, q_0) \le k + 1$ for each $s \in Q$.
2. For each $s \in Q$ there is a $q \in R$ such that $\mathrm{d}_G(s, q) \le k$.

First, check the backward implication. For each $r \in R$, we color by b an edge of the form $r \longrightarrow q_0$ that does not appear in H_1. Then we fix a forest of shortest paths from all the vertices of $Q \backslash R$ into R. Due to the second condition above, the branches have length at most k. We color by a the edges used in the forest. We have completely specified a coloring of edges. Now, for any $s \in Q$ a prefix a^j of $a^k b$ takes us into R, the factor a^{k-j} keeps us inside R, and with the letter b we end up in q_0.

As for the forward implication, the first condition is trivial. For the second one, take any $s \in Q$ and denote $s_j = \delta(s, a^j)$ for $j \ge 0$. Clearly, $s_k \in Q_1$, but we show also that $s_k \in R$, so we can set $q = s_k$ in the last condition. Indeed, whenever $s_j \in Q_1$ for $j \ge k$, we remark that $\delta(s_{j-k+1}, a^k) = q_0$ and thus $s_{j+1} \in Q_1$ as well. Since j can grow infinitely, there is a cycle within Q_1 reachable from s_k. □

Proof for $w \in T_3$. Due to Lemma 1, it is enough to deal with $w = ab^k$ for each $k \ge 2$. For a polynomial-time reduction from W-SAT, take an instance $X = \{x_1, \ldots, x_n\}$, $\Phi = \{C_1, \ldots, C_m\}$, where $C_j = (z_{j,1}, z_{j,2}, z_{j,3}, z_{j,4})$ for each $j = 1, \ldots, m$. We build the graph $G_{k,\phi} = (Q, E)$ defined by Fig. 1. Note that:

- In Fig. 1, states are represented by discs. For each $j = 1, \ldots, m$, the edges outgoing from C'_i and C''_i represent the formula Φ by leading to the states $z_{j,1}, z_{j,2}, z_{j,3}, z_{j,4} \in \{x_1, \ldots, x_n\} \subseteq Q$.
- In the case of $k = 2$ the state $\mathrm{V}_{i,2}$ does not exist, so we set $x_i \longrightarrow \mathrm{D}_0$ and $\mathrm{V}_{i,1} \longrightarrow \mathrm{D}_0$ instead of $x_i \longrightarrow \mathrm{V}_{i,2}$ and $\mathrm{V}_{i,1} \longrightarrow \mathrm{V}_{i,2}$.

We show that $G_{k,\Phi} \in \mathbb{G}_{ab^k}$ if and only if there is an assignment $\xi : X \to \{0, 1\}$ satisfying the conditions given by Φ.

First, let there be a coloring δ of $G_{k,\Phi}$ such that $|\delta(Q, ab^k)| = 1$. Observe that necessarily $\delta(Q, ab^k) = \{\mathrm{D}_0\}$, while there is no loop on D_0. We use this fact to observe that whenever $x_i \in \delta(Q, a)$, the edges outgoing from $x_i, \mathrm{V}_{i,1}, \ldots, \mathrm{V}_{i,k-1}$ must be colored according to Fig. 2, but if $x_i \in \delta(Q, ab)$, then they must be colored according to Fig. 3. Let $\xi(x_i) = 1$ if $x_i \in \delta(Q, ab)$ and $\xi(x_i) = 0$ otherwise. Choose any $j \in \{1, \ldots, m\}$ and observe that

$$\xi(\delta(\mathrm{C}_j, ab)) = 1, \quad \xi(\delta(\mathrm{C}'_j, a)) = 0, \quad \xi(\delta(\mathrm{C}''_j, a)) = 0,$$

thus we can conclude that all the conditions from the definition of W-SAT hold for the clause C_j.

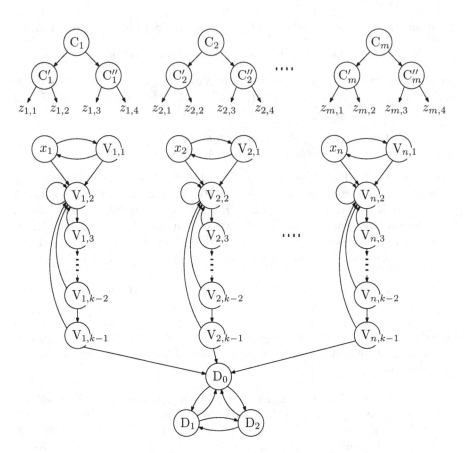

Fig. 1. The graph $G_{k,\Phi}$ reducing W-SAT to $\text{SRCW}_{|I|=2,W=\{ab^k\}}$ for $k \geq 2$

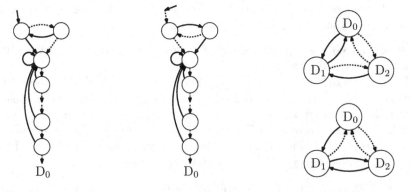

Fig. 2. A coloring corresponding to $\xi(x_i) = \mathbf{0}$

Fig. 3. A coloring corresponding to $\xi(x_i) = \mathbf{1}$

Fig. 4. Colorings for k even (top) and odd (bottom)

On the other hand, let ξ be a satisfying assignment of Φ. For each j we color the edges outgoing from C_j, C_j', C_j'' such that the ab-path from C_j leads to the $z_{j,i}$ with $\xi(z_{j,i}) = \mathbf{1}$ and the a-paths from C_j', C_j'' lead to the $z_{j,i'}$ and $z_{j,i''}$ with $\xi(z_{j,i'}) = \mathbf{0}, \xi(z_{j,i''}) = \mathbf{0}$, where $i' \in \{1, 2\}, i'' \in \{3, 4\}$. For the edges outgoing from $x_i, V_{i,1}, \ldots, V_{i,k-1}$ we use Fig. 2 if $\xi(x_i) = \mathbf{0}$ and Fig. 3 if $\xi(x_i) = \mathbf{1}$. The transitions within D_0, D_1, D_2 are colored according to Fig. 4, depending on the parity of k. Observe that for each $i \in \{1, \ldots, n\}$ we have $x_i \notin \delta(Q, ab)$ if $\xi(x_i) = \mathbf{0}$ and $x_i \notin \delta(Q, a)$ if $\xi(x_i) = \mathbf{1}$. Using this fact we check that $\delta(Q, w) = \{D_0\}$.   □

Proof for $w \in T_4$. Any $w \in T_4$ can be written as $w = va^j b^k a^l$ or $w = vb^j a^k b^l$ for $j, k, l \geq 1$. Due to Lemma 1 it is enough to deal with $w = ab^k a^l$ for each $k, l \geq 1$. Take an instance of W-SAT as above and construct the graph $G_{w,\Phi} = (Q, E)$ defined by Fig. 5. Note that:

– In the case of $l = 1$, the state $Z_{i,1}$ does not exist, so we set $W_i' \longrightarrow D_0$ and $V_{i,k-1} \longrightarrow D_0$ instead of $W_i' \longrightarrow Z_{i,1}$ and $V_{i,k-1} \longrightarrow Z_{i,1}$.
– In the case of $k = 1$, the state $V_{i,1}$ does not exist, so we set $x_i \longrightarrow Z_{i,1}$ (or $x_i \longrightarrow D_0$ if $l = 1$) and $x_i \longrightarrow W_i$ instead of $x_i \rightrightarrows V_{i,1}$.

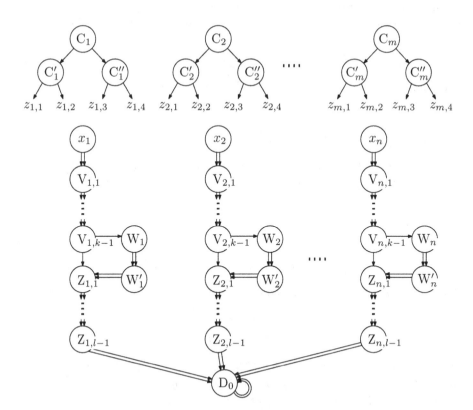

Fig. 5. The graph $G_{w,\Phi}$ reducing W-SAT to $\mathrm{SRCW}^Z_{|I|=2, W=\{ab^k a^l\}}$ for $k, l \geq 1$

Let there be a coloring δ of $G_{w,\Phi}$ such that $|\delta(Q,w)| = 1$. Observe that $\delta(Q,w) = \{D_0\}$. Next, observe that whenever $x_i \in \delta(Q,a)$, then $V_{i,k-1} \xrightarrow{b} Z_{i,1}$, but if $x_i \in \delta(Q,ab)$, then $V_{i,k-1} \xrightarrow{a} Z_{i,1}$. Let $\xi(x_i) = \mathbf{1}$ if $x_i \in \delta(Q,ab)$ and $\xi(x_i) = \mathbf{0}$ otherwise. We choose any $j \in \{1, \ldots, m\}$ and conclude exactly as we did in the case of T_3.

On the other hand, let ξ be a satisfying assignment of Φ. For each j, we color the edges outgoing from C_j, C_j', C_j'' as we did in the case of T_3. For each i, we put $V_{i,k-1} \xrightarrow{a} Z_{i,1}, V_{i,k-1} \xrightarrow{b} W_i$ if $\xi(x_i) = \mathbf{1}$ and the reversed variant if $\xi(x_i) = \mathbf{0}$. $\square$

4   A Partial Classification of Binary Words According to Complexity of $\mathrm{SRCW}^{SC}_{2,\{w\}}$

Clearly, for any $w \in T_1 \cup T_2$ we have $\mathrm{SRCW}^{SC}_{2,\{w\}} \in \mathrm{P}$. In Section 4.1 we show that

$$\mathrm{SRCW}^{SC}_{2,\{abb\}} \in \mathrm{P},$$

which is a surprising result because the general $\mathrm{SRCW}_{2,\{w\}}$ is NP-complete for any $w \in T_3$, including $w = abb$. We are not aware of any other words that witness this difference between SRCW^{SC} and SRCW.

In Section 4.2 we introduce a general method using *sink devices* that allows us to prove the NP-completeness of $\mathrm{SRCW}^{SC}_{2,\{w\}}$ for infinitely many words $w \in T_4$, including any $w \in T_4$ with the first and last letter being the same. However, we are not able to apply the method to each $w \in T_4$.

4.1   A Polynomial-Time Case

A graph $G = (Q, E)$ is said to be k-*lifting* if there exists $q_0 \in Q$ such that for each $s \in Q$ there is an edge leading from s into $V_k(q_0)$. Instead of 2-lifting we just say *lifting*.

Lemma 4. *If G is a k-lifting graph, then $G \in \mathbb{G}_{ab^k}$.*

Lemma 5. *If G is strongly connected, G is not lifting, and $G \in \mathbb{G}_{abb}$ via δ and q_0, then δ has no b-transition ending in $V_2(q_0) \cup V_3(q_0)$. Moreover, $V_3(q_0) = \emptyset$.*

Proof. First, suppose for a contradiction that some $s \in V_2(q_0) \cup V_3(q_0)$ has an incoming b-transition. Together with its outgoing b-transition we have

$$r \xrightarrow{b} s \xrightarrow{b} t,$$

where $s \neq q_0$ and $t \neq q_0$. Due to the strong connectivity there is a shortest path P from q_0 to r (possibly of length 0 if $r = q_0$). The path P is made of b-transitions. Indeed, if there were some a-transitions, let $r' \xrightarrow{a} r''$ be the last one. The abb-path outgoing from r' ends in $\delta(r'', bb)$, which either lies on P or in $\{s, t\}$, so it is different from q_0 and we get a contradiction.

It follows that $\delta(q_0, b) \neq q_0$ and $\delta(q_0, bb) \neq q_0$, so there cannot be any a-transition incoming to q_0. Hence for any $s \in V_1(q_0)$ there is a transition $s \xrightarrow{b} q_0$ and thus there is no a-transition ending in $V_1(q_0)$. Because there is also no a-transition ending in $V_3(q_0)$, all the a-transitions end in $V_2(q_0)$ and thus G is lifting, which is a contradiction.

Second, we show that $V_3(q_0)$ is empty. Suppose that $s \in V_3(q_0)$. No a-transition comes to s since there is no path of length 2 from s to q_0. Thus, s has no incoming transition, which contradicts the strong connectivity.   □

Theorem 6. $\mathrm{SRCW}^{\mathcal{SC}}_{2,\{abb\}}$ *is decidable in polynomial time.*

Proof. As the input we have a strongly connected $G = (Q, E)$. Suppose that q_0 is fixed (we can just try each $q_0 \in Q$) and so we should decide if there is some δ with $\delta(Q, abb) = \{q_0\}$. First we do some preprocessing:

- If G is lifting, according to Lemma 4 we accept.
- If $V_3(q_0) \neq \emptyset$, according to Lemma 5 we reject.
- If there is a loop on q_0, we accept, since due to $V_3(q_0) = \emptyset$ we have $G \in \mathbb{G}_{bb}$.

If we are still not done, we try to find some labeling δ, assuming that none of the three conditions above holds. We deduce two necessary properties of δ. First, Lemma 5 says that we can safely label all the transitions ending in $V_2(q_0)$ by a. Second, we have $q_0 \in \delta(Q, a)$. Indeed, otherwise all the transitions incoming to q_0 are labeled by b, and there cannot be any a-transition ending in $V_1(q_0)$ because we know that the b-transition outgoing from q_0 is not a loop. Thus G is lifting, which is a contradiction.

Let the sets $B_1, \ldots, B_\beta$ denote the connected components (not necessarily strongly connected) of $G[V_1(q_0)]$. Note that maximum out-degree in $G[V_1(q_0)]$ is 1. Let $e = (r, s), e' = (s, t)$ be consecutive edges with $s, t \in V_1(q_0)$ and $r \in Q$. Then the labeling δ has to satisfy

$$e \text{ is labeled by } a \Leftrightarrow e' \text{ is labeled by } b.$$

Indeed:

- The left-to-right implication follows easily from the fact that there is no loop on q_0.
- As for the other one, suppose for a contradiction that both e', e are labeled by b. We can always find a path P (possibly trivial) that starts outside $V_1(q_0)$ and ends in r. Let $\bar{r}$ be the last vertex on P that lies in $\delta(Q, a)$. Such vertex exists because we have $V_2(q_0) \cup \{q_0\} \subseteq \delta(Q, a)$ and $V_3(q_0) = \emptyset$. Now we can deduce that $\delta(\bar{r}, bb) \neq q_0$, which is a contradiction.

It follows that for each B_i there are at most two possible colorings of its inner edges (fix *variant* **0** and *variant* **1** arbitrarily). Moreover, a labeling of any edge incoming to B_i enforces a particular variant for whole B_i.

Let the set A contain the vertices $s \in V_2(q_0) \cup \{q_0\}$ whose outgoing transitions lead both into $V_1(q_0)$. Edges that start in vertices of $(V_2(q_0) \cup \{q_0\}) \setminus A$ have only one possible way of coloring due to Lemma 5, while for each vertex of A there are two possibilities. Now any possible coloring can be described by $|A| + \beta$ Boolean propositions:

$$\mathbf{x}_s \equiv e_s \text{ is labeled by } a$$
$$\mathbf{y}_B \equiv B \text{ is labeled according to variant } \mathbf{1}$$

for each $s \in A$ and $B \in \{B_1, \ldots, B_\beta\}$, where e_s is a particular edge outgoing from s. Moreover, the claim $\delta(Q, abb) = \{q_0\}$ can be equivalently formulated as a conjunction of implications of the form $\mathbf{x}_s \to \mathbf{y}_B$, so we reduce the problem to 2-SAT. □

4.2   NP-Complete Cases

We introduce a method based on *sink devices* to prove the NP-completeness for a wide class of words even under the restriction to strongly connected graphs.

In the proofs below we use the notion of a *partial finite automaton* (*PFA*), which can be defined as a triple $P = (Q, I, \delta)$, where Q is a finite set of states, I is a finite alphabet, and δ is a partial function $Q \times I \to Q$ which can be naturally extended to $Q \times I^\star \to Q$. Again, we write $r \xrightarrow{x} s$ instead of $\delta(r, x) = s$. We say that a PFA is *incomplete* if there is some undefined value of δ. A *sink state* in a PFA has a defined loop for each letter.

Definition 7. *Let* $w \in \{a, b\}^\star$. *We say that a PFA* $B = (Q, \{a, b\}, \delta)$ *is a* sink device *for* w, *if there exists* $q_0 \in Q$ *such that:*

1. $\delta(q_0, u) = q_0$ *for each prefix* u *of* w,
2. $\delta(s, w) = q_0$ *for each* $s \in Q$.

Note that the trivial automaton consisting of a single sink state is a sink device for any $w \in \{a, b\}^\star$. However, we are interested in strongly connected sink devices that are incomplete. In Lemma 8 we show how to prove the NP-completeness using a non-specific sink device in the general case of $w \in T_4$ and after that we construct explicit sink devices for a wide class of words from T_4.

Lemma 8. *Let* $w \in T_4$ *and assume that there exists a strongly connected incomplete sink device* B *for* w. *Then* $\mathrm{SRCW}_{2,\{w\}}^{SC}$ *is NP-complete.*

Proof. We assume that w starts by a and write $w = a^\alpha b^\beta au$ for $\alpha, \beta \geq 1$ and $u \in \{a, b\}^\star$. Denote $B = (Q_B, \{a, b\}, \delta_B)$. For a reduction from W-SAT, take an instance X, Φ with the notation used before, assuming that each $x \in X$ occurs in Φ. We construct a graph $\overline{G}_{w, \Phi} = (\overline{Q}, \overline{E})$ as follows. Let $q_1 \in Q_B$ have an undefined outgoing transition, and let B' be an automaton obtained from B by arbitrarily defining all the undefined transitions except for one transition outgoing from q_1. Let $G_{B'}$ be the underlying graph of B'. By Theorem 3, $\mathrm{SRCW}_{2,\{w\}}^{\mathcal{Z}}$ is NP-complete, so it admits a reduction from W-SAT. Let $G_{w, \Phi} = (Q, E)$ be the graph obtained from such reduction, removing the loop on the sink state $q_0' \in Q$. Let $s_1, \ldots, s_{|Q|-1}$ be an enumeration of all the states of $G_{w, \Phi}$ different

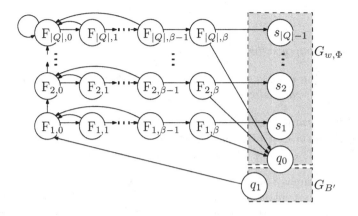

Fig. 6. The graph $\overline{G}_{w,\Phi}$

from q'_0. Then we define $\overline{G}_{w,\Phi}$ as shown in Fig. 6. We merge the state $q'_0 \in Q$ with the state $q_0 \in Q_B$, which is fixed by the definition of a sink device.

First, let there be a coloring $\overline{\delta}$ of $\overline{G}_{w,\Phi}$ such that $\left|\overline{\delta}(\overline{Q}, w)\right| = 1$. It follows easily that $\overline{\delta}$, restricted to Q, encodes a coloring δ of $G_{w,\Phi}$ such that $|\delta(Q, w)| = 1$. The choice of $G_{w,\Phi}$ guarantees that there is a satisfying assignment ξ for Φ.

On the other hand, let ξ be a satisfying assignment of Φ. By the choice of $G_{w,\Phi}$, there is a coloring δ of $G_{w,\Phi}$ such that $|\delta(Q, w)| = 1$. We use the following coloring of $\overline{G}_{w,\Phi}$: The edges outgoing from $s_1, \ldots, s_{|Q|-1}$ are colored according to δ. The edges within $G_{B'}$ are colored according to B'. The edge $q_1 \longrightarrow F_{1,0}$ is colored by b. All the other edges incoming to the states $F_{1,0}, \ldots, F_{|Q|,0}$, together with the edges of the form $F_{i,\beta} \longrightarrow q_0$, are colored by a, while the remaining ones are colored by b. □

For any $w \in \{a,b\}^{\star}$ we construct a strongly connected sink device $\mathbf{D}(w) = (Q_w, \{a,b\}, \delta_w)$. However, for some words $w \in T_4$ (e.g. for $w = abab$) the device $\mathbf{D}(w)$ is not incomplete and thus is not suitable for the reduction above. Take any $w \in \{a,b\}^{\star}$ and let $\mathfrak{C}^P_w, \mathfrak{C}^S_w, \mathfrak{C}^F_w$ be the sets of all prefixes, suffixes and factors of w respectively, including the empty word ϵ. Let

$$Q_w = \left\{[u] \mid u \in \mathfrak{C}^F_w, \, v \notin \mathfrak{C}^S_w \text{ for each nonempty prefix } v \text{ of } u\right\},$$

while the partial transition function δ_w consists of the following transitions:

1. $[u] \xrightarrow{x} [ux]$ whenever $[u], [ux] \in Q_w$,
2. $[u] \xrightarrow{x} [\epsilon]$ whenever $ux \in \mathfrak{C}^S_w$,
3. $[u] \xrightarrow{x} [\epsilon]$ whenever $[ux] \notin Q_w$, $ux \notin \mathfrak{C}^S_w$, and $vx \in \mathfrak{C}^P_w$ for a suffix v of u.

Lemma 9. *For any $w \in \{a,b\}^{\star}$, $\mathbf{D}(w)$ is a strongly connected sink device.*

Lemma 10. *Suppose that $w \in \{a,b\}^{\star}$ starts by x, where $\{x,y\} = \{a,b\}$. If there is $u \in \{a,b\}^{\star}$ satisfying all the following conditions, then $\mathbf{D}(w)$ is incomplete:*

1. $[u] \in Q_w$,
2. $uy \notin \mathfrak{C}_w^{\mathrm{F}}$,
3. *for each nonempty suffix v of uy, $v \notin \mathfrak{C}_w^{\mathrm{P}}$.*

Theorem 11. *If a word $w \in T_4$ satisfies some of the following conditions, then $\mathrm{SRCW}_{2,\{w\}}^{SC}$ is NP-complete:*

1. *w is of the form $w = x\overline{w}x$ for $\overline{w} \in \{a,b\}^\star, x \in \{a,b\}$,*
2. *w is of the form $w = x\overline{w}y$ for $\overline{w} \in \{a,b\}^\star, x,y \in \{a,b\}, x \neq y$, and $x^k y^l x \in \mathfrak{C}_w^{\mathrm{F}}, x^{k+1} \notin \mathfrak{C}_w^{\mathrm{F}}, y^{l+1} \notin \mathfrak{C}_w^{\mathrm{F}}$ for some $k,l \geq 1$.*

Proof. Due to Lemmas 8 and 9, it is enough to show that $\mathbf{D}(w)$ is incomplete. Let $m \geq 1$ be the largest integer such that y^m is a factor of w. It is straightforward to check that $u = y^m$ (in the first case) or $u = x^k y^l$ (in the second case) satisfies the three conditions from Lemma 10.

5   Conclusion and Future Work

We have completely characterized the binary words w that make the computation of road coloring NP-complete if some of them is required to be the reset word for a coloring of a given graph. Except for $w = a^k b$ and $w = a^k$ with $k \geq 1$, each $w \in \{a,b\}^\star$ has this property. We have proved that if we require strong connectivity, the case $w = abb$ becomes solvable in polynomial time. For any w such that the first letter equals to the last one and both a, b occur in w, we have proved that the NP-completeness holds even under this requirement. The main goals of the future research are:

- Complete the classification of binary words in the strongly connected case.
- Give the classifications of words over non-binary alphabets.
- Study SRCW restricted to non-singleton sets of words.

References

1. Adler, R., Goodwyn, L., Weiss, B.: Equivalence of topological Markov shifts. Israel Journal of Mathematics **27**(1), 49–63 (1977)
2. Roman, A., Drewienkowski, M.: A complete solution to the complexity of synchronizing road coloring for non-binary alphabets (2014)
3. Roman, A.: P–NP Threshold for Synchronizing Road Coloring. In: Dediu, A.-H., Martín-Vide, C. (eds.) LATA 2012. LNCS, vol. 7183, pp. 480–489. Springer, Heidelberg (2012)
4. Schaefer, T.J.: The complexity of satisfiability problems. In: Proceedings of the Tenth Annual ACM Symposium on Theory of Computing, STOC 1978, pp. 216–226. ACM, New York (1978)
5. Trahtman, A.N.: The road coloring and Černy conjecture. In: Stringology. pp. 1–12 (2008)
6. Vorel, V., Roman, A.: Parameterized complexity of synchronization and road coloring (2014)

Automata, Logic, and Concurrency

Logics for Unordered Trees with Data Constraints on Siblings

Adrien Boiret[1,2], Vincent Hugot[2,3](✉), Joachim Niehren[2,3], and Ralf Treinen[4]

[1] University Lille 1, Lille, France
[2] Links (INRIA Lille & LIFL, UMR CNRS 8022), Lille, France
vincent.hugot@inria.fr
[3] INRIA, Sophia-antipolis, France
[4] University Paris Diderot Sorbonne Paris Cit PPS UMR 7126, CNRS, 75205
Paris, France

Abstract. We study counting monadic second-order logics (CMso) for unordered data trees. Our objective is to enhance this logic with data constraints for comparing string data values attached to sibling edges of a data tree. We show that CMso satisfiability becomes undecidable when adding data constraints between siblings that can check the equality of factors of data values. For more restricted data constraints that can only check the equality of prefixes, we show that it becomes decidable, and propose a related automaton model with good complexities. This restricted logic is relevant to applications such as checking well-formedness properties of semi-structured databases and file trees. Our decidability results are obtained by compilation of CMso to automata for unordered trees, where both are enhanced with data constraints in a novel manner.

1 Introduction

Logics and automata for unordered trees were studied in the last twenty years mostly for querying Xml documents [5,14,20] and more recently in the context of NoSql databases [2]. They were already studied earlier, for modeling syntactic structures in computational linguistics [16] and records in programming languages [11,12,17]. In our own work, we also find them relevant to the modeling and static verification of file trees, i.e. structures representing directories, files, their contents etcetera, and their transformations, i.e. programs or scripts moving, deleting, or creating files.

Using unordered trees means expressing and evaluating properties on sets – or multisets – of elements, e.g. the data values of the children at the current position. Naturally, this amounts to counting: for instance in a file tree *"there are at least 2 values that match *.txt"* (where * matches any string), or in a bibliographical database *"there are fewer values* inproceedings *than* book". Where the existing approaches differ is in the expressive power available for that counting; for instance, is it possible to compare two variable quantities – as in the second example – or just one variable quantity and a constant – as in the first. In all

© Springer International Publishing Switzerland 2015
A.-H. Dediu et al. (Eds.): LATA 2015, LNCS 8977, pp. 175–187, 2015.
DOI: 10.1007/978-3-319-15579-1_13

```
{ "file.tex" :
    {"\documentclass...":{}},
  "dir" : {
    "x.png" : {"<bin>":{}},
    "y.png" : {"<bin>":{}},
    ...} ... }
```

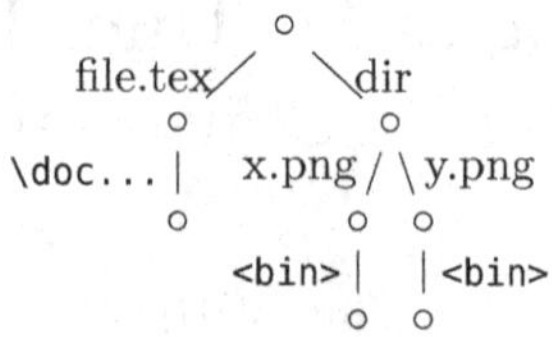

```
               o
       file.tex/   \dir
             o       o
    \doc...|   x.png/ \y.png
         o      o   o
          <bin>|  | <bin>
             o   o
```

Fig. 1. Unordered trees in JSON format, describing a typical file tree

cases, however, each element is considered alone, in isolation from its brothers. We previously studied the complexity of decision problems for automata using various such formalisms as guards for their bottom-up transitions in [3]. The focus was on devising good notions of deterministic machines capable of executing such counting operations, sufficiently expressive but allowing for efficient algorithms. Our present focus, in contrast, is to extend the expressive power of the counting formalisms, while preserving decidability. Since the bottom-up automaton's structure does not play a great role in that, and the yardsticks of expressive power for counting tests are logics, this paper mostly deals directly with second-order logics rather than automata.

Our main goal in this paper is to extend existing formalisms with the ability to express data constraints on unordered data trees, so that each data value may be considered not only in isolation, but also along with sibling values with which it is in relation. Such constraints arise naturally in various circumstances.

By way of example, consider a directory containing LaTeX resources, which may be represented by an edge-labeled tree in the style of Figure 1, given in JSON (JavaScript Object Notation) syntax, where each data value corresponds to a file name or, in the case of leaves, file contents. Suppose that we want to check whether the contents of a LaTeX repository have been properly compiled, which is to say that for every main LaTeX file – i.e. a file whose name has suffix ".tex", and whose contents begin with "\documentclass" – there exists a corresponding PDF file following the version 1.5 of the standard. To express this property, sibling data values – here representing files in the same directory – are put in relation by

$$\theta_{\text{tex2pdf}} = \{ (w".\text{tex}", w".\text{pdf}') \mid w \text{ is a word} \} . \tag{1}$$

Schematically, we express constraints of the form "any value d whose subtree satisfies some property P has a brother $d' = \theta_{\text{tex2pdf}}(d)$ whose subtree satisfies another property P'".

We need to integrate that kind of data constraints in existing formalisms for unordered trees; the two yardsticks of expressive power that have emerged in the literature are the extensions of weak monadic second-order logic (MSO) by horizontal Presburger constraints [14], and by the weaker, but more tractable, counting constraints [8], capable of expressing that the cardinality of a set variable is *less than m* or *equal to m modulo n*, but not of comparing the cardinalities of two set variables directly, unlike Presburger logic. We choose MSO with counting constraints as our starting point, which we write CMSO. We denote by $\Gamma(\Theta)$ and CMSO(Θ) the extensions of counting constraints and CMSO, respectively, with tests on siblings for a certain class Θ of binary relations on data words. Provided that this class contains the relation θ_{tex2pdf} defined above, our example property that, everywhere in the file tree, every TEX file has a corresponding PDF is expressed by the CMSO$(\{\, \theta_{\text{tex2pdf}}\,\})$ formula

$$\forall x \,.\, x \in (\#(*".\text{tex}" \ \wedge \ X_{\text{doc}} \ \wedge \ \neg\theta_{\text{tex2pdf}}.X_{\text{pdf15}}) = 0) \tag{2}$$

where X_{doc} and X_{pdf15} are free set variables assumed to contain the nodes satisfying the "main TEX file" and "valid PDF" properties. Intuitively: "all nodes in the file tree are among the nodes such that the number of their children whose label matches *".tex", which are main TEX documents, and for which there does not exist a corresponding *".pdf" sibling that is a PDF version 1.5, is zero." We shall give the full, closed formula at the end of Sec. 3[p179].

Note that, even for *ordered* data words and in the case of equality tests, – simpler than even the suffix correspondences exemplified by θ_{tex2pdf} – satisfiability, and the emptiness problem in the case of automata, become rapidly intractable or undecidable. This has been studied for register automata, first-order logic, and XPath, [4,9], among others. For instance, satisfiability of $FO^2(=,+1,<)$, i.e. first-order logic with two variables and *successor* and *linear order* relations, while decidable, is not known to be primitive recursive [4]. Unorderedness simplifies matters in this case.

Nevertheless, the choice of the class of string relations Θ to which we have access in our constraints greatly influences the complexity and decidability of the counting constraints using them. We have found that even relatively conservative choices of Θ entail undecidability: merely allowing the replacement of factors of up to three letters, or the addition and deletion of suffixes and prefixes of one letter, suffices. However, we exhibit a relatively large class which *is* decidable, and capable of expressing that the prefixes of two data values are the same, or even in the same regular language, while the suffixes belong to two different languages; this largely covers our envisioned applications. We have also found further restrictions for which the complexities become more reasonable.

Outline:
After a few preliminaries, **Section 3** introduces the logic CMSO(Θ), which is CMSO extended with the ability to put an edge's data value in relation with one of its siblings', the string relation being a member of Θ. In **Section 4**, we show that if relations allow both prefix and suffix manipulations, even restricted to addition or removal of a single letter, CMSO(Θ) becomes undecidable.

After recalling the logic WSkS, which covers a large class of suffix-only manipulations, **Section 5** shows that $\text{CMso}(\Theta_{\text{WS}k\text{S}})$, where allowed relations are WSkS-definable relations, is decidable in non-elementary time, by translating it into automata with horizontal tests in WSkS. **Section 6** presents an algorithm that decides emptiness for the automaton model equivalent to a fragment of the logic where string relations are limited to disjoint suffix replacements. Its complexity is NExpTime – or PSpace if the automaton is deterministic. **Section 7** concludes and hints at possible extensions and different ways of tackling the problem.

2   MSO and Counting Constraints

We recall the definition of MSO and of counting constraints. As models we restrict ourselves to data trees, even though general graph structures could be chosen.

Data Trees.
A *data alphabet* is a finite set $\mathbb{A}$. A *data value* over $\mathbb{A}$ is a string in $\mathbb{A}^*$. The *trees* under consideration are finite, unordered, unranked trees whose edges are labeled by data values in $\mathbb{A}^*$. Formally, a tree t is a multiset $\{\!\{ (d_1, t_1), \ldots, (d_n, t_n) \}\!\}$ where $d_1, \ldots, d_n \in \mathbb{A}^*$ and $t_1, \ldots, t_n$ are trees. d_i is the label of the edge leading into the subtree t_i. For instance, the tree of Fig 1 is $\{\!\{ ("\text{file.tex}", \{\!\{ ("\backslash\text{doc}...", \{\!\{ \}\!\}) \}\!\}), ("\text{dir}", \ldots) \}\!\}$. To simplify the formalisation, we shall not manipulate edges as distinct objects, but instead see an edge label as a property of the node into which the edge leads. Thus we assimilate t to a structure $\langle V_t, \ell_t, \downarrow_t \rangle$, where V_t is the set of nodes of t, $\ell_t(v)$ is the data value labeling the edge leading into the node v – undefined for the root node, – and $v \downarrow_t v'$ holds if v' is a child of v. For our convenience, we also define the "sibling-or-self" relation: $v \curlywedge_t v' \Leftrightarrow \exists v'' . v'' \downarrow_t v \wedge v'' \downarrow_t v'$. By extension of the language of ranked trees, we use the word *arity* to refer to either the multiset of outgoing edge labels of a node, or the set of outgoing edges.

   MSO. Let $\mathbb{A}$ be a data alphabet and $\mathcal{X}$ a countable set of variables of two types, node variables and set variables. A variable assignment I into some tree t will map any node variable $x \in \mathcal{X}$ to a node $I(x) \in V_t$ and any set variable $X \in \mathcal{X}$ to a set of nodes $I(X) \subseteq V_t$.

   As a parameter of our logic we assume a set Ψ of formulæ called *node selectors*, which may contain letters from $\mathbb{A}$ and variables from $\mathcal{X}$. The only assumption we make is that any node selector $\psi \in \Psi$ defines for any tree t and variable assignment I into t a set of nodes $[\![\psi]\!]_{t,I} \subseteq V_t$. For instance, we could choose $\Psi = \Psi_0 = \{\pi \mid \pi \text{ regular expression over } \mathbb{A}\} \cup \{\downarrow x \mid x \in \mathcal{X} \text{ node variable}\}$ such that $[\![\pi]\!]_{t,I} = \{v \mid \ell_t(v) \text{ matches } \pi\}$ is set of all nodes whose incoming edge is labeled by a word in $\mathbb{A}^*$ that matches regular expression π, and $[\![\downarrow x]\!]_{t,I} = \{v \mid v \downarrow_t I(x)\}$ is the set of nodes of which $I(x)$ is a child. Or else, we could also choose $\Psi = \Psi_0 \cup \{\downarrow X \mid X \in \mathcal{X} \text{ set variable}\}$, where a formula $\downarrow X$ requires that some child belongs to $I(X)$. The formulæ of MSO over Ψ are:

$$\xi \in \text{Mso}(\Psi) \quad ::= \quad x \in \psi \quad | \quad x \in X \quad | \quad \exists x . \xi \quad | \quad \exists X . \xi \quad | \quad \xi \wedge \xi \mid \neg \xi \quad ,$$

were $\psi \in \Psi$. Whether a formula is true for a given tree t and variables assignment I into t is defined as follows:

$$t, I \models x \in \psi \Leftrightarrow I(x) \in \llbracket \psi \rrbracket_{t,I} \qquad t, I \models \xi \wedge \xi' \Leftrightarrow t, I \models \xi \text{ and } t, I \models \xi'$$
$$t, I \models x \in X \Leftrightarrow I(x) \in I(X) \qquad t, I \models \neg \xi \Leftrightarrow \text{not } t, I \models \xi$$
$$t, I \models \exists x . \xi \Leftrightarrow t, I[x \mapsto v] \models \xi \text{ for some } v \in V_t$$
$$t, I \models \exists X . \xi \Leftrightarrow t, I[X \mapsto V] \models \xi \text{ for some finite } V \subseteq V_t$$

As syntactic sugar, we will freely use the usual additional logical connectives and set comparisons that can be easily encoded, i.e. formulæ $\forall x.\xi$, $\forall X.\xi$, $\xi \Leftrightarrow \xi'$, $\xi \Rightarrow \xi'$, as well as $X \subseteq X'$, $X = \psi$, and $\psi = \varnothing$.

Children Counting Constraints. A children counting constraint selects a node of a tree by testing the number of its children satisfying some property. Which properties can be tested is defined by the parameter Φ of node selectors. As before, we use as parameter a set of node selectors Φ such that $\llbracket \phi \rrbracket_{t,I} \subseteq V_t$ is defined for all $\phi \in \Phi$, and which may contain variables in $\mathcal{X}$ and letters in $\mathbb{A}$. For instance, we could chose $\Phi = \{\pi \mid \pi \text{ regular expression over } \mathbb{A}\} \cup \mathcal{X}$. A counting constraint over Φ is a formula with the following syntax, where $\phi \in \Phi$ and n, m are natural numbers including 0:

$$\gamma \in \Gamma(\Phi) \quad ::= \quad \#\phi \leqslant n \quad | \quad \#\phi \equiv_m n \quad | \quad \gamma \wedge \gamma \quad | \quad \neg\gamma \ .$$

The first two kinds of formulæ can test whether the number of children satisfying ϕ is less or equal to n or equal to n modulo m. Note that we *cannot* write $\#\phi \leqslant \#\phi'$, which would lead to the richer class of Presburger formulæ.

Any counting constraint γ defines a set of nodes $\llbracket \gamma \rrbracket_{t,I}$ for any variables assignment I to t, so counting constraints themselves can be used as node selectors:

$$\llbracket \#\phi \leqslant n \rrbracket_{t,I} = \{ v \in V_t \mid \text{Card}(\{ v' \mid v \downarrow_t v' \wedge v' \in \llbracket \phi \rrbracket_{t,I} \}) \leqslant n \} \ ,$$
$$\llbracket \#\phi \equiv_m n \rrbracket_{t,I} = \{ v \in V_t \mid \text{Card}(\{ v' \mid v \downarrow_t v' \wedge v' \in \llbracket \phi \rrbracket_{t,I} \}) \equiv_m n \} \ ,$$
$$\llbracket \gamma \wedge \gamma' \rrbracket_{t,I} = \llbracket \gamma \rrbracket_{t,l} \cap \llbracket \gamma' \rrbracket_{t,l} \ , \qquad \llbracket \neg\gamma \rrbracket_{t,I} = V_t \setminus \llbracket \gamma \rrbracket_{t,l} \ .$$

Note that we can define $\#\phi \geqslant n$ as syntactic sugar for $\neg(\#\phi \leqslant n-1)$, and $\#\phi = n$ as syntactic sugar for $\#\phi \geqslant n \wedge \#\phi \leqslant n$.

3   Counting MSO for Data Trees: CMso(Θ)

We now introduce counting MSO for data trees with comparisons of sibling data values. Which precise comparisons are permitted is a parameter of the logic.

As before we assume a set of variables $\mathcal{X}$ and a data alphabet $\mathbb{A}$. In addition, we fix a set Θ of binary relations on $\mathbb{A}^*$ that are called string comparisons. We then define a set of node selectors with regular expressions for matching data values and comparisons of sibling data values from Θ. Such a node selector has the following syntax where $\theta \in \Theta$, π is a regular expression over $\mathbb{A}$, and $x, X \in \mathcal{X}$:

$$
\begin{aligned}
\phi \in \Phi_{\text{rel}}(\Theta) \quad ::= \quad & \pi & & \text{incoming edge label matches } \pi, \\
& \mid x \mid X & & \text{equal to } x \text{ or member of } X, \\
& \mid \theta.\phi & & \exists \text{ sibling satisfying } \phi \text{ with labels related by } \theta, \\
& \mid \phi \wedge \phi \mid \neg\phi & & \text{conjunction and negation.}
\end{aligned}
$$

The sets of selected nodes are defined as follows for formula $\phi \in \Phi_{\text{rel}}$, any tree t and variable assignment I into t:

$$
\begin{aligned}
&[\![\pi]\!]_{t,I} = \{v \mid \ell_t(v) \text{ matches } \pi\} & & [\![\phi \wedge \phi']\!]_{t,I} = [\![\phi]\!]_{t,I} \cap [\![\phi']\!]_{t,I} \\
&[\![x]\!]_{t,I} = \{I(x)\} & & [\![\neg\phi]\!]_{t,I} = \mathsf{V}_t \setminus [\![\phi]\!]_{t,I} \\
&[\![X]\!]_{t,I} = I(X) \\
&[\![\theta.\phi]\!]_{t,I} = \{v \mid \exists v' . v \curlywedge_t v' \wedge (\ell_t(v), \ell_t(v')) \in \theta \wedge v' \in [\![\phi]\!]_{t,I}\}
\end{aligned}
$$

In particular, a node selector $\theta.\phi$ selects all nodes that have a sibling-or-self, so that the data values of these two nodes satisfy comparison θ.

Definition 1. *We define the children counting contraints for data trees with comparisons of data values $\Gamma(\Theta)$ by $\Gamma(\Phi_{rel}(\Theta))$ and the counting MSO for data trees with comparison of sibling data values $\mathrm{CMso}(\Theta)$ by $\mathrm{Mso}(\Gamma(\Theta))$.*

Note that the childhood $x \downarrow x'$ can be defined in $\mathrm{CMso}(\Theta)$ by $x \in (\#x' = 1)$ independently of the choice of Θ. Hence, sibling-or-self contraints $x \curlywedge x'$ can also be defined by $\exists x''. (x'' \downarrow x \wedge x'' \downarrow x')$ for any Θ. The elements of Θ intervene only if one wants to compare the data values of sibling nodes.

Example 1. Recall the TEX compilation example of equation $(2)_{[\text{p177}]}$ and its free variables. There remains to bind X_{doc} and X_{pdf15} to the relevant sets of nodes in a closed formula. A TEX main document (resp. a valid PDF version 1.5) is represented by a node with a single outgoing edge, whose label is prefixed by "\documentclass" (resp. "%PDF-1.5"), leading to a leaf. Thus the closed $\mathrm{CMso}(\{\theta_{\text{tex2pdf}}\})$ formula:

$$
\begin{aligned}
\exists X_{\text{leaf}} . \exists X_{\text{doc}} . \exists X_{\text{pdf15}} . \quad X_{\text{leaf}} \;&=\; (\#(*) = 0) \\
\wedge \; X_{\text{doc}} \;&=\; (\#(*) = 1 \wedge \#(\text{"\documentclass"} * \wedge X_{\text{leaf}}) = 1) \\
\wedge \; X_{\text{pdf15}} \;&=\; (\#(*) = 1 \wedge \#(\text{"\%PDF-1.5"} * \wedge X_{\text{leaf}}) = 1) \\
\wedge \; \forall x . x \;&\in\; (\#(*\text{".tex"} \wedge X_{\text{doc}} \wedge \neg\theta_{\text{tex2pdf}}.X_{\text{pdf15}}) = 0) .
\end{aligned}
$$

Example 2. Another useful thing to require of a data tree is the *feature tree* property, stating that no two sibling edges may share the same label. This property can be used to specify files systems, since one needs to state that no two files in the same directory have the same name. Taking θ_{id} as the identity relation, we can define feature trees in $\mathrm{CMso}(\{\theta_{\text{id}}\})$ as follows:

$$\forall x \, . \, (\#(x \, \wedge \, \theta_{\mathrm{id}} . \neg x) \geqslant 1) = \varnothing \, .$$

Example 3. Consider now a transformation θ_{bck}, which to w associates $w''.\mathrm{bck}''$, thus relating a file's name to that of its automatic backup. Suppose that the system can back up a backup, and so on, up to a certain point, and we need to check that this bound is not overstepped. That is to say, given $n \in \mathbb{N}$, we want to write a formula ξ_n enforcing that there is no chain of backups of length greater than n. Suppose we had a least-fixed point operator μ among our child-selectors, following the syntax – $\mu X.\phi$ – and semantics of μ-calculus. Then we could write ξ_n in $\mathrm{CMso}(\{\theta_{\mathrm{bck}}\})$:

$$\forall x \, . \, (\#\mu X.(x \, \vee \, \theta_{\mathrm{bck}}.X) > n + 1) = \varnothing \, .$$

$\mu X.(x \vee \theta_{\mathrm{bck}}.X)$ intuitively captures the set of nodes related to x by successive iterations of θ_{bck}; we can do the same thing without needing μ by explicitly binding a set variable Y to the least fixpoint of $x \vee \theta_{\mathrm{bck}}.X$, wrt. X:

$$\forall x \, . \, \exists Y \, . \, (\#((x \vee \theta_{\mathrm{bck}}.Y) \wedge \neg Y) \geqslant 1) = \varnothing$$
$$\wedge \, \nexists Y' \, . \, Y' \subseteq Y \wedge (\#((x \vee \theta_{\mathrm{bck}}.Y') \wedge \neg Y') \geqslant 1) = \varnothing$$
$$\wedge \, \forall x \, . \, [\#Y > n + 1] = \varnothing \, .$$

The first line establishes Y as a fixed point, as it means that there are no nodes with a child satisfying $x \vee \theta_{\mathrm{bck}}.Y$ but not Y. The second line states that there is no smaller fixpoint than Y. This encoding can be generalised to any use of μ.

4   Undecidable Instances of $\mathrm{CMso}(\Theta)$

In this section, we exhibit conditions on the expressive power of the class of data constraints Θ sufficient to render satisfiability for $\Gamma(\Theta)$, and therefore for $\mathrm{CMso}(\Theta)$, undecidable. As we shall see, not much is needed. Even merely allowing Θ to express the addition or removal of a single letter at the beginning or end of a word is enough; the argument developed in the next theorem is that even this is sufficient to encode the solution of the Post Correspondence Problem.

Theorem 1. *Let Θ_1 be the set of string relations of the forms $w \mapsto wa$, $w \mapsto aw$, $wa \mapsto w$, or $aw \mapsto w$, with $a \in \mathbb{A}, w \in \mathbb{A}^*$. Then $\mathrm{CMso}(\Theta_1)$ is undecidable.*

Proof. We reduce PCP, with input dominoes $\left[{u_1 \atop v_1} \right], \ldots, \left[{u_n \atop v_n} \right]$. Let us write the relations in Θ_1 as θ_{+a}, θ_{a+}, θ_{-a}, and θ_{a-}, respectively. Given a word $w = a_1 \ldots a_m$, by abuse of notation we abbreviate $\theta_{+a_1}.\ldots.\theta_{+a_m}.\phi$ into $\theta_{+w}.\phi$. Although Θ_1 is not closed by composition, this construction enables us to pretend that it is – the difference is that it requires the existence of siblings for each intermediate step, which does not affect us. $\theta_{w-}.\phi$ is defined likewise. $\theta_{a_m+}.\ldots.\theta_{a_1+}.\phi$ is written $\theta_{w+}.\phi$, and likewise for $\theta_{-w}.\phi$. Let $\$_1, \$_2 \in \mathbb{A}$ be symbols not appearing in any domino, serving as markers for the first and the second phase of the construction. The operation for "placing domino i around previous dominoes" is defined as $\theta_i.\phi \equiv \theta_{\$_1-}.\theta_{\overline{u_i}+}.\theta_{+v_i}.\theta_{\$_1+}.\phi$; "accepting dominoes" is $\theta_{\mathrm{acc}}.\phi \equiv \theta_{\$_1-}.(*_1 \wedge \theta_{\$_2+}.\phi)$, where $*_1$ matches any string of length $\geqslant 1$, to avoid the empty sequence as a trivial solution; "reading a on both ends" is $\theta_a.\phi \equiv \theta_{\$_2-}.\theta_{a-}.\theta_{-a}.\theta_{\$_2+}.\phi$. Abbreviating $\theta.*$ or $\theta.true$ into simply θ, consider now the formula $\gamma \in \Gamma(\Theta_1) =$

$$\#\$_1 = 1 \wedge \#\$_2 = 1 \wedge$$
$$\#(\$_1 * \wedge \neg(\theta_1 \vee \cdots \vee \theta_n \vee \theta_{\mathrm{acc}})) = 0 \quad \wedge \quad \#(\$_2 * \wedge \neg(\bigvee_{a \neq \$_1, \$_2} \theta_a)) = 1 .$$

It is satisfiable iff there is a tree whose arity contains $\$_1, \$_2$, and such that every label beginning with $\$_1$ (i.e. phase one) has a sibling (along with the intermediate siblings) obtained either by placing some domino so that u_i mirrors v_i, staying in phase one, or by moving to phase two. At this point, a label is of the form $\$_2\overline{u_{i_k}} \ldots \overline{u_{i_1}} v_{i_1} \ldots v_{i_k}$. Furthermore, all but one label beginning with $\$_2$ (i.e. all but $\$_2$) have a sibling obtained by removing the same letter at the beginning and the end; all letters must be read until only $\$_2$ remains. Thus, γ is satisfiable iff there are $i_1, \ldots, i_k$ such that $u_{i_1} \ldots u_{i_k} = v_{i_1} \ldots v_{i_k}$. This shows that $\Gamma(\Theta_1)$ is undecidable. This carries over to $\mathrm{CMso}(\Theta_1)$: consider the formula $\exists x . x \in \gamma$. $\square$

Proof.   We reduce PCP, with input dominoes $[\begin{smallmatrix} u_1 \\ v_1 \end{smallmatrix}], \ldots, [\begin{smallmatrix} u_n \\ v_n \end{smallmatrix}]$. Let us write the relations in Θ_1 as θ_{+a}, θ_{a+}, θ_{-a}, and θ_{a-}, respectively. Given a word $w = a_1 \ldots a_m$, by abuse of notation we abbreviate $\theta_{+a_1}.\ldots.\theta_{+a_m}.\phi$ into $\theta_{+w}.\phi$. Although Θ_1 is not closed by composition, this construction enables us to pretend that it is – the difference is that it requires the existence of siblings for each intermediate step, which does not affect us. $\theta_{w-}.\phi$ is defined likewise. $\theta_{a_m+}.\ldots.\theta_{a_1+}.\phi$ is written $\theta_{w+}.\phi$, and likewise for $\theta_{-w}.\phi$. Let $\$_1, \$_2 \in \mathbb{A}$ be symbols not appearing in any domino, serving as markers for the first and the second phase of the construction. The operation for "placing domino i around previous dominoes" is defined as $\theta_i.\phi \equiv \theta_{\$_1-}.\theta_{\overline{u_i}+}.\theta_{+v_i}.\theta_{\$_1+}.\phi$; "accepting dominoes" is $\theta_{\mathrm{acc}}.\phi \equiv \theta_{\$_1-}.(*_1 \wedge \theta_{\$_2+}.\phi)$, where $*_1$ matches any string of length $\geqslant 1$, to avoid the empty sequence as a trivial solution; "reading a on both ends" is $\theta_a.\phi \equiv \theta_{\$_2-}.\theta_{a-}.\theta_{-a}.\theta_{\$_2+}.\phi$. Abbreviating $\theta.*$ or $\theta.true$ into simply θ, consider now the formula $\gamma \in \Gamma(\Theta_1) =$

$$\#\$_1 = 1 \wedge \#\$_2 = 1 \wedge$$
$$\#(\$_1 * \wedge \neg(\theta_1 \vee \cdots \vee \theta_n \vee \theta_{\mathrm{acc}})) = 0 \quad \wedge \quad \#(\$_2 * \wedge \neg(\bigvee_{a \neq \$_1, \$_2} \theta_a)) = 1 .$$

It is satisfiable iff there is a tree whose arity contains $\$_1, \$_2$, and such that every label beginning with $\$_1$ (i.e. phase one) has a sibling (along with the intermediate siblings) obtained either by placing some domino so that u_i mirrors v_i, staying in phase one, or by moving to phase two. At this point, a label is of the form $\$_2\overline{u_{i_k}} \ldots \overline{u_{i_1}} v_{i_1} \ldots v_{i_k}$. Furthermore, all but one label beginning with $\$_2$ (i.e. all but $\$_2$) have a sibling obtained by removing the same letter at the beginning and the end; all letters must be read until only $\$_2$ remains. Thus, γ is satisfiable iff there are $i_1, \ldots, i_k$ such that $u_{i_1} \ldots u_{i_k} = v_{i_1} \ldots v_{i_k}$. This shows that $\Gamma(\Theta_1)$ is undecidable. This carries over to $\text{CMso}(\Theta_1)$: consider the formula $\exists x \,.\, x \in \gamma$. □

5   Satisfiability of $\text{CMso}(\Theta_{\text{WS}k\text{S}})$ is Decidable

We shall now see that, in spite of the bleak picture painted by the previous section, Θ *can* be made rather large and useful without forgoing decidability. Indeed, the most frequent operation in applications, illustrated in particular by the TeX example $(1)_{[\text{p176}]}$, is suffix replacement. The property that we really need is thus decidability of satisfiability for $\text{CMso}(\Theta_{\text{suffix}})$, where the relations of Θ_{suffix} are of the form $\theta_{u,u'} = \{ (wu, wu') \mid w \in \mathbb{A}^* \}$, for $u, u' \in \mathbb{A}^*$. We show decidability for a class that is actually more general: WSkS-definable relations.

The well-known logic Weak Monadic Second-Order Logic with k Successors (WSkS) [6], for any $k \geqslant 1$, is based on first-order variables z, and second-order variables Z. Terms τ and formulæ ω of this logic are defined by

$$\tau ::= \epsilon \mid z \mid \tau i \qquad\qquad\qquad 1 \leqslant i \leqslant k$$
$$\omega ::= \tau = \tau \mid \tau \in Z \mid \omega \wedge \omega \mid \neg \omega \mid \exists z \,.\, \omega \mid \exists Z \,.\, \omega$$

First-order variables range over words in $\{ 1, \ldots, k \}^*$, and second-order variables range over finite subsets of $\{ 1, \ldots, k \}^*$. The constant ϵ denotes the empty word, and each of the functions i, written in postfix notation, denotes appending the symbol i at the end of a word. Validity and satisfiability of formulæ in WSkS are decidable [19], even though with a non-elementary complexity [18].

Some useful relations expressible in WSkS are zz' (prefix partial order on words), $z \leqslant_{\text{lex}} z'$ (lexicographic total order on words), $z \in \pi$ for any regular expression π, $Z \subseteq Z'$, $Z = Z' \cup Z''$, $Z = Z' \cap Z''$, $Z = \overline{Z'}$ (complement), $Z = \varnothing$, $|Z| \equiv_n m$ for any constants n, m. Most of those are shown in [7, p88].

The unary predicates on words definable in WSkS are precisely the regular sets [10,13]. A binary relation $R \subseteq \{ 1, \ldots, k \}^* \times \{ 1, \ldots, k \}^*$ is called *special* if it is of the form $\{ (ab, ac) \mid a \in L, \, b \in M, \, c \in N \}$ for some regular sets L, M, and N. A binary relation on words is definable in WSkS iff it is a finite union of special relations [10]. Some relations which are known *not* to be expressible in WSkS are $z = z'z''$, $z = iz'$, z is a suffix of z', z and z' have the same length, Z and Z' have the same cardinality. Let us note that what is definable largely includes the kinds of suffix manipulations which we need for applications and, conversely, that the dangerous properties highlighted in the previous section are not expressible: one cannot manipulate suffixes and prefixes at the same time.

Let $\Theta_{\text{WS}k\text{S}}$ be the set of WSkS-definable relations, with the letters of $\mathbb{A}$ taken as successor functions, along with a fresh letter \$; we sketch the proof of decidability of $\text{CMso}(\Theta_{\text{WS}k\text{S}})$. Child-selectors ϕ and counting constraints ψ are encoded into WSkS, and thus shown decidable. The Mso layer can then be translated into automata, yielding a model of automata for unordered trees as in [3], for which the emptiness problem is known to be decidable under certain conditions, which are here satisfied.

We encode multisets A of edge labels w as sets of WSkS strings, accounting for multiplicities $A(w)$ by appending different numbers of $ to ws to differentiate them. Let t be a tree and A_v^t the arity – the multiset of labels – of node v; the encoding of A_v^t is denoted by $\overline{A_v^t}$ and that of v by $\overline{v}$, such that

$$\overline{A_v^t} = \{\, w\$^k \mid 1 \leqslant k \leqslant A_v^t(w)\,\} = \{\, \overline{v'} \mid v \downarrow_t v'\,\},$$

where $\overline{v'} = \ell_t(v')\$^i$ for some i. Note that all children sharing the same label must get a different i; while there are several valid encodings depending on that assignment, we simply choose one, indifferently. Taking $\overline{X}$ as fresh WSkS set variables, this encoding extends to interpretations in the obvious way. We can now encode any child-selector ϕ as a WSkS formula $\overline{\phi}$ with free variables z, Z (standing for the current node and its arity), such that for any tree t, interpretation I, and nodes $v' \downarrow_t v$:

$$t, I, v \models \phi \quad \Longleftrightarrow \quad \overline{I}[z \mapsto \overline{v}, Z \mapsto \overline{A_{v'}^t}] \models \overline{\phi}.$$

Our building blocks are: (1) $z \models \pi$, where π is a regular expression, which is known to be WSkS-expressible, (2) $z\theta z'$ is expressible by definition, since θ is a WSkS-expressible relation, and (3) $z - \$$, which removes all the $ at the end of the word, testing its well-formedness at the same time it restitutes the edge-label, and is encoded as

$$z' = z - \$ \quad \equiv \quad z'\$z \quad \wedge \quad z' \models \mathbb{A}^*.$$

Using this, we have the following encodings:

$$\overline{\pi} \equiv (z - \$) \models \pi, \qquad \overline{X} \equiv z \in \overline{X},$$
$$\overline{\theta.\phi} \equiv \exists z' \in Z . (z - \$)\theta(z' - \$) \wedge \overline{\phi}[z \leftarrow z'].$$

There remains to handle counting constraints ψ, which is simply a matter of showing that WSkS can encode the primitives $|Z| \leqslant m$ – which is easy – and $|Z| \equiv_n m$ – which rests on a total order such as the lexicographic one, and on the idea of affecting each element in turn to a second-order variable corresponding to the value of the modulo. (Note that the same cannot be said of Presburger logic's $|X| = |Y|$ tests, which are not expressible in WSkS, and whose addition would make it undecidable.) With this done, all decidability results for WSkS carry over to $\Gamma(\Theta_{\mathrm{WS}k\mathrm{S}})$; in particular:

Lemma 1. *Satisfiability of $\Gamma(\Theta_{WSkS})$ is decidable.*

There now remains to deal with the Mso layer; it could be encoded in WSkS as well (as it is a second order logic with sufficient expressive power), but it is simpler to take an automaton-based viewpoint, similar to [3,15] (with the addition of θs). We summarise the model of automata for our unordered trees, $\mathrm{AUT}(\Theta)$, as bottom-up automata with rules $\psi \rightarrow q$, where ψ are formulæ of $\Gamma(\Theta)$ whose child-selectors have an additional test q determining whether a child node has been evaluated in q previously (this corresponds to an "X_q"

test). A tree language L is said to be $\mathrm{CMSO}(\Theta)$-definable if there exists a closed formula $\xi_L \in \mathrm{CMSO}(\Theta)$ such that $L = \{\, t \mid t \models \xi_L \,\}$. Through straightforward adaptations of the usual encodings [7,19], and further noting that $\mathrm{AUT}(\Theta)$ are effectively closed by all boolean operations, we obtain:

Lemma 2. *A set of trees is* $\mathrm{CMSO}(\Theta)$-*definable iff it is accepted by an* $\mathrm{AUT}(\Theta)$.

Of course, this result is constructive, and we can then adapt the usual reachability algorithm: provided that $\Gamma(\Theta)$ is decidable, so is emptiness for $\mathrm{AUT}(\Theta)$, and, in turn, so is $\mathrm{CMSO}(\Theta)$. In particular:

Theorem 2. *Satisfiability of* $\mathrm{CMSO}(\Theta_{WSkS})$ *is decidable.*

6   More Efficient Fragments

We can further gain in efficiency by further restricting the θ relation. To this end, we consider mutually exclusive suffix replacement: we pick a set of suffixes $L = \{w_1, \ldots, w_n\}$ such that w_i is never a suffix of another w_j. Let Θ_L be the set of string relations θ_{w_i, w_j} linking uw_i to uw_j, we denote $\Gamma_{\mathrm{suf}L}$ the counting formulæ of $\Gamma(\Theta_L)$, with the additional restriction that regular expressions testing labels are of the form $\mathbb{A}^* \cdot w_i$. We use a small-model argument to find an efficient algorithm for satisfiability. We will later use this logic in bottom-up automata of $\mathrm{AUT}(\Theta_L)$ as we did in Part 5.

We consider that our arities are already annotated by set variables $X \in \mathcal{X}$. These variables will later correspond to state labelings of an automaton of $\mathrm{AUT}(\Theta)$. If we consider vertically deterministic automata of $\mathrm{AUT}(\Theta)$ [3], where each tree is evaluated in at most one state, the variables X are mutually exclusive. By restricting ourselves to mutually exclusive suffixes, we only need to consider the edges labeled in uL, i.e. the *orbit* of uw_i under the action of all θ_{w_i, w_j}. This allows us to guessing a valid arity for $\phi \in \Gamma_{\mathrm{suf}L}$ orbit by orbit. All we need then is a small-model theorem: if $\#\phi \leqslant n$ appears in a formula ψ, we need to keep track of how many elements are selected by ϕ in a counter that stops at n. if $\#\phi \equiv_m n$ appears in ψ, we need to keep track of how many elements are selected by ϕ in a counter modulo m. This leads to an exponential number of configurations, which means that, if ψ is satisfiable, then we can find a solution using an exponential number of orbits of exponential size. We finally get:

Lemma 3. *The satisfiability problem for an arity formula of* $\Gamma_{\mathrm{suf}L}$ *is decidable in* NEXPTIME. *Furthermore, if the variables* X *are mutually exclusive, the satisfiability problem for an arity formula of* $\Gamma_{\mathrm{suf}L}$ *is decidable in* PSPACE.

We can then use the techniques of [3,15], to extend our results to a class $\mathrm{AUT}(\Theta_L)$ of bottom-up automata with rules $\psi \to q$, where ψ are formulæ of $\Gamma_{\mathrm{suf}L}$.

Theorem 3. *The emptiness problem for automata in* $\mathrm{AUT}(\Theta_L)$ *is decidable in* NEXPTIME. *Furthermore, for deterministic automata of* $\mathrm{AUT}(\Theta_L)$, *the emptiness problem is decidable in* PSPACE.

7 Conclusions and Future Works

We have introduced the logic $\mathrm{CMSO}(\Theta_{\mathrm{WS}k\mathrm{S}})$ on unordered data trees. It is an extension of CMSO to data trees, where tests on a given child may include enforcing the existence of a sibling whose label is in relation with that child's own label, the relation being WSkS-definable. That logic's expressive power is largely sufficient for concrete applications, such as the verification of common constraints on file trees, which usually involve suffix manipulations, largely captured by WSkS. We have shown that satisfiability for $\mathrm{CMSO}(\Theta_{\mathrm{WS}k\mathrm{S}})$ is decidable. However, we have also shown that any attempt to allow additional data relations for both prefix *and* suffix manipulations, even of the simplest kind, would render the logic undecidable. We have also studied the complexity of the emptiness tests for automata where horizontal counting constraints are restricted to relations that only involve disjoint suffixes, and shown that the test is then NExpTime for alternating automata, and only PSpace for deterministic automata.

There are two main ways in which this work can be extended. One is to find more expressive string relations for which the logic remains decidable; our undecidability results indicate that such an extension may not be very natural. Another is to extend the reach of the string relation from merely the set of siblings to something larger. In a first step towards that, the proof of Thm 3 can be extended to support equality constraints between brother subtrees without changing the NExpTime complexity. Another promising direction is the use of Monadic Datalog on data trees [1], which is capable of expressing relations not only with siblings but also with parents, cousins etcetera, and for which efficient algorithms are known.

References

1. Abiteboul, S., Bourhis, P., Muscholl, A., Wu, Z.: Recursive queries on trees and data trees. In: ICDT, pp. 93–104. ACM (2013)
2. Benzaken, V., Castagna, G., Nguyen, K., Siméon, J.: Static and dynamic semantics of NoSQL languages. In: POPL, pp. 101–114. ACM (2013)
3. Boiret, A., Hugot, V., Niehren, J., Treinen, R.: Deterministic Automata for Unordered Trees. In: Fifth International Symposium on Games, Automata, Logics and Formal Verification (Gandalf) (2014)
4. Bojanczyk, M., David, C., Muscholl, A., Schwentick, T., Segoufin, L.: Two-variable logic on data words. ACM Trans. Comput. Log. **12**(4), 27 (2011)
5. Boneva, I., Talbot, J.-M.: Automata and Logics for Unranked and Unordered Trees. In: Giesl, J. (ed.) RTA 2005. LNCS, vol. 3467, pp. 500–515. Springer, Heidelberg (2005)
6. Büchi, J.R.: Weak second-order arithmetic and finite automata. Mathematical Logic Quarterly **6**(1–6), 66–92 (1960)
7. Comon, H., Dauchet, M., Gilleron, R., Löding, C., Jacquemard, F., Lugiez, D., Tison, S., Tommasi, M.: Tree automata techniques and applications (2007). http://www.grappa.univ-lille3.fr/tata
8. Courcelle, B.: The monadic second-order logic of graphs. i. recognizable sets of finite graphs. Information and computation **85**(1), 12–75 (1990)

9. Figueira, D.: On XPath with transitive axes and data tests. In: ACM Symposium on Principles of Database, System, pp. 249–260 (2013)
10. Läuchli, H., Savioz, C.: Monadic second order definable relations on the binary tree. Journal of Symbol Logic **52**(1), 219–226 (1987)
11. Müller, M., Niehren, J., Treinen, R.: The first-order theory of ordering constraints over feature trees. In: 13th annual IEEE Symposium on Logic in Computer Sience, pp. 432–443 (1998)
12. Niehren, J., Podelski, A.: Feature automata and recognizable sets of feature trees. In: Gaudel, M.-C., Jouannaud, J.-P. (eds.) CAAP 1993, FASE 1993, and TAP-SOFT 1993. LNCS, vol. 668, pp. 356–375. Springer, Heidelberg (1993)
13. Rabin, M.: Automata on Infinite Objects and Church's Problem. Number 13 in Conference Board of the Mathematical Sciences Regional Conference Series in Mathematics. AMS (1972)
14. Seidl, H., Schwentick, T., Muscholl, A.: Numerical document queries. In: ACM Symposium on Principles of Database Systems, pp. 155–166 (2003)
15. Seidl, H., Schwentick, T., Muscholl, A.: Counting in trees. In: Logic and Automata. Texts in Logic and Games, vol. 2, pp. 575–612. Amsterdam University Press (2008)
16. Smolka, G.: Feature constraint logics for unification grammars. Journal of Logic Programming **12**, 51–87 (1992)
17. Smolka, G., Treinen, R.: Records for logic programming. J. Log. Program. **18**(3), 229–258 (1994)
18. Stockmeyer, L., Meyer, A.: Word problems requiring exponential time. In: Symposium on the Theory of Computing. Association for Computing Machinery, Association for Computing Machinery, pp. 1–9 (1973)
19. Thatcher, J.W., Wright, J.B.: Generalized finite automata theory with an application to a decision problem of second-order logic. Mathematical systems theory **2**(1), 57–81 (1968)
20. Zilio, S.D., Lugiez, D.: XML Schema, Tree Logic and Sheaves Automata. In: Nieuwenhuis, R. (ed.) RTA 2003. LNCS, vol. 2706, pp. 246–263. Springer, Heidelberg (2003)

Weak and Nested Class Memory Automata

Conrad Cotton-Barratt[1]([✉]), Andrzej S. Murawski[2], and C.-H. Luke Ong[1]

[1] Department of Computer Science, University of Oxford, Oxford, UK
{conrad.cotton-barratt,luke.ong}@cs.ox.ac.uk
[2] Department of Computer Science, University of Warwick, Warwick, UK
a.murawski@warwick.ac.uk

Abstract. Automata over infinite alphabets have recently come to be studied extensively as potentially useful tools for solving problems in verification and database theory. One popular model of automata studied is the Class Memory Automata (CMA), for which the emptiness problem is equivalent to Petri Net Reachability. We identify a restriction – which we call weakness – of CMA, and show that they are equivalent to three existing forms of automata over data languages. Further, we show that in the deterministic case they are closed under all Boolean operations, and hence have an ExpSpace-complete equivalence problem. We also extend CMA to operate over multiple levels of nested data values, and show that while these have undecidable emptiness in general, adding the weakness constraint recovers decidability of emptiness, via reduction to coverability in well-structured transition systems. We also examine connections with existing automata over nested data.

Keywords: Automata · Data languages · Nested Data

1 Introduction

A data word is a word over a finite alphabet in which every position in the word also has an associated *data value*, from an infinite domain. Data languages provide a useful formalism both for problems in database theory and verification [3,13,16]. For example, data words can be used to model a system of a potentially unbounded number of concurrent processes: the data values are used as identifiers for the processes, and the data word then gives an interleaving of the actions of the processes. Having expressive, decidable logics and automata over data languages then allows properties of the modelled system to be checked.

Class memory automata (CMA) [3] are a natural form of automata over data languages. CMA can be thought of as finite state machines extended with the ability, on reading a data value, to remember what state the automaton was in when it last saw that data value. A run of a CMA is accepting if the following two conditions hold: (i) the run ends in a *globally accepting* state; and (ii) each data

---

C. Cotton-Barratt—Supported by an EPSRC Doctoral Training Grant.
A.S. Murawski—Supported by EPSRC (EP/J019577/1).
C.-H. Luke Ong—Partially supported by Merton College Research Fund.

A.-H. Dediu et al. (Eds.): LATA 2015, LNCS 8977, pp. 188–199, 2015.
DOI: 10.1007/978-3-319-15579-1_14

value read in the run was last seen in a *locally accepting* state. If using data values to distinguish semi-autonomous parts of a system, while the first condition can check the system as a whole has behaved correctly, the second of these conditions can be used to check that each part of the system independently behaved correctly. The emptiness problem for CMA is equivalent to Petri net reachability, and while closed under intersection, union, and concatenation, they are not closed under complementation, and do not have a decidable equivalence problem.

We earlier described how data words can be used to model concurrent systems: each process can be identified by a data value, and CMA can then verify properties of the system. What happens when these processes can spawn subprocesses, which themselves can spawn subprocesses, and so on? In these situations the parent-child relationship between processes becomes important, and a single layer of data values cannot capture this; instead we want a notion of *nested* data values, which themselves contain the parent-child relationship. In fact, such nested data values have applications beyond just in concurrent systems: they are prime candidates for modelling many computational situations in which names are used hierarchically. This includes higher-order computation where intermediate functional values are being created and named, and later used by referring to these names. More generally, this feature is characteristic of numerous encodings into the π-calculus [14].

This paper is concerned with finding useful automata models which are expressive enough to decide properties we may wish to verify, as well as having good closure and decidability properties, which make them easy to abstract our queries to. We study a restriction of class memory automata, which we find leads to improved complexity and closure results, at the expense of expressivity. We then extend class memory automata to a nested data setting, and find a decidable class of automata in this setting.

Contributions. In Section 3 we identify a natural restriction of Class Memory Automata, which we call *weak* Class Memory Automata, in which the local-acceptance condition of CMA is dropped. We show that these weak CMA are equivalent to: (i) Class Counting Automata, which were introduced in [12]; (ii) non-reset History Register Automata, introduced in [18]; and (iii) locally prefix-closed Data Automata, introduced in [5].

These automata have an EXPSPACE-complete emptiness problem. The primary advantage of having this equivalent model as a kind of Class Memory Automaton is that there is a natural notion of determinism, and we show that Deterministic Weak CMA are closed under all Boolean operations (and hence have decidable containment and equivalence problems).

In Section 4 we introduce a new notion of nesting for data languages, based on tree-structured datasets. This notion does not commit all letters to be at the same level of nesting and appears promising from the point of view of modelling scenarios with hierarchical name structure, such as concurrent or higher-order computation. We extend Class Memory Automata to operate over these nested datasets, and show that this extension is Turing-powerful in general, but reintroducing the Weakness constraint recovers decidability. We show how these Nested Data CMA recognise the same string languages as Higher-Order Multicounter

Automata, introduced in [2], and also how the weakness constraint corresponds to a natural weakness constraint on these Higher-Order Multicounter Automata. Finally, we show these automata to be equivalent to the Nested Data Automata introduced in [5].

Related Work. Class memory automata are equivalent to data automata (introduced in [4]), though unlike data automata, they admit a notion of determinism. Data automata (and hence class memory automata) were shown in [4] to be equiexpressive with the two-variable fragment of existential monadic second order logic over data words. Temporal logics have also been studied over data words [6], and the introduction of locally prefix-closed data automata and of nested data automata in [5] is motivated by extensions to BD-LTL, a form of LTL over multiple data values introduced in [10].

Fresh register automata [17] are a precursor to the History Register automata [18] which we examine a restriction of in this paper. Class counting automata, which we show to be equivalent to weak CMA, have been extended to be as expressive as CMA by adding resets and counter acceptance conditions [11,12].

We note that our restriction of class memory automata, which we call weak class memory automata, sound similar to the weak data automata introduced in [9]. However, these are two quite different restrictions, with emptiness problems of different complexities, and the two automata models should not be confused.

In the second part of this paper we examine automata over nested data values. First-order logic over nested data values has been studied in [3], where it was shown that the $<$ predicate quickly led to undecidability, but that only having the $+1$ predicate preserved decidability. They also examined the link between nested data and shuffle expressions. In [5] Decker et al. introduced ND-LTL, extending BD-LTL to nested data values. To show decidability of certain fragments of ND-LTL they extended data automata to run over nested data values, giving the nested data automata we examine in this paper.

Further Work. We would like to understand better whether there is a natural fragment of the π-calculus that corresponds to the new classes of automata. On the logical side, an interesting outstanding question is to characterize languages accepted by our classes of automata with suitable logics.

2   Preliminaries

Let Σ be a finite alphabet, and $\mathcal{D}$ an infinite set of data values. A *data alphabet*, $\mathbb{D}$, is of the form $\Sigma \times \mathcal{D}$. The set of finite data words over $\mathbb{D}$ is denoted $\mathbb{D}^*$.

Class Memory Automata and Data Automata. Given a set S, we write $S_\perp$ to mean $S \cup \{\perp\}$, where $\perp$ is a distinguished symbol (representing a fresh data value). A *Class Memory Automaton* [3] is a tuple $\langle Q, \Sigma, q_I, \delta, F_L, F_G \rangle$ where Q is a finite set of states, Σ is a finite alphabet, $q_I \in Q$ is the initial state, $F_G \subseteq F_L \subseteq Q$ are sets of globally- and locally-accepting sets (respectively), and δ is the transition map $\delta : Q \times \Sigma \times Q_\perp \to \mathcal{P}(Q)$. The automaton is deterministic if each set in the image of the transition function is a singleton. A *class memory function* is a map $f : \mathcal{D} \to Q_\perp$ such that $f(d) \neq \perp$ for only finitely many $d \in \mathcal{D}$.

We view f as a record of the history of computation: it holds the state of the automaton after the data value d was last read, where $f(d) = \bot$ means that d is fresh. A configuration of the automaton is a pair (q, f) where $q \in Q$ and f is a class memory function. The initial configuration is (q_0, f_0) where $f_0(d) = \bot$ for every $d \in \mathcal{D}$. Suppose $(a, d) \in \Sigma \times \mathcal{D}$ is the input. The automaton can transition from configuration (q, f) to configuration (q', f') just if $q' \in \delta(q, a, f(d))$ and $f' = f[d \mapsto q']$. A data word w is *accepted* by the automaton just if the automaton can make a sequence of transitions from the initial configuration to a configuration (q, f) where $q \in F_G$ and $f(d) \in F_L \cup \{\bot\}$ for every data value d.

A *Data Automaton* [4] is a pair $(\mathcal{A}, \mathcal{B})$ where $\mathcal{A}$ is a letter-to-letter string transducer with output alphabet Γ, called the Base Automaton, and $\mathcal{B}$ is a NFA with input alphabet Γ, called the Class Automaton. A data word $w = w_1 \ldots w_n \in \mathbb{D}^*$ is accepted by the automaton if there is a run of $\mathcal{A}$ on the string-projection of w (to Σ) with output $b_1 \ldots b_n$ such that for each maximal set of positions $\{x_1, \ldots, x_k\} \subseteq \{1, \ldots, n\}$ such that $w_{x_1}, \ldots, w_{x_k}$ share the same data value, the word $b_{x_1} \ldots b_{x_k}$ is accepted by $\mathcal{B}$.

CMA and DA are expressively equivalent, with PTIME translation [3]. The emptiness problem for these automata is decidable, and equivalent to Petri Net Reachability [4]. The class of languages recognised by CMA is closed under intersection, union, and concatenation. It is not closed under complementation or Kleene star. Of the above, the class of languages recognised by deterministic CMA is closed only under intersection.

Locally Prefix-Closed Data Automata. A Data Automaton $\mathcal{D} = (\mathcal{A}, \mathcal{B})$ is locally prefix-closed (pDA) [5] if all states in $\mathcal{B}$ are final. The emptiness problem for pDA is EXPSPACE-complete [5].

Class Counting Automata. A *bag* over $\mathcal{D}$ is a function $h : \mathcal{D} \to \mathbb{N}$ such that $h(d) = 0$ for all but finitely many $d \in \mathcal{D}$. Let $C = \{=, \neq, <, >\} \times \mathbb{N}$, which we call the set of constraints. If $c = (\mathrm{op}, e) \in C$ and $n \in \mathbb{N}$ we write $n \vDash c$ iff $n \mathrm{\,op\,} e$. A *Class Counting Automaton* (CCA) [12] is a tuple $\langle Q, \Sigma, \Delta, q_0, F \rangle$ where Q is a finite set of states, Σ is a finite alphabet, q_0 is the initial state, $F \subseteq Q$ is the set of accepting states, and Δ, the transition relation, is a finite subset of $Q \times \Sigma \times C \times \{\uparrow^+, \downarrow\} \times \mathbb{N} \times Q$. A configuration of a CCA, $\mathcal{C} = \langle Q, \Sigma, \Delta, q_0, F \rangle$, is a pair (q, h) where $q \in Q$ and h is a bag. The initial configuration is (q_0, h_0) where h_0 is the zero function. Given a data word $w = (a_1, d_1)(a_2, d_2) \ldots (a_n, d_n)$ a run of w on $\mathcal{C}$ is a sequence of configurations $(q_0, h_0)(q_1, h_1) \ldots (q_n, h_n)$ such that for all $0 \leq i < n$ there is a transition (q, a, c, π, m, q') where $q = q_i$, $q' = q_{i+1}$, $a = a_{i+1}$, $h_i(d_{i+1}) \vDash c$, and $h_{i+1} = h_i[d_{i+1} \mapsto h_i(d_{i+1}) + m]$ if $\pi = \uparrow^+$ or $h_{i+1} = h_i[d_{i+1} \mapsto m]$ if $\pi = \downarrow$ The run is accepting if $q_n \in F$. The emptiness problem for Class Counting Automata was shown to be EXPSPACE-complete in [12].

Non-Reset History Register Automata. For a positive integer k write $[k]$ for the set $\{1, 2 \ldots, k\}$. Fixing a positive integer m, define the set of labels $\mathsf{Lab} = \mathcal{P}([m])^2$. A non-reset History Register Automaton (nrHRA) of type m with initially empty assignment is a tuple $\mathcal{A} = \langle Q, \Sigma, \delta, q_0, F \rangle$ where $q_0 \in Q$ is the initial state, $F \subseteq Q$ is the set of final states, and $\delta \subseteq Q \times \Sigma \times \mathsf{Lab} \times Q$. A configuration of $\mathcal{A}$ is a pair (q, H) where $q \in Q$ and $H : [m] \to$

$\mathcal{P}_{fn}(\mathcal{D})$ where $\mathcal{P}_{fn}(\mathcal{D})$ is the set of finite subsets of $\mathcal{D}$. We call H an *assignment*, and for $d \in \mathcal{D}$ we write $H^{-1}(d)$ for the set $\{i \in [m] : d \in H(i)\}$. The initial configuration is (q_0, H_0), where H_0 assigns every integer in $[m]$ to the empty set. When the automaton is in configuration (q, H), on reading input (a, d) it can transition to configuration (q', H') providing there exists $X \subseteq [m]$ such that $(q, a, (H^{-1}(d), X), d) \in \delta$ and H' is obtained by removing d from $H(i)$ for each i then adding d to each $H(i)$ such that $i \in X$. A run is defined in the usual way, and a run is accepting if it ends in a configuration (q, H) where $q \in F$.

Higher-Order Multicounter Automata. A multiset over a set A is a function $m : A \to \mathbb{N}$. A level-1 multiset over A is a finite multiset over A. A level-$(k+1)$ multiset over A is a finite multiset of level-k multisets over A. We can visualise this with nested set notation: e.g. $\{\{a, a\}, \{\}, \{\}\}$ represents the level-2 multiset containing one level-1 multiset containing two copies of a, and two empty level-1 multisets. A multiset is *hereditarily empty* if, written in nested set notation, it contains no symbols from A.

Higher-Order Multicounter Automata (HOMCA) were introduced in [2], and their emptiness problem was shown to be Turing-complete at level-2 and above. A level-k multicounter automaton is a tuple $\langle Q, \Sigma, A, \Delta, q_0, F \rangle$ where Q is a finite set of states, Σ is the input alphabet, A is the multiset alphabet, q_0 is the initial state, and F is the set of final states. A configuration is a tuple $(q, m_1, m_2, \ldots, m_k)$ where $q \in Q$ and each m_i is either undefined ($\perp$) or a level-i multiset over A. The initial configuration is $(q_0, \perp, \ldots, \perp)$. Δ is the transition relation, and is a subset of $Q \times \Sigma \times ops \times Q$ where ops is the set of possible counter operations. These operations, and meanings, are as follows: (i) new_i ($i \leq k$) turns m_i from $\perp$ into the empty level-i multiset; (ii) inc_a ($a \in A$) adds a to m_1; (iii) dec_a ($a \in A$) removes a from m_1; (iv) $store_i$ ($i < k$) adds m_i to m_{i+1} and sets m_i to $\perp$; (v) $load_i$ ($i < k$) non-deterministically removes an m from m_{i+1} and turns m_i from $\perp$ to m. This can happen only when $m_1 \ldots m_i$ are all $\perp$. The automaton reads the input word from left to right, updating $m_1 \ldots m_k$ as determined by the transitions. A word is accepted by the automaton just if there is a run of the word such that the automaton ends up in configuration $(q, m_1, \ldots, m_k)$ where $q \in F$ and each m_i is hereditarily empty.

3   Weak Class Memory Automata

In this section we introduce a restriction of class memory automata, *weak* class memory automata (WCMA), that have improved closure and complexity properties. We show that WCMA correspond to a natural restriction of data automata, locally-prefix closed data automata, as well as two other independent automata models, class counting automata and non-reset history register automata.

Definition 1. *A class memory automaton $\langle Q, \Sigma, \Delta, q_0, F_L, F_G \rangle$ is weak if all states are locally accepting (i.e. $F_L = Q$).*

When defining a weak CMA (WCMA) we may omit the set of locally accepting states, and just give one set of final states, F.

The emptiness problem for class memory automata is reducible (in fact, equivalent) to emptiness of multicounter automata (MCA) [3,4]. This reduction works by using counters to store the number of data values last seen in each state. The local-acceptance condition is checked by the zero-test of each counter at the end of a run of an MCA. In the weak CMA case, this check is not necessary, and so emptiness is reducible to emptiness of weak MCA. Just as MCA emptiness is equivalent to Petri net reachability, weak MCA emptiness is equivalent to Petri net coverability.

Example 2. We give an example showing how a very simple Petri net reachability query can be reduced to an emptiness of CMA problem, and the small change required to reduce coverability queries to emptiness of WCMA. The idea is to encode tokens in the Petri net using data values: the location of the token is stored by the class memory function's memory for the data value. Transitions in the Petri net will be simulated by sequences of transitions in the automaton, which change class memory function appropriately. Consider the Petri net shown in Figure 1, with initial marking above and target marking below.

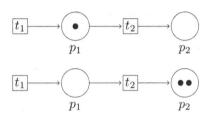

Fig. 1. An example Petri net with initial and target markings

We give the automaton which models this reachability query in Figure 2. The first transitions from the initial state just set up the initial marking. As there is only one token in the initial marking, this just involves reading one fresh data value: this is the transition from s_0 to s_1 below. Once the initial marking has been set up (reaching s_2 below), the automaton can simulate the transitions firing any number of times. Each loop from s_2 back to itself represents one transition in the Petri net firing: the loop above represents t_1 firing, and the loop below represents t_2 firing. For t_1 to fire, no preconditions must be met, and a new data value can be read in state s_3, thus data values last seen in either of states s_1 and s_3 represent tokens in p_1. For t_2 to fire, a token must be removed from p_1, since tokens in p_1 are represented by tokens in either s_1 or s_3, the first transition in this loop – to s_4 – involves reading a data value last seen in one of these states. Thus data values seen in s_4 represent removed tokens, which we do not use again. Then a new token is placed in p_2 by

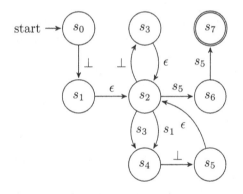

Fig. 2. A class memory automaton simulating the Petri net query shown in Figure 1

reading a fresh data value in s_5. Once back in s_2 these loops can be taken more, or the fact that a marking covering the target marking has been reached can be checked by reading two data values last seen in s_5 to reach a final state. The only globally accepting state is s_7, and all states except those which are used to represent tokens – i.e. all except s_1, s_3, and s_5 – are locally accepting. The local acceptance condition thus checks that no other tokens remain in the simulated Petri net.

If we were interested in a coverability query, the same automaton, but without the local acceptance condition, would obviously suffice. Thus emptiness of WCMA is equivalent to Petri net coverability, which is ExpSpace-complete.

We now give the main observation of this section: that weak CMA are equivalent to three independent existing automata models.

Theorem 3. *Weak CMA, locally prefix-closed DA, class counting automata, and non-reset history register automata are all* PTime-*equivalent.*

Proof. That Weak CMA and pDA are equivalent is a simple alteration of the proof of equivalence of CMA and DA provided in [3].

Recall that CCA use a "bag", which essentially gives a counter for each data value. Weak CMA can easily be simulated by CCA by identifying each state with a natural number; then the bag can easily simulate the class memory function, by setting the data value's counter to the appropriate number when it is read. To simulate a CCA with a WCMA, we first observe that for any CCA, since counter values can only be incremented or reset, there is a natural number, N, above which different counter values are indistinguishable to the automaton. Thus we need only worry about a finite set of values. This means the value for the counter of each data value can be stored in the automaton state, and thereby the class memory function.

In [18] the authors already show that nrHRA can be simulated by CMA. Their construction does not make use of the local-acceptance condition, so the fact that nrHRA can be simulated by WCMA is immediate. In order to simulate a given WCMA with state set $[m]$, one can take a nrHRA of type m, with place i storing the data values last seen in state i.

Data automata, and hence pDA, unlike CMA and WCMA, do not have a natural notion of determinism, nor a natural restriction corresponding to deterministic CMA or WCMA. What about for CCA? We define CCA to be deterministic if for each state q and input letter a, the transitions $(q, a, c, \dots) \in \Delta$ are such that the c's partition $\mathbb{N}$. The same translations given in Theorem 3 also show that deterministic WCMA and deterministic CCA are equivalent. We can ask the same question of non-reset HRA. We find that the natural notion of determinism here is: for each $q \in Q, a \in \Sigma$, and $X \subseteq [m]$ there is precisely one $Y \subseteq [m]$ and $q' \in Q$ such that $(q, a, (X, Y), q') \in \delta$. Similarly, the translations discussed above show deterministic WCMA to be equivalent to deterministic nrHRA.

It follows from the results for CCA in [12] that Weak CMA, like normal CMA, are closed under intersection and union, though these closures can easily be shown directly using product constructions (and these constructions preserve determinism). In fact, Deterministic Weak CMA have even nicer closure properties: the language recognised can be complemented using the same method as for DFA: complementing the final states.

Proposition 4. *Deterministic Weak CMA are closed under all Boolean operations.*

Corollary 5. *The containment and equivalence problems for Deterministic Weak CMA are* ExpSpace*-complete.*

4   Nested Data Class Memory Automata

In Section 1 we discussed how data values fail to provide a good model for modelling computations in which names are used hierarchically, such as a system of concurrent processes which can spawn subprocesses. Motivated by these applications, in this section we introduce a notion of nested data values in which the data set has a forest-structure. This is a stylistically different presentation to earlier work on nested data in that [3,5] require that each position in the words considered have a data value in each of a fixed number of levels. By giving the data set a forest-structure, we can explicitly handle variable levels of nesting within a word. However, we note that there is a natural translation between the two presentations.

Definition 6. *A* rooted tree *(henceforth, just* tree*) is a simple directed graph* $\langle D, pred \rangle$*, where* $pred : D \rightharpoonup D$ *is the predecessor map defined on every node of the tree except the root, such that every node has a unique path to the root. A node* n *of a tree has* level l *just if* $pred^{l-1}(n)$ *is the root (thus the root has level 1). A tree has* bounded level *just if there exists a least* $l \geq 1$ *such that every node has level no more than* l*; we say that such a tree has* level l*.*

We define a nested dataset $\langle \mathcal{D}, pred \rangle$ *to be a forest of infinitely many trees of level* l *which is* full *in the sense that for each data value* d *of level less than* l*,* d *has infinitely many children (i.e. there are infinitely many data values* d' *s.t.* $pred(d') = d$*).*

We now extend CMA to nested data by allowing the nested data class memory automaton (NDCMA), on reading a data value d, to access the class memory function's memory of not only d, but each ancestor of d in the nested data set. Once a transition has been made, the class memory function updates the remembered state not only of d, but also of each of its ancestors. Formally:

Definition 7. *Fix a nested data set of level* l*. A* Nested Data CMA of level l *is a tuple* $\langle Q, \Sigma, \delta, q_0, F_L, F_G \rangle$ *where* Q *is a finite set of states,* $q_0 \in Q$ *is the initial state,* $F_G \subseteq F_L \subseteq Q$ *are sets of globally and locally accepting states respectively, and* δ *is the transition map.* δ *is given by a union* $\delta = \bigcup_{1 \leq i \leq l} \delta_i$ *where each* δ_i *is a function:*

$$\delta_i : Q \times \Sigma \times (\{i\} \times (Q_\perp)^i) \to \mathcal{P}(Q)$$

The automaton is deterministic *if each set in the image of δ is a singleton; and is* weak *if $F_L = Q$. A* configuration *is a pair (q, f) where $q \in Q$, and $f : \mathcal{D} \rightarrow Q_\perp$ is a* class memory function *(i.e. $f(d) = \perp$ for all but finitely many $d \in \mathcal{D}$). The* initial configuration *is (q_0, f_0) where f_0 is the class memory function mapping every data value to $\perp$. A configuration (q, f) is* final *if $q \in F_G$ and $f(d) \in F_L \cup \{\perp\}$ for all $d \in \mathcal{D}$. The automaton can transition from configuration (q, f) to configuration (q', f') on reading input (a, d) just if d is a level-i data value, $q' \in \delta(q, a, (i, f(pred^{i-1}(d)), \ldots, f(pred(d)), f(d)))$, and $f' = f[d \mapsto q, pred(d) \mapsto q, \ldots, pred^{i-1}(d) \mapsto q]$. A* run *$(q_0, f_0), (q_1, f_1), \ldots, (q_n, f_n)$ is* accepting *if the configuration (q_n, f_n) is final. $w \in L(\mathcal{A})$ if there is an accepting run of $\mathcal{A}$ on w.*

It is clear that level-1 NDCMA are equivalent to normal CMA. We know that emptiness of class memory automata is equivalent to reachability in Petri nets; it is natural to ask whether there is any analogous correspondence – to some kind of high-level Petri net – once nested data is used.

Example 8. In Example 2, we showed how CMA (resp. WCMA) can encode Petri net reachability (resp. coverability). A similar technique allows reachability (resp. coverability) of Petri nets with reset arcs to be reduced to emptiness of NDCMA (resp. weak NDCMA). The key idea is to have, for each place in the net, a level-1 data value – essentially as a "bag" holding the tokens for that place. Nested under the level-1 data value, level-2 data values are used to represent tokens just as before. When a reset arc is fired, the corresponding level-1 data value is moved to a "dead" state – from where it and the data values nested under it are not moved again – and a fresh level-1 data value is then used to hold subsequently added tokens to that place.

Theorem 9. *The emptiness problem for NDCMA is undecidable. Emptiness of Weak NDCMA is decidable, but Ackermann-hard.*

Proof. This result follows from Theorem 12 together with results in [5], though we also provide a direct proof.

We show decidability by reduction to a well-structured transition system [8] constructed as follows: a class memory function on a nested data set can be viewed as a labelling of the data set by labels from the set of states. Since we only care about the shape of the class memory function (i.e. up to automorphisms of the nested data set), we can remove the nodes labelled by $\perp$, and view a class memory function as a finite set of labelled trees. The set of finite forests of finite trees of bounded depth with the order given by $F \leq F'$ iff there is a forest homomorphism from F to F' (where a forest is the natural extension of tree homomorphisms to forests) is a well-quasi-order [7], which provides the basis for the well-structured transition system.

Undecidability for NDCMA and Ackermann-hardness for Weak NDCMA follow from the ideas in Example 8: the reachability (resp. coverability) problem for Petri nets with reset arcs is encodable in NDCMA (resp. Weak NDCMA), and this is undecidable [1] (resp. Ackermann-hard [15]).

Weak Nested Data CMA have similar closure properties to Weak CMA (and this can be shown using the same techniques as for DFA).

Proposition 10.   *(i) Weak NDCMA are closed under intersection and union. (ii) Deterministic Weak NDCMA are closed under all Boolean operations.*

Hence, as for weak CMA, the containment and equivalence problems for Deterministic Weak NDCMA are decidable.

4.1   Link with Nested Data Automata

In [5] Decker et al. also examined "Nested Data Automata" (NDA), and showed the locally prefix-closed NDA (pNDA) to have decidable emptiness (via reduction to well-structured transition systems). In fact, these NDA precisely correspond to NDCMA, and again being locally prefix-closed corresponds to weakness. In this section we briefly outline this connection.

   Nested Data Automata. ([5]) A k-nested data automaton (k-NDA) is a tuple $(\mathcal{A}, \mathcal{B}_1, \ldots, \mathcal{B}_k)$ where $(\mathcal{A}, \mathcal{B}_i)$ is a data automaton for each i. Such automata run on words over the alphabet $\Sigma \times \mathcal{D}^k$, where $\mathcal{D}$ is a (normal, unstructured) dataset. As for normal data automata, the transducer, $\mathcal{A}$, runs on the string projection of the word, giving output w. Then for each i the class automaton $\mathcal{B}_i$ runs on each subsequence of w corresponding to the positions which agree on the first i data values. The NDA is locally prefix-closed if each $(\mathcal{A}, \mathcal{B}_i)$ is.

   Since these NDA are defined on a slightly different presentation of nested data, we provide the following presentation of NDCMA over multiple levels of data.

Definition 11. *A Nested Data CMA of level k over the alphabet $\Sigma \times \mathcal{D}^k$ is a tuple $\langle Q, \Sigma, \delta, q_0, F_L, F_G \rangle$ where Q is a finite set of states, $q_0 \in Q$, $F_G \subseteq F_L \subseteq Q$, and $\delta : Q \times \Sigma \times (Q_\perp)^k \to \mathcal{P}(Q)$ is the transition map.*

   A configuration is a tuple $(q, f_1, f_2, \ldots, f_k)$, where each $f_i : \mathcal{D}^i \to Q_\perp$ maps an i-tuple of data values to a state in the automaton (or $\perp$). The initial configuration is $(q_0, f_1^0, \ldots, f_k^0)$ where f_i^0 maps every tuple in the domain to $\perp$. A configuration $(q, f_1, \ldots, f_k)$ is final if each f_i maps into $F_L \cup \{\perp\}$. The automaton can transition from configuration $(q, f_1, \ldots, f_k)$ to configuration $(q', f_1', \ldots, f_k')$ on reading input $(a, d_1, \ldots, d_k)$ just if $q' \in \delta(q, a, (f_1(d_1), f_2(d_1, d_2), \ldots, f_k(d_1, \ldots, d_k)))$, and each $f_i' = f_i[(d_1, \ldots, d_i) \mapsto q']$.

Using ideas from the proof of equivalence between CMA and DA in [3], we can show the following result:

Theorem 12. *NDCMA (resp. weak NDCMA) and NDA (resp. pNDA) are expressively equivalent, with effective translations.*

4.2   Link with Higher-Order Multicounter Automata

In [2] the authors examined a link between nested data values and shuffle expressions. In doing so, they introduced higher-order multicounter automata

(HOMCA). While not explicitly over nested data values, they are closely related to the ideas involved, and we show that, just as multicounter automata and CMA are equivalent, there is a natural translation between HOMCA and the NDCMA we have introduced. Further, just as the equivalence between MCA and CMA descends to one between weak multicounter automata and weak CMA, we find an equivalence between weak NDCMA and "weak" HOMCA in which

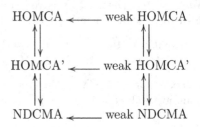

Fig. 3. Translations between HOMCA, HOMCA', NDCMA, and their weak counterparts

the corresponding acceptance condition – hereditary emptiness – is dropped. To show this, we introduce HOMCA', which add restrictions to the $store_i$ and new_i counter operations analogous to the restriction for the $load_i$ operation. We show that these HOMCA' are equivalent to HOMCA, and that HOMCA' are equivalent to NDCMA, with both of these equivalences descending to the weak versions. These equivalences are summarised in Figure 3.

Definition 13. *We define* weak *HOMCA to be just as HOMCA, but without the hereditary-emptiness condition on acceptance, i.e. a run is accepting just if it ends in a final state.*

Definition 14. *We define HOMCA' to be the same as HOMCA, except for the following changes to the counter operations: (i) $store_i$ operations are only enabled when $m_1 = m_2 = \cdots = m_{i-1} = \bot$; and (ii) new_i operations are only enabled when $m_k \neq \bot, m_{k-1} \neq \bot, \ldots, m_{i+1} \neq \bot$ and $m_{i-1} = m_{i-2} = \cdots = m_1 = \bot$.*

As for HOMCA, we define weak *HOMCA' to be HOMCA' without the hereditary-emptiness condition.*

This means that each reachable configuration $(q, m_1, \ldots, m_k)$ is such that there is a unique $0 \leq i \leq k$ such that for all $j \leq i$, $m_j = \bot$ and each $l > i$, $m_l \neq \bot$.

Theorem 15. *HOMCA (respectively weak HOMCA) and HOMCA' (resp. weak HOMCA') are expressively equivalent, with effective translations between them.*

Proof. This requires simulating the HOMCA operations $store_i$ and new_i in HOMCA': which can be difficult if, for instance, the HOMCA is carrying out a $store_i$ operation when it has a current level-$(i-1)$ multiset in memory. The trick is to move the level-$(i-1)$ multiset across to be nested under a new level-i multiset, and this can be done one element at a time in a "folding-and-unfolding" method. The hereditary emptiness condition checks that the each of these movements was completed, i.e. no element was left unmoved. In the weak case some elements not being moved could not change an accepting run to a non-accepting run, so the fallibility of the moving method does not matter.

Theorem 16. *For every (weak) level-k NDCMA, $\mathcal{A}$, there is a (weak) level-k HOMCA', $\mathcal{A}'$, such that $\mathcal{L}(\mathcal{A}')$ is equal to the Σ-projection of $\mathcal{L}(\mathcal{A})$, and vice-versa.*

Proof. This proof rests on the strong similarity between the nesting of data values, and the nesting of level-i multisets in level-$(i+1)$ multisets. For a NDCMA to simulate a HOMCA', we use level-k data values to represent instances of the multiset letters, level-$(k-1)$ data values to represent level-1 multisets, and so on, up to level-1 data values representing level-$(k-1)$ multisets. Since each run of a HOMCA' can have at most one level-k multiset, this does not need to be encoded in data values. Conversely, when simulating a NDCMA with a HOMCA', a level-k data value is represented by an instance of an appropriate multiset letter. The letter contains the information on which state the data value was last seen in.

References

1. Araki, T., Kasami, T.: Some decision problems related to the reachability problem for Petri nets. Theor. Comput. Sci. **3**(1), 85–104 (1976)
2. Björklund, H., Bojańczyk, M.: Shuffle Expressions and Words with Nested Data. In: Kučera, L., Kučera, A. (eds.) MFCS 2007. LNCS, vol. 4708, pp. 750–761. Springer, Heidelberg (2007)
3. Björklund, H., Schwentick, T.: On notions of regularity for data languages. Theor. Comput. Sci. **411**(4–5), 702–715 (2010)
4. Bojanczyk, M., Muscholl, A., Schwentick, T., Segoufin, L., David, C.: Two-variable logic on words with data. In: LICS. pp. 7–16. IEEE Computer Society (2006)
5. Decker, N., Habermehl, P., Leucker, M., Thoma, D.: Ordered Navigation on Multi-attributed Data Words. In: Baldan, P., Gorla, D. (eds.) CONCUR 2014. LNCS, vol. 8704, pp. 497–511. Springer, Heidelberg (2014)
6. Demri, S., Lazic, R.: LTL with the freeze quantifier and register automata. ACM Trans. Comput. Log. 10(3) (2009)
7. Ding, G.: Subgraphs and well-quasi-ordering. J. Graph Theory **16**, 489–502 (1992)
8. Finkel, A., Schnoebelen, P.: Well-structured transition systems everywhere!. Theor. Comput. Sci. **256**(1–2), 63–92 (2001)
9. Kara, A., Schwentick, T., Tan, T.: Feasible Automata for Two-Variable Logic with Successor on Data Words. In: Dediu, A.-H., Martín-Vide, C. (eds.) LATA 2012. LNCS, vol. 7183, pp. 351–362. Springer, Heidelberg (2012)
10. Kara, A., Schwentick, T., Zeume, T.: Temporal logics on words with multiple data values. In: FSTTCS. LIPIcs, vol. 8, pp. 481–492. Schloss Dagstuhl (2010)
11. Manuel, A.: Counter automata and classical logics for data words. Ph.D. thesis, Institute of Mathematical Sciences, Chennai (2011)
12. Manuel, A., Ramanujam, R.: Counting Multiplicity over Infinite Alphabets. In: Bournez, O., Potapov, I. (eds.) RP 2009. LNCS, vol. 5797, pp. 141–153. Springer, Heidelberg (2009)
13. Neven, F., Schwentick, T., Vianu, V.: Finite state machines for strings over infinite alphabets. ACM Trans. Comput. Log. **5**(3), 403–435 (2004)
14. Sangiorgi, D.: Expressing Mobility in Process Algebras: First-Order and Higher-Order Paradigms. Ph.D. thesis, University of Edinburgh (1992)
15. Schnoebelen, P.: Verifying lossy channel systems has nonprimitive recursive complexity. Inf. Process. Lett. **83**(5), 251–261 (2002)
16. Segoufin, L.: Automata and Logics for Words and Trees over an Infinite Alphabet. In: Ésik, Z. (ed.) CSL 2006. LNCS, vol. 4207, pp. 41–57. Springer, Heidelberg (2006)
17. Tzevelekos, N.: Fresh-register automata. In: POPL 2011. pp. 295–306. ACM (2011)
18. Tzevelekos, N., Grigore, R.: History-Register Automata. In: Pfenning, F. (ed.) FOSSACS 2013 (ETAPS 2013). LNCS, vol. 7794, pp. 17–33. Springer, Heidelberg (2013)

Insertion Operations on Deterministic Reversal-Bounded Counter Machines

Joey Eremondi[1], Oscar H. Ibarra[2], and Ian McQuillan[3]($\boxtimes$)

[1] Department of Information and Computing Sciences, Utrecht University,
P.O. Box 80.089, 3508 TB Utrecht, The Netherlands
j.s.eremondi@students.uu.nl
[2] Department of Computer Science, University of California,
Santa Barbara, CA 93106, USA
ibarra@cs.ucsb.edu
[3] Department of Computer Science, University of Saskatchewan,
Saskatoon, SK S7N 5A9, Canada
mcquillan@cs.usask.ca

Abstract. Several insertion operations are studied applied to languages accepted by one-way and two-way deterministic reversal-bounded multicounter machines. These operations are defined by the ideals obtained from relations such as the prefix, infix, suffix and outfix relations. The insertion of regular languages and other languages into deterministic reversal-bounded multicounter languages is also studied. The question of whether the resulting languages can always be accepted by deterministic machines with the same number of turns on the input tape, the same number of counters, and reversals on the counters is investigated. In addition, the question of whether they can always be accepted by increasing either the number of input tape turns, counters, or counter reversals is addressed. The results in this paper form a complete characterization based on these parameters. Towards these new results, we use a technique for simultaneously showing a language cannot be accepted by both one-way deterministic reversal-bounded multicounter machines, and by two-way deterministic machines with one reversal-bounded counter.

Keywords: Automata and logic · Counter machines · Insertion operations · Reversal-bounds · Determinism · Finite automata

1 Introduction

One-way deterministic multicounter machines are deterministic finite automata augmented by a fixed number of counters, which can each be independently increased, decreased or tested for zero. If there is a bound on the number of switches each counter makes between increasing and decreasing, then the

The research of O. H. Ibarra was supported, in part, by NSF Grant CCF-1117708.
The research of I. McQuillan was supported, in part, by the Natural Sciences and Engineering Research Council of Canada.

A.-H. Dediu et al. (Eds.): LATA 2015, LNCS 8977, pp. 200–211, 2015.
DOI: 10.1007/978-3-319-15579-1_15

machine is reversal-bounded [1,8]. The family of languages accepted by one-way deterministic reversal-bounded multicounter machines (denoted by DCM) is interesting as it is more general than regular languages, but still has a decidable emptiness, infiniteness, equivalence, inclusion, universe and disjointness problems [8]. Moreover, these problems remain decidable if the machines operate with two-way input that is finite-crossing in the sense that there is a fixed k such that the number of times the boundary between any two adjacent input cells is crossed is at most k times [4].

Reversal-bounded counter machines (both deterministic and nondeterministic) have been extensively studied. Many generalizations have been investigated, and they have found applications in areas such as verification of infinite-state systems, membrane computing systems, Diophantine equations, etc.

In this paper, we study various insertion operations on deterministic reversal-bounded multicounter languages. Common word and language relations are the prefix, suffix, infix and outfix relations. For example, w is an infix of z, written $w \leq_i z$, if $z = xwy$, for some $x, y \in \Sigma^*$. Viewed as an operation on the first component of the relation, $\leq_i (w) = \{z \mid w \leq_i z, z \in \Sigma^*\}$, which is equal to the set of all words with w as infix, which is $\Sigma^* w \Sigma^*$. If we consider the inverse of this relation, $z \leq_i^{-1} w$, if $z = xwy$, then viewing this as an operation, $\leq_i^{-1} (z) = \{w \mid z \leq_i^{-1} w, w \in \Sigma^*\} = \{w \mid w \leq_i z\}$, the set of all infixes of z. These can be extended to operations on languages. The prefix, suffix, infix and outfix operations can be defined on languages in this way, along with their inverses. This is the approach taken in [10]. Using the more common notation of $\inf(L)$ for the set of infixes of L, then $\inf^{-1}(L) = \Sigma^* L \Sigma^*$, the set of all words having a word in L as an infix. This is the same as what is often called the *two-sided ideal*, or the *infix ideal* [10]. For the suffix operation, $\mathrm{suff}(L) = (\Sigma^*)^{-1} L$, and $\mathrm{suff}^{-1}(L) = \Sigma^* L$, with the latter being called the *left ideal*, or the *suffix ideal*. For prefix, $\mathrm{pref}(L) = L(\Sigma^*)^{-1}$, and $\mathrm{pref}^{-1}(L) = L \Sigma^*$, the *prefix ideal*, or the *right ideal*. The inverse of each operation defines a natural insertion operation.

We will examine the insertion operations defined by the inverse of the prefix, suffix, infix, outfix and embedding relations, and their effects on deterministic reversal-bounded multicounter languages. We will also examine certain standard generalizations of these operations such as left and right concatenation with regular or more general languages. In particular, if we start with a language that can be accepted with a parameterized number of input tape turns, counters, and reversals on the counters, is the result of the various insertion operations always accepted with the same type of machines? And if not, can they always be accepted by increasing either the turns on the input tape, counters, or reversals on the counters? Results in this paper form a complete characterization in this regard, and are summarized in Section 5. Surprisingly, even if we have languages accepted by deterministic 1-reversal bounded machines with either one-way input and 2 counters, or 1 counter and 1 turn on the input, then concatenating Σ^* to the right can result in languages that can neither be accepted by DCM machines (any number of reversal-bounded counters), nor by two-way deterministic reversal-bounded 1-counter machines (2DCM(1), which have no bound on input turns). This is in contrast to deterministic pushdown languages which are closed under right concatenation with regular languages [6]. In addition, concatenating Σ^* to the left of a one-way 1-reversal-bounded one counter machine can create languages that are neither in DCM nor 2DCM(1). Furthermore, as a consequence of the results in this paper, it is evident that the right input end-marker strictly increases the power for even one-way deterministic reversal-bounded multicounter languages when there are at least two counters. This is usually not the case for various classes of one-way machines. To do this, a new mode of acceptance, by *final state without end-marker*, is defined and studied.

Most non-closure results in this paper use a technique that simultaneously shows languages are not in DCM and not in DCM(1). The technique does not rely on any pumping arguments. A similar technique was used in [2] for showing that there is a language accepted by a deterministic pushdown automaton whose stack makes only one reversal (1-reversal DPDA) that cannot be accepted by any one-way nondeterministic reversal-bounded multicounter machine (NCM).

2 Preliminaries

The set of non-negative integers is represented by $\mathbb{N}_0$, and positive integers by $\mathbb{N}$. For $c \in \mathbb{N}_0$, let $\pi(c)$ be 0 if $c = 0$, and 1 otherwise.

We use standard notations for formal languages, referring the reader to [6,7]. The empty word is denoted by λ. We use Σ and Γ to represent finite alphabets, with Σ^* as the set of all words over Σ and $\Sigma^+ = \Sigma^* \setminus \{\lambda\}$. For a word $w \in \Sigma^*$, if $w = a_1 \cdots a_n$ where $a_i \in \Sigma, 1 \le i \le n$, the length of w is denoted by $|w| = n$, and the reversal of w is denoted by $w^R = a_n \cdots a_1$. The number of a's, for $a \in \Sigma$, in w is $|w|_a$. Given a language $L \subseteq \Sigma^*$, the complement of L, $\Sigma^* \setminus L$ is denoted by $\overline{L}$.

Definition 1. *For a language $L \subseteq \Sigma^*$, we define the prefix, inverse prefix, suffix, inverse suffix, infix, inverse infix, outfix and inverse outfix operations, respectively:*

$\mathrm{pref}(L) = \{w \mid wx \in L, x \in \Sigma^*\}$	$\mathrm{pref}^{-1}(L) = \{wx \mid w \in L, x \in \Sigma^*\}$
$\mathrm{suff}(L) = \{w \mid xw \in L, x \in \Sigma^*\}$	$\mathrm{suff}^{-1}(L) = \{xw \mid w \in L, x \in \Sigma^*\}$
$\mathrm{inf}(L) = \{w \mid xwy \in L, x, y \in \Sigma^*\}$	$\mathrm{inf}^{-1}(L) = \{xwy \mid w \in L, x, y \in \Sigma^*\}$
$\mathrm{outf}(L) = \{xy \mid xwy \in L, w \in \Sigma^*\}$	$\mathrm{outf}^{-1}(L) = \{xwy \mid xy \in L, w \in \Sigma^*\}$

We generalize the outfix relation to the notion of embedding [10]:

Definition 2. *The m-embedding of a language $L \subseteq \Sigma^*$ is the following set:* $\mathrm{emb}(L, m) = \{w_0 \cdots w_m \mid w_0 x_1 \cdots w_{m-1} x_m w_m \in L, w_i \in \Sigma^*, 0 \le i \le m, x_j \in \Sigma^*, 1 \le j \le m\}$.

We define the inverse as follows: $\mathrm{emb}^{-1}(L, m) = \{w_0 x_1 \cdots w_{m-1} x_m w_m \mid w_0 \cdots w_m \in L, w_i \in \Sigma^*, 0 \le i \le m, x_j \in \Sigma^*, 1 \le j \le m\}$

Note that $\mathrm{outf}(L) = \mathrm{emb}(L, 1)$ and $\mathrm{outf}^{-1}(L) = \mathrm{emb}^{-1}(L, 1)$.

A language L is called *prefix-free* if, for all words $x, y \in L$, where x is a prefix of y, then $x = y$.

A *one-way k-counter machine* is a tuple $M = (k, Q, \Sigma, \$, \delta, q_0, F)$, where $Q, \Sigma, \$, q_0, F$ are respectively the finite set of states, the input alphabet, the right end-marker, the initial state in Q, and the set of final states, which is a subset of Q. The transition function δ (defined as in [8] except with only a right end-marker since these machines only use one-way inputs) is a mapping from $Q \times (\Sigma \cup \{\$\}) \times \{0, 1\}^k$ into $Q \times \{S, R\} \times \{-1, 0, +1\}^k$, such that if $\delta(q, a, c_1, \dots, c_k)$ contains $(p, d, d_1, \dots, d_k)$ and $c_i = 0$ for some i, then $d_i \geq 0$ to prevent negative values in any counter. The symbols S are R indicate the direction of input tape head movement, either *stay* or *right* respectively. The machine M is *deterministic* if δ is a function. The machine M is *non-exiting* if there are no transitions defined on final states. A *configuration* of M is a $k+2$-tuple $(q, w\$, c_1, \dots, c_k)$ representing the fact that M is in state q, with $w \in \Sigma^*$ still to read as input, and $c_1, \dots, c_k \in \mathbb{N}_0$ are the contents of the k counters. The derivation relation $\vdash_M$ is defined between configurations, where $(q, aw, c_1, \dots, c_k) \vdash_M (p, w', c_1+d_1, \dots, c_k+d_k)$, if $(p, d, d_1, \dots, d_k) \in \delta(q, a, \pi(c_1)), \dots, \pi(c_k))$ where $d \in \{S, R\}$ and $w' = aw$ if $d = S$, and $w' = w$ if $d = R$. We let $\vdash_M^*$ be the reflexive, transitive closure of $\vdash_M$. And, for $m \in \mathbb{N}_0$, let $\vdash_M^m$ be the application of $\vdash_M$ m times. A word $w \in \Sigma^*$ is accepted by M if $(q_0, w\$, 0, \dots, 0) \vdash_M^* (q, \$, c_1, \dots, c_k)$, for some $q \in F$, and $c_1, \dots, c_k \in \mathbb{N}_0$. The language accepted by M, denoted by $L(M)$, is the set of all words accepted by M.

The machine M is l-reversal bounded if, in every accepting computation, the count on each counter alternates between increasing and decreasing at most l times. We will sometimes refer to a multicounter machine as being in $\mathsf{DCM}(k, l)$, if it has k l-reversal bounded counters.

We denote by $\mathsf{NCM}(k, l)$ the family of languages accepted by one-way non-deterministic l-reversal-bounded k-counter machines. We denote by $\mathsf{DCM}(k, l)$ the family of languages accepted by one-way deterministic l-reversal-bounded k-counter machines. The union of the families of languages are denoted by $\mathsf{NCM} = \bigcup_{k,l \geq 0} \mathsf{DCM}(k, l)$ and $\mathsf{DCM} = \bigcup_{k,l \geq 0} \mathsf{DCM}(k, l)$.

Given a DCM machine $M = (k, Q, \Sigma, \$, \delta, q_0, F)$, the language accepted by *final state without end-marker* is the set of words w such that $(q_0, w\$, 0, \dots, 0) \vdash_M^* (q', a\$, c'_1, \dots, c'_k) \vdash_M (q, \$, c_1, \dots, c_k)$, for some $q \in F$, $q' \in Q$, $a \in \Sigma$, $c_i, c'_i \in \mathbb{N}_0, 1 \leq i \leq k$. Such a machine does not "know" when it has reached the end-marker $\$$. The state that the machine is in when the last letter of input from Σ is consumed entirely determines acceptance or rejection. It would be equivalent to require $(q_0, w, 0, \dots, 0) \vdash_M^* (q, \lambda, c_1, \dots, c_k)$, for some $q \in F$, but we continue to use $\$$ for compatibility with the end-marker definition. We use $\mathsf{DCM_{NE}}(k, l)$ to denote the family of languages accepted by these machines when they have k counters that are l-reversal-bounded. We define $\mathsf{DCM_{NE}} = \bigcup_{k,l \geq 0} \mathsf{DCM_{NE}}(k, l)$.

We denote by $\mathsf{2DCM}(1)$ to be the family of languages accepted by two-way deterministic finite automata (with both a left and right input tape end-marker) augmented by one reversal-bounded counter, accepted by final state. A machine of this form is said to be *finite-crossing* if there is a bound on the number of changes of direction on the input tape, and *t-crossing* if it makes at most t changes of direction on the input tape for every computation.

3   Closure for Insertion and Concatenation Operations

Closure under concatenation is difficult for DCM languages because of determinism. However, we show special cases where closure results can be obtained. Additionally, we study the necessity of an end-of-tape marker, showing that it makes DCM languages strictly more powerful, but adding no power to $DCM(1, l)$ languages. To our knowledge, the necessity of the right end-marker for one-way deterministic reversal-bounded multicounter machines has not been documented.

To show that the end-marker is not necessary for $DCM(1, l)$, the proof of the lemma below takes an arbitrary $DCM(1, l)$ machine M and builds another M' that accepts by final state without end-marker and accepts the same language. Before building M', the construction builds an NCM machine for every state q of M. This machine accepts all words of the form a^i where there exists some word x (this word is guessed using nondeterminism) such that M can read x from state q and i on the counter and reach a final state. Although these languages use nondeterminism, they are unary, and all NCM languages are semilinear [8], and all unary semilinear languages are regular [6]. Therefore, a DFA can be build for each such language (for each state of M). Because these languages are unary, the structure of the DFAs are well-known [12]. Every unary DFA is isomorphic to one with states $\{0, \ldots, m-1\}$ where there exists some state k, and a transition from i to $i+1$, for all $0 \leq i < k$ (the "tail"), and a transition from j to $j+1$ for all $k \leq j < m-1$, plus a transition from m to k (the "loop"), and no other transitions. Let t be the maximum tail size, over all DFAs constructed, plus one.

Then, intuitively, the construction of M' involves M' simulating M, and after reading input w, if M has counter value c, M' has counter value $c - t$ if $c > t$, with t stored in the finite control. If $c \leq t$, then M' stores c in the finite control with zero on the counter. This allows M' to know what counter value M would have after reading a given word, but also to know when the counter value is less than t (and the specific value less than t). In the finite control, M' simulates each DFA in parallel. To do this, each time M increases the counter, from i to $i+1$, the state of each DFA switches forward by one letter. Each time M decreases the counter from i to $i-1$, the state of each DFA changes deterministically "going backwards in the loop" if $i > t$, and if $i \leq t$, then the counter of M is stored in the finite control, and thus each DFA can tell when to switch deterministically from loop to tail. Then, when in state q of M, M' can tell if the current counter value would lead to acceptance from q using the appropriate DFA.

The proof is omitted due to space constraints, and can be found online in [3].

Lemma 3. *For any* l, $DCM(1, l) = DCM_{NE}(1, l)$.

We will extend these closure results with a lemma about prefix-free DCM_{NE} languages. It was shown in [5] that a regular language is prefix-free if and only if there is a non-exiting DFA accepting the language. While we omit the proof (see [3]), the same logic gives this result for DCM_{NE} languages.

Lemma 4. *Let* $L \in DCM_{NE}$. *Then* L *is prefix-free if and only if there exists a* DCM-*machine* M *accepting* L *by final state without end-marker which is non-exiting.*

From this, we obtain a special case where DCM is closed under concatenation, if the first language can be both accepted by final state without end-marker, and is prefix-free. The construction considers a non-exiting machine accepting L_1 by final state without end-marker, where transitions into its final state are replaced by transitions into the initial state of the machine accepting L_2. The proof is omitted due to space constraints, and can be found online in [3].

Proposition 5. *Let $L_1 \in \mathsf{DCM_{NE}}(k, l), L_2 \in \mathsf{DCM}(k', l')$, with L_1 prefix-free. Then $L_1 L_2 \in \mathsf{DCM}(k + k', \max(l, l'))$.*

If we remove the condition that L_1 is prefix-free however, the proposition is no longer true, as we will see in the next section that even the regular language Σ^* (which is in $\mathsf{DCM_{NE}}(0, 0)$) concatenated with a DCM language produces a language outside DCM.

Corollary 6. *Let $L \in \mathsf{DCM}(k, l), R \in \mathsf{REG}$, where R is prefix-free. Then $RL \in \mathsf{DCM}(k, l)$.*

In contrast to left concatenation of a regular language with a DCM language (Corollary 6), where it is required that R be prefix-free (the regular language is always in $\mathsf{DCM_{NE}}$), for right concatenation, it is only required that it be a $\mathsf{DCM_{NE}}$ language. We will see in the next section that this is not true if the restriction that L accepts by final state without end-marker is removed.

The following proof takes a DCM machine M_1 accepting by final state without end-marker, and M_2 a DFA accepting R, and builds a DCM machine M' accepting LR by final state without end-marker. Intuitively, M' simulates M_1 while also storing a subset of M_2's states in a second component of the states. Every time it reaches a final state of M_1, it places the initial state of M_2 in the second component. Then, it continues to simulate M_1 while in parallel simulating the DFA M_2 separately on every state in the second component.

Proposition 7. *Let $L \in \mathsf{DCM_{NE}}(k, l)$, $R \in \mathsf{REG}$. Then $LR \in \mathsf{DCM_{NE}}(k, l)$. Also, $\mathrm{pref}^{-1}(L) \in \mathsf{DCM_{NE}}(k, l)$.*

As a corollary, we get that $\mathsf{DCM}(1, l)$ is closed under right concatenation with regular languages. This corollary could also be inferred from the proof in [6] that deterministic context-free languages are closed under concatenation with regular languages.

Corollary 8. *Let $L \in \mathsf{DCM}(1, l)$ and $R \in \mathsf{REG}$. Then $LR \in \mathsf{DCM}(1, l)$.*

Corollary 9. *If $L \in \mathsf{DCM}(1, l)$, then $\mathrm{pref}^{-1}(L) \in \mathsf{DCM}(1, l)$.*

4   Relating (Un)Decidable Properties to Non-closure Properties

In this section, we use a technique that proves non-closure properties using (un)decidable properties. A similar technique was used in [2] for showing that there is a language accepted by a 1-reversal DPDA that cannot be accepted by any NCM. In particular, we use this technique to prove that some languages are not accepted by 2DCM(1)s (i.e., two-way DFAs with one reversal-bounded counter). Since 2DCM(1)s have two-way input and a reversal-bounded counter, it does not seem easy to derive "pumping" lemmas for these machines. 2DCM(1)s are quite powerful, e.g., although the Parikh map of the language accepted by any finite-crossing 2NCM (hence by any NCM) is semilinear [8], 2DCM(1)s can accept non-semilinear languages. For example, $L_1 = \{a^i b^k \mid i, k \geq 2, i \text{ divides } k\}$ can be accepted by a 2DCM(1) whose counter makes only one reversal. However, it is known that $L_2 = \{a^i b^j c^k \mid i, j, k \geq 2, k = ij\}$ cannot be accepted by a 2DCM(1) [9].

We will need the following result (the proof for DCMs is in [8]; the proof for 2DCM(1)s is in [9]):

Theorem 10

1. *The class of languages accepted by* DCMs *is closed under Boolean operations. Moreover, the emptiness problem is decidable.*
2. *The class of languages accepted by* 2DCM(1)s *is closed under Boolean operations. Moreover, the emptiness problem is decidable.*

We note that the emptiness problem for 2DCM(2)s, even when restricted to machines accepting only letter-bounded languages (i.e., subsets of $a_1^* \cdots a_k^*$ for some $k \geq 1$ and distinct symbols $a_1, \ldots, a_k$) is undecidable [8].

We will show that there is a language $L \in$ DCM(1,1) such that $\inf^{-1}(L)$ is not in DCM $\cup$ 2DCM(1).

The proof uses the fact that that there is a recursively enumerable language $L_{re} \subseteq \mathbb{N}_0$ that is not recursive (i.e., not decidable) which is accepted by a deterministic 2-counter machine [11]. Thus, the machine when started with $n \in \mathbb{N}_0$ in the first counter and zero in the second counter, eventually halts (i.e., accepts $n \in L_{re}$).

A close look at the constructions in [11] of the 2-counter machine, where initially one counter has some value d_1 and the other counter is zero, reveals that the counters behave in a regular pattern. The 2-counter machine operates in phases in the following way. The machine's operation can be divided into phases, where each phase starts with one of the counters equal to some positive integer d_i and the other counter equal to 0. During the phase, the positive counter decreases, while the other counter increases. The phase ends with the first counter having value 0 and the other counter having value d_{i+1}. Then in the next phase the modes of the counters are interchanged. Thus, a sequence of configurations corresponding to the phases will be of the form:

$$(q_1, d_1, 0), (q_2, 0, d_2), (q_3, d_3, 0), (q_4, 0, d_4), (q_5, d_5, 0), (q_6, 0, d_6), \ldots$$

where the q_i's are states, with $q_1 = q_s$ (the initial state), and $d_1, d_2, d_3, \ldots$ are positive integers. Note that in going from state q_i in phase i to state q_{i+1} in phase $i + 1$, the 2-counter machine goes through intermediate states. Note that the second component of the configuration refers to the value of c_1 (first counter), while the third component refers to the value of c_2 (second counter).

For each i, there are 5 cases for the value of d_{i+1} in terms of d_i: $d_{i+1} = d_i, 2d_i, 3d_i, d_i/2, d_i/3$. (The division operation is done only if the number is divisible by 2 or 3, respectively.) The case is determined by q_i. Thus, we can define a mapping h such if q_i is the state at the start of phase i, $d_{i+1} = h(q_i)d_i$ (where $h(q_i)$ is 1, 2, 3, 1/2, 1/3).

Let T be a 2-counter machine accepting a recursively enumerable set L_{re} that is not recursive. We assume that $q_1 = q_s$ is the initial state, which is never re-entered, and if T halts, it does so in a unique state q_h. Let T's state set be Q, and 1 be a new symbol.

In what follows, α is any sequence of the form $\#I_1\#I_2\#\cdots\#I_{2m}\#$ (thus we assume that the length is even), where $I_i = q1^k$ for some $q \in Q$ and $k \geq 1$, represents a possible configuration of T at the beginning of phase i, where q is the state and k is the value of counter c_1 (resp., c_2) if i is odd (resp., even).

Define L_0 to be the set of all strings α such that

1. $\alpha = \#I_1\#I_2\#\cdots\#I_{2m}\#$;
2. $m \geq 1$;
3. for $1 \leq j \leq 2m-1$, $I_j \Rightarrow I_{j+1}$, i.e., if T begins in configuration I_j, then after one phase, T is in configuration I_{j+1} (i.e., I_{j+1} is a valid successor of I_j);

Lemma 11. L_0 is not in $\mathsf{DCM} \cup 2\mathsf{DCM}(1)$.

Proof. Suppose L_0 is accepted by a DCM (resp., $2\mathsf{DCM}(1)$). The following is an algorithm to decide, given any n, whether n is in L_{re}.

1. Let $R = \#q_s1^n((\#Q1^+\#Q1^+))^*\#q_h1^+\#$. Clearly R is regular.
2. Then $L' = L_0 \cap R$ is also in DCM (resp., $2\mathsf{DCM}(1)$) by Theorem 10.
3. Check if L' is empty. This is possible, since emptiness of DCM (respectively, $2\mathsf{DCM}(1)$) is decidable by Theorem 10.

The claim follows, since L' is empty if and only if n is not in L_{re}. □

4.1 Non-closure Under Inverse Infix

Theorem 12. *There is a language $L \in \mathsf{DCM}(1,1)$ such that $\inf^{-1}(L)$ is not in $\mathsf{DCM} \cup 2\mathsf{DCM}(1)$.*

Proof. Let T be a 2-counter machine. Let $L = \{\#q1^m\#p1^n\# \mid T$ when started in state q when one counter has value m and the other counter has value 0, does not reach the configuration in the next phase where the first counter becomes zero, the other counter has value n, and the state is $p\}$. Thus, $L = \{\#I\#I'\# \mid I$ and I' are configurations of T, and I' is not a valid successor of $I\}$. Clearly, L can be accepted by a $\mathsf{DCM}(1,1)$.

We claim that $L_1 = \inf^{-1}(L)$ is not in $\mathsf{DCM} \cup 2\mathsf{DCM}(1)$. Otherwise, by Theorem 10, $\overline{L_1}$ (the complement of L_1) is also in $\mathsf{DCM} \cup 2\mathsf{DCM}(1)$, and $\overline{L_1} \cap (\#Q1^+\#Q1^+)^+\# = L_0$ would be in $\mathsf{DCM} \cup 2\mathsf{DCM}(1)$. This contradicts Lemma 11. □

4.2   Non-closure Under Inverse Prefix

Theorem 13. *There exists a language L such that $L \in$ DCM(2,1) and $L \in$ 2DCM(1) (which makes only 1 turn on the input and 1 reversal on the counter) such that* $\text{pref}^{-1}(L) = L\Sigma^* \notin$ DCM $\cup$ 2DCM(1).

Proof. Consider $L = \{\#w\# \mid w \in \{a, b, \#\}^*, |w|_a \neq |w|_b\}$. Clearly, L can be accepted by a DCM(2,1) and by a 2DCM(1) which makes only 1 turn on the input and 1 reversal on the counter.

Suppose to the contrary that $\text{pref}^{-1}(L) \in$ DCM $\cup$ 2DCM(1). Then, $L' \in$ DCM $\cup$ 2DCM(1), where $L' = \text{pref}^{-1}(L) \cap (\#\{a, b, \#\}^* \#) = \{\#w_1 \cdots \#w_n\# \mid \exists i. |w_1 \cdots w_i|_a \neq |w_1 \cdots w_i|_b\}$.

We know that DCM and 2DCM(1) are closed under complement. So we can see that $L'' \in$ DCM $\cup$ 2DCM(1), where we define $L'' = \overline{L'} \cap (\#a^*b^*)^+\# = \{\#a^{k_1}b^{k_1}\# \cdots \#a^{k_m}b^{k_m}\# \mid m > 0\}$.

We will show that L'' is not in DCM $\cup$ 2DCM(1). Suppose L'' is in DCM $\cup$ 2DCM(1). Define two languages:

- $L_1 = \{\#1^{k_1}\#1^{k_1}\# \cdots \#1^{k_m}\#1^{k_m}\# \mid m \geq 1, k_i \geq 1\}$,
- $L_2 = \{\#1^{k_0}\#1^{k_1}\#1^{k_1}\# \cdots \#1^{k_{m-1}}\#1^{k_{m-1}}\#1^{k_m}\# \mid m \geq 1, k_i \geq 1\}$.

Note that L_1 and L_2 are similar. In L_1, the odd-even pairs of 1's are the same, but in L_2, the even-odd pairs of 1's are the same. Clearly, if M'' in DCM $\cup$ 2DCM(1) accepts L'', then we can construct (from M'') M_1 and M_2 in DCM $\cup$ 2DCM(1) to accept L_1 and L_2, respectively.

We now refer to the language L_0 that was shown not to be in DCM$\cup$2DCM(1) in Lemma 11. We will construct a DCM (resp., 2DCM(1)) to accept L_0, which would be a contradiction. Define the languages:

- $L_{odd} = \{\#I_1\#I_2\# \cdots \#I_{2m} \mid m \geq 1, I_1, \cdots, I_{2m}$ are configurations of the 2-counter machine T, for odd i, I_{i+1} is a valid successor of $I_i\}$.
- $L_{even} = \{\#I_1\#I_2\# \cdots \#I_{2m} \mid m \geq 1, I_1, \cdots, I_{2m}$ are configurations of the 2-counter machine T, for even i, I_{i+1} is a valid successor of $I_i\}$.

Clearly, $L_0 = L_{odd} \cap L_{even}$. Since DCM (resp., 2DCM(1)) is closed under intersection, we need only to construct two DCMs (resp., 2DCM(1)s) M_{odd} and M_{even} accepting L_{odd} and L_{even}, respectively. We will only describe the construction of M_{odd}, the construction of M_{even} being similar.

Case: Suppose $L'' \in$ DCM:
First consider the case of DCM. We will construct two machines: a DCM A and a DFA B such that $L(M_{odd}) = L(A) \cap L(B)$.

Let $L_A = \{\#I_1\#I_2\# \cdots \#I_{2m} \mid m \geq 1, I_1, \cdots, I_{2m}$ are configurations of the 2-counter machine T, for odd i, if $I_i = q_i1^{d_i}$, then $d_{i+1} = h(q_i)d_i\}$. We can construct a DCM A to accept L_A by simulating the DCM M_1. For example, suppose $h(q_i) = 3$. Then A simulates M_1 but whenever M_1 moves its input head one cell, A moves its input head 3 cells. If $h(q_i) = 1/2$, then when M_1 moves its head 2 cells, A moves its input head 1 cell. (Note that A does not use the 2-counter machine T.)

Now Let $L_B = \{\#I_1\#I_2\#\cdots\#I_{2m} \mid m \geq 1, I_1, \cdots, I_{2m}$ are configurations of the 2-counter machine, for odd i, if $I_i = q_i 1^{d_i}$, then T in configuration I_i ends phase i in state $q_{i+1}\}$. Clearly, a DFA B can accept L_B by simulating T for each odd i starting in state q_i on 1^{d_i} *without* using a counter, and checking that the phase ends in state q_{i+1}. (Note that the DCM A already checks the "correctness" of d_{i+1}.)

We can then construct from A and B a DCM M_{odd} such that $L(M_{odd}) = L(A) \cap L(B)$. In a similar way, we can construct M_{even}.

Case: Suppose $L'' \in 2DCM(1)$:
The case $2DCM(1)$ can be shown similarly. For this case, the machines M_{odd} and M_{even} are $2DCM(1)$s, and machine A is a $2DCM(1)$, but machine B is still a DFA.   □

From this, we can immediately get the result that the right end-marker is necessary for deterministic counter machines when there are at least two 1-reversal-bounded counters. In fact, without it, no amount of reversal-bounded counters with a deterministic machine could accept even some languages that can be accepted with two 1-reversal-bounded counters could with the end-marker.

Corollary 14. *There are languages in* $DCM(2, 1)$ *that are not in* DCM_{NE}.

Proof. Since DCM_{NE} is closed under concatenation with Σ^*, it follows that $pref^{-1}(L)$ from Theorem 13 is not in DCM_{NE}.   □

4.3   Non-closure for Inverse Suffix, Outfix and Embedding

Proposition 15. *There exists a language* $L \in DCM(1, 1)$ *such that* $suff^{-1}(L) \notin DCM$ *and* $suff^{-1}(L) \notin 2DCM(1)$.

Proof. Let L be as in Theorem 12. We know $DCM(1, 1)$ is closed under $pref^{-1}$ by Corollary 9, so $pref^{-1}(L) \in DCM(1, 1)$. Suppose $suff^{-1}(pref^{-1}(L)) \in DCM$. This implies that $inf^{-1}(L) \in DCM$, but we showed this language was not in DCM. Thus we have a contradiction. A similar contradiction can be reached when we assume $suff^{-1}(pref^{-1}(L)) \in 2DCM(1)$.

□

Corollary 16. *There exists* $L \in DCM(1, 1)$ *and regular languages* R *such that* $RL \notin DCM$ *and* $RL \notin 2DCM(1)$.

This implies that without the prefix-free condition on L_1 in Proposition 5, concatenation closure does not follow.

Corollary 17. *There exists* $L_1 \in DCM_{NE}(0, 0)$ *(regular), and* $L_2 \in DCM(1, 1)$, *where* $L_1 L_2 \notin DCM$ *and* $L_1 L_2 \notin 2DCM(1)$.

The result also holds for inverse outfix.

Proposition 18. *There exists a language $L \in$ DCM$(1,1)$ such that* outf$^{-1}(L) \notin$
DCM *and* outf$^{-1}(L) \notin$ 2DCM(1).

Proof. Consider $L \subseteq \Sigma^*$ where $L \in$ DCM$(1,1)$, and suff$^{-1}(L) \notin$ DCM and
suff$^{-1}(L) \notin$ 2DCM(1). The existence of such a language is guaranteed by Proposition 15. Let $\Gamma = \Sigma \cup \{\%\}$.

Suppose outf$^{-1}(L) \in$ DCM. Then $L' \in$ DCM, where $L' =$ outf$^{-1}(L) \cap \%\Sigma^*$.
We can see $L' = \{\%yx \mid x \in L, y \in \Sigma^*\}$, since the language we intersected with
ensures that the section is always added to the beginning of a word in L.

However, we also have $\%^{-1}L' \in$ DCM because DCM is clearly closed under
left quotient with a fixed word. We can see $\%^{-1}L' = \{yx \mid x \in L, y \in \Sigma^*\}$. This
is just suff$^{-1}(L)$, so suff$^{-1}(L) \in$ DCM, a contradiction.

The result is the same for 2DCM(1), relying on the closure of the family
under left quotient with a fixed word, which is clear. □

Corollary 19. *Let $m \in \mathbb{N}$. There exists a language $L \in$ DCM$(1,1)$ such that*
emb$^{-1}(m, L) \notin$ DCM *and* emb$^{-1}(m, L) \notin$ 2DCM(1).

This is similar to Proposition 18 except starting with $\#^{m-1}$, then

$$\text{emb}^{-1}(\#^{m-1}L) \cap (\#\%)^{m-1}L = \{(\#\%)^{m-1}yx \mid x \in L, y \in \Sigma^*\},$$

and so $L' \in$ DCM.

5 Summary of Results

Assume $R \in$ REG, $L_{\text{DCM}} \in$ DCM, and $L_{\text{DCM}_{\text{NE}}} \in$ DCM$_{\text{NE}}$.

The question: For all $L \in$ DCM(k, l):

Table 1. Summary of results for DCM. When applying the operation in the first column
to any $L \in$ DCM(k, l), is the result necessarily in DCM(k, l) (column 2), and in DCM
(column 3)? This is parameterized in terms of k and l, and the theorems showing each
result is provided.

Operation	is $Op(L) \in$ DCM(k,l)?		is $Op(L) \in$ DCM?	
pref$^{-1}(L)$	Yes if $k = 1, l \geq 1$	Cor 9	Yes if $k = 1, l \geq 1$	Cor 9
	No if $k \geq 2, l \geq 1$	Thm 13	Yes if $L \in$ DCM$_{\text{NE}}$	Prop 7
			No otherwise if $k \geq 2, l \geq 1$	Thm 13
suff$^{-1}(L)$	No if $k, l \geq 1$	Prop 15	No if $k, l \geq 1$	Prop 15
inf$^{-1}(L)$	No if $k, l \geq 1$	Thm 12	No if $k, l \geq 1$	Thm 12
outf$^{-1}(L)$	No if $k, l \geq 1$	Prop 18	No if $k, l \geq 1$	Prop 18
LR	Yes if $k = 1, l \geq 1$	Cor 8	Yes if $k = 1, l \geq 1$	Cor 8
	Yes if $L \in$ DCM$_{\text{NE}}$	Prop 7	Yes if $L \in$ DCM$_{\text{NE}}$	Prop 7
	No otherwise if $k \geq 2, l \geq 1$	Thm 13	No otherwise if $k \geq 2, l \geq 1$	Thm 13
RL	Yes if R prefix-free	Cor 6	Yes if R prefix-free	Cor 6
	No otherwise if $k, l \geq 1$	Cor 16	No otherwise if $k, l \geq 1$	Cor 16
$L_{\text{DCM}}L$	No if $k, l \geq 1$	Cor 17	No if $k, l \geq 1$	Cor 17
$L_{\text{DCM}_{\text{NE}}}L$	No if $k, l \geq 1$	Cor 17	Yes if $L_{\text{DCM}_{\text{NE}}}$ prefix-free	Prop 5
			No otherwise if $k, l \geq 1$	Cor 17

Also, for 2DCM(1), the results are summarized as follows:

- There exists $L \in$ DCM(1, 1) (one-way), s.t. suff$^{-1}(L) \notin$ 2DCM(1) (Prop 15).
- There exists $L \in$ DCM(1, 1) (one-way) , R regular, s.t. $RL \notin$ 2DCM(1) (Cor 16).
- There exists $L \in$ DCM(1, 1) (one-way), s.t. outf$^{-1}(L) \notin$ 2DCM(1) (Prop 18).
- There exists $L \in$ DCM(1, 1) (one-way), s.t. inf$^{-1}(L) \notin$ 2DCM(1) (Thm 12).
- There exists $L \in$ 2DCM(1), 1 input turn, 1 counter reversal, s.t. pref$^{-1}(L) \notin$ 2DCM(1) (Thm 13).
- There exists $L \in$ 2DCM(1), 1 input turn, 1 counter reversal, R regular, s.t. $LR \notin$ 2DCM(1) (Thm 13).

This resolves every open question summarized above, optimally, in terms of the number of counters, reversals on counters, and reversals on the input tape.

References

1. Baker, B.S., Book, R.V.: Reversal-bounded multipushdown machines. Journal of Computer and System Sciences **8**(3), 315–332 (1974)
2. Chiniforooshan, E., Daley, M., Ibarra, O.H., Kari, L., Seki, S.: One-reversal counter machines and multihead automata: Revisited. Theoretical Computer Science **454**, 81–87 (2012)
3. Eremondi, J., Ibarra, O., McQuillan, I.: Insertion operations on deterministic reversal-bounded counter machines. Tech. Rep. 2014–01, University of Saskatchewan (2014). http://www.cs.usask.ca/documents/techreports/2014/TR-2014-01.pdf
4. Gurari, E.M., Ibarra, O.H.: The complexity of decision problems for finite-turn multicounter machines. Journal of Computer and System Sciences **22**(2), 220–229 (1981)
5. Han, Y., Wood, D.: The generalization of generalized automata: Expression automata. International Journal of Foundations of Computer Science **16**(03), 499–510 (2005)
6. Harrison, M.: Introduction to Formal Language Theory. Addison-Wesley Pub. Co., Addison-Wesley series in computer science (1978)
7. Hopcroft, J.E., Ullman, J.D.: Introduction to Automata Theory, Languages, and Computation. Addison-Wesley, Reading (1979)
8. Ibarra, O.H.: Reversal-bounded multicounter machines and their decision problems. Journal of the ACM **25**(1), 116–133 (1978)
9. Ibarra, O.H., Jiang, T., Tran, N., Wang, H.: New decidability results concerning two-way counter machines. SIAM J. Comput. **23**(1), 123–137 (1995)
10. Jürgensen, H., Kari, L., Thierrin, G.: Morphisms preserving densities. International Journal of Computer Mathematics **78**, 165–189 (2001)
11. Minsky, M.L.: Recursive unsolvability of Post's problem of "tag" and other topics in theory of Turing Machines. Annals of Mathematics **74**(3), 437–455 (1961)
12. Nicaud, C.: Average state complexity of operations on unary automata. In: Kutyłowski, M., Pacholski, L., Wierzbicki, T. (eds.) Mathematical Foundations of Computer Science 1999. Lecture Notes in Computer Science, vol. 1672, pp. 231–240. Springer, Berlin Heidelberg (1999)

On the Synchronizing Probability Function and the Triple Rendezvous Time
New Approaches to Černý's Conjecture

François Gonze[(✉)] and Raphaël M. Jungers

ICTEAM Institute UCLouvain, Louvain La Neuve, Belgium
{francois.gonze,raphael.jungers}@uclouvain.be

Abstract. We push further a recently proposed approach for studying synchronizing automata and Černý's conjecture, namely, the *synchronizing probability function*. In this approach, the synchronizing phenomenon is reinterpreted as a Two-Player game, in which the optimal strategies of the players can be obtained through a Linear Program.

Our analysis mainly focuses on the concept of *triple rendezvous time*, the length of the shortest word mapping three states onto a single one. It represents an intermediate step in the synchronizing process, and is a good proxy of its overall length.

Our contribution is twofold. First, using the synchronizing probability function and properties of linear programming, we provide a new upper bound on the triple rendezvous time. Second, we disprove a conjecture on the synchronizing probability function by exhibiting a family of counterexamples. We discuss the game theoretic approach and possible further work in the light of our results.

Keywords: Automata and logic · Synchronization · Černý's conjecture · Game theory · Synchronizing probability function · Triple rendezvous time

1 Synchronizing Automata and Černý's Conjecture

Synchronizing automata have been the source of intense research in the past 50 years. An automaton is called *synchronizing* if there exists a sequence of letters which maps all the states onto a single one (see the next subsection for rigorous definitions). Figure 1 shows an example of such an automaton. The interest for the subject appeared in computers and relay control systems in the 60s. The aim was to restore control over these devices without knowing their current

---

This is a short conference version. For a long version with more details and examples see [12].

R.M. Jungers is a F.R.S.-FNRS Research Associate.

This work was also supported by the communauté francaise de Belgique - Actions de Recherche Concertées and by the Belgian Program on Interuniversity Attraction Poles initiated by the Belgian Federal Science Policy Office.

© Springer International Publishing Switzerland 2015
A.-H. Dediu et al. (Eds.): LATA 2015, LNCS 8977, pp. 212–223, 2015.
DOI: 10.1007/978-3-319-15579-1_16

state. In the 80s and 90s, synchronizing automata found applications in robotics and industry. From a theoretical perspective, the synchronizing property is also linked with active research topics in engineering, like the consensus theory and the primitivity of matrix sets (see [24]).

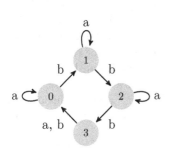

Fig. 1. A synchronizing automaton. The word *abbbabbba* maps any state onto state 0.

In this paper, we will represent automata as sets of matrices. A set of states of an automaton with n possible states will be represented by its $1 \times n$ *characteristic vector*[1], and the letters of the automaton will be represented as $n \times n$ matrices, acting multiplicatively on the characteristic vector of states:

Definition 1. *A (deterministic, finite state, complete) automaton (DFA) is a set of m column-stochastic matrices $\Sigma \subset \{0,1\}^{n \times n}$ (where m, n are respectively the number of letters in the alphabet, and the number of states of the automaton). Each letter corresponds to a matrix $L \in \Sigma$ with binary entries, which* satisfies $Le^T = e^T$, where e is the $1 \times n$ all-ones vector. We write Σ^t for the set of matrices which are products of length t of matrices taken in Σ. We refer to these matrices as words of length t.

Definition 2. *An automaton $\Sigma \subset \{0,1\}^{n \times n}$ is synchronizing if there is an index $1 \leq i \leq n$ and a finite product $W = L_{c_1} \ldots L_{c_s} : L_{c_j} \in \Sigma$ which satisfy*

$$W = e^T e_i,$$

where e_i is the ith standard basis vector $(1 \times n)$.
In this case, the sequence of letters $L_{c_1} \ldots L_{c_s}$ is said to be a synchronizing word.

Jan Černý stated his conjecture on DFA in 1964 [8]. Although it is very simple in its formulation, it has not been proven since then.

Conjecture 3 (Černý's conjecture, 1964 [8]). *Let $\Sigma \subset \{0,1\}^{n \times n}$ be a synchronizing automaton. Then, there is a synchronizing word of length at most $(n-1)^2$.*

In [7], Černý proposes an infinite family of automata attaining this bound, for any number of states. We refer to this family as the Černý family of automata. The automaton in Fig. 1 is the automaton of the family with four states. Synchronizing automata attaining the bound of Conjecture 3 or getting close to it are very infrequent (see [1], [14], [18] for examples).

Since its formulation, Conjecture 3 has been the subject of intense research. On the one hand theoretical research is aiming to prove the conjecture, on the other hand numerical research is aiming to design efficient algorithms to find

---

[1] A vector $x \subset \{0,1\}^n$ for which $x_i = 1$ if state i is in the set, and 0 otherwise.

synchronizing words in automata (see [19], [21]). Černý's Conjecture has been proven to hold for several families of automata [2,3,6,7,9,10,15,22]. However the best known general upper bound is $(n^3 - n)/6$, obtained by Pin and Frankl [11][16]. This bound has been holding for more than 30 years[2]. A state of the art overview is given by Volkov [25].

Recently, several research efforts have tried to shed light on the problem by making use of probabilistic approaches (see [13], [20]). The main tool we will focus on, the *synchronizing probability function* (SPF), was introduced by the second author in 2012 [13]. This tool allows the reformulation of the synchronizing property as a game theoretical problem whose solution can be obtained through convex optimization. This new link between convex optimization (a mature discipline with strong theoretical basis, see [5], [17]) and synchronizing automata is arguably promising towards a better understanding of the synchronizing phenomenon.

The paper is organised as follows. In Section 2 we recall the main properties of the synchronizing probability function. In Section 3, we introduce the concept of *triple rendezvous time*, and, making use of the synchronizing probability function we obtain a new upper bound on this value. In Section 4, we refute a recent conjecture on the synchronizing probability function (Conjecture 2 in [13]) by presenting a particular family of automata which doesn't satisfy it.

2   A Game Theoretical Framework and the Synchronizing Probability Function

In this section, we recall the definition and the properties of the synchronizing probability function needed to develop our results. A more complete introduction to the SPF and the details of the proofs can be found in [13]. This concept is based on a reformulation of the synchronization over an automaton as a Two-Player game with the following rules:

1. The length t is chosen.
2. Player Two secretly chooses a state for the automaton.
3. Player One chooses a word of length at most t. It is applied to the automaton, and changes the state of the automaton accordingly.
4. Player One guesses what the final state of the automaton is. If it is the right final state, he wins. Otherwise, Player Two wins.

The *policy of Player Two* is defined as a probability distribution over the states, that is, any vector $p \in \mathbb{R}^{+n}, ep^T = 1$. Player Two chooses the state i with probability p_i, in which case the automaton will end up at the state corresponding to $e_i A$, where A is the matrix representation of the word chosen by Player One. Since Player One wants to maximize the probability of choosing

---

[2] A bound of $n(7n^2 + 6n - 16)/48$ was proposed by Trahtman [23], but its proof was incomplete.

the right final state, he will pick up the state where the probability for the automaton to end is maximal, that is,

$$\operatorname*{argmax}_{i}(pA)e_i^T.$$

Therefore the probability of winning for Player One is

$$\max_{i,A}(pA)e_i^T. \tag{1}$$

The aim of Player Two is to minimize that probability.

In the following, $\Sigma^{\leq t}$ is the set of products of length at most t of matrices taken in Σ. By convention, and for the ease of notation, it contains the product of length zero, which is the identity matrix.

Definition 4 (SPF, Definition 2 in [13]). *Let $n \in \mathbb{N}$ and $\Sigma \subset \{0,1\}^{n \times n}$ be an automaton. The synchronizing probability function (SPF) of Σ is the function $k_\Sigma : \mathbb{N} \to \mathbb{R}^+$:*

$$k_\Sigma(t) = \min_{p \in \mathbb{R}^{+n},\ ep^T=1} \left\{ \max_{A \in \Sigma^{\leq t}} \left\{ \max_i (pA)e_i^T \right\} \right\}. \tag{2}$$

Conjecture (3) can now be reformulated in terms of the SPF:

Proposition 5 (Proposition 1 in [13]). *The following conjecture is equivalent to Conjecture (3):*
If $\Sigma \subset \{0,1\}^{n \times n}$ is a synchronizing automaton, then,

$$\forall t \geq (n-1)^2, \quad k_\Sigma(t) = 1.$$

If there is no ambiguity on the automaton, we use $k(t)$ for $k_\Sigma(t)$.

In order to use the SPF, we need an explicit algorithmic construction of the optimal strategies for both players, which allows us to compute the SPF value. Each basic strategy of Player One, i.e. the choice of a word and a final state, is equivalent to choosing a column in this word. Therefore, we consider the set of all the different columns reached in words of length at most t.

Definition 6. *We call* reachable columns *the set $A(t)$ of all the different columns in the matrices (words) in $\Sigma^{\leq t}$. We represent $A(t)$ as a $n \times M(t)$ matrix, where $M(t)$ is the number of different columns.*

When there is no ambiguity on t, we use A for $A(t)$. We notice that if $t = 0$, the reachable columns are the columns of the identity matrix, by definition of Σ^0.

It turns out that the SPF can be computed efficiently thanks to the following linear programs[3].

---

[3] The following inequalities are entrywise.

Theorem 7 (Theorem 1 in [13]). *The synchronizing probability function $k_\Sigma(t)$ of Σ is given by*

$$\min_{p,k} \; k \tag{3}$$

$$s.t. \; pA \le ke^T$$
$$ep^T = 1$$
$$p \ge 0.$$

It is also given by:

$$\max_{q,k} \; k \tag{4}$$

$$s.t. \; Aq \ge ke^T$$
$$eq = 1$$
$$q \ge 0.$$

In the equations above, A denotes the set of reachable columns at time t (see Def.6), q is a $M(t) \times 1$ vector, e represents all-ones vectors of the appropriate dimension, 1 is a scalar, and 0 represents zero vectors of the appropriate dimension.

The linear Program (4) is the dual of Program (3). For any primal feasible p and any dual feasible q, the objective value $k(p)$ of Program (3) and the objective value $k(q)$ of Program (4) satisfy $k(q) \le k(p)$. Therefore, if the objective value k is the same for both programs with feasible solutions p and q, this value is the optimum (see [5], [17] for more details on convex optimization and linear programming).

In the following, our main arguments will be based on the dimension of the set of optimal strategies of (4):

Definition 8 (Definition 3 in [13]). *Let Σ represent an automaton and t be a positive integer. The polytopes P_t and Q_t are the sets of optimal solutions of respectively (3) and (4).*

Lemma 9 (Lemma 1 in [13]). *If $k(t) = k(t+1)$ then $P_{t+1} \subset P_t$.*

3 A New Bound on the Triple Rendezvous Time

The *triple rendezvous time* is equal to the length of the shortest word mapping three states of the automaton onto a single one. Although it is a very natural concept, we are not aware of any attempts to bound its value for synchronizing automata. In what follows, the *weight* of a vector is the number of its non-zero elements.

Definition 10. *For a synchronizing automaton Σ, the triple rendezvous time $T_{3,\Sigma}$ is defined as the smallest integer t such that $A(t)$ contains a column of weight superior or equal to 3.*

In other words, it is the length of the shortest word W such that for three of the possible initial states, applying this word to the automaton leaves it in the same final state, i.e. such that there exists states q_i, q_j and q_k with $q_i W = q_j W = q_k W$. In the following, we will use T_3 for $T_{3,\Sigma}$ when there is no ambiguity on the automaton.

Our motivations for studying T_3 are multiple. There are empirical evidences that T_3 is correlated with the length of the shortest synchronizing word. Indeed, for the known automata achieving the bound of Conjecture 3, T_3 is close to n. Moreover, numerical experiments showed that automata with small T_3 have short synchronizing word. In addition, the triple rendezvous time is directly linked with the synchronizing probability function evolution (Proposition 6 and Conjecture 4 in [?]). It is also related to the k-extension property developed in [4][4].

For the known automata achieving the bound of Conjecture 3, the synchronizing probability function is growing close to linearly. This consideration led to the following conjecture:

Conjecture 11 (Conjecture 2 in [13]). *In a synchronizing automaton Σ with n states, for any $1 \leq j \leq n-1$,*

$$k_\Sigma(1 + (j-1)(n+1)) \geq j/(n-1).$$

This conjecture is stronger than Černý's conjecture (Theorem 4 in [?]). Conjecture 11 would also imply that the following conjecture about the triple rendezvous time is true:

Conjecture 12 (Conjecture 4 in [13]). *In a synchronizing automaton Σ with n states,*

$$T_{3,\Sigma} \leq n + 2.$$

In Section 4, we provide a family of automata which are counterexamples for both Conjecture 11 and Conjecture 12.

We now focus on bounding T_3. A first upper bound can be easily obtained without using the SPF:

Proposition 13. *In a synchronizing automaton Σ with n states,*

$$T_{3,\Sigma} \leq \frac{n(n-1)}{2} + 1.$$

Proof. For any positive integer t smaller than the length of the shortest synchronizing word, the matrix $A(t+1)$ must contain columns that are not in $A(t)$ (Lemma 1 in [13]). However, there are only $n(n-1)/2$ possible different columns

---

[4] T_3 is the smallest number such that there is a pair of states in the set of pairs of states which are synchronized by some single letter, which is $(T_3 - 1)$-extendable.

of weight two. As $A(0)$ includes the n columns of weight one, $A(n(n-1)/2+1)$ includes at least $n + n(n-1)/2 + 1$ columns, in which one must be of weight superior or equal to 3.   □

In order to obtain a better upper bound on T_3, we study the evolution of the SPF and $A(t)$ for $t < T_3$. In that setting, $A(t)$ only contains columns of weight one or two. We will associate the graph $G(t)$ with $A(t)$, $A(t)$ being the incidence matrix of $G(t)$ (ignoring columns of weight one in $A(t)$).

In the graph $G(t)$ associated with $A(t)$, we call a *singleton* a vertex which is disconnected from the rest of the graph, a *pair* two vertices which are connected to each other and disconnected from the rest of the graph, and a *cycle* a set of vertices connected between them as a cycle[5] and disconnected from the rest of the graph. We call a cycle *odd* (resp. *even*) if it contains an odd (resp. even) number of vertices.

We also use the reverse correspondence. With a singleton or a pair is associated the column in $A(t)$ corresponding to its characteristic vector, and with a cycle of c vertices is associated the set of c columns in $A(t)$ corresponding to the characteristic vectors of the c edges of the cycle.

In the following, based on this matrix-graph approach, we study the values that $k(t)$ can take with $t < T_3$, and the maximal dimension that P_t could take for each value $k(t)$. To do so, we start from the matrix $A(t)$. We prove that it is possible to extract a matrix A' from $A(t)$ by keeping only some of its columns, satisfying the following properties. The matrix A' is such that its associated graph G' is composed of disjoint singletons, pairs and odd cycles, and such that the optimal objective value for Program (3) and Program (4) is the same if $A(t)$ is replaced by A'. This structure allows us to compute easily the value $k(t)$ and the dimension of the solution set P_t associated with (3) (with A' instead of $A(t)$). Replacing $A(t)$ with A' can only increase the dimension of P_t as it reduces the amount of constraints in (3), while still achieving the same objective value by definition. We call *support* of the strategy q the set of columns in $A(t)$ corresponding to non zero entries in q.

Lemma 14. *If $t < T_3$, there exists an optimal solution q for Program (4) such that its support is associated with a graph composed of disjoint singletons, pairs and odd cycles.*

Proof (sketch). The sketch of the proof is the following. We proceed by induction on the number of variables of Program (4). If there is only one or two states, it is trivially true. Otherwise, we use the fact that the graph $G(t)$ associated with $A(t)$ can either be connected or disconnected.

If $G(t)$ is disconnected, we can define two subprograms with the structure of (4) with less variables, for which we know by induction that there exist optimal solutions satisfying the lemma. From these solutions, we can build an optimal solution of the original program whose support will also be of the right shape.

---

[5] c vertices are forming a *cycle* if we can number them from 1 to c in such a way that node 1 is only connected to nodes 2 and c, each node $1 < i < c$ is only connected to nodes $i-1$ and $i+1$, and the node c is only connected to nodes $c-1$ and 1.

If $G(t)$ is connected, we show that we can either find an optimal solution with a support associated with an odd cycle including all the vertices, or find a solution with a support associated with a disconnected graph. In this latter case, we will again be able to split the program as in the disconnected case.

The full proof is given in [12].   □

For programs with this particular structure, we can compute $k(t)$ and the dimension of P_t:

Lemma 15. *If the graph $G(t)$ associated with $A(t)$ is composed of disjoint odd cycles, pairs and singletons, then the optimum of Program (4) is given by $2/(n + n_1)$, where n_1 is the number of singletons. Moreover, the dimension of P_t is the number of pairs.*

Proof. To make notations concise, we define $K = n + n_1$. Our claim is that $k(t) = 2/K$. We provide an admissible solution for the primal (3), as well as for the dual (4), with the same objective value. Therefore this value is optimal.

A solution of (3) can be built as follows:

$$p_i = \begin{cases} 2/K \text{ if state i corresponds to a singleton,} \\ 1/K \text{ otherwise.} \end{cases} \tag{5}$$

The sum of the coefficients is $\sum_{i=1}^{n}(p_i) = (n - n_1)/K + 2n_1/K = 1$, and p is a feasible solution for (3) with objective value of $2/K$.

For (4), a solution can be built as follows:

$$q_i = \begin{cases} 2/K \text{ if column i corresponds to a pair,} \\ 1/K \text{ if column i corresponds to an edge of an odd cycle,} \\ 2/K \text{ if column i corresponds to a singleton.} \end{cases} \tag{6}$$

The sum of the coefficients is 1, and q is a feasible solution for (4) with objective value of $2/K$. Summarizing, equation (5) describes a solution for (3), and equation (6) describes a solution for (4), achieving the same objective value. As the programs are dual, this implies that both strategies are optimal, and $k(t) = 2/K$.

We can now give an explicit expression for P_t. Reordering the vertices such that the first f indexes correspond to singletons, the next g indexes correspond to vertices in pairs, grouped by pair (two vertices in the same pair have indices $f + 2j - 1$ and $f + 2j$, $1 \leq j \leq g/2$), and the last h indexes correspond to vertices in odd cycles, we have the following set of optimal solutions for (3):

$$P_t = \{(p_1, ..., p_{f+g+h}) | \; p_i = 2/K, \qquad\qquad 1 \leq i \leq f,$$
$$p_{f+2j-1} = 1/K + x_j, \qquad 1 \leq j \leq g/2,$$
$$p_{f+2j} = 1/K - x_j, \qquad -1/K \leq x_j, \leq 1/K, \tag{7}$$
$$p_k = 1/K, \qquad f + g + 1 \leq k \leq f + g + h\}.$$

Indeed, the total value assigned to each pair in the strategy p must be $k(t)$, and it can be split in any way between both vertices as long as none of the values is negative. The value assigned to singletons must be $k(t)$. For the odd cycles, the total value assigned to each pair in the cycle must be $k(t)$ (as it cannot be more than that for any pair, and in the optimal solution this value is effectively reached). However the only way to achieve that is to assign $k(t)/2$ to each vertex of the cycle.

The dimension of P_t is $g/2$, the number of pairs (see [12] for details).   □

We now present the main result of this section, which provides a universal upper bound on the triple rendezvous time for synchronizing automata. The main steps of the reasoning are as follows: starting from any original program obtained from an automaton and a value t, we can from Lemma 14 replace $A(t)$ with an other matrix to obtain a new optimization program with the same objective value, higher dimension for P_t, and the same structure as in Lemma 15. For this program, from Lemma 15, we can easily compute the value of the SPF and an upper bound on the dimension of P_t. Then, making use of the lemma 9 on the evolution of P_t, we then obtain a lower bound on the SPF growing rate before T_3:

Corollary 1. *If $t < T_3$, then $k(t)$ can only take the values $2/(n+s)$, $0 \leq s \leq n-1$, and this value cannot be optimal at more than $\lfloor (n-s)/2 \rfloor + 1$ consecutive values of t.*

Proof. By Lemma 14, for any $t < T_3$, we can replace $A(t)$ with a matrix $\mathcal{A}$ which is a subset of the columns of $A(t)$, such that the graph $\mathcal{G}$ associated is composed of singletons, pairs and odd cycles, and such that Program (4)', based on this matrix, achieves the same optimal objective. The set of optimal solutions P_t' of this program has a dimension superior or equal to the dimension of the set of optimal solution of the original program P_t. Let s be the number of singletons in $\mathcal{G}$. There are $n - s$ vertices which are either in pairs or in odd cycles in $\mathcal{G}$. As the dimension of P_t' is the number of pairs, it is at most $(n - s)/2$. Lemma 9 states that when $k(t)$ does not increase, the dimension of P_t has to decrease. As the dimension of P_t is bounded by the dimension of P_t', $k(t)$ cannot stay at the same value for more than $\lfloor (n - s)/2 \rfloor + 1$ consecutive values of t.   □

With Corollary (1), we can now obtain the main result of this section:

Theorem 16. *In a synchronizing automaton Σ with n states,*

$$T_{3,\Sigma} \leq \frac{n(n+4)}{4} - \frac{n \bmod 2}{4}.$$

Proof. By Corollary 1, the different values that the function $k(t)$ can take before T_3 are of the shape $2/(n+s)$, $0 \leq s \leq n-1$, and this value can be the same for $\lfloor (n-s)/2 \rfloor + 1$ steps at most. Summing over all possible values for $k(t)$, one gets

$$\sum_{s=0}^{n-1} (\lfloor (n-s)/2 \rfloor + 1) = \sum_{s=1}^{n} (\lfloor s/2 \rfloor + 1) = \frac{n(n+4)}{4} - \frac{n \bmod 2}{4}.$$

□

4   A Counterexample to a Conjecture on the Synchronizing Probability Function

In this section, we present an infinite family of automata which are counterexamples to Conjecture 11 and Conjecture 12. This family provides us with a lower bound on the maximum value of the triple rendezvous time for automata with n states for every odd integer $n \geq 9$. The automaton with nine states and two letters in Fig. 2 is the first of the family.

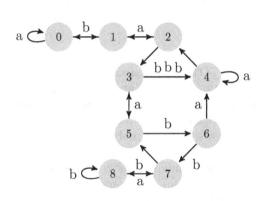

Conjecture 11 would imply that, for $j = 2$ and $n = 9$, $k(11) \geq 2/8$. However, the synchronizing probability function of the automaton in Fig.2 at $t = 11$ is $k(11) = 2/9$, disproving the conjecture. This automaton also has the particularity that its triple rendezvous time equals 12, which is the number of states plus 3. Indeed, on the one hand it can be verified that the matrix $A(11)$ contains only columns of weight two. On the other hand, for the three initial states 0, 4 and 6 of the automaton, the automaton ends at state 4 after application of the word $abbabbababba$,

Fig. 2. Automaton with 9 states and $k(11) = 2/9$

which is twelve letters long. Therefore this automaton is also a counterexample to Conjecture 12.

We can now extend this automaton with 9 states to an infinite family of automata with an odd number of states. Figure 3 shows the automata of this family with 11 and 13 states.

The recursive process to build the automaton of the family with n states from the one with $n - 2$ states is the following:

1. We start from the automaton of the family with $n - 2$ states (with states numbered from 0 to $n - 3$, with $n - 4$ and $n - 3$ added last).
2. We remove the self loops from states $n - 4$ and $n - 3$.
3. We add state $n - 2$ with a self loop labelled as the self loop removed from state $n - 3$, and state $n - 1$ with a self loop with the other label.

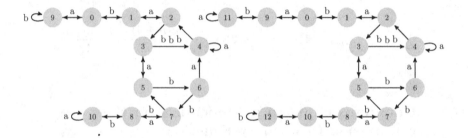

Fig. 3. Automata of the family with 11 and 13 states

4. We add the connections between states $n - 2$ and $n - 4$ in both directions, labelled as the self loop removed from state $n-4$, and we add the connections in both directions between $n - 1$ and $n - 3$, with the other label.

All the automata of this family are such that $T_3 = n+3$, and $k(n+2) = 2/n$.

5   Conclusion

In this paper, we pushed further the study of the synchronizing probability function as a tool to represent the synchronization of an automaton. Our results are twofold and somewhat antagonistic: on the one hand, we managed to prove a non trivial upper bound on the triple rendezvous time thanks to the synchronizing probability function. This result shows that this tool can effectively help in understanding synchronizing automata. On the other hand, we refuted Conjecture 11, formulated in [?], by providing an infinite family of automata for which $T_3 = n+3$ (with n being the number of states of the automaton). Conjecture 11 was stated as a tentative roadmap toward a proof of Cerny's conjecture with the help of the synchronizing probability function, and in that sense our conterexample is a negative result towards that direction.

A natural continuation to this research would be to find non-trivial bounds for T_s, with $3 < s \leq n$ (i.e. the smallest number such that the reachable column set includes a column of weight at least s). Another research question is how to narrow the gap between $n + 3$ and $n^2/4$ for the triple rendezvous time.

References

1. Ananichev, D., Gusev, V., Volkov, M.: Slowly synchronizing automata and digraphs. In: Hliněný, P., Kučera, A. (eds.) MFCS 2010. LNCS, vol. 6281, pp. 55–65. Springer, Heidelberg (2010)
2. Ananichev, D.S., Volkov, M.V.: Synchronizing generalized monotonic automata. Theoretical Computer Science **330**(1), 3–13 (2005)
3. Béal, M.P., Berlinkov, M.V., Perrin, D.: A quadratic upper bound on the size of a synchronizing word in one-cluster automata. International Journal of Foundations of Computer Science **22**(2), 277–288 (2011)

4. Berlinkov, M.V.: On a conjecture by Carpi and D'Alessandro. In: Gao, Y., Lu, H., Seki, S., Yu, S. (eds.) DLT 2010. LNCS, vol. 6224, pp. 66–75. Springer, Heidelberg (2010)
5. Boyd, S., Vandenberghe, L.: Convex optimization. Cambridge University Press, New York (2004)
6. Carpi, A., D'Alessandro, F.: Independent sets of words and the synchronization problem. Advances in Applied Mathematics **50**(3), 339–355 (2013)
7. Černý, J.: Poznámka k homogénnym eksperimentom s konečnými automatami. Matematicko-fysikalny Casopis SAV **14**, 208–216 (1964)
8. Černý, J., Pirická, A., Rosenauerova, B.: On directable automata. Kybernetica **7**, 289–298 (1971)
9. Dubuc, L.: Sur les automates circulaires et la conjecture de Černý. RAIRO Informatique Theorique et Appliquée **32**, 21–34 (1998)
10. Eppstein, D.: Reset sequences for monotonic automata. SIAM Journal on Computing **19**(3), 500–510 (1990). http://www.citeseer.ist.psu.edu/eppstein90reset.html
11. Frankl, P.: An extremal problem for two families of sets. European Journal on Combinatorics **3**, 125–127 (1982)
12. Gonze, F., Jungers, R.M.: On the synchronizing probability function and the triple rendezvous time for synchronizing automata (2014) ArXiv preprint. http://arxiv.org/abs/1410.4034
13. Jungers, R.M.: The synchronizing probability function of an automaton. SIAM Journal on Discrete Mathematics **26**(1), 177–192 (2012)
14. Kari, J.: A counter example to a conjecture concerning synchronizing words in finite automata. EATCS Bulletin **73**, 146 (2001)
15. Kari, J.: Synchronizing finite automata on eulerian digraphs. Theoretical Computer Science **295**, 223–232 (2003)
16. Pin, J.E.: On two combinatorial problems arising from automata theory. Annals of Discrete Mathematics **17**, 535–548 (1983)
17. Rockafellar, R.T.: Convex Analysis. Princeton University Press, Princeton (1970)
18. Roman, A.: A note on Černý conjecture for automata over 3-letter alphabet. Journal of Automata, Languages and Combinatorics **13**(2), 141–143 (2008)
19. Roman, A.: Genetic algorithm for synchronization. In: Dediu, A.H., Ionescu, A.M., Martín-Vide, C. (eds.) LATA 2009. LNCS, vol. 5457, pp. 684–695. Springer, Heidelberg (2009)
20. Steinberg, B.: The averaging trick and the Černý conjecture. In: Gao, Y., Lu, H., Seki, S., Yu, S. (eds.) DLT 2010. LNCS, vol. 6224, pp. 423–431. Springer, Heidelberg (2010)
21. Trahtman, A.N.: An efficient algorithm finds noticeable trends and examples concerning the Černy conjecture. In: Královič, R., Urzyczyn, P. (eds.) MFCS 2006. LNCS, vol. 4162, pp. 789–800. Springer, Heidelberg (2006)
22. Trahtman, A.N.: The Černý conjecture for aperiodic automata. Discrete mathematics and Theoretical Computer Science **9**(2), 3–10 (2007)
23. Trahtman, A.N.: Modifying the upper bound on the length of minimal synchronizing word. In: Owe, O., Steffen, M., Telle, J.A. (eds.) FCT 2011. LNCS, vol. 6914, pp. 173–180. Springer, Heidelberg (2011)
24. Blondel, V., Jungers, R.M., Olshevsky, A.: On primitivity of sets of matrices (2014) ArXiv preprint. http://arxiv.org/abs/1306.0729
25. Volkov, M.V.: Synchronizing automata and the Černý conjecture. In: Martín-Vide, C., Otto, F., Fernau, H. (eds.) LATA 2008. LNCS, vol. 5196, pp. 11–27. Springer, Heidelberg (2008)

On Robot Games of Degree Two

Vesa Halava[1,2], Reino Niskanen[2]([✉]), and Igor Potapov[2]

[1] Department of Mathematics and Statistics, University of Turku,
20014 Turku, Finland
vesa.halava@utu.fi
[2] Department of Computer Science, University of Liverpool, Ashton Building,
Liverpool L69 3BX, UK
{r.niskanen,potapov}@liverpool.ac.uk

Abstract. Robot Game is a two player vector addition game played in integer lattice $\mathbb{Z}^n$. In a degree k case both players have k vectors and in each turn the vector chosen by a player is added to the current configuration vector of the game. One of the players, called Attacker, tries to play the game from the initial configuration to the origin while the other player, Defender, tries to avoid origin. The decision problem is to decide whether or not Attacker has a winning strategy. We prove that the problem is decidable in *polynomial time* for the degree two games in any dimension n.

Keywords: Automata and concurrency · Reachability games · Vector addition game · Decidability · Winning strategy

1 Introduction

There is growing interest in the analysis of two player reachability games defined in the infinite state systems. These games appear in the verification, refinement and compatibility checking of reactive systems [11], in control problems, analysis of programs with recursion [5] and have deep connections with automata theory and logic [9,10,13].

In general two-player reachability games are played on a graph, called an arena, where the set of vertices is partitioned into two subsets to designate which player is in turn to move. Each move consists of picking an edge from the current location and adding its label to a counter vector under some semantics [7,12]. The objective of one player is to reach a given counter value in a given location while the other player tries to avoid it.

The Robot Game (also known as Attacker-Defender game [8]) is the simpler variant of the counter games under $\mathbb{Z}$ semantics (i.e. a counter vector can have any value in $\mathbb{Z}^n$) where the graph consists of only two vertices corresponding to two players: Attacker and Defender. Despite the simplicity of this model, the problem of checking the existence of winning strategy for the Robot Game in

Reino Niskanen - The author was partially supported by Nokia Foundation Grant.

A.-H. Dediu et al. (Eds.): LATA 2015, LNCS 8977, pp. 224–236, 2015.
DOI: 10.1007/978-3-319-15579-1_17

dimension one is EXPTIME-complete [2]. The problem is known to be undecidable for more complex scenario if the game is played on a graph with states of player 1 and states of player 2 with $\mathbb{Z}^2$ or $\mathbb{N}^2$ as the vector space [1,4,12]. In particular it follows from [6,8] that the Robot Game is undecidable in dimension $d \geq 8$ and in dimension $d \geq 2$ if Attacker has internal states.

Obviously, there are two parameters that can be considered in the Robot Game - the *number of vectors* each player has, which we call the *degree* of a game, and the *dimension* of the vectors. The results on complexity and decidability in [2,4,8] for such games were shown when the degree is not fixed.

In this paper we consider the Robot Game in a degree 2 case (i.e. where both players have at most 2 vectors each). We prove that the problem is decidable in *polynomial time* for the degree two games in any dimension n. Also we note that the same problem for counter games of degree 2 on arbitrary graph (i.e. where in each vertex of a graph (arena) only two vectors are available) is undecidable.

The proof of decidability is started with consideration of the dimension one games. The decidability of the problem in dimension one follows clearly from [2], but we shall present our proof, since the method is useful in higher dimensions and leads to a polynomial time algorithm. In particular games of dimension $n \geq 3$ reduce to games of dimension 1 or 2.

2   Notations and Definitions

We denote the set of all integers by $\mathbb{Z}$ and the set of all non-negative integers by $\mathbb{N}$. We denote by $0_n = (0, \ldots, 0)$ n-dimensional zero vector.

A *Counter Reachability Game* (CRG) consists of a directed graph $G = (V, E)$, where set of vertices is partitioned into two parts V_1 and V_2, each edge $e \in E \subseteq V \times \mathbb{Z}^n \times V$ is labeled with vectors in $\mathbb{Z}^n$, and an initial vector $\mathbf{x}_0 \in \mathbb{Z}^n$. *Configuration* of the game is $(v, \mathbf{x})$, successive configuration is $(v', \mathbf{x} + \mathbf{x}')$, where an edge $(v, \mathbf{x}', v') \in E$ is chosen by player 1 if $v \in V_1$ or by player 2 if $v \in V_2$. A *play* is a sequence of successive configurations. The goal of the first player, called *Attacker*, is to reach *final configuration* $(v_f, 0_n)$ for some $v_f \in V$ while the goal of the second player, called *Defender*, is to keep Attacker from reaching $(v_f, 0_n)$. A *strategy* for a player is a function that maps a configuration to an edge that can be applied. We say that Attacker has a *winning strategy* if he can reach the final configuration regardless of the strategies of Defender. On the other hand, we say that Defender has a *spoiling strategy* if there is an infinite play that never reaches the final configuration. In the figures we use $\bigcirc$ for Attacker's states and $\square$ for Defender's states. The *dimension* of the game is clearly the dimension of the integer lattice n and the *degree* is the largest degree of a vertex in the graph.

A *Robot Game* (RG) [6] is special case of Counter Reachability Game, where graph consists of only two vertices, q_0 of Defender and q of Attacker. The goal of the game is the configuration $(q_0, 0_n)$. That is, a Robot Game consists of two players, *Attacker* and *Defender* having a set of vectors U, V over $\mathbb{Z}^n$, respectively, and an *initial vector* $\mathbf{x}_0$. Here the degree is just the upper bound for number of vectors in the sets, that is, $|U|, |V| \geq 2$. Starting from $\mathbf{x}_0$ players add a vector from their set to the current position of the game in turns. As in Counter Rechability Game, Attacker tries to reach the origin while Defender tries to keep Attacker from reaching the origin.

Since by our definition game can only end after Attacker's move, we denote the *configuration* of a game at time t by the vector $\mathbf{w}_t$. Clearly $\mathbf{w}_t = \mathbf{x}_0 + \mathbf{v}_{i_1} + \mathbf{u}_{i_1} + \mathbf{v}_{i_2} + \mathbf{u}_{i_2} + \ldots + \mathbf{v}_{i_t} + \mathbf{u}_{i_t}$ after t rounds of the game where $\mathbf{u}_{i_j} \in U$ and $\mathbf{v}_{i_j} \in V$ for all indices i_j. Note that in each round Defender chooses the vector before Attacker.

In Robot Game Decision Problem, RGDP for short, the task is to determine whether or not there exists a winning strategy for Attacker, i.e., can Attacker reach the origin regardless of the vectors Defender chooses during his turns.

3   Dimension One

In this section we consider a game as an equation describing the configuration vector, where variables represent how many times each vector has been played. The goal of Attacker is to make the equation equal to zero with his choices of variables, while Defender does the opposite. The proof consists of carefully studying this equation and an additional equation stating that total number of vectors played by the players should be equal.

Assume that in the dimension one game that Attacker has numbers $k_1, k_2 \in \mathbb{Z}$ and Defender plays with numbers $\ell_1, \ell_2 \in \mathbb{Z}$, and assume that the initial number or position on the line is $a \in \mathbb{Z}$. Denote by x, y, z, w the times numbers k_1, k_2, ℓ_1, ℓ_2 are played in the game, respectively. Assume by symmetry that $k_2 \geq k_1$ and that $k_2 \geq 0$. If $k_2 < 0$, then $k_1 < 0$ too, and we may multiply all numbers of the game by -1 implying a symmetric game.

If $\ell_1 = \ell_2$, then Defender has no control over the game and after each turn $k_1 + \ell_1$ or $k_2 + \ell_1$ will be added to the configuration, depending on Attacker. In this case winning strategy is equivalent to reachability of the origin by two numbers. In other words, does or does not, the equation

$$x(k_1 + \ell_1) + y(k_2 + \ell_1) + a = 0$$

have a solution with $x, y \in \mathbb{N}$ and a linear diophantine equation can be solved in polynomial time [3]. Indeed, this equivalent to asking, does there exist $x \in \mathbb{N}$ such that y is in $\mathbb{N}$.

$$y = \frac{-x(k_1 + \ell_1) - a}{k_2 + \ell_1.} \tag{1}$$

From now on we assume that $\ell_1 \neq \ell_2$. Now the game is in the origin if and only if the following equations are satisfied:

$$\begin{cases} xk_1 + yk_2 + z\ell_1 + w\ell_2 + a & = 0 \\ x + y - z - w & = 0 \end{cases} \tag{2}$$

under constrain $x, y, z, w \in \mathbb{N}$. We shall look the game from Defender's point of view, and note that Attacker does not have a winning strategy, if Defender can somehow avoid the solutions of the above pair of equations under the constrain.

Assume first that $k_1 = k_2 = k$. In this case, Attacker has no control over the game and Defender will always win. Indeed, each turn either $k + \ell_1$ or $k + \ell_2$ will be added to the configuration. If the configuration is $-(k+\ell_1)$ then Defender will choose vector ℓ_2 and the next configuration is $-(k+\ell_1)+k+\ell_2 = \ell_2 - \ell_1 \neq 0$ or if the configuration is $-(k+\ell_2)$ then Defender will choose ℓ_1 and the configuration is $-(k+\ell_2)+k+\ell_1 = \ell_1 - \ell_2 \neq 0$. Therefore, Attacker does not have a winning strategy.

Assume that $k_2 > k_1$. Let us first express Attacker's variables x, y as follows:

$$x = \frac{(k_2 + \ell_1)z + (k_2 + \ell_2)w + a}{k_2 - k_1},$$
$$y = \frac{(-k_1 - \ell_1)z + (-k_1 - \ell_2)w - a}{k_2 - k_1}. \tag{3}$$

Denote the number $k_2 - k_1 = d$. Consider the game as a sequence of pairs of numbers (z, w), that is, the sequence $((z_i, w_i))_{i=0}^{\infty}$, where $(z_0, w_0) = (0,0)$ and

$$(z_{i+1}, w_{i+1}) = \begin{cases} (z_i + 1, w_i), & \text{or} \\ (z_i, w_i + 1). \end{cases}$$

Indeed, there is no winning strategy for Attacker if and only if Defender can construct a sequence $((z_i, w_i))_{i=0}^{\infty}$ such that Attacker cannot construct a similar sequence $((x_i, y_i))_{i=0}^{\infty}$ starting from $(0,0)$ and defined as $(x_{i+1}, y_{i+1}) = \begin{cases} (x_i + 1, y_i) \text{ or} \\ (x_i, y_i + 1). \end{cases}$ such that, for some i, (x_i, y_i, z_i, w_i) is a solution of the equations (3). First of all, there is an integer solution if and only if $(k_2 + \ell_1)z + (k_2 + \ell_2)w + a$ and $(-k_1 - \ell_1)z + (-k_1 - \ell_2)w - a$ are both divisible by $d = k_2 - k_1$ for some z and w. Therefore, we first rule out the cases where Defender avoids the zero in the play by rationality.

3.1   Avoiding Zero with Rationality

In this section we prove simple conditions that guarantee Defender's victory formulated in Corollary 2. Consider a sequence $x_i = nz_i + mw_i + a$, where $m, n, a \in \mathbb{Z}$, for a play $((z_i, w_i))_{i=0}^{\infty}$.

Lemma 1. *There exists a play $((z_i, w_i))_{i=0}^{\infty}$ such that $x_i \neq 0 \pmod{d}$ for all $i > 0$ if (and only if)*

1. $n \not\equiv m \pmod{d}$, *or*
2. $m \equiv n \equiv b \pmod{d}$ *and* $bj \not\equiv -a \pmod{d}$ *for all* $j \in \mathbb{N}$.

Proof. We prove the claim by constructing sequences for both cases.

Assume that $n \not\equiv m \pmod{d}$. We prove the claim by induction on index i. When $i = 1$, $x_1 = nz_1 + mw_1 + a$ is either $n + a$ or $m + a$. Both of these can't be zero, so Defender can choose the one for which $x_1 \neq 0$. Assume that the claim holds for all indexes smaller than $k+1$. Now $x_{k+1} = nz_{k+1} + mw_{k+1} + a$. By definition of the sequence $((z_i, w_i))_{i=0}^{\infty}$,

$$x_{k+1} = \begin{cases} n(z_k+1) + mw_k + a = x_k + n & \text{or} \\ nz_k + m(w_k+1) = x_k + m. \end{cases}$$

By induction hypothesis $x_k \not\equiv 0 \pmod{d}$ and by assumption $n \not\equiv m \pmod{d}$, thus at most one of these can be zero and Defender can choose the non-zero one.

Assume that $m \equiv n \equiv b \pmod{d}$ and $bj \not\equiv -a \pmod{d}$ for all j. Now $x_i = nz_i + mw_i + a \equiv b(z_i + w_i) + a \pmod{d}$ and by assumption $b(z_i + w_i) \not\equiv -a \pmod{d}$, so $x_i \not\equiv 0 \pmod{d}$.   □

Based on the above, clearly Defender can spoil all games not satisfying

1. $(k_2 + \ell_1) \equiv (k_2 + \ell_2) \equiv b_1 \pmod{k_2 - k_1}$, and $b_1 j_1 \equiv -a \pmod{k_2 - k_1}$, for some j_1, and
2. $(-k_1 - \ell_1) \equiv (-k_1 - \ell_2) \equiv b_2 \pmod{k_2 - k_1}$ and $b_2 j_2 \equiv a \pmod{k_2 - k_1}$, for some j_2.

Obviously $a + \ell_1 \equiv a + \ell_2 \pmod{d}$ if and only if $\ell_1 \equiv \ell_2 \pmod{d}$. On the other hand, for both x and y to be integers in (3), both sequences must be 0 modulo d at the same time, implying that we may assume that $j = j_1 = j_2$, implying that there exists a minimal such j that $j(k_2 + \ell_1) \equiv a \pmod{d}$ and $j(-k_1 - \ell_1) \equiv -a \pmod{d}$.

Corollary 2. *Defender can spoil all games not satisfying*

1. $\ell_1 \equiv \ell_2 \pmod{k_2 - k_1}$, and
2. $j(k_2 + \ell_1) \equiv a \pmod{k_2 - k_1}$ and $j(-k_1 - \ell_1) \equiv -a \pmod{k_2 - k_1}$ for some $j \geq 0$, $j \in \mathbb{N}$.

Next we concentrate on the cases of Corollary 2, and characterize those games where Defender has a spoiling play. Therefore, we assume that the game satisfies cases 1-2 of the previous Corollary. Note that it can be checked in polynomial time whether the game fulfills the conditions of the Corollary 2 or not. Assume next that c is the minimal number satisfying $c(k_2 + \ell_1) \equiv c(-k_1 - \ell_1) \equiv 0 \pmod{k_2 - k_1}$. Now the variables x and y in equations (3) are integers in step $j + tc$ for all $t \in \mathbb{N}$.

3.2   A Case of Positive Factors for z and w in (3)

In this section we prove that if the factors of z and w are positive in (3) and $-\ell_1 = k_1$ and $-\ell_2 = k_2$ then Attacker can only win the game with his first move. Assume first that $(k_2 + \ell_1), (k_2 + \ell_2), (-k_1 - \ell_1), (-k_1 - \ell_2)$ are all non-negative. Then $(k_2 \geq -\ell_1)$, $(k_2 \geq -\ell_2)$, $(k_1 \leq -\ell_1)$, $(k_1 \leq -\ell_2)$ implies that $k_1 \leq -\ell_1, -\ell_2 \leq k_2$. Since $\ell_1 \equiv \ell_2 \pmod{k_2 - k_1}$ and there are $k_2 - k_1 + 1$ numbers between k_2 and k_1, necessarily either $-\ell_1 = -\ell_2$ or $-\ell_1 = k_1$ and $-\ell_2 = k_2$ (or vice versa).

If $\ell_1 = \ell_2$. This case was completely studied already on page 226, but here with the specified conditions of Corollary 2, the decision is even simpler. Indeed, in (3) Attacker may calculate the first t such that $(k_2 + \ell_1)tj + a$ and $(-k_1 - \ell_1)tj - a$ are both positive and play his tj moves corresponding to the numbers x and y in solution of (3). Therefore, in this case Attacker has a winning strategy.

If $-\ell_1 = k_1$ and $-\ell_2 = k_2$. Then Defender can return the game into the same position once and again. Attacker has a winning strategy if and only if he can win the game after the first moves, that is, if

$$(\ell_1 + k_1 + a = 0 \text{ or } \ell_1 + k_2 + a = 0) \text{ and } (\ell_2 + k_1 + a = 0 \text{ or } \ell_2 + k_2 + a = 0).$$

Indeed, assume that $\ell_1 + k_1 + a \neq 0$ and $\ell_1 + k_2 + a \neq 0$. Then Defender plays first ℓ_1 and after that on each turn ℓ_i if Attacker played k_i on the previous round, and Attacker cannot win.

3.3   A Case of Negative or Mixed Factors for z and w in (3)

In this section we find winning conditions for Attacker if at least one of the factors of z and w is negative in (3). In this case extra tools are needed. From Defender's point of view, obvious strategy is to play so that only negative values of x or y satisfy the equation (3).

Assume that at least one of the factors of z or w is negative in (3). Now if $a \leq 0$ and $(k_2 + \ell_1)$ or $(k_2 + \ell_2)$ is negative, then by playing only z or w, respectively, Defender can keep x negative for the whole play and therefore, has a winning play. Similarly, if $-a \leq 0$ and $(-k_1 - \ell_1)$ or $(-k_1 - \ell_2)$ is negative, Attacker does not have a winning strategy.

Next assume that $a > 0$ and $(k_2 + \ell_1)$ or $(k_2 + \ell_2)$ is negative. Note that they both may be negative. It is clear that $-a$ is negative, and the sequence of pairs of rational numbers (x_i, y_i) counted for a play $((z_i, w_i))_{i=0}^{\infty}$ of Defender, initially has $y_0 < 0$. Now this sequence meets integers in steps $j + ct$ for all $t \in \mathbb{N}$. Note also, that y_i becomes positive at least at the same step than x_i turns to negative and not before, since clearly $x_i + y_i = z_i + w_i > 0$ for all $i > 0$.

Now consider two plays by Defender. Assume that Defender has played (z_{m-1}, w_{m-1}) and next step Defender has two choices $(z_{m-1} + 1, w_{m-1})$ or $(z_{m-1}, w_{m-1} + 1)$. Let (x_{m-1}, y_{m-1}) be the pair according to (z_{m-1}, w_{m-1}) and (x', y') and (x'', y'') be the pairs of integers according to the mth choice of Defender. Obviously, there exists an integer d such that

$$(x', y') = (x'' - d, y'' + d). \tag{4}$$

Indeed, changing one w to z makes a difference of

$$\frac{k_2 + \ell_1 - k_2 - \ell_2}{k_2 - k_1} = \frac{\ell_1 - \ell_2}{k_2 - k_1} = -d \tag{5}$$

in x as $\ell_1 \equiv \ell_2 \pmod{k_2 - k_1}$. The number d in (4) is called the *distance* of solutions in equations (3). Note that as $\ell_1 \neq \ell_2$, d is nonzero.

First we show that Defender has a spoiling strategy if the distance $|d| \geq 2$. In this case Defender can play through the solutions of Attacker as the distance between correct pair is too large. Indeed, assume that Defender has played (z_{j+ct-1}, w_{j+ct-1}) for some $t \in \mathbb{N}$ and Attacker (x, y). To have a winning strategy, it must be so that if Defender plays ℓ_2, that is $(z_{j+ct-1}, w_{j+ct-1} + 1)$, then Attacker can play $(x + 1, y)$ or $(x, y + 1)$ according to the correct case in (3). But now if Defender plays ℓ_1, Attacker has to be able to play $(x + 1 - d, y + d)$ or $(x - d, y + d + 1)$ with one move. Since $|d| \geq 2$, this is impossible. Therefore, Defender can choose the play at step $j + ct$ so that Attacker cannot win.

Indeed, in similar fashion, Defender may play through solutions for all cases of $j + ct$ steps where $t \in \mathbb{N}$ having (x, y) integers in (3), and indeed, there is no need for reaching negative x in this case for Defender to win. This holds also in the case of $j = c = 1$, i.e., in a game where x and y are integers for all (z, w) in (3), unless the very first turn of the play leads to a win of Attacker.

Finally, we turn into the case where the distance d has $|d| = 1$. From the equation (5) it follows that $\ell_1 - \ell_2 = \pm(k_2 - k_1)$.

Therefore, either $k_1 + \ell_1 = k_2 + \ell_2$ or $k_1 + \ell_2 = k_2 + \ell_1$. Assume the first equation to hold and denote $m = k_1 + \ell_1$, as there are no assumption on ℓ_1, ℓ_2 except the fact that they are nonequal, we could rename ℓ_1 and ℓ_2 so that the first equation holds. When considering the current state of the game, that is

$$x k_1 + y k_2 + z \ell_1 + w \ell_2 + a = 0,$$

It is obvious that when Defender plays ℓ_i by playing the k_i Attacker can move game by $\ell_i + k_i = m$. On the other hand, the same holds for Defender. As Attacker finishes a turn by k_i, by playing on the next turn ℓ_i, Defender moves the game by m.

We shall now prove that Attacker has winning strategy if and only if Attacker can force the game into position $-t \cdot m$, for some $t \in \mathbb{N}$, on the first round of the game. First of all, it is obvious that if for both $a + \ell_1$ and $a + \ell_2$, there is a $k', k'' \in \{k_1, k_2\}$ such that $a + \ell_1 + k' = -t_1 m$ and $a + \ell_2 + k'' = -t_2 m$, for some $t_1, t_2 \in \mathbb{N}$, then Attacker can win by playing k_i if Defender plays ℓ_i, since then m is added to position on each rounds and then after t_1 or t_2 rounds Attacker can win.

Assume now, $a + \ell_1 + k_i \neq -tm$ for all $t \in \mathbb{N}$. We prove that in this case Defender can avoid the zero position with the following strategy. Denote the play of Attacker by $k_{i_1}, k_{i_2}, \ldots$, where $i_n \in \{1, 2\}$ for all $n = 1, 2, \ldots$. Defender play on round $n \geq 2$ $\ell_{i_{n-1}}$, in other word, after the first round, Defender adds m to the position. Therefore, right before closing move of Attacker on round $n \geq 2$ full rounds, the game is in position $a + \ell_1 + (n - 1)m$. Assume contrary, that Attacker can win the game. Then for some $n \geq 2$ and $k' \in \{k_1, k_2\}$

$$a + \ell_1 + (n - 1)m + k' = 0,$$

which implies that $a + \ell_1 + k = -(n-1)m$, a contradiction. This proves the claim. Therefore, in the $|d| = 1$ games, Attacker has a winning strategy if and only if for both $a + \ell_1$ and $a + \ell_2$, there are $k', k'' \in \{k_1, k_2\}$ such that $a + \ell_1 + k' = -t_1 m$ and $a + \ell_2 + k'' = -t_2 m$, for some $t_1, t_2 \in \mathbb{N}$.

Proposition 3. *Attacker has a winning strategy only if*

1. $\ell_1 = \ell_2$ *and equation* (1) *has solution in* $\mathbb{N}$, *or*
2. *the game satisfies conditions in Corollary 2 and*
   (a) *Factors of* z *and* w *are positive in* (3) *and* $-\ell_1 = k_1$ *and* $-\ell_2 = k_2$ *and Attacker can win the game with the first move, or*
   (b) *At least one of the factors of* z *and* w *is negative in* (3). *Denote by* d *the distance of the game.*
      i. *If* $|d| \geq 2$ *and* $j = 1 = c$ *and Attacker always can win the game after the first round.*
      ii. *If* $|d| = 1$, *if for both* $a + \ell_1$ *and* $a + \ell_2$, *there is a* $k', k'' \in \{k_1, k_2\}$ *such that* $a + \ell_1 + k' = -t_1 m$ *and* $a + \ell_2 + k'' = -t_2 m$, *for some* $t_1, t_2 \in \mathbb{N}$.

Based on above cases it is easy to construct algorithm for the Robot Games of dimension one and degree two. Indeed, since finding all the cases where Attacker has a winning strategy requires first in the case $\ell_1 = \ell_2$ detecting, whether or not there are positive integer solution to equation (1). As mentioned when considering the case, such solutions can be detected in polynomial time. In the case $\ell_1 \neq \ell_2$ we must detect whether the game fulfills the conditions of Corollary 2, which mean solving two modular equations. After that all cases with winning strategy are detected by checking the first round moves. Note that in the case 2b(ii), checking the existence and finding the numbers t_1 and t_2 requires solving four equations of the form $t_i = \frac{a + \ell_i + k}{m} \in \mathbb{N}$. Obviously, this case can be solved in polynomial time too.

Theorem 4. *RGDP in dimension one with* 2 *vectors is in* **P**.

4 Dimension Two

First we approach solving a problem in dimension two in similar manner to the case of dimension one using a system of equations. Let $\mathbf{u}_1 = (\alpha_1, \alpha_2), \mathbf{u}_2 = (\beta_1, \beta_2), \mathbf{v}_1 = (\gamma_1, \gamma_2), \mathbf{v}_2 = (\delta_1, \delta_2)$. Let $U = \{\mathbf{u}_1, \mathbf{u}_2\}$ and $V = \{\mathbf{v}_1, \mathbf{v}_2\}$ be Attacker's and Defender's vector sets, respectively and an initial vector $\mathbf{a} = (a_1, a_2)$.

For Attacker to win, the following equations need to be satisfied

$$x\alpha_1 + y\beta_1 + z\gamma_1 + w\delta_1 + a_1 = 0$$
$$x\alpha_2 + y\beta_2 + z\gamma_2 + w\delta_2 + a_2 = 0 \tag{6}$$

and

$$x + y - z - w = 0 \tag{7}$$

under constrain $x, y, z, w \in \mathbb{N}$.

If the equations in (6) are linearly dependent, then it is enough to solve one of them as in dimension 1. Thus we assume that the equations are linearly independent. If Defender's vectors $\mathbf{v}_1 = \mathbf{v}_2$, then we have a one-player game that is considered in the following proposition.

Proposition 5. *The one-player RG of degree 2 in dimension n is in* **P**.

Proof. Let $U = \{(\alpha_1, \ldots, \alpha_n), (\beta_1, \ldots, \beta_n)\}$ be Attacker's vector set and $\mathbf{a} = (a_1, \ldots, a_n)$ the initial vector in one-player Robot Game of degree 2 in dimension n. Now winning the game is equivalent to whether Attacker can reach the origin from the initial vector. We need to solve system of linear Diophantine equations

$$x\alpha_1 + y\beta_1 - a_1 = 0,$$

$$\vdots$$

$$x\alpha_n + y\beta_n - a_n = 0,$$

with constrain $x, y \in \mathbb{N}$. As mentioned previously, solving linear Diophantine equations is in **P**.   □

Let us consider sets $U' = \{\mathbf{u}_1 - \mathbf{u}_2, (0,0)\} = \{\mathbf{u}', (0,0)\}$ and $V' = \{\mathbf{v}_1 + \mathbf{u}_2, \mathbf{v}_2 + \mathbf{u}_2\} = \{\mathbf{v}'_1, \mathbf{v}'_2\}$, where by default Attacker plays $\mathbf{u}_2$, which is added to each vector. In the following Lemma we prove that only specific type of games can be won by Attacker. The initial point must be on a line defined by one of Defender's vectors and Attacker has to have a way to move back to the line if the second vector is used. This is depicted in Figure 1, where four different games are shown. Note that since one of Attacker's vectors is $(0,0)$, Defender can have consecutive vectors in the play.

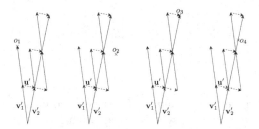

Fig. 1. Four different Robot Games of degree 2 in dimension 2. Here o_i is the origin of each game.

Lemma 6. *Attacker can win a game if and only if* $\mathbf{v}'_1 + \mathbf{u}' = \mathbf{v}'_2$ *and* $\mathbf{a} = -k\mathbf{v}'_2$ *or* $\mathbf{v}'_2 + \mathbf{u}' = \mathbf{v}'_1$ *and* $\mathbf{a} = -k\mathbf{v}'_1$ *for some* $k \in \mathbb{N}$.

Proof. Assume first that $\mathbf{v}'_1 + \mathbf{u}' = \mathbf{v}'_2$ and $\mathbf{a} = -k\mathbf{v}'_2$ for some $k \in \mathbb{N}$. Now if Defender plays $\mathbf{v}'_1$, then Attacker plays $\mathbf{u}'$ and the configuration is $-k\mathbf{v}'_2 + \mathbf{v}'_1 + \mathbf{u}' = (-k+1)\mathbf{v}'_2$. If on the other hand Defender plays $\mathbf{v}'_2$, then Attacker plays $(0,0)$ and the configuration is again $(-k+1)\mathbf{v}'_2$. That is, Attacker can win after k turns. This case is depicted in Figure 1 with origin o_3. The other case is symmetric.

Now we assume the contrary and show that in each case Defender has a spoiling strategy. Up to symmetry, there are three cases to consider:

(i) $\mathbf{v}_1' + \mathbf{u}' = \mathbf{v}_2'$ and $\mathbf{a} = -k\mathbf{v}_1'$ holds,

(ii) only $\mathbf{v}_1' + \mathbf{u}' = \mathbf{v}_2'$ holds,

(iii) neither $\mathbf{v}_2' + \mathbf{u}' = \mathbf{v}_1'$ nor $\mathbf{v}_2' + \mathbf{u}' = \mathbf{v}_1'$ hold.

Consider the first case. This is depicted in Figure 1 with origin being o_1. Clearly Defender spoils by moving away from the line defined by $\mathbf{v}_1'$.

In second case $\mathbf{v}_1' + \mathbf{u}' = \mathbf{v}_2'$ but $\mathbf{a} \neq -k\mathbf{v}_2'$. There are two subcases to consider. Either $\mathbf{a} = -r\mathbf{v}_2'$, where $r \notin \mathbb{Z}$ or $\mathbf{a}$ is not on a line defined by $\mathbf{v}_2'$. This is depicted in Figure 1 with origins being o_4 and o_2 respectively. Apart from one special case, in both of these cases Defender has a spoiling strategy by keeping the play on the line defined by $\mathbf{v}_2'$ that is playing $\mathbf{v}_2'$ until Attacker plays $\mathbf{u}'$. If $\mathbf{u}'$ is played, Defender matches it with $\mathbf{v}_1'$. Clearly the game will not reach origin unless $\mathbf{a} = -k\mathbf{v}_2' - \mathbf{u}'$, but in that case Defender takes the initiative and plays $\mathbf{v}_1'$ until Attacker does not match it with $\mathbf{u}'$, then he plays $\mathbf{v}_2'$. That is, configuration at all times is either

$$\mathbf{a} + \ell\mathbf{v}_2' + \mathbf{v}_1' + \mathbf{u}' = (\ell + 1 - k)\mathbf{v}_2' - \mathbf{u}' \neq (0,0) \text{ or}$$
$$\mathbf{a} + \ell\mathbf{v}_2' + \mathbf{v}_1' + (0,0) = (\ell - k)\mathbf{v}_2' + \mathbf{v}_1' - \mathbf{u}' \neq (0,0), \text{ for some } \ell \in \mathbb{Z}.$$

In the final case, if $\mathbf{v}_1' + \mathbf{u}' \neq \mathbf{v}_2'$ and $\mathbf{v}_2' + \mathbf{u}' \neq \mathbf{v}_1'$, then we consider points from which origin can be reached in one turn. These points are $-\mathbf{v}_1'$, $-\mathbf{v}_1' - \mathbf{u}'$, $-\mathbf{v}_2'$ and $-\mathbf{v}_2' - \mathbf{u}'$. If the point is $-\mathbf{v}_i'$, then Defender will play $-\mathbf{v}_j'$, where $i, j \in \{1, 2\}$ and $i \neq j$. Now if Attacker plays $(0,0)$, then resulting point is not origin, since we assumed that $\mathbf{v}_1 \neq \mathbf{v}_2$, and if Attacker plays $\mathbf{u}'$, then resulting point is $-\mathbf{v}_i' + \mathbf{v}_j' + \mathbf{u}'$ which is nonzero by our assumption. For the other two points we can see by similar reasoning that resulting point is not the origin. That is, Attacker cannot reach origin.   □

Theorem 7. *RGDP of degree two in dimension two is in* **P**.

Proof. We have classified simple conditions for Attacker to have a winning strategy. If one of the equations is a multiple of another, we reduce the game to one-dimensional game. If Defender has only one vector, we use Proposition 5. Or we need to check whether the game satisfies conditions of Lemma 6. Clearly these can be checked in polynomial time.   □

5   Dimension Three or Higher

In this section we consider RG in dimension $n \geq 3$ with 2 vectors. Let $U = \{(\alpha_1, \alpha_2, \ldots, \alpha_n), (\beta_1, \beta_2, \ldots, \beta_n)\}$ and $V = \{(\gamma_1, \gamma_2, \ldots, \gamma_n), (\delta_1, \delta_2, \ldots, \delta_n)\}$ be Attacker's and Defender's sets respectively and starting vector $\mathbf{a} = (a_1, \ldots, a_n)$.

For Attacker to win, the following equations need to be satisfied

$$x\alpha_1 + y\beta_1 + z\gamma_1 + w\delta_1 + a_1 = 0$$

$$\vdots \tag{8}$$

$$x\alpha_n + y\beta_n + z\gamma_n + w\delta_n + a_n = 0 \quad \text{and}$$

$$x + y - z - w = 0 \tag{9}$$

under constrain $x, y, z, w \in \mathbb{N}$.

As there are at least four equations with four variables, we first check by Gaussian elimination for number of linearly independent equations. Note that with Gaussian elimination we can also keep track of which equations are linearly independent. We have the following cases:

(i) There are at least 5 linearly independent equations. In this case there is no solution to the system of equations and obviously Defender always spoils the game.

(ii) There are 4 linearly independent equations. That is, there is a unique solution to the systems of equations. In this case there is no winning strategy since Defender can avoid the (z, w) values of the solution.

(iii) There are 3 linearly independent equations. We have two subcases. In the first subcase, two of the linearly independent equations are from (8) and one from (9). In this case we have a two-dimensional game with these three equations that can be solved by previous section. In the second subcase, all three linearly independent equations are from (8). That is

$$f_i = x\alpha_i + y\beta_i + z\gamma_i + w\delta_i + a_i$$
$$f_j = x\alpha_j + y\beta_j + z\gamma_j + w\delta_j + a_j$$
$$f_k = x\alpha_k + y\beta_k + z\gamma_k + w\delta_k + a_k$$

are the linearly independent equations for some indices i, j, k, and

$$x + y - z - w = af_i + bf_j + cf_k$$

for some coefficients $a, b, c \in \mathbb{Z}$. From this we can express one of the equations with the others, for example $cf_k = x+y-z-w- af_i-bf_j$, and consider 2-dimensional game with equations f_i, f_j and the constrain $x+y-z-w = 0$. Clearly if there is a solution to this game, also f_k will be equal to 0 with the same values x, y, z, w.

(iv) There are 2 linearly independent equations, then either one of them is from (8) and another is (9) and we have to solve a one-dimensional Robot Game. If both linearly independent equations are from (8) then we have two-dimensional Robot Game with the usual constrain equation [1].

(v) There is only 1 linearly independent equation. That is, every equation is a multiple of a single equation. Since (9) has ± 1 as coefficients, it has to be the linearly independent equation and Attacker can win after the first turn by playing any vector regardless of Defender's choice.

Based on Theorem 7 and above considerations, we can prove

Theorem 8. *RGDP of degree two is in* **P** *for all dimensions.*

---

[1] Note that we could have used similar approach to (iii)b and reduced it to one-dimensional game instead.

Corollary 9. *Robot Game Decision Problem of degree two with initial point x_0 and target point x_f is in **P** for all dimensions.*

Proof. The claim follows from Theorem 8 when we consider RG with initial point $x_0 - x_f$ and the origin as target. Clearly if Attacker has a winning strategy in the modified game, the same strategy is winning in the original game.   □

Finally let us consider decidability of Counter Reachability Game. In [12] it was proven that the problem is undecidable in dimension 2 even when all vectors are in $\{-1, 0, 1\}^2$. Although the degree of the game was not considered, it is easy to modify any Counter Reachability Game to be of degree 2 by splitting vertices with higher degree into a chain of vertices.

Proposition 10. *CRG of degree two in dimension two is undecidable.*

Proof. Consider a Counter Reachability Game of any degree. For each vertex q with degree $k > 2$, we construct a chain $q_1, \ldots, q_{k-1}$ of $k - 1$ vertices such that each ith edge $(q, \mathbf{v}, q')$ is $(q_i, \mathbf{v}, q')$. Finally we connect the vertices with edges $(q_i, 0_n, q_{i+1})$ for $i \in \{1, \ldots, k - 1\}$ and $(q, 0_n, q_1)$. Now the modified game is of degree 2. This proves the claim.   □

References

1. Abdulla, P.A., Bouajjani, A., d'Orso, J.: Deciding monotonic games. In: Baaz, M., Makowsky, J.A. (eds.) CSL 2003. LNCS, vol. 2803, pp. 1–14. Springer, Heidelberg (2003)
2. Arul, A., Reichert, J.: The complexity of robot games on the integer line. In: Proceedings of QAPL 2013. EPTCS, vol. 117, pp. 132–148 (2013)
3. Bradley, G.H.: Algorithms for hermite and smith normal matrices and linear diophantine equations. Math. Comp., Amer. Math. Soc. **25**, 897–907 (1971)
4. Brázdil, T., Jančar, P., Kučera, A.: Reachability games on extended vector addition systems with states. In: Abramsky, S., Gavoille, C., Kirchner, C., Meyer auf der Heide, F., Spirakis, P.G. (eds.) ICALP 2010. LNCS, vol. 6199, pp. 478–489. Springer, Heidelberg (2010)
5. Chatterjee, K., Fijalkow, N.: Infinite-state games with finitary conditions. In: Proceedings of CSL 2013. LIPIcs, vol. 23, pp. 181–196 (2013)
6. Doyen, L., Rabinovich, A.: Robot games. Tech. Rep. LSV-13-02, LSV, ENS Cachan (2013)
7. Haase, C., Halfon, S.: Integer vector addition systems with states. In: Ouaknine, J., Potapov, I., Worrell, J. (eds.) RP 2014. LNCS, vol. 8762, pp. 112–124. Springer, Heidelberg (2014)
8. Halava, V., Harju, T., Niskanen, R., Potapov, I.: Weighted automata on infinite words in the context of attacker-defender games. Tech. Rep. 1118, TUCS (2014)
9. Kupferman, O., Vardi, M.Y., Wolper, P.: An automata-theoretic approach to branching-time model checking. J. ACM **47**(2), 312–360 (2000)
10. Rabin, M.O.: Decidability of second-order theories and automata on infinite trees. Bull. Amer. Math. Soc. **74**(5), 1025–1029 (1968)

11. Alur, R., Henzinger, T.A., Kupferman, O.: Alternating-time temporal logic. J. ACM **49**(5), 672–713 (2002)
12. Reichert, J.: On the complexity of counter reachability games. In: Abdulla, P.A., Potapov, I. (eds.) RP 2013. LNCS, vol. 8169, pp. 196–208. Springer, Heidelberg (2013)
13. Walukiewicz, I.: Pushdown processes: Games and model-checking. Inf. Comput. **164**(2), 234–263 (2001)

Time-Bounded Reachability Problem
for Recursive Timed Automata Is Undecidable

Shankara Narayanan Krishna, Lakshmi Manasa$^{(\boxtimes)}$, and Ashutosh Trivedi

Indian Institute of Technology Bombay, Mumbai, India
{krishnas,manasa,trivedi}@cse.iitb.ac.in

Abstract. Motivated by the success of bounded model checking framework for finite state machines, Ouaknine and Worrell proposed a time-bounded theory of real-time verification by claiming that restriction to bounded-time recovers decidability for several key decision problem related to real-time verification. In support of this theory, the list of undecidable problems recently shown decidable under time-bounded restriction is rather impressive: language inclusion for timed automata, emptiness problem for alternating timed automata, and emptiness problem for rectangular hybrid automata. The objective of our study was to recover decidability for general recursive timed automata(RTA)—and perhaps for recursive hybrid automata—under time-bounded restriction in order to provide an appealing verification framework for powerful modeling environments such as Stateflow/Simulink. Unfortunately, however, we answer this question in negative by showing that time-bounded reachability problem stays undecidable for RTA with 5 clocks.

Keywords: Recursive State Machines · Timed Automata · Reachability

1 Introduction

Recursive state machines (RSMs), as introduced by Alur, Etessami, and Yannakakis [2], are a variation on various visual notations to represent hierarchical state machines, notably Harel's statecharts [8] and Object Management Group supported UML diagrams [11], that permits recursion while disallowing concurrency. RSMs closely correspond [2] to pushdown systems [6], context-free grammars, and Boolean programs [4], and provide a natural specification and verification framework to reason with sequential programs with recursive procedure calls. The two fundamental verification questions for RSM, namely reachability and Büchi emptiness checking, are known to be decidable in polynomial time [2,7].

Timed automata [3] extend finite state machines with continuous variables called clocks that permit a natural modeling of timed systems. In a timed automaton the variables continuously flow with uniform rates within each discrete state, while they are allowed to have discontinuous jumps during transitions between states that are guarded by constraints over variables. It is well

© Springer International Publishing Switzerland 2015
A.-H. Dediu et al. (Eds.): LATA 2015, LNCS 8977, pp. 237–248, 2015.
DOI: 10.1007/978-3-319-15579-1_18

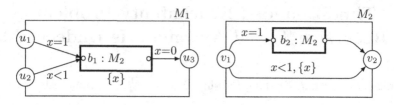

Fig. 1. An example of recursive timed automata with one clock and two components

known that the reachability problem is decidable (PSPACE-complete) for timed automata [3]. Trivedi and Wojtczak [12] introduced *recursive timed automata* (RTAs) as an extension of timed automata with recursion to model real-time software systems. Formally, an RTA is a finite collection of components where each component is a timed automaton that in addition to making transitions between various states, can have transitions to "boxes" that are mapped to other components modeling a potentially recursive call to a subroutine. During such invocation a limited information can be passed through clock values from the "caller" component to the "called" component via two different mechanism: a) *pass-by-value*, where upon returning from the called component a clock assumes the value prior to the invocation, and b) *pass-by-reference*, where upon return clocks reflect any changes to the value inside the invoked procedure.

Example 1 (Visual Presentation). The visual presentation of a recursive timed automaton with two components M_1 and M_2, and one clock variable x is shown in Figure 1 (example taken from [12]), where component M_1 calls component M_2 via box b_1 and component M_2 recursively calls itself via box b_2. Components are shown as thinly framed rectangles with their names written next to upper right corner. Various control states, or "nodes", of the components are shown as circles with their labels written inside them, e.g. see node u_1. Entry nodes of a component appear on the left of the component (see u_1), while exit nodes appear on the right (see u_3).

Boxes are shown as thickly framed rectangles inside components labeled $b :$ M, where b is the label of the box, M is the component it is mapped to. We write the set of clocks passed to M by value just below the box, while we omit this notation if all the clocks are passed by reference. The rest of the clocks are assumed to be passed by reference. For the sake of clarity of presentation, we often abuse the notation and write $\overline{Y}$ to denote the set $\mathcal{X} \setminus Y$. For instance, in the component INC c shown in Figure 3, we pass the clocks $\{x, z_1, z_2, b\}$ by value to the component mapped to the box $F_1{:}\mathrm{UP}_2^y$.

Call ports of boxes are drawn as small circles on the left of the box, while return ports are on the right. We omit labeling the call and return ports as these labels are clear from their position on the boxes. For example, call port (b_1, v_1) is the top small circle on the left-hand side of box b_1, since box b_1 is mapped to M_2 and v_1 is the top node on its left-hand side. Each transition is labeled with a guard and the set of reset variables, (e.g. transition from node v_1 to v_2 can be taken only when variable $x<1$, and after this transition is taken, x is reset).

Related Work. Trivedi and Wojtczak [12] showed that the reachability and termination (reachability with empty calling context) problem is undecidable for RTAs with three or more clocks. Moreover, they considered the so-called glitch-free restriction of RTAs—where at each invocation either all clocks are passed by value or all clocks are passed by reference— and showed that the reachability (and termination) is EXPTIME-complete for RTAs with two or more clocks. In the model of [12] it is compulsory to pass all the clocks at every invocation with either mechanism. Abdulla, Atig, and Stenman [1] studied a related model called timed pushdown automata where they disallowed passing clocks by value. They allowed clocks to be passed either by reference or not passed at all (in that case they are stored in the call context and continue to tick with the uniform rate). It is shown in [1] that the reachability problem for this class remains decidable (EXPTIME-complete). In this paper we restrict ourselves to the recursive timed automata model as introduced in [12]. In particular, we consider time-bounded reachability problem for RTA and show that the problem stays undecidable for RTA with 5 or more clocks. We have also studied two player reachability games on RTA in [9], and showed that time-bounded reachability games are undecidable for RTA with 3 or more clocks.

For a survey of models related to recursive timed automata and dense-time pushdown automata we refer the reader to [12]and [1]. Another closely related model is introduced in [5] where pushdown automata is extended with an additional stack used to store clock valuations. The reachability problem is known to be undecidable for this model. We do not consider this model in the current paper, but we conjecture that time-bounded reachability problem for this model is also undecidable.

2 Preliminaries

2.1 Labeled Transition System

A *labeled transition system* (LTS) is a tuple $\mathcal{L} = (S, A, X)$ where S is the set of *states*, A is the set of *actions*, and $X : S \times A \to S$ is the *transition function*. We say that an LTS $\mathcal{L}$ is *finite* (*discrete*) if both S and A are finite (countable). We write $A(s)$ for the set of actions available at $s \in S$, i.e., $A(s) = \{a : X(s,a) \neq \emptyset\}$.

We say that $(s, a, s') \in S \times A \times S$ is a transition of $\mathcal{L}$ if $s' = X(s,a)$ and a *run* of $\mathcal{L}$ is a sequence $\langle s_0, a_1, s_1, \ldots \rangle \in S \times (A \times S)^*$ such that (s_i, a_{i+1}, s_{i+1}) is a transition of $\mathcal{L}$ for all $i \geq 0$. We write $Runs^{\mathcal{L}}$ ($FRuns^{\mathcal{L}}$) for the sets of infinite (finite) runs and $Runs^{\mathcal{L}}(s)$ ($FRuns^{\mathcal{L}}(s)$) for the sets of infinite (finite) runs starting from state s. For a set $F \subseteq S$ and a run $r = \langle s_0, a_1, \ldots \rangle$ we define $Stop(F)(r) = \inf \{i \in \mathbb{N} : s_i \in F\}$. Given a state $s \in S$ and a set of final states $F \subseteq S$ we say that a final state is reachable from s_0 if there is a run $r \in Runs^{\mathcal{L}}(s_0)$ such that $Stop(F)(r) < \infty$. Given an LTS, an initial state, and a set of final states, the *reachability problem* for LTS is to decide whether a final state is reachable from the given initial state.

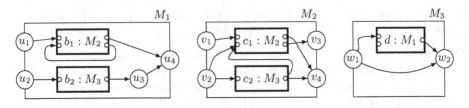

Fig. 2. Example recursive state machine taken from [2]

2.2   Recursive State Machines

Definition 2. *A recursive state machine [2]* $\mathcal{M}$ *is a tuple* $(\mathcal{M}_1, \mathcal{M}_2, \ldots, \mathcal{M}_k)$ *of components, where each component* $\mathcal{M}_i = (N_i, \mathrm{EN}_i, \mathrm{EX}_i, B_i, Y_i, A_i, X_i)$ *for each* $1 \leq i \leq k$ *is such that:*

- N_i *is a finite set of* nodes *including a distinguished set* EN_i *of entry nodes and a set* EX_i *of exit nodes such that* EX_i *and* EN_i *are disjoint sets;*
- B_i *is a finite set of* boxes;
- $Y_i : B_i \rightarrow \{1, 2, \ldots, k\}$ *is a mapping that assigns every box to a component. We associate a set of* call ports *Call(b) and* return ports *Ret(b) to each box* $b \in B_i$: *Call(b)* $= \{(b, en) : en \in \mathrm{EN}_{Y_i(b)}\}$ *and Ret(b)* $= \{(b, ex) : ex \in \mathrm{EX}_{Y_i(b)}\}$. *Let* $Call_i = \cup_{b \in B_i} Call(b)$ *and* $Ret_i = \cup_{b \in B_i} Ret(b)$ *be the set of call and return ports of component* $\mathcal{M}_i$.
  We define the set of vertices Q_i *of component* $\mathcal{M}_i$ *as the union of the set of nodes, call ports and return ports, i.e.* $Q_i = N_i \cup Call_i \cup Ret_i$;
- A_i *is a finite set of* actions; *and*
- $X_i : Q_i \times A_i \rightarrow Q_i$ *is the transition function with a condition that call ports and exit nodes do not have any outgoing transitions.*

For the sake of simplicity, we assume that the set of boxes $B_1, \ldots, B_k$ *and the set of nodes* $N_1, N_2, \ldots, N_k$ *are mutually disjoint. We use symbols* N, B, A, Q, X, *etc. to denote the union of the corresponding symbols over all components.*

An example of a RSM is shown in Figure 2. An execution of a RSM begins at the entry node of some component and depending upon the sequence of input actions the state evolves naturally like a labeled transition system. However, when the execution reaches an entry port of a box, this box is stored on a stack of pending calls, and the execution continues naturally from the corresponding entry node of the component mapped to that box. When an exit node of a component is encountered, and if the stack of pending calls is empty, then the run terminates; otherwise, it pops the box from the top of the stack, and jumps to the exit port of the just popped box corresponding to the just reached exit of the component. We formalize the semantics of a RSM using a discrete LTS, whose states are pairs consisting of a sequence of boxes, called the context, mimicking the stack of pending calls and the current vertex.

Definition 3 (RSM semantics). *Let* $\mathcal{M} = (\mathcal{M}_1, \mathcal{M}_2, \ldots, \mathcal{M}_k)$ *be an RSM where the component* $\mathcal{M}_i$ *is* $(N_i, En_i, Ex_i, B_i, Y_i, A_i, X_i)$. *The semantics of* $\mathcal{M}$ *is the discrete labelled transition system* $[\![\mathcal{M}]\!] = (S_{\mathcal{M}}, A_{\mathcal{M}}, X_{\mathcal{M}})$ *where:*

- $S_{\mathcal{M}} \subseteq B^* \times Q$ *is the set of states;*
- $A_{\mathcal{M}} = \cup_{i=1}^{k} A_i$ *is the set of actions;*
- $X_{\mathcal{M}} : S_{\mathcal{M}} \times A_{\mathcal{M}} \to S_{\mathcal{M}}$ *is the transition relation such that for* $s = (\langle \kappa \rangle, q) \in S_{\mathcal{M}}$ *and* $a \in A_{\mathcal{M}}$, *we have that* $s' = X_{\mathcal{M}}(s, a)$ *if and only if one of the following holds:*
  1. *the vertex* q *is a call port, i.e.* $q = (b, en) \in Call$, *and* $s' = (\langle \kappa, b \rangle, en)$;
  2. *the vertex* q *is an exit node, i.e.* $q = ex \in EX$ *and* $s' = (\langle \kappa' \rangle, (b, ex))$ *where* $(b, ex) \in Ret(b)$ *and* $\kappa = (\kappa', b)$;
  3. *the vertex* q *is any other vertex, and* $s' = (\langle \kappa \rangle, q')$ *and* $q' \in X(q, a)$.

Given $\mathcal{M}$ and a subset $Q' \subseteq Q$ of its nodes we define the set $[\![Q']\!]_{\mathcal{M}}$ as the set $\{(\langle \kappa \rangle, v') : \kappa \in B^* \text{ and } v' \in Q'\}$. We also define the set of terminal configurations $Term_{\mathcal{M}}$ as the set $\{(\langle \varepsilon \rangle, ex) : ex \in EX\}$ with the empty context $\langle \varepsilon \rangle$. Given a recursive state machine $\mathcal{M}$, an initial node v, and a set of *final vertices* $F \subseteq Q$ the *reachability problem* on $\mathcal{M}$ is defined as the reachability problem on the LTS $[\![\mathcal{M}]\!]$ with the initial state $(\langle \varepsilon \rangle, v)$ and final states $[\![F]\!]$. We define *termination problem* as the reachability of one of the exits with the empty context. The following is a well known result.

Theorem 4 ([2]). *The reachability and the termination problem for recursive state machines can be solved in polynomial time.*

3   Recursive Timed Automata

Recursive timed automata (RTAs) extend classical timed automata (TAs) with recursion in a similar way RSMs extend LTSs.

3.1   Syntax

Let $\mathbb{R}$ be the set of real numbers. Let $\mathcal{X}$ be a finite set of real-valued clocks. A *valuation* on $\mathcal{X}$ is a function $\nu : \mathcal{X} \to \mathbb{R}$. We assume an arbitrary but fixed ordering on the clocks and write x_i for the variable with order i. This allows us to treat a valuation ν as a point $(\nu(x_1), \nu(x_2), \ldots, \nu(x_n)) \in \mathbb{R}^{|\mathcal{X}|}$. For a subset of clocks $X \subseteq \mathcal{X}$ and a valuation $\nu' \in \mathcal{X}$, we write $\nu[X := \nu']$ for the valuation where $\nu[X := \nu'](x) = \nu'(x)$ if $x \in X$, and $\nu[X := \nu'](x) = \nu(x)$ otherwise. The valuation $\mathbf{0} \in \mathbb{R}^{|\mathcal{X}|}$ is a special valuation such that $\mathbf{0}(x) = 0$ for all $x \in \mathcal{X}$.

We define a constraint over a set $\mathcal{X}$ as a subset of $\mathbb{R}^{|\mathcal{X}|}$. We say that a constraint is *rectangular* if it is defined as the conjunction of a finite set of constraints of the form $x \bowtie k$, where $k \in \mathbb{Z}$, $x \in \mathcal{X}$, and $\bowtie \in \{<, \leq, =, >, \geq\}$. For a constraint G, we write $[\![G]\!]$ for the set of valuations in $\mathbb{R}^{|\mathcal{X}|}$ satisfying the constraint G. We write $\top$ (resp., $\bot$) for the special constraint that is true (resp., false) in all the valuations, i.e. $[\![\top]\!] = \mathbb{R}^{|\mathcal{X}|}$ (resp., $[\![\bot]\!] = \emptyset$). We write $rect(\mathcal{X})$ for the set of rectangular constraints over $\mathcal{X}$ including $\top$ and $\bot$.

Definition 5. *A recursive timed automaton* $\mathcal{H} = (\mathcal{X}, (\mathcal{H}_1, \mathcal{H}_2, \ldots, \mathcal{H}_k))$ *is a pair of set of clocks* $\mathcal{X}$ *and a collection of components* $(\mathcal{H}_1, \mathcal{H}_2, \ldots, \mathcal{H}_k)$ *where every* $\mathcal{H}_i = (N_i, \mathrm{EN}_i, \mathrm{EX}_i, B_i, Y_i, A_i, X_i, P_i, Inv_i, E_i, J_i, F_i)$ *is such that:*

- N_i *is a finite set of* nodes *including a distinguished set* EN_i *of* entry nodes *and a set* EX_i *of* exit nodes *such that* EX_i *and* EN_i *are disjoint sets;*
- B_i *is a finite set of* boxes;
- $Y_i : B_i \to \{1, 2, \ldots, k\}$ *is a mapping that assigns every box to a component. (Call ports* $Call(b)$ *and return ports* $Ret(b)$ *of a box* $b \in B_i$, *and call ports* $Call_i$ *and return ports* Ret_i *of a component* $\mathcal{H}_i$ *are defined as before. We set* $Q_i = N_i \cup Call_i \cup Ret_i$ *and refer to this set as the set of locations of* $\mathcal{H}_i$.)
- A_i *is a finite set of* actions.
- $X_i : Q_i \times A_i \to Q_i$ *is the transition function with a condition that call ports and exit nodes do not have any outgoing transitions.*
- $P_i : B_i \to 2^{\mathcal{X}}$ *is pass-by-value mapping that assigns every box the set of clocks that are passed by value to the component mapped to the box; (The rest of the clocks are assumed to be passed by reference.)*
- $Inv_i : Q_i \to rect(\mathcal{X})$ *is the* invariant condition;
- $E_i : Q_i \times A_i \to rect(\mathcal{X})$ *is the* action enabledness function;
- $J_i : A_i \to 2^{\mathcal{X}}$ *is the* variable reset function;

We assume that the sets of boxes, nodes, locations, etc. are mutually disjoint across components and we write $(N, B, Y, Q, P, X,$ *etc.) to denote corresponding union over all components.*

We say that a recursive timed automaton is *glitch-free* if for every box either all clocks are passed by value or none is passed by value, i.e. for each $b \in B$ we have that either $P(b) = \mathcal{X}$ or $P(b) = \emptyset$. Any general RTA with one clock is trivially glitch-free. We say that a RTA is *hierarchical* if there exists an ordering over components s.t. a component never invokes another component of higher order or same order.

3.2   Semantics

A *configuration* of an RTA $\mathcal{H}$ is a tuple $(\langle \kappa \rangle, q, \nu)$, where $\kappa \in (B \times \mathbb{R}^{|\mathcal{X}|})^*$ is sequence of pairs of boxes and variable valuations, $q \in Q$ is a location and $\nu \in \mathbb{R}^{|\mathcal{X}|}$ is a variable valuation over $\mathcal{X}$ such that $\nu \in Inv(q)$. The sequence $\langle \kappa \rangle \in (B \times \mathbb{R}^{|\mathcal{X}|})^*$ denotes the stack of pending recursive calls and the valuation of all the variables at the moment that call was made, and we refer to this sequence as the context of the configuration. Technically, it suffices to store the valuation of variables passed by value, because other variables retain their value after returning from a call to a box, but storing all of them simplifies the notation. We denote the the empty context by $\langle \epsilon \rangle$. For any $t \in \mathbb{R}$, we let $(\langle \kappa \rangle, q, \nu) + t$ equal the configuration $(\langle \kappa \rangle, q, \nu + t)$.

Informally, the behavior of an RTA is as follows. In configuration $(\langle \kappa \rangle, q, \nu)$ time passes before an available action is triggered, after which a discrete transition occurs. Time passage is available only if the invariant condition $Inv(q)$ is satisfied while time elapses, and an action a can be chosen after time t elapses only if it is enabled after time elapse, i.e., if $\nu+t \in E(q,a)$. If the action a is chosen then the successor state is $(\langle\kappa\rangle, q', \nu')$ where $q' \in X(q,a)$ and $\nu' = (\nu + t)[J(a) := \mathbf{0}]$. Formally, the semantics of an RTA is given by an LTS which has both an uncountably infinite number of states and transitions.

Definition 6 (RTA semantics). *Let* $\mathcal{H} = (\mathcal{X}, (\mathcal{H}_1, \mathcal{H}_2, \ldots, \mathcal{H}_k))$ *be an RTA where each component is of the form* $\mathcal{H}_i = (N_i, \mathrm{EN}_i, \mathrm{EX}_i, B_i, Y_i, A_i, X_i, P_i, Inv_i, E_i, J_i, F_i)$. *The semantics of* $\mathcal{H}$ *is a labelled transition system* $[\![\mathcal{H}]\!] = (S_\mathcal{H}, A_\mathcal{H}, X_\mathcal{H})$ *where:*

- $S_\mathcal{H} \subseteq (B \times \mathbb{R}^{|\mathcal{X}|})^* \times Q \times \mathbb{R}^{|\mathcal{X}|}$, *the set of states, is s.t.* $(\langle\kappa\rangle, q, \nu) \in S_\mathcal{H}$ *if* $\nu \in Inv(q)$.
- $A_\mathcal{H} = \mathbb{R}_\oplus \times A$ *is the set of* timed actions, *where* $\mathbb{R}_\oplus$ *is the set of non-negative reals;*
- $X_\mathcal{H} : S_\mathcal{H} \times A_\mathcal{H} \to S_\mathcal{H}$ *is the transition function such that for* $(\langle\kappa\rangle, q, \nu) \in S_\mathcal{H}$ *and* $(t,a) \in A_\mathcal{H}$, *we have* $(\langle\kappa'\rangle, q', \nu') = X_\mathcal{H}((\langle\kappa\rangle, q, \nu), (t,a))$ *if and only if the following condition holds:*
  1. *if the location* q *is a call port, i.e.* $q = (b,en) \in Call$ *then* $t = 0$, *the context* $\langle\kappa'\rangle = \langle\kappa, (b,\nu)\rangle$, $q' = en$, *and* $\nu' = \nu$.
  2. *if the location* q *is an exit node, i.e.* $q = ex \in Ex$, $\langle\kappa\rangle = \langle\kappa'', (b,\nu'')\rangle$, *and let* $(b,ex) \in Ret(b)$, *then* $t = 0$; $\langle\kappa'\rangle = \langle\kappa''\rangle$; $q' = (b,ex)$; *and* $\nu' = \nu[P(b):=\nu'']$.
  3. *if location* q *is any other kind of location, then* $\langle\kappa'\rangle = \langle\kappa\rangle$, $q' \in X(q,a)$, *and*
     (a) $\nu + t' \in Inv(q)$ *for all* $t' \in [0,t]$;
     (b) $\nu + t \in E(q,a)$;
     (c) $\nu' = (\nu + t)[J(a) := \mathbf{0}]$.

3.3   Reachability and Time-Bounded Reachability Problems

For a subset $Q' \subseteq Q$ of states of a recursive time automaton $\mathcal{H}$ we define the set $[\![Q']\!]_\mathcal{H}$ as the set $\{(\langle\kappa\rangle, q, \nu) \in S_\mathcal{H} : q \in Q'\}$. We define the terminal configurations as $Term_\mathcal{H} = \{((\langle\varepsilon\rangle, q, \nu) \in S_\mathcal{H} : q \in \mathrm{EX}\}$. Given a recursive timed automaton $\mathcal{H}$, an initial node q and valuation $\nu \in \mathbb{R}^{|\mathcal{X}|}$, and a set of *final locations* $F \subseteq Q$, the *reachability problem* on $\mathcal{H}$ is to decide the existence of a run in the LTS $[\![\mathcal{H}]\!]$ staring from the initial state $(\langle\varepsilon\rangle, q, \nu)$ to some state in $[\![F]\!]_\mathcal{H}$. As with RSMs, we also define *termination problem* as reachability of one of the exits with the empty context. Hence, given an RTA $\mathcal{H}$ and an initial node q and a valuation $\nu \in \mathbb{R}^{|\mathcal{X}|}$, the termination problem on $\mathcal{H}$ is to decide the existence of a run in the LTS $[\![\mathcal{H}]\!]$ from initial state $(\langle\varepsilon\rangle, q, \nu)$ to a final state in $Term_\mathcal{H}$.

Given a run $r = \langle s_0, (t_1, a_1), s_2, (t_2, a_2), \ldots, (s_n, t_n) \rangle$ of an RTA, its time duration $time(r)$ is defined as $\sum_{i=1}^{n} t_i$. Given a recursive timed automaton $\mathcal{H}$, an initial node q, a bound $T \in \mathbb{N}$, and valuation $\nu \in \mathbb{R}^{|\mathcal{X}|}$, and a set of *final locations* $F \subseteq Q$, the *time-bounded reachability problem* on $\mathcal{H}$ is to decide the existence of a run r in the LTS $[\![\mathcal{H}]\!]$ staring from the initial state $(\langle\varepsilon\rangle, q, \nu)$ to some state in $[\![F]\!]_\mathcal{H}$ such that $time(r) \leq T$. Time-bounded termination problem is defined in an analogous manner. The following is the key result of the paper which is proved in the rest of the paper.

Theorem 7. *Time-Bounded Reachability problem is undecidable for unrestricted RTAs with at least 5 clocks.*

4   Undecidability of Time-Bounded Reachability Problem

In this section, we provide a complete proof of Theorem 7 by reducing the halting problem for two counter machines to the reachability problem in an RTA.

A *two-counter machine* M is a tuple (L, CTR) where $L = \{\ell_0, \ell_1, \ldots, \ell_n\}$ is the set of instructions including a distinguished terminal instruction ℓ_n called HALT, and the set $CTR = \{C, D\}$ of two *counters*. The instructions L are of the type:

1. (increment c_j) $\ell_i : c_j := c_j + 1$; goto ℓ_k,
2. (decrement c_j) $\ell_i : c_j := c_j - 1$; goto ℓ_k,
3. (zero-check c_j) $\ell_i :$ if $(c_j > 0)$ then goto ℓ_k else goto ℓ_m,

where $c_j \in CTR$, $\ell_i, \ell_k, \ell_m \in L$. A configuration of a two-counter machine is a tuple (ℓ, c, d) where $\ell \in L$ is an instruction, and $c, d \in \mathbb{N}$ are the values of counters C and D, resp. A run of a two-counter machine is a (finite or infinite) sequence of configurations $\langle k_0, k_1, \ldots \rangle$ where $k_0 = (\ell_0, 0, 0)$ and the relation between subsequent configurations is governed by transitions between respective instructions. The *halting problem* for a two-counter machine asks whether its unique run ends at the terminal instruction ℓ_n. It is well known ([10]) that the halting problem for two-counter machines is undecidable.

In order to prove the results of Theorem 7, we construct a recursive timed automaton whose main components simulate various instructions. In these constructions the reachability of the exit node of each component corresponding to an instruction is due to a faithful simulation of various increment, decrement and zero check instructions of the machine by choosing appropriate delays to adjust the clocks, to reflect changes in counter values.

We specify a main component for each instruction of the two counter machine. The entry node and exit node of a main component corresponding to an increment instruction $\langle \ell_i : c_j := c_j + 1$; goto $\ell_m \rangle$ are respectively ℓ_i and ℓ_m. Similarly, a main component corresponding to a zero check instruction $\langle \ell_i :$ if $(c_j > 0)$ then goto ℓ_m else goto $\ell_n \rangle$, has a unique entry node ℓ_i, and two exit nodes corresponding to ℓ_m and ℓ_n respectively. We get the complete RTA for the two-counter machine when we connect these main components in the same sequence as the corresponding machine. We prove that the problem of reaching a chosen vertex in an RTA within 18 units of total elapsed time is undecidable. In order to get the undecidability result, we use a reduction from the halting problem for two counter machines. Our reduction uses an RTA with 5 clocks.

We maintain three sets of clocks. The first set $X = \{x\}$ encodes correctly the current value of counter C; the second set $Y = \{y\}$ encodes correctly the current value of counter D; while the third set $Z = \{z_1, z_2\}$ of 2 clocks helps in zero-check. An extra clock b is used to enforce urgency in some locations. The clock b is zero at the entry nodes of all the main components. Let $\mathcal{X}$ be the set of all 5 clocks.

To be precise, on entry into a main component simulating the $(k + 1)$th instruction, we have the values of z_1, z_2 as $\nu(Z) = 1 - \frac{1}{2^k}$, the value of x as

$\nu(x) = 1 - \frac{1}{2^{c+k}}$, and the value of y as $\nu(y) = 1 - \frac{1}{2^{d+k}}$, where c, d are the current values of the counters after simulating the first k instructions. If the $(k+1)$th instruction ℓ_{k+1} is an increment counter C instruction, then after the simulation of ℓ_{k+1}, we need $\nu(Z) = 1 - \frac{1}{2^{k+1}}$, $\nu(x) = 1 - \frac{1}{2^{c+k+2}}$ and $\nu(y) = 1 - \frac{1}{2^{d+k+1}}$. Similarly, if ℓ_{k+1} is a decrement C instruction, then after the simulation of ℓ_{k+1}, we need $\nu(Z) = 1 - \frac{1}{2^{k+1}}$, $\nu(x) = 1 - \frac{1}{2^{c+k}}$ and $\nu(y) = 1 - \frac{1}{2^{d+k+1}}$. Likewise, if ℓ_{k+1} is a zero check instruction, then after the simulation of ℓ_{k+1}, we need $\nu(Z) = 1 - \frac{1}{2^{k+1}}$, $\nu(x) = 1 - \frac{1}{2^{c+k+1}}$ and $\nu(y) = 1 - \frac{1}{2^{d+k+1}}$. We show by our construction that, the time taken to simulate the $(k+1)$th instruction is $< \frac{9}{2^k}$. Thus, the time taken to simulate the first instruction is < 9, the second instruction is $< \frac{9}{2} \ldots$, so that the total time taken in simulating the two counter machine is < 18.

Increment Instruction. Let us discuss the case of simulating an increment instruction for counter C. Assume that this is the $(k+1)$th instruction. Figure 3 gives the figure for incrementing counter C. At the entry node en_1 of the component INC c, we have $\nu(x) = 1 - \frac{1}{2^{c+k}}$, $\nu(y) = 1 - \frac{1}{2^{d+k}}$ and $\nu(Z) = 1 - \frac{1}{2^k}$, and $\nu(b) = 0$.

The component INC c has three subcomponents sequentially lined up one after another: Let $\beta = \frac{1}{2^k}, \beta_c = \frac{1}{2^{c+k}}$, and $\beta_d = \frac{1}{2^{d+k}}$.

1. The first subcomponent is UP_2^y. If UP_2^y is entered with $\nu(y) = 1 - \beta_d$, then on exit, we have $\nu(y) = 1 - \frac{\beta_d}{2}$. The values of X, Z are unchanged. Also, the total time elapsed in UP_2^y is $\leq \frac{5\beta}{2}$.
2. The next subcomponent is UP_4^x. If UP_4^x is entered with $\nu(x) = 1 - \beta_c$, then on exit, we have $\nu(x) = 1 - \frac{\beta_c}{4}$. The values of Z, Y are unchanged. Also, the total time elapsed in UP_4^x is $\leq \frac{11\beta}{4}$.
3. The next subcomponent is UP_2^Z which updates the value of Z. If UP_2^Z is entered with $\nu(Z) = 1 - \beta$, then on exit, we have $\nu(Z) = 1 - \frac{\beta}{2}$. The values of X, Y are unchanged. Also, the total time elapsed in UP_2^Z is $\leq \frac{5\beta}{2}$.
4. Thus, at the end of the INC c, we obtain $\nu(Z) = 1 - \frac{1}{2^{k+1}}$, $\nu(x) = 1 - \frac{1}{2^{c+k+2}}$, $\nu(y) = 1 - \frac{1}{2^{d+k+1}}$. Also, the total time elapsed in INC c is $\leq [\frac{5}{2} + \frac{11}{4} + \frac{5}{2}]\beta < 8\beta$.

Consider subcomponent UP_2^y obtained by instantiating a by y and n by 2 in the UP_n^a in the Figure 3. Let us discuss the details of UP_2^y, UP_4^x and UP_2^Z have similar functionality.

1. On entry into the first subcomponent $F_4:D$, we have $\nu(Z) = 1 - \beta$, $\nu(b) = 0$, $\nu(x) = 1 - \beta_c$, $\nu(y) = 1 - \beta_d$. D is called, and clock z_2 is passed by reference and the rest by value. A non-deterministic amount of time t_1 elapses at the entry node en_3 of D. At the return port of $F_4:D$, we have z_2 added by t_1.
2. We are then at the entry node of the subcomponent $F_5:C_{z_2}^{y=}$ with values $\nu(z_2) = 1 - \beta + t_1$, and $\nu(z_1) = 1 - \beta$, $\nu(x) = 1 - \beta_c$, $\nu(y) = 1 - \beta_d$ and $\nu(b) = 0$. $C_{z_2}^{y=}$ is called by passing all clocks by value. The subcomponent $C_{z_2}^{y=}$ ensures that $t_1 = \beta - \beta_d$.
3. To ensure $t_1 = \beta - \beta_d$, at the entry node en_4 of $C_{z_2}^{y=}$, a time β_d elapses. This makes $y = 1$. If z_2 must be 1, then we need $1 - \beta + t_1 + \beta_d = 1$, or the time t_1 elapsed is $\beta - \beta_d$. That is, $C_{z_2}^{y=}$ ensures that z_2 has grown to be equal

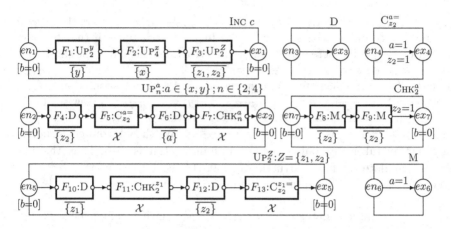

Fig. 3. *TB Term* in RTA: Increment c. Note that $a \in \{x, y\}$. $C_{z_2}^{z_1=}$ is obtained by instantiating $a = z_1$ in $C_{z_2}^{a=}$. The component CHK_4^a is similar to CHK_2^a. It has 4 calls to M inside it each time passing only z_2 by reference.

to y by calling $F_4:D$. At the return port of $F_5:C_{z_2}^{y=}$, we next enter the call port of $F_6:D$ with $\nu(z_2) = \nu(y) = 1 - \beta_d$ and $\nu(z_1) = 1 - \beta$. D is called by passing y by reference, and all others by value. A non-deterministic amount of time t_2 is elapsed in D. At the return port of $F_6:D$, we get $\nu(z_1) = 1 - \beta$, $\nu(z_2) = 1 - \beta_d$, and $\nu(y) = 1 - \beta_d + t_2$.

4. At the call port of $F_7:\text{CHK}_2^y$, we have the same values, since $b = 0$ has to be satisfied at the exit node ex_7 of CHK_2^y. That is, at the call port of $F_7:\text{CHK}_2^y$, we have $\nu(z_1) = 1 - \beta$, $\nu(z_2) = 1 - \beta_d$, and $\nu(y) = 1 - \beta_d + t_2$. F_7 calls CHK_2^y, and passes all clocks by value. CHK_2^y checks that $t_2 = \frac{\beta_d}{2}$.

5. At the entry port en_7 of CHK_2^y, no time elapses. CHK_2^y sequentially calls M twice, each time passing z_2 by reference, and all others by value. In the first invocation of M, we want y to reach 1; thus a time $\beta_d - t_2$ is spent at en_6. This makes $z_2 = 1 - \beta_d + \beta_d - t_2 = 1 - t_2$. After the second invocation, we obtain $z_2 = 1 + \beta_d - 2t_2$ at the return port of $F_9:M$. No time can elapse at the return port of $F_9:M$; for z_2 to be 1, we need $t_2 = \frac{\beta_d}{2}$.

6. No time elapses in the return port of $F_7:\text{CHK}_2^y$, and we are at the exit node ex_2 of UP_2^y. Now, we have $\nu(z_1) = 1 - \beta$, $\nu(z_2) = 1 - \beta_d$ and $\nu(y) = 1 - \beta_d + t_2 = 1 - \frac{\beta_d}{2}$.

7. The time elapsed in UP_2^y is the sum of t_1, t_2 and the times elapsed in $C_{z_2}^{y=}$ and CHK_2^y. That is, $(\beta - \beta_d) + \frac{\beta_d}{2} + \beta_d + 2(\beta_d - t_2) = \beta + \frac{3\beta_d}{2} \le \frac{5\beta}{2}$ since $\beta_d \le \beta$.

At the return port of $F_1 : \text{UP}_2^y$, we thus have $\nu(Z) = 1 - \beta$ ($\nu(z)$ restored to $1 - \beta$ as it was passed by value to UP_2^y), $\nu(x) = 1 - \beta_c$, and $\nu(y) = 1 - \frac{\beta_d}{2}$. No time elapses here, and we are at the call port of $F_2 : \text{UP}_4^x$. The component CHK_4^x is similar to CHK_2^y. It has 4 calls to M inside it, each passing respectively, z_2 by reference to M and x by value. An analysis similar to the above gives that the total time elapsed in UP_4^x is $\le \frac{11\beta}{4}$, and at the return port of $F_2 : \text{UP}_4^x$,

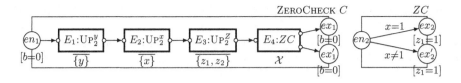

Fig. 4. Unrestricted RTA with 5 clocks: Zero Check if ($C = 0$) then reaches exit ex_1 else reaches ex_1'. Note that $\overline{S} = \mathcal{X} - S$. UP_2^y, UP_2^x and UP_2^Z are in Figure 3.

we get $\nu(x) = 1 - \frac{\beta_c}{4}$, $\nu(y) = 1 - \frac{\beta_d}{2}$ and $\nu(Z) = 1 - \beta$. This is followed by entering $F_3 : \text{UP}_2^Z$, with these values. At the return port of $F_3 : \text{UP}_2^Z$, we obtain $\nu(x) = 1 - \frac{\beta_c}{4}$, $\nu(y) = 1 - \frac{\beta_d}{2}$ and $\nu(Z) = 1 - \frac{\beta}{2}$, with the total time elapsed in UP_2^Z being $\leq \frac{5\beta}{2}$.

From the explanations above, the following propositions can be proved. The same arguments given above will apply to prove this.

Proposition 8. *For any box B and context $\langle \kappa \rangle$, and $\nu(Z) = 1 - \beta$, we have that*
$$(\langle \kappa \rangle, (B, en), (\nu(x), \nu(y), 1 - \beta, \nu(b))) \xrightarrow[\text{UP}_2^Z]{*} (\langle \kappa \rangle, (B, ex), (\nu(x), \nu(y), 1 - \tfrac{\beta}{2}, \nu(b))).$$

Proposition 9. *For any box B and context $\langle \kappa \rangle$, and $\nu(x) = 1 - \beta_c$, we have that*
$$(\langle \kappa \rangle, (B, en), (1 - \beta_c, \nu(y), \nu(Z), \nu(b))) \xrightarrow[\text{UP}_4^x]{*} (\langle \kappa \rangle, (B, ex), (1 - \tfrac{\beta_c}{4}, \nu(y), \nu(Z), \nu(b))).$$

Proposition 10. *For any box B and context $\langle \kappa \rangle$, and $\nu(y) = 1 - \beta_d$, we have*
$$(\langle \kappa \rangle, (B, en), (\nu(x), 1 - \beta_d, \nu(Z), \nu(b))) \xrightarrow[\text{UP}_2^y]{*} (\langle \kappa \rangle, (B, ex), (\nu(x), 1 - \tfrac{\beta_d}{2}, \nu(Z), \nu(b))).$$

Decrement Instruction. Assume that the $(k + 1)$st instruction is decrementing counter C. Then we construct the main component $\text{DEC}\ c$ similar to the component $\text{INC}\ c$ above. The main component $\text{DEC}\ c$ will have the subcomponents UP_2^y and UP_2^Z lined up sequentially. There is no need for any UP_n^x subcomponent here, since the value of x stays unchanged on decrementing c. The total time spent in $\text{DEC}\ c$ is also, less than 8β.

Zero Check Instruction. The main component for checking if counter $C = 0$ is given in Figure 4. It follows the same pattern as $\text{INC}\ c$. Components UP_2^y, UP_2^x and UP_2^Z update the clock values to account for end of the $k + 1$ instruction : changing the value of x from $1 - \beta_c$ to $1 - \frac{\beta_c}{2}$, of y from $1 - \beta_d$ to $1 - \frac{\beta_d}{2}$ and that of Z from $1 - \beta$ to $1 - \frac{\beta}{2}$. Additionally, the final component ZC, to which all clocks are passed by value, checks if $x = z$ by checking $z_1 = 1 \wedge x = 1$ which means $c = 0$ else $z_1 = 1 \wedge x_1 \neq 1$ meaning $c \neq 0$. Time elapsed in ZC component is $\frac{\beta}{2}$ (to make $z_1 = 1$). Thus, total time spent in $\text{ZEROCHECK}\ c$ is $< 9\beta$.

Note that the components for incrementing, decrementing and zero check of counter D can be obtained in a manner similar to the above. The proof that we reach the vertex *Halt* of the RTA iff the two counter machine halts follows: Clearly, the exit node of each main component is reached iff the corresponding instruction is simulated correctly. Thus, if the counter machine halts, we will indeed reach the exit node of the main component corresponding to the last instruction. However, if the machine does not halt, then we keep going between the various main components simulating each instruction, and never reach *Halt*.

Acknowledgments. This work is partially supported by CEFIPRA funded project AVeRTS (Algorithmic Verification of Real-Time Systems). The second author is supported by Microsoft Research Fellowship RSMSRI0033.

References

1. Abdulla, P.A., Atig, M.F., Stenman, J.: Dense-timed pushdown automata. In: Proceedings of the 27th Annual IEEE Symposium on Logic in Computer Science. LICS, pp. 35–44 (2012)
2. Alur, R., Benedikt, M., Etessami, K., Godefroid, P., Reps, T., Yannakakis, M.: Analysis of recursive state machines. ACM Trans. Program. Lang. Syst. **27**(4), 786–818 (2005)
3. Alur, R., Dill, D.: Automata for modeling real-time systems. In: Paterson, M. (ed.) ICALP 1990. LNCS, vol. 443, pp. 322–335. Springer, Heidelberg (1990)
4. Ball, T., Rajamani, S.K.: Bebop: A symbolic model checker for boolean programs. In: Havelund, K., Penix, J., Visser, W. (eds.) SPIN 2000. LNCS, vol. 1885, pp. 113–130. Springer, Heidelberg (2000)
5. Benerecetti, M., Minopoli, S., Peron, A.: Analysis of timed recursive state machines. In: Proceedings of 17th International Symposium on Temporal Representation and Reasoning TIME, pp. 61–68 (2010)
6. Mazurkiewicz, A., Winkowski, J.: Reachability analysis of pushdown automata: Application to model-checking. In: Mazurkiewicz, A., Winkowski, J. (eds.) CONCUR 1997. LNCS, vol. 1243, pp. 135–150. Springer, Heidelberg (1997)
7. Esparza, J., Hansel, D., Rossmanith, P., Schwoon, S.: Efficient algorithms for modelchecking pushdown systems. In: Emerson, E., Sistla, A.P. (eds.) CAV 2000. LNCS, vol. 1855, pp. 232–247. Springer, Heidelberg (2000)
8. Harel, D.: Statecharts: A visual formalism for complex systems. Sci. Comput. Program. **8**(3), 231–274 (1987)
9. Krishna, S.N., Manasa, L., Trivedi, A.: Improved undecidability results for reachability games on recursive timed automata. In: Proceedings of Fifth International Symposium on Games, Automata, Logics and Formal Verification GandALF, pp. 245–259 (2014)
10. Minsky, M.L.: Computation: finite and infinite machines. Prentice-Hall, Inc. (1967)
11. Rumbaugh, J., Jacobson, I., Booch, G.: The unified modeling language reference manual. Addison-Wesley-Longman (1999)
12. Trivedi, A., Wojtczak, D.: Recursive timed automata. In: Bouajjani, A., Chin, W.-N. (eds.) ATVA 2010. LNCS, vol. 6252, pp. 306–324. Springer, Heidelberg (2010)

On Observability of Automata Networks via Computational Algebra

Rui Li[(✉)] and Yiguang Hong

Key Laboratory of Systems and Control, Institute of Systems Science,
Chinese Academy of Sciences, Beijing 100190, China
rui.li@pku.edu.cn, yghong@iss.ac.cn

Abstract. Automata networks have been successfully used as abstract modeling schemes in many different fields. This paper deals with the observability of automata networks, which describes the ability to uniquely infer the network's initial configuration by measuring its output sequence. Simple necessary and sufficient conditions for observability are given. The results employ techniques from symbolic computation and can be easily implemented within the computer algebra environments. Two examples are worked out to illustrate the application of the results.

Keywords: Automata networks · Computational algebra · Discrete dynamical systems · Observability

1 Introduction

Automata networks are discrete dynamical systems originally introduced by McCulloch, Pitts, and von Neumann [13,14]. Roughly speaking, an automata network is defined by a graph with each vertex taking states in a finite set. The vertex variables can change their states at discrete time steps according to some given transition rules that take into account only the states of their neighborhoods in the graph. The configuration of the network is completely specified by the states of the variables at each vertex. In fact, automata networks provide a nice framework for modeling and analyzing various phenomena in areas as diverse as physics, biology, and computer science (see, e.g., [1,4]).

Over the years, much effort has been devoted to the study of automata networks. For example, Saito and Nishio [19] considered the structural equivalence relation induced by the structure of a graph and the behavioral equivalence relation induced by the behavior of each vertex variable, and they further discussed the relationships between these equivalence relations. Paulevé *et al.* [15] considered the reachability of an automata network and proposed an algorithm for identifying sets of state variables whose activity is necessary for reachability. In [3], the problem of modeling asynchronous discrete event systems by a network of input-output automata was studied. A comparison regarding automata networks and input-output automata networks was made in [9]. Finite automata systems as

© Springer International Publishing Switzerland 2015
A.-H. Dediu et al. (Eds.): LATA 2015, LNCS 8977, pp. 249–262, 2015.
DOI: 10.1007/978-3-319-15579-1_19

parallel communicating automata networks were investigated in [10–12]. Negative circuits and sustained oscillations in asynchronous automata networks were analyzed in [17].

Quantitative descriptions of practical systems modeled using automata networks are inherently limited by our ability to estimate the networks' configurations from experimentally available outputs. Although simultaneous measurement of all state variables furnishes a complete description of a network's configuration, in practice the measurement is often limited to a subset or a function of state variables. It is known that observability of automata has been studied for the state estimation or identification [16,23]. Similarly, observability of an automata network also allows us to uniquely infer the network's initial configuration from its outputs. However, to our knowledge, very few works have been done on this topic.

In this paper, we consider the observability of automata networks. The methods to check the observability of automata given in [16,23] are not efficient in automata network situations, though a brute-force enumeration can be certainly carried out to test the observability for a small-size automata network. However, for a large-scale network, brute-force search becomes unfeasible since the number of network configurations grows exponentially with the number of vertices of the graph. For example, in a model defined on a graph consisting of 100 vertices with each vertex taking 3 possible states, there are a total of $3^{100} \approx 5.2 \times 10^{47}$ possible network configurations. Even if a fast computer could check 1 trillion configurations per second, it would still take octillions of years to test the complete space of network configurations. We therefore need other methods to replace model simulation. In this context, the purpose of this paper is to use a symbolic computation approach, which draws from the rich theory of computational algebra, to analyze the observability of automata networks (see Appendix A for some definitions and properties of commutative algebra and finite field algebra). The results obtained provide necessary and sufficient conditions for observability that can be easily implemented in computer algebra systems, and are thus believed to scale well to some large automata networks.

The remainder of this paper is organized as follows. Section 2 gives the definition of an automata network and describes the problem formulation. Section 3 analyzes the observability of automata networks and derives the main results. Two examples are treated in Section 4, and a brief conclusion is drawn in Section 5.

2   Background and Problem Formulation

In this section, we provide the research background and formulate our problem.

2.1   Automata Networks

We first recall that a (finite) directed graph $G = (V, E)$ is a pair consisting of sets of vertices $V = \{1, 2, \ldots, n\}$ and edges $E \subset V \times V$. The vertex i is said to be a neighbor of j if there is an edge from i to j, that is, $(i, j) \in E$. The neighborhood of

j, denoted N_j, is the set of all neighbors of j. Given a directed graph, we can define an automata network by allocating a finite state automaton (also called a cell) to each vertex. More precisely, an automata network is a quintuple

$$\mathscr{A} = (G, \{X_i\}_{i=1}^n, Y, \{f_i\}_{i=1}^n, h), \tag{1}$$

where

- $G = (V = \{1, 2, \ldots, n\}, E)$ is a directed graph,
- X_i is the finite set of states for the automaton placed at the vertex i,
- $Y = \prod_{\lambda=1}^q Y_\lambda$ with each Y_λ finite is the output set,
- $f_i \colon \prod_{j \in N_i} X_j \to X_i$ is the local transition function associated to the vertex i,
- $h \colon \prod_{i=1}^n X_i \to Y$ is the output map.

Let $X = \prod_{i=1}^n X_i$. Without loss of generality it can be assumed that

$$
\begin{aligned}
X_i &= \{0, 1, \ldots, m_i\}, & i &= 1, 2, \ldots, n, \\
Y_\lambda &= \{0, 1, \ldots, l_\lambda\}, & \lambda &= 1, 2, \ldots, q,
\end{aligned}
$$

and that each f_i is defined on X since the graph is finite. An element $x = (x_1, x_2, \ldots, x_n)$ of X is called a configuration of $\mathscr{A}$. The global transition function of $\mathscr{A}$, which transforms a configuration into another one, defines the dynamics of the network, and is constructed with the local transition functions $\{f_i\}_{i=1}^n$ and with some kind of updating scheme. The updating scheme of an automata network can take different forms, among which the two most common ones are the synchronous (or parallel) and the sequential updating schemes [4].

In the synchronous scheme, the states of all automata in the network $\mathscr{A}$ are updated at the same time. So the global transition function of $\mathscr{A}$ is $F = (f_1, f_2, \ldots, f_n)$, and the corresponding dynamics of $\mathscr{A}$ is described by

$$
\begin{aligned}
x(t+1) &= F(x(t)), \\
y(t) &= h(x(t)), & t = 0, 1, 2, \ldots,
\end{aligned} \tag{2}
$$

where $x(t)$ and $y(t)$ denote the configuration and the output of $\mathscr{A}$ at time t, taking values in X and Y, respectively.

In a sequential updating scheme, the automata in the network are assumed to have a prescribed ordering (given by a relation denoted by $<$) and update one by one sequentially. That is, starting from an initial configuration, the first automaton updates first, then the second automaton updates, taking account of the effects of the changes in the first automaton, and so on. For instance, if we assume that

$$\text{the automaton } i < \text{the automaton } j \Leftrightarrow i < j,$$

then the dynamics of the sequential iteration is

$$
\begin{aligned}
x_1(t+1) &= f_1(x_1(t), x_2(t), \ldots, x_n(t)), \\
x_2(t+1) &= f_2(x_1(t+1), x_2(t), \ldots, x_n(t)), \\
&\vdots \\
x_n(t+1) &= f_n(x_1(t+1), x_2(t+1), \ldots, x_{n-1}(t+1), x_n(t)), \quad t = 0, 1, 2, \ldots,
\end{aligned} \tag{3}
$$

where $x_i(t)$ denotes the state of the automaton i at time t.

Remark 1. Given local transition functions $f_1, \ldots, f_n$, there are $n!$ possible sequential modes of operation for $\mathscr{A}$, corresponding to the $n!$ permutations of the set $V = \{1, 2, \ldots, n\}$. It is a known fact that each sequential mode of operation is identical to a synchronous mode of operation [18]. For example, for every $1 \le i \le n$, let F_i be the map $X \to X$, given by

$$x = (x_1, \ldots, x_n) \mapsto F_i(x) = (x_1, \ldots, f_i(x), \ldots, x_n),$$

and put $G = F_n \circ \cdots \circ F_2 \circ F_1$. Then the sequential iteration (3) can be written as

$$x(t+1) = G(x(t)), \qquad t = 0, 1, 2, \ldots,$$

which means that the sequential operation mode given by (3) is identical to the synchronous operation mode defined by G.

2.2   Problem Formulation

Consider an automata network $\mathscr{A}$ of the form (1). For an initial configuration $x(0) \in X$, let $y(t; x(0)) \in Y$ denote the output of $\mathscr{A}$ at the time step t. We say that two different configurations $\xi, \eta \in X$ are indistinguishable if $y(t; \xi) = y(t; \eta)$ for all $t \ge 0$. In this case, it is not possible to uniquely infer the complete initial configuration of $\mathscr{A}$ by observing the output sequence.

Definition 1.   The automata network $\mathscr{A}$ is said to be observable if any two different configurations $\xi, \eta \in X$ are distinguishable.

In other words, observability means that the map sending each initial configuration to its output sequence is injective.

In this paper, we study the observability for the automata network $\mathscr{A}$. The aim is to find necessary and sufficient conditions to analyze the question whether or not an automata network is observable. We will approach this question by using techniques from computational algebra (see Appendix A for notation and terminology). It is clear that, from a theoretical point of view, we only need to consider automata networks that operate in synchronous modes.

3   Main Results

Let $\mathscr{A}$ be an automata network of the form (1). In the sequel, we assume that $\mathscr{A}$ operates in the synchronous mode with dynamics given by (2). To analyze observability, it is useful to impose algebraic structures on the sets X and Y. Choose a prime integer p such that

$$p \ge \max\{m_1 + 1, \ldots, m_n + 1, l_1 + 1, \ldots, l_q + 1\},$$

and let $\mathbb{F}_p = \{0, 1, \ldots, p - 1\}$ be the prime field with p elements. Then we can consider that $X \subset \mathbb{F}_p^n$ and $Y \subset \mathbb{F}_p^q$.

As a consequence, the coordinate functions of F and h can be represented by polynomials over $\mathbb{F}_p$. That is, there exist polynomials $\tilde{f}_1, \ldots, \tilde{f}_n, \tilde{h}_1, \ldots, \tilde{h}_q \in \mathbb{F}_p[x_1, \ldots, x_n]$ such that

$$F(a_1, \ldots, a_n) = (\tilde{f}_1(a_1, \ldots, a_n), \ldots, \tilde{f}_n(a_1, \ldots, a_n)),$$
$$h(a_1, \ldots, a_n) = (\tilde{h}_1(a_1, \ldots, a_n), \ldots, \tilde{h}_q(a_1, \ldots, a_n))$$

for all $(a_1, \ldots, a_n) \in X$. It is worth noting that these polynomial representatives may not be unique. For example, the polynomials

$$g_i(x_1, \ldots, x_n) = \sum_{(a_1, \ldots, a_n) \in X} f_i(a_1, \ldots, a_n)(1 - (x_1 - a_1)^{p-1}) \cdots (1 - (x_n - a_n)^{p-1})$$

$$(4)$$

and

$$g_i'(x_1, \ldots, x_n) = \sum_{(b_1, \ldots, b_n) \in \mathbb{F}_p^n} f_i(\min\{b_1, m_1\}, \ldots, \min\{b_n, m_n\})$$

$$\times (1 - (x_1 - b_1)^{p-1}) \cdots (1 - (x_n - b_n)^{p-1}) \quad (5)$$

define the same local transition function f_i on X. Put $f = (\tilde{f}_1, \ldots, \tilde{f}_n) \in (\mathbb{F}_p[x_1, \ldots, x_n])^n$. We say that f is a proper representative for F if $f(a_1, \ldots, a_n) \in X$ for all $(a_1, \ldots, a_n) \in \mathbb{F}_p^n$. Note that $(g_1, \ldots, g_n)$ and $(g_1', \ldots, g_n')$, with g_i and g_i' given by (4) and (5), respectively, are both proper representatives for F. Thus, every $F \colon X \to X$ can be expressed by a proper representative.

Let $f^{(k)}$ denote the composition of f with itself k times. That is $f^{(k)} = f \circ f \circ \cdots \circ f$ (k times), where by convention $f^{(0)} = (x_1, x_2, \ldots, x_n)$. We define a sequence of ideals in $\mathbb{F}_p[x_1, \ldots, x_n, z_1, \ldots, z_n]$ recursively as follows:

$$J_0 = \langle x_i(x_i - 1) \cdots (x_i - m_i) \colon 1 \le i \le n \rangle + \langle z_i(z_i - 1) \cdots (z_i - m_i) \colon 1 \le i \le n \rangle$$
$$+ \langle \tilde{h}_j(x_1, \ldots, x_n) - \tilde{h}_j(z_1, \ldots, z_n) \colon 1 \le j \le q \rangle,$$
$$J_{k+1} = J_k + \langle \tilde{h}_j(f^{(k+1)}(x_1, \ldots, x_n)) - \tilde{h}_j(f^{(k+1)}(z_1, \ldots, z_n)) \colon 1 \le j \le q \rangle,$$
$$k = 0, 1, 2, \ldots.$$

Observe that for all $k \ge 0$ we have $J_k \subset J_{k+1}$, and so

$$J_0 \subset J_1 \subset J_2 \subset \cdots$$

forms an ascending chain of ideals. Since the ring $\mathbb{F}_p[x_1, \ldots, x_n, z_1, \ldots, z_n]$ satisfies the ascending chain condition (see Appendix A), there exists an integer $N \ge 0$ such that

$$J_N = J_{N+1} = J_{N+2} = \cdots.$$

Write $J = J_N$. The following theorem tells us that the ideal J characterizes the observability of the automata network $\mathscr{A}$.

Theorem 1. *The automata network $\mathscr{A}$ is observable if and only if*

$$V(J) = \{(\zeta, \zeta) \colon \zeta \in X\},\tag{6}$$

where $V(J)$ is the affine variety defined by J (see Appendix A for the definition of an affine variety).

Proof. (*Necessity*). It is clear that $V(J) \supset \{(\zeta, \zeta) \colon \zeta \in X\}$. To establish the reverse inclusion, suppose, on the contrary, that there is $(\xi, \eta) \in V(J)$ which is not in $\{(\zeta, \zeta) \colon \zeta \in X\}$. Write $\xi = (\xi_1, \ldots, \xi_n)$, $\eta = (\eta_1, \ldots, \eta_n)$. Since $J_0 \subset J$, it follows from Lemma A1 that

$$\begin{aligned}
V(J) \subset V(J_0) &= V(\langle x_i(x_i - 1) \cdots (x_i - m_i) \colon 1 \le i \le n \rangle) \\
&\quad \cap V(\langle z_i(z_i - 1) \cdots (z_i - m_i) \colon 1 \le i \le n \rangle) \\
&\quad \cap V(\langle \tilde{h}_j(x_1, \ldots, x_n) - \tilde{h}_j(z_1, \ldots, z_n) \colon 1 \le j \le q \rangle).
\end{aligned}$$

Hence

$$\xi_i(\xi_i - 1) \cdots (\xi_i - m_i) = 0, \quad \eta_i(\eta_i - 1) \cdots (\eta_i - m_i) = 0, \quad i = 1, 2, \ldots, n,$$

so that $\xi_i, \eta_i \in X_i$ for each i. Thus, $\xi, \eta \in X$ and $\xi \ne \eta$.

Since $\mathscr{A}$ is observable, there exists an integer $k \ge 0$ such that

$$\tilde{h}_j(f^{(k)}(\xi_1, \ldots, \xi_n)) \ne \tilde{h}_j(f^{(k)}(\eta_1, \ldots, \eta_n))$$

for some $1 \le j \le q$. Hence $(\xi, \eta) \notin V(J_k)$. However, since $J_k \subset J$, it follows that $(\xi, \eta) \in V(J) \subset V(J_k)$, which is a contradiction. Therefore, $V(J) \subset \{(\zeta, \zeta) \colon \zeta \in X\}$.

(*Sufficiency*). Suppose that there exist distinct configurations $\xi, \eta \in X$ which are indistinguishable. Then for $j - 1, 2, \ldots, q$, $k - 0, 1, 2, \ldots$, we have

$$\tilde{h}_j(f^{(k)}(\xi)) = \tilde{h}_j(f^{(k)}(\eta)).$$

Hence

$$(\xi, \eta) \in V(\langle \tilde{h}_j(f^{(k)}(x_1, \ldots, x_n)) - \tilde{h}_j(f^{(k)}(z_1, \ldots, z_n)) \colon 1 \le j \le q \rangle),$$
$$k = 0, 1, 2, \ldots.$$

It is clear that

$$\begin{aligned}
(\xi, \eta) \in\ & V(\langle x_i(x_i - 1) \cdots (x_i - m_i) \colon 1 \le i \le n \rangle) \\
& \cap V(\langle z_i(z_i - 1) \cdots (z_i - m_i) \colon 1 \le i \le n \rangle).
\end{aligned}$$

Then a straightforward induction on k shows that $(\xi, \eta) \in V(J_k)$ for all $k \ge 0$, so that $(\xi, \eta) \in V(J) = \{(\zeta, \zeta) \colon \zeta \in X\}$. This contradiction proves that every pair of distinct configurations of $\mathscr{A}$ are distinguishable. $\qquad\square$

Theorem 1 asserts that the observability of the automata network $\mathscr{A}$ can be judged by computing the variety $V(J)$. Note that the ideal J satisfies

$$\langle x_1^p - x_1, \ldots, x_n^p - x_n, z_1^p - z_1, \ldots, z_n^p - z_n \rangle \subset J, \tag{7}$$

since in the polynomial ring $\mathbb{F}_p[x_1, \ldots, x_n, z_1, \ldots, z_n]$ one has $x_i^p - x_i = x_i(x_i - 1) \cdots (x_i - p + 1)$, and $z_i^p - z_i = z_i(z_i - 1) \cdots (z_i - p + 1)$ for all $1 \leq i \leq n$. It is easily verified that if $J = \langle x_1 - z_1, \ldots, x_n - z_n, z_1(z_1 - 1) \cdots (z_1 - m_1), \ldots, z_n(z_n - 1) \cdots (z_n - m_n) \rangle$, then (6) and (7) hold. In fact, this is the unique ideal in $\mathbb{F}_p[x_1, \ldots, x_n, z_1, \ldots, z_n]$ that satisfies (6) and (7), as the next proposition shows.

Proposition 1. *If $J \subset \mathbb{F}_p[x_1, \ldots, x_n, z_1, \ldots, z_n]$ is an ideal such that (6) and (7) hold, then*

$$J = \langle x_1 - z_1, \ldots, x_n - z_n, z_1(z_1 - 1) \cdots (z_1 - m_1), \ldots, z_n(z_n - 1) \cdots (z_n - m_n) \rangle.$$

Proof. See Appendix B.   □

As an immediate consequence of Proposition 1, we can deduce the following simpler criterion for observability.

Theorem 2. *The automata network $\mathscr{A}$ is observable if and only if*

$$J = \langle x_1 - z_1, \ldots, x_n - z_n, z_1(z_1 - 1) \cdots (z_1 - m_1), \ldots, z_n(z_n - 1) \cdots (z_n - m_n) \rangle. \tag{8}$$

Remark 2. In [5–7], similar algebraic frameworks were used to address the observability and local observability of polynomial systems over $\mathbb{R}$. There, the results are described in terms of certain ideals and the corresponding affine varieties. In contrast, by exploiting the structure of finite fields, we show that for an automata network, the observability can be checked directly using the ideal J, thereby saving the computation of the variety $V(J)$.

So far we have seen that the ideals J_k play a crucial role in deciding whether or not an automata network is observable. Our next result provides a more convenient way for calculating J_k. (Note that for each integer $k \geq 0$ the ideal J_k is finitely generated; see Appendix A.)

Proposition 2. *Suppose that $f = (\tilde{f}_1, \ldots, \tilde{f}_n)$ is proper. Let $k \geq 0$ be an integer, and let $J_k = \langle g_{k1}, \ldots, g_{kl_k} \rangle$ for some $g_{k1}, \ldots, g_{kl_k} \in \mathbb{F}_p[x_1, \ldots, x_n, z_1, \ldots, z_n]$. Let $A(x_1, \ldots, x_n, z_1, \ldots, z_n) = (f(x_1, \ldots, x_n), f(z_1, \ldots, z_n))$. Then*

$$J_{k+1} = J_0 + \langle g_{k1} \circ A, \ldots, g_{kl_k} \circ A \rangle.$$

Proof. For simplicity of notation, we will write $f(x_1, \ldots, x_n)$ and $f(z_1, \ldots, z_n)$ as $f(x)$ and $f(z)$, respectively.

We proceed by induction on k. For $k = 0$, we have

$$\langle g_{01} \circ A, \ldots, g_{0l_0} \circ A \rangle = \langle \tilde{f}_i(x)(\tilde{f}_i(x) - 1) \cdots (\tilde{f}_i(x) - m_i) \colon 1 \leq i \leq n \rangle$$
$$+ \langle \tilde{f}_i(z)(\tilde{f}_i(z) - 1) \cdots (\tilde{f}_i(z) - m_i) \colon 1 \leq i \leq n \rangle$$
$$+ \langle \tilde{h}_j(f(x)) - \tilde{h}_j(f(z)) \colon 1 \leq j \leq q \rangle.$$

Since f is proper, it follows that

$$\tilde{f}_i(a_1,\ldots,a_n)(\tilde{f}_i(a_1,\ldots,a_n) - 1)\cdots(\tilde{f}_i(a_1,\ldots,a_n) - m_i) = 0$$

for all $(a_1,\ldots,a_n) \in \mathbb{F}_p^n$ and for all $1 \le i \le n$. By Lemma A2,

$$\tilde{f}_i(x)(\tilde{f}_i(x) - 1)\cdots(\tilde{f}_i(x) - m_i) \in \langle x_1^p - x_1,\ldots,x_n^p - x_n\rangle \subset J_0,$$
$$\tilde{f}_i(z)(\tilde{f}_i(z) - 1)\cdots(\tilde{f}_i(z) - m_i) \in \langle z_1^p - z_1,\ldots,z_n^p - z_n\rangle \subset J_0,$$
$$i = 1, 2, \ldots, n.$$

Hence

$$J_0 + \langle g_{01} \circ A, \ldots, g_{0l_0} \circ A\rangle = J_0 + \langle \tilde{h}_j(f(x)) - \tilde{h}_j(f(z)) : 1 \le j \le q\rangle = J_1.$$

Now we assume that $k \ge 1$ and that the statement is true for $k - 1$. Since

$$J_k = \langle g_{k-1\,1},\ldots,g_{k-1\,l_{k-1}}\rangle + \langle \tilde{h}_j(f^{(k)}(x)) - \tilde{h}_j(f^{(k)}(z)) : 1 \le j \le q\rangle,$$

we have

$$\langle g_{k\,1} \circ A, \ldots, g_{k\,l_k} \circ A\rangle = \langle g_{k-1\,1} \circ A, \ldots, g_{k-1\,l_{k-1}} \circ A\rangle$$
$$+ \langle \tilde{h}_j(f^{(k+1)}(x)) - \tilde{h}_j(f^{(k+1)}(z)) : 1 \le j \le q\rangle.$$

Then it follows from the induction hypothesis that

$$J_0 + \langle g_{k1} \circ A, \ldots, g_{kl_k} \circ A\rangle = J_k + \langle \tilde{h}_j(f^{(k+1)}(x)) - \tilde{h}_j(f^{(k+1)}(z)) : 1 \le j \le q\rangle$$
$$= J_{k+1}.$$

This completes the proof of the proposition. □

Proposition 2 also has the following important consequence for determining the ideal J.

Proposition 3. *Suppose that $f = (\tilde{f}_1,\ldots,\tilde{f}_n)$ is proper. If $J_{k+1} = J_k$ for some $k \ge 0$, then $J = J_k$.*

Proof. It suffices to show that $J_{k+2} = J_{k+1}$. Suppose that $J_{k+1} = J_k = \langle g_1,\ldots, g_l\rangle$ for some $g_1,\ldots,g_l \in \mathbb{F}_p[x_1,\ldots,x_n,z_1,\ldots,z_n]$. Then Proposition 2 shows that $J_{k+2} = J_0 + \langle g_1 \circ A, \ldots, g_l \circ A\rangle = J_{k+1}$. □

Remark 3. Note that in order to determine J and to check whether (8) is satisfied, it is required to decide whether or not two sets of polynomials generate the same ideal. This problem can be solved by employing an ideal equality algorithm [2]: simply fix a monomial order and compute the reduced Gröbner bases for the ideals. Then the ideals are equal if and only if the reduced Gröbner bases are the same. The key property used in the algorithm is that for a given monomial order, every nonzero polynomial ideal possesses a unique reduced Gröbner basis. Many computer algebra systems such as Maple and Mathematica have a built-in command for computing a Gröbner basis whose elements are constant multiples of the elements in a reduced Gröbner basis. So the observability condition presented here is easily programmable within the computer algebra environments.

Remark 4. In most cases, polynomial algebra computations do not depend so strongly on the size or complexity of a model [21]. Thus, in practice, it is reasonable to believe that our method can scale gracefully even to some large-scale automata networks.

4   Examples

In this section, we give examples for illustration.

Example 1. Let G be the directed graph on $\mathbb{Z}_{100}$, the set of integers modulo 100, with the neighborhood of the vertex i given by $\{i-1, i, i+1\}$ (modulo 100). Let $\mathscr{A}$ be an automata network defined on G, where each automaton may take on the three possible states $0, 1$, or 2. The local transition function associated to the vertex i is

$$f_i(x_{i-1}, x_i, x_{i+1}) = x_i^2 - x_{i-1} + x_{i+1} \text{ (modulo 3)}.$$

(We interpret the subscripts modulo 100 here and in the remaining part of the example.) We assume that the network operates in the synchronous mode.

Consider now an observability problem for the automata network $\mathscr{A}$. More specifically, we assume that we can monitor a selected subset of automata, which we call sensors, and we are interested in determining the smallest number of sensors from whose measurements the automata network is observable. In other words, observing these sensors allows to determine the network's complete initial configuration. Notice that if $\mathscr{A}$ is observable (resp., not observable) by monitoring the sensors $x_{i_1}, x_{i_2}, \ldots, x_{i_s}$, then it is also observable (resp., not observable) when monitoring $x_{i_1+k}, x_{i_2+k}, \ldots, x_{i_s+k}$, where $1 \leq k \leq 99$.

It is trivial to see that observability holds when taking all automata as sensors. As a next possible step, we consider to monitor $x_0, x_2, x_4, \ldots, x_{98}$, and use the criterion developed above to check the observability of $\mathscr{A}$. The result shows that the corresponding automata network is not observable. So in particular we derive that (i) any sensor set consisting of a single automaton is not sufficient for observability, and (ii) for each $0 \leq i \leq 99$ and each $1 \leq k \leq 49$, the sensor set $\{x_i, x_{i+2k}\}$ does not guarantee observability. We now add another automaton, for example x_1, to the sensor set $\{x_0, x_2, x_4, \ldots, x_{98}\}$ and then check the observability. The result gives that the network is observable. This motivates us to choose to monitor $\{x_0, x_1, x_2\}$ or $\{x_0, x_1\}$. As a consequence, we find that observability holds for both of these choices. Thus, one needs at least two sensors to observe $\mathscr{A}$. For every $0 \leq i \leq 99$, the sensors $\{x_i, x_{i+1}\}$ yield an observable network.

Furthermore, it is easily checked that we cannot obtain observability when monitoring $x_0, x_3, x_5, \ldots, x_{97}$. This implies in particular that sensor sets of the form $\{x_i, x_{i+2k+1}\}$ with $0 \leq i \leq 99$ and $1 \leq k \leq 48$ cannot guarantee observability, so that $\{x_0, x_1\}, \{x_1, x_2\}, \ldots, \{x_{99}, x_0\}$ are exactly all the sensor sets consisting of two automata that ensure observability.   □

Example 2. In this example, we apply the methodology of this paper to an automata network that attempts to model a real genetic system. The network we consider is a discrete model for lactose metabolism in *Escherichia coli* [20]. The global transition function of the network is $F = (f_1, \ldots, f_9) \colon X \to X$, where $X = \{0,1\} \times \{0,1\} \times \{0,1\} \times \{0,1\} \times \{0,1,2\} \times \{0,1,2\} \times \{0,1,2\} \times \{0,1\} \times \{0,1,2\}$. The model is represented in polynomial form (modulo 3) as:

$$f_1 = -x_4^2 x_5^2 + x_4^2,$$
$$f_2 = x_1^2,$$
$$f_3 = x_1^2,$$
$$f_4 = -x_8^2 + 1,$$
$$f_5 = x_6^2 - 1,$$
$$f_6 = -x_3^2 x_7^2 + x_3^2 x_7 + x_7^2,$$
$$f_7 = -x_2^2 x_8^2 x_9^2 - x_2^2 x_8^2 x_9 + x_2 x_8^2 x_9 + x_2^2 x_9^2 + x_8^2 x_9^2 + x_2^2 x_9 - x_8^2 x_9$$
$$\qquad - x_2 x_9 - x_9^2 + x_9,$$
$$f_8 = x_8,$$
$$f_9 = x_9.$$

Here $x_1, x_2, x_3, x_4, x_5, x_6, x_7, x_8, x_9$ denote mRNA, permease, β-galactosidase, CAP, the repressor protein, allolactose, intracellular lactose, glucose, and extracellular lactose, respectively. We refer the interested reader to [20] and [22] for more biological background of the model.

In order to identify the minimum set of sensors that ensures observability, we take 8 out of the 9 nodes as sensors (there are 9 possible combinations), and test each system's observability. For example, consider the case where the sensor set includes all the nodes except for x_1. It is easy to verify that the system is observable, and so x_1 is not necessary for observability. Proceeding in this way, we can conclude that observability holds only for a unique choice of 8 sensor nodes, namely the aforementioned sensor set consisting of all the nodes except for x_1. This indicates in particular that we cannot uncover the system's initial state by measuring less than 8 sensor nodes.   □

5   Conclusion

In many practical systems modeled using automata networks, experimental access is usually limited to a subset or a function of state variables. Hence, one needs to infer network configurations from experimentally accessible outputs. In this paper, we have considered the problem of uniquely determining the initial configuration of an automata network given an output sequence. The discussion is based on the idea of representing the transition functions in polynomial form. We have thus available to us the well-developed theory of computational algebra, with a wide variety of implemented procedures. Necessary and sufficient conditions for observability have been derived by using these powerful tools. Given the fundamental role observability plays, the results provide avenues to systematically explore the dynamical behaviors of automata networks.

A Notation and Mathematical Preliminaries

This appendix briefly summarizes some notation and properties of commutative algebra and finite field algebra used in the paper. Further details can be found in standard texts, such as [2] and [8].

Let p be a prime integer. The ring of integers modulo p forms a field with p elements and is denoted by $\mathbb{F}_p$. The ring of polynomials in $x_1, \ldots, x_n$ over $\mathbb{F}_p$ is denoted by $\mathbb{F}_p[x_1, \ldots, x_n]$. Let $I \subset \mathbb{F}_p[x_1, \ldots, x_n]$ be an ideal. The affine variety defined by I, denoted $V(I)$, is the set

$$\{(a_1, \ldots, a_n) \in \mathbb{F}_p^n \colon f(a_1, \ldots, a_n) = 0 \text{ for all } f \in I\}.$$

Given an affine variety $V \subset \mathbb{F}_p^n$, we define by $I(V)$ the set

$$\{f \in \mathbb{F}_p[x_1, \ldots, x_n] \colon f(a_1, \ldots a_n) = 0 \text{ for all } (a_1, \ldots, a_n) \in V\}.$$

Note that $I(V)$ is actually an ideal in $\mathbb{F}_p[x_1, \ldots, x_n]$, which is called the ideal of V.

Let $f_1, \ldots, f_s$ be polynomials in $\mathbb{F}_p[x_1, \ldots, x_n]$. Denote by $\langle f_1, \ldots, f_s \rangle$ the set $\{\sum_{i=1}^{s} h_i f_i \colon h_1, \ldots, h_s \in \mathbb{F}_p[x_1, \ldots, x_n]\}$. Then $\langle f_1, \ldots, f_s \rangle$ is an ideal of $\mathbb{F}_p[x_1, \ldots, x_n]$. We call $\langle f_1, \ldots, f_s \rangle$ the ideal generated by $f_1, \ldots, f_s$. It follows from the Hilbert Basis Theorem that every ideal $I \subset \mathbb{F}_p[x_1, \ldots, x_n]$ is finitely generated. That is, there exist $f_1, \ldots, f_s \in \mathbb{F}_p[x_1, \ldots, x_n]$ such that $I = \langle f_1, \ldots, f_s \rangle$. This property is equivalent to the statement that $\mathbb{F}_p[x_1, \ldots, x_n]$ satisfies the ascending chain condition. That is, for every chain $I_1 \subset I_2 \subset I_3 \subset \cdots$ of ideals of $\mathbb{F}_p[x_1, \ldots, x_n]$, there is an integer N such that $I_N = I_{N+1} = I_{N+2} = \cdots$.

Let I and J be ideals in $\mathbb{F}_p[x_1, \ldots, x_n]$. The sum of I and J, denoted $I + J$, is the set $\{f + g \colon f \in I, g \in J\}$. Note that $I + J$ is also an ideal in $\mathbb{F}_p[x_1, \ldots, x_n]$.

The following technical lemmas are used in some of our proofs.

Lemma A1 ([2]). *Let I and J be ideals in $\mathbb{F}_p[x_1, \ldots, x_n]$. Then*

(a) $I \subset J \Rightarrow V(J) \subset V(I)$;
(b) $V(I + J) = V(I) \cap V(J)$;
(c) if $I = \langle f_1, \ldots, f_s \rangle$ and $J = \langle g_1, \ldots, g_r \rangle$, then $I + J = \langle f_1, \ldots, f_s, g_1, \ldots, g_r \rangle$.

Lemma A2 ([8]). *For $f, g \in \mathbb{F}_p[x_1, \ldots, x_n]$ we have $f(a_1, \ldots, a_n) = g(a_1, \ldots, a_n)$ for all $(a_1, \ldots, a_n) \in \mathbb{F}_p^n$ if and only if $f - g \in \langle x_1^p - x_1, \ldots, x_n^p - x_n \rangle$.*

B Proof of Proposition 1

Put

$$\tilde{J} = \langle x_1 - z_1, \ldots, x_n - z_n, z_1(z_1 - 1) \cdots (z_1 - m_1), \ldots, z_n(z_n - 1) \cdots (z_n - m_n) \rangle.$$

To prove Proposition 1, we need the following lemmas.

Lemma B1. *Let $r \in \mathbb{F}_p[z_1, \ldots, z_n]$ be a polynomial of degree $\leq m_i$ in z_i. If $r(a_1, \ldots, a_n) = 0$ for all $(a_1, \ldots, a_n) \in \prod_{i=1}^{n} X_i$, then r is the zero polynomial.*

Proof. We shall use induction on the number of indeterminates n. The case $n = 1$ is trivial, since a nonzero polynomial in $\mathbb{F}_p[z]$ of degree m has at most m distinct roots. Now let $n > 1$ and assume by induction that the result holds for $n - 1$. If $r \in \mathbb{F}_p[z_1, \ldots, z_n]$ is of the indicated type, we can write

$$r = \sum_{j=0}^{m_n} r_j(z_1, \ldots, z_{n-1}) z_n^j,$$

where each $r_j(z_1, \ldots, z_{n-1})$ is of degree $\leq m_i$ in the indeterminates z_i. Let $(a_1, \ldots, a_{n-1}) \in \prod_{i=1}^{n-1} X_i$ be fixed. Then the polynomial $r(a_1, \ldots, a_{n-1}, z_n) \in \mathbb{F}_p[z_n]$ vanishes for every $a_n \in X_n$. It follows from the case $n = 1$ that $r(a_1, \ldots, a_{n-1}, z_n)$ is the zero polynomial in $\mathbb{F}_p[z_n]$. Hence $r_j(a_1, \ldots, a_n) = 0$ for all j. Since $(a_1, \ldots, a_{n-1}) \in \prod_{i=1}^{n-1} X_i$ is arbitrary, the induction hypothesis implies that each $r_j = 0$, hence $r = 0$. $\qquad\square$

Lemma B2. $I(\{(\zeta, \zeta) \colon \zeta \in X\}) = \tilde{J}.$

Proof. The inclusion $\tilde{J} \subset I(\{(\zeta, \zeta) \colon \zeta \in X\})$ is obvious. To get the opposite inclusion, suppose that $g \in I(\{(\zeta, \zeta) \colon \zeta \in X\})$. Using lexicographic order with $x_1 \succ \cdots \succ x_n \succ z_1 \succ \cdots \succ z_n$, it is easy to verify that $G = \{x_1 - z_1, \ldots, x_n - z_n, z_1(z_1-1) \cdots (z_1-m_1), \ldots, z_n(z_n-1) \cdots (z_n-m_n)\}$ is a Gröbner basis for $\tilde{J}$ (see, e.g., [2]). Then it follows that [2] there is a polynomial $r \in \mathbb{F}_p[x_1, \ldots, x_n, z_1, \ldots, z_n]$ with the following two properties:

(a) There is $g' \in \tilde{J}$ such that $g = g' + r$.
(b) No term of r is divisible by $x_1, \ldots, x_n, z_1^{m_1+1}, \ldots, z_n^{m_n+1}$.

Property (a) shows that $r = g - g' \in I(\{(\zeta, \zeta) \colon \zeta \in X\})$. Property (b) shows that r is a polynomial only in $z_1, \ldots, z_n$ and of degree $\leq m_i$ in z_i. Then by Lemma B1, we have $r = 0$. Hence $g - g' \subset \tilde{J}$, so that $I(\{(\zeta, \zeta) \colon \zeta \in X\}) \subset \tilde{J}$. $\qquad\square$

Now, we are ready to prove Proposition 1.

Proof of Proposition 1. By the Hilbert Basis Theorem, the ideal J has a finite generating set: $J = \langle g_1, \ldots, g_s \rangle$. Define

$$\alpha = 1 - (1 - g_1^{p-1}) \cdots (1 - g_s^{p-1}),$$
$$\beta_i = (x_i - z_i)\alpha,$$
$$\gamma_i = z_i(z_i - 1) \cdots (z_i - m_i)\alpha, \qquad i = 1, 2, \ldots, n.$$

Since

$$\alpha = g_1^{p-1} + \cdots + g_s^{p-1} - (g_1^{p-1} g_2^{p-1} + \cdots + g_{s-1}^{p-1} g_s^{p-1}) + \cdots$$
$$+ (-1)^{s+1} g_1^{p-1} \cdots g_s^{p-1} \in J,$$

it follows that $\beta_i, \gamma_i \in J$ for each i. Let $\xi, \eta \in \mathbb{F}_p^n$. If $\xi = \eta \in X$, then $(\xi, \eta) \in V(J) = V(\langle g_1, \ldots, g_s \rangle)$, hence $g_j(\xi_1, \ldots, \xi_n, \eta_1, \ldots, \eta_n) = 0$ for all $1 \leq j \leq s$, so that $\alpha(\xi_1, \ldots, \xi_n, \eta_1, \ldots, \eta_n) = 0$. Otherwise, we have $g_l(\xi_1, \ldots, \xi_n, \eta_1, \ldots, \eta_n) \neq 0$ for some $1 \leq l \leq s$, so that $\alpha(\xi_1, \ldots, \xi_n, \eta_1, \ldots, \eta_n) = 1$.

Fix $1 \leq i \leq n$. If $\xi_i \neq \eta_i$, then

$$\beta_i(\xi_1, \ldots, \xi_n, \eta_1, \ldots, \eta_n) = (\xi_i - \eta_i)\alpha(\xi_1, \ldots, \xi_n, \eta_1, \ldots, \eta_n) = \xi_i - \eta_i;$$

if $\xi_i = \eta_i$, then

$$\beta_i(\xi_1, \ldots, \xi_n, \eta_1, \ldots, \eta_n) = \xi_i - \eta_i = 0.$$

Hence $\beta_i(\xi_1, \ldots, \xi_n, \eta_1, \ldots, \eta_n) = \xi_i - \eta_i$ for all $\xi, \eta \in \mathbb{F}_p^n$. By Lemma A2 we have

$$x_i - z_i - \beta_i \in \langle x_1^p - x_1, \ldots, x_n^p - x_n, z_1^p - z_1, \ldots, z_n^p - z_n \rangle \subset J,$$

so that $x_i - z_i \in J$. A similar argument using γ_i instead of β_i shows that $z_i(z_i - 1) \cdots (z_i - m_i) \in J$. Thus, $\tilde{J} \subset J$.

The opposite inclusion follows immediately from Lemma B2 and the fact that $J \subset I(V(J))$. □

Acknowledgments. This work was supported by the National Natural Science Foundation of China under Grants 61333001, 61174071, and 61273111, and by the Guozhi Xu Posdoctoral Research Foundation.

References

1. Choffrut, C. (ed.): Automata Networks. LNCS, vol. 316. Springer, Heidelberg (1988)
2. Cox, D., Little, J., O'Shea, D.: Ideals, Varieties, and Algorithms: An Introduction to Computational Algebraic Geometry and Commutative Algebra, 3rd edn. Springer, New York (2007)
3. Drüppel, S., Lunze, J., Fritz, M.: Modeling of asynchronous discrete-event systems as networks of input-output automata. In: Proceedings of the 17th IFAC World Congress, Seoul, Korea, pp. 544–549 (2008)
4. Goles, E., Martínez, S.: Neural and Automata Networks: Dynamical Behavior and Applications. Kluwer Academic Publishers, Dordrecht (1990)
5. Kawano, Y., Ohtsuka, T.: An algebraic approach to local observability at an initial state for discrete-time polynomial systems. In: Proceedings of the 18th IFAC World Congress, Milano, Italy, pp. 6449–6453 (2011)
6. Kawano, Y., Ohtsuka, T.: Necessary condition for local observability of discrete-time polynomial systems. In: Proceedings of the 2012 American Control Conference, Montréal, Canada, pp. 6757–6762 (2012)
7. Kawano, Y., Ohtsuka, T.: Sufficiency of a necessary condition for local observability of discrete-time polynomial systems. In: Proceedings of the 2013 European Control Conference, Zürich, Switzerland, pp. 1722–1727 (2013)
8. Lidl, R., Niederreiter, H.: Finite Fields, 2nd edn. Cambridge University Press, Cambridge (1997)
9. Lunze, J.: Relations between networks of standard automata and networks of I/O automata. In: Proceedings of the 9th International Workshop on Discrete Event Systems, Göteborg, Sweden, pp. 425–430 (2008)
10. Martín-Vide, C., Mateescu, A., Mitrana, V.: Parallel finite automata systems communicating by states. International Journal of Foundations of Computer Science **13**(5), 733–749 (2002)

11. Martín-Vide, C., Mitrana, V.: Parallel communicating automata systems - A survey. Korean Journal of Computational & Applied Mathematics **7**(2), 237–257 (2000)
12. Martín-Vide, C., Mitrana, V.: Some undecidable problems for parallel communicating finite automata systems. Information Processing Letters **77**(5–6), 239–245 (2001)
13. McCulloch, W.S., Pitts, W.: A logical calculus of the ideas immanent in nervous activity. Bulletin of Mathematical Biology **5**(4), 115–133 (1943)
14. von Neumann, J.: Theory of Self-Reproducing Automata. University of Illinois Press, Urbana (1966)
15. Paulevé, L., Andrieux, G., Koeppl, H.: Under-approximating cut sets for reachability in large scale automata networks. In: Sharygina, N., Veith, H. (eds.) CAV 2013. LNCS, vol. 8044, pp. 69–84. Springer, Heidelberg (2013)
16. Ramadge, P.J.: Observability of discrete event systems. In: Proceedings of the 25th IEEE Conference on Decision and Control, Athens, Greece, pp. 1108–1112 (1986)
17. Richard, A.: Negative circuits and sustained oscillations in asynchronous automata networks. Advances in Applied Mathematics **44**(4), 378–392 (2010)
18. Robert, F.: Discrete Iterations: A Metric Study. Springer, Heidelberg (1986)
19. Saito, T., Nishio, H.: Structural and behavioral equivalence relations in automata networks. Theoretical Computer Science **63**(2), 223–237 (1989)
20. Veliz-Cuba, A.: An algebraic approach to reverse engineering finite dynamical systems arising from biology. SIAM Journal on Applied Dynamical Systems **11**(1), 31–48 (2012)
21. Veliz-Cuba, A., Jarrah, A.S., Laubenbacher, R.: Polynomial algebra of discrete models in systems biology. Bioinformatics **26**(13), 1637–1643 (2010)
22. Veliz-Cuba, A., Stigler, B.: Boolean models can explain bistability in the *lac* operon. Journal of Computational Biology **18**(6), 783–794 (2011)
23. Xu, X., Hong, Y.: Observability analysis and observer design for finite automata via matrix approach. IET Control Theory and Applications **7**(12), 1609–1615 (2013)

Reasoning on Schemas of Formulas: An Automata-Based Approach

Nicolas Peltier[(✉)]

CNRS, LIG, University of Grenoble, 38000 Grenoble, France
Nicolas.Peltier@imag.fr

Abstract. We present a new proof procedure for reasoning on schemas of formulas defined over various (decidable) base logics (e.g. propositional logic, Presburger arithmetic etc.). Such schemas are useful to model sequences of formulas defined by induction on some parameter. The approach works by computing an automaton accepting exactly the values of the parameter for which the formula is satisfiable. Its main advantage is that it is completely modular, in the sense that external tools are used as "black boxes" both to reason on base formulas and to detect cycles in the proof search. This makes the approach more efficient and scalable than previous attempts. Experimental results are presented, showing evidence of the practical interest of the proposed method.

Keywords: Automated reasoning · Satisfiability modulo theories · Inductive reasoning · Schemata of formulas

1 Introduction

Schemas of formulas are useful to model sequences of structurally similar formulas constructed over some base language. A first way of defining such sequences is to use generalized logical connectives of the form $\bigvee_{i=a}^{b} \phi$ or $\bigwedge_{i=a}^{b} \phi$, where a and b denote arithmetic expressions (possibly containing variables). For instance $\phi(n) \stackrel{\text{def}}{=} p(0) \wedge (\bigwedge_{i=0}^{n} p(i) \Rightarrow p(i+1)) \wedge \neg p(n+1)$ is a schema of formulas that is clearly unsatisfiable for every value of n. Note that n denotes an arithmetic variable (called a *parameter*), not a fixed integer. If n is fixed (e.g., $n = 0, 1, 2, \dots$), then $\phi(n)$ can be viewed as a propositional formula, but it is clear that showing that the formula $\phi(n)$ is unsatisfiable for every value of $n \in \mathbb{N}$ requires mathematical induction and is thus out of the scope of propositional and even first-order theorem provers. More generally and more conveniently, schemas can also be defined inductively, e.g., by using convergent systems of rewrite rules. For instance the following rules define a schema $\psi(n) \stackrel{\text{def}}{=} (p \Leftrightarrow (p \Leftrightarrow \dots (p \Leftrightarrow p) \dots))$ that is valid if n is even and equivalent to p if n is odd:

$$\begin{cases} \psi(0) \rightarrow \top \\ \psi(n+1) \rightarrow (p \Leftrightarrow \psi(n)) \end{cases}$$

Schemas can be defined over various base languages and theories. It is also possible to consider schemas defined over algebraic structures different from the natural numbers (in the present paper, we consider schemas defined on words).

© Springer International Publishing Switzerland 2015
A.-H. Dediu et al. (Eds.): LATA 2015, LNCS 8977, pp. 263–274, 2015.
DOI: 10.1007/978-3-319-15579-1_20

Schemas are useful for many applications, for instance to reason on parameterized systems, such as circuits constructed by composing inductively n elementary layers (e.g., an n-bit adder), or programs with loops or recursive functions (the parameter then denotes the number of iterations or recursive calls). They also appear frequently in human-constructed proofs.

In previous papers (see, e.g., [3,4]), we have established decidability and undecidability results for different classes of schemas, defined using various base logics and construction schemes, and we have proposed automated proof procedures for testing satisfiability of schemas. These procedures intertwine usual logical inference rules (used to reason on the base logic) with additional mechanisms intended to simulate a particular form of mathematical induction (based mainly on lazy instantiation schemes and cycle detection rules).

In the present work, we use another approach, which relies more on external tools. It consists to compute an automaton accepting exactly the values of the parameter such that the schema is satisfiable (hence the schema is unsatisfiable iff the accepted language is empty). External provers are used both to reason in the base logic (i.e., to determine whether a state is accepting or not) and to prune the search space by detecting cycles (i.e., to merge equivalent states). The overall approach is similar to that of [1], however the devised procedure is completely different and the considered class of schemas is much larger (the present approach handles schemas defined over arbitrary base theories whereas the procedure in [1] is restricted to propositional logic). The main advantage of the new method is that it is completely modular, in the sense that no modification is required to the proof procedure used to reason on base formulas. This feature allows us to use the most efficient theorem provers available for reasoning on the base logic, instead of having to develop new systems. This makes the approach much more tractable and scalable. The only requirement is that the base language must be decidable and must admit quantifier elimination. Furthermore, the parameter must be unique and has to be interpreted as a finite word on a finite alphabet (or equivalently, as a term defined on a monadic signature).

The rest of the paper is structured as follows. In Section 2 we briefly review some usual definitions. In Section 3 we introduce the syntax and semantics of the class of schemas considered in the present paper. In Section 4, a procedure is described for constructing, given a formula ϕ, an automaton accepting the values of the parameter for which ϕ is satisfiable. Termination results are presented in Section 5 and experimental results are provided in Section 6. Finally, Section 7 briefly concludes the paper and outlines possible directions of future work.

2   Preliminaries

We briefly review usual definitions (see, e.g., [11] for more details).

Let S be a set of *sort symbols*, containing in particular a symbol `bool`, denoting booleans. Let $\mathcal{F}$ be a set of *function symbols*, together with a function *profile* mapping each element of $\mathcal{F}$ to a non-empty sequence of sort symbols. A *constant symbol* is a function whose profile is of length 1. If $profile(f) = (s_1, \ldots, s_n, s)$

then $\mathbf{s}$ is the *range* of f and $(\mathbf{s}_1, \ldots, \mathbf{s}_n)$ is its *domain*. We write $f : \mathbf{s}_1, \ldots, \mathbf{s}_n \to \mathbf{s}$ (resp. $f : \mathbf{s}$) to state that $f \in \mathcal{F}$ and $profile(f) = (\mathbf{s}_1, \ldots, \mathbf{s}_n, \mathbf{s})$ (resp. $profile(f) = (\mathbf{s})$). Let $\mathcal{V}$ be a set of *variables*, associated with a function *sort* mapping each element of $\mathcal{V}$ to a sort symbol. The set $\mathcal{T}_\mathbf{s}$ of *terms of sort* $\mathbf{s}$ is the least set satisfying the following conditions: $x \in \mathcal{V} \wedge sort(x) = \mathbf{s} \Rightarrow x \in \mathcal{T}_\mathbf{s}$ and $f : \mathbf{s}_1, \ldots, \mathbf{s}_n \to \mathbf{s} \in \mathcal{F} \wedge \forall i \in [1, n]\, t_i \in \mathcal{T}_{\mathbf{s}_i} \Rightarrow f(t_1, \ldots, t_n) \in \mathcal{T}_\mathbf{s}$. For any subset $\mathcal{S}'$ of $\mathcal{S}$, the set $\mathcal{T}_{\mathcal{S}'}$ is defined as follows: $\mathcal{T}_{\mathcal{S}'} \overset{\text{def}}{=} \bigcup_{\mathbf{s} \in \mathcal{S}'} \mathcal{T}_\mathbf{s}$. For every term t we denote by $var(t)$ the set of variables occurring in t. A term is *ground* if $var(t) = \emptyset$. For every term t, $head(t)$ denotes the head symbol of t (i.e., $head(f(t_1, \ldots, t_n)) \overset{\text{def}}{=} f$ and $head(x) \overset{\text{def}}{=} x$ if $x \in \mathcal{V}$) and $|t|$ denotes the size of t (not counting variables or constants), defined as follows: $|x| \overset{\text{def}}{=} 0$ if $x \in \mathcal{V} \cup \mathcal{F}$, $|f(t_1, \ldots, t_n)| \overset{\text{def}}{=} 1 + \Sigma_{i=1}^n |t_i|$ if $n \geq 1$.

The set of *formulas* $\mathcal{F}$ is the least set satisfying the following conditions: $\bot, \top \in \mathcal{F}$; $\mathcal{T}_{\text{bool}} \subseteq \mathcal{F}$; $t, s \in \mathcal{T}_\mathbf{s} \wedge \mathbf{s} \neq \text{bool} \Rightarrow t \simeq s \in \mathcal{F}$ and $\phi, \psi \in \mathcal{F}, x \in \mathcal{V} \Rightarrow \phi \vee \psi, \phi \wedge \psi, \phi \Leftrightarrow \psi, \phi \Rightarrow \psi, \neg\phi, \forall x\, \phi, \exists x\, \phi \in \mathcal{F}$. For every formula ϕ, $fvar(\phi)$ is the set of free variables in ϕ. A formula is *closed* if $fvar(\phi) = \emptyset$. An *expression* is a term or a formula.

A *position* is a finite sequence of natural numbers. The empty position is denoted by ε and the concatenation of positions p and q is written $p.q$. A sequence p is a position *in* an expression E if one of the following conditions holds: (i) $p = \varepsilon$; (ii) $E = f(t_1, \ldots, t_n)$, $p = i.q$ (with $i \in [1, n]$), and q is a position in t_i; (iii) $E = \phi_1 \star \phi_2$ (with $\star \in \{\vee, \wedge, \Rightarrow, \Leftrightarrow\}$), $p = i.q$ with $i \in \{1, 2\}$ and q is a position in ϕ_i; or (iv) $p = 1.q$, q is a position in ϕ and $E = Qx\, \phi$ (with $Q \in \{\forall, \exists\}$) or $E = \neg\phi$. The expression $E|_p$ then denotes the expression occurring at position p in E and, for every expression F of the same nature and sort as $E|_p$, $E[F]_p$ denotes the expression obtained by replacing the expression at position p in E by F (the formal definitions are standard hence are omitted for conciseness).

An *interpretation* $\mathcal{I}$ is a function mapping:

- Every sort $\mathbf{s} \in \mathcal{S}$ to a non-empty set of elements $\mathbf{s}^\mathcal{I}$, with $\text{bool}^\mathcal{I} = \{\top, \bot\}$.
- Every function $f : \mathbf{s}_1, \ldots, \mathbf{s}_n \to \mathbf{s} \in \mathcal{S}$ to a function $f^\mathcal{I}$ from $\mathbf{s}_1^\mathcal{I} \times \cdots \times \mathbf{s}_n^\mathcal{I}$ to $\mathbf{s}^\mathcal{I}$ (in particular if $n = 0$ then $f^\mathcal{I}$ is an element of $\mathbf{s}^\mathcal{I}$).
- Every variable x of sort $\mathbf{s}$ to an element of $\mathbf{s}^\mathcal{I}$.

The *value* of an expression E in an interpretation $\mathcal{I}$ is defined as usual and denoted by $[E]^\mathcal{I}$. For every formula ϕ, we write $\mathcal{I} \models \phi$ if $[\forall x_1, \ldots, x_n\, \phi]^\mathcal{I} = \top$, where $\{x_1, \ldots, x_n\} = fvar(\phi)$.

Substitutions are sort-preserving functions mapping variables to terms. For every substitution σ and every expression E, we denote by $E\sigma$ the expression obtained from E by replacing every free occurrence of a variable x by its image by the substitution σ. We let $|\sigma| \overset{\text{def}}{=} \max\{|x\sigma| \mid x \in \mathcal{V}\}$.

We now recall basic notions about word automata.

Definition 1 (Automata). *An* automaton *is a tuple* $\mathcal{A} = (Q, \Sigma, \delta, q_0, A)$ *where* Q *is a set of* states, Σ *is a finite set of* symbols *(*vocabulary*),* δ *is a subset of* $Q \times \Sigma \times Q$ *(*transition function*),* q_0 *is an* initial state *in* Q *and* A *is a set of* accepting states *in* Q. *An automaton is* finite *if* Q *is finite.*

Definition 2 (Accepted Language). *A sequence of states* $q_0, \ldots, q_n$ *is a run of an automaton* $\mathcal{A} = (Q, \Sigma, \delta, q_0, A)$ *on a word* $w = a_1 \ldots a_n \in \Sigma^*$ *if for every* $i \in [1, n]$, $(q_{i-1}, a_i, q_i) \in \delta$. *Word* w *is accepted if there is a run* $q_0, \ldots, q_n$ *on* w *such that* $q_n \in A$. *We denote by* $\mathcal{L}(\mathcal{A})$ *the set of words that are accepted by* $\mathcal{A}$.

3   Schemas of Formulas

Schemas are defined over some *base theory* $\mathcal{T}$, which is defined by giving: a set of base sorts $\mathcal{S}_{base}$; a set of base function symbols $\mathcal{F}_{base}$ of profile $s_1, \ldots, s_n \to s$, with $s_1, \ldots, s_n, s \in \mathcal{S}_{base}$; and a set of interpretations $\mathfrak{I}_{base}$ of the symbols in $\mathcal{S}_{base} \cup \mathcal{F}_{base}$. An interpretation is a $\mathcal{T}$-*interpretation* if its restriction to the symbols in $\mathcal{S}_{base} \cup \mathcal{F}_{base}$ is in $\mathfrak{I}_{base}$. A $\mathcal{T}$-*formula* is a formula containing only function symbols in $\mathcal{F}_{base}$. A formula is $\mathcal{T}$-*satisfiable* (resp. $\mathcal{T}$-*valid*) if it admits a model in $\mathfrak{I}_{base}$ (resp. if it holds in all interpretations in $\mathfrak{I}_{base}$). In the following, we assume that an algorithm exists for testing the $\mathcal{T}$-satisfiability of $\mathcal{T}$-formulas.

Example 3. The base theory of propositional logic is defined by considering the set of sorts $\mathcal{S}_{base} = \{\texttt{bool}\}$, where $\mathcal{F}_{base}$ contains only constant symbols of sort $\texttt{bool}$. The set $\mathfrak{I}_{base}$ contains all possible interpretations of these symbols.

Presburger arithmetic is defined as follows: $\mathcal{S}_{base}$ is $\{\texttt{int}, \texttt{bool}\}$, and $\mathcal{F}_{base}$ contains constant symbols of sort $\texttt{int}$ and the usual functions $0 : \texttt{int}, 1 : \texttt{int}, - :$ $\texttt{int} \to \texttt{int}, + : \texttt{int}, \texttt{int} \to \texttt{int}, \leq: \texttt{int}, \texttt{int} \to \texttt{bool}$. The set $\mathfrak{I}_{base}$ contains all interpretations $\mathcal{I}$ such that $\texttt{int}^{\mathcal{I}} = \mathbb{Z}$, where $0, 1, -, +, <$ have their usual meaning: $0^{\mathcal{I}} \stackrel{\text{def}}{=} 0, 1^{\mathcal{I}} \stackrel{\text{def}}{=} 1, -^{\mathcal{I}}x \stackrel{\text{def}}{=} -x, x +^{\mathcal{I}} y \stackrel{\text{def}}{=} x + y, x \leq^{\mathcal{I}} y = \top$ iff $x \leq y$.

As explained in the Introduction, schemas are defined by induction on some parameters. These parameters must be of some special sort not occurring in the base theory, and must be interpreted as finite words over a finite alphabet, or, equivalently, as terms on a finite signature containing only constants and monadic functions. More formally, let $\mathcal{S}_P \subset \mathcal{S} \setminus \mathcal{S}_{base}$ be a set of sorts. Variables of a sort in $\mathcal{S}_P$ are called *parameters*. Let $\mathcal{C}$ be a finite set of function symbols, called *constructors*, of profile $s' \to s$ or s, with $s, s' \in \mathcal{S}_P$. A term is a *constructor term* if all function symbols occurring in t are in $\mathcal{C}$.

Example 4. In all the considered examples, $\mathcal{S}_P$ will simply refer to the set of natural numbers $\{\texttt{nat}\}$ with the constructors: $0 : \texttt{nat}, s : \texttt{nat} \to \texttt{nat}$. The set of terms constructed on this signature is isomorphic to the natural numbers.

Let $\Omega \subseteq \mathcal{F}$ be a set of *schema symbols*, and let $\prec$ be a well-founded ordering among symbols in Ω. Informally, symbols in Ω are used to denote functions mapping ground constructor terms to base terms. We thus assume that the profile of each symbol $f \in \Omega$ is of the form $s \to s'$ with $s \in \mathcal{S}_P$ and $s' \in \mathcal{S}_{base}$. The set Ω is partitioned into two disjoint sets $\Omega = \mathcal{D} \uplus \mathcal{E}$. Intuitively, symbols in $\mathcal{E}$ will denote individuals in the domain of the base theory depending on the parameter (e.g., the symbol p in the first example in the Introduction is in $\mathcal{E}$), whereas the symbols in $\mathcal{D}$ will denote inductively defined schemas of terms or formulas such as $\bigwedge_{i=0}^{n} p(i) \Rightarrow p(i+1)$ or $\Sigma_{i=0}^{n} i$. Thus the interpretation of a term $e(\alpha)$ with $e \in \mathcal{E}$ will be unspecified (it denotes an arbitrary element of the domain of the base theory), whereas that of a term $d(\alpha)$ with $d \in \mathcal{D}$ will be defined by induction on α. More precisely, the semantics of the symbols in $\mathcal{D}$ will be defined by sets of rewrite rules, reducing every (ground) term of defined head to a base term or formula. In order to formally define these rewrite systems, we need to introduce some definitions.

For every $\mathcal{G} \subseteq \mathcal{F}$, a $\mathcal{G}$-term is a non-variable term of head $f \in \mathcal{G}$. We denote by $\Theta_{\mathcal{G}}(t)$ the set of $\mathcal{G}$-terms occurring in t. In particular, an Ω-term is a term whose head is in Ω. The ordering $\prec$ is extended to Ω-terms as follows: $f(t) \prec g(s)$ if $f \prec g$ or if $f = g$ and t is a proper subterm of s. A term or formula is t-dominated if the relation $s \prec t$ holds for every Ω-term s occurring in it.

A term or formula is $\mathcal{S}_{base}$-compatible if it contains no quantification on parameters and if the terms of a sort in $\mathcal{S}_P$ only occur in the scope of a function in Ω.

Let $\mathcal{R}$ be a set of rewrite rules (see, e.g., [5]). For every symbol $d : \mathbf{s} \to \mathbf{s}' \in \mathcal{D}$, where $\mathbf{s}' \neq \mathtt{bool}$, we assume that the set $\mathcal{R}$ contains the following rules.
- A rule of the form $d(a) \to t$, where t is a ground term of sort $\mathbf{s}'$, for each constant symbol $a : \mathbf{s}$ in $\mathcal{F}$.
- A rule of the form $d(f(x)) \to t$, for each symbol $f : \mathbf{s}'' \to \mathbf{s}$ in $\mathcal{F}$, where t is a term of sort $\mathbf{s}'$ such that $var(t) \subseteq \{x\}$ and x is a variable of sort $\mathbf{s}''$.

Similarly, for every symbol $d : \mathbf{s} \to \mathtt{bool} \in \mathcal{D}$, the set $\mathcal{R}$ contains the following rules.
- A rule of the form $d(a) \to \phi$, where ϕ is a closed formula, for each constant symbol $a : \mathbf{s}$ in $\mathcal{F}$.
- A rule of the form $d(f(x)) \to \phi$ for each symbol $f : \mathbf{s}' \to \mathbf{s}$ in $\mathcal{F}$, where ϕ is a formula with $fvar(\phi) \subseteq \{x\}$ and x is a variable of sort $\mathbf{s}'$.

We also assume that for every rule $l \to r$ occurring in $\mathcal{R}$, r is l-dominated and $\mathcal{S}_{base}$-compatible. The system $\mathcal{R}$ is assumed to be fixed in the rest of the paper.

Proposition 5. *The system $\mathcal{R}$ is convergent, hence every expression E has a unique normal form w.r.t. $\mathcal{R}$, denoted by $E \downarrow_{\mathcal{R}}$.*

Definition 6. *A schema interpretation $\mathcal{I}$ is a base interpretation such that:*

1. *Every sort $\mathbf{s} \in \mathcal{S}_P$ is mapped to the set of ground constructor terms of sort $\mathbf{s}$ (we assume that this set is not empty).*
2. *For every constant symbol $c : \mathbf{s} \in \mathcal{C}$ we have $c^{\mathcal{I}} \stackrel{def}{=} c$, and for every constructor $f : \mathbf{s} \to \mathbf{s}' \in \mathcal{C}$ we have $f^{\mathcal{I}}(x) \stackrel{def}{=} f(x)$ (this implies that $[\mathbf{s}]^{\mathcal{I}} = \mathbf{s}$, for every ground constructor term $\mathbf{s}$).*
3. *For every defined symbol $d : \mathbf{s}' \to \mathbf{s}$ and for every ground constructor term s of sort $\mathbf{s}'$, we have: $d^{\mathcal{I}}(s) \stackrel{def}{=} [d(s) \downarrow_{\mathcal{R}}]^{\mathcal{I}}$.*

We write $\phi \models \psi$ if $\mathcal{I} \models \phi \Rightarrow \mathcal{I} \models \psi$ holds for every schema interpretation $\mathcal{I}$.

Example 7. The schema $p(0) \land \bigwedge_{i=0}^{n} p(i) \Rightarrow p(i+1) \land \neg p(n+1)$ in the Introduction can be encoded as follows: $p(0) \land d(\mathbf{s}(n)) \land \neg p(\mathbf{s}(n))$ with the rules:
$$\begin{cases} d(0) \rightarrow \top \\ d(\mathbf{s}(n)) \rightarrow d(n) \land (p(n) \rightarrow p(\mathbf{s}(n))) \end{cases}$$
The symbol p occurs in $\mathcal{E}$ whereas d occurs in $\mathcal{D}$.

Example 8. The schema $\Sigma_{i=0}^{n} i$ can be encoded as follows: $d(\mathbf{s}(n))$ with the rules:
$$\begin{cases} d(0) \rightarrow 0 \\ d(\mathbf{s}(n)) \rightarrow d'(n) + d(n) \\ d'(0) \rightarrow 0 \\ d'(\mathbf{s}(n)) \rightarrow 1 + d'(n) \end{cases}$$
The term $d'(n)$ denotes the integer n (this is needed because $\mathcal{S}_{base}$-compatible formulas cannot contain parameters outside the scope of a symbol in Ω).

Our goal is to devise a procedure to check the satisfiability of formulas in schema interpretations. To this purpose, we assume that the considered formula contains at most one parameter and that a procedure is available to test the satisfiability of (quantified) formulas in the base theory. Note that the problem of testing the satisfiability of formulas in schema interpretations cannot be encoded directly in the base theory, even if it allows for universal quantifications and uninterpreted symbols, due to the fact that the parameters are interpreted over an inductively defined domain: while the rules in $\mathcal{R}$ can be encoded as first-order axioms (equalities or equivalences), the fact that the parameter n is to be interpreted as a ground term in $\mathcal{T}_{\mathcal{S}_P}$ (e.g., a natural number) cannot be stated in first-order logic.

4   From Schemas to Automata

4.1   Base Abstractions of Schemas

In this section, we devise a simplification operation *simp*, whose principle is to abstract away some of the $\mathcal{E}$-terms occurring in a formula by replacing them by existential variables. As we shall see, this operation preserves satisfiability (under some conditions) and is essential for the termination of the construction of the automaton (in some cases).

Example 9. Consider the formula $p(\mathbf{s}(n)) \land d(n)$, where d is defined as in Example 7. From the rules defining d, it is easy to check that the normal form of a formula $d(\mathbf{s}^k(0))$ (with $k \in \mathbb{N}$) contains no formula of the form $p(\mathbf{s}^l(0)$ with $l > k$. Therefore, the truth value of $d(n)$ does not depend on that of $p(\mathbf{s}(n))$, which entails that $p(\mathbf{s}(n)) \land d(n)$ is satisfiable iff the formula $\exists x \, (x \land d(n))$ (where x is a variable of sort $\mathbf{bool}$) is satisfiable (hence iff $d(n)$ is satisfiable). Note that, in constrast, a formula $p(n) \land d(n)$ *cannot* be reduced to $\exists x \, (x \land d(n))$, because the truth value of $d(n)$ depends on that of $p(n)$.

In order to formally define the simplification operation, we need to introduce some definitions. We first devise a function μ mapping every symbol $f \in \Omega$ to a natural number. This function is defined in such a way that the size of the terms in $\Theta_{\mathcal{E}}$ occurring in the normal form of $f(t)$ is bounded by $1 + \mu(f) + |t|$.

Definition 10. *The function* $\mu : \Omega \to \mathbb{Z}$ *is defined inductively as follows:*
- *If* $f \in \mathcal{E}$ *then* $\mu(f) \overset{def}{=} 0$.
- *If* $f \in \mathcal{D}$ *then* $\mu(f) \overset{def}{=} \max\{\mu(g)+|t|-|l|+1 \mid l \to r \in \mathcal{R}, g(t) \in \Theta_\Omega(r), g \prec f\}$.

We let: $\nu(E) \overset{def}{=} \max\{\mu(f) + |t| \mid f(t) \in \Theta_\Omega(E)\}$ *and* $\nu'(E) \overset{def}{=} \max\{\mu(f) + |t| \mid f(t) \in \Theta_\mathcal{D}(E), \text{ or } f(t) \in \Theta_\mathcal{E}(E) \text{ and } t \text{ is ground}\}$.

Example 11. With the definitions of Example 7, we have $\mu(p) = 0$ (since $p \in \mathcal{E}$) and $\mu(d) = \max\{\mu(p)+|n|-2+1, \mu(p)+|s(n)|-2+1\} = \max\{\mu(p)-1, \mu(p)\} = 0$ (since $|n| = 0$ and $|s(n)| = 1$). Consider now the symbol d' associated with the rules: $\begin{cases} d'(0) \to \top \\ d'(s(n)) \to d(s(s(n))) \end{cases}$

We have $\mu(d') = \mu(d) + 2 - 2 + 1 = 1$.

The next definition introduces the simplification operation.

Definition 12. *Let* ϕ *be a formula, and let* $\{t_1, \ldots, t_n\}$ *be the set of non-ground terms in* $\Theta_\mathcal{E}$ *of size strictly greater than* $\nu'(\phi) + 1$. *We denote by* $simp(\phi)$ *the formula* $\exists x_1, \ldots, x_n \phi'$, *where* $x_1, \ldots, x_n$ *are new pairwise distinct variables not occurring in* ϕ *of the same sorts as* $t_1, \ldots, t_n$ *(respectively) and* ϕ' *is obtained from* ϕ *by replacing each occurrence of* t_i *by* x_i *($1 \leq i \leq n$).*

Example 13. Consider the following schema defined on Presburger arithmetic:
$$\phi \overset{def}{=} a(n) \geq 0 \land d(n), \text{ with } \begin{cases} d(0) \to \top \\ d(s(n)) \to a(s(n)) \leq a(n) \land d(n) \end{cases}$$

Let $\psi \overset{def}{=} \phi\{n \mapsto s(n)\} \downarrow_\mathcal{R} = (a(s(n)) \geq 0 \land d(s(n))) \downarrow_\mathcal{R} = a(s(n)) \geq 0 \land a(s(n)) \leq a(n) \land d(n)$. It is easy to check that $\mu(d) = 0$. The only term in $\Theta_\mathcal{E}$ of size strictly greater than 1 occurring in ψ is $a(s(n))$. The formula $simp(\psi)$ is therefore obtained from ψ by replacing the term $a(s(n))$ by a new existential variable. This yields: $\exists x \, x \geq 0 \land x \leq a(n) \land d(n)$. Note that by the Fourier-Motzkin transformation, this formula is equivalent to $0 \leq a(n) \land d(n)$, i.e. to ϕ.

The following lemma states the soundness of the simplification operation.

Lemma 14. *Let* ϕ *be an* $\mathcal{S}_{base}$*-compatible irreducible formula containing a unique parameter* α *and let* u *be a ground term of the same sort as* α. *The formula* ϕ *has a model* $\mathcal{I}$ *with* $[\alpha]^\mathcal{I} = u$ *iff the same holds for* $simp(\phi)$.

We now devise a sufficient criterion to test the satisfiability of arbitrary formulas in the base theory. The idea is simply to get rid of the symbols not occurring in the base signature by replacing the corresponding terms by variables.

Definition 15. *Let* ϕ *be an* $\mathcal{S}_{base}$*-compatible formula. We denote by* $\lfloor \phi \rfloor_\mathcal{T}$ *the formula* $\forall y_1, \ldots, y_m \phi'$, *where* $\{s_1, \ldots, s_m\} = \Theta_\Omega(\phi)$, $y_1, \ldots, y_m$ *are fresh pairwise distinct variables of the same sorts as* $s_1, \ldots, s_m$ *respectively, and* ϕ' *is obtained from* ϕ *by replacing each occurrence of* s_i *by* y_i *($1 \leq i \leq m$).*

Lemma 16. *For every* $\mathcal{S}_{base}$*-compatible formula* ϕ, $\lfloor \phi \rfloor_\mathcal{T}$ *is a* $\mathcal{T}$*-formula. Moreover,* $\lfloor \phi \rfloor_\mathcal{T} \models \phi$.

Let ϕ, ψ be two formulas. We write $\phi \sim_T \psi$ if ϕ and ψ have the same set of free parameters and if $\lfloor \phi \Leftrightarrow \psi \rfloor_T$ is T-valid.

Example 17. Consider the formula $\phi \stackrel{\text{def}}{=} d(n) \Leftrightarrow \exists x (x \wedge d(n))$. The validity of ϕ cannot be directly established because it is not a T-formula (it contains a symbol $d \in \Omega$). Now consider the formula $\lfloor \phi \rfloor_T = \forall y (y \Leftrightarrow \exists x (x \wedge y))$. $\lfloor \phi \rfloor_T$ is a T-formula hence it can be handled by the decision procedure for checking validity in T. Since $\lfloor \phi \rfloor_T$ is T-valid, we have $d(n) \sim_T \exists x (x \wedge d(n))$.

4.2   Constructing the Automaton

We now describe the procedure for constructing the automaton. Let $s \mapsto \alpha_s$ be a function mapping each sort symbol s to a variable of sort s. For any formula ϕ containing a unique parameter $\alpha = \alpha_{sort(\alpha)}$ and for any symbol $c \in C$ of range $sort(\alpha)$, we denote by $\phi\uparrow^c$ the formula defined as follows.

- If c is a constant symbol then $\phi\uparrow^c \stackrel{\text{def}}{=} simp(\phi\{\alpha \mapsto c\}\downarrow_R)$.
- Otherwise, $\phi\uparrow^c \stackrel{\text{def}}{=} simp(\phi\{\alpha \mapsto c(\alpha_s)\}\downarrow_R)$, where s is the domain of c.

Informally, $\phi\uparrow^c$ is obtained from ϕ by letting $head(\alpha) = c$ (and simplifying).

Definition 18. *Let ϕ be a formula containing at most one parameter. We define an automaton $\mathcal{A}(\phi) = (Q, \Sigma, \delta, q_0, A)$ as follows.*

- *The alphabet is the set of constructors: $\Sigma \stackrel{\text{def}}{=} C$.*
- *States are equivalence classes of formulas w.r.t. the relation $\sim_T$ (the equivalence class of a formula ϕ is denoted by $[\phi]_{\sim_T}$). The set of states Q and the transition relation δ are inductively defined as follows.*
  - *Q contains the initial state: $q_0 \stackrel{\text{def}}{=} [simp(\phi\downarrow_R)]_{\sim_T}$.*
  - *If Q contains a state s, ψ is an arbitrarily chosen representative[1] of s, ψ contains a parameter $\alpha : s$, and c is a constructor of range s then $s' \stackrel{\text{def}}{=} [\psi\uparrow^c]_{\sim_T} \in Q$ and $(s, c, s') \in \delta$.*
- *The set of final states A is the set of states s such that $\neg\lfloor\neg\psi\rfloor_T$ is satisfiable and ψ contains no parameter, where ψ is an arbitrarily chosen representative[1] of s.*

Note that, since every symbol in C is monadic, all terms $t \in T_{S_P}$ can be viewed as words in C^* (of course the converse does not hold). The following theorem states the soundness of the construction.

Theorem 19. *Let ϕ be a formula containing at most one parameter and let u be a word in C^*. The following assertions are equivalent.*

- *ϕ has a model $\mathcal{I}$ and either $u = \varepsilon$ and ϕ contains no parameter or ϕ has a parameter α and $\alpha^{\mathcal{I}} = u$ (i.e., the word corresponding to $\alpha^{\mathcal{I}}$ is u).*
- *There is a run of $\mathcal{A}(\phi)$ that accepts u.*

---

[1] The test is performed only for one particular representative of s (i.e., for one arbitrarily chosen element of s), not for all $\psi \in s$.

Example 20. Consider the schema of Example 7. The initial formula is ϕ : $p(0) \wedge d(\mathbf{s}(n)) \wedge \neg p(\mathbf{s}(n))$. We first compute its normal form w.r.t. $\mathcal{R}$. We get: $\phi \downarrow_{\mathcal{R}}$: $p(0) \wedge d(n) \wedge (p(n) \Rightarrow p(\mathbf{s}(n))) \wedge \neg p(\mathbf{s}(n))$. Next, we compute the formula $simp(\phi \downarrow_{\mathcal{R}})$. As explained in Example 11, we have $\mu(d) = 0$ thus $\nu'(\phi \downarrow_{\mathcal{R}}) = 0$. Consequently, the term $p(\mathbf{s}(n))$, whose size is strictly greater than $\nu'(\phi \downarrow_{\mathcal{R}}) + 1 = 1$, must be replaced by a new existential variable. We get the formula: $simp(\phi \downarrow_{\mathcal{R}})$: $\exists x \, p(0) \wedge d(n) \wedge (p(n) \Rightarrow x) \wedge \neg x)$, e.g., $p(0) \wedge d(n) \wedge \neg p(n)$ (in practice this simplification does not need to be explicitly performed: all theory reasoning will be handled by the external solver).

This yields a first state q_0. This state is not accepting since the formula contains a parameter. There are two constructors 0 and s, which produce new states q_1 and q_2, corresponding respectively to the schemas: $simp(\phi \downarrow_{\mathcal{R}})\{n \mapsto 0\}$: $p(0) \wedge d(0) \wedge \neg p(0)$ and $simp(\phi \downarrow_{\mathcal{R}})\{n \mapsto \mathbf{s}(n)\}$: $p(0) \wedge d(\mathbf{s}(n)) \wedge \neg p(\mathbf{s}(n))$. After normalization, the first formula is reduced to $p(0) \wedge \neg p(0)$. We have $\neg\lfloor \neg(p(0) \wedge \neg p(0)) \rfloor_{\mathcal{T}} = \neg \exists x \neg (x \wedge \neg x) \equiv \bot$ hence the corresponding state is non accepting. The formula contains no parameter, hence no new transition is produced from this state. The second formula is identical to the initial one, hence the state q_2 is actually identical to the initial state q_0 Note that the formulas are not compared syntactically: in practice the equivalence will be detected by the external solver by checking that the base formula $\forall x, y, z \, ((x \wedge y \wedge \neg z) \Leftrightarrow (x \wedge y \wedge \neg z))$ is valid.

We thus obtain the following automata (with no accepting state):

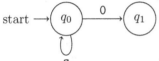

Note that the simplification operation $\phi \mapsto simp(\phi)$ is essential for termination, even in this trivial example. Without it, we would get formulas (hence states) of the form:

$$p(0) \wedge d(n) \wedge (p(n) \Rightarrow p(\mathbf{s}(n))) \wedge \neg p(\mathbf{s}(n))$$
$$p(0) \wedge d(n) \wedge (p(n) \Rightarrow p(\mathbf{s}(n))) \wedge (p(\mathbf{s}(n)) \Rightarrow p(\mathbf{s}(\mathbf{s}(n)))) \wedge \neg p(\mathbf{s}(\mathbf{s}(n)))$$
$$\cdots$$

and the procedure would diverge.

5   Termination

The procedure described in Section 4.2 diverges in general ($\mathcal{A}(\phi)$ is infinite). In this section we identify classes of formulas for which the procedure terminates, thus yielding a decision procedure for testing validity. We start by devising an abstract semantic criterion then we provide concrete examples of syntactic fragments satisfying the required properties.

Definition 21. *Let $\mathfrak{E}$ be a set of expressions. An expression E is $\mathfrak{E}$-supported if it is of the form $E'\sigma$, where $E' \in \mathfrak{E}$ and σ is a substitution mapping every variable in E' to an Ω-term.*

Definition 22. *A class of $\mathcal{T}$-expressions $\mathfrak{E}$ is* admissible *if it satisfies the following properties.*

- *It is* closed under replacement, *i.e.: if $E \in \mathfrak{E}$, p is a position of a variable x in E whose sort is the range of a symbol in $\mathcal{D}$ and F is an expression in $\mathfrak{E}$ of the same nature and sort as x then $E[F]_p \in \mathfrak{E}$.*
- *It is* closed under projection, *i.e.: if ϕ is a formula in $\mathfrak{E}$ and x is a variable in $fvar(\phi)$ whose sort is the range of a symbol in $\mathcal{E}$, then there exists a formula $\psi \in \mathfrak{E}$ that is equivalent to $\exists x\, \phi$.*
- *It is* compatible with $\mathcal{R}$, *i.e., for every rule $l \rightarrow r \in \mathcal{R}$, r is $\mathfrak{E}$-supported.*
- *For every $k \in \mathbb{N}$, the set of formulas $\phi \in \mathfrak{E}$ with $card(var(\phi)) \leq k$ is finite (up to $\mathcal{T}$-equivalence).*

A class of formulas $\mathfrak{F}$ is well-supported *if every formula $\phi \in \mathfrak{F}$ is $\mathfrak{E}$-supported, for some admissible set of expressions $\mathfrak{E}$.*

Theorem 23. *If $\mathfrak{F}$ is well-supported, then for every $\phi \in \mathfrak{F}$, $\mathcal{A}(\phi)$ is finite.*

Examples of Well-Supported Classes

We now provide examples of classes of formulas fulfilling the properties of Definition 22. These classes are defined by imposing conditions on the base theory and on the symbols in the signature (not only on the symbols occurring in the formula itself, but also on those in the rewrite system $\mathcal{R}$).

Definition 24. *A base theory is* well-structured *if for every sort symbol $s \in \mathcal{S}_{base}$, either $s^{\mathcal{I}}$ is infinite in all interpretations $\mathcal{I} \in \mathfrak{I}_{base}$ or there exists a finite set of ground terms $\{c_1, \ldots, c_k\}$ such that $\forall x \bigvee_{i=1}^{k} x \simeq c_i$ holds in all interpretations in $\mathfrak{I}_{base}$ (where x is a variable of sort s). A formula ϕ is:*

- quasi-propositional *if all function symbols are of profile $\mathtt{bool}$ or $s \rightarrow \mathtt{bool}$, with $s \in \mathcal{S}_P$;*
- flat *if all symbols in $\mathcal{D}$ are of range $\mathtt{bool}$ and if the only non-ground terms of a sort $s \in \mathcal{S}_{base} \setminus \{\mathtt{bool}\}$ occurring in ϕ or in the right-hand side of a rule in $\mathcal{R}$ are variables and $\mathcal{E}$-terms.*
- *a $\simeq$-formula if $\mathcal{F}_{base}$ contains no symbol of range $\mathtt{bool}$ (i.e. $\simeq$ is the only predicate symbol);*
- *a $\leq$-formula if the only symbols of range $\mathtt{bool}$ in $\mathcal{F}_{base}$ are symbols $\leq_s\colon s, s \rightarrow \mathtt{bool}$ interpreted as total non-strict orders on s.*

Theorem 25. *The class of quasi-propositional formulas is well-supported. Furthermore, if the base theory is well-structured, then the class of $\simeq$-formulas and the class of $\leq$-formulas are well-supported.*

6   Experimental Results

The procedure has been implemented in SWI Prolog [12] using Z3 [10] as an external tool (more generally any SMT-Lib2 compatible prover [6] can be used).

The programs communicate by exchanging strings. We present some preliminary experimental results. We first run our procedure on schemas of propositional formulas, comparing our approach with the system RegSTAB described in [2] (RegSTAB is based on an extension of semantic tableaux, see, e.g., [9]). The benchmarks are generated by formalizing an n-bit adder circuit as a schema of propositional formulas (parameterized by the number of bits n) and by checking some basic properties of this circuit: associativity and commutativity of $+$, neutrality of 0, equivalence between ripple-carry and carry-propagate adders etc. (see [3] for more details). All the running times for RegSTAB are taken from [3] (for conciseness, easy instances are omitted since they are not really informative). All times are in seconds, unless specified otherwise. The results show significant improvement in favor of the new approach ($\mathcal{A}$ is slower than RegSTAB on some easy instances, which is probably due to the overcost of the communication between the two systems, but faster on hard instances).

Problem	$\mathcal{A}$	RegSTAB
$x + 0 = x$	0.547	0.017
$x + y = y + x$	0.531	0.267
$x + (y + z) = (x + y) + z$	0.687	28.902
Ripple carry $\equiv$ Carry-propagate	0.516	0.194
$x_1 \leq x_2 \wedge y_1 \leq y_2 \Rightarrow x_1 + y_1 \leq x_2 + y_2$	0.641	2.949
$x_1 \leq x_2 \leq x_3 \wedge y_1 \leq y_2 \leq y_3 \Rightarrow x_1 + y_1 \leq x_2 + y_2 \leq x_3 + y_3$	1.328	46m57
$1 \leq x + y \leq 5 \wedge x \geq 3 \wedge y \geq 4$	0.750	7m9

We also provide some results for non propositional schemas (for which there is no available tool, as far as we are aware). The first problems are constructed by formalizing some basic functions on arrays and by stating some of their properties: $insert(T, x)$ denotes the array obtained by inserting a new element x at the appropriate position in a sorted array T, $T \geq 0$ means that all elements in T are positive, $|T|$ is the sum of the elements in T. The next problem states the equivalence between two definitions of the lexicographic ordering: the recursive one and the existential one ($T < S \Leftrightarrow \exists i\, T[i] < S[i] \wedge \bigwedge_{j=0}^{i-1} T[j] = S[j]$). The last problem is defined as the propositional ones, but using a decimal basis instead of bits (we check simultaneously that the adder is associative, commutative and that 0 is neutral).

Problem	$\mathcal{A}$
$sorted(T) \wedge T[0] \geq a \Rightarrow T[n] \geq a$	0.656
$sorted(T) \Rightarrow sorted(insert(T, x))$	0.610
$insert(T, x) \geq 0 \Leftrightarrow (T \geq 0 \wedge x \geq 0))$	0.594
$sorted(T) \wedge sorted(S) \Rightarrow sorted(T + S)$	0.516
$|insert(T, x)| = |T| + x$	0.547
$|T + S| = |T| + |S|$	0.422
Lexicographic ordering	0.541
Decimal adder	17.4

7　Conclusion

We have presented a new method for checking the satisfiability of schemas of formulas defined over arbitrary (but decidable) base logics. The procedure diverges in general, but it is proven terminating on some fragments. The method has the advantage that it is modular, relying on external tools for reasoning in the base theory. The experimental results, although preliminary, are very encouraging. Concerning future work, many refinements of the procedure for constructing the automaton can be considered, for instance devising more elaborate techniques for detecting and pruning redundant nodes, or decomposing formulas into independent parts when possible (each part being associated with a new state of the automaton, instead of considering a single state for the conjunction). The extension of the procedure to more general theories also deserves to be considered, for instance semi-decidable theories such as full first-order logic.

References

1. Aravantinos, V., Caferra, R., Peltier, N.: Complexity of the satisfiability problem for a class of propositional schemata. In: Dediu, A.-H., Fernau, H., Martín-Vide, C. (eds.) LATA 2010. LNCS, vol. 6031, pp. 58–69. Springer, Heidelberg (2010)
2. Aravantinos, V., Caferra, R., Peltier, N.: RegSTAB: A SAT solver for propositional schemata. In: Giesl, J., Hähnle, R. (eds.) IJCAR 2010. LNCS, vol. 6173, pp. 309–315. Springer, Heidelberg (2010)
3. Aravantinos, V., Caferra, R., Peltier, N.: Decidability and undecidability results for propositional schemata. Journal of Artificial Intelligence Research **40**, 599–656 (2011)
4. Aravantinos, V., Peltier, N.: Schemata of SMT-problems. In: Brünnler, K., Metcalfe, G. (eds.) TABLEAUX 2011. LNCS, vol. 6793, pp. 27–42. Springer, Heidelberg (2011)
5. Baader, F., Nipkow, T.: Term Rewriting and All That. Cambridge University Press (1998)
6. Barrett, C., Stump, A., Tinelli, C.: The Satisfiability Modulo Theories Library (SMT-LIB) (2010). http://www.SMT-LIB.org
7. Comon, H., Dauchet, M., Gilleron, R., Jacquemard, F., Lugiez, D., Tison, S., Tommasi, M.: Tree automata techniques and applications (1997). http://www.grappa.univ-lille3.fr/tata
8. Dershowitz, N., Jouannaud, J.: Rewrite systems. In: Handbook of Theoretical Computer Science, Volume B: Formal Models and Sematics (B), pp. 243–320 (1990)
9. Hähnle, R.: Tableaux and related methods. In: Robinson, A., Voronkov, A. (eds.) Handbook of Automated Reasoning, ch. 3, vol. I, pp. 100–178. Elsevier Science (2001)
10. de Moura, L., Bjørner, N.: Z3: An efficient SMT solver. In: Ramakrishnan, C.R., Rehof, J. (eds.) TACAS 2008. LNCS, vol. 4963, pp. 337–340. Springer, Heidelberg (2008)
11. Robinson, A., Voronkov, A. (eds.): Handbook of Automated Reasoning. North-Holland (2001)
12. Wielemakers, J.: SWI-Prolog version 7 extensions. In: WLPE 2014, July 2014

Derivatives for Regular Shuffle Expressions

Martin Sulzmann[1]([⊠]) and Peter Thiemann[2]

[1] Faculty of Computer Science and Business Information Systems,
Karlsruhe University of Applied Sciences, Moltkestrasse 30, 76133 Karlsruhe,
Germany
martin.sulzmann@hs-karlsruhe.de
[2] Faculty of Engineering, University of Freiburg,
Georges-Köhler-Allee 079, 79110 Freiburg, Germany
thiemann@acm.org

Abstract. There is a rich variety of shuffling operations ranging from asynchronous interleaving to various forms of synchronizations. We introduce a general shuffling operation which subsumes earlier forms of shuffling. We further extend the notion of a Brzozowski derivative to the general shuffling operation and thus to many earlier forms of shuffling. This extension enables the direct construction of automata from regular expressions involving shuffles that appear in specifications of concurrent systems.

Keywords: Automata and logic · Shuffle expressions · Derivatives

1 Introduction

We consider an extension of regular expressions with a binary shuffle operation which bears similarity to shuffling two decks of cards. Like the extension with negation and complement, the language described by an expression extended with shuffling remains regular. That is, any expression making use of shuffling can be expressed in terms of basic operations such as choice, concatenation and Kleene star. However, the use of shuffling yields a much more succinct representation of problems that occur in modeling of concurrent systems [4,12].

Our interest is to extend the notion of a Brzozowski derivative [3] to regular expressions with shuffles. Derivatives support the elegant and efficient construction of automata-based word recognition algorithms [10] and are also useful in the development of related algorithms for equality and containment among regular expressions [1,6].

Prior work in the area. To the best of our knowledge, there is almost no prior work which studies the notion of Brzozowski derivatives in connection with shuffling. We are only aware of one work [9] which appears to imply a definition of derivatives for strongly synchronized shuffling [2]. Further work in the area studies the construction of automata for a specific form of shuffling commonly referred to as asynchronous interleaving [5,7]. In contrast, we provide detailed definitions

© Springer International Publishing Switzerland 2015
A.-H. Dediu et al. (Eds.): LATA 2015, LNCS 8977, pp. 275–286, 2015.
DOI: 10.1007/978-3-319-15579-1_21

how to obtain derivatives including formal results for various shuffling operations [2,4,11]. Our results imply algorithms for constructing automata as well as for checking equality and containment of regular expressions with shuffles.

Contributions. After introducing our notation in Section 2 and reviewing existing variants of shuffling in Section 3, we claim the following contributions:

- We introduce a general shuffle operation which subsumes previous forms of shuffling (Section 4).
- We extend the notion of Brzozowski derivatives to the general shuffle operation and are able to re-establish all of its "good" properties (Section 5).
- Based on the general shuffle operation, we provide systematic methods to obtain derivatives for specific variants of shuffling (Section 5.1).

We conclude in Section 6.

2   Preliminaries

Let Σ be a fixed alphabet (i.e., a finite set of symbols). We usually denote symbols by x, y and z. The set Σ^* denotes the set of finite words over Σ. We write Γ, Δ to denote subsets (sub-alphabets) of Σ. We write ϵ to denote the empty word and $v \cdot w$ to denote the concatenation of two words v and w. We generally write $L_1, L_2 \subseteq \Sigma^*$ to denote languages over Σ.

We write $L_2 \backslash L_1$ to denote the left quotient of L_1 with L_2 where $L_2 \backslash L_1 = \{w \mid \exists v \in L_2 . v \cdot w \in L_1\}$. We write $x \backslash L$ as a shorthand for $\{x\} \backslash L$.

We write $\alpha(w)$ to denote the set of symbols which appear in a word. The inductive definition is as follows: (1) $\alpha(\epsilon) = \emptyset$, (2) $\alpha(x \cdot w) = \alpha(w) \cup \{x\}$. The extension to languages is as follows: $\alpha(L) = \bigcup_{w \in L} \alpha(w)$.

We write $\Pi_\Gamma(w)$ to denote the projection of a word w onto a sub-alphabet Γ. The inductive definition is as follows:

$$\Pi_\Gamma(\epsilon) = \epsilon \qquad\qquad \Pi_\Gamma(x \cdot w) = \begin{cases} x \cdot \Pi_\Gamma(w) & x \in \Gamma \\ \Pi_\Gamma(w) & x \notin \Gamma \end{cases}$$

3   Shuffling Operations

Definition 1 (Shuffling).   *The shuffle operator* $\| :: \Sigma^* \times \Sigma^* \to \wp(\Sigma^*)$ *is defined inductively as follows:*

$$\begin{aligned} \epsilon \| w &= \{w\} \\ w \| \epsilon &= \{w\} \\ x \cdot v \| y \cdot w &= \{x \cdot u \mid u \in v \| y \cdot w\} \cup \{y \cdot u \mid u \in x \cdot v \| w\} \end{aligned}$$

We lift shuffling to languages by $L_1 \| L_2 = \{u \mid u \in v \| w \land v \in L_1 \land w \in L_2\}$.

For example, we find that $x \cdot y \| z = \{x \cdot y \cdot z, x \cdot z \cdot y, z \cdot x \cdot y\}$.

While the shuffle operator represents the asynchronous interleaving of two words $v, w \in \Sigma^*$, there are also shuffle operators that include some synchronization. The strongly synchronized shuffle of two words w.r.t. some sub-alphabet Γ imposes the restriction that the traces must synchronize on all symbols in Γ. All symbols not appearing in Γ are shuffled. In its definition, we write $(p) \Rightarrow X$ for: if p then X else $\emptyset$.

Definition 2 (Strong Synchronized Shuffling). *The* synchronized shuffling operator *w.r.t.* $\Gamma \subseteq \Sigma$, $\||_\Gamma :: \Sigma^* \times \Sigma^* \to \wp(\Sigma^*)$, *is defined inductively as follows.*

$$
\begin{aligned}
\epsilon \||_\Gamma w \quad &= (\Gamma \cap \alpha(w) = \emptyset) \Rightarrow \{w\} & \text{(S1)} \\
w \||_\Gamma \epsilon \quad &= (\Gamma \cap \alpha(w) = \emptyset) \Rightarrow \{w\} & \text{(S2)} \\
x \cdot v \||_\Gamma y \cdot w &= (x = y \wedge x \in \Gamma) \Rightarrow \{x \cdot u \mid u \in v \||_\Gamma w\} & \text{(S3)} \\
&\cup (x \notin \Gamma) \Rightarrow \{x \cdot u \mid u \in v \||_\Gamma y \cdot w\} & \text{(S4)} \\
&\cup (y \notin \Gamma) \Rightarrow \{y \cdot u \mid u \in x \cdot v \||_\Gamma w\} & \text{(S5)}
\end{aligned}
$$

We lift strongly synchronized shuffling to languages by

$$ L_1 \||_\Gamma L_2 = \{u \mid u \in v \||_\Gamma w \wedge v \in L_1 \wedge w \in L_2\} $$

The basic cases (S1) and (S2) impose the condition (via $\Gamma \cap \alpha(w) = \emptyset$) that none of the symbols in w shall be synchronized. If the condition is violated we obtain the empty set. For example, $\epsilon \||_{\{x\}} y \cdot z = \{y \cdot z\}$, but $\epsilon \||_{\{x\}} x \cdot y \cdot z = \emptyset$.

In the inductive step, a symbol in Γ appearing on both sides forces synchronization (S3). If the leading symbol on either side does not appear in Γ, then it can be shuffled arbitrarily. See cases (S4) and (S5). These three cases ensure progress until one side is reduced to the empty string. For example, we find that $x \cdot y \||_{\{x\}} x \cdot z = \{x \cdot y \cdot z, x \cdot z \cdot y\}$. On the other hand, $x \cdot y \||_{\{x,y\}} x \cdot z = \emptyset$.

Shuffling and strongly synchronized shuffling correspond to the *arbitrary* synchronized shuffling and strongly synchronized shuffling operations by Beek and coworkers [2]. The inductive definitions that we present simplify our proofs.

Beek and coworkers [2] also introduce a *weak* synchronized shuffling operation. In its definition, we write $L \cdot x$ as a shorthand for $\{w \cdot x \mid w \in L\}$ and $x \cdot L$ as a shorthand for $\{x \cdot w \mid w \in L\}$

Definition 3 (Weak Synchronized Shuffling). *Let* $v, w \subseteq \Sigma^*$ *and* $\Gamma \subseteq \Sigma$. *Then, we define*

$$
\begin{aligned}
v| \sim |_\Gamma w = \{u \mid \exists n \geq 0, & x_i \in \Gamma, v_i \in \Gamma, w_i \in \Gamma. \\
& v = v_1 \cdot x_1 ... x_n \cdot v_{n+1} \wedge w = w_1 \cdot x_1 ... x_n \cdot w_{n+1} \wedge \\
& u \in (v_1 \| w_1) \cdot x_1 ... x_n \cdot (v_{n+1} \| w_{n+1}) \\
& \alpha(v_i) \cap \alpha(w_i) \cap \Gamma = \emptyset\}
\end{aligned}
$$

and for languages: $L_1| \sim |_\Gamma L_2 = \{u \mid u \in v| \sim |_\Gamma w \wedge v \in L_1 \wedge w \in L_2\}$.

The weak synchronized shuffle of two words v and w synchronizes only on those symbols in Γ that occur in both v and w. For example, $x \cdot y| \sim |_{\{x,y\}} x \cdot z = \{x \cdot y \cdot z, x \cdot z \cdot y\}$ because $y \notin \alpha(x \cdot z)$ whereas $x \cdot y \||_{\{x,y\}} x \cdot z = \emptyset$.

There is another variant of synchronous shuffling called *synchronous composition* [4,11]. The difference to strongly synchronized shuffling is that synchronization occurs on symbols common to both operands. Thus, synchronous composition can be defined by projecting onto the symbols of the operands.

Definition 4 (Synchronous Composition). *The synchronous composition operator* $|||$ *is defined by:*

$$L_1|||L_2 = \{w \in (\alpha(L_1) \cup \alpha(L_2))^* \mid \Pi_{\alpha(L_1)}(w) \in L_1 \wedge \Pi_{\alpha(L_2)}(w) \in L_2\}$$

For example, $x \cdot y|||x \cdot z$ equals $\{x \cdot y \cdot z, x \cdot z \cdot y\}$.

It turns out that the strong synchronized shuffling operation $|||_\Gamma$ subsumes synchronous composition due to the customizable set Γ. In Section 4, we show an even stronger result: All of the shuffling variants we have seen can be expressed in terms of a general synchronous shuffling operation.

4   General Synchronous Shuffling

The general synchronous shuffling operation is parameterized by a set of synchronizing symbols, Γ, and two additional sets P_1 and P_2 that keep track of 'out of sync' symbols from Γ. We write $X \cup x$ as a shorthand for $X \cup \{x\}$.

Definition 5 (General Synchronous Shuffling). *Let* $\Gamma, P_1, P_2 \subseteq \Sigma$. *The general synchronous shuffling operator* $^{P_1}||_\Gamma^{P_2} :: \Sigma^* \times \Sigma^* \to \wp(\Sigma^*)$ *is defined inductively as follows.*

$$
\begin{aligned}
\epsilon\,^{P_1}||_\Gamma^{P_2} w \quad &= ((\alpha(w) \cap \Gamma = \emptyset) \vee (P_1 \cap (P_2 \cup \alpha(w)) = \emptyset)) \Rightarrow \{w\} && (G1)\\
w\,^{P_1}||_\Gamma^{P_2} \epsilon \quad &= ((\alpha(w) \cap \Gamma = \emptyset) \vee (P_1 \cup \alpha(w)) \cap P_2 = \emptyset)) \Rightarrow \{w\} && (G2)\\
x \cdot v\,^{P_1}||_\Gamma^{P_2} y \cdot w &= (x \notin \Gamma) \Rightarrow \{x \cdot u \mid u \in v\,^{P_1}||_\Gamma^{P_2} y \cdot w\} && (G3)\\
&\sqcup (y \notin \Gamma) \to \{y \cdot u \mid u \subset x \cdot v\,^{P_1}||_\Gamma^{P_2} w\} && (G4)\\
&\sqcup (x = y \wedge x \in \Gamma \wedge P_1 \cap P_2 = \emptyset) \Rightarrow \{x \cdot u \mid u \in v^{\emptyset}||_\Gamma^{\emptyset} w\} && (G5)\\
&\sqcup (x = y \wedge x \in \Gamma \wedge P_1 \cap P_2 \neq \emptyset) \Rightarrow \{x \cdot u \mid u \in v^{P_1}||_\Gamma^{P_2} w\} && (G6)\\
&\sqcup (x \in \Gamma \wedge (P_1 \cup x) \cap P_2 = \emptyset) \Rightarrow \{x \cdot u \mid u \in v^{P_1 \cup x}||_\Gamma^{P_2} y \cdot w\} && (G7)\\
&\sqcup (y \in \Gamma \wedge P_1 \cap (P_2 \cup y) = \emptyset) \Rightarrow \{y \cdot u \mid u \in x \cdot v^{P_1}||_\Gamma^{P_2 \cup y} w\} && (G8)
\end{aligned}
$$

For $L_1, L_2 \subseteq \Sigma^*$, $\Gamma \subseteq \Sigma$, $P_1, P_2 \subseteq \Sigma$ *we define*

$$L_1\,^{P_1}||_\Gamma^{P_2} L_2 = \{u \mid u \in v^{P_1}||_\Gamma^{P_2} w \wedge v \in L_1 \wedge w \in L_2\}$$

The definition of general synchronous shuffling is significantly more involved compared to the earlier definitions. Cases (G1-8) are necessary to encode the earlier shuffle operations from Section 3. The exact purpose of the individual cases will become clear shortly.

In our first result we observe that $^\Sigma||_\Gamma^\Sigma$ exactly corresponds to strongly synchronized shuffling ($|||_\Gamma$).

Theorem 6. *For any* $L_1, L_2 \subseteq \Sigma^*$ *and* $\Gamma \subseteq \Sigma$: $L_1|||_\Gamma L_2 = L_1{}^\Sigma||_\Gamma^\Sigma L_2$.

Proof. We choose a 'maximal' assignment for P_1 and P_2 such that definition of $P_1||_\Gamma^{P_2}$ reduces to $|||_\Gamma$. We observe that property $P_1 = \Sigma \wedge P_2 = \Sigma$ (SP) is an invariant. For cases (G3-4) and (G6) the invariant property clearly holds. For cases (G5), (G7-8) the preconditions are violated. Hence, for $^\Sigma||_\Gamma^\Sigma$ only cases (G1-4) and (G6) will ever apply.

Under the given assumptions, we can relate the cases in Definition 2 and Definition 5 as follows. Cases (G1-2) correspond to cases (S1-2). Cases (G3-4) correspond to cases (S4-5). Case (G6) corresponds to case (S3). Due to the invariant property (SP) cases (G5) and (G7-8) never apply.

Hence, $^\Sigma||_\Gamma^\Sigma$ and $|||_\Gamma$ yield the same result. $\qquad\square$

Via similar reasoning we can show that for $\Gamma = \emptyset \wedge P_1 = \emptyset \wedge P_2 = \emptyset$ general synchronized shuffling boils down to (arbitrary) shuffling.

Theorem 7. *For any $L_1, L_2 \subseteq \Sigma^*$: $L_1||L_2 = L_1{}^\emptyset||_\emptyset^\emptyset L_2$.*

An immediate consequence from Theorem 6 (set Σ and Γ to $\emptyset$) and Theorem 7 is that shuffling can also be expressed in terms of strong synchronized shuffling.

Corollary 8. *For any $L_1, L_2 \subseteq \Sigma^*$: $L_1||L_2 = L_1|||_\emptyset L_2$.*

Our next result establishes a connection to weak synchronized shuffling [2].

Theorem 9. *For any $L_1, L_2 \subseteq \Sigma^*, \Gamma \subseteq \Sigma$: $L_1| \sim |_\Gamma L_2 = L_1{}^\emptyset||_\Gamma^\emptyset L_2$.*

Proof. Property $P_1, P_2 \subseteq \Gamma \wedge P_1 \cap P_2 = \emptyset$ (WP) is an invariant of $^{P_1}||_\Gamma^{P_2}$. For cases (G3-4) the invariant property clearly holds. More interesting are cases (G7-8) where P_1, resp., P_2 is extended. Under the precondition property (WP) remains invariant. Case (G6) never applies. Case (G5) clearly maintains the invariant.

Recall that for strong synchronized shuffling the roles of (G5) and (G6) are switched. See proof of Theorem 6. This shows that while cases (G5) and (G6) look rather similar both are indeed necessary.

As we can see, under the invariant condition (WP), the purpose of P_1, P_2 is to keep track of "out of sync" symbols from Γ. See cases (G7-8). Case (G5) synchronizes on $x \in \Gamma$. Hence, P_1 and P_2 are (re)set to $\emptyset$.

Thus, we can show (via some inductive argument) that $^\emptyset||_\Gamma^\emptyset$ under the (WP) invariant corresponds to weak synchronized shuffling as defined in Definition 3. $\qquad\square$

It remains to show that synchronous composition is subsumed by general synchronized shuffling. First, we verify that synchronous composition is subsumed by strongly synchronous shuffling.

Theorem 10. *For any $L_1, L_2 \subseteq \Sigma^*$ we find that $L_1|||L_2 = L_1|||_{\alpha(L_1) \cap \alpha(L_2)} L_2$.*

Proof. To establish the direction $L_1|||L_2 \supseteq L_1|||_{\alpha(L_1) \cap \alpha(L_2)} L_2$ we verify that the projection of a strongly synchronizable word w.r.t. $\alpha(L_1) \cap \alpha(L_2)$ yields words in the respective languages L_1 and L_2.

Formally: Let $u, v, w \in \Sigma^*$, $L_1, L_2 \subseteq \Sigma^*$, $\Gamma \subseteq \Sigma$ such that $u \in v|||_\Gamma w$ where $v \in L_1$, $w \in L_2$ and $\Gamma \subseteq \alpha(L_1) \cap \alpha(L_2)$. Then, we find that (1) $\Pi_{\alpha(L_1)}(u) = v$ and (2) $\Pi_{\alpha(L_2)}(u) = w$. The proof of this statement proceeds by induction over u.

Direction $L_1|||L_2 \supseteq L_1|||_{\alpha(L_1) \cap \alpha(L_2)} L_2$ can be verified similarly. We show that if the projection of a word onto $\alpha(L_1)$ and $\alpha(L_2)$ yields words in the respective language, then the word must be strongly synchronizable w.r.t. $\alpha(L_1) \cap \alpha(L_2)$.

Formally: Let $u, v, w \in \Sigma^*$, $L_1, L_2 \subseteq \Sigma^*$ such that $w \in (\alpha(L_1) \cup \alpha(L_2))^*$, $\Pi_{\alpha(L_1)}(w) = u \in L_1$ and $\Pi_{\alpha(L_1)}(w) = v \in L_2$. Then, we find that $w \in u|||_{\alpha(L_1) \cap \alpha(L_2)} v$. The proof proceeds again by induction, this time over w.   □

An immediate consequence of Theorem 6 and Theorem 10 is the following result. Synchronous composition is subsumed by general synchronous shuffling.

Corollary 11. *For any $L_1, L_2 \subseteq \Sigma^*$ we find that $L_1|||L_2 = L_1{}^\Sigma||_{\alpha(L_1) \cap \alpha(L_2)}^\Sigma L_2$.*

5   Derivatives for General Synchronous Shuffling

Brzozowski derivatives [3] are a useful tool to translate regular expressions into finite automata and to obtain decision procedures for equivalence and containment for regular expressions. We show that Brzozowski's results and their applications can be extended to regular expressions that contain shuffle operators. Based on the results of the previous section, we restrict our attention to regular expressions extended with the general synchronous shuffle operator.

Definition 12. *The set R_Σ of regular shuffle expressions is defined inductively by $\phi \in R_\Sigma$, $\epsilon \in R_\Sigma$, $\Sigma \subseteq R_\Sigma$, and for all $r, s \in R_\Sigma$ and $\Gamma, P_1, P_2 \subseteq \Sigma$ we have that $r + s$, $r \cdot s$, r^*, $r^{P_1}||_\Gamma^{P_2} s \in R_\Sigma$.*

Definition 13. *The language $\mathcal{L}() : R_\Sigma \to \Sigma^*$ denoted by a regular shuffle expression is defined inductively as follows. $\mathcal{L}(\phi) = \emptyset$. $\mathcal{L}(\epsilon) = \{\epsilon\}$. $\mathcal{L}(x) = \{x\}$. $\mathcal{L}(r+s) = \mathcal{L}(r) \cup \mathcal{L}(s)$. $\mathcal{L}(r \cdot s) = \{v \cdot w \mid v \in \mathcal{L}(r) \wedge w \in \mathcal{L}(s)\}$. $\mathcal{L}(r^*) = \{w_1...w_n \mid n \geq 0 \wedge w_i \in \mathcal{L}(r) \wedge i \in \{1, ..., n\}\}$. $\mathcal{L}(r^{P_1}||_\Gamma^{P_2} s) = \{u \mid u \in v^{P_1}||_\Gamma^{P_2} w \wedge v \in \mathcal{L}(r) \wedge w \in \mathcal{L}(s)\}$.*

An expression r is *nullable* if $\epsilon \in \mathcal{L}(r)$. The following function $n(\_)$ detects nullable regular expressions.

Definition 14. *We define $n(\_) : R_\Sigma \to Bool$ inductively as follows. $n(\phi) = false$. $n(\epsilon) = true$. $n(x) = false$. $n(r + s) = n(r) \vee n(s)$. $n(r \cdot s) = n(r) \wedge n(s)$. $n(r^*) = true$. $n(r^{P_1}||_\Gamma^{P_2} s) = n(r) \wedge n(s)$.*

Lemma 15. *For all $r \in R_\Sigma$ we have that $\epsilon \in \mathcal{L}(r)$ iff $n(r) = true$.*

Proof. The proof proceeds by induction over r. For brevity, we only consider the shuffle case as the remaining cases are standard. $\epsilon \in \mathcal{L}(r^{P_1}||_\Gamma^{P_2} s)$ iff (by definition) $\epsilon \in v^{P_1}||_\Gamma^{P_2} w$ for some $v \in \mathcal{L}(r)$ and $w \in \mathcal{L}(s)$. By definition of $^{P_1}||_\Gamma^{P_2}$ it must be that $v = \epsilon$ and $w = \epsilon$. By induction, this is equivalent to $n(r) \wedge n(s)$.   □

The derivative of an expression r w.r.t. some symbol x, written $d_x(r)$ yields a new expression where the leading symbol x has been removed. In its definition, we write $(p) \Rightarrow r$ for: if p then r else ϕ.

Definition 16. *The derivative of $r \in R_\Sigma$ w.r.t. $x \in \Sigma$, written $d_x(r)$, is computed inductively as follows.*

$$
\begin{aligned}
d_x(\phi) \quad &= \phi & \text{(D1)}\\
d_x(\epsilon) \quad &= \phi & \text{(D2)}\\
d_y(x) \quad &= (x = y) \Rightarrow \epsilon & \text{(D3)}\\
d_x(r + s) \quad &= d_x(r) + d_x(s) & \text{(D4)}\\
d_x(r \cdot s) \quad &= (d_x(r)) \cdot s + (n(r)) \Rightarrow d_x(s) & \text{(D5)}\\
d_x(r^*) \quad &= (d_x(r)) \cdot r^* & \text{(D6)}\\
d_x(r^{P_1}||_\Gamma^{P_2} s) &= (x \notin \Gamma) \Rightarrow (d_x(r)^{P_1}||_\Gamma^{P_2} s) + (r^{P_1}||_\Gamma^{P_2} d_x(s)) & \text{(D7)}\\
&+ (x \in \Gamma \wedge P_1 \cap P_2 = \emptyset) \Rightarrow (d_x(r)^\emptyset ||_\Gamma^\emptyset d_x(s)) & \text{(D8)}\\
&+ (x \in \Gamma \wedge P_1 \cap P_2 \neq \emptyset) \Rightarrow (d_x(r)^{P_1}||_\Gamma^{P_2} d_x(s)) & \text{(D9)}\\
&+ (x \in \Gamma \wedge (P_1 \cup x) \cap P_2 = \emptyset) \Rightarrow (d_x(r)^{P_1 \cup x}||_\Gamma^{P_2} s) & \text{(D10)}\\
&+ (x \in \Gamma \wedge P_1 \cap (P_2 \cup x) = \emptyset) \Rightarrow (r^{P_1}||_\Gamma^{P_2 \cup x} d_x(s)) & \text{(D11)}
\end{aligned}
$$

The definition extends to words and sets of words. We define $d_\epsilon(r) = r$ and $d_{xw}(r) = d_w(d_x(r))$. For $L \subseteq \Sigma^*$ we define $d_L(r) = \{d_w(r) \mid w \in L\}$.

We refer to the special case $d_{\Sigma^*}(r)$ as the set of descendants *of r. A descendant is either the expression itself, a derivative of the expression, or the derivative of a descendant.*

The first six cases (D1-6) correspond to Brzozowski's original definition [3]. As a minor difference we may concatenate with ϕ when building the derivative of a concatenated expression whose first component is not nullable. The new sub-cases (D7-11) for the general shuffle closely correspond to the sub-cases of Definition 5. For example, compare (D7) and (G3-4), (D8) and (G5), (D9) and (G6), (D10) and (G7), and lastly (D11) and (G8).

An easy induction shows that the derivative of a shuffle expression is again a shuffle expression.

Theorem 17 (Closure). *For any $r \in R_\Sigma$ and $x \in \Sigma$ we have that $d_x(r) \in R_\Sigma$.*

Brzozowski proved that the derivative of a regular expression denotes a left quotient. This result extends to shuffle expressions.

Theorem 18 (Left Quotients). *For any $r \in R_\Sigma$ and $x \in \Sigma$ we have that $\mathcal{L}(d_x(r)) = x \backslash \mathcal{L}(r)$.*

Proof. It suffices to consider the new case of general synchronous shuffling. We consider the direction $\mathcal{L}(d_x(r^{P_1}||_\Gamma^{P_2} s)) \subseteq x \backslash \mathcal{L}(r^{P_1}||_\Gamma^{P_2} s)x = \{w \mid x \cdot w \in \mathcal{L}(r^{P_1}||_\Gamma^{P_2} s)\}$.

Suppose $u \in \mathcal{L}(d_x(r^{P_1}||_\Gamma^{P_2} s))$. We will verify that $x \cdot u \in \mathcal{L}(r^{P_1}||_\Gamma^{P_2} s)$. We proceed by distinguishing among the following cases.

– Case $x \notin \Gamma$:

By definition of the derivative operation, we find that either (D7a) $u \in \mathcal{L}(d_x(r)^{P_1} ||_\Gamma^{P_2} s)$ or (D7b) $u \in \mathcal{L}(r^{P_1} ||_\Gamma^{P_2} d_x(s))$.

  • Case (D7a):

    1. By definition $u \in v^{P_1} ||_\Gamma^{P_2} w$ for some $v \in \mathcal{L}(d_x(r))$ and $w \in \mathcal{L}(s)$.
    2. By induction $x \cdot v \in \mathcal{L}(r)$.
    3. By observing the various cases for w (see (G2-3) in Definition 5) we follow that $x \cdot u \in x \cdot v^{P_1} ||_\Gamma^{P_2} w$.
    4. Hence, $x \cdot u \in \mathcal{L}(r^{P_1} ||_\Gamma^{P_2} s)$ and we are done.

  • Case (D7b): Similar to the above.

– Case $x \in \Gamma$:

By definition of the derivative operation (D8) $u \in \mathcal{L}(d_x(r)^\emptyset ||_\Gamma^\emptyset d_x(s))$ where $P_1 \cap P_2 = \emptyset$, or (D9) $u \in \mathcal{L}(d_x(r)^{P_1} ||_\Gamma^{P_2} d_x(s))$ where $P_1 \cap P_2 \neq \emptyset$, or (D10) $u \in \mathcal{L}(d_x(r)^{P_1 \cup x} ||_\Gamma^{P_2} s)$ where $(P_1 \cup x) \cap P_2 = \emptyset$, or (D11) $u \in \mathcal{L}(r^{P_1} ||_\Gamma^{P_2 \cup x} d_x(s))$ where $P_1 \cap (P_2 \cup x) = \emptyset$.

  • Case (D8):

    1. By definition $u \in v^\emptyset ||_\Gamma^\emptyset w$ for some $v \in \mathcal{L}(d_x(r))$ and $w \in \mathcal{L}(d_x(s))$.
    2. By induction $x \cdot v \in \mathcal{L}(r)$ and $x \cdot w \in \mathcal{L}(s)$.
    3. By case (G5) $x \cdot u \in x \cdot v^{P_1} ||_\Gamma^{P_2} x \cdot w$.
    4. Hence, $x \cdot u \in \mathcal{L}(r^{P_1} ||_\Gamma^{P_2} s)$ and we are done.

  • Case (D9): Similar to the above. Instead of (G5) we can apply (G6).

  • Case (D10):

    1. By definition $u \in v^{P_1 \cup x} ||_\Gamma^{P_2} w$ for some $v \in \mathcal{L}(d_x(r))$ and $w \in \mathcal{L}(s)$.
    2. By induction $x \cdot v \in \mathcal{L}(r)$.
    3. By observing the various cases for w (see (G2) and (G7)) we follow that $x \cdot u \in x \cdot v^{P_1} ||_\Gamma^{P_2} w$.
    4. Hence, $x \cdot u \in \mathcal{L}(r^{P_1} ||_\Gamma^{P_2} s)$ and we are done.

  • Case (D11): Similar to the above.

The other direction $\mathcal{L}(d_x(r^{P_1} ||_\Gamma^{P_2} s)) \supseteq \{x \cdot w \mid w \in \mathcal{L}(r^{P_1} ||_\Gamma^{P_2} s)\}$ follows via similar reasoning. □

Based on the above result, we obtain a simple algorithm for membership testing. Given a word w and expression r, we exhaustively apply the derivative operation and on the final expression we apply the nullable test. That is $w \in \mathcal{L}(r)$ iff $n(d_w(r))$.

In general, it seems wasteful to repeatedly generate derivatives just for the sake of testing a specific word. A more efficient method is to construct a DFA via which we can then test many words. Brzozowski recognized that there is an elegant DFA construction method based on derivatives. Expressions are treated as states. For each expression and its derivative we find a transition.

For this construction to work we must establish that (1) the transitions implied by the derivatives cover all cases, and (2) the set of states remains finite. This is what we will consider next.

First, we establish (1) by verifying that each shuffle expression can be represented as a sum of its derivatives, extending another result of Brzozowski.

Theorem 19 (Representation). *For any* $r \in R_\Sigma$, $\mathcal{L}(r) = \mathcal{L}((n(r)) \Rightarrow \epsilon) \cup \bigcup_{x \in \Sigma} \mathcal{L}(x \cdot d_x(r))$.

Proof. Follows immediately from Lemma 15 and Theorem 18.   □

States are descendants of expression r. Hence, we must verify that the set $d_{\Sigma^*}(r)$ is finite. In general, this may not be the case as shown by the following example

$$
\begin{aligned}
d_x(x^*) &= \epsilon \cdot x^* \\
d_x(\epsilon \cdot x^*) &= \phi \cdot x^* + \epsilon \cdot x^* \\
d_x(\phi \cdot x^* + \epsilon \cdot x^*) &= (\phi \cdot x^* + \epsilon \cdot x^*) + (\phi \cdot x^* + \epsilon \cdot x^*)
\end{aligned}
$$
...

To guarantee finiteness, we need to consider expressions modulo *similarity*.

Definition 20 (Similarity). *We say that two expressions* $r, s \in R_\Sigma$ *are similar, written* $r \approx s$, *if one can be transformed into the other by applying the following identities:*

(I1) $r + s = s + r$ *(I2)* $r + (s + t) = (r + s) + t$ *(I3)* $r + r = r$

For $S \subseteq R_\Sigma$ *we write* $S/\approx$ *to denote the set of equivalence classes of all similar expressions in* S.

For the above example, we find that $(\phi \cdot x^* + \epsilon \cdot x^*) + (\phi \cdot x^* + \epsilon \cdot x^*) \approx \phi \cdot x^* + \epsilon \cdot x^*$ by application of (I3). To show an application of (I2), consider

$$
\begin{aligned}
d_x(x^* + x \cdot x^*) &= = \epsilon \cdot x^* + x^* \\
d_x(\epsilon \cdot x^* + x^*) &= (\phi \cdot x^* + \epsilon x^*) + \epsilon \cdot x^*
\end{aligned}
$$

Clearly, $(\phi \cdot x^* + \epsilon x^*) + \epsilon \cdot x^* \approx \phi \cdot x^* + \epsilon x^*$. Application of (I1) is omitted for brevity.

To verify (dis)similarity among descendants it suffices to apply identities (I1-3) at the top-level, i.e. highest position in the abstract syntax tree representation of expressions. Top-level alternatives are kept in a list and sorted according to the number of occurrences of symbols. Any duplicates in the list are removed. Thus, the set $d_{\Sigma^*}(r)/\approx$ is obtained by generating dissimilar descendants, starting with $\{r\}$. That this generation step reaches a fix-point is guaranteed by the following result.

Theorem 21 (Finiteness). *For any* $r \in R_\Sigma$ *the set* $d_{\Sigma^*}(r)/\approx$ *is finite.*

Proof. It suffices to consider the new case of shuffle expressions. Our argumentation is similar to the case of concatenation in Brzozowski's original result. See proofs of Theorems 4.3(a) and 5.2 in [3]. It suffices to consider the new form $r^{P_1} ||_\Gamma^{P_2} s$ which we will abbreviate by t.

By inspection of the definition of $d()$ on $^{P_1}||_\Gamma^{P_2}$ and application of identity (I2) (associativity) we find that all descendants of t can be represented as a sum of expressions which are either of the shape ϕ or $r'^{P_1'}||_\Gamma^{P_2'} s'$ where r' is a descendant of r and s' is a descendant of s, but P_1' and P_2' are arbitrary subsets of Σ.

Thus, we can apply a similar argument as in Brzozowski's original proof to approximate the number of descendants of $r^{P_1}||_\Gamma^{P_2} s$ by the number of descendants of r and s.

Suppose $\sharp D_r$ denotes the number of all descendants of r and n is the number of elements in Σ. Then, we can approximate $\sharp D_t$ as follows:

$$\sharp D_t \le 2^{\sharp D_r * \sharp D_s * 2^n * 2^n}$$

The exponent counts the number of different factors of the form $r'^{P_1}||_\Gamma^{P_2} s'$ and a sum corresponds to a subset of these factors. The factor $2^n * 2^n$ arises because P_1, P_2 range over subsets of Σ where $n = |\Sigma|$. The factor $\sharp D_s * \sharp D_r$ arises from the variation of r' and s' over the descendants of r and s, respectively.

As $\sharp D_r$ and $\sharp D_s$ are finite, by the inductive hypothesis, we obtain a very large, but finite bound on $\sharp D_t$. $\square$

We summarize.

Definition 22 (Derivative-Based DFA Construction). *For any $r \in R_\Sigma$ we define $\mathcal{D}(r) = (Q, \Sigma, q_0, \delta, F)$ where $Q = d_{\Sigma^*}(r)/\approx$, $q_0 = r$, for each $q \in Q$ and $x \in \Sigma$ we define $\delta(q, x) = d_x(q)$, and $F = \{q \in Q \mid n(q)\}$.*

Theorem 23. *For any $r \in R_\Sigma$ we have that $\mathcal{L}(r) = \mathcal{L}(\mathcal{D}(r))$.*

5.1 Discussion

Derivatives for free. To obtain derivatives for the various shuffling variants in Section 3 we apply the following method. Each shuffling variant is transformed into its general synchronous shuffle representation as specified in Section 4. On the resulting expression we can then apply the derivative construction from Section 5.

For interleaving ($||$), weakly ($| \sim |_\Gamma$) and strongly synchronized shuffling ($|||_\Gamma$) the transformation step is purely syntactic. For example, in case of weak synchronous shuffling expressions $r_1| \sim |_\Gamma r_2$ are exhaustively transformed into $r_1{}^\emptyset||_\Gamma^\emptyset r_2$. The transformation step is more involved for synchronous composition ($|||$) as we must compute the alphabet of expressions (resp. the alphabet of the underlying languages). We define $\alpha(r) = \alpha(\mathcal{L}(r))$.

For plain regular expressions, we can easily compute the alphabet by observing the structure of expressions. For example, $\alpha(r^*) = \alpha(r)$. $\alpha(x) = \{x\}$. $\alpha(\phi) = \{\}$. $\alpha(\epsilon) = \{\}$. $\alpha(r + s) = \alpha(r) \cup \alpha(s)$. $\alpha(r \cdot s) = \alpha(r) \cup \alpha(s)$ if $\mathcal{L}(r), \mathcal{L}(s) \ne \{\}$. Otherwise, $\alpha(r \cdot s) = \{\}$. The test $\mathcal{L}(r) \ne \{\}$ can again be defined by observing the expression structure. We omit the details.

In the presence of shuffle expressions such as synchronous composition, it is not obvious how to appropriately extend the above structural definition. For example, consider $x \cdot x \cdot y |||x \cdot y$. We find that $\alpha(x \cdot x \cdot y) = \{x, y\}$ and $\alpha(x \cdot y)$ but $\alpha(x \cdot x \cdot y |||x \cdot y) = \{\}$ due to the fact that $x \cdot x \cdot y |||x \cdot y$ equals ϕ.

Hence, to compute the alphabet of some $r \in R_\Sigma$ we first convert r into a DFA M using the derivative-based automata construction. To compute the alphabet of M we use a variant of the standard emptiness check algorithm for DFAs. First, we compute all reachable paths from any of the final states to the initial state. To avoid infinite loops we are careful not to visit a transition twice on a path. Then, we obtain the alphabet of M by collecting the set of all symbols on all transitions along these paths.

Thus, the transformation of expressions composed of synchronous composition proceeds as follows. In the to be transformed expression, we pick any subexpression $r_1|||r_2$ where $r_1, r_2 \in R_\Sigma$. If there is none we are done. Otherwise, we must find r_1 and r_2 which have already been transformed, resp., do not contain any shuffling operator. Alphabets $\alpha(r_1)$ and $\alpha(r_2)$ are computed as described and subexpression $r_1|||r_2$ is replaced by $r_1{}^\Sigma||^\Sigma_{\alpha(r_1)\cap\alpha(r_2)}r_2$. This process repeats until all synchronous composition operations have been replaced.

Specialization of derivative method. For shuffling ($||$) and strongly synchronized shuffling ($|||_\Gamma$), it is possible to derive specialized derivative operations. In the following, $R_\Sigma^{|\cdot|}$ denotes the subset of regular expressions restricted to shuffle expressions composed of $|\cdot|$ where $|\cdot|$ stands for any of the shuffling forms we have seen so far.

Theorem 24 (Derivatives Closure for Shuffling). *For any $r \in R_\Sigma^{||}$ and $x \in \Sigma$ we have that $d_x(r) \in R_\Sigma^{||}$.*

Proof. Recall that $||$ can be expressed as $^\emptyset||^\emptyset_\emptyset$. By case analysis of Definition 16. Case (D7) applies only.   □

Theorem 25 (Derivatives Closure for Strongly Synchronized Shuffling). *For any $r \in R_\Sigma^{|||_\Gamma}$ and $x \in \Sigma$ we have that $d_x(r) \in R_\Sigma^{|||_\Gamma}$.*

Proof. Recall that $|||_\Gamma$ can be expressed as $^\Gamma||^\Gamma_\Gamma$. Again by case analysis. This time only cases (D7) and (D9) apply.   □

The above closure results show that the general derivative method in Definition 16 can be specialized for the case of asynchronous and strongly synchronized shuffling. The respective proofs describe the relevant cases.

For weak synchronous shuffling, we can no longer guarantee the closure property. For example, consider weak synchronous shuffling $|\sim|_\Gamma$ which is expressed by $^\emptyset||^\emptyset_\Gamma$. For expression $x \cdot z^\emptyset||^\emptyset_{\{x\}}y$. we find that

$$d_x(x \cdot z^\emptyset||^\emptyset_{\{x\}}y) = z^{\{x\}}||^\emptyset_{\{x\}}y + x \cdot z^\emptyset||^{\{x\}}_{\{x\}}\phi$$

The expression on the right-hand side is *not* part of $R_\Sigma^{|\sim|_\Gamma}$ due to subexpressions of the form $z^{\{x\}}||^\emptyset_{\{x\}}y$. However, these forms are necessary for correctness. Hence, to appropriately define derivatives for weak synchronous shuffling derivatives it is strictly necessary to enrich the expression language with general synchronous shuffling.

A similar observation applies to synchronous composition. For brevity, we omit the details.

6   Conclusion

Thanks to a general form of synchronous shuffling we can extend the notion of Brzozowski derivatives to various forms of shuffling which appear in the literature [2,4,11]. This enables the application of algorithms based on derivatives for shuffle expressions such as automata-based word recognition algorithms [10] and equality/containment checking [1,6].

There are several avenues for future work. For example, it is well-known that associativity does not hold for synchronous composition. The work in [8] identifies sufficient conditions to guarantee associativity and other algebraic laws. It would be interesting to identify such conditions in terms of our general synchronous shuffling operation.

In another direction, it would be interesting to study in detail the impact of the various shuffling variants on the size of the derivative-automata. Earlier work [5] only considers the specific case of (asynchronous) shuffling.

Acknowledgments. We thank the reviewers for their comments.

References

1. Antimirov, V.M.: Rewriting regular inequalities. In: Reichel, H. (ed.) FCT 1995. LNCS, vol. 965, pp. 116–116. Springer, Heidelberg (1995)
2. ter Beek, M.H., Martín-Vide, C., Mitrana, V.: Synchronized shuffles. Theor. Comput. Sci. **341**(1–3), 263–275 (2005)
3. Brzozowski, J.A.: Derivatives of regular expressions. J. ACM **11**(4), 481–494 (1964)
4. Garg, V.K., Ragunath, M.T.: Concurrent regular expressions and their relationship to petri nets. Theor. Comput. Sci. **96**(2), 285–304 (1992)
5. Gelade, W.: Succinctness of regular expressions with interleaving, intersection and counting. Theor. Comput. Sci. **411**(31–33), 2987–2998 (2010)
6. Grabmayer, C.: Using proofs by coinduction to find "traditional" proofs. In: Fiadeiro, J.L., Harman, N.A., Roggenbach, M., Rutten, J. (eds.) CALCO 2005. LNCS, vol. 3629, pp. 175–193. Springer, Heidelberg (2005)
7. Kumar, A., Verma, A.K.: A novel algorithm for the conversion of parallel regular expressions to non-deterministic finite automata. Applied Mathematics & Information Sciences **8**, 95–105 (2014)
8. Latteux, M., Roos, Y.: Synchronized shuffle and regular languages. In: Jewels are Forever, Contributions on Theoretical Computer Science in Honor of arto Salomaa, pp. 35–44. Springer, London (1999)
9. Lodaya, K., Mukund, M., Phawade, R.: Kleene theorems for product systems. In: Holzer, M. (ed.) DCFS 2011. LNCS, vol. 6808, pp. 235–247. Springer, Heidelberg (2011)
10. Owens, S., Reppy, J., Turon, A.: Regular-expression derivatives reexamined. Journal of Functional Programming **19**(2), 173–190 (2009)
11. de Simone, R.: Langages infinitaires et produit de mixage. Theor. Comput. Sci. **31**, 83–100 (1984)
12. Stotts, P.D., Pugh, W.: Parallel finite automata for modeling concurrent software systems. J. Syst. Softw. **27**(1), 27–43 (1994)

From ω-Regular Expressions to Büchi Automata via Partial Derivatives

Peter Thiemann[1]([⊠]) and Martin Sulzmann[2]

[1] Faculty of Engineering, University of Freiburg, Georges-Köhler-Allee 079,
79110 Freiburg, Germany
thiemann@acm.org
[2] Faculty of Computer Science and Business Information Systems,
Karlsruhe University of Applied Sciences, Moltkestrasse 30,
76133 Karlsruhe, Germany
martin.sulzmann@hs-karlsruhe.de

Abstract. We extend Brzozowski derivatives and partial derivatives from regular expressions to ω-regular expressions and establish their basic properties. We observe that the existing derivative-based automaton constructions do not scale to ω-regular expressions. We define a new variant of the partial derivative that operates on linear factors and prove that this variant gives rise to a translation from ω-regular expressions to nondeterministic Büchi automata.

Keywords: Automata and logic · Omega-regular languages · Derivatives

1 Introduction

Brzozowski derivatives [3] and partial derivatives [2] are well-known tools to transform regular expressions to automata and to define algorithms for equivalence and containment on them [1]. Derivatives had quite some impact on the study of algorithms for regular languages on finite words and trees [4,9], but they received less attention in the study of ω-regular languages.

While the extension of Brzozowski derivatives to ω-regular expressions is straightforward, the corresponding automaton construction does not easily extend to ω-automata. This observation leads Park [6] to suggest resorting to a different acceptance criterion based on transitions. Redziejowski [7] remarks that "the automaton constructed from the derivative has, in general, too few transitions as well as too few states." As a remedy, Redziejowski presents a construction of a deterministic automaton where states are certain combinations of derivatives with a non-standard transition-based acceptance criterion. In subsequent work, Redziejowski [8] improves on this construction by lowering the number of states and by simplifying some technical details. To the best of our knowledge, these papers [7,8] are the only attempts to construct ω-automata using derivatives.

© Springer International Publishing Switzerland 2015
A.-H. Dediu et al. (Eds.): LATA 2015, LNCS 8977, pp. 287–298, 2015.
DOI: 10.1007/978-3-319-15579-1_22

In comparison, our construction and proof are much simpler, we gain new insights into the structure of linear factors as a stepping stone to partial derivatives, and we obtain a standard nondeterministic Büchi automaton. Because Brzozowski derivatives invariably lead to deterministic automata, we analyze Antimirov's partial derivatives and identify linear factors as a suitable structure on which we base the construction of a nondeterministic automaton.

Overview

Section 2 reviews the basic definitions for (ω-) regular expressions and (Büchi) automata. Section 3 reviews Brzozowski derivatives, extends them to ω-regular expressions, and demonstrates the failure of the automaton construction based on Brzozowski derivatives. Section 4 introduces Antimirov's linear factors and partial derivatives, extends them to ω-regular expressions, establishes their basic properties, and demonstrates the failure of the automaton construction based on partial derivatives. Section 5 introduces a new notion of partial derivative that operates directly on linear factors of an ω-regular expression, defines a Büchi automaton on that basis, and proves its construction correct.

2 Preliminaries

An alphabet Σ is a finite set of symbols. The set Σ^* denotes the set of finite words over Σ, $\varepsilon \in \Sigma^*$ stands for the empty word; the set Σ^ω denotes the set of infinite words over Σ. For $u \in \Sigma^*$, we write $u \cdot v$ for the concatenation of words; if $v \in \Sigma^*$, then $u \cdot v \in \Sigma^*$; if $v \in \Sigma^\omega$, then $u \cdot v \in \Sigma^\omega$. Concatenation extends to sets of words as usual: $U \cdot V = \{u \cdot v \mid u \in U, v \in V\}$ where $U \subseteq \Sigma^*$ and $V \subseteq \Sigma^*$ or $V \subseteq \Sigma^\omega$.

Given a language $U \subseteq \Sigma^*$ and $W \subseteq \Sigma^*$ or $W \subseteq \Sigma^\omega$, the *left quotient* $U^{-1}W = \{v \mid \exists u \in U : uv \in W\}$. It is a subset of Σ^* or Σ^ω depending on W. For a singleton language $U = \{u\}$, we write $u^{-1}W$ for the left quotient.

Definition 1. *The set R_Σ of regular expressions over Σ is defined inductively by $\mathbf{1} \in R_\Sigma$, $\mathbf{0} \in R_\Sigma$, $\Sigma \subseteq R_\Sigma$, and, for all $r, s \in R_\Sigma$, $(r.s), (r+s), r^* \in R_\Sigma$. The explicit bracketing guarantees unambiguous parsing of regular expressions.*

Definition 2. *The language denoted by a regular expression is defined inductively by $\mathcal{L} : R_\Sigma \to \wp(\Sigma^*)$ as usual. $\mathcal{L}(\mathbf{1}) = \{\varepsilon\}$. $\mathcal{L}(\mathbf{0}) = \{\}$. $\mathcal{L}(a) = \{a\}$ (singleton word) for each $a \in \Sigma$. $\mathcal{L}(r.s) = \mathcal{L}(r) \cdot \mathcal{L}(s)$. $\mathcal{L}(r+s) = \mathcal{L}(r) \cup \mathcal{L}(s)$. $\mathcal{L}(r^*) = \{u_1 \ldots u_n \mid n \in \mathbb{N}, u_i \in \mathcal{L}(r)\}$.*

Definition 3. *The operations $\odot, \oplus : R_\Sigma \times R_\Sigma \to R_\Sigma$ are smart concatenation and smart union constructors for regular expressions.*

$$r \odot s = \begin{cases} 0 & r = 0 \vee s = 0 \\ r & s = 1 \\ s & r = 1 \\ (r.s) & \text{otherwise} \end{cases} \qquad r \oplus s = \begin{cases} r & s = 0 \\ s & r = 0 \\ r & r = s \\ (r+s) & \text{otherwise} \end{cases}$$

Lemma 4. *For all r, s: $\mathcal{L}(r \odot s) = \mathcal{L}(r.s)$; $\mathcal{L}(r \oplus s) = \mathcal{L}(r + s)$.*

Definition 5. *A regular expression r is* nullable *if $\varepsilon \in \mathcal{L}(r)$. The function $N :$ $R_\Sigma \to \{0, 1\}$ detects nullable expressions: $N(\mathbf{1}) = \mathbf{1}$. $N(\mathbf{0}) = \mathbf{0}$. $N(a) = \mathbf{0}$. $N(r.s) = N(r) \odot N(s)$. $N(r + s) = N(r) \oplus N(s)$. $N(r^*) = \mathbf{1}$.*

Lemma 6. *For all $r \in R_\Sigma$. $N(r) = \mathbf{1}$ iff $\varepsilon \in \mathcal{L}(r)$.*

Definition 7. *The set R_Σ^ω of ω-regular expressions over Σ is defined by $\mathbf{0} \in R_\Sigma^\omega$; for all $\alpha, \beta \in R_\Sigma^\omega$, $(\alpha + \beta) \in R_\Sigma^\omega$; for all $r \in R_\Sigma$ and $\alpha \in R_\Sigma^\omega$, $(r.\alpha) \in R_\Sigma^\omega$; for all $s \in R_\Sigma$, if $\varepsilon \notin \mathcal{L}(s)$, then $s^\omega \in R_\Sigma^\omega$.*

Remark 8. Definition 7 is equivalent to an alternative definition often seen in the literature, where an ω-regular-expression has a *sum-of-product form* $\sum_{i=1}^{n}(r_i.s_i^\omega)$ with $\varepsilon \notin \mathcal{L}(s_i)$. An easy induction shows that every α can be rewritten in this form: cases $\mathbf{0}$, $(\alpha + \beta)$, s^ω: immediate; case $(r.\alpha)$: by induction, α can be written as $\sum_{i=1}^{n}(r_i.s_i^\omega)$, distributivity and associativity yield $\sum_{i=1}^{n}(r.r_i).s_i^\omega$ for $(r.\alpha)$. When convenient for a proof, we assume that an expression is in sum-of-product form.

Definition 9. *The language denoted by an ω-regular expression is defined inductively by $\mathcal{L}^\omega : R_\Sigma^\omega \to \wp(\Sigma^\omega)$: $\mathcal{L}^\omega(\mathbf{0}) = \emptyset$. $\mathcal{L}^\omega(\alpha + \beta) = \mathcal{L}^\omega(\alpha) \cup \mathcal{L}^\omega(\beta)$. $\mathcal{L}^\omega(r.\alpha) = \mathcal{L}(r) \cdot \mathcal{L}^\omega(\alpha)$. $\mathcal{L}^\omega(s^\omega) = \{v_1 v_2 \cdots \mid \forall i \in \mathbb{N} : v_i \in \mathcal{L}(s)\}$.*

Definition 10. *A (nondeterministic) finite automaton (NFA) is a tuple $\mathcal{A} = (Q, \Sigma, \delta, q_0, F)$ where Q is a finite set of states, Σ an alphabet, $\delta : Q \times \Sigma \to \wp(Q)$ the transition function, $q_0 \in Q$ the initial state, and $F \subseteq Q$ the set of final states.*

Let $w = a_0 \ldots a_{n-1} \in \Sigma^*$ be a word. A run *of $\mathcal{A}$ on w is a sequence $q_0 \ldots q_n$ such that, for all $0 \leq i < n$, $q_{i+1} \in \delta(q_i, a_i)$. The run is* accepting *if $q_n \in F$. The language $\mathcal{L}(\mathcal{A}) = \{w \in \Sigma^* \mid \exists$ accepting run of $\mathcal{A}$ on $w\}$ is* recognized *by $\mathcal{A}$. The automaton $\mathcal{A}$ is* deterministic *if $|\delta(q, a)| = 1$, for all $q \in Q$, $a \in \Sigma$.*

Definition 11. *A (nondeterministic) Büchi-automaton (NBA) is a tuple $\mathcal{B} = (Q, \Sigma, \delta, Q_0, F)$ where Q is a finite set of states, Σ an alphabet, $\delta : Q \times \Sigma \to \wp(Q)$ the transition function, $Q_0 \subseteq Q$ the set of initial states, and $F \subseteq Q$ the set of accepting states.*

Let $w = (a_i)_{i \in \mathbb{N}} \in \Sigma^\omega$ be an infinite word. A run *of $\mathcal{B}$ on w is an infinite sequence of states $(q_i)_{i \in \mathbb{N}}$ such that $q_0 \in Q_0$ and for all $i \in \mathbb{N}$: $q_{i+1} \in \delta(q_i, a_i)$.*

A run $(q_i)_{i \in \mathbb{N}}$ of $\mathcal{B}$ is accepting *if there exists a strictly increasing sequence $(n_j)_{j \in \mathbb{N}}$ such that $q_{n_j} \in F$, for all $j \in \mathbb{N}$. The language $\mathcal{L}^\omega(\mathcal{B}) = \{w \in \Sigma^\omega \mid \exists$ accepting run of $\mathcal{B}$ on $w\}$ is* recognized *by $\mathcal{B}$. The automaton $\mathcal{B}$ is* deterministic *if $|Q_0| = 1$ and $|\delta(q, a)| = 1$, for all $q \in Q$, $a \in \Sigma$.*

3   Regular Expressions to Finite Automata

The textbook construction to transform a regular expression into a finite automaton is taken from Kleene's work [5]. However, there is an alternative approach based on Brzozowski's idea of derivatives for regular expressions.

Given a regular expression r and a symbol $a \in \Sigma$, the derivative $r' = d_a(r)$ is a regular expression such that $\mathcal{L}(r') = \{w \mid aw \in \mathcal{L}(r)\}$, the left quotient of $\mathcal{L}(r)$ by the symbol a. The derivative can be defined symbolically by induction on regular expressions.

Definition 12 (Brzozowski derivative [3]).

$$
\begin{aligned}
d_a(\mathbf{0}) &= \mathbf{0} & d_a(r.s) &= (d_a(r) \odot s) \oplus (N(r) \odot d_a(s)) \\
d_a(\mathbf{1}) &= \mathbf{0} & d_a(r + s) &= d_a(r) \oplus d_a(s) \\
d_a(b) &= \begin{cases} \mathbf{1} & a = b \\ \mathbf{0} & a \neq b \end{cases} & d_a(r^*) &= d_a(r) \odot r^*
\end{aligned}
$$

Brzozowski proved the following representation theorem that factorizes a regular language into its ε-part and the quotient languages with respect to each symbol of the alphabet.

Theorem 13 (Representation [3]). $\mathcal{L}(r) = \mathcal{L}(N(r)) \cup \bigcup_{a \in \Sigma} \{a\} \cdot \mathcal{L}(d_a(r))$

He further proved that there are only finitely many different regular expressions derivable from a given regular expression. This finiteness result considers expressions modulo a *similarity relation* $\approx$ that contains (at least) associativity, commutativity, and idempotence of the $+$ operator as well as considering $\mathbf{0}$ as the neutral element. We further assume *associativity of concatenation*.

Definition 14 (Similarity). *Similarity* $\approx \subseteq R_\Sigma \times R_\Sigma$ *is the smallest compatible relation that encompasses the following elements for all* $r, s, t \in R_\Sigma$.

$$(r+s)+t \approx r+(s+t) \quad r+s \approx s+r \quad r+r \approx r \quad r+\mathbf{0} \approx r \quad (r.s).t \approx r.(s.t)$$

Similarity extends to $\approx^\omega \subseteq R_\Sigma^\omega \times R_\Sigma^\omega$ *as the smallest compatible relation that contains the following elements for all* $\alpha, \beta, \gamma \in R_\Sigma^\omega$.

$$(\alpha + \beta) + \gamma \approx^\omega \alpha + (\beta + \gamma) \quad \alpha + \beta \approx^\omega \beta + \alpha \quad \alpha + \alpha \approx^\omega \alpha \quad \alpha + \mathbf{0} \approx^\omega \alpha$$

$$(r.s).\alpha \approx^\omega r.(s.\alpha) \quad r \approx s \Rightarrow (r.t^\omega) \approx^\omega (s.t^\omega) \quad s \approx t \Rightarrow (r.s^\omega) \approx^\omega (r.t^\omega)$$

Definition 15. *The derivative operator extends to words* $w \in \Sigma^*$ *by* $d_\varepsilon(r) = r$, $d_{aw}(r) = d_w(d_a(r))$ *and to sets of words* $W \subseteq \Sigma^*$ *by* $d_W(r) = \{d_w(r) \mid w \in W\}$.

Theorem 16 (Finiteness [3]). *For each* $r \in R_\Sigma$, *the set* $d_{\Sigma^*}(r)/_\approx$ *is finite.*

Taken together, these two theorems yield an effective transformation from a regular expression to a deterministic finite automaton.

Theorem 17 (DFA from regular expression [3]). *Define the DFA* $\mathcal{D}(r) = (Q, \Sigma, \delta, q_0, F)$ *where* $Q = d_{\Sigma^*}(r)/_\approx$, *for all* $s \in Q, a \in \Sigma$: $\delta(s, a) = \{d_a(s)\}$, $q_0 = r$, $F = \{s \in Q \mid N(s) = \mathbf{1}\}$. *Then* $\mathcal{D}(r)$ *is a deterministic finite automaton and* $\mathcal{L}(\mathcal{D}(r)) = \mathcal{L}(r)$.

Let's try to apply an analogous construction to ω-regular expressions. We first straightforwardly extend the definition of derivatives [7].

Definition 18 (Brzozowski derivative for ω-regular expressions).

$$d_a(0) \quad = 0 \qquad\qquad d_a(r.\alpha) = (d_a(r) \odot \alpha) \oplus (N(r) \odot d_a(\alpha))$$
$$d_a(\alpha + \beta) = d_a(\alpha) \oplus d_a(\beta) \qquad d_a(s^\omega) = d_a(s) \odot s^\omega$$

Lemma 19. $\mathcal{L}^\omega(d_a(\alpha)) = a^{-1}\mathcal{L}^\omega(\alpha)$

Lemma 20. $\mathcal{L}^\omega(\alpha) = \bigcup_{a \in \Sigma}\{a\} \cdot \mathcal{L}^\omega(d_a(\alpha))$.

The operation $d_w(\Sigma)$ also yields finitely many derivatives modulo similarity (extended to $R_\Sigma^\omega \times R_\Sigma^\omega$ in the obvious way), but applying Brzozowski's automata construction analogously results in a *deterministic* Büchi automaton, which is known to be weaker than its nondeterministic counterpart.

Example 21. Consider the ω-regular expression $(a + b)^*.b^\omega$ that describes the language of infinite words that contain only finitely many as. It is known that this language cannot be recognized with a deterministic Büchi automaton. Applying Brzozowski's automaton construction analogously yields the following:

$$
\begin{aligned}
Q &= \{q_0, q_1\} & \delta(q_0, a) &= q_0 \\
q_0 &= (a + b)^*.b^\omega & \delta(q_0, b) &= q_1 \\
q_1 &= (a + b)^*.b^\omega + b^\omega & \delta(q_1, a) &= q_0 \\
Q_0 &= \{q_0\} & \delta(q_1, b) &= q_1
\end{aligned}
$$

As all states "contain" the looping expression b^ω, it is not clear which states should be accepting. Furthermore, the automaton is deterministic, so it cannot recognize $\mathcal{L}^\omega((a + b)^*.b^\omega)$, regardless.

4 Partial Derivatives

As Brzozowski's construction only results in a deterministic automaton, we next consider a construction that yields a nondeterministic automaton. It is based on Antimirov's *partial derivatives* [2]. The partial derivative $\partial_a(r)$ of a regular expression r with respect to a is a *set* of regular expressions $\{s_1, \ldots, s_n\}$ such that $\bigcup_{i=1}^n \mathcal{L}(s_i) = \{w \mid aw \in \mathcal{L}(r)\}$. As a stepping stone to their definition, Antimirov introduces *linear factors* of regular expressions. A linear factor is a pair of a first symbol that can be consumed by the expression and a "remaining" regular expression. The following definition corresponds to Antimirov's definition [2, Definition 2.4], but we replace the smart constructor $\odot$ for concatenation (that elides ε) by plain concatenation to simplify the finiteness proof.

Definition 22 (Linear factors [2]).

$$
\begin{aligned}
\mathrm{LF}(0) &= \{\} & \mathrm{LF}(r.s) &= \mathrm{LF}(r).s \cup N(r) \odot \mathrm{LF}(s) \\
\mathrm{LF}(1) &= \{\} & \mathrm{LF}(r + s) &= \mathrm{LF}(r) \cup \mathrm{LF}(s) \\
\mathrm{LF}(a) &= \{\langle a, 1 \rangle\} & \mathrm{LF}(r^*) &= \mathrm{LF}(r).r^*
\end{aligned}
$$

where

$$
\begin{aligned}
0 \odot F &= \{\} & 1 \odot F &= F \\
\langle a, r \rangle.s &= \langle a, r.s \rangle & F.s &= \{f.s \mid f \in F\}
\end{aligned}
$$

Defining the language of a linear factor and a set of linear factors F by

$$\mathcal{L}(\langle a, r \rangle) = a \cdot \mathcal{L}(r) \qquad \mathcal{L}(F) = \bigcup \{ \mathcal{L}(f) \mid f \in F \}$$

we can prove the following results about linear factors by induction on r.

Lemma 23. *If $\langle a, r' \rangle \in \mathrm{LF}(r)$, then $a \cdot \mathcal{L}(r') \subseteq \mathcal{L}(r)$.*

Lemma 24. *If $av \in \mathcal{L}(r)$, then there exists $\langle a, r' \rangle \in \mathrm{LF}(r)$ such that $v \in \mathcal{L}(r')$.*

Lemma 25. *For all r, $\mathcal{L}(\mathrm{LF}(r)) = \mathcal{L}(r) \setminus \{ \varepsilon \}$.*

We label the symbol for partial derivative with ^{A} to signify Antimirov's definition. In Section 5, we define a different version of the partial derivative.

Definition 26 (Partial derivative [2]).

$$\partial_a^A(r) = \{ r' \mid \langle a, r' \rangle \in \mathrm{LF}(r), r' \neq \mathbf{0} \}$$

Partial derivatives extend to words and sets of words $W \subseteq \Sigma^*$ in the usual way:

$$\partial_\varepsilon^A(r) = \{ r \} \quad \partial_{aw}^A(r) = \bigcup \{ \partial_w^A(r') \mid r' \in \partial_a^A(r) \} \quad \partial_W^A(r) = \bigcup \{ \partial_w^A(r) \mid w \in W \}$$

Antimirov proves [2, Theorem 3.4] that the set of all partial derivatives of a given regular expression is finite. While his definition of linear factors uses the smart concatenation $\odot$, the finiteness proof does not rely on it: it approximates smart concatenation by the standard concatenation operator.

Theorem 27. *For any $r \in R_\Sigma$, $|\partial_{\Sigma^+}^A(r)| \leq \|r\|$ where $\|r\|$ is the alphabetic width of r (i.e., the number of occurrences of symbols from Σ in r).*

Furthermore, a language can be represented from its partial derivatives.

Lemma 28. $\mathcal{L}(r) = \mathcal{L}(N(r)) \cup \bigcup_{a \in \Sigma} a \cdot \mathcal{L}(\sum \partial_a^A(r))$.

Here, we write $\sum \{ r_i \mid 1 \leq i \leq n \}$ for $r_1 + \cdots + r_n$, if $n > 0$, or for $\mathbf{0}$ if $n = 0$. We also have the following characterization.

Lemma 29. *If $\partial_a^A(r) = \{ s_1, \ldots, s_n \}$, then $\bigcup_{i=1}^n \mathcal{L}(s_i) = \{ w \mid aw \in \mathcal{L}(r) \}$.*

Antimirov defines a *nondeterministic* automaton for $\mathcal{L}(r)$ as follows.

Theorem 30 (NFA from regular expression [2]). *Define the NFA $\mathcal{N}(r) = (Q, \Sigma, \delta, q_0, F)$ where $Q = \partial_{\Sigma^*}^A(r)$, for all $s \in Q$, $a \in \Sigma$: $\delta(s, a) = \partial_a^A(s)$, $q_0 = r$, $F = \{ s \in Q \mid N(s) = 1 \}$. Then $\mathcal{N}(r)$ is an NFA and $\mathcal{L}(r) = \mathcal{L}(\mathcal{N}(r))$.*

Lemma 31. $w \in \mathcal{L}(r)$ *iff* $\varepsilon \in \bigcup N(\partial_w^A(r))$.

Proof. By induction on w.

Base case: $\varepsilon \in \mathcal{L}(r)$ iff $\varepsilon \in N(r)$ by Lemma 6. The claim follows because $N(r) = \bigcup N(\{ r \}) = \bigcup N(\{ \partial_\varepsilon^A(r) \})$.

Inductive case: Suppose that $aw \in \mathcal{L}(r)$ and $\partial_a^A(r) = \{ r_1, \ldots, r_k \}$. By Lemma 29, $\bigcup_i \mathcal{L}(r_i) = \{ v \mid av \in \mathcal{L}(r) \}$ so that $w \in \bigcup_i \mathcal{L}(r_i)$, i.e., $\exists i: w \in \mathcal{L}(r_i)$. By induction, $\varepsilon \in \bigcup N(\partial_w^A(r_i)) \subseteq N(\partial_{aw}^A(r))$.

For the reverse direction, suppose that $\varepsilon \in \bigcup N(\partial_{aw}^A(r)) = N(\bigcup \{ \partial_w^A(r') \mid r' \in \partial_a^A(r), r' \neq \mathbf{0} \})$. Hence, there exists $r' \in \partial_a^A(r)$ such that $\varepsilon \in N(\partial_w^A(r'))$. By induction, $w \in \mathcal{L}(r')$ and thus, by Lemma 29, $aw \in \mathcal{L}(r)$. $\square$

To scale the definition from Theorem 30 to ω-regular expressions we need to extend Definition 22.

Definition 32 (ω-Linear factors). *Define* LF : $R_\Sigma^\omega \to \Sigma \times R_\Sigma^\omega \times \{0,1\}$ *by*

$$\text{LF}(\mathbf{0}) \quad = \emptyset \qquad\qquad \text{LF}(r.\alpha) = \text{LF}(r).\alpha \times \{0\} \cup N(r) \odot \text{LF}(\alpha)$$
$$\text{LF}(\alpha + \beta) = \text{LF}(\alpha) \cup \text{LF}(\beta) \qquad \text{LF}(s^\omega) \quad = \text{LF}(s).s^\omega \times \{1\}$$

Compared to the linear factor of a regular expression, an ω-linear factor is a *triple* of a next symbol, an ω-regular expression, and a bit that indicates whether the factor resulted from unrolling an ω-iteration.

For an ω-linear factor define $\mathcal{L}^\omega(\langle a, \beta, g \rangle) = a \cdot \mathcal{L}^\omega(\beta)$ and for a set F of ω-linear factors accordingly $\mathcal{L}^\omega(F) = \bigcup\{\mathcal{L}^\omega(f) \mid f \in F\}$.

Each ω-regular language can be represented by its set of ω-linear factors. Compared to the finite case (Lemma 25), the empty string need not be considered because it is not an element of Σ^ω.

Lemma 33. *For all* α, $\mathcal{L}^\omega(\alpha) = \mathcal{L}^\omega(\text{LF}(\alpha))$.

Proof. By induction on α. We only show one illustrative case.

Case s^ω: let $w \in \mathcal{L}^\omega(s^\omega)$. By definition, $w = v_0 v_1 \ldots$ with $\varepsilon \neq v_i \in \mathcal{L}(s)$, for all $i \in \mathbb{N}$. Suppose that $w = aw'$. Then $v_0 = av_0'$. Show that there exists $f = \langle a, s', 1 \rangle \in \text{LF}(s^\omega)$ such that $w' \in \mathcal{L}^\omega(s')$.

If $\text{LF}(s^\omega) = \emptyset$, then $\mathcal{L}^\omega(s^\omega) = \emptyset$, which contradicts the existence of w.

Suppose next that all ω-linear factors have the form $\langle b, s', 1 \rangle$ for some $b \neq a$. But then we obtain a contradiction to $av_0' \in \mathcal{L}(s)$.

Thus, we need to examine the ω-linear factors of the form $\langle a, s'.s^\omega, 1 \rangle \in \text{LF}(s).s^\omega \times \{1\} = \text{LF}(s^\omega)$. By Lemma 24, there must be a linear factor $\langle a, s' \rangle \in \text{LF}(s)$ such that $v_0' \in \mathcal{L}(s')$. Hence, $w' = v_0' v_1 \cdots \in \mathcal{L}^\omega(s'.s^\omega)$ and thus $w = aw' \in \mathcal{L}^\omega(\langle a, s'.s^\omega, 1 \rangle) \subseteq \mathcal{L}^\omega(\text{LF}(s^\omega))$.

For the reverse direction, suppose that $w \in \mathcal{L}^\omega(\text{LF}(s^\omega))$. Then there exists $\langle a, s' \rangle \in \text{LF}(s)$ and hence $\langle a, s'.s^\omega, 1 \rangle \in \text{LF}(s).s^\omega \times \{1\} = \text{LF}(s^\omega)$ such that $w \in a \cdot \mathcal{L}^\omega(s'.s^\omega) = a \cdot \mathcal{L}(s') \cdot \mathcal{L}^\omega(s^\omega)$. By Lemma 23, $a \cdot \mathcal{L}(s') \subseteq L(s)$ so that $w \in a \cdot \mathcal{L}(s') \cdot \mathcal{L}^\omega(s^\omega) \subseteq \mathcal{L}(s) \cdot \mathcal{L}^\omega(s^\omega) = \mathcal{L}^\omega(s^\omega)$. $\square$

Using the obvious extension of the partial derivative operator, Lemma 29 extends to the ω-regular case.

Lemma 34. *If* $\partial_a^A(\alpha) = \{\beta_1, \ldots, \beta_n\}$, *then* $\bigcup_{i=1}^n \mathcal{L}^\omega(\beta_i) = \{w \mid aw \in \mathcal{L}^\omega(\alpha)\}$.

However, again it is not clear how to extend Antimirov's automaton construction to Büchi automata. The critical part is to come up with a characterization of the accepting states.

Example 35. Let $\alpha = (a + b)^*.b^\omega$ as in the previous example. Constructing an automaton analogously to Theorem 30 yields

$$
\begin{aligned}
q_0 &= (a+b)^*.b^\omega & \delta(q_0, a) &= \{q_0\} \\
q_1 &= b^\omega & \delta(q_0, b) &= \{q_0, q_1\} \\
Q &= \{q_0, q_1\} & \delta(q_1, a) &= \{\} \\
Q_0 &= \{q_0\} & \delta(q_1, b) &= \{q_1\}
\end{aligned}
$$

Thus, adopting the set of accepting states $F = \{q_1\}$ yields a nondeterministic Büchi automaton that accepts exactly $\mathcal{L}(\alpha)$. Apparently, we may categorize states of the form s^ω as accepting.

While the previous example is encouraging in that the construction leads to a correct automaton, a simple transformation of the ω-regular expression shows that the criterion for accepting states is not sufficient in the general case.

Example 36. Let $\beta = (a + b)^*.(b.b^*)^\omega$. This expression recognizes the same language as the expression of the previous example.

$$
\begin{aligned}
\partial_a(\beta) \quad &= \partial_a((a+b)^*.(b.b^*)^\omega) \\
&= \partial_a((a+b)^*).(b.b^*)^\omega \cup \partial_a(b.b^*) \odot (b.b^*)^\omega \\
&= \{(a+b)^*.(b.b^*)^\omega\} \\
\partial_b(\beta) \quad &= \partial_b((a+b)^*.(b.b^*)^\omega) \\
&= \partial_b((a+b)^*).(b.b^*)^\omega \cup \partial_b(b.b^*) \odot (b.b^*)^\omega \\
&= \{(a+b)^*.(b.b^*)^\omega\} \cup \{b^*.(b.b^*)^\omega\} \\
\partial_b(b^*.(b.b^*)^\omega) &= \partial_b(b^*).(b.b^*)^\omega \cup \partial_b(b.b^*) \odot (b.b^*)^\omega \\
&= \{b^*.(b.b^*)^\omega\} \cup \{b^*.(b.b^*)^\omega\} \\
\partial_a(b^*.(b.b^*)^\omega) &= \{\}
\end{aligned}
$$

Thus, we cannot construct a Büchi automaton for $\mathcal{L}^\omega(\beta)$ by simply classifying the states of the form s^ω as accepting because there are no such states in this automaton: thus, the automaton would accept the empty language.

Alternatively, we might be tempted to consider all expressions of the form $r.s^\omega$ where r is nullable as accepting states. This choice would classify *all states* in the example as accepting, which would cause the automaton to wrongly accept the infinite word a^ω.

5   NBA from ω-Linear Factors

The difficulties with the previous examples demonstrate that Antimirov's partial derivatives cannot be used directly as the states of a Büchi automaton. To fix these problems, we base our construction directly on the ω-linear factors that arise as an intermediate step in Antimirov's work.

Definition 37. *For an ω-linear factor (and a set F of ω-linear factors) define the partial derivative as a set of ω-linear factors:*

$$
\partial_b(\langle a, \beta, g \rangle) = \begin{cases} \{\} & a \neq b \\ \mathrm{LF}(\beta) & a = b \end{cases} \qquad \partial_b(F) = \bigcup_{f \in F} \partial_b(f)
$$

Define further the extension to words $\partial_\varepsilon(F) = F$ and $\partial_{aw}(F) = \partial_w(\partial_a(F))$ and the extension to sets of finite words $W \subseteq \Sigma^*$: $\partial_W(F) = \bigcup\{\partial_w(F) \mid w \in W\}$.

This definition of the derivative serves as the basis for defining the set of states $\mathcal{Q}(\alpha)$ for the NBA, which we are aiming to construct.

Definition 38. *Define* $\mathcal{Q}(\alpha)$ *inductively as the smallest set such that* $\mathrm{LF}(\alpha) \subseteq \mathcal{Q}(\alpha)$ *and, for each* $a \in \Sigma$, $\partial_a(\mathcal{Q}(\alpha)) \subseteq \mathcal{Q}(\alpha)$.

Lemma 39. *If* $\langle a, \beta, g \rangle \in \mathcal{Q}(\alpha)$, *then* $\exists w \in \Sigma^*$ *such that* $\langle a, \beta, g \rangle \in \partial_w(\mathrm{LF}(\alpha))$.

Proof. By induction on the construction of $\mathcal{Q}(\alpha)$.

Base case: $\langle a, \beta, g \rangle \in \mathrm{LF}(\alpha) = \partial_\varepsilon(\mathrm{LF}(\alpha))$.

Inductive case: $\langle a, \beta, g \rangle \in \partial_a(f)$, for some $f \in \mathcal{Q}(\alpha)$ and $a \in \Sigma$. By induction, $f \in \partial_w(\mathrm{LF}(\alpha))$, for some w, and thus $\langle a, \beta, g \rangle \in \partial_a(\partial_w(\mathrm{LF}(\alpha))) = \partial_{aw}(\mathrm{LF}(\alpha))$. □

Proposition 40. *For each* ω-*regular expression* α, $\mathcal{Q}(\alpha)$ *is finite.*

Proof. We prove that $\mathcal{Q}(\alpha) \subseteq \Sigma \times \partial_{\Sigma^+}^A(\alpha) \times \{0, 1\}$.

Suppose that $\langle a, \alpha', g \rangle \in \mathcal{Q}(\alpha)$. There are two cases. If $\langle a, \alpha', g' \rangle \in \mathrm{LF}(\alpha)$, then $a \in \Sigma$ and $\alpha' \in \partial_a^A(\alpha) \subseteq \partial_{\Sigma^+}^A(\alpha)$.

If $\langle a, \alpha', g' \rangle \in \partial_b(\langle b, \beta, g \rangle)$ for some $\langle b, \beta, g \rangle \in \mathcal{Q}(\alpha)$, then there exists some $w \in \Sigma^*$ such that $\beta \in \partial_{wb}^A(\alpha)$ and $\langle a, \alpha', g \rangle \in \mathrm{LF}(\beta)$. By definition, $\alpha' \in \partial_{wba}^A(\alpha) \subseteq \partial_{\Sigma^+}^A(\alpha)$.

By Theorem 27, $|\partial_{\Sigma^+}^A(\alpha)|$ is finite and so is $|\mathcal{Q}(\alpha)| \leq |\Sigma| \cdot |\partial_{\Sigma^+}^A(\alpha)| \cdot 2$. □

Given this finiteness, we construct a non-deterministic Büchi automaton from an ω-regular expression as follows.

Definition 41 (NBA from ω-regular expression). *Define the NBA* $\mathcal{B}(\alpha) = (Q, \Sigma, \delta, Q_0, F)$ *by* $Q = \mathcal{Q}(\alpha)$; $Q_0 = \mathrm{LF}(\alpha)$; $F = \{\langle a, \beta, g \rangle \in Q \mid g = 1\}$; *and* $\delta(f, a) = \partial_a(f)$.

Example 42. Consider (again) $\alpha = (a + b)^*.b^\omega$.

$$\begin{aligned}
\mathrm{LF}(\alpha) &= \mathrm{LF}((a + b)^*).b^\omega \cup \mathrm{LF}(b^\omega) \\
&= \{\langle a, (a + b)^*.b^\omega, 0 \rangle, \langle b, (a + b)^*.b^\omega, 0 \rangle, \langle b, b^\omega, 1 \rangle\} \\
&= Q = Q_0
\end{aligned}$$

$$\begin{aligned}
\delta(\langle b, b^\omega, 1 \rangle, a) &= \{\} \\
\delta(\langle b, b^\omega, 1 \rangle, b) &= \{\langle b, b^\omega, 1 \rangle\} \\
\delta(\langle a, (a + b)^*.b^\omega, 0 \rangle, a) &= \mathrm{LF}((a + b)^*.b^\omega) = Q \\
\delta(\langle a, (a + b)^*.b^\omega, 0 \rangle, b) &= \{\} \\
\delta(\langle b, (a + b)^*.b^\omega, 0 \rangle, a) &= \{\} \\
\delta(\langle b, (a + b)^*.b^\omega, 0 \rangle, b) &= \mathrm{LF}((a + b)^*.b^\omega) = Q
\end{aligned}$$

Accepting states: $F = \{\langle b, b^\omega, 0 \rangle\} = \mathrm{LF}(b^\omega)$.

The resulting automaton properly accepts $\mathcal{L}^\omega(\alpha)$.

Example 43. Next consider $\beta = (a + b)^*.(b.b^*)^\omega$.

$$\begin{aligned}
\mathrm{LF}(\beta) &= \mathrm{LF}((a + b)^*).(b.b^\omega) \times \{0\} \cup \mathrm{LF}((b.b^*)^\omega) \\
&= \mathrm{LF}(a + b).(a + b)^*.(b.b^\omega) \times \{0\} \cup \mathrm{LF}(b.b^*).(b.b^*)^\omega \times \{1\} \\
&= \{\langle a, (a + b)^*.(b.b^\omega), 0 \rangle, \langle b, (a + b)^*.(b.b^\omega), 0 \rangle\} \\
&\quad \cup \mathrm{LF}(b).b^*.(b.b^*)^\omega \times \{1\} \\
&= \{\langle a, (a + b)^*.(b.b^\omega), 0 \rangle, \langle b, (a + b)^*.(b.b^\omega), 0 \rangle \\
&\quad , \langle b, b^*.(b.b^*)^\omega, 1 \rangle\} \\
&= \{\langle a, \beta, 0 \rangle, \langle b, \beta, 0 \rangle, \langle b, b^*.(b.b^*)^\omega, 1 \rangle\}
\end{aligned}$$

$$\begin{aligned}
\delta(\langle a,\beta\rangle,a) &= \text{LF}(\beta)\\
\delta(\langle b,\beta\rangle,b) &= \text{LF}(\beta)\\
\delta(\langle b,b^*.(b.b^*)^\omega\rangle,b) &= \text{LF}(b^*.(b.b^*)^\omega)\\
&= \text{LF}(b^*).(b.b^*)^\omega \times \{1\} \cup \text{LF}((b.b^*)^\omega)\\
&= \text{LF}(b).b^*.(b.b^*)^\omega \times \{1\} \cup \text{LF}(b.b^*).(b.b^*)^\omega \times \{1\}\\
&= \text{LF}(b).b^*.(b.b^*)^\omega \times \{1\} \cup \text{LF}(b).b^*.(b.b^*)^\omega \times \{1\}\\
&= \{\langle b,b^*.(b.b^*)^\omega,1\rangle\}\\
&= \text{LF}((b.b^*)^\omega)
\end{aligned}$$

Accepting states:

$$F = \{\langle b,b^*.(b.b^*)^\omega,1\rangle\} = \text{LF}((b.b^*)^\omega)$$

The resulting automaton properly accepts $\mathcal{L}^\omega(\beta)$ with the same number of states as in the previous example.

It remains to prove the correctness of the construction in Definition 41.

Theorem 44. *For all $\alpha \in R_\Sigma^\omega$: $\mathcal{L}^\omega(\alpha) = \mathcal{L}^\omega(\mathcal{B}(\alpha))$.*

We start with some technical lemmas.

Lemma 45. *For all $v \neq \varepsilon$, $\partial_v(\text{LF}(s^\omega)) = \partial_v(\text{LF}(s.s^\omega))$.*

Proof. By definition of ω-regular expressions, $\varepsilon \notin \mathcal{L}(s)$ that is $N(s) = \mathbf{0}$.
    Observe that $\text{LF}(s^\omega) = \text{LF}(s).s^\omega \times \{1\}$,
    whereas $\text{LF}(s.s^\omega) = \text{LF}(s).s^\omega \times \{0\} \cup N(s) \odot \text{LF}(s^\omega) = \text{LF}(s).s^\omega \times \{0\}$.
    Because $v \neq \varepsilon$, it must be that $v = av'$, for some a.
    Hence, $\partial_a(\text{LF}(s^\omega)) = \bigcup\{\text{LF}(s'.s^\omega) \mid \langle a,s'\rangle \in \text{LF}(s)\} = \partial_a(\text{LF}(s.s^\omega))$.
    Hence, $\partial_{av'}(\text{LF}(s^\omega)) = \partial_{av'}(\text{LF}(s.s^\omega))$ □

The next lemma is our workhorse in proving that $\mathcal{L}^\omega(\alpha)$ is contained in the language of $\mathcal{B}(\alpha)$.

Lemma 46. *If $u \in \mathcal{L}(r)$, then $\text{LF}(\alpha) \subseteq \partial_u(\text{LF}(r.\alpha))$.*

Proof. Induction on r.
    Case $r = \mathbf{0}$: contradiction because $\mathcal{L}(\mathbf{0}.\alpha) = \{\}$.
    Case $r = \mathbf{1}$: Then $u = \varepsilon$ and $\partial_\varepsilon(\text{LF}(\mathbf{1}.\alpha)) = \text{LF}(\mathbf{1}.\alpha) = \text{LF}(\alpha)$.
    Case $r = a$: Then $u = a$ and $\partial_a(\text{LF}(a.\alpha)) = \partial_a(\langle a,\alpha,0\rangle) = \text{LF}(\alpha)$.
    Case $r = r_1.r_2$: Then $u = u_1 u_2$ with $u_1 \in \mathcal{L}(r_1)$ and $u_2 \in \mathcal{L}(r_2)$.
By similarity (cf. Definition 14), $\text{LF}((r_1.r_2).\alpha) = \text{LF}(r_1.(r_2.\alpha))$.
By induction on r_1, $\text{LF}(r_2.\alpha) \subseteq \partial_{u_1}(\text{LF}(r_1.(r_2.\alpha)))$.
By induction on r_2,

$$\text{LF}(\alpha) \subseteq \partial_{u_2}(\text{LF}(r_2.\alpha)) \subseteq \partial_{u_2}(\partial_{u_1}(\text{LF}(r_1.(r_2.\alpha)))) = \partial_u(\text{LF}(r.\alpha))$$

    Case $r = r_1 + r_2$: Assume that $u \in \mathcal{L}(r_1) \subseteq \mathcal{L}(r)$. By induction, $\text{LF}(\alpha) \subseteq \partial_u(\text{LF}(r_1.\alpha)) \subseteq \partial_u(\text{LF}(r.\alpha))$. The case for r_2 is analogous.

Case $r = r_1^*$: Consider

$$\text{LF}(r_1^*.\alpha) = \text{LF}(r_1^*).\alpha \cup N(r_1^*) \odot \text{LF}(\alpha) = \text{LF}(r_1).r_1^*.\alpha \cup \text{LF}(\alpha)$$

For $u \in \Sigma^*$, $\partial_u(\text{LF}(r_1^*.\alpha)) = \partial_u(\text{LF}(r_1).r_1^*.\alpha) \cup \partial_u(\text{LF}(\alpha))$.
If $u \in \mathcal{L}(r)$, then $u = u_1 \ldots u_n$, for some $n \in \mathbb{N}$, where all $u_i \neq \varepsilon$. Continue by induction on n.
If $n = 0$, $u = \varepsilon$, then clearly $\text{LF}(\alpha) \subseteq \partial_\varepsilon(\text{LF}(r_1^*.\alpha))$.
Otherwise,

$$
\begin{aligned}
&\partial_u(\text{LF}(r_1^*.\alpha)) \\
&= \partial_{u_1 \ldots u_n}(\text{LF}(r_1^*.\alpha)) \\
&= \partial_{u_2 \ldots u_n}(\partial_{u_1}(\text{LF}(r_1^*.\alpha))) \\
&= \partial_{u_2 \ldots u_n}(\partial_{u_1}(\text{LF}(r_1).r_1^*.\alpha) \cup \partial_{u_1}(\text{LF}(\alpha))) \\
&\supseteq \partial_{u_2 \ldots u_n}(\partial_{u_1}(\text{LF}(r_1).r_1^*.\alpha)) \\
&\supseteq \partial_{u_2 \ldots u_n}(\text{LF}(r_1^*.\alpha)) \\
&\qquad \text{by induction} \\
&\supseteq \text{LF}(\alpha)
\end{aligned}
$$

$\square$

The next, final lemma is our workhorse in proving that the language of $\mathcal{B}(\alpha)$ is contained in $\mathcal{L}^\omega(\alpha)$. The proof requires the extra bit in the ω-linear factors.

Lemma 47. *Let $q_0 q_1 \ldots q_n$ be a prefix of an accepting run of $\mathcal{B}(r.s^\omega)$ on $uw = a_1 \ldots a_n w$ where $q_n \in \text{LF}(s^\omega)$, but $q_i \notin \text{LF}(s^\omega)$, for $0 \leq i < n$. Then $u \in \mathcal{L}(r)$.*

Proof. Induction on n.
    Case 0; $u = \varepsilon$: $q_0 \in \text{LF}(s^\omega) \cap \text{LF}(r.s^\omega)$ because $q_0 \in Q_0$. Now $\text{LF}(s^\omega) = \text{LF}(s).s^\omega \times \{1\}$ and $\text{LF}(r.s^\omega) = \text{LF}(r).s^\omega \times \{0\} \cup N(r) \odot \text{LF}(s).s^\omega \times \{1\}$.
    If $N(r) = \mathbf{1}$, then $q_0 \in \text{LF}(s^\omega) \subseteq \text{LF}(r.s^\omega)$ and $u = \varepsilon \in \mathcal{L}(r)$.
    If $N(r) = \mathbf{0}$, then $q_0 \in \text{LF}(s).s^\omega \times \{1\} \cap \text{LF}(r).s^\omega \times \{0\} = \emptyset$ so that this case is not possible. (Without the extra bit in LF, there may be common linear factors if $\mathcal{L}(r) \cap \mathcal{L}(s^*) \neq \emptyset$.)
    Case $n > 0$: $u = au'$ and $q_1 \in \partial_a(q_0)$. As $q_0 \in Q_0 = \text{LF}(r.s^\omega) = \text{LF}(r).s^\omega \times \{0\} \cup N(r) \odot \text{LF}(s^\omega)$ but $q_0 \notin \text{LF}(s^\omega)$, it must be that $q_0 \in \text{LF}(r).s^\omega \times \{0\}$.
    Thus, $q_1 \in \partial_a(\text{LF}(r).s^\omega \times \{0\})$, so that there is a linear factor $\langle a, r' \rangle \in \text{LF}(r)$ such that $q_1 \in \text{LF}(r'.s^\omega)$.
    Thus, $q_1 \ldots q_n$ is a prefix of an accepting run of $\mathcal{B}(r'.s^\omega)^1$ on $u'w = a_2 \ldots a_n w$ where $q_n \in \text{LF}(s^\omega)$, but $q_i \notin \text{LF}(s^\omega)$, for $1 \leq i < n$. By induction, $u' \in \mathcal{L}(r')$ so that $u = au' \in \mathcal{L}(r)$ by Lemma 23. $\square$

Proof (of Theorem 44). It is sufficient to consider $\alpha = r.s^\omega$.
    Case "$\subseteq$": Let $w \in \mathcal{L}^\omega(r.s^\omega)$. Then $w = uv_0 v_1 \ldots$ where $u \in \mathcal{L}(r)$ and $\varepsilon \neq v_i \in \mathcal{L}(s)$, for $i \in \mathbb{N}$.
    Let $Q_0 = \text{LF}(r.s^\omega)$. By Lemma 46, $\text{LF}(s^\omega) \subseteq \partial_u(\text{LF}(r.s^\omega)) = \delta(Q_0, u)$.

---

[1] While the set Q' of states of $\mathcal{B}(r'.s^\omega)$ is a subset of the states Q of $\mathcal{B}(r.s^\omega)$, it is easy to see that the states $q_1 \ldots q_n$ as well as the remaining states $q_{n+1} q_{n+2} \ldots$ of the accepting run are all elements of Q'.

Furthermore, for each $i \in \mathbb{N}$, by Lemmas 45 and 46,

$$\partial_{v_i}(\mathrm{LF}(s^\omega)) = \partial_{v_i}(\mathrm{LF}(s.s^\omega)) \supseteq \mathrm{LF}(s^\omega)$$

Hence, there exists a run of $\mathcal{B}(\alpha)$ which visits states from $F = \mathrm{LF}(s^\omega)$ infinitely often.

Case "$\supseteq$": Suppose that $a_0 a_1 \cdots \in \mathcal{L}^\omega(\mathcal{B}(\alpha))$. Hence, there is a run $q_0 q_1 \cdots \in Q^\omega$ and a strictly increasing sequence $(n_i)_{i \in \mathbb{N}} \in \mathbb{N}^\omega$ such that, for all $j \in \mathbb{N}$, $q_j \in F$ iff $\exists i : j = n_i$.

Let $q = q_{n_0}$ be the first accepting state in the run and let $u = a_0 \dots a_{n_0 - 1}$. By construction of $\mathcal{B}(\alpha)$, $q \in \delta(Q_0, u)$ and $q \in \mathrm{LF}(s^\omega) = F$. By Lemma 47, $u \in \mathcal{L}(r)$.

Next, for each $i \in \mathbb{N}$, define $v_i = a_{n_i} \dots a_{n_{i+1}}$ so that $w = u v_0 v_1 \dots$.

For each i, $q_{n_i} \in F$ and $\varepsilon \neq v_i = b_i v_i'$. By construction $q_{n_i+1} \in \delta(q_{n_i}, b_i)$ so that $q_{n_i+1} \dots q_{n_{i+1}} \dots$ is a prefix of an accepting run of $\mathcal{B}(q_{n_i+1})$ where $q_{n_i+1} = \langle b_i, s'.s^\omega, 1 \rangle$, for some $\langle b_i, s' \rangle \in \mathrm{LF}(s)$. By Lemma 47, $v_i' \in \mathcal{L}(s')$ so that $v_i = b_i v_i' \in \mathcal{L}(s)$ by Lemma 23.

Taken together, we have shown that $w \in \mathcal{L}(r) \cdot \{v_0 v_1 \cdots \mid v_i \in \mathcal{L}(s)\} = \mathcal{L}^\omega(r.s^\omega)$.   □

We believe that it is possible to reduce the number of states of $\mathcal{B}(\alpha)$ by a factor of $|\Sigma|$ by merging suitable linear factors, but we leave this for future work.

References

1. Antimirov, V.M.: Rewriting regular inequalities. In: Reichel, H. (ed.) FCT 1995. LNCS, vol. 965, pp. 116–125. Springer, Heidelberg (1995)
2. Antimirov, V.M.: Partial derivatives of regular expressions and finite automaton constructions. Theoretical Computer Science **155**(2), 291–319 (1996)
3. Brzozowski, J.A.: Derivatives of regular expressions. J. ACM **11**(4), 481–494 (1964)
4. Caron, P., Champarnaud, J.-M., Mignot, L.: Partial derivatives of an extended regular expression. In: Dediu, A.-H., Inenaga, S., Martín-Vide, C. (eds.) LATA 2011. LNCS, vol. 6638, pp. 179–191. Springer, Heidelberg (2011)
5. Kleene, S.C.: Representation of events in nerve nets and finite automata. Automata Studies (1956)
6. Park, D.: Concurrency and automata on infinite sequences. In: Deussen, P. (ed.) Theoretical Computer Science. LNCS, vol. 104, pp. 167–183. Springer, Heidelberg (2003)
7. Redziejowski, R.R.: Construction of a deterministic ω-automaton using derivatives. Informatique Théorique et Applications **33**(2), 133–158 (1999)
8. Redziejowski, R.R.: An improved construction of deterministic omega-automaton using derivatives. Fundam. Inform. **119**(3–4), 393–406 (2012)
9. Roşu, G., Viswanathan, M.: Testing extended regular language membership incrementally by rewritin. In: Nieuwenhuis, R. (ed.) RTA 2003. LNCS, vol. 2706, pp. 499–514. Springer, Heidelberg (2003)

Quotient of Acceptance Specifications Under Reachability Constraints

Guillaume Verdier$^{(\boxtimes)}$ and Jean-Baptiste Raclet

IRIT/CNRS, 118 Route de Narbonne, 31062 Toulouse Cedex 9, France
{verdier,raclet}@irit.fr

Abstract. The quotient operation, which is dual to the composition, is crucial in specification theories as it allows the synthesis of missing specifications and thus enables incremental design. In this paper, we consider a specification theory based on marked acceptance specifications (MAS) which are automata enriched with variability information encoded by acceptance sets and with reachability constraints on states. We define a sound and complete quotient for MAS hence ensuring reachability properties by construction.

1 Introduction

Component-based design aims at building complex reactive systems by assembling components, possibly taken off-the-shelf. This approach can be supported by a specification theory in which requirements correspond to specifications while components are models of the specifications. Such theories come equipped with a set of operations enabling modular system design.

Several recent specification theories are based on modal specifications [9,14, 16] including in timed [7,11] or quantitative [1] contexts and with data [2]. In this paper, we introduce *marked acceptance specifications* (MAS): they are based on an extension of modal specifications, called acceptance specifications, which we enrich with marked states to model reachability objectives. This last addition is needed to model session terminations, component checkpoints or rollbacks.

A crucial feature in a specification theory is the operation of quotient. Let S_1 be the specification of a target system and S_2 be the specification of an available black-box component. The specification S_1/S_2 characterizes all the components that, when composed with any model of S_2, conform with S_1. In other words, S_1/S_2 tells what remains to be implemented to realize S_1 while reusing a component doing S_2. By allowing to characterize missing specifications, quotient thus enables incremental design and component reuse.

The quotient of specifications also plays a central role in contract-based design. In essence, a contract describes what a system should guarantee under some assumptions about its context of use. It can be modeled as a pair of specifications (A, G) for, respectively, the assumptions and the guarantees. Satisfiability of a contract then corresponds to the satisfiability of the specification G/A (see [6] for more explanations on contract satisfaction).

A full version with proofs is available online [17].

© Springer International Publishing Switzerland 2015
A.-H. Dediu et al. (Eds.): LATA 2015, LNCS 8977, pp. 299–311, 2015.
DOI: 10.1007/978-3-319-15579-1_23

Our contribution. Firstly, we define MAS and their semantics. The included marked states allow to specify reachability objectives that must be fulfilled by any model of the MAS. A MAS then characterizes a set of automata called *terminating* as they satisfy the reachability property telling that a marked state can always be reached.

Secondly, we study the compositionality of MAS. We define a compatibility criterion such that two MAS S_1 and S_2 are compatible if and only if the product of any models of S_1 and S_2 is terminating. Further, given two incompatible MAS S_1 and S_2, we propose a construction to refine S_1 into the most general S_1' such that S_1' and S_2 become compatible.

Last, we define the quotient of MAS. This is a two-step construction that makes use of the previous cleaning construction. The operation is shown to be sound and complete.

Related work. Modal specifications [13] enriched with marked states (MMS) have been introduced in [10] for the supervisory control of services. Product of MMS has been investigated in [8]. These papers did not show the need for the more expressive framework of MAS as quotient was not considered. Acceptance specifications have first been proposed in [15] based on [12]. Their non-deterministic version is named Boolean MS in [5]. The LTL model checking of MS has been studied in [4]. However, the reachability considered in this paper can be stated in CTL by `AG(EF(final))` and cannot be captured in LTL.

Quotient of modal and acceptance specifications has been studied in [9,15] and in [3] for the non-deterministic case. It has also been defined for timed [7,11] and quantitative [1] extensions of modal specifications. None of these works consider reachability constraints.

Outline of the paper. We recall some definitions about automata and introduce MAS in Sec. 2. Then, we define the pre-quotient operation in Sec. 3 which only partially solves the problem as it does not ensure the reachability of marked states. In Sec. 4, we give a criterion of compatible reachability telling whether the product of the models of two MAS is always terminating. When this condition of compatible reachability is not met, it is possible to impose some constraints on one of the specifications in order to obtain it, as shown in Sec. 5. Based on this construction, Sec. 6 finally defines the quotient operation on MAS.

2   Modeling with Marked Acceptance Specifications

2.1   Background on Automata

A (deterministic) *automaton* over an alphabet Σ is a tuple $M = (R, r^0, \lambda, G)$ where R is a finite set of states, $r^0 \in R$ is the initial state, $\lambda : R \times \Sigma \rightharpoonup R$ is the labeled transition map and $G \subseteq R$ is the set of marked states. The set of *fireable* actions from a state r, denoted ready(r), is the set of actions a such that $\lambda(r, a)$ is defined.

Given a state r, we define pre$^*(r)$ and post$^*(r)$ as the smallest sets such that $r \in$ pre$^*(r)$, $r \in$ post$^*(r)$ and for any r', a and r'' such that $\lambda(r', a) = r''$, $r' \in$

$\mathrm{pre}^*(r)$ if $r'' \in \mathrm{pre}^*(r)$ and $r'' \in \mathrm{post}^*(r)$ if $r' \in \mathrm{post}^*(r)$. We also define $\mathrm{pre}^+(r)$ as the union of $\mathrm{pre}^*(r')$ for all r' such that $\exists a : \lambda(r', a) = r$ and $\mathrm{post}^+(r)$ as the union of $\mathrm{post}^*(\lambda(r, a))$ for all $a \in \mathrm{ready}(r)$. Let $\mathrm{Loop}(r) = \mathrm{pre}^+(r) \cap \mathrm{post}^+(r)$.

Two automata M_1 and M_2 are bisimilar iff there exists a simulation relation $\pi : R_1 \times R_2$ such that $(r_1^0, r_2^0) \in \pi$ and for all $(r_1, r_2) \in \pi$, $\mathrm{ready}(r_1) = \mathrm{ready}(r_2) = Z$, $r_1 \in G_1$ iff $r_2 \in G_2$ and for any $a \in Z$, $(\lambda(r_1, a), \lambda(r_2, a)) \in \pi$.

The *product* of two automata M_1 and M_2, denoted $M_1 \times M_2$, is the automaton $(R_1 \times R_2, (r_1^0, r_2^0), \lambda, G_1 \times G_2)$ where $\lambda((r_1, r_2), a)$ is defined as the pair $(\lambda_1(r_1, a), \lambda_2(r_2, a))$ when both $\lambda_1(r_1, a)$ and $\lambda_2(r_2, a)$ are defined.

Given an automaton M and a state r of M, r is a *deadlock* if $r \notin G$ and $\mathrm{ready}(r) = \emptyset$; r belongs to a *livelock* if $\mathrm{Loop}(r) \neq \emptyset$, $G \cap \mathrm{Loop}(r) = \emptyset$ and there is no transition $\lambda(r', a) = r''$ such that $r' \in \mathrm{Loop}(r)$ and $r'' \notin \mathrm{Loop}(r)$. An automaton is *terminating* if it is deadlock-free and livelock-free.

2.2 Marked Acceptance Specification

We now enrich acceptance specifications [15] with marked states to model reachability constraints. The resulting formalism allows to specify a (possibly infinite) set of terminating automata called models.

Definition 1 (MAS). *A* marked acceptance specification *(MAS) over an alphabet Σ is a tuple $S = (Q, q^0, \delta, \mathrm{Acc}, F)$ where Q is a finite set of states, $q^0 \in Q$ is the initial state, $\delta : Q \times \Sigma \to Q$ is the labeled transition map, $\mathrm{Acc} : Q \to 2^{2^{\Sigma}}$ associates to each state its acceptance set and $F \subseteq Q$ is a set of marked states.*

Basically, an acceptance set is a set of sets of actions a model of the specification is ready to engage in. The *underlying* automaton associated to S is $\mathrm{Un}(S) = (Q, q^0, \delta, F)$. We only consider MAS such that $\mathrm{Un}(S)$ is deterministic.

Definition 2 (Satisfaction). *A* terminating automaton M satisfies *a MAS S, denoted $M \models S$, iff there exists a simulation relation $\pi \subseteq R \times Q$ such that $(r^0, q^0) \in \pi$ and for all $(r, q) \in \pi$: $\mathrm{ready}(r) \in \mathrm{Acc}(q)$; if $r \in G$ then $q \in F$; and, for any a and r' such that $\lambda(r, a) = r'$, $(r', \delta(q, a)) \in \pi$. M is called a model of S.*

Example 1. A MAS is depicted in Fig. 1(a). Marked states are double-circled while the acceptance sets are indicated near their associated state. The terminating automata M' and M'' in Fig. 1(b) and Fig. 1(c) are models of S_1 because of the respective simulation relation $\pi' = \{(0', 0), (1', 1)\}$ and $\pi'' = \{(0'', 0), (1'', 0), (2'', 1)\}$. Observe that the transitions labeled by b and c are optional in state 0 from the MAS S_1 as these actions are not present in all sets in $\mathrm{Acc}(0)$ and thus may not be present in any model of the specification. Moreover, state 1 in S_1 is marked to encode the constraint that it must be simulated in any model. As a result, although the actions b and c are optional, at least one of the two must be present in any model of S_1. This kind of constraint entails that MAS are more expressive than MS.

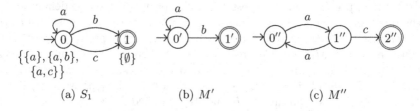

      (a) S_1 (b) M' (c) M''

Fig. 1. Example of MAS with two models

A MAS is said to be in normal form if the following holds for any state q:
- $\text{post}^*(q) \cap F \neq \emptyset$ and $\text{Acc}(q) \neq \emptyset$;
- when $\emptyset \in \text{Acc}(q)$ then $q \in F$;
- for any $a \in \Sigma$, $\delta(q,a)$ is defined iff $a \in \bigcup \text{Acc}(q)$

Any MAS S can be transformed into a MAS in normal form $\rho(S)$ without altering its set of models. The cleaning operation ρ is presented in [17]. In particular, if S has no model then $\rho(S)$ returns the empty specification. As a result, from now on, we always suppose that MAS are in normal form.

3   Pre-Quotient Operation of MAS

We first define an operation called *pre-quotient*. Given two MAS S_1 and S_2, it returns a MAS $S_1 /\!/ S_2$ such that the product of any of its models with any model of S_2, *if terminating*, will be a model of S_1. Another operation, defined in Sec. 5, will then be used in Sec. 6 to remove the *"if terminating"* assumption.

Definition 3 (Pre-quotient).   *The* pre-quotient *of two MAS S_1 and S_2, denoted $S_1 /\!/ S_2$, is the MAS $(Q_1 \times Q_2, (q_1^0, q_2^0), \delta, \text{Acc}, F)$ with:*
- *$\text{Acc}(q_1, q_2) = \{X \mid (\forall X_2 \in \text{Acc}_2(q_2) : X \cap X_2 \in \text{Acc}_1(q_1)) \wedge X \subseteq (\bigcup \text{Acc}_1(q_1))$*
  $\cap (\bigcup \text{Acc}_2(q_2))\}$;
- *$\forall a \in \Sigma : \delta((q_1, q_2), a)$ is defined if and only if there exists $X \in \text{Acc}(q_1, q_2)$ such that $a \in X$ and then $\delta((q_1, q_2), a) = (\delta_1(q_1, a), \delta_2(q_2, a))$;*
- *$(q_1, q_2) \in F$ if and only if $q_1 \in F_1$ or $q_2 \notin F_2$.*

Theorem 1 (Correctness).   *Given two MAS S_1 and S_2 and an automaton $M \models S_1 /\!/ S_2$, for any $M_2 \models S_2$ such that $M \times M_2$ is terminating, $M \times M_2 \models S_1$.*

The specification returned by the quotient is also expected to be complete, ie., to characterize all the possible automata whose product with a model of S_2 is a model of S_1. However, such a specification may become very large as it will, in particular, have to allow from a state (q_1, q_2) all the transitions which are not fireable from q_2 in S_2. As these transitions will always be removed by the product with models of S_2, they serve no real purpose for the quotient. We propose to return a compact specification for the quotient, without these transitions which we then call *unnecessary*.

An automaton M is said to have no *unnecessary transitions* regarding a MAS S, denoted $M \sim_\mathcal{U} S$, if and only if there exists a simulation relation $\pi \subseteq R \times Q$ such that $(r^0, q^0) \in \pi$ and for all $(r, q) \in \pi$, $\text{ready}(r) \subseteq \bigcup \text{Acc}(q)$ and for every a and r' such that $\lambda(r, a) = r'$, $(r', \delta(q, a)) \in \pi$.

When an automaton M has unnecessary transitions regarding a MAS S, it is possible to remove these transitions. Let $\rho_u(M, S)$ be the automaton $M' = (R \times Q, (r^0, q^0), \lambda', G \times Q)$ with:

$$\lambda'((r, q), a) = \begin{cases} (\lambda(r, a), \delta(q, a)) & \text{if } a \in \bigcup \text{Acc}(q) \\ \text{undefined} & \text{otherwise} \end{cases}$$

This automaton has no unnecessary transitions regarding S and for any $M_S \models S$, the automata $M \times M_S$ and $\rho_u(M, S) \times M_S$ are bisimilar (see [17] for proofs).

We can then prove that our pre-quotient is complete for automata without unnecessary transitions. Given an arbitrary automaton, it suffices to remove these transitions with ρ_u before checking if it is a model of the quotient.

Theorem 2. *Given two MAS S_1 and S_2 and an automaton M such that $M \sim_\mathcal{U} S_2$ and for all $M_2 \models S_2$ we have $M \times M_2 \models S_1$, then $M \models S_1 /\!\!/ S_2$.*

Corollary 1 (Completeness). *Given two MAS S_1 and S_2 and an automaton M such that for all $M_2 \models S_2$ we have $M \times M_2 \models S_1$, then $\rho_u(M, S_2) \models S_1 /\!\!/ S_2$.*

Observe now that the pre-quotient $S_1 /\!\!/ S_2$ may admit some models whose product with some models of S_2 may not be terminating. Consider indeed the specifications S_1 and S_2 of Fig. 1(a) and 2(a) and their pre-quotient in Fig. 2(b). The product of the models M_1^1 of $S_1 /\!\!/ S_2$ (Fig. 2(c)) and M' of S_2 (Fig. 1(b)) has a livelock and thus is not terminating. One may think that there is an error in the pre-quotient computation and that it should not allow to realize only $\{a, c\}$, without b in $\text{Acc}(0, 0')$. Indeed, it would forbid the model M_1^1, but it would also disallow some valid models such as M_1^2 (Fig. 2(d)), which realizes $\{a, c\}$ in a state and $\{a, b\}$ in another, thus synchronizing on b with any model of S_2 and allowing the joint reachability of the marked states.

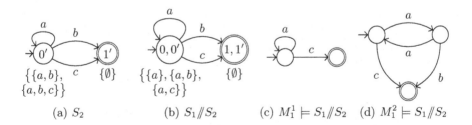

(a) S_2 (b) $S_1 /\!\!/ S_2$ (c) $M_1^1 \models S_1 /\!\!/ S_2$ (d) $M_1^2 \models S_1 /\!\!/ S_2$

Fig. 2. Example of pre-quotient

In the next section, we define a criterion allowing to test whether the product of any models of two MAS is terminating or not. On this basis, we will then refine the pre-quotient in Sec. 6 in order to guarantee the reachability property.

4 Compatible Reachability of MAS

By definition, the product of some models of two MAS may not terminate due to two different causes, namely deadlock and livelock. We consider separately the two issues to derive a compatible reachability criterion on MAS.

4.1 Deadlock-Free Specifications

In this section, we propose a test to check if two MAS S_1 and S_2 have some models M_1 and M_2 such that $M_1 \times M_2$ has a deadlock. To do so, we characterize deadlock-free pairs of states, from which no deadlock may arise in the product of any two models of S_1 and S_2.

Given two acceptance sets A_1 and A_2, let $\text{Compat}(A_1, A_2)$ be true iff for all $X_1 \in A_1$ and $X_2 \in A_2$, $X_1 \cap X_2 \neq \emptyset$. Now a pair of states (q_1, q_2) is said to be deadlock-free, denoted $\text{DeadFree}(q_1, q_2)$, if $\text{Acc}_1(q_1) = \text{Acc}_2(q_2) = \{\emptyset\}$ or $\text{Compat}(\text{Acc}_1(q_1), \text{Acc}_2(q_2))$.

Definition 4 (Deadlock-free MAS). *Two MAS S_1 and S_2 are deadlock-free when all the reachable pairs of states in $\text{Un}(S_1) \times \text{Un}(S_2)$ are deadlock-free.*

Theorem 3. *Two MAS S_1 and S_2 are deadlock-free if and only if for any $M_1 \models S_1$ and $M_2 \models S_2$, $M_1 \times M_2$ is deadlock-free.*

4.2 Livelock-Free Specifications

In this section, we explain how we can check if two MAS S_1 and S_2 have some models M_1 and M_2 such that $M_1 \times M_2$ has a livelock. We identify the cycles shared between S_1 and S_2 along with the transitions leaving them. We check if at least one of these transitions is preserved in the product of any two models of S_1 and S_2. Before studying these common cycles, a first step consists in unfolding S_1 and S_2 so as possible synchronizations become unambiguous.

Unfolding. Given two specifications S_1 and S_2, we define the *partners* of a state q_1 as $Q_2(q_1) = \{q_2 \mid (q_1, q_2) \text{ is reachable in } \text{Un}(S_1) \times \text{Un}(S_2)\}$; the set $Q_1(q_2)$ is defined symmetrically. As a shorthand, if we know that a state q_1 has exactly one partner, we will also use $Q_2(q_1)$ to denote this partner.

If some states of S_2 have several partners, it is possible to transform S_2 so that each of its states has at most one partner, while preserving the set of models of the specification. The unfolding of S_2 in relation to S_1 is the specification $((Q_1 \cup \{q^?\}) \times Q_2, (q_1^0, q_2^0), \delta_u, \text{Acc}_u, (Q_1 \cup \{q^?\}) \times F_2)$ where:
- $q^?$ is a fresh state ($q_1^?$ denotes a state in $Q_1 \cup \{q^?\}$);
- $\delta_u((q_1^?, q_2), a)$ is defined if and only if $\delta_2(q_2, a)$ is defined and then:

$$\delta_u((q_1, q_2), a) = \begin{cases} (\delta_1(q_1, a), \delta_2(q_2, a)) & \text{if } \delta_1(q_1, a) \text{ is defined} \\ (q^?, \delta_2(q_2, a)) & \text{otherwise} \end{cases}$$
$$\delta_u((q^?, q_2), a) = (q^?, \delta_2(q_2, a))$$

– $\mathrm{Acc}_u((q_1^?, q_2)) = \mathrm{Acc}_2(q_2)$.

Two MAS S_1 and S_2 have *single partners* if and only if for all $q_1 \in Q_1$, we have $|Q_2(q_1)| \leq 1$ and for all $q_2 \in Q_2$, we also have $|Q_1(q_2)| \leq 1$.

Given two MAS S_1 and S_2, there exists some MAS S_1' and S_2', called *unfoldings* of S_1 and S_2, with single partners and which have the same models as S_1 and S_2. These two MAS can be computed by unfolding S_1 in relation to S_2 and then S_2 in relation to the unfolding of S_1 (see [17] for proofs).

Cycles. In order to detect livelocks, we need to study the cycles that may be present in the models of a specification. Intuitively, a cycle is characterized by its states and the transitions between them.

Given a MAS S, the partial map $C : Q \rightharpoonup 2^{\Sigma}$ represents a *cycle* in S if and only if for any $q \in \mathrm{dom}(C)$, (a) $C(q) \neq \emptyset$, (b) $\exists X \in \mathrm{Acc}(q)$ such that $C(q) \subseteq X$, (c) $\mathrm{dom}(C) \subseteq \mathrm{post}^*(q)$ and (d) $\forall a \in C(q) : \delta(q, a) \in \mathrm{dom}(C)$.

A model M of a MAS S *implements* a cycle C if and only if there exists a set $\mathcal{R}$ of states of M such that each $q \in \mathrm{dom}(C)$ is implemented by at least one state of $\mathcal{R}$ and for each $r \in \mathcal{R}$ and for each q such that $(r, q) \in \pi$, (a) $q \in \mathrm{dom}(C)$, (b) $C(q) \subseteq \mathrm{ready}(r)$, (c) $\forall a \in C(q) : \lambda(r, a) \in \mathcal{R}$ and (d) $\forall a \in \mathrm{ready}(r) \backslash C(q) : \lambda(r, a) \notin \mathcal{R}$. A cycle is said to be *implementable* if there exists a model M of S implementing the cycle.

We define in Algo. 1 an operation, $\mathrm{Loop}_{\models}$-rec, which computes the cycles of a MAS passing by a given state. However, some of these cycles may not be implementable. For instance, the cycle $C = \{0 \mapsto \{a\}\}$ is not implementable in the MAS depicted in Figure 3, as all the models have to eventually realize the transition by b to reach the marked state and then are not allowed to simultaneously realize the transition by a.

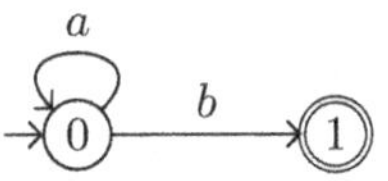

$$\mathrm{Acc}(0) = \{\{a\}, \{b\}\}$$
$$\mathrm{Acc}(1) = \{\emptyset\}$$

Fig. 3. A MAS over $\{a,b\}$ with no implementable cycle

In order to be implementable, a cycle has to contain a marked state or it must be possible to realize a transition that is not part of the cycle in addition to the transitions of the cycle. Thus, the set of implementable cycles of a MAS S, denoted $\mathrm{Loop}_{\models}(S)$, is $\bigcup_{q \in Q} \{C \in \mathrm{Loop}_{\models}\text{-rec}(S, q, \emptyset) \mid \mathrm{dom}(C) \cap F \neq \emptyset \vee \exists q_C \in \mathrm{dom}(C) : \exists X \in \mathrm{Acc}(q_C) : C(q_C) \subset X\}$.

Livelock-freeness. We can now analyze the cycles of two MAS with single partners in order to detect if there may be a livelock in the product of some of their models. To do so, we distinguish two kinds of transitions: those, denoted $\mathcal{A}$, which are always realized when the cycle is implemented and those, denoted $\mathcal{O}$, which may (or may not) be realized when the cycle is implemented. These sets

---

Algorithm 1. Loop$_\models$-rec (S: MAS, q: State, cycle: Cycle): Set Cycle

---

1: **if** $q \in$ dom(cycle) **then return** {cycle}
2: res $\leftarrow \emptyset$
3: **for all** $A \in$ Acc(q):
4:     cycle_acc $\leftarrow \{a \mid a \in A \wedge q \in \text{post}^*(\delta(q,a))\}$
5:     **for all** $C \in 2^{\text{cycle\_acc}} \setminus \{\emptyset\}$:
6:         current $\leftarrow$ {cycle}
7:         **for all** $a \in C$:
8:             current $\leftarrow \bigcup_{cycle \in \text{current}}$ Loop$_\models$ -rec($S, \delta(q,a), cycle \cup \{q \mapsto C\}$)
9:         res $\leftarrow$ res $\cup$ current
10: **return** res

---

are represented by partial functions from a state to a set of sets of actions and given, for a particular cycle $\mathcal{C}$, by the following formulae:

$$\mathcal{A} = \{q \mapsto \text{leaving}(q, A) \mid A \notin \text{Acc}(q), (q, A) \in \mathcal{C}\}$$

$$\mathcal{O} = \{q \mapsto \text{leaving}(q, A) \mid A \in \text{Acc}(q), (q, A) \in \mathcal{C} \wedge \text{leaving}(q, A) \neq \emptyset\}$$

$$\text{where leaving}(q, A) = \{X \setminus A \mid X \in \text{Acc}(q) \wedge A \subset X\}$$

Definition 5. *Given two MAS S_1 and S_2 with single partners and a cycle $\mathcal{C}_1$ in S_1 such that all its states have a partner, $\mathcal{C}_1$ is livelock-free in relation to S_2, denoted* LiveFree($\mathcal{C}_1, S_2$), *if and only if, when the cycle $\mathcal{C}_2 = \{Q_2(q) \mapsto \mathcal{C}_1(q) \mid q \in \text{dom}(\mathcal{C}_1)\}$ is in* Loop$_\models$(S_2):
1. $\mathcal{A}_{\mathcal{C}_1} \neq \emptyset$, $\mathcal{A}_{\mathcal{C}_2} \neq \emptyset$ *and there exists* $q_1' \in \text{dom}(\mathcal{A}_{\mathcal{C}_1})$ *such that* $Q_2(q_1') \in \text{dom}(\mathcal{A}_{\mathcal{C}_2})$ *and* Compat($\mathcal{A}_{\mathcal{C}_1}(q_1'), \mathcal{A}_{\mathcal{C}_2}(Q_2(q_1'))$), *or*
2. $\mathcal{A}_{\mathcal{C}_1} \neq \emptyset$, $\mathcal{A}_{\mathcal{C}_2} = \emptyset$, dom($\mathcal{C}_2$) $\cap F_2 = \emptyset$ *and* $\forall q_2' \in \text{dom}(\mathcal{O}_{\mathcal{C}_2})$: $Q_1(q_2') \in \text{dom}(\mathcal{A}_{\mathcal{C}_1})$ *and* Compat($\mathcal{A}_{\mathcal{C}_1}(Q_1(q_2')), \mathcal{O}_{\mathcal{C}_2}(q_2')$), *or*
3. $\mathcal{A}_{\mathcal{C}_1} = \emptyset$, $\mathcal{A}_{\mathcal{C}_2} \neq \emptyset$, dom($\mathcal{C}_1$) $\cap F_1 = \emptyset$ *and* $\forall q_1' \in \text{dom}(\mathcal{O}_{\mathcal{C}_1})$: $Q_2(q_1') \in \text{dom}(\mathcal{A}_{\mathcal{C}_2})$ *and* Compat($\mathcal{O}_{\mathcal{C}_1}(q_1'), \mathcal{A}_{\mathcal{C}_2}(Q_2(q_1'))$).

Definition 6 (Livelock-free specifications). *Two MAS S_1 and S_2 with single partners are* livelock-free *if all the implementable cycles of S_1 are livelock-free in relation to S_2.*

This definition only tests the implementable cycles of S_1. It is not necessary to do the symmetrical test (checking that the implementable cycles of S_2 verify LiveFree) because we only compare the cycle of S_1 with the same cycle in S_2 and the tests of Def. 6 are symmetric.

The previous definition offers a necessary and sufficient condition to identify MAS which can have two respective models whose product has a livelock:

Theorem 4. *Two MAS S_1 and S_2 with single partners are livelock-free if and only if for any $M_1 \models S_1$ and $M_2 \models S_2$, $M_1 \times M_2$ is livelock-free.*

Specifications with Compatible Reachability. By combining the tests for deadlock-free and livelock-free specifications, we can define a criterion checking if two MAS S_1 and S_2 have some models M_1 and M_2 such that $M_1 \times M_2$ is not terminating.

Definition 7 (Compatible reachability). *Two MAS S_1 and S_2 have a compatible reachability, denoted $S_1 \sim_T S_2$, if and only if they are deadlock-free and their unfoldings are livelock-free. They have an incompatible reachability otherwise.*

Theorem 5. *Given two MAS S_1 and S_2, $S_1 \sim_T S_2$ if and only if for any $M_1 \models S_1$ and $M_2 \models S_2$, $M_1 \times M_2$ is terminating.*

This theorem allows *independent implementability* of MAS: given two MAS with compatible reachability, each specification may be implemented independently from the other while keeping the guarantee that the composition of the resulting implementations will be terminating and thus satisfy by construction a reachability property.

5   Correction of MAS with Incompatible Reachability

We now define an operation that, given two MAS S_1 and S_2 with incompatible reachability, returns a MAS refining S_1 with a compatible reachability with S_2.

5.1   Deadlock Correction

First, given two non-deadlock-free MAS S_1 and S_2, we propose to refine S_1 such that the obtained MAS S_1' is deadlock-free with S_2. For this, we iterate through all the non-deadlock-free pairs of states (q_1, q_2) and remove the elements of the acceptance set of q_1 which may cause a deadlock, as described in Algo. 2. Note that it may return an empty specification, because of ρ, which then means that for any model M_1 of S_1, there exists a model M_2 of S_2 such that $M_1 \times M_2$ has a deadlock.

Theorem 6 (Deadlock correction). *Given two MAS S_1 and S_2, $M_1 \models S_1$ is such that for any $M_2 \models S_2$, $M_1 \times M_2$ is deadlock-free if and only if $M_1 \models dead\_correction(S_1, S_2)$.*

Algorithm 2. dead_correction (S_1: MAS, S_2: MAS): MAS

1: $S_1' \leftarrow S_1$
2: **for all** (q_1, q_2) such that $\neg \mathrm{DeadFree}(q_1, q_2)$:
3:     **if** $\mathrm{Acc}_2(q_2) = \{\emptyset\}$:
4:         **if** $\emptyset \in \mathrm{Acc}_1'(q_1)$ **then** $\mathrm{Acc}_1'(q_1) \leftarrow \{\emptyset\}$ **else** $\mathrm{Acc}_1'(q_1) \leftarrow \emptyset$
5:     **else:**
6:         $\mathrm{Acc}_1'(q_1) \leftarrow \{X_1 \mid X_1 \in \mathrm{Acc}_1'(q_1) \land \forall X_2 \in \mathrm{Acc}_2(q_2) : X_1 \cap X_2 \neq \emptyset\}$
7: **return** $\rho(S_1')$

5.2 Livelock Correction

Secondly, given S_1 and S_2 two deadlock-free MAS, we propose to refine S_1 such that the obtained specification S_1' is livelock-free with S_2.

There are two ways to prevent livelocks from occuring in the product of the models of two MAS: removing some transitions so that states from which it is not possible to guarantee termination will not be reached and forcing some transitions to be eventually realized in order to guarantee that it will be possible to leave cycles without marked states. For this last method, we introduce marked acceptance specifications with priorities that are MAS with some priority transitions which have to be eventually realized.

Definition 8 (MAS with priorities). *A marked acceptance specification with priorities (MASp) is a tuple* $(Q, q^0, \delta, \mathrm{Acc}, P, F)$ *where* $(Q, q^0, \delta, \mathrm{Acc}, F)$ *is a MAS and* $P : 2^{2^{Q \times \Sigma}}$ *is a set of* priorities.

Definition 9 (Satisfaction). *An automaton M implements a MASp S if M implements the underlying MAS and for all* $\mathcal{P} \in P$, *either* $\forall (q, a) \in \mathcal{P} : \forall r : (r, q) \notin \pi$ *or* $\exists (q, a) \in \mathcal{P} : \exists r : (r, q) \in \pi \wedge a \in \mathrm{ready}(r)$.

Intuitively, P represents a conjunction of disjunctions of transitions: at least one transition from each element of P must be implemented by the models of the specification.

Let S_1 and S_2 be two MAS and q_1 a state of S_1 such that q_1 belongs to a livelock. Then, there exists a cycle $\mathcal{C}_1$ in S_1 and its partner $\mathcal{C}_2$ in S_2 such that the conditions given in Def. 5 are false. Given these cycles, Algo. 3 ensures that the possible livelock will not happen, either by adding some priorities or by removing some transitions.

Algorithm 3. live_corr_cycle (S_1: MASp, $\mathcal{C}_1$: Cycle, S_2: MAS, $\mathcal{C}_2$: Cycle): MASp

1: **if** $\mathcal{A}_{\mathcal{C}_2} \neq \emptyset$:
2:     $Q_A \leftarrow \{q_1 \mid Q_2(q_1) \in \mathrm{dom}(\mathcal{A}_{\mathcal{C}_2}) \wedge \forall A \in \mathcal{A}_{\mathcal{C}_2}(Q_2(q_1)) : A \cap \mathrm{ready}(q_1) \neq \emptyset\}$
3:     **if** $Q_A \neq \emptyset$:
4:         $P \leftarrow \{\bigcup_{1 \leq i \leq |Q_A|} \{(q_i, a) \mid a \in X_i\} \mid X_i \in \{A \cap \mathrm{ready}(q_i) \mid A \in \mathcal{A}_{\mathcal{C}_2}(Q_2(q_i))\}\}$
5:         **return** $(Q_1, q_1^0, \delta_1, \mathrm{Acc}_1, P_1 \cup P, F_1)$
6: **else if** $\mathrm{dom}(\mathcal{C}_2) \cap F_2 = \emptyset$:
7:     $\mathrm{Acc}' \leftarrow \mathrm{Acc}_1$
8:     **for all** $q_1 \in \{Q_1(q_2) \mid q_2 \in \mathrm{dom}(\mathcal{O}_{\mathcal{C}_2})\}$:
9:         $\mathrm{Acc}'(q_1) \leftarrow \{X \mid X \in \mathrm{Acc}_1(q_1) \wedge \forall O \in \mathcal{O}_{\mathcal{C}_2}(Q_2(q_1)) : X \cap O \neq \emptyset\}$
10:     **return** $\rho((Q_1, q_1^0, \delta_1, \mathrm{Acc}', P_1, F_1))$
11: $\mathrm{Acc}' \leftarrow \mathrm{Acc}_1$
12: **for all** $q_1 \in Q_1$:
13:     $\mathrm{Acc}'(q_1) \leftarrow \{X \mid X \in \mathrm{Acc}_1(q_1) \wedge \forall a \in X : \delta(q_1, a) \notin \mathrm{dom}(\mathcal{C}_1)\}$
14: **return** $\rho((Q_1, q_1^0, \delta_1, \mathrm{Acc}', P_1, F_1))$

We then iterate over the possible cycles, fixing those which may cause a livelock (see [17] for a detailed explanation of the construction of live_correction).

Theorem 7 (Livelock correction). *Given two MAS S_1 and S_2, $M_1 \models S_1$ is such that for any $M_2 \models S_2$, $M_1 \times M_2$ is livelock-free if and only if $M_1 \models$ live_correction(S_1, S_2).*

As a result, by applying successively dead_correction and live_correction, we can define the following operation ρ_T:

$$\rho_T(S_1, S_2) = \text{live\_correction}(\text{dead\_correction}(S_1, S_2), S_2)$$

Given two MAS S_1 and S_2, it refines the set of models of S_1 as precisely as possible so that their product with any model of S_2 is terminating.

Theorem 8 (Incompatible reachability correction). *Given two MAS S_1 and S_2, $M \models \rho_T(S_1, S_2)$ if and only if $M \models S_1$ and for any $M_2 \models S_2$, $M \times M_2$ is terminating.*

6   Quotient Operation of MAS

We can now combine the pre-quotient and cleaning operations to define the quotient of two MAS.

Definition 10. *Given two MAS S_1 and S_2, their quotient S_1/S_2 is given by $\rho_T(S_1 /\!/ S_2, S_2)$.*

Theorem 9 (Soundness). *Given two MAS S_1 and S_2 and an automaton $M \models S_1/S_2$, for any $M_2 \models S_2$, $M \times M_2 \models S_1$.*

Theorem 10 (Completeness). *Given two MAS S_1 and S_2 and an automaton M such that $\forall M_2 \models S_2 : M \times M_2 \models S_1$, then $\rho_u(M, S_2) \models S_1/S_2$.*

These theorems indicate that each specification S_2 and S_1/S_2 may be implemented independently from the other and that the composition of the resulting implementations will eventually be terminating and will also satisfy S_1.

7   Conclusion

In this paper, we have introduced marked acceptance specifications. We have developed several compositionality results ensuring a reachability property by construction and, in particular, a sound and complete quotient. Note that this framework can almost immediately be enriched with a refinement relation, parallel product and conjunction by exploiting the constructions available in [15] and [8], hence providing a complete specification theory as advocated in [16].

Considering an acceptance setting instead of a modal one offers a gain in terms of expressivity as MAS provide more flexibility than the marked extension of modal specifications [8]. This benefit becomes essential for the quotient as may/must modalities are not rich enough to allow for a complete operation [17]. Observe also that quotient of two MAS is heterogeneous in the sense that its result may be a MASp. By definition, MASp explicitly require to *eventually* realize some transitions fixed in the priority set P. By bounding the delay before the implementation of the transitions, a MASp could become a standard MAS and the quotient would then become homogeneous. Algorithms for bounding MASp are left for future investigations.

References

1. Bauer, S.S., Fahrenberg, U., Juhl, L., Larsen, K.G., Legay, A., Thrane, C.R.: Weighted modal transition systems. Formal Methods in System Design **42**(2), 193–220 (2013)
2. Bauer, S.S., Larsen, K.G., Legay, A., Nyman, U., Wasowski, A.: A modal specification theory for components with data. Sci. Comput. Program. **83**, 106–128 (2014)
3. Beneš, N., Delahaye, B., Fahrenberg, U., Křetínský, J., Legay, A.: Hennessy-Milner Logic with Greatest Fixed Points as a Complete Behavioural Specification Theory. In: D'Argenio, P.R., Melgratti, H. (eds.) CONCUR 2013 – Concurrency Theory. LNCS, vol. 8052, pp. 76–90. Springer, Heidelberg (2013)
4. Beneš, N., Černá, I., Křetínský, J.: Modal Transition Systems: Composition and LTL Model Checking. In: Bultan, T., Hsiung, P.-A. (eds.) ATVA 2011. LNCS, vol. 6996, pp. 228–242. Springer, Heidelberg (2011)
5. Beneš, N., Křetínský, J., Larsen, K.G., Møller, M.H., Srba, J.: Parametric Modal Transition Systems. In: Bultan, T., Hsiung, P.-A. (eds.) ATVA 2011. LNCS, vol. 6996, pp. 275–289. Springer, Heidelberg (2011)
6. Benveniste, A., Raclet, J.B., Caillaud, B., Nickovic, D., Passerone, R., Sangiovanni-Vincentelli, A., Henzinger, T., Larsen, K.G.: Contracts for the design of embedded systems - part II: theory (2012)
7. Bertrand, N., Legay, A., Pinchinat, S., Raclet, J.B.: Modal event-clock specifications for timed component-based design. Sci. Comput. Program. **77**(12), 1212–1234 (2012)
8. Caillaud, B., Raclet, J.-B.: Ensuring Reachability by Design. In: Roychoudhury, A., D'Souza, M. (eds.) ICTAC 2012. LNCS, vol. 7521, pp. 213–227. Springer, Heidelberg (2012)
9. Chen, T., Chilton, C., Jonsson, B., Kwiatkowska, M.: A Compositional Specification Theory for Component Behaviours. In: Seidl, H. (ed.) Programming Languages and Systems. LNCS, vol. 7211, pp. 148–168. Springer, Heidelberg (2012)
10. Darondeau, P., Dubreil, J., Marchand, H.: Supervisory control for modal specifications of services. In: WODES. pp. 428–435 (2010)
11. David, A., Larsen, K.G., Legay, A., Nyman, U., Wasowski, A.: Timed I/O automata: A complete specification theory for real-time systems. In: HSCC. pp. 91–100. ACM (2010)
12. Hennessy, M.: Acceptance trees. J. ACM **32**(4), 896–928 (1985)
13. Larsen, K.G., Thomsen, B.: A modal process logic. In: LICS. pp. 203–210. IEEE (1988)

14. Lüttgen, G., Vogler, W.: Modal interface automata. Logical Methods in Computer Science 9(3) (2013)
15. Raclet, J.B.: Residual for component specifications. FACS. Electr. Notes Theor. Comput. Sci. **215**, 93–110 (2008)
16. Raclet, J.B., Badouel, E., Benveniste, A., Caillaud, B., Legay, A., Passerone, R.: A modal interface theory for component-based design. Fun. Informaticae **108**(1–2), 119–149 (2011)
17. Verdier, G., Raclet, J.B.: Quotient of acceptance specifications under reachability constraints. arXiv:1411.6463 (2014)

Codes, Semigroups, and Symbolic Dynamics

Structure and Measure of a Decidable Class of Two-dimensional Codes

Marcella Anselmo[1], Dora Giammarresi[2], and Maria Madonia[3(✉)]

[1] Dipartimento di Informatica, Università di Salerno, Via Giovanni Paolo II,
132-84084 Fisciano (SA), Italy
anselmo@dia.unisa.it
[2] Dipartimento di Matematica, Università Roma "Tor Vergata",
via della Ricerca Scientifica, 00133 Roma, Italy
giammarr@mat.uniroma2.it
[3] Dipartimento di Matematica e Informatica, Università di Catania,
Viale Andrea Doria 6/a, 95125 Catania, Italy
madonia@dmi.unict.it

Abstract. A two-dimensional code is defined as a set $X \subseteq \Sigma^{**}$ such that any picture over Σ is tilable in at most one way with pictures in X. It is in general undecidable whether a set X of pictures is a code also in the finite case. Very recently in [3] strong prefix picture codes were defined as a decidable subclass that generalizes prefix string codes. Here a characterization for strong prefix codes that results in an effective procedure to construct them is presented. As a consequence there are also proved interesting results on the measure of strong prefix codes and a connection with the family of string prefix codes.

Keywords: Two-dimensional languages · Codes · Prefix codes

1 Introduction

A two-dimensional word, or picture, is a rectangular array of symbols taken from a finite alphabet Σ. The set of all pictures over Σ is usually denoted by Σ^{**}: a two-dimensional language is thus a subset of Σ^{**}. Extending the theory of formal (string) languages to two dimensions is a very challenging task. The two-dimensional structure in fact imposes its intrinsic difficulties to all the theory: the two concatenation operations (horizontal and vertical) are only partial operations and do not induce a monoid structure to the set Σ^{**}. Moreover, if we cut out a "prefix" from a picture (i.e. delete a rectangular portion in top-left corner) the remaining part is not in general a picture.

During the last fifty years, and still intensively nowadays, many researchers investigated how the notion of finite state recognizability can be transferred into a two-dimensional (2D) world (e.g. [8,11,13,14,17–20]). Among the most accredited generalizations to 2D of regular string languages, there is the family REC of picture languages recognized by tiling systems that extend a characterization

© Springer International Publishing Switzerland 2015
A.-H. Dediu et al. (Eds.): LATA 2015, LNCS 8977, pp. 315–327, 2015.
DOI: 10.1007/978-3-319-15579-1_24

of finite automata ([13]). A crucial difference with string language theory is that REC is intrinsically non-deterministic. In [2,6] deterministic and unambiguous tiling systems, together with the corresponding subfamilies DREC and UREC, are defined; it is proved that all the inclusions among REC, UREC and DREC are proper. Moreover the problem whether a given tiling system is unambiguous is undecidable.

In the theoretical study of formal string languages, string codes have been always a relevant subject of research, also because of their applications to practical problems. Theoretical results on string codes are related to combinatorics on words, automata theory and semigroup theory (see [10] for complete references).

The notion of code can be intuitively and naturally transposed to two-dimensional objects by exploiting the notion of unique tiling decomposition. Several attempts of developing a formal theory of *two-dimensional codes* have been done by using polyominoes (connected two-dimensional figures). Unfortunately, most of the published results show that in the 2D context we loose important properties. In [9] D. Beauquier and M. Nivat proved that the problem whether a finite set of polyominoes is a code is undecidable, and that the same result holds also for dominoes. Other variants of two-dimensional codes are also studied in [1,12,15] still proving undecidability results. It is worthwhile to remark that all mentioned results consider 2D codes independently from a 2D formal language theory.

Very recently, in [4,5], a new definition for picture codes was introduced in connection with the family REC of picture languages recognized by finite tiling systems. Codes are defined by using the formal operation of tiling star as defined in [20]: the tiling star of a set of pictures X is the set X^{**} of all pictures that are tilable (in the polyominoes style) by elements of X. Then X is a code if any picture in X^{**} is tilable in a unique way. Remark that if $X \in$ REC then X^{**} is also in REC. Similarly to the string case, it holds that if X is a finite picture code then, starting from pictures in X, we can construct an unambiguous tiling system for X^{**}. Unfortunately, despite this nice connection to the string code theory, it is still undecidable whether a given set of pictures is a code.

Definitions of *two-dimensional prefix code* and *strong prefix code* are introduced in [3,4]: they are introduced as the two-dimensional counterpart of prefix string codes and they seem to be the first non-trival decidable classes of two-dimensional codes. Remark that, at this stage, the aim of this study is purely theoretical.

The formal definition of prefix sets of pictures (see [4]) results in a decidable family that includes interesting examples: nevertheless the definition is quite involved and it is based on special kind of polyominoes that have straight top border. The strong prefix sets (proposed in [3]) have a simpler definition based on the notion of overlapping. Main results show that strong prefix sets are a decidable family of picture codes with a simple polynomial decoding algorithm. It is also proved that it is decidable whether a given finite strong prefix set is maximal and that it is possible to construct a finite maximal strong prefix code containing a given strong prefix code.

In this paper we deeply investigate strong prefix codes and prove that they have a recursive structure. This allows us to describe an effective procedure to construct *all* (maximal) finite strong prefix codes of pictures, starting from the "singleton" pictures containing only one alphabet symbol. The construction in some sense extends the literal representation of prefix codes of strings. Using such construction we prove some results on the measure of these codes that generalize known results from the string code theory. In the last part of the paper we show an interesting connection between strong prefix picture codes and prefix string codes that gives a deeper comprehension of the structure of strong prefix picture codes and relates them to a conjecture on commutative equivalence that holds in the string code theory (see [10, 16]).

2 Preliminaries

2.1 Two-Dimensional Languages

A *picture* over a finite alphabet Σ is a two-dimensional rectangular array of elements of Σ. Given a picture p, $|p|_{row}$ and $|p|_{col}$ denote the number of rows and columns, respectively while $size(p) = (|p|_{row}, |p|_{col})$ denotes the picture *size*. The set of all pictures over Σ of fixed size (m, n) is denoted by $\Sigma^{m,n}$, while Σ^{m*} and Σ^{*n} denote the set of all pictures over Σ with fixed number of rows m and columns n, respectively. The empty pictures, referred to as $\lambda_{m,0}$ and $\lambda_{0,n}$, for all $m, n \geq 0$, correspond to all pictures of size $(m, 0)$ or $(0, n)$. The set of all pictures over Σ is denoted by Σ^{**}, while Σ^{++} denotes the set of all non-empty pictures over Σ. A *two-dimensional language*, or *picture language*, over Σ is a subset of Σ^{**}.

The set of coordinates $dom(p) = \{1, 2, \ldots, |p|_{row}\} \times \{1, 2, \ldots, |p|_{col}\}$ is referred to as the *domain* of a picture p. We let $p(i, j)$ denote the symbol in p at coordinates (i, j). The top-left corner (*tl-corner*) of p refers to position $(1, 1)$. We also fix the scanning direction for a picture from the top-left corner toward the bottom-right one (*tl2br*).

A *subdomain* of $dom(p)$ is a set d of the form $\{i, i + 1, \ldots, i'\} \times \{j, j + 1, \ldots, j'\}$, where $1 \leq i \leq i' \leq |p|_{row}$, $1 \leq j \leq j' \leq |p|_{col}$, also specified by the pair $[(i, j), (i', j')]$. The portion of p corresponding to positions in subdomain $[(i, j), (i', j')]$ is denoted by $p[(i, j), (i', j')]$. Then a non-empty picture x is *subpicture of p* if $x = p[(i, j), (i', j')]$, for some $1 \leq i \leq i' \leq |p|_{row}$, $1 \leq j \leq j' \leq |p|_{col}$.

Definition 1. *Given pictures* $x, p \in \Sigma^{++}$, *with* $|x|_{row} \leq |p|_{row}$ *and* $|x|_{col} \leq |p|_{col}$, *picture x is a* prefix *of p, denoted by* $x \trianglelefteq p$, *if x is the subpicture of p of size* $size(x)$ *corresponding to the top-left portion of p, i.e.* $x = p[(1, 1), (|x|_{row}, |x|_{col})]$.

Two concatenation products are usually considered. Let $p, q \in \Sigma^{**}$ be pictures of size (m, n) and (m', n'), respectively, the *column concatenation* of p and q ($p \oslash q$) and the *row concatenation* of p and q ($p \ominus q$) are partial operations, defined only if $m = m'$ and if $n = n'$, respectively, as:

These row and column concatenations can be extended to languages and to define *row* and *column stars* denoted by $X^{\ominus*}$ and $X^{\oplus*}$, respectively [13]. Another star operation for picture languages, introduced by D. Simplot in [20], is the tiling star. The idea is to compose pictures in a way to cover a rectangular area, as for example in the following figure.

Definition 2. *The* tiling star *of X, denoted by X^{**}, is the set that contains all the empty pictures together with the non-empty pictures p whose domain can be partitioned in disjoint subdomains $\{d_1, d_2, \ldots, d_k\}$ such that any subpicture p_h of p associated with the subdomain d_h belongs to X, for all $h = 1, \ldots, k$.*

The language X^{**} is called the set of all tilings by X in [20]. Denote X^{++} the set X^{**} without the empty pictures. In the sequel, if $p \in X^{++}$, we say that p is *tilable* in X while the partition $t = \{d_1, d_2, \ldots, d_k\}$ of $dom(p)$, together with the corresponding pictures $\{p_1, p_2, \ldots, p_k\}$, is called a *tiling decomposition* of p in X.

2.2   One-Dimensional Codes and Measure

A set of strings S over an alphabet Σ is a *code* if every word $w \in \Sigma^*$ can be obtained in at most one way as concatenation of strings in S. A set of strings S is a *prefix set* if any string in S is not a prefix of another one in S. It holds that any prefix set of non-empty strings is a code. The construction of prefix codes exploits the one-to-one correspondence between prefix codes and trees and it is referred to as the *literal representation* of prefix codes.

Important properties of string codes are connected to the notion of measure of a language. We give below some definitions and list some major results (see [10]). Intuitively the results state that a set S is not a code only if there are "too many too short words".

Given an alphabet Σ, a *probability distribution* π on Σ is a map $\pi : \Sigma \to [0, 1]$ such that $\sum_{a \in \Sigma} \pi(a) = 1$. As a particular case, the *uniform probability distribution* is defined on Σ by $\pi(a) = \frac{1}{Card(\Sigma)}$ for any $a \in \Sigma$.

Given an alphabet Σ and a probability distribution $\pi : \Sigma \to [0, 1]$ on Σ, for any $s = a_1 \ldots a_n \in \Sigma^*$ the *probability* of s is $\pi(s) = \Pi_{i=1}^{n} \pi(a_i)$. Given a probability distribution π on an alphabet Σ and $S \subseteq \Sigma^*$, the *measure* of S relative to π, denoted $\mu_\pi(S)$, is $\mu_\pi(S) = \sum_{s \in S} \pi(s)$.

Theorem 3. *If $S \subseteq \Sigma^*$ is a code, then $\mu_\pi(S) \leq 1$ for any probability distribution π on Σ.*

Theorem 4. *[Kraft-McMillan] Given a sequence $(u_n)_{n \geq 1}$ of integers, there exists a code S over an alphabet Σ of k symbols such that $u_n = Card(S \cap \Sigma^n)$ if and only if $\sum_{n \geq 1} u_n k^{-|n|} \leq 1$. Moreover, the code S can be chosen to be prefix.*

Theorem 5. *A finite code $S \subseteq \Sigma^*$ is a maximal code if and only if $\mu_\pi(S) = 1$, for any probability distribution π on Σ.*

2.3   Two-Dimensional Codes

We directly refer to the results in [4] where two-dimensional codes are introduced in the setting of the theory of recognizable two-dimensional languages and coherently to the notion of language unambiguity as in [2,6,7]. The following definition is based on the operation of tiling star recalled in Definition 2.

Definition 6. *Let Σ be a finite alphabet. $X \subseteq \Sigma^{++}$ is a code if any $p \in \Sigma^{++}$ has at most one tiling decomposition in X.*

Example 7. Let $\Sigma = \{a, b\}$ be the alphabet and let $X = \left\{ \boxed{a\ b},\ \boxed{\begin{smallmatrix} a \\ b \end{smallmatrix}},\ \boxed{\begin{smallmatrix} a\ a \\ a\ a \end{smallmatrix}} \right\}$.
The set X is a code. Any picture $p \in X^{++}$ can be univocally decomposed in X starting at the tl-corner, checking the subpicture $p[(1,1), (2,2)]$ and then proceeding similarly for the next contiguous subpictures of size $(2,2)$.

In this paper we consider the smaller family of *strong prefix* codes introduced in [3]. To get in the formal definition, we specialize the definition of "picture p prefix of picture q" when p and q have the same number of rows (columns, resp.).

Definition 8. *Let $p, q \in \Sigma^{++}$. The picture p is a horizontal prefix of q, denoted by $p \trianglelefteq_h q$, if there exists $x \in \Sigma^{**}$ such that $q = p \oslash x$. The picture p is a vertical prefix of q, denoted by $p \trianglelefteq_v q$, if there exists $y \in \Sigma^{**}$ such that $q = p \ominus y$.*

Definition 9. *Let $p, q \in \Sigma^{++}$. Pictures p and q overlap if either $p \trianglelefteq q$ or there exists a picture $x \in \Sigma^{++}$ such that $x \trianglelefteq_h p$ and $x \trianglelefteq_v q$. Pictures p and q strictly overlap if there exists a picture $x \in \Sigma^{++}$, $x \neq p, q$, such that $x \trianglelefteq_h p$ and $x \trianglelefteq_v q$.*

For example, in the following figure, the pictures p and q strictly overlap:

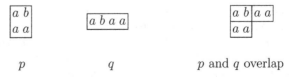

p	q	p and q overlap

Definition 10. *Let $X \subseteq \Sigma^{++}$. X is strong prefix if there are no two pictures $p, q \in X$, with $p \neq q$, such that p and q overlap.*

Remark that the previous Definition 10 can be stated by referring directly to picture domains by imposing that any two distinct pictures in X differ in the common part of the domain, i.e. they do not overlap if we let their tl-corners coincide. Let us give some examples.

Example 11 (Running Example). The following language X_{run} is strong prefix. Indeed, no two pictures in X_{run} overlap.

$$X_{run} = \left\{ \boxed{a\,b\,a}, \boxed{a\,b\,b}, \begin{array}{|c|}\hline b \\\hline b \\\hline\end{array}, \begin{array}{|cc|}\hline a & a \\ a & a \\\hline\end{array}, \begin{array}{|cc|}\hline a & a \\ a & b \\\hline\end{array}, \begin{array}{|cc|}\hline a & a \\ b & a \\\hline\end{array}, \begin{array}{|cc|}\hline a & a \\ b & b \\\hline\end{array}, \begin{array}{|cc|}\hline b & a \\ a & a \\\hline\end{array}, \begin{array}{|cc|}\hline b & a \\ a & b \\\hline\end{array}, \begin{array}{|cc|}\hline b & b \\ a & a \\\hline\end{array}, \begin{array}{|cc|}\hline b & b \\ a & b \\\hline\end{array} \right\}.$$

In [3], it is proved that strong prefix sets are codes and that they form a decidable family of picture codes. A strong prefix code $X \subseteq \Sigma^{++}$ is *maximal strong prefix* over Σ if it is not properly contained in any other strong prefix code over Σ; that is, $X \subseteq Y \subseteq \Sigma^{++}$ and Y strong prefix imply $X = Y$. Some results on maximality can be found in [3].

3   Structure of Finite Strong Prefix Codes

Strong prefix codes of pictures have many valuable properties that make them a valid counterpart of prefix codes of strings (see [3]). In this section we exhibit a recursive construction of all finite maximal strong prefix codes. We start by introducing the notion of "extension of a picture" and proving some properties of finite maximal strong prefix codes related to these extensions. Let us fix an order between pairs of integers as follows:

- $(m, n) \leq (m', n')$ if $m \leq m'$ and $n \leq n'$;
- $(m, n) < (m', n')$ if $(m, n) \leq (m', n')$ and $(m \neq m'$ or $n \neq n')$.

Definition 12. *Let Σ be an alphabet, $p \in \Sigma^{++}$, $m, n \geq 0$ be positive integers with $size(p) < (m, n)$. The set of extensions of p to size (m, n) is $E_{(m,n)}(p) = \{q \in \Sigma^{m,n} \mid q[(1,1), (|p|_{row}, |p|_{col})] = p\}$.*

Proposition 13. *Let $X \subseteq \Sigma^{++}$ be a finite maximal strong prefix code, $X \neq \Sigma^{1,1}$. Then there exist $p \in \Sigma^{++}$, $m, n \in \mathbb{N}$, with $size(p) < (m, n)$, such that $E_{(m,n)}(p) \subseteq X$.*

Proof. Let $X \subseteq \Sigma^{++}$ be a finite maximal strong prefix code and let r_X be the maximum number of rows of some picture in X, that is $r_X = max\{|x|_{row}, x \in X\}$. Consider a picture $\bar{x} \in X$ with a maximal number $\bar{c}$ of columns, among all pictures with r_X rows. Suppose that $|\bar{x}|_{row} \neq 1$ and $|\bar{x}|_{col} \neq 1$. The goal is to show that there exists a prefix p of $\bar{x}$ such that $E_{(r_X, \bar{c})}(p) \subseteq X$. Suppose that this is not the case. Consider the prefix $\bar{x}_r$ obtained by deleting the last row of $\bar{x}$. By contradiction there exists $t \in \Sigma^{**}$ such that $\bar{t}_r = \bar{x}_r \ominus t \notin X$. Furthermore the maximality of X implies that $X \cup \{\bar{t}_r\}$ is no longer a strong prefix code. From the maximality of the size of $\bar{t}_r$, neither $\bar{t}_r$ can be a prefix of another picture in X, nor $\bar{t}_r$ and another picture in X can strictly overlap. Therefore the unique possibility is that there exists $y \in X$ that is a prefix of $\bar{t}_r$; more precisely $y \leq_h \bar{t}_r$ (otherwise y would be a prefix of $\bar{x}$ too). In a dual way, considering the picture $\bar{x}_c$ obtained from $\bar{x}$ by deleting its last column, one can show that there exists $y' \in X$ such that $y' \trianglelefteq_v \bar{t}_r$. Then y and y' are two pictures in X that overlapp: this is a contradiction. The cases $|\bar{x}|_{row} = 1$ or $|\bar{x}|_{col} = 1$ can be similarly handled. □

Observe that the proof of Proposition 13 identifies a set $E_{(m,n)}(p)$ of pictures in X such that m is the maximum number of rows of a picture in X and n is the maximum number of columns of a picture with m rows in X. Note that some other sets $E_{(m',n')}(p')$, with $m' \leq m$ and $n' \leq n$, can be found as subsets of X. The idea in the next proposition is that, given a maximal strong prefix code X, we can always find a subset of the form $E_{(m,n)}(p)$ and *reduce* the set X by taking out pictures in $E_{(m,n)}(p)$ and replacing them by p, in such a way to keep a maximal strong prefix code.

Proposition 14. *Let $X \subseteq \Sigma^{++}$ be a finite maximal strong prefix code, $X \neq \Sigma^{1,1}$. Let $Y \subseteq X$ with $Y = E_{(m,n)}(p)$, for $p \in \Sigma^{++}$ and positive integers m, n, with $size(p) < (m,n)$. Then $X_{red} = (X \setminus Y) \cup \{p\}$ is a maximal strong prefix code.*

Proof. (Sketch) To prove that X_{red} is a maximal strong prefix code, we show that, for any picture $q \in \Sigma^{++}$, q and p overlap if and only if q and some picture in Y overlap. The proof considers all possible relation cases between p and q. □

Using Proposition 13 and 14 we give the following recursive characterization of finite maximal strong prefix codes. Remark that $\Sigma^{1,1}$ is a maximal strong prefix code.

Proposition 15. *Any finite maximal strong prefix code X on alphabet Σ either is equal to $\Sigma^{1,1}$ or can be constructed from $\Sigma^{1,1}$ by a finite number of subsequent replacements of $p \in X$ with $E_{m,n}(p)$, for $m, n \in \mathbb{N}$, with $size(p) < (m,n)$.*

Example 16. Consider the language of our running example:

$$X_{run} = \left\{ \boxed{a\,b\,a}, \boxed{a\,b\,b}, \boxed{\begin{matrix} b \\ b \end{matrix}}, \boxed{\begin{matrix} a\,a \\ a\,a \end{matrix}}, \boxed{\begin{matrix} a\,a \\ a\,b \end{matrix}}, \boxed{\begin{matrix} a\,a \\ b\,a \end{matrix}}, \boxed{\begin{matrix} a\,a \\ b\,b \end{matrix}}, \boxed{\begin{matrix} b\,a \\ a\,a \end{matrix}}, \boxed{\begin{matrix} b\,a \\ a\,b \end{matrix}}, \boxed{\begin{matrix} b\,b \\ a\,a \end{matrix}}, \boxed{\begin{matrix} b\,b \\ a\,b \end{matrix}} \right\}.$$

By Proposition 15 we get the following construction procedure.
- Start with $X_1 = \{\boxed{a}, \boxed{b}\}$.
- Replace picture $p_1 = \boxed{a}$ with $E_{1,2}(p_1)$, yielding $X_2 = \{\boxed{a\,a}, \boxed{a\,b}, \boxed{b}\}$.
- Replace $p_2 = \boxed{b}$ with $E_{2,1}(p_2)$, yielding $X_3 = \left\{ \boxed{a\,a}, \boxed{a\,b}, \boxed{\begin{matrix}b\\a\end{matrix}}, \boxed{\begin{matrix}b\\b\end{matrix}} \right\}$.
- Replace $p_3 = \boxed{a\,a}$ with $E_{2,2}(p_3)$, yielding

$$X_4 = \left\{ \boxed{\begin{matrix} a\,a \\ a\,a \end{matrix}}, \boxed{\begin{matrix} a\,a \\ a\,b \end{matrix}}, \boxed{\begin{matrix} a\,a \\ b\,a \end{matrix}}, \boxed{\begin{matrix} a\,a \\ b\,b \end{matrix}}, \boxed{a\,b}, \boxed{\begin{matrix}b\\a\end{matrix}}, \boxed{\begin{matrix}b\\b\end{matrix}} \right\}.$$

- Replace $p_4 = \boxed{a\,b}$ with $E_{1,3}(p_4)$, yielding

$$X_5 = \left\{ \boxed{\begin{matrix} a\,a \\ a\,a \end{matrix}}, \boxed{\begin{matrix} a\,a \\ a\,b \end{matrix}}, \boxed{\begin{matrix} a\,a \\ b\,a \end{matrix}}, \boxed{\begin{matrix} a\,a \\ b\,b \end{matrix}}, \boxed{a\,b\,a}, \boxed{a\,b\,b}, \boxed{\begin{matrix}b\\a\end{matrix}}, \boxed{\begin{matrix}b\\b\end{matrix}} \right\}.$$

- Finally replace $p_5 = \boxed{\begin{matrix}b\\a\end{matrix}}$, with $E_{2,2}(p_5)$, and obtain X_{run}.

It is interesting to represent the construction in the following tree. Note the similarities with the literal representation of a maximal prefix string code.

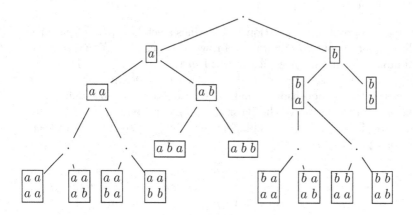

4   Measure of Two-Dimensional Languages and Codes

In the theory of string codes important results are stated in connection with the notion of measure of a language. We show that some of them hold also for strong prefix picture codes.

Definition 17. *Let Σ be an alphabet and π be a probability distribution on Σ. The* probability *of a picture $p \in \Sigma^{**}$ is defined as*

$$\pi(p) = \prod_{1 \leq i < m, 1 \leq j \leq n} \pi(p(i,j)).$$

The measure *of a language $X \subseteq \Sigma^{**}$ relative to π is $\mu_\pi(X) = \sum_{p \in X} \pi(p)$.*

Particular interest is devoted to the *uniform distribution*, that associates to every symbol a in alphabet Σ of cardinality k, the probability $\pi_u(a) = \frac{1}{k}$. Then the *uniform probability* of a picture $p \in \Sigma^{**}$ is $\pi_u(p) = \frac{1}{k^{area(p)}}$. The *uniform measure* of a language $X \subseteq \Sigma^{**}$, is $\mu_u(X) = \sum_{p \in X} \pi_u(p)$.

Example 18. Let $\Sigma = \{a, b\}$ and consider language $X = \left\{ \boxed{\begin{matrix} a & b \end{matrix}}, \boxed{\begin{matrix} a \\ b \end{matrix}}, \boxed{\begin{matrix} a & a \\ a & a \end{matrix}}, \boxed{\begin{matrix} a & a \\ a & b \end{matrix}} \right\}$ on Σ. Its uniform measure is $\mu_u(X) = 5/8 < 1$. Indeed for any probability distribution $\pi(a) = p$, $\pi(b) = 1 - p$, $0 < p < 1$, then $\mu_\pi(X) = p^3 - 2p^2 + 2p < 1$. Note that X is a code (see [4] for a proof).

Example 19. The uniform measure of the strong prefix code X_{run} on $\Sigma = \{a, b\}$, as defined in Example 11, is $\mu_u(X_{run}) = 1$. Moreover, by some calculations, one can show that the measure of X_{run} is 1, for any probability distribution on Σ.

Example 20. Let $X = X_{run} \cup \{p\}$ where $p =$
$$\begin{vmatrix} b & b & a & b & b & a & a & a & a \\ b & b & b & b & b & a & a & a & a \\ a & b & b & a & b & a & a & a & a \\ b & b & b & b & b & a & a & a & a \\ a & a & a & a & a & a & a & a & a \\ a & a & a & a & a & a & a & a & a \\ b & a & a & a & a & b & a & a & a \end{vmatrix}.$$ For any probability distribution π on $\Sigma = \{a, b\}$, the relative measure of X is $\mu_\pi(X) = \mu_\pi(X_{run}) + \pi(p) > 1$ since $\mu_\pi(X_{run}) = 1$, as shown in Example 19. Note that X is a code, as can be argued from [4].

The previous example shows that, contrarily to the one dimensional case, there exist finite picture codes of measure greater than 1. Hence the results on the measure of string codes cannot be transferred to the whole class of picture codes. Our goal is now to show that they hold for the subfamily of finite strong prefix codes. The proof will be based on the characterization of maximal strong prefix codes.

Proposition 21. *Let $X \subseteq \Sigma^{++}$ be a finite maximal strong prefix code and π be a probability distribution on Σ. Then $\mu_\pi(X) = 1$.*

Proof. Let π be a probability distribution on Σ, and μ simply denote the measure relative to π. The proof is by induction on $Card(X)$. The basis is the case $X = \Sigma^{1,1}$ and $Card(X) = 2$, where $\mu(X) = 1$. Suppose now that any maximal strong prefix code with cardinality strictly less than n has measure 1, and prove it for X with $Card(X) = n$. As remarked in the Proposition 13 there exists $p \in \Sigma^{++}$ and positive integers m, n, with $size(p) < (m, n)$, such that $Y = E_{(m,n)}(p) \subseteq X$, and $X_{red} = (X \setminus Y) \cup \{p\}$ is a maximal strong prefix code. Note that $Card(X_{red}) < Card(X)$; thus by the inductive hypothesis $\mu(X_{red}) = 1$. The proof is completed by $\mu(X_{red}) = \mu(X)$. Indeed $\mu(X_{red}) = \mu(X) - \mu(Y) + \pi(p)$ and $\pi(p) = \sum_{q \in E_{(m,n)}(p)} \pi(q) = \mu(Y)$. □

Next theorem completes the previous proposition and it is the analogous of Theorems 3 and 5 for strong prefix picture codes.

Theorem 22. *Let $X \subseteq \Sigma^{++}$ be a finite strong prefix code and μ be a measure. Then $\mu(X) \leq 1$. Moreover $\mu(X) = 1$ if and only if X is a finite maximal strong prefix code.*

Proof. Let $X \subseteq \Sigma^{++}$ be a finite strong prefix code. Let us show that if X is not maximal then $\mu(X) < 1$. This claim together with Proposition 21 proves the (whole) statement. If X is not maximal as strong prefix code, then, as shown in [3], there exists a finite language Y such that Y is maximal strong prefix and $X \subsetneq Y$. Since, for any $x \in \Sigma^{++}$, $\pi(x) > 0$ it follows $\mu(X) < \mu(Y)$. Moreover, from Proposition 21 we have $\mu(Y) = 1$ and, therefore, $\mu(X) < \mu(Y) = 1$. □

The previous theorem provides in particular a simple algorithm to test whether a finite strong prefix set of pictures is a maximal strong prefix code, in a different way from [3]. We conclude the section by showing an analogous to the Kraft-McMillan inequality (Theorem 4) for finite strong prefix picture codes.

Theorem 23. *Given a finite sequence* $(u_n)_{1 \leq n \leq r}$ *of positive integers, there exists a strong prefix code X over an alphabet Σ of k symbols, such that* $u_n = Card(\{p \in X \mid area(p) = n\})$ *if and only if* $\sum_{1 \leq n \leq r} u_n k^{-n} \leq 1$.

Proof. Let X be a finite strong prefix code over an alphabet Σ of k symbols, and denote by u_n its area distribution: $u_n = Card(\{p \in X \mid area(p) = n\})$. One can easily observe that $\sum_{1 \leq n \leq r} u_n k^{-n} = \mu_u(X)$ and hence $\sum_{1 \leq n \leq r} u_n k^{-n} \leq 1$, applying Theorem 22 to X, in the particular case of the uniform measure. Vice versa, given a finite sequence $(u_n)_{1 \leq n \leq r}$ of positive integers, such that $\sum_{1 \leq n \leq r} u_n k^{-n} \leq 1$, the Kraft-McMillan inequality (Theorem 4) states that there exists a finite prefix code of strings S over an alphabet Σ of k symbols, such that $u_n = Card(\{w \in S \mid |w| = n\})$. Replacing each string in $S \subseteq \Sigma^+$ by its corresponding one-row picture, one can obtain a picture language $X \subseteq \Sigma^{**}$, and the length distribution of S coincides with the area distribution of X. Moreover X is a strong prefix code. $\qquad\square$

5   A Correspondence with Prefix String Codes

Strong prefix codes of pictures behave as prefix string codes from many points of view including their measure. This section aims to enucleate a special relation between strong prefix codes and prefix string codes, that underlies such connection. A simple way to associate a string language with a picture language is to consider the symbols sequences when reading each picture. One can read a picture row by row or, alternatively, column by column. Indeed many other scanning strategies can be used. Let us start with some examples.

Example 24. Let $\Sigma = \{a, b\}$ and let $X \subseteq \Sigma^{3*}$ be the following strong prefix code:

$$X = \left\{ \begin{array}{|cc|} \hline a & a \\ b & b \\ b & b \\ \hline \end{array}, \begin{array}{|ccc|} \hline a & b & a \\ a & b & a \\ b & b & b \\ \hline \end{array}, \begin{array}{|cccc|} \hline a & b & a & b \\ a & b & a & a \\ b & b & a & b \\ \hline \end{array}, \begin{array}{|cccc|} \hline b & a & b & b \\ a & a & b & b \\ b & a & a & b \\ \hline \end{array} \right\}.$$ If we scan the pictures in X column by

column, we obtain the following string language on Σ:
$L_c(X) = \{abbabb, aabbbbaab, aabbbbaaabab, babaaabbabbb\}$. It is easy to check that $L_c(X)$ is a prefix code (of strings). Notice that also by scanning X row by row the resulting language $L_r(X)$ is prefix.

It is possible to verify that for any strong prefix code of the form $X \subseteq \Sigma^{m*}$, with $m > 0$, the column by column scanning strategy yields a prefix code of strings (similarly for $X \subseteq \Sigma^{*n}$ and the row by row scanning strategy). The next example takes back our running language X_{run}.

Example 25. Let X_{run} be as in Example 11. Both languages $L_r(X_{run})$ and $L_c(X_{run})$ (obtained reading the pictures by rows or by columns, respectively) are not prefix codes. In fact in $L_r(X_{run})$ there is bb that is a prefix of $bbaa$, while in $L_c(X_{run})$ there is aba that is a prefix of $abaa$. However if we scan the pictures of X_{run} in a different (more complex) way, we obtain a prefix code. Saying differently, symbols of strings in $L_r(X_{run})$ can be commuted to obtain the following prefix code
$P_{run} = \{aba, abb, bb, aaaa, aaab, aaba, aabb, baaa, baab, baba, babb\}$.

In the rest of the section we prove that, what is shown for X_{run} holds in general for all finite strong prefix codes. Given a finite strong prefix code X, there exists a particular scanning of its pictures (not necessarily the row-by-row, or column-by-column one) that yields a prefix string code. It can be obtained following the construction of X by replacements described in Section 3. The notations in the next definitions refer directly to the ones in Propositions 13, 14, 15.

Definition 26. *Let $X \subseteq \Sigma^{++}$ be a finite maximal strong prefix code. The* stringing *of X is the language $P(X) \subseteq \Sigma^+$ recursively defined as follows:*

- *if $X = \Sigma^{1,1}$ then $P(X) = \Sigma$*
- *if $X = (X_{red} \setminus \{p\}) \cup E_{(m,n)}(p)$ then $P(X) = (P(X_{red}) \setminus \{y_p\}) \cup y_p\Sigma^k$, where $y_p \in P(X_{red})$ is the string corresponding to p in $P(X_{red})$ and $k = m \times n - size(p)$.*

Example 27. Let X_{run} be the running example. Referring to the proof of Proposition 14, p, $Y = E_{(m,n)}(p)$, and X_{red} are respectively: p_5, $Y = E_{(2,2)}(p_5)$, and X_5, as in Example 16. Then $P(X_{red})$ can be recursively obtained as $P(X_{red}) = \{aaaa, aaab, aaba, aabb, aba, abb, ba, bb\}$ and $P(X_{run}) = (P(X_{red}) \setminus \{ba\}) \cup ba\Sigma^2$, is: $P(X_{run}) = \{aaaa, aaab, aaba, aabb, aba, abb, baaa, baab, baba, babb, bb\}$. Notice that language $P(X_{run})$ is a maximal prefix code. Its corresponding literal representation is given below in order to point out the resemblance of such tree with the tree representing the construction of X_{run} in Example 16. Strings in brackets in some nodes of the tree, denote strings that do not correspond to intermediate steps in the recursive construction of X_{run}.

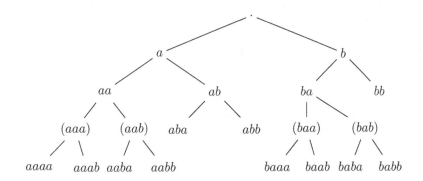

Proposition 28. *Let $X \subseteq \Sigma^{++}$ be a finite maximal strong prefix code of pictures. Then the stringing $P(X)$ of X is a maximal prefix code of strings.*

Proof. The proof is by induction on the cardinality of X. The basis is that $\{a, b\}$ is a prefix code of strings. Then suppose the statement holds for any cardinality less than $|X|$. Let p, $Y = E_{(m,n)}(p) \subseteq X$, and X_{red} as claimed in Proposition 14. X_{red} is a a maximal strong prefix code and $|X_{red}| < |X|$. Then, by the inductive hypothesis, $P(X_{red})$ is a maximal prefix code. Let $y_p \in P(X_{red})$ the string corresponding to p in $P(X_{red})$. Then, $P(X) = P(X_{red}) \setminus \{y_p\} \cup y_p \Sigma^k$, with $k = m \times n - size(p)$, is a prefix code. Moreover $P(X)$ is maximal prefix (see [10] for results on string codes).   □

Remark that the property stated in Proposition 28 for any finite strong prefix code does not hold for general picture codes. In the theory of string codes there are indeed examples of codes whose strings cannot be rearranged in any way to give a prefix code [10]. Further any string code $S \subseteq \Sigma^+$ can be considered as a picture code, where any string is viewed as one-row picture.

In [3] it is proved that any strong prefix code X can be embedded in a maximal one: this guarantees that Proposition 28 holds also by removing the maximality hypothesis for X. Then the stringing $P(X)$ can be defined also for (non-maximal) strong prefix codes. Note that in this case, $P(X)$ will be a prefix code, but not a maximal one.

Corollary 29. *Let X be a finite strong prefix code. Then its stringing $P(X)$ is a prefix code and X is maximal strong prefix if and only if $P(X)$ is a maximal prefix code.*

The correspondence between strong prefix codes of pictures and prefix codes of strings sheds light on the results on the measure, presented in Section 4. Remark that the measure of any finite strong prefix code X, is equal to the measure of its corresponding prefix code $P(X)$. That is actually the intrinsic reason for which results on string codes that involve the measure can be generalized also to strong prefix codes.

As conclusion of the paper, we mention a conjecture on string codes, still open since the 80's and referred to as "the commutative equivalence conjecture", that is related to other famous conjectures in the theory of codes (see [10]). Given two (string) languages S_1 and S_2, we say that they are commutatively equivalent if S_1 can be obtained by rearranging the order of the symbols in each string of S_2. The mentioned conjecture claims that every finite maximal code is commutatively equivalent to a finite maximal prefix code. Researchers, with the aim of solving the conjecture, have detected families of commutatively prefix codes [16]. The stringing of a strong prefix picture code X can be regarded to as a way to rearrange the symbols of each picture and obtain a prefix string code. Therefore the results in this section have shown that also finite strong prefix codes of pictures are a family of codes that are in some sense commutatively prefix.

References

1. Aigrain, P., Beauquier, D.: Polyomino tilings, cellular automata and codicity. Theoretical Computer Science **147**, 165–180 (1995)
2. Anselmo, M., Giammarresi, D., Madonia, M.: Deterministic and unambiguous families within recognizable two-dimensional languages. Fund. Inform **98**(2–3), 143–166 (2010)

3. Anselmo, M., Giammarresi, D., Madonia, M.: Strong Prefix Codes of Pictures. In: Muntean, T., Poulakis, D., Rolland, R. (eds.) CAI 2013. LNCS, vol. 8080, pp. 47–59. Springer, Heidelberg (2013)

4. Anselmo, M., Giammarresi, D., Madonia, M.: Two Dimensional Prefix Codes of Pictures. In: Béal, M.-P., Carton, O. (eds.) DLT 2013. LNCS, vol. 7907, pp. 46–57. Springer, Heidelberg (2013)

5. Anselmo, M., Giammarresi, D., Madonia, M.: Prefix picture codes: a decidable class of two-dimensional codes. Int. Jou. of Foundations of Computer Science (to appear, 2015)

6. Anselmo, M., Giammarresi, D., Madonia, M., Restivo, A.: Unambiguous recognizable two-dimensional languages. RAIRO -ITA **40**(2), 227–294 (2006)

7. Anselmo, M., Madonia, M.: Deterministic and unambiguous two-dimensional languages over one-letter alphabet. Theoretical Computer Science **410**(16), 1477–1485 (2009)

8. Anselmo, M., Jonoska, N., Madonia, M.: Framed Versus Unframed Two-Dimensional Languages. In: Nielsen, M., Kučera, A., Miltersen, P.B., Palamidessi, C., Tůma, P., Valencia, F. (eds.) SOFSEM 2009. LNCS, vol. 5404, pp. 79–92. Springer, Heidelberg (2009)

9. Beauquier, D., Nivat, M.: A codicity undecidable problem in the plane. Theoret. Comp. Sci. **303**, 417–430 (2003)

10. Berstel, J., Perrin, D., Reutenauer, C.: Codes and Automata. Cambridge University Press (2009)

11. Blum, M., Hewitt, C.: Automata on a 2-dimensional tape. In: SWAT (FOCS). pp. 155–160 (1967)

12. Bozapalidis, S., Grammatikopoulou, A.: Picture codes. RAIRO - ITA **40**(4), 537–550 (2006)

13. Giammarresi, D., Restivo, A.: Two-dimensional languages. In: Rozenberg, G., (ed.) Handbook of Formal Languages, Vol. III, pp. 215–268. Springer (1997)

14. Kari, J., Salo, V.: A Survey on Picture-Walking Automata. In: Kuich, W., Rahonis, G. (eds.) Algebraic Foundations in Computer Science. LNCS, vol. 7020, pp. 183–213. Springer, Heidelberg (2011)

15. Kolarz, M., Moczurad, W.: Multiset, Set and Numerically Decipherable Codes over Directed Figures. In: Smyth, B. (ed.) IWOCA 2012. LNCS, vol. 7643, pp. 224–235. Springer, Heidelberg (2012)

16. Mauceri, S., Restivo, A.: A family of codes commutatively equivalent to prefix codes. Inf. Process. Lett. **12**(1), 1–4 (1981)

17. Otto, F., Mráz, F.: Extended Two-Way Ordered Restarting Automata for Picture Languages. In: Dediu, A.-H., Martín-Vide, C., Sierra-Rodríguez, J.-L., Truthe, B. (eds.) LATA 2014. LNCS, vol. 8370, pp. 541–552. Springer, Heidelberg (2014)

18. Pradella, M., Cherubini, A., Crespi-Reghizzi, S.: A unifying approach to picture grammars. Inf. Comput. **209**(9), 1246–1267 (2011)

19. Průša, D., Mráz, F., Otto, F.: New Results on Deterministic Sgraffito Automata. In: Béal, M.-P., Carton, O. (eds.) DLT 2013. LNCS, vol. 7907, pp. 409–419. Springer, Heidelberg (2013)

20. Simplot, D.: A characterization of recognizable picture languages by tilings by finite sets. Theoretical Computer Science **218**(2), 297–323 (1991)

On Torsion-Free Semigroups Generated by Invertible Reversible Mealy Automata

Thibault Godin, Ines Klimann, and Matthieu Picantin[✉]

Université Paris Diderot, Sorbonne Paris Cité, LIAFA, UMR 7089 CNRS,
Paris, France
{godin,klimann,picantin}@liafa.univ-paris-diderot.fr

Abstract. This paper addresses the torsion problem for a class of automaton semigroups, defined as semigroups of transformations induced by Mealy automata, aka letter-by-letter transducers with the same input and output alphabet. The torsion problem is undecidable for automaton semigroups in general, but is known to be solvable within the well-studied class of (semi)groups generated by invertible bounded Mealy automata. We focus on the somehow antipodal class of invertible reversible Mealy automata and prove that for a wide subclass the generated semigroup is torsion-free.

Keywords: Automaton semigroup · Reversible mealy automaton · Labeled orbit tree · Torsion-free semigroup

1 Introduction

In this paper we address the torsion problem for a class of automaton semigroups.

In a (semi)group, a *torsion*—or *periodic*—element is an element of finite order, that is an element generating a finite monogenic sub(semi)group. In particular, a (semi)group is *torsion-free (resp. torsion)* if its only torsion element is its possible identity element (*resp.* if all its elements are torsion elements). Like most of the major group or semigroup theoretical decision problems, the word, torsion and finiteness problems are undecidable in general [8].

Automaton (semi)groups, that is (semi)groups generated by Mealy automata, were formally introduced a half century ago (for details, see [9] and references therein). Two decades later, important results started revealing their full potential. In particular, contributing to the so-called Burnside problem, the articles [2,15] construct particularly simple Mealy automata generating infinite finitely generated torsion groups, and, answering the so-called Milnor problem, the articles [6,16] describe Mealy automata generating the first examples of (semi)groups with intermediate growth. Since these pioneering works, a substantial theory continues to develop using various methods, ranging from finite

M. Picantin—The authors are partially supported by the French *Agence Nationale pour la Recherche*, through the Project **MealyM** ANR-JCJC-12-JS02-012-01.

A.-H. Dediu et al. (Eds.): LATA 2015, LNCS 8977, pp. 328–339, 2015.
DOI: 10.1007/978-3-319-15579-1_25

automata theory to geometric group theory and never ceases to show that automaton (semi)groups possess multiple interesting and sometimes unusual features.

For automaton (semi)groups, the word problem is solvable using standard minimization techniques [7,12,19]. The torsion problem and the finiteness problem for automaton semigroups have been proven to be undecidable [14] but remain open for automaton groups. However there exist various criteria for recognizing whether such a (semi)group or one of its element has finite order, see for instance [1,3,4,9–11,18,20,22,23,25,27,28]. In particular, there are many partial methods to find elements of infinite order in such (semi)groups. Their efficiency may vary significantly. By contrast, the class of so-called invertible bounded Mealy automata, which has received considerable attention, admits an effective solution to both problems of torsion and finiteness [5,10,27]. This class happens to correspond to some tight restriction on the underlying automata: the non-trivial cycles are disjoint and none can be reached from another.

Here we tackle the torsion problem, focusing on a very different class of Mealy automata, namely reversible Mealy automata, in which each connected component turns out to be strongly connected. This class was known as the class for which most of the existing partial methods do not work or perform poorly. We prove that for a wide subclass of invertible reversible Mealy automata—roughly the non-bireversible ones—the generated semigroup is torsion-free. It is worth mentioning that the class of bounded Mealy automata and the class of reversible Mealy automata are somehow at the opposite ends of the spectrum.

The proof of torsion-freeness relies on deep structural properties of the so-called *labeled orbit tree* which happens to capture the behavior of the (strongly) connected components during the exponentiation of a reversible Mealy automaton, and it gives hopefully a new insight even in the still mysterious subclass of bireversible Mealy automata (see [7,20,24] and the references therein).

The paper is organized as follows. In Section 2, we set up notation, provide well-known definitions and facts concerning Mealy automata and automaton semigroups. Some results concerning connected components of reversible Mealy automata are given in Section 3. In Section 4 we introduce a crucial construction, namely the labeled orbit tree of a Mealy automaton, and define the notion of a self-liftable path, especially relevant for investigating torsion-freeness. Finally, Section 5 contains the proof of our main result.

2 Mealy Automata

We first recall the formal definition of an automaton. A *(finite, deterministic, and complete) automaton* is a triple $\big(Q, \Sigma, \delta = (\delta_i \colon Q \to Q)_{i \in \Sigma}\big)$, where the *stateset* Q and the *alphabet* Σ are non-empty finite sets, and where the δ_i are functions.

A *Mealy automaton* is a quadruple $\big(Q, \Sigma, \delta = (\delta_i \colon Q \to Q)_{i \in \Sigma}, \rho = (\rho_x \colon \Sigma \to \Sigma)_{x \in Q}\big)$, such that both (Q, Σ, δ) and (Σ, Q, ρ) are automata. In other terms, a Mealy automaton is a complete, deterministic, letter-to-letter transducer with the same input and output alphabet.

The graphical representation of a Mealy automaton is standard, see Figures 1 and 2.

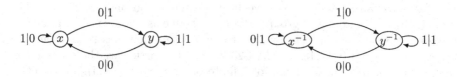

Fig. 1. An invertible reversible non-bireversible Mealy automaton $\mathcal{L}$ (left) and its inverse $\mathcal{L}^{-1}$ (right), both generating the lamplighter group $\mathbb{Z}_2 \wr \mathbb{Z}$ (see [17])

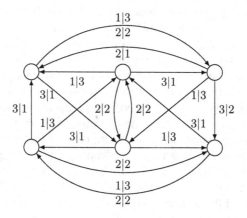

Fig. 2. A 3-letter 6-state inv. reversible non-bireversible Mealy automaton $\mathcal{J}$

In a Mealy automaton $\mathcal{A} = (Q, \Sigma, \delta, \rho)$, the sets Q and Σ play dual roles. So we may consider the *dual (Mealy) automaton* defined by $\mathfrak{d}(\mathcal{A}) = (\Sigma, Q, \rho, \delta)$. Alternatively, we can define the dual Mealy automaton via the set of its transitions:

$$x \xrightarrow{i|j} y \in \mathcal{A} \quad \Longleftrightarrow \quad i \xrightarrow{x|y} j \in \mathfrak{d}(\mathcal{A}).$$

Definition 1. *A Mealy automaton $(Q, \Sigma, \delta, \rho)$ is said to be* invertible *if the functions $(\rho_x)_{x \in Q}$ are permutations of Σ and* reversible *if the functions $(\delta_i)_{i \in \Sigma}$ are permutations of Q.*

Consider a Mealy automaton $\mathcal{A} = (Q, \Sigma, \delta, \rho)$. Let $Q^{-1} = \{x^{-1}, x \in Q\}$ be a disjoint copy of Q. The *inverse* $\mathcal{A}^{-1}$ of $\mathcal{A}$ is defined by the set of its transitions:

$$x \xrightarrow{i|j} y \in \mathcal{A} \quad \Longleftrightarrow \quad x^{-1} \xrightarrow{j|i} y^{-1} \in \mathcal{A}^{-1}.$$

If $\mathcal{A}$ is invertible, then its inverse $\mathcal{A}^{-1}$ is a Mealy automaton, see for instance Figure 1.

Definition 2. *A Mealy automaton is* bireversible *if it is invertible, reversible and its inverse is reversible.*

The terms "invertible", "reversible", and "bireversible" are standard since [21]. Figure 3 gives characterizations of invertibility and reversibility in terms of forbidden configurations in a Mealy automaton.
Here we define a new class:

Definition 3. *A Mealy automaton is* coreversible *whenever Configuration* (c) *in Figure 3 does not occur. This means that each output letter induces a permutation on the stateset.*

The bireversible Mealy automata are those which are simultaneously invertible, reversible, and coreversible. We emphasize that an invertible reversible Mealy automaton is bireversible if and only if it is coreversible.

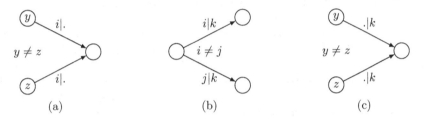

$$\text{(a)} \qquad\qquad\qquad \text{(b)} \qquad\qquad\qquad \text{(c)}$$

Fig. 3. Configuration (a) is forbidden for reversible automata, Configuration (b) for invertible ones, and Configuration (c) for coreversible ones

We view $\mathcal{A} = (Q, \Sigma, \delta, \rho)$ as an automaton with an input and an output tape, thus defining mappings from input words over Σ to output words over Σ. Formally, for $x \in Q$, the map $\rho_x \colon \Sigma^* \to \Sigma^*$, extending $\rho_x \colon \Sigma \to \Sigma$, is defined recursively by:

$$\forall i \in \Sigma, \ \forall \mathbf{s} \in \Sigma^*, \qquad \rho_x(i\mathbf{s}) = \rho_x(i)\rho_{\delta_i(x)}(\mathbf{s}) . \qquad (1)$$

Equation (1) can be easier to understood if depicted by a *cross-diagram* (see [1]):

$$
\begin{array}{ccc}
 & i & \mathbf{s} \\
x \ \longmapsto\!\!+ & \delta_i(x) & \longmapsto\!\!+ \ \ \delta_\mathbf{s}(\delta_i(x)) \\
 \rho_x(i) & \rho_{\delta_i(x)}(\mathbf{s})
\end{array}
$$

By convention, the image of the empty word is itself. The mapping ρ_x for each $x \in Q$ is length-preserving and prefix-preserving. We say that ρ_x is the *production function* associated with $(\mathcal{A}, x)$. For $\mathbf{x} = x_1 \cdots x_n \in Q^n$ with $n > 0$, set $\rho_\mathbf{x} \colon \Sigma^* \to \Sigma^*, \rho_\mathbf{x} = \rho_{x_n} \circ \cdots \circ \rho_{x_1}$. Denote dually by $\delta_i \colon Q^* \to Q^*, i \in \Sigma$, the production functions associated with the dual automaton $\mathfrak{d}(\mathcal{A})$. For $\mathbf{s} = s_1 \cdots s_n \in \Sigma^n$ with $n > 0$, set $\delta_\mathbf{s} \colon Q^* \to Q^*, \ \delta_\mathbf{s} = \delta_{s_n} \circ \cdots \circ \delta_{s_1}$.

The semigroup of mappings from Σ^* to Σ^* generated by $\{\rho_x, x \in Q\}$ is called the *semigroup generated by* $\mathcal{A}$ and is denoted by $\langle \mathcal{A} \rangle_+$. When $\mathcal{A}$ is invertible, its production functions are permutations on words of the same length and thus we may consider the group of mappings from Σ^* to Σ^* generated by $\{\rho_x, x \in Q\}$. This group is called the *group generated by* $\mathcal{A}$ and is denoted by $\langle \mathcal{A} \rangle$.

It is know from [1] that the possible behaviors of invertible reversible non-bireversible Mealy automata provide less variety than those of bireversible automata whenever finiteness is concerned:

Proposition 4. ([1, Corollary 22]) *Any invertible reversible non-bireversible Mealy automaton generates an infinite group.*

Note that the ratio of these invertible reversible non-bireversible Mealy automata tends to supersede the bireversible one, when the size of alphabet and/or stateset increases.

3   On the Behavior of Connected Components

In this section, we gather some properties satisfied by the connected components of the underlying graph of a reversible Mealy automaton and we focus on those properties preserved when making products. We use the following crucial property: any connected component of a reversible Mealy automaton is strongly connected. Our main tool, described in the next section, captures the behavior of the connected components of the successive powers of a given reversible Mealy automaton, allowing a much finer analysis.

Definition 5. *Let $\mathcal{A} = (Q, \Sigma, \delta, \rho)$ and $\mathcal{B} = (Q', \Sigma, \delta', \rho')$ be two Mealy automata acting on the same alphabet. Their* product *is the Mealy automaton $\mathcal{A} \times \mathcal{B} = (Q \times Q', \Sigma, \gamma, \pi)$ with transition*

$$xy \xrightarrow{\;i \mid \rho'_y(\rho_x(i))\;} \delta_i(x)\delta'_{\rho_x(i)}(y) \, ,$$

which can be seen in terms of cross-diagram as:

Note that the product of two reversible (*resp.* invertible) Mealy automata is still a reversible (*resp.* invertible) Mealy automaton. Let us consider the coreversibility property.

Lemma 6. *Let A and B be Mealy automata on the same alphabet with A connected and reversible. Then, for any connected component C of $A \times B$, every state of A occurs as a prefix of some state of C.*

Proof. Let $A = (Q, \Sigma, \delta, \rho)$ and let C be a connected component of $A \times B$. Let $xx' \in C$ and $y \in Q$. Since A is connected and reversible, there exists $\mathbf{s} \in \Sigma^*$ satisfying $y = \delta_{\mathbf{s}}(x)$, hence y is a prefix of the state $\delta_{\mathbf{s}}(xx')$ in C.   □

Proposition 7. *Let A and B be reversible Mealy automata on the same alphabet. If A is connected and non-coreversible, then every connected component of $A \times B$ is reversible and non-coreversible.*

Proof. Let Q be the stateset of A and let C be a connected component of $A \times B$. As A and B are reversible, so is C.
Since A is not coreversible, there exist two states $x \neq y \in Q$ leading to the same state z, when producing the same letter j:

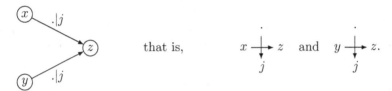

that is, $x \xrightarrow{\ \ } z$ and $y \xrightarrow{\ \ } z$.
$\quad\ \ j\qquad\qquad\ \ j$

By Lemma 6, C admits a state prefixed with x, say xx'. Let

$$x' \overset{j}{\underset{k}{\xrightarrow{\ \ }}} z'$$

be a transition of B, then the following configuration occurs in C:

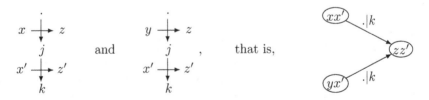

which means that C cannot be coreversible.   □

A convenient and natural operation is to raise a Mealy automaton to some power. The *n-th power* of $A = (Q, \Sigma, \delta, \rho)$ is recursively defined. By convention, A^0 is the trivial Mealy automaton with only one state, which acts like identity on Σ. For $n > 0$, A^n is the Mealy automaton

$$A^n = \left(Q^n, \Sigma, (\delta_i : Q^n \to Q^n)_{i \in \Sigma}, (\rho_{\mathbf{u}} : \Sigma \to \Sigma)_{\mathbf{u} \in Q^n} \right).$$

Corollary 8. *If a Mealy automaton is (invertible) reversible without coreversible connected component, then every connected component of any of its powers is (invertible) reversible and non-coreversible.*

Definition 9. *[17, 24] The action of a Mealy automaton $\mathcal{A}$ is said to be spherically transitive or level-transitive whenever all the powers of $\mathfrak{d}(\mathcal{A})$ are connected.*

4   The Labeled Orbit Tree

There exist strong links between the successive sizes of the connected components of the powers of a reversible Mealy automaton and some finiteness properties of the generated semigroup, as emphasized by the two following results. Such links can be captured by a suitable tree, playing a fundamental role in the sequel.

Proposition 10. *A reversible Mealy automaton generates a finite semigroup if and only if the sizes of the connected components of its powers are bounded.*

The latter is proven in [20] within the framework of invertible reversible Mealy automata, but the invertibility is not invoked in the proof. We need the following result, also from [20].

Proposition 11. *Let $\mathcal{A} = (Q, \Sigma, \delta, \rho)$ be an invertible reversible Mealy automaton. For any $\mathbf{u} \in Q^+$, the following are equivalent:*

(i) the action $\rho_{\mathbf{u}}$ induced by $\mathbf{u}$ has finite order;
(ii) the sizes of the connected components of $(\mathbf{u}^n)_{n \in \mathbb{N}}$ are bounded.

A direct consequence of Proposition 10 provides a simple yet interesting result concerning torsion-freeness.

Corollary 12. *Let $\mathcal{A}$ be a reversible Mealy automaton. Whenever the action of $\mathfrak{d}(\mathcal{A})$ is spherically transitive, the semigroup $\langle \mathcal{A} \rangle_+$ is torsion-free.*

Proof. Let $\mathcal{A}$ be a Mealy automaton with stateset Q such that all its powers are connected. By Proposition 10, $\mathcal{A}$ generates an infinite semigroup.
Assume that there exists $\mathbf{u} \in Q^+$ whose action has finite order, say $\rho_{\mathbf{u}^p} = \rho_{\mathbf{u}^q}$ with $p < q$. By reversibility of $\mathcal{A}$, every state of $\mathcal{A}^q$ is equivalent to some state of $\mathcal{A}^p$, hence $\mathcal{A}$ generates a finite semigroup, which is a contradiction.   □

Corollary 12 applies for instance to the Mealy automaton $\mathcal{L}$ on Figure 1(left): the subsemigroup of the lamplighter group generated by x and y is torsion-free.
    We are now ready to introduce our main tool.

Definition 13. *Let $\mathcal{A}$ be a reversible Mealy automaton with stateset Q. Rooted in $\mathcal{A}^0$, the labeled orbit tree $\mathfrak{t}(\mathcal{A})$ is constructed as the graph of the (strongly) connected components of the powers of $\mathcal{A}$, with an edge between two nodes $\mathcal{C}, \mathcal{D}$ whenever there is $\mathbf{u} \in \mathcal{C}$ with $\mathbf{u}x \in \mathcal{D}$ and $x \in Q$, such an edge being labeled by the (integer) ratio $|\mathcal{D}|/|\mathcal{C}|$.*

Such a tree $t(\mathcal{A})$ is more precisely named the labeled orbit tree of the dual $\mathfrak{d}(\mathcal{A})$ since it can be seen as the tree of the orbits of Q^* under the action of the group $\langle \mathfrak{d}(\mathcal{A}) \rangle$ (see [13, 20]).

Figure 4 displays the labeled orbit tree $t(\mathcal{J})$, where $\mathcal{J}$ is the Mealy automaton defined in Figure 2.

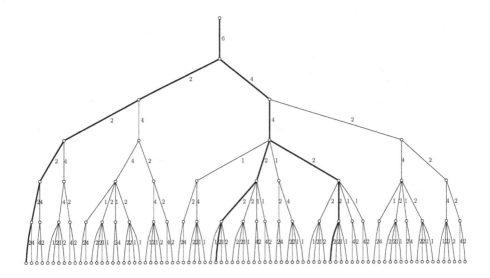

Fig. 4. The labeled orbit tree $t(\mathcal{J})$ (up to level 6) with $\mathcal{J}$ defined on Figure 2 (the thickened edges emphasize the 1-self-liftable paths defined below in Def. 17)

Since the orbit trees are rooted, we choose the classical orientation where the root is the higher vertex and the tree grows from top to bottom. A path is a (possibly infinite) sequence of adjacent edges without backtracking. The initial vertex of an edge e is denoted by $\top(e)$ and its terminal vertex by $\bot(e)$; by extension, the initial vertex of a non-empty path $\mathbf{e}$ is denoted by $\top(\mathbf{e})$ and its terminal vertex by $\bot(\mathbf{e})$ whenever the path is finite. The label of a (possibly infinite) path is the ordered sequence of labels of its edges.

Definition 14. *In a tree, a (possibly infinite) path $\mathbf{e}$ is said to be* initial *if $\top(\mathbf{e})$ is the root of the tree.*

Definition 15. *Let $\mathcal{A}$ be a reversible Mealy automaton and let e, f be two edges in the orbit tree $t(\mathcal{A})$. We say that e is* liftable *to f if each word of $\bot(e)$ admits some word of $\bot(f)$ as a suffix.*

One can notice that this condition is not as strong as it seems:

Lemma 16. *Let $\mathcal{A}$ be a reversible Mealy automaton and let e, f be two edges in the orbit tree $t(\mathcal{A})$. If there exists a word of $\bot(e)$ which admits a word of $\bot(f)$ as suffix, then e is liftable to f.*

Proof. Assume $\mathbf{uv} \in \perp(e)$ with $\mathbf{v} \in \perp(f)$. By reversibility, for any word $\mathbf{w}$ in the connected component $\perp(e)$, there exists $\mathbf{s} \in \Sigma^*$ satisfying $\mathbf{w} = \delta_{\mathbf{s}}(\mathbf{uv})$, which can also be written $\mathbf{w} = \delta_{\mathbf{s}}(\mathbf{u})\delta_{\mathbf{t}}(\mathbf{v})$ with $\mathbf{t} = \rho_{\mathbf{u}}(\mathbf{s})$. Hence the suffix $\delta_{\mathbf{t}}(\mathbf{v})$ of $\mathbf{w}$ belongs to the connected component $\perp(f)$ of $\mathbf{v}$.   □

Definition 17. *Let $\mathcal{A}$ be a reversible Mealy automaton and let e be a (possibly infinite) initial path in the orbit tree $\mathfrak{t}(\mathcal{A})$. We say that $\mathbf{e} = e_0 e_1 \cdots$ is 1-self-liftable whenever every edge e_{i+1} is liftable to its predecessor e_i, for $i \geq 0$.*

This important notion generalizes that of an *e-liftable path* used in [20] where the liftability is required with respect to a uniquely specified edge e.
Using thickened edges, Figure 4 highlights each of the 1-self-liftable paths in the orbit tree $\mathfrak{t}(\mathcal{J})$, where the Mealy automaton $\mathcal{J}$ is displayed on Figure 2.

Definition 18. *Let $\mathcal{A}$ be a reversible Mealy automaton with stateset Q. The path of a word $\mathbf{u} \in Q^* \cup Q^\omega$ is the unique initial path in $\mathfrak{t}(\mathcal{A})$ going from the root through the connected components of the prefixes of $\mathbf{u}$.*

Lemma 19. *Let $\mathcal{A}$ be a reversible Mealy automaton. For any state x of $\mathcal{A}$, the path of x^ω in $\mathfrak{t}(\mathcal{A})$ is 1-self-liftable.*

Proof. By Lemma 16, x^n being a suffix of x^{n+1}, such a path is 1-self-liftable.   □

Lemma 19 guarantees the existence of 1-self-liftable paths in any orbit tree.

5   Main Result

Assume that $\mathcal{A}$ is an invertible reversible Mealy automaton without bireversible component. Our aim is to prove that every element of $\langle \mathcal{A} \rangle_+$ has infinite order. We first prove this property for the states of $\mathcal{A}$, whenever $\mathcal{A}$ is connected, by looking at some 1-self-liftable paths in $\mathfrak{t}(\mathcal{A})$ (defined in Section 4). Then we extend it to arbitrary elements of $\langle \mathcal{A} \rangle_+$ by using the properties of products of Mealy automata (established in Section 3).

Proposition 20. *Let $\mathcal{A}$ be some connected invertible reversible non-bireversible Mealy automaton. A 1-self-liftable path in $\mathfrak{t}(\mathcal{A})$ cannot contain an edge labeled by 1.*

Proof. Let $\mathcal{A} = (Q, \Sigma, \delta, \rho)$, $\mathbf{e}$ be a 1-self-liftable path in $\mathfrak{t}(\mathcal{A})$ and e be an edge of $\mathbf{e}$. Let T (*resp.* L) denote the set of states of $\top(e)$ (*resp.* of $\perp(e)$). For any word $\mathbf{w}$, $L_{\mathbf{w}} = \{\mathbf{u} \mid \mathbf{wu} \in L\}$ is the *left quotient* of L by $\mathbf{w}$ (see for instance [26]). As $\mathcal{A}$ is connected and reversible, according to Lemma 6, for any $x \in Q$, the left quotient L_x is non-empty. Hence we have

$$L = \bigsqcup_{x \in Q} x L_x \quad \text{(disjoint union)}.$$

The hypotheses that the path $\mathbf{e}$ is 1-self-liftable and that $\mathcal{A}$ is invertible yield

$$T = \bigcup_{x \in Q} L_x.$$

Indeed, let $x\mathbf{u} \in L$ with $x \in Q$ (and $\mathbf{u} \in T$ by 1-self-liftability) and let $\mathbf{v} \in T$. By reversibility, there exists $\mathbf{s} \in \Sigma^*$ verifying $\delta_{\mathbf{s}}(\mathbf{u}) = \mathbf{v}$. Now, by invertibility, there exists $\mathbf{t} \in \Sigma^*$ with $\rho_x(\mathbf{t}) = \mathbf{s}$:

$$
\begin{array}{c}
\mathbf{t} \\
x \xrightarrow{} \delta_{\mathbf{t}}(x) = x' \\
\downarrow \mathbf{s} \\
\mathbf{u} \xrightarrow{} \delta_{\mathbf{s}}(\mathbf{u}) = \mathbf{v} \\
\rho_{\mathbf{u}}(\mathbf{s}).
\end{array}
$$

Therefore, $\mathbf{v}$ is a suffix of $\delta_{\mathbf{t}}(x\mathbf{u})$, hence $\mathbf{v} \in L_{x'}$ for $x' = \delta_{\mathbf{t}}(x) \in Q$. Since $\mathcal{A}$ is not coreversible, there exist $y \neq y', z \in Q$ and $i, j, k \in \Sigma$ satisfying

$$
\begin{array}{cc}
i & k \\
y \xrightarrow{} z \quad \text{and} \quad y' \xrightarrow{} z. \\
\downarrow j & \downarrow j
\end{array}
$$

So, in the connected component $\perp(e)$, we have

$$\delta_i(yL_y) = zL_z \quad \text{and} \quad \delta_k(y'L_{y'}) = zL_z.$$

From reversibility of $\mathcal{A}$, δ_j is injective and we deduce $L_y = L_{y'}$. Therefore the union $T = \bigcup_{x \in Q} L_x$ is not disjoint and we find $|T| < |L|$ which implies that the label of e is greater than 1. □

An easy but interesting first consequence is the following.

Corollary 21. *A connected 3-state invertible reversible non-bireversible Mealy automaton generates a free semigroup.*

Proof. We deduce from Proposition 20 that any connected 3-state invertible reversible non-bireversible Mealy automaton sees its dual to be spherically transitive and the result follows from [18, Proposition 14]. □

Let us go back to our main purpose.

Proposition 22. *Let $\mathcal{A}$ be some connected invertible reversible non-bireversible Mealy automaton. Then any state of $\mathcal{A}$ induces an action of infinite order.*

Proof. Let x be a state of $\mathcal{A}$. The path of x^ω is 1-self-liftable by Lemma 19. So by Proposition 20, this path has no edge labeled with 1, which means that the sizes of the connected components of $(x^n)_{n \in \mathbb{N}}$ are unbounded. By Proposition 11, the action induced by x has infinite order. □

We can now state our main result by extending Proposition 22:

Theorem 23. *Any invertible reversible Mealy automaton without bireversible component generates a torsion-free semigroup.*

Proof. Let $\mathcal{A}$ be an invertible reversible Mealy automaton without bireversible component with stateset Q. Let $\mathbf{u} \in Q^+$ and let $\mathcal{C}$ its connected component in $\mathfrak{t}(\mathcal{A})$. From Corollary 8, $\mathcal{C}$ is a connected invertible reversible non-bireversible Mealy automaton with $\mathbf{u}$ as a state. Hence by Proposition 22, $\mathbf{u}$ induces an action of infinite order.   □

Note that Theorem 23 cannot provide extra information on the torsion-freeness of the generated group. Take for instance the Mealy automaton $\mathcal{L}$ of Figure 1 (left): the action induced by yx^{-1} has order 2. However, Theorem 23 ensures that an invertible reversible Mealy automaton without bireversible component cannot generate an infinite Burnside group (see [24] for background on the Burnside problem).

All these results and constructions emphasize the relevance of the reversibility property and question us further on those (semi)groups structures generated by bireversible automata that, despite the tightness of the hypothesis on them, reveal more complex to study.

Acknowledgments. The authors thank an anonymous referee, whose relevant comments improved the paper.

References

1. Akhavi, A., Klimann, I., Lombardy, S., Mairesse, J., Picantin, M.: On the finiteness problem for automaton (semi) groups. Internat. J. Algebra Comput. **22**(6), 26 (2012)
2. Alešin, S.V.: Finite automata and the Burnside problem for periodic groups. Mat. Zametki **11**, 319–328 (1972)
3. Antonenko, A.S.: On transition functions of Mealy automata of finite growth. Matematychni Studii. **29**(1), 3–17 (2008)
4. Antonenko, A.S., Berkovich, E.L.: Groups and semigroups defined by some classes of Mealy automata. Acta Cybernetica **18**(1), 23–46 (2007)
5. Bartholdi, L., Kaimanovich, V.A., Nekrashevych, V.V.: On amenability of automata groups. Duke Math. J. **154**(3), 575–598 (2010)
6. Bartholdi, L., Reznykov, I.I., Sushchanskiĭ, V.I.: The smallest Mealy automaton of intermediate growth. J. Algebra **295**(2), 387–414 (2006)
7. Bartholdi, L., Silva, P.V.: Groups defined by automata. In: Handbook AutoMathA, ArXiv:cs.FL/1012.1531, ch. 24 (2010)
8. Baumslag, G., Boone, W.W., Neumann, B.H.: Some unsolvable problems about elements and subgroups of groups. Math. Scand. **7**, 191–201 (1959)
9. Bondarenko, I., Grigorchuk, R.I., Kravchenko, R., Muntyan, Y., Nekrashevych, V., Savchuk, D., Šunić, Z.: On classification of groups generated by 3-state automata over a 2-letter alphabet. Algebra Discrete Math. (1), 1–163 (2008)

10. Bondarenko, I.V., Bondarenko, N.V., Sidki, S.N., Zapata, F.R.: On the conjugacy problem for finite-state automorphisms of regular rooted trees. Groups Geom. Dyn. **7**(2), 323–355 (2013). with an appendix by Raphaël M. Jungers
11. Cain, A.J.: Automaton semigroups. Theor. Comput. Sci. **410**, 5022–5038 (2009)
12. Eilenberg, S.: Automata, languages, and machines, vol. A. Academic Press (A subsidiary of Harcourt Brace Jovanovich, Publishers), New York (1974)
13. Gawron, P.W., Nekrashevych, V.V., Sushchansky, V.I.: Conjugation in tree automorphism groups. Internat. J. Algebra Comput. **11**(5), 529–547 (2001)
14. Gillibert, P.: The finiteness problem for automaton semigroups is undecidable. Internat. J. Algebra Comput. **24**(1), 1–9 (2014)
15. Grigorchuk, R.I.: On Burnside's problem on periodic groups. Funktsional. Anal. i Prilozhen. **14**(1), 53–54 (1980)
16. Grigorchuk, R.I.: On the Milnor problem of group growth. Dokl. Akad. Nauk SSSR **271**(1), 30–33 (1983)
17. Grigorchuk, R.I., Nekrashevich, V.V., Sushchanskiĭ, V.I.: Automata, dynamical systems, and groups. Tr. Mat. Inst. Steklova **231**, 134–214 (2000)
18. Klimann, I.: The finiteness of a group generated by a 2-letter invertible-reversible Mealy automaton is decidable. In: Proc. 30th STACS. LIPIcs, vol. 20, pp. 502–513 (2013)
19. Klimann, I., Mairesse, J., Picantin, M.: Implementing computations in automaton (semi) groups. In: Moreira, N., Reis, R. (eds.) CIAA 2012. LNCS, vol. 7381, pp. 240–252. Springer, Heidelberg (2012)
20. Klimann, I., Picantin, M., Savchuk, D.: A connected 3-state reversible Mealy automaton cannot generate an infinite Burnside group. arXiv:1409.6142 (2014)
21. Macedonska, O., Nekrashevych, V.V., Sushchansky, V.I.: Commensurators of groups and reversible automata. Dopov. Nats. Akad. Nauk Ukr., Mat. Pryr. Tekh. Nauky (12), 36–39 (2000)
22. Maltcev, V.: Cayley automaton semigroups. Internat. J. Algebra Comput. **19**(1), 79–95 (2009)
23. Mintz, A.: On the Cayley semigroup of a finite aperiodic semigroup. Internat. J. Algebra Comput. **19**(6), 723–746 (2009)
24. Nekrashevych, V.: Self-similar groups. Mathematical Surveys and Monographs, vol. 117. American Mathematical Society, Providence (2005)
25. Russyev, A.: Finite groups as groups of automata with no cycles with exit. Algebra and Discrete Mathematics **9**(1), 86–102 (2010)
26. Sakarovitch, J.: Elements of Automata Theory. Cambridge University Press (2009)
27. Sidki, S.N.: Automorphisms of one-rooted trees: growth, circuit structure, and acyclicity. J. Math. Sci. (New York) **100**(1), 1925–1943 (2000). algebra, 12
28. Silva, P.V., Steinberg, B.: On a class of automata groups generalizing lamplighter groups. Internat. J. Algebra Comput. **15**(5–6), 1213–1234 (2005)

Coding Non-orientable Laminations

Luis-Miguel Lopez[1] and Philippe Narbel[2](✉)

[1] Tokyo University of Social Welfare, Gunma 372-0831, Japan
[2] LaBRI, University of Bordeaux, 33405 Talence, France
narbel@labri.fr

Abstract. Surface laminations are classic closed sets of disjoint curves in surfaces. We give here a full description of how to obtain codings of such laminations when they are non-orientable by using lamination languages, i.e. specific linear complexity languages of two-way infinite words. We also compare lamination languages with symbolic laminations, i.e. the coding counterparts of algebraic laminations.

1 Introduction

A *surface lamination* is essentially a closed set of pairwise disjoint closed or two-way infinite curves (leaves) rolling around a surface, a notion related to *foliations* of surfaces [6,18]. The curves of a lamination can always be continuously deformed onto paths of labeled embedded graphs – closely related to classic *train-tracks* [16,18] –, that we call *train-track like (ttl) graphs*. This deformation of curves onto a graph is called *carrying*, and the set of labels of the paths the curves of a lamination are deformed onto is called a *lamination language* [14]. Lamination languages not only reflect some of the geometric behavior of the laminations they represent, they are also a family of languages with specific combinatorial properties. They are shifts, characterized by their sequences of embedded *Rauzy graphs*, and by their sets of *bispecial factors* [14] (as an extension of results about *interval exchange transformations* obtained in [1, 10]). Lamination languages always have *linear factor complexity*, whose possible functional forms have been described in [15]. The fact that the non-orientable case – that is, when laminations and their carrier train-tracks are non-orientable –, reduces to the orientable case [18] was traditionally taken for granted. However, recent advances in the non-orientable case through a detailed symbolic study of *linear involutions* [2], together with the general framework in [8] readily including the non-orientable case and using a representation called *symbolic laminations*, show that this reduction to the orientable case is not as straightforward as it seems. The purpose of this paper is to describe how this reduction goes, and in particular we prove:

Proposition A. *Every ttl graph Γ admits a canonically oriented ttl graph $\tilde{\Gamma}$ as an orientation covering space, and any lamination carried by Γ, orientable or not, can be coded by $\tilde{\Gamma}$, yielding a lamination language.*

© Springer International Publishing Switzerland 2015
A.-H. Dediu et al. (Eds.): LATA 2015, LNCS 8977, pp. 340–352, 2015.
DOI: 10.1007/978-3-319-15579-1_26

Lamination languages are mostly included into the set of symbolic laminations, which is a much larger set of shifts. In fact, symbolic laminations are in bijection with *algebraic laminations*, a notion defined for free groups of finite type, and among the algebraic laminations are those which describe geodesic laminations on compact surfaces, called *algebraic surface laminations* [8]. With this respect, the second main result we prove here is the following, making a link possible between the results of [8] and [14]:

Proposition B. *The set of symbolic laminations Λ_{surf} corresponding to algebraic surface laminations is equal to the set* $\mathbf{L}_d$ *of lamination languages obtained through Proposition A.*

2   Context and Basic Definitions

We first give basic definitions (details can be found in [14]). Let $\Gamma = (V, E)$ be a graph with V as a set of vertices and E as a set of edges. Γ is **connected** if any two of its vertices are linked by a path. It is **directed** (or **oriented**) if an orientation is given to each of its edges, thus every edge has an initial vertex and a terminal vertex. A directed graph is **connected** if any two of its vertices are linked by a **(directed) admissible path**, i.e. a sequence of consecutive edges with the same orientation. Now, a classic notion in lamination theory is *train-tracks* [16,18], usually defined as *1D branched differentiable submanifolds* of surfaces, but which can also be defined as a kind of graphs embedded in surfaces:

Definition 1. *A graph $\Gamma = (V, E)$ is said to be* **train track-like** *(ttl for short) if it is embedded in a surface Σ, and for each vertex $v \in V$ of degree > 1, the set of incident edges at v has been non-trivially partitioned into two sets $E_{1,v}, E_{2,v}$, both formed of incident edges which are next to each other around v.*

A ttl graph can be thought of as a "semi-directed" graph: at each vertex v, the sets $E_{1,v}$, $E_{2,v}$ induce two possible local orientations, where the edges in $E_{1,v}$ are considered incoming at v while the ones of $E_{2,v}$ outgoing from v, or the other way around. Thus a **(ttl) admissible path** (or *trainpath*) in a ttl graph Γ is a sequence of consecutive edges such that at each vertex v crossed, the path enters v by an edge in $E_{1,v}$ (resp. $E_{2,v}$) and leaves it by an edge in $E_{2,v}$ (resp. $E_{1,v}$). This semi-directedness reflects the idea of "railroad switches" in a train-track.

Definition 2. *A curve γ in a surface Σ is said to be* **carried** *by a ttl graph Γ in Σ if there is a continuous deformation – a homotopy – of γ onto a ttl admissible path of Γ.*

A ttl graph Γ is **orientable** if at each vertex v of Γ, there is a choice of one of the local orientations among the two induced by $E_{1,v}$ and $E_{2,v}$, so that these orientations consistently match on each edge in Γ. A ttl graph Γ is **directed** (or **oriented**) when such orientation choices for Γ have been made, and Γ becomes then a plain embedded directed graph.

Lemma 1 (Two orientations). *Let Γ be a connected and orientable ttl graph. Then Γ has only two possible orientations, opposite of each other on each edge.*

Example 1 (A non-orientable graph). *One of the simplest non-orientable ttl graph is the "yin-yang graph" which is made of two vertices v_1, v_2, three edges e_1, e_2, e_3 linking them, and such that $E_{1,v_1} = \{e_1, e_2\}$, $E_{2,v_1} = \{e_3\}$, $E_{1,v_2} = \{e_1\}$, $E_{2,v_2} = \{e_2, e_3\}$. This graph is shown on the left of the next figure. In the middle are represented the four possible sets of local orientations derived from the $E_{\cdot,\cdot}$ sets; each of them induces some non-consistently oriented edge, hence the non-orientability of the yin-yang graph:*

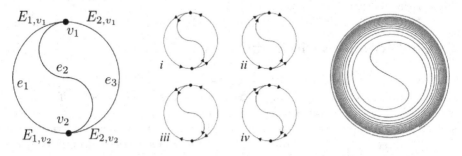

A **geodesic lamination** $\mathcal{L}$ on a surface Σ (with a Poincaré metric) is a non-empty closed subset of Σ forming a union of simple and pairwise disjoint geodesics [18]. Another definition, equivalent up to continuous deformation, goes essentially as follows [14]: A **(topological) lamination** $\mathcal{L}$ in Σ is a set of simple closed or two-way infinite curves in Σ, all pairwise disjoint and non-homotopic, such that there exists an embedded ttl graph Γ which carries $\mathcal{L}$ in a maximal way with respect to inclusion (no other curve carried by Γ can be added to $\mathcal{L}$ while preserving all its curve set properties). The right part of the above figure shows an example of a lamination, carried by the yin-yang graph, and made of two curves: one looks like a Fermat's spiral rescaled so as to accumulate at the other one, a surrounding circle. Orientability for a lamination, can be defined via the notion of carrying [4]: A lamination $\mathcal{L}$ is **orientable** if all its curves can be carried by a directed ttl graph Γ, using only directed admissible paths of Γ. A lamination is **oriented** when it is carried by a given directed graph Γ.

An embedded directed graph is **coherent** if at each vertex v of Γ all the incoming edges are next to each other around v (hence the outgoing edges are too). Following Definition 1 a directed ttl graph Γ is always coherent. Conversely, an embedded directed graph Γ can become a ttl graph only when it is coherent so that the sets $E_{1,v}$, $E_{2,v}$ at each vertex v can be accordingly defined. Nevertheless, when the focus is on curve carrying by directed graphs, non-coherent carrier graphs can be considered too. These graphs have directed admissible paths as any other directed graphs, and Definition 2 applies to them as well. Thus we can drop the constraint in Definition 1 that $E_{1,v}, E_{2,v}$ need to be formed by consecutive incident edges around v. A non-necessarily coherent directed ttl graph is an embedded directed graph, and conversely. In fact, a non-coherent graph Γ is always the result of *edge contractions* applied to a coherent graph while preserving its carried laminations (see Sections 3 and 5.4 in [14]). Non-coherent ttl graphs can indeed be seen as contracted train tracks.

Let A be a finite alphabet, and let $A^{\mathbb{Z}}$ denote the set of the two-way infinite words over A. Any subset of $A^{\mathbb{Z}}$ is a **language** of two-way infinite words. $A^{\mathbb{Z}}$ can be endowed with the topology coming from the *Cantor metric*, and the **shift map** σ is a continuous map on $A^{\mathbb{Z}}$ which sends $...a_{-1}a_0a_1...$ to $...a'_{-1}a'_0a'_1...$ where $a'_i = a_{i+1}$ for $i \in \mathbb{Z}$. A **shift** is a closed σ-invariant language in $A^{\mathbb{Z}}$. Now, an embedded directed graph Γ is here said to be **labeled** by A if its edges are bijectively labeled by A, and the **label** of a directed admissible path of Γ is the word obtained by concatenating the labels of its edges. If γ is a curve carried by Γ, and if it is carried by a unique path (up to indexing), its **coding** is the label of this path. In this case, we also say that γ is **coded** by this label, or **coded** by Γ. The coding of a carried closed curve γ is the two-way infinite periodic word $^{\omega}u^{\omega}$, where u is the label of the closed directed path in Γ carrying γ.

Definition 3. *A* **lamination language** *is the σ-closure in $A^{\mathbb{Z}}$ of the codings of all the curves of a lamination $\mathcal{L}$ coded by a directed graph Γ labeled by A.*

Embedded directed graphs not only occur in the definition of lamination languages, they are also useful when analyzing the combinatorial properties of these languages [14]. But then, the assumption that laminations should be orientable seems necessary. The next sections show that the non-orientable case reduces to the orientable one, thus making possible to stick to directed graphs.

3   Orientation Coverings for Carrier Graphs

First of all, to have a more combinatorial description of laminations together with carrier graphs less dependent on the embedding surfaces, the idea in [14] was to consider ttl graphs including their embedding information, that is, to consider *ribbon graphs* (close to *combinatorial maps* [12]). This technical move is possible since the coding of lamination curves only depends on the way curves are carried by a graph, not on the *genus* or the *punctures* of their embedding surfaces.

Definition 4. *A* **ribbon graph** *(or fat graph) Γ is a quintuple (V, H, h, i, ξ) where: V is a set of vertices; H is a set of half-edges; $h : H \to H$ is an involution without fixed points, which exchanges the pairs of half-edges, thus inducing a set E of (full) edges (giving Γ an associated usual graph structure); $i : H \to V$ is an incidence map, which indicates the vertex of Γ each half-edge is incident with; ξ is a permutation on H defined as the product of the cyclic orderings defined on each subset $i^{-1}(v)$, with $v \in V$.*

An embedding of a graph in a surface naturally endows it with a unique ribbon structure. Also, similarly to a graph, a ribbon graph Γ is **connected** if its associated graph is connected; it is **orientable** if for each half-edge in H, there is a choice of orientation so that these orientations consistently match on each edge in E; it is **directed** (or **oriented**) when such orientation choices for Γ have been made; it is **coherent**, if ξ implies that at each vertex in V, all the incoming edges are next to each other around. A ttl structure is assigned to a ribbon graph by partitioning the half-edges of H incident at each $v \in V$ into sets $E_{1,v}$ and $E_{2,v}$.

Example 2 (A non-orientable coherent ttl ribbon graph). *Reconsidering the yin-yang graph of Example 1 with its embedding in the plane, and denoting by $e_{j,s}$, $e_{j,t}$ the two half-edges of the edge e_j, we get the following ribbon graph: $V = \{v_1, v_2\}$, $H = \{e_{1,s}, e_{1,t}, e_{2,s}, e_{2,t}, e_{3,s}, e_{3,t}\}$, $h(e_{j,s}) = e_{j,t}$ and $h(e_{j,t}) = e_{j,s}$ for $j = 1,...3$, $i(e_{j,s}) = v_1$ and $i(e_{j,t}) = v_2$ for $j = 1,...3$, and $\xi = (e_{1,s}\ e_{2,s}\ e_{3,s})(e_{1,t}\ e_{3,t}\ e_{2,t})$. We can also define the same ttl structure as in Example 1: $E_{1,v_1} = \{e_{1,s}, e_{2,s}\}$, $E_{2,v_1} = \{e_{3,s}\}$, $E_{1,v_2} = \{e_{1,t}\}$, $E_{2,v_2} = \{e_{2,t}, e_{3,t}\}$.*

Let us give other instances of embedded graphs describable as ribbon graphs. First of all, a ttl graph Γ is **recurrent** if for every edge e in Γ there exists a closed admissible path which includes e. Then Γ is recurrent iff its edges can be weighted by a positive map $\mu : E \to \mathbb{R}_+^*$ such that at every vertex v of Γ the branch equation $\sum_{e \in E_{1,v}} \mu(e) = \sum_{e \in E_{2,v}} \mu(e)$ holds [16]. A ttl graph Γ endowed with such a map μ is said to be **weighted**. A **bouquet of circles** is a graph having only one vertex v. Now, **interval exchange transformations** [7] (**iets** for short) are orientation-preserving and piecewise isometric maps of bounded intervals, and they are characterized by the fact their dynamics correspond to laminations carried by orientable (thus trivially recurrent), coherent and weighted ttl ribbon bouquets of circles [4]. When these laminations are coded by their bouquets, they define a subset of lamination languages that we call **iet languages**, corresponding to the usual way of coding iets, and characterized in [1,10] (note that the languages coding iets over two subintervals are the *Sturmian languages*). **Non-classical iets** can be defined similarly to iets, as they are also characterized by the fact their dynamics correspond to laminations carried by coherent and weighted (hence recurrent) ttl ribbon bouquets of circles, but non-orientable ones [11]. Note that for a ttl bouquet, non-orientability is equivalent to the presence of at least one **reversing loop**, that is, an edge whose both half-edges lie either in $E_{1,v}$ or in $E_{2,v}$. A non-orientable ttl bouquet is recurrent iff there are reversing loops of both kinds, with half-edges in $E_{1,v}$ and $E_{2,v}$. Iets and non-classical iets dynamics are thus represented by weighted ttl ribbon bouquets of circles, and they characterize what is called **linear involutions** [2,9].

Example 3 (Some ttl ribbon bouquets of circles). *In the following figure, three recurrent ttl ribbon bouquets of circles are shown; their embeddings induce a ribbon structure at their unique vertex v, and their ttl structures are indicated by the fact that every half-edge starting above the dotted line belongs to $E_{1,v}$, while every one below belongs to $E_{2,v}$. The bouquet (i) corresponds to an iet over three intervals; (ii) and (iii) corresponds to non-classical iets (linear involutions), respectively given in [11, Section 2.4], and in [2, Section 2.2]:*

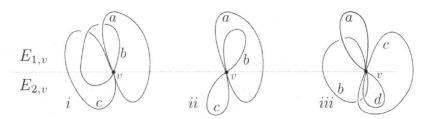

Ribbon graphs allow us to make abstraction of the embedding surfaces. We can show then directly how to associate an oriented ttl ribbon graph with any non-orientable one, reflecting the two possible edge orientations in their admissible paths, and using a construction based on the classic method [13,17]:

Definition 5. *Let $\Gamma = (V, H, h, i, \xi)$ be a ttl ribbon graph. Let (V^+, H^+, i^+, ξ^+) and (V^-, H^-, i^-, ξ^-) be two copies of (V, H, i, ξ), that is, copies of Γ without half-edges linking information. For every $v \in V$, let $v^+ \in V^+$, $v^- \in V^-$ denote its corresponding copies, and define the local orientation at $v^+ \in V^+$ and $v^- \in V^-$ so that the half-edges in E_{1,v^+} are outgoing (and those in E_{2,v^+} incoming), and the half-edges in E_{1,v^-} are incoming (and those in E_{2,v^-} outgoing).*

*Then the **orientation covering** of Γ is a ribbon graph $\tilde{\Gamma} = (V^+ \sqcup V^-, H^+ \sqcup H^-, \tilde{h}, i^+ \sqcup i^-, \xi^+ \sqcup \xi^-)$, where $\tilde{h}$ is an involution without fixed point defined as follows. For each $e \in H$, let $e^+ \in H^+$, $e^- \in H^-$ denote its copies, and:*

a) *If the orientation of e^+ is consistent with that of $h(e)^+$, $\tilde{h}(e^+) = h(e)^+$ (defining an oriented edge linking vertices in V^+); and similarly for its copy, $\tilde{h}(e^-) = h(e)^-$ (linking vertices in V^-).*

b) *If the orientation of e^+ is consistent with that of $h(e)^-$, $\tilde{h}(e^+) = h(e)^-$ (defining an oriented edge linking a vertex in V^+ to a vertex in V^-); similarly for its copy, $\tilde{h}(e^-) = h(e)^+$ (linking a vertex in V^- to a vertex in V^+).*

The ribbon graph $\tilde{\Gamma}$ comes with a **canonical orientation**, since Cases (a) and (b) above fix an orientation for each edge. With respect to the classic theory [13,17], $\tilde{\Gamma}$ is just the two-sheeted (two-fold) **orientation covering space** of Γ coming with a **covering map** $\psi : \tilde{\Gamma} \to \Gamma$ defined by sending vertices and half-edges of $\tilde{\Gamma}$ to the corresponding ones of Γ in the obvious way. Also, since $\tilde{\Gamma}$ is a two-sheeted covering, it has a covering involution s, which is the map exchanging V^+ and V^- together with their incident half-edges.

Example 4 (Some orientation covering graphs). *Considering again the yin-yang graph of Example 1, and the non-orientable bouquet of circles (ii) in the above figure, here are their respective orientation coverings following Definition 5:*

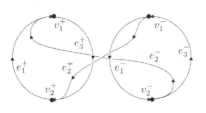
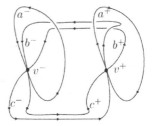

The next lemma is the analogous to a classic result in manifold theory [13]:

Lemma 2 (Connectedness and Orientation). *Let Γ be a connected ttl ribbon graph. Then $\tilde{\Gamma}$ is connected iff Γ is non-orientable.*

Lemma 3 (Uniqueness of $\tilde{\Gamma}$). *Let Γ be a non-orientable ttl ribbon graph. Then the possible orientations of $\tilde{\Gamma}$ yield isomorphic directed graphs.*

4   Orientation Coverings for Laminations

In order to make a ttl ribbon graph Γ as an effective carrier graph for laminations, there are ways of building surfaces which embed Γ and laminations carried by Γ in a unique way. A **ribbon graph surface** $\Sigma(\Gamma)$, unique up to isometry, can be obtained as follows [14]: each vertex of Γ with degree ≤ 2 and each half-edge in H is replaced by an Euclidean square having side length 1, and each vertex in V with degree $d > 2$ is replaced by an Euclidean regular polygon with d sides of length 1; these polygons are then glued together by Euclidean positive isometries according to the patterns given by h, i and ξ. Laminations can then be defined on $\Sigma(\Gamma)$ so as to be carried by Γ. To obtain surfaces without boundary embedding $\Sigma(\Gamma)$, the boundary components of $\Sigma(\Gamma)$ can be capped off, e.g. with disks, but as we already said, we mostly focus here at how a lamination is carried by Γ, not at the specifics of their embedding surfaces. Also, any lamination $\mathcal{L}$ on a surface Σ carried by an embedded graph Γ can always be moved on Σ so as to lie in a regular neighborhood $N(\Gamma)$ of Γ in Σ. The embedding of Γ in Σ endows it then with a ribbon structure, and $\Sigma(\Gamma)$ is homeomorphic to $N(\Gamma)$ by uniqueness of regular neighborhoods. Thus, to be able to speak about laminations, there is no loss of generality using only surfaces of the form $\Sigma(\Gamma)$.

Now, considering the orientation covering graph $\tilde{\Gamma}$ of Γ, we get $\Sigma(\tilde{\Gamma})$, where ψ extends to a covering map to $\Sigma(\Gamma)$, whereto corresponding laminations can be then defined consistently:

Definition 6. *Let $\mathcal{L}$ be a lamination carried by Γ in $\Sigma(\Gamma)$. Let Γ^+ and Γ^- be two copies of Γ, and let $\mathcal{L}^+$ and $\mathcal{L}^-$ be two copies of $\mathcal{L}$ embedded in $\Sigma(\Gamma^+)$ and $\Sigma(\Gamma^-)$, respectively. Then, the **orientation covering lamination** $\tilde{\mathcal{L}}$ of $\mathcal{L}$ in $\Sigma(\tilde{\Gamma})$ carried by $\tilde{\Gamma}$ is built after Definition 5: the surface $\Sigma(\tilde{\Gamma})$ is obtained by gluing together the polygons making $\Sigma(\Gamma^+)$ and $\Sigma(\Gamma^-)$, using $\tilde{h}$, $i^+ \sqcup i^-$, and $\xi^+ \sqcup \xi^-$ of $\tilde{\Gamma}$. These polygons contain pieces of curves of $\mathcal{L}^+$ and $\mathcal{L}^-$, fitting together on $\Sigma(\Gamma^+)$ by construction. The result is $\tilde{\mathcal{L}}$ and, by extension $\psi(\tilde{\mathcal{L}}) = \mathcal{L}$.*

A technical point is that if we put some *Poincaré metric* on $\Sigma(\Gamma)$ (with geodesic boundary), then $\Sigma(\tilde{\Gamma})$ inherits one too in such a way that ψ becomes a local isometry, and in this case the covering involution s becomes an isometry.

The main properties of carrier graphs and laminations are preserved by their orientation coverings, showing that the tools described in [14] remain effective on them. First, a carrying of a lamination $\mathcal{L}$ by Γ is said to be **full** if every edge of Γ is used to carry $\mathcal{L}$:

Lemma 4 (Fullness of the carrying). *Let $\mathcal{L}$ be a lamination fully carried by Γ. Then $\tilde{\mathcal{L}}$ is fully carried by $\tilde{\Gamma}$.*

Lemma 5 (Coherence). *Let Γ be a coherent ttl ribbon graph. Then $\tilde{\Gamma}$ is coherent.*

In reference to the topological definition of laminations, a carrying of a lamination $\mathcal{L}$ by Γ is said to be **maximal** if no other curve carried by Γ can be added to $\mathcal{L}$ while at the same time preserving its nature of lamination (the definition of carrying does not require maximality, only the definition of laminations does):

Lemma 6 (Maximality). *Let $\mathcal{L}$ be a lamination maximally carried by a coherent graph Γ. Then $\tilde{\mathcal{L}}$ is maximally carried by $\tilde{\Gamma}$.*

A lamination $\mathcal{L}$ is said to be **minimal** if it does not contain any lamination as a proper non-empty subset:

Lemma 7 (Minimality). *Let $\mathcal{L}$ be a minimal lamination carried by a graph Γ. Then $\tilde{\mathcal{L}}$ is minimal iff $\mathcal{L}$ is non-orientable.*

Example 5 (An orientation covering non-minimal lamination). *Considering the yin-yang graph Γ and the lamination $\mathcal{L}$ of Example 1, together with the orientation covering graph $\tilde{\Gamma}$ shown in Example 4, the corresponding orientation covering lamination $\tilde{\mathcal{L}}$ can be seen to be made of four curves. We show them on the right of the next figure in two pairs, each pair containing one curve with its two spiralling ends and one surrounding limit cycle:*

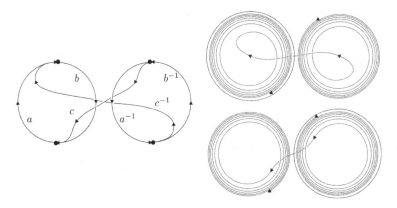

Using the above labeling of $\tilde{\Gamma}$, the lamination language coding $\tilde{\mathcal{L}}$ is the σ-closure of $\{^{\omega}(b^{-1}a^{-1})c^{-1}(ba)^{\omega},\ ^{\omega}(a^{-1}b^{-1})c(ab)^{\omega},\ ^{\omega}(ab)^{\omega},\ ^{\omega}(a^{-1}b^{-1})^{\omega}\}$.

5   Coding Laminations

Given a lamination $\mathcal{L}$ carried by a ttl ribbon graph Γ, we are now ready for:

Proof (of Proposition A). First, we can consider their respective orientation coverings $\tilde{\Gamma}$ and $\tilde{\mathcal{L}}$ (cf. Definitions 5 and 6). As a consequence of Definition 5, $\tilde{\Gamma}$ is directed, and by Lemma 3, it is uniquely determined. Now, since $\tilde{\mathcal{L}}$ is carried by $\tilde{\Gamma}$, it is oriented too, thus it can be coded by $\tilde{\Gamma}$ as a lamination language following Definition 3.   □

As a consequence, here are all the possible ways of coding $\mathcal{L}$ by Γ into a lamination language considering its orientability status:

  i. If $\mathcal{L}$ is oriented, Γ is too, and the coding process is just based on taking the labels of the admissible paths in Γ carrying the curves of $\mathcal{L}$.
 ii. If $\mathcal{L}$ is orientable but does not come with an orientation:
      a. Either we fix one of the two orientations (see Lemma 1) and $\mathcal{L}$ is coded through this orientation (as in Case (i)),
      b. Or we consider both orientations, that is, $\mathcal{L}$ is coded through the coding of $\tilde{\mathcal{L}}$ by $\tilde{\Gamma}$ (see Proposition A), and we are in Case (i) again. Recall that $\tilde{\Gamma}$ is then made of two disjoint copies of Γ (see Lemma 2), respectively having one of the two possible orientations of Γ (see Lemma 3).
iii. If $\mathcal{L}$ is non-orientable, $\mathcal{L}$ is coded through the coding of $\tilde{\mathcal{L}}$ by $\tilde{\Gamma}$ (see Proposition A), and we are in Case (i) again. Here, $\tilde{\Gamma}$ is a connected graph.

As an example of a combinatorial property, the (factor) **complexity** [5] of a language L is the map $p_L : \mathbb{N}^* \to \mathbb{N}^*$, where $p_L(n)$ is the number of distinct factors (*subblocks*) in the words of L, and as a consequence of [14, 4.1.1]:

Remark 1 (Factor complexity of non-orientable laminations). *Let $\mathcal{L}$ be a non-orientable lamination, maximally carried by a coherent ttl graph Γ with a set V of vertices and a set E of edges. Let L be the lamination language obtained by coding $\mathcal{L}$ by $\tilde{\Gamma}$. Then we have:* $p_L(n) = 2(|E| - |V|)n + 2|V|, \forall n > 0$.

For instance, the complexity of the lamination language L of Example 5 is $p_L(n) = 2n + 4$. An additional remark is that for a lamination language L coming from some linear involution *without connections*, i.e. without simplifications so that maximality of the corresponding laminations holds, we have $p_L(n) = 2(|E| - 1)n + 2, \forall n > 0$ (in accordance with [2]).

Note that the context of lamination languages is mostly the one of free monoids for their sets of factors. Thus the way the carrier graphs are labeled is of no importance (in particular from a word combinatorics point-of-view) as long as we use as many distinct letters as there are edges. Nevertheless, considering the orientation covering graphs allows one to extend the context to free groups, as these graphs include edges with both orientations. In order to indicate these orientations in the coding, there is a way of labeling an orientation covering graph $\tilde{\Gamma}$, so that the edge directions taken by an admissible path γ in Γ as a ttl graph are reflected in $\psi^{-1}(\gamma)$:

Definition 7. *Let Γ be a non-orientable graph, and $\tilde{\Gamma}$ its orientation covering. Let $A = \{a_1, ..., a_n\}$ and $A^{-1} = \{a_1^{-1}, ..., a_n^{-1}\}$, for which s is defined as $s(a_i) = a_i^{-1}$ and $s(a_i^{-1}) = a_i$, for every $i = 1...n$, and where n is the number of edges of Γ. Then, following Definition 5, a* **natural labeling** *of $\tilde{\Gamma}$ is given as:*

a. *For every edge e built by Case (a) of Definition 5 of $\tilde{\Gamma}$, we assign a distinct letter $a \in A$, and a^{-1} to $s(e)$ (its corresponding copy);*
b. *For every edge e built by Case (b) of Definition 5 and its copy $s(e)$, we assign letters $a \in A$ and a^{-1}, with an arbitrary choice (since they correspond to edges linking vertices in V^+ and V^-).*

For instance, the orientation covering of the yin-yang graph shown in Example 5 is labeled according to a natural labeling.

6   Lamination Languages and Symbolic Laminations

Let $w = ...w_{-2}w_{-1}w_0w_1w_2...$ be a two-way infinite word over an alphabet $A \sqcup A^{-1}$ equipped with the involution s exchanging a and a^{-1} (see Definition 7), then the **symmetric** word of w is $\nu(w) = ...s(w_2)s(w_1)s(w_0)s(w_{-1})s(w_{-2})....$ By extension, the symmetric language $\nu(L)$ of a language L over $A \sqcup A^{-1}$ is the language made of the symmetric words of all the words in L. The word w is said to be **reduced** if $w_i \neq s(w_{i+1})$ for all i (neither $a_i a_i^{-1}$ nor $a_i^{-1} a_i$ occurs).

Lemma 8. *Let $\mathcal{L}$ be a lamination carried by a ttl ribbon graph Γ. Let $\tilde{\Gamma}$ be labeled with a natural labeling. Let L be the lamination language of $\mathcal{L}$ coded by $\tilde{\Gamma}$. Then $L = \nu(L)$, and all the words in L are reduced.*

Now, there exists another symbolic way of dealing with laminations, which is defined in a very general setting in [8] by:

Definition 8. *A* **symbolic lamination** *is a symmetric-invariant shift over an alphabet $A \sqcup A^{-1}$ made of reduced words.*

Let us denote the set of all symbolic laminations by Λ, the set of lamination languages by $\mathbf{L}$, and the subset of all lamination languages obtained by coding laminations using the orientation coverings of their carrier graphs (cf. Cases (ii)(b) and (iii) in p. 348) by $\mathbf{L}_d \subsetneq \mathbf{L}$. According to Lemma 8, all the words of these sets of languages coded over alphabets of the form $A \sqcup A^{-1}$ are reduced, thus these sets are comparable, combinatorially fitting in the context of free monoids:

Lemma 9. $\mathbf{L} \not\subset \Lambda$ *and* $\mathbf{L}_d \subsetneq \Lambda$.

An important case illustrating the preceding result is given by the set of all the iet languages (cf. p. 344), that we denote by $\mathbf{L}_{iet}$:

Lemma 10. $\mathbf{L}_{iet} \subsetneq \mathbf{L}$, $\mathbf{L}_{iet} \not\subset \Lambda$, *and* $\mathbf{L}_{iet} \cap \Lambda \neq \emptyset$.

In [8], a symbolic lamination in Λ is said to be *orientable* if it can be written as a disjoint union $L \sqcup \nu(L)$, where L and $\nu(L)$ are both closed shifts, and Λ is called *positive* if either L or $\nu(L)$ uses letters from A only (the other one from A^{-1} only). This situation corresponds to the case when orientable laminations are coded with lamination languages using their orientation covering graphs (cf. Case (ii)(b) in p. 348), made then of two copies of the carrier graphs. This situation is also the one of orientable linear involutions, which can be seen as a pair of mutually inverse iets [2]. Now, similarly to iets, let us denote by $\mathbf{L}_{invol}$ all the codings of the dynamics of the linear involutions by the orientation coverings of their carrying ttl bouquets of circles (that is, the codings studied in [2]):

Lemma 11 $\mathbf{L}_{invol} \subsetneq \mathbf{L}_d$, *and* $(\mathbf{L}_{invol} \cap \mathbf{L}_{iet}) = (\mathbf{L}_{iet} \cap \Lambda)$.

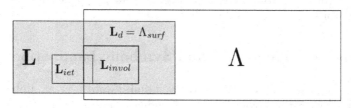

In order to explain some of the differences in the above language sets, note first that the choices underlying the notion of lamination language were made so that these languages code laminations according to the way their curves roll up on a surface, and to their carrier graphs, thus yielding specific combinatorial properties [14]. As for orientability, the concreteness of lamination languages makes them to include the codings of orientable laminations without their symmetric counterpart, using their directed carrier graphs, and not necessarily their orientation coverings. Instead, symbolic laminations always preserve the link with group theory, so that in particular free groups can be exploited together with the methods used to study their *outer automorphisms* [8, Remark 4.5].

Also, in order to have a more precise view at the inclusion $\mathbf{L}_d \subsetneq \Lambda$ (cf. Lemma 9), i.e. how the lamination languages coded by the orientation covering graphs are included into the set of symbolic laminations, we now introduce some more notions from [8]. Let F_n denote the free group on a set A of n generators, where A^{-1} denote the set of A's inverses, and whose elements can be seen to be the reduced finite words over $A \sqcup A^{-1}$. The group F_n can be represented by a labeled tree $\mathcal{T}_n$ (a *Cayley graph* of F_n), defined by associating one vertex with each reduced word in F_n, and one edge, with label $x \in A \sqcup A^{-1}$, from $v \in F_n$ to $v' \in F_n$ if $v' = vx$ without reduction. The vertex associated with the empty word is taken as the origin of $\mathcal{T}_n$. It is then possible to consider the limit language $\partial^2 F_n$ made of all the possible pairs (w, w'), $w \neq w'$, where w, w' are right-infinite labels of admissible paths in $\mathcal{T}_n$ starting at the origin. The **flip involution** is the map on $\partial^2 F_n$ which sends (w, w') to (w', w). Also, F_n acts on $\partial^2 F_n$ as $v \cdot (w, w') = (vw, vw')$, $v \in F_n$, where vw, vw' are reduced if necessary. These pairs (w, w') give rise to two-way infinite words $w'^{-1}w$, where all the reductions (if any) have been applied, and the action of F_n preserves the set of these two-way infinite words. Geometrically, it is then known that the words

$w'^{-1}w$ correspond to geodesics in some surfaces (the geodesics are determined by two limit points on the boundary of what is called the *universal covering*, where $\mathcal{T}_n$ embeds, of each of these surfaces), leading to the following definition:

Definition 9. *[8]. An* **algebraic lamination** *is a non-empty, closed set in $\partial^2 F_n$, invariant by the flip and by F_n's action.*

The set of algebraic laminations is denoted by Λ^2. By construction there is a bijection ρ between Λ^2 and Λ [8, Proposition 4.4] (the flip corresponds to the symmetry property, and the F_n's action to the shift-closure, see Definition 8).

Now, accordingly, an algebraic lamination determines a closed set of geodesics on a surface, but this set is generally not a geodesic lamination since these geodesics intersect on the surface (it is even very rare that they do not intersect [3]). As a consequence, symbolic laminations form a much larger set of languages than lamination languages, since lamination languages only code geodesic laminations. When an algebraic lamination actually determines a geodesic lamination, it is called an **algebraic surface lamination** [8]. Their set is denoted by $\Lambda^2_{surf} \subsetneq \Lambda^2$, and let $\Lambda_{surf} = \rho(\Lambda^2_{surf})$ denote their corresponding symbolic laminations. We are now in position to prove Proposition B, i.e. $\Lambda_{surf} = \mathbf{L}_d$ (also completing the figure of the set inclusions shown in p. 350):

Proof (of Proposition B, sketch). That $\Lambda_{surf} \subseteq \mathbf{L}_d$ comes from the fact that any algebraic surface lamination is related to a lamination carried by a bouquet of n circles corresponding to the quotient of $\mathcal{T}_n$ by the action of F_n. The converse $\Lambda_{surf} \supseteq \mathbf{L}_d$ is essentially obtained by the fact that any ttl carrier graph can be continuously transformed into a ttl bouquet of n circles by identifying all its vertices onto one single vertex. □

References

1. Belov, A.Y., Chernyatiev, A.L.: Describing the set of words generated by interval exchange transformation. Comm. Algebra **38**(7), 2588–2605 (2010)
2. Berthé, V., Delecroix, V., Dolce, F., Perrin, D., Reutenauer, C., Rindone, G.: Natural coding of linear involutions (2014), arXiv:1405.3529 (see also the abstracts of the 15th Mons Theoretical Computer Science Days)
3. Birman, J., Series, C.: Geodesics with bounded intersection number on surfaces are sparsely distributed. Topology **24**(2), 217–225 (1985)
4. Bonahon, F.: Geodesic laminations on surfaces. In: Laminations and foliations in dynamics, geometry and topology, Contemp. Math., vol. 269, pp. 1–37. Amer. Math. Soc. (2001)
5. Cassaigne, J., Nicolas, F.: Factor complexity. In: Combinatorics, automata and number theory, Encyclopedia Math. Appl., vol. 135, pp. 163–247. Cambridge Unversity Press, Cambridge (2010)
6. Casson, A., Bleiler, S.: Automorphisms of surfaces after Nielsen and Thurston, London Mathematical Society Student Texts, vol. 9. Cambridge Unversity Press, Cambridge (1988)

7. Cornfeld, I.P., Fomin, S.V., Sinaĭ, Y.G.: Ergodic theory, Grundlehren der Mathematischen Wissenschaften [Fundamental Principles of Mathematical Sciences], vol. 245. Springer, New York (1982)
8. Coulbois, T., Hilion, A., Lustig, M.: R-trees and laminations for free groups I: Algebraic laminations. J. Lond. Math. Soc. **78**(2), 723–736 (2008)
9. Danthony, C., Nogueira, A.: Measured foliations on nonorientable surfaces. Ann. Sci. École Norm. Sup. (4) 23(3), 469–494 (1990)
10. Ferenczi, S., Zamboni, L.Q.: Languages of k-interval exchange transformations. Bull. Lond. Math. Soc. **40**(4), 705–714 (2008)
11. Gadre, V.S.: Dynamics of non-classical interval exchanges. Ergodic Theory Dynam. Systems **32**(6), 1930–1971 (2012)
12. Lando, S.K., Zvonkin, A.K.: Graphs on surfaces and their applications, Encyclopaedia of Mathematical Sciences, vol. 141. Springer, Berlin (2004)
13. Lee, J.M.: Introduction to smooth manifolds, Graduate Texts in Mathematics, vol. 218, 2nd edn. Springer, New York (2013)
14. Lopez, L.M., Narbel, P.: Lamination languages. Ergodic Theory Dynam. Systems **33**(6), 1813–1863 (2013)
15. Narbel, P.: Bouquets of circles for lamination languages and complexities. RAIRO, Theoretical Informatics and Applications **48**(4), 391–418 (2014)
16. Penner, R., Harer, J.: Combinatorics of train tracks, Annals of Mathematics Studies, vol. 125. Princeton University Press, Princeton (1992)
17. Spanier, E.H.: Algebraic topology. Springer, New York (1981)
18. Thurston, W.: The geometry and topology of three-manifolds (Princeton University Lecture Notes) (Electronic version 1.1 - March 2002). http://library.msri.org/books/gt3m (1980) (accessed September 2014)

Preset Distinguishing Sequences and Diameter of Transformation Semigroups

Pavel Panteleev[(✉)]

Faculty of Mechanics and Mathematics, Lomonosov Moscow State University,
GSP-1, Leninskiye Gory, Moscow 119991, Russian Federation
panteleev@intsys.msu.ru

Abstract. We investigate the length $\ell(n, k)$ of a shortest preset distinguishing sequence (PDS) in the worst case for a k-element subset of an n-state Mealy automaton. It was mentioned by Sokolovskii [18] that this problem is closely related to the problem of finding the maximal subsemigroup diameter $\ell(\mathbf{T}_n)$ for the full transformation semigroup $\mathbf{T}_n$ of an n-element set. We prove that $\ell(\mathbf{T}_n) = 2^n \exp\{\sqrt{\frac{n}{2}\ln n}(1 + o(1))\}$ as $n \to \infty$ and, using approach of Sokolovskii, find the asymptotics of $\log_2 \ell(n, k)$ as $n, k \to \infty$ and $k/n \to a \in (0, 1)$.

Keywords: Automata · Finite-state machine · Preset distinguishing sequence · Transformation semigroup · Diameter

1 Introduction

Finite state machines are widely used models for systems in a variety of areas, including sequential circuits [8] and communication protocols [10]. The study of finite automata testing is motivated by applications in the verification of these systems. One of the basic tasks in the verification of finite automata is to identify the state of the automaton under investigation. Once the state is known, the behavior of the automaton becomes predictable and it is possible to force the automaton into the desirable mode of operation. Suppose we have a finite deterministic Mealy automaton $\mathfrak{A}$ whose transition and output functions are available and we know that its initial state q_0 is in some subset S of its set of states Q. The *state-identification* problem is to find an input sequence called a *preset distinguishing sequence* (PDS) for S in $\mathfrak{A}$ that produces different outputs for different states from S. Before we give a formal definition of a PDS and state the results of the paper we need to fix notations and recall some standard definitions from automata theory.

A *finite deterministic Mealy automaton* (an *automaton* for short) is a quintuple $\mathfrak{A} = (A, Q, B, \delta, \lambda)$, where: A, Q, B are finite nonempty sets called the *input alphabet*, the *set of states*, and the *output alphabet*, respectively; $\delta \colon Q \times A \to Q$ and $\lambda \colon Q \times A \to B$ are total functions called the *transition function* and the *output function*, respectively.

© Springer International Publishing Switzerland 2015
A.-H. Dediu et al. (Eds.): LATA 2015, LNCS 8977, pp. 353–364, 2015.
DOI: 10.1007/978-3-319-15579-1_27

If we omit in the definition of automaton the output alphabet B and the output function λ we obtain an object $\mathfrak{A} = (A, Q, \delta)$ called *finite semiautomaton*. If the output function δ is partial then the semiautomaton is also called *partial*.

Let Σ be an arbitrary alphabet. By Σ^* we denote the set of all words over the alphabet Σ. Denote by $|\alpha|$ the length of a word $\alpha \in \Sigma^*$. Denote by ε the *empty* word, i.e., $|\varepsilon| = 0$.

As usual, we extend functions δ and λ to the set $Q \times A^*$ in the following way: $\delta(q, \varepsilon) = q$, $\delta(q, \alpha a) = \delta(\delta(q, \alpha), a)$, $\lambda(q, \varepsilon) = \varepsilon$, $\lambda(q, \alpha a) = \lambda(q, \alpha)\lambda(\delta(q, \alpha), a)$, where $q \in Q, a \in A, \alpha \in A^*$. Moreover, if $S \subseteq Q$ is a subset of states, then we let $\delta(S, \alpha) = \{\delta(q, \alpha) \mid q \in S\}$.

We say that two states $q_1, q_2 \in Q$ of an automaton $\mathfrak{A}$ are *distinguishable* by an input word $\alpha \in A^*$ if $\lambda(q_1, \alpha) \neq \lambda(q_2, \alpha)$. If there are no such words we say that the states q_1, q_2 are *indistinguishable* or *equivalent*. An automaton is called *reduced* or *minimal* if it does not have equivalent states.

Definition.   *Let S be a subset of states of an automaton $\mathfrak{A}$. We say that an input word α is a preset distinguishing sequence (PDS) for S in $\mathfrak{A}$ if α pairwise distinguishes the states in the set S, i.e., $\lambda(q_1, \alpha) \neq \lambda(q_2, \alpha)$ for all $q_1, q_2 \in S$, $q_1 \neq q_2$.*

Denote by $\ell(\mathfrak{A}, S)$ the length of a shortest PDS for S in $\mathfrak{A}$, or 0 if such a PDS does not exist. It is a well known fact [12] that there are reduced automata that do not have a PDS for some k-element subsets of states when $k \geq 3$. Consider the function

$$\ell(n, k) = \max_{\mathfrak{A} \in \mathscr{A}_n, |S| = k} \ell(\mathfrak{A}, S),$$

where $\mathscr{A}_n$ is the class of all n-state automata. This function can be interpreted as the length of a shortest PDS in the worst case for a k-element subset of states in an n-state automaton.

The function $\ell(n, k)$ was studied by many authors. In his seminal paper [12] Moore proves that $\ell(n, 2) = n - 1$. Gill [4] gives the upper bound $\ell(n, k) \leq (k - 1)n^k$. Sokolovskii finds the lower bounds in [17]:

$$\ell(n, k) \geq \binom{n-1}{k-1} \text{ if } 1 \leq k \leq n/2, \tag{1}$$

$$\ell(n, k) \geq \binom{n-2}{\lfloor (n-2)/2 \rfloor} \text{ if } n/2 < k < n. \tag{2}$$

In [14] Rystsov shows that $\log_3 \ell(n, n) \sim n/6$ as $n \to \infty$. The result is proved reducing the problem of estimating $\ell(n, n)$ to the problem of estimating the function $T(n)$ that is equal to the length of a shortest irreducible word in the worst case for a partial n-state semiautomaton. An *irreducible word* for a partial semiautomaton $\mathfrak{A} = (A, Q, \delta)$ is a word $\alpha \in A^*$ such that its action is defined on all states and for any word $\beta \in A^*$ such that its action is defined on the set $\delta(Q, \alpha)$ we have $|\delta(Q, \alpha)| = |\delta(Q, \alpha\beta)|$. In [14] it is proved that $\log_3 T(n) \sim n/3$ as $n \to \infty$. It is interesting to note that $T(n)$ coincides with the function $d_3(n)$

studied by several authors [3,11] which is equal to the length of a shortest carefully synchronizing word[1] in the worst case for a partial n-state semiautomaton. This is due to the fact that every carefully synchronizing word is also irreducible and the worst case irreducible word is always carefully synchronizing[2]. Thus we have $\log_3 d_3(n) \sim n/3$ which was conjectured in [3].

In the paper [18] Sokolovskii investigate the relationship between the function $\ell(n,k)$ and the maximum of a subsemigroup diameter in the full transformation semigroup of an n-element set.

Definition. *Let Ω_n be an n-element set. The full transformation semigroup of Ω_n (also called the symmetric semigroup of Ω_n) is the set $\mathbf{T}_n$ of all transformations of Ω_n.*

The set $\mathbf{T}_n$ contains the proper subset $\mathbf{S}_n$ of all bijections on the set Ω_n called the *symmetric group* on Ω_n. We see that $\mathbf{T}_n$ is a monoid and $\mathbf{S}_n$ is a group with function composition as the multiplication operation. In this paper, by the *composition* fg of transformations $f, g \in \mathbf{T}_n$ we mean the *left* composition $x \mapsto g(f(x))$.

Consider $\mathcal{B} \subseteq \mathbf{T}_n$. By $\langle \mathcal{B} \rangle$ denote the *closure* of the set $\mathcal{B}$, i.e., the set $\{f_1 \ldots f_\ell \mid f_1, \ldots, f_\ell \in \mathcal{B}\}$. Let $f \in \langle \mathcal{B} \rangle$ and ℓ be the minimum natural number such that $f = f_1 \ldots f_\ell$ for some $f_1, \ldots, f_\ell \in \mathcal{B}$. Then ℓ is called the *complexity* of the function f over the *basis* $\mathcal{B}$ and is denoted by $\ell_\mathcal{B}(f)$. We should also mention that the same function was considered in the paper [16] under the name *depth*.

For any subset $\mathcal{C} \subseteq \mathbf{T}_n$ we define the following function:

$$\ell(\mathcal{C}) = \max_{\mathcal{B} \subseteq \mathcal{C}, f \in \langle \mathcal{B} \rangle} \ell_\mathcal{B}(f). \tag{3}$$

The function $\ell(\mathcal{C})$ can be interpreted as the worst-case complexity of the functions from $\mathcal{C}$. In the paper [18] Sokolovskii shows that:

$$\binom{n-1}{\lfloor \frac{n-1}{2} \rfloor} < \ell(\mathbf{T}_n) < n^{\frac{n}{2}(1+o(1))},$$

$$e^{\sqrt{n \ln n}(1+o(1))} < \ell(\mathbf{S}_n) < n!^{\frac{1}{2}(1+o(1))},$$

as $n \to \infty$. It is worth mentioning that the lower bound for $\ell(\mathbf{S}_n)$ follows from the asymptotic estimate of the maximum order of the permutations from $\mathbf{S}_n$ called Landau's function [9]. The stronger result for $\ell(\mathbf{S}_n)$ follows from [1]. The author considers only *closed* sets $\mathcal{C}$ (i.e., $\langle \mathcal{C} \rangle = \mathcal{C}$), which are subgroups of $\mathbf{S}_n$. For any subgroup G of $\mathbf{S}_n$ the *directed diameter* $\mathrm{diam}^+(G)$ of the group G is defined as follows:

$$\mathrm{diam}^+(G) = \max_{f \in G, \langle \mathcal{B} \rangle = G} \ell_\mathcal{B}(f).$$

---

[1] A word α is a *carefully synchronizing* for a partial semiautomaton $\mathfrak{A} = (A, Q, \delta)$ if the action of α is defined on all states and $|\delta(Q, \alpha)| = 1$.

[2] If α is a shortest irreducible word for a partial semiautomaton $\mathfrak{A} = (A, Q, \delta)$ and $|\delta(Q, \alpha)| > 1$ then we can always add a new input symbol a to $\mathfrak{A}$ and obtain $\mathfrak{A}' = (A \cup \{a\}, Q, \delta')$ such that αa is a shortest irreducible word for $\mathfrak{A}'$ and $|\delta'(Q, \alpha a)| = 1$.

It is easily shown that $\ell(\mathbf{S}_n) = \max_G \text{diam}^+(G)$, where G ranges over all subgroups of $\mathbf{S}_n$. From the results of [1] it follows that

$$\ell(\mathbf{S}_n) = e^{\sqrt{n \ln n}(1+o(1))} \text{ as } n \to \infty. \tag{4}$$

We are now ready to state the first of the two main results of this paper.

Theorem 1. *We have* $\ell(\mathbf{T}_n) = 2^n e^{\sqrt{\frac{n}{2} \ln n}(1+o(1))}$ *as* $n \to \infty$.

As we mentioned before, Sokolovskii discovered (see [18]) the relationship between functions $\ell(\mathbf{T}_n)$ and $\ell(n,k)$. He proved in particular that

$$\ell(n,k) \le (k-1)\ell(\mathbf{T}_n). \tag{5}$$

The *binary entropy* function denoted by $H_2(x)$ is defined as follows:

$$H_2(x) = -x \log_2 x - (1-x) \log_2(1-x), \text{ where } x \in (0,1).$$

Combining inequalities (1), (2), and (5) with theorem 1 the second main result of the paper can be proved.

Theorem 2. *We have* $\log_2 \ell(n,k) \sim \varphi(a)n$ *as* $n \to \infty$ *and* $k/n \to a \in (0,1)$, *where* $\varphi(a) = H_2(a)$ *if* $a < 1/2$ *and* $\varphi(a) = 1$ *if* $a \ge 1/2$.

2   Proofs of the Main Results

Before we proceed to the formal proofs of the main results, let us give some definitions and state some useful lemmas first. Consider the set $\mathbf{T}_n^{(k)}$ of all bijections $f: D \to D'$ such that $D, D' \subseteq \Omega_n$ and $|D| = |D'| = k$. Suppose $\mathcal{B} \subseteq \mathbf{T}_n$, $f \in \mathbf{T}_n^{(k)}$, $f: D \to D'$, and there is a map $g \in \langle \mathcal{B} \rangle$ such that $f = g|_D$. Then we denote by $\ell_\mathcal{B}(f)$ the minimum of $\ell_\mathcal{B}(g)$ over all such maps g, or 0 if there are no such maps. The value $\ell_\mathcal{B}(f)$ is also called the *complexity* of f over $\mathcal{B}$. Consider the following function:

$$\ell(\mathbf{T}_n^{(k)}) = \max_{\mathcal{B} \subseteq \mathbf{T}_n, f \in \mathbf{T}_n^{(k)}} \ell_\mathcal{B}(f).$$

If $f(D) = D'$, then we say that f *transforms* D *into* D' and write $D \xrightarrow{f} D'$. For any set of maps $\mathcal{B} \subseteq \mathbf{T}_n$ the k-*graph over* $\mathcal{B}$ is the directed graph G (loops and multiple edges are permitted[3]) with the set of vertices

$$V(G) = \{D \mid D \subseteq \Omega_n, |D| = k\},$$

the set of arcs

$$E(G) = \{f \in \mathbf{T}_n^{(k)} \mid f = g|_D \text{ for some } g \in \mathcal{B} \text{ and } D \in V(G)\},$$

---

[3] Sometimes such graphs are called *pseudographs*.

and every arc f goes from the vertex D to the vertex D' whenever $D \xrightarrow{f} D'$.

A *walk* from the vertex D to the vertex D' in the k-graph G is a sequence of vertices and arcs $\mathbf{w} = D_0, f_1, D_1, \ldots, f_\ell, D_\ell$ such that $D_0 = D$, $D_\ell = D'$ and the arc f_i goes from the vertex D_{i-1} to the vertex D_i for $i = 1, \ldots, \ell$, or, in terms of maps,

$$D_0 \xrightarrow{f_1} D_1 \xrightarrow{f_2} \cdots \xrightarrow{f_\ell} D_\ell.$$

We often omit vertices in walks and write simply $\mathbf{w} = f_1, \ldots, f_\ell$. The number ℓ is called the *length* of the walk $\mathbf{w}$ and is denoted by $\ell(\mathbf{w})$. By a *subwalk* of $\mathbf{w}$ we mean a subsequence $f_i, f_{i+1}, \ldots, f_j$, $1 \leq i < j \leq \ell$.

For any walk $\mathbf{w} = f_1, \ldots, f_\ell$ from D to D' consider the map $[\mathbf{w}]: D \to D'$, where $[\mathbf{w}] = f_1 \ldots f_\ell$ (the composition of the maps $f_1, \ldots, f_\ell$). Two walks $\mathbf{w}$ and $\mathbf{w}'$ are called *equivalent* if $[\mathbf{w}] = [\mathbf{w}']$. For a *closed* walk $\mathbf{w}$, which starts and ends in the same vertex D, the map $[\mathbf{w}]$ is a permutation of D. For any closed walk $\mathbf{w}$, by definition, put

$$\mathbf{w}^k = \underbrace{\mathbf{w}, \ldots, \mathbf{w}}_{k}, \quad k \in \mathbb{N}.$$

It is readily seen that $\mathbf{w}^k$ is also a closed walk and $[\mathbf{w}^k] = [\mathbf{w}]^k$.

The next lemma is an immediate consequence of the previous definitions.

Lemma 3. *Given a basis $\mathcal{B} \subseteq \mathbf{T}_n$ and a map $f \in \mathbf{T}_n^{(k)}$ such that $f: D \to D'$ is a restriction of some map from $\langle \mathcal{B} \rangle$. Consider the k-graph G over $\mathcal{B}$. Then $\ell_\mathcal{B}(f)$ is the length of a shortest walk $\mathbf{w}$ in G from D to D' such that $[\mathbf{w}] = f$.*

We say that a vertex D is *reachable from* a vertex D' in a k-graph G if there is a walk in G from D to D'. Vertices D and D' are called *mutually reachable* if D is reachable from D' and D' is reachable from D. A k-graph is called *strongly connected* if all its vertices are mutually reachable. Obviously, mutual reachability is an equivalence relation on vertices and it partitions the set of vertices $V(G)$ into equivalence classes $V(G) = V_1 \cup \ldots \cup V_r$. Subgraphs $G_1, \ldots, G_r$ induced by $V_1, \ldots, V_r$ are called *strongly connected components* of G. Evidently, every strongly connected component is strongly connected.

Lemma 4. *For any walk $\mathbf{w}$ in a strongly connected k-graph G over $\mathcal{B} \subseteq \mathbf{T}_n$ there is an equivalent walk $\mathbf{w}'$ such that $\ell(\mathbf{w}') < 2|V(G)| \cdot (\ell(\mathbf{S}_k) + 1) - 1$.*

Proof. Given a walk $\mathbf{w} = D_0, f_1, D_1, \ldots, f_\ell, D_\ell$ in the k-graph G. For any vertex D in G we define a walk $\mathbf{w}_D$, equivalent to $\mathbf{w}$, called a *D-saturation* of $\mathbf{w}$, as follows. For every vertex D_i, $0 \leq i \leq \ell$, we consider two paths[4]: $\mathbf{p}_{D_i \to D}$ from D_i to D and $\mathbf{p}_{D \to D_i}$ from D to D_i. Connecting them, we obtain the closed walk $\mathbf{c}_i = \mathbf{p}_{D_i \to D}, \mathbf{p}_{D \to D_i}$. Since G is strongly connected, it follows that these two paths exist and $\ell(\mathbf{p}_{D \to D_i}), \ell(\mathbf{p}_{D \to D_i})$ are bounded by $|V(G)| - 1$. For every closed walk $\mathbf{c}_i$ we consider the permutation $\pi_i = [\mathbf{c}_i]$ of the set D_i. Let m_i be

---

[4] A *path* is a walk in which all vertices and edges are distinct.

the *order* of c_i, i.e., the smallest positive integer m such that $\pi_i^m = e_{D_i}$ (where e_M denotes the identity map on M). Finally, by definition, put

$$\mathbf{w}_D = \mathbf{c}_0^{m_0}, f_1, \mathbf{c}_1^{m_1}, \ldots, f_\ell, \mathbf{c}_\ell^{m_\ell}. \tag{6}$$

It now follows that

$$[\mathbf{w}_D] = \pi_0^{m_0} f_1 \pi_1^{m_1} \ldots f_\ell \pi_\ell^{m_\ell} = e_{D_0} f_1 e_{D_1} f_2 \ldots f_\ell e_{D_\ell} = f_1 f_2 \ldots f_\ell = [\mathbf{w}],$$

and we see that the D-saturation $\mathbf{w}_D$ is equivalent to the walk $\mathbf{w}$.

Consider all the occurrences of the vertex D in the walk $\mathbf{w}_D$. These occurrences partition $\mathbf{w}_D$ into subwalks, i.e., $\mathbf{w}_D = \mathbf{w}_0, \mathbf{w}_1, \ldots, \mathbf{w}_s, \mathbf{w}_{s+1}$, where $\mathbf{w}_0$ is the subwalk from the begin to the first occurrence of D, $\mathbf{w}_{s+1}$ is the subwalk from the last occurrence of D to the end, and the closed subwalks $\mathbf{w}_1, \ldots, \mathbf{w}_s$ connect successive occurrences of D. Using (6) and recalling that $\mathbf{c}_i = \mathbf{p}_{D_i \to D}, \mathbf{p}_{D \to D_i}$, where $\ell(\mathbf{p}_{D \to D_i})$ and $\ell(\mathbf{p}_{D \to D_i})$ are bounded by $|V(G)| - 1$, we have $\ell(\mathbf{w}_0) \le |V(G)| - 1$, $\ell(\mathbf{w}_{s+1}) \le |V(G)| - 1$, and $\ell(\mathbf{w}_i) \le 2|V(G)| - 1$ for $i = 1, \ldots, s$. Let $\pi_1 = [\mathbf{w}_1], \ldots, \pi_s = [\mathbf{w}_s]$. Consider the set $\mathcal{B} = \{\pi_1, \ldots, \pi_s\}$ and the permutation $\pi = \pi_1 \ldots \pi_s \in \langle \mathcal{B} \rangle$. Now note that $\pi, \pi_1, \ldots, \pi_s$ are permutations of the same k-element set D. Thus, taking into account (3), we obtain $\pi = \pi_{i_1} \ldots \pi_{i_r}$, where $r \le \ell(\mathbf{S}_k)$.

Finally, let $\mathbf{w}' = \mathbf{w}_0, \mathbf{w}_{i_1}, \ldots, \mathbf{w}_{i_r}, \mathbf{w}_{s+1}$. Then we get

$$[\mathbf{w}'] = [\mathbf{w}_0]\pi_{i_1} \ldots \pi_{i_r}[\mathbf{w}_{s+1}] = [\mathbf{w}_0]\pi_1 \ldots \pi_r[\mathbf{w}_{s+1}] = [\mathbf{w}_D] = [\mathbf{w}],$$

i.e., $\mathbf{w}'$ is equivalent to $\mathbf{w}$. Moreover, we have

$$\ell(\mathbf{w}') \le 2(|V(G)| - 1) + (2|V(G)| - 1) \cdot \ell(\mathbf{S}_k) < 2|V(G)| \cdot (\ell(\mathbf{S}_k) + 1) - 1.$$

The lemma is proved. ∎

In the previous lemma we deal with strongly connected k-graphs only. More general case is considered in the next lemma.

Lemma 5. *For any walk $\mathbf{w}$ in a k-graph G over $\mathcal{B} \subseteq \mathbf{T}_n$ there is an equivalent walk $\mathbf{w}'$ such that $\ell(\mathbf{w}') < 2|V(G)| \cdot (\ell(\mathbf{S}_k) + 1)$.*

Proof. Consider an arbitrary walk $\mathbf{w}$ in G. It is readily seen that this walk can be represented as $\mathbf{w} = \mathbf{w}_1, f_1, \mathbf{w}_2, \ldots, f_{s-1}\mathbf{w}_s$, where every subwalk $\mathbf{w}_i$ belongs completely to one strong component G_i of G and all the components $G_1, \ldots, G_s$ are different. On the other hand, from Lemma 4 it follows that for any walk $\mathbf{w}_i$, $1 \le i \le s$, there exists an equivalent walk $\mathbf{w}_i'$ such that $\ell(\mathbf{w}_i') < 2|V(G_i)| \cdot (\ell(\mathbf{S}_k) + 1) - 1$. Then we let $\mathbf{w}' = \mathbf{w}_1', f_1, \mathbf{w}_2', \ldots, f_{s-1}, \mathbf{w}_s'$ and obtain that $[\mathbf{w}'] = [\mathbf{w}]$. Moreover, we have

$$\ell(\mathbf{w}') = \sum_{i=1}^{s} \ell(\mathbf{w}_i') + s - 1 < \left(\sum_{i=1}^{s} 2|V(G_i)|\right) \cdot (\ell(\mathbf{S}_k) + 1) \le 2|V(G)| \cdot (\ell(\mathbf{S}_k) + 1).$$

This proves the lemma. ∎

Consider an arbitrary basis $\mathcal{B} \subseteq \mathbf{T}_n$. It is clear that for the k-graph G over $\mathcal{B}$ we have $|V(G)| = \binom{n}{k}$. Therefore from Lemmas 3 and 5 it follows that

$$\ell(\mathbf{T}_n^{(k)}) < 2\binom{n}{k}(\ell(\mathbf{S}_k) + 1). \tag{7}$$

Combining this fact with equality (4), we obtain the following

Lemma 6. *We have* $\ell(\mathbf{T}_n^{(k)}) < \binom{n}{k}e^{\sqrt{k \ln k}(1+o(1))}$ *as* $n, k \to \infty$.

Consider a basis $\mathcal{B} \subseteq \mathbf{T}_n$ and a map $f \in \langle \mathcal{B} \rangle$. Let $f = f_1 \ldots f_\ell$ be a shortest representation of f over $\mathcal{B}$, i.e., $\ell = \ell_\mathcal{B}(f)$. Thus we have

$$D_0 \xrightarrow{f_1} D_1 \xrightarrow{f_2} \ldots \xrightarrow{f_\ell} D_\ell,$$

where $D_i = \Omega_n f_1 \ldots f_i$. Suppose $k_i = |D_i|$ for $i = 0, \ldots, \ell$; then we obtain

$$k_0 = \ldots = k_{i_1} = r_0 > k_{i_1+1} = \ldots = k_{i_2} = r_1 > \ldots > k_{i_s+1} = \ldots = k_\ell = r_s.$$

Therefore we get $f = g_0 f_{i_1} g_1 \ldots g_{s-1} f_{i_s} g_s$, where $g_i \in \mathbf{T}_n^{(r_i)}$, $0 \le i \le s$. Thus it is easily shown that $\ell_\mathcal{B}(f) = s + \ell_\mathcal{B}(g_0) + \cdots + \ell_\mathcal{B}(g_s)$. It is clear that $s \le n - 1$. Therefore, we have

$$\ell_\mathcal{B}(f) \le n - 1 + \ell(\mathbf{T}_n^{(r_0)}) + \cdots + \ell(\mathbf{T}_n^{(r_s)}) < n \max_{0 \le i \le s} \left\{ \ell(\mathbf{T}_n^{(r_i)}) + 1 \right\}.$$

Finally, we obtain

$$\ell(\mathbf{T}_n) < n \max_{1 \le k \le n} \left\{ \ell(\mathbf{T}_n^{(k)}) + 1 \right\}. \tag{8}$$

Lemma 7. *We have* $\ell(\mathbf{T}_n) \le 2^n e^{\sqrt{\frac{n}{2} \ln n}(1+o(1))}$ *as* $n \to \infty$.

Proof. Using (4), (7), and (8), we obtain $\ln \ell(\mathbf{T}_n) \le \ln n + \ell(n)$, where

$$\ell(n) = \max_{1 \le k \le n} \left\{ \ln \binom{n}{k} + \varphi(k) \right\},$$

and $\varphi(k)$ is a function such that $\varphi(k) \sim \sqrt{k \ln k}$ as $k \to \infty$.

Recall that the function $\ln \binom{n}{k}$ achieves its maximum value at the point $k_n = \lceil n/2 \rceil$ when n is fixed. Suppose $\ln \binom{n}{k} + \varphi(k)$ achieves its maximum value at $k_n' = k_n + h_n$, i.e., $\ell(n) = \ln \binom{n}{k_n'} + \varphi(k_n')$. We claim that $h_n/n \to 0$ as $n \to \infty$. Indeed, in the converse case, we can take $\varepsilon \in (0, 1/2)$ such that $|h_n/n| \ge \varepsilon$ holds for an infinite sequence of indexes n. Further, since we have $|h_n/n| \in [\varepsilon, 1/2]$ for this sequence; then it has an infinite subsequence $n_1, n_2, \ldots, n_i, \ldots$ such that $h_{n_i}/n_i \to a \in [-1/2, -\varepsilon] \cup [\varepsilon, 1/2]$ as $i \to \infty$. On the other hand, it is well known that

$$\frac{\ln \binom{n}{m}}{n} \to H(p) \tag{9}$$

as $n, m \to \infty$ and $m/n \to p \in [0,1]$, where[5]

$$H(p) = -p \ln p - (1-p) \ln(1-p)$$

is the *entropy function*. Since $\varphi(k_{n_i}) = o(n_i), \varphi(k'_{n_i}) = o(n_i), k'_{n_i}/n_i \to 1/2 + a$, and $k_{n_i}/n_i \to 1/2$ as $i \to \infty$, we obtain:

$$\lim_{i \to \infty} \frac{\ln \binom{n_i}{k'_{n_i}} + \varphi(k'_{n_i})}{n_i} = H\left(\frac{1}{2} + a\right) \geq \lim_{i \to \infty} \frac{\ln \binom{n_i}{k_{n_i}} + \varphi(k_{n_i})}{n_i} = H\left(\frac{1}{2}\right).$$

The latter contradicts the fact that the function $H(p)$ achieves its maximum value at the point $1/2$ only (see Fig. 1). This contradiction proves that $h_n/n \to 0$ as $n \to \infty$.

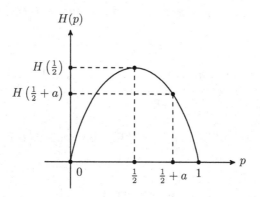

Fig. 1. Entropy function $H(p)$

Further, since the function $\ln \binom{n}{k} + \varphi(k_n)$ achieves its maximum value at $k = k'_n$ and the function $\ln \binom{n}{k}$ at $k = k_n$ when n is fixed; then we obtain

$$0 \leq \ln \binom{n}{k'_n} + \varphi(k'_n) - \left(\ln \binom{n}{k_n} + \varphi(k_n) \right) \leq \varphi(k'_n) - \varphi(k_n). \qquad (10)$$

Since $h_n/n \to 0$, we see that $k_n \sim k'_n$ and $\sqrt{k_n \ln k_n} \sim \sqrt{k'_n \ln k'_n}$ as $n \to \infty$. From $\varphi(k_n) \sim \sqrt{k_n \ln k_n}$ and $\varphi(k'_n) \sim \sqrt{k'_n \ln k'_n}$ it follows that $\varphi(k_n) \sim \varphi(k'_n)$. Hence $\varphi(k'_n) - \varphi(k_n) = o(\sqrt{k_n \ln k_n}) = o(\sqrt{n \ln n})$ as $n \to \infty$.

Thus, recalling that $\binom{n}{\lceil n/2 \rceil} \sim \sqrt{\frac{2}{\pi n}} \cdot 2^n$ as $n \to \infty$, from (10) it follows that

$$\ln \binom{n}{k'_n} + \varphi(k'_n) = \ln \binom{n}{k_n} + \varphi(k_n) + o(\sqrt{n \ln n}) = n \ln 2 + \sqrt{\frac{n}{2} \ln n} + o(\sqrt{n \ln n}).$$

---

[5] Here we assume that $0 \cdot \ln 0 = 0$.

Therefore

$$\ell(n) = n \ln 2 + \sqrt{\frac{n}{2} \ln n} \cdot (1 + o(1))$$

and recalling that $\ln \ell(\mathbf{T}_n) \leq \ln n + \ell(n)$, we obtain

$$\ell(\mathbf{T}_n) \leq n e^{\ell(n)} = 2^n e^{\sqrt{\frac{n}{2} \ln n}(1+o(1))} \text{ as } n \to \infty.$$

This completes the proof. □

Each semiautomaton $\mathfrak{A} = (A, Q, \delta)$ induces the transformation semigroup $\mathbf{T}(\mathfrak{A})$ acting on the set of states Q in the following way. For every word $\alpha \in A^*$, let $T_\alpha \colon Q \to Q$ be the map $q \mapsto \delta(q, \alpha)$. Then by definition, put

$$\mathbf{T}(\mathfrak{A}) = \{T_\alpha \mid \alpha \in A^*\}.$$

It is obvious that $\mathbf{T}(\mathfrak{A}) = \langle \mathcal{B} \rangle$, where $\mathcal{B} = \{T_a \mid a \in A\}$. Moreover, if $f \in \mathbf{T}(\mathfrak{A})$, then $\ell_{\mathcal{B}}(f)$ is equal to the length of a shortest word $\alpha \in A^*$ such that $f = T_\alpha$.

In Lemma 7 we obtain an upper bound on the function $\ell(\mathbf{T}_n)$. The next lemma shows that this bound is in some sense exact.

Lemma 8. *We have* $\ell(\mathbf{T}_n) \geq 2^n e^{\sqrt{\frac{n}{2} \ln n}(1+o(1))}$ *as* $n \to \infty$.

Proof. For each n and $k < n$ consider a semiautomaton $\mathfrak{A} = (A, Q, \delta)$ such that $Q = \{1, \dots, n\}$, $A = \{1, \dots, m\}$, where $m = \binom{n-1}{k}$; and the transition function δ is defined as follows. First we take in some order all k-element subsets of the set $\{1, \dots, n-1\} \subseteq Q$:

$$D_1 = \{q_1^{(1)}, \dots, q_k^{(1)}\}, \dots, D_m = \{q_1^{(m)}, \dots, q_k^{(m)}\}.$$

Further, we choose a permutation $\pi \in \mathbf{S}_k$ of the maximum order, and define the transition function such that $\delta(q_j^{(i)}, i) = q_j^{(i+1)}$ and $\delta(q_j^{(m)}, m) = q_{\pi(j)}^{(1)}$ for $i = 1, \dots, m-1$; $j = 1, \dots, k$. Moreover, we let $\delta(q, i) = n$ whenever $q \notin D_i$ for $i = 1, \dots, m$.

It is not hard to see that we have

$$D_1 \xrightarrow{1} D_2 \xrightarrow{2} \cdots \xrightarrow{m-1} D_m \xrightarrow{m} D_1. \tag{11}$$

Furthermore, we claim that we have $\delta(D_1, \alpha) = D_1$ iff $\alpha = (12 \dots m)^s$, $s \geq 0$. Indeed, if $\alpha = (12 \dots m)^s$; then from (11) we obtain $\delta(D_1, \alpha) = D_1$. Suppose we have $\delta(D_1, \alpha) = D_1$ for some word $\alpha = a_1 \dots a_\ell \in A^*$. Consider the sequence $D_1', \dots, D_\ell'$, where $D_1' = D_\ell' = D_1$ and $D_{i+1}' = \delta(D_i', a_i)$ for $i = 1, \dots, \ell - 1$. Let us show that

$$a_1 = 1, a_2 = 2, \dots, a_m = m, a_{m+1} = 1, a_{m+2} = 2, \dots, a_\ell = m. \tag{12}$$

Assume the converse, and let i be the smallest index such that condition (12) does not hold for a_i. Then it is readily seen that

$$D_1' = D_1, D_2' = D_2, \dots, D_i' = D_i \text{ and } n \in D_{i+1}' \neq D_{i+1}.$$

Further, since $\delta(n, a) = n$ for all $a \in A$; then we get $n \in \delta(D'_i, a_{i+1} \ldots a_\ell) = D'_\ell$, and hence $n \in D_1 \subseteq \{1, \ldots, n-1\}$. This contradiction proves condition (12) and we obtain $\alpha = (12 \ldots m)^s$ for some $s \geq 0$.

Let r_k be the order of the previously defined permutation $\pi \in \mathbf{S}_k$. Hence r_k is Landau's function [9], i.e., the maximum order of an element of $\mathbf{S}_k$, and we get $r_k = e^{\sqrt{k \ln k}(1+o(1))}$ as $k \to \infty$. Consider the map $f \colon q \mapsto \delta(q, (12 \ldots m)^{r_k-1})$. Since $f(D_1) = D_1$; then for each word $\alpha \in A^*$ such that $f = T_\alpha$ we have $\alpha = (12 \ldots m)^s$, and it is not hard to see that $f|_{D_1} = \pi^s$. Therefore we have $s \geq r_k - 1$ and $\ell_\mathcal{B}(f) \geq |\alpha| = ms \geq \binom{n-1}{k}(r_k - 1)$, where $\mathcal{B} = \{T_a \mid a \in A\}$. Finally, if we let $k = \lfloor n/2 \rfloor$, then we obtain the inequality

$$\ell_\mathcal{B}(f) \geq \binom{n-1}{k}(r_k - 1) = 2^n e^{\sqrt{\frac{n}{2} \ln n}(1+o(1))} \text{ as } n \to \infty.$$

This completes the proof of the lemma. $\square$

Now we can prove the first of the two main results of this paper.

Proof (of Theorem 1). The result follows from Lemmas 7 and 8. $\square$

Before we give the proof of Theorem 2, we introduce the following definitions and notions.

A *partition* of a set S is a set $\pi = \{B_1, \ldots, B_m\}$ of pairwise disjoint non-empty subsets $B_i \subseteq S$ (called *blocks*) such that $\cup_i B_i = S$. We say that a partition π' is a *refinement* of a partition π and write $\pi' \leq \pi$ if every element of π' is a subset of some element of π. It is easily shown that the set of all partitions of S is a partially ordered set with respect to the relation "$\leq$". It has the least element (called *discrete* partition), which contains $|S|$ singleton blocks, and the greatest element (called *trivial* partition), which contains one $|S|$-element block.

Given a finite automaton $\mathfrak{A} = (A, Q, B, \delta, \lambda)$ and a subset of states $S \subseteq Q$, the *initial state uncertainty* (with respect to $\mathfrak{A}$ and S) after applying input word α is a partition π_α of S such that two states $q, q' \in S$ are in the same block iff $\lambda(q, \alpha) = \lambda(q', \alpha)$. Informally speaking, the initial state uncertainty describes what we know about the initial state $q_0 \in S$ of the automaton $\mathfrak{A}$ after applying the input word α. From the definition of a PDS for S in $\mathfrak{A}$ it follows that an input word α is a PDS iff the partition π_α is discrete. Moreover, it is easy to prove that for every $\alpha, \beta \in A^*$ the partition $\pi_{\alpha\beta}$ is a refinement of π_α.

Proof (of Theorem 2). Given an n-state automaton $\mathfrak{A}$ and a k-element subset S of its states. Let $\alpha = a_1 \ldots a_\ell$ be a minimum length PDS for the subset S. For each $i \in \{0, 1, \ldots, \ell\}$ we consider the two values $k_i = |\delta(S, a_1 \ldots a_i)|$ and $r_i = |\pi_{a_1 \ldots a_i}|$. It is clear that $k_0 \geq k_1 \geq \cdots \geq k_l$ and $r_0 \leq r_1 \leq \cdots \leq r_l$. Let $i_1, \ldots, i_m$ be the increasing sequence of all indexes $i \in \{1, \ldots, \ell\}$ such that $k_{i-1} > k_i$ or $r_{i-1} < r_i$. Since α is a minimum length PDS, then $r_{\ell-1} < r_\ell$ and $i_m = \ell$. Let also $i_0 = 0$. Hence the word α can be represented as $\alpha = \alpha_1 a_{i_1} \ldots \alpha_m a_{i_m}$. Moreover, it is readily seen that $m \leq 2(n-1)$.

Further, for each $j \in \{1, \ldots, m\}$ there exists an input word α'_j such that $T_{\alpha_j}|_{S_{j-1}} = T_{\alpha'_j}|_{S_{j-1}}$ and $|\alpha'_j| \leq \ell(\mathbf{T}_n^{(p_j)})$, where $S_j = \delta(S, a_1 \ldots a_{i_j})$, $p_j = k_{i_{j-1}} =$

$k_{i_j-1+1} = \cdots = k_{i_j-1}$. We claim that the word $\alpha' = \alpha_1' a_{i_1} \ldots \alpha_m' a_{i_m}$ is also a PDS for S. Indeed, in the converse case, there exist two states $q_1, q_2 \in S$ such that $\lambda(q_1, \alpha') = \lambda(q_2, \alpha')$. Since α is a PDS, we obtain $\lambda(q_1, \alpha) \neq \lambda(q_2, \alpha)$. Let j be the minimum index such that

$$\lambda(q_1, \alpha_1 a_{i_1} \ldots \alpha_j a_{i_j}) \neq \lambda(q_2, \alpha_1 a_{i_1} \ldots \alpha_j a_{i_j}).$$

Then from $r_{i_j} > r_{i_j-1} = \cdots = r_{i_{j-1}}$ and the minimality of the index j it follows that

$$\lambda(q_1, \alpha_1 a_{i_1} \ldots \alpha_j) = \lambda(q_2, \alpha_1 a_{i_1} \ldots \alpha_j).$$

Therefore for the states $q_1' = \delta(q_1, \alpha_1 a_{i_1} \ldots \alpha_j)$, $q_2' = \delta(q_2, \alpha_1 a_{i_1} \ldots \alpha_j)$ we get $\lambda(q_1', a_{i_j}) \neq \lambda(q_2', a_{i_j})$. At the same time since $T_{\alpha_1 a_{i_1} \ldots \alpha_j}|_S = T_{\alpha_1' a_{i_1} \ldots \alpha_j'}|_S$, we have $q_1' = \delta(q_1, \alpha_1' a_1 \ldots \alpha_j')$, $q_2' = \delta(q_2, \alpha_1' a_1 \ldots \alpha_j')$ and we finally obtain

$$\lambda(q_1, \alpha_1' a_1 \ldots \alpha_j' a_{i_j}) \neq \lambda(q_2, \alpha_1' a_1 \ldots \alpha_j' a_{i_j}).$$

Therefore the word α' distinguishes the states q_1, q_2 and hence is a PDS for S. Moreover, we have $|\alpha'| \leq m + |\alpha_1'| + \cdots + |\alpha_m'| \leq m + \ell(\mathbf{T}_n^{(p_1)}) + \cdots + \ell(\mathbf{T}_n^{(p_m)})$ and therefore

$$\ell(n, k) < m \max_{1 \leq p \leq k} \left\{ \ell(\mathbf{T}_n^{(p)}) + 1 \right\}.$$

Since the function $\ell(\mathbf{S}_k)$ is increasing; then from $k \leq n$, $m \leq 2(n-1)$, asymptotic equality (4), and inequality (7) it follows that

$$\ell(n, k) < \binom{n}{k} e^{\sqrt{n \ln n}(1+o(1))} \quad \text{if } k \leq \frac{n}{2}; \tag{13}$$

$$\ell(n, k) < 2^n e^{\sqrt{n \ln n}(1+o(1))} \quad \text{if } k > \frac{n}{2}. \tag{14}$$

To conclude the proof, it remains to use inequalities (1) and (2) with asymptotic equality (9). $\qquad\square$

3 Remarks and Related Work

Despite the fact that the length of a shortest PDS is exponential in the worst case in the class of all Mealy automata there are a number of natural automata classes where it is much smaller. For example, for the class of linear automata it is only logarithmic [2] and for the class of automata with finite memory it is linear [15] in the number of states. Moreover, if in a reduced automaton $\mathfrak{A}$ for each input symbol a and for each pair of different states q, q' such that $\delta(q, a) = \delta(q', a)$ we always have $\lambda(q, a) \neq \lambda(q', a)$ then every preset homing sequence (PHS) for $\mathfrak{A}$ is also a PDS for $\mathfrak{A}$ [15]. Hence using the classical result of Hibbard [6] for PHSs it immediately follows that for any such n-state automaton a PDS always exists, can be efficiently computed, and the length of a shortest PDS is upper bounded by $\frac{n(n-1)}{2}$. Moreover, this upper bound is tight [6,7]. The class of such automata was investigated by the author in [13] under the name multiply reduced automata. It is interesting to note that exactly the same class was considered in a recent paper [5] under the name DMFSM where an $O(n^3)$ upper bound on the PDS length was obtained and an $O(n^2)$ upper bound was only conjectured.

References

1. Babai, L.: On the diameter of eulerian orientations of graphs. In: SODA 2006: Proceedings of the Seventeenth Annual ACM-SIAM Symposium on Discrete algorithms, pp. 822–831. ACM, New York (2006)
2. Cohn, M.: Properties of linear machines. J. ACM 11(3), 296–301 (1964), http://doi.acm.org/10.1145/321229.321233
3. Gazdag, Z., Iván, S., Nagy-György, J.: Improved upper bounds on synchronizing nondeterministic automata. Information Processing Letters 109(17), 986–990 (2009), http://www.sciencedirect.com/science/article/pii/S0020019009001811
4. Gill, A.: State-identification experiments in finite automata. Inform. Control 4(2-3), 132–154 (1961), http://www.sciencedirect.com/science/article/pii/S001999586180003X
5. Güniçen, C., İnan, K., Türker, U.C., Yenigün, H.: The relation between preset distinguishing sequences and synchronizing sequences. Formal Aspects of Computing, 1–15 (2014), http://dx.doi.org/10.1007/s00165-014-0297-8
6. Hibbard, T.N.: Least upper bounds on minimal terminal state experiments for two classes of sequential machines. J. ACM 8(4), 601–612 (1961)
7. Karacuba, A.A.: Solution to a problem in the theory of finite automatons. Uspehi Mat. Nauk 15(3) (93), 157–159 (1960)
8. Kohavi, Z.: Switching and finite automata theory. McGraw-Hill (1970)
9. Landau, E.: Über die maximalordnung der permutationen gegebenes grades. Archiv der Math. und Phys. 5, 92–103 (1903)
10. Lee, D., Yannakakis, M.: Principles and methods of testing finite state machines – a survey. Proceedings of the IEEE 84(8), 1090–1123 (1996)
11. Martyugin, P.: A lower bound for the length of the shortest carefully synchronizing words. Russian Mathematics 54(1), 46–54 (2010), http://dx.doi.org/10.3103/S1066369X10010056
12. Moore, E.F.: Gedanken experiments on sequential machines. In: Shannon, C., McCarthy, J. (eds.) Automata Studies, pp. 129–153. Princeton U. (1956)
13. Panteleev, P.A.: On the distinguishability of states of an automaton under distortions at the input. Intellekt. Sist. 11(1–4), 653–678 (2007) (in Russian)
14. Rystsov, I.K.: Asymptotic estimate of the length of a diagnostic word for a finite automaton. Cybernetics and Systems Analysis 16, 194–198 (1980)
15. Rystsov, I.K.: Diagnostic words for automata having a finite memory. Cybernetics 9(6), 927–928 (1973), http://dx.doi.org/10.1007/BF01071671
16. Salomaa, A.: Composition sequences for functions over a finite domain. Theoret. Comput. Sci. 292, 263–281 (2003)
17. Sokolovskii, M.N.: Diagnostic experiments with automata. Cybernetics and Systems Analysis 7, 988–994 (1971)
18. Sokolovskii, M.N.: The complexity of the generation of transformations, and experiments with automata. In: Discrete analysis methods in the theory of codes and schemes, vol. 29, pp. 68–86. Institute of Mathematics, Siberian. Branch USSR Acad. Sci. (1976) (in Russian)

Hierarchy and Expansiveness
in 2D Subshifts of Finite Type

Charalampos Zinoviadis[1,2]([⊠])

[1] TUCS-Turku Center for Computer Science, University of Turku,
20014 Turku, Finland
[2] Department of Mathematics and Statistics, University of Turku,
20014 Turku, Finland
chzino@utu.fi

Abstract. We present a hierarchical construction of 2D subshifts of finite type with a unique direction of non-expansiveness. Our construction combines various techniques that were developed in previous self-similar constructions of SFTs and cellular automata.

Keywords: Subshift of finite type · Multidimensional symbolic dynamics · Expansiveness · Dynamical systems · Self-similarity

Introduction

Two-dimensional subshifts of finite type (2D SFT) are a well-known model of dynamical systems that can also perform universal computation [14]. As multidimensional dynamical systems, they can be studied from the point of view of their subdynamics, defined by Boyle and Lind in [1], and especially with respect to their expansive subdynamics. The study of the expansive directions of 2D SFTs was initiated by Kari in [9]. Even though the SFT constructed in that paper did not have any direction of expansiveness, it had a direction of determinism, which is a weaker notion that can be seen as "semi"-expansiveness. Soon, Kari and Papazoglu gave an example of an aperiodic 2D SFT with all but two directions expansive, see [10]. This left open only the case of a unique direction of non-expansiveness (extremely expansive), since it is known that any SFT that does not have any direction of non-expansiveness must be finite, and hence periodic.

In [3], Durand, Romashchenko and Shen gave a very general construction of aperiodic 2D SFTs, which they call fixed-point tile sets. Their method finds inspiration in Gács' construction of reliable cellular automata, see [4]. However, their SFTs are not expansive in any direction. In [7], Hochman constructed 2D subshifts with a unique direction of non-expansiveness. However, his subshifts are not of finite type. Our construction can be seen as a blending of these two

The author was supported by TUCS (Turku Center for Computer Science) and the Academy of Finland Grant 131558.

© Springer International Publishing Switzerland 2015
A.-H. Dediu et al. (Eds.): LATA 2015, LNCS 8977, pp. 365–377, 2015.
DOI: 10.1007/978-3-319-15579-1_28

constructions, which keeps the good properties of each one. As a result, we obtain aperiodic 2D SFTs with a unique direction of non-expansiveness. Furthermore, the same methods can be used to obtain new results about extremely expansive 2D SFTs.

In this paper, we give an informal presentation of the construction, referring to the papers [3] and [7] and mentioning the places where we do things differently and why we have to do this. A complete description from scratch needs a long article, so we also assume some familiarity with these papers and their ideas.

In Section 1 we give the basic definitions and some technical notions that we need . In Section 2, we describe the construction of a rule that can simulate any RPCA. In section 3, we use this RPCA to construct an aperiodic extremely expansive 2D SFT. Finally, as an application of our method, in Section 4 we construct extremely-expansive SFT covers for the limit sets of 2D substitutions.

1   Preliminaries

We will denote by $\mathbb{Z}, \mathbb{N}$ and $\mathbb{R}$ the set of integers, non-negative integers and real numbers,respectively. $[i, j[$ and $[i, j]$ denote the integer intervals $\{i, \ldots, j - 1\}$ and $\{i, \ldots, j\}$, respectively, while $[x, y]$ denotes an interval of real numbers.

An **alphabet** is any finite set, whose elements are called **letters**. A special alphabet that we will often use is $\mathcal{C} = \{0, 1, \$, \sharp\}$. If $\mathcal{A}$ is an alphabet, $\mathcal{A}^* = \bigcup_{l \in \mathbb{N}} \mathcal{A}^l$ denotes the set of finite **words** over $\mathcal{A}$ and $\mathcal{A}^{**} = \bigcup (\mathcal{A}^*)^n$ the set of finite tuples of words. If $w \in \mathcal{A}^n$, we write $w = w_0 \cdots w_{n-1}$, and call $\|w\| = n$ **length** of w. If $u \in (\mathcal{A}^*)^l$, we write $u = (u_1, \ldots, u_l)$, and call l the number of **fields** of u. For every vector $\boldsymbol{k} \in \mathbb{N}^l$, where $l \in \mathbb{N}$, let $\mathcal{B}_{\boldsymbol{k}} = \mathcal{C}^{k_1} \times \ldots \times \mathcal{C}^{k_l}$. If $\mathcal{B} \subseteq \mathcal{B}_{\boldsymbol{k}}$, for some vector $\boldsymbol{k}$, then we say that $\mathcal{B}$ has **constant lengths**. The notion of alphabets with constant lengths is a technicality which allows us to implement the exchange of information between so-called colonies in an efficient way.

If $n \in \mathbb{N}$, then we denote by $\|n\|$ the length of its quaternary representation.If $w \in \mathcal{C}^*$, $[w]$ is the number represented by w in the quaternary representation system, where $\$$ is interpreted as 2 and $\sharp$ as 3.

For every $i \in \mathbb{N}$, $\pi_i \colon \mathcal{C}^{**} \to \mathcal{C}^*$ is the **projection** onto the i'th coordinate. We can see π_i as a partial function of $\mathcal{C}^{**}$, by leaving it undefined over those tuples that have less than i components. A **field** is a projection π_i together with a label `Field`, written in typewriter form. We denote $\pi_i(u) = u.\texttt{Field}$ the value of `Field` in u. The notion of fields makes it possible to talk about the components of a tuple in a more intuitive way.

1.1   Symbolic Dynamics and Cellular Automata

$\mathcal{A}^{\mathbb{Z}^d}$ is the set of d-dimensional **configurations**, endowed with the product of the discrete topology, and with the **shift** dynamical system σ, defined as the action of $\mathbb{Z}^d$ by $(\sigma^i)_{i \in \mathbb{Z}^d}$, where for any configuration $x \in \mathcal{A}^{\mathbb{Z}^d}$ and any $k \in \mathbb{Z}^d$, $\sigma^i(x)_k = x_{i+k}$.

A **pattern** over a *finite* **support** $D \subset \mathbb{Z}^d$ is a map $p \in \mathcal{A}^D$. A (d-dimensional) **subshift** is defined as the set of configurations that avoid a set of forbidden patterns: There exists a family of patterns $\mathcal{F} \subset \bigcup_{D \subset_{\text{finite}} \mathbb{Z}^d} \mathcal{A}^D$ such that $X_{\mathcal{F}} = \left\{ x \in \mathcal{A}^{\mathbb{Z}^d} \,\middle|\, \forall i \in \mathbb{Z}^d, \forall D \subset_{\text{finite}} \mathbb{Z}^d, \sigma^i(x)_{|D} \notin \mathcal{F} \right\}$. If $\mathcal{F}$ is a finite family of patterns, we say that $X_{\mathcal{F}}$ is a **subshift of finite type (SFT)**. A subshift $Y \subseteq \mathcal{B}^{\mathbb{Z}^d}$ is called **sofic** if it is the image of an SFT X through an alphabet projection. X is called an **SFT cover** of Y.

A configuration $c \in \mathcal{A}^{\mathbb{Z}^d}$ is called **periodic** if there exists $i \in \mathbb{Z}^d$ such that $\sigma^i(c) = c$. A subshift X is called **aperiodic** if it does not contain any periodic configurations.

A (1D) **partial cellular automaton** (PCA) is a partial map $F : \mathcal{A}^{\mathbb{Z}} \to \mathcal{A}^{\mathbb{Z}}$ defined by a partial local map: There exists $r \in \mathbb{N}$, called the **radius** of F, and a partial map $f : \mathcal{A}^{[-r,r]} \to \mathcal{A}$ such that for all $c \in \mathcal{A}^{\mathbb{Z}}$ and $i \in \mathbb{Z}$, $F(c)_i = f(c_{|i+[-r,r]})$; $F(c)$ is not defined if and only if $f(c_{|i+[-r,r]})$ is not defined for some $i \in \mathbb{Z}$. f is called the **local rule** of F, and F is called the **global map** induced by f. The **domain** $dom(F)$ of F is the set of those configuration c for which $F(c)$ is defined.

A PCA is called **reversible** (RPCA) if it is injective. In this case, it is known that there exists another RPCA, denoted by F^{-1}, such that $F^{-1}F$ and FF^{-1} are restrictions of Id, and $dom(F^{-1}) = F(\mathcal{A}^{\mathbb{Z}})$ [6]. In particular, there exist so-called inverse radius and inverse local rule. If r is both the radius and the inverse radius of an RPCA F, we call it the **bi-radius** of F. In the rest of the paper, we only need RPCA of bi-radius 1.

If F is an RPCA, then we say that $c \in \mathcal{A}^{\mathbb{Z}}$ is **ultimately rejected** if there exists $m \in \mathbb{Z}$ such that $F^m(c)$ is not defined. There is a very natural way to associate a 2D SFT $\mathcal{X}_F$ to an RPCA F: it consists of the space-time diagrams of the configurations that are not ultimately rejected. These are called the **valid** configurations of F. Formally,

$$\mathcal{X}_F = \left\{ x \in \mathcal{A}^{\mathbb{Z}^2} \,\middle|\, \forall t \in \mathbb{Z}, x_{|\mathbb{Z} \times \{t\}} = F(x_{|\mathbb{Z} \times \{t-1\}}) \right\} \ .$$

A **partitioned PCA** (PPCA) is a PCA $F = (\sigma^{i_1} \times \ldots \times \sigma^{i_l}) \circ \mu$ over some alphabet $\mathcal{A} = \mathcal{A}_1 \times \ldots \times \mathcal{A}_l$, where μ is a partial permutation of $\mathcal{A}$, and $\sigma^{i_1} \times \ldots \times \sigma^{i_l}$ is a collection of shift maps on the components of $\mathcal{A}$. Integer i_j is called the **direction** of the j'th field. We describe PPCA by giving a list of their fields, with the directions of the fields in parentheses, and by describing their partial permutation.

PPCA are reversible, and conversely every RPCA F is essentially partitioned (see for instance [8, propsps53]). We will use PPCA rather than general RPCA because of their nice structure. The bi-radius of a PPCA is equal to $\max\{|i_1|, \ldots, |i_l|\}$. In this article, each $i_j \in \{-1, 0, 1\}$.

When we want to define families of PCA, it is convenient to do it by allowing the alphabet of the PCA to be infinite. In other words, a **PCA with infinite alphabet** (IPCA) is defined by a local function $f : \mathcal{A}^{[-r,r]} \to \mathcal{A}$, where $\mathcal{A}$

is an infinite set. If $(\mathcal{A}_i)_{i \in I}$ is a family of finite subalphabets of $\mathcal{A}$ satisfying $f(\mathcal{A}_i^{\mathbb{Z}}) \subseteq \mathcal{A}_i^{\mathbb{Z}}$, then the family of local rules $(f_{|\mathcal{A}_i})_{i \in I}$ defines a family of PCA.

Some definition similar to IPCA is always, even though implicitly, used in self-similar constructions. The idea is to give a uniform construction of a family of SFTs (or CA) that depend on a set of parameters. Then, for suitable values of these parameters, one can prove that the corresponding SFT (or CA) behaves in a certain desirable way. This is done explicitly in [4], even though it is lost amidst the general complexity of that construction. Using IPCA, we can make formal, precise (even though usually complicated) statements that can be useful to anyone who understands the ideas of fixed-point constructions, but still wants to check the precise details.

In order to define useful families of RPCA with infinite alphabet (IRPCA), we will use the corresponding version of PPCA. Let $\mathcal{A}_1, \ldots, \mathcal{A}_l$ be infinite sets and μ be a partial permutation of $\mathcal{A} = \mathcal{A}_1 \times \ldots \times \mathcal{A}_l$. Then the function $F = (\sigma^{i_1} \times \ldots \times \sigma^{i_l}) \circ \mu$ is an **infinite PPCA** (IPPCA). Any restriction of F to a stable finite subalphabet is a PPCA, and hence an RPCA. A similar approach has already been used in [2].

1.2 Expansiveness

Definition 1. *Let X be a 2D subshift, $l \in \mathbb{R} \sqcup \{\infty\}$ be a **direction**, $\mathbf{l} \subseteq \mathbb{R}^2$ the line of slope l passing through the origin and $\mathbf{v} \in \mathbb{R}^2$ a unit vector orthogonal to $\mathbf{l}$. We say that direction l is **expansive** for X if there exists $b \in \mathbb{R}$, called the **radius of expansiveness**, such that:*

$$\forall x, y \in X, x_{|(\mathbf{l}+[-b,b]\mathbf{v}) \cap \mathbb{Z}^2} = y_{|(\mathbf{l}+[-b,b]\mathbf{v}) \cap \mathbb{Z}^2} \Rightarrow x = y \ .$$

We denote by $\mathcal{E}(X)$ the set of expansive directions of X and by $\mathcal{N}(X)$ its complement, which is called the set of non-expansive directions.

Expansive directions were first introduced by Boyle and Lind [1] in a more general setting. In the case of subshifts, l is an expansive direction of X if every configuration of X is uniquely determined by values of the cells in the infinite strip of slope l and width $2b$ that passes through the origin.

The following is a particular case of [1, Theorem 3.7].

Proposition 2. *Let X be a 2D subshift. Then, $\mathcal{N}(X)$ is closed with respect to the one-point compactification of the euclidean topology of $\mathbb{R}$. In addition, $\mathcal{N}(X)$ is empty if and only if X is finite.*

We say that X is **extremely expansive** if $|\mathcal{N}(X)| = 1$, which is, according to Proposition 2, the most constrained nontrivial case. A sofic shift Y is called **extremely-expansively sofic** if it has an extremely expansive SFT cover. Note that an extremely-expansively sofic shift does not have to be extremely expansive itself, or even have an expansive direction.

It is straightforward to see that if F is an RPCA, then the horizontal direction is expansive for $\mathcal{X}_F$.

1.3 Simulation

Let F and G be RPCA over alphabets $\mathcal{A}$ and $\mathcal{B}$ and $S, T \in \mathbb{N}$. We say that F (S,T)-**simulates** G if there exist alphabet projections $\pi_{\text{Addr}} : \mathcal{A} \to [\![0, S[\![, \pi_{\text{Age}} : \mathcal{A} \to [\![0, T[\![$ and an injection $\psi : \mathcal{B} \to \mathcal{A}^S$ with "bulked" global map $\Psi : \mathcal{B}^{\mathbb{Z}} \to \mathcal{A}^{\mathbb{Z}}$ (*i.e.*, $\forall y \in \mathcal{B}^{\mathbb{Z}}, i \in \mathbb{Z}, \Psi(y)_{[\![iS,(i+1)S[\![} = \psi(y_i))$ such that:

1. For all $x \in \mathcal{X}_F$, there exist $s \in [\![0, S[\![$ and $t \in [\![0, T[\![$ such that $\pi_{\text{Addr}}(x_{i,j}) = s + i \bmod S$ and $\pi_{\text{Age}}(x_{i,j}) = t + j \bmod T$, for all $i, j \in \mathbb{Z}$.
2. For all $e \in \mathcal{B}^{\mathbb{Z}}$, $F^T \Psi(e) = \Psi G(e)$.
3. If $c \in F^{-T}(\mathcal{A}^{\mathbb{Z}})$, $\pi_{\text{Addr}}(c_0) = 0$ and $\pi_{\text{Age}}(c_i) = 0$ for all i, then $c = \Psi(e)$, for some $e \in G^{-1}(\mathcal{B}^{\mathbb{Z}})$.

Addr and Age partition every (2D) valid configuration into **macro-tiles**, *i.e.*, rectangles whose lower-left corner has Addr and Age equal to 0. Similarly, every (1D) configuration is divided into **colonies**, *i.e.*, segments whose leftmost-point has Addr 0. All the cells of a colony have the same Age. Every colony belongs in a unique macro-tile (in the corresponding space-time diagram). When Age = 0, every colony of F encodes one letter of $\mathcal{B}$, which we call the **simulated letter** of the colony (or macro-tile) and it takes T time steps of F to simulate one time-step of G. Intuitively, inside every macro-tile, F computes the local rule of G using T time steps.

The following Proposition is a corollary of Theorem 5.4 proven in [7]:

Proposition 3. *Let* $F_1, F_2, \ldots$ *be a sequence of RPCA with the following properties:*

1. *For all* $i \in \mathbb{N}$ *there exist* S_i, T_i *such that* F_i (S_i, T_i)-*simulates* F_{i+1}.
2. *For all* $i \in \mathbb{N}$, $\mathcal{X}_{F_i}$ *contains infinitely many points.*
3. $\lim_i \frac{\prod S_i}{\prod T_i} = 0$.

Then, for all $i \in \mathbb{N}$, $\mathcal{X}_{F_i}$ *is aperiodic and* $\mathcal{N}(\mathcal{X}_{F_i}) = \{\infty\}$.

1.4 Simulating Turing Machines with IPPCA

We define a model of Turing machine (TM) with some additional technical conditions, which, however, do not change the class of computable and, more importantly, polynomially computable functions.

In the definition of simulation, every configuration that is not ultimately rejected is divided into macro-tiles. Inside the macro-tiles, we want to perform some computation, in order to compute the new value of the simulated letter. This can be done more or less as in all previous self-similar constructions, but there are two things that are worth mentioning, because they have to do with reversibility (which was not a concern in [3,4]) and polynomial computability (which was not a concern in [7]; in fact, computability in general was not a concern in that paper).

When simulating a TM with a PCA, one source of non-reversibility could be that the simulated TM is not reversible. It is not difficult to cope with this problem. Morita, in [12], has shown that every TM can be simulated by a PPCA. In our case, we want to construct a single IPPCA that can simulate *any* TM when restricted to the appropriate subalphabet.

Another source of irreversibility is that inside the macro-tiles, we do not know when to stop simulating the TM. It is certainly not enough to stop as soon as the TM halts, because then, when going back in time from the top of the macro-tile to the bottom, we do not know when we should start running the computation backwards. In order to deal with this problem, we use the field `Schedule`. This is a field that is horizontally and vertically constant in valid configurations. We only simulate the TM when the age is between 0 and `Schedule`, thus preserving reversibilty, because we know exactly during which period to simulate the TM.

1.5   The Programming Language

We define a programming language that defines functions $\mathcal{C}^{**} \to \mathcal{C}^{**}$. We can prove that every function defined in this programming language is a partial permutation and computable in polynomial time. This allows to define all the needed IPPCA with this language.

A programming language was also explicitly defined in [4]. In [5], there is a very clear explanation of the reason that we need to use rules defined by a programming language.

2   Construction of a Universal IPPCA

In this section, we will describe the basic building blocks of the construction.

First of all, we define the rule `Grid`. This is the rule that imposes the periodic structure `Addr` and `Age`. Unlike [3], the `Addr` and `Age` not are sufficient to achieve this. We have to use some extra fields because of the constant lengths restrictions and because we want to implement this behaviour with IPPCA. However, we can easily write a program that defines a permutation called `Grid` which has 6 fields, among them `Addr` and `Age`, and which imposes the following:

For all $S, T \in \mathbb{N}$, there exists a subalphabet $\mathcal{A}_{\mathrm{Grid},S,T}$ of $\mathcal{A}_{\mathrm{Grid}}$ such that if $c \in \mathcal{A}_{\mathrm{Grid},S,T}^{\mathbb{Z}}$ is not ultimately rejected by `Grid`, then in every valid configuration the fields `Addr` and `Age` partition the plane into $S \times T$ macro-tiles, as described in Subsection 1.3. Furthermore, $\mathcal{A}_{\mathrm{Grid},S,T}$ has constant lengths and $\|\mathcal{A}_{\mathrm{Grid},S,T}\| = \mathcal{O}(\log(ST))$.

Let c be a configuration that is not ultimately rejected, `Field` be a field of the alphabet and B_i^c be the i'th colony (we arbitrarily choose the 0'th colony to be the one that contains the origin) We denote $B_i^c.\mathtt{Field} = \pi_{\mathtt{Field}}(c_{|B_i^c})$. $B_i^c.\mathtt{Field}$ is the word of length S that we obtain when we project $c_{|B_i^c}$ onto `Field`. The previous discussion implies that $B_i^c.\mathtt{Addr} = 01\ldots(S-1)$ and $B_i^c.\mathtt{Age} = t^S$, for some $t \in [\![0, T[\![$.

Here is the prototype of the method that we use: First, we define (in the programming language) a partial permutation of $(\mathcal{C}^*)^l$, for some l, then we give directions to the fields, then for some set of parameters and for all possible values of these parameters we define a subalphabet of $(\mathcal{C}^*)^l$, then we prove that for all, or at least for all sufficiently large, values of the parameters the IPPCA behaves nicely when restricted to the corresponding subalphabet, and, finally, we observe that for all values of the parameters, the subalphabet has constant lengths and we give a bound on its size.

Next, we describe the rule `MacroRule`. This rule defines a permutation over $(\mathcal{C}^*)^{14}$. Among its fields, there are `Age`, `Addr`, `Tape`, `NewTape`, `Prog` and `Schedule` plus some more fields that are not needed for an informal presentation of the behaviour of `MacroRule`. We give some fixed directions to the fields, so that together with `MacroRule` we obtain an IPPCA. Then, for all $S, T, r \in \mathbb{N}$ and $p \in 2^*$, we can define a subalphabet $\mathcal{A}_{\texttt{MacroRule},p,S,T,r}$ with constant lengths and size $\mathcal{O}(\log(STr) + \|p\|)$ such that the following holds:

Let $c \in \mathcal{A}_{\texttt{MacroRule},p,S,T,r}^{\mathbb{Z}}$ be a configuration whose colonies have `Age` $= 0$. Assume, further, that $B_i^c.\texttt{Tape}$ is a letter $b_i \in \mathcal{B}$, where $\mathcal{B}$ is a finite alphabet with constant lengths. Assume, finally, that p is the program of a rule μ of the programming language. Then, if S, T, r and p satisfy a set of inequalities, $\texttt{MacroRule}^{4r}(c)$ exists if and only if $\mu(b_i)$ exists, for all i. In this case, $B_i^{\texttt{MacroRule}^{4r}(c)}.\texttt{Tape} = \mu(b_i)$.

In fact, this statement is not totally precise, but it gives a good idea of what we do. Let us now describe how the implementation of the computation phase differs from those in [3] and [7]. `Prog` is the field where we write p, the program of the permutation μ of the simulated PPCA. Contrary to [3], we do not write the program in the beginning of the simulated letter, but instead we include it in the state of every letter of the simulating PPCA. This makes the exchange of information phase easier. First of all, when `Age` $= 0$, `MacroRule` checks that some initialization constraints are satisfied, like, for example, that its working tape is empty or that the head of the TM is on the leftmost position of the colony. Then, using the program p in `Prog`, it simulates p onto the input written on `Tape` for r steps. If it halts during the computation, then c is rejected. Also, when `Age` $= r$, it checks that the accepting state is at the leftmost position of the colony. In the model of TM that we use, this implies that $\mu(b_i)$ exists and that at `Age` $= r$ it is written in the `Tape` track of colony i. However, during the computation, the TM has written information on various tapes and we need to delete this information. For this we use the Bennett trick. We copy `Tape` onto the field `NewTape` which was initially empty and run the computation backwards. Thus, at `Age` $= 2r$, in the i'th colony, the `Tape` track contains b_i and the `NewTape` track contains $\mu(b_i)$. Then, we use the same trick backwards to delete b_i from `Tape` and we move $\mu(b_i)$ from `NewTape` to `Tape`, while at the same time deleting all information that the TM writes during the computation. Thus, at `Age` $= 4r$, `Tape` contains $\mu(b_i)$. We can also prove that when `Age` $= 4r$, the fields satisfy the initialization constraints, so that when `Age` becomes 0 again, a new computation phase can start.

Using `MacroRule`, we can simulate any permutation, provided that S, T and r are large enough. If we want to simulate any PPCA, we also have to simulate shifts. This is accomplished with by the rule `ParShift`.

`ParShift` has 11 fields, among which `Age, Addr, Tape, Dir, Schedule, RMail` and `LMail`. The `Tape` track contains the simulated letter. `Dir` is a field whose values are $-1, 0$ and $+1$ and it says whether the letter in the `Tape` will go to the colony on the left, stay at its place, or go the colony on the right, respectively. `RMail` is a field with direction $+1$ that is used to carry the letter to the right colony, while `LMail` is a letter with direction -1. For all $S, T, r \in \mathbb{N}$, we define a subalphabet $\mathcal{A}_{\texttt{ParShift}, S, T, r}$ with constant lengths and size $\mathcal{O}(\log(STr))$. Let us describe the behaviour of `ParShift`.

Let $\mathcal{B}$ be an alphabet with constant lengths and l fields. Let $\boldsymbol{d} \in \{-1, 0, +1\}^l$ be a vector of *directions*. We want to simulate the shift $\sigma_{\boldsymbol{d}} = \sigma^{d_1} \times \ldots \times \sigma^{d_l} : \mathcal{B}^{\mathbb{Z}} \to \mathcal{B}^{\mathbb{Z}}$. Let $c \in \mathcal{A}_{\texttt{ParShift}, S, T, r}$ be a configuration whose colonies have `Age` $= 4r$. Assume, further, that $\mathcal{B}_i^c.\texttt{Tape}$ is a letter $b_i \in \mathcal{B}$. Then, if the values of `Dir` are well-chosen and $T > S + 4r$, $B_i^{\texttt{ParShift}^S(c)}.\texttt{Tape} = b_i'$, where $b' = \sigma_{\boldsymbol{d}}(b)$. In other words, S steps of `ParShift` simulate one step of $\sigma_{\boldsymbol{d}}$.

At this point, we have to say something about the encoding function. Formally speaking, it does not make sense to say that the word in the info track is equal to some letter of $\mathcal{C}^{**}$, because the elements of $\mathcal{C}^{**}$ are products of words and not words. What we do instead is that we *embed* $\mathcal{C}^{**}$ into $\mathcal{C}^*$ with the encoding function χ. The details of this embedding are not important. It essentially consists of writing the fields of the letter one after the other, separating them with markers. In addition, the length of the encoding of a field depends only on the length of the field, so that letters of an alphabet with constant lengths have their markers in the same positions. Finally, in some sense, χ can be defined in the programming language.

In order to explain how this works, we need to say what it means that the `Dir` are well-chosen. If a letter is part of the encoding of the j'th field of b_i, then its `Dir` has to be equal to d_j. Otherwise, it is 0. At `Age` $= 4r$, every letter is moved to the `LMail` if `Dir` $= -1$, to `RMail` if `Dir` $= +1$ and stays at its place if `Dir` $= 0$. During the next S time steps, nothing else happens except that the letter in `LMail` travel to the left and the letters in `RMail` travel to the right. After exactly S steps, at `Age` $= 4r + S$, the letters in the mail fields have reached their destinations and are moved back to `Tape`.

We remark here that the condition of constant lengths is necessary so that the above procedure is successful. If the alphabet does not have constant lengths, then when we move the letters back from the mail fields to `Tape`, we try to move some letter onto some already occupied position, or we will have empty space between the encodings of the fields. Having the constant length position, we know that after S steps, everything fits in exactly.

By combining `MacroRule` and `ParShift`, we immediately obtain an IPPCA `Universal` over the infinite alphabet $\mathcal{A}_{\texttt{Universal}}$ that can simulate any PPCA in the following sense:

Proposition 4. *Let $\mu\colon (\mathcal{C}^*)^l \to (\mathcal{C}^*)^l$ be a permutation defined by program p, $\boldsymbol{d} \in \mathbb{N}^l$ be a vector of directions and $\mathcal{B} \subseteq (\mathcal{C}^*)^l$ be an alphabet with constant lengths. Let $F = \sigma_{\boldsymbol{d}} \circ \mu$. For all $S, T, r \in \mathbb{N}$, there exists an alphabet $\mathcal{A}_{\mathtt{Universal},p,S,T,r}$ with constant lengths and size $\mathcal{O}(\log(STr))$, and an SFT $X_{\mathcal{B}} \subseteq \mathcal{A}_{\mathtt{Universal},p,S,T,r}^{\mathbb{Z}}$, such that if S, T, r and p satisfy that*

1. $T \geq 4r + S$,
2. $S > \max(2r, \|\mathcal{B}\|)$,
3. $r \geq t_p(\|\mathcal{B}\|)$, *(where t_p is the time complexity of the program p),*

then $\mathtt{Universal}_{|X_{\mathcal{B}}}$ (S, T)-simulates $F_{|\mathcal{B}}$.

Formally speaking, we prove something different, because there are some small technical details that need to be taken care of, but we have focused on giving the basic ideas only. $\mathtt{Universal}$ is a very useful IPPCA that is used in all of the self-similar and hierarchical constructions. All we have to do is to make sure that a set of (simple) inequalities are satisfied, and then we immediately obtain a simulation simply by restricting $\mathtt{Universal}$ to a suitable 1D SFT.

3   Self-simulation and Hierarchical Simulation

Up to this point, we have constructed the IPPCA $\mathtt{Universal}$ that can simulate any PPCA $F\colon \mathcal{B}^{\mathbb{Z}} \to \mathcal{B}^{\mathbb{Z}}$ when restricted to an appropriate alphabet of constant lengths and an appropriate 1D SFT. The appropriate subalphabet is defined through a finite set of parameters that must satisfy some inequalities, while the subshift is essentially obtained by saying that the simulated letter of every colony comes from the alphabet of $\mathcal{B}$. Notice that in the statements about $\mathtt{MacroRule}$ and $\mathtt{ParShift}$ there is an assumption that the word read on the $\mathtt{Tape}$ track of every colony is a letter of $\mathcal{B}$, which is the alphabet of the PPCA we want to simulate. In general, the simulated letters can belong to other alphabets, too, so without restricting $\mathtt{Universal}$ to $X_{\mathcal{B}}$, we would not simulate F, but rather a disjoint union of all those PPCA for which the inequalities are satisfied for the given values of the parameters.

If we try to use $\mathtt{Universal}$ to construct a self-simulating PPCA, then we encounter the following problem: let p^0 be the program of $\mathtt{Universal}$. On the one hand, since $\mathcal{A}_{\mathtt{Universal},p^0,S,T,r}$ has constant lengths and size $\mathcal{O}(\log(STr))$, it is certainly possible to satisfy the inequalities, when S, T, r are sufficiently large, and by restricting $\mathtt{Universal}$ to $X_{\mathcal{A}_{\mathtt{Universal},p^0,S,T,r}}$, we obtain a PPCA F_1 that simulates $\mathtt{Universal}_{|\mathcal{A}_{\mathtt{Universal},p^0,S,T,r}}$, but not $\mathtt{Universal}_{|X_{\mathcal{A}_{\mathtt{Universal},p^0,S,T,r}}}$. To achieve this, we have to impose further restrictions, which means that we construct a PPCA F_2 such that F_2 simulates F_1, while F_1 simulates $\mathtt{Universal}_{|\mathcal{A}_{\mathtt{Universal},p^0,S,T,r}}$. We can keep on doing this for any finite number of steps, but not to the infinity. Therefore, this approach fails. This problem is explained in greater detail in [3].

In that same paper, it is said that the way to overcome this problem is to construct a new rule that checks that the simulated letter belongs to the same alphabet as the simulating ones and then do the simulation. This is achieved by "hard-wiring" the information about the simulating alphabet and then having a TM check that this information is the same for the simulated letters. This can be done in polynomial time. The imporant thing is that the TM does not "know" that it is checking that the simulated letters come from the same alphabet (since this would make the construction circular). Instead, it is just checking that some information that is "hard-wired" in its state is the same as the information appearing on the Tape track. It is only us, the constructors of the SFT that know that the TM is actually checking that, because we have set up the computation in a clever way.

First of all, by "hardwiring" some information onto the simulating alphabet, for example the width of the colonies S, we mean that the alphabet has a field MaxAddr, such that whenever we use the parameter S to define a subalphabet, MaxAddr $= S$. Similarly, we can "hard-wire" the program p onto Prog and so on. Second, thanks to the good properties of the encoding of the letters of C^{**}, for every l, we can explicitly construct two functions $v_l \colon C^{**} \to \mathbb{N}$ and $v'_l \colon C^{**} \to \mathbb{N}$ such that for every $a \in C^{**}$, the encoding of the l'th field of a is written between the position $v_l(a)$ and $v'_l(a)$. What is more important, this function can be expressed in the programming language.

This means that we can write a rule in the programming language that checks the following things: First, check that the simulated letter has the same number of fields as the simulating alphabet. This is doable, because the letters of the simulating alphabet have a fixed number of fields, so we can check that without circularity problems. Then, check that the length of every field of the simulated letter is the same as the length of the same field in the simulating alphabet. This is doable, because the simulating alphabet has constant lengths, so, in some sense, the information about its lengths is also "hard-wired" in its letters. Then, for some of its fields (we do not have to do this for all of the fields because the alphabet $\mathcal{A}_{\text{Universal},p,S,T,r}$ is defined by restricting only some of its fields), check that the value of the simulated field is the same. This is a rule that enforces that the simulated letters belong in $\mathcal{A}_{\text{Universal},p,S,T,r}$ and the configuration in $X_\mathcal{B}$. Now, if we combine this rule with Universal, we obtain the rule Self. Let p^1 be the program of Self. There is a set of inequalities, similar to the ones we had for Universal, such that when they are satisfied, we have that $\text{Self}|_{\mathcal{A}_{\text{Universal},p^1,S,T,r}}$ simulates itself. It is also not difficult to prove that the inequalities are satisfied for all large enough values of S, T, r, using the fact that $\|\mathcal{A}_{\text{Universal},p^1,S,T,r}\| = \mathcal{O}(\log(STr))$.

Corollary 5. *There exist a (S,T)- self-simulating RPCA.*

It is also possible to prove that $X_{\text{Self}|_{\mathcal{A}_{\text{Universal},p^1,S,T,r}}}$ is infinite, so that using Proposition 3 and the fact that a self-simulating RPCA gives rise to an aperiodic SFT, see [3], we obtain our main result:

Corollary 6. *There exists an aperiodic, extremely expansive 2D SFT.*

According to Proposition 2, this is the best possible result concerning expansive directions of SFTs, since an SFT without any directions of expansiveness is finite, and hence all of its points are periodic. Our next goal is to use the methods developed up to here to prove that in various cases, extremely expansive SFTs are as powerful as ordinary SFTs. In the next section, we give an example of this.

4   Extremely Expansive SFT Covers of Substitutions

A (deterministic, $S \times T$) substitution is a function $\tau \colon \mathcal{A} \to \mathcal{A}^{I_{S,T}}$, where $I_{S,T} = [\![0, S[\![\times [\![0, T[\![$. If $S = T = n$, we call τ a **square**, $n \times n$ substitution.

τ can be extended to configurations in a natural way: if $c \in \mathcal{A}^{\mathbb{Z}^2}$ and $(x, y) = (Sk + i, Tl + j)$, where $(i, j) \in I_{S,T}$, then $\tau(c)_{x,y} = \tau(c_{k,l})_{i,j}$.

Let $\Lambda_\tau^0 = \mathcal{A}^{\mathbb{Z}^2}$ and $\Lambda_\tau^{n+1} = \{\sigma^v(\tau(c)) \colon c \in \Lambda_\tau^n$ and $v \in I_{S,T}\}$. The **limit set** of τ is the set $\Lambda_\tau = \bigcap_{n \in \mathbb{N}} \Lambda_\tau^n$.

Using a geometrical construction, Mozes in [13] proved that the limit set of every substitution is sofic. The same thing was reproved in [3] using the fixed-point method, while [11] gives a geometrical construction of an SFT cover with 2 non-expansive directions. By adapting one of their ideas to our method, we are actually able to reduce the number of non-expansive directions by 1, which is the best possible.

Proposition 7. *Let τ be a $n \times n$ substitution. Then, Λ_τ has an extremely expansive SFT cover.*

Our construction is a blend of [3] and [11]. From [3], we use the hierarchical structure provided by the fixed-point method. However, there are two problems that we need to deal with. First, in [3], it is assumed that the macro-tiles have size $n \times n$ and that τ can be "computed quickly" within the macro-tiles. In our construction, it is not possible to assume that the rectangles are actually squares, because we always have that $T > S$. However, it is not difficult to circumvent this problem and also provide a concrete definition of what "compute quickly" means. The real problem lies everywhere.

In [3], every letter of the simulating alphabet has two fields, CurSon and CurFath, with the following behaviour: CurFath is constant within macro-tiles. CurSon is determined by the value of the CurFath and the Age and Addr fields, in such a way that the projection of a macro-tile onto CurSon is the image through τ of the letter in CurFath. In this way, we ensure that the projection of a valid configuration onto CurSon belongs in $\tau(\mathcal{A}^{\mathbb{Z}^2})$. Then, we only need to check that inside every colony the CurSon of the simulated letter is equal to the CurFath of the simulating letters in order to hierarchically deduce that the projection onto the CurSon actually belongs in Λ_τ. This approach will not work for SFTs defined by RPCA without some modifications for the following reason:

Inside a macro-tile, CurFath is constant and CurSon is updated according to a rule that depends on CurFath, Addr and Age. Since CurFath is constant within a macro-tile and Addr and Age are updated according to a reversible rule, CurSon can be updated in a reversible way, as long as the coordinates are not on the bottom or lower edge of the macro-tile. But in these cases, in order to update CurSon, we have to know the CurFath of the macro-tile that lies above (or below) the current macro-tile. In [3], this was done by "guessing" what the next CurFath is, and then checking that the guess was a good one. In RPCA, guessing is not allowed, so we have to do something different.

Unfortunately, due to space restrictions, we cannot describe in detail how to deal with this problem. Let us just mention that we use the idea of [11] to obtain, at the right moment, the CurFath of the macro-tile above the current one from "arbitrarily far in the past". The basic idea is that we obtain the next CurFath from the simulated letter *after* the simulation and at the right time we copy it onto every field of the colony. Of course, we have to do everything in a reversible way, which makes the construction slightly complex, but this is the basic idea that makes the construction work.

Acknowledgments. The author would like to thank P. Guillon and J. Kari for a very careful reading of various drafts of this work and for many useful comments, I. Törmä and V. Salo for many interesting discussions and T. Jolivet for sharing his knowledge of Latex.

References

1. Boyle, M., Lind, D.: Expansive subdynamics. Transactions of the American Mathematical Society **349**(1), 55–102 (1997).
   http://www.math.umd.edu/~mmb/papers/subdynamics.pdf
2. Dennunzio, A., Formenti, E., Weiss, M.: Multidimensional cellular automata: Closing property, quasi-expansivity, and (un)decidability issues. Theor. Comput. Sci. **516**, 40–59 (2014)
3. Durand, B., Romashchenko, A.E., Shen, A.: Fixed-point tile sets and their applications. J. Comput. Syst. Sci. **78**(3), 731–764 (2012)
4. Gács, P.: Reliable cellular automata with self-organization. Journal of Statistical Physics **102**(1–2), 45–267 (2001).
   http://www.cs.bu.edu/fac/gacs/recent-publ.html
5. Gray, L.F.: A reader's guide to Gacs's positive rates paper. Journal of Statistical Physics **103**(1–2), 1–44 (2001)
6. Hedlund, G.: Endomorphisms and automorphisms of the shift dynamical system. Mathematical systems theory **3**(4), 320–375 (1969).
   http://dx.doi.org/10.1007/BF01691062
7. Hochman, M.: Non-expansive directions for z2 actions. Ergodic Theory and Dynamical Systems **31**, 91–112 (2011).
   http://journals.cambridge.org/article_S0143385709001084
8. Kari, J.: Lecture notes on cellular automata. http://users.utu.fi/jkari/ca/part4.pdf
9. Kari, J.: The nilpotency problem of one-dimensional cellular automata. SIAM Journal on Computing **21**(3), 571–586 (1992)
10. Kari, J., Papasoglu, P.: Deterministic aperiodic tile sets. Geometric and Functional Analysis GAFA **9**(2), 353–369 (1999). http://dx.doi.org/10.1007/s000390050090
11. Le Gloannec, B., Ollinger, N.: Substitutions and strongly deterministic tilesets. In: Cooper, S.B., Dawar, A., Löwe, B. (eds.) CiE 2012. LNCS, vol. 7318, pp. 462–471. Springer, Heidelberg (2012)

12. Morita, K.: Reversible computing and cellular automata a survey. Theoretical Computer Science **395**(1), 101–131 (2008).
http://www.sciencedirect.com/science/article/pii/S030439750800100X
13. Mozes, S.: Tilings, substitution systems and dynamical systems generated by them. Journal d'analyse mathématique **53**, 139–186 (1988)
14. Robinson, R.M.: Undecidability and nonperiodicity for tilings of the plane. Inventiones Mathematicæ 12(3) (1971)

Combinatorics on Words

On the Number of Closed Factors in a Word

Golnaz Badkobeh[1], Gabriele Fici[2($\boxtimes$)], and Zsuzsanna Lipták[3]

[1] Department of Computer Science, University of Sheffield, Sheffield, UK
g.badkobeh@sheffield.ac.uk
[2] Dipartimento di Matematica e Informatica, Università di Palermo, Palermo, Italy
gabriele.fici@unipa.it
[3] Dipartimento di Informatica, Università di Verona, Verona, Italy
zsuzsanna.liptak@univr.it

Abstract. A closed word (a.k.a. periodic-like word or complete first return) is a word whose longest border does not have internal occurrences, or, equivalently, whose longest repeated prefix is not right special. We investigate the structure of closed factors of words. We show that a word of length n contains at least $n + 1$ distinct closed factors, and characterize those words having exactly $n + 1$ closed factors. Furthermore, we show that a word of length n can contain $\Theta(n^2)$ many distinct closed factors.

Keywords: Combinatorics on words · Closed word · Complete return · Rich word · Bitonic word

Introduction

It is known (see for example [8]) that any word w of length n contains at most $n + 1$ palindromic factors. Triggered by this result, several researchers initiated a study to characterize words that can accommodate a maximal number of palindromes, called *rich* (or *full*) *words* (see, for example, [2,4,5,10,12]).

In this paper, we consider the notion of *closed word* (a.k.a. periodic-like word or complete first return). A word w is closed if and only if it is empty or has a factor $v \neq w$ occurring exactly twice in w, as a prefix and as a suffix of w. We also say in this case that w is a complete return to v. For example, aaa, $ababa$, $ccabcc$ are all closed words (they are complete returns to aa, aba and cc, respectively), while ab and $abaabab$ are not. As shown in Proposition 4, any word whose exponent is at least two is closed.

The *closed factors* of a word are its factors that are closed words. In contrast to the case of palindromic factors, we show that a word of length n contains at least $n + 1$ closed factors (Lemma 8). Inspired by this property, we study the class of words that contain the smallest number of closed factors, and we call them *CR-poor words*.

G. Fici—Partially supported by Italian MIUR Project PRIN 2010LYA9RH, "Automi e Linguaggi Formali: Aspetti Matematici e Applicativi".

A.-H. Dediu et al. (Eds.): LATA 2015, LNCS 8977, pp. 381–390, 2015.
DOI: 10.1007/978-3-319-15579-1_29

As an example, *abca* is a CR-poor word, since it has length 4 and exactly 5 closed factors, namely ε, a, b, c and *abca*, whereas the word *ababa* is not CR-poor since it has length 5 but contains 8 closed factors: ε, *a*, *b*, *aba*, *bab*, *abab*, *baba* and *ababa*.

However, there is some relation between rich words and CR-poor words. Bucci, de Luca and De Luca [5] showed that a palindromic word is rich if and only if all of its palindromic factors are closed. We show, in Proposition 10, that if a word *w* has the property that all of its closed factors are palindromes, then *w* is a CR-poor word, and it is also rich. CR-poor words are also connected to some problems on *privileged words* (see [11]).

While having only palindromic closed factors is a necessary and sufficient condition for a binary word to be CR-poor (Theorem 23), we prove that in a word *w* over an alphabet Σ of arbitrary cardinality, the set of closed factors and the set of palindromic factors of *w* coincide if and only if *w* is both rich and CR-poor (Proposition 20).

In Theorem 17, we give a combinatorial characterization of CR-poor words over an alphabet Σ of cardinality greater than one: A word over Σ is CR-poor if and only if it does not contain any closed factor that is a complete return to *xy*, for *x, y* different letters in Σ. In other words, CR-poor words are exactly those words having as their closed factors only complete returns to powers of a single letter. As a consequence, the language of CR-poor words over Σ is a regular language. In contrast, the language of closed words is not regular (Proposition 5).

We give some further characterizations of CR-poor words in the case of the binary alphabet (Theorem 23). One of them is that the binary CR-poor words are the *bitonic words*, i.e., the conjugates to words in a^*b^*. We therefore have that binary CR-poor words form a regular subset of the language of rich words.

Finally, we show that a word of length *n* can contain $\Theta(n^2)$ many distinct closed factors (Theorem 25).

1   Closed Words

A *word* is a finite sequence of elements from a finite set Σ. We refer to the elements of Σ as *letters* and to Σ as the *alphabet*. The *i*-th letter of a word *w* is denoted by w_i. Given a word $w = w_1 w_2 \cdots w_n$, with $w_i \in \Sigma$ for $1 \leq i \leq n$, the nonnegative integer *n* is the *length* of *w*, denoted by $|w|$. The empty word has length zero and is denoted by ε. The set of all words over Σ is denoted by Σ^*. Any subset of Σ^* is called a *language*. A language is *regular* (or *rational*) if it can be recognized by a finite state automaton.

A *prefix* (resp. a *suffix*) of a word *w* is any word *u* such that $w = uz$ (resp. $w = zu$) for some word *z*. A *factor* of *w* is a prefix of a suffix (or, equivalently, a suffix of a prefix) of *w*. The set of prefixes, suffixes and factors of the word *w* are denoted by $\mathrm{Pref}(w)$, $\mathrm{Suff}(w)$ and $\mathrm{Fact}(w)$ respectively. A *border* of a word *w* is any word in $\mathrm{Pref}(w) \cap \mathrm{Suff}(w)$ different from *w*. From the definitions, we have that ε is a prefix, a suffix, a border and a factor of any word. An *occurrence* of a factor *u* in *w* is a factorization $w = vuz$. An occurrence of *u* is *internal* if both *v* and *z* are non-empty.

The word $\tilde{w} = w_n w_{n-1} \cdots w_1$ is called the *reversal* (or *mirror image*) of w. A *palindrome* is a word w such that $\tilde{w} = w$. In particular, the empty word is a palindrome. A *conjugate* of a word w is any word of the form vu such that $uv = w$, for some $u, v \in \Sigma^*$. A conjugate of a word w is also called a *rotation* of w.

A *period* for the word w is a positive integer p, with $0 < p \leq |w|$, such that $w_i = w_{i+p}$ for every $i = 1, \ldots, |w| - p$. Since $|w|$ is always a period for w, we have that every non-empty word has at least one period. We can unambiguously define *the* period of the word w as the smallest of its periods. The *exponent* of a word w is the ratio between its length and its smallest period. A *power* is a word whose exponent is an integer greater than 1. A word that is not a power is called *primitive*

We denote by $\mathrm{PAL}(w)$ the set of factors of w that are palindromes. A word w of length n is *rich* [10] (or *full* [2]) if $|\mathrm{PAL}(w)| = n + 1$, i.e., if it contains the largest number of palindromes a word of length n can contain.

A language L is called *factorial* if $L = \mathrm{Fact}(L)$, i.e., if L contains all the factors of its words. A language L is *extendible* if for every word $w \in L$, there exist letters $a, b \in \Sigma$ such that $awb \in L$. The language of rich words over a fixed alphabet Σ is an example of a factorial and extendible language.

We recall the definition of closed word given in [9]:

Definition 1. *A word w is* closed *if and only if it is empty or has a factor $v \neq w$ occurring exactly twice in w, as a prefix and as a suffix of w.*

The word aba is a closed, since its factor a appears in it only as a prefix and as a suffix. The word $abaa$, on the contrary, is not closed. Note that for any letter $a \in \Sigma$ and for any integer $n > 0$, the word a^n is closed, a^{n-1} being a factor occurring only as a prefix and as a suffix in it (this includes the special case of single letters, for which $n = 1$ and $a^{n-1} = \varepsilon$).

Remark 2. The notion of closed word is equivalent to that of *periodic-like* word [6]. A word w is periodic-like if its longest repeated prefix does not have two occurrences in w followed by different letters, i.e., if its longest repeated prefix is not right special.

The notion of closed word is also closely related to the concept of *complete return* to a factor, as considered in [10]. A complete return to the factor u in a word w is any factor of w having exactly two occurrences of u, one as a prefix and one as a suffix. Hence a non-empty word w is closed if and only if it is a complete return to one of its factors; such a factor is clearly both the longest repeated prefix and the longest repeated suffix of w (i.e., the longest border of w).

Remark 3. Let w be a non-empty word over Σ. The following characterizations of closed words follow easily from the definition:

1. w has a factor $v \neq w$ occurring exactly twice in w, as a prefix and as a suffix of w;
2. the longest repeated prefix (resp. suffix) of w does not have internal occurrences in w, i.e., occurs in w only as a prefix and as a suffix;

3. the longest repeated prefix (resp. suffix) of w does not have two occurrences in w followed (resp. preceded) by different letters;
4. w has a border that does not have internal occurrences in w;
5. the longest border of w does not have internal occurrences in w;
6. w is a complete return to its longest repeated prefix;
7. w is a complete return to its longest border.

For more details on closed words and related results see [1,3,5–7,9,13].
We end this section by exhibiting some properties of closed words.

Proposition 4. *Any word whose exponent is at least 2 is closed.*

Proof. Let $w = v^n v'$ for $n \geq 2$, v a primitive word, and v' a prefix of v such that the exponent of w is equal to $n + |v'|/n$. Then $v^{n-1}v'$ is a border of w. If $v^{n-1}v'$ has an internal occurrence in w, then there exists a proper prefix u of v such that $uv = vu$, and it is a basic result in Combinatorics on Words that two words commute if and only if they are powers of the same word, in contradiction with our hypotheses on u and v. □

Moreover, it is easy to see that for any rational number x between 1 and 2, there exists a closed word having exponent x (it is sufficient to take a word over $\{a, b\}$ ending with b and with only one other occurrence of b, placed in the first half of the word).

Proposition 5. *Let Σ be an alphabet of cardinality $|\Sigma| \geq 2$. The language of closed words over Σ is not regular.*

Proof. Let L be the language of closed words over Σ and let $a, b \in \Sigma$ be different letters. Let us assume that L is regular. This implies that also $L \cap a^*b^*a^*$ is regular, since $a^*b^*a^*$ is a regular language and the intersection of two regular languages is regular. We claim that $L \cap a^*b^*a^* = \{a^n b^m a^n \mid n, m \geq 0\}$, which is not a regular language, and so we have a contradiction.

Clearly, every word in $\{a^n b^m a^n \mid n, m \geq 0\}$ is closed. Suppose now that w belongs to $a^*b^*a^*$. Hence, $w = a^n b^m a^k$, for some $n, m, k \geq 0$. If $n \neq k$, say $n < k$, then the longest repeated prefix of w is a^n and it has at least one internal occurrence in w. By Remark 3, w is not closed. The case $n > k$ is symmetric. □

Finally, we recall two results from [7].

Lemma 6. *[7, Lemma 4] Let w be a non-empty word over Σ. Then there exists at most one letter $x \in \Sigma$ such that wx is closed.*

Lemma 7. *[7, Lemma 5] Let w be a closed word. Then wx, $x \in \Sigma$, is closed if and only if wx has the same period of w.*

2   Closed Factors

Let w be a word. A factor of w that is a closed word is called a *closed factor* of w. The set of closed factors of the word w is denoted by $C(w)$.

Lemma 8. *For any word w of length n, one has $|C(w)| \geq n + 1$.*

Proof. We show that every position of w is the ending position of an occurrence of a distinct closed factor of w. Thus w contains at least n non-empty closed factors, and the claim follows. Indeed, let v be the longest non-empty closed factor ending in position i, so that $w_{i-|v|+1} \cdots w_i = v$. Since a is closed for every $a \in \Sigma$, such a factor always exists. If v did not occur before in w, then we are done. Otherwise, let j be the largest position smaller than i such that $w_{j-|v|+1} \cdots w_j = v$. Set $v' = w_{j-|v|+1} \cdots w_i$ and observe that v' is a closed factor ending in i, with longest border v. But $|v'| > |v|$, in contradiction to the choice of v. □

Lemma 9. *For any words u, v one has $|C(u)| + |C(v)| \leq |C(uv)| + 1$.*

Proof. Clearly, $C(u) \subseteq C(uv)$. In order to prove the statement, it is sufficient to prove that for any non-empty z in $C(v)$, there exists an $f(z)$ in $C(uv) \setminus C(u)$ and f is injective. So let $z \in C(v)$, $uv = w = w_1 \cdots w_n$, and let j be the smallest integer greater than $|u|$ such that $z = w_j \cdots w_{j+|z|-1}$. If j is the smallest integer such that $z = w_j \cdots w_{j+|z|-1}$, then set $f(z) = z$. Otherwise, there is in w a closed z' to z ending in position $w_{j+|z|-1}$. If this is the first occurrence of z' in w, then set $f(z) = z'$, otherwise repeat the construction for z'. Eventually, we will find a closed factor $f(z) = z^{(k)}$ whose first occurrence in w ends in position $w_{j+|z|-1}$.

By construction, f has the desired properties. □

Proposition 10. *Let w be a word of length n. If $C(w) \subseteq \mathrm{PAL}(w)$, then $C(w) = \mathrm{PAL}(w)$ and $|C(w)| = |\mathrm{PAL}(w)| = n + 1$. In particular, w is a rich word.*

Proof. On the one hand, from Lemma 8, one has $|C(w)| \geq n + 1$. On the other hand, one has $|\mathrm{PAL}(w)| \leq n + 1$. Hence, if $C(w) \subseteq \mathrm{PAL}(w)$, then it must be $C(w) = \mathrm{PAL}(w)$ and $|C(w)| = |\mathrm{PAL}(w)| = n + 1$, and so w is a rich word. □

Bucci et al. [5, Proposition 4.3] showed that a word w is rich if and only if every closed factor v of w has the property that the longest palindromic prefix (or suffix) of v is unrepeated in v. Moreover, they proved the following remarkable result:

Theorem 11 (Bucci et al. [5, Corollary 5.2]). *A palindromic word w is rich if and only if $\mathrm{PAL}(w) \subseteq C(w)$.*

In Section 4, we will prove that the condition $\mathrm{PAL}(w) = C(w)$ characterizes the CR-poor words over a binary alphabet.

3   CR-poor Words

By Lemma 8, we have that $n+1$ is a lower bound on the number of closed factors of a word of length n. We introduce the following definition:

Definition 12. *A word* $w \in \Sigma^*$ *is CR-poor if* $|C(w)| = |w| + 1$. *We also set*

$$\mathcal{L}_\Sigma = \{w \in \Sigma^* : |C(w)| = |w| + 1\}$$

the language of CR-poor words over the alphabet Σ.

Remark 13. If $|\Sigma| = 1$, then $\mathcal{L}_\Sigma = \Sigma^*$. So in what follows we will suppose $|\Sigma| \geq 2$.

Note that, for any alphabet Σ, the language $\mathcal{L}_\Sigma$ of CR-poor words over Σ is closed under reversal. Indeed, it follows from the definition that a word $w \in \Sigma^*$ is closed if and only if its reversal $\tilde{w}$ is closed.

Proposition 14. *The language* $\mathcal{L}_\Sigma$ *of CR-poor words over* Σ *is a factorial language.*

Proof. We have to prove that for any word CR-poor w and any factor v of w, v is a CR-poor word. Suppose by contradiction that there exists a CR-poor word w containing a factor v that is not a CR-poor word, i.e., $w \in \mathcal{L}_\Sigma$, $w = uvz$ and $|C(v)| > |v| + 1$. By Lemma 9, $|C(w)| \geq |C(u)| + |C(v)| + |C(z)| - 2 > |u| + |z| + |v| + 1 = |w| + 1$ and therefore w cannot be a CR-poor word. □

The following technical lemma will be used in the proof of the next theorem.

Lemma 15. *Let* w *be a CR-poor word over the alphabet* Σ *and* $x \in \Sigma$. *The word* wx *(resp.* xw*) is CR-poor if and only if it has a unique suffix (resp. prefix) that is closed and is not a factor of* w.

Proof. We prove the statement for wx, the one for xw will follow by symmetry. The "if" part is straightforward. For the "only if" part, recall from the proof of Lemma 8 that there is at least one new closed factor ending in every position, so in particular wx has at least one suffix that is closed and is not a factor of w. □

Remark 16. Suppose that a word w contains as a factor a complete return to some word u. Then for every factor u' of u, the word w contains as a factor a complete return to u'.

We now give a characterization of CR-poor words.

Theorem 17. *A word* w *over* Σ *is CR-poor if and only if for any two different letters* $a, b \in \Sigma$, w *does not contain any complete return to* ab. *In other words,*

$$\mathcal{L}_\Sigma = \Sigma^* \setminus \bigcup_{a \neq b} \Sigma^* ab \Sigma^* ab \Sigma^*.$$

Proof. Let u be a complete return to ab for $a, b \in \Sigma$ different letters. We claim that u is not CR-poor. Since by Proposition 14, a CR-poor word cannot contain a factor that is not CR-poor, once the claim is proved the "only if" part of the theorem follows. So let u' be the longest suffix of u that is closed and starts with the letter b. Such a suffix exists since u contains at least two occurrences of b. Then u' is unioccurrent in u, and since u is a closed suffix of itself we have, by Lemma 15, that u is not CR-poor.

Conversely, suppose that the word w is not CR-poor. Then, analogously as in the proof of Lemma 3, it follows that there is a position i of w such that there are at least two different closed factors u and u' of w that end in position i and do not occur in $w_1 \cdots w_{i-1}$. If both u and u' are complete returns to a power of the letter w_i, then one of them must occur in $w_1 \cdots w_{i-1}$, so this situation is not possible, and we can therefore suppose that there is a factor ending in position i that is a complete return to a word containing at least two different letters. The statement then follows from Remark 16. $\square$

Corollary 18. *A word w over Σ is CR-poor if and only if every closed factor of w is a complete return to a power of a single letter.*

Corollary 19. *The language $\mathcal{L}_\Sigma$ of CR-poor words over Σ is a regular language.*

We can now state the following result:

Proposition 20. *Let w be a word over Σ. Then $C(w) = \mathrm{PAL}(w)$ if and only if w is rich and CR-poor.*

Proof. If $C(w) = \mathrm{PAL}(w)$, then $|C(w)| = |\mathrm{PAL}(w)|$, and since $|C(w)| \geq |w| + 1$ (by Lemma 8) and $|\mathrm{PAL}(w)| \leq |w| + 1$, then it must be $|C(w)| = |\mathrm{PAL}(w)| = |w| + 1$, and hence by definition w is rich and CR-poor.

Conversely, suppose that w is rich and CR-poor. Let $v \in C(w)$. By Corollary 18, v is a complete return to a power of a single letter, so v is a complete return to a palindrome. It is known (see [10, Theorem 2.14]) that a word is rich if and only if all of its factors that are complete returns to a palindrome are palindromes themselves. Therefore, v is a palindrome, and hence we proved that $C(w) \subseteq \mathrm{PAL}(w)$. By Proposition 10, $C(w) = \mathrm{PAL}(w)$ and we are done. $\square$

4   The Case of Binary Words

In this section we fix the alphabet $\Sigma = \{a, b\}$. For simplicity of exposition, we will denote the language of CR-poor words over $\{a, b\}$ by $\mathcal{L}$ rather than by $\mathcal{L}_{\{a,b\}}$. We first recall the definition of bitonic word.

Definition 21. *A word $w \in \{a, b\}^*$ is bitonic if it is a conjugate of a word in $a^* b^*$, i.e., if it is of the form $a^i b^j a^k$ or $b^i a^j b^k$ for some integers $i, j, k \geq 0$.*

By Theorem 17, it is easy to see that a binary word is in $\mathcal{L}$ if and only if it is bitonic.

Lemma 22. *Let w be a bitonic word. Then $C(w) \subseteq PAL(w)$.*

Proof. Since w is bitonic, a closed factor of w can only be the complete return to a power of a single letter. So a closed factor u of w is of the form $u = a^n$, $u = b^n$, $u = a^n b^m a^n$ or $u = b^n a^m b^n$, for some $n, m > 0$, and these words are all palindromes.   □

Thus, by Proposition 10, any bitonic word w of length $n > 0$ contains exactly $n + 1$ closed factors and so is a CR-poor word. We therefore have the following characterizations of CR-poor binary words.

Theorem 23. *Let $w \in \{a, b\}^*$. The following are equivalent:*

1. $w \in \mathcal{L}$;
2. w does not contain any complete return to ab or ba;
3. $C(w) \subseteq PAL(w)$;
4. $C(w) = PAL(w)$;
5. w is a bitonic word.

Notice that the condition $C(w) \subseteq PAL(w)$ does not hold in general for CR-poor words over alphabets larger than two. As an example, the word $abca$ is CR-poor but contains a closed factor ($abca$) that is not a palindrome. In view of Theorem 11, a natural question would be that of establishing whether a palindrome w is CR-poor if and only if $C(w) = PAL(w)$, i.e., whether the characterization in Theorem 23 can be generalized to larger alphabets at least for palindromes. However, the answer to this question is negative since, for example, the word $w = abcacba$ is a CR-poor palindrome and contains the non-palindromic closed factor $abca$. Note that, coherently with Theorem 11 (and with Proposition 20), w is not rich. However, in the case of a binary alphabet, we have, by Theorem 23 and Proposition 10, that every CR-poor word is rich. Since by Theorem 17 it follows that the language $\mathcal{L}_\Sigma$ is extendible for any alphabet Σ, the language $\mathcal{L}$ is therefore a factorial and extendible subset of the language of (binary) rich words.

In the following proposition we exhibit a closed enumerative formula for the language $\mathcal{L}$.

Proposition 24. *For every $n > 0$, there are exactly $n^2 - n + 2$ distinct words in $\mathcal{L}$.*

Proof. Each of the $n - 1$ words of length $n > 0$ in $a^+ b^+$ has n distinct rotations, while for the words a^n and b^n all the rotations coincide. Thus, there are $n(n - 1) + 2$ bitonic words of length n, and the statement follows from Theorem 23.   □

5   How Many Closed Factors Can a Word Contain?

We showed in Lemma 8 that any word of length n contains at least $n + 1$ distinct closed factors. But how many closed factors, at most, can a word contain? We provide an answer in the following theorem.

Theorem 25. *For every $n > 4$, there exists a word $w \in \{a, b\}^n$ with quadratically many closed factors.*

Proof. Let $n > 4$ be fixed. We construct a word w of length n such that $|C(w)| \geq (k+1)(k+2)/2$, where $k = \lfloor n/4 \rfloor$.

Let $w = a^k b^k a^k b^k a^{n-4k}$. Clearly $|w| = n$. Let $v_{i,j} = w_i \cdots w_j$, $1 \leq i \leq j \leq n$, be a factor of w. We claim that for every $i = 1, 2, \ldots, k - 1$, every factor $v_{i,j}$, with $3k - 1 + i \leq j \leq 4k$, is closed. Indeed, fixed i between 1 and $k - 1$, the factor $v_{i,3k-1+i}$, of length $3k$, is equal to $a^{k-i+1} b^k a^k b^{i-1}$, and therefore it is closed since it is a complete return to $a^{k-i+1} b^{i-1}$. Then, for every j such that $3k - 1 + i \leq j \leq 4k$, the factor $v_{i,j}$ has the same period of $v_{i,3k-1+i}$, and therefore is closed by Lemma 7.

Finally, notice that whenever (i', j') is different from (i, j), for i' and j' in the same range of i and j, respectively (that is, $1 \leq i \leq k-1$ and $3k-1+i \leq j \leq 4k$), the factor $v_{i',j'}$ is different from the factor $v_{i,j}$.

Therefore we conclude that w contains at least $(k+1)(k+2)/2 = \Theta(n^2)$ many different closed factors, and we are done. $\square$

It is possible to exhibit a formula for the precise value of the maximal number of closed factors in a word of length n, but we think this adds nothing to the general picture provided by Theorem 25. Moreover, the words realizing the upper bound do not have a nice characterization, contrarily to the case of words realizing the lower bound, discussed in the previous sections. However, for completeness, we report in Table 1 the first values of the sequence of the maximum number of closed factors for binary words.

Table 1. The sequence of the maximum number of closed factors in a binary word

n	1	2	3	4	5	6	7	8	9	10	11	12	13	14	15	16	17	18	19	20
max	2	3	4	6	8	10	12	15	18	21	25	29	33	37	42	47	52	58	64	70

6   Conclusion and Open Problems

This paper is a first attempt to study the set of closed factors of a finite word. In particular, we investigated the words with the smallest number of closed factors, which we referred to as CR-poor words. We provided a combinatorial characterization of these words and exhibited some relations with rich words.

An enumerative formula for rich words is not known, not even in the binary case. A possible approach to this problem is to separate rich words in subclasses to be enumerated separately. Our enumerative formula for binary CR-poor words given in Proposition 24 could constitute a step towards this direction.

The set of closed factors could be investigated for specific (finite or infinite) words or classes of words, and could be a tool to derive new combinatorial properties of words.

Finally, the notion of closed factor has recently found applications in string algorithms [1], hence a better understanding of the structure of closed factors of a word could lead to some applications.

References

1. Badkobeh, G., Bannai, H., Goto, K., I, T., Iliopoulos, C.S., Inenaga, S., Puglisi, S.J., Sugimoto, S.: Closed factorization. In: Proceedings of the Prague Stringology Conference 2014, pp. 162–168 (2014)
2. Brlek, S., Hamel, S., Nivat, M., Reutenauer, C.: On the palindromic complexity of infinite words. Internat. J. Found. Comput. Sci. 15, 293–306 (2004)
3. Bucci, M., De Luca, A., Fici, G.: Enumeration and Structure of Trapezoidal Words. Theoret. Comput. Sci. 468, 12–22 (2013)
4. Bucci, M., De Luca, A., Glen, A., Zamboni, L.: A new characteristic property of rich words. Theoretical Computer Science 410(30), 2860–2863 (2009)
5. Bucci, M., de Luca, A., De Luca, A.: Rich and periodic-like words. In: Diekert, V., Nowotka, D. (eds.) DLT 2009. LNCS, vol. 5583, pp. 145–155. Springer, Heidelberg (2009)
6. Carpi, A., de Luca, A.: Periodic-like words, periodicity and boxes. Acta Inform. 37, 597–618 (2001)
7. De Luca, A., Fici, G.: Open and closed prefixes of sturmian words. In: Karhumäki, J., Lepistö, A., Zamboni, L. (eds.) WORDS 2013. LNCS, vol. 8079, pp. 132–142. Springer, Heidelberg (2013)
8. Droubay, X., Justin, J., Pirillo, G.: Episturmian words and some constructions of de Luca and Rauzy. Theoret. Comput. Sci. 255(1–2), 539–553 (2001)
9. Fici, G.: A classification of trapezoidal words. In: WORDS 2011, 8th International Conference on Words. Electronic Proceedings in Theoretical Computer Science, vol. 63, pp. 129–137 (2011)
10. Glen, A., Justin, J., Widmer, S., Zamboni, L.Q.: Palindromic richness. European J. Combin. 30, 510–531 (2009)
11. Peltomäki, J.: Introducing privileged words: Privileged complexity of Sturmian words. Theoret. Comput. Sci. 500, 57–67 (2013)
12. Restivo, A., Rosone, G.: Burrows-Wheeler transform and palindromic richness. Theoret. Comput. Sci. 410(30), 3018–3026 (2009)
13. Sloane, N.J.A.: The On-Line Encyclopedia of Integer Sequences. Available electronically at http://oeis.org, Sequence A226452: Number of closed binary words of length n

Online Computation of Abelian Runs

Gabriele Fici[1]([✉]), Thierry Lecroq[2],
Arnaud Lefebvre[2], and Élise Prieur-Gaston[2]

[1] Dipartimento di Matematica e Informatica, Università di Palermo, Palermo, Italy
Gabriele.Fici@unipa.it
[2] Normandie Université, LITIS EA4108, NormaStic CNRS FR 3638, IRIB,
Université de Rouen, 76821 Mont-Saint-Aignan Cedex, France
{Thierry.Lecroq,Arnaud.Lefebvre,Elise.Prieur}@univ-rouen.fr

Abstract. Given a word w and a Parikh vector $\mathcal{P}$, an abelian run of period $\mathcal{P}$ in w is a maximal occurrence of a substring of w having abelian period $\mathcal{P}$. We give an algorithm that finds all the abelian runs of period $\mathcal{P}$ in a word of length n in time $O(n \times |\mathcal{P}|)$ and space $O(\sigma + |\mathcal{P}|)$.

Keywords: Combinatorics on words · Text algorithms · Abelian period · Abelian run

1 Introduction

Computing maximal (non-extendable) repetitions in a string is a classical topic in the area of string algorithms (see for example [7] and references therein). Detecting maximal repetitions of substrings, also called *runs*, gives information on the repetitive regions of a string, and is used in many applications, for example in the analysis of genomic sequences.

Kolpakov and Kucherov [5] gave a linear time algorithm for computing all the runs in a word and conjectured that any word of length n contains less than n runs. Bannai et al. [1] recently proved this conjecture using the notion of Lyndon root of a run.

Here we deal with a generalization of this problem to the commutative setting. Recall that an abelian power is a concatenation of two or more words that have the same Parikh vector, i.e., that have the same number of occurrences of each letter of the alphabet. For example, *aababa* is an abelian square, since *aab* and *aba* both have 2 *a*'s and 1 *b*. When an abelian power occurs within a string, one can search for its "maximal" occurrence by extending it to the left and to the right character by character without violating the condition on the number of occurrences of each letter. Following the approach of Constantinescu and Ilie [2], we say that a Parikh vector $\mathcal{P}$ is an abelian period for a word w over a finite ordered alphabet $\Sigma = \{a_1, a_2, \ldots, a_\sigma\}$ if w can be written as $w = u_0 u_1 \cdots u_{k-1} u_k$ for some $k > 2$ where for $0 < i < k$ all the u_i's have the same Parikh vector $\mathcal{P}$ and the Parikh vectors of u_0 and u_k are contained in $\mathcal{P}$. Note that the factorization above is not necessarily unique. For example,

© Springer International Publishing Switzerland 2015
A.-H. Dediu et al. (Eds.): LATA 2015, LNCS 8977, pp. 391–401, 2015.
DOI: 10.1007/978-3-319-15579-1_30

$a \cdot bba \cdot bba \cdot \varepsilon$ and $\varepsilon \cdot abb \cdot abb \cdot a$ (ε denotes the empty word) are two factorizations of the word $abbabba$ both corresponding to the abelian period $(1, 2)$. Moreover, the same word can have different abelian periods.

In this paper we define an *abelian run* of period $\mathcal{P}$ in a word w as an occurrence of a substring v of w such that v has abelian period $\mathcal{P}$ and this occurrence cannot be extended to the left nor to the right by one letter into a substring having the same abelian period $\mathcal{P}$.

For example, let $w = ababaaa$. Then the prefix $ab \cdot ab \cdot a = w[1..5]$ has abelian period $(1, 1)$ but it is not an abelian run since the prefix $a \cdot ba \cdot ba \cdot a = w[1..6]$ has also abelian period $(1, 1)$. This latter, instead, is an abelian run of period $(1, 1)$ in w.

Looking for abelian runs in a string can be useful to detect those regions in a string in which there is some kind of non-exact repetitiveness, for example regions in which there are several consecutive occurrences of a substring or its reverse.

Matsuda et al. [6] recently presented an offline algorithm for computing all abelian runs of a word of length n in $O(n^2)$ time. Notice that, however, the definition of abelian run in [6] is slightly different from the one we consider here. We will comment on this in Section 3.

We present an online algorithm that, given a word w of length n over an alphabet of cardinality σ, and a Parikh vector $\mathcal{P}$, returns all the abelian runs of period $\mathcal{P}$ in w in time $O(n \times |\mathcal{P}|)$ and space $O(\sigma + |\mathcal{P}|)$.

2   Definitions and Notation

Let $\Sigma = \{a_1, a_2, \ldots, a_\sigma\}$ be a finite ordered alphabet of cardinality σ and let Σ^* be the set of finite words over Σ. We let $|w|$ denote the length of the word w. Given a word $w = w[0..n-1]$ of length $n > 0$, we write $w[i]$ for the $(i+1)$-th symbol of w and, for $0 \leqslant i \leqslant j < n$, we write $w[i..j]$ for the substring of w from the $(i+1)$-th symbol to the $(j+1)$-th symbol, both included. We let $|w|_a$ denote the number of occurrences of the symbol $a \in \Sigma$ in the word w.

The *Parikh vector* of w, denoted by $\mathcal{P}_w$, counts the occurrences of each letter of Σ in w, that is, $\mathcal{P}_w = (|w|_{a_1}, \ldots, |w|_{a_\sigma})$. Notice that two words have the same Parikh vector if and only if one word is a permutation (i.e., an anagram) of the other.

Given the Parikh vector $\mathcal{P}_w$ of a word w, we let $\mathcal{P}_w[i]$ denote its i-th component and $|\mathcal{P}_w|$ its norm, defined as the sum of its components. Thus, for $w \in \Sigma^*$ and $1 \leqslant i \leqslant \sigma$, we have $\mathcal{P}_w[i] = |w|_{a_i}$ and $|\mathcal{P}_w| = \sum_{i=1}^{\sigma} \mathcal{P}_w[i] = |w|$.

Finally, given two Parikh vectors $\mathcal{P}, \mathcal{Q}$, we write $\mathcal{P} \subset \mathcal{Q}$ if $\mathcal{P}[i] \leqslant \mathcal{Q}[i]$ for every $1 \leqslant i \leqslant \sigma$ and $|\mathcal{P}| < |\mathcal{Q}|$.

Definition 1 (Abelian period [2]). *A Parikh vector $\mathcal{P}$ is an abelian period for a word w if $w = u_0 u_1 \cdots u_{k-1} u_k$, for some $k > 2$, where $\mathcal{P}_{u_0} \subset \mathcal{P}_{u_1} = \cdots = \mathcal{P}_{u_{k-1}} \supset \mathcal{P}_{u_k}$, and $\mathcal{P}_{u_1} = \mathcal{P}$.*

Note that since the Parikh vector of u_0 and u_k cannot be equal to $\mathcal{P}$ it implies that $|u_0|, |u_k| < |\mathcal{P}|$. We call u_0 and u_k respectively the *head* and the *tail* of the abelian period. Note that in [2] the abelian period is characterized by $|u_0|$ and $|\mathcal{P}|$ thus we will sometimes use the notation (h, p) for an abelian period of norm p and head length h of a word w. Notice that the length t of the tail is uniquely determined by h, p and $n = |w|$, namely $t = (n - h) \bmod p$.

Definition 2 (Abelian repetition). *A substring $w[i..j]$ is an abelian repetition with period length p if $i - j + 1$ is a multiple of p, $i - j + 1 \geq 2p$ and there exists a Parikh vector $\mathcal{P}$ of norm p such that $\mathcal{P}_{w[i+kp..i+(k+1)p-1]} = \mathcal{P}$ for every $0 \leq k \leq p/(i - j + 1)$.*

An abelian repetition $w[i..j]$ with period length p such that $i - j + 1 = 2p$ is called an abelian square. An abelian repetition $w[i..j]$ of period length p of a string w is maximal if:

1. $\mathcal{P}_{w[i-p..i-1]} \neq \mathcal{P}_{w[i..i+p-1]}$ or $i - p < 0$;
2. $\mathcal{P}_{w[j-p+1..j]} \neq \mathcal{P}_{w[j+1..j+p]}$ or $j + p \geq n$.

We now give the definition of an abelian run. Let $v = w[b..e]$, $0 \leq b \leq e \leq |w| - 1$, be an occurrence of a substring in w and suppose that v has an abelian period $\mathcal{P}$, with head length h and tail length t. Then we denote this occurrence by the tuple (b, h, t, e).

Definition 3. *Let w be a word. An occurrence (b, h, t, e) of a substring of w starting at position b, ending at position e, and having abelian period $\mathcal{P}$ with head length h and tail length t is called **left-maximal** (resp. **right maximal**) if there does not exist an occurrence of a substring $(b - 1, h', t', e)$ (resp. $(b, h', t', e + 1)$) with the same abelian period $\mathcal{P}$. An occurrence (b, h, t, e) is called **maximal** if it is both left-maximal and right-maximal.*

This definition leads to the one of abelian run.

Definition 4. *An **abelian run** is a maximal occurrence (b, h, t, e) of a substring with abelian period $\mathcal{P}$ of norm p such that $(e - b - h - t + 1) \geq 2p$ (see Fig. 1).*

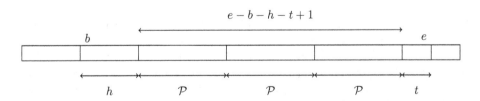

Fig. 1. The tuple (b, h, t, e) denotes an occurrence of a substring starting at position b, ending at position e, and having abelian period $\mathcal{P}$ with head length h and tail length t

The next result limits the number of abelian runs starting at each position in a word.

Lemma 5. *Let w be a word. Given a Parikh vector $\mathcal{P}$, there is at most one abelian run with abelian period $\mathcal{P}$ starting at each position of w.*

Proof. If two abelian runs start at the same position, the one with the shortest head cannot be maximal. □

Corollary 6. *Let w be a word. Given a Parikh vector $\mathcal{P}$, for every position i in w there are at most $|\mathcal{P}|$ abelian runs with period $\mathcal{P}$ overlapping at i.*

The next lemma shows that a left-maximal abelian substring at the right of another left-maximal abelian substring starting at position i in a word w cannot begin at a position smaller than i.

Lemma 7. *If $(b_1, h_1, 0, e_1)$ and $(b_2, h_2, 0, e_2)$ are two left-maximal occurrences of substrings with the same abelian period $\mathcal{P}$ of a word v such that $e_1 < e_2$ and $b_1 > e_1 - 2 \times |\mathcal{P}| + 1$ and $b_2 > e_2 - 2 \times |\mathcal{P}| + 1$, then $b_1 \leq b_2$.*

Proof. If $b_2 < b_1$ then since $e_2 > e_1$, $w[b_1..b_1 + h_1 - 1]$ is a substring of $w[b_2..b_2 + h_2 - 1]$. Thus $\mathcal{P}_{w[b_1..b_1+h_1-1]} \subset \mathcal{P}_{w[b_2..b_2+h_2-1]} \subset \mathcal{P}$ which implies that $\mathcal{P}_{w[b_1-1..b_1+h_1-1]} \subset \mathcal{P}$ meaning that $(b_1, h_1, 0, e_1)$ is not left-maximal: a contradiction. □

We recall the following proposition, which shows that if we can extend the abelian period with the longest tail of a word w when adding a symbol a, then we can extend all the other abelian periods with shorter tail.

Proposition 8. *[4] Suppose that a word w has s abelian periods $(h_1, p_1) < (h_2, p_2) < \cdots < (h_s, p_s)$ such that $(|w| - h_i) \bmod p_i = t > 0$ for every $1 \leq i \leq s$. If for a letter $a \in \Sigma$, (h_1, p_1) is an abelian period of wa, then $(h_2, p_2), \ldots, (h_s, p_s)$ are also abelian periods of wa.*

We want to give an algorithm that, given a string w and a Parikh vector $\mathcal{P}$, returns all the abelian runs of w having abelian period $\mathcal{P}$.

3   Previous Work

In [6], the authors presented an algorithm that computes all the abelian runs of a string w of length n in $O(n^2)$ time and space complexity. They consider that a substring $w[i - h..j + t]$ is an abelian run if $w[i..j]$ is a maximal abelian repetition with period length p and $h, t \geq 0$ are the largest integers satisfying $\mathcal{P}_{w[i-h..i-1]} \subset \mathcal{P}_{w[i..i+p-1]}$ and $\mathcal{P}_{w[j+1..j+t]} \subset \mathcal{P}_{w[i..i+p-1]}$. Their algorithm works as follows. First, it computes all the abelian squares using the algorithm of [3]. For each $0 \leq i \leq n - 1$, it computes a set L_i of integers such that

$$L_i = \{j \mid \mathcal{P}_{w[i-j..i]} = \mathcal{P}_{w[i+1..i+j+1]}, 0 \leq j \leq \min\{i + 1, n - i\}\}.$$

The L_i's are stored in a two-dimensional boolean array L of size $\lfloor n/2 \rfloor \times (n - 1)$: $L[j, i] = 1$ if $j \in L_i$ and $L[j, i] = 0$ otherwise. An example of array L is given in

Figure 2. All entries in L are initially unmarked. Then, for each $1 \leq j \leq \lfloor n/2 \rfloor \times$ all maximal abelian repetitions of period length j are computed in $O(n)$. The j-th row of L is scanned in increasing order of the column index. When an unmarked entry $L[j, i] = 1$ is found then the largest non-negative integer k such that $L[j, i + pj + 1] = 1$, for $1 \leq p \leq k$, is computed. This gives a maximal abelian repetition with period length j starting at position $i - j + 1$ and ending at position $i + (k + 1)j$. Meanwhile all entries $L[j, i + pj + 1]$, for $-1 \leq p \leq k$, are marked. Thus all abelian repetitions are computed in $O(n^2)$ time. It remains to compute the length of their heads and tails. This cannot be done naively otherwise it would lead to a $O(n^3)$ time complexity overall. Instead, for each $0 \leq i \leq n - 1$, let T_i be the set of positive integers such that for each $j \in T_i$ there exists a maximal abelian repetition of period j and starting at position $i - j + 1$. Elements of T_i are processed in increasing order. Let j_k denote the k-th smallest element of T_i. Let h_k denote the length of the head of the abelian run computed from the abelian repetition of period j_k. Then h_k can be computed from h_{k-1}, j_{k-1} and j_k as follows. Two cases can arise:

1. If $k = 0$ or $j_{k-1} + h_{k-1} \leq j_k$, then h_k can be computed by comparing the Parikh vector $\mathcal{P}_{w[i-j_k-p..i-j_k]}$ for increasing values of $\mathcal{P}$ from 0 up to $h_k + 1$, with the Parikh vector $\mathcal{P}_{w[i-j_k+1..i]}$.
2. If $j_{k-1} + h_{k-1} > j_k$, then
   $\mathcal{P}_{w[i-j_{k-1}-h_k..i-j_k]}$ can be computed from $\mathcal{P}_{w[i-j_{k-1}-h_{k-1}+1..i-j_{k-1}]}$. Then, h_k is computed by comparing the Parikh vector $\mathcal{P}_{w[i-j_{k-1}-h_{k-1}+1-p..i-j_k]}$ for increasing values of p from 0 up to $h_k + j_k - h_{k-1} - j_{k-1} + 1$.

This can be done in $O(n)$ time. The lengths of the tails can be computed similarly. Overall, all the runs can be computed in time and space $O(n^2)$.

	a	b	a	a	b	a	b	a	a	b	b	b
	0	1	2	3	4	5	6	7	8	9	10	11
1	0	0	1	0	0	0	0	1	0	1	1	0
2	0	0	0	0	1	1	0	1	0	0	0	0
3	0	0	1	0	1	1	0	0	0	0	0	0
4	0	0	0	0	0	0	1	0	0	0	0	0
5	0	0	0	0	1	0	0	0	0	0	0	0
6	0	0	0	0	0	0	0	0	0	0	0	0

Fig. 2. An example of array L for $w = $ abaababaabbb. $L_{4,6} = 1$ which means that $\mathcal{P}_{w[3..6]} = \mathcal{P}_{w[7..10]}$.

This previous method works offline: it needs to know the whole string before reporting any abelian run. We will now give what we call an online method meaning that we will be able to report the abelian runs ending at position $i - 1$ of a string w when processing position i. However, this method is restricted to a given Parikh vector.

4 A Method for Computing Abelian Runs of a Word with a Given Parikh Vector

4.1 Algorithm

Positions of w are processed in increasing order. Assume that when processing position i we know all the, at most $|\mathcal{P}|$, abelian substrings ending at position $i-1$. At each position i we checked if $\mathcal{P}_{w[i-|\mathcal{P}|+1..i]} = \mathcal{P}$ then all abelian substrings ending at position $i - 1$ can be extended and thus become abelian substrings ending at position i. Otherwise, if $\mathcal{P}_{w[i-|\mathcal{P}|+1..i]} \neq \mathcal{P}$ then abelian substrings ending at positions $i-1$ are processed in decreasing order of tail length. When an abelian substring cannot be extended it is considered as an abelian run candidate. As soon as an abelian substring ending at position $i - 1$ can be extended then all the others (with smaller tail length) can be extended: they all become abelian substrings ending at position i. At most one candidate (with the smallest starting position) can be output at each position.

4.2 Implementation

The algorithm RUNS$(\mathcal{P}, w)$ given below computes all the abelian runs with Parikh vector $\mathcal{P}$ in the word w. It uses:

- function FIND$(\mathcal{P}, w)$, which returns the ending position of the first occurrence of Parikh vector $\mathcal{P}$ in w or $|w| + 1$ if such an occurrence does not exist;
- function FINDHEAD$(w, i, \mathcal{P})$, which returns the leftmost position $j < i$ such that $\mathcal{P}_{w[j..i-1]} \subset \mathcal{P}$ or i is such a substring does not exist;
- function MIN(B) that returns the smallest element of the integer array B.

Positions of w are processed in increasing order (Lines 4–21). We will now describe the situation when processing position i of w:

- array B stores the starting positions of abelian substrings ending at position $i - 1$ for the different $|\mathcal{P}|$ tail lengths (B is considered as a circular array);
- t_0 is the index in B of the possible abelian substring with a tail of length 0 ending at position i.

All the values of the array B are initially set to $|w|$. Then, when processing position i of w, for $0 \leq k < |\mathcal{P}|$ and $k \neq t_0$, if $B[k] = b < |w|$ then $w[b..i - 1]$ is an abelian substring with Parikh vector $\mathcal{P}$ with tail length $((t_0 - k + |\mathcal{P}|) \bmod |\mathcal{P}|) - 1$. Otherwise, if $B[k] = |w|$ then it means that there is no abelian substring in w ending at position $i - 1$ with tail length $((t_0 - k + |\mathcal{P}|) \bmod |\mathcal{P}|) - 1$.

The algorithm RUNS$(\mathcal{P}, w)$ uses two other functions:

- function GETTAIL$(tail, t_0, p)$, which returns $(t_0 - tail + p) \bmod p$ which is the length of the tail for the abelian substring ending at position $i - 1$ and starting in $B[tail]$;

- function GETRUN($B, tail, t_0, e, p$), which returns the abelian substring ($B[tail], h, t, e$).

If $\mathcal{P}_{w[i-|\mathcal{P}|+1..i]} = \mathcal{P}$ (Line 6) then all abelian substrings ending at position $i - 1$ can be extended (see Fig. 3). Either this occurrence does not extend a previous occurrence at position $i - |\mathcal{P}|$ (Line 7): the starting position has to be stored in B (Line 8) or this occurrence extends a previous occurrence at position $i - |\mathcal{P}|$ and the starting position is already stored in the array B.

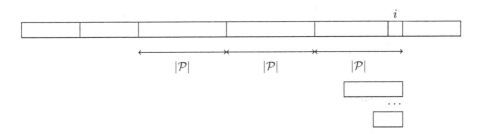

Fig. 3. If $\mathcal{P}_{w[i-|\mathcal{P}|+1..i]} = \mathcal{P}$ then $\mathcal{P}_{w[j..i]} \subset \mathcal{P}$ for $i - |\mathcal{P}| + 1 < j < i$

If $\mathcal{P}_{w[i-|\mathcal{P}|+1..i]} \neq \mathcal{P}$ (Lines 9-21) then abelian substrings ending at position $i - 1$ are processed in decreasing order of tail length. To do that, the circular array B is processed in increasing order of index starting from t_0 (Lines 11-19).

Let $tail$ be the current index in array B. At first, $tail$ is set to t_0 (Line 10). In this case there is no need to check if there is an abelian substring with tail length 0 ending at position i (since it has been detected in Line 6) and thus ($B[t_0], h, |\mathcal{P}| - 1, i - 1$) is considered as an abelian substring candidate (Line 15) and array B is updated (Line 16) since ($B[t_0], h, 0, i$) is not an abelian substring.

When $tail \neq t_0$, let $t =$ GETTAIL($tail, t_0, |\mathcal{P}|$). If $\mathcal{P}_{w[i-t+1..i]} \not\subset \mathcal{P}$ thus ($B[tail], h, t, i - 1$) is considered as an abelian substring candidate (Line 15) and array B is updated (Line 16) since ($B[tail], h, t + 1, i$) is not an abelian substring. If $\mathcal{P}_{w[i-t+1..i]} \subset \mathcal{P}$ then, for $tail \leq k \leq (t_0 - 1 + |\mathcal{P}|) \mod |\mathcal{P}|$, $\exists h'_k, t'_k$ such that ($B[k], h'_k, t'_k, i$) is an abelian substring. It comes directly from Prop. 8.

At each iteration of the loop in Lines 11-19 b is either equal to $|w|$ or to the position of the leftmost abelian run ending at position $i - 1$. Thus a new candidate is found if its starting position is smaller than b (Lines 14-15). It comes directly from Lemma 7.

Algorithm 1. GETTAIL($tail, t_0, p$)

---

1 **return** ($t_0 - tail + p$) mod p

Algorithm 2. GETRUN$(B, tail, t_0, e, p)$

---

1   $b \leftarrow B[tail]$
2   **if** $tail = t_0$ **then**
3      $t \leftarrow p - 1$
4   $t \leftarrow \text{GETTAIL}(tail, t_0, p) - 1$
5   $h \leftarrow (e - t - b + 1) \bmod p$
6   **return** (b, h, t, e)

Algorithm 3. RUNS$(\mathcal{P}, w)$

---

1   $j \leftarrow \text{FIND}(\mathcal{P}, w)$
2   $(B, t_0) \leftarrow (|w|^{|\mathcal{P}|}, 0)$
3   $B[t_0] \leftarrow \text{FINDHEAD}(w, j - |\mathcal{P}| + 1, \mathcal{P})$
4   **for** $i \leftarrow j + 1$ *to* $|w|$ **do**
5      $t_0 \leftarrow (t_0 + 1) \bmod |\mathcal{P}|$
6   **if** $i < |w|$ **and** $\mathcal{P}_{w[i-|\mathcal{P}|+1..i]} = \mathcal{P}$ **then**
7   **if** $B[t_0] = |w|$ **then**
8            $B[t_0] \leftarrow \text{FINDHEAD}(w, i - |\mathcal{P}| + 1, \mathcal{P})$
9   **else**
10       $(b, tail) \leftarrow (|w|, t_0)$
11   **repeat**
12   **if** $B[tail] \neq |w|$ **then**
13   **if** $tail = t_0$ **or** $i = |w|$ **or** $\mathcal{P}_{w[i-\text{GETTAIL}(tail, t_0, |\mathcal{P}|)+1..i]} \not\subset \mathcal{P}$ **then**
14   **if** $B[tail] \leqslant b$ **then**
15               $(b, h, t, e) \leftarrow \text{GETRUN}(B, tail, t_0, i - 1, |\mathcal{P}|)$
16               $B[tail] \leftarrow |w|$
17   **else break**
18         $tail \leftarrow (tail + 1) \bmod |\mathcal{P}|$
19   **until** $tail = t_0$
20   **if** $\text{MIN}(B) > b$ **and** $e - t - h - b + 1 > |\mathcal{P}|$ **then**
21         $\text{OUTPUT}(b, h, t, e)$

Example

Let us see the behaviour of the algorithm on $\Sigma = \{\mathsf{a}, \mathsf{b}\}$, $w = \mathsf{abaababaabbb}$ and $\mathcal{P} = (2, 2)$:

    $j = 4, B = (12, 12, 12, 12), t_0 = 0$
    $B[0] = 0, B = (0, 12, 12, 12)$
    $i = 5$
        $t_0 = 1$
        $\mathcal{P}_{w[2..5]} \neq \mathcal{P}$
        $(b, tail) = (12, 1)$
        $tail = 3, 2, 1, 0$
    $i = 6$
        $t_0 = 2$

$$\mathcal{P}_{w[3..6]} = \mathcal{P}$$
$$B[2] = 0, B = (0, 12, 0, 12)$$

$i = 7$

$$t_0 = 3$$
$$\mathcal{P}_{w[4..7]} = \mathcal{P}$$
$$B[3] = 1, B = (0, 12, 0, 1)$$

$i = 8$

$$t_0 = 0$$
$$\mathcal{P}_{w[5..8]} \neq \mathcal{P}$$
$$(b, tail) = (12, 0)$$
$$(b, h, t, e) = (0, 1, 3, 7)$$
$$B[0] = 12, B = (12, 12, 0, 1)$$
$$tail = 1$$
$$tail = 2$$

$i = 9$

$$t_0 = 1$$
$$\mathcal{P}_{w[6..9]} = \mathcal{P}$$
$$B[1] = 3, B = (12, 3, 0, 1)$$

$i = 10$

$$t_0 = 2$$
$$\mathcal{P}_{w[7..10]} = \mathcal{P}$$
$$B[2] \neq 12$$

$i = 11$

$$t_0 = 3$$
$$\mathcal{P}_{w[8..11]} \neq \mathcal{P}$$
$$(b, tail) = (12, 3)$$
$$(b, h, t, e) = (1, 3, 3, 10)$$
$$B[3] = 12, B = (12, 3, 0, 12)$$
$$tail = 0$$
$$tail = 1$$

$i = 12$

$$t_0 = 0$$
$$i \geq 12$$
$$(b, tail) = (12, 0)$$
$$tail = 1$$
$$(b, h, t, e) = (3, 3, 2, 11)$$
$$B[1] = 12, B = (12, 12, 0, 12)$$
$$tail = 2$$
$$(b, h, t, e) = (0, 3, 1, 11)$$
$$B[1] = 12, B = (12, 12, 12, 12)$$
$$tail = 3$$
$$tail = 0$$
$$\text{OUTPUT}((0, 3, 1, 11)$$

4.3   Correctness and Complexity

Theorem 9. *The algorithm* $\mathrm{RUN}(\mathcal{P}, w)$ *computes all the abelian runs with Parikh vector* $\mathcal{P}$ *in a string* w *of length* n *in time* $O(n \times |\mathcal{P}|)$ *and additional space* $O(\sigma + |\mathcal{P}|)$.

Proof. The correctness of the algorithm comes from Corollary 6, Lemma 7 and Prop. 8. The loop in lines 4-21 iterates at most n times. The loop in lines 11-19 iterates at most $|\mathcal{P}|$ times. The instructions in lines 6, 8 and 13 regarding the comparison of Parikh vectors can be performed in $O(n)$ time overall, independently from the alphabet size, by maintaining the Parikh vector of a sliding window of length $|\mathcal{P}|$ on w and a counter r of the number of differences between this Parikh vector and $\mathcal{P}$. At each sliding step, from $w[i - |\mathcal{P}|..i - 1]$ to $w[i - |\mathcal{P}| + 1..i]$ the counters of the characters $w[i - |\mathcal{P}|]$ and $w[i]$ are updated, compared to their counterpart in $\mathcal{P}$ and r is updated accordingly. The additional space comes from the Parikh vector and from the array B, which has $|\mathcal{P}|$ elements.   □

5   Conclusions

We gave an algorithm that, given a word w of length n and a Parikh vector $\mathcal{P}$, returns all the abelian runs of period $\mathcal{P}$ in w in time $O(n \times |\mathcal{P}|)$ and space $O(\sigma + |\mathcal{P}|)$. The algorithm works in an online manner. To the best of our knowledge, this is the first algorithm solving the problem of searching for all the abelian runs having a given period.

We believe that further combinatorial results on the structure of the abelian runs in a word could lead to new algorithms.

One of the reviewers of this submission pointed out that our algorithm can be modified in order to achieve time complexity $O(n)$. Due to the limited time we had for preparing the final version of this paper, we did not include such improvement here. We will provide the details in a forthcoming full version of the paper. By the way, we warmly thank the reviewer for his comments.

References

1. Bannai, H., I, T., Inenaga, S., Nakashima, Y., Takeda, M., Tsuruta, K.: A new characterization of maximal repetitions by Lyndon trees. CoRR abs/1406.0263 (2014). http://arxiv.org/abs/1406.0263
2. Constantinescu, S., Ilie, L.: Fine and Wilf's theorem for abelian periods. Bulletin of the European Association for Theoretical Computer Science **89**, 167–170 (2006)
3. Cummings, L.J., Smyth, W.F.: Weak repetitions in strings. Journal of Combinatorial Mathematics and Combinatorial Computing **24**, 33–48 (1997)

4. Fici, G., Lecroq, T., Lefebvre, A., Prieur-Gaston, É.: Algorithms for computing abelian periods of words. Discrete Applied Mathematics **163**, 287–297 (2014)

5. Kolpakov, R., Kucherov, G.: Finding maximal repetitions in a word in linear time. In: Proceedings of the 1999 Symposium on Foundations of Computer Science (FOCS 1999), 17–19 October 1999, pp. 596–604. IEEE Computer Society, New York (1999)

6. Matsuda, S., Inenaga, S., Bannai, H., Takeda, M.: Computing abelian covers and abelian runs. In: Prague Stringology Conference 2014, p. 43 (2014)

7. Smyth, W.F.: Computing regularities in strings: a survey. European J. Combinatorics **34**(1), 3–14 (2013)

Coverability in Two Dimensions

Guilhem Gamard[1] and Gwenaël Richomme[1,2]([✉])

[1] LIRMM, (CNRS, Univ. Montpellier 2) UMR 5506, CC 477, 161 rue Ada,
34095 Montpellier Cedex 5, France
guilhem.gamard@lirmm.fr
[2] Dpt MIAp, Univ. Paul-Valéry Montpellier 3, Route de Mende,
34199 Montpellier Cedex 5, France
gwenael.richomme@lirmm.fr

Abstract. A word is *quasiperiodic* (or *coverable*) if it can be covered by occurrences of another finite word, called its *quasiperiod*. This notion was previously studied in the domains of text algorithms and combinatorics of right infinite words. We extend several results to two dimensions. We also characterize all rectangular words that cover non-periodic two-dimensional infinite words. Then we focus on two-dimensional words with infinitely many quasiperiods. We show that such words have zero entropy. However, contrarily to the one-dimensional case, they may not be uniformly recurrent.

Keywords: Combinatorics on words · Patterns

1 Introduction

At the beginning of the 1990's, in the area of text algorithms, Apostolico and Ehrenfeucht introduced the notion of *quasiperiodicity* [1]. Their definition is as follows: "a string w is quasiperiodic if there is a second string $u \neq w$ such that every position of w falls within some occurrence of u in w". The word w is also said to be u-quasiperiodic, and u is called a *quasiperiod* (or a *cover*) of w. For instance, the string:

$$ababaabababaababababaababa$$

is *aba*-quasiperiodic and *ababa*-quasiperiodic.

In 2004, Marcus extended this notion to right-infinite words and observed some basic facts about this new class. He opened several questions [10], most of them related to Sturmian words and the subword complexity. First answers were given in [7]. A characterization of right-infinite quasiperiodic Sturmian words was given in [8] and extended to episturmian words in [5]. More details on the complexity function were given in [11,12].

In [11], Marcus and Monteil showed that quasiperiodicity is independent from several other classical notions in combinatorics on words. They also introduced a stronger notion of quasiperiodicity, namely *multi-scale quasiperiodicity*, with better properties.

© Springer International Publishing Switzerland 2015
A.-H. Dediu et al. (Eds.): LATA 2015, LNCS 8977, pp. 402–413, 2015.
DOI: 10.1007/978-3-319-15579-1_31

Finally, in [4], the authors introduced a two-dimensional version of quasiperiodicity. In particular, they gave a linear-time algorithm computing all square quasiperiods of a given square matrix of letters. Our approach is to continue the study of two-dimensional quasiperiodicity by generalizing the results from [11] to infinite two-dimensional words.

- First, we recall definitions of some classical notions from combinatorics on words in a two-dimensional context. Then, to illustrate these notions and quasiperiodicty, we check that independence between these notions and quasiperiodicity is still true in two dimensions (Section 2).
- We determine a necessary and sufficient condition for a word to be a quasiperiod of non-periodic two-dimensional word. Given a quasiperiod q, we construct a substitution allowing to forge q-quasiperiodic words with various properties, in particular aperiodicity (Section 3).
- We define multi-scale quasiperiodicity in two dimensions. Then we study how multi-scale quasiperiodicity is linked to other classical notions from combinatorics on words. (Section 4).

Warning. Note that in some contexts, most notably in the field of tilings, "quasiperiodic" means "uniformly recurrent". Hence we refer to quasiperiodic words as *coverable* words; each quasiperiod is a *cover* (or *covering pattern*).

2   Coverability

Let Σ be a finite alphabet. A *two-dimensional word* (or $\mathbb{Z}^2$-word) is a function from $\mathbb{Z}^2$ to Σ. Unless otherwise stated, those functions are assumed to be total. When clarification will be needed, we will note $\text{dom}(\mathbf{w})$ the domain of $\mathbf{w}$, i.e. the set of coordinates where it has defined letters.

A *rectangular word* is a word w such that $\text{dom}(w) = \{i, \ldots, i+n\} \times \{j, \ldots, j+m\}$, for $i, j \in \mathbb{Z}$ and $n, m \in \mathbb{N}$. In that case, let $\text{width}(w) = n+1$ and $\text{height}(w) = m+1$. The set of rectangular words of dimension $n \times m$ is $\Sigma^{n \times m}$. More generally, if u is a rectangular word, then $u^{n \times m}$ denotes the $n\,\text{width}(u) \times m\,\text{height}(u)$-rectangle which consists only in occurrences of u.

Let $\mathcal{C}_{\Sigma,n}$ denote the set of n-columns over Σ, i.e. $1 \times n$-rectangular words over Σ. Those columns are concatenated horizontally. Likewise, let $\mathcal{L}_{\Sigma,m}$ denote the set of m-lines over Σ, concatenated vertically. We will occasionally view rectangular words as finite one-dimensional words over $\mathcal{C}_{\Sigma,n}$ or $\mathcal{L}_{\Sigma,m}$, considered as alphabets.

In what follows, let $\mathbf{w}$ be a bidimensional word and let u, v be rectangular words. We say that u is a *cover* (or a *covering pattern*) of $\mathbf{w}$ if, for all $(x, y) \in \mathbb{Z}^2$, there exists $(i, j) \in \mathbb{N}^2$ with $0 \le i < \text{width}(u)$ and $0 \le j < \text{height}(u)$ such that $\mathbf{w}[x - i \ldots x - i + \text{width}(u) - 1;\ y - j \ldots y - j + \text{height}(u) - 1]$ is equal to u up to shift. Intuitively, u is a cover of $\mathbf{w}$ when each position of $\mathbf{w}$ belongs to an occurrence of u.

Now we recall some classical notions from combinatorics on words, adapted to the two-dimensional case. Then we will check that coverability is independent from these notions. This will generalize the first part of [11] to two-dimensional words.

Let $\mathbf{w}[i \ldots i + n;\ j \ldots j + m]$ denote the restriction of $\mathbf{w}$ to the rectangle $\{i \ldots i+n\} \times \{j \ldots j+m\}$, for $i, j \in \mathbb{Z}$ and $n, m \in \mathbb{N}$. If $u = \mathbf{w}[i \ldots i+n;\ j \ldots j+m]$ for some i, j, n and m, then u is a *block* of $\mathbf{w}$.

A two-dimensional word $\mathbf{w}$ is *uniformly recurrent* if, for all $k \in \mathbb{N}$, there exists some $\ell \in \mathbb{N}$ such that all $k \times k$-blocks of $\mathbf{w}$ appear in all $\ell \times \ell$-blocks of $\mathbf{w}$. Intuitively, this means that any block of $\mathbf{w}$ appears infinitely often with bounded gaps.

Moreover, a two-dimensional word $\mathbf{w}$ has a *vector of periodicity* $\overrightarrow{x} \in \mathbb{Z}^{2*}$ if, for all vectors $\overrightarrow{y} \in \mathbb{Z}^2$, we have $\mu(\overrightarrow{x}) = \mu(\overrightarrow{x} + \overrightarrow{y})$. We say that $\mathbf{w}$ is *periodic* if it has at least two non-colinear vectors of periodicity. Links between periodicity and others notion defined in this section (most notably the block complexity function, see below) are currently investigated, see e.g. [3].

Let $c_{\mathbf{w}}(n, m)$ be the number of $n \times m$-blocks of $\mathbf{w}$ ($c_{\mathbf{w}}$ is known as the *block complexity function* of $\mathbf{w}$). Then the the *topological entropy* of $\mathbf{w}$ is the following quantity:

$$H(\mathbf{w}) = \lim_{n \to \infty} \frac{\log_{|\Sigma|} c_{\mathbf{w}}(n, n)}{n^2}$$

Intuitively, if $c_{\mathbf{w}}(n, n) \simeq |\Sigma|^{\epsilon n^2}$, then $H(\mathbf{w}) \simeq \epsilon$. In other words, when the complexity function of $\mathbf{w}$ is polynomial, $\mathbf{w}$ has zero entropy. This is a classical regularity property on words, often used in the context of dynamical systems.

Let $|u|_v$ denote the number of occurrences of v in u. The *frequency* of u in $\mathbf{w}$ is the following quantity:

$$f_{\mathbf{w}}(u) = \lim_{n \to \infty} \frac{|\mathbf{w}[-n \cdots + n, -n \cdots + n]|_u}{n^2}$$

when it exists. If $f_{\mathbf{w}}(u)$ exists for all blocks u of $\mathbf{w}$, then u is said to *have frequencies*. This is another common regularity property coming from dynamical systems.

Proposition 1. *Coverability is independent from uniform recurrence, subword complexity and existence of frequencies.*

Proof. For uniform recurrence, observe that $q = \begin{smallmatrix} b & b & a \\ b & b & b \\ a & b & b \end{smallmatrix}$ is a cover of the non-uniformly recurrent word displayed on Figure 1. With the same value of q, the q-periodic two-dimensional word is uniformly recurrent.

Let $\mathbf{w}$ be a two-dimensional word over $\{a, b\}$ with polynomial (resp. exponential) complexity. Consider the following function:

$$\nu(a) = ababaaba$$
$$\nu(b) = abaababa$$

The image $\nu(\mathbf{w})$ has polynomial with the same degree (resp. exponential) complexity and is *aba*-coverable (viewing *aba* as a 3×1-rectangle).

Finally, the word $\nu(a^{\mathbb{Z}^2})$ has frequencies for all its blocks. By contrast, if **w** is a word having no frequencies for any block, then $\nu(\mathbf{w})$ has no frequencies either. □

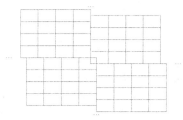

Fig. 1. A coverable, non-uniformly recurrent word

As a conclusion, coverability is a weak notion: it does not bring much information about two-dimensional words it characterizes.

3   Aperiodic Coverings

In this section, we determine under which conditions a rectangular word q can be a cover of an aperiodic $\mathbb{Z}^2$-word w. First, let us consider the question for $\mathbb{N}$-words. Recall that, in this context, a border is a block of q which is both a proper prefix and a suffix of q. (A word u is a proper block of v if it is a block of v and $u \neq v$).

Lemma 2. *A finite one-dimensional word q is a cover of an aperiodic coverable $\mathbb{N}$-word if and only if the primitive root of q has a non-empty border.*

Sketch of proof. If q is a cover of an aperiodic infinite word, then so is its primitive root. If the primitive root r of q has no non-empty borders, then two occurrences of r never properly overlap. Hence any r-covered word must be periodic.

Conversely, if r has a non-empty border and if $q = r^k$ for some positive integer k, then $r = uvu$. Let h be the morphism defined by $h(a) = (uvu)^k$ and $h(b) = (uvu)^{k-1}vu$. The image of any aperiodic word by h is an aperiodic, r-coverable word. The proof is omitted by lack of space, but proof of Theorem 5 works quite the same. □

The previous result also holds for $\mathbb{Z}$-words. For $\mathbb{N}^2$-words, one can prove similarly: a finite rectangular word q is a cover of some aperiodic infinite word if and only if the primitive root of q has a non-empty horizontal border or a non-empty vertical border. In this context, a horizontal (resp. vertical) border is a rectangular word which has the same width (resp. height) as q and which occurs both at the top and the bottom (resp. left and right) of q.

The proof of Lemma 2 used a morphism. Given a word w, readers can check that $h(w)$ shares a lot of common properties with w. From now on, we focus on the $\mathbb{Z}^2$ case. Theorem 5 below generalizes the previous result for $\mathbb{Z}^2$-words. However, our deeper goal is to to construct a substitution over $\mathbb{Z}^2$ allowing to obtain q-coverable $\mathbb{Z}^2$-words with various properties.

3.1   Primitive Roots of Rectangular Words

We need some simple definitions to state our characterization. Let q and r be rectangular words. By definition, r is a *root* of q if $q = r^{n \times m}$, for some positive integers n and m. If q has no roots except itself, it is said to be *primitive*.

These notions initially came from combinatorics on one-dimensional words. The following lemma is a classical result about roots in one dimension. It shows that any one-dimensional finite word has a smallest root, called its *primitive root*.

Lemma 3. (See, e.g., [9], Prop. 1.3.1 and 1.3.2.)
Given any finite one-dimensional words u and v, the following statements are equivalent:

1. *there exist integers $n, m \leq 0$ with $(n, m) \neq (0, 0)$, such that $u^n = v^m$;*
2. *there exist a word t and positive integers k and ℓ such that $u = t^k$ and $v = t^\ell$;*
3. *$uv = vu$.*

Let us show that primitive roots are also well-defined on rectangular words.

Lemma 4. *Let q be a rectangular word. Suppose that q has two distinct roots r_1 and r_2. Then there exists a rectangular word r_3 such that r_3 is a root of both r_1 and r_2.*

Proof. Let r_1^k (resp. r_2^k) denote k occurrences of r_1 (resp. r_2) concatenated vertically. Since r_1 and r_2 are roots of q, there exist integers n and m such that both r_1^n and r_2^m are roots of q, with height$(q) = $ height$(r_1^n) = $ height(r_2^m). Consider q, r_1^n and r_2^m as words over $\mathcal{C}_{\Sigma, \text{height}(q)}$; by Lemma 3, there exists a word c over $\mathcal{C}_{\Sigma, \text{height}(q)}$ such that c is a root of both r_1^n and r_2^m.

Let r_3 (resp. r_4) be the horizontal prefix of r_1 (resp. r_2) of length width(c). Both r_3 and r_4 are prefixes of q, hence $r_3^n = r_4^m$ (the power is still taken for vertical concatenation). Now view r_3 and r_4 as words over $\mathcal{L}_{\Sigma, \text{width}(c)}$. By Lemma 3, there exists a word r over $\mathcal{L}_{\Sigma, \text{width}(c)}$ which is a common root of r_3 and r_4.

As r_1 (resp. r_2) is obtained by horizontal concatenations of occurrences of r_3 (resp. r_4), we deduce that r is a root of r_1 and of r_2.   □

The *primitive root* of a rectangular word q is the root minimal for the "is a root of" relation. By Lemma 4, it is the only root of q (possibly itself) which is primitive.

We need one last definition before stating our first theorem. Let q be a rectangular word. Following [4], a proper block b of q is a *diagonal border* of q if b occurs in two opposite corners of q. Note that it is possible to have either width$(b) = $ width(q) (horizontal border) or height$(b) = $ height(q) (vertical border), but not both.

3.2   Patterns Covering Aperiodic Bidimensional Words

Now we can state the condition under which a rectangular word can be the covering pattern of a non-periodic $\mathbb{Z}^2$-word.

Theorem 5. *Let q be a finite, rectangular word. Then there exists a q-coverable, non-periodic $\mathbb{Z}^2$-word if and only if the primitive root of q has a non-empty diagonal border.*

This subsection is entirely dedicated to the proof of Theorem 5.

Proof of the "only if" part. First, suppose that $\mathbf{w}$ is a $\mathbb{Z}^2$-word which is both q-coverable and non-periodic. There exists at least two overlapping occurrences of q in $\mathbf{w}$. Moreover, the overlapping part is not a power of the primitive root of q: if all overlappings are powers of some root of q, then $\mathbf{w}$ is periodic. Therefore, q must have at least one border which is not a power of its primitive root. Hence its primitive root has a non-empty border.

Proof of the "if" part. Suppose that q's primitive root has a non-empty diagonal border. Let us build an infinite $\mathbb{Z}^2$-word which is q-coverable, but not periodic.

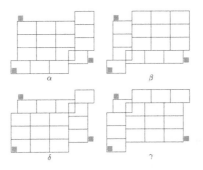

Fig. 2. Four tiles to build a q-coverable word. Each rectangle is an occurrence of q.

Let r be the primitive root of q and b be a non-empty diagonal border of r. Consider the four tiles α, β, δ and γ displayed on Figure 2. Each rectangle is an occurrence of q. The overlapping zones are all occurrences of b and the shifts on tile borders are sized accordingly. If the border b is on the opposite corner, all tiles are built symmetrically.

Let $A = \{a_1, a_2, a_3, a_4\}$ and μ be the function from $A^{\mathbb{Z}^2}$ to $\Sigma^{\mathbb{Z}^2}$, defined by $\mu(a_1) = \alpha$, $\mu(a_2) = \beta$, $\mu(a_3) = \gamma$ and $\mu(a_4) = \delta$. If its input is regular enough, μ behaves more or less like a morphism, with the following concatenation rules.

On Figure 2, each tile has three *anchors*, i.e. letters marked by a small square. Concatenate two tiles horizontally by merging the right-anchor of the first one with the left-anchor of the second one. Concatenate two tiles vertically by merging the bottom-anchor of the first one with the top-anchor of the second one.

More formally, we have:

$$\mu(a_i \cdot u) = \mu(a_i) \cup S_{(4\,\mathrm{width}(q);\mathrm{height}(b))} \circ \mu(u)$$

$$\mu\begin{pmatrix} u \\ v \end{pmatrix} = \mu(v) \cup S_{(\mathrm{width}(b);4\,\mathrm{height}(q))} \circ \mu(u)$$

where the operator $\cup$ denotes the superposition of two finite words. Recall that we view two-dimensional words as (possibly partial) functions from $\mathbb{Z}^2$ to the alphabet. These functions have domains which may be strictly included in $\mathbb{Z}^2$. If w_1 and w_2 are two words with disjoints domains, then $(w_1 \cup w_2)[x, y] = w_1[x, y]$ where w_1 is defined and $w_2[x, y]$ where w_2 is defined. In what follows, we will only consider superpositions where no position (x, y) is defined in both $w_1[x, y]$ and $w_2[x, y]$.

If u is a rectangular word, the leftmost bottom anchor of $\mu(u[i, j])$ has coordinates:

$$(i \times 4 \times \mathrm{width}(q) + j \times \mathrm{width}(b);\ j \times 4 \times \mathrm{height}(q) + i \times \mathrm{height}(b))$$

in $\mu(u)$. Figure 3 gives an example of how μ works.

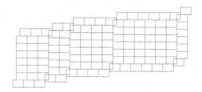

Fig. 3. $\mu\left(\begin{smallmatrix} a_3 & a_4 & a_4 & a_3 \\ a_1 & a_2 & a_2 & a_1 \end{smallmatrix}\right)$, each rectangle is an occurrence of q

A word over A is *suitable* when it satisfies the following conditions:

1. each line is either on alphabet $\{a_1, a_2\}$ or on alphabet $\{a_3, a_4\}$;
2. each column is either on alphabet $\{a_1, a_3\}$ or on alphabet $\{a_2, a_4\}$.

First, we check that if $\mathbf{w}$ is suitable, then each letter of $\mu(\mathbf{w})$ belongs to the image of exactly one letter of $\mathbf{w}$. This essentially means that all tiles "fit together" with no overlaps.

By construction, tiles α and δ fit together vertically, and tiles β and γ fit as well. Hence $\mu\left(\begin{smallmatrix} a_1 \\ a_3 \end{smallmatrix}\right)$ and $\mu\left(\begin{smallmatrix} a_2 \\ a_4 \end{smallmatrix}\right)$ are well-defined. Likewise, tiles α and β fit together horizontally, and tiles δ and γ fit as well. Hence $\mu(a_1 a_2)$ and $\mu(a_3 a_4)$ and are well-defined. Iterating this argument, we deduce that the image of any suitable word is well-defined.

Moreover, we let readers check that $\mu(\mathbf{w})$ has no "holes". More precisely, if if $\mathbf{w}$ is a suitable rectangular word, $\mu(w)$ satisfies the following weak convexity properties:

- for all $i, j, j_1, j_2 \in \mathbb{N}$ with $j_1 \leq j \leq j_2$, if (i, j_1) and (i, j_2) are in $\mathrm{dom}(\mu(\mathbf{w}))$, then (i, j) is in $\mathrm{dom}(\mu(\mathbf{w}))$ as well;

– for all $i, j, i_1, i_2 \in \mathbb{N}$ with $i_1 \leq i \leq i_2$, if (i_1, j) and (i_2, j) are in $\mathrm{dom}(\mu(\mathbf{w}))$, then (i, j) is in $\mathrm{dom}(\mu(\mathbf{w}))$ as well.

As a consequence, the definition of μ can be extended to suitable $\mathbb{Z}^2$-words. If $\mathbf{w}$ is a suitable $\mathbb{Z}^2$-word, then $\mu(\mathbf{w})$ is a well-defined $\mathbb{Z}^2$-word as well.

We will now prove that $\mu(\mathbf{w})$ is aperiodic for any aperiodic bidimensional word $\mathbf{w}$. First, we need a technical lemma about our tiles.

Lemma 6. *Let x and y be different tiles from $\{\alpha, \beta, \gamma, \delta\}$. Then an occurrence of x and an occurrence of y cannot overlap when their anchor points coincide.*

This essentially means that situations from Figure 4 cannot occur.

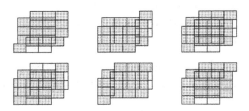

Fig. 4. All other possible overlappings

Proof. There are six possibilities for the set $\{x, y\}$. All proofs are quite similar, so we only provide a proof when $x = \alpha$ and $y = \beta$ (illustrated by the top left-hand case of Figure 4). In what follows, q refers to the rectangular word used for the construction of the tiles, r to its primitive root and b to a diagonal border of r.

There are three occurrences of q, named q_1, q_2 and q_3, such that q_1 is covered by q_2 and q_3 and all three are horizontally aligned. (See for instance the top second column of q's in the figure). View q_1, q_2 and q_3 as one-dimensional words over the alphabet $\mathcal{L}_{\Sigma, \mathrm{width}(q)}$. There exist words x and x' over $\mathcal{L}_{\Sigma, \mathrm{width}(q)}$ such that $q_1 = xx'$ and $q_2 = q_3 = x'x$ (where words are concatenated from bottom to top).

By Lemma 3, x and x' (and q) are powers of a same word s over $\mathcal{L}_{\Sigma, \mathrm{width}(q)}$. Notice that $\mathrm{height}(x') = \mathrm{height}(b)$ and $\mathrm{height}(x) = \mathrm{height}(q) - \mathrm{height}(b)$. It follows that $\mathrm{height}(s)$ divides $\mathrm{height}(x)$ and $\mathrm{height}(q) - \mathrm{height}(b)$.

Observe that s is a vertical prefix of both q and x. Thence one can find three occurrences of s, named s_1, s_2 and s_3, such that s_1 is covered by s_2 and s_3 and all three are vertically aligned. (See for instance the second line of q's in the figure).

Now view s as a one-dimensional word on the alphabet $\mathcal{C}_{\Sigma, \mathrm{height}(s)}$. There exist words y, y' such that $s_1 = yy'$, $s_2 = s_3 = y'y$ and $\mathrm{width}(y') = \mathrm{width}(b)$. By Lemma 3, we deduce that there exists a word t over $\mathcal{C}_{\Sigma, \mathrm{height}(s)}$ such that y and y' (and s) are powers of t.

Let $k \leq 1$ be the integer such that $q = s^k$ (for vertical concatenation) and let $\ell \geq 1$ be the integer such that $s = t^\ell$ (for horizontal concatenation). We have that $q = t^{\ell \times k}$. Therefore t is a root of q such that width$(t) \leq width(y') = width(b)$ and height$(t) =$ height$(s) \leq$ height(b). Thus width$(s) \times$ height$(s) \leq$ width$(b) \times$ height(b) which is a contradiction with the definition of b. Indeed, recall that b is a border (hence a proper block) of the primitive root of q, which is smallest (in number of letters) roots of q.   □

In the proof of next lemma, Lemma 6 helps to establish a correspondence between the letters of the $\mathbb{Z}^2$-word $\mu(\mathbf{w})$ and the "tiling" consisting of occurrences of α, β, δ and γ. We need this correspondence to prove that some $\mu(\mathbf{w})$ can always be made aperiodic.

Lemma 7. *Let q be a rectangular word, r its primitive root and b one non-empty diagonal border of r. Let $\mathbf{w}$ be an aperiodic, suitable $\mathbb{Z}^2$-word. Then $\mu(\mathbf{w})$ is an aperiodic, q-coverable $\mathbb{Z}^2$-word.*

Proof. By construction, $\mu(\mathbf{w})$ is q-coverable for all $\mathbf{w}$. Suppose that $\mu(\mathbf{w})$ has a non-null vector of periodicity $\overrightarrow{p} \in \mathbb{Z}^2$. Let us prove that, under this assumption, $\mathbf{w}$ is periodic.

Let $a \in \mathbb{Z}^2$ be the coordinates of the anchor point of some tile in $\mu(\mathbf{w})$. For any $i \in \mathbb{Z}$, let $t_i = a + i \times \overrightarrow{p}$. Since tiles have at most $16 \times$ width$(q) \times$ height(q) letters, by pigeonhole principle, there are two pairs of coordinates t_i and t_j which have the same offset to the anchor points of their respective tiles (i.e. the tiles covering their respective positions). Hence the difference between these anchor points is a multiple of the vector of periodicity $\overrightarrow{p}$.

Let T_i (resp. T_j) be the tile covering position t_i (resp. t_j). Since T_i is the $(j - i) \times \overrightarrow{p}$-translation of T_j, they are both occurrences of a same tile. Moreover, the right-neighbours of T_i and T_j are both occurrences of a same tile, otherwise we would have a configuration forbidden by Lemma 6. Likewise, the top-neighbour, bottom-neighbour and left-neighbour of T_i and T_j are also equal. By iterating this argument over the neighbours' neighbours, and so on, we conclude that the tiling itself is periodic. Hence, $\mathbf{w}$ is periodic.   □

This ends the proof of Theorem 5. From any rectangular word q with at least one non-empty diagonal border in its primitive root, we can build $\mu(\mathbf{w})$ for any aperiodic, suitable $\mathbb{Z}^2$-word $\mathbf{w}$.

3.3   Lifting other Properties to Coverable Words

Notice how we "lifted" aperiodicity from an arbitrary $\mathbb{Z}^2$-word to a q-coverable word. This technique can be used to lift other properties, such as existence of frequencies, uniform recurrence, block complexity or topological entropy. The proof is as in Proposition 1, using μ instead of ν.

Hence, for any rectangular word q, there exist q-coverable $\mathbb{Z}^2$-words with or without uniform recurrence, with or without frequencies, and with any complexity function. Any rectangle which is the cover of a $\mathbb{Z}^2$-word is also the cover of

$\mathbb{Z}^2$-words *with any properties*. In other terms, the contents of a covering pattern do not bring any information about the covered word.

As a conclusion of this section, remark that our substitution preserves various kind of non-periodic properties. In Lemma 3.5, we could have assumed that there exists no-sequence $(x_i, y_i)_{i \in \mathbb{N}}$ of coordinates in w with $(x_{i+1} - x_i, y_{i+1} - y_i)$ constant. We would have obtained exactly the same property for $\mu(\mathbf{w})$.

4   Multi-scale Coverability in Two Dimensions

In [11], Monteil and Marcus called *multi-scale coverable* any $\mathbb{N}$-word having infinitely many covers. We want to exclude cases where coverability is obtained on groups of lines (or columns) stacked all over $\mathbb{Z}^2$. Hence our generalization is more specific. A $\mathbb{Z}^2$-word (or a $\mathbb{N}^2$-word) is called *multi-scale coverable* if, for each $n \in \mathbb{N}$, it has a $k \times \ell$ cover with both $k \geq n$ and $\ell \geq n$.

In [11], Monteil and Marcus prove that multi-scale coverable right-infinite words have zero entropy and are uniformly recurrent. We study these results for $\mathbb{Z}^2$-words.

4.1   Topological Entropy

Let $\mathbf{w}$ be a $\mathbb{Z}^2$-word. Recall that $c_\mathbf{w}(n, m)$ is the number of rectangles of size $n \times m$ which occur in $\mathbf{w}$ and that the topological entropy of $\mathbf{w}$ is the following quantity:

$$H(w) = \lim_{n \to +\infty} \frac{\log_{|\Sigma|} c_w(n, n)}{n^2} \tag{1}$$

This sequence converges since $\frac{\log_{|\Sigma|} c_w(n,n)}{n}$ is sub-additive (thanks to the Fekete's Subadditive Lemma, see e.g. [13]).

Proposition 8. *Any multi-scale coverable, $\mathbb{Z}^2$-word $\mathbf{w}$ has zero entropy.*

Proof. Consider a covering pattern q of $\mathbf{w}$ with size $n \times m$. Suppose without loss of generality that $n \leq m$. Let s be a $m \times m$-square of $\mathbf{w}$. The square s is covered with occurrences of q (which may spill out of s). The relative position of s and of occurrences of q completely defines s.

We need at most $4m$ occurrences of q to define a covering of s. Each occurrence of q must have at least one of its corners in s. If some occurrence of q has its bottom right-hand corner in s, then no other occurrence of q may have their bottom right-hand corners on the same line of s. Otherwise, one of these occurrences would supersede the other one, which would be "useless" in the covering.

Proceed the same way for the other corners and deduce that at most $4m$ occurrences of q (4 per line) uniquely define s. Each of these occurrences is uniquely determined by its position of its corner on a line of s. There are at most m possibilities for each. Therefore, there are at most m^{4m} q-coverings which define all possible squares s.

This bound on $c_{\mathbf{w}}(m, m)$ allows us to compute the entropy of $\mathbf{w}$. Observe that:

$$\lim_{m \to \infty} \frac{\log m^{4m}}{m^2} = \frac{4m \log m}{m^2} \to 0 \qquad (2)$$

Since there are infinitely many covering patterns of $\mathbf{w}$ with growing sizes, there are infinitely many integers m such that $c_{\mathbf{w}}(m, m) \le m^{4m}$. Hence equation (2) shows that then topological entropy of $\mathbf{w}$ converges to zero. $\qquad \square$

Note that since the Kolmogorov complexity is bound by the entropy (see [2]), this result also shows that the Kolmogorov complexity of multi-scale coverable words is zero as well.

4.2   Uniform Recurrence

Recall that a $\mathbb{Z}^2$-word $\mathbf{w}$ is *uniformly recurrent* when all its blocks occur infinitely often with bounded gaps. In $\mathbb{N}$-words, multi-scale coverability implies uniform recurrence. Quite surprisingly, this is not true for infinite two-dimensional words. Consider $q = \begin{smallmatrix} b & b & a \\ b & b & b \\ a & b & b \end{smallmatrix}$ and the word displayed on Figure 1. The central block

$$\begin{array}{ccccc} b & b & b & b & a \\ b & b & b & b & b \\ b & b & a & b & b \\ b & b & b & b & b \\ a & b & b & b & b \end{array}$$

occurs only once, hence this word is not uniformly recurrent.

Actually, the problem does not lie in the dimension two, but in the absence of origin. The statement "multi-scale coverability implies uniform recurrence" is true for $\mathbb{N}$-words (see [11]) and $\mathbb{N}^2$-words, and false for $\mathbb{Z}$-words and $\mathbb{Z}^2$-words.

Here is an example of a $\mathbb{Z}$-word which is multi-scale coverable, but not uniformly recurrent:

$$^\omega(ab)a(ab)^\omega = \ldots bababab a\, a\, bababab a \ldots$$

Any word matching the $aba(ba)^*$ regular expression is a covering pattern of this word. However, the pattern aa only occur once, hence it is not uniformly recurrent.

Proposition 9. *Any multi-scale, $\mathbb{N}^2$-word $\mathbf{w}$ is uniformly recurrent.*

Proof. This is an adaptation of the proof from [11]. Consider a rectangle r occurring in $\mathbf{w}$. Since $\mathbf{w}$ has arbitrarily large covering patterns and all these patterns occur at the origin, one of these patterns contains r entirely. Hence r occurs whenever the covering patterns occurs, and the latter occurs infinitely many times with bounded gaps. $\qquad \square$

As a conclusion, uniform recurrence from multi-scale coverability does not generalize to $\mathbb{Z}^2$-words. However, the situation as a whole generalizes to two dimensions: the implication is true on words "with origins" ($\mathbb{N}, \mathbb{N}^2$), and false on words "without" ($\mathbb{Z}, \mathbb{Z}^2$).

5   Conclusion and Future Work

As a conclusion, let us point out several questions on which we are currently working.

In [11], it is shown that all multi-scale coverable words have uniform frequencies. Although the result seems still true for two-dimensional words, the proof appears to be not directly generalizable.

Moreover, we have the feeling that non-uniformly recurrent coverable $\mathbb{Z}^2$-words are pathological cases. We suspect that they are all similar to the one displayed on Figure 1. We are currently working on a full characterization of those words.

Finally, one-dimensional coverable words may be decomposed to a *normal form* (see [6]). This allows to view one-dimensional coverable coverable words as images of arbitrary words by some morphisms (which depend on the cover). However, there does not seem to exist such normal form for coverable two-dimensional words. Is any q-coverable word an image by some kind of substitution?

References

1. Apostolico, A., Ehrenfeucht, A.: Efficient detection of quasiperiodicities in strings. Theor. Comput. Sci. **119**(2), 247–265 (1993)
2. Brudno, A.A.: Entropy and complexity of the trajectories of a dynamical system. Tr. Mosk. Mat. O.-va **44**, 124–149 (1982)
3. Cassaigne, J.: Subword complexity and periodicity in two or more dimensions. In: Rozenberg, G., Thomas, W. (eds.) Developments in Language Theory, Foundations, Applications, and Perspectives 1999, pp. 14–21. World Scientific, Aachen (1999)
4. Crochemore, M., Iliopoulos, C.S., Korda, M.: Two-dimensional prefix string matching and covering on square matrices. Algorithmica **20**(4), 353–373 (1998)
5. Glen, A., Levé, F., Richomme, G.: Quasiperiodic and lyndon episturmian words. Theor. Comput. Sci. **409**(3), 578–600 (2008)
6. Iliopoulos, C.S., Mouchard, L.: Quasiperiodicity: From detection to normal forms. Journal of Automata, Languages and Combinatorics **4**(3), 213–228 (1999)
7. Levé, F., Richomme, G.: Quasiperiodic infinite words: Some answers (column: Formal language theory). Bulletin of the EATCS **84**, 128–138 (2004)
8. Levé, F., Richomme, G.: Quasiperiodic sturmian words and morphisms. Theor. Comput. Sci. **372**(1), 15–25 (2007)
9. Lothaire, M.: Combinatorics on Words. Cambridge University Press. Cambridge Mathematical Library (1997)
10. Marcus, S.: Quasiperiodic infinite words (columns: Formal language theory). Bulletin of the EATCS **82**, 170–174 (2004)
11. Monteil, T., Marcus, S.: Quasiperiodic infinite words: multi-scale case and dynamical properties. http://arxiv.org/abs/math/0603354
12. Polley, R., Staiger, L.: The maximal subword complexity of quasiperiodic infinite words. In: McQuillan, I., Pighizzini, G. (eds.) Proceedings Twelfth Annual Workshop on Descriptional Complexity of Formal Systems, DCFS 2010, vol. 31, pp. 169–176. EPTCS, Saskatoon (2010)
13. Schechter, E.: Handbook of Analysis and its Foundations. Elsevier Science (1996)

Equation $x^i y^j x^k = u^i v^j u^k$ in Words

Jana Hadravová$^{(\boxtimes)}$ and Štěpán Holub

Department of Algebra Faculty of Mathematics and Physics, Charles University,
186 75 Praha 8, Sokolovská 83, Prague, Czech Republic
hadravova@ff.cuni.cz, holub@karlin.mff.cuni.cz

Abstract. We will prove that the word $a^i b^j a^k$ is periodicity forcing if
$j \geq 3$ and $i + k \geq 3$, where i and k are positive integers. Also we will
give examples showing that both bounds are optimal.

Keywords: Combinatorics on words · Word equations · Periodicity

1 Introduction

Periodicity forcing words are words $w \in A^*$ such that the equality $g(w) = h(w)$
is satisfied only if $g = h$ or both morphisms $g, h : A^* \to \Sigma^*$ are periodic. The first
analysis of short binary periodicity forcing words was published by J. Karhumäki
and K. Culik II in [2]. Besides proving that the shortest periodicity forcing
words are of length five, their work also covers the research on the non-periodic
homomorphisms agreeing on the given small word w over a binary alphabet.
What in their work attracts attention the most, is the fact, that even short word
equations can be quite difficult to solve. The intricacies of the equation $x^2 y^3 x^2 =
u^2 v^3 u^2$, proved to have only periodic solution [3], nothing but reinforced the
perception of difficulty. Not frightened, we will extend the result and prove that
the word $a^i b^j a^k$ is periodicity forcing if $j \geq 3$ and $i + k \geq 3$, where i and k
are positive integers. Also we will give examples showing that both bounds are
optimal.

2 Preliminaries

Standard notation of combinatorics on words will be used: $u \leq_p v$ ($u \leq_s v$ resp.)
means that u *is a prefix of* v (u *is a suffix of* v resp.). The maximal common prefix
(suffix resp.) of two word $u, v \in A^*$ will be denoted by $u \wedge v$ ($u \wedge_s v$ resp.). By the
length of a word u we mean the number of its letters and we denote it by $|u|$. A
(one-way) *infinite word* composed of infinite number of copies of a word u will be
denoted by u^ω. It should be also mentioned that the *primitive root* of a word u,
denoted by p_u, is the shortest word r such that $u = r^k$ for some positive k. A word
u is *primitive* if it equals to its primitive root. Words u, v are *conjugate* if there are
words α, β such that $u = \alpha\beta$ and $v = \beta\alpha$. For further reading, please consult [6].

We will briefly recall a few basic and a few more advanced concepts which
will be needed in the proof of our main theorem. Key role in the proof will be
played by the Periodicity lemma (see [6, Chap. 6, Theorem 6.1]):

© Springer International Publishing Switzerland 2015
A.-H. Dediu et al. (Eds.): LATA 2015, LNCS 8977, pp. 414–423, 2015.
DOI: 10.1007/978-3-319-15579-1_32

Lemma 1 (Periodicity Lemma). *Let p and q be primitive words. If p^ω and q^ω have a common factor of length at least $|p|+|q|-1$, then p and q are conjugate. If, moreover, p and q are prefix (or suffix) comparable, then $p = q$.*

Reader should also recall that if two words satisfy an arbitrary non-trivial relation, then they have the same primitive root. Another well-known result is the fact that the maximal common prefix (suffix resp.) of any two different words from a binary code is bounded (see [6, Chap. 6, Lemma 3.1]). We formulate it as the following lemma:

Lemma 2. *Let $X = \{x, y\} \subseteq A^*$ and let $\alpha \in xX^*$, $\beta \in yX^*$ be words such that $\alpha \wedge \beta \geq |x| + |y|$. Then x and y commute.*

The previous lemma can be formulated also for the maximal common suffix:

Lemma 3. *Let $X = \{x, y\} \subseteq A^*$ and let $\alpha \in X^*x$, $\beta \in X^*y$ be words such that $\alpha \wedge_s \beta \geq |x| + |y|$. Then x and y commute.*

The most direct and most well-known case is the following.

Lemma 4. *Let $s = s_1 s_2$ and let $s_1 \leq_s s$ and $s_2 \leq_p s$. Then s_1 and s_2 commute.*

Proof. Directly, we obtain $s = s_1 s_2 = s_2 s_1$.

Next, let us remind the following property of conjugate words:

Lemma 5. *Let $u, v, z \in A^*$ be words such that $uz = zv$. Then u and v are conjugate and there are words $\sigma, \tau \in A^*$ such that $\sigma\tau$ is primitive and*

$$u \in (\sigma\tau)^*, \qquad\qquad z \in (\sigma\tau)^*\sigma, \qquad\qquad v \in (\tau\sigma)^*.$$

We will also need not so well-know, but interesting, result by A. Lentin and M.-P. Schützenberger [4].

Lemma 6. *Suppose that $x, y \in A^*$ do not commute. Then $xy^+ \cup x^+y$ contains at most one imprimitive word.*

We now introduce some more terminology. Suppose that x and y do not commute and let $X = \{x, y\}$, i.e., we suppose that X is a binary code. We say that a word $u \in X^*$ is X-*primitive* if $u = v^i$ with $v \in X^*$ implies $u = v$. Similarly, $u, v \in X^*$ are X-*conjugate*, if $u = \alpha\beta$ and $v = \beta\alpha$ and the words α and β are from X^*.

In the following lemma, first proved by J.-C. Spehner [7], and consequently by E. Barbin-Le Rest and M. Le Rest [1], we will see that all words that are imprimitive and X-primitive are X-conjugate of a word from the set $x^*y \cup xy^*$. Source of the inspiration of both articles was an article by A. Lentin and M.-P. Schützenberger [4] with its weaker version stating that if the set of X-primitive words contains some imprimitive words, then so does the set $x^*y \cup xy^*$. As a curiosity, we mention that Lentin and Schützenberger formulated the theorem for $x^*y \cap y^*x$ instead of $x^*y \cup y^*x$ (for which they proved it). Also, the Le Rests did not include in the formulation of the theorem the trivial possibility that the word x or the word y is imprimitive.

Lemma 7. *Suppose that $x, y \in A^*$ do not commute and let $X = \{x, y\}$. If $w \in X^*$ is a word that is X-primitive and imprimitive, then w is X-conjugate of a word from the set $x^* y \cup y^* x$. Moreover, if $w \notin \{x, y\}$, then primitive roots of x and y are not conjugate.*

Putting together Lemma 6 with Lemma 7, we get the following result:

Lemma 8. *Suppose that $x, y \in A^*$ do not commute and let $X = \{x, y\}$. Let $\mathcal{C}$ be the set of all X-primitive words from $X^+ \setminus X$ that are not primitive. Then either $\mathcal{C}$ is empty or there is $k \geq 1$ such that*

$$\mathcal{C} = \{x^i y x^{k-i}, 0 \leq i \leq k\} \text{ or } \mathcal{C} = \{y^i x y^{k-i}, 0 \leq i \leq k\}.$$

The previous lemma finds its interesting application when solving word equations. For example, we can see that an equation $x^i y^j x^k = z^\ell$, with $\ell \geq 2$, $j \geq 2$ and $i + k \geq 2$ has only periodic solutions. (This is a slight modification of a well known result of Lyndon and Schützenberger [5]). Notice, that we can use the previous lemma also with equations which would generate notable difficulties if solved "by hand". E.g. equation

$$(yx)^i yx(xxy)^j xy(xy)^k = z^m,$$

with $m \geq 2$, has only periodic solutions.

We formulate it as a special lemma:

Lemma 9. *Suppose that $x, y \in A^*$ do not commute and let $X = \{x, y\}$. If there is an X-primitive word $\alpha \in X^*$ and a word $z \in A^*$, such that*

$$\alpha = z^i,$$

with $i \geq 2$, then $\alpha = x^k y x^\ell$ or $\alpha = y^k x y^\ell$, for some $k, \ell \geq 0$.

We finish this preliminary part with the following useful lemmas:

Lemma 10. *Let $u, v, z \in A^*$ be words such that $z \leq_s v$ and $uv \leq_p zv^i$, for some $i \geq 1$. Then $uv \in zp_v^*$.*

Proof. Let $0 \leq j < i$ be the largest exponent such that $zv^j \leq_p uv$ and let $r = (zv^j)^{-1} uv$. Then r is a prefix of v. Our assumption that $z \leq_s v$ yields that $v \leq_s vr$ and

$$r(r^{-1}v) = v = (r^{-1}v)r.$$

From the commutativity of words $r^{-1}v$ and r, it follows that they have the same primitive root, namely p_v. Since $uv = (zv^j)r$ we have $uv \in zp_v^*$, which concludes the proof. □

Lemma 10 has the following direct corollary.

Lemma 11. *Let* $w, v, t \in A^*$ *be words such that* $|t| \leq |w|$ *and* $wv \leq_p tv^i$, *for some* $i \geq 1$. *Then* $w \in tp_v^*$.

Proof. Lemma 10 with $u = t^{-1}w$ and z empty yields that $uv \in p_v^*$. Then $wv \in tp_v^*$ and from $|t| \leq |w|$, we obtain that $w \in tp_v^*$. □

Lemma 12. *Let* $u, v \in A^*$ *be words such that* $|u| \geq |v|$. *If* αu *is a prefix of* v^i *and* $u\beta$ *is a suffix of* v^i, *for some* $i \geq 1$, *then* $\alpha u\beta$ *and* v *commute*.

Proof. Since $\alpha u \leq_p v^i$ and $|u| \geq |v|$ we have

$$\alpha^{-1} v\alpha \leq_p u \leq_p u\beta.$$

Our assumption that $u\beta$ is a suffix of v^i yields that $u\beta$ has a period $|v|$. Then, $u\beta \leq_p (\alpha^{-1} v\alpha)^i$ and, consequently, $\alpha u\beta \leq_p v^i$. From $v \leq_s u\beta$ and Lemma 10, it follows that $\alpha u\beta \in p_v^*$, which concludes the proof. □

Lemma 13. *Let* $u, v \in A^*$ *be words such that* $|u| \geq |v|$. *If* αu *and* βu *are prefixes of* v^i, *for some* $i \geq 1$, *and* $|\alpha| \leq |\beta|$, *then* α *is a suffix of* β, *and* $\beta\alpha^{-1}$ *commutes with* v.

Proof. Since αu is a prefix of v^+ and $|u| \geq |v|$, we have $\alpha^{-1} v\alpha \leq_p u$. Similarly, $\beta^{-1} v\beta \leq_p u$. Therefore,

$$\alpha^{-1} v\alpha = \beta^{-1} v\beta,$$

and $|\alpha| \leq |\beta|$ yields $\alpha \leq_s \beta$. From $\beta\alpha^{-1} v = v\beta\alpha^{-1}$ we obtain commutativity of v and $\beta\alpha^{-1}$. □

Notice that the previous result can be reformulated for suffixes of v^i:

Lemma 14. *Let* $u, v \in A^*$ *be words such that* $|u| \geq |v|$. *If* $u\alpha$ *and* $u\beta$ *are suffixes of* v^i, *for some* $i \geq 1$, *and* $|\alpha| \leq |\beta|$, *then* α *is a prefix of* β, *and* $\alpha^{-1}\beta$ *commutes with* v.

3 Solutions of $x^i y^j x^k = u^i v^j u^k$

Theorem 15. *Let* $x, y, u, v \in A^*$ *be words such that* $x \neq u$ *and*

$$x^i y^j x^k = u^i v^j u^k, \tag{1}$$

where $i + k \geq 3$, $ik \neq 0$ *and* $j \geq 3$. *Then all words* x, y, u *and* v *commute*.

Proof. First notice that, by Lemma 9, theorem holds in case that either of the words x, y, u or v is empty. In what follows, we suppose that x, y, u and v are non-empty. By symmetry, we also suppose, without loss of generality, that $|x| > |u|$ and $i \geq k$; in particular, $i \geq 2$. Recall that p_x (p_y, p_u, p_v resp.) denote the primitive root of x (y, u, v resp.).

We first prove the theorem for some special cases.

(A) Let $p_x = p_u$.

Then $p_x^{in} y^j p_x^{kn} = v^j$ for some $n \geq 1$, and we are done by Lemma 9.

Notice that the solution of case (A) allows us to assume the useful inequality

$$(i + k - 1)|u| < |p_x|, \qquad (*)$$

since otherwise p_x^ω and u^ω have a common factor of length at least $|p_x| + |u|$, and u and x commute by the Periodicity lemma.

From

$$(u^{-i+1} p_x u^{-k})u = u(u^{-i} p_x u^{-k+1})$$

and Lemma 5 we see that there are words σ and τ such that $\sigma\tau$ is primitive and

$$u^{-i+1} p_x u^{-k} \in (\sigma\tau)^m, \qquad u = (\sigma\tau)^\ell \sigma, \qquad u^{-i} p_x u^{-k+1} \in (\tau\sigma)^m,$$

for some $m \geq 1$ and $\ell \geq 0$. Then we have

$$u = (\sigma\tau)^\ell \sigma, \qquad p_x = u^i(\tau\sigma)^m u^{k-1} = u^{i-1}(\sigma\tau)^m u^k, \qquad (**)$$

for some $m \geq 1$ and $\ell \geq 0$.

(B) Let p_y and p_v be conjugate.

Let α and β be such that $p_y = \alpha\beta$ and $p_v = \beta\alpha$. Since $x^i p_y$ is a prefix of $u^i p_v^+$, we can see that $u^{-i} x^i \alpha\beta \leq_p \beta(\alpha\beta)^+$. From Lemma 10 we infer that and $u^{-i} x^i \in \beta(\alpha\beta)^*$. Similarly, by the mirror symmetry, $p_y x^k \leq_s p_v^+ u^k$ yields that $x^k u^{-k} \in (\alpha\beta)^* \alpha$. Then

$$x^{i+k} = u^i p_v^n u^k,$$

for some $n \geq 1$. From $|v| > |y|$, it follows that $|v| \geq |y| + |p_v|$ and, consequently,

$$(i + k)(|x| - |u|) = j(|v| - |y|) \geq 3|p_v|.$$

Then $n \geq 3$ and we are done by Lemma 9.

(C) Let p_x and p_v be conjugate.

Let α and β be such that $p_x = \alpha\beta$ and $p_v = \beta\alpha$. From $(*)$ and $i \geq 2$, it follows that $u^i p_v$ is a prefix of p_x^2. Then $u^i(\beta\alpha) \leq_p \alpha(\beta\alpha)^+$ and Lemma 10 yields that $u^i \in \alpha(\beta\alpha)^*$. From $i|u| < |p_x|$, it follows $u^i = \alpha$. Since p_x is a suffix of $\alpha\beta\alpha u^k = p_x u^{i+k}$ and u is a prefix of p_x, we have

$$u(u^{-1} p_x) = u(u^{-1} p_x) \wedge_s u(u^{-1} p_x)u^{i+k}.$$

We then deduce from Lemma 3 that x and u commute, that implies $p_x = p_u$, i.e., case (A).

Fig. 1. Case $|x| \geq |v|$

We will now discuss separately cases when $|x| \geq |v|$ and $|x| < |v|$.

1. Suppose that $|x| \geq |v|$.

If $i \geq 3$ or $x \neq p_x$, then $(u^{-i}x)x^{i-1}$ is a prefix of v^j that is longer than $|p_x| + |x|$ by (∗). By the Periodicity lemma, p_x is a conjugate of p_v and we are in case (C). The remaining cases deal with $i = k = 2$ and $i = 2$, $k = 1$.

1a) First suppose that $i = k = 2$. Since $(u^{-i}x)x$ is a prefix of v^j and $x(xu^{-k})$ is a suffix of v^j, we get, by Lemma 12, that $(u^{-i}x)x(xu^{-k})$ commutes with v. Then

$$x^3 = u^i p_v^n u^k,$$

for some $n \geq 0$. From $(i + k - 1)|u| < |p_x| \leq |x|$ and $|p_v| \leq |v| \leq |x|$ we infer that $n \geq 2$. Therefore, $p_u = p_x$ holds by Lemma 9, and we have case (A).

1b) Suppose now that $i = 2$ and $k = 1$. We will have a look at the words u and $x = p_x$ expressed by (∗∗). Let $h = (\sigma\tau)^m$ and $h' = (\tau\sigma)^m$. Then (∗∗) yields

$$u = (\sigma\tau)^\ell \sigma, \qquad\qquad x = u^2 h' = uhu.$$

1b.i) Suppose now that $|p_v| \leq |uh|$. Since $h'uh$ is a prefix of v^j and uh is a suffix of v^j, we obtain by Lemma 12 that $h'uh = p_v^n$. From $|p_v| \leq |uh|$, we infer $n \geq 2$ and, according to Lemma 9, σ and τ commute. Then also x and u commute and we have case (A).

1b.ii) Suppose that $|p_v| > |uh|$. From $|x| \geq |v| \geq |p_v|$, it follows that $p_v = h'uu_1$ for some prefix u_1 of u. We can suppose that u_1 is a proper prefix of u, otherwise x and v are conjugate and we have case (C). Then $u_1 h' \leq_p uh' \leq_p (\sigma\tau)^+$ and, by Lemma 13, we obtain $uu_1^{-1} \in (\sigma\tau)^+$. Therefore, $u_1 \in (\sigma\tau)^*\sigma$. Since $h \leq_s p_v$, we can see that $\sigma\tau \leq_s \tau\sigma^+$. Lemma 3 then implies commutativity of σ and τ. Therefore, the words x and u also commute and we are in case (A).

2. Suppose that $|x| < |v|$ and $i|x| = i|u| + |v|$.

From $x \leq_s v$, we have $x \leq_s xu^k$. Since $u \leq_p x$ we deduce from Lemma 3 that x and u commute, thus we have case (A).

3. Suppose that $|x| < |v|$ and $i|x| > i|u| + |v|$.

Let r be a non-empty word such that $u^i vr = x^i$. Notice that $|r| < |p_x|$ otherwise the words p_x and p_v are conjugate and we have case (C). Considering the words u and p_x expressed by (∗∗), we can see that $(\tau\sigma)^m u^{k-1} u^i$ is a prefix of v and $u^{i-1}(\sigma\tau)^m$ is a suffix of v. Notice also that we have case (A) if σ and τ commute.

3a) Consider first the special case when $r = u^k$.

3a.i) If $i = k$, then $v^{j-2} = u^i y^j u^i$. If $j \geq 4$, we have case (B) by Lemma 9. If $j = 3$, then the equality $u^i vr = x^i$ implies $x^i = u^{2i} y^j u^{2i}$ and we get case (A) again by Lemma 9.

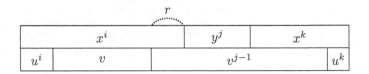

Fig. 2. Case $|x| < |v|$ and $i|x| > i|u| + |v|$

3a.ii) Suppose therefore that $k < i$. Notice that $u = \sigma$, otherwise, from $\tau\sigma \leq_p v$ and $u^k = r \leq_p v$, we get commutativity of σ and τ. Therefore,

$$v \in (\tau\sigma)^m \sigma^{k-1} p_x^* \sigma^{i-1} (\sigma\tau)^m.$$

We have

$$vu^k x^{-k} = vrx^{-k} = u^{-i} x^{i-k}.$$

From $i > k$ and $(*)$ we get $|u^{-i} x^{i-k}| > 0$ and, consequently, $|vu^k| > |x^k|$. Let v' dente the word $vu^k x^{-k}$. Then $v^{j-2} v' = ry^j$, and $j \geq 3$ together with $|v| > |x| > |u^k| = |r|$ yields that v' is a suffix of y^j. According to $(**)$, $v' = u^{-i} x^{i-k} \in (\tau\sigma)^m \sigma^{k-1} p_x^*$. Then, σ^k is a suffix of y^j and we have

$$(\sigma^k y \sigma^{-k})^j = \sigma^k y^j \sigma^{-k} = v^{j-2} v' \sigma^{-k}.$$

This is a point where Lemma 9 turns out to be extremely useful. Direct inspection yields that $v^{j-2} v' \sigma^{-k}$ is not a jth power of a word from $\{\sigma, \tau\}^*$. One can verify, for example, that the expression of $v^{j-2} v' \sigma^{-k}$ in terms of σ and τ contains exactly $j-2$ occurrences of τ^2. Therefore, Lemma 9 yields that σ and τ commute, a contradiction.

3b) We first show that $r = u^k$ holds if $k \geq 2$. Indeed, if $k \geq 2$ then $u^k p_x u^{-k}$ is a suffix of v and, consequently, $u^k p_x u^{-k} r$ is a suffix of x^i. Since $u^k p_x u^{-k} u^k$ is also a suffix of x^i, we can use Lemma 14 and get commutativity of x with one of the words $u^{-k} r$ or $r^{-1} u^k$. From $|r| < |p_x|$ and $|u^k| < |p_x|$, we get $r = u^k$.

3c) Suppose that $k = 1$ and $r \neq u$.

3c.i) If $|r| < |u|$, then r is a suffix of u and $|xr^{-1} u| > |x|$. Since $xr^{-1} \leq_s v$ and $k = 1$, the word $x = xr^{-1} r$ is a suffix of $xr^{-1} u$. Therefore, xr^{-1} is a suffix of $(ur^{-1})^+$. Since $u^2 \leq_p x$ and $|xr^{-1}| \geq |u| + (|u| - |r|)$, the Periodicity lemma implies that the primitive root of ur^{-1} is a conjugate of p_u. But since p_u is prefix comparable with ur^{-1}, we obtain that $ur^{-1} \in p_u^+$. Then also $r \in p_u^+$ and $xr^{-1} \in p_u^+$. Consequently, x and u commute, and we have case (A).

3c.ii) Suppose therefore that $|r| > |u|$. Then u is a suffix of r. Since r is a suffix of p_x and $p_x = u^i(\tau\sigma)^m$, the word r is a suffix of $u^i(\tau\sigma)^m$. From $|v| > |x|$ we obtain $u^{-i} x u^i \leq_p v$. Consequently, from $p_x = u^i(\tau\sigma)^m$ and $r \leq_s v$, it follows that r is a prefix of $(\tau\sigma)^m u^i$.

Consider first the special case when $r \in (\tau\sigma)^m p_u^*$. If $r \in (\tau\sigma)^m p_u^+$, then $r \leq_s u^i(\tau\sigma)^m$ yields that $(\tau\sigma)^m$ and u commute by Lemma 3. Consequently, σ and τ commute, and we have case (A). Therefore, $r = (\tau\sigma)^m$, $p_x = u^i r$ and $v = u^{-i} x^i r^{-1} \in (ru^i)^+$. We have proved that x and v have conjugate primitive roots, which yields case (C). Consider now the general case.

If $m \leq \ell$, then $(\tau\sigma)^m$ is a suffix of u. Since r is a prefix of $(\tau\sigma)^m u^i$, and $u \leq_s r$, we get from Lemma 10 the case $r \in (\tau\sigma)^m p_u^*$.

Suppose that $m > \ell$. Then u is a suffix of $(\tau\sigma)^m$. Let s denote the word $(\tau\sigma)^m u^{-1} = (\tau\sigma)^{m-\ell-1}\tau$.

If $|r| \geq |(\tau\sigma)^m|$, then $r = s'su$ for some s'. From $r \leq_p (\tau\sigma)^m u^i$, it follows that $s'su$ is a prefix of su^{i+1}. Lemma 11 then yields $s's \in sp_u^*$. Therefore $r \in sup_u^*$ and from $su = (\tau\sigma)^m$, we have the case $r \in (\tau\sigma)^m p_u^*$.

Let $|r| < |(\tau\sigma)^m|$. From $|r| > |u|$ and $(\tau\sigma)^m = su$, we obtain that there are words s_1, s_2 such that $s = s_1 s_2$, $r = s_2 u \leq_p v$ and $s_1 \leq_s v$. Since s is both a prefix and a suffix of v, Lemma 4 implies that s_1 and s_2 have the same primitive root, namely p_s.

Note that $p_x = u^i su$. We now have

$$u^i s_2 s_1 = u^i s \leq_s v \leq_s x^i r^{-1} = (u^i su)^{ni-1} u^i s_1,$$

for some $n \geq 1$. From $i \geq 2$, it follows that $u^i s_2$ is a suffix of su^{i+1}. Lemma 3 then yields commutativity of s and u. Hence, words x and u also commute and we are in case (A).

4. Suppose now that $|x| < |v|$ and $i|x| < i|u| + |v|$.

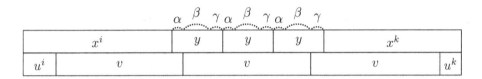

Fig. 3. Case $|x| < |v|$ and $i|x| < i|u| + |v|$

First notice that in this case also $k|x| < k|u| + |v|$. If $j|y| \geq |v| + |p_y|$, then, by the Periodicity lemma, p_v and p_y are conjugate, and theorem holds by (B). Assume that $j|y| < |v| + |p_y|$. Then, since $i|x| < i|u| + |v|$ and $k|x| < k|u| + |v|$, we can see that $j = 3$ and there are non-empty words α, β and γ for which $y = \alpha\beta\gamma$ and $v = (\beta\gamma)(\alpha\beta\gamma)(\alpha\beta)$, with $|\alpha\gamma| < |p_y|$.

4a) Suppose first that $|u^i\gamma| \leq |x|$. Notice that also $|\alpha u^k| \leq |x|$ since $k \leq i$ and $|\gamma| = (i-k)(|x| - |u|) + |\alpha|$. Then $|\gamma x| \leq |v|$ and $u^i\gamma x$ is a prefix of x^2. Therefore, by Lemma 10, $u^i\gamma$ commutes with x. We obtain the following equalities:

$$v = \gamma p_x^n \alpha, \qquad\qquad y^j = \alpha v\gamma = (\alpha\gamma)p_x^n(\alpha\gamma),$$

where $n \geq 1$. If $n \geq 2$, then x and y commute by Lemma 9. If $n = 1$, then $p_x = x$ and $i = 2$. Since $\gamma x^k = vu^k = \gamma x\alpha u^k$ and $|\alpha u^k| \leq |x|$, also $k = 2$ and $\alpha u^k = x$. Then $|\alpha| = |\gamma|$ and $u^2\gamma = x = \alpha u^2$. If $|u| \geq |\gamma|$, then u and γ commute, a contradiction with $p_x = x$. Therefore, $|x| < 3|\gamma|$ and $|v| = |\gamma x\alpha| < 5|\gamma|$. Since γ is a suffix of x and α is a prefix of x, $(\gamma\alpha\beta)^3\gamma\alpha$ is a factor of v^3 longer than $|y| + |v|$. Therefore, by the Periodicity lemma, words y and v are conjugate, and we have case (B).

4b) Suppose that $|u^i\gamma| > |x|$, denote $z = x^{-1}u^i\gamma$ and $z' = \gamma^{-1}v\alpha^{-1} = x^k u^{-k}\alpha^{-1}$. From

$$|y| + |\gamma| + |\alpha| < |v| = |\gamma z'\alpha|,$$

we deduce $|y| < |z'|$. Since $x^{i-1} = zz'$ and z' is a prefix of x^k, the word zz' has a period $|z| < |\gamma|$. Since zz' is a factor of v greater than $|z| + |y|$ and v has a period $|p_y|$, the Periodicity lemma implies $|p_y| \le |z| < |\gamma|$, a contradiction with $|\gamma| < |p_y|$.   □

4   Conclusion

The minimal bounds for i, j, k in the previous theorem are optimal. In case that $i = k$ and j is even, Eq. (1) splits into two separate equations, which have a solution if and only if either $j = 2$, or $i = k = 1$ (see [2]).

Apart from these solutions, we can find non-periodic solutions also in case that $i \neq k$. Namely, for $j = 2$ and $i = k + 1$, we have

$$x = \alpha^{2k+1}(\beta\alpha^k)^2, \qquad u = \alpha,$$
$$y = \beta\alpha^k, \qquad v = (\alpha^k\beta)^2(\alpha^{3k+1}\beta\alpha^k\beta)^k.$$

So far this seems to be the only situation when the equation

$$x^i y^2 x^k = u^i v^2 u^k \tag{2}$$

with $i > k$ has a non-periodic solution. We conjecture that if $|i - k| \ge 2$, then Eq. (2) has only periodic solutions.

If $i = k = 1$ and j is odd, then Eq. (1) has several non-periodic solutions, for example:

$$x = \alpha\beta\alpha, \qquad u = \alpha,$$
$$y = \gamma, \qquad v = \alpha\gamma^j\alpha,$$

where $\beta^2 = v^{j-1}$.

References

1. Barbin-Le Rest, E., Le Rest, M.: Sur la combinatoire des codes à deux mots. Theor. Comput. Sci. **41**, 61–80 (1985)
2. Culik II, K., Karhumäki, J.: On the equality sets for homomorphisms on free monoids with two generators. RAIRO ITA **14**(4), 349–369 (1980)

3. Czeizler, E., Holub, Š., Karhumäki, J., Laine, M.: Intricacies of simple word equations: an example. Internat. J. Found. Comput. Sci. **18**(6), 1167–1175 (2007)
4. Lentin, A., Schützenberger, M.-P.: A combinatorial problem in the theory of free monoids. In: Pollak, G. (ed.) Algebraic Theory of Semigroups. North-Holland Pub. Co, Amsterdam (1979)
5. Lyndon, R.C., Schützenberger, M.-P.: The equation $a^m = b^n c^p$ in a free group. The Michigan Mathematical Journal **9**(4), 289–298 (1962)
6. Rozenberg, G., Salomaa, A. (eds.): Handbook of formal languages, vol. 1: word, language, grammar. Springer-Verlag New York Inc., USA (1997)
7. Spehner, J.-C.: Quelques problèmes d'extension, de conjugaison et de presentation des sous-monoïdes d'un monoïde libre. Ph.D. thesis, Université Paris VII, Paris (1976)

Square-Free Words over Partially Commutative Alphabets

Łukasz Mikulski[1,2]([✉]), Marcin Piątkowski[1,3], and Wojciech Rytter[1,4]

[1] Faculty of Mathematics and Computer Science, Nicolaus Copernicus University,
Toruń, Poland
{lukasz.mikulski,marcin.piatkowski}@mat.umk.pl
[2] School of Computing Science, Newcastle University, Newcastle upon Tyne, UK
[3] Department of Computer Science, University of Helsinki, Helsinki, Finland
[4] Faculty of Mathematics, Informatics and Mechanics,
Warsaw University, Warsaw, Poland
rytter@mimuw.edu.pl

Abstract. There exist many constructions of infinite words over three-letter alphabet avoiding squares. However, the characterization of the lexicographically minimal square-free word is an open problem. Efficient construction of this word is not known. We show that the situation changes when some letters commute with each other. We give two characterizations (morphic and recursive) of the lexicographically minimal square-free word $\tilde{v}$ in the case of a partially commutative alphabet Θ of size three. We consider the only non-trivial relation of partial commutativity, for which $\tilde{v}$ exists: there are two commuting letters, while the third one is blocking (does not commute at all). We also show that the n-th letter of $\tilde{v}$ can be computed in time logarithmic with respect to n.

1 Introduction

Problems related to repetitions are crucial in the combinatorics on words due to many practical application, for instance in data compression, pattern matching, text indexing and so on (see [16]). On the other hand, in some cases it is important to consider words avoiding regularities and repetitions. Example applications can be found in such research areas as cryptography and bioinformatics. Languages of words over partially commutative alphabets are fundamental tools for concurrent systems investigation, see [9]. Therefore, the study of repetitions and their avoidability in such languages is significant.

The simplest form of repetition is a square – the factor of the form $x \cdot x$, where x is not empty. Therefore, to show that a word w contains no repetitions, it is sufficient to show that w does not contain squares. Another interesting type of repetition is the abelian square – a factor of the form $x \cdot y$, where x can be obtained from y by permutation of the letters. For example, $baca \cdot caab$ is an abelian square, whereas $bcca \cdot cbba$ is not. A word that contains no abelian squares is called abelian square-free.

© Springer International Publishing Switzerland 2015
A.-H. Dediu et al. (Eds.): LATA 2015, LNCS 8977, pp. 424–435, 2015.
DOI: 10.1007/978-3-319-15579-1_33

Square-free and abelian square-free words have been extensively studied. In 1906 Thue showed that squares are avoidable over three-letter alphabet (see [19]), i.e. there exist infinitely many ternary words without a square. In 1961 Erdös raised the question whether abelian squares are avoidable (see [13]). First attempt to answer this question was made by Evdokimov in 1968 (see [14]), who showed that abelian squares are avoidable over alphabets consisting of at least 25 letters. Then, in 1970, the required size of the alphabet was decreased to 5 by Pleasants (see [18]), and finally, in 1992, to 4 by Keränen (see [15]). Moreover, it can be easily shown that abelian squares cannot be avoided over three-letter alphabet.

A one step further is to study repetitions and their avoidability in words over partially commutative alphabets, see for instance [6–8,10,11]. In contrast to the abelian case, only some fixed pairs of letters from the alphabet are allowed to commute. It complicates considerably the analysis of repetitions in such classes of words.

Our Results. In this paper we deal with the avoidability of repetitions in words over three-letter alphabet Θ with one pair of commuting letters. Then we describe an infinite language of length-increasing square-free words and investigate their combinatorial properties. We use this language, utilising the results of [10], to define the infinite language of partially abelian square-free words over Θ.

As a final result, we give two characterizations of the infinite lexicographically minimal Θ-square-free word $\tilde{\mathbf{v}}$ and give an efficient construction of this word. The n-th letter of $\tilde{\mathbf{v}}$ can be computed in logarithmic time with respect to n. The first 176 letters of $\tilde{\mathbf{v}}$ are:

$$\tilde{\mathbf{v}} = abacabcbacabacbcbabacabcbacbcbabcbacabacbcbabacabcbacabacbcbabcb$$
$$acbcabacabcbacabacbcbabacabcbacbcbabcbacabacbcbabcbacbcbabacabcb$$
$$acbcabcbacabacbcbabacabcbacabacbcbabcbacbcbabacabcbacabacbc\ldots$$

Due to the page limitation, the proofs of some facts were omitted. The full version of this paper, including all proofs, is available as [17].

2   Basic Notions

Throughout the paper we use the standard notions of the formal language theory (see [16] for a more detailed introduction). By Σ we denote a finite set, called the *alphabet*. Elements of the alphabet are called *letters*. A finite word over Σ is a finite sequence of letters. The length of a word w is defined as the number of its letters and denoted $|w|$. The set of all finite words over Σ is denoted by Σ^* and is equipped with a binary associative concatenation operation $\cdot$, where $a_1 \ldots a_n \cdot b_1 \ldots b_m$ is simply $a_1 \ldots a_n b_1 \ldots b_m$. An empty sequence of letters, called the *empty word* and denoted by ε, is the neutral element of the concatenation operation. Thus for any word w we have $\varepsilon \cdot w = w \cdot \varepsilon = w$. An infinite word over

Σ is a sequence of letters indexed by non-negative integers. On the other hand, it can be also defined as a limit of infinite sequence of finite words.

A word u is called a *factor* of a word w if there exist words x and y such that $w = xuy$. If $y = \varepsilon$ then u is called a *prefix* of w and if $x = \varepsilon$ then u is called a *suffix* of w. For a word $w = a_1 a_2 \ldots a_n$ and $1 \leq i, j \leq n$ by $w[i..j]$ we denote its factor of the form $a_i a_{i+1} \ldots a_j$.

We assume that the alphabet Σ is given together with a strict total order $<$, called the *lexicographical order*. This notion is extended in a natural way to the level of words. For any two words x and y we have $x < y$ if x is a proper prefix of y or we have $x = uav_1$ and $y = ubv_2$, where a, b are letters and $a < b$.

A mapping $\phi : \Sigma_1^* \to \Sigma_2^*$ is called a *morphism* if we have $\phi(u \cdot v) = \phi(u) \cdot \phi(v)$ for every $u, v \in \Sigma_1^*$. A morphism ϕ is uniquely determined by its values on the alphabet. Moreover, ϕ maps the neutral element of Σ_1^* into the neutral element of Σ_2^*.

A *partially commutative alphabet* is a pair $\Theta = (\Sigma, ind)$, where Σ is an ordered alphabet and $ind \subseteq \Sigma \times \Sigma$ is a symmetric *commutation* relation. Such an alphabet defines an equivalence relation $\equiv_\Theta$ identifying words, which differ only by the ordering of commuting letters. Two words $w, v \in \Sigma^*$ satisfy $w \equiv_\Theta v$ if there exists a finite sequence of commutations of adjacent commuting letters transforming w into v. For example let us consider the partially commutative alphabet $\Theta = (\{a, b, c\}, \{(b, c), (c, b)\})$. Then the word $w = acbcacb$ is equivalent to $v = accbabc$, but is not equivalent to $u = baccacb$. Words over a partially commutative alphabet $\Theta = (\Sigma, ind)$ are called *partially commutative words*. Note that it is usually assumed that for each $a \in \Sigma$ we have $(a, a) \notin ind$, but in the case of this paper such an assumption is not essential and it does not affect the presented results.

A square in a word w is a factor of the form $x \cdot x$, where x is not empty. A word w is called *square-free* if none of its factors is a square. If we consider a partially commutative alphabet $\Theta = (\Sigma, ind)$ a square is called a partially commutative square or a Θ-square in short.

Definition 1 ([10]). *Let $\Theta = (\Sigma, ind)$ be a partially commutative alphabet. A Θ-square is a factor of the form $u \cdot v$ such that $u \equiv_\Theta v$. A word w is Θ-square-free if it does not contain a nonempty Θ-square.*

There are possible other (nonequivalent) definitions of a partially commutative square-free words, see [8]. Moreover, in the case of the full commutation relation (i.e. any pair of letters can commute) Θ-squares are called the *abelian squares*, and words avoiding them – the *abelian square-free* words.

Example 1. Let $\Theta = (\{a, b, c\}, \{(b, c), (c, b)\})$ be a partially commutative alphabet. The word $w_1 = abc \cdot acb$ is a Θ-square, but it is not an ordinary square. On the other hand, $w_2 = abc \cdot bac$ is an abelian square, which is neither a Θ-square nor an ordinary square. Therefore, w_1 is a square-free word, which it is neither Θ-square-free nor abelian square-free, while w_2 is square-free and Θ-square-free, but not abelian square-free.

3   Partially Abelian Square-Free Words over Three-Letter Alphabets

It is easy to see that any binary word consisting of at least four letters must contain a square. In 1906 Thue shown that three letters are sufficient to construct an infinite square-free word (see [19]). Moreover, in 1992 Keränen proved that to avoid abelian squares (i.e. factors of the form $x \cdot y$, where x and y differ only by permutation of their letters) four letters are sufficient (see [15]). It follows immediately that any four-letter alphabet with more restricted commutation relation also allows to avoid partially commutative squares. Therefore, the alphabets of size three are the most interesting boundary case.

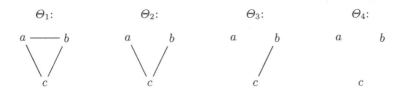

Fig. 1. The possible shapes of the commutation relation over three-letter alphabet. The pairs of letters connected by an edge can commute.

In partially commutative alphabets consisting of three letters one can consider four distinct commutation relations as depicted on Fig. 1. We start with the most restricted case of Θ_4. Observe that the concepts of Θ_4-square freeness and ordinary square-freeness are equivalent. Due to the results of Thue (see [19]), the number of Θ_4-square-free words is infinite. Moreover, the number of finite square-free words of a given length is exponential with respect to this length (see [5] for more details).

On the other hand, the concepts of abelian square-freeness and Θ_1-square-freeness are equivalent. We have only 117 words without Θ_1-square and the longest of them consists of 7 letters. Similarly, the number of Θ_2-square-free words is finite. In this case we have 289 such words with the longest having 15 letters (see [10] for more details).

The most interesting case is the remaining alphabet Θ_3. Similarly as in the case of Θ_4, the number of finite Θ_3-square-free words is infinite, however it is polynomially proportional to the length of the word (see [11]). In what follows, we focus on this alphabet and investigate the combinatorial structure of Θ_3-square-free words in more details.

Remark 1. From now on we only consider the alphabet Θ_3 and denote it by Θ. Thus, by Θ-square and Θ-square-freeness we mean the Θ_3-square and Θ_3-square-freeness.

Conditions for Θ-square-freeness

We start with giving some necessary and sufficient conditions for a word to be Θ-square-free. The more detailed study of their combinatorial structure is presented in the subsequent sections. We follow here the results of [10] presented below. Initially we present a statement, which is used further to formulate the conditions for Θ-square-freeness.

Definition 2 (Condition (F), see [10]). *The word $v \in \Sigma^*$ satisfies the condition (F) if neither abca nor acba is a factor of v, where $a, b, c \in \Theta = (\Sigma, ind)$.*

The possible structure of a finite word containing a Θ-square is established by the following fact (see Proposition 3.2 in [10]).

Proposition 1 (see [10]). *Let w be a finite square-free word satisfying the condition (F) and containing a Θ-square as a factor. Then w admits one of the following decompositions:*

(i) $w = w_1 bcw_2 bcbw_2 bw_3$ (ii) $w = w_1 cbw_2 cbcw_2 cw_3$
(iii) $w = w_1 bw_2 bcbw_2 cbw_3$ (iv) $w = w_1 cw_2 cbcw_2 bcw_3$

where $w_1, w_2, w_3 \in \Sigma^*$. Moreover in such a decomposition one of the factors w_1 or w_3 is of length at most 1.

As a corollary to Proposition 1 we can formulate the following fact characterizing the possible building blocks of Θ-square-free words (see the proof of Proposition 2.1 and Proposition 3.2 in [10]). It will be utilized further in construction an infinite Θ-square-free word.

Corollary 1. *Any infinite Θ-square-free word w starting with a consists of the factors belonging to the following: $\mathcal{B} = \{aba, aca, abcba, acbca, abca, acba\}$. Moreover, the factors acba and abca can appear only as a prefix of w.*

Remark 2. Note that no two different words created by concatenating factors (without ending a) from the set $\mathcal{B}$ defined in Corollary 1 are equivalent under the relation $\equiv_\Theta$.

Finally, the following theorem gives a sufficient condition for the Θ-square-freeness of an infinite word, see Corollary 3.3 in [10] for the proof.

Theorem 1 (see [10]). *Any infinite square-free word over Σ starting with a and satisfying the (F) condition is Θ-square-free.*

4   The Structure of Θ-square-free Words

In the preceding section we presented the necessary and sufficient conditions for a word to be Θ-square-free. Below we investigate the combinatorial structure of such words in more detail.

Recall that due to Corollary 1 any infinite Θ-square-free word w consists only of the factors aba, aca, $abcba$, $acbca$, $acba$ and $abca$, where the last two can appear only as a prefix w. It can be easily proven that neither abc nor acb could be a prefix of the lexicographically minimal infinite Θ-square-free word. Therefore we have to consider only the factors aba, aca, $abcba$, $acbca$.

The above observations are the basis of the idea of encoding the possible building blocks of Θ-square-free words as the symbols of a four-letter meta-alphabet $\Delta = \{A, B, C, D\}$.

Definition 3. Let $\Sigma=\{a,b,c\}$ (alphabet) and $\Delta=\{A,B,C,D\}$ (meta-alphabet). We define a morphism $M : \Delta^* \longrightarrow \Sigma^*$ as follows:

$$M = \begin{cases} A \longrightarrow ab & B \longrightarrow ac \\ C \longrightarrow abcb & D \longrightarrow acbc \end{cases}.$$

It is worth to note that the morphism defined above is a code with finite deciphering delay. This fact is utilized in operations described further in this paper.

In what follows, if a word w over Σ is an image of a word u over Δ we call u an M-reduction of w and w is called M-reducible[1].

The alphabet Δ consists of four letters, hence it allows us to construct words without repetitions. However, not all such words over Δ lead to words with no repetitions over Θ. Since we are interested in Θ-square-free words, the considered words over Δ must satisfy additional conditions presented further.

Lemma 1. *Let w be an infinite, M-reducible and Θ-square-free word starting with abacabcbaca. Then M-reduction of w does not contain any of the factors: AC, CA, BD, DB, ABA, BAB, CBC, DAD, ADCB, BCDA.*

Remark 3. Let $w \in \Sigma^*$ be a Θ-square-free word satisfying the condition (F) stated in Definition 2. Then w consists of blocks, which are images of letters from the alphabet Δ by the morphism M defined above, hence it is always M-reducible and we can apply the inverse mapping M^{-1} (M-reduction) to w. Moreover, the obtained result is a square-free word. On the other hand, the image by M of a square-free word over Δ does not have to be Θ-square-free word. For instance AC is a square-free word over Δ, but $M(AC) = ababcb$ is not Θ-square-free.

As a corollary to Lemma 1 we can describe the structure of Θ-square-free words in the terms of meta-alphabet.

Corollary 2. *Each M-reduction of a Θ-square-free word starting with aba is an element of the set defined by a following regular expression $\Upsilon = \big((A|C)(B|D)\big)^*$.*

---

[1] In the approach presented in this paper the morphism M is in fact used as a translation of an infinite word over four-letter alphabet into an infinite word over three-letter alphabet.

Remark 4. The inverse of Corollary 2 is not true, i.e. there exist words whose M-reductions are in the set Υ, but they are not Θ-square-free, for instance $abacabacbc$ which is an image of $ABAD$.

Lemma 2. *Let $w < v \in \Sigma^*$ be two M-reducible words starting with aba, which M-reductions have equal length and are contained in Υ, and let $u \in \Sigma^*$ be the longest M-reducible word such that $M^{-1}(w) = M^{-1}(u)M^{-1}(w')$ and $M^{-1}(v) = M^{-1}(u)M^{-1}(v')$. Then $M^{-1}(v')$ starts with C and $M^{-1}(w')$ starts with A or $M^{-1}(v')$ starts with D and $M^{-1}(w')$ starts with B.*

Theorem 2. *An infinite word w starting with the letter a and not starting with abca or acba is Θ-square-free if and only if w is square-free and M-reducible.*

5   Two Equivalent Characterizations of the Infinite Word $\tilde{v}$

In this section we present two alternative definitions of the language of square-free words over the meta-alphabet $\Delta = \{A, B, C, D\}$ introduced in the previous section. We start with a definition using a morphism, then we show a recurrent procedure, which generates the same class of words.

5.1   Morphic Characterization

Let us define the sequence of words $\{X_i\}_{i \geq 0}$ over Δ together with the languages L_m and L_{dep} similar to $M(L_m)$. Both of them are based on the following morphism.

Definition 4. *We define a morphism $m : \{A, B, C, D\}^* \longrightarrow \{A, B, C, D\}^*$ as:*

$$m = \begin{cases} A \longrightarrow BCB \\ B \longrightarrow ADA \\ C \longrightarrow BCDCB \\ D \longrightarrow ADCDA \end{cases}.$$

Definition 5. *Let $\{X_i\}_{i>0}$ be defined as:*

$$X_i = \begin{cases} AB & \text{for} \quad i = 0 \\ A \cdot m(X_{i-1}) \cdot B & \text{for } i > 0 \end{cases}.$$

We define the languages

$$L_m = \left\{ m^i(AB) : \ i \geq 0 \right\}; \quad \text{and} \quad L_{dep} = \left\{ M(X_i) : \ i \geq 0 \right\}.$$

The subsequent fact describes the combinatorial structure of words contained in the language L_{dep}.

Lemma 3. *For every two words* $u, v \in L_{dep}$ *either* u *is the prefix of* v *or* v *is the prefix of* u.

Note that L_{dep} is an infinite set of words with strictly growing lengths. Therefore, for any $k > 0$ there exists a word $w \in L_{dep}$ such that $|w| > k$. This observation, together with Lemma 3, constitutes the correctness of the definition of the infinite word $\tilde{v}$.

Id Z_i is a sequence of length increasing words, such that Z_i is a prefix of Z_{i+1} for each i then $lim_{i \to \infty} Z_i$ denotes the infinite word containing all Z_i as its prefixes.

Definition 6. *Define* $\tilde{v} = lim_{i \to \infty} M(X_i)$, *or equivalently as* $\tilde{v} = \sup(L_{dep})$.

We show that $\tilde{v}$ is the lexicographically least word over our partially commutative alphabet.

5.2   Recurrent Characterization of $\tilde{v}$

In this subsection we define two sequences of words using recurrence. Furthermore, at the end of this section, we show that one of them is equivalent to the sequence $\{X_i\}_{i \geq 0}$ defined previously. We start with defining the operation of so-called complement for letters.

Definition 7. *We define the operation* $\hat{\ }$: $\{A, B, C, D\} \to \{A, B, C, D\}$ *as follows:*

$$\hat{\ } = \begin{cases} A \to B, & B \to A \\ C \to D, & D \to C \end{cases}.$$

The mapping defined above is in a natural way extended to the level of words. It allows us to define two recurrent sequences of words.

Definition 8. *We define the sequences of words* $\{Y_i\}_{i \geq -1}$ *and* $\{S_i\}_{i \geq 0}$ *over the alphabet* $\Delta = \{A, B, C, D\}$ *as follows:*

$$Y_{-1} = \varepsilon, \qquad Y_0 = AB, \qquad S_0 = C,$$
$$Y_{n+1} = Y_n S_n \widehat{Y}_n \widehat{S}_n Y_n, \qquad S_{n+1} = S_n \widehat{Y}_{n-1} \widehat{S}_n Y_{n-1} S_n.$$

Example 2. The first few elements of sequences defined above are as follows:

$$Y_0 = AB, \quad S_0 = C, \qquad Y_1 = ABCBADAB, \quad S_1 = CDC,$$
$$Y_2 = ABCBADABCDCBADABCBADCDABCBADAB,$$
$$S_2 = CDCBADCDABCDC.$$

The following facts describe some of the combinatorial properties of the sequences defined above.

Lemma 4. *For each* i, *the word* S_i *is a palindrome and the word* Y_i *is a pseudo-palindrome, i.e. for each* $1 \leq k \leq l$ *we have* $Y[l - k + 1] = \widehat{Y}[k]$, *where* $l = |Y_i|$.

Lemma 5. *Let X_i be as in Definition 5. Then*
(1) for each $i \geq 0$ we have $X_i = Y_i$;
(2) $\tilde{v} = \lim_{i \to \infty} M(Y_i)$.

Taking into account the images of the words Y_i and S_i by morphism M we can formulate the following fact, which will be very useful further.

Proposition 2 (Block lengths). *Let Y_i and S_i be as defined above. Then for each $i \geq 1$ we have:*

$$\left| M(Y_i) \right| = \frac{4(4^{i+1} - 1)}{3} \quad \text{and} \quad \left| M(S_i) \right| = \frac{4(2 \cdot 4^i + 1)}{3}.$$

6   Combinatorial Properties of the Word $\tilde{v}$

In this section we formulate and prove the main results of the paper. Namely, we show the Θ-square-freeness (Theorem 3) and lexicographical minimality (Theorem 4) of the word $\tilde{v}$ and the time complexity of the computing the n-th letter of $\tilde{v}$ (Theorem 5). The proof of the latter yields in fact a very efficient procedure.

We start with a series of facts which lead to the proof of Θ-square-freeness of $\tilde{v}$. Let us recall languages L_m and $L_{dep} = M(L_m)$ from Definition 5.

Proposition 3. *The language L_m consists of square-free words only.*

Lemma 6. $L_m \subseteq \Upsilon = \left((A|C)(B|D) \right)^*.$

Proposition 4. *Let $v \in L_m$. Then v does not contain a factor of the form $wxwy$, where $w \in \Delta^*$, and $(x = A \wedge y = C)$ or $(x = B \wedge y = D)$.*

Theorem 3. *The languages $M(L_m)$ and L_{dep} consists of square-free words only. The word $\tilde{v}$ is an infinite Θ-square-free word.*

Theorem 4. *The word $\tilde{v}$ is the lexicographically minimal infinite Θ-square-free word.*

Proof. The Θ-square-freeness of $\tilde{v}$ follows from Theorem 3.

Suppose that there exists an infinite Θ-square-free word $\tilde{w}$ that is lexicographically smaller than $\tilde{v}$. Then, by the analysis of short Θ-square-free words, $\tilde{w}$ has to start with $abacabcbaca$. Moreover, due to Theorem 2 the word $\tilde{w}$ is M-reducible. Let us consider $u \in \Sigma^*$ – the longest common M-reducible prefix of $\tilde{v}$ and $\tilde{w}$. Moreover, let $X, Y \in \Delta$ be such that $v = uM(X)$ and $w = uM(Y)$ are prefixes of $\tilde{v}$ and $\tilde{w}$, respectively. We have that $M(X) > M(Y)$. Potentially there are 6 cases for X, Y, however from our previous results it follows that the only cases to consider are:

$$(X, Y) = (C, A), \quad \text{or} \ (X, Y) = (D, B).$$

Precisely, w, v and u satisfy all assumptions of Lemma 2, hence indeed, one of the following conditions holds:

1. $M^{-1}(v) = M^{-1}(u)C$ and $M^{-1}(w) = M^{-1}(u)A$,
2. $M^{-1}(v) = M^{-1}(u)D$ and $M^{-1}(w) = M^{-1}(u)B$.

Without the loss of generality, we can assume the first case. According to Lemma 1, the last meta-letter of $M^{-1}(u)$ is either B or D. We deal with each of those cases separately.

1° ($M^{-1}(u)$ ends with B): Due to the morphic definition of $\widetilde{v}$, the last but one letter of $M^{-1}(u)$ is A. Hence, the word $M^{-1}(w)$ contains a forbidden factor ABA. Therefore, by Lemma 1, the infinite word $\widetilde{w}$ cannot be Θ-square-free.

2° ($M^{-1}(u)$ ends with D): Following similar reasoning as above, we obtain that $M^{-1}(u)$ ends with BCD. Hence, the word $M^{-1}(w)$ contains a forbidden factor $BCDA$. Therefore, due to Lemma 1, the infinite word $\widetilde{w}$ cannot be Θ-square-free.

The contradictions obtained above prove that the initial assumption concerning the existence of $\widetilde{w}$ was wrong. Therefore, $\widetilde{v}$ is indeed the lexicographically minimal infinite Θ-square-free word.

Theorem 5. *For each $n > 0$ the n-th letter of the word $\widetilde{v}$ can be determined in time $O(\log n)$.*

Proof. To prove the above lemma we present a simple recurrent procedure.

For given $n \geq 0$ we find the shortest word Y_i of length $l \geq n$, where Y_i's are as in Definition 8. Due to Proposition 2, both the index i and length l are given by simple arithmetic formulas. By Definition 8, the word Y_i consists of five factors (either Y_{i-1} or S_{i-1} or their complements) with lengths given by Proposition 2. Thus, we can determine the factor $F \in \{Y_{i-1}, \widehat{Y}_{i-1}, S_{i-1}, \widehat{S}_{i-1}\}$ containing the considered position k in a constant time. It remains to determine (using a recurrent call) the letter in F on a position n', which is obtained by subtracting from k the starting position of F. The recurrence stops when $i = 1$ and we have F equal to one of the words $M(Y_1)$, $M(S_1)$, $M(\widehat{Y_1})$ or $M(\widehat{S_1})$.

Note that at each call of the recurrence it is necessary to memorize whether we are looking for a letter in one of words ($M(Y_1)$ or $M(S_1)$) or their complements ($M(\widehat{Y_1})$ or $M(\widehat{S_1})$). It could be done by using a single boolean variable.

It is easy to see that the number of iterations performed by the procedure described above is logarithmic with respect to n. Moreover, the required computations on each level of recurrence can be performed in a constant time.

7   Final Remarks

In partially commutative alphabets of size three with one pair of commuting letters one can consider three commutation relations:

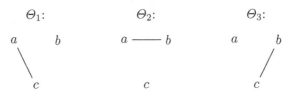

From the point of view of the partially abelian square-freeness all above alphabets are equivalent. We considered in this paper square-free words over the partially commutative alphabet Θ_3, but any Θ_3-square-free word could be transformed by a morphism (precisely an isomorphism) to a Θ_1-square-free or Θ_2-square-free word and almost all the results follow.

However, if we are interested in construction of the lexicographically minimal partially abelian square-free word, the choice of the alphabet is very important. Such a choice determines the blocking letter and the structure of the lexicographically minimal word. In the case of alphabets Θ_1 and Θ_2 it requires further investigation.

In [2] Allouche and Shallit presented an open problem of characterizing the lexicographically minimal square-free word over three-letter alphabet without any commutation allowed. The construction of Thue (see [19]) leads to a word which is not lexicographically minimal.

On the other hand, there is a procedure, proposed by Currie [12], which allows to determine if a given finite word is a prefix of an infinite word avoiding some repetitions. It immediately gives an algorithm computing arbitrary long prefix of the lexicographically least infinite word. However, generating n-th letter is definitely not logarithmic with respect to n. Moreover, in the case of square-freeness, it seems to be directly applicable for alphabets consisting more than four letters.

Another problem related to square-freeness is the overlap-freeness (i.e., avoiding pattern $axaxa$, where a is a letter and x is a word). Berstel proved [3], (see also [4]), that the lexicographically *greatest* infinite overlap-free word on the binary alphabet $\Sigma = \{0, 1\}$ that begins with 0 is the Thue-Morse overlap-free sequence τ.

Moreover, it has been shown in [1] that the lexicographically least infinite overlap-free binary word is $001001\bar{\tau}$, where $\bar{\tau}$ is the negation of overlap-free Thue-Morse word τ. This makes the problem of extremal cases for overlap-freeness closed. However, its solution relies on Thue-Morse word, which is a fix point of a morphism. This supports the claim that there is also an efficient construction of the lexicographically least square-free word over a ternary alphabet without commutation. We believe that the techniques utilized in this paper might be helpful in finding such a construction.

Acknowledgements. We would like to thank the anonymous referees for their comments and useful suggestions.

This research was supported by Polish National Science Center under the grant No.2013/09/D/ST6/03928 and by research fellowship within project "Enhancing Educational Potential of Nicolaus Copernicus University in the Disciplines of Mathematical and Natural Sciences" (project no. POKL.04.01.01-00-081/10).

References

1. Allouche, J., Currie, J.D., Shallit, J.: Extremal infinite overlap-free binary words. Electr. J. Comb. 5 (1998)
2. Allouche, J., Shallit, J.O.: Automatic Sequences - Theory, Applications, Generalizations. Cambridge University Press (2003)
3. Berstel, J.: A rewriting of Fife's theorem about overlap-free words. In: Karhumäki, J., Maurer, H., Rozenberg, G. (eds.) Results and Trends in Theoretical Computer Science, pp. 19–29. Springer, Heidelberg (1994)
4. Berstel, J.: Axel Thue's work on repetitions in words. Séries formelles et combinatoire algébrique 11, 65–80 (1992)
5. Brandenburg, F.J.: Uniformly growing k-th power-free homomorphisms. Theoretical Computer Science 23(1), 69–82 (1983)
6. Carpi, A.: On the number of abelian square-free words on four letters. Discrete Applied Mathematics 81(1–3), 155–167 (1998)
7. Carpi, A.: On abelian squares and substitutions. Theoretical Computer Science 218(1), 61–81 (1999)
8. Carpi, A., de Luca, A.: Square-free words on partially commutative free monoids. Information Processing Letters 22(3), 125–131 (1986)
9. Cartier, P., Foata, D.: Problèmes combinatoires de commutation et réarrangements. LNM, vol. 85. Springer, Heidelberg (1969)
10. Cori, R., Formisano, M.R.: ITA. Partially abelian squarefree words 24, 509–520 (1990)
11. Cori, R., Formisano, M.R.: On the number of partially abelian square-free words on a three-letter alphabet. Theoretical Computer Science 81(1), 147–153 (1991)
12. Currie, J.D.: On the structure and extendibility of k-power free words. European Journal of Combinatorics 16(2), 111–124 (1995)
13. Erdös, P.: Some unsolved problems. Magyar Tud. Akad. Mat. Kutató 6, 221–254 (1961)
14. Evdokimov, A.: Strongly asymmetric sequences generated by a finite number of symbols. Dokl. Akad. Nauk. SSSR 179, 1268–1271 (1968)
15. Keränen, V.: Abelian squares are avoidable on 4 letters. In: Kuich, W. (ed.) ICALP 1992. LNCS, vol. 623. Springer, Heidelberg (1992)
16. Lothaire : Algebraic Combinatorics on Words. Encyclopedia of mathematics and its application, vol. 90. Cambridge University Press (2002)
17. Mikulski, Ł., Piątkowski, M., Rytter, W.: Square-free words over partially commutative alphabets. CS-TR 1439, Newcstle University (2014)
18. Pleasants, P.A.B.: Non-repetitive sequences. Mathematical Proceedings of the Cambridge Philosophical Society 68, 267–274 (1970)
19. Thue, A.: Über unendliche Zeichenreihen. Videnskabsselskabets Skrifter, I Mat.-nat. K1, 1–22 (1906)

On the Language of Primitive Partial Words

Ananda Chandra Nayak$^{(\boxtimes)}$ and Kalpesh Kapoor

Department of Mathematics,
Indian Institute of Technology Guwahati, Guwahati, India
{n.ananda,kalpesh}@iitg.ernet.in

Abstract. A partial word is a word which contains some holes known as *do not know* symbols and such places can be replaced by any letter from the underlying alphabet. We study the relation between language of primitive partial words with the conventional language classes viz. regular, linear and deterministic context-free in Chomsky hierarchy. We give proofs to show that the language of primitive partial words over an alphabet having at least two letters is not regular, not linear and not deterministic context free language. Also we give a 2DPDA automaton that recognizes the language of primitive partial words.

Keywords: Combinatorics on words · Partial word · Primitive word · Regular · Linear · Deterministic Context-free · 2DPDA

1 Introduction

Let Σ be a finite alphabet. We assume that Σ is a nontrivial alphabet, which means that it has at least two distinct symbols. A total word (referred to as simply a word) $u = a_0 a_1 a_2 \ldots a_{n-1}$ of length n can be defined by a total function $u : \{0, \ldots, n-1\} \rightarrow \Sigma$ where each $a_i \in \Sigma$. We use string and word interchangeably. The set Σ^* is the free monoid generated by Σ which contains all the strings. The length of a string u is the number of symbols contained in it and denoted by $|u|$. The string with length zero (also referred to as empty string) is denoted as λ. The set of all words of length n over Σ is denoted by Σ^n. We define, $\Sigma^* = \bigcup_{n \in N} \Sigma^n$ where $\Sigma^0 = \{\lambda\}$ and, $\Sigma^+ = \Sigma^* \setminus \{\lambda\}$. A language L over Σ is a subset of Σ^*.

A partial word u of length n over alphabet Σ can be defined by a partial function $u : \{0, \ldots, n-1\} \rightarrow \Sigma$. The partial word u contains some do not know symbols known as holes along with the usual symbols. For $0 \leq i < n$, if $u(i)$ is defined, then we say $i \in D(u)$ (the domain of u), otherwise $i \in H(u)$ (the set of holes). A word is a partial word without any holes. If u is a partial word of length n over Σ, then the companion of u is the total function $u_\diamond : \{0, \ldots, n-1\} \rightarrow \Sigma \cup \{\diamond\}$ defined by $u_\diamond(i) = u(i)$ if $i \in D(u)$ and, $u_\diamond(i) = \diamond$ otherwise [4]. We denote the language of partial words with arbitrary number of holes as Σ_p^*. The set Σ_p^* is also a monoid under the operation of concatenation where λ serves as the identity. Peter Leupold in [14] has given some connection between partial words and languages in Chomsky hierarchy [17].

© Springer International Publishing Switzerland 2015
A.-H. Dediu et al. (Eds.): LATA 2015, LNCS 8977, pp. 436–445, 2015.
DOI: 10.1007/978-3-319-15579-1_34

The outline of the rest of the paper is as follows. The next section contains some basic concepts and describes the hierarchy of languages of partial words. In Section 3, we prove that the language of primitive partial words is not regular, not linear and not deterministic context-free language. In Section 4, we give a 2DPDA automaton that accepts the language of primitive partial words and partial Lyndon words.

2   The Language of Primitive Partial Words

A word u is said to be primitive if it cannot be represented as nontrivial power of another word, that is, if $u = v^n$ then it implies $u = v$ and $n = 1$. The language containing all primitive words over an alphabet is represented as Q. The language of primitive words plays a vital role in various fields such as coding theory, formal languages and applications and combinatorics on words [20].

If $w = x^n$ and x is a primitive, then x is called as the primitive root of w. Thus, for every total word there is a word which is considered as its root.

Lemma 1 ([20]). *Every nonempty word w can be expressed uniquely in the form $w = x^n$, where $n \geq 1$ and x is primitive.*

Similarly, for each partial word a set of words can be considered as root. For example, $a\Diamond = \{ab, a^2\}$. Thus, $a\Diamond \subset ab$ and $a\Diamond \subset a^2$. Hence $\sqrt{a\Diamond} = \{ab, a\}$.

The language of primitive words Q has been extensively studied and many facts have been proved about relation of Q with conventional formal language classes. The language Q is known to be not regular [7], not linear [12], not DCFL [16] but context sensitive language [13]. However, it is still an open question whether the language of primitive words Q is context-free or not [8,9,15,19].

We briefly mention two basic concepts, *containment* and *compatibility*, that are required to extend the definition of primitivity to partial words [3]. If u and v are two partial words of equal length then u is said to be contained in v, if all elements in $D(u)$ are also in $D(v)$ and $u(i) = v(i)$ for all $i \in D(u)$. It is denoted by $u \subset v$.

The partial words u and v are called *compatible* if there exists a partial word w such that $u \subset w$ and $v \subset w$. The compatibility of u and v is denoted by $u \uparrow v$. Note that *containment* is not a symmetric relation where as *compatibility* is a symmetric relation.

A partial word u is said to be primitive if there does not exists a word v such that $u \subset v^n, n \geq 2$. Note that if u is primitive and $u \subset v$, then v is primitive as well[3]. We denote $W_i(\Sigma)$ as the set of words over alphabet Σ with at most i holes [4]. It is easy to see the following relation.

$$W_0(\Sigma) \subset W_1(\Sigma) \subset W_2(\Sigma) \subset \cdots \subset W_i(\Sigma) \subset \cdots$$

where $W_0(\Sigma) = \Sigma^*$. We put $W(\Sigma) = \bigcup_{i \geq 0} W_i(\Sigma)$. Let us denote the language of primitive partial words with at most i holes as Q_p^i. We know that $Q \cup \overline{Q} = \Sigma^*$ where $\overline{Q}$ is the set of all non-primitive words. We denote the language of partial words with at most i holes as Σ_i^* and the language of partial words with arbitrary number of holes as Σ_p^* over the alphabet $\Sigma \cup \{\Diamond\}$. Thus, we define

$$\Sigma_p^* = \Sigma_0^* \cup \Sigma_1^* \cup \Sigma_2^* \cup \Sigma_3^* \cup \cdots$$

The root of a partial word w is defined as follows.

$$\sqrt{w} = \{p \mid p \text{ is primitive and total and there exists } k \text{ such that } w \subset p^k\}$$

For a language L of partial words, we define $\sqrt{L} = \{\sqrt{w} \mid w \in L\}$.

Let us recall the relation between the finiteness of the set of root terms with the set of regular languages.

Theorem 2 ([14]). *A regular language has finite root if and only if it can be described by a root term.*

Theorem 3 ([14]). *All context-free languages with finite root are regular.*

We denote the language of primitive partial words as Q_p. Therefore,

$$Q_p = Q_p^0 \cup Q_p^1 \cup Q_p^2 \cup Q_p^3 \cup \cdots .$$

The language of partial words with at most i holes can be viewed in a similar way as Chomsky hierarchy of formal languages. The following figure shows the hierarchy of language of partial words.

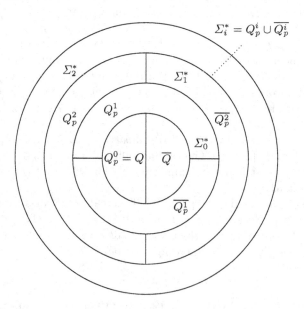

Fig. 1. The hierarchy of language of partial words

In [5], a special language class of partial words $CF \subseteq \Sigma_p^*$ is defined, where

$$CF = \{w \mid w \text{ is a partial word over } \{a, b\} \text{ that has a critical factorization}\}$$

It has been proved that CF is not regular [2] but CF is context sensitive [6]. It is an open problem whether CF is context-free or not [5]. In this paper we study a different language viz. the language of primitive partial words Q_p and its relation with the languages in Chomsky hierarchy.

3   Properties of the Language of Primitive Partial Words

In this section, we shall prove that the language of primitive partial words Q_p is not regular, not linear as well as not Deterministic Context-free Language (DCFL). First we recall the pumping lemma for regular languages and linear languages and also some of the properties of DCFL which are required to prove our result.

Lemma 4 (Pumping Lemma for Regular Languages [10]). *For a regular language L, there exists an integer $n > 0$ such that for every word $w \in L$ with $|w| \geq n$, there exist a decomposition of w as $w = xyz$ such that the following conditions holds.*

(i) $|y| > 0$,
(ii) $|xy| \leq n$, and
(iii) $xy^i z \in L$ for all $i \geq 0$.

It is easy to observe that the language of primitive partial words with at most i holes, Q_p^i, is not regular. This is because pumping up a hole in a word having at most i holes will give a word which is not in Q_p^i. The following result proves the claim in general. We use the similar idea as used in [7].

Lemma 5 ([7]). *For any fixed integer k, there exist a positive integer m such that the equation system $(k - j)x_j + j = m$, $j = 0, 1, 2, \ldots, k - 1$ has a nontrivial solution with appropriate positive integers $x_1, x_2, \ldots, x_j > 1$.*

Theorem 6. Q_p *is not regular.*

Proof. Suppose that the language of primitive partial words Q_p is regular. So there exist a natural number $n > 0$ depending upon the number of states of finite automaton for Q_p.

Consider the partial word $w = a^n \Diamond a^m \Diamond a^m b, m > n$. Note that w is a primitive partial word over $\Sigma \cup \{\Diamond\}$, where $|\Sigma| \geq 2$ and $a \neq b$. Since $w \in Q_p$ and $|w| \geq n$, then it must satisfy the other conditions of pumping Lemma for regular languages. So there exist a decomposition of w into x, y, z such that $w = xyz, |y| > 0$ and $xy^i z \in Q_p$ for all $i \geq 0$.

Let $x = a^k$, $y = a^{(n-j)}$, $z = a^{j-k} \Diamond a^m \Diamond a^m b$. Now choose $i = x_j$ and since we know by Lemma 5 that for every $j \in \{0, 1, \ldots, n-1\}$, there exists a positive integer $x_j > 1$ such that $xy^{x_j} z = a^k a^{(n-j)x_j} a^{j-k} \Diamond a^m \Diamond a^m b = a^{(n-j)x_j+j} \Diamond a^m \Diamond a^m b = a^m \Diamond a^m \Diamond a^m b \subset (a^m b)^3 \notin Q_p$ which is a contradiction. Hence the language of primitive partial words Q_p is not regular. □

Next, we prove that the language of primitive partial words Q_p is not linear. Let us recall the pumping lemma for linear languages.

Lemma 7 (Pumping Lemma for Linear Languages [11]). *Let L be a linear language. There exists an integer n such that any word $p \in L$ with $|p| \geq n$, admits a factorization $p = uvwxy$ satisfying the following conditions.*

(i) $uv^m wx^m y \in L$ $\forall m \in \mathbb{N}$,
(ii) $|vx| > 0$, and
(iii) $|uvxy| \leq n$.

Theorem 8. *The language of primitive partial words Q_p is not linear.*

Proof. Suppose that the language Q_p is linear. Let $n > 0$ be an integer. Consider $s = a^n b^m a^m \lozenge^n \in Q_p$ be a partial word and $m > n$ and $a \neq b$. Since $|s| \geq n$, so there exists a factorization of s into u, v, w, x, y such that it satisfies the conditions of pumping Lemma for linear languages.

Let $u = a^i, v = a^j, x = b^k, y = b^l$ such that $i + j + k + l \leq n, j + k > 0, w = a^r b^m a^m \lozenge^q$ and $i, j, k, l, r, q \geq 0$. So $0 < i + j + k + l \leq n$. Also, $i + j + r = s + k + l = n$. Now it must be that $uv^t wx^t y \in Q_p$ for all $t \in \mathbb{N}$. However, $uv^t wx^t y = a^{i+tj+r} b^m a^m \lozenge^{q+tk+l} = a^m b^m a^m \lozenge^m \subset (a^m b^m)^2 \notin Q_p$. It will happen that $i + tj + r = m = q + tk + l$ because if we consider the left hand side equality we get,

$$i + tj + r = m$$
$$\Rightarrow i + j + r + (t-1)j = m$$
$$\Rightarrow n + (t-1)j = m$$
$$\Rightarrow t = \frac{m-n}{j} + 1$$

It is true that there will be some integers m, n and j such that t is an integer. So, $uv^t wx^t y = a^m b^m a^m \lozenge^m \subset (a^m b^m)^2$ for some t. Hence Q_p is not linear. $\square$

Next we prove that the language of primitive partial words is not deterministic context-free language. We will use the closure properties of DCFL. In particular, the set of DCFLs is closed under complementation.

Theorem 9. Q_p *is not deterministic context-free.*

Proof. Suppose that the language of primitive partial words Q_p is deterministic context-free. As the set of DCFLs is closed under complementation, the complement of Q_p, that is, $\overline{Q_p}$ (set of partial periodic words) is also a DCFL.

Also, we know that the intersection of a DCFL with a regular language is also DCFL. Therefore, $\overline{Q_p} \cap \{a^* b^* a^* b^*\} = \{a^n b^m a^n b^m \mid m, n \in N\}$ is also a DCFL. But the language $\{a^n b^m a^n b^m \mid m, n \in N\}$ is not a Context Free Language (CFL) which can be proved by using pumping lemma for CFLs. Hence it is a contradiction that the language of primitive partial words Q_p is a DCFL. Therefore the language Q_p is not deterministic context-free. $\square$

4   2DPDA for Q_p

In this section we define *Partial Lyndon* words denoted by L_p and prove that the language of primitive partial words with 1-hole Q_p^1 is accepted by Two-way Pushdown Automaton (2DPDA). Next, we show that the language of primitive partial words with at least two holes is accepted by a 2DPDA. Let us recall the basic definition of 2DPDA and some of the results which are required in the proof.

A 2DPDA is the same as ordinary DPDA but with an additional ability to move its input head in both directions. We use Z_0 as the bottom of stack and one left end and one right end symbol as $\vdash$ and $\dashv$, respectively.

Definition 10 ([1]). *A 2DPDA, P, consists of 7-tuple*

$$P = \langle S, I, T, \delta, s_0, Z_0, s_t \rangle,$$

where

(a) S is the states of the finite control.
(b) I is the input alphabet (excluding $\vdash$ and $\dashv$).
(c) T is the pushdown list alphabet (excluding Z_0).
(d) δ is a mapping on $(S - \{s_t\}) \times (I \cup \{\vdash, \dashv\}) \times (T \cup Z_0)$. The value of $\delta(s, a, A)$ is, if defined, is of one of the forms $(s', d, push\ B), (s', d), \ or\ (s', d, pop)$ where $s' \in S, B \in T$ and $d \in \{-1, 0, +1\}$. We assume a 2DPDA makes no moves from the final state s_t and certain other states may have no moves defined.
(e) $s_0 \in S$ is the initial state of the finite control.
(f) Z_0 is the special symbol that indicates the bottom of the pushdown list.
(g) s_t is one of the designated final state.

In the above definition $\delta(s, a, A)$ indicates the transition function of the machine when it is in the state s, the input head scans a symbol a, and the pushdown list has the symbol A on top. There are three possible actions.

$$\delta(s, a, A) = \begin{cases} (s', d, push\ B) & \text{provided } B \neq Z_0 \\ (s', d); \\ (s', d, pop) & \text{if } A \neq Z_0 \end{cases}$$

In these transitions, the machine enters in state s' and moves its input head in the direction d, where $d = -1, +1$ or 0, to indicate to move its head to left, right or remain stationary, respectively. The operation *push B* means add the symbol B on the top of pushdown list and *pop* means remove the topmost symbol from the pushdown list.

Definition 11 ([3]). *Let k and l be two positive integers satisfying $k \leq l$. For $0 \leq i < k + l$, we define the sequence of i relative to k, l as $seq_{k,l}(i) = (i_0, i_1, \ldots, i_{n+1})$ where* *(a)* $i_0 = i = i_{n+1}$

(b) For $1 \leq j \leq n$, $i_j \neq i$

(c) For $1 \leq j \leq n+1$, i_j *is defined as* $i_j = \begin{cases} i_{j-1} + k & \text{if } \neq i_{j-1} < l, \\ i_{j-1} - l & \text{otherwise.} \end{cases}$

Definition 12 ([4]). *Let* k *and* l *be two positive integers satisfying* $k \leq l$ *and let* z *be a partial word of length* $(k + l)$. *We say that* z *is* **(k,l)-special** *if there exists* $0 \leq i \leq \gcd(k, l)$ *such that* $seq_{k,l}(i) = (i_0, i_1, \ldots, i_{n+1})$ *contains at least two positions that are holes of* z *while* $z(i_0)z(i_1) \ldots z(i_{n+1})$ *is not 1-periodic.*

Lemma 13 ([4]). *Let* u *and* v *be two nonempty partial words such that* uv *contains at most one hole. The words* u *and* v *commute if and only if they are contained in powers of the same words, that is,* $uv \uparrow vu$ *if there exists a word* w *such that* $u \subset w^m$ *and* $v \subset w^n$ *for some integer* m, n.

Lemma 14 ([4]). *Let* u *and* v *be two nonempty partial words such that* $|u| \leq |v|$. *If* $uv \uparrow vu$ *and* uv *is not* $(|u|, |v|)$*-special, then there exists a word* w *such that* $u \subset w^m$ *and* $v \subset w^n$ *for some integer* m, n.

Definition 15. *A partial word* $w \in \Sigma^+$ *is primitive if and only if it is not contained in two non-empty commuting words, that is,* $w \in Q_p \Leftrightarrow w \neq \lambda$ *and* $\forall u, v \in \Sigma^* : (w \subset uv$ *and* $w \subset vu \Rightarrow \lambda \in \{u, v\})$.

Several facts about primitive partial words are known; we recall some of them which will be useful below.

Lemma 16 ([4]). *Let* u *be a partial word with one hole. Then* u *is primitive if and only if* $uu \uparrow xuy$ *for some partial words* x, y *implies* $x = \epsilon$ *or* $y = \epsilon$.

Lemma 17 ([4]). *Let* u *be a partial word with at least two holes.*

I. If $uu \uparrow xuy$ *for some partial words* x, y *implies* $x = \epsilon$ *or* $y = \epsilon$, *then* u *is primitive.*

II. If $uu \uparrow xuy$ *for some nonempty partial words* x *and* y *satisfying* $|x| \leq |y|$, *then the following hold:*

(a) If $|x| = |y|$, *then* u *is not primitive.*

(b) If u *is not* $(|x|, |y|)$*-special, then* u *is not primitive (it is contained in a power of a word of length* $|x|$).

(c) If u *is* $(|x|, |y|)$*-special, then* u *is not contained in a power of a word of length* $|x|$.

Definition 18. *A primitive partial word* w *is a partial Lyndon word if and only if it is minimal in its conjugate class (with respect to lexicographic order where we assume* $a < b < \cdots < \Diamond$).

Lemma 19. *The cardinality of the class of conjugates of a primitive partial word* w *of length* n *is* n.

Proof. Let w be a primitive partial word over the alphabet $\Sigma \cup \{\Diamond\}$ of length n. So every new partial word which is generated by cyclic permutation of w is a conjugate of w. Since w is primitive and we know that the primitive word is closed under cyclic permutation [20], then each of the conjugate of w is also primitive. So the number of such conjugates of w is n. This proves our claim. $\qquad\Box$

In [18], a 2DPDA automaton for the language of primitive partial words Q_p is presented. We show that a 2DPDA can also be constructed for Q_p^1 by using the same idea used in [18].

Theorem 20. Q_p^1 *is accepted by 2DPDA.*

Proof. The informal idea is as follows: Let P be a 2DPDA. Let $\vdash w \dashv$ be the input partial word with at most one hole augmented with two end markers. If $w = \lambda$, P rejects. If not, then P moves its head towards $\dashv$. P skips the last symbol of w and pushes the remaining symbols of w onto its stack. Again P moves to $\dashv$, pushes all symbols of w onto the stack and pops one symbol. If we write the input $w = xw'y$ with $x, y \in \Sigma \cup \{\Diamond\}$ (assuming $|w| \geq 2$), the contents of the stack will be:

$$w' y x w' Z_0$$

The automaton P compares w with the pushdown contents one symbol at a time. If the symbols match (assuming that $a \neq b$ and $\Diamond = a$ for any $a \in \Sigma$), the head moves right and P pops the pushdown. If in this way the entire word w is completely scanned and is compatible to a factor of $w' y x w'$ P rejects (by assuming that the hole $\Diamond$ is compatible to any of the symbol in Σ).

If a mismatch is encountered P moves the input head back to $\vdash$ and pushes the symbols scanned during this move. Then P pops the topmost symbol from the pushdown and repeats the process. If the pushdown becomes empty then P accepts.

Hence the 2DPDA accepts w if and only if w is not compatible to a factor of $w' y x w'$ by using Lemma 16. $\qquad\Box$

Observe that the above proof method does not work for the set of primitive partial words with at least two holes. A counter example is given below.

Example 21. Let $w = a \Diamond b \Diamond$. So by the method described in Theorem 20 we have $x = a, y = \Diamond$ and $w' = \Diamond b$. Now $w' y x w' = \Diamond b \Diamond a \Diamond b$ and w is compatible to a factor of $w' y x w'$. We know that if a partial word u is contained in another word v and u is primitive then v is also primitive, that is, if v is not primitive then u is also not primitive. We can see that a partial word is contained in a set of words. If $w = a \Diamond b \Diamond$, then $w \subset \{aaba, aabb, abba, abbb\}$.

Next we extend the Theorem 20 to show that the language Q_p with at least two holes is also recognizable by a 2DPDA by using a similar idea as in proof of Theorem 20.

Theorem 22. *The language of primitive partial words Q_p with at least two holes is accepted by a 2DPDA.*

Proof. Let A be a 2DPDA. Let $\vdash w \dashv$ be a partial word augmented with two end markers which is the input to A. The informal idea is as follows: since w is a partial word, first compute the set of words in which w is contained. Let w is contained in the set $\{w_1, w_2, \ldots, w_k\}$. Now we will give input each of the w_i, $1 \leq i \leq k$ to A one at a time and if at least one of the w_i is rejected by A, then the partial word w is rejected and stop the process; otherwise continue the same process.

If $w = \lambda$ then A rejects.

Otherwise it computes the set of words $\{w_1, w_2, \ldots, w_k\}$. Suppose $\vdash w_1 \dashv$ is the input to A. Next, A advances its input head to $\dashv$. Then A skips the last symbol of w_1 and pushes the remainder of w_1 onto its pushdown store. Then A moves its head to $\dashv$ again and pushes all of w_1 onto its store and pop one symbol. If we write $w_1 = xw''y$ where $x, y \in \Sigma$ with $|w_1| \geq 2$ the contents of pushdown are of the form:

$$w''yxw''Z_0$$

The automaton A compares w_1 with the pushdown contents symbol by symbol. If the symbols match, the head moves right and A pops the pushdown. If in this way w_1 is completely scanned A rejects. If a mismatch occurs A moves the input head back to $\vdash$ and pushes the symbols scanned during this move. Then A pops one symbol and repeats the process. If the pushdown is empty then A chooses the next word w_2 from the set and continue.

The automaton A accepts the partial word w if and only if each of the word in the set $\{w_1, w_2, \ldots, w_k\}$ is accepted by A. □

Lemma 23. *The language of partial Lyndon words L_p is accepted by 2DPDA.*

Proof. The key idea is based on that a partial Lyndon word w is smaller than each of its proper right factors. Let us describe an automaton B that accepts L_p. Similar to the previous theorem, let $\vdash w \dashv$ be the input where $w \in \Sigma \cup \{\Diamond\}$. B scans the input w and push the symbols of w into the pushdown while scanning. Then B compares the topmost symbol of stack with w. If B encounters a symbol a in the stack and symbol b in w with $a < b$ (assuming the comparison $a < b < \ldots < \Diamond$) or reach the bottom most symbol Z_0, then it rejects.

If the symbols in w and the topmost symbol in stack matches, in particular a don't know symbol matches only with a don't know symbol, then B pops the top symbol of stack and moves its head one position right. If the symbol on the stack is greater than the symbol in the input then B returns to $\vdash$ and push the symbols onto stack during the move. If B reached the symbol $\vdash$ in the input and some symbols are left in the pushdown then B accepts. The procedure is then repeated. □

References

1. Aho, A.V., Hopcroft, J.E.: Design & Analysis of Computer Algorithms. Pearson Education India (1974)

2. Blanchet-Sadri, F., Zhang, J.: On the critical factorization theorem (preprint)
3. Blanchet-Sadri, F.: Primitive partial words. Discrete Applied Mathematics **148**(3), 195–213 (2005)
4. Blanchet-Sadri, F.: Algorithmic combinatorics on partial words. CRC Press (2007)
5. Blanchet-Sadri, F.: Open problems on partial words. In: Bel-Enguix, G., Dolores Jiménez-López, M., Martín-Vide, C. (eds.) New Developments in Formal Languages and Applications. SCI, vol. 113, pp. 11–58. Springer, Heidelberg (2008)
6. Blanchet-Sadri, F., Davis, C., Dodge, J., Mercaş, R., Moorefield, M.: Unbordered partial words. Discrete Applied Mathematics **157**(5), 890–900 (2009)
7. Dömösi, P., Horváth, G.: The language of primitive words is not regular: two simple proofs. Bulletin of European Association for Theoretical Computer Science **87**, 191–194 (2005)
8. Dömösi, P., Horváth, S., Ito, M., Kászonyi, L., Katsura, M.: Formal languages consisting of primitive words. In: Ésik, Zoltán (ed.) FCT 1993. LNCS, vol. 710, pp. 194–203. Springer, Heidelberg (1993)
9. Dömösi, P., Ito, M., Marcus, S.: Marcus contextual languages consisting of primitive words. Discrete Mathematics **308**(21), 4877–4881 (2008)
10. Hopcroft, J.E., Motwani, R., Ullman, J.D.: Introduction to automata theory, languages, and computation. ACM SIGACT News **32**(1), 60–65 (2001)
11. Horváth, G., Nagy, B.: Pumping lemmas for linear and nonlinear context-free languages. arXiv preprint arXiv:1012.0023 (2010). http://arxiv.org/abs/1012.0023
12. Horváth, S.: Strong interchangeability and nonlinearity of primitive words. In: Proc. Workshop AMAST Workshop on Algebraic Methods in Language Processing, vol. 95, pp. 173–178 (1995)
13. Kunimochi, Y.: A context sensitive grammar generating the set of all primitive words. In: Algebras, Languages, Computations and their Applications, vol. 1562, pp. 143–145 (2007)
14. Leupold, P.: Languages of partial words. Grammars **7**, 179–192 (2004)
15. Leupold, P.: Primitive words are unavoidable for context-free languages. In: Dediu, A.-H., Fernau, H., Martín-Vide, C. (eds.) LATA 2010. LNCS, vol. 6031, pp. 403–413. Springer, Heidelberg (2010)
16. Lischke, G.: Primitive words and roots of words. arXiv preprint arXiv:1104.4427 (2011). http://arxiv.org/abs/1104.4427
17. Miller, G.A., Chomsky, N., Luce, D.R., Bush, R.R., Galanter, E.: Handbook of mathematical psychology (1963)
18. Petersen, H.: The ambiguity of primitive words. In: Enjalbert, P., Mayr, Ernst W., Wagner, K.W. (eds.) STACS 1994. LNCS, vol. 775, pp. 679–690. Springer, Heidelberg (1994)
19. Petersen, H.: On the language of primitive words. Theoretical Computer Science **161**(1), 141–156 (1996)
20. Shallit, J.: A second course in formal languages and automata theory. Cambridge University Press (2008)

Complexity and Recursive Functions

Complexity of Regular Functions

Eric Allender[(✉)] and Ian Mertz

Department of Computer Science, Rutgers University,
Piscataway, NJ 08854, USA
allender@cs.rutgers.edu, iwmertz@gmail.com

Abstract. We give complexity bounds for various classes of functions computed by cost register automata.

Keywords: Computational complexity · Transducers · Weighted automata

1 Introduction

We study various classes of *regular functions*, as defined in a recent series of papers by Alur *et al.* [7–9]. In those papers, the reader can find pointers to work describing the utility of regular functions in various applications in the field of computer-aided verification. Additional motivation for studying these functions comes from their connection to classical topics in theoretical computer science; we describe these connections now.

The class of functions computed by *two-way* deterministic finite transducers is well-known and widely-studied. Engelfriet and Hoogeboom studied this class [16] and gave it the name of *regular string transformations*. They also provided an alternative characterization of the class in terms of monadic second-order logic. It is easy to see that this is a strictly larger class than the class computed by *one-way* deterministic finite transducers, and thus it was of interest when Alur and Černý [4] provided a characterization in terms of a new class of *one-way* deterministic finite automata, known as *streaming string transducers*; see also [5]. Streaming string transducers are traditional deterministic finite automata, augmented with a finite number of *registers* that can be updated at each time step, as well as an output function for each state. Each register has an initial value in Γ^* for some alphabet Γ, and at each step receives a new value consisting of the concatenation of certain other registers and strings. (There are certain other syntactic restrictions, which will be discussed later, in Section 2.)

The model that has been studied in [7–9], known as *cost register automata* (CRAs), is a generalization of streaming string transducers, where the register update functions are not constrained to be the concatenation of strings, but instead may operate over several other algebraic structures such as monoids, groups and semirings. Stated another way, streaming string transducers are cost register automata that operate over the monoid $(\Gamma^*, \circ)$ where $\circ$ denotes concatenation. Another important example is given by the so-called "tropical semiring",

© Springer International Publishing Switzerland 2015
A.-H. Dediu et al. (Eds.): LATA 2015, LNCS 8977, pp. 449–460, 2015.
DOI: 10.1007/978-3-319-15579-1_35

where the additive operation is min and the multiplicative operation is $+$; CRAs over $(\mathbb{Z}, \min, +)$ can be used to give an alternative characterization of the class of functions computed by weighted automata [7].

The cost register automaton model is the main machine model that was advocated by Alur et al. [7] as a tool for defining and investigating various classes of "regular functions" over different domains. Their definition of "regular functions" does not always coincide exactly with the CRA model, but does coincide in several important cases. In this paper, we will focus on the functions computed by (various types of) CRAs.

Although there have been papers examining the complexity of several decision problems dealing with some of these classes of regular functions, there has not previously been a study of the complexity of computing the functions themselves. There was even a suggestion [3] that these functions might be difficult or impossible to compute efficiently in parallel. Our main contribution is to show that most of the classes of regular functions that have received attention lie in certain low levels of the NC hierarchy.

2   Preliminaries

The reader should be familiar with some common complexity classes, such as L (deterministic logspace), and P (deterministic polynomial time). Many of the complexity classes we deal with are defined in terms of families of circuits. A language $A \subseteq \{0,1\}^*$ is accepted by circuit family $\{C_n : n \in \mathbb{N}\}$ if $x \in A$ iff $C_{|x|}(x) = 1$. Our focus in this paper will be on *uniform* circuit families; by imposing an appropriate uniformity restriction (meaning that there is an algorithm that describes C_n, given n) circuit families satisfying certain size and depth restrictions correspond to complexity classes defined by certain classes of Turing machines.

For more detailed definitions about the following standard circuit complexity classes (as well as for motivation concerning the standard choice of the U_E-uniformity), we refer the reader to [20, Section 4.5].

- $NC^i = \{A : A$ is accepted by a U_E-uniform family of circuits of bounded fan-in AND, OR and NOT gates, having size $n^{O(1)}$ and depth $O(\log^i n)\}$.
- $AC^i = \{A : A$ is accepted by a U_E-uniform family of circuits of unbounded fan-in AND, OR and NOT gates, having size $n^{O(1)}$ and depth $O(\log^i n)\}$.
- $TC^i = \{A : A$ is accepted by a U_E-uniform family of circuits of unbounded fan-in MAJORITY gates, having size $n^{O(1)}$ and depth $O(\log^i n)\}$.

We remark that, for constant-depth classes such as AC^0 and TC^0, U_E-uniformity coincides with U_D-uniformity, which is also frequently called DLOGTIME-uniformity.) We use these same names to refer to the associated classes of *functions* computed by the corresponding classes of circuits.

We also need to refer to certain classes defined by families of *arithmetic* circuits. Let $(S, +, \times)$ be a semiring. An *arithmetic circuit* consists of input gates, $+$ gates, and $\times$ gates connected by directed edges (or "wires"). One gate is designated as an "output" gate. If a circuit has n input gates, then it computes a function from $S^n \to S$ in the obvious way. In this paper, we consider only arithmetic circuits where all gates have bounded fan-in.

- $\#\text{NC}^1{}_S$ is the class of functions $f : \bigcup_n S^n \to S$ for which there is a U_E-uniform family of arithmetic circuits $\{C_n\}$ of logarithmic depth, such that C_n computes f on S^n.
- By convention, when there is no subscript, $\#\text{NC}^1$ denotes $\#\text{NC}^1{}_N$, with the additional restriction that the functions in $\#\text{NC}^1$ are considered to have domain $\bigcup_n \{0,1\}^n$. That is, we restrict the inputs to the Boolean domain. (Boolean negation is also allowed at the input gates.)
- GapNC^1 is defined as $\#\text{NC}^1 - \#\text{NC}^1$; that is: the class of all functions that can be expressed as the difference of two $\#\text{NC}^1$ functions. It is the same as $\#\text{NC}^1{}_Z$ restricted to the Boolean domain. See [1, 20] for more on $\#\text{NC}^1$ and GapNC^1.

The following inclusions are known:

$$\text{NC}^0 \subseteq \text{AC}^0 \subseteq \text{TC}^0 \subseteq \text{NC}^1 \subseteq \#\text{NC}^1 \subseteq \text{GapNC}^1 \subseteq \text{L} \subseteq \text{AC}^1 \subseteq \text{P}.$$

All inclusions are straightforward, except for $\text{GapNC}^1 \subseteq \text{L}$ [17].

2.1   Cost-Register Automata

A *cost-register automaton* (CRA) is a deterministic finite automaton (with a read-once input tape) augmented with a fixed finite set of *registers* that store elements of some algebraic domain $\mathcal{A}$. At each step in its computation, the machine

- consumes the next input symbol (call it a),
- moves to a new state (based on a and the current state (call it q)),
- based on q and a, updates each register r_i using updates of the form $r_i \leftarrow f(r_1, r_2, \ldots, r_k)$, where f is an expression built using the registers $r_1, \ldots, r_k$ using the operations of the algebra $\mathcal{A}$.

There is also an "output" function μ defined on the set of states; μ is a partial function – it is possible for $\mu(q)$ to be undefined. Otherwise, if $\mu(q)$ is defined, then $\mu(q)$ is some expression of the form $f(r_1, r_2, \ldots, r_k)$, and the output of the CRA on input x is $\mu(q)$ if the computation ends with the machine in state q.

More formally, here is the definition as presented by Alur *et al.* [7].

A cost-register automaton M is a tuple $(\Sigma, Q, q_0, X, \delta, \rho, \mu)$, where

- Σ is a finite input alphabet.
- Q is a finite set of states.
- $q_0 \in Q$ is the initial state.
- X is a finite set of *registers*.
- $\delta : Q \times \Sigma \to Q$ is the state-transition function.

– $\rho : Q \times \Sigma \times X \to E$ is the register update function (where E is a set of algebraic expressions over the domain $\mathcal{A}$ and variable names for the registers in X).

– $\mu : Q \to E$ is a (partial) final cost function.

A *configuration* of a CRA is a pair (q, ν), where ν maps each element of X to an algebraic expression over $\mathcal{A}$. The *initial configuration* is (q_0, ν_0), where ν_0 assigns the value 0 to each register. Given a string $w = a_1 \ldots a_n$, the *run* of M on w is the sequence of configurations $(q_0, \nu_0), \ldots (q_n, \nu_n)$ such that, for each $i \in \{1, \ldots, n\}$ $\delta(q_{i-1}, a_i) = q_i$ and, for each $x \in X$, $\nu_i(x)$ is the result of composing the expression $\rho(q_{i-1}, a_i, x)$ to the expressions in ν_{i-1} (by substituting in the expression $\nu_{i-1}(y)$ for each occurrence of the variable $y \in X$ in $\rho(q_{i-1}, a_i, x)$). The output of M on w is undefined if $\mu(q_n)$ is undefined. Otherwise, it is the result of evaluating the expression $\mu(q_n)$ (by substituting in the expression $\nu_n(y)$ for each occurrence of the variable $y \in X$ in $\mu(q_n)$).

It is frequently useful to restrict the algebraic expressions that are allowed to appear in the transition function $\rho : Q \times \Sigma \times X \to E$. One restriction that is important in previous work [7] is the "copyless" restriction.

A CRA is *copyless* if, for every register $r \in X$, for each $q \in Q$ and each $a \in \Sigma$, the variable "r" appears at most once in the multiset $\{\rho(q, a, s) : s \in X\}$. In other words, for a given transition, no register can be used more than once in computing the new values for the registers. Following [8], we refer to copyless CRAs as CCRAs. Over many algebras, unless the copyless restriction is imposed, CRAs compute functions that can not be computed in polynomial time. For instance, CRAs that can concatenate string-valued registers and CRAs that can multiply integer-valued registers can perform "repeated squaring" and thereby obtain results that require exponentially-many symbols to write down.

3   CRAs over Monoids

In this section, we study CRAs operating over algebras with a single operation. We focus on two canonical examples:

– CRAs operating over the commutative monoid $(\mathbb{Z}, +)$.
– CRAs operating over the noncommutative monoid $(\Gamma^*, \circ)$.

3.1   CRAs over the Integers

Additive CRAs (ACRAs) are CRAs that operate over commutative monoids. They have been studied in [7–9]; in [9] the ACRAs that were studied operated over $(\mathbb{Z}, +)$, and thus far no other commutative monoid has received much attention, in connection with CRAs.

Theorem 1. *All functions computable by CCRAs over $(\mathbb{Z}, +)$ are computable in* NC^1. *(This bound is tight, since there are regular sets that are complete for* NC^1 *under projections [10].)*

Proof. It was shown in [7] that CCRAs (over any commutative semiring) have equivalent power to CRAs that are not restricted to be copyless, but that have another restriction: the register update functions are all of the form $r \leftarrow r' + c$ for some register r' and some semiring element c. Thus assume that the function f is computed by a CRA M of this form. Let M have k registers $r_1, \ldots, r_k$.

It is straightforward to see that the following functions are computable in NC^1:

- $(x, i) \mapsto q$, such that M is in state q after reading the prefix of x of length i.
- $(x, i) \mapsto G_i$, where G_i is a labeled bipartite graph on $[k] \times [k]$, with the property that there is an edge labeled c from j on the left-hand side to ℓ on the right hand side, if the register update operation that takes place when M consumes the i-th input symbol includes the update $r_\ell \leftarrow r_j + c$. If the register update operation includes the update $r_\ell \leftarrow c$, then vertex ℓ on the right hand side is labeled c. (To see that this is computable in NC^1, note that by the previous item, in NC^1 we can determine the state q that M is in as it consumes the
  i-th input symbol. Thus G_i is merely a graphical representation of the register update function corresponding to state q.) Note that the indegree of each vertex in G_i is at most one. (The *outdegree* of a vertex may be as high as k.)

Now consider the graph G that is obtained by concatenating the graphs G_i (by identifying the right-hand side of G_i with the left-hand side of G_{i+1} for each i). This graph shows how the registers at time $i + 1$ depend on the registers at time i. G is a constant-width graph, and it is known that reachability in constant-width graphs is computable in NC^1. Note that we can determine in NC^1 the register that provides the output when the last symbol of x is read. By tracing the edges back from that vertex in G (following the unique path leading back toward the left, using the fact that each vertex has indegree at most one) we eventually encounter a vertex of indegree zero. In NC^1 we can determine which edges take part in this path, and add the labels that occur along that path. This yields the value of $f(x)$. □

We remark that the NC^1 upper bound holds for any commutative monoid where iterated addition of monoid elements can be computed in NC^1.

A related bound holds, when the copyless restriction is dropped:

Theorem 2. *All functions computable by CRAs over* $(\mathbb{Z}, +)$ *are computable in* GapNC^1. *(This bound is tight, since there is one such function that is hard for* GapNC^1 *under* AC^0 *reductions.)*

Proof. We use a similar approach as in the proof of the preceding theorem. We build a bipartite graph G_i that represents the register update function that is executed while consuming the i-th input symbol, as follows. Each register update operation is of the form $r_\ell \leftarrow a_0 + r_{i_1} + r_{i_2} + \ldots r_{i_m}$. Each register r_j appears, say, a_j times in this sum, for some nonnegative integer a_j. If $r_\ell \leftarrow a_0 + \sum_{j=1}^{k} a_j \cdot r_j$

is the update for r_ℓ at time i, then if $a_j > 0$, then G_i will have an edge labeled a_j from j on the left-hand side to ℓ on the right-hand side, along with an edge from 0 to ℓ labeled a_0, and an edge from 0 to 0. Let the graph G_i correspond to matrix M_i. An easy inductive argument shows that $(\sum_{j=0}^{k}(\prod_{i=1}^{t} M_i))_{j,\ell}$ gives the value of register ℓ after time t. The upper bound now follows since iterated multiplication of $O(1) \times O(1)$ integer matrices can be computed in GapNC^1 [15].

For the lower bound, observe that it is shown in [15], building on [12], that computing the iterated product of 3×3 matrices with entries from $\{0, 1, -1\}$ is complete for GapNC^1. More precisely, taking a sequence of such matrices as input and outputting the (1,1) entry of the product is complete for GapNC^1. Consider the alphabet Γ consisting of such matrices. There is a CRA taking input from Γ^* and producing as output the contents of the $(1, 1)$ entry of the product of the matrices given as input. (The CRA simulates matrix multiplication in the obvious way.) □

3.2   CRAs over $(\Gamma^*, \circ)$

Unless we impose the copyless restriction, CRAs over this monoid can generate exponentially-long strings. Thus in this subsection we consider only CCRAs.

CCRAs operating over the algebraic structure $(\Gamma^*, \circ)$ are precisely the so-called *streaming string transducers* that were studied in [5], and shown there to compute precisely the functions computed by two-way deterministic finite transducers (2DFAs). This class of functions is very familiar, and it is perhaps folklore that such functions can be computed in NC^1, but we have found no mention of this in the literature. Thus we present the proof here.

Theorem 3. *All functions computable by CCRAs over $(\Gamma^*, \circ)$ are computable in NC^1. (This bound is tight, since there are regular sets that are complete for NC^1 under projections [10].)*

Proof. Let M be a 2DFA computing a (partial) function f, and let x be a string of length n. If $f(x)$ is defined, then M halts on input x, which means that M visits no position i of x more than k times, where k is the size of the state set of M.

Define the *visit sequence at i* to be the sequence $q_{(i,1)}, q_{(i,2)}, \cdots q_{(i,\ell_i)}$ of length $\ell_i \leq k$ such that $q_{(i,j)}$ is the state that M is in the j-th time that it visits position i. Denote this sequence by V_i.

We will show that the function $(x, i) \mapsto V_i$ is computable in NC^1. Assume for the moment that this is computable in NC^1; we will show how to compute f in NC^1.

Note that there is a planar directed graph G of width at most k having vertex set $\bigcup_i V_i$, where all edges adjacent to vertices V_i go to vertices in either V_{i-1} or V_{i+1}, as follows: Given V_{i-1}, V_i and V_{i+1}, for any $q_{(i,j)} \in V_i$, it is trivial to compute the pair (i', j') such that, when M is in state $q_{(i,j)}$ scanning the i-th symbol of the input, then at the next step it will be in state $q_{(i',j')}$ scanning the i'-th symbol of the input. (Since this depends on only $O(1)$ bits, it is computable in U_E-uniform NC^0.) The edge set of G consists of these "next move" edges from $q_{(i,j)}$ to $q_{(i',j')}$. It is immediate that no edges cross when embedded in the plane in the obvious way (with the vertex sets $V_1, V_2, \ldots$ arranged in vertical columns with V_1 at the left end, and V_{i+1} immediately to the right of V_i, and with the vertices $q_{(i,1)}, q_{(i,2)}, \ldots q_{(i,\ell_i)}$ arranged in order within the column for V_i).

Let us say that (i,j) *comes before* (i',j') if there is a path from $q_{(i,j)}$ to $q_{(i',j')}$ in G. Since reachability in constant-width planar graphs is computable in AC^0 [11], it follows that the "comes before" predicate is computable in AC^0.

Thus, in TC^0, one can compute the size of the set $\{(i',j') : (i',j')$ comes before (i,j) and M produces an output symbol when moving from $q_{(i',j')}\}$. Call this number $m_{(i,j)}$. Hence, in TC^0 one can compute the function $(x,m) \mapsto (i,j)$ such that $m_{(i,j)} = m$. But this allows us to determine what symbol is the m-th symbol of $f(x)$. Hence, given the sequences V_i, $f(x)$ can be computed in $\mathsf{TC}^0 \subseteq \mathsf{NC}^1$.

It remains to show how to compute the sequences V_i.

It suffices to show that the set $B = \{(x,i,V) : V = V_i\} \in \mathsf{NC}^1$. To do this, we will present a nondeterministic constant-width branching program recognizing B; such branching programs recognize only sets in NC^1 [10]. Our branching program will guess each V_j in turn; note that each V_j can be described using only $O(k \log k) = O(1)$ bits, and thus there are only $O(1)$ choices possible at any step. When guessing V_{j+1}, the branching program rejects if V_{j+1} is inconsistant with V_j and the symbols being scanned at positions j and $j+1$. When $i = j$ the branching program rejects if V is not equal to the guessed value of V_i. When $j = |x|$ the branching program halts and accepts if all of the guesses $V_1, \ldots, V_n$ have been consistent. It is straightforward to see that the algorithm is correct. $\qquad\square$

4   CRAs over Semirings

In this section, we study CRAs operating over algebras with two operations satisfying the semiring axioms. We focus on three such structures:

- CRAs operating over the commutative ring $(\mathbb{Z}, +, \times)$.
- CRAs operating over the commutative semiring $(\mathbb{N} \cup \{\infty\}, \min, +)$: the so-called "tropical" semiring.
- CRAs operating over the noncommutative semiring $(\Gamma^* \cup \{\bot\}, \max, \circ)$.

There is a large literature dealing with *weighted automata* operating over semirings. It is shown in [7] that the functions computed by weighted automata operating over a semiring $(S, +, \times)$ is exactly equal to the class of functions computed by CRAs operating over $(S, +, \times)$, where the only register operations involving $\times$ are of the form $r \leftarrow r' \times c$ for some register r' and some semiring element c. Thus for each structure, we will also consider CRAs satisfying this restriction.

We should mention the close connection between iterated matrix product and weighted automata operating over commutative semirings. As in the proof of Theorem 2, when a CRA is processing the i-th input symbol, each register update function is of the form $r_\ell \leftarrow a_0 + \sum_{j=1}^{k} a_j \cdot r_j$, and thus the register updates for position i can be encoded as a matrix. Thus the computation of the machine on an input x can be encoded as an instance of iterated matrix multiplication. In fact, some treatments of weighted automata essentially *define* weighted automata in terms of iterated matrix product. (For instance, see [18, Section 3].) Thus, since iterated product of $k \times k$ matrices lies in $\#\mathsf{NC}^1{}_S$ for any commutative semiring S, the functions computed by weighted automata operating over S all lie in $\#\mathsf{NC}^1{}_S$. (For the case when $S = \mathbb{Z}$, iterated matrix product of $k \times k$ matrices is *complete* for GapNC^1 for all $k \geq 3$ [12,15].)

4.1   CRAs Over the Integers

First, we consider the copyless case:

Theorem 4. *All functions computable by CCRAs over $(\mathbb{Z}, +, \times)$ are computable in GapNC^1. (Some such functions are hard for NC^1, but we do not know if any are hard for GapNC^1.)*

Proof. Consider a CCRA M computing a function f, operating on input x. There is a function computable in NC^1 that maps x to an encoding of an arithmetic circuit that computes $f(x)$, constructed as follows: The circuit will have gates $r_{j,i}$ computing the value of register j at time i. The register update functions dictate which operations will be employed, in order to compute the value of $r_{j,i}$ from the gates $r_{j',i-1}$. Due to the copyless restriction, the outdegree of each gate is at most 1 (which guarantees that the circuit is a formula).

It follows from Lemma 5 below that $f \in \mathsf{GapNC}^1$.   □

Lemma 5. *If there is a function computable in NC^1 that takes an input x and produces an encoding of an arithmetic formula that computes $f(x)$ when evaluated over the integers, then $f \in \mathsf{GapNC}^1$.*

Proof. By [14], there is a logarithmic-depth arithmetic-Boolean formula over the integers, that takes as input an encoding of a formula F and outputs the integer represented by F. An arithmetic-Boolean formula is a formula with Boolean gates AND, OR and NOT, and arithmetic gates $+, \times$, as well as *test* and *select* gates that provide an interface between the two types of gates. Actually, the construction given in [14] does not utilize any *test* gates [13], and thus we need not concern ourselves with them. (Note that this implies that there is no path in the circuit from an arithmetic gate to a Boolean gate.)

A *select* gate takes three inputs (y, x_0, x_1) and outputs x_0 if $y = 0$ and outputs x_1 otherwise. In the construction given in [14], *select* gates are only used when y is a Boolean value. When operating over the integers, then, $select(y, x_0, x_1)$ is equivalent to $y \times x_1 + (1 - y) \times x_0$. But since Boolean NC^1 is contained in $\#\mathsf{NC}^1 \subseteq \mathsf{GapNC}^1$ (see, e.g., [1]), the Boolean circuitry can all be replaced by arithmetic circuitry. (When operating over algebras other than $\mathbb{Z}$, it is not clear that such a replacement is possible.)   □

We cannot entirely remove the copyless restriction while remaining in the realm of polynomial-time computation, since repeated squaring allows one to obtain numbers that require exponentially-many bits to represent in binary. However, as noted above, if the multiplicative register updates are all of the form $r \leftarrow r' \times c$, then again the GapNC^1 upper bound holds (and in this case, some of these CRA functions are complete for GapNC^1, just as was argued in the proof of Theorem 2).

4.2   CRAs over the Tropical Semiring

Again, we first consider the copyless case.

Theorem 6. *All functions computable by CCRAs over the tropical semiring are computable in* L, *and in* $\mathsf{NC}^1(\#\mathsf{NC}^1_{trop})$.

Here, $\mathsf{NC}^1(\#\mathsf{NC}^1_{trop})$ refers to the class of functions expressible as $g(f(x))$ for some functions $f \in \mathsf{NC}^1$ and $g \in \#\mathsf{NC}^1_{trop}$. No inclusion relation is known between L and $\mathsf{NC}^1(\#\mathsf{NC}^1_{trop})$.

Proof. The L upper bound follows easily, because the only operation that increases the value of a register is a $+$ operation, and because of the copyless restriction the value of a register after i computation steps can be expressed as a sum of $i^{O(1)}$ values that are present as constants in the program of the CRA. Thus, in particular, the value of a register at any point during the computation on input x can be represented using $O(\log|x|)$ bits. Thus a logspace machine can simply simulate a CRA directly, storing the value of each of the $O(1)$ registers, and computing the updates at each step.

For the $\mathsf{NC}^1(\#\mathsf{NC}^1_{trop})$ upper bound, first note that there is a function h computable in NC^1 that takes x as input, and outputs a description of an arithmetic formula over the tropcial semiring that computes $f(x)$. This is exactly as in the first paragraph of the proof of Theorem 4.

Next, as in the proof of Lemma 5, recall that, by [14], there is a uniform family of *logarithmic-depth* arithmetic-Boolean formulae $\{C_n\}$ over the tropical semiring, that takes as input an encoding of a formula F and outputs the integer represented by F. Furthermore, each arithmetic-Boolean formula C_n has Boolean gates AND, OR and NOT, and arithmetic gates min, $+$, as well as *select* gates, and there is no path in C_n from an arithmetic gate to a Boolean gate.

Let $\{D_n\}$ be the uniform family of arithmetic circuits, such that D_n is the connected subcircuit of C_n consisting only of arithmetic min and $+$ gates. We now have the following situation: The NC^1 function h (which maps x to an encoding of a formula F having some length m) composed with the circuit C_m (which takes F as input and produces $f(x)$ as output) is identical with some NC^1 function h' (computed by the NC^1 circuitry in the composed hardware for $C_m(h(x))$) feeding into the arithmetic circuitry of D_m. This is precisely what is needed, in order to establish our claim that $f \in \mathsf{NC}^1(\#\mathsf{NC}^1_{trop})$. $\square$

Unlike the case of CRAs operating over the integers, CRAs over the tropical semiring without the copyless restriction compute only functions that are computable in polynomial time (via a straightforward simulation). We know of no better upper bound than P in this case, and we also have no lower bounds.

As noted above at the beginning of Section 4, if the "multiplicative" register updates (i.e., $+$ in the tropical semiring) are all of the form $r \leftarrow r' + c$, then even without the copyless restriction, the computation of a CRA function f reduces to iterated matrix multiplication of $O(1) \times O(1)$ matrices over the tropical semiring. Again, it follows easily that the contents of any register at any point in the computation can be represented using $O(\log n)$ bits. Thus the upper bound of L holds also in this case.

4.3   CRAs over the Max-Concat Semiring

As in Section 3.2, we consider only CCRAs.

Theorem 7. *All functions computable by CCRAs over $(\Gamma^*, \max, \circ)$ are computable in AC^1.*

Proof. Let f be computed by a CCRA M operating over $(\Gamma^*, \max, \circ)$.

We first present a logspace-computable function h with the property that $h(1^n)$ is a description of a circuit C_n computing f on inputs of length n. The input convention is slightly different for this circuit family. For each input symbol a and each $i \leq n$ there is an input gate $g_{i,a}$ that evaluates to λ (the empty string) if $x_i = a$, and evaluates to $\bot$ otherwise. (This provides an "arithmetical" answer to the Boolean query "is the i-th input symbol equal to a?")

Assume that there are gates $r_{1,i}, r_{2,i}, \ldots, r_{k,i}$ storing the values of each of the registers at time i. For $i = 0$ these gates are constants. For each input symbol a and each $j \leq k$, let $E_{a,j}(r_{1,i}, \ldots, r_{k,i})$ be the expression that describes how register j is updated if the $i + 1$-st symbol is a. Then the value $r_{j,i+1} = \max_a\{g_{i,a} \circ E_{a,j}(r_{1,i}, \ldots, r_{k,i})\}$. This yields a very uniform circuit family, since the circuit for inputs of length n consists of n identical blocks of this form connected in series. That is, there is a function computable in NC^1 that takes 1^n as input, and produces an encoding of circuit C_n as output.

Although the depth of circuit C_n is linear in n, its *algebraic degree* is only polynomial in n. (Recall that the additive operation of the semiring is max and the multiplicative operation is $\circ$. Thus the degree of a max gate is the maximum of the degrees of the gates that feed into it, and the degree of a $\circ$ gate is the sum of the degrees of the gates that feed into it.) This degree bound follows from the copyless restriction. (Actually, the copyless restriction is required only for the $\circ$ gates; inputs to the max gates could be re-used without adversely affecting the degree.)

By [2, Proposition 5.2], arithmetic circuits of polynomial size and algebraic degree over $(\Gamma^*, \max, \circ)$ characterize exactly the complexity class OptLogCFL. OptLogCFL was defined by Vinay [19] as follows: f is in OptLogCFL if there is a nondeterministic logspace-bounded auxiliary pushdown automaton M running in polynomial time, such that, on input x, $f(x)$ is the lexicographically largest string that appears on the output tape of M along any computation path. The proof of Proposition 5.2 in [2], which shows how an auxiliary pushdown automaton can simulate the computation of a max-concat circuit, also makes it clear that an auxiliary pushdown machine, operating in polynomial time, can take a string x as input, use its logarithmic workspace to compute the bits of $h(1^{|x|})$ (i.e., to compute the description of the circuit $C_{|x|}$), and then to produce $C_{|x|}(x) = f(x)$ as the lexicographically-largest string that appears on its output tape along any computation path. That is, we have $f \in \mathsf{OptLogCFL}$.

By [2, Lemma 5.5], $\mathsf{OptLogCFL} \subseteq \mathsf{AC}^1$, which completes the proof. □

Acknowledgments. This work was supported by NSF grant CCF-1064785 and an REU supplement.

References

1. Allender, E.: Arithmetic circuits and counting complexity classes. In: Krajíček, J. (ed.) Complexity of Computations and Proofs, Quaderni di Matematica, vol. 13, pp. 33–72. Seconda Università di Napoli (2004)
2. Allender, E., Jiao, J., Mahajan, M., Vinay, V.: Non-commutative arithmetic circuits: Depth reduction and size lower bounds. Theoretical Computer Science **209**(1–2), 47–86 (1998)
3. Alur, R.: Regular functions (2013); lecture presented at Horizons in TCS: A Celebration of Mihalis Yannakakis's 60th Birthday. Center for Computational Intractability, Princeton (2013)
4. Alur, R., Cerný, P.: Expressiveness of streaming string transducers. In: Conference on Foundations of Software Technology and Theoretical Computer Science (FST&TCS). LIPIcs, vol. 8, pp. 1–12. Schloss Dagstuhl - Leibniz-Zentrum fuer Informatik (2010)
5. Alur, R., Cerný, P.: Streaming transducers for algorithmic verification of single-pass list-processing programs. In: 38th ACM SIGPLAN-SIGACT Symposium on Principles of Programming Languages (POPL), pp. 599–610 (2011)
6. Alur, R., D'Antoni, L., Deshmukh, J.V., Raghothaman, M., Yuan, Y.: Regular functions, cost register automata, and generalized min-cost problems. CoRR abs/1111.0670 (2011)
7. Alur, R., D'Antoni, L., Deshmukh, J.V., Raghothaman, M., Yuan, Y.: Regular functions and cost register automata. In: 28th Annual ACM/IEEE Symposium on Logic in Computer Science (LICS). pp. 13–22 (2013), see also the expanded version, [6]
8. Alur, R., Freilich, A., Raghothaman, M.: Regular combinators for string transformations. In: Joint Meeting of the Twenty-Third EACSL Annual Conference on Computer Science Logic and the Twenty-Ninth Annual ACM/IEEE Symposium on Logic in Computer Science, (CSL-LICS), p. 9. ACM (2014)
9. Alur, R., Raghothaman, M.: Decision problems for additive regular functions. In: Fomin, F.V., Freivalds, R., Kwiatkowska, M., Peleg, D. (eds.) ICALP 2013, Part II. LNCS, vol. 7966, pp. 37–48. Springer, Heidelberg (2013)
10. Barrington, D.A.: Bounded-width polynomial-size branching programs recognize exactly those languages in NC^1. Journal of Computer and System Sciences **38**, 150–164 (1989)

11. Mix Barrington, D.A., Lu, C.-J., Miltersen, P.B., Skyum, S.: Searching constant width mazes captures the AC^0 hierarchy. In: Meinel, C., Morvan, M. (eds.) STACS 1998. LNCS, vol. 1373, pp. 73–83. Springer, Heidelberg (1998)
12. Ben-Or, M., Cleve, R.: Computing algebraic formulas using a constant number of registers. SIAM Journal on Computing **21**(1), 54–58 (1992)
13. Buss, S.: Comment on formula evaluation (2014); personal communication
14. Buss, S.R., Cook, S., Gupta, A., Ramachandran, V.: An optimal parallel algorithm for formula evaluation. SIAM Journal on Computing **21**(4), 755–780 (1992)
15. Caussinus, H., McKenzie, P., Thérien, D., Vollmer, H.: Nondeterministic NC^1 computation. Journal of Computer and System Sciences **57**(2), 200–212 (1998)
16. Engelfriet, J., Hoogeboom, H.J.: MSO definable string transductions and two-way finite-state transducers. ACM Trans. Comput. Log. **2**(2), 216–254 (2001)
17. Hesse, W., Allender, E., Barrington, D.A.M.: Uniform constant-depth threshold circuits for division and iterated multiplication. Journal of Computer and System Sciences **65**, 695–716 (2002)
18. Kiefer, S., Murawski, A.S., Ouaknine, J., Wachter, B., Worrell, J.: On the complexity of equivalence and minimisation for Q-weighted automata. Logical Methods in Computer Science 9(1) (2013)
19. Vinay, V.: Counting auxiliary pushdown automata. In: Proceedings of the Sixth Annual Structure in Complexity Theory Conference, June 30-July 3, Chicago, Illinois, USA, pp. 270–284 (1991). http://dx.doi.org/10.1109/SCT.1991.160269
20. Vollmer, H.: Introduction to Circuit Complexity: A Uniform Approach. Springer-Verlag New York Inc. (1999)

A Nonuniform Circuit Class with Multilayer of Threshold Gates Having Super Quasi Polynomial Size Lower Bounds Against NEXP

Kazuyuki Amano and Atsushi Saito[(✉)]

Department of Computer Science, Gunma University,
Tenjin 1-5-1, Kiryu, Gunma 376-8515, Japan
amano@cs.gunma-u.ac.jp, saito@amano-lab.cs.gunma-u.ac.jp

Abstract. Recently, Williams [STOC '14] proved a separation between NEXP and ACC ∘ THR, where an ACC ∘ THR circuit has a single layer of threshold gates at the bottom and an ACC circuit at the top. Two main ideas of his strategy are a closure property of circuit class and an algorithm for counting satisfying assignments of circuits.

In this paper, we show that this general scheme based on these two ideas can be applied for a certain class of circuits with *multi layer* of threshold gates. The circuit class we give has the symmetric gate at the top and poly-log layers of threshold gates to which an extra condition on the *dependency* is imposed. Two gates in a circuit are dependent, if the output of the one is always greater than or equal to the other one. An independent gate set is a set of gates in which two arbitrary gates are *not* dependent. We show that, if the size of a maximum independent gate set of each layer of threshold gates is at most n^γ for sufficiently small $\gamma > 0$, then two key ingredients needed to apply his strategy can be established. Namely, (i) we can efficiently find a circuit in our class being equivalent to the AND of two input circuits in our class, and (ii) we can construct a faster than brute-force algorithm for counting satisfying assignments for this class by introducing a partial order to represent the dependency of gates. As a result, we give super quasi-polynomial size lower bounds for our class against NEXP.

Keywords: Nonuniform circuit class · Satisfiability · Lower bounds

1 Introduction

Boolean circuit is one of the most popular and natural computation models. cience. For example, proving the existence of some NP problem having super polynomial size circuits led us to P ≠ NP. The best general boolean circuit lower bounds for NP problems are, however, $5n - o(n)$ by Iwama and Morizumi [9]. Various restricted circuit classes are studied. Bounded depth circuit class is one of the most successful restricted classes with a lot of remarkable results [4,6,10,11]. Williams established a landmark in the circuit complexity theory with the separation between NEXP and ACC^0 [14]. He incorporated many known results [2,3,7]

© Springer International Publishing Switzerland 2015
A.-H. Dediu et al. (Eds.): LATA 2015, LNCS 8977, pp. 461–472, 2015.
DOI: 10.1007/978-3-319-15579-1_36

into a perspective between algorithms and lower bounds [12]. The class TC^0, which is a class of constant depth polynomial size threshold circuits, is a well known natural circuit class larger than ACC^0. Current understanding of bounded depth threshold circuits is extremely inadequate [5,8]. Recently, Williams [13] proved a separation between NEXP and $\mathsf{ACC} \circ \mathsf{THR}$, where an $\mathsf{ACC} \circ \mathsf{THR}$ circuit has single layer of threshold gates at the bottom and an ACC circuit at the top. Two main ideas of his strategy are a closure property of circuit class and an algorithm for counting satisfying assignments of circuits. Thus it is a plausible direction to consider the usefulness of the framework based on these ideas.

In this paper, we show that this general framework based on these two ideas can be applied for some restricted class of circuits with *multi layer* of threshold gates. The circuit class we give has the symmetric gate at the top and at most poly-log layers of threshold gates to which an extra condition on the *dependency* is imposed. Two gates in a circuit are dependent, if the output of the one is always greater than or equal to the output of the other one. An independent gate set is a set of gates in which two arbitrary gates are *not* dependent. Each layer of threshold gates in our class has independent gate sets of size at most n^γ for sufficiently small $\gamma > 0$. We show that two main ideas in [13] are workable for our circuit class. It is notable that our circuit class is universal even if there is no two independent gates and that the general framework can be applied for poly-log depth circuits. As far as we know, no restricted poly-log depth subclass having super linear lower bounds for even NEXP had been found. First, we show that we can efficiently find a circuit in our class being equivalent to the AND of two input circuits in our class. Thus our class has a closure property (Lemma 1). Second, we design an algorithm for counting satisfying assignments for our circuit class (Lemma 2). We connect dependency to a structure of a partial order on the gate set. This connection make counting assignments easier than general settings. By pluging them into William's schema (Theorem 1), we obtain super quasi-polynomial size lower bounds for our circuit class against NEXP (Theorem 3).

2   Preliminaries

In this section, we give several definitions for stating our work.

Definition 1. Let $x_1, ..., x_n$ be boolean variables. Let $w_1, ..., w_n, t$ be real numbers.

(1) We define a threshold gate as a gate computing a boolean function $THR_{w_1,...,w_n,t}(x_1, ..., x_n)$ such that $THR_{w_1,...,w_n,t}(x_1, ..., x_n) = 1 \iff \Sigma_{i=1}^n w_i x_i \geq t$. The real numbers $w_1, ..., w_n$ and t are called weights and threshold value, respectively. When all weights are one and the threshold value is the half of fan-in wires, the threshold gate is called *majority gate*.

(2) We define a symmetric gate $SYM_S(x_1, ..., x_n)$ as a gate computing a boolean function $SYM_S(x_1, ..., x_n) = 1 \iff \Sigma_{i=1}^n x_i \in S$ for a subset $S \subseteq \{0, 1, ..., n\}$. We call S the characteristic set of the symmetric gate SYM_S.

Remark of Definition 1. In this paper, we suppose that the absolute value of any weight in threshold gates is at most $2^{poly(n)}$ and any weight are coded by a binary string of length $poly(n)$. We will use the term *source* (*sink*, resp.) to represent an input variable or a gate which is connected to the input terminal (the output terminal, resp.) of a wire.

We give a notion of "dependency" of gates which was introduced in [1].

Definition 2. Let C be an arbitrary circuit. For a gate G in C, let $G(x) \in \{0,1\}$ denote the output value of G when we feed an input string x to the circuit C.
(1) Two gates G_1, G_2 in C have *dependency*, if one of two preimages $G_1^{-1}(1)$ and $G_2^{-1}(1)$ is a subset of the other one. In other words, $\forall x \in \{0,1\}^n [G_1(x) \leq G_2(x)] \vee \forall x \in \{0,1\}^n [G_2(x) \leq G_1(x)]$.
(2) A subset of gates in C is called *independent gate set*, if any two gates in the set do not have dependency. A circuit may contain several independent gate sets.

One may think that circuits with bounded size independent gate sets seems to have very weak computational ability and seems to compute only boolean functions in a narrow class. However, for example, the class of depth two threshold circuits is universal, even if there is *no* pair of independent gates. We will prove this in section 3.

Definition 3. Let $\mathcal{L}_i$ be a type of gates for each $i = 1, 2, ..., d$. Let C be a circuit, and let V_0 and V be respectively the set of input variables of C and the set of gates of C.
(1) A circuit C is a $\mathcal{L}_d \circ \mathcal{L}_{d-1} \circ \cdots \circ \mathcal{L}_1$ circuit, if there exists some partition $V_1, ..., V_d$ of the set V such that **(i)** any wire from $G \in V_i$ to $G' \in V_j$ satisfies that $i < j$ and **(ii)** a type of all gates in V_i is $\mathcal{L}_i$ for each i. We call this partition $V_1, ..., V_d$ a *layering partition* of C. We also call each V_i *the i-th layer*.
(2) We assume that any gate G of a $\mathcal{L}_d \circ \cdots \circ \mathcal{L}_1$ circuit has an integer label i such that G is in the i-th layer, where $1 \leq i \leq d$. We call such labels *layering labels*.

Remark of Definition 3. Note that there is no wire connecting two gates belonging to the same layer. For example, any AC^0 circuit C is in $\mathcal{L}_d \circ \mathcal{L}_{d-1} \circ \cdots \circ \mathcal{L}_1$ for some constant d, where for each i ($1 \leq i \leq d$) $\mathcal{L}_i \in \{\mathsf{AND}, \mathsf{OR}, \mathsf{NOT}\}$.

Definition 4. Let C be a $\mathcal{L}_d \circ \mathcal{L}_{d-1} \circ \cdots \circ \mathcal{L}_1$ circuit. Let $V_1, ..., V_d$ be a layering partition of C.
(1) A set V_i is called *the i-th k-$\mathcal{L}_i$ layer* in C, if the maximum size of an independent gate set $I \subseteq V_i$ in C is at most k.
(2) We call C a k-$\mathcal{L}_d \circ k$-$\mathcal{L}_{d-1} \circ \cdots \circ k$-$\mathcal{L}_1$ circuit, if there exists some layering partition $V_1, ..., V_d$ such that each V_i is the i-th k-$\mathcal{L}_i$ layer in C. Let $(\mathcal{L})^d$ denote an abbreviation of $\mathcal{L}_d \circ \mathcal{L}_{d-1} \circ \cdots \circ \mathcal{L}_1$, if $\mathcal{L}_i$ is the same type $\mathcal{L}$ for all i.
(3) We define $\mathcal{C}_k[d]$ as a class of $\mathsf{SYM} \circ (k\text{-}\mathsf{THR})^d$ circuits.

Remark of Definition 4. **(1)** We usually use the term "circuit" as single output circuits, and we particularly mention the use of multi-output circuits.

When we consider a class of single output circuits, we write $\mathcal{C}_d \circ k\text{-}\mathcal{C}_{d-1} \circ \cdots \circ k\text{-}\mathcal{C}_1$ instead of $k\text{-}\mathcal{C}_d \circ k\text{-}\mathcal{C}_{d-1} \circ \cdots \circ k\text{-}\mathcal{C}_1$. **(2)** In this paper, we may use the words *a k-THR layer, a layer of k-THR gates* or just *a layer of threshold gates* to mention one of the above sets $V_1, ..., V_d$.

Let $\mathcal{A}_t$ be an $\mathcal{A}$ gate with at most t fan-in wires, where $\mathcal{A}$ is a gate type.

3   Prior Work and Our Results

In this section, we firstly review the proof of super quasi-polynomial lower bounds for ACC $\circ$ THR by Williams [13] since it is closely related to our work in subsection 3.1. In subsection 3.2, we define a complexity class to separate from NEXP and give a formal statement of our result. In subsection 3.3, we give an intuitive explanation about our proof strategy and formally state notions to understand our proof methods.

3.1   Prior Work

In [13], the following property of circuit classes plays an important role.

Definition 5. Let $\mathcal{C}$ be a circuit class. The class $\mathcal{C}$ is *weakly closed under AND*, if there is a polynomial time procedure such that for given the AND of two $\mathcal{C}$ circuits the procedure produces an equivalent $\mathcal{C}$ circuit.

Remark of Definition 5. Note that the time complexity of a procedure in the above definition is a function in the size of a code of two circuits.

We state a meta theorem in [13].

Theorem 1 ([13]). Let $\mathcal{C}$ be a circuit class *weakly closed under AND*. Suppose for any $c \geq 1$, there is an $\varepsilon > 0$ and an algorithm for counting the satisfying assignments in time $2^{n-\Omega(n^\varepsilon)}$ on $\mathcal{C}$ circuits with n inputs and $n^{\log^c n}$ size. Then NEXP does not have quasi-polynomial size $\mathcal{C}$ circuits.

Note that ACC $\circ$ THR clearly satisfies the closure property under AND. By this general theorem, we can derive super quasi-polynomial ACC $\circ$ THR lower bounds against NEXP, if we construct a faster counting algorithm. The following theorem achieves this.

Theorem 2 ([13]). For every $m > 1$ and $d > 0$, there is an $\varepsilon > 0$ such that counting satisfying assignments to ACC $\circ$ THR circuits of size 2^{n^ε}, depth d and modulus m gates can be solved in 2^{n-n^ε} time.

Remark 2. Williams actually constructed a counting algorithm for the class of circuits ACC $\circ$ SYM. He showed that there is a transformation from an arbitrary threshold gate to a constant depth circuit with single layer of symmetric gates. Using this transformation we can transform an arbitrary ACC $\circ$ THR circuit to an ACC $\circ$ SYM circuit. Thus, Theorem 2 can be proved by giving a counting algorithm for ACC $\circ$ SYM circuits. Below we formally state this transformation since we will also use this in the proof of our result.

Claim 1 ([13]). An arbitrary THR gate G can be replaced with $AC^0 \circ MAJ$ circuit C such that the size of C is at most polynomial in the input size of G. Moreover, time complexity for the replacement is at most polynomial time.

3.2 Lower Bounds Against NEXP for a Circuit Class with Multi Layers of Threshold Gates

In this subsection, we firstly define a class of circuits and a relating class of languages.

Definition 6. We define $\tilde{\mathcal{C}}_k[d]$ as $\bigcup\limits_{c>0} \mathcal{C}_{ck}[d]$.

We note that $\tilde{\mathcal{C}}_k[d]$ is a class of circuits with n input variables, where c does not depend on n and k.

Definition 7. Let $\mathcal{A}$ be an arbitrary circuit class. We define $\mathcal{A}\text{-SIZE}[S(n)]$ as the class of languages having a family of $\mathcal{A}$ circuits with n inputs and of size $O(S(n))$.

We mention our primal goal.

Theorem 3. Let $d = poly \log n$ for the number of input variables n. There is some $\gamma > 0$ such that $NEXP \not\subseteq \tilde{\mathcal{C}}_k[d]\text{-SIZE}[2^{poly(\log n)}]$ for $k \leq n^\gamma$.

Before proving lower bounds for the class $\tilde{\mathcal{C}}_k[d]$, we see that $\mathcal{C}_k[d]$ is able to compute all boolean functions even when $k = 1$ and $d = 1$. This means that $NEXP \subseteq \tilde{\mathcal{C}}_1[1]\text{-SIZE}[2^n]$ and gives the motivation for studying the complexity of circuits $\tilde{\mathcal{C}}_k[d]$ with small values of k and d. The following claim says that $SYM \circ k\text{-THR}$ as well as $THR \circ k\text{-THR}$ is universal for $k = 1$.

Claim 2. For any positive integer n and for any boolean function $f : \{0,1\}^n \rightarrow \{0,1\}$ there are a $THR \circ 1\text{-THR}$ circuit and a $SYM \circ 1\text{-THR}$ circuit with at most $2^n + 1$ gates.

Proof. Let $f(x_{n-1}, ..., x_1, x_0)$ be a boolean function on n variables. For simplicity, we assume $f(0, 0, ..., 0) = 0$. First, we give a construction of a $THR \circ 1\text{-THR}$ circuit for f. For $0 \leq j \leq 2^n - 1$, let y_j denote the binary representation of j of length n, i.e., $y_j := (x_{n-1}, ..., x_1, x_0)$ with $\sum_{i=0}^{n-1} x_i 2^i$. Let G_j be the threshold gate whose output is 1 iff $\sum_{i=0}^{n-1} 2^i x_i \geq j$. The bottom level of a circuit is consisting of $\mathcal{G} = \{G_j : f(y_j) \neq f(y_{j-1}) \ (1 \leq y \leq 2^n - 1)\}$. Obviously, there is no pair of independent gates in $\mathcal{G}$. The top gate outputs 1 iff $\sum_{G_j \in \mathcal{G}} w_j G_j \geq 1$ where the weight w_j is $f(y_j) - f(y_{j-1})$ which is 1 or -1. In fact, the value of f is *equal* to $\sum_{G_j \in \mathcal{G}} w_j G_j$.

It is easy to observe that the top gate can be replaced by a symmetric gate that outputs 1 iff $\sum_{G_j \in \mathcal{G}} G_j$ is odd. This says that $SYM \circ 1\text{-THR}$ is also universal. $\square$

We prove Theorem 3 by applying Theorem 1. Apparently, it is sufficient to prove the following two lemmas.

Lemma 1. The class of $\tilde{C}_k[d]$ circuits with n inputs and of quasi-polynomial $2^{\log^{O(1)} n}$ size is weakly closed under AND.

Lemma 2. Let d be *poly* $\log n$ in the number of input variables n. There exist some $\varepsilon > 0$ and $\gamma > 0$ such that counting satisfying assignments to $C_k[d]$ circuits of size $S(n) = 2^{n^{o(1)}}$ can be solved in $2^{n - \Omega(n^\varepsilon)}$ time for $k \leq n^\gamma$.

We will give a proof outline of the lemma about closure property in section 4. We will give a counting algorithm in section 5. The algorithm in Theorem 2 is incorporated to our counting algorithm as a subroutine.

3.3   Restrictions to Output of Threshold Gates

In this subsection, we give several notions to understand the reason why we can construct a faster satisfiability or counting algorithm for circuits with bounded size independent gate sets. We will give extensions of these notions to multilayer setting in section 5. Suppose that $\mathcal{G}$ is a set of gates which has independent gate sets of size at most k and has *no pair of equivalent gates*. Then, we can form a partial ordered set $(\mathcal{G}, \preceq)$ by defining the order $\preceq$ on $\mathcal{G}$ so that $G_1 \preceq G_2$ iff $G_1^{-1}(1) \subseteq G_2^{-1}(1)$ for $G_1, G_2 \in \mathcal{G}$. It is easy to observe that the size of an independent vertex set of the Hasse diagram of $(\mathcal{G}, \preceq)$ is at most k. We call this Hasse diagram an Induced Hasse Diagram (I.H.D, in short) of $\mathcal{G}$. In [1], depth two threshold circuits are considered, and I.H.D is defined for the set of bottom gates. In this paper, we will extend the notion of I.H.D for k-THR layer in section 5.

Definition 8.
(1) Let V be a set of gates and let $H = (V, E)$ be I.H.D of V. A map $\chi : V \mapsto \{0, 1\}$ is called a *validly ordered restriction*, if $\forall (u, v) \in E, \chi(u) \leq \chi(v)$.
(2) Let χ be a validly ordered restriction for an arbitrary I.H.D $H = (V, E)$. We define the *min-set of H for χ* as the set $\{u_{min} \in V \cap \chi^{-1}(1) : \forall v \in V \setminus \{u_{min}\}[v \preceq u_{min} \Rightarrow \chi(v) = 0]\}$.
We define the *max-set of H for χ* as the set $\{u_{max} \in V \cap \chi^{-1}(0) : \forall v \in V \setminus \{u_{max}\}[u_{max} \preceq v \Rightarrow \chi(v) = 1]\}$.

Threshold gates in min-sets and max-sets are regarded as some critical local information about all k-THR layers in a $C_k[d]$ circuit. We give the following notion stating min-sets and max-sets from a viewpoint of graph theory.

Definition 9. Let H be an I.H.D and χ be a validly ordered restriction of H. Let I_1, I_0 be independent sets in H. The pair of independent sets (I_1, I_0) satisfies the *covering condition* for H, if the following condition holds.

Condition: For any $v \in V \setminus (I_1 \cup I_0)$ in H, either $\exists u_1 \in I_1, u_1 \preceq v$ or $\exists u_0 \in I_0, v \preceq u_0$ according to the order $\preceq$ of H.

We define X'_H as the set $\{\chi : \chi$ is a validly ordered restriction of $H\}$. We also define $\mathcal{I}_H$ as follows: $\mathcal{I}_H := \{(I_1, I_0) \subseteq V \times V : I_1, I_0$ are independent sets satisfying the covering condition in $H\}$. The following lemma is in [1].

Lemma 3 ([1]). There is a bijection $\mu_H : X'_H \ni \chi \mapsto (I_1, I_0) \in \mathcal{I}_H$ such that if $\mu_H(\chi) = (I_1, I_0)$ then I_1 is the min-set of H for χ and I_0 is the max-set of H for χ.

Observation 1. For boolean functions $P_1(x), ..., P_m(x) : \{0,1\}^n \to \{0,1\}$, the following statements hold.

(1) If $\forall x \forall i [P_i(x) \leq P_{i-1}(x)]$, then $\forall x [\bigwedge_{i=1}^m P_i(x) = 1 \iff P_1(x) = 1]$.

(2) If $\forall x \forall i [P_i(x) \leq P_{i-1}(x)]$, then $\forall x [\bigvee_{i=1}^m P_i(x) = 0 \iff P_m(x) = 0]$.

We explain the reason why this observation and Lemma 3 is useful to prove Lemma 2. Intuitively, the output of an arbitrary gate of a circuit is determined by fixing gates in the union of min-set and max-set by   Observation 1 . By Lemma 3, it is also determined by fixing the output of threshold gates in $I_1 \cup I_0$, where $(I_1, I_0) \in \mathcal{I}_H$. Note that for given pair of subsets of the vertex set of H we can efficiently decide whether $(I_1, I_0) \in \mathcal{I}_H$ holds or not. If the maximum size of independent gate sets is small, then the number of threshold gates we have to fix is not so many. We will give a more detailed description of the algorithm in subsection 5.2.

4   Closure Property Under AND

In this section, we state how to prove the following lemma.

Lemma 1(restated). The class of $\tilde{\mathcal{C}}_k[d]$ circuits with n inputs and of quasi-polynomial $2^{\log^{O(1)} n}$ size is weakly closed under AND.

We assume that the input size is measured by the number of wires of circuits. The following claim is in [13].

Claim 3 ([13]). There is a procedure such that for given $\mathsf{AND}_u \circ \mathsf{SYM}$ circuit C with n input variables and $N = N(n)$ wires, where each input variable can have more than one wire connecting to a same symmetric gate in C, the procedure converts the circuit C to a *single* SYM *gate* with $O(N^u)$ wires.   □

We directly apply the above claim to our setting as follows.

Claim 4. There is a procedure such that for given AND of two $\tilde{\mathcal{C}}_k[d]$ circuits C_1 and C_2, where C_1 and C_2 have n input variables and at most $t = t(n)$ top fan-in and at most $N = N(n)$ wires, the procedure transforms the input to a $\mathsf{SYM} \circ (\mathsf{THR})^d$ circuit C_3 having at most $O(t^2 N)$ wires. Moreover, any gate in the l-th layer in C_3 is in either the l-th layer of C_1 or the l-th layer of C_2.   □

We omit the proofs of these two claims because of the page limitation. We give an outline of the proof of Claim 4 as follows.

Proof outline of Claim 4. Remove the top symmetric gate from C_1, and regard fan-ins of the top symmetric gate as output wires. Let C'_1 be the multi output circuit obtained by this operation. Let C_2' be the circuit which is obtained from C_2 in the same way. *We take B copies of C_2'* and let these copies be $C_{2,1}', ..., C_{2,B}'$, where B is at most t. Layering labels are also copied. We take a new symmetric gate. Connect the output wires of $C_{2,1}', ..., C_{2,B}', C_1'$ to the new symmetric gate. Then, we obtain C_3. □

In the proof outline of Claim 4, taking copies of a multi output circuit is particularly important to prove Lemma 1. We omit the poof of Lemma 1, because it is not hard to show Lemma 1. We give the outline of the proof as follows.

Proof outline of Lemma 1. Suppose that C_1, C_2 are respectively a $\mathcal{C}_{c_1 k}[d]$ circuit and a $\mathcal{C}_{c_2 k}[d]$ circuit. We show that there is a polynomial time procedure such that for given the AND of a $\mathcal{C}_{c_1 k}[d]$ circuit C_1 and a $\mathcal{C}_{c_2 k}[d]$ circuit C_2, where C_1, C_2 have n input variables and at most N wires and c_1, c_2 do not depend on n and k, the procedure outputs an equivalent $\mathcal{C}_{(c_1 + c_2) k}[d]$ circuit C_3 with $poly(N)$ wires. Note that top fan-in of a circuit is at most the number of wires in the circuit. Hence, the number of wires in the output circuit C_3 in Claim 4 is at most $poly(N)$. All that we prove is that the output circuit C_3 in Claim 4

is a $\mathcal{C}_{(c_1 + c_2) k}[d]$ circuit. For each $i = 1, 2$, let C_i' be a $(c_i k\text{-THR})^d$ circuit with n input variables and at most $t = t(n)$ output wires such that C_i' is obtained by removing the top symmetric gate from C_i and by regarding input wires of the top symmetric gate as output wires. *We can construct C_3 by copying C_2'.* Let $\mathcal{G}_{2,l}$ be the set $\{G_{i,j} : G_{i,j}$ is the i-th gate in the j-th copy of the l-th $c_2 k\text{-THR}$ layer in $C_2\}$, and let $\mathcal{G}_{1,l}$ be the l-th threshold layer in C_1. By contradiction, we prove that $\mathcal{G}_{2,l}$ has *no* independent gate set of size greater than $c_2 k$ for each fixed l. □

5 Transforming of Circuits and a Counting Algorithm

In this section, our goal is to prove the following lemma.

Lemma 2(restated). Let d be $poly \log n$ in the number of input variables n. There exist some $\varepsilon > 0$ and $\gamma > 0$ such that counting satisfying assignments to $\mathcal{C}_k[d]$ circuits of size $S(n) = 2^{n^{o(1)}}$ can be solved in $2^{n - \Omega(n^\varepsilon)}$ time for $k \leq n^\gamma$.

5.1 Notions for Bottom Up Procedures

Let $f : X \to Y$ be a map for finite sets X, Y. For an arbitrary $A \subseteq X$, let $f|_A$ denote the map satisfying that $\forall x \in A, f|_A(x) = f(x)$.

Definition 10. Let $\mathcal{C}$ be a circuit class $\mathcal{C}' \circ (k\text{-THR})^d$ for an arbitrary $\mathcal{C}'$ gate at the top level. We call a $\mathcal{C}' \circ (k\text{-THR})^d$ circuit an *abbreviated circuit*, if for any threshold layer in the circuit there is *no* pair of equivalent gates in the threshold layer.

Definition 11. Let $\mathcal{C}$ be a circuit class $\mathcal{C}' \circ (k\text{-THR})^d$ for an arbitrary $\mathcal{C}'$ gate at the top level. Let C be an abbreviated $\mathcal{C}$ circuit. Let V_j be the j-th k-THR layer in C for each $1 \le j \le d$.

We call a family of directed graphs $\mathcal{F}_i = \{H_j = (V_j, E_j) : 1 \le j \le i\}$ an i-th *Induced Hasse Diagram Family* of C, if for each j the directed graph H_j is the Hasse diagram of the partial ordered set $(V_j, \preceq)$ defined as follows:

$$\forall G_1 \in V_j, \forall G_2 \in V_j [G_1 \preceq G_2 \iff G_1^{-1}(1) \subseteq G_2^{-1}(1)].$$

We call a map $\rho : V_1 \cup \cdots \cup V_d \to \{0, 1\}$ a *validly ordered restriction for a family* $\mathcal{F}$, if $\rho|_{V_i}$ is a validly ordered restriction to V_i for any i. We also call a d-th induced Hasse diagram family an *induced Hasse diagram family* of C, and $\mathcal{F}_d$ is simply denoted by $\mathcal{F}$.

5.2   Proof of Lemma 2

Let $\{\mathcal{A}, \mathcal{B}\}$ be a layer of gates which contains $\mathcal{A}$ gates or $\mathcal{B}$ gates, where $\mathcal{A}, \mathcal{B}$ are types of gates. We state the following lemma, but we omit the proof of this lemma because of page limitation.

Lemma 4. Let $\mathcal{C}$ be a circuit class $\mathcal{C}' \circ (k\text{-THR})^d$, where $\mathcal{C}'$ is either SYM or THR. There is a procedure such that for given abbreviated $\mathcal{C}$ circuit C of size $S(n)$ and $\mathcal{F}$ which is the induced Hasse diagram family of C, the procedure outputs an $AC^0 \circ$ SYM circuit C' of size at most $k^{2d} \left(\frac{S(n)}{O(k)}\right)^d poly(S(n))$ such that C is equivalent to C'. Moreover, this procedure runs in $O\left(k^{2d} \left(\frac{S(n)}{O(k)}\right)^d poly(S(n))\right)$ time. $\square$

It is not hard to prove this lemma using Lemma 3. We give the proof outline as follows.

Proof outline. Let $\mathcal{F}$ be $\{H_1 = (V_1, E_1), ..., H_d = (V_d, E_d)\}$. We note that $V_1 \cup \cdots \cup V_d$ is the set of threshold gates of C. We define $\mathcal{I}_i$ as follows: $\mathcal{I}_i :=\{(I_i, J_i) \subseteq V_i \times V_i : (I_i, J_j)$ is a pair of independent sets satisfying the covering condition in $H_i\}$. By Lemma 3, the definition of min-set and max-set in Definition 8 and Observation 1 , for each $1 \le i \le d$ and for any $(I_i, J_i) \in \mathcal{I}_i$ there uniquely exists $\mu_i : V_i \to \{0, 1\}$ such that (A) μ_i is a validly ordered restriction to V_i and (B) the two images $\mu_i(I_i)$ and $\mu_i(J_i)$ are respectively $\{1\}$ and $\{0\}$, and (C) we can fix all outputs of threshold gates in the i-th k-THR layer according to μ_i. Suppose that $1 \le \forall i \le d, (I_i, J_i) \in \mathcal{I}_i$. Then, for each validly ordered restriction to $V_1 \cup \cdots \cup V_d$, there is an integer linear programming (ILP, in short) instance such that the feasible space of the instance is equal to the set of all inputs agreeing with the restriction. Therefore, we obtain the $OR \circ AND \circ \{SYM, THR\}$ circuit by taking OR of the $AND \circ \{SYM, THR\}$ circuits for all validly ordered restrictions to $V_1 \cup \cdots \cup V_d$. Intuitively, the number of fan-ins of the OR gate is bounded above by the number of small subsets of threshold gates. We note that there are efficient ways for listing all pairs of independent gates and implementing Lemma 3 than the trivial brute-force. The former can be executed by solving an ILP instance with two linear constraints for each pair of independent gates, and the later can be done by checking all small size subsets of threshold gates.   □

Lemma 5. Let $d = poly \log n$. There is a procedure such that for given $\mathcal{C}_k[d]$ circuit C of size $S(n) = 2^{n^{o(1)}}$ it outputs an abbreviated circuit C' and an Induced Hasse Diagram Family $\mathcal{F}$ of C' such that C is equivalent to C'. Moreover, there exist some $\varepsilon > 0$ and some $\gamma > 0$ such that it runs in time $2^{n-\Omega(n^\varepsilon)}$ for $k \le n^\gamma$.

Proof overview. We give an outline of our algorithm. We first explain a simple procedure which is incorporated to our algorithm. For given $\mathcal{C}_k[d]$ circuit C and two gates $G_1, G_2 \in V_i$, where V_i is the i-th threshold layer in C and $G_1(x) = G_2(x)$ for any input x, it outputs a $\mathcal{C}_k[d]$ circuit which is equivalent to C. Essentially, this procedure replaces the gate G_2 with G_1. We call $V_{i+1} \cup \cdots \cup V_d$ the *upper layers than the i-th threshold layer*. The following is a description of this procedure.

1. For each threshold gate T in the upper layers than the i-th threshold layer, if there is an input wire from G_2 then the label $w_{G_2}y_{G_2} + \sum\limits_{U \ne G_2} w_U y_U \ge t_T$ in the gate T is replaced with $w_{G_2}y_{G_1} + \sum\limits_{U \ne G_2} w_U y_U \ge t_T$, and *a wire is drawn from G_1 to T*, where each U is a source of T.
2. For the top SYM gate S, if there is an input wire from G_2 then the label $y_{G_2} + \sum\limits_{U \ne G_2} y_U \in S_1$ in the gate S is replaced with $y_{G_1} + \sum\limits_{U \ne G_2} y_U \in S_1$, and *a wire is drawn from the gate G_1 to S*, where S_1 is the characteristic set of the symmetric gate S.
3. Remove G_2 and all input and output wires of G_2 from C.

We call this procedure *abbreviation procedure*. We note that eliminating threshold gates in a threshold layer does not increase the size of maximum independent gate sets. We explain our approach to make our algorithm. Our algorithm progresses from the bottom layer to the top layer step by step. For each i, the $(i + 1)$-th threshold layer is abbreviated by using an i-th induced Hasse diagram family. We give the complete proof of Lemma 5 as follows.

Proof of Lemma 5. We consider the following procedure about a bottom up construction of Induced Hasse Diagram Family.

1. Let $\mathcal{F}$ be $\emptyset$. For $i = 1, 2, ..., d$, let V_i be the i-th k-THR layer in C, and let E_i be $\emptyset$, and do the following steps **2.**, **8.**, and **9.**.
   2. For any $G_1, G_2 \in V_i$, do the following steps **3.**, **4.**, **5.**, **6.**, and **7.**.
      3. Let C_1, C_2 be THR $\circ$ (k-THR)$^{i-1}$ sub-circuits in C whose top gates are G_1 and G_2, respectively, if $i \ge 2$. Let C_1, C_2 be threshold gates G_1, G_2, respectively, if $i = 1$.
      4. For $b = 1, 2$, transform the circuit C_b to an $AC^0 \circ$ SYM circuit C_b', by running the procedure in Lemma 4 on the input C_b and

$\mathcal{F} = \{H_1, ..., H_{i-1}\}$ for any $i \geq 2$ or by running the procedure in Claim 1 on the input threshold gate C_b for $i = 1$.

5. Call the counting algorithm in Remark 2 to check the satisfiability of the two $\mathsf{AC}^0 \circ \mathsf{SYM}$ circuits $A_1{:}\neg C_1' \wedge C_2'$ and $A_2{:}\neg C_2' \wedge C_1'$.
6. If it outputs "Unsatisfiable" for A_2 (i.e. $C_1'(x) \leq C_2'(x)$ for any input string x) then $E_i := E_i \cup \{(G_1, G_2)\}$.
7. Else if it outputs "Unsatisfiable" for A_1 (i.e. $C_2'(x) \leq C_1'(x)$ for any input string x) then $E_i := E_i \cup \{(G_2, G_1)\}$.
8. For each G_1, G_2 in C, if both (G_1, G_2) and (G_2, G_1) are in E_i then run the abbreviation procedure on C, G_1, and G_2. Let C be the resulting circuit (with no pair of equivalent gates in the i-th k-THR layer).
9. Let H_i be (V_i, E_i) and let $\mathcal{F}$ be $\mathcal{F} \cup \{H_i\}$.
10. Output C (with no pair of equivalent threshold gates in any k-THR layer) and $\mathcal{F} = \{H_i : H_i = (V_i, E_i) \ (1 \leq i \leq d)\}$.

Running time analysis is as follows. The most dominant contribution to the entire running time is in the step testing dependency of two circuits. Note that we can construct an $\mathsf{AC}^0 \circ \mathsf{SYM}$ circuit with n inputs and of size $S_1(n) = O\left(k^{2d}\left(\frac{S(n)}{O(k)}\right)^d poly(S(n))\right)$ by Lemma 4. By Theorem 2 and Remark 2, there is some $\varepsilon > 0$ such that an algorithm can count the satisfying assignments to $\mathsf{AC}^0 \circ \mathsf{SYM}$ circuits of size 2^{n^ε} and runs in 2^{n-n^ε} time. We can take sufficiently small constant $\gamma > 0$ such that $S_1(n) \leq 2^{n^\varepsilon}$ for $k \leq n^\gamma$. Thus the running time in step 5. is at most 2^{n-n^ε}. The entire running time is at most

$$d \cdot \left(O\left(\binom{S(n)}{2}\right)\right) \cdot \left(poly(S(n)) + k^{2d}\left(\frac{S(n)}{O(k)}\right)^d poly(S(n)) + 2^{n-n^\varepsilon}\right),$$

for some constant $\varepsilon > 0$. Note that $S(n) = 2^{n^{o(1)}}$. Therefore, there exist some $\varepsilon > 0, \gamma > 0$ such that the entire running time is at most $2^{n-\Omega(n^\varepsilon)}$ for $k \leq n^\gamma$. □

Finally, we give the proof of Lemma 2.

Proof of Lemma 2. By the procedure in Lemma 5, for given input circuit C with depth $d = poly \log n$ and size $S(n) = 2^{n^{o(1)}}$, we compute an Induced Hasse Diagram Family $\mathcal{F}$ and an abbreviated circuit C_1 such that C and C_1 are equivalent. By Lemma 4, we obtain an $\mathsf{AC}^0 \circ \mathsf{SYM}$ circuit C_2 with size $S_2(n) = O\left(k^{2d}\left(\frac{S(n)}{O(k)}\right)^d poly(S(n))\right)$. By $S(n) = 2^{n^{o(1)}}$ and $d = poly \log n$, we have $S_2(n) = 2^{n^{o(1)}}$. Thus, there exist $\varepsilon_1, \gamma > 0$ such that this transformation from a $\mathcal{C}_k[d]$ circuit C to C_2 runs in time $2^{n-\Omega(n^{\varepsilon_1})}$ for $k \leq n^\gamma$. Finally, run the algorithm in Theorem 2 on C_2. There is $\varepsilon_2 > 0$ such that the running time of this algorithm is $2^{n-\Omega(n^{\varepsilon_2})}$. There exist $\varepsilon = \min_{i=1,2} \varepsilon_i$ and $\gamma > 0$ such that counting satisfying assignments to given $\mathcal{C}_k[d]$ circuit C can be done in time $2^{n-\Omega(n^\varepsilon)}$ for $k \leq n^\gamma$. □

6   Concluding Remark

In Lemma 4, we implicitly prove that any boolean function computed by restricted circuits of our form can be computed by $\mathrm{OR} \circ \mathrm{AND} \circ \{\mathrm{SYM}, \mathrm{THR}\}$ circuits of exponential size. Currently, there are no known exponential size lower bounds for $\mathrm{OR} \circ \mathrm{AND} \circ \{\mathrm{SYM}, \mathrm{THR}\}$ circuits, and a direct application of [13] is not workable because any exponential function is not *sub-half-exponential* (see [14] for more details). We hope that our proof method can be applied to prove results for the class of quasi-poly size $\mathrm{ACC} \circ \mathrm{THR} \circ (k\text{-}\mathrm{THR})^d$ circuits, extending results of [13]. At this moment, it is not clear that threshold gates are essential for our lower bounds. It would be an also interesting future work to extend our method to be applicable for circuit classes with more general boolean functions as gates.

References

1. Amano, K., Saito, A.: A satisfiability algorithm for some class of dense depth two threshold circuits, IEICE Trans. Inf. Sys., to appear
2. Beigel, R., Tarui, J.: On ACC. Computational Complexity **4**, 350–366 (1994)
3. Coppersmith, D.: Rapid multiplication of rectangular matrices. SIAM J. Comput. **11**(3), 467–471 (1982)
4. Furst, M., Saxe, J., Sipser, M.: Parity, circuits, and the polynomial time hierarchy. Mathematical Systems Theory **17**, 13–27 (1984)
5. Hajnal, A., Maass, W., Pudlak, P., Szegedy, M., Turan, G.: Threshold circuits of bounded depth. In: FOCS 1987, pp. 99–110 (1987)
6. Håstad, J.: Almost optimal lower bounds for small depth circuits. Advances in Computing Research **5**, 143–170 (1989)
7. Impagliazzo, R., Kabanets, V., Wigderson, A.: In search of an easy witness: exponential time versus probabilistic polynomial time. J. Comput. and Sys. Sci. **65**(4), 672–694 (2002)
8. Impagliazzo, R., Paturi, R., Saks, M.E.: Size-depth tradeoffs for threshold circuits. SIAM J. Comput. **26**(3), 693–707 (1997)
9. Iwama, K., Morizumi, H.: An explicit lower bound of $5n\text{-}o(n)$ for boolean circuits. In: Diks, K., Rytter, W. (eds.) MFCS 2002. LNCS, vol. 2420, pp. 353–364. Springer, Heidelberg (2002)
10. Razborov, A.: Lower bounds on the size of bounded depth networks over a complete basis with logical addition. Mathematical Notes of Academy of Sciences USSR **41**(4), 598–607 (1987)
11. Smolensky, R.: Algebraic methods in the theory of lower bounds for boolean circuit complexity. In: STOC 1987, pp. 77–82 (1987)
12. Williams, R.: Improving exhaustive search implies superpolynomial lower bounds. In: STOC 2010, pp. 231–240 (2010)
13. Williams, R.: New algorithms and lower bounds for circuits with linear threshold gates. In: STOC 2014, pp. 194–102 (2014)
14. Williams, R.: Non-uniform ACC circuit lower bounds. Journal of the ACM **61**(1), Article 22 (January 2014)

Finite Automata for the Sub- and Superword Closure of CFLs: Descriptional and Computational Complexity

Georg Bachmeier, Michael Luttenberger, and Maximilian Schlund[(✉)]

Technische Universität München, Boltzmannstrae 3,
85748 Garching Bei Mnchen, Germany
{bachmeie,luttenbe,schlund}@in.tum.de

Abstract. We answer two open questions by (Gruber, Holzer, Kutrib, 2009) on the state-complexity of representing sub- or superword closures of context-free grammars (CFGs): (1) We prove a (tight) upper bound of $2^{\mathcal{O}(n)}$ on the size of nondeterministic finite automata (NFAs) representing the subword closure of a CFG of size n. (2) We present a family of CFGs for which the minimal deterministic finite automata representing their subword closure matches the upper-bound of $2^{2^{\mathcal{O}(n)}}$ following from (1). Furthermore, we prove that the inequivalence problem for NFAs representing sub- or superword-closed languages is only NP-complete as opposed to PSPACE-complete for general NFAs. Finally, we extend our results into an approximation method to attack inequivalence problems for CFGs.

Keywords: Descriptional complexity · Subword closure · Nfa equivalence · Language approximation

1 Introduction

Given a (finite) word $w = w_1 w_2 \ldots w_n$ over some alphabet Σ, we say that u is a *(scattered) subword or subsequence* of w if u can be obtained from w by erasing some letters of w. We denote the fact that u is a subword of w by $u \preccurlyeq w$, and alternatively say that w is a *superword* of u. As shown by Higman [11] in 1952 $\preccurlyeq$ is a well-quasi-order on Σ^*, implying that *every* language $L \subseteq \Sigma^*$ has a finite set of $\preccurlyeq$-minimal elements. This proves that both the subword (also: downward) closure $\nabla L := \{u \in \Sigma^* \mid \exists w \in L \colon u \preccurlyeq w\}$ and the superword (also: upward) closure $\Delta L := \{w \in \Sigma^* \mid \exists u \in L \mid u \preccurlyeq w\}$ are regular for *any* language L. While in general, we cannot effectively construct a finite automaton accepting ∇L resp. ΔL, for specific classes of languages effective constructions are known.

It is well-known that this is the case when L is given as a context-free grammar (CFG). This was first shown by van Leeuwen [13] in 1978. Later, Courcelle

This work was partially funded by the DFG project "Polynomial Systems on Semirings: Foundations, Algorithms, Applications".

© Springer International Publishing Switzerland 2015
A.-H. Dediu et al. (Eds.): LATA 2015, LNCS 8977, pp. 473–485, 2015.
DOI: 10.1007/978-3-319-15579-1_37

gave an alternative proof of this result in [6]. Section 3 builds up on these results by Courcelle. We also mention that for Petri-net languages an effective construction is known thanks to Habermehl, Meyer, and Wimmel [10].

These results can be used to tackle undecidable questions regarding the ambiguity, inclusion, equivalence, universality or emptiness of languages by over-approximating one or both languages by suitable regular languages [8,10,14,15]: For instance, consider the scenario where we are given a procedural program whose runs can be described as a pushdown automaton resp. a CFG G_1 and a context-free specification G_2 of all safe executions, and we want to check whether all runs of the system conform to the safety specification $\mathcal{L}(G_1) \subseteq \mathcal{L}(G_2)$. As $\mathcal{L}(G_1) \cap \overline{\nabla\mathcal{L}(G_2)} \neq \emptyset \Rightarrow \mathcal{L}(G_1) \not\subseteq \mathcal{L}(G_2)$, we can obtain at least a partial answer to the otherwise undecidable question. Of course, in the case $\mathcal{L}(G_1) \subseteq \nabla\mathcal{L}(G_2)$ no information is gained, and one needs to refine the problem e.g. by using some sort of counter-example guided abstraction refinement as done e.g. in [14].

Contributions and Outline. Our first results (Sections 3 and 4) concern the blow-up incurred when constructing a (non-)deterministic finite automaton (NFA resp. DFA) for the subword closure of a language given by a context-free grammar G where we improve the results of [9]: For a CFG G of size n, [9] shows that an NFA recognizing $\nabla\mathcal{L}(G)$ has at most $2^{2^{\mathcal{O}(n)}}$ states, and there are CFGs requiring at least $2^{\Omega(n)}$ states. (For linear CFGs the upper and lower bounds are both single exponential.) The upper bound of [9] is established by analyzing the inductive construction of [13]. We improve this result in Section 3 to $2^{\mathcal{O}(n)}$ by slightly adapting Courcelle's construction [6] (we also briefly discuss that naively applying Courcelle's construction cannot do better than $2^{\Omega(n \log n)}$ in general). This result of course yields immediately an upper bound of $2^{2^{\mathcal{O}(n)}}$ on the size of minimal DFA representing $\nabla\mathcal{L}(G)$. In Section 4 we show this bound is tight already over a binary alphabet. To the best of our knowledge, so far only examples were known which showcase the single-exponential blow-up when constructing an NFA accepting the subword closure of a context-free grammar [9] resp. a DFA accepting the subword closure of a DFA or NFA [17]. We then study in Section 5 the equivalence problem for NFAs recognizing subword- resp. supword-closed languages. While for general NFAs this problem is PSPACE-complete, we show that it becomes coNP-complete under this restriction. We combine these results in Section 6 to derive a conceptual simple semi-decision procedure for checking language-inequivalence of two CFGs G_1, G_2: we first construct NFAs for $\nabla\mathcal{L}(G_1)$ and $\nabla\mathcal{L}(G_2)$, and check language-inequivalence of these NFAs; if the NFAs are inequivalent, we construct a witness of the language-inequivalence of G_1 and G_2; otherwise we refine the grammars, and repeat the test on the so obtained new grammars. This approach is motivated by the abstraction-refinement approach of [14] for checking if the intersection of two context-free languages is empty. We experimentally evaluate our approach by comparing it to *cfg-analyzer* of [2] which uses incremental SAT-solving to tackle the language-inequivalence problem. Missing proofs can be found in the extended version of the paper [3].

2   Preliminaries

By Σ we denote a finite alphabet. For every natural number n, let $\Sigma^{\leq n}$ denote the words of length at most n over Σ. The empty word is denoted by ε; the set of all finite words by Σ^*.

We measure the *size* $|G|$ of a CFG G as the total number of symbols on the right hand sides of all productions. The size of an NFA is simply measured as the number of states (this is an adequate measure for a constant alphabet, since the number of transitions is at most quadratic in the number of states).

Throughout the paper we will always assume that all CFGs are reduced, i.e. do not contain any unproductive or unreachable nonterminals (any CFG can be reduced in polynomial time). Let X be a nonterminal in a CFG G. We define $\mathcal{L}(X)$ as the set of all words $w \in \Sigma^*$ derivable from X. If S is the start symbol of G, then $\mathcal{L}(G) := \mathcal{L}(S)$. Moreover, $\Sigma_X \subseteq \Sigma$ denotes the set of all terminals reachable from X. Overloading notation we sometimes write ∇X for $\nabla \mathcal{L}(X)$.

The dependency graph of a CFG G is the finite graph with nodes the nonterminals of G where there is an edge from X to Y if there is a production $X \to \alpha Y \beta$ in G. We say that X *depends directly on* Y (written as $X \rhd Y$) if $X \neq Y$ and there is an edge from X to Y. The reflexive and transitive closure of $\rhd$ is denoted by $\rhd^*$. We write $X \equiv Y$ if $X \rhd^* Y \wedge Y \rhd^* X$, i.e. if X and Y are located in a common strongly-connected component of the dependency graph. We say that G is strongly connected if the dependency graph is strongly connected.

From [6] we recall some useful facts concerning the subword closure:

Lemma 1. *For any nonterminals X, Y, Z in a CFG G it holds that:*

1. $\nabla(\mathcal{L}(X) \cup \mathcal{L}(Y)) = \nabla \mathcal{L}(X) \cup \nabla \mathcal{L}(Y)$
2. $\nabla(\mathcal{L}(X) \cdot \mathcal{L}(Y)) = \nabla \mathcal{L}(X) \cdot \nabla \mathcal{L}(Y)$
3. $X \equiv Y \Rightarrow \nabla X = \nabla Y$
4. *If $X \to^* \alpha Y \beta Z \gamma$ for $Y, Z \equiv X$ then $\nabla X = \Sigma_X^*$*

3   Computing the Subword Closure of CFGs

In this section we describe an optimized version of the construction in [6] to compute an NFA for the subword closure of a CFG G of size $2^{\mathcal{O}(|G|)}$, which is asymptotically optimal. We first illustrate the construction by a simple example.

As explained at the end of the next section, a naive implementation of the construction of [6] leads to an automaton of size $2^{\Omega(n)} n! = 2^{\Omega(n \log n)}$ whereas our approach achieves the (optimal) bound of $2^{\mathcal{O}(n)}$.

3.1   Construction by Example

Consider the grammar G with start symbol S defined by the productions:

$$
\begin{aligned}
S &\rightarrow XaU \mid UaU \mid X & X &\rightarrow ZbY \mid \varepsilon \\
Y &\rightarrow XYa \mid b & U &\rightarrow VZ \mid acb \\
V &\rightarrow ZU \mid \varepsilon & Z &\rightarrow cZ \mid bc
\end{aligned}
$$

On the right-hand side, the dependency graph is shown where an edge $x \rightarrow y$ stands for $x \unrhd y$. To simplify the construction, we first transform the grammar G into a certain normal form G' (with $\nabla \mathcal{L}(G) = \nabla \mathcal{L}(G')$) and then construct an NFA from G'.

In the first step we compute the strongly connected components (SCCs) of G, here $\{X, Y\}$ and $\{U, V\}$. Since $Y \rightarrow XYa$ (with $Y \equiv X$ and $X \equiv X$), we know that $\nabla Y = \nabla X = \Sigma_X^* = \{a, b, c\}^*$. We therefore can replace any occurrence of Y by X (thereby removing Y from the grammar) and redefine the rules for X to $X \rightarrow aX \mid bX \mid cX \mid \varepsilon$. In case of the SCC $\{U, V\}$ the grammar is linear w.r.t. U and V, i.e. starting from either of the two we can never produce sentential forms in which the total number of occurrences of U and V exceeds one. Hence, we can identify U and V without changing the subword closure. Finally, we introduce unique nonterminals for each terminal symbol and restrict the right-hand side of each production to at most two symbols by introducing auxiliary nonterminals W and T:

$$
\begin{aligned}
S &\rightarrow XW \mid UW \mid X & W &\rightarrow A_a U \\
X &\rightarrow A_a X \mid A_b X \mid A_c X \mid A_\varepsilon & U &\rightarrow UZ \mid ZU \mid A_a T \mid A_\varepsilon \\
T &\rightarrow A_c A_b & Z &\rightarrow A_c Z \mid A_b A_c \\
A_a &\rightarrow a & A_b &\rightarrow b \\
A_c &\rightarrow c & A_\varepsilon &\rightarrow \varepsilon
\end{aligned}
$$

Note that the dependency graph of this transformed grammar is now acyclic apart from self-loops. Because of this, we can directly transform the grammar into an *acyclic* equation system (or straight-line program, or algebraic circuit) whose solution is a regular expression for ∇S:

$$
\begin{aligned}
\nabla A_a &= (a + \varepsilon) & \nabla A_b &= (b + \varepsilon) \\
\nabla A_c &= (c + \varepsilon) & \nabla A_\varepsilon &= \varepsilon \\
\nabla Z &= c^*(\nabla A_b \nabla A_c) & \nabla T &= \nabla A_c \nabla A_b \\
\nabla U &= \Sigma_Z^*(\nabla A_a \nabla T)\Sigma_Z^* & \nabla W &= \nabla A_a \nabla U \\
\nabla X &= \Sigma_X^* & \nabla S &= \nabla X \nabla W + \nabla U \nabla W + \nabla X
\end{aligned}
$$

In order to obtain an NFA for ∇S, we evaluate this equation system from bottom to top while re-using as many of the already constructed automata as possible. For instance, consider the equation: $\nabla S = \nabla X \nabla W + \nabla U \nabla W + \varepsilon \cdot \nabla X$. Because of acyclicity of the equation system, we may assume inductively that we have already constructed NFAs $A_{\nabla X}$, $A_{\nabla W}$, and $A_{\nabla U}$ for ∇X, ∇W, and ∇U, respectively. To construct the NFA for ∇S, we first make two copies $A^{(1)}$, $A^{(2)}$ of each of these automata. Automata with superscript (1) will be used exclusively for variable occurrences to the left of the concatenation operator, while automata with superscript (2) will be used for the remaining occurrences. We then read quadratic monomials, like $\nabla X \nabla W$, as an ε-transition connecting $A^{(1)}_{\nabla X}$ with $A^{(2)}_{\nabla W}$ as shown in Figure 1 where all edges represent ε-transitions.

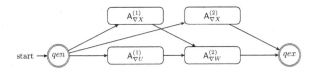

Fig. 1. Efficient re-use of re-occurring NFAs in Courcelle's construction

We do not claim that this construction yields the smallest NFA, but it is easy to describe and yields an NFA of sufficiently small size in order to deduce in the following subsections an asymptotically tight upper bound on the number of states. We recall that using a CFG of size $3n + 2$ to succinctly represent the singleton language $\{a^{2^n}\}$, the bound of $2^{\Theta(n)}$ follows [9].

In [1] it is remarked that a straight-forward implementation of Courcelle's construction yields an NFA "single exponential" size w.r.t. $|G|$. However, no detailed complexity analysis is given. Consider the CFG with start-symbol A_n and consisting of the rules $A_0 \to a$ and for all $1 \le k \le n$: $\quad A_k \to A_i A_j \quad \forall 0 \le i, j \le (k-1)$. If we compute an NFA for ∇A_n via the straight-forward bottom-up construction it will have size $a_n := |A_{\nabla A_n}|$ with $a_n = 2 + \sum_{0 \le i,j \le (n-1)} (a_i + a_j)$. It is easy to show that $a_n \ge 2^n n! \in 2^{\Omega(n \log n)}$. Hence, the crucial part to achieve the optimal bound of $2^{\mathcal{O}(n)}$ is to reuse already computed automata. We just remark that one can also achieve similar savings by factoring out common terms in the right hand side of the acyclic equations. A subsequent bottom-up construction leads to an NFA of size $2^{\mathcal{O}(n)}$ as well but the constant hidden in the $\mathcal{O}$ is larger and the analysis is more involved. Note that this also shows that we can construct a regular expression of size $2^{\mathcal{O}(n)}$ representing the subword closure.

3.2 Normal Form for Computing the Subword Closure

To simplify our construction, we will assume that our grammar has a special form which is similar to CNF but with unary rules allowed. Any CFG can be transformed into this form with at most linear blowup in size preserving its subword closure (but not its language).

Definition 2. *A CFG G is in quadratic normal form (QNF) if for every terminal $x \in \Sigma \cup \{\varepsilon\}$ there is a unique nonterminal A_x with the only production $A_x \to x$ and every other production is in one of the following forms:*

- $X \to YX$ or $X \to XY$ (with $Y \neq X$)
- $X \to Y$ or $X \to YZ$ (with $Y, Z \neq X$)

A grammar in QNF is called simple if

- for all $X \to YX$ or $X \to XY$, we have $X \rhd Y$
- for all $X \to Y$ or $X \to YZ$, we have $X \rhd Y, Z$.

Note that the dependency graph associated with a grammar in simple QNF is acyclic with the exception of self-loops.

First, we need a small lemma that allows us to eliminate all linear productions "within" some SCC, i.e. productions of the form $X \to \alpha Y \beta$ such that $X \neq Y$ but $Y \unrhd^* X$.

Lemma 3. *Let G be a strongly connected linear CFG with nonterminals $\mathcal{X} = \{X_1, \ldots, X_n\}$ so that every production is either of the form $X \to \alpha Y \beta$ or $X \to \alpha$ for $\alpha, \beta \in \Sigma^*$. Consider the grammar G' which we obtain from G by replacing in every production of G every occurrence of a nonterminal X_i by Z. We then have that $\nabla\mathcal{L}(Z) = \nabla\mathcal{L}(X_i)$ for all $i \in [n]$.*

Using the preceding lemma, we can show that it suffices to consider only CFG in simple QNF in the following.

Theorem 4. *Every CFG G can be transformed into a CFG G' in simple QNF such that $\nabla\mathcal{L}(G) = \nabla\mathcal{L}(G')$ and $|G'| \in \mathcal{O}(|G|)$.*

Proof (sketch). First, we use Lemma 1 to simplify all productions involving an X with $X \Rightarrow^* \alpha X \beta X \gamma$. Then we apply Lemma 3 to contract SCCs to a single non-terminal. Finally, we introduce auxiliary variables for the terminals and we binarize the grammar (keeping unary rules like [12]).

Theorem 5. *For any CFG G in simple QNF with n nonterminals there is an NFA A with at most $2 \cdot 3^{n-1}$ states which recognizes the subword closure of G, i.e. $\nabla\mathcal{L}(G) = \mathcal{L}(\mathsf{A})$.*

Proof (sketch). Since the dependency graph of a grammar in simple QNF is a DAG (if we ignore self-loops), we can order the nonterminals according to a topological ordering of this graph. We proceed bottom-up to inductively build an NFA for $\nabla\mathcal{L}(G) = \nabla S$ as in section 3.1. Since our grammar is in QNF, at each stage we only have to produce at most two copies of every automaton representing the subword-closure of a "lower" nonterminal Y. Inductively, for each of these Y we can build an NFA with at most $2 \cdot 3^i$ many states where i is Y's position in the topological ordering. Using the "bipartite wiring" sketched in Figure 1 the size of the automaton for X can then be estimated as

$$|\mathsf{A}_S| \leq 2 + \sum_{Y:\, S \rhd Y} 2 \cdot |\mathsf{A}_Y| \leq 2 + 4 \cdot \sum_{i=0}^{n-2} 3^i = 2 \cdot 3^{n-1}.$$

Corollary 6. *For every CFG G of size n there is an NFA A of size $2^{\mathcal{O}(n)}$ and a DFA D of size $2^{2^{\mathcal{O}(n)}}$ with $\nabla\mathcal{L}(G) = \mathcal{L}(\mathsf{A}) = \mathcal{L}(D)$.*

4   CFG $\rightarrow$ DFA: Double-Exponential Blowup

As seen in the preceding section, moving from a context-free grammar G representing a subword-closed language to a language-equivalent NFA $\mathcal{A}$, the size of the automaton is bounded from above by $2^{O(|G|)}$. For superword closures [9] prove the same upper bound for the size of the NFA. From both results we immediately obtain the upper bound $2^{2^{O(|G|)}}$ on the size of the minimal language-equivalent DFA recognizing the sub- or superword closure of a CFG G. This bound is essentially tight as witnessed by the family of finite languages

$$L_k = \bigcup_{j=1}^{k} \{0,1\}^{j-1}\{0\}\{0,1\}^k\{0\}\{0,1\}^{k-j}.$$

L_k contains exactly all those words $w \in \{0,1\}^{2k+1}$ which contain two 0s which are separated by exactly k letters. Using the idea of iterated squaring in order to succinctly encode the language $\{a^{2^n}\}$ as a context-free grammar (resp. straight-line program) of size $\mathcal{O}(n)$, the language L_{2^n} can be represented by a context-free grammar of size $\mathcal{O}(n)$ as well. One then easily shows that the Myhill-Nerode relation w.r.t. L_{2^n}, ∇L_{2^n}, and ΔL_{2^n}, respectively, has at least 2^{2^n} equivalence classes:

Theorem 7. *There exists a family of CFGs G_n of size $\mathcal{O}(n)$ (generating finite languages) such that the minimal DFAs accepting either $L(G_n)$, or $\nabla L(G_n)$, or $\Delta L(G_n)$, have at least 2^{2^n} states.*

5   Equivalence of NFAs Modulo Sub-/Superword Closure

As hinted at in the introduction, one application of the sub- resp. superword closure is (in-)equivalence checking of CFGs by regular over-approximation. For this, we must solve the equivalence problems for NFAs representing sub/sup-word closed languages. Naturally, the question arises how hard this is.

Let A and B denote NFAs over the common alphabet Σ, having n_A and n_B many states, respectively. Recall that the universality problem for NFAs, i.e. $\mathcal{L}(A) \stackrel{?}{=} \Sigma^*$, and hence also the equivalence problem $\mathcal{L}(A) \stackrel{?}{=} \mathcal{L}(B)$ are PSPACE-complete. Only recently, it was shown in [18] that these problems *stay* PSPACE-complete even when restricted to NFAs representing languages which are closed w.r.t. either prefixes or suffixes or factors. However, in [18] it was also shown that for subword-closed NFAs (i.e. $\nabla \mathcal{L}(A) = \mathcal{L}(A)$), universality is decidable in linear time as $\mathcal{L}(A) = \Sigma^*$ holds if and only if there is an SCC in A whose labels cover all of Σ. It is easily shown that a similar result also holds for superword-closed NFAs (i.e. $\Delta \mathcal{L}(A) = \mathcal{L}(A)$): We have $\mathcal{L}(A) = \Sigma^*$ if and only if $\varepsilon \in \mathcal{L}(A)$.

In this section we show that both equivalence problems, i.e. $\nabla \mathcal{L}(A) \stackrel{?}{=} \nabla \mathcal{L}(B)$ and $\Delta \mathcal{L}(A) \stackrel{?}{=} \Delta \mathcal{L}(B)$, are coNP-complete, hence are easier than in the general case (unless NP = PSPACE). In the following, we write more succinctly $A \stackrel{?}{\equiv}_\nabla B$ and $A \stackrel{?}{\equiv}_\Delta B$ for these two problems. The following lemma is easy to prove:

Lemma 8. *Let* A *be an NFA. Define* A^∇ *as the NFA we obtain from* A *by adding for every transition* $q \xrightarrow{a} q'$ *of* A *the* ε*-transition* $q \xrightarrow{\varepsilon} q'$*. Similarly, define* $A^\triangle$ *to be the NFA we obtain by adding the loops* $q \xrightarrow{a} q$ *for every state* q *and every terminal* $a \in \Sigma$ *to* A*. Then* $\nabla \mathcal{L}(A) = \mathcal{L}(A^\nabla)$ *and* $\triangle \mathcal{L}(A) = \mathcal{L}(A^\triangle)$.

To prove that both $A \overset{?}{\equiv}_\triangle B$ and $A \overset{?}{\equiv}_\nabla B$ are coNP-complete we will give a polynomial bound on the length of a *separating word*, i.e. a word w in the symmetric difference of $\mathcal{L}(A^\nabla)$ and $\mathcal{L}(B^\nabla)$ resp. of $\mathcal{L}(A^\triangle)$ and $\mathcal{L}(B^\triangle)$.

We first show that the DFA obtained from A^∇ resp. $A^\triangle$ using the powerset construction has a particular simple structure (this was also observed in [17]).

Lemma 9. *Let* A *be an NFA. Let* D_A^∇ *(resp.* $D_A^\triangle$*) be the DFA we obtain from* A^∇ *(resp.* $A^\triangle$*) by means of the powerset construction. For any transition* $S \xrightarrow{a} T$ *of* D_A^∇ *(*$D_A^\triangle$*) it holds that* $S \supseteq T$ *(resp.* $S \subseteq T$*).*

Thus, the transition relation of D_A^∇ (disregarding self-loops) can be "embedded" into the lattice of subsets of the states of A, which has height $n_A - 1$.

Corollary 10. *With the assumptions of the preceding lemma: The length of the longest simple path in* D_A^∇ *(resp.* $D_A^\triangle$*) is at most* $n_A - 1$.

It now immediately follows that a shortest separating word for sub- resp. supword closed NFAs – if one exists – has at most length linear in the size of the two NFAs.

Lemma 11. *Let* A *and* B *be two NFAs. If* $A \not\equiv_\nabla B$ *(resp.* $A \not\equiv_\triangle B$*), then there exists a separating word of length at most* $n_A + n_B - 2$.

Theorem 12. *The decision problems* $A \overset{?}{\equiv}_\nabla B$ *and* $A \overset{?}{\equiv}_\triangle B$ *are in* coNP.

To show coNP-hardness, recall the proof that the equivalence problem for starfree regular expressions is coNP-hard by reduction from TAUT: Given a formula ϕ in propositional calculus, we build a regular expression ρ (without Kleene stars) over $\Sigma = \{0, 1\}$ that enumerates exactly the satisfying assignments of ϕ. Hence, $\phi \in$ TAUT iff $\mathcal{L}(\rho) = \Sigma^n$ iff $\nabla \mathcal{L}(\rho) = \Sigma^{\leq n}$, since the subword closure can only add new words of length less than n (analogously for $\triangle$).

Theorem 13. *The decision problems* $A \overset{?}{\equiv}_\nabla B$ *and* $A \overset{?}{\equiv}_\triangle B$ *are* coNP-hard.

6   Application to Grammar Problems

We apply our results to devise an approximation approach for the well-known undecidable problem whether $\mathcal{L}(G_1) = \mathcal{L}(G_2)$ for two CFGs G_1, G_2. Possible attacks on this problem include exhaustive search for a word in the symmetric difference $w \in (L_1 \oplus L_2) \cap \Sigma^{\leq n}$ w.r.t. some increasing bound n e.g. by using incremental SAT-solving [2]. Unfortunately, this quickly becomes infeasible for large problems. Previous work has successfully applied regular approximation for ambiguity detection [5,19] or intersection non-emptiness of CFGs [14].

1. Compute NFAs A_1 and A_2 for the subword closures of G_1 and G_2, respectively.
2. Check, if $\mathcal{L}(A_1) = \mathcal{L}(A_2)$.
   (a) Case "Not equal": Generate a witness $w \in \mathcal{L}(G_1) \oplus \mathcal{L}(G_2)$.
   (b) Case "Equal": Refine the grammars and restart at **1**.

Fig. 2. Equivalence checking via subword closure approximation

A high-level description of our approach to (in-)equivalence-checking is given in Figure 2.

Of course the procedure will not terminate if $\mathcal{L}(G_1) = \mathcal{L}(G_2)$, so in practice a timeout will be used after which the algorithm will terminate itself and output "Maybe equal". Steps (1) and (2) might take time (at most) double exponential in the size of the grammars G_1 and G_2: Recall that the construction of Section 3 yields in the worst-case an NFA A_i whose number of states is exponential in the size of the given CFG G_i. To check if $\nabla \mathcal{L}(G_1) = \nabla \mathcal{L}(G_2)$, an on-the-fly construction of the power-set automaton for $A_1 \times A_2$ can be used which terminates as soon as a set of states is reached which contains at least one accepting state of, say, A_1 but no accepting state of A_2. Using Lemma 11, we can safely terminate the exploration of simple paths if their length exceeds the bound stated in Lemma 11. In the worst case this might take time exponential in the size of A_1 and A_2, so at most double exponential in the size of G_1 and G_2.

In the following, we describe in greater detail how we generate a separating word w' in $\mathcal{L}(G_1)$ or $\mathcal{L}(G_2)$ if we find a separating word $w \in \nabla \mathcal{L}(G_1) \oplus \nabla \mathcal{L}(G_2)$, resp. how we refine G_1 and G_2 if $\nabla \mathcal{L}(G_1) = \nabla \mathcal{L}(G_2)$.

6.1 Witness Generation for $\mathcal{L}(G_1) \neq \mathcal{L}(G_2)$

If our check in step (2) returns "Not equal" we know that $\nabla \mathcal{L}(G_1) \neq \nabla \mathcal{L}(G_2)$ and we obtain a word $w \in \nabla \mathcal{L}(G_1) \oplus \nabla \mathcal{L}(G_2)$, w.l.o.g. assume in the following $w \in \nabla \mathcal{L}(G_1) \setminus \nabla \mathcal{L}(G_2)$. This word has length linear in $|A_1|$ and $|A_2|$, i.e. at most exponential w.r.t. $|G_1|$ and $|G_2|$.

To obtain a (direct) certificate for the fact that $\mathcal{L}(G_1) \neq \mathcal{L}(G_2)$, we construct a superword $w' \succcurlyeq w$ with $w' \in \mathcal{L}(G_1)$ – such a w' is guaranteed to exist as it is the reason for $w \in \nabla \mathcal{L}(G_1)$. Straight-forward induction on w shows:

Lemma 14. *For $w \in \Sigma^*$ a DFA recognizing $\nabla \mathcal{L}(\{w\})$ resp. $\Delta \mathcal{L}(\{w\})$ and having at most $|w| + 2$ states can be constructed in time polynomial in $|w|$.*

We can therefore intersect G_1 with a DFA accepting $\Delta \mathcal{L}(\{w\})$, to obtain a new CFG G_1' whose size is at most cubic in $|w|$[4,16], i.e. exponential in the size of G_1. From this grammar, we can obtain in time linear in $|G_1'|$ a shortest word w' in $\mathcal{L}(G_1') = \mathcal{L}(G_1) \cap \Delta \mathcal{L}(\{w\})$. The length of w' is at most exponential in $|G_1'|$, i.e. at most double exponential in $|G_1|$.

In practice, shorter witnesses are preferable, so we construct the shortest word in $\overline{\mathcal{L}(A_2)} \cap \mathcal{L}(G_1)$. In theory this might incur a triple exponential blow-up resulting from complementing A_2, but this way we can find a separating word w' which is *not* a superword of w and hence is usually shorter.

6.2   Refinement

In case that the test in step (2) returns "Equal", we refine both grammars such that subsequent subword-approximations may find a counterexample to equality. Assume that our equivalence check yields $\nabla\mathcal{L}(G_1) = \nabla\mathcal{L}(G_2)$. A possible refinement strategy is to cover $L := \nabla\mathcal{L}(G_1)$ using a finite number of regular languages $L \subseteq L' := L_0 \cup L_1 \cup \cdots \cup L_k$ and then to repeat the equivalence check for all pairs of refined languages $\mathcal{L}(G_1) \cap L_i$ and $\mathcal{L}(G_2) \cap L_i$ for all i. The requirement $L' \supseteq L$ protects the refinement from cutting off potential witnesses.

A simple method is covering using prefixes: Here we generate all prefixes $p_1, \ldots, p_k$ of words in L of increasing length (up to some small bound d called the *refinement depth*) and set $L_i := p_i \Sigma^*$ and $L_0 = \nabla\{p_i \mid i \in [k]\}$. Since $\bigcup_i L_i \supseteq L$ this strategy preserves potential witnesses and since any counterexample eventually appears as a prefix, this yields a semi-decision procedure for grammar inequivalence. In our experiments we disregard the finite language L_0 (which can also be checked by enumeration) and only check refinement using the infinite sets $p_i \Sigma^*$ with the goal of quickly finding *some* (not the shortest) distinguishing word. This strategy is often able to tell apart different CFLs after few iterations as shown in the following.

6.3   Implementation and Experiments

We implemented the inequivalence check in an extension[1] of the FPSOLVE tool [7]. The additional code comprises roughly 1800 lines of C++ and uses libfa[2] to handle finite automata.

Our worst-case descriptional complexity results for the subword closure of CFGs (exponential sized NFA, double-exponential sized DFA) and our remarks on the length of possible counterexamples might suggest that our inequivalence checking procedure is merely of academic interest. Here we briefly show that this is not the case, and that overapproximation via subword closures is actually quite fast in practice.

The paper [2] presents cfg-analyzer, a tool that uses SAT-solving to attack several undecidable grammar problems by exhaustive enumeration. We demonstrate the feasibility of our approximation approach on several slightly altered grammars (cf. [20]) for the PASCAL programming language[3]. The altered grammars were obtained by adding, deleting, or mutating a single rule from the original grammar [20]. We used FPSOLVE and cfg-analyzer to check equivalence of the altered grammar with the original. Both tools were given a timeout of 30 seconds. We want to stress that we do not strive to replace enumeration-based tools like cfg-analyzer, but rather envision a combined approach: Use overapproximations like the subword closure (with small refinement depth) as a quick check and resort to more computationally demanding techniques like

---

[1] The fork is available from https://github.com/regularApproximation/newton

[2] http://augeas.net/libfa/

[3] Available from https://github.com/nvasudevan/experiment/tree/master/grammars/mutlang/acc

SAT-solving for a thorough test. Also note that it is not too hard to find examples where enumeration-based tools cannot detect inequivalence anymore, e.g. by considering grammars with large alphabet (like C# or Java) for which the shortest word in the language is already longer than 20 tokens. Here we just showcase an example where both approaches can be fruitfully combined.

Table 1 demonstrates that even if our tool uses the very simple prefix-refinement (which is the main bottleneck in terms of speed), we can successfully solve 100 cases where cfg-analyzer has to give up after 30 seconds and even in cases where both tools find a difference, FPSOLVE does so much faster

Table 1. Numbers of solved instances for different scenarios and respective average times: #CA: solved by cfg-analyzer, #FP: solved by FPSOLVE, $\#(CA \wedge FP)$: solved by both tools, $t_{\text{tool}}^{\wedge}$: time needed by *tool* on instances from $(CA \wedge FP)$

scenario	# instances	# CA	t_{CA}	#FP	t_{FP}	$\#(CF \wedge FP)$	$t_{\text{CA}}^{\wedge}$	$t_{\text{FP}}^{\wedge}$
add	700	190	17.9	18	2.43	8	10.7	4.97
delete	284	61	17.8	34	0.424	10	14.4	0.464
empty	69	32	18.7	1	1.35	1	5.62	1.35
mutate	700	167	19.1	100	1.3	36	15.8	2.87
switchadj	187	16	20.5	2	5.46	1	9.68	0.34
switchany	328	35	18	9	3.72	8	9.09	2.84
$\sum$	2268	501	–	164	–	64	–	–

7 Discussion and Future Work

Motivated by the language-equivalence problem for context-free languages, we have studied the problems of the space requirements of representing the subword closure of CFGs by NFAs and DFAs, and the computational complexity of the equivalence problem of subword-closed NFAs. We have shown how to construct from a context-free grammar G an NFA accepting $\nabla\mathcal{L}(G)$ consisting of at most $2^{\mathcal{O}(|G|)}$ states – a small gap between the lower bound of $\Omega(2^{|G|})$ and our upper bound of $\mathcal{O}(3^{|G|})$ for grammars in QNF remains for future work. A further question is if this bound can be improved in the case of languages given as deterministic pushdown automata. We have further shown that the upper bound on the size of a DFA accepting $\nabla\mathcal{L}(G)$ of $2^{2^{\mathcal{O}(|G|)}}$ is tight. Interestingly, a binary alphabet suffices for the presented language family L_k: for instance the worst-case example of [17], which showcases the exponential blow-up suffered when constructing an DFA for the subword closure of a language given as DFA or NFA, requires an unbounded alphabet. We note that a unary context-free language cannot lead to this double exponential blow-up – this follows from the proof of Theorem 3.14 in [9] (see also Lemma 14 here). Regarding the language-equivalence problem, we have shown that it becomes coNP-complete when restricted to sub- resp. superword-closed NFAs. This is somewhat surprising given the fact that it stays PSPACE-complete for many related families (e.g. for prefix-, suffix-, or factor-closed languages). Finally, we have briefly described an approach to tackle the equivalence problem for CFGs using the presented results, though much work remains to turn our current implementation into a mature tool: In particular, since the intersection of two regular overapproximations is again a regular overapproximation, it could be fruitful to combine the subword closure (or variants like [14]) with other regular approximation techniques like [15]. We also need to improve the refinement of the approximations when scaling the problem size.

References

1. Atig, M.F., Bouajjani, A., Touili, T.: On the Reachability Analysis of Acyclic Networks of Pushdown Systems. In: van Breugel, F., Chechik, M. (eds.) CONCUR 2008. LNCS, vol. 5201, pp. 356–371. Springer, Heidelberg (2008)
2. Axelsson, R., Heljanko, K., Lange, M.: Analyzing Context-Free Grammars Using an Incremental SAT Solver. In: Aceto, L., Damgård, I., Goldberg, L.A., Halldórsson, M.M., Ingólfsdóttir, A., Walukiewicz, I. (eds.) ICALP 2008, Part II. LNCS, vol. 5126, pp. 410–422. Springer, Heidelberg (2008)
3. Bachmeier, G., Luttenberger, M., Schlund, M.: Finite Automata for the Sub- and Superword Closure of CFLs: Descriptional and Computational Complexity. CoRR abs/1410.2737 (2014). http://arxiv.org/abs/1410.2737
4. Bar-Hillel, Y., Perles, M., Shamir, E.: On Formal Properties of Simple Phrase Structure Grammars. Zeitschrift für Phonetik, Sprachwissenschaft und Kommunikationsforschung 14, 143–172 (1961)
5. Brabrand, C., Giegerich, R., Møller, A.: Analyzing Ambiguity of Context-Free Grammars. Sci. Comput. Program. 75(3), 176–191 (2010)
6. Courcelle, B.: On Constructing Obstruction Sets of Words. Bulletin of the EATCS 44, 178–186 (1991)
7. Esparza, J., Luttenberger, M., Schlund, M.: FPSOLVE: A Generic Solver for Fixpoint Equations over Semirings. In: Holzer, M., Kutrib, M. (eds.) CIAA 2014. LNCS, vol. 8587, pp. 1–15. Springer, Heidelberg (2014)
8. Ganty, P., Majumdar, R., Monmege, B.: Bounded underapproximations. Formal Methods in System Design 40(2), 206–231 (2012)
9. Gruber, H., Holzer, M., Kutrib, M.: More on the Size of Higman-Haines Sets: Effective Constructions. Fundam. Inf. 91(1), 105–121 (2009)
10. Habermehl, P., Meyer, R., Wimmel, H.: The Downward-Closure of Petri Net Languages. In: Abramsky, S., Gavoille, C., Kirchner, C., Meyer auf der Heide, F., Spirakis, P.G. (eds.) ICALP 2010. LNCS, vol. 6199, pp. 466–477. Springer, Heidelberg (2010)
11. Higman, G.: Ordering by Divisibility in Abstract Algebras. Proc. London Math. Soc. s3-2(1), 326–336 (Jan 1952)
12. Lange, M., Leiß, H.: To CNF or not to CNF? An Efficient Yet Presentable Version of the CYK Algorithm. Informatica Didactica 8 (2009)
13. van Leeuwen, J.: Effective constructions in well-partially-ordered free monoids. Discrete Mathematics 21(3), 237–252 (1978)
14. Long, Z., Calin, G., Majumdar, R., Meyer, R.: Language-Theoretic Abstraction Refinement. In: de Lara, J., Zisman, A. (eds.) Fundamental Approaches to Software Engineering. LNCS, vol. 7212, pp. 362–376. Springer, Heidelberg (2012)
15. Mohri, M., Nederhof, M.J.: Regular Approximation of Context-Free Grammars through Transformation. In: Junqua, J.C., van Noord, G. (eds.) Robustness in Language and Speech Technology. Text, Speech and Language Technology, vol. 17, pp. 153–163. Springer, Netherlands (2001)

16. Nederhof, M., Satta, G.: New Developments in Formal Languages and Applications. Studies in Computational Intelligence, pp. 229–258. Springer, Heidelberg (2008)
17. Okhotin, A.: On the State Complexity of Scattered Substrings and Superstrings. Fundam. Inform. **99**(3), 325–338 (2010)
18. Rampersad, N., Shallit, J., Xu, Z.: The Computational Complexity of Universality Problems for Prefixes, Suffixes, Factors, and Subwords of Regular Languages. Fundam. Inform. **116**(1–4), 223–236 (2012)
19. Schmitz, S.: Conservative Ambiguity Detection in Context-Free Grammars. In: Arge, L., Cachin, C., Jurdziński, T., Tarlecki, A. (eds.) ICALP 2007. LNCS, vol. 4596, pp. 692–703. Springer, Heidelberg (2007)
20. Vasudevan, N., Tratt, L.: Detecting Ambiguity in Programming Language Grammars. In: Erwig, M., Paige, R.F., Van Wyk, E. (eds.) SLE 2013. LNCS, vol. 8225, pp. 157–176. Springer, Heidelberg (2013)

A Game Characterisation of Tree-like
Q-resolution Size

Olaf Beyersdorff[1]([✉]), Leroy Chew[1], and Karteek Sreenivasaiah[2]

[1] School of Computing, University of Leeds, Leeds, UK
o.beyersdorff@leeds.ac.uk
[2] The Institute of Mathematical Sciences, Chennai, India

Abstract. We provide a characterisation for the size of proofs in tree-like Q-Resolution by a Prover-Delayer game, which is inspired by a similar characterisation for the proof size in classical tree-like Resolution [10]. This gives the first successful transfer of one of the lower bound techniques for classical proof systems to QBF proof systems. We confirm our technique with two previously known hard examples. In particular, we give a proof of the hardness of the formulas of Kleine Büning et al. [20] for tree-like Q-Resolution.

1 Introduction

Proof complexity is a well established field that has rich connections to fundamental problems in computational complexity and logic [14,21]. In addition to these foundational contributions, proof complexity provides the main theoretical approach towards an understanding of the performance of SAT solvers, which have gained a wide range of applications for the efficient solution of practical instances of NP-hard problems. As most modern SAT solvers employ CDCL-based methods, they correspond to Resolution. Lower bounds to the size and space of Resolution proofs therefore imply sharp bounds for running time and memory consumption of SAT algorithms. Consequently, Resolution has received key attention in proof complexity; and many ingenious techniques have been devised to understand the complexity of Resolution proofs (cf. [13,26] for surveys).

There has been growing interest and research activity to extend the success of SAT solvers to the more expressive *quantified boolean formulas (QBF)*. Due to its PSPACE completeness, QBF is far more expressive than SAT and thus applies to further fields such as formal verification or planning [6,25]. As for SAT solvers, runs of QBF solvers produce witnesses (respectively proofs) of unsatisfiability, and there has been great interest in trying to understand which formal system would correspond to the solvers. In particular, a number of

---

This work was supported by the EU Marie Curie IRSES grant CORCON, grant no. 48138 from the John Templeton Foundation, EPSRC grant EP/L024233/1, and a Doctoral Training Grant from EPSRC (2nd author).

© Springer International Publishing Switzerland 2015
A.-H. Dediu et al. (Eds.): LATA 2015, LNCS 8977, pp. 486–498, 2015.
DOI: 10.1007/978-3-319-15579-1_38

Resolution-based proof systems have been developed for QBF, most notably Q-Resolution, introduced by Kleine Büning et al. [20], long-distance Q-Resolution [2], QU-Resolution [27], and ∀Exp+Res [19]. Designing two further calculi IR-calc and IRM-calc, a unifying framework for most of these systems has recently been suggested in [7].

Understanding the sizes of proofs in these systems is important as lower bounds for proof size directly translate into lower bounds for running time of the corresponding QBF-solvers. However, in contrast to classical proof complexity we do not yet have many established methods that could be employed for this task. Very recently, the paper [8] introduces a general proof technique for QBF systems based on strategy extraction, that transfers circuit lower bounds to proof size lower bounds. However, no technique for classical Resolution is known to be effective for QBF systems. Except for recent results shown by the new strategy extraction method [8] all present lower bounds for QBF proof systems are either shown ad hoc (e.g. [18] or the lower bound for KBKF(t) in [8]) or are obtained by directly lifting known classical lower bounds to QBF (e.g. [15]).

Our contribution in this paper is to transfer one of the main game methods from classical proof complexity to QBF. Game techniques have a long tradition in proof complexity, as they provide intuitive and simplified methods for lower bounds in Resolution, e.g. for Haken's exponential bound for the pigeonhole principle in dag-like Resolution [23], or the optimal bound in tree-like Resolution [9] , and even work for strong systems [4] and other measures such as proof space [17] and width [1]. A unified game approach to hardness measures was recently established in [12]. Building on the classic game of Pudlák and Impagliazzo [24] for tree-like Resolution, the papers [9,11] devise an asymmetric Prover-Delayer game, which was shown in [10] to even characterise tree-like Resolution size. Thus, in contrast to the classic symmetric Prover-Delayer game of [24], the asymmetric game in principle allows to always obtain the optimal lower bounds, which was demonstrated in [9] for the pigeonhole principle.

Inspired by these games, we develop here a Prover-Delayer game which tightly characterises the proof size in tree-like Q-Resolution. The idea behind this game is that a Delayer claims to know a satisfying assignment to a false formula, while a Prover asks for values of variables until eventually finding a contradiction. In the course of the game the Delayer scores points proportional to the progress the Prover makes towards reaching a contradiction. By an information-theoretic argument we show that the optimal Delayer will score exactly logarithmically many points in the size of the smallest tree-like Q-Resolution proof of the formula. Thus exhibiting clever Delayer strategies gives lower bounds to the proof size, and in principle these bounds are guaranteed to be optimal. In comparison to the game of [9–11], our formulation here needs a somewhat more powerful Prover, who can forget information as well as freely set universal variables. This is necessary as the Prover needs to simulate more complex Q-Resolution proofs involving universal variables and ∀-reductions.

We illustrate this new technique with two examples. The first was used by Janota and Marques-Silva [18] to separate Q-Resolution from the system

∀Exp+Res defined in [19]. We use these separating formulas as an easy first illustration of our technique. Our Delayer strategy as well as the analysis here are quite straightforward; in fact, a simple symmetric game in the spirit of [24] would suffice to get the lower bound. Our second example are the well-known KBKF(t)-formulas of Kleine Büning, Karpinski and Flögel [20]. In the same work [20], where Q-Resolution was introduced, these formulas were suggested as hard formulas for the system. Very recently, the formulas KBKF(t) were even shown to be hard for IR-calc, a system stronger than Q-Resolution [8]. In fact, a number of further separations of QBF proof systems builds on the hardness of KBKF(t) [3,16] (cf. also [8]). Here we use our new technique to show that these formulas require exponential-size proofs in tree-like Q-Resolution. In terms of the lower bound, this result is weaker than the result obtained in [8]. However, it provides an interesting example for our new game technique. In contrast to the first example, both the Delayer strategy as well as the scoring analysis is technically involved. Here we need the refined asymmetric game. The formulas KBKF(t) have very unbalanced proofs, so we cannot use a symmetric Delayer, as symmetric games only yield a lower bound according to the largest full binary tree embeddable into the proof tree (cf. [10]).

The remaining part of this paper is organised as follows. We start in Section 2 with setting up notation and reviewing Q-Resolution. Section 3 contains our characterisation of tree-like Q-Resolution in terms of the Prover-Delayer game. The two mentioned examples for this lower bound technique follow in Sections 4 and 5, the latter of which contains the hardness proof for KBKF(t). We conclude with some open directions for future research in Section 6.

2   Preliminaries

A *literal* is a Boolean variable or its negation; we say that the literal x is *complementary* to the literal $\neg x$ and vice versa. If l is a literal, $\neg l$ denotes the complementary literal, i.e. $\neg\neg x = x$. A *clause* is a disjunction of zero or more literals. The empty clause is denoted by $\bot$, which is semantically equivalent to false. A formula in *conjunctive normal form* (CNF) is a conjunction of clauses. Whenever convenient, a clause is treated as a set of literals and a CNF formula as a set of clauses. For a literal $l = x$ or $l = \neg x$, we write var(l) for x and extend this notation to var(C) for a clause C and var(ψ) for a CNF ψ.

Quantified Boolean Formulas (QBFs) extend propositional logic with quantifiers with the standard semantics that $\forall x.\Psi$ is satisfied by the same truth assignments as $\Psi[0/x] \land \Psi[1/x]$ and $\exists x.\Psi$ as $\Psi[0/x] \lor \Psi[1/x]$. Unless specified otherwise, we assume that QBFs are in *closed prenex* form with a CNF *matrix*, i.e., we consider the form $\mathcal{Q}_1 X_1 \ldots \mathcal{Q}_k X_k. \phi$, where X_i are pairwise disjoint sets of variables; $\mathcal{Q}_i \in \{\exists, \forall\}$ and $\mathcal{Q}_i \neq \mathcal{Q}_{i+1}$. The formula ϕ is in CNF and is defined only on variables $X_1 \cup \cdots \cup X_k$. The propositional part ϕ of a QBF is called the *matrix* and the rest the *prefix*. If a variable x is in the set X_i, we say that x is at *level* i and write lev(x) = i; we write lev(l) for lev(var(l)). A closed QBF is *false* (resp. *true*), iff it is semantically equivalent to the constant 0 (resp. 1).

Often it is useful to think of a QBF $Q_1 X_1 \ldots Q_k X_k. \, \phi$ as a *game* between the *universal* and the *existential player*. In the i-th step of the game, the player Q_i assigns values to the variables X_i. The existential player wins the game iff the matrix ϕ evaluates to 1 under the assignment constructed in the game. The universal player wins iff ϕ evaluates to 0. A QBF is false iff there is a *winning strategy* for the universal player, i.e. if the universal player can win any game.

Q-Resolution, by Kleine Büning et al. [20], is a resolution-like calculus that operates on QBFs in prenex form where the matrix is a CNF. The rules are given in Figure 1. All proofs in Q-Resolution are refutations, deriving $\emptyset$. Q-Resolution derivations can be associated with a graph where vertices are the clauses of the proof and each resolution inference $\frac{C \quad D}{E}$ gives rise to two directed edges (C, E) and (D, E). Likewise a universal reduction $\frac{C}{D}$ yields an edge (C, D). We speak of *tree-like Q-Resolution* if we only allow Q-Resolution proofs which have trees as its associated graphs. This means that intermediate clauses cannot be used more than once and have to be rederived otherwise. There are exponential separations known between tree-like and dag-like Resolution in the classical case (cf. [26]), that carry over between tree-like and dag-like Q-Resolution.

$$\frac{}{C} \text{ (Axiom)} \qquad \frac{C_1 \cup \{x\} \qquad C_2 \cup \{\neg x\}}{C_1 \cup C_2} \text{ (Res)}$$

C is a clause in the matrix. Variable x is existential. If $z \in C_1$, then $\neg z \notin C_2$.

$$\frac{D \cup \{u\}}{D} \text{ (}\forall\text{-Red)} \qquad \begin{array}{l} \text{Literal } u \text{ is universal. For existential} \\ x \text{ in clause } D, \, \text{lev}(x) < \text{lev}(u). \end{array}$$

Fig. 1. The rules of Q-Res [20]

3   Prover-Delayer Game

In this section, we present a two player game along with a scoring system. The players will be called Prover and Delayer (referred by pronouns 'she' and 'he' respectively). The game is played on a QBF F. The Delayer tries to maximise the score. The Prover tries to win the game by falsifying the formula (which ends the game) and giving the Delayer as small a score as possible. The game proceeds in rounds. Each round of the game has the following phases:

1. *Setting universal variables:* The Prover can assign values to any number of universal variables of her choice that are not blocked, i.e., a universal variable u can be assigned a value by the Prover if all the existential variables with higher quantification level than u are currently unassigned.
2. *Declare Phase:* The Delayer can choose to assign values to any unassigned existential variables of his choice. The Delayer does not score from this.

3. *Query Phase:* This phase has three stages:
   (a) Prover queries any one existential variable x that is currently unassigned.
   (b) Delayer replies with positive weights w_0 and w_1 such that $w_0 + w_1 = 1$.
   (c) Prover assigns a value for x. If she assigns $x = b$ for some $b \in \{0, 1\}$, the Delayer scores $\lg(\frac{1}{p_b})$ points.
4. *Forget Phase:* The Prover can forget values of any of the assigned variables of her choice. Any variable chosen in this phase will become unassigned.

The Prover wins the game if any clause in F is falsified. In every round, we check if the Prover has won the game after each phase.

We will now show that our game characterizes tree like Q-Resolution.

Theorem 1. *If ϕ has a tree-like Q-Resolution proof of size at most s, then there exists a Prover strategy such that any Delayer scores at most $\lg\lceil \frac{s}{2} \rceil$ points.*

Proof. We take a similar approach as in [10]. Let Π be a tree-like Q-Resolution refutation of ϕ. Informally, the Prover plays according to Π, starting at the empty clause and following a path in the tree to one of the axioms. At a Resolution inference the Prover will query the resolved variable and at a universal reduction she will set the universal variable. The Prover will keep the invariant that at each moment in the game, the current assignment α assigns exactly all literals from the current clause C on the path in Π, and moreover α falsifies C. This invariant holds in the beginning at the empty clause, and in the end, Prover wins by falsifying an axiom.

We will now elaborate and describe a randomized Prover strategy. Let the Prover be at a node in Π labelled with clause C. We describe what she does in the four stages.

Setting universal variables: If the current clause C was derived in the proof Π by a $\forall$-reduction $\frac{C \vee z}{C}$, then Prover sets $z = 0$. This is possible as the current assignment contains only variables from C and therefore z is not blocked. Prover then moves to the clause $C \vee z$. The Prover repeats this till arriving at a clause derived by the Resolution rule (or winning the game).

Query phase: Prover is now at a clause in Π that was derived by a Resolution step $\frac{C_1 \vee x \quad C_2 \vee \neg x}{C_1 \vee C_2}$. If the Delayer already set the value of x in his Declare phase, then Prover follows this choice and moves on in the proof tree, possibly setting further universal variables. She does this until she reaches a clause derived by Resolution, where resolved variable x is unassigned. She queries x. On Delayer replying with weights w_0 and w_1, she chooses $x = i$ with probability w_i.

If $x = 0$, then Prover defines S to be the set of all variables not in $C_1 \vee x$ and proceeds down to the subtree rooted at that clause. Else, she defines S to be all variables not in $C_2 \vee \neg x$ and proceeds down to the corresponding subtree.

Forget Phase: The Prover forgets all variables in the set S.

For a fixed Delayer D, let $q_{D,\ell}$ denote the probability (over all random choices made within the game) that the game ends at leaf ℓ. Let π_D be the corresponding distribution induced on the leaves.

For the Prover strategy described above, we have the following claim:

Claim. If the game ends at a leaf ℓ, then the Delayer scores exactly $\alpha_\ell = \lg\left(\frac{1}{q_{D,\ell}}\right)$ points.

Proof. Note that since Π is a tree-like Q-Resolution proof, there is exactly one path from the root of Π to ℓ. Let p be the unique path that leads to the leaf ℓ and let the number of random choices made along p be m. Then, we have $q_{D,\ell} = \prod_{i=1}^{m} q_i$ where q_i is the probability for the ith random choice made along p. Since p is the unique path that leads to ℓ, the number of points α_ℓ scored by the Delayer when the game ends at ℓ is exactly the number of points scored when the game proceeds along the path p. The number of points scored by the Delayer along p is given by: $\alpha_\ell = \sum_{i=1}^{m} \lg\left(\frac{1}{q_i}\right) = \lg\left(\prod_i \frac{1}{q_i}\right) = \lg\left(\frac{1}{q_{D,\ell}}\right)$ □

The Prover strategy we described is randomized. The expected score over all leaves ℓ is the following expression: $\sum_{\text{leaves } \ell \in \Pi} q_{D,\ell}\alpha_\ell = \sum_{\text{leaves } \ell \in \Pi} q_{D,\ell} \lg\frac{1}{q_{D,\ell}}$.

But this quantity is exactly the Shannon entropy $\mathcal{H}(\pi_D)$. Since D is fixed, this entropy will be maximum when π_D is the uniform distribution; i.e., $\mathcal{H}(\pi_D)$ is maximum when, for all leaves ℓ, the probability that the game ends at ℓ is the same. A tree like Q-Resolution proof of size s has at most $\lceil s/2\rceil$ leaves. So the support of the distribution π_D has size at most $\lceil s/2\rceil$ and hence $\mathcal{H}(q_{D,\ell}) \leq \lg\lceil s/2\rceil$.

If the expected score with the randomised Prover is $\leq \lg\lceil s/2\rceil$, then there is a deterministic Prover who restricts the scores to at most $\lg\lceil s/2\rceil$. Now we derandomise the Prover by just fixing her random choices accordingly. If the Delayer is optimal she can pick arbitrarily if not she can pick to exploit this.   □

To obtain the characterisation of Q-Resolution we also need to show the opposite direction, exhibiting an optimal Delayer:

Theorem 2. *Let ϕ be an unsatisfiable QBF formula and let s be the size of a shortest tree-like Q-Resolution proof for ϕ. Then there exists a Delayer who scores at least $\lg\lceil s/2\rceil$ points against any Prover.*

Proof. For any unsatisfiable QBF formula ϕ, let $L(\phi)$ denote the number of leaves in the shortest tree-like Q-Resolution proof of ϕ. For a partial assignment α to variables in ϕ, let $\phi|_\alpha$ denote ϕ restricted to the partial assignment α.

The Delayer starts with the empty assignment α and changes α throughout the game. On receiving a query for an existential variable x, the Delayer does the following:

1. Updates α to reflect any changes made by the Prover to any of the variables. These changes include assignments made to both universal variables as well as existential variables.
2. Computes the quantities $\ell_0 = L(\phi|_{\alpha,x=0})$ and $\ell_1 = L(\phi|_{\alpha,x=1})$.
3. Replies with weights $w_0 = \frac{\ell_0}{\ell_0+\ell_1}$ and $w_1 = \frac{\ell_1}{\ell_0+\ell_1}$.

We show by induction on the number of existential variables n in ϕ that the Delayer always scores at least $\lg L(\phi)$ points: Base case $n = 0$, $L(\phi) = 0$ and the Delayer scores at least 0 points. Assume the statement is true for all $n < k$. Now for $n = k$, consider the first query by the Prover, after she possibly made some universal choices according to the partial assignment α. Let the queried variable be x. If the Prover chose $x = b$ where $b \in \{0, 1\}$, then the Delayer scores $\lg \frac{1}{w_b}$ for this step alone. After assigning $x = b$, the formula $\phi|_{\alpha, x=b}$ has $k - 1$ existential variables and hence we use induction hypothesis to conclude that the remaining rounds in the game give the Delayer at least $\lg L(\phi|_{\alpha, x=b})$. Hence the total score is evaluated as: $\lg (L(\phi|_{\alpha, x=0}) + L(\phi|_{\alpha, x=1})) \geq \lg L(\phi|_\alpha) \geq \lg L(\phi)$.

The last inequality holds, because if $\phi|_\alpha$ is unsatisfiable, we can refute ϕ by deriving a clause with no existential literals, just containing all variables in the domain of α and then $\forall$-reduce. The theorem follows since for any binary tree of size s, the number of leaves is $\lceil s/2 \rceil$. □

4   A First Example

We consider the following formulas studied by Janota and Marques-Silva [18]:

$$F_n = \exists e_1 \forall u_1 \exists c_1^1 c_1^2 \cdots \exists e_i \forall u_i \exists c_i^1 c_i^2 \cdots \exists e_n \forall u_n \exists c_n^1 c_n^2 :$$

$$\bigwedge_{i=1}^{n} (e_i \to c_i^1) \wedge (u_i \to c_i^1) \wedge (\neg e_i \to c_i^2) \wedge (\neg u_i \to c_i^2) \wedge \bigvee_{i=1}^{n} (\neg c_i^1 \vee \neg c_i^2)$$

These formulas were used in [18] to show that $\forall$Exp+Res does not simulate Q-Resolution, i.e., F_n requires exponential-size proofs in $\forall$Exp+Res, but has polynomial-size Q-Resolution proofs. Janota and Marques-Silva [19] also show that $\forall$Exp+Res p-simulates tree-like Q-resolution, and hence it follows that F_n is also hard for the latter system. We reprove this result using our characterisation.

Let $\mathcal{U} = \{u_1, u_2, \ldots, u_n\}$ be the set of all universal variables. In the following, we show a Delayer strategy that scores at least n points against any Prover. For the Declare phase, the Delayer executes Algorithm 1 till reaching a fixed point. For any variable queried by Prover, Delayer responds with weights $(\frac{1}{2}, \frac{1}{2})$. For $i \in [n]$, let $T_i = \{e_i, c_i^1, c_i^2\}$. Let $\mathcal{C} = \bigvee_{i=1}^{n} (\neg c_i^1 \vee \neg c_i^2)$. Note that except for $\mathcal{C}$, all other clauses have only two literals.

Lemma 3. *Algorithm 1 never falsifies a clause that has only two literals.*

Lemma 4. *If the Delayer uses the strategy outlined above, then for any winning Prover strategy, the clause falsified is $\mathcal{C}$.*

Proof. Suppose the clause falsified was D. We will show that if $D \neq \mathcal{C}$, then the Delayer did not use our strategy. We consider the following cases:

Algorithm 1. Declare Routine

---

for all clauses $(\ell_1 \to \ell_2)$ in F_n **do**
    if $\ell_1 = 1$ **then** Declare $\ell_2 = 1$.
    if $\ell_2 = 0$ and $\mathrm{var}(\ell_1) \notin \mathcal{U}$ **then** Declare $\ell_1 = 0$.
end for

---

1. D involves variable u_i for some $i \in [n]$:

   Note that u_i appears in clauses with either c_i^1 or c_i^2. Since both c_i^1 and c_i^2 block u_i, it has to be the case that when u_i was set by the Prover, the variables c_i^1 and c_i^2 were unassigned. Now it is straightforward to see that if the Delayer indeed used the declare routine described in Algorithm 1, then all clauses involving u_i become satisfied after u_i is set by the Prover.

2. D is $(e_i \rightarrow c_i^1)$ or $(\neg e_i \rightarrow c_i^2)$:

   Suppose w.l.o.g. that $D = (e_i \rightarrow c_i^1)$. As a consequence of Lemma 3, it must be the case that D was falsified because of the Prover choosing a value for either e_i or c_i^1. So we have two cases:

   - Prover chose a value for e_i to falsify D: So e_i was unassigned just before the query phase began. But if Algorithm 1 left e_i unassigned, then this means c_i is unassigned or $c_i^1 \neq 0$. Hence if the Delayer indeed used Algorithm 1, D could not have been falsified.
   - Prover chose a value for c_i^1 to falsify D: Following an argument just like the previous case, if the Delayer indeed used Algorithm 1, then c_i would be unassigned at the start of the query phase only if $e_i = 0$ or unassigned. In both these cases D cannot be falsified by choosing a value for c_i^1. $\square$

Theorem 5. *Delayer scores at least n points against any Prover strategy.*

Proof. From Lemma 4, it is sufficient to show that any Prover strategy that falsifies $\mathcal{C}$ will give the Delayer a score of at least n. $\mathcal{C}$ can be falsified only if all variables c_i^1, c_i^2 have been assigned to 1. We observe that for any $i \in [n]$, the Prover can get at most one of c_i^1 or c_i^2 to be declared for free by setting u_i appropriately. To assign the other c_i to 1, the Prover can either query c_i directly and set it to 1 or query e_i and set it appropriately. Both these ways give the Delayer 1 point. Hence for every $i \in [n]$, the Delayer scores at least 1 point. $\square$

With Theorem 1 this reproves the hardness of F_n for tree-like Q-Resolution, already implicitly established in [18,19]:

Corollary 6. *Formulas F_n require tree-like Q-Resolution proofs of size $\Omega(2^n)$.*

This bound is tight as tree-like Q-Resolution refutations of size $O(2^n)$ exist.

5   Hardness of the Formulas of Kleine Büning et al.

In our second example we look at a family of formulas first defined by Kleine Büning, Karpinski and Flögel [20]. The formulas are known to be hard for Q-Resolution and indeed for the stronger system IR-calc [8]. Here we use our technique to give an independent proof of their hardness in tree-like Q-Resolution.

Definition 7 (Kleine Büning, Karpinski and Flögel [20]). *Consider the clauses*

$$
\begin{aligned}
C_- &= \{\neg y_0\} & C_0 &= \{y_0, \neg y_1^0, \neg y_1^1\} \\
C_i^0 &= \{y_i^0, x_i, \neg y_{i+1}^0, \neg y_{i+1}^1\} & C_i^1 &= \{y_i^1, \neg x_i, \neg y_{i+1}^0, \neg y_{i+1}^1\} & i &\in [t-1] \\
C_t^0 &= \{y_t^0, x_t, \neg y_{t+1}, \ldots, \neg y_{t+t}\} & C_t^1 &= \{y_t^1, \neg x_t, \neg y_{t+1}, \ldots, \neg y_{t+t}\} \\
C_{t+i}^0 &= \{x_t, y_{t+i}\} & C_{t+i}^1 &= \{\neg x_i, y_{t+i}\} & i &\in [t]
\end{aligned}
$$

The KBKF(t) formulae are defined as the union of these clauses under the quantifier prefix $\exists y_0, y_1^0, y_1^1 \, \forall x_1 \, \exists y_2^0, y_2^0 \, \forall x_2, \ldots, \forall x_{t-1} \, \exists y_t^0, y_t^1 \, \forall x_t \, \exists y_{t+1} \ldots \exists y_{t+t}.$

We now want to show an exponential lower bound on proof size for the KBKF(t) formulas via our game. We will assume throughout that $t > 2$. We start with an informal description of the Delayer strategy.

Delayer Strategy – Informal Description

At any point of time during a run of the game, there is a partial assignment that has been constructed by the Prover and Delayer. We define the following:

Definition 8. *For any partial assignment α to the variables, we define z_α to be the index of the highest subscript such that an α assigns a 0 to one or more existential variables with that subscript. If no such subscript exists, then $z = 0$.*

For convenience, we will drop the subscript and just say z when the partial assignment is clear from context. We usually mention the time during a run of the game by referring to z instead of explicitly mentioning the induced partial assignment. The idea behind the Delayer strategy is the following: We observe that for all $i < t - 2$ and $j \in \{0, 1\}$, to falsify the clause C_i^j, it is necessary that y_i^j is set to 0 and both y_{i+1}^0 and y_{i+1}^1 are set to 1. The strategy we design will not let the Prover win on clauses C_-, C_0, C_i^0, or C_i^1 for any $i < (t - 2)$. We do this by declaring either y_{i+1}^0 or y_{i+1}^1 to 0 at a well chosen time. Furthermore, we will show : (1) When the game ends, $z \geq t$ and (2) After any round in the game, the Delayer has a score of at least $O(z)$ It is easy to see that the lower bound of $\Omega(t)$ for the score of the Delayer follows from statements (1) and (2).

Delayer Strategy – Details

Declare Phase: The Delayer sets y_0 to 0 in the declare phase of the first round.

Let F be the set of all existential variables that were chosen to be forgotten by the Prover in the forget phase of the previous round. The Delayer first does the following "Reset Step": For all variables y in F that had value 0 just before the forget phase of the previous round, the Delayer declares $y = 0$. After the reset step, the Delayer executes Algorithm 2 repeatedly until reaching a fixed point. The notation $y \leftarrow b$ means that the Delayer declares $y = b$ if and only if y is an unassigned variable. Also, we assume that z is updated automatically to be the highest subscript for existential literals set to 0. We observe the following about the reset step:

Observation 9. *The reset step ensures that z always increases monotonically (when z is measured at the beginning of each query phase).*

Line 14 of Algorithm 2 gives us the following observation:

Observation 10. *After the declare phase, for all $i < z$, the existential variables y_i^0 and y_i^1 has been assigned a value.*

Algorithm 2. Declare Routine

1: $y_z^0 \leftarrow 1$, $y_z^1 \leftarrow 1$, $z' := z$
2: **if** $y_z^{x_z} \neq 0$ or x_z unassigned **then**
3:     **for all** $i > z$ **do** $y_i^0 \leftarrow 1$; $y_i^1 \leftarrow 1$
4: **end if**
5: **for** $i = t - 1$ to 1 **do**
6:     **for** $j = 0$ to 1 **do**
7:         **if** C_i^j is not satisfied with only one literal l that is unassigned **then** Satisfy C_i^j with that literal (if existential).
8:     **end for**
9: **end for**
10: **if** $z \leq t - 2$ and either $y_{z+2}^0 = 1$ or $y_{z+2}^1 = 1$ **then** $y_{z+1}^{1-x_{z+1}} \leftarrow 0$
11: **if** $z \neq z'$, x_z assigned and $y_z^{x_z} = 0$ **then**
12:     **if** x_{z+1} unassigned **then** $y_{z+1}^0 \leftarrow 0$ **else** $y_{z+1}^{1-x_z} \leftarrow 0$
13: **end if**
14: **for all** $i < z$ **do** $y_i^0 \leftarrow 0$, $y_i^1 \leftarrow 0$

Observation 11. *For all $i > z$, Algorithm 2 assigns all y_i^0 and y_i^1 to 1 before assigning any of them to 0.*

Query Phase:
Let the variable queried be y_i^b. From Observation 10, it is easy to see that $i \geq z$. We have the following cases:

- If $i > t$, then the Delayer replies with weights $w_0 = 2^{z-t-1}$ and $w_1 = 1 - w_0$.
- Else $z \leq i \leq t$. We have two cases:
  - If x_i is unassigned, then the Delayer replies with weights $w_0 = 2^{z-i}$ and $w_1 = 1 - w_0$.
  - Else x_i holds a value. Then we have the following cases:
    * If $b = \neg x_i$, then the Delayer replies with weights $w_0 = 2^{z-i}$ and $w_1 = 1 - w_0$.
    * Else $b = x_i$ and Delayer replies with weight $w_0 = 2^{z-j}$, where j is the largest index such that $\forall k : z < k \leq j, x_k$ is assigned and $y_k^{1-x_k} = 1$. Weight $w_1 = 1 - w_0$.

We now analyze the above strategy: We start with the following lemma:

Lemma 12. *If the Delayer uses the strategy outlined above, then against any Prover, at the end of the game on $KBKF(t)$, $z \geq t$ (where z is defined as in Definition 8).*

Remark 13. If the Prover choses to assign 1 to a variable queried in the query phase on turn k, then by the query phase on turn $k + 1$, the value of z (index of the rightmost zero) increments by at most 1. For the increase by 1 it is required that $y_z^{x_z} = 0$ and that for all $c \in \{0, 1\}$, y_{z+1}^c and y_{z+2}^c are unassigned before the query phase on turn k. If the Prover chose to assign 1 to the variable queried and it results in a change of z, then it must cause any of y_{z+1}^0, y_{z+1}^1, y_{z+2}^0 or y_{z+2}^1 to be set to 1, incrementing z be at most one.

For all $i \in [t]$, and $z < t-1$, let $s_z(y_i^c)$ denote the minimum (over all possible Prover strategies) Delayer score when y_i^c is assigned 1 by the Prover for the first time starting from a partial assignment where the right most zero is in column z and every variable to the right of column z is unassigned.

Combining Observation 9 with the fact that at the start of the game $z = 0$, Lemma 12 implies that the Prover increases z by at least t in the process of winning the game. We will now measure the scores that the Delayer accumulates.

Lemma 14. *For all $z < t-1$ and $i < t$, each of $s_z(y_i^0)$ and $s_z(y_i^1)$ is at least* $2^{t-i} \lg \frac{2^{t-z}}{2^{t-z}-1}$.

During a run of the game, z increases from 0 to t. Now we show that the Delayer scores $\Omega(z)$ points during any run of the game on KBKF(t) for large enough t:

Lemma 15. *There exists constants $t_0 > 0$ and $\alpha > 0$ such that for all $t > t_0$, at any point of time during a run of the game on KBKF(t), the Delayer has a score of at least αz.*

To show this lemma, we argue that the Delayer scores $\Omega(1)$ points for every increment of z during the game. This immediately gives us the required claim. Combining Lemma 12 and Lemma 15, we have:

Theorem 16. *There exists a Delayer strategy that scores $\Omega(t)$ against any Prover in the Prover-Delayer game on KBKF(t).*

Corollary 17. *KBKF(t) require tree-like Q-Resolution proofs of size $2^{\Omega(t)}$.*

6   Conclusion

In this paper we have shown that lower bound techniques from classical proof complexity can be transferred to the more complex setting of QBF. We have demonstrated this with respect to prover-delayer games, even obtaining a characterisation of tree-like size in Q-Resolution. Although tree-like (Q-)Resolution is a weak system, it is an important one as it corresponds to runs of the plain DLL algorithm, which serves as the basis of most SAT and QBF-solvers.

A very interesting question for further research is to understand how far this transfer of techniques can be extended. In particular, it seems likely that the very general game-theoretic approaches of [23] can also be utilised for QBF systems. Two other seminal techniques that have found wide-spread applications for classical Resolution are feasible interpolation [22], which also applies to many further systems, and the size-width method of Ben-Sasson and Wigderson [5]. Is it possible to use analogous methods for Q-Resolution and its extensions?

References

1. Atserias, A., Dalmau, V.: A combinatorial characterization of resolution width. Journal of Computer and System Sciences **74**(3), 323–334 (2008)

2. Balabanov, V., Jiang, J.-H.R.: Unified QBF certification and its applications. Formal Methods in System Design **41**(1), 45–65 (2012)
3. Balabanov, V., Widl, M., Jiang, J.-H.R.: QBF resolution systems and their proof complexities. In: Sinz, C., Egly, U. (eds.) SAT 2014. LNCS, vol. 8561, pp. 154–169. Springer, Heidelberg (2014)
4. Ben-Sasson, E., Harsha, P.,: Lower bounds for bounded depth Frege proofs via Buss-Pudlák games. ACM Trans. on Computational Logic **11**(3) (2010)
5. Ben-Sasson, E., Wigderson, A.: Short proofs are narrow - resolution made simple. Journal of the ACM **48**(2), 149–169 (2001)
6. Benedetti, M., Mangassarian, H.: QBF-based formal verification: Experience and perspectives. JSAT **5**(1–4), 133–191 (2008)
7. Beyersdorff, O., Chew, L., Janota, M.: On unification of QBF resolution-based calculi. In: Csuhaj-Varjú, E., Dietzfelbinger, M., Ésik, Z. (eds.) MFCS 2014, Part II. LNCS, vol. 8635, pp. 81–93. Springer, Heidelberg (2014)
8. Beyersdorff, O., Chew, L., Janota, M.: Proof complexity of resolution-based QBF calculi. ECCC **21**, 120 (2014)
9. Beyersdorff, O., Galesi, N., Lauria, M.: A lower bound for the pigeonhole principle in tree-like resolution by asymmetric prover-delayer games. Information Processing Letters **110**(23), 1074–1077 (2010)
10. Beyersdorff, O., Galesi, N., Lauria, M.: A characterization of tree-like resolution size. Information Processing Letters **113**(18), 666–671 (2013)
11. Beyersdorff, O., Galesi, N., Lauria, M.,: Parameterized complexity of DPLL search procedures. ACM Trans. on Computational Logic **14**(3) (2013)
12. Beyersdorff, O., Kullmann, O.: Unified characterisations of resolution hardness measures. In: Sinz, C., Egly, U. (eds.) SAT 2014. LNCS, vol. 8561, pp. 170–187. Springer, Heidelberg (2014)
13. Buss, S.R.: Towards NP-P via proof complexity and search. Ann. Pure Appl. Logic **163**(7), 906–917 (2012)
14. Cook, S.A., Nguyen, P.: Logical Foundations of Proof Complexity. Cambridge University Press (2010)
15. Egly, U.: On sequent systems and resolution for QBFs. In: Cimatti, A., Sebastiani, R. (eds.) SAT 2012. LNCS, vol. 7317, pp. 100–113. Springer, Heidelberg (2012)
16. Egly, U., Lonsing, F., Widl, M.: Long-distance resolution: Proof generation and strategy extraction in search-based QBF solving. In: McMillan, K., Middeldorp, A., Voronkov, A. (eds.) LPAR-19 2013. LNCS, vol. 8312, pp. 291–308. Springer, Heidelberg (2013)
17. Esteban, J.L., Torán, J.: A combinatorial characterization of treelike resolution space. Information Processing Letters **87**(6), 295–300 (2003)
18. Janota, M., Marques-Silva, J.: ∀Exp+Res does not p-simulate Q-resolution. International Workshop on Quantified Boolean Formulas (2013)
19. Janota, M., Marques-Silva, J.: On propositional QBF expansions and Q-resolution. In: Järvisalo, M., Van Gelder, A. (eds.) SAT 2013. LNCS, vol. 7962, pp. 67–82. Springer, Heidelberg (2013)
20. Büning, H.K., Karpinski, M., Flögel, A.: Resolution for quantified Boolean formulas. Inf. Comput. **117**(1), 12–18 (1995)
21. Krajíček, J.: Bounded Arithmetic, Propositional Logic, and Complexity Theory. Cambridge University Press, Cambridge (1995)
22. Krajíček, J.: Interpolation theorems, lower bounds for proof systems and independence results for bounded arithmetic. J. Symb. Log. **62**(2), 457–486 (1997)
23. Pudlák, P.: Proofs as games. American Math. Monthly, pp. 541–550 (2000)

24. Pudlák, P., Impagliazzo, R.: A lower bound for DLL algorithms for SAT. In: Proc. 11th Symposium on Discrete Algorithms, pp. 128–136 (2000)
25. Rintanen, J.: Asymptotically optimal encodings of conformant planning in QBF. In: AAAI, pp. 1045–1050. AAAI Press, (2007)
26. Segerlind, N.: The complexity of propositional proofs. Bulletin of Symbolic Logic **13**(4), 417–481 (2007)
27. Van Gelder, A.: Contributions to the theory of practical quantified Boolean formula solving. In: CP, pp. 647–663 (2012)

Recurrence Relations, Succession Rules, and the Positivity Problem

Stefano Bilotta[1][✉], Elisa Pergola[1], Renzo Pinzani[1], and Simone Rinaldi[2]

[1] Dipartimento di Matematica e Informatica "Ulisse Dini",
University of Florence, Viale Morgagni 65, 50134 Firenze, Italy
{stefano.bilotta,elisa.pergola,renzo.pinzani}@unifi.it
[2] Dipartimento di Ingegneria Dell'Informazione e Scienze Matematiche,
University of Siena, Via Roma 56, 53100 Siena, Italy
rinaldi@unisi.it

Abstract. In this paper we present a method which can be used to investigate on the positivity of a number sequence defined by a recurrence relation having constant coefficients (in short, a C-recurrence).

Keywords: C-recurrences · Positive numbers sequence

1 Introduction

Succession rules (sometimes called *ECO-systems*) have been proved to be an efficient tool in order to solve several combinatorial problems. The concept of a succession rule was introduced in [6] to study reduced Baxter permutations, and only later this has been recognized as an extremely useful tool for the ECO method, a methodology applied for the enumeration of various combinatorial structures [2].

An *(ordinary) succession rule* Ω is a system constituted by an *axiom* and a set of *productions*. A production constructs the *successors* of any given label (k). The rule Ω can be represented by means of a *generating tree* having the axiom as the label of the root and each node labelled (k) at level n has k sons at level $n+1$.

A succession rule Ω defines a sequence of positive integers $\{f_n\}_{n\geq 0}$ where f_n is the number of the nodes at level n in the generating tree defined by Ω. By convention the root is at level 0, so $f_0 = 1$. The function $f_\Omega(x) = \sum_{n\geq 0} f_n x^n$ is the *generating function* determined by Ω.

More recently, there have been some efforts in developing methods to pass from a recurrence relation defining an integer sequence to a succession rule defining the same sequence; in this case we say that the succession rule and the recurrence relation are *equivalent*.

Our work fits into this research line, and tries to deepen the relations between succession rules and recurrence relations.

It is worth mentioning that almost all studies realized until now on this topic have regarded linear recurrence relations with a finite number of integer

© Springer International Publishing Switzerland 2015
A.-H. Dediu et al. (Eds.): LATA 2015, LNCS 8977, pp. 499–510, 2015.
DOI: 10.1007/978-3-319-15579-1_39

coefficients [4, 7]. We will address to these ones as *C-finite recurrence relations*, and to the defined sequences as *C-finite sequences* [18].

Accordingly, our work will start considering C-finite recurrences. Compared with the methods presented in [4, 7], our approach is completely different.

To achieve this goal, we first translate the given C-finite recurrence relation into an *extended succession rule*, which differs from the ordinary succession rules since it admits both jumps and marked labels. Then we recursively eliminate jumps and marked labels from such an extended succession rule, thus obtaining an ordinary succession rule equivalent to the previous one. We need to point out that this translation is possible only if a certain condition – called *positivity condition* – is satisfied. Such a condition ensures that all the labels of the generating tree are non marked, hence the sequence defined by the succession rule has all positive terms.

If the recurrence relation has degree k with coefficients $a_1, \ldots, a_k$, such a condition can be expressed in terms of a set of k inequalities which can be obtained from a set of quotients and remainders given by the coefficients. To the authors' knowledge, such a condition is completely new in literature. It directly follows that our positive condition provides a sufficient condition for testing the positivity of a C-finite sequence, then it is related to the so called *positivity problem*.

Positivity Problem: given a C-finite sequence $\{f_n\}_{n \geq 0}$, establish if all its terms are positive.

This problem was originally proposed as an open problem in [3], and then re-presented in [16] (Theorems 12.1-12.2, pages 73-74), but no general solution has been found yet.

It is worth mentioning that the positivity problem can be solved for a large class of C-finite sequences, precisely those whose generating function is a N-rational series. We also recall that the class of N-rational series is precisely the class of the generating functions of regular languages, and that a Soittola's Corollary in [17] states that the problem of establishing whether a rational generating function is N-rational is decidable.

N-rational series have been recently revisited using modern combinatorial techniques in [4, 15], using different approaches and some algorithms to pass from an N-rational series to a regular expression enumerated by such a series have been proposed [1, 13]. However, none of these techniques provides a method to face C-finite recurrence relations which are not N-rational.

Following the attempt of enlightening some questions on positive sequences, some researches have recently focused on determining sufficient conditions to establish the possible positivity of a given C-finite recurrence relation, as interestingly described in [11]. As a matter of fact, up to now, we only know that the positivity problem is decidable for C-finite recurrences of two [12] or three terms [14]. Another approach to tackle the positivity problem is to develop algorithms to test possible positivity of recursively defined sequences (and, in particular, C-finite sequences) by means of computer algebra, as in [10].

Our work fits into this research line, since the positive condition we propose is a sufficient condition for testing the positivity of a C-finite sequence.

2   Basic Definitions and Notations

In this section we present some basic definitions and notations related to the concept of succession rule. For further definitions and examples we address the reader to [6].

Two succession rules are *equivalent* if they have the same generating function. A succession rule is *finite* if it has a finite number of labels and productions.

For example, the two succession rules:

$$\begin{cases} (2) \\ (2) \rightsquigarrow (2)(2) \end{cases} \qquad \begin{cases} (2) \\ (k) \rightsquigarrow (1)^{k-1}(k+1) \end{cases}$$

are equivalent rules, and define the sequence $f_n = 2^n$. The one on the left is a finite rule, since it uses only the label (2), while the one on the right is an infinite rule.

According to the technique of *colored label* [8], in a succession rule there can be labels $(k)_1$ and $(k)_2$ having the same number k of sons but having different productions, in this case we refer to *colored succession rules*.

A slight generalization of the concept of ordinary succession rule is provided by the so called *jumping succession rule* [9]. Roughly speaking, the idea is to consider a set of succession rules acting on the objects of a class and producing sons at different levels.

The usual notation to indicate a jumping succession rule Ω is the following:

$$\begin{cases} (a) \\ (k) \overset{j_1}{\rightsquigarrow} (e_{11}(k))(e_{12}(k))\dots(e_{1k}(k)) \\ (k) \overset{j_2}{\rightsquigarrow} (e_{21}(k))(e_{22}(k))\dots(e_{2k}(k)) \\ \quad\vdots \\ (k) \overset{j_m}{\rightsquigarrow} (e_{m1}(k))(e_{m2}(k))\dots(e_{mk}(k)) \end{cases}$$

The generating tree associated with Ω has the property that each node labelled (k) lying at level n produces m sets of sons at level $n + j_1, n + j_2, \dots, n+j_m$, respectively and each of such set has labels $(e_{i1}(k)), (e_{i2}(k)), \dots, (e_{ik}(k))$ respectively, $1 \leq i \leq m$.

We need to point out that a node labelled (k) has precisely k sons, according to the above definitions. A rule having this property is said to be *consistent*. However, in many cases we can relax this constraint and consider rules, where the number of sons is a function of the label k.

Another generalization is used in [5], where the authors deal with *jumping and marked succession rules*. In this case the labels appearing in a jumping succession rule can be marked, and the *marked* labels are considered together with the unmarked ones.

A *jumping and marked generating tree* is a rooted labelled tree where there appear marked and unmarked labels according to the corresponding succession rule. The main property is that in the generating tree a marked label $(\overline{k})$ kills or annihilates the unmarked label (k) lying on the same level n. In particular, the enumeration of the combinatorial objects in a class is the difference between the number of unmarked and marked labels lying on a given level.

For any label (k), we introduce the following notation for generating tree specifications:

$$(\overline{\overline{k}}) = (k); \quad (k)^n = \underbrace{(k)\ldots(k)}_{n} \ n > 0; \quad (k)^{-n} = \underbrace{(\overline{k})\ldots(\overline{k})}_{n} \ n > 0.$$

3   A Method to Translate C-Sequences into Succession Rules

The main purpose of our research is to develop a general formal method to translate a given recurrence relation into a succession rule defining the same number sequence. In this case we will say that the recurrence relation and the succession rules are *equivalent* by abuse of language.

This section is organized as follows.

i) We deal with C-finite recurrences of the form

$$f_n = a_1 f_{n-1} + a_2 f_{n-2} + \cdots + a_k f_{n-k} \qquad a_i \in \mathbb{Z}, 1 \le i \le k \qquad (1)$$

with *default* initial conditions, i.e. $f_0 = 1$ and $f_h = 0$ for all $h < 0$. First, we translate the given C-finite recurrence relation into an extended succession rule, possibly using both jumps and marked labels (Section 3.1).

ii) Then, we recursively eliminate jumps and marked labels from such an extended succession rule, thus obtaining a finite succession rule equivalent to the previous one (Section 3.2). We remark that steps i) and ii) can be applied independently of the positivity of $\{f_n\}_{n\ge0}$, but at this step we cannot be sure that all the labels of the obtained rule are nonnegative integers.

iii) We state a condition to ensure that the labels of the obtained succession rule are all nonnegative. If such a condition holds, then the sequence $\{f_n\}_{n\ge0}$ has all positive terms, thus we refer to this as *positivity condition* (Section 3.3).

3.1   C-Sequences with Default Initial Conditions

Let us consider a C-finite recurrence relation expressed as in (1), with default initial conditions and the related C-finite sequence $\{f_n\}_{n\ge0}$. We recall that the generating function of $\{f_n\}_{n\ge0}$ is rational, and precisely it is $f(x) = \sum_{n\ge0} f_n x^n = \frac{1}{1-a_1 x - a_2 x^2 - \cdots - a_k x^k}$.

The first step of our method consists into translating the C-finite recurrence relation (1) into an extended succession rule. The translation is rather straightforward, since in practice it is just an equivalent way to represent the recurrence relation.

Proposition 1. *The recurrence relation (1) with default initial conditions is equivalent to the following extended succession rule:*

$$\begin{cases} (a_1) \\ (a_1) \overset{1}{\rightsquigarrow} (a_1)^{a_1} \\ (a_1) \overset{2}{\rightsquigarrow} (a_1)^{a_2} \\ \quad \vdots \\ (a_1) \overset{k}{\rightsquigarrow} (a_1)^{a_k} \end{cases} \tag{2}$$

For example, the recurrence relation $f_n = 3f_{n-1} + 2f_{n-2} - f_{n-3}$ with default initial conditions, defines the sequence $1, 3, 11, 38, 133, 464, 1620, 5655, \ldots$, and it is equivalent to the following extended succession rule:

$$\begin{cases} (3) \\ (3) \overset{1}{\rightsquigarrow} (3)^3 \\ (3) \overset{2}{\rightsquigarrow} (3)^2 \\ (3) \overset{3}{\rightsquigarrow} (\overline{3}) \end{cases} \tag{3}$$

Figure 1 shows the first few levels of the associated generating tree.

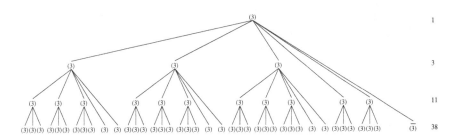

Fig. 1. Four levels of the generating tree associated with the succession rule (3)

3.2 Elimination of Jumps and Marked Labels

The successive step of our method consists into recursively eliminating jumps from the extended succession rule (2) in order to obtain a finite succession rule which is equivalent to the previous one. Once jumps have been eliminated we will deal with marked labels.

Proposition 2. *The succession rule:*

$$\begin{cases} (a_1) \\ (a_1) \quad \rightsquigarrow (a_1 + a_2)(a_1)^{a_1 - 1} \\ (a_1 + a_2) \rightsquigarrow (a_1 + a_2 + a_3)(a_1)^{a_1 + a_2 - 1} \\ \quad \vdots \\ (\sum_{l=1}^{k-1} a_l) \rightsquigarrow (\sum_{l=1}^{k} a_l)(a_1)^{(\sum_{l=1}^{k-1} a_l) - 1} \\ (\sum_{l=1}^{k} a_l) \rightsquigarrow (\sum_{l=1}^{k} a_l)(a_1)^{(\sum_{l=1}^{k} a_l) - 1} \end{cases} \quad (4)$$

is equivalent to the recurrence relation $f_n = a_1 f_{n-1} + a_2 f_{n-2} + \cdots + a_k f_{n-k}$, $a_i \in \mathbb{Z}, 1 \le i \le k$, *with default initial conditions.*

Please notice that the numbers inside a label are the coefficients of the recurrence relation and their algebraic sum gives the number of successors of that label. Obviously, the labels $(a_1), (a_1 + a_2), \cdots, (\sum_{l=1}^{k} a_l)$ are different labels even if the algebraic sums of the numbers inside labels gives the same value. For example, given the recurrence relation $f_n = 3f_{n-1} + 4f_{n-3}$ with default initial conditions, in this case $a_1 = 3, a_2 = 0, a_3 = 4$, we have the following colored succession rule:

$$\begin{cases} (3)_1 \\ (3)_1 \rightsquigarrow (3)_2 (3)_1^2 \\ (3)_2 \rightsquigarrow (7)(3)_1^2 \\ (7) \rightsquigarrow (7)(3)_1^6 \end{cases}$$

Proof. Let $A_k(x)$ be the generating function of the label $(\sum_{l=1}^{k} a_l)$ related to the succession rule (4). We have:

$$A_1(x) = 1 + (a_1 - 1)xA_1(x) + (a_1 + a_2 - 1)xA_2(x) + \ldots$$

$$\cdots + (a_1 + a_2 + \cdots + a_k - 1)xA_k(x);$$

$$A_2(x) = xA_1(x);$$

$$A_3(x) = xA_2(x) = x^2 A_1(x);$$

$$\vdots$$

$$A_{k-1}(x) = xA_{k-2}(x) = x^{k-2}A_1(x);$$

$$A_k(x) = xA_{k-1}(x) + xA_k(x) = \frac{x^{k-1}}{1-x}A_1(x).$$

Therefore,

$$A_1(x) = 1 + x(a_1 - 1)A_1(x) + x^2(a_1 + a_2 - 1)A_1(x) + \ldots$$

$$\cdots + \frac{x^k}{1-x}(a_1 + a_2 + \cdots + a_k - 1)A_1(x),$$

and we obtain the generating function $A_1(x) = \frac{1-x}{1 - a_1 x - a_2 x^2 - \cdots - a_k x^k}$.

At this point we can consider the generating function determined by the succession rule (4) as following:

$$\sum_{i=1}^{k} A_i(x) = A_1(x) + A_2(x) + \cdots + A_{k-1}(x) + A_k(x) =$$

$$= A_1(x) + xA_1(x) + \cdots + x^{k-2}A_1(x) + \frac{x^{k-1}}{1-x}A_1(x) =$$

$$= \frac{(1-x)+x(1-x)+\cdots+x^{k-2}(1-x)+x^{k-1}}{1-a_1x-a_2x^2-\cdots-a_kx^k} =$$

$$= \frac{1}{1-a_1x-a_2x^2-\cdots-a_kx^k} .$$

Following the previous statement, the extended succession rule (3) – determined in the previous section – can be translated into the following succession rule:

$$\begin{cases} (3) \\ (3) \rightsquigarrow (5)(3)^2 \\ (5) \rightsquigarrow (4)(3)^4 \\ (4) \rightsquigarrow (4)(3)^3 \end{cases}$$

We observe that the previously obtained succession rule is an ordinary finite succession rule, but it may happen that the value of the label $(\sum_{l=1}^{i} a_l)$ is negative, for some i with $i \le k$, then the succession rule (4) contains marked labels.

For example, the recurrence relation $f_n = 5f_{n-1} - 6f_{n-2} + 2f_{n-3}$, with default initial conditions, which defines the sequence 1,5,19,67,231,791,2703, ..., (sequence A035344 in the The On-Line Encyclopedia of Integer Sequences) is equivalent to the following succession rule:

$$\begin{cases} (5) \\ (5) \rightsquigarrow (-1)(5)^4 \\ (-1) \rightsquigarrow (1)(\overline{5})^2 \\ (1) \rightsquigarrow (1) \end{cases}$$

Therefore our next goal is to remove all possible marked labels from the succession rule. We observe that in order to obtain this goal, the recurrence relation $f_n = a_1f_{n-1} + a_2f_{n-2} + \cdots + a_kf_{n-k}$ with default initial conditions needs $a_1 > 0$. We assume that this condition holds throughout the rest of the present section.

In order to furnish a clearer description of our method, we start considering the case $k = 2$.

Proposition 3. *The C-finite recurrence $f_n = a_1f_{n-1} + a_2f_{n-2}$, with default initial conditions, and having $a_1 > 0$, is equivalent to*

$$\begin{cases} (a_1) \\ (a_1) \rightsquigarrow (0)^{q_2}(r_2)(a_1)^{a_1-(q_2+1)} \\ (r_2) \rightsquigarrow \left((0)^{q_2}(r_2)\right)^{q_2}(0)^{q_2}(r_2)(a_1)^{r_2-(q_2+1)^2} \end{cases} \tag{5}$$

where, by convention, the label (0) *does not produce any son, and* q_2, r_2 *are defined as follows:*
- *if* $a_1 + a_2 \leq 0$ *then* $q_2, r_2 > 0$ *such that* $|a_1 + a_2| = q_2 a_1 - r_2$;
- *otherwise* $q_2 = 0$, $r_2 = a_1 + a_2$.

Proof. We have to distinguish two cases: in the first one $a_1 + a_2 \leq 0$ and in the second one $a_1 + a_2 > 0$.

If $a_1 + a_2 \leq 0$, we have to prove that the generating tree associated to the succession rule (5) is obtained by performing some actions on the generating tree associated to the extended succession rule (6) which is obviously equivalent to the recurrence $f_n = a_1 f_{n-1} + a_2 f_{n-2}$ having $a_1 > 0$ and $a_2 < 0$, with $f_0 = 1$ and $f_h = 0$ for each $h < 0$.

$$\begin{cases} (a_1) \\ (a_1) \overset{1}{\rightsquigarrow} (a_1)^{a_1} \\ (a_1) \overset{2}{\rightsquigarrow} (a_1)^{a_2} \end{cases} \tag{6}$$

The proof consists in eliminating jumps and marked labels at each level of the generating tree associated with succession rule (6), sketched in Figure 2, by modifying the structure of the generating tree, still maintaining f_n nodes at level n, for each n.

Let (a_1) be a label at a given level n. We denote by B_1 the set of a_1 labels (a_1) at level $n+1$ and by B_2 the set of a_2 labels (a_1) at level $n+2$, see Figure 2. We remark that $(a_1)^{a_2} = \underbrace{\overline{(a_1)} \dots \overline{(a_1)}}_{-a_2}$.

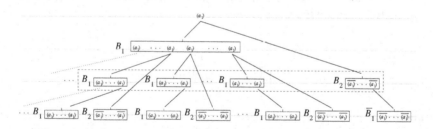

Fig. 2. Step 1

In order to eliminate both jumps and marked labels in B_2 at level 2 produced by the root (a_1) at level 0, we have to consider the set of a_1 labels (a_1) in B_1 at level 2 obtained by (a_1) which lie at level 1. At level 2, each label (a_1) in a given set B_1 kills one and only one marked label $\overline{(a_1)}$ in B_2. At this point $|a_1 + a_2|$ labels $\overline{(a_1)}$ in B_2 always exist at level 2.

In order to eliminate such marked labels we have to consider more than a single set B_1 of label (a_1) at level 2. Let q_2 be a sufficient number of sets B_1 at level 2 able to kill all the labels $\overline{(a_1)}$ in B_2 at level 2. Therefore $|a_1 + a_2| = q_2 a_1 - r_2$ with $q_2, r_2 > 0$.

We have the desired number of labels (a_1) at level 2 by setting q_2 labels (a_1) at level 1 equal to (0) and one more label (a_1) to (r_2). Note that the marked labels at level 2 are not generated and the labels (a_1) at level 1 are revised in order to have the right number of labels at level 2, see Figure 3.

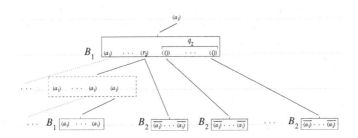

Fig. 3. Step 2

Note that, when a label (a_1) kills a marked label $\overline{(a_1)}$ at a given level n, then the subtree, having such label (a_1) as its root, kills the subtree having $\overline{(a_1)}$ as its root. So, when a label (a_1) of B_1 kills a label $\overline{(a_1)}$ of B_2 at level 2, then the two subtrees having such labels as their roots are eliminated too, see Figure 3.

On the other hand, the $q_2 + 1$ sets B_2 at level 3 obtained by the $q_2 + 1$ labels at level 1, once labelled with (a_1) and now having value $r_2, 0, \ldots, 0$, respectively, are always present in the tree, see Figure 3. In order to eliminate such undesired marked labels we can only set the production of (r_2). As a set B_2 at a given level is eliminated by using $q_2 + 1$ labels at previous level then (r_2) must give $(r_2) \underbrace{(0) \ldots (0)}_{q_2}$ exactly $q_2 + 1$ times. This explains the first part of the production rule of the label (r_2) in succession rule (5). Since (r_2) has r_2 sons then the remaining $r_2 - (q_2 + 1)^2$ labels are set to be equal to (a_1) as in the previous case, see Figure 4.

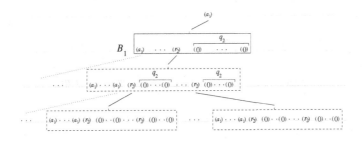

Fig. 4. Step 3

By the way, the modified q_2+1 labels having value $r_2, 0, \ldots, 0$, respectively, at a given level n, produce the labels $\left((0)^{q_2}(r_2)\right)^{q_2+1}(a_1)^{r_2-(q_2+1)^2}$ at level $n+1$. Just as obtained for levels 1 and 2, the labels $\left((0)^{q_2}(r_2)\right)^{q_2+1}$ automatically annihilate the remaining $q_2 + 1$ sets B_2 of marked labels at level $n + 2$, once obtained by the modified $q_2 + 1$ labels at level n, see Figure 4.

Till now we have modified a portion P of the total generating tree in a way that it does not contain any marked label. Note that, the remaining labels (a_1) will be the roots of subtrees which are all isomorphic to P.

The value f_n defined by the tree associated to the extended succession rule (6), is given by the difference between the number of non-marked and marked labels. The just described algorithm modifies the number of generated non-marked labels and sets to 0 the number of marked ones in a way that f_n is unchanged, for each n, so the succession rule (5) is equivalent to the recurrence $f_n = a_1 f_{n-1} + a_2 f_{n-2}$.

In the case $a_1 + a_2 > 0$ we have marked labels only if $a_2 < 0$. In this case a single set B_1 is sufficient to kill all the marked labels in B_2 at level 2. By the way, both in the case $a_2 < 0$ and $a_2 > 0$ we have that $q_2 = 0$ and $r_2 = a_1 + a_2$, and the succession rule (5) has the same form of the rule (4) which is equivalent to the recurrence $f_n = a_1 f_{n-1} + a_2 f_{n-2}$ having $a_1 > 0$ and $a_2 \in \mathbb{Z}$, with $f_0 = 1$ and $f_h = 0$ for each $h < 0$.

The statement of Proposition 3 can be naturally extended to the general case $k > 2$.

Proposition 4. *The C-finite sequence $\{f_n\}_n$ satisfying $f_n = a_1 f_{n-1} + a_2 f_{n-2} + \cdots + a_k f_{n-k}$, with default initial conditions and $a_1 > 0$ is equivalent to*

$$
\begin{cases}
(a_1) \\
(a_1) \rightsquigarrow (0)^{q_2}(r_2)(a_1)^{a_1-(q_2+1)} \\
(r_2) \rightsquigarrow \left((0)^{q_2}(r_2)\right)^{q_2}(0)^{q_3}(r_3)(a_1)^{r_2-(q_2(q_2+1)+q_3+1)} \\
\quad \vdots \\
(r_i) \rightsquigarrow \left((0)^{q_2}(r_2)\right)^{q_i}(0)^{q_{i+1}}(r_{i+1})(a_1)^{r_i-(q_i(q_2+1)+q_{i+1}+1)} \\
\quad \vdots \\
(r_k) \rightsquigarrow \left((0)^{q_2}(r_2)\right)^{q_k}(0)^{q_k}(r_k)(a_1)^{r_k-(q_k(q_2+1)+q_k+1)}
\end{cases}
\tag{7}
$$

where the parameters q_i and r_i, with $2 \leq i \leq k$, can be determined in the following way:
- if $\sum_{l=1}^{i} a_l \leq 0$ then $q_i, r_i > 0$ such that $|\sum_{l=1}^{i} a_l| = q_i a_1 - r_i$,
- otherwise $q_i = 0$ and $r_i = \sum_{l=1}^{i} a_l$.

Proof. It is omitted for brevity sake.

We can translate the previously considered recurrence relation $f_n = 5f_{n-1} - 6f_{n-2} + 2f_{n-3}$, with default initial conditions, into the following ordinary succession rule by using Proposition 4:

$$\begin{cases} (5) \\ (5) \rightsquigarrow (0)(4)(5)^3 \\ (4) \rightsquigarrow (0)(4)(1)(5) \\ (1) \rightsquigarrow (1) \end{cases}$$

being $q_2 = 1$, $r_2 = 4$, $q_3 = 0$ and $r_3 = 1$.

3.3 Positivity Condition

The statement of Proposition 4 is indeed a tool to translate C-finite recurrences into finite succession rules. However this property turns out to be effectively applicable only when the labels of the succession rule are all positive, and the reader can easily observe that Proposition 4 does not give us an instrument to test whether this happens or not.

In particular, if the labels of the succession rule are all positive then the terms of the C-finite sequence are all positive. It is then interesting to relate our problem with the so called *positivity problem*, which we have already mentioned in the Introduction.

Corollary 1. *Let us consider the recurrence relation* $f_n = a_1 f_{n-1} + a_2 f_{n-2} + \cdots + a_k f_{n-k}$ *having* $a_1 > 0$ *and* $a_i \in \mathbb{Z}$, $2 \leq i \leq k$, *with* $f_0 = 1$ *and* $f_h = 0$ *for each* $h < 0$. *If*

$$\begin{cases} a_1 - (q_2 + 1) \geq 0 \\ r_2 - (q_2(q_2 + 1) + q_3 + 1) \geq 0 \\ \vdots \\ r_i - (q_i(q_2 + 1) + q_{i+1} + 1) \geq 0 \ , \ \ 3 \leq i \leq k - 1 \\ \vdots \\ r_k - (q_k(q_2 + 1) + q_k + 1) \geq 0 \end{cases} \tag{8}$$

then $f_n > 0$ *for all* n.

As previously mentioned, condition (8) ensures that all the labels of the succession rules equivalent to the given C-finite recurrence are positive, hence all the terms f_n are positive. Thus it can be viewed as a sufficient condition to test the positivity of a given C-finite sequence.

Note that our criterion deals with a subclass of C-finite recurrence relations as it requires that $a_1 > 0$.

4 Conclusions and Further Developments

In this paper we have presented a general method to translate a given C-finite recurrence into an ordinary succession rule and we have proposed a sufficient condition for testing the positivity of a given C-finite sequence.

A further development could take into consideration the average complexity necessary to prove the positivity of a given C-finite sequence.

Afterwards, it should be interesting to develop the study concerning the C-finite recurrences with generic initial conditions in order to examine in depth the potentiality of our method.

References

1. Barcucci, E., Del Lungo, A., Frosini, A., Rinaldi, S.: A technology for reverse-engineering a combinatorial problem from a rational generating function. Advances in Applied Mathematics **26**(2), 129–153 (2001)
2. Barcucci, E., Del Lungo, A., Pergola, E., Pinzani, R.: ECO: a methodology for the Enumeration of Combinatorial Objects. Journal of Difference Equations and Applications **5**, 435–490 (1999)
3. Berstel, J., Mignotte, M.: Deux propriétés décidables des suites récurrentes linéaires. Bulletin de la Société Mathématique de France **104**(2), 175–184 (1976)
4. Berstel, J., Reutenauer, C.: Another proof of Soittola's Theorem. Theoretical Computer Science **393**, 196–203 (2008)
5. Bilotta, S., Grazzini, E., Pergola, E., Pinzani, R.: Avoiding cross-bifix-free binary words. ACTA Informatica **50**, 157–173 (2013)
6. Chung, F.R.K., Graham, R.L., Hoggatt, V.E., Kleimann, M.: The number of Baxter permutations. Journal of Combinatorial Theory Series A **24**, 382–394 (1978)
7. Duchi, E., Frosini, A., Pinzani, R., Rinaldi, S.: A note on rational succession rules. Journal of Integer Sequences 6, Article 03.1.7 (2003)
8. Ferrari, L., Pergola, E., Pinzani, R., Rinaldi, S.: An algebraic characterization of the set of succession rules. Theoretical Computer Science **281**, 351–367 (2002)
9. Ferrari, L., Pergola, E., Pinzani, R., Rinaldi, S.: Jumping succession rules and their generating functions. Discrete Mathematics **271**, 29–50 (2003)
10. Gerhold, S.: Sequences: non-holonomicity and inequalities, ph.D. Thesis
11. Gessel, I.: Rational functions with nonnegative integer coefficients. In The 50th seminaire Lotharingien de Combinatoire, page Domaine Saint-Jacques. Unpublished, available at Gessels homepage March 2003
12. Halava, V., Harju, T., Hirvensalo, M.: Positivity of second order linear recurrent sequences. Discrete Applied Mathematics **154**, 447–451 (2006)
13. Koutschan, C.: Regular languages and their generating functions: The inverse problem. Theoretical Computer Science **391**, 65–74 (2008)
14. Laohakosol, V., Tangsupphathawat, P.: Positivity of third order linear recurrence sequences. Discrete Applied Mathematics **157**, 3239–3248 (2009)
15. Perrin, D.: On positive matrices. Theoretical Computer Science **94**(2), 357–366 (1992)
16. Salomaa, A., Soittola, M.: Automata-Theoretic Aspects of Formal Power Series. Springer-Verlag (1978)
17. Soittola, M.: Positive rational sequences. Theoretical Computer Science **2**(3), 317–322 (1976)
18. Zeilberger, D.: A holonomic systems approach to special functions identities. Journal of Computational and Applied Mathematics **32**, 321–368 (1990)

On the Complexity of Fragments of the Modal Logic of Allen's Relations over Dense Structures

Davide Bresolin[1], Dario Della Monica[2]([⊠]), Angelo Montanari[3],
Pietro Sala[4], and Guido Sciavicco[5]

[1] Department of Computer Science and Engineering,
University of Bologna, Bologna, Italy
davide.bresolin@unibo.it

[2] ICE-TCS, School of Computer Science, Reykjavik University, Reykjavik, Iceland
dariodm@ru.is

[3] Department of Mathematics and Computer Science,
University of Udine, Udine, Italy
angelo.montanari@uniud.it

[4] Department of Computer Science, University of Verona, Verona, Italy
pietro.sala@univr.it

[5] Department of Information, Engineering and Communications,
University of Murcia, Murcia, Spain
guido@um.es

Abstract. Interval temporal logics provide a natural framework for temporal reasoning about interval structures over linearly ordered domains, where intervals are taken as the primitive ontological entities. Their computational behaviour and expressive power mainly depend on two parameters: the set of modalities they feature and the linear orders over which they are interpreted. In this paper, we consider all fragments of Halpern and Shoham's interval temporal logic HS with a decidable satisfiability problem over the class of all dense linear orders, and we provide a complete classification of them in terms of their complexity and expressiveness by solving the last two open cases.

Keywords: Computational complexity · Interval temporal logics · Satisfiability · Expressiveness · Decidability

1 Introduction

Most temporal logics proposed in the literature assume a point-based structure of time. They have been successfully applied in a variety of fields, ranging from the specification and verification of communication protocols to temporal data mining. However, a number of relevant application domains, such as, for instance, those of planning and synthesis of controllers, are often characterized by advanced features like durative actions (and their temporal relationships), accomplishments, and temporal aggregations, which are neglected or dealt with in an unsatisfactory way by point-based formalisms. The distinctive features of

© Springer International Publishing Switzerland 2015
A.-H. Dediu et al. (Eds.): LATA 2015, LNCS 8977, pp. 511–523, 2015.
DOI: 10.1007/978-3-319-15579-1_40

interval temporal logics turn out to be useful in these domains. As an example, they allow one to model telic statements [18], that is, statements that express goals or accomplishments, like the statement: "The airplane flew from Venice to Toronto" (see [8, Sect. II.B]). Temporal logics with interval-based semantics have also been proposed as suitable formalisms for the specification and verification of hardware [15] and of real-time systems [10]. Finally, successful implementations of interval-based systems can be found in the areas of learning (the adaptive learning system TERENCE [11], that provides a support to poor comprehenders and their educators, is based on the so-called Allen's interval algebra [3]) and real-time data systems (the algorithm RISMA [13], for performance and behaviour analysis of real-time data systems, is based on Halpern and Shoham's modal logic of Allen's relations [12]).

The variety of binary relations between intervals in a linear order was first studied by Allen [3], who investigated their use in systems for time management and planning. In [12], Halpern and Shoham introduced and systematically analyzed the (full) modal logic of Allen's relations (HS for short), that features one modality for each Allen relation. In particular, they showed that HS is highly undecidable over most classes of linear orders. This result motivated the search for (syntactic) fragments of HS offering a good balance between expressiveness and computational complexity. During the last decade, a systematic analysis has been carried out to characterize the complexity of the satisfiability problem for HS fragments [4,5,16], as well as their relative expressive power [1,2,5]. Such an analysis pointed out that such characterizations also depend on the class of linearly ordered set over which formulae are interpreted.

This paper aims at completing the classification of decidable HS fragments with respect to both their complexity and expressiveness, relative to the class of (all) dense linear orders. For our purposes, the class of dense linear orders and the linear order of the rational numbers $\mathbb{Q}$ are indistinguishable. Thus, all the results presented here directly apply to $\mathbb{Q}$ as well. The paper is organized as follows. In Section 2, we introduce syntax and semantics of (fragments of) HS. Next, in Section 3 we summarize known results about dense linear orders. In Section 4 and Section 5, we solve the last two open problems, thus completing the picture for the class of dense linear structures. It is worth mentioning that an analogous classification has been provided in [5] for the class of finite linear orders, the class of discrete linear orders, the linear order of the natural numbers $\mathbb{N}$, and the linear order of the integers $\mathbb{Z}$.

2   The Modal Logic of Allen's Relations

Let us consider a linearly ordered set $\mathbb{D} = \langle D, < \rangle$, where D is an element domain and $<$ is a total ordering on it. An *interval* over $\mathbb{D}$ is an ordered pair $[x, y]$, where $x, y \in D$ and $x \leq y$. An interval is called a *point interval* if $x = y$ and a *strict interval* if $x < y$. In this paper, we assume the *strict semantics*, that is, we exclude point intervals and only consider strict intervals. The adoption of the strict semantics, excluding point intervals, instead of the *non-strict semantics*,

HS modalities	Allen's relations	Graphical representation
$\langle A \rangle$	$[x,y]R_A[x',y'] \Leftrightarrow y = x'$	
$\langle L \rangle$	$[x,y]R_L[x',y'] \Leftrightarrow y < x'$	
$\langle B \rangle$	$[x,y]R_B[x',y'] \Leftrightarrow x = x', y' < y$	
$\langle E \rangle$	$[x,y]R_E[x',y'] \Leftrightarrow y = y', x < x'$	
$\langle D \rangle$	$[x,y]R_D[x',y'] \Leftrightarrow x < x', y' < y$	
$\langle O \rangle$	$[x,y]R_O[x',y'] \Leftrightarrow x < x' < y < y'$	

Fig. 1. Allen's interval relations and the corresponding HS modalities

which includes them, conforms to the definition of interval adopted by Allen in [3], but differs from the one given by Halpern and Shoham in [12]. It has at least two strong motivations: first, a number of representation paradoxes arise when the non-strict semantics is adopted, due to the presence of point intervals, as pointed out in [3]; second, when point intervals are included there seems to be no intuitive semantics for interval relations that makes them both pairwise disjoint and jointly exhaustive. If we exclude the identity relation, there are 12 different relations between two strict intervals in a linear order, often called *Allen's relations* [3]: the six relations R_A (*meets* or *adjacent*), R_L (*after* or *later*), R_B (*starts* or *begins*), R_E (*finishes* or *ends*), R_D (*during*), and R_O (*overlaps*), depicted in Fig. 1, and their inverses, that is, $R_{\overline{X}} = (R_X)^{-1}$, for each $X \in \{A, L, B, E, D, O\}$.

We interpret interval structures as Kripke structures with Allen's relations playing the role of the accessibility relations. Thus, we associate a modality $\langle X \rangle$ with each Allen relation R_X. For each $X \in \{A, L, B, E, D, O\}$, the *transpose* of modality $\langle X \rangle$ is modality $\langle \overline{X} \rangle$, corresponding to the inverse relation $R_{\overline{X}}$ of R_X. Halpern and Shoham's logic HS [12] is a multi-modal logic with formulae built from a finite, non-empty set $\mathcal{AP}$ of atomic propositions (also referred to as proposition letters), the propositional connectives $\vee$ and $\neg$, and a modality for each Allen relation. With every subset $\{R_{X_1}, \ldots, R_{X_k}\}$ of these relations, we associate the fragment $X_1 X_2 \ldots X_k$ of HS, whose formulae are defined by the grammar:

$$\varphi ::= p \mid \neg\varphi \mid \varphi \vee \varphi \mid \langle X_1 \rangle \varphi \mid \ldots \mid \langle X_k \rangle \varphi,$$

where $p \in \mathcal{AP}$. The other propositional connectives and constants (e.g., $\wedge$, $\rightarrow$, and $\top$), as well as the dual modalities (e.g., $[A]\varphi \equiv \neg\langle A \rangle \neg\varphi$), can be derived in the standard way.

The (strict) semantics of HS is given in terms of *interval models* $M = \langle \mathbb{I}(\mathbb{D}), V \rangle$, where $\mathbb{D}$ is a linear order, $\mathbb{I}(\mathbb{D})$ is the set of all (strict) intervals over $\mathbb{D}$, and V is a *valuation function* $V : \mathcal{AP} \rightarrow 2^{\mathbb{I}(\mathbb{D})}$, which assigns to each atomic proposition $p \in \mathcal{AP}$ the set of intervals $V(p)$ on which p holds. The *truth* of a formula on a given interval $[x, y]$ in an interval model M is defined by structural induction on formulae as follows:

- $M, [x, y] \Vdash p$ if and only if $[x, y] \in V(p)$, for each $p \in \mathcal{AP}$;
- $M, [x, y] \Vdash \neg\psi$ if and only if it is not the case that $M, [x, y] \Vdash \psi$;
- $M, [x, y] \Vdash \varphi \vee \psi$ if and only if $M, [x, y] \Vdash \varphi$ or $M, [x, y] \Vdash \psi$;
- $M, [x, y] \Vdash \langle X \rangle \psi$ if and only if there exists $[x', y']$ such that $[x, y] R_X [x', y']$ and $M, [x', y'] \Vdash \psi$, for each modality $\langle X \rangle$.

Formulae of HS can be interpreted over a given class of interval models; we identify the class of interval models over linear orders in $\mathcal{C}$ with the class $\mathcal{C}$ itself. Thus, we will use, for example, the expression 'formulae of HS are interpreted over the class $\mathcal{C}$ of linear orders' instead of the extended one 'formulae of HS are interpreted over the class of interval models over linear orders in $\mathcal{C}$'. Among others, we mention the following important classes of linear orders: *(i)* the class of *all* linear orders Lin; *(ii)* the class of all *dense* linear orders Den, that is, those in which for every pair of different points there exists at least one point in between them; *(iii)* the class of all *weakly discrete* linear orders WDis, that is, those in which every element, apart from the greatest one, if it exists, has an immediate successor, and every element, other than the least one, if it exists, has an immediate predecessor; *(iv)* the class of all *strongly discrete* linear orders Dis, that is, those in which for every pair of different points there are only finitely many points in between them; *(v)* the class of all *finite* linear orders Fin, that is, those having only finitely many points; *(vi)* the singleton classes consisting of the standard linear orders over $\mathbb{R}$, $\mathbb{Q}$, $\mathbb{Z}$, and $\mathbb{N}$. The *mirror image* (or, simply, *mirror*) of a fragment $\mathcal{F}$ is obtained by simultaneously substituting $\langle A \rangle$ with $\langle \overline{A} \rangle$, $\langle B \rangle$ with $\langle E \rangle$, $\langle \overline{B} \rangle$ with $\langle \overline{E} \rangle$, $\langle O \rangle$ with $\langle \overline{O} \rangle$, $\langle L \rangle$ with $\langle \overline{L} \rangle$, and the other way around. When interpreted over left/right symmetric classes of structures (i.e., classes $\mathcal{C}$ such that if $\mathcal{C}$ contains a linear order $\mathbb{D} = \langle D, \prec \rangle$, then it also contains a linear order isomorphic to its dual linear order $\mathbb{D}^d = \langle D, \succ \rangle$, where $\succ$ is the inverse of $\prec$), such as Den, all computational properties of a fragment are preserved for its mirror one; thanks to this observation, we can safely deal with only one fragment for each pair of mirror fragments.

3   Known and Unknown Results

It has been proved in [1] that there are precisely 9 different optimal definabilities that hold among HS modalities in the dense case. As a consequence, only 966 HS fragments are expressively different (out of 4096 different subsets of 12 modalities). Of those, 146 are decidable, thanks to the following results:

Undecidability: we know from [4] that each fragment containing (as definable) O, AD, or $A\overline{D}$ is undecidable;

Non-primitive recursive: the decidability of $A\overline{A}B\overline{B}$ has been proved in [14], where it has also been shown that each fragment containing $A\overline{A}B$ or $A\overline{A}B$ is non-primitive recursive;

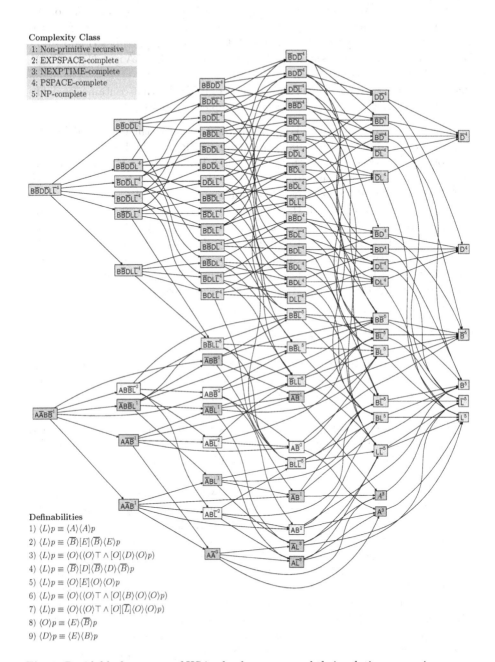

Complexity Class
1: Non-primitive recursive
2: EXPSPACE-complete
3: NEXPTIME-complete
4: PSPACE-complete
5: NP-complete

Definabilities

1) $\langle L\rangle p \equiv \langle A\rangle\langle A\rangle p$
2) $\langle L\rangle p \equiv \langle\overline{B}\rangle[E]\langle\overline{B}\rangle\langle E\rangle p$
3) $\langle L\rangle p \equiv \langle O\rangle(\langle O\rangle\top \wedge [O]\langle D\rangle\langle O\rangle p)$
4) $\langle L\rangle p \equiv \langle\overline{B}\rangle[D]\langle\overline{B}\rangle\langle D\rangle\langle\overline{B}\rangle p$
5) $\langle L\rangle p \equiv \langle O\rangle[E]\langle O\rangle\langle O\rangle p$
6) $\langle L\rangle p \equiv \langle O\rangle(\langle O\rangle\top \wedge [O]\langle B\rangle\langle O\rangle\langle O\rangle p)$
7) $\langle L\rangle p \equiv \langle O\rangle(\langle O\rangle\top \wedge [O][\overline{L}]\langle O\rangle\langle O\rangle p)$
8) $\langle O\rangle p \equiv \langle E\rangle\langle\overline{B}\rangle p$
9) $\langle D\rangle p \equiv \langle E\rangle\langle B\rangle p$

Fig. 2. Decidable fragments of HS in the dense case and their relative expressive power

ExpSpace-completeness: as a consequence of the results presented in [8], we know that ABB̄L̄ is in ExpSpace, and each fragment containing AB or AB̄ is ExpSpace-hard (in particular, the hardness result given in [8] for ABB̄ can be suitably rephrased to deal with the smaller fragments AB and AB̄);

NExpTime-completeness: it has been proved in [7] that AĀ is in NExpTime, and both A and Ā are NExpTime-hard;

PSpace-completeness: each sub-fragment of BB̄DD̄L̄L̄ that contains (as definable) D or D̄ is shown to be PSpace-complete in [6,16].

The purpose of this paper is to fill in the few gaps still uncovered by this collection of results. Here, we shall prove that: *(i)* BB̄LL̄ and all its fragments are NP-complete (observe that each fragment is NP-hard, given that it is at least as expressive as propositional logic), and *(ii)* all the fragments that contain ĀB or ĀB̄ are non-primitive recursive. The aforementioned results allow us to draw a picture that encompasses all HS fragments, ordered according to their relative expressive power and grouped by computational complexity. We show here such a picture (see Fig. 2), limited to all and only decidable HS fragments (for the sake of readability, we omit fragments that are expressively equivalent or mirror image of another fragment featured in the picture). In Fig. 2 we also show the 9 definabilities that hold among HS modalities over dense linear orders.

4   NP-Complete Fragments

In this section we show that the fragment BB̄LL̄ is in NP (NP-completeness immediately follows as propositional logic is embedded into BB̄LL̄). By defining a suitable notion of pseudo-model for formulae of BB̄LL̄ we can show that each satisfiable formula admits a pseudo-model of size at most $P(|\varphi|)$ for some polynomial P. For lack of space, in this paper we only give the intuition behind the concept of pseudo-model and the main ideas behind the small pseudo-model theorem. A detailed account of the proof can be found in [9].

We start the discussion by considering the fragment LL̄. The semantics of the interval modalities implies that intervals with the same ending point agree on the truth of $\langle L \rangle$-formulae (i.e., formulae of the kind $\langle L \rangle \varphi$); symmetrically, intervals with the same beginning point agree on $\langle \overline{L} \rangle$-formulae. Hence, given a model M for a formula φ, we can associate to every point x the set of its LL̄-*requests*, defined as the pair of sets $(L_x, \overline{L}_x)$, where L_x contains all formulae ψ in the *closure of* φ (that is, the set of all sub-formulae of φ and their negations) such that $\langle L \rangle \psi$ is true over all intervals $[y, x]$, and $\overline{L}_x$ contains all formulae ψ in the closure of φ such that $\langle \overline{L} \rangle \psi$ is true over all intervals $[x, y]$. Since the closure of a formula is a finite set, we can partition the domain of the model into a finite number of clusters of points with the same set of LL̄-requests. Moreover, by the transitivity of both $\langle L \rangle$ and $\langle \overline{L} \rangle$, we have that the set of LL̄-requests is monotone with respect to the ordering of points, that is, for every pair of points $x < y$ we have $L_x \supseteq L_y$ and $\overline{L}_x \subseteq \overline{L}_y$. This implies that every cluster is either a single *point* or a *segment* of D, and that the number of clusters is at most $4|\varphi|$.

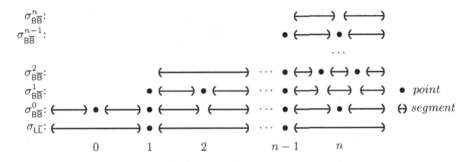

Fig. 3. A pseudo-model for $B\overline{B}L\overline{L}$

A pseudo-model for $L\overline{L}$ is an abstract representation of the partitioning. It is formally defined as a finite $L\overline{L}$-*sequence* of triples:

$$(L_0, \overline{L}_0, Type_0), (L_1, \overline{L}_1, Type_1), \ldots, (L_n, \overline{L}_n, Type_n),$$

where each *Type* is either *point* or *segment*, and such that: *(i)* the monotonicity of $L\overline{L}$-requests is respected; *(ii)* the first and the last triple of the sequence are of type *segment*; *(iii)* clusters of type *point* cannot be adjacent. To represent a well-formed model for φ, an $L\overline{L}$-sequence must respect the following additional constraints:

- it must be *consistent*: for every pair of indexes $i < j$ there must exists an *atom* F (that is, a maximally consistent subset of the closure) that contains the formula $\langle L \rangle \psi$ for every $\psi \in L_j$, the formula $\neg \xi$ for every $\xi \notin L_i$, the formula $\langle \overline{L} \rangle \eta$ for every $\eta \in \overline{L}_i$ and the formula $\neg \zeta$ for every $\zeta \notin \overline{L}_j$;
- it must be L-*fulfilling*: for every index i and every formula $\psi \in L_i$ there must exists a pair of indexes $i < j < k$ and atom F containing ψ and consistent with the clusters j and k;
- it must be $\overline{L}$-*fulfilling*, which is defined analogously.

The consistency condition guarantees that $[L]$- and $[\overline{L}]$-formulae are satisfied, while the fulfillment conditions guarantee that $\langle L \rangle$- and $\langle \overline{L} \rangle$-formulae are satisfied as well. We have already observed that the number of clusters (and thus, the length of an $L\overline{L}$-sequence) is bounded by $4|\varphi|$. Hence, by guessing a $L\overline{L}$-sequence and then checking it for consistency and fulfillment we can easily obtain an NP procedure for deciding the satisfiability of a formula in $L\overline{L}$.

The extension of the above result to the full $B\overline{B}L\overline{L}$ language is based on the following observation. Given a model for the formula and an interval $[x, y]$ we define the set of $B\overline{B}$-*requests* of the interval as the pair $(B_{[x,y]}, \overline{B}_{[x,y]})$, where $B_{[x,y]}$ contains all formulae ψ in the closure of φ such that $\langle B \rangle \psi$ is true on $[x, y]$, and $\overline{B}_{[x,y]}$ contains all formulae ξ in the closure of φ such that $\langle \overline{B} \rangle \xi$ is true on $[x, y]$. Fixed a point x in the model, we have that the sets of $B\overline{B}$-requests of the intervals $[x, y]$ with begin point x respect the same monotonicity property as for $L\overline{L}$-requests: for every pair of points $y < z$ we have $B_{[x,y]} \subseteq B_{[x,z]}$ and

$\overline{B}_{[x,y]} \supseteq \overline{B}_{[x,z]}$. Hence, it is possible to partition the intervals starting in any given point x into at most $4|\varphi|$ "points" and "segments". A pseudo-model for B$\overline{\text{B}}$L$\overline{\text{L}}$ is then made of the following components (see Fig. 3 for a graphical account):

– an L$\overline{\text{L}}$-sequence $\sigma_{\text{L}\overline{\text{L}}} = (L_0, \overline{L}_0, Type_0), (L_1, \overline{L}_1, Type_1), \ldots, (L_n, \overline{L}_n, Type_n)$ defining the partitioning of L$\overline{\text{L}}$-requests;
– for every cluster $(L_i, \overline{L}_i, Type_i)$ of the L$\overline{\text{L}}$-sequence, a B$\overline{\text{B}}$-sequence $\sigma^i_{\text{B}\overline{\text{B}}} = (B_i, \overline{B}_i, Type_i), (B_{i+1}, \overline{B}_{i+1}, Type_{i+1}), \ldots, (B_m, \overline{B}_m, Type_m)$ representing all intervals $[x, y]$ such that x belongs to the ith cluster $(L_i, \overline{L}_i, Type_i)$. $\sigma^i_{\text{B}\overline{\text{B}}}$ must be a refinement of the partitioning $(L_i, \overline{L}_i, Type_i) \ldots (L_n, \overline{L}_n, Type_n)$.

The consistency and the fulfillment condition are suitably extended to guarantee satisfiability of B$\overline{\text{B}}$-formulae. Since the size of a B$\overline{\text{B}}$L$\overline{\text{L}}$ pseudo-model is quadratic in the size of the formula, we can easily obtain an NP decision procedure that guesses a pseudo-model and checks the satisfiability of a formula in B$\overline{\text{B}}$L$\overline{\text{L}}$.

Theorem 1. *The satisfiability problem for the logic* B$\overline{\text{B}}$L$\overline{\text{L}}$ *and each one of its fragments, interpreted over the class of dense linear orders, is NP-complete.*

5   Non-Primitive Recursive Fragments

As we have mentioned, the last piece needed to complete the picture in Fig. 2 concerns the non-primitive recursive fragments. In [14] the non-primitive recursiveness of A$\overline{\text{A}}$B and A$\overline{\text{AB}}$ has been proved. We shall prove here that, in actuality, every fragment that contains $\overline{\text{AB}}$ or $\overline{\text{A}}$B is non-primitive recursive.

Lossy counter machines are a variant of Minsky counter automata where transitions may non-deterministically decrease the values of counters. A comprehensive survey on faulty machines and on the relevant complexity, decidability, and undecidability results can be found in [17]. Formally, a *counter automaton* is a tuple $\mathcal{A} = (Q, q_0, C, \Delta)$, where Q is a finite set of control *states*, $q_0 \in Q$ is the initial state, $C = \{c_1, \ldots, c_k\}$ is the set of *counters*, whose values range over $\mathbb{N}$, and Δ is a *transition relation*. The relation Δ is a subset of $Q \times L \times Q$, where L is the *instruction set* $L = \{inc, dec, ifz\} \times \{1, \ldots, k\}$. A *configuration* of $\mathcal{A}$ is a pair $(q, \bar{v})$, where $q \in Q$ and $\bar{v}$ is the vector of counter values. A *run* of a Minsky (i.e., with no error) counter automaton is a finite or infinite sequence of configurations such that, for every pair of consecutive configurations $(q, \bar{v}), (q', \bar{v}')$, a *transition* $(q, \bar{v}) \xrightarrow{l} (q', \bar{v}')$ has been taken (for some $(q, l, q') \in \Delta$). The value of $\bar{v}'$ is obtained from the value of $\bar{v}$ by performing instruction l, where $l = (dec, i)$ requires $v_i > 0$ and $l = (ifz, i)$ requires $v_i = 0$. In lossy machines, which is the type in which we are interested, once a faulty transition has been taken, counter values may have been decreased nondeterministically before or after the execution of the exact transition by an arbitrary natural number. We use the notation $(q, \bar{v}) \xrightarrow{l}_\dagger (q', \bar{v}')$ to denote that there exist $\bar{v}_\dagger, \bar{v}'_\dagger$ such that $\bar{v} \geq \bar{v}_\dagger$, $(q, \bar{v}_\dagger) \xrightarrow{l} (q', \bar{v}'_\dagger)$, and $\bar{v}'_\dagger \geq \bar{v}'$, where the ordering $\leq$ is defined component-wise in the obvious way. We are interested here in the *non-termination problem* for lossy machines, defined as the problem of deciding whether $\mathcal{A}$ has at least one infinite run starting with the *initial configuration* $(q_0, \bar{0})$. This problem is non-primitive recursive [17].

Lemma 2. *There exists a reduction from the non-termination problem for lossy counter machines to the satisfiability problem for $\overline{\text{AB}}$ over the class of all dense linear orders.*

Proof. Let $\mathcal{A} = (Q, q_0, C, \Delta)$ be a lossy counter machine. We write an $\overline{\text{AB}}$-formula $\varphi_{\mathcal{A}}$ which is satisfiable over a dense linear order if and only if $\mathcal{A}$ has at least one infinite run starting with the initial configuration. The computation is encoded left-to-right over a dense domain $\mathbb{D}$, by choosing an evaluation interval $[x, y]$ that works as the "last" one, and taking into account that, given any $x_0 < x$, there are infinitely many intervals between x_0 and x. We shall make use of the propositional letters u (*units*), q_i (*states*, where i ranges from 0 to $|Q|$), *conf* (*configurations*), c_i (*counters' instances*, where i ranges from 1 to $|C|$), and *corr*, *corr_i* (*corresponds*; i ranges from 1 to $|C|$). Counters' instances, or simply *counters*, allow us to encode the counters of $\mathcal{A}$: given a configuration where the value of the i-th counter is n, the corresponding *conf*-interval will contain precisely n c_i-intervals. (By *p-interval* we denote those intervals that satisfy p, for every propositional letter p.) Additional propositional letters will be used in the reduction for technical reasons.

Let $[G]$ (*universal modality*) be the following shortcut:

$$[G]\varphi = \varphi \wedge [B]\varphi \wedge [\overline{A}]\varphi \wedge [\overline{A}][\overline{A}]\varphi.$$

The first step in our construction consists in *discretizing* the domain, making use of a propositional variable u. In doing so, we also set the first configuration:

$$\varphi_{u\text{-}chain} = \begin{cases} \langle \overline{A} \rangle \langle \overline{A} \rangle (u \wedge conf \wedge start \wedge q_0) \wedge [\overline{A}](\langle \overline{A} \rangle u \to \langle B \rangle u) \\ [G](u \to [B]\neg u) \wedge [G](u \to [B]u_b) \wedge [G](u \to [\overline{A}]\neg u_b) \\ [G](start \to u) \wedge [G](start \to [\overline{A}](\neg u \wedge [\overline{A}]\neg u)) \end{cases}$$

Consider an interval $[x, y]$ over which the formula of our reduction is evaluated. The sense of the above formula $\varphi_{u\text{-}chain}$ is to generate an infinite discrete chain $x_0, x_1, \ldots$ such that $x_0 < x_1 < \ldots < x < y$, and that each $[x_k, x_{k+1}]$ is labeled by u. With the above formulae we also guarantee that *start* is unique and no u-interval overlaps a u-interval in the chain.

With the next formulae we make sure that there is a infinite sequence of configurations. The first one (*start*) coincides with the unit $[x_0, x_1]$, and contains the starting state q_0 only. This is consistent with our requirement that all counters start with the value 0. Moreover, we guarantee that configurations' endpoints coincide with endpoints of elements of the u-chain, that every configuration contains a state, and that *start* is unique. In our reduction, the state is placed on the last unit of every configuration.

$$\varphi_{conf\text{-}chain} = \begin{cases} [G](conf \to (u \vee \langle B \rangle u)) \wedge [G](\langle \overline{A} \rangle conf \to \langle \overline{A} \rangle u) \\ [\overline{A}](\langle \overline{A} \rangle conf \to \langle B \rangle conf) \wedge [G](conf \to [B]conf_b \wedge [B]\neg conf) \\ [G](conf \to [\overline{A}]\neg conf_b) \wedge [G](\langle \overline{A} \rangle conf \leftrightarrow \langle \overline{A} \rangle (\bigvee_{i=0,\ldots,|Q|} q_i)) \end{cases}$$

Notice that states (q_i-intervals) occur exactly as last u-intervals of configurations. Since configurations do not overlap, this implies that each configuration contains exactly one state.

Configurations also contain counters' instances c_i for each counter i whose value is greater than zero. Besides, a special placeholder c_i^+ or c_i^- may be placed in a configuration, in order to make it possible to deal with increment and decrement operations. States, counters' instances, and placeholders may only hold over units, which, in turn, all have to contain one of the above. A placeholder must be placed over the counter to which it refers. Moreover, counters and states are mutually incompatible, and there cannot be more than one per type on a given unit. These requirements are guaranteed by the following formula:

$$\varphi_{units} = \begin{cases} [G](\bigwedge_{i=0,\ldots,|Q|}(q_i \to u) \wedge \bigwedge_{i=1,\ldots,|C|}((c_i \vee c_i^+ \vee c_i^-) \to u)) \\ [G](u \to ((\bigvee_{i=0,\ldots,|Q|} q_i) \vee (\bigvee_{i=1,\ldots,|C|} c_i))) \\ [G]\bigwedge_{i=0,\ldots,|Q|}(q_i \to (\bigwedge_{j=i+1,\ldots,|Q|} \neg q_j)) \\ [G]\bigwedge_{i=0,\ldots,|Q|}(q_i \to (\bigwedge_{j=1,\ldots,|C|} \neg c_j)) \\ [G]\bigwedge_{i=1,\ldots,|C|}((c_i \to (\bigwedge_{j=i+1,\ldots,|C|} \neg c_j)) \wedge (c_i^- \to c_i) \wedge (c_i^+ \to c_i)) \end{cases}$$

Before we can actually encode the transition function Δ, we have to axiomatize the properties of $corr$ and $corr_i$ for each i. In a perfect (non-faulty) machine, when a counter is not modified by any operation from a configuration to the next one its value is preserved. Since we are encoding a lossy machine, it suffices to guarantee that no counter's value is ever incremented, except for the special case of an incrementing operation. To this end, we use the propositional letter $corr$ as a basis for correspondence, and the proposition $corr_i$ to identify the correspondence for the i-th counter:

$$\varphi_{corr} = \begin{cases} [G]\bigwedge_{i=1,\ldots,|C|}(((c_i \wedge \neg c_i^+) \to \langle \overline{A} \rangle corr_i) \wedge (c_i^+ \to \neg \langle \overline{A} \rangle corr_i)) \\ [G]\bigwedge_{i=1,\ldots,|C|}(corr_i \to corr) \\ [G]\bigwedge_{i=1,\ldots,|C|}(corr_i \to \langle \overline{A} \rangle (c_i \wedge \neg c_i^-)) \wedge [G](corr \to [B] corr_b) \\ [G](((\bigvee_{i=0,\ldots,|Q|} q_i) \wedge corr_b) \to corr_b{}^*) \\ [G]((\bigvee_{i=0,\ldots,|Q|} q_i) \to [\overline{A}](corr_b \to corr_b{}^*)) \\ [G](corr \to [B] \neg corr) \wedge [G](corr_b{}^* \to [B] \neg corr_b{}^*) \\ [G](\langle \overline{A} \rangle corr_b{}^* \to \langle \overline{A} \rangle u) \wedge [G](corr \to \langle B \rangle corr_b{}^*) \\ [G]((u \wedge \neg(\bigvee_{i=0,\ldots,|Q|} q_i)) \to [\overline{A}] \neg corr_b{}^*) \end{cases}$$

To finalize the reduction, we now take care of incrementing and decrementing operations, as well as of the zero test. For each $(q, l, q') \in \Delta$, let $conf_{(q,l,q')}$ be a special propositional letter holding on a configuration and carrying information on which transition produced that configuration. Clearly, every configuration but $start$ is the result of precisely one transition. Therefore, we have:

$$\varphi_{conf} = \begin{cases} [G]((conf \wedge \neg start) \leftrightarrow (\bigvee_{(q,l,q') \in \Delta} conf_{(q,l,q')})) \\ [G](\bigwedge_{(q,l,q') \in \Delta}(conf_{(q,l,q')} \to (\bigwedge_{(q'',l',q''') \neq (q,l,q')} \neg conf_{(q'',l',q''')}))) \end{cases}$$

We can now implement the actual transitions. To deal with the increment (resp., decrement) operation we make use of the symbol c_i^+ (resp., c_i^-), as follows:

$$\varphi_{inc} = \begin{cases} [G](\bigwedge_{(q,(inc,i),q')\in\Delta}(conf_{(q,(inc,i),q')} \to (\langle\overline{A}\rangle q \wedge \langle B\rangle c_{i,b}^+))) \\ [G](\bigwedge_{(q,(inc,i),q')\in\Delta}(\langle\overline{A}\rangle conf_{(q,(inc,i),q')} \to \langle\overline{A}\rangle q')) \\ [G](\bigwedge_{i=1,\ldots,|C|}(\langle\overline{A}\rangle c_{i,b}^+ \leftrightarrow \langle\overline{A}\rangle c_i^+)) \\ [G](\bigwedge_{i=1,\ldots,|C|}(c_{i,b}^+ \to (\langle\overline{A}\rangle conf \wedge [B]\neg conf))) \\ [G](\bigwedge_{i,j=1,\ldots,|C|}(c_{i,b}^+ \to [B]\neg c_{j,b}^+)) \\ [G](\bigwedge_{i=1,\ldots,|C|}((conf \wedge \langle B\rangle c_{i,b}^+) \to (\bigvee_{q,q'\in Q} conf_{(q,(inc,i),q')}))) \end{cases}$$

$$\varphi_{dec} = \begin{cases} [G](\bigwedge_{(q,(dec,i),q')\in\Delta}(conf_{(q,(dec,i),q')} \to (\langle\overline{A}\rangle q \wedge [\overline{A}](conf \to \langle B\rangle c_{i,b}^-)))) \\ [G](\bigwedge_{(q,(dec,i),q')\in\Delta}(\langle\overline{A}\rangle conf_{(q,(dec,i),q')} \to \langle\overline{A}\rangle q')) \\ [G](\bigwedge_{i=1,\ldots,|C|}(\langle\overline{A}\rangle c_{i,b}^- \leftrightarrow \langle\overline{A}\rangle c_i^-)) \\ [G](\bigwedge_{i=1,\ldots,|C|}(c_{i,b}^- \to (\langle\overline{A}\rangle conf \wedge [B]\neg conf))) \\ [G](\bigwedge_{i,j=1,\ldots,|C|}(c_{i,b}^- \to [B]\neg c_{j,b}^-)) \\ [G](\bigwedge_{i=1,\ldots,|C|}((conf \wedge \langle\overline{A}\rangle\langle B\rangle c_{i,b}^-) \to (\bigvee_{q,q'\in Q} conf_{(q,(dec,i),q')}))) \end{cases}$$

$$\varphi_{ifz} = \begin{cases} [G](\bigwedge_{(q,(ifz,i),q')\in\Delta}(conf_{(q,(ifz,i),q')} \to (\langle\overline{A}\rangle q \wedge [\overline{A}](conf \to [B]c_{i,b}^z)))) \\ [G](\bigwedge_{(q,(ifz,i),q')\in\Delta}(\langle\overline{A}\rangle conf_{(q,(ifz,i),q')} \to \langle\overline{A}\rangle q')) \\ [G](\bigwedge_{i=1,\ldots,|C|}((\langle\overline{A}\rangle c_i \to [\overline{A}]c_{i,b}^z) \wedge (\neg\langle\overline{A}\rangle c_i \to [\overline{A}]\neg c_{i,b}^z))) \end{cases}$$

The formula $\varphi_{u-chain} \wedge \varphi_{conf-chain} \wedge \varphi_{units} \wedge \varphi_{corr} \wedge \varphi_{conf} \wedge \varphi_{inc} \wedge \varphi_{dec} \wedge \varphi_{ifz}$ is satisfiable if and only if $\mathcal{A}$ has at least one infinite run. $\square$

Since it is possible to construct a similar reduction using the fragment $\overline{A}B$, we can conclude the following theorem.

Theorem 3. *The complexity of the satisfiability problem for the fragments $A\overline{B}$ and $\overline{A}B$ over the class of dense linear orders is non-primitive recursive.*

6   Conclusions

In this paper, we solved the last open problems about the complexity of HS fragments whose satisfiability problem is decidable when interpreted over the class of dense linear orders (equivalently, $\mathbb{Q}$). If we look at the emerging picture, we notice that such a class turns out to be the best one from the point of view of computational complexity. The satisfiability problem for any HS fragment over the class of finite (resp, discrete) linear orders, as well as over $\mathbb{N}$ and $\mathbb{Z}$, is indeed at least as complex as over the class of dense linear orders. Moreover, there are some fragments, like the logic of subintervals D, for which the problem is decidable (in fact, PSPACE) over the latter class and undecidable over the former ones. The same relationships hold between the class of dense linear orders and the class of all linear orders (resp., $\mathbb{R}$) with respect to the known fragments.

Acknowledgments. The authors acknowledge the support from the Spanish fellowship program *'Ramon y Cajal' RYC-2011-07821* (G. Sciavicco), the project *Processes and Modal Logics* (project nr. 100048021) and the project *Decidability and Expressiveness for Interval Temporal Logics* (project nr. 130802-051) of the Icelandic Research Fund (D. Della Monica), and the Italian GNCS project *Automata, Games, and Temporal Logics for the verification and synthesis of controllers in safety-critical systems* (D. Bresolin, A. Montanari).

References

1. Aceto, L., Della Monica, D., Goranko, V., Ingólfsdóttir, A., Montanari, A., Sciavicco, G.: A complete classification of the expressiveness of interval logics of Allen's relations: the dense and the general case. Acta Informatica 2014, (in press)
2. Aceto, L., Della Monica, D., Ingólfsdóttir, A., Montanari, A., Sciavicco, G.: On the expressiveness of the interval logic of allen's relations over finite and discrete linear orders. In: Fermé, E., Leite, J. (eds.) JELIA 2014. LNCS, vol. 8761, pp. 267–281. Springer, Heidelberg (2014)
3. Allen, J.F.: Maintaining knowledge about temporal intervals. Communications of the ACM **26**(11), 832–843 (1983)
4. Bresolin, D., Della Monica, D., Goranko, V., Montanari, A., Sciavicco, G.: The dark side of interval temporal logic: marking the undecidability border. Annals of Mathematics and Artificial Intelligence **71**(1–3), 41–83 (2014)
5. Bresolin, D., Della Monica, D., Montanari, A., Sala, P., Sciavicco, G.: Interval temporal logics over strongly discrete linear orders: Expressiveness and complexity. Theoretical Computer Science 560, 269–291 (2014)
6. Bresolin, D., Goranko, V., Montanari, A., Sala, P.: Tableaux for logics of subinterval structures over dense orderings. Journal of Logic and Computation **20**(1), 133–166 (2010)
7. Bresolin, D., Montanari, A., Sala, P., Sciavicco, G.: Optimal tableau systems for propositional neighborhood logic over all, dense, and discrete linear orders. In: Brünnler, K., Metcalfe, G. (eds.) TABLEAUX 2011. LNCS, vol. 6793, pp. 73–87. Springer, Heidelberg (2011)
8. Bresolin, D., Montanari, A., Sala, P., Sciavicco, G.: What's decidable about Halpern and Shoham's interval logic? The maximal fragment $A\overline{BBL}$. In: Proc. of the 26th LICS. pp. 387–396 (2011)
9. Bresolin, D., Della Monica, D., Montanari, A., Sala, P., Sciavicco, G.: On the complexity of fragments of the modal logic of Allen's relations over dense structures (extended version) (2014), available at. http://www.icetcs.ru.is/dario/techrep/lata15ext.pdf
10. Chaochen, Z., Hansen, M.R.: Duration calculus: A formal approach to real-time systems. Monographs in Theoretical Computer Science, Springer, EATCS (2004)
11. Gennari, R., Tonelli, S., Vittorini, P.: An AI-based process for generating games from flat stories. In: Proc. of the 33th SGAI, pp. 337–350 (2013)
12. Halpern, J., Shoham, Y.: A propositional modal logic of time intervals. Journal of the ACM **38**(4), 935–962 (1991)
13. Laban, S., El-Desouky, A.: RISMA: A rule-based interval state machine algorithm for alerts generation, performance analysis and monitoring real-time data processing. In: Proc. of the European Geosciences Union General Assembly. Geophysical Research Abstracts, vol. 15 (2013)

14. Montanari, A., Puppis, G., Sala, P.: Decidability of the interval temporal logic
AĀBB̄ over the rationalsd. In: Csuhaj-Varjú, E., Dietzfelbinger, M., Ésik, Z. (eds.)
MFCS 2014, Part I. LNCS, vol. 8634, pp. 451–463. Springer, Heidelberg (2014)
15. Moszkowski, B.: Reasoning about digital circuits. Tech. rep. stan-cs-83-970, Dept.
of Computer Science. Stanford University, Stanford, CA (1983)
16. Sala, P.: Decidability of Interval Temporal Logics. Ph.D. thesis, Department of
Mathematics and Computer Science, University of Udine, Udine, Italy (2010)
17. Schnoebelen, P.: Lossy counter machines decidability cheat sheet. In: Proc. of the
4th International Workshop (RP 2010), pp. 51–75 (2010)
18. Terenziani, P., Snodgrass, R.T.: Reconciling point-based and interval-based semantics in temporal relational databases: A treatment of the telic/atelic distinction.
IEEE Transactions on Knowledge and Data Engineering **16**(5), 540–551 (2004)

Parameterized Enumeration for Modification Problems

Nadia Creignou[1]([⊠]), Raïda Ktari[1], Arne Meier[2], Julian-Steffen Müller[2],
Frédéric Olive[1], and Heribert Vollmer[2]

[1] CNRS, LIF UMR 7279, Aix-Marseille Université, 163 av. de Luminy,
13288 Marseille, France
{Nadia.Creignou,Raida.Ktari,Frederic.Olive}@lif.univ-mrs.fr
[2] Institut für Theoretische Informatik, Leibniz Universität Hannover, Appelstrasse 4,
30167 Hannover, Germany
{meier,mueller,vollmer}@thi.uni-hannover.de

Abstract. Recently the class DelayFPT has been introduced into parameterized complexity in order to capture the notion of efficiently solvable parameterized enumeration problems. In this paper we propose a framework for parameterized *ordered* enumeration and will show how to obtain DelayFPT enumeration algorithms in the context of graph modification problems. We study these problems considering two different orders of solutions, lexicographic and by size. We present generic algorithmic strategies: The first one is based on the well-known principle of self-reducibility in the context of lexicographic order. The second one shows that the existence of some neighborhood structure among the solutions implies the existence of a DelayFPT algorithm which outputs all solutions ordered non-decreasingly by their size.

Keywords: Parameterized complexity · Enumeration · Bounded search tree · Parameterized enumeration · Enumeration with ordering

1 Introduction

Enumeration problems, the task of generating all solutions of a given computational problem, find applications, e.g., in query answering in databases and web search engines, bioinformatics and computational linguistics. From a complexity-theoretic viewpoint, the notion of Delay-P, the class of problems whose instance solutions can be output in such a way that the delay between two outputs is bounded by a polynomial, is of utmost importance [14].

For many enumeration problems it is of high interest that the output solutions obey some given ordering. In particular in many applications it is interesting

Supported by a Campus France/DAAD Procope grant, Campus France Projet No 28292TE, DAAD Projekt-ID 55892324.

This work has received support from the French Agence Nationale dela Recherche, AGGREG project reference ANR-14-CE25-0017-01.

© Springer International Publishing Switzerland 2015
A.-H. Dediu et al. (Eds.): LATA 2015, LNCS 8977, pp. 524–536, 2015.
DOI: 10.1007/978-3-319-15579-1_41

to get the solutions with the smallest "cost" at the beginning. Enumerating all solutions in non-decreasing order allows to determine not only the smallest solution, but also the kth-smallest one. Also with such a generation algorithm it is possible to find the smallest solution satisfying some additional constraints in checking at each generation step whether these constraints are satisfied. The disadvantage of this method is that it cannot guarantee fast results, however it has the advantage to be applicable to any additional constraint (see, e.g., [18]). Let us illustrate this with some examples.

The question for which classes of propositional CNF formulas an enumeration of all satisfying solutions is possible in Delay-P, above defined, was studied in [4]. In terms of the well-known Schaefer framework for classification of Boolean constraint satisfaction problems, it was shown that for the classes of Horn, anti-Horn, affine or bijunctive formulas, such an algorithm exists. For other classes of formulas the existence of a Delay-P algorithm implies P = NP. It is interesting to note that the result hinges on the self-reducibility of the propositional satisfiability problem. Since variables systematically are tried first with an assignment 0 and then 1, it can be observed that the given enumeration algorithms output all satisfying assignments in lexicographic order.

In [6] the enumeration of satisfying assignments for propositional formulas was studied under a different order, namely in non-decreasing weight, and it was shown that under this new requirement, enumeration with polynomial delay is only possible for Horn formulas and width-2 affine formulas (i.e., affine formulas with at most 2 literals per clause). One of the main ingredients of these algorithms is the use of a priority queue to ensure enumeration in order (as is the case already in [14]).

While parameterized enumeration has already been considered (see, e.g., [7–9]), the notion of fixed-parameter tractable delay was introduced only recently in this context, leading to the definition of the complexity class DelayFPT [5]. The "polynomial time" in the definition of DelayP here is simply replaced by a time-bound of the form $p(n) \cdot f(k)$, where n denotes the input length and k the input parameter, p is an arbitrary polynomial, and f is an arbitrary computable function. By this the notion of efficiency in the context of the parameterized world, i.e., fixed parameter tractability (FPT), has been combined with the enumeration framework. A number of problems from propositional logic were studied in [5] and enumeration algorithms based on self-reducibility and on the technique of kernelization were developed. In particular it was shown that membership of an enumeration problem in DelayFPT can be characterized by a certain tailored form of kernelizability, very much as in the context of usual decision problems.

In the present paper we study *ordered enumeration* in the context of *parameterized complexity*. First we develop a formal framework for enumeration with *any* order. Then we consider the special context of graph modification problems where we are interested in ordered enumeration for the two mostly studied orders, namely *lexicographic* and by non-decreasing *size* (where the size is the number of modifications that have to be made). We use two algorithmic strategies, depending on the order: Based on the principle of self-reducibility we obtain DelayFPT (and polynomial-space) enumeration algorithms for lexicographic order, as soon as the decision problem is efficiently solvable. Second, we present a DelayFPT enumeration algorithm for order by size as soon as a certain FPT-computable neighborhood function on the solutions set exists (see Proposition 11). In order to take care of the order, we use a priority queue that may require exponential space.

We prove the wide scope of applicability of our method by presenting FPT-delay ordered enumeration algorithms for a large variety of problems, such as cluster editing, triangulation, triangle deletion, closest-string, and backdoor sets.

2   Preliminaries

We start by defining parameterized enumeration problems with a specific ordering and their corresponding enumeration algorithms. Most definitions in this section transfer those of [14,16] from the context of enumeration and those of [5] from the context of parameterized enumeration to the context of parameterized *ordered* enumeration.

Definition 1. *A* parameterized enumeration problem with ordering *is a quadruple* $E = (I, \kappa, \mathrm{Sol}, \preceq)$ *such that the following holds: I is the set of* instances*; $\kappa \colon I \to \mathbb{N}$ is the* parameterization function *and κ is required to be polynomial-time computable; Sol is a function such that for all $x \in I$, $\mathrm{Sol}(x)$ is a finite set, the set of* solutions *of x, further we write $\mathcal{S} = \bigcup_{x \in I} \mathrm{Sol}(x)$; $\preceq$ is a quasiorder (or* preorder*, i.e., a reflexive and transitive binary relation) on $\mathcal{S}$.*

We will write I_E, κ_E, etc. to denote that we talk about instance set, parameterization function, etc. of problem E.

Definition 2. *Let $E = (I, \kappa, \mathrm{Sol}, \preceq)$ be a parameterized enumeration problem with ordering. Then an algorithm $\mathcal{A}$ is an* enumeration algorithm *for E if the following holds: For every $x \in I$, $\mathcal{A}(x)$ terminates after a finite number of steps. For every $x \in I$, $\mathcal{A}(x)$ outputs exactly the elements of $\mathrm{Sol}(x)$ without duplicates. For every $x \in I$ and $y, z \in \mathrm{Sol}(x)$, if $y \preceq z$ and $z \not\preceq y$ then $\mathcal{A}(x)$ outputs solution y before solution z.*

Before we define complexity classes for parameterized enumeration, we need the notion of delay for enumeration algorithms.

Definition 3 (Delay). *Let $E = (I, \kappa, \mathrm{Sol}, \preceq)$ be a parameterized enumeration problem with ordering and $\mathcal{A}$ be an enumeration algorithm for E. Let $x \in I$ be an instance. The i-th delay of $\mathcal{A}$ is the time between outputting the i-th and $(i+1)$-st solution in $\mathrm{Sol}(x)$. The 0-th delay is the* precalculation time *which is the time from the start of the computation to the first output statement. Analogously, the n-th delay, for $n = |\mathrm{Sol}(x)|$, is the* postcalculation time *which is the time needed after the last output statement until $\mathcal{A}$ terminates. The delay of $\mathcal{A}$ is then defined as the maximum over all $0 \leq i \leq n$ of the i-th delay of $\mathcal{A}$.*

Now we are able to define two different complexity classes for parameterized enumeration following the notion of [5].

Definition 4. *Let $E = (I, \kappa, \mathrm{Sol}, \preceq)$ be a parameterized enumeration problem. We say E is FPT-enumerable if there exists an enumeration algorithm $\mathcal{A}$, a computable function $f \colon \mathbb{N} \to \mathbb{N}$, and a polynomial p such that for every $x \in I$, $\mathcal{A}$ outputs all solutions of $\mathrm{Sol}(x)$ in time $f(\kappa(x)) \cdot p(|x|)$.*

An enumeration algorithm $\mathcal{A}$ is a DelayFPT algorithm if there exists a computable function $f \colon \mathbb{N} \to \mathbb{N}$, and a polynomial p such that for every $x \in I$, $\mathcal{A}$ outputs all solutions of $\mathrm{Sol}(x)$ with delay of at most $f(\kappa(x)) \cdot p(|x|)$.

The class DelayFPT consists of all parameterized enumeration problems that admit a DelayFPT enumeration algorithm.

Some of our enumeration algorithms will make use of the concept of *priority queues* to enumerate all solutions in the correct order and to avoid duplicates. We will follow the approach of Johnson et al. [14]. A priority queue Q stores a potentially exponential number of elements. Let x be an instance. The *insert operation* of Q requires $O(|x| \cdot \log |\mathrm{Sol}(x)|)$ time. The *extract minimum operation* requires $O(|x| \cdot \log |\mathrm{Sol}(x)|)$ time, too. It is important, however, that the computation of the order between two elements takes at most $O(|x|)$ time. As pointed out by Johnson et al. the required queue can be implemented with the help of standard balanced tree schemes.

3   Graph Modification Problems

Graph modifications problems have been studied for a long time in computational complexity theory. Already in the monograph by Garey and Johnson [11], among the graph-theoretic problems considered, many fall into this problem class. To the best of our knowledge, graph modification problems were studied in the context of parameterized complexity for the first time in [3]. Given some graph property $\mathcal{P}$ and some graph G, we write $G \models \mathcal{P}$ if the graph G obeys the property $\mathcal{P}$. A *(graph) operation* for G is either removing a vertex, or adding/removing an edge. Two operations are *dependent* if one removes a vertex v and the other removes or adds an edge incident to v, or if one removes an edge e and the other adds the same edge e again. A set of operations is *consistent* if it does not contain two dependent operations. Given such a consistent set of operations S, the graph obtained from G by applying the operations in S is denoted by $S(G)$. Notice that for two consistent sets of operations S and S' such that $S \neq S'$, we have $S(G) \neq S'(G)$.

Definition 5. *Given some graph property $\mathcal{P}$, a graph G, $k \in \mathbb{N}$, and a set of operations O, we say that S is a solution for (G, k, O) with respect to $\mathcal{P}$ if the following three properties hold: (1) $S \subseteq O$ is a consistent set of operations ; (2) $|S| \leq k$; (3) $S(G) \models \mathcal{P}$. A solution S is minimal if there is no solution S' such that $S' \subsetneq S$.*

Cai was interested in the following parameterized graph modification decision problem w.r.t. some given graph property $\mathcal{P}$:

Problem: $\mathcal{M_P}$
Input: (G, k, O), G undirected graph, $k \in \mathbb{N}$, O set of operations on G.
Parameter: k
Question: Does there exist a solution for (G, k, O) with respect to $\mathcal{P}$?

Some of the most important examples of graph modification problems are presented now.

A *chord* in a graph $G = (V, E)$ is an edge between two vertices of a cycle C in G which is not part of C. A given graph $G = (V, E)$ is *triangular* (or *chordal*) if each of its induced cycles of 4 or more nodes has a chord. The problem TRIANGULATION then asks, given an undirected graph G and $k \in \mathbb{N}$, whether there exists a set of at most k edges such that adding this set of edges to G makes it triangular. Yannakakis showed that this problem is NP-complete [20].

A *cluster* is a graph such that all its connected components are cliques. In order to transform (or modify) a graph G we allow here only two kinds of operations: adding or removing an edge. CLUSTER-EDITING asks, given a graph G and a parameter k, whether there exists a consistent set of operations of cardinality at most k such that $S(G)$ is cluster. It was shown by Shamir et al. that the problem is NP-complete [17].

The problem TRIANGLE-DELETION asks whether a given graph can be transformed into a triangle-free graph by deletion of at most k vertices. Yannakakis has shown that the problem is NP-complete [19].

Analogous problems can be defined for many other classes of graphs, e.g., line graphs, claw-free graphs, Helly circular-arc graphs, etc., see [2].

Now turn towards the main focus of the paper. Here we are interested in corresponding enumeration problems with ordering. In particular, we will focus on two well-known preorders, lexicographic and by size. Since our solutions are subsets of an ordered set of operations, they can be encoded as binary strings in which the ith bit from right indicates whether the ith operation is in the subset. We define lexicographic ordering of solutions as the lexicographic ordering of these strings. We define the size of a solution simply as its cardinality.

Problem: ENUM-$\mathcal{M}_{\mathcal{P}}^{\text{LEX}}$
Instance: (G, k, O), G undir. graph, $k \in \mathbb{N}$, O ordered set of oper. on G.
Parameter: k
Output: All solutions of (G, k, O) w.r.t. $\mathcal{P}$ in lexicographic order.

Problem: ENUM-$\mathcal{M}_{\mathcal{P}}^{\text{SIZE}}$
Instance: (G, k, O), G undir. graph, $k \in \mathbb{N}$, O set of operations on G.
Parameter: k
Output: All solutions of (G, k, O) w.r.t. $\mathcal{P}$ in non-decreasing size.

If the context is clear we omit the subscript $\mathcal{P}$ for the graph modification problem and simply write $\mathcal{M}$. We write $\mathrm{Sol}_{\mathcal{M}}(x)$ for the function associating solutions to a given instance, and also $\mathcal{S}_{\mathcal{M}}$ for the set of all solutions of $\mathcal{M}$.

4 Enumeration of Graph Modification Problems with Ordering

4.1 Lexicographic Order

We first prove that, for any graph modification problem $\mathcal{M}_{\mathcal{P}}$, if the decision problem is in FPT then there is an efficient enumeration algorithm for ENUM-$\mathcal{M}_{\mathcal{P}}^{\mathrm{LEX}}$.

Theorem 6. *Let $\mathcal{M}_{\mathcal{P}}$ be a graph modification problem. If $\mathcal{M}_{\mathcal{P}}$ is in* FPT *then* ENUM-$\mathcal{M}_{\mathcal{P}}^{\mathrm{LEX}} \in$ DelayFPT *with polynomial space.*

Proof. We present an algorithm enumerating all solutions of an instance of a given modification problem $\mathcal{M}_{\mathcal{P}}$ by the method of self-reducibility, see Algorithm 1. The algorithm uses a function $\mathtt{ExistsSol}(G, k, O)$ that tests if the instance (G, k, O) of the modification problem $\mathcal{M}_{\mathcal{P}}$ has a solution. By assumption, this test is in FPT. We use calls to this function to avoid exploration of branches of the recursion tree that do not lead to any output.   □

Algorithm 1. Enumerate all solutions of $\mathcal{M}_{\mathcal{P}}$ in lexicographic order

---

Input: a graph G, $k \in \mathbb{N}$, an ordered set of operations $O = \{o_1, \dots, o_n\}$
Output: all consistent sets $S \subseteq O$ s.t. $|S| \leq k$, $S(G) \models \mathcal{P}$ in lexicographic order
1 **if** $\mathtt{ExistsSol}(G, k, O)$ **then** $\mathtt{Generate}(G, k, O, \emptyset)$.

 Procedure $\mathtt{Generate}(G, k, O, S)$:
1 **if** $O = \emptyset$ *or* $k = 0$ **then return** S;
2 **else**
3 $\quad$ let o_p be the last operation in O, let $O := O \setminus \{o_p\}$;
4 $\quad$ **if** $\mathtt{ExistsSol}(S(G), k, O)$ **then** $\mathtt{Generate}(S(G), k, O, S)$;
5 $\quad$ **if** $S \cup \{o_p\}$ *is consistent and* $\mathtt{ExistsSol}((S \cup \{o_p\})(G), k - 1, O)$ **then**
6 $\quad\quad$ $\mathtt{Generate}((S \cup \{o_p\})(G), k - 1, O, S \cup \{o_p\})$.

Corollary 7. ENUM-TRIANGULATION$^{\mathrm{LEX}} \in$ DelayFPT *with polynomial space.*

Proof. Kaplan et al. [15] and Cai [3] showed that TRIANGULATION $\in$ FPT. Now by applying Theorem 6 we get the result.   □

In [3], Cai identified a class of graph properties whose associated modification problems belong to FPT. Let us introduce some terminology. Given two graphs $G = (V, E)$ and $H = (V', E')$, we write $H \trianglelefteq G$ if H is an induced subgraph of G, i.e., $V' \subseteq V$ and $E' = E \cap (V' \times V')$. Let $\mathcal{F}$ be a set of graphs and $\mathcal{P}$ be some graph property. We say that $\mathcal{F}$ *is a forbidden set characterization of* $\mathcal{P}$ *if for any graph* G *it holds that*: $G \models \mathcal{P}$ iff for all $H \in \mathcal{F}, H \not\trianglelefteq G$.

Among the problems presented in the previous section (see page 5) TRIANGLE-DELETION and CLUSTER-EDITING have a finite forbidden set characterization, namely by triangles and paths of length two. In contrast, TRIANGULATION has a forbidden set characterization which is not finite, since cycles of arbitrary length are problematic.

Actually, for properties having a finite forbidden set characterization, the corresponding modification problem is fixed-parameter tractable. Together with Theorem 6, this provides a positive result in terms of enumeration.

Proposition 8 ([3]). *If a property* $\mathcal{P}$ *has a* finite *forbidden set characterization then the problem* $\mathcal{M}_\mathcal{P}$ *is in* FPT.

Proposition 9. *For any graph modification problem, if* $\mathcal{P}$ *has a finite forbidden set characterization then* ENUM-$\mathcal{M}_\mathcal{P}^{\text{LEX}} \in$ DelayFPT *with polynomial space.*

4.2   Size Ordering

A common strategy in enumeration context consists of defining a notion of neighborhood that allows to compute a new solution from a previous one with small amounts of computation time (see, e.g., the work of Avis and Fukuda [1]). We introduce the notion of a neighborhood function, which, roughly speaking, generates some initial solutions from which all solutions can be produced. Taking care of the order and avoiding duplicates is then handled by a priority queue, which may require exponential space. For the graph modification problems, we show that if the inclusion-minimal solutions can be generated in FPT, then such a neighborhood function exists, thus providing a DelayFPT enumeration algorithm. In the following $\mathbb{O}$ (the "seed") is a technical symbol that will be used in order to generate the initial solutions.

Definition 10. *Let* $\mathcal{M}$ *be some graph modification problem. A neighborhood function for* $\mathcal{M}$ *is a (partial) function* $\mathcal{N}_\mathcal{M} \colon I_\mathcal{M} \times (\mathcal{S}_\mathcal{M} \cup \{\mathbb{O}\}) \to 2^{\mathcal{S}_\mathcal{M}}$ *such that the following holds:*

1. *For all* $x = (G, k, O) \in I_\mathcal{M}$ *and* $S \in \mathrm{Sol}_\mathcal{M}(x) \cup \{\mathbb{O}\}$, $\mathcal{N}_\mathcal{M}(x, S)$ *is defined.*
2. *For all* $x \in I_\mathcal{M}$, $\mathcal{N}_\mathcal{M}(x, \mathbb{O}) = \emptyset$ *if* $\mathrm{Sol}_\mathcal{M}(x) = \emptyset$, *and* $\mathcal{N}_\mathcal{M}(x, \mathbb{O})$ *is an arbitrary set of solutions otherwise.*
3. *For all* $x \in I_\mathcal{M}$ *and* $S \in \mathrm{Sol}_\mathcal{M}(x)$, *if* $S' \in \mathcal{N}_\mathcal{M}(x, S)$ *then* $|S| < |S'|$.
4. *For all* $x \in I_\mathcal{M}$ *and all* $S \in \mathrm{Sol}_\mathcal{M}(x)$, *there exists* $p > 0$ *and* $S_1, \dots, S_p \in \mathrm{Sol}_\mathcal{M}(x)$ *such that (i)* $S_1 \in \mathcal{N}_\mathcal{M}(x, \mathbb{O})$, *(ii)* $S_{i+1} \in \mathcal{N}_\mathcal{M}(x, S_i)$ *for* $1 \le i < p$, *and (iii)* $S_p = S$.

Furthermore, we say that $\mathcal{N}_\mathcal{M}$ *is* FPT*-computable, when* $\mathcal{N}_\mathcal{M}(x, S)$ *is computable in time* $f(k) \cdot poly(|x|)$ *for any* $x \in I_\mathcal{M}$ *and* $S \in \mathrm{Sol}_\mathcal{M}(x)$.

Thus, a neighborhood function for a problem $\mathcal{M}$ is a function that in a first phase computes from scratch some initial set of solutions (see Definition 10(2)). In many of our applications below, $\mathcal{N}_\mathcal{M}(x, \mathbb{O})$ will be the set of all minimal solutions for x. In a second phase these solutions are iteratively enlarged (see condition (3)), where condition (4) guarantees that we do not miss any solution, as we will see in the next theorem.

Proposition 11. *Let $\mathcal{M}$ be a graph modification problem. If $\mathcal{M}$ admits a neighborhood function $\mathcal{N}_\mathcal{M}$ that is FPT-computable, then* ENUM-$\mathcal{M}^{\text{SIZE}} \in$ DelayFPT.

Proof. Algorithm 2 outputs all solutions in DelayFPT. By the definition of the priority queue (recall in particular that insertion of an element is done only if the element is not yet present in the queue) and by the fact that all elements of $\mathcal{N}_\mathcal{M}((G, k, O), S)$ are of bigger size than S by Definition 10(3), it is easily seen that *the solutions are output in the right order* and that *no solution is output twice*.

Besides, *no solution is omitted*. Indeed, given $S \in \text{Sol}_\mathcal{M}(G, k, O)$ and $S_1, \ldots, S_p$ associated with S by Definition 10(4), we prove by induction that each S_i is inserted in Q during the run of the algorithm: For $i = 1$, this proceeds from line 2 of the algorithm; for $i > 1$, the solution S_{i-1} is inserted in Q by induction hypothesis and hence all elements of $\mathcal{N}_\mathcal{M}((G, k, O), S_{i-1})$, including S_i, are inserted in Q (line 6). Thus, each S_i is inserted in Q and then output during the run. In particular, this holds for $S = S_p$.

Finally, we claim that *Algorithm 2 runs in* DelayFPT. Indeed, the delay between the output of two consecutive solutions is bounded by the time required to compute a neighborhood of the form $\mathcal{N}_\mathcal{M}((G, k, O), \mathbb{O})$ or $\mathcal{N}_\mathcal{M}((G, k, O), S)$ and to insert all its elements in the priority queue. This is in FPT due to the assumption on $\mathcal{N}_\mathcal{M}$ being FPT-computable and as there is only a single extraction and FPT-many insertion operations on the queue. □

Algorithm 2. DelayFPT algorithm for ENUM-$\mathcal{M}$

Input: (G, k, O), G is an undirected graph, $k \in \mathbb{N}$, and O is a set of operations.
1   compute $\mathcal{N}_\mathcal{M}((G, k, O), \mathbb{O})$;
2   insert all elements of $\mathcal{N}_\mathcal{M}((G, k, O), \mathbb{O})$ into priority queue Q (ordered by size);
3   **while** Q *is not empty* **do**
4   **extract** the minimum solution S of Q and output it;
5   **insert** all elements of $\mathcal{N}_\mathcal{M}((G, k, O), S)$ into Q;

A natural way to provide a neighborhood function for a graph modification problem $\mathcal{M}$ is to consider the *inclusion minimal* solutions of $\mathcal{M}$. Let us denote by MIN-$\mathcal{M}$ the problem of enumerating all inclusion minimal solutions of $\mathcal{M}$.

Theorem 12. *Let $\mathcal{M}$ be a graph modification problem. If* MIN-$\mathcal{M}$ *is FPT-enumerable then* ENUM-$\mathcal{M}^{\text{SIZE}} \in$ DelayFPT.

Proof. Let $\mathcal{A}$ be an FPT-algorithm for MIN-$\mathcal{M}$. Because of Proposition 11, it is sufficient to build an FPT neighborhood function for $\mathcal{M}$. For an instance (G, k, O) of $\mathcal{M}$ and for $S \in \mathrm{Sol}_{\mathcal{M}}(G, k, O) \cup \{\mathbb{O}\}$, we define $\mathcal{N}_{\mathcal{M}}((G, k, O), S)$ as the result of Algorithm 3.

Algorithm 3. Procedure for computing $\mathcal{N}_{\mathcal{M}}((G, k, O), S)$

Input: (G, k, O), G is an undirected graph, $k \in \mathbb{N}$, and O is a set of operations.
1 **if** $S = \mathbb{O}$ **then return** $\mathcal{A}(G, k, O)$;
2 res := $\emptyset$;
3 **forall the** $t \in O$ **do**
4 　　**forall the** $S' \in \mathcal{A}((S \cup \{t\})(G), k - |S| - 1, O \setminus \{t\})$ **do**
5 　　　　**if** $S \cup S' \cup \{t\}$ *is consistent* **then** res := res $\cup \{S \cup S' \cup \{t\}\}$;

6 **return** res;

The function $\mathcal{N}_{\mathcal{M}}$ thus defined clearly fulfills conditions 2 and 3 of Definition 10. We prove by induction that it also satisfies condition 4 (that is, each solution T of size k comes with a sequence $T_1, \ldots, T_p = T$ such that $T_1 \in \mathcal{N}_{\mathcal{M}}((G, k, O), \mathbb{O})$ and $T_{i+1} \in \mathcal{N}_{\mathcal{M}}((G, k, O), T_i)$ for each i). If T is a minimal solution for (G, k, O), then $T \in \mathcal{N}_{\mathcal{M}}((G, k, O), \mathbb{O})$ and the expected sequence (T_i) reduces to $T_1 = T$. Otherwise, there exists an $S \in \mathrm{Sol}_{\mathcal{M}}(G, k, O)$ and a non-empty set of transformations, say $S' \cup \{t\}$, such that $T = S \cup S' \cup \{t\}$ and there is no solution for G between S and $S \cup S' \cup \{t\}$. This entails that S' is a minimal solution for $((S \cup \{t\})(G), k - |S| - 1)$ and hence $T \in \mathcal{N}_{\mathcal{M}}((G, k, O), S)$ (see lines 4–5 of Algorithm 3). The conclusion follows from the induction hypothesis that guarantees the existence of solutions $S_1, \ldots, S_q$ such that $S_1 \in \mathcal{N}_{\mathcal{M}}((G, k, O), \mathbb{O})$, $S_{i+1} \in \mathcal{N}_{\mathcal{M}}((G, k, O), S_i)$ and $S_q = S$. The expected sequence $T_1, \ldots, T_p$ for T is nothing but $S_1, \ldots, S_q, T$. To conclude, it remains to see that Algorithm 3 is FPT. This follows from the fact that $\mathcal{A}$ is an FPT-algorithm (lines 1 and 4). □

Corollary 13. ENUM-TRIANGULATION[SIZE] $\in$ DelayFPT.

Proof. All minimal k-triangulations can be output in time $O(2^{4k} \cdot |E|)$ for a given graph G and $k \in \mathbb{N}$ [15, Thm. 2.4]. This immediately yields the expected result, by help of Theorem 12. □

Proposition 14. *For any property $\mathcal{P}$ that has a finite forbidden set characterization, the problem ENUM-$\mathcal{M}_{\mathcal{P}}^{\text{SIZE}}$ is in DelayFPT.*

Proof. The algorithm developed by Cai [3] for the decision problem is based on a bounded search tree, whose exhaustive examination provides all minimal solutions in FPT. Theorem 12 yields the conclusion. □

Corollary 15. *The enumeration problems* ENUM-CLUSTER-EDITING[SIZE] *and* ENUM-TRIANGLE-DELETION[SIZE] *are in DelayFPT.*

Proof. As we already mentioned, these two properties have a finite forbidden set characterization. □

5 Generalization

Modification problems can be defined for other structures than graphs, thus providing similar complexity results. We give two different examples in this section.

5.1 Closest String

Given a set of binary strings I we want to find a string s whose maximum Hamming distance $\max\{d_H(s, s') \mid s' \in I\} \leq d$ for some $d \in \mathbb{N}$. This problem is NP-hard [10]. Given a string $w = w_1 \cdots w_n$ with $w_i \in \{0, 1\}, n \in \mathbb{N}$, and a set $S \subseteq \{1, \ldots, n\}$, $S(w)$ denotes the string obtained from w in flipping the bits indicated by S. The corresponding parameterized problem is the following.

Problem:	CLOSEST-STRING
Input:	$(s_1, \ldots, s_k, n, d)$, where $s_1, \ldots, s_k$ is a sequence of strings over $\{0, 1\}$ of length $n \in \mathbb{N}$, and an integer $d \in \mathbb{N}$.
Parameter:	d
Question:	Does there exist $S \subseteq \{1, \ldots, n\}$ such that $d_H(S(s_1), s_i) \leq d$ for all $1 \leq i \leq k$?

Gramm et al. have shown that this problem is in FPT [13]. Moreover, an exhaustive examination of a bounded search tree constructed from the idea of Gramm et al. [13, Fig.1] allows to produce all minimal solutions of this problem in FPT. Therefore, we get the following result for the corresponding enumeration problems.

Theorem 16. ENUM-CLOSEST-STRING$^{\text{LEX}} \in$ DelayFPT *with polynomial space and* ENUM-CLOSEST-STRING$^{\text{SIZE}} \in$ DelayFPT.

5.2 Backdoors

In the following, let $\mathcal{C}$ be some class of CNF-formulas, and φ be a propositional CNF formula. If X is a set of propositional variables we denote with $\Theta(X)$ the set of all assignments over the variables in X. For some $\theta \in \Theta(X)$ the expression $\theta(\varphi)$ is the formula obtained by applying the assignment θ to φ, i.e., clauses with a satisfied literal are removed, and falsified literals are removed. A set S of variables from φ is a *weak $\mathcal{C}$-backdoor of φ* if there exists an assignment $\theta \in \Theta(S)$ such that $\theta(\varphi) \in \mathcal{C}$ and $\theta(\varphi)$ is satisfiable. The set S is a *strong $\mathcal{C}$-backdoor of φ* if for all $\theta \in \Theta(S)$ the formula $\theta(\varphi)$ is in $\mathcal{C}$.

Now we can define the corresponding parameterized problems:

Problem:	WEAK/STRONG-$\mathcal{C}$-BACKDOORS
Input:	A formula φ in 3CNF, $k \in \mathbb{N}$.
Parameter:	k
Question:	Does there exist a weak/strong $\mathcal{C}$-backdoor of size $\leq k$?

The class C is a *base class* if it can be recognized in P, satisfiability of its formulas is in P, and the class is closed under isomorphisms w.r.t. variable names. We say C is *clause defined* if for every CNF-formula φ it holds: $\varphi \in C$ iff $\{C\} \in C$ for all clauses C from φ. Gaspers and Szeider [12] investigated a specific type of C-formulas, namely the clause-defined base classes C, and showed that for any such class C, the detection of weak C-backdoors is in FPT for input formulas in 3CNF. They describe in [12, Prop. 2] that a bounded search tree technique allows to solve the detection of weak C-backdoors in FPT time. This technique results in obtaining all minimal solutions in FPT time.

Theorem 17. *For every clause-defined base class C and input formulas in 3CNF,* ENUM-WEAK-C-BACKDOORS^{LEX} $\in$ DelayFPT *with polynomial space and* ENUM-WEAK-C-BACKDOORS^{SIZE} $\in$ DelayFPT.

Proof. Given an instance $x = (\varphi, k)$ we first compute all its minimal backdoors. Then, given some backdoor S we define $\mathcal{N}(x, S)$ as the set of the pairwise unions of all minimal weak C-backdoors of $(\theta(\varphi), k - |S| - 1)$ together with $S \cup \{x_i\}$ for each $\theta \in \Theta(S \cup \{x_i\})$ for $x_i \in Vars(\varphi) \setminus S$. Observe that there are only FPT-many assignments for which the minimal weak C-backdoors have to be computed and as the satisfiability test for the formulas is in P this yields a DelayFPT algorithm. □

Let φ be a CNF-formula and $V \subseteq Vars(\varphi)$ be a subset of its variables. Then $\varphi - V$ is the formula where all literals over variables from V have been removed from all clauses in φ. Now we want to consider strong C-backdoors for clause-defined base classes C. Note that in this case the notion of strong C-backdoors coincides with the notion of deletion C-backdoors, i.e., a set $V \subseteq Vars(\varphi)$ is a strong C-backdoor of φ if and only if $\varphi - V \in C$.

Theorem 18. *For every clause-defined base class C and input formulas in 3CNF,* ENUM-STRONG-C-BACKDOORS^{LEX} $\in$ DelayFPT *with polynomial space and* ENUM-STRONG-C-BACKDOORS^{SIZE} $\in$ DelayFPT.

Proof. For every clause-defined base class C and input formulas in 3CNF, the problem MIN-STRONG-C-BACKDOORS is FPT-enumerable. Indeed, we only need to branch on the variables from a clause $C \notin C$ and remove the corresponding variable from φ. The size of the branching tree is at most 3^k. As for base classes the satisfiability test is in P, this yields an FPT-algorithm. The neighbourhood function $\mathcal{N}(x, S)$ for $x = (\varphi, k)$ is defined to be the set of the pairwise unions of all minimal strong C-backdoors of $(\varphi - (S \cup \{x_i\}), k - |S| - 1)$ together with $S \cup \{x_i\}$ for all variables $x_i \notin S$. □

6 Conclusion

We presented FPT-delay ordered enumeration algorithms for a large variety of problems, such as cluster editing, triangulation, closest-string, and backdoors. An important point of our paper is that we propose an algorithmic strategy for efficient enumeration. This is rather rare in the literature, where usually algorithms are devised individually for specific problems. In particular, our scheme yields DelayFPT algorithms for all graph modification problems that are characterized by a finite set of forbidden patterns.

We would like to mention that the DelayFPT algorithms for size order presented in this paper require exponential space (in the size of the input) due to the use of the priority queues. An interesting question is whether there is a method which requires less space but uses a comparable delay between the output of solutions and still obeys the underlying order on solutions.

References

1. Avis, D., Fukuda, K.: Reverse search for enumeration. Discrete Applied Mathematics **65**, 21–46 (1996)
2. Brandtstädt, A., Le, V.B., Spinrad, J.P.: Graph Classes: A Survey. Monographs on Discrete Applied Mathematics. SIAM, Philadelphia (1988)
3. Cai, L.: Fixed-parameter tractability of graph modification problems for hereditary properties. Information Processing Letters **58**(4), 171–176 (1996)
4. Creignou, N., Hébrard, J.J.: On generating all solutions of generalized satisfiability problems. Theoretical Informatics and Applications **31**(6), 499–511 (1997)
5. Creignou, N., Meier, A., Müller, J.-S., Schmidt, J., Vollmer, H.: Paradigms for parameterized enumeration. In: Chatterjee, K., Sgall, J. (eds.) MFCS 2013. LNCS, vol. 8087, pp. 290–301. Springer, Heidelberg (2013)
6. Creignou, N., Olive, F., Schmidt, J.: Enumerating all solutions of a boolean CSP by non-decreasing weight. In: Sakallah, K.A., Simon, L. (eds.) SAT 2011. LNCS, vol. 6695, pp. 120–133. Springer, Heidelberg (2011)
7. Damaschke, P.: Parameterized enumeration, transversals, and imperfect phylogeny reconstruction. Theoretical Computer Science **351**(3), 337–350 (2006)
8. Fernau, H.: On parameterized enumeration. In: Ibarra, O.H., Zhang, L. (eds.) COCOON 2002. LNCS, vol. 2387, pp. 564–573. Springer, Heidelberg (2002)
9. Fomin, F.V., Saurabh, S., Villanger, Y.: A polynomial kernel for proper interval vertex deletion. SIAM Journal Discrete Mathematics **27**(4), 1964–1976 (2013)
10. Frances, M., Litman, A.: On covering problems of codes. Theory of Computing Systems **30**(2), 113–119 (1997)
11. Garey, M.R., Johnson, D.S.: Computers and Intractability: A Guide to the Theory of NP-Completeness. W. H. Freeman & Co., New York (1990)
12. Gaspers, S., Szeider, S.: Backdoors to Satisfaction. In: Bodlaender, H.L., Downey, R., Fomin, F.V., Marx, D. (eds.) Fellows Festschrift 2012. LNCS, vol. 7370, pp. 287–317. Springer, Heidelberg (2012). http://dx.doi.org/10.1007/978-3-642-30891-8_15
13. Gramm, J., Niedermeier, R., Rossmanith, P.: Fixed-parameter algorithms for CLOSEST STRING and related problems. Algorithmica **37**(1), 25–42 (2003)
14. Johnson, D.S., Papadimitriou, C.H., Yannakakis, M.: On generating all maximal independent sets. Information Processing Letters **27**(3), 119–123 (1988)
15. Kaplan, H., Shamir, R., Tarjan, R.E.: Tractability of parameterized completion problems on chordal, strongly chordal, and proper interval graphs. SIAM Journal on Computing **28**(5), 1906–1922 (1999)
16. Schmidt, J.: Enumeration: Algorithms and Complexity. Master's thesis, Leibniz Universität Hannover (2009)

17. Shamir, R., Sharan, R., Tsur, D.: Cluster graph modification problems. Discrete Applied Mathematics **114**(1–2), 173–182 (2004)
18. Sörensen, K., Janssens, G.K.: An algorithm to generate all spanning trees of a graph in order of increasing cost. Pesquisa Operacional **25**(2), 219–229 (2005)
19. Yannakakis, M.: Node- and edge-deletion NP-complete problems. In: Proc. STOC, pp. 253–264 (1978)
20. Yannakakis, M.: Computing the minimum fill-in is NP-complete. SIAM Journal on Algebraic Discrete Methods **2**(1), 77–79 (1981)

Preimage Problems for Reaction Systems

Alberto Dennunzio[1], Enrico Formenti[2],
Luca Manzoni[1], and Antonio E. Porreca[1](✉)

[1] Dipartimento di Informatica, Sistemistica e Comunicazione,
Università Degli Studi di Milano-Bicocca, Viale Sarca 336/14, 20126 Milano, Italy
{dennunzio,luca.manzoni,porreca}@disco.unimib.it
[2] CNRS, I3S, UMR 7271, Université Nice Sophia Antipolis,
06900 Sophia Antipolis, France
formenti@unice.fr

Abstract. We investigate the computational complexity of some problems related to preimages and ancestors of states of reaction systems. In particular, we prove that finding a minimum-cardinality preimage or ancestor, computing their size, or counting them are all intractable problems, with complexity ranging from $\mathbf{FP}^{\mathbf{NP}[\log n]}$ to $\mathbf{FPSPACE}$(poly).

Keywords: Reaction systems · Computational complexity

1 Introduction

Recently many new computational models have been introduced. Most of them are inspired by natural phenomena. This is also the case of Reaction Systems (RS), proposed by Ehrenfeucht and Rozenberg in [2], which are a metaphor for basic chemical reactions. Informally, a reaction system is made of a (finite) set of entities (molecules) and a (finite) set of admissible reactions. Each reaction is a triple of sets: *reactants*, *inhibitors* and *products* (clearly the set of reactants and the one of inhibitors are disjoint). Given a set of reactants T, a reaction (R, I, P) is applied if $R \subseteq T$ and if there are no inhibitors (*i.e.* $T \cap I$ is empty); the result is the replacement of T by the set of products P. Given a set of reactants T, all admissible reactions are applied in parallel. The final set of products is the union of all single sets of products of each reaction which is admissible for T.

Studying RS is interesting for a number of reasons, not only as a clean computational model allowing precise formal analysis but also as a reference *w.r.t.* other computing systems. For example, in [5], the authors showed an embedding of RS into Boolean automata networks (BAN), a well-known model used in a

This work has been supported by Fondo d'Ateneo (FA) 2013 of Università degli Studi di Milano-Bicocca: "Complessità computazionale in modelli di calcolo bioispirati: Sistemi a membrane e sistemi a reazioni", by the Italian MIUR PRIN 2010–2011 grant "Automata and Formal Languages: Mathematical and Applicative Aspects" H41J12000190001, and by the French National Research Agency project EMC (ANR-09-BLAN-0164).

© Springer International Publishing Switzerland 2015
A.-H. Dediu et al. (Eds.): LATA 2015, LNCS 8977, pp. 537–548, 2015.
DOI: 10.1007/978-3-319-15579-1_42

number of application domains. Remark that for BAN the precise complexity of only for a bunch of problems about the dynamical behaviour is known. Via the embedding of RS into BAN, all the complexity results about RS are indeed lower bounds for the corresponding ones for BAN.

In this paper, we continue the exploration of the computational complexity of properties of RS. The focus is on preimages and ancestors of minimal size. In more practical terms, this could be useful when minimising the number of chemical entities necessary to obtain a target compound. Indeed, given a current state T, the minimal pre-image (resp., n-th-ancestor) problem or MPP (resp., MAP) consists in finding the minimal set ($w.r.t.$ cardinality) of reactants which produces T in one step (resp., n steps). Variants of MPP and MAP consider counting the pre-images (#MPP); counting the ancestors (#MAP); or computing the size of the minimal pre-image (SMPP) or of the minimal ancestor (SMAP). We prove that (see Section 2 for the precise definition of the complexity classes):

- MPP $\in$ **FP^{NP}** and it is **FP$_{\|}^{NP}$**-hard under metric reductions;
- #MPP is in **#P$^{NP[\log n]}$** and it is **#P**-hard under parsimonious reductions;
- SMPP is **FP$^{NP[\log n]}$**-complete under metric reductions;
- MAP and #MAP are complete for **FPSPACE**(poly) under metric reductions;
- SMAP is **FPSPACE**(log)-complete under metric reductions.

These results are important for further understanding the computational capabilities of RS but they also provide clean new items to the (relatively) short list of examples of problems in high functional complexity classes. Finally, remark that the problem of pre-image existence has been proved to be in **NP** by Salomaa [10]. However, here the complexity is higher because of the minimality requirement.

2   Basic Notions

We briefly recall the basic notions about RS [3]. In this paper we require the sets of reactants and inhibitors of a reaction to be nonempty, as is sometimes enforced in the literature; our results also hold when empty sets are allowed.

Definition 1. *Consider a finite set S, whose elements are called entities. A reaction a over S is a triple (R_a, I_a, P_a) of nonempty subsets of S. The set R_a is the set of reactants, I_a the set of inhibitors, and P_a is the set of products. The set of all reactions over S is denoted by $\text{rac}(S)$.*

Definition 2. *A Reaction System (RS) is a pair $\mathcal{A} = (S, A)$ where S is a finite set, called the background set, and $A \subseteq \text{rac}(S)$.*

Given a *state* $T \subseteq S$, a reaction a is said to be *enabled* in T when $R_a \subseteq T$ and $I_a \cap T = \varnothing$. The *result function* $\text{res}_a : 2^S \to 2^S$ of a, where 2^S denotes the power set of S, is defined as $\text{res}_a(T) = P_a$ if a is enabled in T, and $\text{res}_a(T) = \varnothing$ otherwise. The definition of res_a naturally extends to sets of reactions: given $T \subseteq S$ and $A \subseteq \text{rac}(S)$, define $\text{res}_A(T) = \bigcup_{a \in A} \text{res}_a(T)$. The result function $\text{res}_\mathcal{A}$ of a RS $\mathcal{A} = (S, A)$ is res_A, i.e., it is the result function on the whole set of reactions.

Definition 3. *Let* $\mathcal{A} = (S, A)$ *be a RS. For any* $T \subseteq S$, *an element* $U \subseteq S$ *is an* ancestor *of* T *if* $\mathrm{res}_A^t(U) = T$ *for some* $t \in \mathbb{N}$. *If* $t = 1$, U *is called* preimage *of* T. *An ancestor (resp., preimage)* U *of* T *is* minimal *if* $|U| \leq |V|$ *for all ancestors (resp., preimages)* V *of* T.

A state always admits at least itself as ancestor but might not have a preimage.

We describe the complexity of preimage and ancestor problems for RS with complexity classes of functions problems (see [7,8] for further details). Let Σ be an alphabet. The class **FP** (resp., $\mathbf{FP^{NP}}$) consists of all binary relations R over $\Sigma^\star$ having a "choice function" $f \subseteq R$ with $\mathrm{dom}\, f = \mathrm{dom}\, R$ that can be computed in polynomial time by a deterministic Turing machine (TM) without access to oracles (resp., with access to an oracle for an **NP** decision problem). Additional requirements on the oracle queries define the subclasses $\mathbf{FP^{NP[\log n]}}$ and $\mathbf{FP_{\parallel}^{NP}}$ of $\mathbf{FP^{NP}}$, where the number of allowed oracle queries is $O(\log n)$ and the queries are performed in parallel (i.e., every current query string does not depend on the results of previous queries), respectively. We have $\mathbf{FP} \subseteq \mathbf{FP^{NP[\log n]}} \subseteq \mathbf{FP_{\parallel}^{NP}} \subseteq \mathbf{FP^{NP}}$. The class **#P** consists of all functions $f \colon \Sigma^\star \to \mathbb{N}$ with a polynomial-time nondeterministic TM having exactly $f(x)$ accepting computations on every input x. If in addition $O(\log n)$ queries to an **NP** oracle are allowed, the class $\mathbf{\#P^{NP[\log n]}}$ is defined. Clearly, $\mathbf{FP} \subseteq \mathbf{\#P} \subseteq \mathbf{\#P^{NP[\log n]}}$. In this paper we will also refer to **FPSPACE**, i.e., the collection of binary relations having a choice function computable in polynomial space, and its two subclasses **FPSPACE**(poly) and **FPSPACE**(log), in which the output is limited to polynomial length and logarithmic length, respectively, rather than exponential length. Remark that **FPSPACE**(poly) is just ♮**PSPACE**, i.e., the class of functions $f \colon \Sigma^\star \to \mathbb{N}$ such that there exists a polynomial-space nondeterministic TM performing only a polynomial number of nondeterministic choices and having exactly $f(x)$ accepting computations on every input x.

Hardness for these classes is defined in terms of two kinds of reductions. The first one is the many-one reduction, also called *parsimonious* when dealing with counting problems: a function f is many-one reducible to g if there exists a function $h \in \mathbf{FP}$ such that $f(x) = g(h(x))$ for every input x. A generalisation is the *metric reduction* [6]: a function f is metric reducible to g if there exist functions $h_1, h_2 \in \mathbf{FP}$ such that $f(x) = h_2(x, g(h_1(x)))$ for every input x. These notions of reduction can be generalised to reductions between binary relations.

Since we do not deal with sublinear space complexity, without loss of generality, throughout this paper we assume that all TM computing functions use a unique tape, both for input and work. We also assume that they move their tape head to the leftmost cell before entering a final state.

3 Preimage Problems

First of all, inspired by the algorithm described in [9], we show a tight relation between the MPP and the problem of finding a minimal unary travelling salesman tour (TSP) [8], where the edge weights are encoded in unary and, hence, bounded by a polynomial in the number of vertices.

Lemma 4. *Finding a minimal unary TSP tour is metric reducible to* MPP.

Proof. Suppose we are given any set of vertices V, with $|V| = n$, and any unary-encoded weight function $w\colon V^2 \to \mathbb{N}$. We build a RS $\mathcal{A} = (S, A)$ admitting a state whose preimages encode the weighted simple cycles over V. The background set is given by $S = E \cup W \cup \{\heartsuit, \spadesuit\}$, where

$$E = \{(u, v)_t : u, v \in V \text{ and } 0 \le t < n\}$$
$$W = \{\Diamond_{(u,v),i} : u, v \in V \text{ and } 1 \le i \le w(u, v)\}$$

while the reactions in A, with u, v, u_1, u_2, v_1, v_2 ranging over V and t, t_1, t_2 ranging over $\{0, \ldots, n-1\}$, are

$$(\{(u_1, v_1)_t, (u_2, v_2)_t, \heartsuit\}, \{\spadesuit\}, \{\spadesuit\}) \qquad \text{if } (u_1, v_1) \ne (u_2, v_2) \quad (1)$$
$$(\{(u, v_1)_{t_1}, (u, v_2)_{t_2}, \heartsuit\}, \{\spadesuit\}, \{\spadesuit\}) \qquad \text{if } v_1 \ne v_2 \text{ or } t_1 \ne t_2 \quad (2)$$
$$(\{(u_1, v)_{t_1}, (u_2, v)_{t_2}, \heartsuit\}, \{\spadesuit\}, \{\spadesuit\}) \qquad \text{if } u_1 \ne u_2 \text{ or } t_1 \ne t_2 \quad (3)$$
$$(\{(u_1, v_1)_t, (u_2, v_2)_{(t+1) \bmod n}, \heartsuit\}, \{\spadesuit\}, \{\spadesuit\}) \qquad \text{if } v_1 \ne u_2 \quad (4)$$
$$(\{\heartsuit\}, \{(u, v)_t : u, v \in V\} \cup \{\spadesuit\}, \{\spadesuit\}) \qquad\qquad\qquad (5)$$
$$(\{(u, v)_t, \heartsuit\}, \{\Diamond_{(u,v),i}, \spadesuit\}, \{\spadesuit\}) \qquad \text{for } 1 \le i \le w(u, v) \quad (6)$$
$$(\{\Diamond_{(u,v),i}, \heartsuit\}, \{(u, v)_t : 0 \le t < n\} \cup \{\spadesuit\}, \{\spadesuit\}) \qquad \text{for } 1 \le i \le w(u, v) \quad (7)$$
$$(\{\heartsuit\}, \{\spadesuit\}, \{\heartsuit\}) \qquad\qquad\qquad\qquad\qquad\qquad\qquad (8)$$

The meaning of an element $(u, v)_t$ in a state of $\mathcal{A}$ is that edge (u, v) is the t-th edge (for $0 \le t < n$) of a simple cycle over V, i.e., of a candidate solution for the TSP. The edge weights of the cycle must also appear, in unary notation: if $(u, v)_t$ is part of a state and $w(u, v) = k$, then also $\Diamond_{(u,v),1}, \ldots, \Diamond_{(u,v),k}$ must be part of the state. Hence, a length-n simple cycle $\mathbf{c} = (v_0, \ldots, v_{n-1})$ over V is encoded as a state $T(\mathbf{c}) = E(\mathbf{c}) \cup W(\mathbf{c}) \cup \{\heartsuit\}$, where

$$E(\mathbf{c}) = \{(v_0, v_1)_0, (v_1, v_2)_1, \ldots, (v_{n-2}, v_{n-1})_{n-2}, (v_{n-1}, v_0)_{n-1}\}$$

is the set of edges traversed by the cycle, indexed in order of traversal, and

$$W(\mathbf{c}) = \{\Diamond_{(v_t, v_{(t+1) \bmod n}), i} : 0 \le t < n, 1 \le i \le w(v_t, v_{(t+1) \bmod n})\}$$

contains the elements encoding the weights of the edges in $E(\mathbf{c})$.

Moreover, consider a state $T \subseteq S$. If $\heartsuit \notin T$, then necessarily $\mathrm{res}_{\mathcal{A}}(T) = \varnothing$, since all reactions have $\heartsuit$ as a reactant. Similarly, $\spadesuit \in T$ implies $\mathrm{res}_{\mathcal{A}}(T) = \varnothing$, since $\spadesuit$ inhibits all reactions. Now suppose $\heartsuit \in T$ and $\spadesuit \notin T$. Reactions (1)–(5) produce $\spadesuit$ from T when any of the following conditions (implying that $T \cap E$ does *not* encode a simple cycle over V) occur:

(1) the state T contains two distinct elements denoting edges occurring as the t-th edge of the candidate cycle;
(2) one of the vertices of the candidate cycle has outdegree greater than 1;
(3) one of the vertices of the candidate cycle has indegree greater than 1;

(4) two consecutive edges do not share an endpoint;
(5) no edge is the t-th edge of the candidate cycle, for some $0 \leq t < n$.

Any set $T \cap E$ where *none* of the above apply encodes a valid simple cycle over V.

Reaction (6) produces ♠ if an edge $(u, v)_t$ occurs, but some element $\Diamond_{(u,v),i}$, encoding a unit of the weight of the edge, is missing. Conversely, reaction (7) produces ♠ if a unit of the weight of a missing edge occurs. These reactions are all simultaneously disabled exactly when $T \cap W$ contains the weights of the edges in $T \cap E$. Finally, reaction (8) preserves the ♡. The result function of $\mathcal{A}$ is thus

$$
\mathrm{res}_{\mathcal{A}}(T) = \begin{cases} \{\heartsuit\} & \text{if } T \text{ encodes a weighted simple cycle over } V \\ \{\heartsuit, \spadesuit\} & \text{if } \heartsuit \in T, \spadesuit \notin T, \text{ but } T \text{ fails to encode} \\ & \quad \text{a weighted simple cycle over } V \\ \varnothing & \text{if } \heartsuit \notin T \text{ or } \spadesuit \in T \end{cases}
$$

Hence, the preimages of $\{\heartsuit\}$ are exactly the weighted simple cycles over V. Since each preimage T of $\{\heartsuit\}$ has size $|T| = n + 1 + \sum_{t=0}^{n-1} w(v_t, v_{(t+1) \bmod n})$, a preimage T of $\{\heartsuit\}$ of minimum size corresponds to a shortest tour over V, which can be extracted from T in polynomial time just by listing the elements in $T \cap E$, ordered by their subscript. Since the mapping $(V, w) \mapsto \mathcal{A}$ described by the above construction can be computed in polynomial time, the thesis follows. □

Lemma 5. *For RS MPP is metric reducible to the problem of finding, among the possible output strings of a polynomial-time nondeterministic TM, a string having the minimum number of 1s.*

Proof. Given any instance $(\mathcal{A} = (S, A), T)$ of the RS minimal preimage problem, let M be the nondeterministic TM which behaves as follows. M guesses a state $U \subseteq S$ and checks whether $\mathrm{res}_{\mathcal{A}}(U) = T$; if this is the case, then M outputs U as a binary string in $\{0, 1\}^{|S|}$; M outputs $1^{|S|+1}$, otherwise. Clearly, M works in time $p(n)$ for some polynomial p, and its outputs are all the preimages of T (together with an easily distinguishable dummy output if no preimage exists); in particular, the outputs of M that are minimal with respect to the number of 1s correspond to the smallest preimages of T. □

The following is proved similarly to the equivalence of binary TSP and finding the maximum binary output of a polynomial-time nondeterministic TM [8].

Lemma 6. *Finding a string with minimal number of 1s among those output by a nondeterministic polynomial-time TM is metric reducible to the unary TSP.* □

We can now provide lower and upper bounds to the complexity of MPP.

Theorem 7. *MPP for RS is equivalent to the unary TSP under metric reductions. Hence, MPP $\in$ **FP$^{\mathbf{NP}}$** and it is **FP$^{\mathbf{NP}}_{\|}$**-hard under metric reductions.*

Proof. The equivalence is a consequence of Lemmata 4, 5, and 6. Moreover, TSP belongs to **FP$^{\mathbf{NP}}$** which is closed under metric reductions [8]. The travelling salesman with 0/1 weights is hard for **FP$^{\mathbf{NP}}_{\|}$** (see [1]) and can be reduced to the minimal RS preimage problem. □

The complexity of finding the *size* of minimal preimages is given by:

Corollary 8. SMPP *is* $\mathbf{FP}^{\mathbf{NP}[\log n]}$*-complete under metric reductions.*

Proof. The size of the minimal preimage can be found by binary search, using an oracle answering the question "Is there a preimage of size at most k?", which belongs to **NP**. Hence, the problem is in $\mathbf{FP}^{\mathbf{NP}[\log n]}$. The hardness of the problem follows from the fact that the size of a minimal preimage is just the length of the shortest unary travelling salesman tour increased by $n + 1$ in the reduction of Lemma 4, and that the unary TSP is $\mathbf{FP}^{\mathbf{NP}[\log n]}$-complete [8]. ☐

Finally, we can also prove lower and upper bounds to the complexity of finding *the number of* minimal preimages or #MPP in short.

Theorem 9. #MPP *is in* $\#\mathbf{P}^{\mathbf{NP}[\log n]}$ *and it is* #P-*hard under parsimonious reductions.*

Proof. The following algorithm shows the membership in $\#\mathbf{P}^{\mathbf{NP}[\log n]}$: given $(\mathcal{A}, T)$, compute the size k of the smallest preimage of T by binary search using $\log n$ queries to the oracle (as in the proof of Corollary 8); then, guess a state $U \subseteq S$ with $|U| = k$, and accept if and only if $\mathrm{res}_{\mathcal{A}}(U) = T$. The number of accepting computations corresponds to the number of minimal preimages of T.

In order to prove the #P-hardness of the problem, we perform a reduction from #SAT (a variant of [5, Theorem 4]). Let $\varphi = \varphi_1 \wedge \cdots \wedge \varphi_m$ be a Boolean formula in conjunctive normal form over the variables $V = \{x_1, \ldots, x_n\}$. Let $\mathcal{A} = (S, A)$ be a RS with $S = V \cup \overline{V} \cup C \cup \{\spadesuit\}$, where $\overline{V} = \{\bar{x}_1, \ldots, \bar{x}_n\}$ and $C = \{\varphi_1, \ldots, \varphi_m\}$, and A consisting of the following reactions:

$$(\{x_i\}, \{\spadesuit\}, \{\varphi_j\}) \qquad \text{for } 1 \leq i \leq n,\, 1 \leq j \leq m, \text{ if } x_i \text{ occurs in } \varphi_j \qquad (9)$$

$$(\{\bar{x}_i\}, \{\spadesuit\}, \{\varphi_j\}) \qquad \text{for } 1 \leq i \leq n,\, 1 \leq j \leq m, \text{ if } \bar{x}_i \text{ occurs in } \varphi_j \qquad (10)$$

$$(\{s\}, \{x_i, \bar{x}_i, \spadesuit\}, \{\spadesuit\}) \qquad \text{for } 1 \leq i \leq n,\, s \notin \{x_i, x_i, \spadesuit\} \qquad (11)$$

$$(\{x_i, \bar{x}_i\}, \{\spadesuit\}, \{\spadesuit\}) \qquad \text{for } 1 \leq i \leq n. \qquad (12)$$

A state $T \subseteq S$ encodes a valid assignment for φ if, for each $1 \leq i \leq n$, it contains either x_i or $\bar{x}_i$ (denoting the truth value of variable x_i), but not both, and no further element. The reactions of type (9) (resp., (10)) produce the set of elements representing the clauses satisfied when x_i is assigned a true (resp., false) value. Hence, the formula φ has as many satisfying truth assignments as the number of states $T \subseteq S$ encoding valid assignment such that $\mathrm{res}_A(T) = C$, the whole set of clauses. Any such T contains exactly n elements.

If a state T has n elements or less, but it is not a valid assignment to φ, then there is at least one literal x_i or $\bar{x}_i$ missing in T: thus, either $T = \varnothing$, or reaction (11) is enabled and $\spadesuit \in \mathrm{res}_A(T)$; in both cases $\mathrm{res}_A(T) \neq C$. If T has strictly more than n elements, then either it is an inconsistent assignment, containing both x_i and $\bar{x}_i$ for some i, and in that case $\spadesuit \in \mathrm{res}_A(T) \neq C$ by reaction (12), or it has a subset $T' = T \cap (V \cup \overline{V})$ with $|T'| = n$ such that $\mathrm{res}_A(T') = \mathrm{res}_A(T)$.

Thus, the number of *minimal* preimages of T in $\mathcal{A}$ is exactly the number of assignments satisfying φ, as required. ☐

4   Ancestor Problems

We now turn our attention to the ancestor problems, which will show a (supposedly) higher complexity. First of all, we need a few technical results, providing us with a **FPSPACE**(poly)-complete function suitable for reductions.

Lemma 10. *The "universal" function $U(M, 1^m, x)$, defined as $M(x)$ if the TM M halts in space m, and undefined otherwise, is* **FPSPACE**(poly)-*complete under many-one reductions.*

Proof. We have $U \in$ **FPSPACE**(poly), since there exist universal TMs having only a polynomial space overhead [8]. Let $R \in$ **FPSPACE**(poly), and let M be a polynomial-space TM for R. Then, there exists a polynomial p bounding both the working space and the output length of M. The mapping $f(x) = (M, 1^{p(|x|)}, x)$ can be computed in polynomial time, and $R(x, U(f(x)))$ holds for all $x \in$ dom R. This proves the **FPSPACE**(poly)-hardness of U.   □

Lemma 11. *Let the binary relation $R((M, 1^m, y), x)$ hold if and only if the deterministic TM M, on input x, halts with output y on its tape, $|y| \leq m$, and M does not exceed space m during the computation. Then, the relation R is* **FPSPACE**(poly)-*complete under many-one reductions.*

Proof. The relation R is in **FPSPACE**(poly): a polynomial-space deterministic TM can try all strings x of length at most m, one by one, and simulate M on input x (within space m) until the output y is produced or the strings have been exhausted. We now reduce the function U from Lemma 10 to R. Given an instance $(M, 1^m, x)$ of U, consider the instance $(M', 1^k, 1)$ of R, where

- M' is the TM which on any input y, first simulates M on input x within space k; if $M(x) = y$, then M' outputs 1; otherwise, M' outputs 0.
- $k = p(m)$, where p is the polynomial space overhead needed by M' in order to simulate M.

We have $U(M, 1^m, x) = y$ if and only if M' outputs 1 in space k on input y, that is, if and only if $R((M', 1^k, 1), y)$. Since the mapping $(M, 1^m, x) \mapsto (M', 1^k, 1)$ can be computed in polynomial time, the relation R is **FPSPACE**(poly)-hard.   □

By exploiting the ability of RS to simulate polynomial-space TMs [4], we obtain the following result.

Theorem 12. MAP *is complete for* **FPSPACE**(poly) *under metric reductions.*

Proof. The problem is in **FPSPACE**(poly), since a polynomial-space TM can enumerate all states of a RS in order of size, and check whether they lead to the target state (the reachability problem for RS is known to be in **PSPACE** [4]).

In order to prove the **FPSPACE**(poly)-hardness of the problem, we describe a variant of the simulation of polynomial-space TMs by means of RS from [4], where a distinguished state T, encoding the final configuration of the TM, has as minimal ancestors all states encoding the inputs of the TM leading to that configuration. Given a TM M and a space bound m, let Q and Γ be the set of states and the tape alphabet of M, including a symbol $\sqcup$ for "blank" cells. We define a RS $\mathcal{A}$ having background set

$$S = \{q_i : q \in Q, -1 \leq i \leq m\} \cup \{a_i : a \in \Gamma, 0 \leq i < m\} \cup \{\spadesuit\}.$$

We encode the configurations of M as states of $\mathcal{A}$ as follows: if M is in state q, the tape contains the string $w = w_0 \cdots w_{m-1}$, and the tape head is located on the i-th cell, then the corresponding state of $\mathcal{A}$ is $\{q_i, w_{0,0}, \ldots, w_{m-1,m-1}\}$, with an element q_i representing TM state q and head position i, and m elements corresponding to the symbols on the tape indexed by their position.

A transition $\delta(q, a) = (r, b, d)$ of M is implemented by the following reactions

$$(\{q_i, a_i\}, \{\spadesuit\}, \{r_{i+d}, b_i\}) \qquad \text{for } 0 \leq i < m$$

which update state, head position, and symbol under the tape head. The remaining symbols on the tape (which have an index different from the tape head position) are instead preserved by the following reactions:

$$(\{a_i\}, \{q_i : q \in Q\} \cup \{\spadesuit\}, \{a_i\}) \qquad \text{for } a \in \Gamma, 0 \leq i < m$$

If an element encoding TM state and position is not part of the current RS state, the element representing the initial state $s \in Q$ of M, with tape head on the first cell, is produced by the following reactions

$$(\{a_i\}, \{q_j : q \in Q, 0 \leq j < m\} \cup \{\spadesuit\}, \{s_0\}) \qquad \text{for } a \in \Gamma, 0 \leq i < m \qquad (13)$$

and the simulation of M by $\mathcal{A}$ begins in the next time step with the same tape contents. If M exceeds its space bound, by moving the tape head to the left of position 0 or to the right of position m, the following reactions become enabled

$$(\{q_{-1}\}, \{\spadesuit\}, \{\spadesuit\}) \qquad \text{for } q \in Q$$
$$(\{q_m\}, \{\spadesuit\}, \{\spadesuit\}) \qquad \text{for } q \in Q$$

and produce the universal inhibitor $\spadesuit$, which halts the simulation in the next time step. The universal inhibitor is also produced by the following reactions when the state of $\mathcal{A}$ is not a valid encoding of a configuration of M, namely, when multiple state elements appear:

$$(\{q_i, r_j\}, \{\spadesuit\}, \{\spadesuit\}) \qquad \text{for } q, r \in Q, q \neq r, 0 \leq i < m, 0 \leq j < m$$

or when multiple symbols are located on the same tape cell:

$$(\{a_i, b_i\}, \{\spadesuit\}, \{\spadesuit\}) \qquad \text{for } a, b \in \Gamma, a \neq b, 0 \leq i < m$$

or when a tape cell does not contain any symbol (recall that a blank cell contains a specified symbol from Γ):

$$(\{s\}, \{a_i : a \in \Gamma\} \cup \{\spadesuit\}, \{\spadesuit\}) \qquad \text{for } s \notin \{a_i : a \in \Gamma\} \cup \{\spadesuit\}, 0 \leq i < m \qquad (14)$$

The result function of $\mathcal{A}$ can thus be described as follows:

$$
\mathrm{res}_{\mathcal{A}}(T) = \begin{cases}
T' & \text{if } T \text{ encodes a configuration of } M \\
& \text{and } T' \text{ its next configuration} \\
T \cup \{s_0\} & \text{if } T \text{ encodes a tape of } M \\
T' \cup \{\spadesuit\} & \text{for some } T' \subseteq S, \text{ if } \spadesuit \notin T \text{ but } T \text{ does} \\
& \text{not encode a configuration of } M \\
\varnothing & \text{if } \spadesuit \in T \text{ or } T = \varnothing
\end{cases}
$$

Notice that all states of $\mathcal{A}$ encoding configurations of M have $m + 1$ elements, and that either $\mathrm{res}_{\mathcal{A}}(T) = \varnothing$ or $\spadesuit \in \mathrm{res}_{\mathcal{A}}(T)$ if $|T| < m$.

Given an instance $(M, 1^m, y)$ of relation R from Lemma 11, we can ask for a minimal state X of $\mathcal{A}$ leading to Y, where Y encodes the configuration of M in its final state, with the string y on the tape.

If there exists a string x such that $M(x) = y$ and the space bound m is never exceeded by M during its computation, then there exists a state $X \subseteq S$ encoding a tape for M containing the input x (padded to length m with blanks) and such that $\mathrm{res}_{\mathcal{A}}^t(X) = Y$ for some $t \geq 0$. We have $|X| = m$, and X is minimal with respect to size among all states leading to Y, since all smaller states lead to $\varnothing$ in at most two steps. Furthermore, from X we can easily recover a string x with $M(x) = y$. Conversely, any state $X \subseteq S$ with $|X| = m$ and $\mathrm{res}_{\mathcal{A}}^t(X) = Y$ necessarily encodes a string x such that $M(x) = y$ within space m.

If $M(x) \neq y$ for all strings x (or all such computations exceed the space bound m), then all states T such that $\mathrm{res}_{\mathcal{A}}^t(T) = Y$, and in particular the minimal ones, contain $m + 1$ elements. By observing this fact, we can infer that no input of M produces the output y in space m.

Since the mapping $(M, 1^m, y) \mapsto (\mathcal{A}, Y)$ described by the above construction can be computed in polynomial time, and the answer for R can be extracted from the answer to the minimal RS ancestor search problem in polynomial time, the latter problem is **FPSPACE**(poly)-hard under metric reductions. $\square$

We now deal with the problem of finding the size of a minimal ancestor.

Lemma 13. *Let* $f(M, 1^m) = \min\{|x| : \text{the TM } M \text{ accepts } x \text{ in space } m\}$, *undefined if no such x exists. Then f is* **FPSPACE**(log)-*complete under many-one reductions.*

Proof. Given $g \in$ **FPSPACE**(log), let G be a deterministic TM computing g in space $p(n)$ for some polynomial p, and let $x \in \Sigma^\star$. Let M be a deterministic TM that, on input y, first simulates $G(x)$, then accepts if and only if $|y| \geq G(x)$. Hence, we have $g(x) = f(M, 1^{q(|x|)})$, where q is a polynomial bound on the space needed by M to simulate G. This proves the hardness of f. The function can be computed in **FPSPACE**(log) by simulating M on all strings of length at most m until one is accepted in space m, then outputting its length. $\square$

Theorem 14. SMAP *is* **FPSPACE**(log)-*complete under metric reductions.*

Proof. The problem is in **FPSPACE**(log), since a polynomial-space TM can find an ancestor U of a state of a RS as in the proof of Theorem 12, outputting the size of U rather than U itself. In order to show the hardness of the problem, we reduce the function f from Lemma 13 to it. Given $(M, 1^m)$, let $\mathcal{A}$ be the RS of Theorem 12, simulating M in space 1^m, and let $\mathcal{A}'$ be $\mathcal{A}$ modified as follows. First of all, we add $\heartsuit$ to S, and we also add it as a reactant to all reactions. Then, the reactions of type (13) are replaced by

$$(\{\heartsuit\}, \{a_i : a \in \Gamma\} \cup \{q_j : q \in Q, -1 \leq j \leq m\} \cup \{\spadesuit\}, \{\sqcup_i\}) \quad \text{for } 0 \leq i < m \tag{15}$$

$$(\{\heartsuit\}, \{q_j : q \in Q, -1 \leq j \leq m\} \cup \{\spadesuit\}, \{s_0\}) \tag{16}$$

The reactions of type (15) complete the tape of the TM by producing a blank symbol in position i if no symbol a_i and no state q_j occur. Reaction (16) produces the initial state of M in position 0 when no other state element occurs.

The reactions of type (14) are replaced by

$$(\{q_j\}, \{a_i : a \in \Gamma\} \cup \{\spadesuit\}, \{\spadesuit\}) \quad \text{for } q \in Q, 0 \leq i < m, -1 \leq j \leq m \tag{17}$$

which give an error (producing $\spadesuit$) when a tape symbol is missing, but only if a state element is already present. Finally, we add the reaction $(\{\heartsuit\}, \{\spadesuit\}, \{\heartsuit\})$, which preserves the $\heartsuit$ element.

The behaviour of $\mathcal{A}'$ differs from $\mathcal{A}$ in the following ways. The input string $x = x_0 \cdots x_{n-1}$ of M is provided as a state $X = \{x_{0,0}, \ldots, x_{n-1,n-1}, \heartsuit\}$. In the first step of $\mathcal{A}'$, the tape is completed by adding blanks and the initial state of M (reactions (15)–(16)); this produces the state $X \cup \{\sqcup_n, \ldots, \sqcup_{m-1}, s_0\}$, which encodes the initial configuration of M. The simulation of M then proceeds as for $\mathcal{A}$ (with the additional element $\heartsuit$ always present).

The ancestors of state T describing an accepting configuration of M (empty tape, accepting state, head in position 0) and of minimal size encode the initial input x of M together with $\heartsuit$, if M accepts at least one string in space m (hence, such ancestors have size at most $m + 1$); if no string is accepted, the minimal ancestors of T all have size at least $m + 2$, by the same reasoning as in the proof of Theorem 12. Hence, $f(M, 1^m) + 1$ is the size of a minimal ancestor of T, if the latter is at most $m + 1$, and $f(M, 1^m)$ is undefined otherwise: this defines a metric reduction of f to this problem. □

Remark 15. The problem of Theorem 14 is actually complete under metric reductions that only increase linearly the length of the output; the class **FPSPACE**(log) is closed under such reductions, but not under general metric reductions.

Finally, we show that counting the number of minimal ancestors has the same complexity as finding one of them.

Lemma 16. *Given a TM M, a unary integer 1^m, and a string y, computing the number of strings x of length at most m such that $M(x) = y$ in space m is* **FPSPACE**(poly)-*complete under many-one reductions.*

Proof. Recalling that **FPSPACE**(poly) = ♮**PSPACE**, the following nondeterministic polynomial-space TM has the required number of accepting computations: on input $(M, 1^m, y)$, it guesses a string x of length at most m (this requires polynomially many guesses); then it simulates M on x, accepting if M outputs y without exceeding space m, and rejecting otherwise.

Given $f \in$ ♮**PSPACE**, let N be a nondeterminstic, polynomial-space TM with $f(x)$ accepting computations on input x; let $p(n)$ be both a space bound for N and bound on the number of nondeterministic choices it makes. Consider the following polynomial-space TM M: on input $z \in \{0, 1\}^\star$, it simulates a computation of N on input x, but replaces the nondeterministic choices of N with deterministic lookups to successive bits of z. If N exceeds space $m = p(|x|)$, or halts without having made exactly $|z|$ nondeterministic choices, or the simulated computation of N rejects, then M writes 0 as output; otherwise, M outputs 1.

Hence, M outputs 1 once for each accepting computation of N, that is, for exactly $f(x)$ input strings. Since the mapping $x \mapsto (M, 1^{p(|x|)}, 1)$ can be computed in polynomial time, the **FPSPACE**(poly)-hardness of the problem follows. □

Theorem 17. #MAP *is* **FPSPACE**(poly)-*complete under metric reductions.*

Proof. The problem is in **FPSPACE**(poly), since a polynomial-space TM can compute the size of a minimal ancestor of a state T of a RS, then enumerate all states of the same size and count how many of them lead to T.

Let $(M, 1^m, y)$ be an instance of the problem of Lemma 16, and let $\mathcal{A}$ be the RS simulating M as in the proof of Theorem 12. Let $Y = \{f_0, z_{0,0}, \ldots, z_{m-1,m-1}\}$ be the state of $\mathcal{A}$ encoding the final configuration of M with output y, where $z = y_0 \cdots y_{k-1} \sqcup^{m-k}$ is y padded to length m with blanks and $f \in Q$ is the final state of M. In order to distinguish the presence or absence of at least a string x such that $M(x) = y$, we add a large number of ancestors of Y having size $m + 1$, ensuring that are minimal only if no such string exists. Let $\mathcal{A}'$ be $\mathcal{A}$ augmented with the following reactions:

$$(\{a_i, \spadesuit\}, \{q_j : q \in Q, -1 \leq j \leq m\}, \{z_{i,i}, f_0\}) \quad \text{for } a \in \Gamma, 0 \leq i < m. \quad (18)$$

When $\spadesuit$ is present, and all q_j are missing, these reactions map each element representing a symbol in tape cell i, to the symbol z_i in tape cell i, together with the final state f of M in position 0. In particular, when at least one symbol per position i is present, the whole target state Y is produced. Hence, these reactions introduce exactly $|\Gamma|^m$ new ancestors of Y of size $m + 1$. The state Y then has at least $|\Gamma|^m + 1$ ancestors of size $m + 1$, including Y itself. From the proof of Theorem 12 we may infer that the maximum number of ancestors of Y of size m is $|\Gamma|^m$. The number of strings x of length at most m such that $M(x) = y$ in space m is then equivalent to the number of minimal ancestors of Y for $\mathcal{A}'$, if and only if this number is at most $|\Gamma|^m$. If the number is larger than $|\Gamma|^m$, then the minimal ancestors have size $m + 1$, indicating that no such string x exists. This defines a metric reduction, proving the **FPSPACE**(poly)-hardness of #MAP. □

5   Conclusions

We investigated the problem of finding the minimal preimage of a state of a RS and proved that this problem is equivalent, under metric reductions, to finding a minimal TSP tour when the weights are expressed in unary. We also studied the complexity of finding a minimal ancestor of a given state and showed that it is as hard as simulating a polynomial-space TM (with polynomial-length output). Furthermore, we have investigated the complexity of other problems related to preimages (resp., ancestors): finding the size and the number of minimal preimages (resp., ancestors). All these problems were proved to be intractable.

In the future we plan to continue the exploration of problems related to preimages and ancestors of RS. In the more general model [3], RSs behave as interactive processes, where new entities are introduced at every time step by means of a context sequence. Under which conditions does the presence of a context sequence increase the complexity of the problems we considered? We are also interested in questions related to the approximability of the aforementioned problems and the complexity of finding a minimal ancestor that is not "too far" from the target state.

References

1. Chen, Z.Z., Toda, S.: On the complexity of computing optimal solutions. International Journal of Foundations of Computer Science **2**(3), 207–220 (1991)
2. Ehrenfeucht, A., Rozenberg, G.: Basic notions of reaction systems. In: Calude, C.S., Calude, E., Dinneen, M.J. (eds.) DLT 2004. LNCS, vol. 3340, pp. 27–29. Springer, Heidelberg (2004)
3. Ehrenfeucht, A., Rozenberg, G.: Reaction systems. Fundamenta Informaticae **75**, 263–280 (2007)
4. Formenti, E., Manzoni, L., Porreca, A.E.: Cycles and global attractors of reaction systems. In: Jürgensen, H., Karhumäki, J., Okhotin, A. (eds.) DCFS 2014. LNCS, vol. 8614, pp. 114–125. Springer, Heidelberg (2014)
5. Formenti, E., Manzoni, L., Porreca, A.E.: Fixed points and attractors of reaction systems. In: Beckmann, A., Csuhaj-Varjú, E., Meer, K. (eds.) CiE 2014. LNCS, vol. 8493, pp. 194–203. Springer, Heidelberg (2014)
6. Krentel, M.W.: The complexity of optimization problems. Journal of Computer and System Sciences **36**, 490–509 (1988)
7. Ladner, R.E.: Polynomial space counting problems. SIAM Journal on Computing **18**(6), 1087–1097 (1989)
8. Papadimitriou, C.H.: Computational Complexity. Addison-Wesley (1993)
9. Porreca, A.E., Murphy, N., Pérez-Jiménez, M.J.: An optimal frontier of the efficiency of tissue P systems with cell division. In: García-Quismondo, M., Macías-Ramos, L.F., Păun, Gh., Valencia-Cabrera, L. (eds.) Tenth Brainstorming Week on Membrane Computing, vol. II, pp. 141–166. Fénix Editora (2012)
10. Salomaa, A.: Minimal and almost minimal reaction systems. Natural Computing **12**(3), 369–376 (2013)

Parameterized Complexity of CTL
A Generalization of Courcelle's Theorem

Martin Lück, Arne Meier[(✉)], and Irina Schindler

Institut für Theoretische Informatik, Leibniz Universität Hannover,
Appelstrasse 4, 30167 Hannover, Germany
{lueck,meier,schindler}@thi.uni-hannover.de

Abstract. We present an almost complete classification of the parameterized complexity of all operator fragments of the satisfiability problem in computation tree logic CTL. The investigated parameterization is the sum of temporal depth and structural pathwidth. The classification shows a dichotomy between W[1]-hard and fixed-parameter tractable fragments. The only real operator fragment which is confirmed to be in FPT is the fragment containing solely AX. Also we prove a generalization of Courcelle's theorem to infinite signatures which will be used to proof the FPT-membership case.

Keywords: Parameterized complexity · Temporal logic · Computation tree logic · Courcelle's theorem

1 Introduction

Temporal logic is the most important concept in computer science in the area of program verification and is a widely used concept to express specifications. Introduced in the late 1950s by Prior [16] a large area of research has been evolved up to today. Here the most seminal contributions have been made by Kripke [9], Pnueli [14], Emerson, Clarke, and Halpern [2,6] to name only a few. The maybe most important temporal logic so far is the computation tree logic CTL due to its polynomial time solvable model checking problem which influenced the area of program verification significantly. However the satisfiability problem, i.e., the question whether a given specification is consistent, is beyond tractability, i.e., complete for deterministic exponential time. One way to attack this intrinsic hardness is to consider restrictions of the problem by means of operator fragments leading to a trichotomy of computational complexity shown by Meier [12]. This landscape of intractability depicted completeness results for nondeterministic polynomial time, polynomial space, and (of course) deterministic exponential time showing how combinations of operators imply jumps in computational complexity of the corresponding satisfiability fragment.

For more than a decade now there exists a theory which allows us to better understand the structure of intractability: 1999 Downey and Fellows developed

Irina Schindler: Supported in part by DFG ME 4279/1-1.

© Springer International Publishing Switzerland 2015
A.-H. Dediu et al. (Eds.): LATA 2015, LNCS 8977, pp. 549–560, 2015.
DOI: 10.1007/978-3-319-15579-1_43

the area of parameterized complexity [4] and up to today this field has grown vastly. Informally the main idea is to detect a specific part of the problem, the *parameter*, such that the intractability of the problems complexity vanishes if the parameter is assumed to be constant. Through this approach the notion of *fixed parameter tractability* has been founded. A problem is said to be fixed parameter tractable (or short, FPT) if there exists a deterministic algorithm running in time $f(k) \cdot poly(n)$ for all input lengths n, corresponding parameter values k, and a recursive function f. As an example, the usual propositional logic satisfiability problem SAT (well-known to be NP-complete) becomes fixed parameter tractable under the parameter number of variables.

In this work we almost completely classify the parameterized complexity of all operator fragments of the satisfiability problem for the computation tree logic CTL under the parameterization of formula pathwidth and temporal depth. Only the case for AF resisted a full classification. We will explain the reasons in the conclusion. For all other fragments we show a dichotomy consisting of two fragments being fixed parameter tractable and the remainder being hard for the complexity class **W**[1] under fpt-reductions. **W**[1] can be seen as an analogue of intractability in the decision case in the parameterized world. To obtain this classification we prove

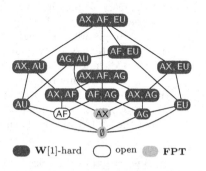

Fig. 1. Parameterized complexity of CTL-SAT($\mathcal{T}$) parameterized by formula pathwidth and temporal depth (see Theorem 1)

a generalization of Courcelle's theorem [3] for *infinite* signatures which may be of independent interest.

Related work. Similar research for modal logic has been done by Praveen and influenced the present work in some parts [15]. Other applications of Courcelle's theorem have been investigated by Meier et al. [11] and Gottlob et al. [8]. In 2010 Elberfeld et al. proved that Courcelle's theorem can be extended to give results in XL as well [5] wherefore the results of Theorem 4 can be extended to this class, too.

2   Preliminaries

We assume familiarity with standard notions of complexity theory as Turing machines, reductions, the classes **P** and **NP**. For an introduction into this field we confer the reader to the very good textbook of Pippenger [13].

2.1   Complexity Theory

Let Σ be an alphabet. A pair $\Pi = (Q, \kappa)$ is a *parameterized problem* if $Q \subseteq \Sigma^*$ and $\kappa: \Sigma^* \to \mathbb{N}$ is a function. For a given instance $x \in \Sigma^*$ we refer to x as the *input*. A function $\kappa\colon \Sigma^* \to \mathbb{N}$ is said to be a *parameterization of Π* or the *parameter of Π*. We say a parameterized problem Π is *fixed-parameter tractable* (or in the class **FPT**) if there exists a deterministic algorithm deciding Π in time $f(\kappa(x)) \cdot |x|^{O(1)}$ for every $x \in \Sigma^*$ and a recursive function f. Note that the notion of fixed-parameter tractability is easily extended beyond decision problems.

If $\Pi = (Q, \kappa), \Pi' = (Q', \kappa')$ are parameterized problems over alphabets Σ, Δ then an *fpt-reduction from Π to Π'* (or in symbols $\Pi \leq^{fpt} \Pi'$) is a mapping $r\colon \Sigma^* \to \Delta^*$ with the following three properties:

(1) For all $x \in \Sigma^*$ it holds $x \in Q$ iff $r(x) \in Q'$. (2) r is fixed-parameter tractable, i.e., r is computable in time $f(\kappa(x)) \cdot |x|^{O(1)}$ for a recursive function $f\colon \mathbb{N} \to \mathbb{N}$. (3) There exists a recursive function $g\colon \mathbb{N} \to \mathbb{N}$ such that for all $x \in \Sigma^*$ it holds $\kappa'(r(x)) \leq g(\kappa(x))$.

The class **W**[1] is a parameterized complexity class which plays a similar role as **NP** in the sense of intractability in the parameterized world. The class **W**[1] is a superset of **FPT** and a hierarchy of other **W**-classes are build above of it: **FPT** $\subseteq$ **W**[1] $\subseteq$ **W**[2] $\subseteq \cdots \subseteq$ **W**[P]. All these classes are closed under fpt-reductions. It is not known whether any of these inclusions is strict. For further information on this topic we refer the reader to the text book of Flum and Grohe [7].

2.2 Tree- and Pathwidth

Given a structure $\mathcal{A}$ we define a *tree decomposition of $\mathcal{A}$* (with universe A) to be a pair (T, X) where $X = \{B_1, \ldots, B_r\}$ is a family of subsets of A (the set of *bags*), and T is a tree whose nodes are the bags B_i satisfying the following conditions:

1. Every element of the universe appears in at least one bag: $\bigcup X = A$.
2. Every Tuple is contained in a bag: for each $(a_1, \ldots, a_k) \in R$ where R is a relation in $\mathcal{A}$, there exists a $B \in X$ such that $\{a_1, \ldots, a_k\} \in B$.
3. For every element a the set of bags containing a is connected: for all $a \in A$ the set $\{B \mid a \in B\}$ forms a connected subtree in T.

The *width* of a decomposition (T, X) is $\mathrm{width}(T, X) := \max\{|B| \mid B \in X\} - 1$ which is the size of the largest bag minus 1. The *treewidth* of a structure $\mathcal{A}$ is the minimum of the widths of all tree decompositions of $\mathcal{A}$. Informally the treewidth of a structure describes the tree-likeliness of it. The closer the value is to 1 the more the structure is a tree.

A *path decomposition* of a structure $\mathcal{A}$ is similarly defined to tree decompositions however T has to be a path. Here $\mathrm{pw}(\mathcal{A})$ denotes the *pathwidth* of $\mathcal{A}$. Likewise the size of the pathwidth describes the similarity of a structure to a path. Observe that pathwidth bounds treewidth from above.

2.3 Logic

Let Φ be a finite set of propositional letters. A *propositional formula* ($\mathcal{PL}$ formula) is inductively defined as follows. The constants $\top, \bot$, (true, false) and any *propositional letter* (or *proposition*) $p \in \Phi$ are $\mathcal{PL}$ formulas. If ϕ, ψ are $\mathcal{PL}$ formulas then so are $\phi \wedge \psi, \neg\phi, \phi \vee \psi$ with their usual semantics (we further use the shortcuts $\rightarrow, \leftrightarrow$). Temporal logic extends propositional logic by introducing four *temporal operators*, i.e., *next* X, *future* F, *globally* G, and *until* U. Together with the two *path quantifiers*, *exists* E and *all* A, they fix the set of *computation tree logic formulas* (*CTL* formulas) as follows. If $\phi \in \mathcal{PL}$ then $\mathsf{PT}\phi, \mathsf{P}[\phi\mathsf{U}\psi] \in \mathcal{CTL}$ and if $\phi, \psi \in \mathcal{CTL}$ then $\mathsf{PT}\phi, \mathsf{P}[\phi\mathsf{U}\psi], \phi \vee \psi, \neg\psi, \phi \wedge \psi \in \mathcal{CTL}$ hold, where $\mathsf{P} \in \{\mathsf{A}, \mathsf{E}\}$ is a path quantifier and $\mathsf{T} \in \{\mathsf{X}, \mathsf{F}, \mathsf{G}\}$ is a temporal operator. The pair of a single path quantifier and a single temporal operator is referred to as a CTL-operator. If T is a set of CTL-operators then $\mathcal{CTL}(T)$ is the restriction of $\mathcal{CTL}$ to formulas that are allowed to use only CTL-operators from T.

Let us turn to the notion of Kripke semantics. Let Φ be a finite set of propositions. A *Kripke structure* $K = (W, R, V)$ is a finite set of *worlds* W, a *total successor relation* $R: W \rightarrow W$ (i.e., for every $w \in W$ there exists a $w' \in W$ with wRw'), and an *evaluation function* $V: W \rightarrow 2^\Phi$ labeling sets of propositions to worlds. A *path* π in a Kripke structure $K = (W, R, V)$ is an infinite sequence of worlds $w_0, w_1, \ldots$ such that for every $i \in \mathbb{N}$ $w_i R w_{i+1}$. With $\pi(i)$ we refer to the i-th world w_i in π. Denote with $\mathfrak{P}(w)$ the set of all paths starting at w. For $\mathcal{CTL}$ formulas we define the semantics of $\mathcal{CTL}$ formulas ϕ, ψ for a given Kripke structure $K = (W, R, V)$, a world $w \in W$, and a path π as

$$
\begin{aligned}
K, w \models \mathsf{AT}\phi \quad &\Leftrightarrow \quad \text{for all } \pi \in \mathfrak{P}(w) \text{ it holds } K, \pi \models \mathsf{T}\phi, \\
K, w \models \mathsf{ET}\phi \quad &\Leftrightarrow \quad \text{there exists a } \pi \in \mathfrak{P}(w) \text{ it holds } K, \pi \models \mathsf{T}\phi, \\
K, \pi \models \mathsf{X}\phi \quad &\Leftrightarrow \quad K, \pi(1) \models \phi, \\
K, \pi \models \mathsf{F}\phi \quad &\Leftrightarrow \quad \text{there exists an } i \geq 0 \text{ such that } K, \pi(i) \models \phi, \\
K, \pi \models \mathsf{G}\phi \quad &\Leftrightarrow \quad \text{for all } i \geq 0 \ K, \pi(i) \models \phi, \\
K, \pi \models \phi\mathsf{U}\psi \quad &\Leftrightarrow \quad \exists i \geq 0 \forall j < i \ K, \pi(j) \models \phi \text{ and } K, \pi(i) \models \psi.
\end{aligned}
$$

For a formula $\phi \in \mathcal{CTL}$ we define the satisfiability problem CTL-SAT asking if there exists a Kripke structure $K = (W, R, V)$ and $w \in W$ such that $K, w \models \phi$. Then we also say that M is a *model of* ϕ. Similar to before CTL-SAT(T) is the restriction of CTL-SAT to formulas in $\mathcal{CTL}(T)$ for a set of CTL-operators T. A formula $\phi \in \mathcal{CTL}$ is said to be in *negation normal form* (*NNF*) if its negation symbols $\neg$ occur only in front of propositions; we will use the symbol $\mathcal{CTL}_{\text{NNF}}$ to denote the set of CTL-formulas which are in NNF only.

Given $\phi \in \mathcal{CTL}$ we define SF(ϕ) as the *set of all subformulas of* ϕ (containing ϕ itself). The *temporal depth of* ϕ, in symbols td(ϕ), is defined inductively as follows. If Φ is a finite set of propositional symbols and $\phi, \psi \in \mathcal{CTL}$ then

$$
\begin{aligned}
\text{td}(p) &:= 0, & \text{td}(\phi \circ \psi) &:= \max\{\text{td}(\phi), \text{td}(\psi)\}, \\
\text{td}(\top) &:= 0, & \text{td}(\neg\phi) &:= \text{td}(\phi), \\
\text{td}(\bot) &:= 0, & \text{td}(\mathsf{PT}\phi) &:= \text{td}(\phi) + 1, \\
& & \text{td}(\mathsf{P}[\phi\mathsf{U}\psi]) &:= \max\{\text{td}(\phi), \text{td}(\psi)\} + 1,
\end{aligned}
$$

where $\circ \in \{\wedge, \vee, \rightarrow, \leftrightarrow\}$, $\mathsf{P} \in \{\mathsf{A}, \mathsf{E}\}$, and $\mathsf{T} \in \{\mathsf{X}, \mathsf{F}, \mathsf{G}\}$. If $\psi \in \text{SF}(\phi)$ then the *temporal depth of* ψ *in* ϕ is $\text{td}_\phi(\psi) := \text{td}(\phi) - \text{td}(\psi)$.

Vocabularies are *finite* sets of *relation symbols* (or *predicates*) of finite arity $k \geq 1$ (if $k = 1$ then we say the predicate is *unary*) which are usually denoted with the symbol τ. Later we will also refer to similar objects of infinite size wherefore we prefer to denote them with the term *signature* which usually is an countable infinite sized set of symbols. A *structure* $\mathcal{A}$ over a vocabulary (or signature) τ consists of a *universe* A which is a non-empty set, and a relation $P^{\mathcal{A}} \subseteq A^k$ for each predicate P of arity k. Monadic second order logic (MSO) is the restriction of second order logic (SO) in which only quantification over unary relations is allowed (elements of the universe can still be quantified existentially or universally). If P is a unary predicate then $P(x)$ is true if and only if $x \in P$ holds (otherwise it is false).

3   Parameterized Complexity of CTL-SAT(T)

In this section we investigate all operator fragments of CTL-SAT parameterized by temporal depth and formula pathwidth with respect to its parameterized complexity. This means, we the given formulas from $\mathcal{CTL}$ as input are represented by relational structures as follows.

Let $\varphi \in \mathcal{CTL}$ be a $\mathcal{CTL}$ formula. The vocabulary of our interest is τ being defined as $\tau := \{\text{const}_f^1 \mid f \in \{\top, \bot\}\} \cup \{\text{conn}_{f,i}^2 \mid f \in \{\wedge, \vee, \neg\}, 1 \leq i \leq \text{ar}(f)\} \cup \{\text{var}^1, \text{repr}^1, \text{repr}_{\text{PL}}^1\} \cup \{\text{repr}_{\text{C}}^1, \text{body}_{\text{C}}^2 \mid \text{C is a unary CTL-operator}\} \cup \{\text{repr}_{\text{C}}^1, \text{body}_{\text{C}}^3 \mid \text{C is a binary CTL-operator}\}$. We then associate the vocabulary τ with the structure $\mathcal{A}_\varphi$ where its universe consists of elements representing subformulas of φ. The predicates are defined as follows

- $\text{var}(x)$ holds iff x represents a variable,
- $\text{repr}(x)$ holds iff x represents the formula φ,
- $\text{repr}_{\text{PL}}(x)$ holds iff x represents a propositional formula,
- $\text{repr}_{\text{C}}(x)$ holds iff x represents a formula $\text{C}\psi$ where C is a CTL-operator,
- $\text{body}_{\text{C}}(y, x)$ (resp., $\text{body}_{\text{C}}(y, z, x)$) holds iff x represents a formula $\text{C}\psi$ (resp., $\text{C}(\psi, \chi)$) and ψ is represented by y where C is a unary CTL-operator (resp., ψ / χ is represented by y / z where C is a binary CTL-operator),
- $\text{const}_f(x)$ holds iff x represents the constant of f,
- $\text{conn}_{f,i}(x, y)$ holds iff x represents the ith argument of the function f at the root of the formula tree represented by y.

Now we consider the problem CTL-SAT parameterized by the pathwidth of its instance structures $\mathcal{A}_\varphi$ (for the instances φ) as well as the temporal depth of the formula. Hence the parameterization function κ maps, given an instance formula $\varphi \in \mathcal{CTL}$ to the pathwidth of the structures $\mathcal{A}_\varphi$ plus the temporal depth of φ, i.e., $\kappa(\varphi) = \text{pw}(\mathcal{A}_\varphi) + \text{td}(\varphi)$.

The following theorem summarizes the collection of results we have proven in the upcoming lemmas. The subsection on page 554 contains the **FPT** result together with the generalization of Courcelle's theorem to infinite signatures.

Theorem 1. CTL-SAT(T) *parameterized by formula pathwidth and temporal depth is*

1. *in* **FPT** *if* $T = \{AX\}$ *or* $T = \emptyset$, *and*
2. **W**[1]-*hard if* $AG \in T$, *or* $AU \in T$, *or* $\{AX, AF\} \subseteq T$.

Proof. (1.) is witnessed by Corollary 4. The proof of (2.) is split into Lemmas 5 to 7. □

One way to prove the containment of a problem parameterized in that way in the class **FPT** is to use the prominent result of Courcelle [Thm. 6.3 (1)][3]. Informally, satisfiability of CTL-formulas therefore has to be formalized in monadic second order logic. The other ingredient of this approach is expressing formulas by relational structures as described before. Now the crux is that our case requires a family of MSO formulas which depend on the instance. This however seems to be a serious issue at first sight as this prohibits the application of Courcelle's theorem. Fortunately we are able to generalize Courcelle's theorem in a way to circumvent this problem. Moreover we extended it to work with infinite sized signatures under specific restrictions which allows us to state the desired **FPT** result described as follows.

A Generalized Version of Courcelle's Theorem

Assume we are able to express a problem Q in MSO. If instances $x \in Q$ can be modeled via some relational structure $\mathcal{A}_x$ over some finite vocabulary τ and we see Q as a parameterized problem (Q, κ) where κ is the treewidth of $\mathcal{A}_x$ then by Courcelle's theorem we immediately obtain that (Q, κ) is in **FPT** [3]. If we do not have a fixed MSO formula (which is independent of the instance) then we are not able to use the mentioned result. However the following theorem shows how it is possible even with infinite signatures to apply the result of Courcelle. For this, we assume that the problem can be expressed by an infinite family $(\phi_n)_{n \in \mathbb{N}}$ of MSO-formulas along with the restriction that $(\phi_n)_{n \in \mathbb{N}}$ is uniform, i.e., there is a recursive function $f : n \to \phi_n$.

Let κ be a parameterization. Call a function $f : \Sigma^* \to \Sigma^*$ κ-*bounded* if there is a computable function h such that for all x it holds that $|f(x)| \leq h(\kappa(x))$.

Theorem 2. *Let* (Q, κ) *be a parameterized problem such that instances* $x \in \Sigma^*$ *can be expressed via relational structures* $\mathcal{A}_x$ *over a (possibly infinite) signature* τ *and* $tw(\mathcal{A}_x)$ *is* κ-*bounded. If there exists a uniform MSO-formula family* $(\phi_n)_{n \in \mathbb{N}}$ *and a fpt-computable,* κ-*bounded function* f *such that for all* $x \in \Sigma^*$ *it holds* $x \in Q \Leftrightarrow \mathcal{A}_x \models \phi_{|f(x)|}$ *then* $(Q, \kappa) \in$ **FPT**.

Proof. Let (Q, κ), $(\phi_n)_{n \in \mathbb{N}}$, κ and f be given as in the conditions of the theorem. Let $(\phi_n)_{n \in \mathbb{N}}$ be computed by a w.l.o.g. non-decreasing and computable function g. The following algorithm correctly decides Q in fpt-time w.r.t. κ. First compute $i := |f(x)|$ in **FPT** for the given instance x. Since $(\phi_n)_{n \in \mathbb{N}}$ is uniform and f is κ-bounded we can construct ϕ_i in time $g(n) = g(|f(x)|) \leq g(h(\kappa(x)))$, hence in **FPT**. Now we are able to solve the model checking problem instance $(\mathcal{A}_x, \phi_i)$ in time $f'(tw(\mathcal{A}_x), |\phi_i|) \cdot |\mathcal{A}_x|$ for a recursive f' due to Courcelle's theorem. As both tw and $|\phi_i|$ are κ-bounded, the given algorithm then runs in **FPT** time. □

Note that the infinitely sized signature is required to describe the structures from the set of all structures $\mathfrak{A}$ which occur with respect to the corresponding *family* of MSO-formulas $(\phi_n)_{n \in \mathbb{N}}$. Every subset $T \subset \mathfrak{A}$ of structures with respect to each ϕ_i then possess (as desired and required by Courcelle's theorem) a finite signature, i.e., a vocabulary.

Praveen [15] shows the fixed-parameter tractability of ML-SAT (parameterized by pathwidth and modal depth) by applying Courcelle's theorem, using for each modal formula an MSO-formula whose length is linear in the modal depth. This can be seen as a special case of Theorem 2 using a **P**-uniform MSO family that partitions the instance set according to the modal depth.

Again we want to stress that formula pathwidth of φ refers to the pathwidth of the corresponding structures $\mathcal{A}_\varphi$ as defined above.

Lemma 3. *Let* $\varphi \in \mathcal{CTL}_{\mathrm{NNF}}(\{\mathsf{AX}, \mathsf{EX}\}, B)$ *given by the structure* $\mathcal{A}_\varphi$ *over* τ. *Then there exists an MSO formula* $\theta(\varphi)$ *such that* $\varphi \in$ CTL-SAT$(\{\mathsf{AX}\})$ *iff* $\mathcal{A}_\varphi \models \theta(\varphi)$ *and* $\theta(\varphi)$ *depends only on* $\mathrm{td}(\varphi)$.

Proof. The first step is to show that a formula $\varphi \in \mathcal{CTL}_{\mathrm{NNF}}(\{\mathsf{AX}, \mathsf{EX}\})$ is satisfiable if and only if it is satisfied by a Kripke structure of depth $\mathrm{td}(\varphi)$, where the depth of a structure (M, w_0) is the maximal distance in M from w_0 to another state from M. This can be similar proven as the *tree model property* of modal logic [1, p. 269, Lemma 35].

Let φ be the given formula in $\mathcal{CTL}_{\mathrm{NNF}}(\{\mathsf{AX}, \mathsf{EX}\})$. The following formula θ_{struc} describes the properties of the structure $\mathcal{A}_\varphi$. At first it takes care of the uniqueness of the formula representative. If an element x does not represent a formula then it has to be a subformula. Additionally if x it is not a variable it has to be either a constant, or a Boolean function $f \in B$ with the corresponding arity $\mathrm{ar}(f)$, or an AX-, or an EX-formula respectively. Furthermore the distinctness of the representatives has to be ensured which together with the previous constraints implies acyclicity of the relation structure graph.

In the following $f_1(u, v, w, x)$ corresponds to the operator of the function which is true if exactly one of its arguments is true.

$$\theta_{\mathrm{struc}} := \forall x \forall y (\mathrm{repr}(x) \wedge \mathrm{repr}(y) \rightarrow x = y) \wedge$$

$$\forall x \left(\neg \mathrm{repr}(x) \rightarrow \exists y \left(\neg \mathrm{var}(y) \wedge \bigvee_{\substack{f \in \{\wedge, \vee, \neg\}, \\ 1 \leq i \leq \mathrm{ar}(f)}} \mathrm{conn}_{f,i}(x, y) \right) \right) \wedge$$

$$\forall x \, f_1 \Bigg(\mathrm{var}(x), \bigvee_{f \in \{\top, \bot\}} \mathrm{const}_f(x),$$

$$\bigvee_{\substack{f \in B, \\ \mathrm{ar}(f) \geq 1}} \bigwedge_{1 \leq i \leq \mathrm{ar}(f)} \exists y \big(\mathrm{conn}_{f,i}(y, x) \wedge \forall z (\mathrm{conn}_{f,i}(z, x) \rightarrow z = y) \big),$$

$$\exists y \big(\mathrm{body}_{\mathsf{AX}}(y, x) \wedge \forall z (\mathrm{body}_{\mathsf{AX}}(z, x) \rightarrow z = y) \big),$$

$$\left. \exists y \big(\mathrm{body}_{\mathsf{EX}}(y,x) \wedge \forall z \big(\mathrm{body}_{\mathsf{EX}}(z,x) \to z = y \big) \big) \right) \wedge$$

$$\forall x \forall y \big((\mathrm{body}_{\mathsf{AX}}(y,x) \to \mathrm{repr}_{\mathsf{AX}}(x)) \wedge (\mathrm{body}_{\mathsf{EX}}(y,x) \to \mathrm{repr}_{\mathsf{EX}}(x)) \big).$$

The previous formula is a modification of the formula used in the proof of Lemma 1 in [11].

The next formulas will quantify sets M_i which represent sets of satisfied subformulas at worlds in the Kripke structure at depth i. Here the formulas with propositional connectives, resp., all constants, have a valid assignment obeying their function value in the model M_i. The AX- and EX-formulas are processed as expected: the EX-formulas branch to different worlds and the AX-formulas have to hold in all possible next worlds. Now we are ready to define $\theta^i_{\mathrm{assign}}$ in an inductive way. At depth 0 we want to consider only propositional formulas. Here it ensures that all Boolean functions obey the model:

$$\theta^0_{\mathrm{assign}}(M_0) := \forall x, y_1, \dots, y_n \in M_0 : \mathrm{repr}_{\mathsf{PL}}(x) \wedge$$

$$\bigwedge_{f \in B} \left(\bigwedge_{\mathrm{ar}(f)=0} \mathrm{const}_f(x) \to f \wedge \bigwedge_{1 \leq i \leq \mathrm{ar}(f)} \mathrm{conn}_{f,i}(y_i, x) \to f(M_0(y_1), \dots, M_0(y_{\mathrm{ar}(f)})) \right).$$

In the general definition of $\theta^i_{\mathrm{assign}}$ we utilize for convenience two subformulas, $\theta^i_{\mathrm{branchEX}}$ and $\theta^i_{\mathrm{stepAX}}$. The first is defined for an element x representing an EX-formula, a set of elements M_i representing to be satisfied formulas, and a set of elements M_{AX} representing the AX-formulas which are satisfied in the current world. The formula enforces that the formula $\mathsf{EX}\psi$ represented by x has to hold in the next world together with all bodies of the AX-formulas:

$$\theta^i_{\mathrm{branchEX}}(M_i, M_{\mathsf{AX}}, x) := \exists y \Big(\mathrm{body}_{\mathsf{EX}}(y,x) \wedge \exists M_{i-1} \big(M_{i-1}(y) \wedge \forall z \in M_{\mathsf{AX}}$$

$$(\exists w \, \mathrm{body}_{\mathsf{AX}}(w,z) \wedge M_{i-1}(w)) \wedge \theta^{i-1}_{\mathrm{assign}}(M_{i-1})\big) \Big).$$

The second formula is crucial when there are no EX-formulas represented in M_i. Then the AX-formulas still have to be satisfied eventually wherefore we proceed with a single next world (without any branching required):

$$\theta^i_{\mathrm{stepAX}}(M_{\mathsf{AX}}) := \exists M_{i-1} \forall z \in M_{\mathsf{AX}} (\exists w \, \mathrm{body}_{\mathsf{AX}}(w,z) \wedge M_{i-1}(w)) \wedge \theta^{i-1}_{\mathrm{assign}}(M_{i-1}).$$

Now we turn towards the complete inductive definition step where we need to differentiate between the two possible cases for representatives: either a propositional or a temporal formula is represented. The first part is similar to the induction start and the latter follows the observation that for every EX-preceded formula we want to branch. In each such branch all not yet satisfied AX-preceded formulas have to hold. The set M_{AX} contains all AX-formulas which are satisfied in the current world. If we do not have any EX-formulas then we enforce a single next world for the remaining AX-formulas:

$$\theta^i_{\mathrm{assign}}(M_i) := \forall x, y_1, \ldots, y_n \in M_i \bigwedge_{f \in B} \left(\bigwedge_{\mathrm{ar}(f)=0} \mathrm{const}_f(x) \to (M_i(x) \leftrightarrow f) \wedge \right.$$

$$\left. \bigwedge_{1 \le i \le \mathrm{ar}(f)} \mathrm{conn}_{f,i}(y_i, x) \to \left(M_i(x) \leftrightarrow f(M_i(y_1), \ldots, M_i(y_{\mathrm{ar}(f)})) \right) \right) \wedge$$

$$\exists M_{\mathsf{AX}} \subseteq M_i \left(\forall x \left(M_{\mathsf{AX}}(x) \leftrightarrow \left(\mathrm{repr}_{\mathsf{AX}}(x) \wedge M_i(x) \right) \right) \right) \wedge$$

$$\forall x \in M_i \left(\mathrm{repr}_{\mathsf{EX}}(x) \to \theta^i_{\mathrm{branchEX}}(M_i, M_{\mathsf{AX}}, x) \right) \wedge$$

$$\left(\forall x \in M_i (\neg \mathrm{repr}_{\mathsf{EX}}(x)) \right) \to \theta^i_{\mathrm{stepAX}}(M_{\mathsf{AF}}) \right).$$

Through the construction we get that φ is satisfiable iff $\mathcal{A}_\varphi \models \theta_{\mathrm{struc}} \wedge \exists M(\theta^{\mathrm{td}(\varphi)}_{\mathrm{assign}}(M)) =: \theta(\varphi)$. □

Corollary 4. CTL-SAT({AX}) *parameterized by formula pathwidth and temporal depth is fixed-parameter tractable.*

Proof. Assume that the given formula φ is in NNF since such a transformation is possible in linear time. As pathwidth is an upper bound for treewidth, we apply Theorem 2 in the following way. For $|f(\varphi)| = \mathrm{td}(\varphi)$ the function f is κ-bounded and computes the appropriate MSO formula from the uniform family given by Lemma 3. □

Intractable Fragments of CTL-SAT

In the following section we consider fragments of CTL for which their models cannot be bounded by the temporal depth of the formula. Therefore the framework used for the AX case cannot be applied. Instead we prove **W**[1]-hardness.

Lemma 5. CTL-SAT(T) *parameterized by formula pathwidth and temporal depth is* **W**[1]*-hard if* $\{\mathsf{AX}, \mathsf{AF}\} \subseteq T$.

Proof (Sketch). Due to space constraints the full proof is omitted and we refer to the technical report [10]. We will modify the construction in the proof of Praveen [15, Lemma A.3] and thereby state an fpt-reduction from the parameterized problem p-PW-SAT whose input is $(\mathcal{F}, part : \Phi \to [k], tg : [k] \to \mathbb{N})$, where $\mathcal{F}$ is a propositional CNF formula, *part* is a function that partitions the set of propositional variables of $\mathcal{F}$ into k parts, and tg is a function which maps to each part a natural number. The task is to find a satisfying assignment of $\mathcal{F}$ such that in each part $p \in [k]$ exactly $tg(p)$ variables are set to true. A generalization of this

$$determined := \mathsf{AG} \bigwedge_{i=1}^{n} \big((q_i \Rightarrow \mathsf{AX}q_i) \wedge (\neg q_i \Rightarrow \mathsf{AX}\neg q_i)\big)$$

$$depth := \bigwedge_{i=0}^{n-2} \big((d_i \wedge \neg d_{i+1}) \Rightarrow \mathsf{AX}(d_{i+1} \wedge \neg d_{i+2})\big)$$

$$setCounter := (q_1 \Rightarrow t_{\uparrow part(1)}) \wedge (\neg q_1 \Rightarrow f_{\uparrow part(1)}) \wedge$$
$$\mathsf{AG} \bigwedge_{i=2}^{n} \big((d_{i-1} \wedge \neg d_i) \Rightarrow [(q_i \Rightarrow t_{\uparrow part(i)}) \wedge (\neg q_i \Rightarrow f_{\uparrow part(i)})]\big)$$

$$incCounter := \big((t_{\uparrow part(1)} \Rightarrow \mathsf{AX}tr^1_{part(1)}) \wedge (f_{\uparrow part(1)} \Rightarrow \mathsf{AX}fl^1_{part(1)})\big) \wedge$$
$$\mathsf{AG} \bigwedge_{p=1}^{k} \bigwedge_{j=0}^{n[p]-1} \Big[\big(t_{\uparrow p} \Rightarrow (tr^j_p \Rightarrow tr^{j+1}_p \wedge \mathsf{AX}tr^{j+1}_p)\big) \wedge$$
$$\big(f_{\uparrow p} \Rightarrow (fl^j_p \Rightarrow fl^{j+1}_p \wedge \mathsf{AX}fl^{j+1}_p)\big)\Big]$$

$$targetMet := \mathsf{AG} \bigwedge_{p=1}^{k} \big(d_n \Rightarrow tr^{tg(p)}_p \wedge \neg tr^{tg(p)+1}_p \wedge fl^{n[p]-tg(p)}_p \wedge \neg tr^{n[p]-tg(p)+1}_p\big)$$

$$determined' := \mathsf{AG} \bigwedge_{p=1}^{k} \big((tr^0_p \Rightarrow tr^0_p) \wedge (fl^0_p \Rightarrow fl^0_p)\big)$$

$$countInit := d_0 \wedge \neg d_1 \wedge \bigwedge_{p=1}^{k} (\neg tr^1_p \wedge \neg fl^1_p \wedge tr^0_p \wedge fl^0_p)$$

$$depth' := \mathsf{AG} \bigwedge_{p=1}^{k} \bigwedge_{j=0}^{n[p]} \Big[\big(tr^j_p \Rightarrow tr^j_p\big) \wedge \big(fl^j_p \Rightarrow (fl^j_p)\big)\Big]$$

$$countMonotone := \mathsf{AG} \Big(\bigwedge_{i=1}^{n} \big((d_i \Rightarrow d_{i-1})\big) \wedge \bigwedge_{p=1}^{k} \bigwedge_{l=2}^{n[p]} \big[(tr^j_p \Rightarrow tr^{j-1}_p) \wedge (fl^j_p \Rightarrow fl^{j-1}_p)\big]\Big)$$

Fig. 2. Reduction from p-PW-SAT to CTL-SAT($\{\mathsf{AX}, \mathsf{AG}\}$)

problem to arbitrary formulas $\mathcal{F}$ (i.e., the CNF constraint is dropped) is **W**[1]-hard when parameterized by k and the pathwidth of the structural representation $\mathcal{A}_{\mathcal{F}}$ of $\mathcal{F}$ which is similar proven as in [15, Lemma 7.1].

The further idea is to construct a $\mathcal{CTL}$-formula $\phi_{\mathcal{F}}$ in which we are able to verify the required targets. The formula enforces a Kripke structure $K = (W, R, V)$ where in each world $w \in W$ the value of $V(w)$ coincides with a satisfying assignment f of $\mathcal{F}$ together with the required targets. Each such K contains as a substructure a chain $w_0 R w_1 R \cdots R w_n$ of worlds and all variables q_i in $\mathcal{F}$ are labeled to each w_j if $f(q_i)$ holds.

The formula $\phi_{\mathcal{F}}$ that is the conjunction of subformulas (Figure 2) similar to [15, Lemma A.3] states the reduction from p-PW-SAT to CTL-SAT($\{\mathsf{AX}, \mathsf{AG}\}$) parameterized by temporal depth and pathwidth. With respect to Praveens approach we explain how to obtain a formula consisting of only one single AG operator leading to a formula $\phi_{\mathcal{F}} = \psi \wedge \mathsf{AG}\chi$, where ψ is purely propositional and $\chi \in \mathcal{CTL}(\{\mathsf{AX}\})$. Then AG can be replaced by EG and the proof stays valid since there is only one instance of an existential temporal operator and it occurs at temporal depth zero. As $\mathsf{AG}(\alpha) \wedge \mathsf{AG}(\beta) \equiv \mathsf{AG}(\alpha \wedge \beta)$ we can modify the formula $\phi_{\mathcal{F}}$ which is a conjunction of the formulas from above to the desired form containing only a single AG. This is then replaced by EG and the argumentation follows. The correctness of the reduction is similarly proven as in [15, Lemma A.3]. □

Lemma 6. CTL-SAT(T) *parameterized by formula pathwidth and temporal depth is* $\mathbf{W}[1]$-*hard if* $\mathsf{AG} \in T$.

Proof. Now we consider the case were $T = \{\mathsf{AG}\}$. As $\mathsf{AG}\varphi$ is equivalent to $\neg\mathsf{EF}\neg\varphi$ we can simply substitute in the constructed formula $\phi_{\mathcal{F}}$ from [15, Lemma A.3] the occurrence of EX with EF. By this the possible "steps" invoked by the EX-operator become "jumps" through EF. This however allows consecutive worlds to be labeled identically, counting a variable duplicately. We refer to the technical report for the necessary construction to prevent such behaviour [10].

Lemma 7. CTL-SAT(T) *parameterized by formula pathwidth and temporal depth is* $\mathbf{W}[1]$-*hard if* $\mathsf{AU} \in T$.

Proof. Due to space constraints the proof is omitted and we refer to the technical report [10].

4   Conclusion

In this work we present an almost complete classification with respect to parameterized complexity of all possible CTL-operator fragments of the satisfiability problem in computation tree logic CTL parameterized by formula pathwidth and temporal depth. Only the case for the fragment containing solely AF remains open. Currently we are working on a classification of this fragment which aims for an **FPT** result and uses the "full version" of Theorem 2; the main goal is to bound the model depth of an AF-formula in the full parameter, i.e. not only in the temporal depth of the formula. This requires finding lower bounds for the treewidth of the considered structures when the formula enforces a deep model. Then we can construct a family of MSO formulas similar to the AX case. The classified results form a dichotomy with two fragments in **FPT** and the remainder being $\mathbf{W}[1]$-hard.

Comparing our results to the situation in usual computational complexity for the decision case they do not behave as expected. Surprisingly the fragment $\{\mathsf{AX}\}$ is **FPT** whereas on the decision side this fragment is **PSPACE**-complete. For the other classified fragments the rule of thumb is the following: The **NP**-complete fragments are **FPT** whereas the **PSPACE**- and **EXPTIME**-complete fragments are $\mathbf{W}[1]$-hard. For the shown $\mathbf{W}[1]$-hardness results an exact classification with matching upper bounds is open for further research. Similarly a complete classification with respect to all possible Boolean fragments in the sense of Post's lattice is one of our next steps.

Furthermore we constructed a generalization of Courcelle's theorem to infinite signatures for parameterized problems (Q, κ) with $Q \subseteq \Sigma^*$ such that the treewidth of the relational structures $\mathcal{A}_x$ corresponding to instances $x \in \Sigma^*$ is κ-bounded under the existence of a computable family of MSO-formulas (cf. Theorem 2). Previously such a general result for infinite signatures was not known to the best of the authors knowledge and is of independent interest.

Another consequent step will be the classification of other temporal logics fragments, e.g., of linear temporal logic LTL and the full branching time logic CTL* with respect to their parameterized complexity. Also the investigation of other parameterizations beyond the usual considered measures of pathwidth or treewidth and temporal depth may lead to a better understanding of intractability in the parameterized sense.

References

1. Blackburn, P., de Rijke, M., Venema, Y.: Modal logic. Cambridge University Press, New York (2001)
2. Clarke, E.M., Emerson, E.A.: Desing and synthesis of synchronisation skeletons using branching time temporal logic. In: Kozen, D. (ed.) Logic of Programs 1981. LNCS, vol. 131, pp. 52–71. Springer, Heidelberg (1982)
3. Courcelle, B., Engelfriet, J.: Graph structure and monadic second-order logic, a language theoretic approach. Cambridge University Press (2012)
4. Downey, R.G., Fellows, M.R.: Parameterized Complexity, p. 530. Springer (1999)
5. Elberfeld, M., Jakoby, A., Tantau, T.: Logspace versions of the theorems of bodlaender and courcelle. In: Proc. 51th Annual IEEE Symposium on Foundations of Computer Science. IEEE Computer Society (2010)
6. Emerson, E.A., Halpern, J.Y.: Decision procedures and expressiveness in the temporal logic of branching time. Journal of Computer and System Sciences **30**(1), 1–24 (1985)
7. Flum, J., Grohe, M.: Parameterized Complexity Theory. Springer (2006)
8. Gottlob, G., Pichler, R., Wei, F.: Bounded treewidth as a key to tractability of knowledge representation and reasoning. Artificial Intelligence **174**(1), 105–132 (2010)
9. Kripke, S.: Semantical considerations on modal logic. Acta Philosophica Fennica **16**, 84–94 (1963)
10. Lück, M., Meier, A., Schindler, I.: Parameterized Complexity of CTL: A Generalization of Courcelle's Theorem. arXiv 1410.4044 (2014)
11. Meier, A., Schmidt, J., Thomas, M., Vollmer, H.: On the parameterized complexity of default logic and autoepistemic logic. In: Dediu, A.-H., Martín-Vide, C. (eds.) LATA 2012. LNCS, vol. 7183, pp. 389–400. Springer, Heidelberg (2012)
12. Meier, A.: On the Complexity of Modal Logic Variants and their Fragments. Ph.D. thesis, Gottfried Wilhelm Leibniz Universität Hannover (2011)
13. Pippenger, N.: Theories of Computability. Cambridge University Press (1997)
14. Pnueli, A.: The temporal logic of programs. In: Proc. 18th Symposium on Foundations of Computer Science, pp. 46–57. IEEE Computer Society Press (1977)
15. Praveen, M.: Does treewidth help in modal satisfiability? ACM Transactions on Computational Logic **14**(3), 18:1–18:32 (2013)
16. Prior, A.N.: Time and Modality. Clarendon Press, Oxford (1957)

Compression, Inference, Pattern Matching, and Model Checking

Single-Pass Testing Automata
for LTL Model Checking

Ala Eddine Ben Salem[⊠]

LRDE, EPITA, Le Kremlin-Bicêtre, France
ala@lrde.epita.fr

Abstract. *Testing Automaton* (TA) is a new kind of ω-automaton introduced by Hansen et al. [6] as an alternative to the standard Büchi Automata (BA) for the verification of stutter-invariant LTL properties. Geldenhuys and Hansen [5] shown later how to use TA in the automata-theoretic approach to LTL model checking. They propose a TA-based approach using a verification algorithm that requires two searches (two passes) and compare its performance against the BA approach.

This paper improves their work by proposing a transformation of TA into a normal form (STA) that only requires a single one-pass verification algorithm. The resulting automaton is called *Single-pass Testing Automaton* (STA). We have implemented the STA approach in Spot model checking library. We are thus able to compare it with the BA and TA approaches. These experiments show that STA compete well on our examples.

1 Introduction

The automata-theoretic approach [11] to LTL model checking relies on ω-automata (i.e., an extension of finite automata to infinite words). It starts by converting the negation of the LTL formula φ into an ω-automaton $A_{\neg\varphi}$, then composing that automaton with the state-space of a model M given as a Kripke structure $\mathcal{K}_M$ (a variant of ω-automaton), and finally checking the language emptiness of the resulting product automaton $A_{\neg\varphi} \otimes \mathcal{K}_M$. This operation tells whether $A_{\neg\varphi} \otimes \mathcal{K}_M$ accepts an infinite word, and can return such a word as a counterexample. The model M satisfies φ iff $\mathscr{L}(A_{\neg\varphi} \otimes \mathcal{K}_M) = \emptyset$.

As for any model checking process, the automata-theoretic approach suffers from the well known state explosion problem. In practice, it is the product automaton that can be very large, its size can reach $(|A_{\neg\varphi}| \times |\mathcal{K}_M|)$ states, which can make it impossible to be handled using the resources of modern computers.

The ω-automaton representing $A_{\neg\varphi}$ is usually a Büchi Automaton (BA). This paper focuses on improving another kind of ω-automaton called *Testing Automaton (TA)*. TA is a variant of an "extended" Büchi automaton introduced by Hansen et al. [6]. Instead of observing the valuations on states or transitions, the TA transitions only record the changes between these valuations. In addition, TA are less expressive than BA since they are able to represent only stutter-invariant [3] properties. Also they are often larger than their equivalent

© Springer International Publishing Switzerland 2015
A.-H. Dediu et al. (Eds.): LATA 2015, LNCS 8977, pp. 563–576, 2015.
DOI: 10.1007/978-3-319-15579-1_44

BA, but their high degree of determinism [6] often leads to a smaller product size [5].

In a previous work [1], we evaluated the use of TA for the model checking of stutter-invariant LTL properties. We have shown that the TA approach is efficient when the formula to be verified is violated (i.e., a counterexample exists). This is not the case when the property is satisfied since the product $(A_{\neg\varphi} \otimes \mathcal{K}_M)$ has to be visited twice during the the emptiness check. In this work, we improve the TA approach in order to avoid the second pass of the emptiness check algorithm. To achieve this goal, we propose a transformation of TA into a normal form that does not require such a second pass, called *Single-pass Testing Automata* (STA). We have implemented the algorithms of STA approach in Spot [10] library. Our experimental comparisons between BA, TA and STA approaches show that the STA approach is statistically more efficient when no counterexample is found (i.e., the property is satisfied) because it does not require a second pass.

2   Existing Approaches

Let AP a set of atomic propositions, a valuation ℓ over AP is an assignment of truth value to each atomic proposition. We denote by $\Sigma = 2^{AP}$ the set of all valuations over AP, where a valuation $\ell \in \Sigma$ is interpreted either as the set of atomic propositions that are true, or as a Boolean conjunction. For instance, if $AP = \{a, b\}$, then $\Sigma = 2^{AP} = \{\{a, b\}, \{a\}, \{b\}, \emptyset\}$ or $\Sigma = \{ab, a\bar{b}, \bar{a}b, \bar{a}\bar{b}\}$.

2.1   Büchi Automata (BA)

A Büchi Automaton (BA) is an ω-automaton [4] with valuations on transitions and acceptance conditions on states. Any LTL formula φ can be converted into a BA that accepts the same executions that satisfy φ [11].

Definition 1 (BA). *A Büchi Automaton (BA) over the alphabet* $\Sigma = 2^{AP}$ *is a tuple* $\mathcal{B} = \langle \mathcal{Q}, \mathcal{I}, \delta, \mathcal{F} \rangle$ *where:*
- $\mathcal{Q}$ *is a finite set of states,* $\mathcal{I} \subseteq \mathcal{Q}$ *is a finite set of initial states,*
- $\mathcal{F} \subseteq \mathcal{Q}$ *is a finite set of accepting states* ($\mathcal{F}$ *is called the accepting set*),
- $\delta \subseteq \mathcal{Q} \times \Sigma \times \mathcal{Q}$ *is the transition relation where each transition is labeled by a letter* ℓ *of* Σ, *i.e., each element* $(q, \ell, q') \in \delta$ *represents a transition from state* q *to state* q' *labeled by a valuation* $\ell \in 2^{AP}$.

A run of $\mathcal{B}$ *over an infinite word* $\sigma = \ell_0 \ell_1 \ell_2 \ldots \in \Sigma^\omega$ *is an infinite sequence of transitions* $r = (q_0, \ell_0, q_1)(q_1, \ell_1, q_2)(q_2, \ell_2, q_3) \ldots \in \delta^\omega$ *such that* $q_0 \in \mathcal{I}$ *(i.e., the infinite word is recognized by the run). Such a run is said to be accepting if* $\forall i \in \mathbb{N}, \exists j \geq i, q_j \in \mathcal{F}$ *(at least one accepting state is visited infinitely often). The infinite word* σ *is accepted by* $\mathcal{B}$ *if there exists an accepting run of* $\mathcal{B}$ *over* σ.

Figure 1 shows a BA recognizing the LTL formula ($a \cup G\ b$). In this BA, the Boolean conjunctions labeling each transition are valuations over $AP = \{a, b\}$.

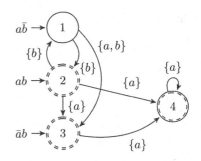

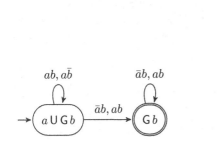

Fig. 1. A BA $\mathcal{B}$ for the LTL formula $a \cup G\,b$, with accepting states shown as double circles

Fig. 2. A TA $\mathcal{T}$ for the LTL formula $a \cup G\,b$

Fig. 3. SCC search stack and how the SCCs are merged

The LTL formulas labeling each state represent the property accepted starting from this state of the automaton: they are shown for the reader's convenience but not used for model checking. As an illustration of Definition 1, the infinite word $ab; a\bar{b}; \bar{a}b; ab; \bar{a}b; ab; \ldots$ is accepted by the BA of Figure 1. A run over such infinite word must start in the initial state labeled by the formula $(a \cup G\,b)$ and remains in this state for the first two valuations $ab; a\bar{b}$, then it changes the value of a, so it has to take the transition labeled by the valuation $\bar{a}b$ to move to the second state labeled by the formula $(G\,b)$. Finally, to be accepted, it must stay on this accepting state by executing infinitely the transitions labeled by $\{\bar{a}b, ab\}$.

Model Checking Using BA. The synchronous product of a BA $\mathcal{B}$ with a Kripke structure $\mathcal{K}$ is a BA $\mathcal{K} \otimes \mathcal{B}$ whose language is the intersection of both languages. Testing this product automaton (i.e., a BA) for emptiness amounts to the search of an accepting cycle that contains at least one accepting states.

Algorithm 1 presented below is an iterative version of the Couvreur's SCC-based algorithm [2] adapted to the emptiness check of BA. Algorithm 1 computes on-the-fly the Maximal Strongly Connected Components (MSCCs) of the BA representing the product $\mathcal{K} \otimes \mathcal{B}$: it performs a Depth-First Search (DFS) for SCC detection and then merges the SCCs belonging to the same Maximal SCC into a single SCC. After each merge, if the merged SCC contains an accepting state from $\mathcal{F}_\otimes$ (line 16), then an accepting run (i.e., a counterexample) is found (line 16) and the $\mathscr{L}(\mathcal{K} \otimes \mathcal{B})$ is not empty. *todo* is the DFS stack. It is used by the procedure **DFSpush** to push the states of the current DFS path and the set of their successors that have not yet been visited. H maps each visited state to its rank in the DFS order, and $H[s] = 0$ indicates that s is a dead state (i.e., s belongs to a maximal SCC that has been fully explored). The SCC stack stores a chain of partial SCCs found during the DFS. For each SCC the attribute *root* is the DFS rank (H) of the first state of the SCC, *acc* is the set of accepting states belonging to the SCC, and *rem* contains the fully explored states of the SCC.

```
1  Input: A BA K ⊗ B = ⟨S⊗, I⊗, δ⊗, F⊗⟩
2  Result: ⊤ if and only if ℒ(K ⊗ B) = ∅
3  Data: todo: stack of ⟨state ∈ S⊗, succ ⊆ δ⊗⟩, H: map of S⊗ ↦ ℕ
           SCC: stack of ⟨root ∈ ℕ, acc ⊆ F⊗, rem ⊆ S⊗⟩, max ← 0
4  begin
5  │   foreach s⁰ ∈ I⊗ do
6  │   │   DFSpush(s⁰)
7  │   │   while ¬todo.empty() do
8  │   │   │   if todo.top().succ = ∅ then
9  │   │   │   │   DFSpop()
10 │   │   │   else
11 │   │   │   │   pick one ⟨s, _, d⟩ off todo.top().succ
12 │   │   │   │   if d ∉ H then
13 │   │   │   │   │   DFSpush(d)
14 │   │   │   │   else if H[d] > 0 then
15 │   │   │   │   │   merge(H[d])
16 │   │   │   │   │   if SCC.top().acc ≠ ∅ then  return ⊥
17 │   return ⊤
18 DFSpush(s ∈ S⊗)
19 │   max ← max + 1; H[s] ← max;
20 │   SCC.push(⟨max, ({s} ∩ F⊗), ∅⟩)
21 │   todo.push(⟨s, {⟨q, l, d⟩ ∈ δ⊗ | q = s}⟩)
22 DFSpop()
23 │   ⟨s, _⟩ ← todo.pop()
24 │   SCC.top().rem.insert(s)
25 │   if H[s] = SCC.top().root then
26 │   │   foreach s ∈ SCC.top().rem do
27 │   │   │   H[s] ← 0
28 │   │   SCC.pop()
29 merge(t ∈ ℕ)
30 │   acc ← ∅; r ← ∅;
31 │   while t < SCC.top().root do
32 │   │   acc ← acc ∪ SCC.top().acc
33 │   │   r ← r ∪ SCC.top().rem
34 │   │   SCC.pop()
35 │   SCC.top().acc ← SCC.top().acc ∪ acc
36 │   SCC.top().rem ← SCC.top().rem ∪ r
```

Algorithm 1. Emptiness check algorithm for BA

1. The algorithm 1 begins by pushing in SCC each state s visited for the first time (line 12), as a trivial SCC with the set $acc = \{s\} \cap \mathcal{F}_\otimes$ (line 20).
2. Then, when the DFS explores a transition t between two states s and d, if d is in the SCC stack (line 14), then t closes a cycle passing through s and d in the product automaton. This cycle "strongly connects" all SCCs pushed in the SCC stack between $SCC[i]$ and $SCC[n]$: the two SCCs that respectively contains the states d and s ($SCC[n]$ is the top of the SCC stack).
3. All the SCCs between $SCC[i]$ and $SCC[n]$ are merged (line 15) into $SCC[i]$. This merging is illustrated by Figure 3: a "back" transition t is found between $SCC[n]$ and $SCC[i]$, therefore the latest SCCs (from i to n) are merged.
4. The set of accepting states of the merged SCC is equal to the union of $SCC[i].acc \cup SCC[i+1].acc \cup \cdots \cup SCC[n].acc$. If this union contains an accepting state of $\mathcal{F}_\otimes$, then the merged SCC is accepting and the algorithm return $false$ (line 16): the product is not empty.

2.2 Testing Automata (TA)

Testing Automata were introduced by Hansen et al. [6] to represent stutter-invariant [3] properties. While a Büchi automaton observes the value of the atomic propositions, the basic idea of TA is to only detect the *changes* in these values, making TA particularly suitable for stutter-invariant properties; if a valuation of AP does not change between two consecutive valuations of an execution, the TA stay in the same state, this kind of transitions are called stuttering transitions. To detect infinite executions that end stuck in the same state because they are stuttering, a new kind of accepting states is introduced: *livelock-accepting states*.

$A \oplus B$ denotes the symmetric set difference between two valuations A and B, i.e., the atomic propositions that differ (e.g., $a\bar{b} \oplus ab = \{b\}$).

Definition 2 (TA). *A Testing Automaton (TA) over the alphabet $\Sigma = 2^{AP}$ is a tuple $\mathcal{T} = \langle Q, \mathcal{I}, U, \delta, \mathcal{F}, \mathcal{G} \rangle$, where:*

- *Q is a finite set of states, $\mathcal{I} \subseteq Q$ is a finite set of initial states,*
- *$U : \mathcal{I} \to 2^\Sigma$ is a function mapping each initial state to a set of valuations (set of possible initial configurations),*
- *$\mathcal{F} \subseteq Q$ is a set of Büchi-accepting states,*
- *$\mathcal{G} \subseteq Q$ is a set of livelock-accepting states,*
- *$\delta \subseteq Q \times (\Sigma \setminus \emptyset) \times Q$ is the transition relation where each transition (s, k, d) is labeled by a changeset: $k \in \Sigma$ is interpreted as a non empty set of atomic propositions whose value must change between states s and d.*

An infinite word $\sigma = \ell_0\ell_1\ell_2\ldots \in \Sigma^\omega$ is accepted by $\mathcal{T}$ iff there exists a sequence $(q_0, \ell_0 \oplus \ell_1, q_1)(q_1, \ell_1 \oplus \ell_2, q_2)\ldots(q_i, \ell_i \oplus \ell_{i+1}, q_{i+1})\ldots \in (Q \times \Sigma \times Q)^\omega$ such that:

- *$q_0 \in \mathcal{I}$ with $\ell_0 \in U(q_0)$,*
- *$\forall i \in \mathbb{N}$, either $(q_i, \ell_i \oplus \ell_{i+1}, q_{i+1}) \in \delta$ (the execution progresses), or $(\ell_i = \ell_{i+1}) \wedge (q_i = q_{i+1})$ (the execution is stuttering and the TA does not progress),*
- *either, $\forall i \in \mathbb{N}$, $(\exists j \geq i, \ell_j \neq \ell_{j+1}) \wedge (\exists l \geq i, q_l \in \mathcal{F})$ (the TA is progressing in a Büchi-accepting way), or, $\exists n \in \mathbb{N}$, $(q_n \in \mathcal{G} \wedge (\forall k \geq n, q_k = q_n \wedge \ell_k = \ell_n))$ (the sequence reaches a livelock-accepting state and then stays on that state because the execution is stuttering).*

The construction of a TA from a BA is detailed in [1,5]. To illustrate Definition 2, let us consider Figure 2, representing a TA $\mathcal{T}$ for $a\mathsf{U}Gb$. In this figure, the initial states 1, 2 and 3 are labeled respectively by the set of valuations $U(1) = \{a\bar{b}\}$, $U(2) = \{ab\}$ and $U(3) = \{\bar{a}b\}$. Each transition of $\mathcal{T}$ is labeled with a changeset over the set of atomic propositions $AP = \{a, b\}$. In a TA, states with a double enclosure belong to either $\mathcal{F}$ or $\mathcal{G}$: states in $\mathcal{F} \setminus \mathcal{G}$ have a double solid line, states in $\mathcal{G} \setminus \mathcal{F}$ have a double dashed line (states 2 and 3 of $\mathcal{T}$), and states in $\mathcal{F} \cap \mathcal{G}$ use a mixed dashed/solid style (state 4).

- The infinite word $ab; \bar{a}b; ab; \bar{a}b; ab; \bar{a}b; \ldots$ is accepted by a Büchi accepting run of $\mathcal{T}$. A run recognizing such word must start in state 2, then it always changes the value of a, so it has to take transitions labeled by $\{a\}$. For instance it could be the run $2 \xrightarrow{\{a\}} 4 \xrightarrow{\{a\}} 4 \xrightarrow{\{a\}} 4 \cdots$ or the run $2 \xrightarrow{\{a\}} 3 \xrightarrow{\{a\}} 4 \xrightarrow{\{a\}} 4 \cdots$ Both visit the state $4 \in \mathcal{F}$ infinitely often, so they are Büchi accepting.

- The infinite word $ab; \bar{a}b; \bar{a}b; \bar{a}b; \ldots$ is accepted by a livelock accepting run of $\mathcal{T}$. An accepting run starts in state 2, then moves to state 4, and stutters on this livelock-accepting state. Another possible accepting run goes from state 2 to state 3 and stutters in $3 \in \mathcal{G}$.

- The infinite word $ab; a\bar{b}; ab; a\bar{b}; ab; a\bar{b}; \ldots$ is not accepted. It would correspond to a run alternating between states 2 and 1, but such a run is neither Büchi accepting (does not visit any $\mathcal{F}$ state) nor livelock-accepting (it passes through state $2 \in \mathcal{G}$, but does not stay into this state continuously).

Model Checking Using TA. The product of a Kripke and a TA is not a TA: while a TA execution is allowed to stutter on any state, the product must execute an explicit stuttering transition.

Definition 3 (Synchronous Product of a TA with a Kripke structure).
For a Kripke structure $\mathcal{K} = \langle \mathcal{S}, \mathcal{S}_0, \mathcal{R}, l \rangle$ and a TA $\mathcal{T} = \langle \mathcal{Q}, \mathcal{I}, U, \delta, \mathcal{F}, \mathcal{G} \rangle$, the product $\mathcal{K} \otimes \mathcal{T}$ is a tuple $\langle \mathcal{S}_\otimes, \mathcal{I}_\otimes, U_\otimes, \delta_\otimes, \mathcal{F}_\otimes, \mathcal{G}_\otimes \rangle$ where
- $\mathcal{S}_\otimes = \mathcal{S} \times \mathcal{Q}$, $\mathcal{F}_\otimes = \mathcal{S} \times \mathcal{F}$, $\mathcal{G}_\otimes = \mathcal{S} \times \mathcal{G}$,
- $\mathcal{I}_\otimes = \{(s, q) \in \mathcal{S}_0 \times \mathcal{I} \mid l(s) \in U(q)\}$ *with* $\forall (s, q) \in \mathcal{I}_\otimes$, $U_\otimes((s, q)) = \{l(s)\}$,
- $\delta_\otimes = \{((s, q), k, (s', q')) \mid (s, s') \in \mathcal{R}, (q, k, q') \in \delta, k = l(s) \oplus l(s')\}$.
   $\cup \{((s, q), \emptyset, (s', q')) \mid (s, s') \in \mathcal{R}, q = q', l(s) = l(s')\}$

An execution $\sigma = \ell_0 \ell_1 \ell_2 \ldots \in \Sigma^\omega$ is accepted by $\mathcal{K} \otimes \mathcal{T}$ if there exists a sequence $(s_0, \ell_0 \oplus \ell_1, s_1)(s_1, \ell_1 \oplus \ell_2, s_2) \ldots (s_i, \ell_i \oplus \ell_{i+1}, s_{i+1}) \ldots \in (\mathcal{S}_\otimes \times \Sigma \times \mathcal{S}_\otimes)^\omega$ where:
- $s_0 \in \mathcal{S}_\otimes^0$ *with* $\ell_0 \in U_\otimes(s_0)$,
- $\forall i \in \mathbb{N}, (s_i, \ell_i \oplus \ell_{i+1}, s_{i+1}) \in \delta_\otimes$ *(we are always progressing in the product)*
- *Either,* $\forall i \in \mathbb{N}, (\exists j \geq i, \ell_j \neq \ell_{j+1}) \wedge (\exists l \geq i, s_l \in \mathcal{F}_\otimes)$ *(the automaton is progressing in a Büchi-accepting way), or,* $\exists n \in \mathbb{N}, \forall k \geq n, (\ell_k = \ell_n) \wedge (s_k \in \mathcal{G}_\otimes)$ *(a suffix of the execution stutters in $\mathcal{G}_\otimes$).*
We have $\mathscr{L}(\mathcal{K} \otimes \mathcal{T}) = \mathscr{L}(\mathcal{K}) \cap \mathscr{L}(\mathcal{T})$ by construction.

Figure 4 shows an example of a synchronous product between a Kripke structure $\mathcal{K}$ and a TA $\mathcal{T}$ recognizing the LTL formula $\mathsf{FG}p$. Each state of $\mathcal{K}$ is numbered and labeled with a valuation of atomic propositions (over $AP = \{p\}$) that

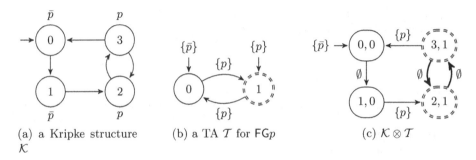

(a) a Kripke structure $\mathcal{K}$ (b) a TA $\mathcal{T}$ for $\mathsf{FG}p$ (c) $\mathcal{K} \otimes \mathcal{T}$

Fig. 4. Example of a product between a Kripke structure $\mathcal{K}$ and a TA $\mathcal{T}$ of $\mathsf{FG}p$. The bold cycle of $\mathcal{K} \otimes \mathcal{T}$ is livelock-accepting.

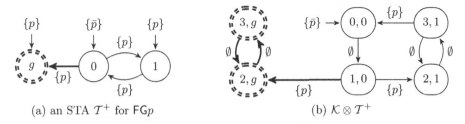

(a) an STA $\mathcal{T}^+$ for $\mathsf{FG}p$ (b) $\mathcal{K} \otimes \mathcal{T}^+$

Fig. 5. Impact on the product of using STA $\mathcal{T}^+$ instead of TA $\mathcal{T}$. Bold states and transitions are addition relative to Figure 4.

hold in this state. In the product $\mathcal{K} \otimes \mathcal{T}$, states are labeled with a pairs of the form (s, q) where s is a state of $\mathcal{K}$ and q of $\mathcal{T}$, and the livelock accepting states are denoted by a double dashed circle.

A Two-Pass Emptiness Check Algorithm. In this section, we present a two-pass algorithm for the emptiness check of the synchronous product between a TA and a Kripke structure. In model checking approach using TA, the emptiness check requires a dedicated algorithm because according to the Definition 3, there are two ways to detect an accepting cycle in the product:

- Büchi accepting: a cycle containing at least a Büchi-accepting state $(\mathcal{F}_\otimes)$ and at least one non-stuttering transition (i.e., a transition (s, k, s') with $k \neq \emptyset$),
- livelock accepting: a cycle composed only by stuttering transitions and livelock accepting states $(\mathcal{G}_\otimes)$.

A straightforward emptiness check would have two passes: a first pass to detect Büchi accepting cycles and a second pass to detect livelock accepting cycles.

The `first-pass` of Algorithm 2 is similar to Algorithm 1, it detects all Büchi-accepting cycles, and with line 18 included in this algorithm, it detects also some livelock-accepting cycles. Since in certain cases it may fail to report some livelock-accepting cycles, a second pass is required to look for possible livelock-accepting cycles. However, if no livelock-accepting state is visited during the first pass, then the second pass can be disabled: this is the purpose of variable *Gseen* of

```
 1 Input: K ⊗ T = ⟨S_⊗, I_⊗, U_⊗, δ_⊗, F_⊗, G_⊗⟩
 2 Result: ⊤ if and only if ℒ(K ⊗ T) = ∅
 3 Data:  todo: stack of ⟨state ∈ S_⊗, succ ⊆ δ_⊗⟩
          SCC: stack ⟨root ∈ ℕ, lk ∈ 2^{AP}, k ∈ 2^{AP}, acc ⊆ F_⊗, rem ⊆ S_⊗⟩
          H: map of S_⊗ ↦ ℕ, max ← 0, Gseen ← false
 4 begin
 5 │   if ¬ first-pass() then return ⊥ if Gseen then return second-pass()

 6 first-pass()
 7 │    foreach s^0 ∈ I_⊗ do
 8 │    │   DFSpush1(∅, s^0)
 9 │    │   while ¬todo.empty() do
10 │    │   │   if todo.top().succ = ∅ then
11 │    │   │   │   DFSpop()
12 │    │   │   else
13 │    │   │   │   pick one ⟨s, k, d⟩ off todo.top().succ
14 │    │   │   │   if d ∉ H then
15 │    │   │   │   │   DFSpush1(k, d)
16 │    │   │   │   else if H[d] > 0 then
17 │    │   │   │   │   merge1(k, H[d])
18 │    │   │   │   │   if (SCC.top().acc ≠ ∅) ∧ (SCC.top().k ≠ ∅) then return ⊥
          │   │   │   │   if (d ∈ G_⊗) ∧ (SCC.top().k = ∅) then return ⊥

19 │    return ⊤
20 DFSpush1(lk ∈ 2^{AP}, s ∈ S_⊗)
21 │    max ← max + 1; H[s] ← max;
22 │    SCC.push(⟨max, lk, ∅, ({s} ∩ F_⊗), ∅⟩)
23 │    todo.push(⟨s, {⟨q, k, d⟩ ∈ δ_⊗ | q = s}⟩)
24 │    if s ∈ G_⊗ then Gseen ← true

25 merge1(lk ∈ 2^{AP}, t ∈ ℕ)
26 │    acc ← ∅; r ← ∅; k ← lk;
27 │    while t < SCC.top().root do
28 │    │   acc ← acc ∪ SCC.top().acc
29 │    │   k ← k ∪ SCC.top().k ∪ SCC.top().lk
30 │    │   r ← r ∪ SCC.top().rem
31 │    │   SCC.pop()
32 │    SCC.top().acc ← SCC.top().acc ∪ acc
33 │    SCC.top().k ← SCC.top().k ∪ k
34 │    SCC.top().rem ← SCC.top().rem ∪ r
```

Algorithm 2. The first-pass of the Emptiness check algorithm for TA products

Algorithm 2 (line 5), where *Gseen* is a flag that records if a livelock-accepting state is detected during the exploration of the product by the first pass (line 24). This **first-pass** is based on the BA emptiness check algorithm presented in Algorithm 1 with the following changes:

– In each item *scc* of the *SCC* stack: the new field *scc.lk* stores the *change-set* labeling the transition coming from the previous SCC, and *scc.k* contains the union of all *change-sets* in *scc* (lines 29 and 33).
– After each merge, *SCC*.top() is checked for Büchi-acceptance (line 18) or livelock-acceptance (line 18) depending on the emptiness of *SCC*.top().*k*.

Figure 4 illustrates how the **first-pass** of Algorithm 2 can fail to detect the livelock accepting cycle in a product $\mathcal{K} \otimes \mathcal{T}$ as defined in Definition 3. In this example, $\mathcal{G}_{\mathcal{T}} = \{1\}$ therefore $(3, 1)$ and $(2, 1)$ are livelock-accepting states, and $C_2 = [(3, 1) \rightarrow (2, 1) \rightarrow (3, 1)]$ is a livelock-accepting cycle.

However, the **first-pass** may miss this livelock-accepting cycle depending on the order in which it processes the outgoing transitions of $(3, 1)$. If the transition $t_1 = ((3, 1), \{p\}, (0, 0))$ is processed before $t_2 = ((3, 1), \emptyset, (2, 1))$, then the cycle $C_1 = [(0, 0) \rightarrow (1, 0) \rightarrow (2, 1) \rightarrow (3, 1) \rightarrow (0, 0)]$ is detected and the four states are merged in the same SCC before exploring t_2. After this merge (line 17), this SCC is at the top of the SCC stack. Subsequently, when the DFS explores t_2, the merge caused by the cycle C_2 does not add any new state to the SCC, and the *SCC* stack remains unchanged. Therefore, the test line 18 still return false because the union *SCC.top()*.*k* of all change-sets labeling the transitions of the SCC is not empty (it includes for example t_1's label: $\{p\}$). Finally, **first-pass** algorithm terminates without reporting any accepting cycle, missing C_2.

In general, to report a livelock-accepting cycle, the first-pass computes the union of all change-sets of the SCC containing this cycle. However, this union may include non-stuttering transitions belonging to other cycles of the SCC. In this case, the **second-pass** is required to search for livelock-accepting cycles, ignoring the non-stuttering transitions that may belong to the same SCC. In the next section, we propose a Single-pass Testing Automata STA, which allows to obtain a synchronous product in which such mixing of non-stuttering and stuttering transitions will never occur in SCCs containing livelock-accepting cycles, making the **second-pass** unnecessary. It is important to say that in the experiments presented in the sequel, we implemented Algorithm 2 including an heuristic proposed by Geldenhuys and Hansen [5] to detect more livelock-accepting cycles during the first pass. However, when properties are satisfied, the second pass is always required because this heuristic fails to report some livelock-accepting cycles [5]. We don't present the details of this heuristic because we show in the next sections other solutions that allow to detect all the livelock-accepting cycles during the first pass and therefore remove the second pass (in all cases).

3   Converting a TA into a Single-Pass Testing Automaton

In this section, we introduce STA, a transformation of TA into a normal form such that livelock-accepting states have no successors, and therefore STA approach does not need the second pass of the emptiness check of TA approach. This improves the efficiency of the model checking (experimentally evaluated in section 4). STA also simplify the implementation (and the optimization) of the emptiness check algorithm as it renders unnecessary the implementation of the second pass.

Definition 4 (STA). *A Single-pass Testing Automaton (STA) is a Testing Automaton* $T = \langle Q, I, U, \delta, F, G \rangle$ *over* Σ *such that* $\delta \cap (G \times \Sigma \times Q) = \emptyset$. *In other words, an STA is a TA in which every livelock-accepting state has no successors.*

3.1   Construction of an STA from a TA

Property 1 formalizes the construction of an STA from a TA. We can transform a TA into an STA by adding an unique livelock-accepting state g (i.e., in STA, $G = \{g\}$), and adding a transition (q, k, g) for any transition (q, k, q') that goes into a livelock-accepting state $q' \in G$ of the original automaton. In addition, if q' has no successors then q' can be removed, since it is bisimilar to the new state g.

Property 1. *Let* $T = \langle Q, I, U, \delta, F, G \rangle$ *be a TA, we construct an equivalent STA* $T' = \langle Q', I', U', \delta', F, \{g\} \rangle$ *such that* $\mathscr{L}(T') = \mathscr{L}(T)$ *by the following:*

 - $Q' = (Q \setminus G_\emptyset) \cup \{g\}$ *where* $G_\emptyset = \{q \in G \mid (\{q\} \times \Sigma \times Q) \cap \delta = \emptyset\}$ *is the set of states of* G *that have no successors, and* $g \notin Q$ *is a new state,*
 - $I' = I \cup \{g\}$ *if* $G \cap I \neq \emptyset$, $I' = I$ *otherwise,*
 - $\delta' = (\delta \setminus (Q \times \Sigma \times G_\emptyset)) \cup \{(q, k, g) \mid (q, k, q') \in \delta, q' \in G\}$,
 - $\forall q \in I, U'(q) = U(q)$ *and* $U'(g) = \bigcup\limits_{q \in (G \cap I)} U(q)$.

Figure 5a shows how the TA from Figure 4b was transformed into an STA using Property 1. The idea behind this transformation is that any livelock-accepting execution of T will be mapped to an execution of T^+ that is captured by the new state g. The new g state has an impact on the product (Figure 5b): the strongly connected components of this new product no longer mix non-stuttering transitions and livelock-accepting cycles: this renders the `second-pass` useless. The objective of STA is to isolate in the product the exploration of the parts that are composed only by livelock-accepting states and stuttering transitions, like the bold part of the product represented in the Figure 5b.

The STA emptiness check algorithm is the `first-pass` of the TA emptiness check algorithm without the `second-pass` procedure. In other words, in STA approach, the emptiness check is only Algorithm 2 (page 570) without line 5.

3.2   Correctness of the One-Pass Emptiness Check Using STA

In the following, $\mathcal{K}$, T, T^+ denote respectively a Kripke, a TA and an STA.

The `first-pass` is an SCC-based algorithm, it computes the set of all MSCCs (i.e., Maximal SCCs) of the product automaton. Therefore, in order to prove that the `first-pass` is sufficient to detect all livelock-accepting cycles, we prove that in $\mathcal{K} \otimes T^+$, searching for all livelock-accepting cycles is equivalent to searching for all MSCCs that are only composed of stuttering transitions and livelock-accepting states. In Algorithm 2, line 18 allows to detect this kind of MSCCs.

Lemma 1. *In a product $\mathcal{K} \otimes \mathcal{T}$: **if** one MSCC M contains a product state (s, q) such that q is a livelock-accepting state that has no successors in $\mathcal{T}$, **then** M is only composed of stuttering transitions and livelock-accepting states.*

Proof. q has no successors in the TA $\mathcal{T}$, therefore from q, a run of $\mathcal{T}$ can only execute stuttering transitions: it stays in the same livelock-accepting state q. Consequently, all product states of M are connected by stuttering transitions. In addition, they have the same livelock-accepting state as TA component (q), therefore by Definition 3 all states of M are livelock-accepting.

Lemma 2. *In a product $\mathcal{K} \otimes \mathcal{T}^+$: one MSCC M contains a livelock-accepting state **iff** M is only composed of stuttering transitions and livelock-accepting states.*

Proof. ($\Longrightarrow$) If an MSCC M contains a livelock-accepting state (s, q) of $\mathcal{K} \otimes \mathcal{T}^+$, then q is a livelock-accepting state that has no successors in $\mathcal{T}^+$ because in STA every livelock-accepting state has no successors. The proof follows from Lemma 1 applied to $\mathcal{K} \otimes \mathcal{T}^+$. ($\Longleftarrow$) Any state of M is livelock-accepting.

The difference between Lemma 1 and Lemma 2 is that the livelock-accepting states of STA have no successors, while those of TA can.

Lemma 3. *In the product $\mathcal{K} \otimes \mathcal{T}^+$: there exists at least one livelock-acceptance cycle C **if and only if** there exists at least one non trivial MSCC M such that M is only composed of stuttering transitions and livelock-accepting states.*

Proof. ($\Longrightarrow$): The cycle C contains at least one livelock-accepting state, therefore applying Lemma 2 with M is the MSCC containing C allows us to conclude. ($\Longleftarrow$): M is non-trivial (it contains at least one state with a self-loop), therefore M contains at least one non-trivial cycle only composed of stuttering transitions and livelock-accepting states. This cycle is the livelock-accepting cycle C.

In Algorithm 2, the `first-pass` computes all MSCCs and line 18 allows to detect only the MSCCs satisfying Lemma 3. Thus, the STA emptiness check algorithm reports one cycle **iff** this cycle is a livelock-accepting or a Büchi-accepting cycle.

STA Optimization. The goal of this optimization is to reduce the number of transitions in STA, by exploiting the fact that the livelock-accepting states $(q' \in \mathcal{G})$ that are also Büchi-accepting $(q' \in \mathcal{F})$ do not require the second pass. Indeed, during the TA to STA transformation described by Property 1, it was unnecessary to add artificial transitions (q, k, g) for any transition (q, k, q') where $q' \in (\mathcal{G} \cap \mathcal{F})$, because any MSCC containing q' is necessarily an accepting MSCC and it is detected by the `first-pass` of Algorithm 2.

4   Experimental Evaluation of STA

This section presents our experimentation conducted under the same conditions as our previous work [1]: within the same tools Spot and CheckPN and we

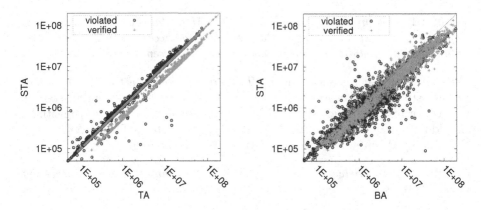

Fig. 6. Performance (transitions visited by the emptiness check) of STA vs. TA and BA

selected some Petri net models and formulas to compare BA, TA and STA approaches. The models are from the Petri net literature [9], we selected two instances of each of the following models: the Flexible Manufacturing System (4/5), the Kanban system (4/5), the Peterson algorithm (4/5), the slotted-ring system (6/5), the dining philosophers (9/10) and the Round-robin mutex (14/15). We also used two models from actual case studies: **PolyORB** [8] and **MAPK** [7]. For each selected model instance, we generated 200 verified formulas (no counterexample in the product) and 200 violated formulas (a counterexample exists): 100 random (length 15) and 100 weak-fairness [1] (length 30) of the two cases of formulas. Since generated formulas are very often trivial to verify (the emptiness check needs to explore only a handful of states), we selected only those formulas requiring more than one second of CPU for the emptiness check in all approaches.

4.1 Results

Figure 6 compares the number of visited transitions when running the emptiness check; plotting STA against TA and BA. This gives an idea of their relative performance. Each point corresponds to one of the 5600 evaluated formulas (2800 violated with counterexample as black circles, and 2800 verified having no counterexample as grey crosses). Each point below the diagonal is in favor of STA while others are in favor of the other approach. Axes are displayed using a logarithmic scale.

4.2 Discussion

On verified properties, the results are very straightforward to interpret when looking at the number transitions explored by the emptiness check in Figure 6. STA significantly improve TA in all cases where a second pass was necessary.

In these cases, the STA approach, with its single-pass emptiness check, is a clear improvement over TA. These cases where the STA approach is twice faster than TA's, appear as a linear cloud of grey crosses below the diagonal in the scatter plot of Figure 6. Otherwise, they have the same performance because if no livelock-acceptance states are detected in the product then the TA and STA approaches explore exactly the same product (these cases correspond to the grey crosses on the diagonal). In the scatter plot comparing STA against BA, in most cases the grey crosses appear below the diagonal, i.e., the points where STA is better. Therefore, STA outperform BA for verified properties.

On violated properties, it is harder to interpret the results because they depend on the order in which non-deterministic transitions of the property automaton are explored. In the best case, the order of transitions leads the emptiness check straight to a counterexample; in the worst case, the algorithm explores the whole product until it finally finds a counterexample. BA, TA and STA provide different orders of transitions and therefore change the number of states and transitions to be explored by the emptiness check before a counterexample is found.

5   Conclusion

In a preliminary work presented in [1], we experiment LTL model checking of stuttering-insensitive properties with various techniques: Büchi automata (BA), Transition-based Generalized Büchi Automata and Testing Automata (TA) [5]. At this time, conclusions were that TA has good performance for violated properties (i.e. when a counterexample was found). However, this was not the case when no counterexample was computed since the entire product had to be visited twice to check for each acceptance mode of a TA (Büchi acceptance or livelock-acceptance).

This paper extends the above work to avoid the second pass of the emptiness check algorithm in TA approach. It proposes a transformation of TA into STA, a Single-pass Testing Automata that avoids the need for a second pass. The STA approach have been implemented in Spot library and used on several benchmark models including large models issued from case studies. Experimentation with Spot reported that, STA remain good for violated properties, and also beat TA and BA in most cases when properties exhibit no counterexample.

References

1. Ben Salem, A.E., Duret-Lutz, A., Kordon, F.: Generalized Büchi automata versus testing automata for model checking. In: Proc. of SUMo 2011, vol. 726, pp. 65–79. CEUR (June 2011)
2. Couvreur, J.-M.: On-the-fly verification of linear temporal logic. In: Wing, J.M., Woodcock, J., Davies, J. (eds.) FM 1999. LNCS, vol. 1708, pp. 253–271. Springer, Heidelberg (1999)

3. Etessami, K.: Stutter-invariant languages, ω-automata, and temporal logic. In: Halbwachs, N., Peled, D.A. (eds.) CAV 1999. LNCS, vol. 1633, pp. 236–248. Springer, Heidelberg (1999)
4. Farwer, B.: ω-automata. In: Grädel, E., Thomas, W., Wilke, T. (eds.) Automata, Logics, and Infinite Games. LNCS, vol. 2500, pp. 3–21. Springer, Heidelberg (2002)
5. Geldenhuys, J., Hansen, H.: Larger automata and less work for LTL model checking. In: Valmari, A. (ed.) SPIN 2006. LNCS, vol. 3925, pp. 53–70. Springer, Heidelberg (2006)
6. Hansen, H., Penczek, W., Valmari, A.: Stuttering-insensitive automata for on-the-fly detection of livelock properties. In: Proc. of FMICS 2002. ENTCS, vol. 66(2). Elsevier (July 2002)
7. Heiner, M., Gilbert, D., Donaldson, R.: Petri nets for systems and synthetic biology. In: Bernardo, M., Degano, P., Zavattaro, G. (eds.) SFM 2008. LNCS, vol. 5016, pp. 215–264. Springer, Heidelberg (2008)
8. Hugues, J., Thierry-Mieg, Y., Kordon, F., Pautet, L., Barrir, S., Vergnaud, T.: On the formal verification of middleware behavioral properties. In: Proc. of FMICS 2004. ENTCS, vol. 133, pp. 139–157. Elsevier, September 2004
9. Kordon, F., Linard, A., Buchs, D., Colange, M., Evangelista, S., Lampka, K., Lohmann, N., Paviot-Adet, E., Thierry-Mieg, Y., Wimmel, H.: Report on the model checking contest at petri nets 2011. T. Petri Nets and Other Models of Concurrency **6**, 169–196 (2012)
10. MoVe/LRDE: The Spot home page (2014). http://spot.lip6.fr
11. Vardi, M.Y.: An automata-theoretic approach to linear temporal logic. In: Moller, F., Birtwistle, G. (eds.) Logics for Concurrency. LNCS, vol. 1043, pp. 238–266. Springer, Heidelberg (1996)

Compressed Data Structures for Range Searching

Philip Bille, Inge Li Gørtz, and Søren Vind$^{(\boxtimes)}$

DTU Compute, Technical University of Denmark, 2800 Kongens Lyngby, Denmark
{phbi,inge,sovi}@dtu.dk

Abstract. We study the orthogonal range searching problem on points that have a significant number of *geometric repetitions*, that is, subsets of points that are identical under translation. Such repetitions occur in scenarios such as image compression, GIS applications and in compactly representing sparse matrices and web graphs. Our contribution is twofold. First, we show how to compress geometric repetitions that may appear in standard range searching data structures (such as K-D trees, Quad trees, Range trees, R-trees, Priority R-trees, and K-D-B trees), and how to implement subsequent range queries on the compressed representation with only a constant factor overhead. Secondly, we present a compression scheme that efficiently identifies geometric repetitions in point sets, and produces a hierarchical clustering of the point sets, which combined with the first result leads to a compressed representation that supports range searching.

Keywords: Data and image compression · Range searching · Relative tree · DAG compression · Hierarchical clustering

1 Introduction

The *orthogonal range searching* problem is to store a set of axis-orthogonal k-dimensional objects to efficiently answer *range queries*, such as reporting or counting all objects inside a k-dimensional query range. Range searching is a central primitive in a wide range of applications and has been studied extensively over the last 40 years [1,3–6,10,11,14,16,19,21–24,26,28,29] (Samet presents an overview in [30]).

In this paper we study range searching on points that have a significant number of *geometric repetitions*, that is, subsets of points that are identical under translation. Range searching on points sets with geometric repetitions arise naturally in several scenarios such as data and image analysis [12,27,32], GIS applications [12,20,31,33], and in compactly representing sparse matrices and web graphs [7,9,17,18].

Supported by a grant from the Danish National Advanced Technology Foundation.
P. Bille and I.L. Gørtz—Supported by a grant from the Danish Council for Independent Research | Natural Sciences.

A.-H. Dediu et al. (Eds.): LATA 2015, LNCS 8977, pp. 577–586, 2015.
DOI: 10.1007/978-3-319-15579-1_45

Our contribution is twofold. First, we present a simple technique to effectively compress geometric repetitions that may appear in standard range searching data structures (such as K-D trees, Quad trees, Range trees, R-trees, Priority R-trees, and K-D-B trees). Our technique replaces repetitions within the data structures by a single copy, while only incurring an $O(1)$ factor overhead in queries (both in standard RAM model and I/O model of computation). The key idea is to compress the underlying tree representation of the point set into a corresponding minimal DAG that captures the repetitions. We then show how to efficiently simulate range queries directly on this DAG. This construction is the first solution to take advantage of geometric repetitions. Compared to the original range searching data structure the time and space complexity of the compressed version is never worse, and with many repetitions the space can be significantly better. Secondly, we present a compression scheme that efficiently identifies translated geometric repetitions. Our compression scheme guarantees that if point set P_1 is a translated geometric repetition of point set P_2 and P_1 and P_2 are at least a factor 2 times their diameter away from other points, the repetition is identified. This compression scheme is based on a hierarchical clustering of the point set that produces a tree of height $O(\log D)$, where D is the diameter of the input point set. Combined with our first result we immediately obtain a compressed representation that supports range searching.

1.1   Related Work

Several succinct data structures and entropy-based compressed data structures for range searching have recently been proposed, see e.g., [2,8,15,25]. While these significantly improve the space of the classic range searching data structure, they all require at least a $\Omega(N)$ *bits* to encode N points. In contrast, our construction can achieve exponential compression for highly compressible point sets (i.e. where there is a lot of geometric repetitions).

A number of papers have considered the problem of compactly representing web graphs and tertiary relations [7,9,18]. They consider how to efficiently represent a binary (or tertiary) quad tree by encoding it as bitstrings. That is, their approach may be considered compact storage of a (sparse) adjacency matrix for a graph. The approach allows compression of quadrants of the quad tree that only contain zeros or ones. However, it does not exploit the possibly high degree of geometric repetition in such adjacency matrices (and any quadrant with different values cannot be compressed).

To the best of our knowledge, the existence of geometric repetitions in the point sets has not been exploited in previous solutions for neither compression nor range searching. Thus, we give a new perspective on those problems when repetitions are present.

1.2   Outline

We first present a general model for range searching, which we call a *canonical range searching data structure*, in Section 2. We show how to compress such data structures efficiently and how to support range searching on the compressed data structure in the same asymptotic time as on the uncompressed data structure in Section 3. Finally, we present a *similarity clustering* algorithm in Section 4, guaranteeing that geometric repetitions are clustered such that the resulting canonical range searching data structure is compressible.

2   Canonical Range Searching Data Structures

We define a *canonical range searching data structure* T, which is an ordered, rooted and labeled tree with N vertices. Each vertex $v \in T$ has an associated k-dimensional axis-parallel range, denoted r_v, and an arbitrary label, denoted $label(v)$. We let $T(v)$ denote the subtree of T rooted at vertex v and require that ranges of vertices in $T(v)$ are contained in the range of v, so for every vertex $u \in T(v)$, $r_u \subseteq r_v$. Leafs may store either points or ranges, and each point or range may be stored in several leafs. The data structure supports *range queries* that produce their result after evaluating the tree through a (partial) traversal starting from the root. In particular, we can only access a node after visiting all ancestors of the node. Queries can use any information from visited vertices. A similar model for showing lower bounds for range searching appeared was used by Kanth and Singh in [21].

Geometrically, the children of a vertex v in a canonical range searching data structure divide the range of v into a number of possibly overlapping ranges. At each level the tree divides the k-dimensional regions at the level above into smaller regions. Canonical range searching data structures directly capture most well-known range searching data structures, including Range trees, K-D trees, Quad trees and R-trees as well as B-trees, Priority R-trees and K-D-B trees.

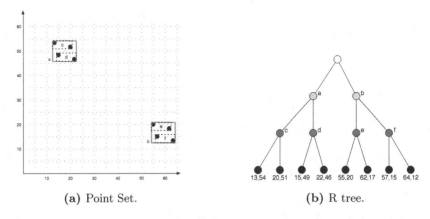

(a) Point Set.   (b) R tree.

Fig. 1. A two-dimensional point set with R tree ranges overlaid, and the resulting R tree. Blue ranges are children of the root in the tree, red ranges are at the second level. A vertex label (a - h) in the R tree identifies the range. We have omitted the precise coordinates for the ranges, but e.g. range a spans the range $[13, 22] \times [46, 54]$.

Example: Two-dimensional R tree. The two-dimensional R tree is a canonical range searching data structure since a vertex covers a range of the plane that contains the ranges of all vertices in its subtree. The range query is a partial traversal of the tree starting from the root, visiting every vertex having a range that intersects the query range and reporting all vertices with their range fully contained in the query range. Figure 1 shows an R tree for a point set, where each vertex is labeled with the range that it covers. The query described for R trees can be used on any canonical range searching data structure, and we will refer to it as a *canonical range query*.

3   Compressed Canonical Range Searching

We now show how to compress geometric repetitions in any canonical range searching data structure T while incurring only a constant factor overhead in queries. To do so we convert T into a *relative tree* representation, which we then compress into a minimal DAG representation that replaces geometric repetitions by single occurrences. We then show how to simulate a range query on T with only constant overhead directly on the compressed representation. Finally, we extend the result to the I/O model of computation.

3.1   The Relative Tree

A *relative tree* R is an ordered, rooted and labeled tree storing a relative representation of a canonical range searching data structure T. The key idea is we can encode a range or a point $r = [x_1, x_1'] \times \ldots \times [x_k, x_k']$ as two k-dimensional vectors $position(r) = (x_1, \ldots, x_k)$ and $extent(r) = (x_1' - x_1, \ldots, x_k' - x_k')$ corresponding to an *origin position* and an *extent* of r. We use this representation in the relative tree, but only store extent vectors at vertices explicitly. The origin position vector for the range r_v of a vertex $v \in R$ is calculated from offset vectors stored on the path from the root of R to v, denoted $path(v)$.

Formally, each vertex $v \in R$ stores a label, $label(v)$, and a k-dimensional extent vector $extent(r_v)$. Furthermore, each edge $(u, v) \in R$ stores an offset vector $offset(u, v)$. The position vector for r_v is calculated as $position(r_v) = \sum_{(a,b) \in path(v)} offset(a, b)$. We say that two vertices $v, w \in R$ are *equivalent* if the subtrees rooted at the vertices are isomorphic, including all labels and vectors. That is, v and w are equivalent if the two subtrees $R(v)$ and $R(w)$ are equal.

It is straightforward to convert a canonical range searching data structure into the corresponding relative tree.

Lemma 1. *Given any canonical range searching data structure T, we can construct the corresponding relative tree R in linear time and space.*

Proof. First, note that a relative tree allows each vertex to store extent vectors and labels. Thus, to construct a relative tree R representing the canonical range searching data structure T, we can simply copy the entire tree including extent

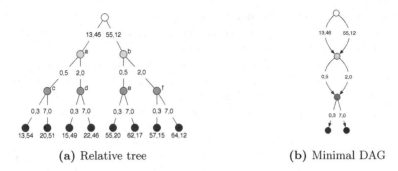

(a) Relative tree   (b) Minimal DAG

Fig. 2. The relative tree obtained from the R tree from Figure 1 and the resulting minimal DAG G generating the tree. Only coordinates of the lower left corner of the ranges in the R tree are shown. In the relative tree, the absolute coordinates for the points are only shown for illustration, in order to see that the relative coordinates sum to the absolute coordinate along the root-to-leaf paths.

vectors and vertex labels. So we only need to show how to store offset vectors in R to ensure that the ranges for each pair of copied vertices are equal.

Consider a vertex $v \in T$ and its copy $v_R \in R$ and their parents $w \in T$ and $w_R \in R$. Since the extent vector and vertex labels are copied, $extent(r_v) = extent(r_{v_R})$ and $label(v) = label(v_R)$. The offset vector for the (w_R, v_R) edge is $offset(w_R, v_R) = position(r_v) - position(r_w)$. We assume the offset for the root is the 0-vector. Observe that summing up all the offset vectors on $path(v)$ is exactly $position(r_v)$, and so $position(r_{v_R}) = position(r_v)$.

Since each vertex and edge in T is only visited a constant number of times during the mapping, the construction time for R is $O(N)$. The total number of labels stored by R is asymptotically equal to the number of labels stored by T. Finally, the degrees of vertices does not change from T to R. Thus, if $v \in T$ is mapped to $v_R \in R$ and v requires s space, v_R requires $\Theta(s)$ space.

3.2 The Data Structure

The compressed canonical data structure is the minimal DAG G of the relative tree R for T. By Lemma 1 and [13] we can build it in $O(N)$ time. Since G replaces equivalent subtrees in R by a single subtree, geometric repetitions in T are stored only once in G. For an example, see Figure 2.

Now consider a range query Q on the canonical range searching data structure T. We show how to simulate Q efficiently on G. Assuming $v_G \in G$ generates $v_R \in R$, we say that v_G generates $v \in T$ if v_R is the relative tree representation of v. When we visit a vertex $v_G \in G$, we calculate the origin position $position(r_{v_G})$ from the sum of the offset vectors along the root-to-v_G path. The origin position for each vertex can be stored on the way down in G, since we may only visit a vertex after visiting all ancestors (meaning that we can only arrive at v_G from a root-to-v_G path in G). Thus, it takes constant time to maintain the origin position for each visited vertex. Finally, a visit to a child of $v \in T$ can be simulated in constant additional time by visiting a child of $v_G \in G$. So we can simulate a visit to $v \in T$ by visiting the vertex $v_G \in G$ that generates v and in constant time calculate the origin position for v_G.

Any label comparison takes the same time on G and T since the label must be equal for $v_G \in G$ to generate $v \in T$. Now, since there is only constant overhead in visiting a vertex and comparing labels, it follows that if Q uses t time we can simulate it in $O(t)$ time on G. In summary, we have the following result.

Theorem 1. *Given a canonical range searching data structure T with N vertices, we can build the minimal DAG representation G of T in linear time. The space required by G is $O(n)$, where n is the size of the minimal DAG for a relative representation of T. We can support any query Q on T that takes time t on G in time $O(t)$.*

As an immediate corollary, we get the following result for a number of concrete range searching data structures.

Corollary 1. *Given a K-D tree, Quad tree, R tree or Range tree, we can in linear time compress it into a data structure using space proportional to the size of the minimal relative DAG representation which supports canonical range searching queries with $O(1)$ overhead.*

3.3   Extension to the I/O Model

We now show that Theorem 1 extends to the I/O model of computation. We assume that each vertex in T require $\Theta(B)$ space, where B is the size of a disk block. To allow for such vertices, we relax the definition of a canonical range searching data structure to allow it to store B k-dimensional ranges. From Lemma 1 and [13], if a vertex $v \in T$ require $\Theta(B)$ space, then so does the corresponding vertex $v_G \in G$. Thus, the layout of the vertices on disk does not asymptotically influence the number of disk reads necessary to answer a query, since only a constant number of vertices can be retrieved by each disk read. This means that visiting a vertex in either case takes a constant number of disk blocks, and so the compressed representation does not asymptotically increase the number of I/Os necessary to answer the query. Hence, we can support any query Q that uses p I/Os on T using $O(p)$ I/Os on G.

4   Similarity Clustering

We now introduce the *similarity clustering* algorithm. Even if there are significant geometric repetitions in the point set P, the standard range searching data structures may not be able to capture this and may produce data structures that are not compressible. The similarity clustering algorithm allows us to create a canonical range searching data structure for which we can guarantee good compression using Theorem 1.

4.1 Definitions

Points and point sets We consider points in k-dimensional space, assuming k is constant. The distance between two points p_1 and p_2, denoted $d(p_1, p_2)$, is their euclidian distance. We denote by $P = \{p_1, p_2, \ldots, p_r\}$ a point set containing r points. We say that two point sets P_1, P_2 are *equivalent* if P_2 can be obtained from P_1 by translating all points with a constant k-dimensional offset vector.

The minimum distance between a point p_q and a point set P, $mindist(P, p_q) = \min_{p \in P} d(p, p_q)$, is the distance between p_q and the closest point in P. The minimum distance between two point sets P_1, P_2 is the distance between the two closest points in the two sets, $mindist(P_1, P_2) = \min_{p_1 \in P_1, p_2 \in P_2} d(p_1, p_2)$. These definitions extend to maximum distance in the natural way, denoted $maxdist(P, p_q)$ and $maxdist(P_1, P_2)$. The diameter of a point set P is the maximum distance between any two points in P, $diameter(P) = \max_{p_1, p_2 \in P} d(p_1, p_2) = maxdist(P, P)$.

A point set $P_1 \subset P$ is *lonely* if the distance from P_1 to any other point is more than twice $diameter(P_1)$, i.e. $mindist(P_1, P \setminus P_1) > 2 \times diameter(P_1)$.

Clustering. A hierarchical clustering of a point set P is a tree, denoted $C(P)$, containing the points in P at the leaves. Each node in the tree $C(P)$ is a cluster containing all the points in the leaves of its subtree. The root of $C(P)$ is the cluster containing all points. We denote by $points(v)$ the points in cluster node $v \in C(P)$. Two cluster nodes $v, w \in C(P)$ are equivalent if $points(v)$ is equivalent to $points(w)$ and if the subtrees rooted at the nodes are isomorphic such that each isomorphic pair of nodes are equivalent.

4.2 Hierarchical Clustering Algorithm for Lonely Point Sets

Order P in lexicographically increasing order according to their coordinates in each dimension, and let $\Delta(P)$ denote the ordering of P. The similarity clustering algorithm performs a greedy clustering of the points in P in levels $i = 0, 1, \ldots, \log D + 1$, where $D = diameter(P)$. Each level i has an associated clustering distance threshold d_i, defined as $d_0 = 0$ and $d_i = 2^{i-1}$ for all other i.

The clustering algorithm proceeds as follows, processing the points in order $\Delta(P)$ at each level. If a point p is not clustered at level $i > 0$, create a new cluster C_i centered around the point p (and its cluster C_{i-1} at the previous level). Include a cluster C_{i-1} from level $i - 1$ in C_i if $maxdist(points(C_{i-1}), p) \leq d_i$. The clusters at level 0 contain individual points and the cluster at level $\log D + 1$ contains all points.

Lemma 2. *Given a set of points P, the similarity clustering algorithm produces a clustering tree containing equivalent clusters for any pair of equivalent lonely point sets.*

Proof. Let P_1 and P_2 be two lonely point sets in P such that P_1 and P_2 are equivalent, and let $d = diameter(P_1) = diameter(P_2)$. Observe that a cluster formed at level i has at most diameter $2d_i = 2^i$. Thus, since all points are clustered at every level and all points outside P_1 have a distance greater than

$2d$ to any point in P_1, there is a cluster $c \in C(P)$ formed around point $a \in P_1$ at level $j = \lceil \log d \rceil$ containing no points outside P_1. Now, assume some point $p \in P_1$ is not in $points(c)$. As all unclustered points within distance $2^j \geq d$ from a are included in c, this would mean that p was clustered prior to creating c. This contradicts the assumption that P_1 is lonely, since it can only happen if some point outside P_1 is closer than $2d$ to p. Concluding, c contains exactly the points in P_1. The same argument naturally extends to P_2.

Now, let C_1, C_2 be the clusters containing the points from P_1, P_2, respectively. Observe that $points(C_1)$ and $points(C_2)$ are equivalent. Furthermore, because each newly created cluster process candidate clusters to include in the same order, the resulting trees for C_1 and C_2 are isomorphic and have the same ordering. Thus, the clusters C_1 and C_2 are equivalent.

Because the clustering proceeds in $O(\log D)$ levels, the height of the clustering tree is $O(\log D)$. Furthermore, by considering all points and all of their candidates at each level, the clustering can be implemented in time $O(N^2 \log D)$. Observe that the algorithm allows creation of paths of clusters with only a single child cluster. If such paths are contracted to a single node to reduce the space usage, the space required is $O(N)$ words. In summary, we have the following result.

Theorem 2. *Given a set of N points with diameter D, the similarity clustering algorithm can in $O(N^2 \log D)$ time create a tree representing the clustering of height $O(\log D)$ requiring $O(N)$ words of space. The algorithm guarantees that any pair of equivalent lonely point sets results in the same clustering, producing equivalent subtrees in the tree representing the clustering.*

Since the algorithm produces equivalent subtrees in the tree for equivalent lonely point sets, the theorem gives a compressible canonical range searching data structure for point sets with many geometric repetitions.

5   Open Problems

The technique described in this paper for generating the relative tree edge labels only allows for translation of the point sets in the underlying subtrees. However, the given searching technique and data structure generalizes to scaling and rotation (if simply storing a parent-relative scaling factor and rotation angle in each node, along with the nodes parent-relative translation vector). We consider it an open problem to efficiently construct a relative tree that uses such transformations of the point set.

Another interesting research direction is if it is possible to allow for small amounts of noise in the point sets. That is, can we represent point sets that are almost equal (where few points have been moved a little) in a compressed way? An even more general question is how well one can do when it comes to compression of higher dimensional data in general.

Finally, the $O(N^2 \log D)$ time bound for generating the similarity clustering is prohibitive for large point sets. So an improved construction would greatly benefit the possible applications of the clustering method and is of great interest.

References

1. Arge, L., Berg, M.D., Haverkort, H., Yi, K.: The Priority R-tree: A practically efficient and worst-case optimal R-tree. ACM TALG **4**(1), 9 (2008)
2. Barbay, J., Claude, F., Navarro, G.: Compact rich-functional binary relation representations. In: López-Ortiz, A. (ed.) LATIN 2010. LNCS, vol. 6034, pp. 170–183. Springer, Heidelberg (2010)
3. Bayer, R., McCreight, E.: Organization and maintenance of large ordered indexes. Acta Informatica **1**(3), 173–189 (1972)
4. Bentley, J.L.: Multidimensional binary search trees used for associative searching. Comm. ACM **18**(9), 509–517 (1975)
5. Bentley, J.L.: Multidimensional binary search trees in database applications. IEEE Trans. Softw. Eng. **4**, 333–340 (1979)
6. Bentley, J.L., Saxe, J.B.: Decomposable searching problems I. Static-to-dynamic transformation. J. Algorithms **1**(4), 301–358 (1980)
7. de Bernardo, G., Álvarez-García, S., Brisaboa, N.R., Navarro, G., Pedreira, O.: Compact querieable representations of raster data. In: Kurland, O., Lewenstein, M., Porat, E. (eds.) SPIRE 2013. LNCS, vol. 8214, pp. 96–108. Springer, Heidelberg (2013)
8. Bose, P., He, M., Maheshwari, A., Morin, P.: Succinct orthogonal range search structures on a grid with applications to text indexing. In: Dehne, F., Gavrilova, M., Sack, J.-R., Tóth, C.D. (eds.) WADS 2009. LNCS, vol. 5664, pp. 98–109. Springer, Heidelberg (2009)
9. Brisaboa, N.R., Ladra, S., Navarro, G.: k^2-trees for compact web graph representation. In: Karlgren, J., Tarhio, J., Hyyrö, H. (eds.) SPIRE 2009. LNCS, vol. 5721, pp. 18–30. Springer, Heidelberg (2009)
10. Clarkson, K.L.: Fast algorithms for the all nearest neighbors problem. In: Proc. 24th FOCS, vol. 83, pp. 226–232 (1983)
11. Comer, D.: Ubiquitous B-tree. ACM CSUR **11**(2), 121–137 (1979)
12. Dick, C., Schneider, J., Westermann, R.: Efficient geometry compression for gpu-based decoding in realtime terrain rendering. CGF **28**(1), 67–83 (2009)
13. Downey, P.J., Sethi, R., Tarjan, R.E.: Variations on the common subexpression problem. J. ACM **27**(4), 758–771 (1980)
14. Eppstein, D., Goodrich, M.T., Sun, J.Z.: Skip quadtrees: Dynamic data structures for multidimensional point sets. IJCGA **18**(01n02), 131–160 (2008)
15. Farzan, A., Gagie, T., Navarro, G.: Entropy-bounded representation of point grids. CGTA **47**(1), 1–14 (2014)
16. Gaede, V., Günther, O.: Multidimensional access methods. ACM CSUR **30**(2), 170–231 (1998)
17. Galli, N., Seybold, B., Simon, K.: Compression of sparse matrices: Achieving almost minimal table size. In: Proc. ALEX, pp. 27–33 (1998)
18. Alvarez Garcia, S., Brisaboa, N.R., de Bernardo, G., Navarro, G.: Interleaved k2-tree: indexing and navigating ternary relations. In: Proc. DCC, pp. 342–351 (2014)
19. Guttman, A.: R-trees: a dynamic index structure for spatial searching. In: Proc. 1984 ACM SIGMOD, vol. 14, pp. 47–57 (1984)
20. Haegler, S., Wonka, P., Arisona, S.M., Van Gool, L., Mueller, P.: Grammar-based encoding of facades. CGF **29**(4), 1479–1487 (2010)
21. Kanth, K.V.R., Singh, A.K.: Optimal dynamic range searching in non-replicating index structures. In: Beeri, C., Bruneman, P. (eds.) ICDT 1999. LNCS, vol. 1540, pp. 257–276. Springer, Heidelberg (1998)

22. van Kreveld, M.J., Overmars, M.H.: Divided k-d trees. Algorithmica **6**(1–6), 840–858 (1991)

23. Lee, D., Wong, C.: Quintary trees: a file structure for multidimensional datbase sytems. ACM TODS **5**(3), 339–353 (1980)

24. Lueker, G.S.: A data structure for orthogonal range queries. In: Proc. 19th FOCS, pp. 28–34 (1978)

25. Mäkinen, V., Navarro, G.: Rank and select revisited and extended. TCS **387**(3), 332–347 (2007)

26. Orenstein, J.A.: Multidimensional tries used for associative searching. Inform. Process. Lett. **14**(4), 150–157 (1982)

27. Pajarola, R., Widmayer, P.: An image compression method for spatial search. IEEE Trans. Image Processing **9**(3), 357–365 (2000)

28. Procopiuc, O., Agarwal, P.K., Arge, L., Vitter, J.S.: Bkd-tree: a dynamic scalable kd-tree. In: Proc. 8th SSTD, pp. 46–65 (2003)

29. Robinson, J.T.: The KDB-tree: a search structure for large multidimensional dynamic indexes. In: Proc. 1981 ACM SIGMOD, pp. 10–18 (1981)

30. Samet, H.: Applications of spatial data structures. Addison-Wesley (1990)

31. Schindler, G., Krishnamurthy, P., Lublinerman, R., Liu, Y., Dellaert, F.: Detecting and matching repeated patterns for automatic geo-tagging in urban environments. In: CVPR, pp. 1–7 (2008)

32. Tetko, I.V., Villa, A.E.: A pattern grouping algorithm for analysis of spatiotemporal patterns in neuronal spike trains. J. Neurosci. Meth. **105**(1), 1–14 (2001)

33. Zhu, Q., Yao, X., Huang, D., Zhang, Y.: An Efficient Data Management Approach for Large Cyber-City GIS. ISPRS Archives, 319–323 (2002)

Average Linear Time and Compressed Space Construction of the Burrows-Wheeler Transform

Alberto Policriti[1,2], Nicola Gigante[1], and Nicola Prezza[1(✉)]

[1] Department of Mathematics and Informatics, University of Udine, Udine, Italy
alberto.policriti@uniud.it, {gigante.nicola,prezza.nicola}@spes.uniud.it
[2] Istituto di Genomica Applicata, Udine, Italy

Abstract. The Burrows-Wheeler Transform is a text permutation that has revolutionized the fields of pattern matching and text compression, bridging the gap existing between the two. In this paper we approach the BWT-construction problem generalizing a well-known algorithm—based on backward search and dynamic strings manipulation—to work in a context-wise fashion, using automata on words. Let n, σ, and H_k be the text length, the alphabet size, and the k-th order empirical entropy of the text, respectively. Moreover, let $H_k^* = \min\{H_k + 1, \lceil \log \sigma \rceil\}$. Under the word RAM model with word size $w \in \Theta(\log n)$, our algorithm builds the BWT in average $\mathcal{O}(nH_k^*)$ time using $nH_k^* + o(nH_k^*)$ bits of space, where $k = \log_\sigma(n/\log^2 n) - 1$. We experimentally show that our algorithm has very good performances (essentially linear time) on DNA sequences, using about 2.6 bits per input symbol in RAM.

1 Introduction

The Burrows-Wheeler Transform of a text T [4], is the (unique) permutation of $T\$$—where $\$$ is a character not appearing in T and lexicographically smaller than all the characters in T—obtained concatenating the characters preceding its sorted circular permutations. Since its discovery [4], this text transform revolutionized text compression [6] and text indexing [7], merging the two in the promising field of compressed self indexing (a complete and accurate survey on the subject is [16]). Despite the intriguing properties of compressed self-indexes, there still exists a bottleneck in their construction, consisting in the *construction* of the BWT. Let n be the text length, σ the alphabet size and H_k the k-th order empirical entropy of the text. To date, none of the solutions in the literature is able to guarantee simultaneously both $\mathcal{O}(n)$ construction time and $nH_k + o(n \log \sigma)$ bits of space. The most time-efficient (also in practice) $\mathcal{O}(n)$ techniques to date, rely on the construction of suffix arrays (see for example [19]), which however require $\mathcal{O}(n \log n)$ bits of space. Very recently, it has been shown that the space requirements can be reduced to $\mathcal{O}(n \log \sigma)$, while maintaining the optimal construction time $\mathcal{O}(n)$ [1]. Despite compact space being asymptotically optimal in the uncompressed domain, the hidden constant in practice could be high and makes this kind of algorithms impractical, especially for large texts (e.g. big genomes). This problem motivates the search for more space-efficient

© Springer International Publishing Switzerland 2015
A.-H. Dediu et al. (Eds.): LATA 2015, LNCS 8977, pp. 587–598, 2015.
DOI: 10.1007/978-3-319-15579-1_46

algorithms, being able either to work in external memory or to exploit the compressibility of the text in order to reduce RAM requirements. External and semi-external solutions on genomic data include [2], which requires about 1 byte per input symbol and works in linear-time, and [21], which requires about 2 bits per input symbol and works in $\mathcal{O}(n \log^2 \log n)$ average time. Of particular interest is the more general implementation described in [5], which requires a *constant* (i.e. text independent) amount of RAM working space. Building a compressed BWT is another common solution in order to save working space. Usually, this is done by inserting the text characters backwards in a compressed dynamic string data structure. The complexity of this approach is deeply influenced by the inherent complexity of dynamic string data structures, which have been proved to have a $\Theta(\log n/\log \log n)$ lower [8] and (amortized) upper [17] bound for queries and updates. In particular, the result in [17] has as direct consequence (clearly mentioned in that paper) that the BWT can be constructed in $nH_k + o(n \log \sigma)$ bits of space and $\mathcal{O}(n \log n/\log \log n)$ worst-case time.

In this scenario, we propose a compressed-space solution able to reach average linear time on semi-uniform inputs (e.g. DNA). Our result makes two (reasonable) assumptions: that the RAM word size is $w \in \Theta(\log n)$ and that the alphabet size is bounded by $\sigma \in \mathcal{O}(polylog(n))$. Table 1 summarizes the above discussed results, comparing them with our bounds. Our algorithm relies on the fact that BWT characters can be partitioned in *contexts*. On the grounds of this observation we generalize the classic backward insertion algorithm to work with multiple dynamic strings (one per context), instead of only one for the whole BWT. The classic algorithm becomes, then, a particular case of ours when the context length k is 0. This strategy reduces considerably the average size of internal data structures, thus leading to better performances. To our knowledge, ours is the first result reaching both linear average-case time and compressed working space. We call our algorithm cw-bwt (context-wise BWT).

Table 1. Comparison among some of the most interesting space-time tradeoffs presented in literature. H_k^* stands for $\min\{H_k + 1, \lceil \log \sigma \rceil\}$, and stems from the use of Huffman encoding.

Space (bits)	average-case time	worst-case time	reference
$\mathcal{O}(n \log n)$	-	$\mathcal{O}(n)$	[19]
$\mathcal{O}(n \log \sigma)$	$\mathcal{O}(n)$	$\mathcal{O}(n)$	[1]
$nH_k + o(n \log \sigma)$	-	$\mathcal{O}(n \log n/\log \log n)$	[17]
$nH_k^* + o(nH_k^*)$	$\mathcal{O}(nH_k^*)$	$\mathcal{O}\left(nH_k^*(\log n/\log \log n)^2\right)$	This work

We implemented cw-bwt in the BWTIL library, freely downloadable at https://github.com/nicolaprezza/BWTIL. cw-bwt has also been integrated in the short-string alignment package ERNE, to be used in DNA analysis (http://erne.sourceforge.net). Our software relies on the bitvector library, freely downloadable at https://github.com/nicola-gigante/bitvector.

2   Notation

Throughout this paper we will work with an alphabet Σ ordered by $<$ and of size $|\Sigma| = \sigma$; we will denote by c a character in Σ. With $T \in \Sigma^n$ we will denote the *text* to be processed, and $\$ \notin \Sigma$ will be a character (terminator) lexicographically smaller than any character in Σ. We will assume to work under the word RAM model with word size $w \in \Theta(\log n)$ and that the alphabet size is $\sigma \in \mathcal{O}(polylog(n)) = \mathcal{O}(poly(w))$. Concatenation of strings $u, v \in (\Sigma \cup \$)^*$ will be denoted by uv. By T_i, $i \leq n$, we will denote $T[i, ..., n-1]$, i.e. the i-th suffix of T. When dealing with dynamic strings/bitvectors, with the term *queries* we will denote *rank* and *access* operations (we do not consider *select*), while by the term *updates* we will denote *insertions* or *substitutions* (we do not consider *deletions*). With u we will denote the length of a generic dynamic bitvector/string, which could be much smaller than n. Given a dynamic string/bitvector S on the alphabet Σ, with $S.rank(c, i)$ we will denote the number of characters equal to c in the substring $S[0, ..., i-1]$. The operation $S.insert(c, i)$ will instead denote character insertion, turning S into $S[0, ..., i-1]cS[i, ..., |S|-1]$. $S[i]$ is the i-th character of S. By $S[i] \leftarrow c$ we will denote character substitution in position i of the string. By $S.F(c)$ we denote the quantity $\sum_{x<c} S.rank(x, |S|)$, i.e. the number of characters lexicographically smaller than c in the whole string S. If S is the BWT of some text, then $S.F(c)$ has a direct interpretation as the first column (usually called F) in the matrix representation of the BWT. By c^k we will denote the string "$cc...c$" (k times). With the term k-*context* we will denote a k-mer on the extended alphabet $\Sigma \cup \{\$\}$, i.e. a string in $(\Sigma \cup \{\$\})^k$. Logarithms are taken in base 2, unless differently specified.

3   The Burrows-Wheeler Transform

The Burrows-Wheeler transform of $T\$$ can be obtained by sorting all circular permutations of $T\$$, representing them in "conceptual" matrix M (see Table 2), and then taking the last column $M^n = L$ of M (the first column will be denoted $M^0 = F$), where M^i, $0 \leq i \leq n$ is the i-th column of M. We will often refer to the BWT matrix M in our algorithm's description. Notice that length-k contexts appear lexicographically sorted in the first k columns of M, and thus they induce a partition of the BWT rows (as depicted in Table 2). A fundamental property of the BWT matrix is the *LF property*: the i-th occurrence of c ($c \in \Sigma \cup \{\$\}$) in the last column corresponds to the i-th occurrence of c in the first column (i.e. they represent the same text position). This property can be generalized.

Given a h-context $s \in (\Sigma \cup \{\$\})^h$, for some h, let $\mathcal{C}^s(M^i)$ be the class of the partition of M^i induced by s.

Lemma 1. *(Context-wise LF property) If $T[j]$ is i-th occurrence of c in $\mathcal{C}^s(L)$, then $T[j]$ is the i-th occurrence of c in $\mathcal{C}^{cs}(F)$.*

The classical *LF*-property is the special case for $k = 0$ of the context-wise *LF* property.

Table 2. Conceptual BWT matrix M of the text mississippi$. Last column is the BWT (ipssm$pissii). There are 9 different contexts of length $k = 2$, corresponding to a partitioning of the BWT in 9 substrings. With our strategy, we will keep one dynamic string data structure for each of these substrings.

$	m	i	s	s	i	s	s	i	p	p	i
i	$	m	i	s	s	i	s	s	i	p	p
i	p	p	i	$	m	i	s	s	i	s	s
i	s	s	i	p	p	i	$	m	i	s	s
i	s	s	i	s	s	i	p	p	i	$	m
m	i	s	s	i	s	s	i	p	p	i	$
p	i	$	m	i	s	s	i	s	s	i	p
p	p	i	$	m	i	s	s	i	s	s	i
s	i	p	p	i	$	m	i	s	s	i	s
s	i	s	s	i	p	p	i	$	m	i	s
s	s	i	p	p	i	$	m	i	s	s	i
s	s	i	s	s	i	p	p	i	$	m	i

4 Data Structures

Our main structure will be, essentially, a de Bruijn automaton: a labeled subgraph of a de Bruijn graph [3], having k-contexts as states. For each automaton's state s we will store a compressed dynamic string encoding the class $\mathcal{C}^s(L)$ and a partial sum data structure encoding the class $\mathcal{C}^s(M^k)$. These data structures will be better specified below. Then, the main algorithm will proceed by reading text's characters (right to left) while navigating automaton's states and updating the corresponding dynamic strings and partial sums structures. Correctness follows from Lemma 1.

de Bruijn Automata

With the term *de Bruijn automaton* we indicate a labeled subgraph of a de Bruijn graph [3], having k-mers appearing in $T\$^k$ as nodes. More in detail, our de Bruijn automaton is $\mathcal{A} = \langle Q, \Sigma, \delta, \$^k \rangle$, where $Q = \{q \mid q$ is a k-mer in $T\$^k\}$ is the set of states, Σ is the set of input symbols, $\delta : Q \times \Sigma \to Q$ is the transition function defined as $\delta(u, c) = v$ iff $v = cu[0, ..., k - 2]$, and $\$^k$ is the start state. Accepting states are not relevant for our application, so we omit them. Using array indexes as k-digit integers in base σ, the representation of the automaton will turn out to be implicit in our data structures. This choice gives the additional benefit that automaton's states can be visited in lexicographic order without overhead, a feature that we will use in our algorithm.

In order to refer to the structures associated to each automaton state, we will use the following notation. Each automaton's state $s \in Q$ will carry a dynamic string denoted (using an object-oriented like notation) with $\mathcal{A}[s].DS$ and a partial sum data structure denoted with $\mathcal{A}[s].PS$. The function $\mathcal{A}.GOTO(s, c) \in Q$, $s \in Q$, $c \in \Sigma$, encodes the automaton's transition function: $\mathcal{A}.GOTO(s, c) = \delta(s, c)$.

Succinct Dynamic Bitvectors

The problem of designing a lightweight bitvector offering efficient query and update operations has been extensively discussed in the literature [10,11,17,18]. Let u be the bitvector length. Given the lower bound $\Omega\left(\log u/\log\log u\right)$ on the maximum of update and queries [8] and the most recent optimal-time and compressed $\mathcal{O}(uH_0)$ space solutions [17,18], this problem can be considered essentially solved for a general bitvector size u. However, under the word RAM model with word size w, better solutions can be found for a small enough bitvector size u (e.g. $u \in \mathcal{O}(poly(w))$).

The core of our bitvector data structure is a packed B-tree. Each leaf of the tree stores $W = p \cdot w$ bits, for a suitable integer $p > 0$. Internal nodes are composed by $3p$ words each and store one rank (number of 1's) and one size (number of bits) counter for each child, in addition to pointers to the children (totalling $3pw$ bits for each internal node). The size of each pointer, rank, and size counter is of $\mathcal{O}(\log u)$ bits, thus the maximum number of children per node is $d = \mathcal{O}(W/\log u)$. With h we denote the height of the tree. Access and rank operations are implemented in $\mathcal{O}(h \cdot p)$ time: size counters guide the search from the root to the leaves, and rank counters give partial rank information (of the subtrees) while searching the leaf ($\mathcal{O}(p)$ operations for each node on the path). The main novelty resides in the insertion algorithm, which is studied to maximize leaf usage and minimize expensive re-arrangement operations. While inserting a bit, if a leaf/node is not full, then we simply insert the bit/key in the right place, updating accordingly the counters. If a leaf/node is full, 4 cases can appear. Let $b = \sqrt{d}$. If a leaf is full, then we count the number m of bits in b adjacent leaves, including the current full leaf (apart from being adjacent, the way leaves are chosen is not relevant for the analysis). If $m > b(W - b)$, then we create a new leaf and redistribute uniformly the m bits in the resulting $b+1$ leaves. If $m \leq b(W - b)$, then we redistribute uniformly the m bits among the b leaves without creating a new one. If an internal node is full, then we choose b adjacent nodes (included the current full node) and we count the total number m of children. If $\lfloor m/(b+1) \rfloor \geq b$, then we create a new internal adjacent node and we uniformly redistribute the children in the resulting $b + 1$ nodes. If $\lfloor m/(b + 1) \rfloor < b$, then we uniformly redistribute the children among the b adjacent nodes. All the above redistributions can be implemented in $\mathcal{O}(p \cdot b)$ time using shifts and masks. In all 4 the cases, it can be easily shown that after a redistribution the number of free bits/positions in the manipulated leaves/nodes is always $\Omega(b)$: as a consequence, if a redistribution takes place, then at least b "easy" ($\mathcal{O}(h \cdot p)$ time) insertions have been made beforehand, resulting in $\mathcal{O}(h \cdot p)$ amortized cost for the insertion. Due to the internal node redistribution policy, the minimum number of children per node is b. It follows that the height of the tree is $h \in \mathcal{O}(\log_b(u))$. We will study space/time complexities in 2 cases: $u \in \mathcal{O}(poly(w))$, and $u \in \mathcal{O}(2^w)$ (intuitively, in our algorithm these situations will represent the average and worst-case, respectively). In order to keep the space always succinct, we choose $p = \log u/\log w$. We obtain $d = \mathcal{O}(w/\log w)$ and $b = \mathcal{O}(\sqrt{w/\log w})$. If $u \in \mathcal{O}(w^c)$, $c \in \mathcal{O}(1)$, then $h \in \mathcal{O}(c) = \mathcal{O}(1)$. Otherwise, if

$u \in \mathcal{O}(2^w)$, then $h \in \mathcal{O}(\log u / \log \log u) = \mathcal{O}(w / \log w)$. The minimum number of used bits per leaf is (following the redistribution policy) $b(W - b)/(b+1)$. From this fact, it can be shown that the maximum overhead (total number of bits allocated in the leaves but not used) is of $\mathcal{O}(u/b)$ bits. Moreover, the maximum number of leaves is $n_L \in \mathcal{O}(u/W)$, thus the maximum number of internal nodes is $(n_L - 1)/(b-1) \in \mathcal{O}(n_L/b) = \mathcal{O}(u/(bW))$, totalling $W \cdot \mathcal{O}(u/(bW)) = \mathcal{O}(u/b)$ bits of space occupancy for the internal nodes. In both cases $u \in \mathcal{O}(poly(w))$ and $u \in \mathcal{O}(2^w)$, the extra space required by the tree is thus $o(u)$, so the whole structure occupies $u + o(u)$ bits of space. Finally, from the particular value chosen for p, it can be easily shown (see above) that all operations have cost $\mathcal{O}(h \cdot p) = \mathcal{O}(1)$ or $\mathcal{O}((w / \log w)^2)$ (in the amortized sense for insertion) if the bitvector size is $u \in \mathcal{O}(poly(w))$ or $u \in \mathcal{O}(2^w)$, respectively.

Compressed Dynamic Strings

Given the bitvector data structure discussed above, a generalization to dynamic strings can be easily made using wavelet trees (see [15] for a complete survey on the subject). Assuming that the frequency of each character to be inserted in the string is known beforehand, we implement this structure using a Huffman-shaped wavelet tree. Huffman encoding requires, on average, at most $H_0 + 1$ bits per symbol, and never more than $\lceil \log \sigma \rceil$. For compactness of notation, we denote by H_0^* the value $\min\{H_k + 1, \lceil \log \sigma \rceil\}$. The total space of the structure is then $nH_0^* + o(nH_0^*) + \mathcal{O}(\sigma \log n) + \mathcal{O}(\sigma \log \sigma)$ bits [12,15], where the last two terms come from the tree topology and the codebook, respectively. Queries and update operations are supported in average $\mathcal{O}(H_0^*)$ or $\mathcal{O}(H_0^*(w / \log w)^2)$ time if the string size is $u \in \mathcal{O}(poly(w))$ or $u \in \mathcal{O}(2^w)$, respectively.

Partial Sums

Our algorithm will require, for each automaton's state s, one partial sum structure of length σ encoding the class $\mathcal{C}^s(M^k)$. A partial sum data structure PS of length j is a list of values $PS[0], ..., PS[j-1]$ offering efficient partial sum and update queries. With $PS.sum(i)$ we will denote the quantity $\sum_{k=0}^{i-1} PS[k]$, and $PS.increment(i)$ will denote the operation $PS[i] \leftarrow PS[i] + 1$.

Efficient solutions offering optimal worst-case space and time bounds, appeared in literature [20]. For self-containedness, here we describe a simple structure based on packed B-trees, optimized for the particular case where $\sigma \in \mathcal{O}(poly(w))$. The main idea is to use a packed B-tree to store temporary partial sums information, plus an array $S[0, ..., \sigma - 1]$ initialized at $S[i] = 0$, $0 \leq i < \sigma$. The packed B-tree has σ leaves (each storing a counter) and internal nodes store the partial sums of the corresponding subtrees. Each counter in the tree is composed by $\log \sigma = \mathcal{O}(\log(w))$ bits, so the height of the tree is $\log_{w / \log w} \sigma \in \mathcal{O}(\log_{w / \log w} poly(w)) = \mathcal{O}(1)$. Updates are implemented as follows. When incrementing a counter (i.e. $PS.increment(i)$), the packed B-tree is updated accordingly (i.e. by incrementing counters from the i-th leaf up to the root). At steps of σ increment operations, $S[i]$ is incremented with the content of the i-th leaf ($0 \leq i < \sigma$) of the packed B-tree, and the tree is re-initialized (i.e. each counter is reset to 0). Since a single update on the packed B-tree takes constant time and the tree is re-built every

σ operations, updates on the whole structure PS take constant amortized time. Finally, a query $PS.sum(i)$ is implemented in $\mathcal{O}(1)$ time by returning the sum between $S[i]$ and the i-th partial sum stored in the packed B-tree. Given that each counter will store a number less than or equal to n (the text length), the described data structure occupies $\mathcal{O}(\sigma \log n) = \mathcal{O}(\sigma w)$ bits of space.

5   The cw-bwt Algorithm

Standard dynamic string based approaches use one string to represent the whole BWT and proceed by inserting the text characters backwards as follows. The construction starts at $i = 0$ with $BWT = $ "$\$$" and at the, generic, i-th step, character $T[n - i - 1]$ is inserted in it. Letting j be such that $BWT[j] = \$$ and $r = BWT.rank(T[n - i - 1], j)$, we update BWT by:

1) $BWT[j] \leftarrow T[n - i - 1]$ and
2) $BWT.insert(\$, BWT.F(T[n - i - 1]) + r)$.

Here we describe how to improve the above algorithm by allowing it to work *context-wise*. We name our algorithm *cw-bwt*.

Optimal k and Space Requirements

First of all notice that, since we partition the BWT using length k contexts and each partition is Huffman-compressed, as a by-product we obtain that, overall, the dynamic string data structures require globally $nH_k^* + o(nH_k^*)$ bits of space in memory [14] (excluding wavelet tree topologies and codebooks, see below), where $H_k^* = \min\{H_k + 1, \lceil \log \sigma \rceil\}$. The number of states of the automaton is bounded by $|Q| \leq \sigma^k + k \in \mathcal{O}(\sigma^k)$. For each automaton state we store a partial sum data structure of size $\mathcal{O}(\sigma w) = \mathcal{O}(\sigma \log n)$ bits. This is, asymptotically, the same space required to store a single wavelet tree topology ($\mathcal{O}(\sigma \log n)$) and a codebook ($\mathcal{O}(\sigma \log \sigma) \subseteq \mathcal{O}(\sigma \log n)$). The choice $k = \log_\sigma \left(n/\sigma \log^2 n\right) = \log_\sigma(n/\log^2 n) - 1$ results in a total space occupancy of all the above discussed structures of $\mathcal{O}(\sigma \log n)\mathcal{O}(\sigma^k) = \mathcal{O}\left(\sigma \log n \cdot n/(\sigma \log^2 n)\right) = \mathcal{O}(n/\log n) = o(n)$ bits. Summing up, the total space occupancy of the cw-bwt algorithm is $nH_k^* + o(nH_k^*)$ bits, where $k = \log_\sigma(n/\log^2 n) - 1$.

Main Algorithm

Our algorithm is reported as Algorithm 1. See below for a detailed discussion of the pseudocode.

In line 3 the de Bruijn automaton is constructed and data structures are initialized. As mentioned in Section 4, a simple direct-hashing strategy permits to perform the automaton construction implicitly and with no overhead. The initialization of dynamic string data structures requires the frequency of each character to be computed for each class $C^s(L)$ (for Huffman encoding). This step can be easily done in linear $\mathcal{O}(n)$ time and $\mathcal{O}(\sigma^k \sigma \log n) = o(n)$ bits of space using, again, direct hashing.

Algorithm 1. cw-bwt(T)

---

input : Text $T \in \Sigma^n$, without terminator appended at the end.
output: BWT of $T\$$.

1 $n \leftarrow |T|$;

2 $k \leftarrow max(\lceil \log_\sigma(n/\log^2 n) - 1 \rceil, 0)$;   /* optimal k */
3 $\mathcal{A} \leftarrow init\_automaton(T, k)$;   /* init automaton and data structures */

4 $s \leftarrow \$^k$;   /* current state */
5 $t \leftarrow 0$;   /* position of the insertion in current state */

6 **for** $i = n - 1$ **downto** 0 **do**

7 | $head \leftarrow T[i]$;   /* symbol entering in the context */
8 | $tail \leftarrow s[k-1]$;   /* symbol exiting from the context */
9 | $s' \leftarrow \mathcal{A}.GOTO(s, head)$;   /* next state */
10 | $\mathcal{A}[s].DS.insert(head, t)$;   /* insert current character */
11 | $\mathcal{A}[s'].PS.increment(tail)$;   /* update partial sums */
12 | $t \leftarrow \mathcal{A}[s'].PS.sum(tail) + \mathcal{A}[s].DS.rank(head, t)$;   /* update t */
13 | $s \leftarrow s'$;   /* update state */

14 $\mathcal{A}[s].DS.insert(\$, t)$;   /* insert terminator */

15 $BWT \leftarrow \epsilon$;   /* the BWT of $T\$$ */
16 **for** $s \in Q$ **in lexicographic order do**
17 | $BWT.append(\mathcal{A}[s].DS)$;   /* append dynamic strings to BWT */

18 **return** BWT;

The *for* loop in line 6 scans backwards all text characters, starting from the rightmost. Variables *head* and *tail* at lines 7 and 8 store the current text character and the rightmost symbol of the current context, respectively. t is the position where *head* has to be inserted in the current state. The new automaton state s' is computed (line 9) by appending *head* at the beginning of the current state and by removing *tail* from its rightmost end. The subsequent 4 lines represent the core of our algorithm. First of all, the current text character $head = T[i]$ is inserted at position t in the dynamic string associated with the class $\mathcal{C}^s(L)$ (line 10). In the BWT matrix, this operation corresponds to the substitution of the terminator $\$$ character with *head* (notice that $\$$ is not explicitly inserted at each step: we just remember its coordinates $\langle s, t \rangle$). The next operations correspond to the insertion in the BWT matrix of the current text suffix T_i, having as prefix $head \cdot s = s' \cdot tail$ and ending with $\$$. Since we will need information about the first $k + 1$ columns of M (see Lemma 1), we need to keep track of the fact that in $\mathcal{C}^{s'}(M^k)$ a new symbol *tail* has been added. Since symbols in $\mathcal{C}^{s'}(M^k)$ appear in lexicographic order, this task is accomplished simply by incrementing a partial sum counter (line 11). In line 12 the new position of $\$$ in $\mathcal{C}^{s'}(L)$ is computed (remember that $\$$ is not explicitly inserted). The operation $\mathcal{A}[s].DS.rank(head, t)$ returns the number of characters equal to *head* before the position that contained $\$$.

Since we are computing this value in $\mathcal{C}^s(L)$, this is the number of text suffixes starting with $head \cdot s$ and lexicographically smaller than the current text suffix T_i. Lemma 1 implies that, in order to compute the new position of \$ in $\mathcal{C}^{s'}(L)$ (i.e. the lexicographic position of the new text suffix in the BWT matrix), we need to add to this value the number of characters smaller than $tail$ in $\mathcal{C}^{s'}(M^k)$, i.e. $\mathcal{A}[s'].PS.sum(tail)$ (line 12).

As mentioned above, we never explicitly insert the terminator character \$ during construction. For this reason, at the end of the first *for* loop, \$ is explicitly inserted (line 14). Finally, the *for* loop at line 16 scans lexicographically the automaton states in order to reconstruct the BWT in the correct order. Notice that states can be scanned in lexicographic order with no overhead if the automaton has been implemented using direct hashing, as described in section 4. Moreover, in this step the BWT can be stored directly to disk, for no additional RAM consumption. The operation $BWT.append(\mathcal{A}[s].DS)$ at line 17 is implemented by appending the characters of $\mathcal{A}[s].DS$ one by one to the string BWT ($|\mathcal{A}[s].DS|$ access operations).

Time Complexity

The most expensive operations in the *for* loops are those at lines 10, 12, and 17 (insert, rank, and access, respectively). All other operations have cost $\mathcal{O}(1)$ (see Section 4). Assuming a uniform text distribution, the expected length of each dynamic string is $\mathcal{O}(n/\sigma^k) = \mathcal{O}(\sigma \log^2 n)$. This value, under the other two assumptions $\sigma \in \mathcal{O}(poly(w))$ and $w \in \Theta(\log n)$, is equal to $\mathcal{O}(polylog(n)) = \mathcal{O}(poly(w))$. These observations imply (Section 4) that the amortized cost of queries/updates on the dynamic strings is of $\mathcal{O}(H_k^*)$, so the following theorem holds:

Theorem 2. *The cw-bwt algorithm builds the BWT of a length n text in average time $\mathcal{O}(nH_k^*)$ using $nH_k^* + o(nH_k^*)$ bits of space, where $k = \log_\sigma(n/\log^2 n) - 1$ and $H_k^* = \min\{H_k + 1, \lceil \log \sigma \rceil\}$.*

The worst case scenario, on the other hand, is represented by a highly repetitive text T in which one or more k-contexts appear $\Theta(n)$ times. This results in the length of the corresponding dynamic strings being $\Theta(n) = \Theta(2^w)$, thus (see Section 4) in $\mathcal{O}(H_k^*(w/\log w)^2) = \mathcal{O}(H_k^*(\log n/\log \log n)^2)$ cost for queries/updates. The following holds:

Theorem 3. *The cw-bwt algorithm builds the BWT of a length n text in worst-case time $\mathcal{O}(nH_k^*(\log n/\log \log n)^2)$ using $nH_k^* + o(nH_k^*)$ bits of space, where $k = \log_\sigma(n/\log^2 n) - 1$ and $H_k^* = \min\{H_k + 1, \lceil \log \sigma \rceil\}$.*

6   Experiments

We tested our tool on datasets from the pizza&chili repository[1] and on the Human genome, build hg19 ($n \approx 3.2 \cdot 10^9$, $\Sigma_{DNA} = \{A, C, T, G, N\}$). In order

---

[1] http://pizzachili.dcc.uchile.cl/

Table 3. Performances of the tested tools on data coming from the pizza&chili repository. The column *Working space* accounts for both RAM and disk working space (input and output excluded). All input files have size 100MB. divsufsort required always 502 MB of working space while having essentially the same speed of dbwt, so it is not reported here. The RAM space of bwte was always forced to be upper-bounded by that of cw-bwt.

Tool	File	Alphabet size	RAM (MB)	Working space (MB)	Time (s)
cw-bwt	english	215	85	85	1376
	proteins	25	69	69	1162
	XML	96	74	74	1184
	sources	227	92	92	1337
	DNA	16	41	41	581
dbwt	english	215	243	243	18
	proteins	25	276	276	27
	XML	96	234	234	15
	sources	227	237	237	16
	DNA	16	197	197	15
bwte	english	215	85	123	193
	proteins	25	69	153	217
	XML	96	74	142	180
	sources	227	92	135	164
	DNA	16	41	121	313
SE-SAIS	english	215	135	614	75
	proteins	25	169	662	78
	XML	96	129	560	65
	sources	227	144	566	61
	DNA	16	129	538	66

to compare performances with other state-of-the-art tools, we also tested an implementation of the semi-external algorithm SE-SAIS[2], the suffix array construction algorithm divsufsort (both implemented in the SDSL library[9]), the external algorithm bwte [5], and a direct-bwt (dbwt) construction tool by Sadakane[2]. All tests were conducted on a intel i7 core, 2.4GHz machine running Ubuntu 14.04 operating system. Table 3 shows running times and memory requirements of all the tested tools. As expected, cw-bwt wins no prize in speed, due to its complicated data structures. On the other hand, cw-bwt is the only tool (among the ones tested) able to operate in *sublinear* working space (disk and RAM), making it useful in situations where both RAM and disk space are at a premium. The results obtained with bwte show that external-memory algorithms can beat cw-bwt running times, while using less RAM space (albeit more total working space). In order to prove the linear time complexity of our tool on semi-uniform inputs, we executed cw-bwt on several prefixes of the Human genome (build hg19), see Figure 1. As predicted by theory (Theorem 2), time complexity grows locally as $\mathcal{O}(n \log n)$, dropping down when k is automatically increased

---

[2] http://researchmap.jp/muuw41s7s-1587/#_1587

cw–bwt running times – Human genome

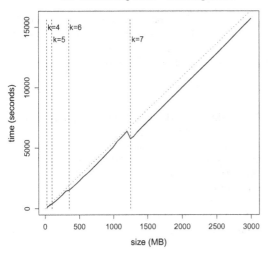

Fig. 1. Running times on genomic data (Human genome)

(which happens at exponential steps). This behaviour keeps the complexity always under a linear function of n—in Figure 1, the dotted line interpolating the first 4 peaks. cw-bwt completed the BWT of the Human genome in 4 hours and 37 minutes using only 994 MB of RAM (about 2.6 bits per input symbol, less than a plain encoding of the alphabet Σ_{DNA}). This improves the performances of state-of-the-art internal-memory tools: the bbwt tool published in [13] has a declared memory consumption of 1.4 GB, and dbwt and divsufsort required 5.8 GB and 14.6 GB, respectively. When compared also with external-memory algorithms, cw-bwt improved upon SE-SAIS, which required 3.62 GB of RAM and 13 GB of disk working space. Finally, bwte was executed allowing 1 GB of RAM and required additional 2.6 GB of disk working space, with times comparable to those of cw-bwt (3 hours).

7   Conclusions

Building the Burrows-Wheeler transform in small space is important in applications such as bioinformatics, due to the huge sizes of the genomes involved. In this paper we showed that, exploiting the semi-uniform distribution of genomic sequences, it is indeed possible to accomplish this goal in average linear time and compressed working space. Our implementation is general and we showed that also on other commonly used texts (english, code sources, XML), our tool reaches sublinear (compressed) space, showing no significant slowdown with respect to semi-uniform genomic texts. Despite being linear in text size (and acceptable for many practical applications), due to the use of more complex data structures, running times of cw-bwt are however still much higher than those of more memory-consuming tools. To overcome this problem, we are considering the possibility to parallelise our algorithm. From a theoretical point of view, we are also investigating the effect that using other kinds of automata on words would have on the data structures load distribution and, consequently, on the worst-case complexity of our algorithm.

References

1. Belazzougui, D.: Linear time construction of compressed text indices in compact space. arXiv preprint arXiv:1401.0936 (2014)
2. Beller, T., Zwerger, M., Gog, S., Ohlebusch, E.: Space-efficient construction of the burrows-wheeler transform. In: Kurland, O., Lewenstein, M., Porat, E. (eds.) SPIRE 2013. LNCS, vol. 8214, pp. 5–16. Springer, Heidelberg (2013)
3. de Bruijn, N.G., Erdos, P.: A combinatorial problem. Koninklijke Nederlandse Akademie v. Wetenschappen 49(49), 758–764 (1946)
4. Burrows, M., Wheeler, D.J.: A block-sorting lossless data compression algorithm (1994)
5. Ferragina, P., Gagie, T., Manzini, G.: Lightweight data indexing and compression in external memory. Algorithmica 63(3), 707–730 (2012)
6. Ferragina, P., Giancarlo, R., Manzini, G., Sciortino, M.: Boosting textual compression in optimal linear time. Journal of the ACM (JACM) 52(4), 688–713 (2005)
7. Ferragina, P., Manzini, G.: Opportunistic data structures with applications. In: Proceedings of the 41st Annual Symposium on Foundations of Computer Science, pp. 390–398. IEEE (2000)
8. Fredman, M., Saks, M.: The cell probe complexity of dynamic data structures. In: Proceedings of the Twenty-First Annual ACM Symposium on Theory of Computing, pp. 345–354. ACM (1989)
9. Gog, S., Beller, T., Moffat, A., Petri, M.: From theory to practice: plug and play with succinct data structures. In: Gudmundsson, J., Katajainen, J. (eds.) SEA 2014. LNCS, vol. 8504, pp. 326–337. Springer, Heidelberg (2014)
10. González, R., Navarro, G.: Rank/select on dynamic compressed sequences and applications. Theoretical Computer Science 410(43), 4414–4422 (2009)
11. He, M., Munro, J.I.: Succinct representations of dynamic strings. In: Chavez, E., Lonardi, S. (eds.) SPIRE 2010. LNCS, vol. 6393, pp. 334–346. Springer, Heidelberg (2010)
12. Huffman, D.A., et al.: A method for the construction of minimum redundancy codes. Proc. IRE 40(9), 1098–1101 (1952)
13. Lippert, R.A., Mobarry, C.M., Walenz, B.P.: A space-efficient construction of the burrows-wheeler transform for genomic data. Journal of Computational Biology 12(7), 943–951 (2005)
14. Manzini, G.: An analysis of the burrows-wheeler transform. Journal of the ACM (JACM) 48(3), 407–430 (2001)
15. Navarro, G.: Wavelet trees for all. Journal of Discrete Algorithms 25, 2–20 (2014)
16. Navarro, G., Mäkinen, V.: Compressed full-text indexes. ACM Computing Surveys (CSUR) 39(1), 2 (2007)
17. Navarro, G., Nekrich, Y.: Optimal dynamic sequence representations. SIAM Journal on Computing 43(5), 1781–1806 (2014)
18. Navarro, G., Sadakane, K.: Fully functional static and dynamic succinct trees. ACM Transactions on Algorithms (TALG) 10(3), 16 (2014)
19. Nong, G., Zhang, S., Chan, W.H.: Linear suffix array construction by almost pure induced-sorting. In: Data Compression Conference, DCC 2009, pp. 193–202. IEEE (2009)
20. Raman, R., Raman, V., Rao, S.S.: Succinct dynamic data structures. In: Dehne, F., Sack, J.-R., Tamassia, R. (eds.) WADS 2001. LNCS, vol. 2125, pp. 426–437. Springer, Heidelberg (2001)
21. Tischler, G.: Faster average case low memory semi-external construction of the burrows-wheeler transform. In: CEUR Workshop Proceedings, vol. 1146, pp. 61–68 (2014)

Backward Linearised Tree Pattern Matching

Jan Trávníček[1]([✉]), Jan Janoušek[1], Bořivoj Melichar[1], and Loek Cleophas[2,3]

[1] Faculty of Information Technology, Department of Theoretical Computer Science,
Czech Technical University in Prague, Thákurova 9, Praha 6,
160 00 Prague, Czech Republic
{Jan.Travnicek,Jan.Janousek,melichar}@fit.cvut.cz

[2] FASTAR Research Group, Department of Information Science,
Stellenbosch University, Private Bag X1 Matieland,
Stellenbosch 7602, Republic of South Africa
loek@fastar.org

[3] Natural and Formal Languages Research Group,
Department of Computer Science, Umeå University, 901 87 Umeå, Sweden

Abstract. We present a new backward tree pattern matching algorithm for ordered trees. The algorithm finds all occurrences of a single given tree pattern which match an input tree. It makes use of linearisations of both the given pattern and the input tree. The algorithm preserves the properties and advantages of standard backward string pattern matching approaches. The number of symbol comparisons in the backward tree pattern matching can be sublinear in the size of the input tree. As in the case of backward string pattern matching, the size of the bad character shift table used by the algorithm is linear in the size of the alphabet. We compare the new algorithm with best performing previously existing algorithms based on (non-linearised) tree pattern matching using finite tree automata or stringpath matchers and show that it outperforms these for single pattern matching.

Keywords: Tree pattern matching · Backward pattern matching · Tree processing · Tree linearisation

1 Introduction

Trees are one of the fundamental data structures used in Computer Science and the theory of formal tree languages has been extensively studied and developed since the 1960s [8,12]. Tree pattern matching on node-labelled trees is an important algorithmic problem with applications in many tasks such as compiler code selection, interpretation of nonprocedural languages, implementation of rewriting systems, or XML processing. Tree patterns are trees whose leaves can be labelled by a special wildcard, the nullary symbol S, which serves as a

This research has been supported in part by the Czech Science Foundation (GAČR) as project No. 13-03253S. This work is based on the research supported in part by the National Research Foundation of South Africa (Grant Number 88214).

© Springer International Publishing Switzerland 2015
A.-H. Dediu et al. (Eds.): LATA 2015, LNCS 8977, pp. 599–610, 2015.
DOI: 10.1007/978-3-319-15579-1_47

placeholder for any subtree. Since the linear notation of a subtree of a tree is a substring of the linear notation of that tree, the subtree matching and tree pattern matching problems are in many ways similar to the string pattern matching problem. We note that the tree pattern matching problem is more complex than the string matching one because there can be at most n^2 distinct substrings of a string of size n, whereas there can be at most $2^{n-1} + n$ distinct tree patterns which match a tree of size n.

As mentioned, trees can be linearised into strings. Such a linear notation can be obtained by a corresponding tree traversal. Moreover, every sequential algorithm on a tree traverses its nodes in a sequential order, which corresponds to some linear notation. Such a linear representation need not be built explicitly. Many algorithms have been proposed for exact string matching [3,9,10,16]. Among the most efficient of them are those based on backward string pattern matching, represented by the Boyer-Moore and Boyer-Moore-Horspool algorithms. Although backward string pattern matching's time complexity is generally $O(n*m)$ (for text and pattern size n and m respectively) in the worst case, due to such algorithms' ability to skip text parts, they often perform sublinearly in practice.

Many tree pattern matching algorithms exist as well [4,6,11,13]. Many of them use some kind of tree automata [6]. Cole et al. [7] use a subset matching approach, but at the cost of large auxiliary data structures. For unrestricted tree pattern sets, among the fastest algorithms in practice are algorithms based on deterministic frontier-to-root (bottom-up) tree automata (DFRTAs) [4,6,13] and on Hoffmann-O'Donnell-style stringpath matchers [1,13]. A few of these tree pattern matching algorithms use principles of matching patterns backwards: Hoffmann and O'Donnell refer to work by Lang et al. [15] that applies such an approach to leftmost stringpaths of trees and involves complications when dealing with nodes of arity greater than 2. [18] compares symbols of a pattern and an input subject tree upwards, with subsequent shifting of the pattern in the subject tree. That algorithm can skip nodes when it is known that no occurrence is skipped, but the tree is not linearised and therefore skipping is somewhat complicated. Our algorithm uses a linear representation of the subject tree where random access to symbols/positions is possible.

While modifying backward string pattern matching to backward subtree matching (searching for occurrences of given subtrees) is straightforward, this is not the case for backward tree pattern matching, where complications arise due to the use of nullary symbol S and matched subtrees being possibly recursively nested. In this paper, a new backward tree pattern matching algorithm is presented. The presented backward tree pattern matching algorithm preserves the properties and the advantages of the standard backward string pattern matching: the number of symbol comparisons in the backward tree pattern matching can be sublinear in n, the size of the subject tree. Based on the Boyer-Moore-Horspool algorithm, a modified bad character shift heuristic is used. As in the case of backward string pattern matching, the size of the bad character shift table used by the algorithm is linear with the size of the alphabet. Our experimental results confirm the properties of the algorithm and show that it outperforms the aforementioned DFRTAs and stringpath matchers.

2   Basic Notions

An *alphabet* is a finite nonempty set of *symbols*. A *ranked alphabet* is a finite nonempty set of symbols each of which has a unique nonnegative *arity* (or *rank*). Given a ranked alphabet $\mathcal{A}$, the arity of a symbol $a \in \mathcal{A}$ is denoted $Arity(a)$. The set of symbols of arity p is denoted by $\mathcal{A}_p$. Elements of arity $0, 1, 2, \ldots, p$ are called nullary (constants), unary, binary, ..., p-ary symbols, respectively. We assume that $\mathcal{A}$ contains at least one constant. In the examples we use numbers at the end of identifiers for a short declaration of symbols with arity. For instance, $a2$ is a short declaration of a binary symbol a.

A *string* x is a sequence of i symbols $s_1 s_2 s_3 \ldots s_i$ from a given alphabet, where i is the size of the string. A sequence of zero symbols is called the empty string. The empty string is denoted by symbol ε.

Based on concepts and notations from graph theory [2], a *rooted tree* t is an acyclic connected directed graph $t = (N, R)$ with a special node $r \in N$, called the *root*, such that (1) r has in-degree 0, (2) all other nodes of t have in-degree 1, and (3) there is just one path from the root r to every $f \in N$, where $f \neq r$. Nodes of a tree with out-degree 0 are called *leaves*. A *labelled and rooted tree* is a tree with the additional property that every node $f \in N$ is labelled by a symbol $a \in \mathcal{A}$, where $\mathcal{A}$ is an alphabet. A node g is a *direct descendant* of node f if a pair $(f, g) \in R$. A *ranked, labelled and rooted tree* is a tree labelled by symbols from a ranked alphabet and where the out-degree of a node f labelled by symbol $a \in \mathcal{A}$ equals $Arity(a)$. Nodes labelled by nullary symbols (constants) are leaves. An *ordered, ranked, labelled and rooted tree* is a tree where direct descendants $a_{f1}, a_{f2}, \ldots, a_{fn}$ of a node a_f having an $Arity(a_f) = n$ are ordered.

The *prefix notation* $pref(t)$ of a tree t is defined as follows:

1. $pref(a) = a0$ if a is a leaf,
2. $pref(t) = an\; pref(b_1)\; pref(b_2) \ldots pref(b_n)$, where a is the root of tree t, $n = Arity(a)$ and $b_1, b_2, \ldots b_n$ are direct descendants of a.

The *prefix bar notation* $pref\_bar(t)$ of a tree t is defined as follows:

1. $pref\_bar(a) = a \uparrow$ if a is a leaf,
2. $pref\_bar(t) = a\; pref\_bar(b_1)\; pref\_bar(b_2) \ldots pref\_bar(b_n) \uparrow$, where a is the root of tree t and $b_1, b_2, \ldots b_n$ are direct descendants of a.

Example 1. Consider a ranked alphabet $\mathcal{A} = \{a2, a1, a0\}$. Consider an ordered, ranked, labelled and rooted tree $t_{1r} = (\{a2_1, a2_2, a0_3, a1_4, a0_5, a1_6, a0_7\}, R_{1r})$ over alphabet $\mathcal{A}$, where $R_{1r} = \{(a2_1, a2_2), (a2_1, a1_6), (a2_2, a0_3), (a2_2, a1_4), (a1_4, a0_5), (a1_6, a0_7)\}$. Tree t_{1r} in prefix notation is $pref(t_{1r}) = a2\; a2\; a0\; a1\; a0\; a1\; a0$. Trees can be represented graphically, as is done for tree t_{1r} in Figure 1(a).   □

Example 2. Consider an unranked alphabet $\mathcal{A} = \{a\}$. Consider an ordered, labelled and rooted tree $t_{1u} = (\{a_1, a_2, a_3, a_4, a_5, a_6, a_7\}, R_{1u})$ over an alphabet $\mathcal{A}$, where $R_{1u} = \{(a_1, a_2), (a_1, a_6), (a_2, a_3), (a_2, a_4), (a_4, a_5), (a_6, a_7)\}$. Tree t_{1u} in prefix bar notation is $pref\_bar(t_{1u}) = a\; a\; a \uparrow\; a\; a \uparrow \uparrow \uparrow\; a\; a \uparrow \uparrow \uparrow$. The tree t_{1u} is illustrated in Figure 1(b).   □

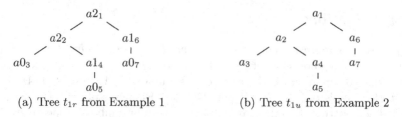

(a) Tree t_{1r} from Example 1   (b) Tree t_{1u} from Example 2

Fig. 1. Tree t_{1r} over a ranked alphabet (left), and the same tree t_{1u} over an unranked alphabet (right) from Examples 1 and 2

To define a *tree pattern*, we use a special wildcard symbol $S \notin \mathcal{A}$, $Arity(S) = 0$, which serves as a placeholder for any subtree. A tree pattern is defined as a labelled ordered tree over an alphabet $\mathcal{A} \cup \{S\}$. We will assume that the tree pattern contains at least one node labelled by a symbol from $\mathcal{A}$. A tree pattern containing at least one symbol S will be called a *tree template*. A tree pattern p with $k \geq 0$ occurrences of the symbol S *matches* a subject tree t at node n if there exist subtrees $t_1, t_2, \ldots, t_k$ (not necessarily the same) of t such that the tree p', obtained from p by substituting the subtree t_i for the i-th occurrence of S in p, $i = 1, 2, \ldots, k$, is equal to the subtree of t rooted at n.

Example 3. Consider a tree $t_{1r} = (\{a2_1, a2_2, a0_3, a1_4, a0_5, a1_6, a0_7\}, R_{1r})$ from Example 1, which is illustrated in Figure 1(a). Consider a subtree p_{1r} over alphabet $\mathcal{A}, p_{1r} = (\{a2_1, a0_2, a1_3, a0_4\}, R_{p1})$. Subtree p_{1r} in prefix notation is $pref(p_{1r}) = a2\ a0\ a1\ a0$ and $R_{p1} = \{((a2_1, a0_2), (a2_1, a1_3)), ((a1_3, a0_4))\}$. Consider a tree pattern p_{2r} over alphabet $\mathcal{A} \cup \{S\}$, $p_{2r} = (\{a2_1, S_2, a1_3, S_4\}, R_{p2})$. Tree pattern p_{2r} in prefix notation is $pref(p_{2r}) = a2\ S\ a1\ S$ and $R_{p2} = \{(a2_1, S_2), (a2_1, a1_3), (a1_3, S_4)\}$. Tree patterns p_{1r} and p_{2r} are illustrated in Figure 2. Tree pattern p_{1r} occurs once in tree t_{1r} — it matches at node 2 of t_{1r}. Tree pattern p_{2r} occurs twice in t_{1r} — it matches at nodes 1 and 2 of t_{1r}. $\square$

(a) Subtree p_{1r} from Example 3   (b) Tree pattern p_{2r} from Example 3

Fig. 2. Subtree and tree pattern over ranked alphabet from Example 3

In backward string pattern matching the symbols of the pattern and the text are compared in opposite direction to the shifting of the pattern.

Instead of a shift by 1 (as per line 8 of Alg. 1), lager shifts can often be made. One heuristic for larger shifts is that of the Boyer-Moore-Horspool algorithm [14], computing the length of the shift based on one symbol aligned to the end of the pattern. This shift, a simplification of the one used by the original Boyer-Moore algorithm, has turned out to perform very well in practice. The shifts are stored in a *bad character shift table*. Given a pattern of size m ($pattern[1..m]$) over an alphabet $\mathcal{A}$, the bad character shift table $BCS(pattern[1..m])[a] = \min(\{m\} \cup \{j : pattern[m - j] = a\ and\ j > 1\})$ for each $a \in \mathcal{A}$.

Name: Basic backward PM.

Input: A string *text* of size n and a string *pattern* of size m.

Output: Locations of the *pattern* in the *text*.

```
1  begin
2  |   i := 0
3  |   while i < n − m do
4  |   |   j := m
5  |   |   while j > 0 and pattern[j] = text[i + j] do
6  |   |   |   j := j − 1
7  |   |   end
8  |   |   if j = 0 then output(i + 1) i := i + 1 {Length of the shift.}
9  |   end
10 end
```

Algorithm 1. Basic backward string pattern matching algorithm

3   Backward Tree Pattern Matching Algorithm

The problem of tree pattern matching can be seen as matching connected subgraphs in trees. Tree patterns in a linear notation are represented by substrings of trees in the linear notation but they can contain gaps given by special wildcards S, which serve as placeholders for any subtrees. The basic idea of backward tree pattern matching for tree patterns is the same as in the string case: moving the pattern in one direction and matching symbols of tree pattern and subject tree in the opposite direction. Wildcard S occurrences must be handled in a special way. For this purpose we use a prefix ranked bar notation of the tree.

Definition 4. *The* prefix ranked bar notation *$pref\_ranked\_bar(t)$ of a tree t is defined as follows:*

1. $pref\_ranked\_bar(S) = S \uparrow S$

2. $pref\_ranked\_bar(a) = a0 \uparrow 0$ if a is a leaf,

3. $pref\_ranked\_bar(t) = an \; pref\_ranked\_bar(b_1) \; pref\_ranked\_bar(b_2) \ldots$
   $pref\_ranked\_bar(b_n) \uparrow n$, where a is the root of the tree t, $n = Arity(a)$ and $b_1, b_2, \ldots b_n$ are direct descendants of a.

Definition 5. *Let $\uparrow n$, where $n \geq 0$ be bar symbols of arity n. The* bar set $\mathcal{A}_\uparrow$ *is the set of all bar symbol $\uparrow n$.*   □

Definition 6. *Let $pattern[1..m]$ be a $pref\_ranked\_bar$ notation of a tree pattern p over an alphabet $\mathcal{A}$. The* bad character shift table $BCS(pattern[1..m])$ *for backward tree pattern matching is defined for each $a \in \mathcal{A}$:*

$BCS(pattern[1..m])[a] = \min($

$\quad \{m\} \cup \{j : pattern[m - j] = a \text{ and } m > j > 0\} \cup$

$\quad \{j + Arity(a) * 2 : pattern[m - j] = S \text{ and } m > j > 0 \text{ and } a \notin \mathcal{A}_\uparrow\} \cup$

$\quad \{j - 1 : pattern[m - j] = S \text{ and } m > j > 1 \text{ and } a \in \mathcal{A}_\uparrow\})$ □

Note that there is no value for the wildcard S in shift table BCS because this symbol cannot occur in the subject tree.

Items of the BCS table are computed as the minimum value from four formulas, see Def. 6 where the formulas are separated by the union operation. The first formula makes sure that the shift is not longer than the size of the pattern m. In this case the size of a subtree corresponding to wilcard S is considered to be the smallest possible one, i.e. 2 (the size of subtree consisting of one nullary symbol $a0 \uparrow 0$). The second formula defines the minimal safe shift for symbols that occur in the pattern. The minimal safe shift for a symbol a is the distance j of the closest occurrence of the symbol a from the end of the pattern. Nullary symbol S is considered to correspond to the smallest possible subtree again.

The third and fourth formulas define the shift for cases when a symbol a is expected to be in a subtree t_e that corresponds to wildcard S. The location of the last wildcard S from the end of the pattern is used to define the base shift length j and this shift can be prolonged by some number depending on the arity of the symbol a, see the second part of the definition. The smallest subtree t_e that contains the symbol a is rooted by a and its direct descendants are nullary symbols $b0$. For each symbol $b0$ in the subtree t_e there is also one symbol $\uparrow 0$. The base shift j is then prolonged by $2 * Arity(a)$. Any symbol from the set $\mathcal{A}_\uparrow$ can occur as the last symbol of a subtree t_e, ie. it can be matched with $\uparrow S$. Therefore, the base shift of each bar is shortened by 1, see fourth part of the definition. The shift cannot be zero and in that case the base shift is not shortened. Note that this case would occur only for pattern $S \uparrow S$.

Firstly, Alg. 7 for the construction of BCS table finds the location of the last wildcard S. Then, the BCS table for all symbols of the alphabet is initialised

---

Name: ConstructBCS.

Input: Tree *pattern* in prefix ranked bar notation $pref\_ranked\_bar(pattern)$ of size m over alphabet $\mathcal{A}$ of the *subject* tree.

Output: The bad character shift table $BCS(pref\_ranked\_bar(pattern))$.

```
 1 begin
 2 │   s := m
 3 │   for i := 1 to m do
 4 │   │   if pref_ranked_bar(pattern)[i] = S then s = m − i
 5 │   end
 6 │   foreach x ∈ A do BCS[x] = m
 7 │   foreach x ∈ A do
 8 │   │   if x ∉ A↑ then shift := s + Arity(x) * 2 else if s >= 2 then
 9 │   │   shift := s − 1 else shift := s if BCS[x] > shift then BCS[x] := shift
   │   end
10 │   for i := 1 to m − 1 do
11 │   │   if pref_ranked_bar(pattern)[i] ∉ {S, ↑S} and
   │   │   BCS[pref_ranked_bar(pattern)[i]] > (m − i) then
   │   │   BCS[pref_ranked_bar(pattern)[i]] := m − i
12 │   end
13 end
```

Algorithm 2. Construction of BCS table

to the size of the pattern. The length of the shift for all symbols of the alphabet is possibly shortened with the use of the information on the position of the last wildcard S. The arity of symbols is used to make this part of the shift function longer according to Def. 6. Finally, the length of the shift is again possibly shortened by the actual positions of symbols in the pattern.

Example 7. Consider a tree pattern p_{3r} in prefix ranked bar notation $pref\_ranked\_bar(p_{3r}) = a2\ a1\ S\ {\uparrow}S\ {\uparrow}1\ a1\ a0\ {\uparrow}0\ {\uparrow}1\ {\uparrow}2$ over an alphabet $\mathcal{A} = \{a3, a2, a1, a0, S, {\uparrow}3, {\uparrow}2, {\uparrow}1, {\uparrow}0, {\uparrow}S\}$. Alg. 7 constructs the following items of the BCS table.
$BCS[a3] = \min(\{10\} \cup \emptyset \cup \{13\}) = 10$, $BCS[a2] = \min(\{10\} \cup \{9\} \cup \{11\}) = 9$, $BCS[a1] = \min(\{10\} \cup \{4, 8\} \cup \{9\}) = 4$, $BCS[a0] = \min(\{10\} \cup \{3\} \cup \{7\}) = 3$, $BCS[{\uparrow}3] = \min(\{10\} \cup \emptyset \cup \{6\}) = 6$, $BCS[{\uparrow}2] = \min(\{10\} \cup \emptyset \cup \{6\}) = 6$, $BCS[{\uparrow}1] = \min(\{10\} \cup \{1, 5\} \cup \{6\}) = 1$, $BCS[{\uparrow}0] = \min(\{10\} \cup \{2\} \cup \{6\}) = 2$. □

The backward tree pattern matching algorithm uses an additional structure subtree jump table (SJT) to efficiently skip subtrees corresponding to S. This structure contains two kinds of positions for each subtree r of a tree t. The first kind of position is the position of the first symbol of the subtree r in $pref\_ranked\_bar(r)$ notation in the $pref\_ranked\_bar(t)$ notation of the tree t as an index and the position one after the last symbol of the subtree r in $pref\_ranked\_bar(r)$ notation as a value. The second one is the position of the last symbol of the subtree r in $pref\_ranked\_bar(r)$ notation in the $pref\_ranked\_bar(t)$ notation of the tree t as an index and the position one before the first symbol of the subtree r in $pref\_ranked\_bar(r)$ notation as a value. This structure has the same size as the $pref\_ranked\_bar$ notation and its construction by Alg. 3 is very fast without performing any time–consuming comparisons of symbols (labels).

Definition 8. Let t and $pref\_ranked\_bar(t)$ of length n be a tree and its prefix ranked bar notation, respectively. A *subtree jump table* $SJT(pref\_ranked\_bar(t))$ is defined as a mapping from set of integers $\{1..n\}$ into a set of integers $\{0..n + 1\}$. If $pref\_ranked\_bar(t)\ [i..j]$ is the $pref\_ranked\_bar$ notation of a subtree of tree t, then $SJT(pref\_ranked\_bar(t))[i] = j + 1$ and $SJT(pref\_ranked\_bar(t))[j] = i - 1$, $1 \leq i < j \leq n$.

Lemma 9. Given $pref\_ranked\_bar(t)$ and $rootIndex$ equal to 1, Algorithm 3 constructs subtree jump table $SJT(pref\_ranked\_bar(t))$. □

Example 10. Consider a tree t_{2r} in prefix ranked bar notation $pref\_ranked\_bar$ $(t_{2r}) = a2\ a2\ a0\ {\uparrow}0\ a0\ {\uparrow}0\ {\uparrow}2\ a2\ a0\ {\uparrow}0\ a0\ {\uparrow}0\ {\uparrow}2\ {\uparrow}2$ over alphabet $\mathcal{A} = \{a3, a2, a1, a0, {\uparrow}3, {\uparrow}2, {\uparrow}1, {\uparrow}0\}$. Table 1 shows the $SJT(pref\_ranked\_bar(t_{2r}))$. □

Our backward tree pattern matching algorithm, shown in Alg. 4, is an extension of the string backward pattern matching algorithm, shown in Alg. 1.

The modification of the string backward matching algorithm is based on the principle that the algorithm performs also tests for wildcards S in the pattern.

Name: ConstructSJT

Input: Tree t in prefix notation $pref\_ranked\_bar(t)$ of length n, index of current node $rootIndex$ default is 1, reference to an empty subtree jump table $SJT(pref\_ranked\_bar(t))$ of length n

Output: index $exitIndex$, subtree jump table $SJT(pref\_ranked\_bar(t))$

```
 1 begin
 2 |   index := rootIndex + 1
 3 |   for i = 1 to Arity(pref_ranked_bar(t)[rootIndex]) do
 4 |   |   index :=
   |   |   ConstructSJT(pref_ranked_bar(t), index, SJT(pref_ranked_bar(t)))
 5 |   end
 6 |   index := index + 1
 7 |   SJT(pref_ranked_bar(t))[rootIndex] = index
 8 |   SJT(pref_ranked_bar(t))[index] = rootIndex - 1
 9 |   return index
10 end
```

Algorithm 3. Construction of subtree jump table

Table 1. Subtree jumping table $SJT(pref\_ranked\_bar(t_{2r}))$ of tree t_{2r}

1	2	3	4	5	6	7	8	9	10	11	12	13	14
$a2$	$a2$	$a0$	$\uparrow0$	$a0$	$\uparrow0$	$\uparrow2$	$a2$	$a0$	$\uparrow0$	$a0$	$\uparrow0$	$\uparrow2$	$\uparrow2$
15	8	5	2	7	4	1	14	11	8	13	10	7	0

Name: BackwardLTPM.

Input: The $subject$ tree in $pref\_ranked\_bar(subject)$ notation of size n, the tree $pattern$ in $pref\_ranked\_bar(pattern)$ notation of size m, $SJT(pref\_ranked\_bar(subject))$, and $BCS(pref\_ranked\_bar(pattern))$.

Output: Locations of occurrences of the pattern $pattern$ in the tree $subject$.

```
 1 begin
 2 |   i := 0
 3 |   while i <= (n - m) do
 4 |   |   j := m
 5 |   |   position := i + j
 6 |   |   while j > 0 and position > 0 do
 7 |   |   |   if pref_ranked_bar(subject)[position] = pref_ranked_bar(pattern)[j]
   |   |   |   then
 8 |   |   |   |   position := position - 1
 9 |   |   |   else if pref_ranked_bar(pattern)[j] = ↑S and
   |   |   |   pref_ranked_bar(subject)[position] ∈ A↑ then
10 |   |   |   |   position := SJT(pref_ranked_bar(subject))[position]
11 |   |   |   |   j = j - 1 {Subtree skip}
12 |   |   |   else break j := j - 1
13 |   |   end
14 |   |   if j = 0 then output(position + 1)
   |   |   i := i + BCS[pref_ranked_bar(subject)[i + m]]
15 |   end
16 end
```

Algorithm 4. Backward tree pattern matching algorithm

The modification is in line 10 of Alg. 4, where a part of the subject tree representing a subtree is skipped when a wildcard S, represented as $S \uparrow S$, is processed. Also, two indexes, one to the pattern and the other one to the text, are needed because subtrees (which need to be skipped) are often longer than two symbols.

Theorem 11. *Given a tree pattern p in prefix ranked bar notation and shift table $BCS(pref\_ranked\_bar(p))$ constructed by Alg. 2, Alg. 4 correctly computes the locations of all occurrences of the pattern p in an input tree t.*

Proof. Backward tree pattern matching algorithm is an extension of the backward string pattern matching algorithm. It is to be proved that shifting using $BCS(pref\_ranked\_bar(t))$ cannot skip any occurrence of the tree pattern p. Let $c \in \mathcal{A}$. Assume that there is an occurrence of p located at position i, $0 < i < BCS(pref\_ranked\_bar(t))[c]$. A symbol c must then be located at some position i either directly or as part of a subtree that corresponds to a wildcard S. According to Def. 6, the $BCS(pref\_ranked\_bar(p))[c]$ is derived from the last occurrence of symbol c in the prefix ranked bar notation of the pattern $pref\_ranked\_bar(p)$, hence we get a contradiction, and from the last occurrence of symbol S and its bar $\uparrow S$. If the symbol c is located in the subtree that corresponds to a wildcard S, then the the shift is already computed from the smallest possible subtree containing the symbol c. Hence, pattern p cannot occur at position i. Therefore, no occurrence of p can be skipped by the algorithm.   □

Example 12. Consider a tree pattern p_{4r} in the prefix ranked bar notation $pref\_ranked\_bar(p_{4r}) = a2\ S\ \uparrow S\ S\ \uparrow S\ \uparrow 2$ over an alphabet $\mathcal{A} = \{a3, a2,$ $a1, a0, S, \uparrow 3, \uparrow 2, \uparrow 1, \uparrow 0, \uparrow S\}$ and a tree t_{2r} in the prefix ranked bar notation $pref\_ranked\_bar\ (t_{2r}) = a2\ a2\ a0\ \uparrow 0\ a0\ \uparrow 0\ \uparrow 2\ a2\ a0\ \uparrow 0\ a0\ \uparrow 0\ \uparrow 2\ \uparrow 2$ over an alphabet $\mathcal{A} = \{a3, a2, a1, a0, \uparrow 3, \uparrow 2, \uparrow 1, \uparrow 0\}$. The $BCS[a3] = 6$, $BCS[a2] = 5$, $BCS[a1] = 4$, $BCS[a0] = 2$, $BCS[\uparrow 3] = 1$, $BCS[\uparrow 2] = 1$, $BCS[\uparrow 1] = 1$, $BCS[\uparrow 0] = 1$. A run of Alg. 4 is depicted in Table 2. Longer subtrees in place of wildcards S are denoted by $S\rightarrow \leftarrow S$.   □

The run of Alg. 4 for Example 12 starts at position 6 of the $pref\_ranked\_bar$ (t_{2r}). Mismatch of $\uparrow 2$ and $\uparrow 0$ results in subsequent shift by 1 symbol to align $\uparrow 0$

Table 2. Trace of the run of Alg. 4 for subject tree t_{2r} and tree pattern p_{4r}

1	2	3	4	5	6	7	8	9	10	11	12	13	14	
a2	a2	a0	↑0	a0	↑0	↑2	a2	a0	↑0	a0	↑0	↑2	↑2	$pref\_ranked\_bar(t_{2r})$
14	6	2	2	2	2	6	6	2	2	2	2	6	14	$subtree\_sizes(t_{2r})$
					↑2									$\uparrow 0 \neq \uparrow 2, shift = 1$
	a2	S→	←S	S→	←S	↑2								match, $shift = 1$
						↑2								$a2 \neq \uparrow 0, shift = 5$
							a2	S→	←S	S→	←S	↑2		match, $shift = 1$
a2	S→				←S	S→						←S	↑2	match, $shift = 1$

with position of the end of the last wildcard S in the $pref\_ranked\_bar(p4r)$. The algorithm recognises pattern match on positions 2 to 7 and shift is by 1 symbol to align $\uparrow 2$ again with the end of the last wildcard S in $pref\_ranked\_bar(p4r)$. Mismatch of $\uparrow 2$ and $a2$ results in a shift by 5 symbol where $a2$ is not only aligned with $a2$ but also with position closes to the end of the pattern where $a2$ can be as a part of the last wildcard S. Another match is recognised and the shift is by 1 symbol where another occurrence is recognised and subsequent shift is to outside of the $pref\_ranked\_bar(t_{2r})$ resulting in the end of the run of the Alg. 4.

The BCS table is the only data structure needed for the algorithm and its size is $\Theta(\mathcal{A})$, where $\mathcal{A}$ is the alphabet size. The preprocessing time is $O(m + \mathcal{A})$, where m is the pattern length and $\mathcal{A}$ is the alphabet size.

Backward string pattern matching is known to perform sublinear number of comparisons of symbols on average. The modification to backward tree pattern matching requires the input tree to be read in prefix ranked bar notation. However, the algorithm still performs $\Omega(\frac{n}{m})$ comparisons of symbols, where n is the size of the input subject tree and m is the size of the given tree pattern and $O(n * m)$ comparisons of symbols as in the case of the backward string pattern matching. The lengths of the shifts depend on the position of the last wildcard S in the pattern p – the closer to the end of the pattern the last occurrence of symbol S is, the longer are the shifts performed.

4   Some Empirical Results

We have implemented our algorithm by extending the existing Forest FIRE toolkit and accompanying FIRE Wood graphical user interface [5,17]. This toolkit already implemented many tree pattern matching algorithms and constructions of automata used in them, but no algorithms based on linearisations of both pattern tree(s) and subject tree. Constructions included in Forest FIRE include ones described in [1,4,6,13] and others. We compared our algorithm's performance to some of the best-performing ones in Forest FIRE, according to the results in [6]. We compared the running times of the search phase of the following algorithms: 1) our new backward tree pattern matching algorithm based on linearisations of pattern and subject tree (BLTPM); 2) an algorithm based on the use of a *deterministic frontier-to-root (bottom-up) tree automaton* constructed for the pattern (DFRTA); and 3) an algorithm based on the use of a *Aho-Corasick automaton* constructed for the pattern's stringpath set (AC).

The comparison was done using a pattern set previously used for benchmarking Forest FIRE. This pattern set was obtained by taking the Mono project's X86 instruction set grammar and, for each grammar production, taking the tree in the production's right hand side, and replacing any nonterminal occurrences by wildcard symbol occurrences. The resulting pattern set consists of 460 tree patterns of varying sizes. Since our pattern matching algorithm is a single-pattern one, we ran each of the algorithms for each pattern individually, and sequentially ran it over each subject tree from two sets of subject trees: a set of 150 trees of approximately 500 nodes each, and a set of 500 trees of approximately 150 nodes each. Both of these sets had previously been used for benchmarking Forest FIRE. Benchmarking was conducted on a 2 GHz Intel Core i7 with 16 GB of RAM running OpenSUSE GNU/Linux version 13.1 using Java SE 7.

Linearised versions of the tree patterns and subject trees were constructed from the in-memory tree representations, using additional memory. However, this is linear in the size of the tree representations, while the shift tables used will typically be much smaller than the automata used in the other algorithms. Because of this and because search time was our primary concern, we do not consider memory use. Figures 3(a) and 3(b) show the search times as boxplots, clearly showing that on average, our new algorithm considerably outperforms existing ones for the single-pattern case (note the logarithmic scale).

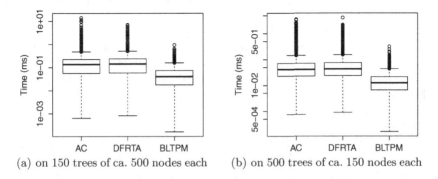

(a) on 150 trees of ca. 500 nodes each   (b) on 500 trees of ca. 150 nodes each

Fig. 3. Distributions of pattern matching times for the respective algorithms

5   Concluding Remarks

We presented a new backward tree pattern matching algorithm that uses a modified bad character shift table and works on trees in prefix ranked bar notation. The algorithm may perform sublinear number of comparisons of symbols (labels) in respect to size of the subject tree and performs well in practice. Future work should involve the investigation of other shift heuristics that might result in longer shifts. Furthermore, we are working on extensions of our algorithm to the case of multiple tree patterns.

References

1. Aho, A.V., Ganapathi, M., Tjiang, S.W.K.: Code generation using tree matching and dynamic programming. ACM Trans. Program. Lang. Syst., 491–516 (1989)
2. Aho, A.V., Ullman, J.D.: The theory of parsing, translation, and compiling. Prentice-Hall (1972)

3. Charras, C., Lecroq, T.: Handbook of exact string matching algorithms. Kings College Publications (2004)

4. Chase, D.R.: An improvement to bottom-up tree pattern matching. In: POPL, pp. 168–177. ACM Press (1987)

5. Cleophas, L.: Forest FIRE and FIRE wood: tools for tree automata and tree algorithms. In: Piskorski, J., Watson, B.W., Yli-Jyrä, A. (eds.) FSMNLP. Frontiers in Artificial Intelligence and Applications, vol. 19, pp. 191–198. IOS Press (2008)

6. Cleophas, L.: Tree Algorithms: Two Taxonomies and a Toolkit. Ph.D. thesis, Department of Mathematics and Computer Science, Eindhoven University of Technology (April 2008)

7. Cole, R., Hariharan, R., Indyk, P.: Tree pattern matching and subset matching in deterministic o($nlog^3n$) time. In: Proceedings of the 10th ACM-SIAM Symposium on Discrete Algorithms, pp. 245–254 (1999)

8. Comon, H., Dauchet, M., Gilleron, R., Jacquemard, F., Lugiez, D., Tison, S., Tommasi, M.: Tree automata: Techniques and applications (2007). http://www.grappa.univ-lille3.fr/tata/ (release October 12, 2007)

9. Crochemore, M.A., Rytter, W.: Jewels of Stringology. World Scientific Publishing Company (2003)

10. Faro, S., Lecroq, T.: The exact online string matching problem: A review of the most recent results. ACM Comput. Surv. **45**(2), 13 (2013)

11. Flouri, T., Janoušek, J., Melichar, B., Iliopoulos, C.S., Pissis, S.P.: Tree template matching in ranked ordered trees by pushdown automata. In: Bouchou-Markhoff, B., Caron, P., Champarnaud, J.-M., Maurel, D. (eds.) CIAA 2011. LNCS, vol. 6807, pp. 273–281. Springer, Heidelberg (2011)

12. Gécseg, F., Steinby, M.: Tree languages. In: Handbook of Formal Languages, vol. 3, pp. 1–68. Springer (1997)

13. Hoffmann, C.M., O'Donnell, M.J.: Pattern matching in trees. Journal of the ACM **29**(1), 68–95 (1982)

14. Horspool, R.N.: Practical fast searching in strings. Software Practice and Experience **10**(6), 501–506 (1980)

15. Lang, H.W., Schimmler, M., Schmeck, H.: Matching tree patterns sublinear on the average. Christian-Albrechts-Universität, Tech. rep. (1980)

16. Smyth, W.F.: Computing Patterns in Strings. Addison-Wesley-Pearson Education Limited (2003)

17. Strolenberg, R.: ForestFIRE & FIREWood, A Toolkit & GUI for Tree Algorithms. Master's thesis, Department of Mathematics and Computer Science, Eindhoven University of Technology (June 2007). http://alexandria.tue.nl/extra1/afstversl/wsk-i/strolenberg2007.pdf

18. Watson, B.W.: A boyer-moore (or watson-watson) type algorithm for regular tree pattern matching. In: Stringology, pp. 33–38 (1997)

BFS-Based Symmetry Breaking Predicates for DFA Identification

Vladimir Ulyantsev$^{(\boxtimes)}$, Ilya Zakirzyanov, and Anatoly Shalyto

ITMO University, Saint-Petersburg, Russia
{ulyantsev,zakirzyanov}@rain.ifmo.ru, shalyto@mail.ifmo.ru

Abstract. It was shown before that the NP-hard problem of deterministic finite automata (DFA) identification can be translated to Boolean
satisfiability (SAT). Modern SAT-solvers can efficiently tackle hard DFA
identification instances. We present a technique to reduce SAT search
space by enforcing an enumeration of DFA states in breadth-first search
(BFS) order. We propose symmetry breaking predicates, which can be
added to Boolean formulae representing various DFA identification problems. We show how to apply this technique to DFA identification from
both noiseless and noisy data. The main advantage of the proposed approach is that it allows to exactly determine the existence or non-existence
of a solution of the noisy DFA identification problem.

Keywords: Grammatical inference · Boolean satisfiability · Learning
automata · Symmetry breaking techniques

1 Introduction

Deterministic finite automata (DFA) are models that recognize regular languages [1], therefore the problem of DFA identification (induction, learning) is
one of the best studied [2] in grammatical inference. The identification problem
consists of finding a DFA with minimal number of states that is consistent with
a given set of strings with language attribution labels. This means that such
a DFA rejects the negative example strings and accepts the positive example
strings. It was shown in [3] that finding a DFA with a given upper bound on its
size (number of states) is an NP-complete problem. Besides, in [4] it was shown
that this problem cannot be approximated within any polynomial.

Despite this theoretical difficulty, several efficient DFA identification algorithms exist [2]. The most common approach is the evidence driven state-merging
(EDSM) algorithm [5]. The key idea of this algorithm is to first construct an augmented prefix tree acceptor (APTA), a tree-shaped automaton, from the given
labeled strings, and then to iteratively apply a state-merging procedure until
no valid merges are left. Thus EDSM is a polynomial-time greedy method that
tries to find a good local optimum. EDSM participated in the Abbadingo DFA
learning competition [5] and won it (in a tie). To improve the EDSM algorithm
several specialized search procedures were proposed, see, e.g., [6,7]. One of the

© Springer International Publishing Switzerland 2015
A.-H. Dediu et al. (Eds.): LATA 2015, LNCS 8977, pp. 611–622, 2015.
DOI: 10.1007/978-3-319-15579-1_48

most successful approaches is the EDSM algorithm in the red-blue framework [5], also called the Blue-fringe algorithm.

The second approach for DFA learning is based on evolutionary computation; early work includes [8,9]. Later the authors of [10] presented an effective scheme for evolving DFA with a multi-start random hill climber, which was used to optimize the transition matrix of the identified DFA. A so-called smart state labeling scheme was applied to choose the state labels optimally, given the transition matrix and the training set. Authors emphasized that smart selection of state labels gives the evolutionary method a significant boost which allowed it to compete with the EDSM. Authors find that the proposed evolutionary algorithm (EA) outperforms the EDSM algorithm on small target DFAs when the training set is sparse. For larger DFAs with 32 states, the hill climber fails and EDSM then clearly outperforms it.

The challenge of the GECCO 2004 Noisy DFA competition [11] was to learn the target DFA when 10 percent of the given training string labels had been randomly flipped. In [12] Lucas and Reynolds show that within limited time EA with smart state labeling is able to identify the target DFA even at such high noise level. Authors compared their algorithm with the results of the GECCO competition and found that EA clearly outperformed all the entries. Thereby it is the state-of-the-art technique for learning DFA from noisy training data.

In several cases the best solution for noiseless DFA identification is the *translation-to-SAT* technique [13], which was altered to suit the *StaMInA* (State Machine Inference Approaches) competition [14] and ultimately won. The main idea of that algorithm is to translate the DFA identification problem to Boolean satisfiability (SAT). Thus we are able to use highly optimized modern DPLL-style SAT solving techniques [15]. The translation-to-SAT approach was also used to efficiently tackle problems such as bounded model checking [16], solving SQL constraints by incremental translation [17], analysis of JML-annotated Java sequential programs [18], extended finite-state machine induction [19].

Many optimization problems exhibit symmetries – groups of solutions which can be obtained from each other via some simple transformations. To speed up the solution search process we can reduce the problem search space by performing *symmetry breaking*. In DFA identification problems the most straightforward symmetries are groups of isomorphic automata. The idea of avoiding isomorphic DFAs by fixing state numbers in breadth-first search (BFS) order was used in the state-merging approach [20] (function NatOrder) and in the genetic algorithm from [21] (*Move To Front* reorganization). Besides, in [13] symmetry breaking was performed by fixing some colors of the APTA vertices from a clique provided by a greedy *max-clique* algorithm was applied in a preprocessing step of translation-to-SAT technique.

In this paper we propose new symmetry breaking predicates [15] which can be added to Boolean formulae representing various DFA identification problems. These predicates enforce DFA states to be enumerated in BFS order. Proposed predicates cannot be applied with the max-clique technique [13] at the same time, but our approach is more flexible. To show the flexibility of the approach, we draw our attention to the case of noisy DFA identfication. Therefore we propose a modification of the noiseless translation-to-SAT for the noisy case (Section 3). We show that the previously proposed max-clique technique is not applicable in this case while our BFS-based approach is. The main advantage of our approach is that we can determine existence or non-existence of a solution in this case. Experiments showed that using BFS-based symmetry breaking predicates can significantly reduce the time of algorithm execution. Also we show that our strategy outperforms the current state-of-the-art EA from [12] if the number of the target DFA states, noise level and number of strings are small.

2   Encoding DFA Identfication into SAT

The goal of DFA identification is to find a smallest DFA A such that every string from S_+, a set of positive examples, is accepted by A, and every string from S_-, a set of negative examples, is rejected. The size of A is defined as the number of states C it contains. The alphabet $\Sigma = \{l_1, \ldots, l_L\}$ of the sought DFA A is the set of all L symbols from S_+ and S_-. The example of the smallest DFA for $S_+ = \{ab, b, ba, bbb\}$ and $S_- = \{abbb, baba\}$ is shown in Fig. 1. In this work we assume that DFA states are numbered from 1 to C and the start state has number 1.

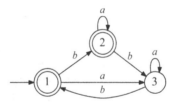

Fig. 1. An example of a DFA

In [13] Heule and Verwer proposed a compact translation of DFA identifi-cation problem into SAT. Here we briefly review the proposed technique, since our symmetry breaking predicates supplement it. The first step of both state-merging and translation-to-SAT techniques is augmented prefix tree acceptor (APTA) construction from the given examples S_+ and S_-. APTA is a tree-shaped automaton such that paths corresponding to two strings reach the same state v if and only if these strings share the same prefix in which the last symbol corresponds to v. We denote by V the set of all APTA states; by v_r – the APTA root; by V_+ – the set of accepting states; and by V_- – the set of rejecting states. Moreover, for state v (except v_r) we denote its incoming symbol as $l(v)$ and its parent as $p(v)$. The APTA for S_+ and S_- mentioned above is shown in Fig. 2a.

The second step of the technique proposed in [13] is the construction of the *consistency graph* (CG) for the obtained APTA. The set of nodes of the CG is identical to the set of APTA states. Two CG nodes v and w are connected

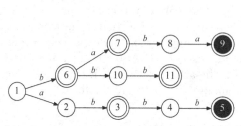 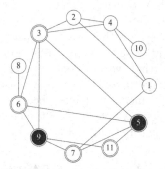

(a) An example of an APTA for $S_+ = \{ab, b, ba, bbb\}$ and $S_- = \{abbb, baba\}$

(b) The consistency graph for APTA from Fig. 2a

Fig. 2. An example of APTA and its consistency graph

with an edge (and called inconsistent) if merging v and w in APTA results in an inconsistency: an accepting state is merged with a rejecting state. Let E denote the set of CG edges. The CG for APTA of Fig. 2a is shown in Fig. 2b.

The key part of the algorithm is translating the DFA identification problem into a Boolean folmula in conjunctive normal form (CNF) and using a SAT solver to find a satisfying assignment. For a given set of examples and fixed DFA size C the solver returns a satisfying assignment (that defines a DFA with C states that is compliant with S_+ and S_-) or a message that it does not exist. The main idea of this translation is to use a distinct color for every state of the identified DFA and to find a consistent mapping of APTA states to colors. Three types of variables were used in the proposed compact translation:

1. *color* variables $x_{v,i} = 1$ ($v \in V$; $1 \leqslant i \leqslant C$) iff APTA state v has color i;
2. *parent relation* variables $y_{l,i,j} \equiv 1$ ($l \in \Sigma$; $1 \leqslant i, j \leqslant C$) iff DFA transition with symbol l from state i ends in state j;
3. *accepting color* variables $z_i \equiv 1$ ($1 \leqslant i \leqslant C$) iff DFA state i is accepting.

Direct encoding, described in [13], uses only variables $x_{v,i}$; variables $y_{l,i,j}$ and z_i are auxiliary and are used in compact encoding predicates, which are described below.

The compact translation proposed in [13] uses nine types of clauses:

1. $x_{v,i} \Rightarrow z_i$ ($v \in V_+$; $1 \leqslant i \leqslant C$) – definitions of z_i values for accepting states $(\neg x_{v,i} \lor z_i)$;
2. $x_{v,i} \Rightarrow \neg z_i$ ($v \in V_-$; $1 \leqslant i \leqslant C$) – definitions of z_i values for rejecting states $(\neg x_{v,i} \lor \neg z_i)$;
3. $x_{v,1} \lor x_{v,2} \lor \ldots \lor x_{v,C}$ ($v \in V$) – each state v has at least one color;
4. $x_{p(v),i} \land x_{v,j} \Rightarrow y_{l(v),i,j}$ ($v \in V \setminus \{v_r\}$; $1 \leqslant i, j \leqslant C$) – a DFA transition is set when a state and its parent are colored $(y_{l(v),i,j} \lor \neg x_{p(v),i} \lor \neg x_{v,j})$;
5. $y_{l,i,j} \Rightarrow \neg y_{l,i,k}$ ($l \in \Sigma$; $1 \leqslant i, j, k \leqslant C$; $j < k$) – each DFA transition can target at most one state $(\neg y_{l,i,j} \lor \neg y_{l,i,k})$;

6. $\neg x_{v,i} \vee \neg x_{v,j}$ $(v \in V; 1 \leqslant i < j \leqslant C)$ – each state has at most one color;
7. $y_{l,i,1} \vee y_{l,i,2} \vee \ldots \vee y_{l,i,C}$ $(l \in \Sigma; 1 \leqslant i \leqslant C)$ – each DFA transition must target at least one state;
8. $y_{l(v),i,j} \wedge x_{p(v),i} \Rightarrow x_{v,j}$ $(v \in V \setminus \{v_r\}; 1 \leqslant i, j \leqslant C)$ – state color is set when DFA transition and parent color are set $(\neg y_{l(v),i,j} \vee \neg x_{p(v),i} \vee x_{v,j})$;
9. $x_{v,i} \Rightarrow \neg x_{w,i}$ $((v,w) \in E; 1 \leqslant i \leqslant C)$ – the colors of two states connected with an edge in the consistency graph must be different $(\neg x_{v,i} \vee \neg x_{w,i})$.

Thus, the constructed formula consists of $\mathcal{O}(C^2|V|)$ clauses and, if the SAT solver finds a solution, we can identify the DFA.

To find a minimal DFA, authors use iterative SAT solving. Initial DFA size C is equal to the size of a *large clique* found in the CG. To find that clique, a greedy algorithm proposed in [13] can be applied. Then the minimal DFA is found by iterating over the DFA size C until the formula is satisfied.

The found clique was also used to perform symmetry breaking: in any valid coloring of a graph, all states in a clique must have a different color. Thus, we can fix the state colors in the clique in a preprocessing step. Later we will see that the max-clique symmetry breaking is not compatible with the one proposed in this paper.

To significantly reduce the SAT search space, the authors applied several EDSM steps before translation to SAT. Since EDSM cannot guarantee the minimality of solution, we will omit the consideration of this step in our paper.

3   Learning DFA from Noisy Samples

The translation described in the previous section deals with exact DFA identification. In this section we show how to modify the translation in order to apply it to noisy examples. We assume that not more than K attribution labels of the given training strings were randomly flipped. Solving this problem was the goal of the GECCO 2004 Noisy DFA competition [11] (with K equal to 10 percent of the number of the given training strings). An EA with smart state labeling was later proposed in [12], and since that time it is, to the best of our knowledge, the state-of-the-art technique for learning DFA from noisy training data.

In noisy case we cannot use APTA node consistency: we cannot determine whether an accepting state is merged with a rejecting state because correct string labels are unknown. Thus we cannot use CG and the max-clique symmetry breaking.

The idea of our modification is rather simple: for each labeled state of APTA we define a variable which states whether the label can be flipped. The number of flips is limited by K. Formally, for each $v \in V_{\pm} = V_{+} \cup V_{-}$ we define f_v which is true if and only if the label of state v can (but does not have to) be incorrect (**f**lipped). Using these variables, we can modify the translation proposed in [13] to take into account mistakes in string labels. To do this, we change the z_i definition clauses (items 1 and 2 from list in Section 2): because of mistake possibility they hold in case f_v is false. Thus, new z_i value definitions are expressed in the following way: $\neg f_v \Rightarrow (x_{v,i} \Rightarrow z_i)$ for $v \in V_{+}$; $\neg f_v \Rightarrow (x_{v,i} \Rightarrow \neg z_i)$ for $v \in V_{-}$.

To limit the number of corrections to K we use an auxiliary array of K integer variables. This array stores the numbers of the APTA states for which labels can be flipped. Thus, f_v is true if and only if the array contains v. To avoid consideration of isomorphic permutations we enforce the array to be sorted in the increasing order.

To represent the auxiliary array as a Boolean formula we define variables $r_{i,v}$ for $1 \leqslant i \leqslant K$ and $v \in V_\pm = \{v_1, \ldots, v_W\}$. $r_{i,v}$ is true if and only if v is stored in the i-th position of the array. To connect variables f_v with $r_{i,v}$ we add so-called channeling constrains: $f_v \Leftrightarrow (r_{1,v} \vee \ldots \vee r_{K,v})$ for each $v \in V_\pm$.

We have to state that exactly one $r_{i,v}$ is true for each position i in the auxiliary array. To achieve that we use the order encoding method [22]. We add auxiliary order variables $o_{i,v}$ for $1 \leqslant i \leqslant K$ and $v \in V_\pm = \{v_1, \ldots, v_W\}$. We assume that $o_{i,v}$ for $v \in \{v_1, \ldots, v_j\}$ and $\neg o_{i,v}$ for $v \in \{v_{j+1}, \ldots, v_W\}$ for some j. This can be expressed by the following constraint: $o_{i,v_{j+1}} \Rightarrow o_{i,v_j}$ for $1 \leqslant j < W$. Now we define that $r_{i,v_j} \Leftrightarrow o_{i,v_j} \wedge \neg o_{i,v_{j+1}}$. Also we add clauses $o_{i,v_j} \Rightarrow o_{i+1,v_{j+1}}$ (for $1 \leqslant i < K$ and $1 \leqslant j < W$) to store corrections in increasing order.

The proposed constraints in CNF are listed in Table 1; there are $\mathcal{O}(C|V_\pm| + K|V_\pm|)$ clauses. Thus, to modify the translation for the noiseless case to deal with noise we can replace the z_i value definition and inconsistency clauses (items 1, 2 and 9 from list in Section 2) with the ones listed in Table 1.

Table 1. Clauses for noisy DFA identification

Clauses	CNF representation	Range
$\neg f_v \Rightarrow (x_{v,i} \Rightarrow z_i)$	$\neg x_{v,j} \vee z_j \vee f_v$	$1 \leqslant j \leqslant C; v \in V_+$
$\neg f_v \Rightarrow (x_{v,i} \Rightarrow \neg z_i)$	$\neg x_{v,j} \vee \neg z_j \vee f_v$	$1 \leqslant j \leqslant C; v \in V_-$
$f_v \Rightarrow (r_{1,v} \vee \ldots \vee r_{K,v})$	$\neg f_v \vee r_{1,v} \vee \ldots \vee r_{K,v}$	$v \in V_\pm$
$r_{i,v} \Rightarrow f_v$	$\neg r_{i,v} \vee f_v$	$1 \leqslant i \leqslant K; v \in V_\pm$
$r_{i,v_j} \Rightarrow o_{i,v_j}$	$\neg r_{i,v_j} \vee o_{i,v_j}$	$1 \leqslant i \leqslant K; 1 \leqslant j \leqslant W$
$r_{i,v_j} \Rightarrow \neg o_{i,v_{j+1}}$	$\neg r_{i,v_j} \vee \neg o_{i,v_{j+1}}$	$1 \leqslant i \leqslant K; 1 \leqslant j < W$
$o_{i,v_j} \wedge \neg o_{i,v_{j+1}} \Rightarrow r_{i,v_j}$	$\neg o_{i,v_j} \vee o_{i,b_{j+1}} \vee r_{i,v_j}$	$1 \leqslant i \leqslant K; 1 \leqslant j < W$
$o_{K,v_W} \Rightarrow r_{K,v_W}$	$\neg o_{K,v_W} \vee r_{K,v_W}$	
$o_{i,v_{j+1}} \Rightarrow o_{i,v_j}$	$\neg o_{i,v_{j+1}} \vee o_{i,v_j}$	$1 \leqslant i \leqslant K; 1 \leqslant j < W$
$o_{i,v_j} \Rightarrow o_{i+1,v_{j+1}}$	$\neg o_{i,v_j} \vee o_{i+1,v_{j+1}}$	$1 \leqslant i < K; 1 \leqslant j < W$

4   Symmetry Breaking Predicates

In this section we propose a way to fix automata state enumeration to avoid consideration of isomorphic DFAs during SAT solving. The main idea of our symmetry breaking is to enforce DFA states to be enumerated in breadth-first search (BFS) order. That idea was also used in function NatOrder in the state-merging approach described in [20] and the Move To Front reorganization algorithm used in the genetic algorithm [21].

BFS uses the *queue* data structure to store intermediate results as it traverses the graph. First we enqueue the initial DFA state (in this paper state number 1). While the queue is not empty we deque a state i and enqueue any direct child states j that have not yet been discovered (enqueued before). Since our transitions are labeled with symbols from Σ, we enqueue child states in alphabetical order of symbols l on transitions $i \xrightarrow{l} j$. We call DFA *BFS-enumerated* if its states are enumerated in dequeuing (equals to enqueuing) order. An example of a BFS-enumerated DFA with six states shown in Fig. 3a (BFS-tree transitions that were used to enqueue states are marked bold); BFS enqueues are shown in Fig. 3b. The DFA shown in Fig. 1 is not BFS-enumerated – BFS first dequeues state 3 rather than state 2 (we consider $1 \xrightarrow{a} 3$ before $1 \xrightarrow{b} 2$).

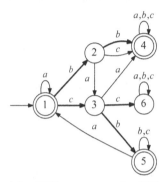

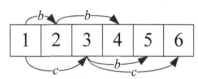

(a)   BFS-enumerated   DFA with bolded BFS-tree edges

(b) BFS queue. Cells correspond to DFA states, transitions correspond to enqueues

Fig. 3. An example of BFS-numerated DFA and its BFS queue

We propose constraints that enforce DFA to be BFS-enumerated. We assume that translation of a given DFA identification problem to SAT deals with Boolean variables $y_{l,i,j}$ ($l \in \Sigma$; $1 \leqslant i, j \leqslant C$) to set the DFA transition function: $y_{l,i,j} \equiv 1$ iff transition with symbol l from state i ends in state j.

The main idea is to determine each state's parent in the BFS-tree and set constrains between states' parents. We store **parents** in values $p_{j,i}$ (for each $1 \leqslant i < j \leqslant C$). $p_{j,i}$ is true if and only if state i is the parent of j in the BFS-tree. Each state except the initial one must have a parent with a smaller number, thus

$$\bigwedge_{2 \leqslant j \leqslant C} (p_{j,1} \vee p_{j,2} \vee \ldots \vee p_{j,j-1}).$$

Moreover, in BFS-enumeration states' parents must be ordered. State j must be enqueued before the next state $j + 1$, thus the next state's parent k cannot be less than current state's parent i (see Fig. 4):

$$\bigwedge_{1 \leqslant k < i < j < C} (p_{j,i} \Rightarrow \neg p_{j+1,k}).$$

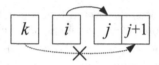

Fig. 4. Part of the queue illustrating the parent ordering predicates. Transitions show parent relations. The dotted transition is not allowed due to BFS-enumeration.

We set parents variables $p_{j,i}$ through $y_{l,i,j}$ using auxiliary variables $t_{i,j}$. In BFS-enumeration state j was enqueued while processing the state with minimal number i among states that have a transition to j:

$$\bigwedge_{1\leqslant i<j\leqslant C} (p_{j,i} \Leftrightarrow t_{i,j} \wedge \neg t_{i-1,j} \wedge \ldots \wedge \neg t_{1,j}),$$

where $t_{i,j} \equiv 1$ iff there is a transition between i and j; we define these auxiliary variables using $y_{l,i,j}$:

$$\bigwedge_{1\leqslant i<j\leqslant C} (t_{i,j} \Leftrightarrow y_{l_1,i,j} \vee \ldots \vee y_{l_L,i,j}).$$

Now to enforce DFA to be BFS-enumerated we have to order children in alphabetical order of symbols on transitions. We consider two cases: alphabet Σ consists of two symbols $\{a, b\}$ and more than two symbols $\{l_1, \ldots, l_L\}$. In the case of two symbols only two states can have the same parent i and they are forced by ordering constraints to have consecutive numbers j and $j+1$. In this case we force the transition that starts in state i labeled with symbol a to end in state j instead of $j+1$:

$$\bigwedge_{1\leqslant i<j<C} (p_{j,i} \wedge p_{j+1,i} \Rightarrow y_{a,i,j}).$$

In the second case we have to introduce a third type of variables in our symmetry breaking predicates. We store the alphabetically minimal symbol on transitions between states: $m_{l,i,j}$ is true if and only if there is a transition $i \overset{l}{\to} j$ and there is no such transition with an alphabetically smaller symbol. We connect these variables with DFA transitions by adding the following channeling predicates:

$$\bigwedge_{1\leqslant i<j\leqslant C} \bigwedge_{1\leqslant n\leqslant L} (m_{l_n,i,j} \Leftrightarrow y_{l_n,i,j} \wedge \neg y_{l_{n-1},i,j} \wedge \ldots \wedge \neg y_{l_1,i,j}).$$

Now it remains to arrange consecutive states j and $j+1$ with the same parent i in the alphabetically order of minimal symbols on transitions between them and i (see Fig. 5):

$$\bigwedge_{1\leqslant i<j<C} \bigwedge_{1\leqslant k<n\leqslant L} (p_{j,i} \wedge p_{j+1,i} \wedge m_{l_n,i,j} \Rightarrow \neg m_{l_k,i,j+1}).$$

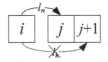

Fig. 5. Illustration of alphabetical ordering predicates. If i is the parent of j and $j+1$, l_n (l_k) is the alphabetically minimal symbol on transitions between i and j (i and $j+1$) then l_k cannot be alphabetically smaller than l_n

Thus we propose symmetry breaking predicates that are composed by listed constraints. Predicates (for three or more symbols case) translated into $\mathcal{O}(C^3 + C^2 L^2)$ CNF clauses are listed in Table 2. Our implementation of proposed predicates and all algorithms can be found on the our labaratory github repository (https://github.com/ctlab/DFA-Inductor).

Table 2. BFS-based symmetry breaking clauses

Clauses	CNF representation	Range
$t_{i,j} \Rightarrow (y_{l_1,i,j} \vee \ldots \vee y_{l_L,i,j})$	$\neg t_{i,j} \vee y_{l_1,i,j} \vee \ldots \vee y_{l_L,i,j}$	$1 \leqslant i < j \leqslant C$
$y_{i,j,l} \Rightarrow t_{i,j}$	$\neg y_{l,i,j} \vee t_{i,j}$	$1 \leqslant i < j \leqslant C; l \in \Sigma$
$m_{l,i,j} \Rightarrow y_{l,i,j}$	$\neg m_{l,i,j} \vee y_{l,i,j}$	$1 \leqslant i < j \leqslant C; l \in \Sigma$
$m_{l_n,i,j} \Rightarrow \neg y_{l_k,i,j}$	$\neg m_{l_n,i,j} \vee \neg y_{l_k,i,j}$	$1 \leqslant i < j \leqslant C; 1 \leqslant k < n \leqslant L$
$(y_{l_n,i,j} \wedge \neg y_{l_{n-1},i,j} \wedge \ldots$	$\neg y_{l_n,i,j} \vee y_{l_{n-1},i,j} \vee \ldots$	$1 \leqslant i < j \leqslant C; 1 \leqslant n \leqslant L$
$\neg y_{l_1,i,j}) \Rightarrow m_{l_n,i,j}$	$\vee y_{l_1,i,j} \vee m_{l_n,i,j}$	
$p_{j,1} \vee p_{j,2} \vee \ldots \vee p_{j,j-1}$	$p_{j,1} \vee p_{j,2} \vee \ldots \vee p_{j,j-1}$	$2 \leqslant j \leqslant C$
$p_{j,i} \Rightarrow t_{i,j}$	$\neg p_{j,i} \vee t_{i,j}$	$1 \leqslant i < j \leqslant C$
$p_{j,i} \Rightarrow \neg t_{k,j}$	$\neg p_{j,i} \vee \neg t_{k,j}$	$1 \leqslant k < i < j \leqslant C$
$(t_{i,j} \wedge \neg t_{i-1,j} \wedge \ldots \wedge \neg t_{1,j}) \Rightarrow p_{j,i}$	$\neg t_{i,j} \vee t_{i-1,j} \vee \ldots \vee t_{1,j} \vee p_{j,i}$	$1 \leqslant i < j \leqslant C$
$p_{j,i} \Rightarrow \neg p_{j+1,k}$	$\neg p_{j,i} \vee \neg p_{j+1,k}$	$1 \leqslant k < i < j < C$
$(p_{j,i} \wedge p_{j+1,i} \wedge m_{l_n,i,j}) \Rightarrow \neg m_{l_k,i,j+1}$	$\neg p_{j,i} \vee \neg p_{j+1,i} \vee \neg m_{l_n,i,j} \vee \neg m_{l_k,i,j+1}$	$1 \leqslant i < j \leqslant C; 1 \leqslant k < n \leqslant L$

5   Experiments

All experiments were performed using a machine with an AMD Opteron 6378 2.4 GHz processor running on Ubuntu 14.04. All algorithms were implemented in Java, the *lingeling* SAT-solver was used. Our own algorithm was used for generating problem instances for all experiments based on randomly generated data sets. This algorithm builds a set of strings with the following parameters: size of DFA N which has to be generated, alphabet size A, number of strings S which have to be generated, noise level K (percent of attribution labels of generated strings which have to be randomly flipped).

In the exact case the max-clique method clearly outperforms BFS-based strategy.

For noisy DFA identification we used randomly generated instances. First we considered the case when the target DFA exists and the Boolean formula is satisfiable. We used following parameters: $N \in [5; 10]$, $A = 2$, $S \in \{10N, 25N, 50N\}$. We compared SAT approach without any symmetry breaking predicates, our solution using BFS-based symmetry breaking predicates and the current state-of-the-art EA from [12]. Each experiment was repeated 100 times. The time limit was set to 1800 seconds. Initial experiments showed that EA clearly outperforms our method when $K > 4\%$. Therefore we set this parameter to $1\% - 4\%$. Results are listed in Table 3. We left only instances which were solved within the time limit. These results indicate that the BFS-based strategy finds the solution faster than the current state-of-the-art EA only when N is small (< 7), noise level is small ($1\% - 4\%$) and the number of strings is also small ($< 50N$). But BFS-based strategy finds the solution extremely faster than SAT approach without symmetry breaking strategy.

Table 3. Mean times of solving noisy DFA identification with count of strings in the each instance set to $10N$, $25N$ and $50N$ respectively

N	K	BFS, sec	SAT, sec	EA, sec
5	2	0.22	0.38	1.22
5	4	0.59	0.9	1.1
6	2	1.05	2.44	2.94
6	4	3.34	7.82	2.85
7	1	4.34	10.83	21.36
7	3	17.22	143.66	19.16
8	1	17.89	31.58	30.29
8	2	163.92	225.31	19.8

N	K	BFS, sec	SAT, sec	EA, sec
5	1	0.54	0.64	2.77
5	2	2.42	4.33	1.80
6	1	6.3	11.95	11.65
6	2	13.3	43.54	4.80
7	1	31.01	114.95	17.24
7	2	286.76	TL	13.11
8	1	239.46	404.32	21.73

N	K	BFS, sec	SAT, sec	EA, sec
5	1	4.2	7.59	6.07
5	2	12.87	22.36	3.05
6	1	20.76	52.5	20.39
6	2	107.94	309.22	11.28

The last experiment considered the case when the target DFA does not exist and the Boolean formula is unsatisfiable. Random dataset was also used here. We tried to find the target DFA using the following parameters: $N \in [5; 7]$, $A = 2$, $S = 50N$, $K \in [1; 2]$ percent. The input set of strings was generated from a $(N + 1)$-sized DFA. It should be noted that the EA from [12] cannot determine that an automaton consistent with a given set of strings does not exist. On the other hand, all SAT-based methods are capable of that. Therefore we compared our implementation of compact SAT encoding without using symmetry breaking predicates and the same with BFS-based predicates. Each experiment was repeated 100 times and the time limit was set to 1800 seconds again. Results are listed in Table 4. It can be seen from the table that BFS-based strategy significantly reduces the mean time of determination that an automaton does not exist.

Table 4. Mean times and percent of passed solutions of solving noisy DFA identification when the target DFA does not exist

N	K	BFS, sec	WO, sec	passed BFS, %	passed WO, %
5	1	11.57	257.13	100	100
5	2	46.42	1296.71	100	30
6	1	110.05	TL	100	0
6	2	581.73	TL	100	0
7	1	995.27	TL	89	0
7	2	TL	TL	0	0

6   Conclusions and Future Work

We proposed symmetry breaking predicates which can be added to the Boolean formula representing various DFA identification problems. By adding the predicates we can reduce SAT search space through enforcing DFA states to be enumerated in breadth-first search (BFS) order.

We drew our attention to the case of noisy DFA identfication. We proposed a modification of the noiseless translation-to-SAT [13] for the noisy case. To achieve compact encoding for that case we used the order encoding method. We showed that the previously proposed max-clique technique for symmetry breaking is not applicable in the noisy case while our BFS-based approach is. We showed that the BFS-based strategy can be applied in the noisy case when an automaton which is consistent with a given set of strings does not exists. The current state-of-the-art EA from [12] cannot determine that. In experimental results, we showed that our approach with BFS-based symmetry breaking predicates clearly outperforms algorithm without any predicates. Also we showed that our strategy outperform EA if the number of the target DFA states is small, noise level is small and number of strings is small either.

We plan to translate noisy DFA identification to Max-SAT in order to limit the number of corrections without using an auxiliary array of integer variables. Also we plan to experiment with alternative integer encoding methods. In the future we would like to solve a problem of finding all solutions (instead of a single DFA) using our predicates.

Acknowledgements. Authors would like to thank Daniil Chivilikhin, Igor Buzhinsky and Andrey Filchenkov for useful comments. This work was financially supported by the Government of Russian Federation, Grant 074-U01, and also partially supported by RFBR, research project No. 14-07-31337 mol_a.

References

1. Hopcroft, J., Motwani, R., Ullman, J.: Introduction to Automata Theory, Languages, and Computation. Addison-Wesley (2006)
2. De La Higuera, C.: A bibliographical study of grammatical inference. Pattern Recognition **38**(9), 1332–1348 (2005)
3. Gold, E.M.: Complexity of automaton identification from given data. Information and Control **37**(3), 302–320 (1978)
4. Pitt, L., Warmuth, M.K.: The minimum consistent DFA problem cannot be approximated within any polynomial. Journal of the ACM **40**(1), 95–142 (1993)
5. Lang, K.J., Pearlmutter, B.A., Price, R.A.: Results of the abbadingo one DFA learning competition and a new evidence-driven state merging algorithm. In: Honavar, V.G., Slutzki, G. (eds.) ICGI 1998. LNCS (LNAI), vol. 1433, pp. 1–112. Springer, Heidelberg (1998)
6. Lang, K.J.: Faster Algorithms for Finding Minimal Consistent DFAs. Technical report (1999)
7. Bugalho, M., Oliveira, A.L.: Inference of regular languages using state merging algorithms with search. Pattern Recognition **38**(9), 1457–1467 (2005)

8. Dupont, P.: Regular grammatical inference from positive and negative samples by genetic search: the GIG method. In: Carrasco, R.C., Oncina, J. (eds.) ICGI 1994. LNCS, vol. 862, pp. 236–2445. Springer, Heidelberg (1994)

9. Luke, S., Hamahashi, S., Kitano, H.: Genetic programming. In: Proceedings of the Genetic and Evolutionary Computation Conference, vol. 2, pp. 1098–1105 (1999)

10. Lucas, S.M., Reynolds, T.J.: Learning DFA: evolution versus evidence driven state merging. In: The 2003 Congress on Evolutionary Computation, CEC 2003, vol. 1, pp. 351–358. IEEE (2003)

11. Lucas, S.: GECCO 2004 noisy DFA results. In: GECCO Proc. (2004)

12. Lucas, S.M., Reynolds, T.J.: Learning deterministic finite automata with a smart state labeling evolutionary algorithm. IEEE Transactions on Pattern Analysis and Machine Intelligence 27(7), 1063–1074 (2005)

13. Heule, M.J.H., Verwer, S.: Exact DFA identification using SAT solvers. In: Sempere, J.M., García, P. (eds.) ICGI 2010. LNCS, vol. 6339, pp. 66–79. Springer, Heidelberg (2010)

14. Walkinshaw, N., Lambeau, B., Damas, C., Bogdanov, K., Dupont, P.: STAMINA: a competition to encourage the development and assessment of software model inference techniques. Empirical Software Engineering 18(4), 791–824 (2013)

15. Biere, A., Heule, M., van Maaren, H.: Handbook of satisfiability, vol. 185. IOS Press (2009)

16. Amla, N., Du, X., Kuehlmann, A., Kurshan, R.P., McMillan, K.L.: An analysis of SAT-based model checking techniques in an industrial environment. In: Borrione, D., Paul, W. (eds.) CHARME 2005. LNCS, vol. 3725, pp. 254–268. Springer, Heidelberg (2005)

17. Lohfert, R., Lu, J.J., Zhao, D.: Solving SQL constraints by incremental translation to SAT. In: Nguyen, N.T., Borzemski, L., Grzech, A., Ali, M. (eds.) IEA/AIE 2008. LNCS (LNAI), vol. 5027, pp. 669–676. Springer, Heidelberg (2008)

18. Galeotti, J.P., Rosner, N., Lopez Pombo, C.G., Frias, M.F.: TACO: Efficient SAT-Based Bounded Verification Using Symmetry Breaking and Tight Bounds. IEEE Transactions on Software Engineering 39(9), 1283–1307 (2013)

19. Ulyantsev, V., Tsarev, F.: Extended finite-state machine induction using SAT-solver. In: Proc. of ICMLA 2011, vol. 2, pp. 346–349. IEEE (2011)

20. Lambeau, B., Damas, C., Dupont, P.E.: State-merging DFA induction algorithms with mandatory merge constraints. In: Clark, A., Coste, F., Miclet, L. (eds.) ICGI 2008. LNCS (LNAI), vol. 5278, pp. 139–153. Springer, Heidelberg (2008)

21. Chambers, L.D.: Practical handbook of genetic algorithms: complex coding systems, vol. 3. CRC Press (2010)

22. Barahona, P., Hölldobler, S., Nguyen, V.: Efficient SAT-encoding of linear csp constraints. In: 13th International Symposium on Artificial Intelligence and Mathematics-ISAIM, Fort Lauderdale, Florida, USA (2014)

Learning Conjunctive Grammars and Contextual Binary Feature Grammars

Ryo Yoshinaka[✉]

Graduate School of Informatics, Kyoto University, Kyoto, Japan
ry@i.kyoto-u.ac.jp

Abstract. Approaches based on the idea generically called distributional learning have been making great success in the algorithmic learning of context-free languages and their extensions. We in this paper show that conjunctive grammars are also learnable by a distributional learning technique. Conjunctive grammars are context-free grammars enhanced with conjunctive rules to extract the intersection of two languages. We also compare our result with the closely related work by Clark et al. (JMLR 2010) on contextual binary feature grammars (CBFGs). Our learner is stronger than theirs. In particular our learner learns every *exact* CBFG, while theirs does not. Clark et al. emphasized the importance of exact CBFGs but they only conjectured there should be a learning algorithm for exact CBFGs. This paper shows that their conjecture is true.

Keywords: Grammatical inference · Algorithmic learning · Distributional learning

1 Introduction

Approaches based on the idea generically called *distributional learning* have been making great success in the algorithmic learning of context-free languages (CFLs). Distributional learning algorithms exploit information on the *distribution* of strings in contexts with respect to a learning target language: that is, we observe which combination of a string $u \in \Sigma^*$ and a context $\langle l, r \rangle \in \Sigma^* \times \Sigma^*$ forms a string $lur \in \Sigma^*$ belonging to the concerned language $L \subseteq \Sigma^*$. In the distributional learning of context-free grammars (CFGs), nonterminal symbols of a grammar are associated with a context, a string, or sets of those and have semantics determined by those associated objects. For example, a nonterminal indexed with a context $\langle l, r \rangle$ is supposed to derive strings u such that lur belongs to the learning target language (cf. [3]). A learner constructs a hypothesis grammar based on the semantics of nonterminals. Approaches that define the semantics by strings are called *primal* and those by contexts are *dual*.

Clark et al. [4] have proposed a grammar formalism called *contextual binary feature grammars* (CBFGs) as "distributionally" learnable representations. The formalism is strong enough to generate all CFLs and some other context-sensitive

© Springer International Publishing Switzerland 2015
A.-H. Dediu et al. (Eds.): LATA 2015, LNCS 8977, pp. 623–635, 2015.
DOI: 10.1007/978-3-319-15579-1_49

languages. CBFGs use contexts as *features*, which play a similar role of nonterminal symbols of CFGs. This idea gives a guide to design a learning algorithm. They proposed a learning algorithm that identifies a subclass of CFLs in the limit from positive data and membership queries.

However, their discussion has a strange gap between the formalism and the learning target. Their learner constructs a CBFG as a hypothesis while the learning target is a CFL. The learnable subclass of CFLs are defined in terms of CFGs rather than CBFGs. Certainly their algorithm learns some non-context-free languages, but no characterization of those languages are given except that they are learned by their method. Actually, features of CBFGs play *only* the role as a guide for learning. Indeed, in the definition of CBFGs, no semantical requirement is imposed on features. One can use arbitrary contexts as features of a grammar, which may be completely irrelevant of the language defined by the grammar. On the other hand, Clark et al. [4] are very much interested in the property called *exactness*, which establishes a clear relation between features and their semantics. In exact CBFGs, a string has a feature if and only if the string can occur in the feature (= context). Indeed this property is the guide for learning. Their learner constructs its hypothesis so that it can be exact. However, they present no rigorous mathematical relation between the exactness and learnability. In fact, there are exact CBFGs which generate CFLs that their algorithm does not learn. They only conjectured that there should exist an algorithm that learns all exact CBFGs.

Another mismatch in their algorithm is found in their grammar construction. While features of a CBFG are contexts, their learner constructs rules based on strings. Their strategy can be seen as a compromise of primal and dual approaches. Their algorithm requires target CFLs to satisfy two conditions which we call the *finite kernel property* (FKP) and *finite fiducial set property* (FFP). Actually later Yoshinaka [9] showed that a primal type algorithm learns CFLs with the FKP, whose hypotheses are standard CFGs.

We in this paper take a closer look at those mismatches in Clark et al.'s result and close the gaps by designing a distributional learning algorithm for a subclass of *conjunctive grammars* [7]. Conjunctive grammars are generalization of CFGs that have conjunctive rules to extract the intersection of the languages derived from nonterminal symbols. CBFGs can be seen as notational variants of conjunctive grammars with some minor restriction. We treat conjunctive grammars for the generality in this paper. The subclass we target has the *finite context property* (FCP) [1,2]. Actually the technique used in our learning algorithm is a straightforward generalization of an existing dual type algorithm for CFGs with the FCPs [9]. We show that exact CBFGs satisfy this property. As an important consequence of this, the conjecture by Clark et al. [4] is shown to be true. After we explain the algorithm and its correctness briefly, we discuss and compare the learning algorithms by Clark et al. and by us. We will also show that if a CFL satisfies the two conditions required by Clark et al.'s algorithm, the FKP and the FFP, then it has the FCP. The FKP is a property with which a primal approach works and the FCP is favorable for a dual one. This explains how their primal-dual mixed strategy works.

2   Distribution of Strings in Contexts

Let Σ be a nonempty finite set of letters. We denote the empty string by λ. We define $\Sigma_\lambda = \Sigma \cup \{\lambda\}$. Any element of $\Sigma^* \times \Sigma^*$ is called a *context*. The empty context $\langle \lambda, \lambda \rangle$ is denoted by Λ. For a string $v \in \Sigma^*$ and a context $\langle u_1, u_2 \rangle \in \Sigma^* \times \Sigma^*$, the *composition* of them is $\langle u_1, u_2 \rangle \odot v = u_1 v u_2 \in \Sigma^*$. The composition operation is naturally generalized to be applied to sets $W \subseteq \Sigma^* \times \Sigma^*$ and $V \subseteq \Sigma^*$ as $W \odot V = \{ w \odot v \mid w \in W,\, v \in V \}$. For a language $L \subseteq \Sigma^*$, we let

$$\mathrm{Sub}(L) = \{ v \in \Sigma^* \mid w \odot v \in L \text{ for some } w \in \Sigma^* \times \Sigma^* \},$$
$$\mathrm{Con}(L) = \{ w \in \Sigma^* \times \Sigma^* \mid w \odot v \in L \text{ for some } v \in \Sigma^* \}.$$

We also have an operation dual to $\odot$. We denote the set of contexts that admit every string in a set $V \subseteq \Sigma^*$ with respect to a language $L \subseteq \Sigma^*$ by

$$L \oslash V = \{ w \in \Sigma^* \times \Sigma^* \mid w \odot v \in L \text{ for all } v \in V \}.$$

Similarly, the set of strings that every context in $W \subseteq \Sigma^* \times \Sigma^*$ accepts is

$$L \oslash W = \{ v \in \Sigma^* \mid w \odot v \in L \text{ for all } w \in W \}.$$

Note that $L \oslash \{\Lambda\} = L$. By definition, $W \odot V \subseteq L$ iff $W \subseteq L \oslash V$ iff $V \subseteq L \oslash W$. When L is understood, particularly when our learning target is L, we denote $L \oslash W$ for $W \subseteq \Sigma^* \times \Sigma^*$ by $W^\dagger$ and $L \oslash V$ for $V \subseteq \Sigma^*$ by $V^\ddagger$. It is easy to see that $V \subseteq (V^\ddagger)^\dagger$, $W \subseteq (W^\dagger)^\ddagger$, $V^\ddagger = ((V^\ddagger)^\dagger)^\ddagger$ and $W^\dagger = ((W^\dagger)^\ddagger)^\dagger$. Moreover, we define $W^\ddagger = (W^\dagger)^\ddagger$ and $V^\dagger = (V^\ddagger)^\dagger$. In both cases where $X \subseteq \Sigma^*$ and where $X \subseteq \Sigma^* \times \Sigma^*$, we have $X^\dagger \subseteq \Sigma^*$ and $X^\ddagger \subseteq \Sigma^* \times \Sigma^*$. For $V_1, V_2 \subseteq \Sigma^*$, we write $V_1 \approx V_2$ if $V_1^\dagger = V_2^\dagger$. For sets $V \subseteq \Sigma^*$ and $W \subseteq \Sigma^* \times \Sigma^*$, we define $W^{(V)} = W^\dagger \cap V$ and $V^{(W)} = V^\ddagger \cap W$.

3   Learning Conjunctive Grammars with Finite Context Property

This section presents an algorithm that learns conjunctive grammars with a special property called the *finite context property* (FCP). This property is first proposed for CFGs by Clark [1,2].

Definition 1 (Okhotin [7,8]). A *conjunctive grammar* is a quadruple $G = \langle \Sigma, V, P, S \rangle$ where Σ is a finite set of *terminal symbols*, V is a finite set of *nonterminal symbols*, $S \in V$ is a special nonterminal called the *initial symbol* and P is a finite set of *rules* each of which has the form

$$X \leftarrow \alpha_1 \& \dots \& \alpha_k$$

where $X \in V$, $\alpha_i \in (\Sigma \cup V)^*$ for each $i \in \{1, \dots, k\}$ for some $k \geq 1$.

The language $\mathcal{L}(G, \alpha)$ of $\alpha \in (\Sigma \cup V)^*$ is recursively defined as follows:

- $a \in \mathcal{L}(G, a)$ for all $a \in \Sigma_\lambda$,
- if $u_i \in \mathcal{L}(G, X_i)$ with $X_i \in \Sigma \cup V$ for $i = 1, \ldots, k$, then $u_1 \ldots u_k \in \mathcal{L}(G, X_1 \ldots X_k)$,
- if $u \in \mathcal{L}(G, \alpha_i)$ with $\alpha_i \in (\Sigma \cup V)^*$ for $i = 1, \ldots, k$ and $X \leftarrow \alpha_1 \& \ldots \& \alpha_k \in P$, then $u \in \mathcal{L}(G, X)$,
- nothing else is in $\mathcal{L}(G, \alpha)$.

The language of G is $\mathcal{L}(G) = \mathcal{L}(G, S)$.

We say that two grammars are *equivalent* if they define the same language. We say that a rule is *useless* if there is no string in $\mathcal{L}(G)$ that can be derived using that rule. For example, the rule $Y \leftarrow b$ is useless in the following grammar.

$$S \leftarrow X \& Y, \ X \leftarrow a, \ Y \leftarrow a, \ Y \leftarrow b.$$

We in this paper assume without loss of generality that no rule is useless.

Conjunctive grammars can define the following context-sensitive languages:

$$\{ a^n b^n c^n \mid n \geq 0 \}, \quad \{ ucu \mid u \in \{a, b\}^* \}.$$

One can show that every conjunctive grammar admits an equivalent grammar whose rules have one of the following forms (cf. [7]):

- $X \leftarrow a$ for some $X \in V$ and $a \in \Sigma_\lambda$,
- $X \leftarrow YZ$ for some $X, Y, Z \in V$,
- $X \leftarrow Y \& Z$ for some $X, Y, Z \in V$.

Hereafter we assume that grammars are in this binary form for simplicity.

Definition 2. Let $X \in V$ be a nonterminal symbol of a conjunctive grammar $G = \langle \Sigma, V, P, S \rangle$. A context set C_X is said to be a *characterizing context set* of X if $\mathcal{L}(G, X) = \mathcal{L}(G) \oslash C_X$. We say that G has *the k-finite context property* (k-FCP) if every nonterminal admits a finite characterizing context set of cardinality at most k. A grammar has the FCP if it has the k-FCP for some k.

In general, it is not necessarily the case that $\mathcal{L}(G, X) \subseteq \mathrm{Sub}(\mathcal{L}(G))$ for a conjunctive grammar G and a nonterminal X, even when G has no useless rules. The FCP requires the languages of a nonterminal to be "observable" in the sense that we have $\mathcal{L}(G, X) \subseteq \mathrm{Sub}(\mathcal{L}(G))$.

Yoshinaka [9] and Leiß [6][1] proposed learning algorithms for CFGs with the *weak* FCP, which requires $\mathcal{L}(G, X) \approx \mathcal{L}(G) \oslash C_X$ rather than the exact equality. It is an open problem whether the classes of languages generated by CFGs (conjunctive grammars) with the stronger (Clark's original) FCP and of those with the weak FCP coincide. This paper requires learning targets to satisfy the stronger FCP. The learning algorithm that will be presented in this section is a

---

[1] Leiß [6] pointed out an error of the original algorithm by Clark [2].

straightforward generalization of Yoshinaka's algorithm, but a learner based on Leiß's algorithm is also possible (see Sec. 3.2).

We remark that a characterizing context set C_X of X is a subset of $\mathrm{Con}(\mathcal{L}(G))$. Otherwise $C_X \nsubseteq \mathrm{Con}(\mathcal{L}(G))$, we have $\mathcal{L}(G) \oslash C_X = \varnothing = \mathcal{L}(G, X)$. The nonterminal X is useless.

3.1 Learner

Our learning paradigm is *identification in the limit from positive data and membership queries*. A *positive presentation* of a language L_* over Σ is an infinite sequence $u_1, u_2, \cdots \in \Sigma^*$ such that $L_* = \{ u_i \mid i \geq 1 \}$. A learner is given a positive presentation of the language $L_* = \mathcal{L}(G_*)$ of the target grammar G_* and each time a new example u_i is given, it outputs a grammar G_i (called a *hypothesis*) computed from $u_1, \ldots, u_i$ with the aid of a *membership oracle*. One may query the oracle whether an arbitrary string u is in L_*, and the oracle answers in constant time. We say that a learning algorithm *identifies G_* in the limit from positive data and membership queries* if for any positive presentation $u_1, u_2, \ldots$ of $\mathcal{L}(G_*)$, there is an integer n such that $G_n = G_m$ for all $m \geq n$ and $\mathcal{L}(G_n) = \mathcal{L}(G_*)$. Trivially every grammar admits a successful learning algorithm. An algorithm should learn a rich class of grammars in a uniform way. We say that a learning algorithm *identifies a class* $\mathbb{G}$ *of grammars in the limit from positive data and membership queries* iff it identifies all $G_* \in \mathbb{G}$.

Since the nonterminals of a target grammar G_* are characterized by contexts, our learner's hypothesis grammar $\hat{G}$ will use nonterminals indexed by context sets. We denote a nonterminal indexed by a context set C by $[\![C]\!]$. We would like $[\![C]\!]$ to be characterized by the set C, i.e., $\mathcal{L}(\hat{G}, [\![C]\!]) = C^\dagger$. If a context set C characterizes a nonterminal X of the target grammar, we want $[\![C]\!]$ to simulate X. Our learning algorithm constructs a grammar $\hat{G} = \mathcal{G}^k(F, K) = \langle \Sigma, \hat{V}, \hat{P}, [\![\{\Lambda\}]\!]\rangle$ from two finite sets $F \subseteq \Sigma^* \times \Sigma^*$ and $K \subseteq \Sigma^*$.

- $\hat{V} = \{ [\![C]\!] \mid C \subseteq F \text{ and } |C| \leq k \}$,
- $\hat{P} = \hat{P}_0 \cup \hat{P}_1 \cup \hat{P}_2$ where
  - $\hat{P}_0 = \{ [\![C]\!] \leftarrow a \mid a \in C^{(\Sigma_\lambda)} \}$,
  - $\hat{P}_1 = \{ [\![C]\!] \leftarrow [\![C_1]\!][\![C_2]\!] \mid C \odot (C_1^{(K)} C_2^{(K)}) \subseteq L_* \}$,
  - $\hat{P}_2 = \{ [\![C]\!] \leftarrow [\![C_1]\!]\&[\![C_2]\!] \mid C \odot (C_1^{(K)} \cap C_2^{(K)}) \subseteq L_* \}$.

Those rules are constructed in polynomial time in $|F|$ and $|K|$ (but not in k) by the aid of the membership oracle.

Lemma 3. *If $E \subseteq F$ then $\mathcal{L}(\mathcal{G}^k(E, K)) \subseteq \mathcal{L}(\mathcal{G}^k(F, K))$.*
If $J \subseteq K$ then $\mathcal{L}(\mathcal{G}^k(F, K)) \subseteq \mathcal{L}(\mathcal{G}^k(F, J))$.

Definition 4. We say that a rule is *correct* if it is compatible with the semantics of the nonterminals in it. That is,

- $[\![C]\!] \leftarrow a$ is correct if $a \in C^\dagger$,
- $[\![C]\!] \leftarrow [\![C_1]\!][\![C_2]\!]$ is correct if $C^\dagger \supseteq C_1^\dagger C_2^\dagger$,

- $[\![C]\!] \leftarrow [\![C_1]\!] \& [\![C_2]\!]$ is correct if $C^\dagger \supseteq C_1^\dagger \cap C_2^\dagger$.

If a rule is not correct, it is called *incorrect*.

Lemma 5. *For any F and K, all the correct rules constructible from the non-terminals of $\hat{V}$ are present in $\mathcal{G}^k(F, K)$.*

Lemma 6. *Let L_* be generated by a conjunctive grammar G_* with the k-FCP and C_X be a characterizing context of each nonterminal X of G_* with $|C_X| \leq k$. If $F \supseteq C_X$ for all nonterminals X, then $L_* \subseteq \mathcal{L}(\mathcal{G}^k(F, K))$.*

Proof. Let $\hat{G} = \mathcal{G}^k(F, K)$. One can easily see that for every rule ρ of G_*, $\phi(\rho)$ is a correct rule where ϕ replaces each nonterminal X of G_* by $[\![C_X]\!]$. By Lemma 5, $\phi(\rho)$ is present in $\hat{G}$ and thus $\hat{G}$ can simulate G_*. We prove this claim for conjunctive rules only. For a rule $X \leftarrow Y \& Z$ of G_*, the fact

$$C_Y^\dagger \cap C_Z^\dagger = \mathcal{L}(G_*, Y) \cap \mathcal{L}(G_*, Z) \subseteq \mathcal{L}(G_*, X) = C_X^\dagger$$

means that the rule $[\![C_X]\!] \leftarrow [\![C_Y]\!] \& [\![C_Z]\!]$ is correct by definition. $\quad\square$

Note that here the weak FCP is not enough to establish that $\phi(\rho)$ will be correct, because $S_i \approx S_i'$ where $S_i \subseteq \Sigma^*$ for $i = 1, 2$ does not imply $S_1 \cap S_2 \approx S_1' \cap S_2'$.

Every rule in $\hat{P}_0$ of $\mathcal{G}^k(F, K)$ is correct for any F and K. However it is not necessarily the case for rules in $\hat{P}_1$ and $\hat{P}_2$.

Lemma 7. *Every F admits a finite set $K \subseteq \mathrm{Sub}(L_*)$ consisting of at most $2|\hat{V}|^3$ strings such that all rules of $\mathcal{G}^k(F, K)$ are correct.*

Proof. We construct a finite set K from F as follows.

For each triple $[\![C]\!], [\![C_1]\!], [\![C_2]\!]$ that forms an incorrect rule $[\![C]\!] \leftarrow [\![C_1]\!][\![C_2]\!]$, there is $u_1 u_2 \in C_1^\dagger C_2^\dagger - C^\dagger$ where $u_i \in C_i^\dagger$ for $i = 1, 2$. Let K contain u_1 and u_2. Then the incorrect rule is suppressed.

For each triple $[\![C]\!], [\![C_1]\!], [\![C_2]\!]$ that forms an incorrect rule $[\![C]\!] \leftarrow [\![C_1]\!] \& [\![C_2]\!]$, there is $u \in C_1^\dagger \cap C_2^\dagger - C^\dagger$. Let K contain u. Then we have $u \in C_i^{(K)}$ and $C \odot (C_1^{(K)} \cap C_2^{(K)}) \nsubseteq L_*$, so the incorrect rule is suppressed. $\quad\square$

Lemma 8. *If $\hat{G} = \mathcal{G}^k(F, K)$ has no incorrect rules, then $\mathcal{L}(\hat{G}) \subseteq L_*$.*

Proof. It is easy to show by induction on derivation that $u \in \mathcal{L}(\hat{G}, [\![C]\!])$ implies $u \in C^\dagger$. In particular for the initial symbol $[\![\Lambda]\!]$, we establish $\mathcal{L}(\hat{G}, [\![\Lambda]\!]) \subseteq \Lambda^\dagger = L_*$. $\quad\square$

Our learner Algorithm 1 expands F to augment the conjecture grammar when we know we do not yet have enough contexts, while keeping expanding K to exclude incorrect rules.

Theorem 9. *Algorithm 1 identifies conjunctive grammars with the k-FCP in the limit from positive data and membership queries.*

Algorithm 1 Learning conjunctive grammars with k-FCP

Data: A positive presentation $u_1, u_2, \ldots$ of L_*; membership oracle $\mathcal{O}$ for L_*;
Result: A sequence of conjunctive grammars $G_1, G_2, \ldots$
let $D := \varnothing$; $F := \varnothing$; $K := \varnothing$; $\hat{G} := \mathcal{G}^k(F, K)$;
for $n = 1, 2, \ldots$ **do**
  let $D := D \cup \{u_n\}$; $K := \mathrm{Sub}(D)$;
  if $D \nsubseteq \mathcal{L}(\hat{G})$ **then**
    let $F := \mathrm{Con}(D)$;
  end if
  output $\hat{G} = \mathcal{G}^k(F, K)$ as G_n;
end for

Proof. Let $D_n = \{u_1, \ldots, u_n\}$. Lemma 6 ensures that Algorithm 1 does not update F infinitely many times. Let $F_{m_0} = \mathrm{Con}(D_{m_0})$ be the limit of F. There is a point n_0 such that $\mathcal{G}^k(F_{m_0}, K_{n_0})$ has no incorrect rules for $K_{n_0} = \mathrm{Sub}(D_{n_0})$ by Lemma 7. For any $n \geq \max\{m_0, n_0\}$, Algorithm 1 outputs $\hat{G}_n = \mathcal{G}^k(F_{m_0}, K_{n_0})$, which contains all and only correct rules. By Lemma 8, $\mathcal{L}(\hat{G}_n) \subseteq L_*$. By the choice of m_0, it is impossible that $\mathcal{L}(\hat{G}_n) \subsetneq L_*$. $\square$

We remark on the efficiency of our algorithm. Algorithm 1 updates its conjecture in polynomial time in the data size. Moreover, to get characterizing contexts of all nonterminals in V_* of G_*, $k|V_*|$ examples are enough. To suppress incorrect rules, $\mathrm{O}(|\hat{V}|^3)$ substrings are enough by Lemma 7.

3.2 Hypothesis Grammar with Closed Nonterminals

Our hypothesis grammars $\mathcal{G}^k(F, K)$ may have many nonterminals playing the same role. Observing that two nonterminals $[\![C_1]\!]$ and $[\![C_2]\!]$ with $C_1^{(K)} = C_2^{(K)}$ occur exactly at the same positions on the right-hand side of rules, we can pick only one of those equivalent nonterminals. We call C *closed* if $C^{(K)(F)} = C$, which is the maximum element of the equivalence class $\{ B \mid B^{(K)} = C^{(K)} \}$. An alternative construction of a hypothesis $\mathcal{H}^k(F, K)$ which uses only closed sets for nonterminals is obtained by merging all nonterminals $[\![C]\!]$ of $\mathcal{G}^k(F, K)$ into $[\![C^{(K)(F)}]\!]$. It is obvious that $\mathcal{L}(\mathcal{H}^k(F, K), [\![C]\!]) = \bigcup\{ \mathcal{L}(\mathcal{G}^k(F, K), [\![B]\!]) \mid B^{(K)} = C^{(K)} \}$, which implies $\mathcal{L}(\mathcal{H}^k(F, K)) = \mathcal{L}(\mathcal{G}^k(F, K))$.

Another idea to construct a hypothesis grammar with nonterminals indexed with closed sets is given by Leiß [6] for learning CFGs with the FCP. His algorithm updates its hypothesis only when $(C_1^{(K)} C_2^{(K)})^{(F)} = (C_1^{(K)} C_2^{(K)})^{(F)(K)(F)}$ or $(C_1^{(K)} C_2^{(K)})^{(F)} = \varnothing$ for all $C_1, C_2 \subseteq F$. We can take this idea where the rule construction need not be altered from $\mathcal{G}^k(F, K)$ except that all nonterminal symbols are indexed with closed sets.

3.3 Non-binary Form

Just for simplicity we assume that target languages are generated by conjunctive grammars in the binary form. Although every conjunctive grammar can be converted into this form preserving its language, the FCP is not necessarily preserved. For example, the language $L_{\mathrm{CEH}} = \{\, a^m b \mid m > 0 \,\} \cup \{\, a^m c^n \mid 0 < m < n \,\}$ [4] can be generated by a CFG with the FCP[2] but it cannot be in the binary form. To overcome this problem, we may allow grammar rules to have a more general form

$$[\![C]\!] \leftarrow \alpha_1 \& \ldots \& \alpha_m \text{ with } \alpha_i \in (\Sigma \cup \hat{V})^* \text{ for each } i$$

where

- if $m \geq 2$ then each α_i contains at least one nonterminal symbol,
- for $\alpha_i = u_0 [\![C_1]\!] u_1 \ldots [\![C]\!]_{n_i} u_{n_i}$, each u_j occurs in some elements of F,
- and some restriction should be satisfied so that polynomially many rules are possible in a hypothesis grammar.

An example of the third condition is to bound the number of nonterminals that occur in the right-hand side of each rule (cf. [10]).

Under this modification, Leiß's condition for updating hypotheses should also be modified accordingly.

4   Contextual Binary Feature Grammars

Clark et al. [4] introduced *contextual binary feature grammars* (CBFGs) as distributionally learnable representations of languages. They are essentially notational variants of conjunctive grammars with some minor restriction.

Definition 10 (Clark et al. [4]). A *contextual binary feature grammar* (CBFG) over Σ is a quadruple $G = \langle F, P_0, P_1, \Sigma \rangle$ where

- $F \subseteq \Sigma^* \times \Sigma^*$ is a finite set of contexts,
- P_0 is a finite set of rules of the form $C \leftarrow a$ for $C \subseteq F$ and $a \in \Sigma$,
- P_1 is a finite set of rules of the form $C \leftarrow C_1 C_2$ for $C, C_1, C_2 \subseteq F$.

A CBFG assigns a context set to each string recursively by

$$f_G(\lambda) = \varnothing,$$
$$f_G(a) = \bigcup \{\, C \mid C \leftarrow a \in P_0 \,\} \text{ for } a \in \Sigma,$$
$$f_G(u) = \bigcup \{\, C \mid C \leftarrow C_1 C_2 \in P_1, \, C_i \subseteq f_G(u_i) \text{ for } i = 1, 2$$
$$\text{for some } u_1, u_2 \in \Sigma^+ \text{ with } u = u_1 u_2 \,\} \text{ for } |u| > 1 .$$

The language defined by G is

$$\mathcal{L}(G) = \{\, u \in \Sigma^+ \mid \langle \lambda, \lambda \rangle \in f_G(u) \,\}.$$

---

[2] The grammar with the following rules have the FCP and generate L_{CEH}.

$$S \leftarrow Xb \mid aYc, \ X \leftarrow aX \mid a, \ Y \leftarrow aYc \mid Yc \mid c.$$

Clark et al. [4] showed that for every conjunctive grammar G one can construct a CBFG G' generating almost the same language: $\mathcal{L}(G') = \mathcal{L}(G)\#$ for a special end marker $\# \notin \Sigma$. The converse is trivial.

Proposition 11. *Every* CBFG *G has an equivalent conjunctive grammar G'.*

Proof. For $G = \langle F, P_0, P_1, \Sigma \rangle$, define $G' = \langle \Sigma, F_0' \cup F_1', P_0' \cup P_1' \cup P_2', (\lambda, \lambda) \rangle$ where

$$F_0' = \{\, [\![w]\!] \mid w \in F \,\},$$
$$F_1' = \{\, [\![C]\!] \mid C \text{ occurs on the right hand side of some rule of } P_1 \,\},$$
$$P_0' = \{\, [\![w]\!] \leftarrow a \mid w \in C \text{ and } C \leftarrow a \in P_0 \text{ with } a \in \Sigma \text{ for some } C \subseteq F \,\},$$
$$P_1' = \{\, [\![w]\!] \leftarrow [\![C_1]\!][\![C_2]\!] \mid w \in C \text{ and } C \leftarrow C_1 C_2 \in P_1 \text{ for some } C \subseteq F \,\},$$
$$P_2' = \{\, [\![C]\!] \leftarrow [\![w_1]\!] \& \ldots \& [\![w_n]\!] \mid C = \{w_1, \ldots, w_n\} \in F' \,\}.$$

It is easy to see that $u \in \mathcal{L}(G', [\![w]\!])$ if and only if $w \in f_G(u)$. $\square$

4.1 Comparison of Clark et al.'s Learner and Ours

Clark et al. [4] proposed a learning algorithm for CFGs with special conditions. The hypotheses output by the algorithm are CBFGs. Although their main theorem mentions the learnability of the specific kind of CFGs, certainly their algorithm learns some CBFGs generating non-context-free languages. In this regard, Clark et al. have already shown a positive result on the learnability of some conjunctive grammars. We show that our result is stronger than Clark et al.'s. Clark et al. [4] showed the learnability of the class of CFLs that are generated by CFGs with the *finite kernel property* (FKP) and that have the *finite fiducial set property* (FFP). Those properties are defined as follows. In what follows, we allow CFGs to have multiple initial symbols. The language of a CFG is defined to be the union of all the languages defined by those initial symbols.

Definition 12 (Yoshinaka [9]). A CFG G with multiple initial symbols has the FKP if every nonterminal X has a string $u_X \in \mathcal{L}(G, X)$ such that $\mathcal{L}(G, X)^\dagger = u_X^\dagger$.

Definition 13. A language L has the FFP if every string $u \in \Sigma^*$ has a finite context set C_u such that $C_u^\dagger = u^\dagger$.

The above definitions might look different from the original ones. The equivalence to Clark et al.'s original definitions is shown in the appendix. Now the following proposition is trivial.

Proposition 14. *If a grammar G satisfies the FKP and its language $\mathcal{L}(G)$ has the FFP, then G satisfies the weak FCP.*

Proof. Each nonterminal X of G has a string u_X such that $\mathcal{L}(G, X)^\dagger = u_X^\dagger$ by the FKP. By the FFP, u_X has a finite context set C_X such that $u_X^\dagger = C_X^\dagger$. That is, $\mathcal{L}(G, X)^\dagger = C_X^\dagger$. $\square$

If Definition 12 is strengthened so that $\mathcal{L}(G, X) = u_X^\dagger$, then the conclusion of Proposition 14 becomes that G satisfies the (stronger) FCP.

Yoshinaka [9] showed that actually either the weak FCP or FKP is enough to make CFGs learnable, whereas Proposition 14 clarifies that Clark et al. [4] require both properties. Their learner's hypothesis grammars have a branching rule $C \leftarrow C_1 C_2$ if there are $u_1, u_2 \in K$ such that $C = (u_1 u_2)^{(F)}$ and $C_i = u_i^{(F)}$ for $i = 1, 2$. This rule will be correct, in the sense that $C \odot C_1^\dagger C_2^\dagger \subseteq L_*$, if $C_i^\dagger = u_i^\dagger$. This is the reason why the FFP is required in addition to the FKP. Although they use contexts as features, rules are determined by strings. This inconsistent strategy demands learning targets to satisfy both the FCP and FKP.

We remark that there is a CFG G with the FCP whose language $\mathcal{L}(G)$ does not satisfy the FFP. An example is L_{CEH} (Sec. 3.3). The string b has no finite context set C such that $C_b^\dagger = b^\dagger$. On the other hand, there is a CFL with the FFP that has no CFG satisfying the FCP, e.g., $L_2 = \{ a^m \mid m \geq 0 \} \cup \{ a^m b^{m'} c b^{n'} a^n \mid m \leq m' \text{ and } n \leq n' \}$. One can easily verify that L_2 has the FFP. If a CFG generates L_2, it must have a nonterminal X whose language L_X is an infinite subset of a^*. We have $L_X^\dagger = \{ \langle a^m, a^n \rangle \mid m, n \geq 0 \}$ but no finite subset of $L_X^\dagger$ characterizes X. For any finite subset F of $L_X^\dagger$, one can find n such that $F \subseteq (b^n c b^n)^{(F)}$, for which $\langle a^{n+1}, a^{n+1} \rangle \in L_X^\dagger - (b^n c b^n)^{(F)}$.

4.2   Exact Contextual Binary Feature Grammars

Although CBFGs use contexts to control derivations, choice of contexts is arbitrary. They may be completely irrelevant of the language defined by the grammar. A restriction that establishes a clear and strong relation between features and their semantics is the *exactness*.

Definition 15 (Clark et al. [4]). A CBFG is said to be *exact* if for all $u \in \Sigma^+$, $f_G(u) = \{u\}^{(F)}$.

Clark et al. [4] emphasize the importance of this property. Indeed their and our learners construct a hypothesis grammar so that it can be exact. However, they found no relation between exactness and learnability of CBFGs, while conjecturing that there should be a learning algorithm for exact CBFGs Actually the exactness entails the FCP through the straightforward translation of CBFGs into conjunctive grammars.

Lemma 16. *If a CBFG G is exact, then the conjunctive grammar G' obtained by the straightforward translation has the FCP.*

Proof. Let G' be the conjunctive grammar obtained by the method of the proof of Proposition 11 from an exact CBFG G. It is obvious that nonterminals $[\![w]\!] \in F_0'$ and $[\![C]\!] \in F_1'$ of G' are characterized by $\{w\}$ and C, respectively. $\square$

Our learning algorithm can be translated for CBFGs accordingly. Actually conjunctive grammars $\mathcal{G}^k(F, K)$ output by our learner are almost CBFGs as nonterminals are indexed with sets of contexts. A translation of $\mathcal{G}^k(F, K)$ will be obtained by removing conjunctive rules and replacing all nonterminals $[\![C]\!]$ by the context sets C. Let us call a CBFG a k-CBFG if every context set used in rules has cardinality at most k.

Corollary 17. *Exact k-CBFGs are polynomial time identifiable in the limit from positive data and membership queries.*

5   Primal Approach

Existing distributional learning algorithms are classified into two types. A *primal* approach uses strings to define the semantics of nonterminals, and a *dual* approach uses contexts. Our approach taken in this paper is dual in this sense. Yoshinaka [9,10] showed a neat symmetry between primal and dual approaches and gave an algorithm that integrates the two. However it does not seem straightforward to design a correct learner of primal type for conjunctive grammars with the FKP. Another formalism that only a dual approach is known to work for is *parallel (multiple)* CFGs [5]. The current state of the art might suggest dual approaches have an advantage over primal approaches, but further investigation is needed to clarify the relation of the two types of approaches.

A   Equivalence of Clark et al.'s and Our Definitions

The original definition of the FKP by Clark et al. [4] is as follows.

Definition 18 (Clark et al. [4]). A CFG G has the FKP if every nonterminal X admits a finite set $S_X \subseteq \mathcal{L}(G, X)$ such that

- $a \in S_X$ if $a \in \mathcal{L}(G, X)$,
- for every $v \in \mathcal{L}(G, X)$ there is $u \in S_X$ such that $u^\ddagger \subseteq v^\ddagger$.

Obviously every grammar satisfying Definition 12 also satisfies Definition 18. We show the converse.

Proposition 19. *Every grammar G satisfying Definition 18 has an equivalent grammar G' satisfying Definition 12.*

Proof. Let S_X be a finite string set for each nonterminal X of a grammar G satisfying Definition 18. Suppose K includes S_X for each nonterminal X. For simplicity we assume that every rule of G is either $X \leftarrow a$ with $a \in \Sigma_\lambda$ or $X \leftarrow YZ$ for some $X, Y, Z \in V$. Let us define $G' = \langle \Sigma, V_K, P_K, I_K \rangle$, where $I_K \subseteq V_K$ is the set of initial symbols, by

- $V_K = \{ [\![u]\!] \mid u \in K \}$,
- $P_K = \{ [\![a]\!] \leftarrow a \mid a \in \Sigma_\lambda \} \cup \{ [\![u]\!] \leftarrow [\![u_1]\!][\![u_2]\!] \mid u^\ddagger \odot u_1 u_2 \subseteq \mathcal{L}(G) \}$,
- $I_K = \{ [\![u]\!] \mid u \in \mathcal{L}(G) \}$.

$[\mathcal{L}(G') \subseteq \mathcal{L}(G)]$ One can show by induction on derivation that $v \in \mathcal{L}(G', \llbracket u \rrbracket)$ implies $u^{\ddagger} \odot v \subseteq \mathcal{L}(G)$. This implies $u^{\ddagger} \odot \mathcal{L}(G', \llbracket u \rrbracket) \subseteq \mathcal{L}(G)$. Applying this claim to the initial symbols $\llbracket u \rrbracket$ of G', we obtain that $\mathcal{L}(G', \llbracket u \rrbracket) \subseteq \mathcal{L}(G)$.

$[\mathcal{L}(G) \subseteq \mathcal{L}(G')]$ For every $u \in S_X$ of every initial symbol X of G, the fact $u \in \mathcal{L}(G)$ implies $\llbracket u \rrbracket \in I_K$. It is enough to show that if $v \in \mathcal{L}(G, X)$ then $v \in \mathcal{L}(G', \llbracket u \rrbracket)$ for some $u \in S_X$ such that $u^{\ddagger} \subseteq v^{\ddagger}$. For X an initial symbol of G, This holds true for $v \in \Sigma_\lambda$ with $u = v$ by construction. Suppose $v = v_1 v_2 \in \mathcal{L}(G, X)$, $v_i \in \mathcal{L}(G, Y_i)$ for $i = 1, 2$ and $X \leftarrow YZ \in P$. We have $\llbracket u_i \rrbracket \in V_K$ such that $v_i \in \mathcal{L}(G', \llbracket u_i \rrbracket)$ and $u_i^{\ddagger} \subseteq v_i^{\ddagger}$ for $i = 1, 2$. Since $u_1 u_2 \in \mathcal{L}(G, Y_1 Y_2) \subseteq \mathcal{L}(G, X)$, one can find $u \in S_X$ be such that $u^{\ddagger} \subseteq (u_1 u_2)^{\ddagger}$. Then G' has the the rule $\llbracket u \rrbracket \leftarrow \llbracket u_1 \rrbracket \llbracket u_2 \rrbracket$ and $v \in \mathcal{L}(G', \llbracket u \rrbracket)$. The fact $v_i^{\dagger} \subseteq u_i^{\dagger}$ implies $v^{\dagger} = (v_1 v_2)^{\dagger} \subseteq (u_1 u_2)^{\dagger} \subseteq u^{\dagger}$.

[FKP] In the proof for $\mathcal{L}(G') \subseteq \mathcal{L}(G)$, we have already seen that $u^{\ddagger} \odot \mathcal{L}(G', \llbracket u \rrbracket) \subseteq \mathcal{L}(G)$, which means $\mathcal{L}(G', \llbracket u \rrbracket) \subseteq u^{\dagger}$. The fact $u \in \mathcal{L}(G', \llbracket u \rrbracket)$ implies $u^{\dagger} \subseteq \mathcal{L}(G', \llbracket u \rrbracket)^{\dagger}$. Hence G' satisfies Definition 12. $\qquad\square$

The original definition of the FFP by Clark et al. [4] is as follows.[3]

Definition 20 (Clark et al. [4]). A language L has the FFP if every string u admits a finite context set (called a *fiducial set*) F such that for any $v \in \Sigma^*$, $u^{(F)} \subseteq v^{(F)}$ iff $u^{\ddagger} \subseteq v^{\ddagger}$.

Proposition 21. *Definitions 13 and 20 are equivalent.*

Proof. If F is a fiducial set of $u \in \Sigma^*$, then $u^{(F)}$ is satisfies Definition 13, since

$$u^{\dagger} = \{\, v \mid u^{\ddagger} \subseteq v^{\ddagger} \,\} = \{\, v \mid u^{(F)} \subseteq v^{(F)} \,\} = (u^{(F)})^{\dagger}.$$

On the other hand, if C satisfies Definition 13 for u, it is a fiducial set, since for any v,

$$C = u^{(C)} \subseteq v^{(C)} \iff C \odot v \subseteq L \iff u^{\ddagger} \odot v \subseteq L \iff u^{\ddagger} \subseteq v^{\ddagger}. \qquad \square$$

Acknowledgements. The authors are grateful to the anonymous reviewers for valuable comments that have improved the quality of this paper. This work was supported in part by JSPS KAKENHI.

References

1. Clark, A.: A learnable representation for syntax using residuated lattices. In: de Groote, P., Egg, M., Kallmeyer, L. (eds.) Formal Grammar. LNCS, vol. 5591, pp. 183–198. Springer, Heidelberg (2011)
2. Clark, A.: Learning context free grammars with the syntactic concept lattice. In: Sempere, J.M., García, P. (eds.) ICGI 2010. LNCS, vol. 6339, pp. 38–51. Springer, Heidelberg (2010)

---

[3] In the original paper [4], this is called *finite context-property* (FCP). In this paper we reserve the term FCP for the meaning defined by Clark [1] rather than by [4].

3. Clark, A.: Towards general algorithms for grammatical inference. In: Hutter, M., Stephan, F., Vovk, V., Zeugmann, T. (eds.) Algorithmic Learning Theory. LNCS, vol. 6331, pp. 11–30. Springer, Heidelberg (2010)
4. Clark, A., Eyraud, R., Habrard, A.: Using contextual representations to efficiently learn context-free languages. Journal of Machine Learning Research **11**, 2707–2744 (2010)
5. Clark, A., Yoshinaka, R.: Distributional learning of parallel multiple context-free grammars. Machine Learning **96**(1–2), 5–31 (2014)
6. Leiß, H.: Learning context free grammars with the finite context property: a correction of A. Clark's Algorithm. In: Morrill, G., Muskens, R., Osswald, R., Richter, F. (eds.) Formal Grammar. LNCS, vol. 8612, pp. 121–137. Springer, Heidelberg (2014)
7. Okhotin, A.: Conjunctive grammars. Journal of Automata, Languages and Combinatorics **6**(4), 519–535 (2001)
8. Okhotin, A.: The dual of concatenation. Theoretical Computer Science **345**(2–3), 425–447 (2005)
9. Yoshinaka, R.: Towards dual approaches for learning context-free grammars based on syntactic concept lattices. In: Mauri, G., Leporati, A. (eds.) DLT 2011. LNCS, vol. 6795, pp. 429–440. Springer, Heidelberg (2011)
10. Yoshinaka, R.: Integration of the dual approaches in the distributional learning of context-free grammars. In: Dediu, A.-H., Martín-Vide, C. (eds.) LATA 2012. LNCS, vol. 7183, pp. 538–550. Springer, Heidelberg (2012)

Graphs, Term Rewriting, and Networks

Recognizable Series on Hypergraphs

Raphaël Bailly[1], François Denis[2], and Guillaume Rabusseau[2](✉)

[1] Université de Technologie de Compiègne, Heudiasyc, France
`baillyra@utc.fr`
[2] LIF - CNRS UMR, Aix-Marseille Université, 7279 Marseille, France
{`francois.denis,guillaume.rabusseau`}`@lif.univ-mrs.fr`

Abstract. We introduce the notion of *Hypergraph Weighted Model* (HWM) that generically associates a tensor network to a hypergraph and then computes a value by tensor contractions directed by its hyperedges. A series r defined on a hypergraph family is said to be recognizable if there exists a HWM that computes it. This model generalizes the notion of recognizable series on strings and trees. We present some properties of the model and study at which conditions finite support series are recognizable.

1 Introduction

Real-valued functions whose domains are composed of syntactical structures, such as strings, trees or graphs, are widely used in computer science. One way to handle them is by means of rational series that use automata devices to jointly analyze the structure of the input and compute its image. Rational series have been defined for strings and trees, but their extension to graphs is challenging.

On the other hand, rational series have an equivalent algebraic characterization by means of linear (or multi-linear) representations. We show in this paper that this last formalism can be naturally extended to graphs (and hypergraphs) by associating tensors to the vertices of the graph.

More precisely, we define the notion of *Hypergraph Weighted Model* (HWM), a computational model that generically associates a tensor network to a hypergraph and that computes a value by successive generalized tensor contractions directed by its hyperedges. We say that a series r defined on a hypergraph family is HWM-recognizable if there exists a HWM M that computes it: we then denote r by r_M. We first show that HWM-recognizable series defined on strings or trees exactly recover the classical notion of recognizable series. We present two closure properties: if r and s are two recognizable series defined on a family $\mathcal{H}$ of connected hypergraphs, then $r + s$ and $r \cdot s$, respectively defined for all graphs $G \in \mathcal{H}$ by $(r + s)(G) = r(G) + s(G)$ and $(r \cdot s)(G) = r(G)s(G)$ (the Hadamard product), are HWM-recognizable.

Raphaël Bailly is funded through a Google Faculty Research Award, and this work has been carried out thanks to the support of the ARCHIMEDE Labex (ANR-11-LABX- 0033) and the A*MIDEX project (ANR-11-IDEX-0001-02) funded by the "Investissements d'Avenir" French government program managed by the ANR.

© Springer International Publishing Switzerland 2015
A.-H. Dediu et al. (Eds.): LATA 2015, LNCS 8977, pp. 639–651, 2015.
DOI: 10.1007/978-3-319-15579-1_50

Recognizable series on strings and trees include polynomials, i.e. finite support series. This is not always the case for recognizable series defined on more general families of hypergraphs. For example, we show that finite support series are not recognizable on the family of circular strings. The main reason is that if a recognizable series is not null on some hypergraph G, it must be also different from zero on *tilings* of G, i.e. connected graphs made of copies of G. We show that if a graph family is tiling-free, then finite support series are recognizable. Strings and trees, as any family of rooted hypergraphs, are tiling-free.

String rational series and weighted automaton have their roots in automata theory [9,16] and their study can be found in [4,8,11,14,15]. The extension of rational series and weighted automaton to trees is presented in [3,8]. Spectral methods for inference of stochastic languages of strings/trees have been developed upon the notion of linear representation of a rational series ([2,7] for example). Tensor networks emerged in the theory of brain functions [13], they have been used in quantum theory (see for example [12]), and the interest for these objects has recently been growing in other fields (e.g. data mining [5]).

We recall notions on tensors and hypergraphs in Section 2, we introduce the Hypergraph Weighted Model and present some of its properties in section 3, we introduce the notion of tilings and we study the recognizability of finite support series in Section 4, we provide some examples in Section 5 and we then propose a short conclusion.

Most of the proofs have been omitted for brevity but can be found in [1].

2   Preliminaries

2.1   Rational Series on Strings and Trees

We refer to [3,4,6,8,14] for notions about recognizable series on strings and trees, and we briefly recall below some basic definitions.

Let Σ be a finite *alphabet*, and Σ^* be the set of strings on Σ. A *series* on Σ^* is a mapping $r : \Sigma^* \to \mathbb{K}$ where $\mathbb{K}$ is a semiring. A series r is *recognizable* if there exists a tuple $\langle V, \iota, \{\mathbf{M}_x\}_{x \in \Sigma}, \boldsymbol{\tau} \rangle$ where $V = \mathbb{K}^d$ for some integer $d \geq 1$, $\iota, \boldsymbol{\tau} \in V$ and $\mathbf{M}_x \in \mathbb{K}^{d \times d}$ for each symbol $x \in \Sigma$, such that for any $u_1 \ldots u_n \in \Sigma^*$, $r(u_1 \ldots u_n) = \iota^\top \mathbf{M}_{u_1} \ldots \mathbf{M}_{u_n} \boldsymbol{\tau}$. In this paper, we will only consider the case where $\mathbb{K} = \mathbb{R}$ or $\mathbb{C}$.

A *ranked* alphabet $\mathcal{F}$ is a tuple $(\Sigma, \sharp)$ where Σ is a finite alphabet and where $\sharp$ maps each symbol x of Σ to an integer $\sharp x$ called its *arity*; for any $k \in \mathbb{N}$, let us denote $\mathcal{F}_k = \sharp^{-1}(\{k\})$. A ranked alphabet is *positive* if $\sharp$ takes its values in $\mathbb{N}_+$.

The set of trees over a ranked alphabet $\mathcal{F}$ is denoted by $T(\mathcal{F})$. A *tree series* on $T(\mathcal{F})$ is a mapping $r : T(\mathcal{F}) \to \mathbb{K}$. A series r is *recognizable* if there exists a tuple $\langle V, \mu, \boldsymbol{\lambda} \rangle$, where $V = \mathbb{K}^d$ for some integer $d \geq 1$, μ maps each $f \in \mathcal{F}_p$ to a p-multilinear mapping $\mu(f) \in \mathcal{L}(V^p; V)$ for each $p \geq 0$ and $\boldsymbol{\lambda} \in V$, such that $r(t) = \boldsymbol{\lambda}^\top \mu(t)$ for all t in $T(\mathcal{F})$, where $\mu(t) \in V$ is inductively defined by $\mu(f(t_1, \ldots, t_p)) = \mu(f)(\mu(t_1), \ldots, \mu(t_p))$.

2.2   Tensors

Let $d \geq 1$ be an integer, $V = \mathbb{K}^d$ where $\mathbb{K} = \mathbb{R}$ or $\mathbb{C}$ and let $(\mathbf{e}_1, \ldots, \mathbf{e}_d)$ be the canonical basis of V. A tensor $\mathcal{T} \in \bigotimes^k V = V \otimes \cdots \otimes V$ (k times) can uniquely be expressed as a linear combination $\mathcal{T} = \sum_{i_1, \ldots, i_k \in [d]} \mathcal{T}_{i_1 \ldots i_k} \mathbf{e}_{i_1} \otimes \cdots \otimes \mathbf{e}_{i_k}$ (where $[d] = \{1, \cdots, d\}$) of *pure tensors* $\mathbf{e}_{i_1} \otimes \cdots \otimes \mathbf{e}_{i_k}$ which form a basis of $\bigotimes^k V$ [10]. Hence, the tensor $\mathcal{T}$ can be represented as the multi-array $(\mathcal{T}_{i_1 \ldots i_k})$.

Definition 1. *The* tensor product *of $\mathcal{T} \in \bigotimes^p V$ and $\mathcal{U} \in \bigotimes^q V$ is the tensor $\mathcal{T} \otimes \mathcal{U} \in \bigotimes^{p+q} V$ defined by $(\mathcal{T} \otimes \mathcal{U})_{i_1 \cdots i_p j_1 \cdots j_q} = \mathcal{T}_{i_1 \cdots i_p} \mathcal{U}_{j_1 \cdots j_q}$. For any $\mathbf{v} \in \mathbb{K}^d$, let $\mathbf{v}^{\otimes k} = \mathbf{v} \otimes \cdots \otimes \mathbf{v}$ (k times) denote its k-th tensor power.*

Let $\odot : V \times V \to V$ be an associative and symmetric bilinear mapping: $\forall u, v, w \in V, u \odot v = v \odot u$ and $u \odot (v \odot w) = (u \odot v) \odot w$. The mapping $\odot$ is called a *product*.

Remark 2. Let $\mathbf{1} = (1, \ldots, 1)^{\top}$ and let $\odot_{id}$ be defined by $\mathbf{e}_i \odot_{id} \mathbf{e}_j = \delta_{ij} \mathbf{e}_i$, where δ is the Kronecker symbol: $\odot_{id}$ is called the *identity product*. The operation of applying the linear form $\mathbf{v} \mapsto \mathbf{1}^{\top} \mathbf{v}$ to the identity product $\mathbf{a} \odot_{id} \mathbf{b}$ of two vectors is related to the notions of *generalized trace* and *contraction*: if $\mathcal{A} = \sum_{i,j \in [d]} \mathcal{A}_{i,j} \mathbf{e}_i \otimes \mathbf{e}_j$ is a 2-order tensor over $\mathbb{K}^d$ (i.e. a square matrix), $\mathbf{v} = \sum_{i,j \in [d]} \mathcal{A}_{i,j} \mathbf{e}_i \odot_{id} \mathbf{e}_j$ is the diagonal vector of $\mathcal{A}$ and $\mathbf{1}^{\top} \mathbf{v}$ is its trace. Furthermore, if $\mathcal{A} = \sum_{i,j \in [d]} \mathcal{A}_{i,j} \mathbf{e}_i \otimes \mathbf{e}_j$ and $\mathcal{B} = \sum_{i,j \in [d]} \mathcal{B}_{i,j} \mathbf{e}_i \otimes \mathbf{e}_j$ are 2-order tensors over $\mathbb{K}^d$, then $\sum_{i,j,k,l} \mathcal{A}_{i,j} \mathcal{B}_{k,l} \mathbf{e}_i \otimes \mathbf{1}^{\top} (\mathbf{e}_j \odot_{id} \mathbf{e}_k) \otimes \mathbf{e}_l$ is the tensor form of the matrix product $\mathcal{A} \cdot \mathcal{B}$ (i.e. the contraction of the tensor $\mathcal{A} \otimes \mathcal{B}$ along its 2nd and 3rd modes).

2.3   Hypergraphs

Definition 3. *A* hypergraph *$G = (V, E, l)$ over a positive ranked alphabet $(\Sigma, \sharp)$ is given by a non empty finite set V, a mapping $l : V \to \Sigma$ and a partition $E = (h_k)_{1 \leq k \leq n_E}$ of $P_G = \{(v, j) : v \in V, 1 \leq j \leq \sharp v\}$ where $\sharp v = \sharp l(v)$.*

V is the set of *vertices*, P_G is the set of *ports* and E is the set of *hyperedges* of G. The arity of a symbol x is equal to the number of ports of any vertex labelled by x. We will sometimes use the notation $v^{(i)}$ for the port $(v, i) \in P_G$. A hypergraph G can be represented as a bipartite graph where vertices from one partite set represent the vertices of G, and vertices from the other represent its hyperedges (see Figure 4). A hypergraph is *connected* if for any partition $V = V_1 \cup V_2$, there exists a hyperedge $h \in E$ and ports $v_1^{(i)}, v_2^{(j)} \in h$ s.t. $v_1 \in V_1$ and $v_2 \in V_2$. A hypergraph is a *graph* if $|h| \leq 2$ for all $h \in E$, and a hypergraph is *closed* if $|h| \geq 2$ for all $h \in E$.

Example 4. Over the ranked alphabet $\{(a, 3), (b, 2)\}$, let $V = \{v_1, v_2, v_3\}$, $l(v_1) = l(v_3) = a$, $l(v_2) = b$, $E = \{h_1, h_2, h_3, h_4\}$ where $h_1 = \{v_1^{(1)}, v_3^{(3)}\}$, $h_2 = \{v_1^{(2)}, v_2^{(1)}, v_3^{(2)}\}$, $h_3 = \{v_1^{(3)}, v_2^{(2)}\}$ and $h_4 = \{v_3^{(1)}\}$ (see Figure 1).

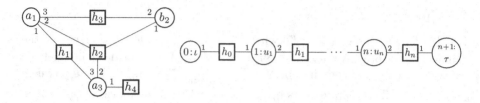

Fig. 1. (left) The hypergraph G from example 4. (right) Graph associated with a string $u = u_1 \cdots u_n$ (where $i : x$ means that $\ell(i) = x$).

Example 5. A string $u = u_1 \ldots u_n$ over an alphabet Σ can be seen as a (hyper)graph over the ranked alphabet $(\Sigma \cup \{\iota, \tau\}, \sharp)$ where $\sharp x = 2$ for any $x \in \Sigma$ and $\sharp \iota = \sharp \tau = 1$. Let $V = \{0, \cdots, n + 1\}$, $l(0) = \iota$, $l(n + 1) = \tau$ and $l(i) = u_i$ for $1 \leq i \leq n$. Let $E = \{h_0, h_1, \ldots, h_n\}$ where $h_0 = \{(0, 1), (1, 1)\}$ and $h_i = \{(i, 2), (i + 1, 1)\}$ for $1 \leq i \leq n$ (see Figure 1). The set of strings Σ^* gives rise to a family of hypergraphs.

Example 6. Similarly, we can associate any tree t over a ranked alphabet $(\Sigma, \sharp)$ with a graph G_t on the ranked alphabet $(\Sigma \cup \{\lambda\}, \sharp')$ where $\sharp'(f) = \sharp f + 1$ for any $f \in \Sigma$, and where the special symbol λ of arity 1 is connected to the free port of the vertex corresponding to the root of t. The explicit construction of G_t can be found in [1], and the graph associated with the tree $t = f(a, f(a, a))$ is shown as an example in Figure 2.

Example 7. Given a finite alphabet Σ, let $\mathcal{F} = (\Sigma, \sharp)$ be the ranked alphabet where $\sharp x = 2$ for each $x \in \Sigma$. We say that a hypergraph $G = (V, E)$ on $\mathcal{F}$ is a circular string if and only if G is connected and every hyperedge $h \in E$ is of the form $h = \{(v, 2), (w, 1)\}$ for $v, w \subset V$ (scc Figure 2).

Example 8. An other interesting extension of strings (naturally modeled by graphs) is the set of 2D-words $w \in \Sigma^{M \times N}$ on a finite alphabet Σ, see Section 5 (Crosswords) for details.

3   Hypergraph Weighted Models

3.1   Definition

In this section, we give the formal definition of Hypergraph Weighted Models. We then explain how to compute its value for a given hypergraph.

Definition 9. *A Hypergraph Weighted Model (HWM) of dimension d on a ranked alphabet $(\Sigma, \sharp)$ is a tuple $M = \langle V_M, \{\mathcal{T}^x\}_{x \in \Sigma}, \odot, \boldsymbol{\alpha} \rangle$ where $V_M = \mathbb{K}^d$, $\odot$ is a product on V_M, $\boldsymbol{\alpha} \in V_M$, and $\{\mathcal{T}^x\}_{x \in \Sigma}$ is a family of tensors where each $\mathcal{T}^x \in \bigotimes^{\sharp x} V_M$.*

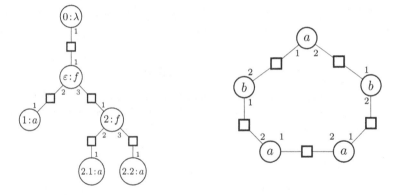

Fig. 2. (left) Hypergraph G_t associated with the tree $t = f(a, f(a, a))$. (right) Example of circular string on the alphabet $\{a, b\}$.

Let $G = (V, E, l)$ be a hypergraph and let $\Gamma = [d]^{P_G}$ be the set of mappings from P_G to $[d]$. The series r_M computed by the HWM M is defined by

$$r_M(G) = \sum_{\gamma \in \Gamma} \mathcal{T}_\gamma \prod_{h \in E} \boldsymbol{\alpha}^\top \underset{i \in \gamma(h)}{\bigodot} \mathbf{e}_i$$

where $\mathcal{T}_\gamma = \prod_{v \in V} \mathcal{T}^v_{\gamma(v^{(1)}) \dots \gamma(v^{(\sharp v)})}$ (using the notation $\mathcal{T}^v = \mathcal{T}^{l(v)}$).

Let $V = \{v_1, \cdots, v_n\}$. The tensor $\mathcal{T}^{v_1} \otimes \mathcal{T}^{v_2} \otimes \cdots \otimes \mathcal{T}^{v_n}$ is of order $|P_G|$, and any element $\gamma \in \Gamma$ can be seen as a multi-index of $[d]^{|P_G|}$. Thus, $\mathcal{T}_\gamma$ is the $\left(\gamma(v_1^{(1)}), \cdots, \gamma(v_1^{(\sharp v_1)}), \cdots, \gamma(v_n^{(1)}), \cdots, \gamma(v_n^{(\sharp v_n)})\right)$-coordinate of the tensor $\bigotimes_{i=1}^n \mathcal{T}^{v_i}$. Hence, the computation consists in forming the tensor product defined over all the vertices and reducing it along the hyperedges.

Example 10. Consider the hypergraph G from Example 4. We have

$$r_M(G) = \sum_{i_1, \cdots, i_8} \mathcal{T}^a_{i_1 i_2 i_3} \mathcal{T}^b_{i_4 i_5} \mathcal{T}^a_{i_6 i_7 i_8} \boldsymbol{\alpha}^\top (\mathbf{e}_{i_1} \odot \mathbf{e}_{i_8}) \boldsymbol{\alpha}^\top (\mathbf{e}_{i_2} \odot \mathbf{e}_{i_4} \odot \mathbf{e}_{i_7}) \boldsymbol{\alpha}^\top (\mathbf{e}_{i_3} \odot \mathbf{e}_{i_5}) \boldsymbol{\alpha}^\top \mathbf{e}_{i_6}.$$

Remark 11. If $\odot = \odot_{id}$ and if $\boldsymbol{\alpha} = \mathbf{1}$, then $r_M(G) = \sum_{\gamma \in \Gamma_{Id}} \mathcal{T}_\gamma$ where $\Gamma_{Id} = \{\gamma \in \Gamma : \forall h \in E, p, q \in h \Rightarrow \gamma(p) = \gamma(q)\}$.

For the hypergraph G from Example 4, this would lead to the following contractions of the tensor $\mathcal{T}^a \otimes \mathcal{T}^b \otimes \mathcal{T}^a$: $r_M(G) = \sum_{i_1, i_2, i_3, i_6} \mathcal{T}^a_{i_1 i_2 i_3} \mathcal{T}^b_{i_2 i_3} \mathcal{T}^a_{i_6 i_2 i_1}$.

Remark 12. Let Σ be a finite alphabet, let $\mathbf{M}_\sigma \in \mathbb{K}^{d \times d}$ for $\sigma \in \Sigma$ and let $A = \langle \mathbb{K}^d, \{\mathbf{M}_\sigma\}_{\sigma \in \Sigma}, \odot_{id}, \mathbf{1}\rangle$ be a HWM. For any non empty word $w = w_1 \cdots w_n \in \Sigma^*$ and its corresponding circular string G_w, it follows from Remark 2 that $r_A(G_w) = Tr(\mathbf{M}_{w_1} \cdots \mathbf{M}_{w_n})$ (where $Tr(\mathbf{M})$ is the trace of the matrix $\mathbf{M}$).

Remark 13. Let $A = \langle \mathbb{R}^d, \{\mathcal{A}^x\}_{x \in \Sigma}, \odot, \boldsymbol{\alpha}\rangle$ be a HWM. Each tensor $\mathcal{A}^x$ can be decomposed as a sum of rank one tensors $\mathcal{A}^x = \sum_{r=1}^R \mathbf{a}_r^{(x,1)} \otimes \cdots \otimes \mathbf{a}_r^{(x,\sharp x)}$ where R

is the maximum rank of the tensors $\boldsymbol{\mathcal{T}}^x$ for $x \in \Sigma$. The computation of A on $G = (V, E, \ell)$ can then be written as $r(G) = \prod_{h \in E} \boldsymbol{\alpha}^\top \left[\bigodot_{(v,i) \in h} \left(\sum_{r=1}^R \mathbf{a}_r^{(\ell(v),i)} \right) \right]$.

Remark 14. If G is a hypergraph with two connected components G_1 and G_2, we have $r_M(G) = r_M(G_1) \cdot r_M(G_2)$ for any HWM M.

Definition 15. *Let $\mathcal{H}$ be a family of hypergraphs on a ranked alphabet $(\Sigma, \sharp)$. We say that a hypergraph series $r : \mathcal{H} \to \mathbb{K}$ is (HWM-)recognizable if and only if there exists a HWM M such that $r_M(G) = r(G)$ for all $G \in \mathcal{H}$.*

3.2   Properties

In this section, we show that HWMs satisfy some basic properties which are desirable for a model extending the notion of recognizable series to hypergraphs. The following propositions (whose proofs can be found in [1]) show that the proposed model naturally generalizes the notion of linear representation of recognizable series on strings and trees.

Proposition 16. *Let $r = \langle V, \iota, \{M^\sigma\}_{\sigma \in \Sigma}, \tau \rangle$ be a recognizable series on Σ^*. For any word $w \in \Sigma^*$, let G_w be the associated hypergraph on the ranked alphabet $(\Sigma \cup \{\iota, \tau\}, \sharp)$, whose construction is described in Example 5. Consider the HWM $M = \langle V, \{\boldsymbol{\mathcal{T}}^x\}_{x \in \Sigma \cup \{\iota, \tau\}}, \odot_{id}, 1 \rangle$ where $\boldsymbol{\mathcal{T}}^\tau = \tau$, $\boldsymbol{\mathcal{T}}^\iota = \iota$ and $\boldsymbol{\mathcal{T}}^\sigma = M^\sigma$ for all $\sigma \in \Sigma$. Then, $r(w) = r_M(G_w)$ for all strings $w \in \Sigma^*$.*

Proposition 17. *Let $r = \langle V, \mu, \boldsymbol{\lambda} \rangle$ be a recognizable series on trees on the ranked alphabet $\mathcal{F} = (\Sigma, \sharp)$. For any tree t over $\mathcal{F}$, let G_t be the associated hypergraph on the ranked alphabet $(\Sigma \cup \{\lambda\}, \sharp')$ (see Example 6).*
There exists a HWM M such that $r_M(G_t) = r(t)$ for any tree t over $\mathcal{F}$.

The following propositions show that the set of HWMs is closed under addition and Hadamard product.

Proposition 18. *Let $A = \langle \mathbb{K}^m, \{\boldsymbol{\mathcal{A}}^x\}_{x \in \Sigma}, \odot_A, \boldsymbol{\alpha} \rangle$, $B = \langle \mathbb{K}^n, \{\boldsymbol{\mathcal{B}}^x\}_{x \in \Sigma}, \odot_B, \boldsymbol{\beta} \rangle$ be two HWMs. Let r_A (resp. r_B) be the series computed by A (resp. by B).*
There exists a HWM of dimension $m + n$ computing the series r_{A+B} defined by $r_{A+B}(G) = r_A(G) + r_B(G)$, for any connected hypergraph G.

Proposition 19. *Let $A = \langle \mathbb{K}^m, \{\boldsymbol{\mathcal{A}}^x\}_{x \in \Sigma}, \odot_A, \boldsymbol{\alpha} \rangle$, $B = \langle \mathbb{K}^n, \{\boldsymbol{\mathcal{B}}^x\}_{x \in \Sigma}, \odot_B, \boldsymbol{\beta} \rangle$ be two HWMs. Let r_A (resp. r_B) be the series computed by A (resp. by B).*
There exists a HWM of dimension mn computing the series $r_{A \cdot B}$ defined by $r_{A \cdot B}(G) = r_A(G) r_B(G)$, for any hypergraph G.

The proofs of these propositions use a construction similar to the one used to show the closure of recognizable series on strings under addition and Hadamard product. For the addition, the two HWMs are juxtaposed in the vector space $\mathbb{K}^{m+n}$, for the Hadamard product, they are combined in the vector space $\mathbb{K}^m \otimes \mathbb{K}^n$ (see [1] for details).

Finally, the next proposition (whose proof can be found in [1]) shows that any recognizable real valued series on closed graphs can be computed by a HWM with coefficients in $\mathbb{C}$ using the identity product $\odot_{id}$ and the vector $\mathbf{1}$.

Proposition 20. *Let* $A = \langle \mathbb{R}^d, \{\mathcal{A}^x\}_{x \in \Sigma}, \odot_A, \alpha \rangle$ *be a HWM. There exists a HWM* $B = \langle \mathbb{C}^d, \{\mathcal{B}^x\}_{x \in \Sigma}, \odot_{id}, 1 \rangle$ *such that* $r_B(G) = r_A(G)$ *for any closed graph* G.

4   Recognizability of Finite Support Series

In this section, we show that finite support series (or *polynomials*: series for which the set of hypergraphs with non-zero value is finite) are not recognizable in general, but we exhibit a wide class of families of hypergraphs for which they are.

First, we show on a simple example why polynomials are not recognizable for all families of hypergraphs. Consider the family of circular strings over a one letter alphabet $\Sigma = \{a\}$ introduced in Example 7 and Remark 12. The following lemma (whose proof can be found in [1]) implies that the series r, defined by $r(G_a) = 1$ and $r(G_{a^k}) = 0$ for all integer $k > 1$, is not recognizable. Indeed, r would be such that $r(G_{a^k}) = Tr(\mathbf{M}_a^k) = 0$ for all $k \geq 2$, but it then follows from Lemma 21 that $r(G_a) = Tr(\mathbf{M}_a) = 0$.

Lemma 21. *Let* $\mathbf{M} \in \mathbb{R}^{n \times n}$. *If* $Tr(\mathbf{M}^k) = 0$ *for all* $k \geq 2$, *then* $Tr(\mathbf{M}) = 0$.

This example illustrates the fact that the computation of a HWM on a hypergraph G is done independently on each hyperedge of G. This implies that if two hypergraphs are not distinguishable by just looking at the ports involved in their hyperedges, the computations of a HWM on these two hypergraphs are strongly dependent. This is clear if we consider a hypergraph G_1 made of two copies of a hypergraph G_2 (i.e. G_1 has two connected components, which are both isomorphic to G_2): we have $r(G_1) = r(G_2)^2$ for any HWM r (see Remark 14).

The following section formally introduces the notion of *tiling* of a hypergraph G and show how this relation between hypergraphs relates to the question of the recognizability of polynomials.

4.1   Tilings

A *tiling* of a hypergraph $\widehat{G}$ is a hypergraph G, built on the same alphabet and made of copies of $\widehat{G}$. More precisely,

Definition 22. *Let* $\widehat{G} = (\widehat{V}, \widehat{E}, \widehat{l})$ *be a hypergraph over a ranked alphabet* $(\Sigma, \sharp)$. *A hypergraph* $G = (V, E, l)$ *on the same alphabet* $(\Sigma, \sharp)$ *is a tiling of* $\widehat{G}$ *if and only if there exists a mapping* $f : V \to \widehat{V}$ *such that*

(i) $l(v) = \widehat{l}(f(v))$ *for any* $v \in V$
(ii) the mapping $g : P_G \to P_{\widehat{G}}$ *defined by* $g(v, i) = (f(v), i)$ *is such that for all* $h \in E$: $g(h) \in \widehat{E}$ *and the restriction* $g_{|h}$ *of* g *to* h *is bijective.*

The following proposition shows that for a connected hypergraph, this formal definition of tiling is equivalent to the intuition of a hypergraph made of copies of the original one.

Let $G = (V, E, l)$ be a tiling of the connected hypergraph $\widehat{G} = (\widehat{V}, \widehat{E}, \widehat{l})$, let $\sim_V$ be the equivalence relation defined on V by $v \sim_V v'$ iff $f(v) = f(v')$, and let $\sim_E$ be the equivalence relation defined on E by $h \sim_E h'$ iff $g(h) = g(h')$ where f and g are the mappings defined above. Clearly, $v \sim_V v'$ entails that $l(v) = l(v')$ and it can easily be shown that $h \sim_E h'$ iff $\exists v^{(i)} \in h, v'^{(i)} \in h'$ such that $v \sim_V v'$. We can thus define the quotient hypergraph $\overline{G} = (V/\sim_V, E/\sim_E, l)$.

Proposition 23. *If $G = (V, E, l)$ is a tiling of a connected hypergraph $\widehat{G} = (\widehat{V}, \widehat{E}, \widehat{l})$, then $\overline{G} = (V/\sim_V, E/\sim_E, l)$ is isomorphic to $\widehat{G}$ and moreover, the cardinal of $f^{-1}(\{\hat{v}\})$ is the same for every $\hat{v} \in \widehat{V}$.*

Proof. We will prove the last part of the proposition, which entails the surjectivity of f. This will be enough since if f is surjective, then $\overline{G}$ is isomorphic to $\widehat{G}$.

Let m be the maximal cardinality of the sets $f^{-1}(\{\hat{v}\})$ and suppose that they have different cardinalities. Let $V_1 = \{\hat{v} \in \widehat{V} : Card(f^{-1}(\{\hat{v}\})) = m\}$ and $V_2 = \widehat{V} \setminus V_1$. Since $\widehat{G}$ is connected, there exists a hyperedge $\hat{h}$ and $\hat{v}_1^{(i)}, \hat{v}_2^{(j)} \in \hat{h}$ such that $\hat{v}_1 \in V_1$ and $\hat{v}_2 \in V_2$. Let $f^{-1}(\{\hat{v}_1\}) = \{v_1, \ldots, v_m\}$ and let $h_1, \ldots, h_m \in E$ be the hyperedges containing $v_1^{(i)}, \ldots, v_m^{(i)}$, respectively. Since each $g_{|h_i}$ is injective and since the vertices $v_1, \ldots, v_m$ are distinct, the hyperedges $h_1, \ldots, h_m$ are also distinct and therefore disjoint. Let $w_1^{(j)} = g_{|h_1}^{-1}(\hat{v}_2^{(j)}), \ldots, w_m^{(j)} = g_{|h_m}^{-1}(\hat{v}_2^{(j)})$. These ports are distinct and therefore, the vertices $w_1, \ldots, w_m$ are also distinct. Since, $f(w_1) = \cdots = f(w_m) = \hat{v}_2$, we obtain a contradiction. $\qquad\square$

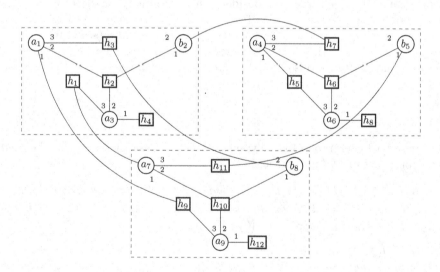

Fig. 3. A tiling made of three copies of the hypergraph from Example 4

4.2   Finite Support Series and Tilings

We end this section with the main result of this paper. We show that we can construct a HWM which assigns a nonzero value to a specific hypergraph over some ranked alphabet and all of its tilings, and zero to any other hypergraph on the same alphabet. This result leads to a sufficient condition on families of hypergraphs for the recognizability of finite support series.

Theorem 24. *Given a hypergraph $\widehat{G} = (\widehat{V}, \widehat{E}, \widehat{l})$ over $(\Sigma, \sharp)$, there exists a recognizable series $r_{\widehat{G}}$ such that $r_{\widehat{G}}(G) \neq 0$ if and only if G is a tiling of $\widehat{G}$.*

Proof.   Let $P_{\widehat{G}}$ be the set of ports of $\widehat{G}$. For any symbol $x \in \Sigma$, we note $\widehat{V}(x)$ the set of vertices in $\widehat{V}$ labeled by x.

Let $\mathcal{S} = 2^{P_{\widehat{G}}}$ bet the set of subsets of $P_{\widehat{G}}$ and let $d = |\mathcal{S}|$. Instead of indexing the canonical basis of $\mathbb{K}^d$ with integers in $[d]$, we will index it with elements of $\mathcal{S}$. For example, for each port $(\widehat{v}, i) \in P_{\widehat{G}}$, the singleton $\{(\widehat{v}, i)\}$ is in $\mathcal{S}$, thus $\mathbf{e}_{\{(\widehat{v}, i)\}}$ is a basis vector (which we will note $\mathbf{e}_{(\widehat{v}, i)}$ for convenience).

Define the HWM $M = \langle \mathbb{K}^d, \{\mathcal{T}^x\}_{x \in \Sigma}, \odot, \boldsymbol{\alpha} \rangle$ by

$$
\mathcal{T}^x = \begin{cases} \mathbf{e}_\emptyset^{\otimes \sharp x} & \text{if } \widehat{V}(x) = \emptyset \\ \sum_{\widehat{v} \in \widehat{V}(x)} \mathbf{e}_{(\widehat{v}, 1)} \otimes \cdots \otimes \mathbf{e}_{(\widehat{v}, \sharp \widehat{v})} & \text{otherwise} \end{cases}
$$

$$
\mathbf{e}_S \odot \mathbf{e}_T = \begin{cases} \mathbf{e}_{S \cup T} & \text{if } S \neq \emptyset, T \neq \emptyset \text{ and } S \cap T = \emptyset \\ \mathbf{e}_\emptyset & \text{otherwise} \end{cases}
$$

$$
\boldsymbol{\alpha}_S = \begin{cases} 1 & \text{if } S \in \widehat{E} \quad \text{(note that } \emptyset \notin \widehat{E}) \\ 0 & \text{otherwise} \end{cases}
$$

for any $x \in \Sigma$ and $S, T \in \mathcal{S}$. Let r be the series computed by M, we claim that r satisfies the property of the theorem.

For any hypergraph $G = (V, E, l)$ with $V = \{v_1, \cdots, v_N\}$, we have $r(G) = \sum_{\gamma \in \Gamma} \mathcal{T}_\gamma \prod_{h \in E} \boldsymbol{\alpha}^\top \bigodot_{S \in \gamma(h)} \mathbf{e}_S$ where $\Gamma = \mathcal{S}^{P_G}$ and $\mathcal{T}_\gamma = \prod_{i=1}^N \mathcal{T}_{\gamma(v_i,1), \cdots, \gamma(v_i, \sharp v_i)}^{v_i}$. Let $\gamma \in \Gamma$. If there exists a port $p \in P_G$ s.t. $\gamma(p) = \emptyset$, then $\prod_{h \in E} \boldsymbol{\alpha}^\top \bigodot_{S \in \gamma(h)} \mathbf{e}_S = 0$; otherwise, it follows from the definition of the tensors $\mathcal{T}^x$ that $\mathcal{T}_\gamma$ is different from 0 (and furthermore equal to 1) if and only if for all $v \in V$, there exists $\widehat{v} \in \widehat{V}(l(v))$ s.t. $\gamma(v, i) = (\widehat{v}, i)$ for all $i \in [\sharp v]$, hence

$$
r(G) = \sum_{\widehat{v}_1 \in \widehat{V}(l(v_1))} \cdots \sum_{\widehat{v}_N \in \widehat{V}(l(v_N))} \prod_{i=1}^N \mathcal{T}_{(\widehat{v}_i, 1) \cdots (\widehat{v}_i, \sharp \widehat{v}_i))}^{v_i} \prod_{h \in E} \boldsymbol{\alpha}^\top \bigodot_{(v_j, i_j) \in h} \mathbf{e}_{(\widehat{v}_j, i_j)}
$$

We have $r(G) \neq 0$ if and only if there exist N vertices $\widehat{v}_i \in \widehat{V}(l(v_i))$ for $i \in [N]$ such that (i) $\boldsymbol{\alpha}^\top \bigodot_{(v_j, i_j) \in h} \mathbf{e}_{(\widehat{v}_j, i_j)} \neq 0$ for all $h \in E$. Let $f : V \to \widehat{V}$ and $g : P_G \to P_{\widehat{G}}$ be the mappings defined by $f(v_i) = \widehat{v}_i$ and $g(v_i, j) = (\widehat{v}_i, j)$ for all $i \in [N]$. It follows from the definitions of $\boldsymbol{\alpha}$ and $\odot$ that (i) is true if and only if $g(h) \in \widehat{E}$ for all $h \in E$, and there are no distinct $(v_j, i_j), (v_k, i_k)$ in a hyperedge $h \in E$ such that $(\widehat{v}_j, i_j) = (\widehat{v}_k, i_k)$, i.e. the restriction $g_{|h}$ is injective for any $h \in E$. $\square$

A family $\mathcal{H}$ of hypergraphs is *tiling-free* if and only if for any $G \in \mathcal{H}$, there are no (non-trivial) tiling of G in $\mathcal{H}$.

Corollary. *For any tiling-free family of hypergraphs $\mathcal{H}$, finite support series on $\mathcal{H}$ are recognizable.*

Remark 25. A simple example of tiling-free family is the family of rooted hypergraphs: hypergraphs on a ranked alphabet $(\Sigma \cup \{\lambda\}, \sharp)$, where the special *root* symbol λ appears exactly once.

5   Examples

In this section, we present some examples of recognizable hypergraph series to give some insight on the expressiveness of HWMs and on how their computation relates to the usual notion of recognizable series on strings and trees.

Rooted Circular Strings. First note that the family of circular strings is not tiling-free: the circular string $abab$ is a tiling of ab. Instead of the construction described in Example 5, we can map each string w on a finite alphabet Σ to a *rooted circular string*. Let $w = w_1 \cdots w_n \in \Sigma^*$, we will consider the circular string G_w on the ranked alphabet $(\Sigma \cup \{\lambda\}, \sharp)$ where $\sharp x = 2$ for any $x \in \Sigma \cup \{\lambda\}$, with vertices $V = \{0, \cdots, n\}$, labels $l(0) = \lambda$ and $l(i) = w_i$ for $i \in [n]$, and edges $\{(n,2),(0,1)\}$ and $\{(i,2),(i+1,1)\}$ for $i \in \{0, \cdots, n-1\}$ (see Figure 4).

Let $r = \langle \mathbb{R}^d, \iota, \{\mathbf{M}_\sigma\}_{\sigma \in \Sigma}, \tau \rangle$ be a rational series on Σ^*. We define the HWM $A = \langle \mathbb{R}^d, \{\mathcal{A}^x\}_{x \in \Sigma}, \odot_{id}, 1 \rangle$ where $\mathcal{A}^\sigma = \mathbf{M}_\sigma$ for all $\sigma \in \Sigma$ and $\mathcal{A}^\lambda = \tau \iota^\top$. We have $r_A(G_w) = Tr(\tau \iota^\top \mathbf{M}_{w_1} \cdots \mathbf{M}_{w_n}) = Tr(\iota^\top \mathbf{M}_{w_1} \cdots \mathbf{M}_{w_n} \tau) = r(w)$ for all $w \in \Sigma^*$.

Now consider m rational series on Σ^* with d-dimensional linear representations $\langle \iota_i, \{\mathbf{M}_\sigma\}_{\sigma \in \Sigma}, \tau_i \rangle$ for $i \in [m]$. The string series $r = r_1 + \cdots + r_m$ is rational and its dimension can be as high as dm. However, the HWM $A = \langle \mathbb{R}^d, \{\mathcal{A}^x\}_{x \in \Sigma}, \odot_{id}, 1 \rangle$ where $\mathcal{A}^\sigma = \mathbf{M}_\sigma$ for all $\sigma \in \Sigma$ and $\mathcal{A}^\lambda = \sum_{i=1}^m \tau_i \iota_i^\top$ is such that $r_A(G_w) = r(w)$ for all $w \in \Sigma^*$, and is of dimension d.

Recognizing $a^n b^n$. Given an even length string on the alphabet $\{a,b\}$, we can enrich the construction described in Example 5 by associating a vertex of arity 3 to each letter, and adding extra edges connecting letters in the first half of the string to letters in the second half. Formally, given a word $w_1 \cdots w_{2n}$ on Σ, we consider the *3-ary graph representation* of w given by the graph $G = (V, E, l)$ on the ranked alphabet $(\Sigma \cup \{\iota, \tau\}, \sharp)$ where $\sharp \iota = \sharp \tau = 1$, $\sharp \sigma = 3$ for all $\sigma \in \Sigma$, $V = \{0 \cdots 2n+1\}$, $l(0) = \iota$, $l(2n+1) = \tau$, $l(i) = w_i$ for $1 \le i \le 2n$, and the set of edges is composed of $\{(0,1),(1,1)\}$, $\{(i,2),(i+1,1)\}$ for $1 \le i \le 2n$ and $\{(i,3),(n+i,3)\}$ for $1 \le i \le n$. The 3-ary graph representation of the string $abaa$ is shown in Figure 4.

Using this construction, it is easy to show that there exists a HWM (on the family of 3-ary graph representations of even length strings) whose support is the set of 3-ary graph representations of the language $\{a^n b^n : n \ge 1\}$.

Crosswords. Let $(\Sigma, \sharp)$ be a ranked alphabet where all symbols have arity 4. An (M, N)-crossword w on Σ is an array of symbols $[w_{ij}]_{ij} \in \Sigma^{M \times N}$. The graph $G_w = (V, E, l)$ associated to the crossword w is the graph with vertices $V = [M] \times [N]$, $l(m, n) = w_{mn}$, and edges $E = E_H \cup E_V$, where the ports are labeled by W, E, N, S, and where

$$E_H = \bigcup_{m \in [M], n \in [N-1]} \{\{(m, n)^E, (m, n+1)^W\}\} \bigcup_{m \in [M]} \{\{(m, 1)^W\}, \{(m, N)^E\}\},$$

$$E_V = \bigcup_{n \in [N], m \in [M-1]} \{\{(m, n)^S, (m+1, n)^N\}\} \bigcup_{n \in [N]} \{\{(1, n)^N\}, \{(M, n)^S\}\}.$$

An example of graph associated to a crossword is shown in Figure 4. The computation of a HWM on a crossword can be independently done on the rows and columns. Since the computation of a rational string series can be simulated by a HWM, we can construct a HWM that will associate to any crossword the product of a rational string series applied to its rows, and another one applied to its columns (this result can easily be generalized to N-dimensional words).

Proposition 26. *Let*

$$\mathcal{A} = \langle \mathbb{R}^{d_1}, \{\mathbf{A}^\sigma\}_{\sigma \in \Sigma}, \alpha_0, \alpha_\infty \rangle, \mathcal{B} = \langle \mathbb{R}^{d_2}, \{\mathbf{B}^\sigma\}_{\sigma \in \Sigma}, \beta_0, \beta_\infty \rangle$$

be two rational string series on Σ^*. There exists a HWM $\mathcal{C} = \langle \mathbb{C}^{d_1+d_2}, \{\mathcal{C}^\sigma\}_{\sigma \in \Sigma}, \odot, \gamma \rangle$ such that $r_{\mathcal{C}}(G_w) = \prod_{m \in [M]} r_{\mathcal{A}}(w_{m:}) \prod_{n \in [N]} r_{\mathcal{B}}(w_{:n})$ for any (M, N)-crossword w (where $w_{m:}$ denotes the m-th row of w and $w_{:n}$ its n-th column).

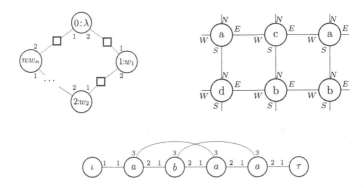

Fig. 4. (left) Rooted circular string, (right) graph associated to the 2D-word $\frac{aca}{dbb}$ and (bottom) graph associated to the string $abaa$ with extra edges connecting u_i to u_{2+i}

6   Conclusion

The model we propose naturally generalizes recognizable series on strings and trees. It satisfies closure properties by sum and Hadamard product. We have analysed why finite support series on some families of hypergraphs are not recognizable, and we exhibit a sufficient condition on families of hypergraph for the recognizability of finite support series.

Since many data over a variety of fields naturally present a graph structure (images, secondary structure of RNA in bioinformatics, dependency graphs in NLP, etc.), this computational model offers a broad range of applications.

The next theoretical step will be to study how learning can be achieved within this framework, i.e. how the tensor components of the model M can be recovered or estimated from samples of the form $(G_1, \widehat{r_M(G_1)}), \ldots, (G_n, \widehat{r_M(G_n)})$. Preliminary results on circular strings indicate that this is a promising direction. General learning algorithms should rely on tensor decomposition techniques, which generalize the spectral methods used for learning rational series on strings and trees. We also plan to tackle algorithmic issues and to study how techniques and methods developed in the field of graphical models, such as message passing, variational methods, etc., could be adapted to the setting of HWMs.

References

1. Bailly, R., Denis, F., Rabusseau, G.: Recognizable series on hypergraphs. arXiv preprint (2014) arXiv:1404.7533
2. Bailly, R., Denis, F., Ralaivola, L.: Grammatical inference as a principal component analysis problem. In: Proceedings of the 26th Annual International Conference on Machine Learning (2009)
3. Berstel, J., Reutenauer, C.: Recognizable formal power series on trees. Theoretical Computer Science **18**(2) (1982)
4. Berstel, J., Reutenauer, C.: Rational series and their languages. Monographs in Theoretical Computer Science. An EATCS Series, vol. 12. Springer, Heidelberg (1988)
5. Cichocki, A.: Era of big data processing: a new approach via tensor networks and tensor decompositions. In: International Workshop on Smart Info-Media Systems in Asia (2013)
6. Comon, H., Dauchet, M., Gilleron, R., Löding, C., Jacquemard, F., Lugiez, D., Tison, S., Tommasi, M.: Tree automata techniques and applications (2007)
7. Denis, F., Habrard, A.: Learning Rational Stochastic Tree Languages. In: Hutter, M., Servedio, R.A., Takimoto, E. (eds.) ALT 2007. LNCS (LNAI), vol. 4754, pp. 242–256. Springer, Heidelberg (2007)
8. Droste, M., Kuich, W., Vogler, H.: Handbook of weighted automata. Springer (2009)
9. Eilenberg, S.: Automata, languages, and machines. Academic press (1974)
10. Hackbusch, W.: Tensor spaces and numerical tensor calculus. Springer Series in Computational Mathematics, vol. 42. Springer, Heidelberg (2012)
11. Kuich, W., Salomaa, A.: Semirings, automata, languages. Springer, Berlin (1986)
12. Orus, R.: A practical introduction to tensor networks: Matrix product states and projected entangled pair states. arXiv preprint arXiv:1306.2164 (2013)

13. Pellionisz, A., Llinas, R.: Brain modeling by tensor network theory and computer simulation. the cerebellum: Distributed processor for predictive coordination. Neuroscience **4**(3) (1979)
14. Sakarovitch, J.: Elements of automata theory. Cambridge University Press (2009)
15. Salomaa, A., Soittola, M., Bauer, F.L., Gries, D.: Automata-Theoretic Aspects of Formal Power Series. Monographs in Computer Science. Springer, New York (1978)
16. Schützenberger, M.P.: On the definition of a family of automata. Information and control (2) (1961)

Towards More Precise Rewriting Approximations

Yohan Boichut[✉], Jacques Chabin, and Pierre Réty

LIFO - Université d'Orléans, B.P. 6759, 45067 Orléans Cedex 2, France
{yohan.boichut,jacques.chabin,pierre.rety}@univ-orleans.fr

Abstract. To check a system, some verification techniques consider a set of terms I that represents the initial configurations of the system, and a rewrite system R that represents the system behavior. To check that no undesirable configuration is reached, they compute an over-approximation of the set of descendants (successors) issued from I by R, expressed by a tree language. Their success highly depends on the quality of the approximation. Some techniques have been presented using regular tree languages, and more recently using non-regular languages to get better approximations: using context-free tree languages [16] on the one hand, using synchronized tree languages [2] on the other hand. In this paper, we merge these two approaches to get even better approximations: we compute an over-approximation of the descendants, using synchronized-context-free tree languages expressed by logic programs. We give several examples for which our procedure computes the descendants in an exact way, whereas the former techniques compute a strict over-approximation.

Keywords: Term rewriting · Tree languages · Logic programming · Reachability

1 Introduction

To check systems like cryptographic protocols or Java programs, some verification techniques consider a set of terms I that represents the initial configurations of the system, and a rewrite system R that represents the system behavior [1,13,14]. To check that no undesirable configuration is reached, they compute an over-approximation of the set of descendants[1] (successors) issued from I by R, expressed by a tree language. Let $R^*(I)$ denote the set of descendants of I, and consider a set Bad of *undesirable* terms. Thus, if a term of Bad is reached from I, i.e. $R^*(I) \cap Bad \neq \emptyset$, it means that the protocol or the program is flawed. In general, it is not possible to compute $R^*(I)$ exactly. Instead, one computes an over-approximation App of $R^*(I)$ (i.e. $App \supseteq R^*(I)$), and checks that $App \cap Bad = \emptyset$, which ensures that the protocol or the program is correct.

However, I, Bad and App have often been considered as regular tree languages, recognized by finite tree automata. In the general case, $R^*(I)$ is not

---

[1] I.e. terms obtained by applying arbitrarily many rewrite steps on the terms of I.

© Springer International Publishing Switzerland 2015
A.-H. Dediu et al. (Eds.): LATA 2015, LNCS 8977, pp. 652–663, 2015.
DOI: 10.1007/978-3-319-15579-1_51

regular, even if I is. Moreover, the expressiveness of regular languages is poor. Then the over-approximation App may not be precise enough, and we may have $App \cap Bad \neq \emptyset$ whereas $R^*(I) \cap Bad = \emptyset$. In other words, the protocol is correct, but we cannot prove it. Some work has proposed CEGAR-techniques (Counter-Example Guided Approximation Refinement) to conclude as often as possible [1,4,6]. However, in some cases, no regular over-approximation works [5].

To overcome this theoretical limit, the idea is to use more expressive languages to express the over-approximation, i.e. non-regular ones. However, to be able to check that $App \cap Bad = \emptyset$, we need a class of languages closed under intersection and whose emptiness is decidable. Actually, if we assume that Bad is regular, closure under intersection with a regular language is enough. The class of context-free tree languages has these properties, and an approximation technique using context-free tree languages has been proposed in [16]. On the other hand, the class of synchronized tree languages [17] also has these properties, and an approximation technique using synchronized tree languages has been proposed in [2]. Both classes include regular languages, but they are incomparable. Context-free tree languages cannot express dependencies between different branches, except in some cases, whereas synchronized tree languages cannot express vertical dependencies.

We want to use a more powerful class of languages that can express the two kinds of dependencies together: the class of synchronized-context-free tree-(tuple) languages [21,22], which has the same properties as context-free languages and as synchronized languages, i.e. closure under union, closure under intersection with a regular language, decidability of membership and emptiness.

In this paper, we propose a procedure that always terminates and that computes an over-approximation of the descendants obtained by a linear rewrite system, using synchronized-context-free tree-(tuple) languages expressed by logic programs. Compared to our previous work [2], we introduce "input arguments" in predicates, which is a major technical change that highly improves the quality of the approximation, and that requires new results and new proofs. This work is a first step towards a verification technique offering more than regular approximations. Some on-going work is discussed in Section 5 in order to make this technique be an accepted verification technique.

The paper is organized as follows: classical notations and notions manipulated throughout the paper are introduced in Section 2. Our main contribution, i.e. computing approximations, is explained in Section 3. Finally, in Section 4 our technique is applied to examples, in particular when $R^*(I)$ can be expressed in an exact way neither by a context-free language, nor by a synchronized language. For lack of space, all proofs are in [3].

Related Work: The class of tree-tuples whose overlapping coding is recognized by a tree automaton on the product alphabet [7] (called "regular tree relations" by some authors), is strictly included in the class of rational tree relations [19]. The latter is equivalent to the class of non-copying[2] synchronized languages [20], which is strictly included in the class of synchronized languages.

---

[2] Clause heads are assumed to be linear.

Context-free tree languages (i.e. without assuming a particular strategy for grammar derivations) [23] are equivalent to OI (outside-in strategy) context-free tree languages, but are incomparable with IO (inside-out strategy) context-free tree languages [11,12]. The IO class (and not the OI one) is strictly included in the class of synchronized-context-free tree languages. The latter is equivalent to the "term languages of hyperedge replacement grammars", which are equivalent to the tree languages definable by attribute grammars [9,10]. However, we prefer to use the synchronized-context-free tree languages, which use the well known formalism of pure logic programming, for its implementation ease.

Much other work computes the descendants in an exact way using regular tree languages (in particular the recent paper [8]), assuming strong restrictions.

2   Preliminaries

Consider a *finite ranked alphabet* $\Sigma = \{a, b, f, g, h, \ldots\}$ and a set of variables $Var = \{x, y, z, \ldots\}$. Each symbol $f \in \Sigma$ has a unique arity, denoted by $ar(f)$. The notions of *first-order term*, *position* and *substitution* are defined as usual. Given σ and σ' two substitutions, $\sigma \circ \sigma'$ denotes the substitution such that for any variable x, $\sigma \circ \sigma'(x) = \sigma(\sigma'(x))$. T_Σ denotes the set of ground terms (without variables) over Σ. For a term t, $Var(t)$ is the set of variables of t, $Pos(t)$ is the set of positions of t. For $p \in Pos(t)$, $t(p)$ is the symbol of $\Sigma \cup Var$ occurring at position p in t, and $t|_p$ is the subterm of t at position p. The term t is *linear* if each variable of t occurs only once in t. The term $t[t']_p$ is obtained from t by replacing the subterm at position p by t'. $PosVar(t) = \{p \in Pos(t) \mid t(p) \in Var\}$, $PosNonVar(t) = \{p \in Pos(t) \mid t(p) \notin Var\}$.

A *rewrite rule* is an oriented pair of terms, written $l \rightarrow r$. We always assume that l is not a variable, and $Var(r) \subseteq Var(l)$. A *rewrite system* R is a finite set of rewrite rules. *lhs* stands for left-hand-side, *rhs* for right-hand-side. The rewrite relation $\rightarrow_R$ is defined as follows: $t \rightarrow_R t'$ if there exist a position $p \in PosNonVar(t)$, a rule $l \rightarrow r \in R$, and a substitution θ s.t. $t|_p = \theta(l)$ and $t' = t[\theta(r)]_p$. $\rightarrow_R^*$ denotes the reflexive-transitive closure of $\rightarrow_R$. t' is a *descendant* of t if $t \rightarrow_R^* t'$. If E is a set of ground terms, $R^*(E)$ denotes the set of descendants of elements of E. The rewrite rule $l \rightarrow r$ is *left (resp. right) linear* if l (resp. r) is linear. R is *left (resp. right) linear* if all its rewrite rules are left (resp. right) linear. R is *linear* if R is both left and right linear.

In the following, we consider the framework of *pure logic programming*, and the class of synchronized-context-free tree-tuple[3] languages [21,22], which is presented as an extension of the class of synchronized tree-tuple languages defined by CS-clauses [17,18]. Given a set *Pred* of *predicate* symbols; *atoms*, *goals*, *bodies* and *Horn-clauses* are defined as usual. Note that both *goals* and *bodies* are sequences of atoms. We will use letters G or B for sequences of atoms, and A for atoms.

Definition 1. *The tuple of terms* $(t_1, \ldots, t_n)$ *is* flat *if* $t_1, \ldots, t_n$ *are variables. The sequence of atoms B is* flat *if for each atom $P(t_1, \ldots, t_n)$ of B, $(t_1, \ldots, t_n)$*

---

[3] For simplicity, "tree-tuple" is sometimes omitted.

is flat. B is linear *if each variable occurring in B (possibly at subterm position) occurs only once in B. Note that the empty sequence of atoms (denoted by $\emptyset$) is flat and linear.*

A Horn clause $P(t_1, \ldots, t_n) \leftarrow B$ is normalized *if $\forall i \in \{1, \ldots, n\}$, t_i is a variable or contains only one occurrence of function-symbol. A program is* normalized *if all its clauses are normalized.*

Example 2. Let x, y, z be variables. The sequence of atoms $P_1(x, y), P_2(z)$ is flat, whereas $P_1(x, f(y))$, $P_2(z)$ is not flat. The clause $P(x, y) \leftarrow Q(x, y)$ is normalized (x, y are variables). The clause $P(f(x), y) \leftarrow Q(x, y)$ is normalized whereas $P(f(f(x)), y) \leftarrow Q(x, y)$ is not.

Definition 3. A *logic program with modes* is a logic program such that a mode-tuple $m \in \{I, O\}^n$ is associated to each predicate symbol P (n is the arity of P). In other words, each predicate argument has mode I (Input) or O (Output). To distinguish them, **output arguments will be covered by a hat.**

Notation: Let P be a predicate symbol. $ArIn(P)$ is the number of input arguments of P, and $ArOut(P)$ is the number of output arguments. Let B be a sequence of atoms (possibly containing only one atom). $In(B)$ is the input part of B, i.e. the tuple composed of the input arguments of B. $ArIn(B)$ is the arity of $In(B)$. $Var^{in}(B)$ is the set of variables that appear in $In(B)$. $Out(B)$, $ArOut(B)$, and $Var^{out}(B)$ are defined in a similar way. We also define $Var(B) = Var^{in}(B) \cup Var^{out}(B)$.

Example 4. Let $B = P(\widehat{t_1}, \widehat{t_2}, t_3), Q(\widehat{t_4}, t_5, t_6)$. Then, $Out(B) = (t_1, t_2, t_4)$ and $In(B) = (t_3, t_5, t_6)$.

Definition 5. Let $B = A_1, \ldots, A_n$ be a sequence of atoms. We say that $A_j \succ A_k$ (possibly $j = k$) if $\exists y \in Var^{in}(A_j) \cap Var^{out}(A_k)$. In other words an input of A_j depends on an output of A_k. We say that B has a *loop* if $A_j \succ^+ A_j$ for some A_j ($\succ^+$ is the transitive closure of $\succ$).

Example 6. $Q(\widehat{x}, s(y)), R(\widehat{y}, s(x))$ (where x, y are variables) has a loop because $Q(\widehat{x}, s(y)) \succ R(\widehat{y}, s(x)) \succ Q(\widehat{x}, s(y))$.

Definition 7. A *Synchronized-Context-Free (S-CF) program Prog* is a logic program with modes, whose clauses $H \leftarrow B$ satisfy:

- $In(H).Out(B)$ (. is the tuple concatenation) is a linear tuple of variables, i.e. each tuple-component is a variable, and each variable occurs only once,
- and B does not have a loop.

A clause of an S-CF program is called *S-CF clause*.

Example 8. $Prog = \{P(\widehat{x}, y) \leftarrow P(\widehat{s(x)}, y)\}$ is not an S-CF program because $In(H).Out(B) = (y, s(x))$ is not a tuple of variables. $Prog' = \{P'(\widehat{s(x)}, y) \leftarrow P'(\widehat{x}, s(y))\}$ is an S-CF program because $In(H).Out(B) = (y, x)$ is a linear tuple of variables, and there is no loop in the clause body.

Definition 9. Let *Prog* be an S-CF program. Given a predicate symbol P without input arguments, the tree-(tuple) language generated by P is $L_{Prog}(P) = \{t \in (T_\Sigma)^{ArOut(P)} \mid P(t) \in Mod(Prog)\}$, where T_Σ is the set of ground terms over the signature Σ and $Mod(Prog)$ is the least Herbrand model of *Prog*. $L_{Prog}(P)$ is called *Synchronized-Context-Free language (S-CF language)*.

Example 10. $Prog = \{S(\widehat{c(x,y)}) \leftarrow P(\widehat{x}, \widehat{y}, a, b).$
$P(\widehat{f(x)}, \widehat{g(y)}, x', y') \leftarrow P(\widehat{x}, \widehat{y}, h(x'), i(y')).\ P(\widehat{x}, \widehat{y}, x, y) \leftarrow\}$ is an S-CF program. The language generated by S is $L_{Prog}(S) = \{c(f^n(h^n(a)), g^n(i^n(b))) \mid n \in \mathbb{N}\}$, which is not synchronized (there are vertical dependencies) nor context-free.

Definition 11. The S-CF clause $H \leftarrow B$ is *non-copying* if the tuple $Out(H)$. $In(B)$ is linear. A program is *non-copying* if all its clauses are non-copying.

Example 12. The clause $P(\widehat{d(x,x)}, y) \leftarrow Q(\widehat{x}, p(y))$ is copying whereas $P(\widehat{c(x)}, y) \leftarrow Q(\widehat{x}, p(y))$ is non-copying.

Remark 13. An S-CF program without input arguments is actually a CS-program (composed of CS-clauses) [17], which generates a synchronized language[4]. A non-copying CS-program such that every predicate symbol has only one argument generates a regular tree language[5]. Conversely, every regular tree language can be generated by a non-copying CS-program.

Definition 14. *Given an S-CF program Prog and a sequence of atoms G,*

- *G derives into G' by a resolution step if there exists a clause[6] $H \leftarrow B$ in Prog and an atom $A \in G$ such that A and H are unifiable by the most general unifier σ (then $\sigma(A) = \sigma(H)$) and $G' = \sigma(G)[\sigma(A) \leftarrow \sigma(B)]$. It is written $G \rightsquigarrow_\sigma G'$. We consider the transitive closure $\rightsquigarrow^+$ and the reflexive-transitive closure $\rightsquigarrow^*$ of $\rightsquigarrow$. If $G_1 \rightsquigarrow_{\sigma_1} G_2$ and $G_2 \rightsquigarrow_{\sigma_2} G_3$, we write $G_1 \rightsquigarrow^*_{\sigma_2 \circ \sigma_1} G_3$.*
- *G rewrites into G' (possibly in several steps) if $G \rightsquigarrow^*_\sigma G'$ s.t. σ does not instantiate the variables of G. It is written $G \rightarrow^*_\sigma G'$.*

Example 15. Let $Prog = \{P(\widehat{x_1}, \widehat{g(x_2)}) \leftarrow P'(\widehat{x_1}, \widehat{x_2}).\ P(\widehat{f(x_1)}, \widehat{x_2}) \leftarrow P''(\widehat{x_1}, \widehat{x_2}).\}$, and consider $G = P(f(x), y)$. Thus, $P(f(x), y)) \rightsquigarrow_{\sigma_1} P'(f(x), x_2)$ with $\sigma_1 = [x_1/f(x), y/g(x_2)]$ and $P(f(x), y)) \rightarrow_{\sigma_2} P''(x, y)$ with $\sigma_2 = [x_1/x, x_2/y]$.

In the remainder of the paper, given an S-CF program *Prog* and two sequences of atoms G_1 and G_2, $G_1 \rightsquigarrow^*_{Prog} G_2$ (resp. $G_1 \rightarrow^*_{Prog} G_2$) also denotes that G_2 can be derived (resp. rewritten) from G_1 using clauses of *Prog*. Note that for any atom A, if $A \rightarrow B$ then $A \rightsquigarrow B$. On the other hand, $A \rightsquigarrow_\sigma B$ implies $\sigma(A) \rightarrow B$. Consequently, if A is ground, $A \rightsquigarrow B$ implies $A \rightarrow B$.

It is well known that resolution is complete.

Theorem 16. *Let A be a ground atom. $A \in Mod(Prog)$ iff $A \rightsquigarrow^*_{Prog} \emptyset$.*

---

[4] Initially, synchronized languages were presented using constraint systems (sorts of grammars) [15], and later using logic programs. CS stands for "Constraint System".

[5] In this case, the S-CF program can easily be transformed into a finite tree automaton.

[6] We assume that the clause and G have distinct variables.

3   Computing Descendants

To make the understanding easier, we first give the completion algorithm in Definition 17. Given a normalized S-CF program $Prog$ and a linear rewrite system R, we propose an algorithm to compute a normalized S-CF program $Prog'$ such that $R^*(Mod(Prog)) \subseteq Mod(Prog')$, and consequently $R^*(L_{Prog}(P)) \subseteq L_{Prog'}(P)$ for each predicate symbol P. Some notions, as strong coherence, will be explained later.

Definition 17 (comp). *Let arity-limit and predicate-limit be positive integers. Let R be a linear rewrite system, and $Prog$ be a finite, normalized and non-copying S-CF program strongly coherent with R. The completion process is defined by:*

Function $\mathsf{comp}_R(Prog)$
   $Prog = \mathsf{removeCycles}(Prog)$
   while *there exists a non-convergent critical pair* $H \leftarrow B$ *in* $Prog$ **do**
      $Prog = \mathsf{removeCycles}(Prog \cup \mathsf{norm}_{Prog}(H \leftarrow B))$
   end while
   return $Prog$

Let us explain this algorithm.

The notion of critical pair is at the heart of the technique. Given an S-CF program $Prog$, a predicate symbol P and a rewrite rule $l \rightarrow r$, a critical pair, explained in details in Section 3.1, is a way to detect a possible rewriting by $l \rightarrow r$ for a term t in a tuple of $L_{Prog}(P)$. A convergent critical pair means that the rewrite step is already handled i.e. if $t \rightarrow_{l \rightarrow r} s$ then s is in a tuple of $L_{Prog}(P)$. Consequently, the language of a normalized CS-program involving only convergent critical pairs is closed by rewriting.

To summarize, a non-convergent critical pair gives rise to an S-CF clause. Adding the resulting S-CF clause to the current S-CF program makes the critical pair convergent. But, let us emphasize on the main problems arising from Definition 17, i.e. the computation may not terminate and the resulting S-CF clause may not be normalized. Concerning the non-termination, there are mainly two reasons. Given a normalized S-CF program $Prog$, 1) the number of critical pairs may be infinite and 2) even if the number of critical pairs is finite, adding the critical pairs to $Prog$ may create new non-convergent critical pairs, and so on.

Actually, as in [2], there is a function called removeCycles whose goal is to get finitely many critical pairs from a given finite S-CF program. For lack of space, many details on this function are given in [3]. Basically, given an S-CF program $Prog$ having infinitely many critical pairs, $\mathsf{removeCycles}(Prog)$ is another S-CF program that has finitely many critical pairs, and such that for any predicate symbol P, $L_{Prog}(P) \subseteq L_{\mathsf{removeCycles}(Prog)}(P)$. The normalization process presented in Section 3.2 not only preserves the normalized nature of the computed S-CF programs but also allows us to control the creation of new non-convergent critical pairs. Finally, in Section 3.3, our main contribution, i.e. the computation of an over-approximating S-CF program, is fully described.

3.1   Critical Pairs

Definition 18. *Let Prog be a non-copying S-CF program and $l \rightarrow r$ be a left-linear rewrite rule. Let $x_1, \ldots, x_n$ be distinct variables such that $\{x_1, \ldots, x_n\} \cap Var(l) = \emptyset$. If there are P and k s.t. the kth argument of P is an output, and $P(x_1, \ldots, x_{k-1}, l, x_{k+1}, \ldots, x_n) \leadsto_\theta^+ G$ where*[7]

1. resolution steps are applied only on atoms whose output is not flat,
2. $Out(G)$ is flat and
3. the clause $P(t_1, \ldots, t_n) \leftarrow B$ used in the first step of this derivation satisfies t_k is not a variable[8]

then the clause $\theta(P(x_1, \ldots, x_{k-1}, r, x_{k+1}, \ldots, x_n)) \leftarrow G$ is called critical pair. *Moreover, if θ does not instantiate the variables of $In(P(x_1, \ldots, x_{k-1}, l, x_{k+1}, \ldots, x_n))$ then the critical pair is said* strict.

Example 19. Let *Prog* be the S-CF program defined by:
$Prog = \{P(\widehat{x}) \leftarrow Q(\widehat{x}, a).\ Q(\widehat{f(x)}, y) \leftarrow Q(\widehat{x}, g(y)).\ Q(\widehat{x}, x) \leftarrow .\}$ and consider the rewrite system: $R = \{f(x) \rightarrow x\}$. Note that $L(P) = \{f^n(g^n(a)) \mid n \in \mathbb{N}\}$.

We have $Q(\widehat{f(x)}, y) \leadsto_{Id} Q(\widehat{x}, g(y))$ where Id denotes the substitution that leaves every variable unchanged. Since $Out(Q(\widehat{x}, g(y)))$ is flat, this generates the strict critical pair $Q(\widehat{x}, y) \leftarrow Q(\widehat{x}, g(y))$.

Lemma 20. *A strict critical pair is an S-CF clause. In addition, if $l \rightarrow r$ is right-linear, a strict critical pair is a non-copying S-CF clause.*

Definition 21. *A critical pair $H \leftarrow B$ is said* convergent *if $H \rightarrow_{Prog}^* B$.*

The critical pair of Example 19 is not convergent.

Let us recall that the completion procedure is based on adding the non-convergent critical pairs into the program. In order to preserve the nature of the S-CF program, the computed non-convergent critical pairs are expected to be strict. So we define a sufficient condition on R and *Prog* called *strong coherence*.

Definition 22. Let R be a rewrite system. We consider the smallest set of *consuming* symbols, recursively defined by: $f \in \Sigma$ is *consuming* if there exists a rewrite rule $f(t_1, \ldots, t_n) \rightarrow r$ in R s.t. some t_i is not a variable, or r contains at least one consuming symbol.

The S-CF program *Prog* is *strongly coherent* with R if 1) for all $l \rightarrow r \in R$, the top-symbol of l does not occur in input arguments of *Prog* and 2) no consuming symbol occurs in clause-heads having input arguments.

In $R = \{f(x) \rightarrow g(x),\ g(s(x)) \rightarrow h(x)\}$, g is consuming and so is f. Thus $Prog = \{P(\widehat{f(x)}, x) \leftarrow .\}$ is not strongly coherent with R. Note that a CS-program (no input arguments) is strongly coherent with any rewrite system.

---

[7] Here, we do not use a hat to indicate output arguments because they may occur anywhere depending on P.

[8] In other words, the overlap of l on the clause head $P(t_1, \ldots, t_n)$ is done at a non-variable position.

Lemma 23. *If Prog is a normalized S-CF program strongly coherent with R, then every critical pair is strict.*

So, we come to our main result that ensures to get the rewriting closure when every computable critical pair is convergent.

Theorem 24. *Let R be a linear rewrite system, and Prog be a non-copying normalized S-CF program strongly coherent with R. If all strict critical pairs are convergent, then for every predicate symbol P without input arguments, $L(P)$ is closed under rewriting by R, i.e. $(t \in L(P) \wedge t \to_R^* t') \implies t' \in L(P)$.*

3.2 Normalizing Critical Pairs – norm$_{Prog}$

If a critical pair is not convergent, we add it into *Prog*, and the critical pair becomes convergent. However, in the general case, a critical pair is not normalized, whereas all clauses in *Prog* should be normalized. In the case of CS-clauses (i.e. without input arguments), a procedure that transforms a non-normalized clause into normalized ones has been presented [2]. For example, $P(\widehat{f(g(x))}, \widehat{b}) \leftarrow Q(\widehat{x})$ is normalized into $\{P(\widehat{f(x_1)}, \widehat{b}) \leftarrow P_1(\widehat{x_1}). \ P_1(\widehat{g(x_1)}) \leftarrow Q(\widehat{x_1}).\}$ (P_1 is a new predicate symbol). Since only output arguments should be normalized, this procedure still works even if there are also input arguments. As new predicate symbols are introduced, possibly with bigger arities, the procedure may not terminate. To make it terminate in every case, two positive integers are used: *predicate-limit* and *arity-limit*. If the number of predicate symbols having the same arity as P_1 (including P_1) exceeds *predicate-limit*, an existing predicate symbol (for example Q) must be used instead of the new predicate P_1. This may enlarge $Mod(Prog)$ in general and may lead to a strict over-approximation. If the arity of P_1 exceeds *arity-limit*, P_1 must be replaced in the clause body by several predicate symbols[9] whose arities are less than or equal to *arity-limit*. This may also enlarge $Mod(Prog)$. See [2] for more details.

In other words norm$_{Prog}(H \leftarrow B)$ builds a set of normalized S-CF clauses N such that $Mod(Prog \cup \{H \leftarrow B\}) \subseteq Mod(Prog \cup N)$.

However, when starting from a CS-program (i.e. without input arguments), it could be interesting to normalize by introducing input arguments, in order to profit from the bigger expressiveness of S-CF programs, and consequently to get a better approximation of the set of descendants, or even an exact computation, like in Examples 26 and 27 presented in Section 4.

3.3 Completion

At the very beginning of Section 3, we have presented in Definition 17 the completion algorithm i.e. comp$_R$. In Sections 3.1 and 3.2, we have described how to

---

[9] For instance, if P_1 is binary and *arity-limit* $= 1$, then $P_1(t_1, t_2)$ should be replaced by the sequence of atoms $P_2(t_1), P_3(t_2)$. Note that the dependency between t_1 and t_2 is lost, which may enlarge $Mod(Prog)$. Symbols P_2 and P_3 are new if it is compatible with *predicate-limit*. Otherwise former predicate symbols should be used instead of P_2 and P_3.

detect non-convergent critical pairs and how to convert them into normalized clauses using norm$_{Prog}$.

Theorem 25 illustrates that our technique leads to a finite S-CF program whose language over-approximates the descendants obtained by a linear rewrite system R.

Theorem 25. *Function* comp *always terminates, and all critical pairs are convergent in* comp$_R(Prog)$. *Moreover, for each predicate symbol P without input arguments, $R^*(L_{Prog}(P)) \subseteq L_{\text{comp}_R(Prog)}(P)$.*

4   Examples

In this section, our technique is applied on several examples. I is the initial set of terms and R is the rewrite system. Initially, we define an S-CF program $Prog$ that generates I and that satisfies the assumptions of Definition 17. For lack of space, the examples should be as short as possible. To make the procedure terminate shortly, we suppose that *predicate-limit*=1, which means that for all i, there is at most one predicate symbol having i arguments, except for $i = 1$ we allow two predicate symbols having one argument.

When the following example is dealt with synchronized languages, i.e. with CS-programs [2, Example 42], we get a strict over-approximation of the descendants. Now, thanks to the bigger expressive power of S-CF programs, we compute the descendants in an exact way.

Example 26. Let $I = \{f(a,a)\}$ and $R = \{f(x,y) \rightarrow u(f(v(x),w(y)))\}$. Intuitively, the exact set of descendants is $R^*(I) = \{u^n(f(v^n(a), w^n(a))) \mid n \in \mathbb{N}\}$ where u^n means that u occurs n times. We define $Prog = \{P_f(\widehat{f(x,y)}) \leftarrow P_a(\widehat{x}), P_a(\widehat{y})., P_a(\widehat{a}) \leftarrow .\}$. Note that $L_{Prog}(P_f) = I$. The run of the completion is given in Fig. 1. The reader can refer to [3] for a detailed explanation. In Fig. 1, the left-most column reports the detected non-convergent critical pairs and the right-most column describes how they are normalized. Note that for the resulting program $Prog$, i.e. clauses appearing in the right-most column, $L_{Prog}(P_f) = R^*(I)$ indeed.

The previous example could probably be dealt in an exact way using the technique of [16] as well, since $R^*(I)$ is a context-free language. It is not the case for the following example, whose language of descendants $R^*(I)$ is not context-free (and not synchronized). It can be handled by S-CF programs in an exact way thanks to their bigger expressive power.

Example 27. Let $I = \{d_1(a,a,a)\}$ and

$$R = \left\{ \begin{array}{ll} d_1(x,y,z) \xrightarrow{1} d_1(h(x),i(y),s(z)), & d_1(x,y,z) \xrightarrow{2} d_2(x,y,z) \\ d_2(x,y,s(z)) \xrightarrow{3} d_2(f(x),g(y),z), & d_2(x,y,a) \xrightarrow{4} c(x,y) \end{array} \right\}$$

Detected non-convergent critical pairs	New clauses obtained by norm_{Prog}
	Starting S-CF program $\newline P_f(f(\widehat{x,y})) \leftarrow P_a(\widehat{x}), P_a(\widehat{y}). \newline P_a(\widehat{a}) \leftarrow .$
$P_f(u(f(v(\widehat{x}), w(y)))) \leftarrow P_a(\widehat{x}), P_a(\widehat{y}).$	$P_f(\widehat{z}) \leftarrow P_1(\widehat{z}, x, y), P_a(\widehat{x}), P_a(\widehat{y}). \newline P_1(\widehat{u(z)}, x, y) \leftarrow P_1(\widehat{z}, v(x), w(y)). \newline P_1(f(\widehat{x,y}), x, y) \leftarrow .$
$\emptyset$	

Fig. 1. Run of comp_R on Example 26

$R^*(I)$ is composed of all terms appearing in the following derivation:

$$d_1(a, a, a) \xrightarrow{1}{}^n d_1(h^n(a), i^n(a), s^n(a)) \xrightarrow{2} d_2(h^n(a), i^n(a), s^n(a))$$
$$\xrightarrow{3}{}^k d_2(f^k(h^n(a)), g^k(i^n(a)), s^{n-k}(a)) \xrightarrow{4} c(f^n(h^n(a)), g^n(i^n(a))) .$$

Note that the last rewrite step by rule 4 is possible only when $k = n$. The run of the completion on this example is given in Fig. 2. Black arrows means that the non-convergent critical pair is directly added to $Prog$ since it is already normalized. The reader can find a full explanation of this example in [3].

Note that the subset of descendants $d_2(f^k(h^n(a)), g^k(i^n(a)), s^{n-k}(a))$ can be seen (with $p = n - k$) as $d_2(f^k(h^{k+p}(a)), g^k(i^{k+p}(a)), s^p(a))$. Let $Prog'$ be the S-CF program composed of all the clauses except the blue one occurring in the right-most column in Fig. 2. Thus, the reader can check by himself that $L_{Prog'}(P_d)$ is exactly $R^*(I)$.

5 Further Work

Computing approximations more precise than regular approximations is a first step towards a verification technique. However, there are at least two steps before claiming this technique as a verification technique: 1) automatically handling the choices done during the normalization process and 2) extending our technique to any rewrite system. The quality of the approximation is closely related to those choices. On one hand, it depends on the choice of the predicate symbol to be reused when *predicate-limit* is reached. On the other hand, the choice of generating function-symbols as output or as input is also crucial. According to the verification context, some automated heuristics will have to be designed in order to obtain well-customized approximations.

Ongoing work tends to show that the linear restriction concerning the rewrite system can be tackled. A non right-linear rewrite system makes the computed S-CF program copying. Consequently, Theorem 24 does not hold anymore. To get rid of the right-linearity restriction, we are studying the transformation of a copying S-CF clause into non-copying ones that will generate an over-approximation.

Detected non-convergent critical pairs	New clauses obtained by norm_{Prog}
	Starting S-CF program $P_d(d_1(\widehat{x,y,z})) \leftarrow P_a(\widehat{x}), P_a(\widehat{y}), P_a(\widehat{z}).$ $P_a(\widehat{a}) \leftarrow .$
$P_d(d_1(h(x),\widehat{i(y)},s(z))) \leftarrow P_a(\widehat{x}), P_a(\widehat{y}), P_a(\widehat{z})$	$P_d(d_1(\widehat{x,y,z})) \leftarrow P_1(\widehat{x},\widehat{y},\widehat{z}).$ $P_1(\widehat{h(x)},\widehat{i(y)},\widehat{s(z)}) \leftarrow P_a(\widehat{x}), P_a(\widehat{y}), P_a(\widehat{z}).$
$P_d(d_2(\widehat{x,y,z})) \leftarrow P_a(\widehat{x}), P_a(\widehat{y}), P_a(\widehat{z}).$	$\longrightarrow$
$P_d(d_1(h(x),\widehat{i(y)},s(z))) \leftarrow P_1(\widehat{x},\widehat{y},\widehat{z})$	$P_1(\widehat{h(x)},\widehat{i(y)},\widehat{s(z)}) \leftarrow P_1(\widehat{x},\widehat{y},\widehat{z}).$
$P_d(d_2(\widehat{x,y,z})) \leftarrow P_1(\widehat{x},\widehat{y},\widehat{z}).$	$\longrightarrow$
$P_d(c(\widehat{x,y})) \leftarrow P_a(\widehat{x}), P_a(\widehat{y}).$	$\longrightarrow$
$P_d(d_2(f(\widehat{h(x)}),g(i(y))),z)) \leftarrow P_a(\widehat{x}), P_a(\widehat{y}), P_a(\widehat{z})$	$P_d(d_2(\widehat{x,y,z})) \leftarrow P_2(\widehat{x},\widehat{y},\widehat{z},x',y',z'), P_a(\widehat{x'}), P_a(\widehat{y'}), P_a(\widehat{z'}).$ $P_2(\widehat{f(x)},\widehat{g(y)},\widehat{z},x',y',z') \leftarrow P_2(\ \widehat{x},\widehat{y},\widehat{z},h(x'),i(y'),z')$ $P_2(\widehat{x},\widehat{y},\widehat{z},x,y,z) \leftarrow .$
A cycle is detected – **removeCycles** replaces the blue clause by the red one.	$P_2(\widehat{f(x)},\widehat{g(y)},\widehat{z},\ x',y',z') \leftarrow P_2(\widehat{x},\widehat{y},\widehat{z_1},h(x').i(y').z'_1),$ $\qquad\qquad\qquad P_2(\widehat{x_1},\widehat{y_1},\widehat{z},\ h(x'_1).i(y'_1),\ z')$
$P_d(d_2(f(\widehat{h(x)}),g(i(y)),z)) \leftarrow P_1(\widehat{x},\widehat{y},\widehat{z})$	$P_d(d_2(\widehat{x,y,z})) \leftarrow P_2(\widehat{x},\widehat{y},\widehat{z},x',y',z'), P_1(\widehat{x'},\widehat{y'},\widehat{z'}).$
$P_d(c(f(\widehat{x}),g(y))) \leftarrow P_2(\widehat{x},\widehat{y},\widehat{z},h(x'),i(y'),z'),$ $\qquad\qquad P_a(\widehat{x'}), P_a(\widehat{y'}).$	$P_3(\widehat{f(x)},\widehat{g(y)}) \leftarrow P_2(\widehat{x},\widehat{y},\widehat{z},h(x'),i(y'),z'),\ P_a(\widehat{x'}), P_a(\widehat{y'}).$ $P_d(c(\widehat{x,y})) \leftarrow P_3(\widehat{x},\widehat{y}).$

Fig. 2. Run of comp_R on Example 27

On the other hand, to get rid of the left-linearity restriction, we are studying a technique based on the transformation of any Horn clause into CS-clauses [17]. However, the method of [17] does not always terminate. We want to ensure termination thanks to an additional over-approximation.

References

1. Boichut, Y., Boyer, B., Genet, T., Legay, A.: Equational Abstraction Refinement for Certified Tree Regular Model Checking. In: Aoki, T., Taguchi, K. (eds.) ICFEM 2012. LNCS, vol. 7635, pp. 299–315. Springer, Heidelberg (2012)
2. Boichut, Y., Chabin, J., Réty, P.: Over-approximating descendants by synchronized tree languages. RTA. LIPIcs **21**, 128–142 (2013)
3. Boichut, Y., Chabin, J., Réty, P.: Towards more precise rewriting approximations (full version). Tech. Rep. RR-2014-02, LIFO, Université d'Orléans (2014)
4. Boichut, Y., Courbis, R., Héam, P.-C., Kouchnarenko, O.: Finer Is Better: Abstraction Refinement for Rewriting Approximations. In: Voronkov, A. (ed.) RTA 2008. LNCS, vol. 5117, pp. 48–62. Springer, Heidelberg (2008)

5. Boichut, Y., Héam, P.C.: A Theoretical Limit for Safety Verification Techniques with Regular Fix-point Computations. IPL **108**(1), 1–2 (2008)
6. Bouajjani, A., Habermehl, P., Rogalewicz, A., Vojnar, T.: Abstract Regular (Tree) Model Checking. STTT **14**(2), 167–191 (2012)
7. Comon, H., Dauchet, M., Gilleron, R., Lugiez, D., Tison, S., Tommasi, M.: Tree Automata Techniques and Applications (TATA)
8. Durand, I., Sylvestre, M.: Left-linear bounded trss are inverse recognizability preserving. RTA. LIPIcs **10**, 361–376 (2011)
9. Engelfriet, J., Heyker, L.: Context-free Hypergraph Grammars have the same Term-generating Power as Attribute Grammars. Acta Informatica 29 (1992)
10. Engelfriet, J., Vereijken, J.: Context-free Grammars and Concatenation of Graphs. Acta Informatica **34**, 773–803 (1997)
11. Engelfriet, J., Schmidt, E.M.: IO and OI (I). Journal of Computer and System Sciences **15**(3), 328–353 (1977)
12. Engelfriet, J., Schmidt, E.M.: IO and OI (II). Journal of Computer and System Sciences **16**(1), 67–99 (1978)
13. Genet, T.: Decidable Approximations of Sets of Descendants and Sets of Normal Forms. In: Nipkow, T. (ed.) RTA 1998. LNCS, vol. 1379, pp. 151–165. Springer, Heidelberg (1998)
14. Genet, T., Klay, F.: Rewriting for cryptographic protocol verification. In: McAllester, D. (ed.) Automated Deduction - CADE-17. LNCS, vol. 1831, pp. 271–290. Springer, Heidelberg (2000)
15. Gouranton, V., Réty, P., Seidl, H.: Synchronized Tree Languages Revisited and New Applications. In: Honsell, F., Miculan, M. (eds.) FOSSACS 2001. LNCS, vol. 2030, pp. 214–229. Springer, Heidelberg (2001)
16. Kochems, J., Ong, C.H.L.: Improved Functional Flow and Reachability Analyses Using Indexed Linear Tree Grammars. RTA. LIPIcs **10**, 187–202 (2011)
17. Limet, S., Salzer, G.: Proving Properties of Term Rewrite Systems via Logic Programs. In: van Oostrom, V. (ed.) RTA 2004. LNCS, vol. 3091, pp. 170–184. Springer, Heidelberg (2004)
18. Limet, S., Salzer, G.: Tree Tuple Languages from the Logic Programming Point of View. Journal of Automated Reasoning **37**(4), 323–349 (2006)
19. Raoult, J.: Rational Tree Relations. Bulletin of the Belgian Mathematical Society Simon Stevin **4**, 149–176 (1997)
20. Réty, P.: Langages synchronisés d'arbres et applications. Habilitation Thesis (in French). LIFO, Université d'Orléans. Tech. rep., June 2001
21. Réty, P., Chabin, J., Chen, J.: R-Unification thanks to Synchronized-Contextfree Tree Languages. In: UNIF (2005)
22. Réty, P., Chabin, J., Chen, J.: Synchronized ContextFree Tree-tuple Languages. Tech. Rep. RR-2006-13 (LIFO, 2006)
23. Rounds, W.C.: Context-free grammars on trees. In: Fischer, P.C., Ginsburg, S., Harrison, M.A. (eds.) STOC. ACM (1969)

Sorting Networks: The End Game

Michael Codish[1], Luís Cruz-Filipe[2(✉)], and Peter Schneider-Kamp[2]

[1] Department of Computer Science, Ben-Gurion University of the Negev, PoB 653,
84105 Beersheva, Israel
mcodish@cs.bgu.ac.il

[2] Department of Mathematics and Computer Science, University of Southern
Denmark, Campusvej 55, 5230 Odense M, Denmark
{lcf,petersk}@imada.sdu.dk

Abstract. This paper studies properties of the back end of a sorting
network and illustrates the utility of these in the search for networks
of optimal size or depth. All previous works focus on properties of the
front end of networks and on how to apply these to break symmetries in
the search. The new properties help shed understanding on how sorting
networks sort and speed-up solvers for both optimal size and depth by
an order of magnitude.

Keywords: Sorting networks · SAT solving · Symmetry breaking

1 Introduction

In the last year, new results were obtained regarding optimality of sorting networks, concerning both the optimal depth of sorting networks on 11 to 16 channels [2] and the optimal size of sorting networks on 9 and 10 channels [3]. Both
these works apply symmetry-breaking techniques that rely on analyzing the
structure at the front of a sorting network in order to reduce the number of
candidates to test in an exhaustive proof by case analysis.

In this work, we focus on the dual problem: what does the *end* of a sorting
network look like? To the best of our knowledge, this question has never been
studied in much detail. Batcher [1] characterizes a particular class of networks
that can be completed to a sorting network in a systematic way, but his work
only applies to the search for efficient sorting networks. Parberry [8] establishes
a necessary condition to avoid examining the last two layers of a candidate prefix
in his proof of optimality of the depth 6 sorting network on 9 channels, but its
application requires fixing the previous layers (although it has similarities to the
idea behind our proof of Theorem 11 below).

We show that the comparators in the last layer of a sorting network are of a
very particular form, and that the possibilities for the penultimate layer are also
limited. Furthermore, we show how to control redundancy of a sorting network

Supported by the Israel Science Foundation, grant 182/13 and by the Danish Council
for Independent Research, Natural Sciences.

in a very precise way in order to restrict its last two layers to a significantly smaller number of possibilities, and we study the impact of this construction in the SAT encodings used in the proofs of optimality described in [2,3].

The analysis, results, and techniques in this paper differ substantially from the work done on the first layers: that work relies heavily on symmetries of sorting networks to show that the comparators in those layers *may* be restricted to be of a particular form. Our results show that the comparators in the last layers *must* have a particular form. When working with the first layers it suffices to work up to renaming of the channels, as there are very general results on how to apply permutations to the first layers of any sorting network and obtain another sorting network of the same depth and size. On the last layer, this is not true: permuting the ending of a sorting network will not, in general, yield the ending of another sorting network. We formalize the fact that, as inputs go through a sorting network, the number of channels between pairs of unsorted values gets smaller, until, at the last layer, all occurrences of unsorted pairs of values are on adjacent channels. To the best of our knowledge, this surprising fact has never been observed before, and it influences the possible positions of comparators in the last layers. This intuition about the mechanism of sorting networks is formally expressed by the notion of k-block and Theorem 11, which is the main contribution of this paper.

2   Preliminaries on Sorting Networks

A *comparator network* C with n channels and depth d is a sequence $C = L_1; \ldots; L_d$ where each *layer* L_k is a set of comparators (i, j) for pairs of channels $i < j$. At each layer, every channel may occur in at most one comparator. The *depth* of C is the number of layers d, and the *size* of C is the total number of comparators in its layers. If C_1 and C_2 are comparator networks, then $C_1; C_2$ denotes the comparator network obtained by concatenating the layers of C_1 and C_2; if C_1 has m layers, it is an *m-layer prefix* of $C_1; C_2$.

An input $\bar{x} \in \{0, 1\}^n$ propagates through C as follows: $\bar{x}_0 = \bar{x}$, and for $0 < k \leq d$, $\bar{x}_k$ is the permutation of $\bar{x}_{k-1}$ obtained as follows: for each comparator $(i, j) \in L_k$, the values at positions i and j of $\bar{x}_{k-1}$ are reordered in $\bar{x}_k$ so that the value at position i is not larger than the value at position j. The output of the network for input $\bar{x}$ is $C(\bar{x}) = \bar{x}_d$, and $\mathsf{outputs}(C) = \{ C(\bar{x}) \mid \bar{x} \in \{0, 1\}^n \}$. The comparator network C is a *sorting network* if all elements of $\mathsf{outputs}(C)$ are sorted (in ascending order). The zero-one principle (e.g. [6]) implies that a sorting network also sorts any other totally ordered set, e.g. integers.

Optimal sorting network problems are about finding the smallest depth and the smallest size of a sorting network for a given number of channels n. Figure 1 shows a sorting network on 5 channels that has optimal size (9 comparators) and optimal depth (5 layers). It also shows how the network sorts the input 10101.

In order to determine the minimal depth of an optimal sorting network on n channels, one needs to consider all possible ways in which such a network can be built. Parberry [8] shows that the first layer of a depth-optimal sorting

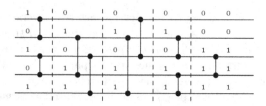

Fig. 1. An optimal-depth, optimal-size sorting network on 5 channels, operating on the input 10101. The channels are numbered from top to bottom, with a comparator (i, j) represented as a vertical line between two channels; each comparator moves its smallest input to its top channel. The layers are separated by a vertical dashed line.

network on n channels can be assumed to consist of the comparators $(2k - 1, 2k)$ for $1 \leq k \leq \lfloor \frac{n}{2} \rfloor$. Parberry and later Bundala and Závodný pursued the study of the possibilities for the second layer and demonstrated the impact of this on the search for optimal sorting networks.

The following two observations will be be instrumental for proofs in later sections. We write $\bar{x} \leq \bar{y}$ to denote that every bit of x is less than or equal to the corresponding bit of y, and $\bar{x} < \bar{y}$ for $\bar{x} \leq \bar{y}$ and $x \neq y$.

Lemma 1. *Let C be a comparator network and $\bar{x}$ be a sorted sequence. Then $\bar{x}$ is unchanged by every comparator in C.*

Lemma 2 (Theorem 4.1 in [1]). *Let C be a comparator network and $\bar{x}, \bar{y} \in \{0, 1\}^n$ be such that $\bar{x} \leq \bar{y}$. Then $C(\bar{x}) \leq C(\bar{y})$.*

3   The Last Layers of a Sorting Network

In this section we analyze the last two layers of a sorting network and derive some structural properties that will be useful both for restricting the search space in proofs of optimality, and as a tool to understand how a sorting network works.

We begin by recalling the notion of redundant comparator (Exercise 5.3.4.51 of [6], credited to R.L. Graham). Let $C; (i, j); C'$ be a comparator network. The comparator (i, j) is *redundant* if $x_i \leq x_j$ for all sequences $x_1 \ldots x_n \in \mathsf{outputs}(C)$. If D' is a comparator network obtained by removing every redundant comparator from D, then D' is a sorting network iff D is a sorting network: from the definition it follows that $D(\bar{x}) = D'(\bar{x})$ for every input $\bar{x} \in \{0, 1\}^n$. This result was already explored in the proof of optimality of the 25-comparator sorting network on 9 channels [3]. We will call a sorting network without redundant comparators *non-redundant*. In this section we focus on non-redundant sorting networks.

Lemma 3. *Let C be a non-redundant sorting network on n channels. Then all comparators in the last layer of C are of the form $(i, i + 1)$.*

Proof. Let C be as in the premise with a comparator $c = (i, i + 2)$ in the last layer. We can assume it is the last comparator. Since c is not redundant, there is an input $\bar{x}$ such that channels i to $i + 2$ before applying c look like (a) or (b) on the right.

1	0
0	0
0	1

(a)

1	0
1	1
0	1

(b)

Suppose $\bar{x}$ is a word yielding case (a), and let $\bar{y}$ be any word obtained by replacing one 0 in $\bar{x}$ by a 1. Since C is a sorting network, $C(\bar{y})$ is sorted, but since $\bar{x} < \bar{y}$ the value in channel i before applying c must be a 1 (Lemma 2), hence $\bar{y}$ yields situation (b). Dually, given $\bar{y}$ yielding (b), we know that any $\bar{z}$ obtained by replacing one 1 in $\bar{y}$ by a 0 will yield (a).

Thus all inputs with the same number of zeroes as $\bar{x}$ or $\bar{y}$ must yield either (a) or (b), in particular sorted inputs, contradicting Lemma 1. The same reasoning shows that c cannot have the form $(i, i + k)$ with $k > 2$, thus it has to be of the form $(i, i + 1)$. □

Corollary 4. *Suppose that C is a sorting network with no redundant comparators that contains a comparator (i, j) at layer d, with $j > i + 1$. Then at least one of channels i and j is used in a layer d' with $d' > d$.*

Proof. If neither i nor j are used after layer d, then the comparator (i, j) can be moved to the last layer without changing the function computed by C. By the previous lemma C can therefore not be a sorting network. □

Lemma 3 restricts the number of possible comparators in the last layer in a sorting network on n channels to $n - 1$, instead of $n(n - 1)/2$ in the general case.

Theorem 5. *The number of possible last layers in an n-channel sorting network with no redundancy is $L_n = F_{n+1} - 1$, where F_n denotes the Fibonacci sequence.*

Proof. Denote by L_n^+ the number of possible last layers on n channels, where the last layer is allowed to be empty (so $L_n = L_n^+ - 1$). There is exactly one possible last layer on 1 channel, and there are two possible last layers on 2 channels (no comparators or one comparator), so $L_1^+ = F_2$ and $L_2^+ = F_3$.

Given a layer on n channels, there are two possibilities. Either the first channel is unused, and there are L_{n-1}^+ possibilities for the remaining $n - 1$ channels; or it is connected to the second channel, and there are L_{n-2}^+ possibilities for the remaining $n - 2$ channels. So $L_n^+ = L_{n-1}^+ + L_{n-2}^+$, whence $L_n^+ = F_{n+1}$. □

Even though L_n grows quickly, it grows slower than the number G_n of possible layers in general [2]; in particular, $L_{17} = 2583$, whereas $G_{17} = 211,799,312$.

To move (backwards) beyond the last layer, we introduce an auxiliary notion.

Definition 6. *et C be a depth d sorting network without redundant comparators, and let $k < d$. A k-block of C is a set of channels B such that $i, j \in B$ if and only if there is a sequence of channels $i = x_0, \ldots, x_\ell = j$ where (x_i, x_{i+1}) or (x_{i+1}, x_i) is a comparator in a layer $k' > k$ of C.*

Note that for each k the set of k-blocks of C is a partition of the set of channels.

Given a comparator network of depth d, we will call its $(d-1)$-blocks simply *blocks* – so Lemma 3 states that a block in a sorting network C is either a channel unused at the last layer of C or two adjacent channels connected by a comparator at the last layer of C.

Example 7. Recall the sorting network shown in Figure 1. Its 4-blocks, or simply blocks, are $\{1\}$, $\{2\}$, $\{3,4\}$ and $\{5\}$, its 3-blocks are $\{1\}$, $\{2,3,4,5\}$, and for $k < 3$ there is only the trivial k-block $\{1,2,3,4,5\}$.

Lemma 8. *Let C be a sorting network of depth d on n channels, and $k < d$. For each input $\bar{x} \in \{0,1\}^n$, there is at most one k-block that receives a mixture of 0s and 1s as input.*

Proof. From the definition of k-block, there is no way for values to move from one k-block to another. Therefore, if there is an input for which two distinct k-blocks receive both 0s and 1s as inputs, the output will not be sorted. □

Lemma 9. *Let C be a depth d sorting network on n channels without redundant comparators. Then all comparators in layer $d-1$ connect adjacent blocks of C.*

Proof. The proof is similar to that of Lemma 3, but now considering blocks instead of channels. Let c be a comparator in layer $d-1$ of C that does not connect adjacent blocks of C. Since c is not redundant, there must be some input $\bar{x}$ that provides c with input 1 on its top channel and 0 on its bottom channel. The situation is depicted below, where A and C are blocks, and B is the set of channels in between. According to Lemma 8, there are five possible cases for A, B and C, depending on the number of 0s in $\bar{x}$.

	A	all 0	all 0	all 0	all 0	mixed
	B	all 0	all 0	mixed	all 1	all 1
	C	mixed	all 1	all 1	all 1	all 1
		(a)	(b)	(c)	(d)	(e)

(To the left: channels labeled, with 1 at A, B in between, 0 at C, with comparator.)

Suppose that $\bar{x}$ yields (a). By changing the appropriate number of 0s in $\bar{x}$ to 1s, we can find a word $\bar{y}$ yielding case (b), since again by monotonicity of C this cannot bring a 0 to the top input of c. Likewise, we can reduce (e) to (d). But now we can move between (b), (c) and (d) by changing one bit of the word at a time. By Lemma 2, this must keep either the top 1 input of c or the lower 0, while the other input is kept by the fact that C is a sorting network. Again this proves that this configuration occurs for all words with the same number of 0s, which is absurd since it cannot happen for the sorted input. □

Combining this result with Lemma 3 we obtain the explicit configurations that can occur in a sorting network.

Corollary 10. *Let C be a depth d sorting network on n channels without redundant comparators. Then every comparator (i,j) in layer $d-1$ of C satisfies $j - i \leq 3$. Furthermore, if $j = i + 2$, then either $(i, i+1)$ or $(i+1, i+2)$ occurs in the last layer; and if $j = i + 3$, then both $(i, i+1)$ and $(i+2, i+3)$ occur in the last layer.*

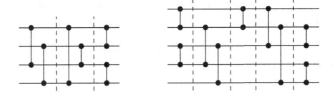

Fig. 2. Sorting networks containing a comparator $(i, i+3)$ in their penultimate layer

The sorting networks in Figure 2 show that the bound $j - i \leq 3$ is tight. We can also state a more general form of Lemma 9, proved exactly in the same way.

Theorem 11. *If C is a sorting network on n channels without redundant comparators, then every comparator at layer k of C connects adjacent k-blocks of C.*

When considering the last n comparators instead of the last k layers, induction on n using Theorem 11 yields the following result.

Corollary 12. *Every k-block with n comparators of a sorting network without redundant comparators uses at most $n + 1$ channels.*

4   Co-Saturation

The results in the previous sections allow us to reduce the search space of all possible sorting networks of a given depth simply by identifying necessary conditions on the comparators those networks may have. However, the successful strategies in [2,3,8] all focus on finding *sufficent* conditions on those comparators: identifying a (smaller) set of networks that must contain one sorting network of depth d (or size k), if such a network exists at all.

We now follow this idea pursuing the idea of saturation in [2]: how many (redundant) comparators can we safely add to the last layers of a sorting network? We will show how to do this in a way that reduces the number of possibilities for the last two layers to a minimum. Note that we are again capitalizing on the observation that redundant comparators do not change the function represented by a comparator network and can, thus, be removed or added at will.

Lemma 13. *Let C be a sorting network on n channels. There is a sorting network N of the same depth whose last layer: (i) only contains comparators between adjacent channels; and (ii) does not contain two adjacent unused channels.*

Proof. We first eliminate all redundant comparators from C to obtain a sorting network S. By Lemma 3 all comparators in the last layer of S are then of the form $(i, i+1)$. Let j be such that j and $j+1$ are unused in the last layer of S; since S is a sorting network, this means that the comparator $(j, j + 1)$ is redundant and we can add it to the last layer of S. Repeating this process for $j = 1, \ldots, n$ we obtain a sorting network N that satisfies both desired properties. $\qquad \square$

We say that a sorting network satisfying the conditions of Lemma 13 is in *last layer normal form* (llnf).

Theorem 14. *The number of possible last layers in llnf on n channels is $K_n = P_{n+5}$, where P_n denotes the Padovan sequence, defined as $P_0 = 1$, $P_1 = P_2 = 0$ and $P_{n+3} = P_n + P_{n+1}$.*

Proof. Let K_n^+ be the number of layers in llnf that begin with the comparator $(1,2)$, and K_n^- the number of those where channel 1 is free. Then $K_n = K_n^+ + K_n^-$. Let $n > 3$. If a layer in llnf begins with a comparator, then there are K_{n-2} possibilities for the remaining channels; if it begins with a free channel, then there are K_{n-1}^+ possibilities for the remaining channels. Therefore $K_n = K_n^+ + K_n^- = K_{n-2} + K_{n-1}^+ = K_{n-2} + K_{n-3}$. There exist one last layer on 1 channel (with no comparator), one on 2 channels (with one comparator between them) and two on 3 channels (one comparator between either the top two or the bottom two channels), so $K_1 = P_6$, $K_2 = P_7$ and $K_3 = P_8$. From the recurrence it follows that $K_n = P_{n+5}$. □

Note that K_n grows much slower than the total number L_n of non-redundant last layers identified in Theorem 5. For example, $K_{17} = 86$ instead of $L_{17} = 2583$.

If the last layer is required to be in llnf, we can also study the previous layer. By Lemma 9, we know that every block can only be connected to the adjacent ones; again we can *add* redundant comparators to reduce the number of possibilities for the last two layers.

Lemma 15. *Let C be a sorting network of depth d in llnf. Let $i < j$ be two channels that are unused in layer $d-1$ and that belong to different blocks. Then adding the comparator (i,j) to layer $d-1$ of C still yields a sorting network.*

Proof. Suppose there is an input $\bar{x}$ such that channel i carries a 1 at layer $d-1$, and channel j carries a 0 at that same layer. Since neither channel is used, their corresponding blocks will receive these values. But then $C(\bar{x})$ has a 1 in a channel in the block containing i and a 0 in the block containing j, and since $i < j$ this sequence is not sorted by C. Therefore the comparator (i,j) at layer $d-1$ of C is redundant, and can be added to this network. □

Lemma 16. *Let C be a sorting network of depth d in llnf. Suppose that there is a comparator $(i, i+1)$ in the last layer of C, that channel $i+2$ is used in layer $d-1$ but not in layer d, and that channels i and $i+1$ are both unused in layer $d-1$ of C (see Figure 3, left). Then there is a sorting network C' of depth d in llnf such that channels $i+1$ and $i+2$ are both used in layers $d-1$ and d.*

Proof. Since channels i and $i+1$ are unused in layer $d-1$, comparator $(i, i+1)$ can be moved to that layer without changing the behaviour of C; then the redundant comparator $(i+1, i+2)$ can be added to layer d, yielding the sorting network C' (Figure 3, left). If $i > 1$ and channel $i-1$ is not used in the last layer of C, then C' must also contain a comparator $(i-1, i)$ in its last layer (Figure 3, right). □

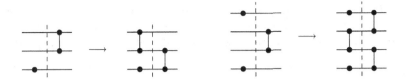

Fig. 3. Transformations in the proof of Lemma 16

Lemma 16 can also be applied if channel $i - 1$ (instead of $i + 2$) is used at layer $d - 1$ and unused in layer d.

Definition 17. *A sorting network of depth d is* co-saturated *if: (i) its last layer is in llnf, (ii) no two consecutive blocks at layer $d - 1$ have unused channels, and (iii) if $(i, i + 1)$ is a comparator in layer d and channels i and $i + 1$ are unused in layer $d - 1$, then channels $i - 1$ and $i + 2$ (if they exist) are used in layer d.*

Theorem 18. *If C is a sorting network on n channels with depth d, then there is a co-saturated sorting network N on n channels with depth d.*

Proof. Assume C is given. Apply Theorem 14 to find a sorting network S in llnf, containing no redundant comparators except possibly in the last layer.

Let $B_1, \ldots, B_k$ be the $(d - 1)$-blocks in S. For $i = 1, \ldots, k - 1$, if blocks B_i and B_{i+1} have a free channel, add a comparator between them. (Note that it may be possible to add *two* comparators between these blocks, namely if they both have two channels and none is used in layer $d - 1$.) Let N be the resulting network. By Lemma 15, all the comparators added from S to N are redundant, so N is a sorting network; by construction, N satisfies (ii).

If N does not satisfy (iii), then applying Lemma 16 transforms it into another sorting network N' that does. □

Table 1 shows the number of possibilities for the last two layers of a co-saturated sorting network on n channels for $n \leq 17$, obtained by a representation of these suffixes similar to the one described in [4].

In the next sections, we show how we can capitalize on these results to improve the proofs of optimal depth and optimal size of sorting networks.

5   Implications for Optimal Depth SAT Encodings

In this section, we describe how SAT encodings in the spirit of [2] can profit from the results in Sections 3 and 4. We detail the boolean variables in the model of the

Table 1. Number of distinct co-saturated two-layer suffixes on n channels, for $n \leq 17$

n	3	4	5	6	7	8	9	10	11	12	13	14	15	16	17
#	4	4	12	26	44	86	180	376	700	1,440	2,892	5,676	11,488	22,848	45,664

encoding, and express our contribution in terms of those. The remaining details of this construction are immaterial to this paper. The encoding represents an n-channel comparator network of depth d by $d \times n(n-1)$ Boolean variables

$$\mathsf{V}_n^d = \left\{ \, c_{i,j}^\ell \mid 1 \le \ell \le d, 1 \le i < j \le n \, \right\}$$

where the intention is that $c_{i,j}^\ell$ is true if and only if the network contains a comparator between channels i and j at depth ℓ. Further, to facilitate a concise and efficient encoding of our new results, we introduce an additional set of $d \cdot n$ Boolean variables capturing which channels are "used" at a given layer

$$\mathsf{U}_n^d = \left\{ \, u_k^\ell \mid 1 \le \ell \le d, 1 \le k \le n \, \right\}$$

where the intention is that u_k^ℓ is true if and only if there is some comparator on channel k at level ℓ. Using these variables, previous work describes how the search for an n-channel sorting network of depth d is encoded by a formula φ_0 satisfiable if and only if there is such a network. If φ_0 is satisfiable, the network found can be reconstructed from the assignment of the variables V_n^d.

5.1 Encoding Necessary Conditions

The results of Section 3 represent necessary conditions for non-redundant sorting networks. Thus, we can just add them to the SAT encoding as further restrictions of the search space without losing solutions. We start by looking at the last layer, i.e., the layer at depth d, and then continue to consider layer $d-1$.

Consider first Lemma 3, which states that non-redundant comparators in the last layer have to be of the form $(i, i+1)$. Seen negatively, we can simply forbid all comparators (i, j) where $j > i+1$, that connect non-adjacent channels. This restriction can be encoded straightforwardly by adding the following $(n-1)(n-2)$ unit clauses φ_1 to the SAT encoding:

$$\varphi_1 = \left\{ \, \neg c_{i,j}^d \mid 1 \le i, i+1 < j \le n \, \right\}$$

The restriction from Lemma 3 is generalized by Corollary 4, which states that whenever a comparator at any layer connects two non-adjacent channels, necessarily one of these channels is used at a later layer. Similarly to φ_1 we can encode this by adding one clause for each of the $(n-1)(n-2)/2$ non-adjacent comparator at any given depth ℓ using $\varphi_1(\ell)$:

$$\varphi_1(\ell) = \left\{ \, c_{i,j}^\ell \rightarrow \bigvee_{\ell < k \le d} u_i^k \vee u_j^k \;\middle|\; 1 \le i, i+1 < j \le n \, \right\}$$

Note that indeed $\varphi_1(d) = \varphi_1$, as there is no depth k with $\ell < k \le d$.

We now move on to consider the penultimate layer $d-1$. According to Corollary 10, no comparator at this layer can connect two channels more than 3 channels apart. Similar to Lemma 3, we encode this restriction by adding unit clauses for each of the $(n-3)(n-4)/2$ comparators more than 3 channels apart:

$$\varphi_2 = \left\{ \, \neg c_{i,j}^{d-1} \mid 1 \le i, i+3 < j \le n \, \right\}$$

Corollary 10 also states that the existence of a comparator $(i, i+3)$ on the penultimate layer implies the existence of the two comparators $(i, i+1)$ and $(i+2, i+3)$ on the last layer. This is straightforwardly encoded using additional $2(n-3)$ implication clauses:

$$\varphi_3 = \left\{ c_{i,i+3}^{d-1} \rightarrow c_{i,i+1}^{d} \right) \wedge \left(c_{i,i+3}^{d-1} \rightarrow c_{i+2,i+3}^{d} \mid 1 \leq i \leq n-3 \right\}$$

Finally, Corollary 10 also states that the existence of a comparator $(i, i+2)$ on the penultimate layer implies the existence of either of the comparators $(i, i+1)$ or $(i+1, i+2)$ on the last layer. This can be encoded using $n-2$ clauses:

$$\varphi_4 = \left\{ c_{i,i+2}^{d-1} \rightarrow c_{i,i+1}^{d} \vee c_{i+1,i+2}^{d} \mid 1 \leq i \leq n-2 \right\}$$

Empirically, we have found that using $\varphi = \varphi_0 \wedge \varphi_1 \wedge \varphi_2 \wedge \varphi_3 \wedge \varphi_4$ instead of just φ_0 decreases SAT solving times dramatically. In contrast, adding $\varphi_1(\ell)$ for $\ell < d$ has not been found to have a positive impact.

5.2 Symmetry Breaking Using Sufficient Conditions

The restrictions encoded so far were necessary conditions for non-redundant sorting networks. In addition, we can break symmetries by using the sufficient conditions from Section 4, essentially forcing the SAT solver to *add redundant comparators*.

According to Lemma 13 (ii) we can break symmetries by requiring that there are no adjacent unused channels in the last layer, i.e., that the network is in llnf.

$$\psi_1 = \left\{ u_i^d \vee u_{i+1}^d \mid 1 \leq i < n \right\}$$

Essentially, this forces the SAT solver to add a (redundant) comparator between any two adjacent unused channels on the last layer.

The next symmetry break is based on a consideration of two adjacent blocks. There are four possible cases: two adjacent comparators, a comparator followed by an unused channel, an unused channel followed by a comparator, and two unused channels. The latter is forbidden by the symmetry break ψ_1 (and thus not regarded further).

The case of two adjacent comparators is handled by formula ψ_2^a:

$$\psi_2^a = \left\{ c_{i,i+1}^d \wedge c_{i+2,i+3}^d \rightarrow \left(u_i^{d-1} \wedge u_{i+1}^{d-1} \right) \vee \left(u_{i+2}^{d-1} \wedge u_{i+3}^{d-1} \right) \mid 1 \leq i \leq n-3 \right\}$$

This condition essentially forces the SAT solver to add a (redundant) comparator on layer $d-1$, if both blocks have an unused channel in that layer.

The same idea of having to add a comparator at layer $d-1$ is enforced for the two remaining cases of a comparator followed by an unused channel or its dual by ψ_2^b and ψ_2^c, respectively:

$$\psi_2^b = \left\{ c_{i,i+1}^d \wedge \neg u_{i+2}^d \rightarrow \left(u_i^{d-1} \wedge u_{i+1}^{d-1} \right) \vee u_{i+2}^{d-1} \mid 1 \leq i \leq n-2 \right\}$$
$$\psi_2^c = \left\{ \neg u_i^d \wedge c_{i+1,i+2}^d \rightarrow u_i^{d-1} \vee \left(u_{i+1}^{d-1} \wedge u_{i+2}^{d-1} \right) \mid 1 \leq i \leq n-2 \right\}$$

Table 2. SAT solving for n-channel, depth 8 sorting networks with $|R_n|$ 2-layer filters. The table shows the impact of the restrictions on the last two layers in the size of the encoding and the solving time (in seconds) for the slowest unsatisfiable instance, as well as the total time for all $|R_n|$ instances.

		unrestricted last layer: φ_0				restricted last layer: ψ			
		slowest instance			total	slowest instance			total
n	$\|R_n\|$	#clauses	#vars	time	time	#clauses	#vars	time	time
15	262	278312	18217	754.74	**130551.42**	335823	25209	148.35	**19029.26**
16	211	453810	27007	1779.14	**156883.21**	314921	22901	300.07	**24604.53**

The final symmetry break is based on Lemma 16, i.e., on the idea of moving a comparator from the last layer to the second last layer. We encode that such a situation cannot occur, i.e., that whenever we have a comparator on the last layer d following or followed by an unused channel, one of the channels of the comparator is used on layer $d - 1$:

$$\psi_3^a = \left\{ \ c_{i,i+1}^d \wedge \neg u_{i+2}^d \rightarrow u_i^{d-1} \vee u_{i+1}^{d-1} \mid 1 \leq i \leq n - 2 \right\}$$
$$\psi_3^b = \left\{ \ c_{i,i+1}^d \wedge \neg u_{i-1}^d \rightarrow u_i^{d-1} \vee u_{i+1}^{d-1} \mid 2 \leq i \leq n - 1 \right\}$$

Empirically, we found that $\psi = \varphi \wedge \psi_1 \wedge \psi_2^a \wedge \psi_2^b \wedge \psi_2^c \wedge \psi_3^a \wedge \psi_3^b$ further improves the performance of SAT solvers. In order to show optimality of the known depth 9 sorting networks on 15 and 16 channels, it is enough to show that there is no sorting network on those numbers of channels with a depth of 8. Previous work [4] introduces the notion of complete set of prefixes: a set R_n such that if there exists a sorting network on n channels with depth d, then there exists one extending a prefix in R_n. Using this result, it suffices to show that there are no sorting networks of depth 8 that extend an element of R_{15} or R_{16}. Table 2 shows the improvement of using ψ instead of φ_0, detailing for both cases the number of clauses, the number of variables and the time to solve the slowest of the $|R_n|$ instances (which are solved in parallel). We also specify the total solving time (both compilation and SAT-solving) for all $|R_n|$ instances together. The new encodings are larger per same instance (the slowest instances, showed in the table, are different), but, as indicated in the table, the total time required in order to show that the formulas are unsatisfiable is reduced by a factor of around 6.

6   Conclusion

This paper presents the first systematic exploration of what happens at the *end* of a sorting network, as opposed to at the beginning. We present properties of the last layers of sorting networks. In order to assess the impact of our contribution, we show how to integrate them into SAT encodings that search for sorting networks of a given depth [2]. Here, we see an order of magnitude improvement in solving times, bringing us closer to being able to solve the next open instance of the optimal depth problem (17 channels).

While the paper presents detailed results on the end of sorting networks in the context of proving optimal depth of sorting networks, the necessary properties of the last layers can also be used to prove optimal size. We experimented on adding constraints similar to those in Section 5 for the last three comparators, as well as constraints encoding Corollary 12, to the SAT encoding presented in [3]. Preliminary results based on uniform random sampling of more than 10% of the cases indicate that we can reduce the total computational time used in the proof that 25 comparators are optimal for 9 channels from 6.5 years to just over 1.5 years. On the 288-thread cluster originally used for that proof, this corresponds to reducing the actual execution time from over 8 days to just 2 days.

These results can also be used to improve times for the *search* for sorting networks. In a recent paper [7], the authors introduce an incremental approach to construct sorting networks (iterating between two different SAT problems). They show that, using the first three layers of a Green filter [5], their approach finds a sorting network with 17 channels and depth 10, thus improving the previous best upper bound on the depth of a 17-channel sorting network. Using the same prefix, together with the constraints on the last two layers described in Section 5, we can find a depth 10 sorting network in under one hour of computation. Without these last layer constraints, this procedure times out after 24 hours.

References

1. Baddar, S.W.A.H., Batcher, K.E.: Designing Sorting Networks: A New Paradigm. Springer (2011)
2. Bundala, D., Závodný, J.: Optimal sorting networks. In: Dediu, A.-H., Martín-Vide, C., Sierra-Rodríguez, J.-L., Truthe, B. (eds.) LATA 2014. LNCS, vol. 8370, pp. 236–247. Springer, Heidelberg (2014)
3. Codish, M., Cruz-Filipe, L., Frank, M., Schneider-Kamp, P.: Twenty-five comparators is optimal when sorting nine inputs (and twenty-nine for ten). In: Proceedings of ICTAI 2014. IEEE (2014) (accepted for publication)
4. Codish, M., Cruz-Filipe, L., Schneider-Kamp, P.: The quest for optimal sorting networks: Efficient generation of two-layer prefixes. In: Proceedings of SYNASC 2014. IEEE (2014) (accepted for publication)
5. Coles, D.: Efficient filters for the simulated evolution of small sorting networks. In: Proceedings of GECCO 2012, pp. 593–600. ACM (2012)
6. Knuth, D.E.: The Art of Computer Programming. Sorting and Searching, vol. III. Addison-Wesley (1973)
7. Müller, M., Ehlers, T.: Faster sorting networks for 17, 19 and 20 inputs. CoRR abs/1410.2736 (2014). http://arxiv.org/abs/1410.2736
8. Parberry, I.: A computer-assisted optimal depth lower bound for nine-input sorting networks. Mathematical Systems Theory **24**(2), 101–116 (1991)

Bounding Clique-Width via Perfect Graphs

Konrad Kazimierz Dabrowski[1]([✉]), Shenwei Huang[2], and Daniël Paulusma[1]

[1] School of Engineering and Computing Sciences, Durham University, Science
Laboratories, South Road, Durham DH1 3LE, UK
{konrad.dabrowski,daniel.paulusma}@durham.ac.uk
[2] School of Computing Science, Simon Fraser University, 8888 University Drive,
Burnaby, BC V5A 1S6, Canada
shenweih@sfu.ca

Abstract. Given two graphs H_1 and H_2, a graph G is (H_1, H_2)-free if
it contains no subgraph isomorphic to H_1 or H_2. We continue a recent
study into the clique-width of (H_1, H_2)-free graphs and present three
new classes of (H_1, H_2)-free graphs that have bounded clique-width. We
also show the implications of our results for the computational complexity of the COLOURING problem restricted to (H_1, H_2)-free graphs. The
three new graph classes have in common that one of their two forbidden
induced subgraphs is the diamond (the graph obtained from a clique on
four vertices by deleting one edge). To prove boundedness of their clique-
width we develop a technique based on bounding clique covering number
in combination with reduction to subclasses of perfect graphs.

Keywords: Clique-width · Forbidden induced subgraphs · Graph class

1 Introduction

Clique-width is a well-known graph parameter and its properties are well studied;
see for example the surveys of Gurski [20] and Kamiński, Lozin and Milanič [22].
Computing the clique-width of a given graph is NP-hard, as shown by Fellows,
Rosamond, Rotics and Szeider [18]. Nevertheless, many NP-complete graph problems are solvable in polynomial time on graph classes of *bounded* clique-width,
that is, classes in which the clique-width of each of its graphs is at most c for some
constant c. This follows by combining the fact that if a graph G has clique-width
at most c then a so-called $(8^c - 1)$-expression for G can be found in polynomial
time [28] together with a number of results [13,23,30], which show that if a
q-expression is provided for some fixed q then certain classes of problems can
be solved in polynomial time. A well-known example of such a problem is the
COLOURING problem, which is that of testing whether the vertices of a graph

The research in this paper was supported by EPSRC (EP/K025090/1). The second
author is grateful for the generous support of the Graduate (International) Research
Travel Award from Simon Fraser University and Dr. Pavol Hell's NSERC Discovery
Grant.

© Springer International Publishing Switzerland 2015
A.-H. Dediu et al. (Eds.): LATA 2015, LNCS 8977, pp. 676–688, 2015.
DOI: 10.1007/978-3-319-15579-1_53

can be coloured with at most k colours such that no two adjacent vertices are coloured alike. Due to these algorithmic implications, it is natural to research whether the clique-width of a given graph class is bounded.

It should be noted that having bounded clique-width is a more general property than having bounded tree-width, that is, every graph class of bounded treewidth has bounded clique-width but the reverse is not true [11]. Cliquewidth is also closely related to other graph width parameters, e.g. for any class, having bounded clique-width is equivalent to having bounded rank-width [29] and also equivalent to having bounded NLC-width [21]. Moreover, clique-width has been studied in relation to graph operations, such as edge or vertex deletions, edge subdivisions and edge contractions. For instance, a recent result of Courcelle [12] solved an open problem of Gurski [20] by proving that if $\mathcal{G}$ is the class of graphs of clique-width 3 and $\mathcal{G}'$ is the class of graphs obtained from graphs in $\mathcal{G}$ by applying one or more edge contraction operations then $\mathcal{G}'$ has unbounded clique-width.

The classes that we consider in this paper consist of graphs that can be characterized by a family $\{H_1, \ldots, H_p\}$ of forbidden induced subgraphs (such graphs are said to be $(H_1, \ldots, H_p)$-free). The clique-width of such graph classes has been extensively studied in the literature (e.g. [1–9, 14, 16, 19, 24–27]). It is straightforward to verify that the class of H-free graphs has bounded cliquewidth if and only if H is an induced subgraph of the 4-vertex path P_4 (see also [17]). Hence, Dabrowski and Paulusma [17] investigated for which pairs (H_1, H_2) the class of (H_1, H_2)-free graphs has bounded clique-width. In this paper we solve a number of the open cases. The underlying research question is:

What kind of properties of a graph class ensure that its clique-width is bounded?

As such, our paper is to be interpreted as a further step towards this direction. Rather than coming up with ad hoc techniques for solving specific cases, we aim to develop more general techniques for attacking a number of the open cases simultaneously. Our technique in this paper is obtained by generalizing an approach followed in the literature. In order to illustrate this approach with some examples, we first need to introduce some notation (see Section 2 for all other terminology).

Notation. The disjoint union $(V(G) \cup V(H), E(G) \cup E(H))$ of two vertex-disjoint graphs G and H is denoted by $G + H$ and the disjoint union of r copies of a graph G is denoted by rG. The complement of a graph G, denoted by $\overline{G}$, has vertex set $V(\overline{G}) = V(G)$ and an edge between two distinct vertices if and only if these vertices are not adjacent in G. The graphs C_r, K_r and P_r denote the cycle, complete graph and path on r vertices, respectively. The graph $\overline{2P_1 + P_2}$ is called the *diamond*. The graph $K_{1,3}$ is the 4-vertex star, also called the *claw*. For $1 \leq h \leq i \leq j$, let $S_{h,i,j}$ be the *subdivided claw* whose three edges are subdivided $h - 1$, $i - 1$ and $j - 1$ times, respectively; note that $S_{1,1,1} = K_{1,3}$.

Our Technique. Dabrowski and Paulusma [16] determined all graphs H for which the class of H-free bipartite graphs has bounded clique-width. Such a classification turns out to also be useful for proving boundedness of the clique-width for other graph classes. For instance, in order to prove that $(\overline{P_1 + P_3}, P_1 + S_{1,1,2})$-free graphs have bounded clique-width, the given graphs were first reduced to $(P_1 + S_{1,1,2})$-free bipartite graphs [17]. In a similar way, Dabrowski, Lozin, Raman and Ries [15] proved that $(K_3, K_{1,3} + K_2)$-free graphs and $(K_3, S_{1,1,3})$-free have bounded clique-width by reducing to a subclass of bipartite graphs. Note that bipartite graphs are perfect graphs. This motivated us to develop a technique based on perfect graphs that are not necessarily bipartite. In order to so, we need to combine this approach with an additional tool. This tool is based on the following observation. If the vertex set of a graph can be partitioned into a small number of cliques and the edges between them are sufficiently sparse, then the clique-width is bounded (see also Lemma 10). Our technique can be summarized as follows.

1. Reduce the input graph to a graph that is in some subclass of perfect graphs;
2. While doing so, bound the clique covering number of the input graph.

Another well-known subclass of perfect graphs is the class of chordal graphs. We show that besides the class of bipartite graphs, the class of chordal graphs and the class of perfect graphs itself may be used for Step 1. We explain Steps 1-2 of our technique in detail in Section 3.

Our Results. In this paper, we investigate whether our technique can be used to find new pairs (H_1, H_2) for which the clique-width of (H_1, H_2)-free graphs is bounded. We show that this is indeed the case. By applying our technique, we are able to present three new classes of (H_1, H_2)-free graphs of bounded clique-width.[1] Namely, it enables us to prove the following result, which we prove in Section 4.

Theorem 1. *The class of (H_1, H_2)-free graphs has bounded clique-width if*

(i) $H_1 = \overline{2P_1 + P_2}$ *and* $H_2 = 3P_1 + P_2$;
(ii) $H_1 = \overline{2P_1 + P_2}$ *and* $H_2 = 2P_1 + P_3$;
(iii) $H_1 = \overline{2P_1 + P_2}$ *and* $H_2 = P_2 + P_3$.

Structural Consequences. Theorem 1 reduces the number of open cases in the classification of the boundedness of the clique-width for (H_1, H_2)-free graphs to 13 open cases, up to some equivalence relation, see also [17]. Note that the graph H_1 is the diamond in each of the three results in Theorem 1. Out of the 13 remaining cases, there are still three cases in which H_1 is the diamond, namely when $H_2 \in \{P_1 + P_2 + P_3, P_1 + 2P_2, P_1 + P_5\}$. However, for each of these graphs H_2, it is not even known whether the clique-width of the corresponding smaller subclasses of (K_3, H_2)-free graphs is bounded. Of particular note is the class of $(K_3, P_1 + 2P_2)$-free graphs, which is contained in all of the above open

---

[1] We do not specify our upper bounds as this would complicate our proofs for negligible gain. This is because in our proofs we apply graph operations that exponentially increase the upper bound of the clique-width, which means that the bounds that could be obtained from our proofs would be very large and far from being tight.

cases and for which the boundedness of clique-width is unknown. Settling this case is a natural next step in completing the classification. Note that for K_3-free graphs the clique covering number is proportional to the size of the graph. Another natural research direction is to determine whether the clique-width of $(\overline{P_1 + P_4}, H_2)$-free graphs is bounded for $H_2 = P_2 + P_3$ (the clique-width is known to be unbounded for $H_2 \in \{3P_1 + P_2, 2P_1 + P_3\}$).

Dabrowski, Golovach and Paulusma [14] showed that COLOURING restricted to $(sP_1 + P_2, \overline{tP_1 + P_2})$-free graphs is polynomial-time solvable for all pairs of integers s, t. They justified their algorithm by proving that the clique-width of the class of $(sP_1, \overline{tP_1 + P_2})$-free graphs is bounded only for small values of s and t, namely only for $s \leq 2$ or $t \leq 1$ or $s + t \leq 6$. In the light of these two results it is natural to try to classify the clique-width of the class of $(sP_1 + P_2, \overline{tP_1 + P_2})$-free graphs for all pairs (s, t). Theorem 1, combined with the aforementioned classification of the clique-width of $(sP_1, \overline{tP_1 + P_2})$-free graphs and the fact that any class of (H_1, H_2)-free graphs has bounded clique-width if and only if the class of $(\overline{H_1}, \overline{H_2})$-free graphs has bounded clique-width, immediately enables us to do this.

Corollary 2. *The class of* $(sP_1 + P_2, \overline{tP_1 + P_2})$*-free graphs has bounded clique-width if and only if* $s \leq 1$ *or* $t \leq 1$ *or* $s + t \leq 5$*.*

Algorithmic Consequences. Our research was (partially) motivated by a study into the computational complexity of the COLOURING problem for (H_1, H_2)-free graphs. As mentioned, COLOURING is polynomial-time solvable on any graph class of bounded clique-width. Of the three classes for which we prove boundedness of clique-width in this paper, only the case of $(\overline{2P_1 + P_2}, 3P_1 + P_2)$-free (and equivalently $(2P_1 + P_2, \overline{3P_1 + P_2})$-free) graphs was previously known to be polynomial-time solvable [14]. Hence, Theorem 1 gives us four new pairs (H_1, H_2) with the property that COLOURING is polynomial-time solvable when restricted to (H_1, H_2)-free graphs, namely if

- $H_1 = 2P_1 + P_2$ and $H_2 \in \{\overline{2P_1 + P_3}, \overline{P_2 + P_3}\}$;
- $H_1 = \overline{2P_1 + P_2}$ and $H_2 \in \{2P_1 + P_3, P_2 + P_3\}$.

As such, there are still 15 potential classes of (H_1, H_2)-free graphs left for which both the complexity of COLOURING and the boundedness of their clique-width is unknown [17].

2   Preliminaries

Below we define some graph terminology used throughout our paper. Let G be a graph. For $u \in V(G)$, the set $N(u) = \{v \in V(G) \mid uv \in E(G)\}$ is the *neighbourhood* of u in G. The *degree* of a vertex in G is the size of its neighbourhood. The *maximum degree* of G is the maximum vertex degree. For a subset $S \subseteq V(G)$, we let $G[S]$ denote the *induced* subgraph of G, which has vertex set S and edge set $\{uv \mid u, v \in S, uv \in E(G)\}$. If $S = \{s_1, \dots, s_r\}$ then, to simplify notation, we may also write $G[s_1, \ldots, s_r]$ instead of $G[\{s_1, \ldots, s_r\}]$. Let H be another graph. We write $H \subseteq_i G$ to indicate that H is an induced subgraph of G. Let $X \subseteq V(G)$. We write $G \setminus X$ for the graph obtained from G after removing X. A set $M \subseteq E(G)$ is a *matching* if no two edges in M share an end-vertex. We say that two disjoint sets $S \subseteq V(G)$ and $T \subseteq V(G)$ are *complete* to each other if every vertex of S is adjacent to every vertex of T. If no vertex of S is joined to a vertex of T by an edge, then S and T are *anti-complete* to each other. Similarly, we say that a vertex u and a set S not containing u may be complete or anti-complete to each other. Let $\{H_1, \ldots, H_p\}$ be a set of graphs. Recall that G is $(H_1, \ldots, H_p)$-*free* if G has no induced subgraph isomorphic to a graph in $\{H_1, \ldots, H_p\}$; if $p = 1$, we may write H_1-free instead of (H_1)-free.

The *clique-width* of a graph G, denoted by $\mathrm{cw}(G)$, is the minimum number of labels needed to construct G by using the following four operations:

(i) creating a new graph consisting of a single vertex v with label i;
(ii) taking the disjoint union of two labelled graphs G_1 and G_2;
(iii) joining each vertex with label i to each vertex with label j ($i \neq j$);
(iv) renaming label i to j.

A class of graphs $\mathcal{G}$ has *bounded* clique-width if there is a constant c such that the clique-width of every graph in $\mathcal{G}$ is at most c; otherwise the clique-width of $\mathcal{G}$ is *unbounded*.

Let G be a graph. We say that G is *bipartite* if its vertex set can be partitioned into two (possibly empty) independent sets B and W. We say that (B, W) is a *bipartition* of G.

Let G be a graph. We define the following two operations. For an induced subgraph $G' \subseteq_i G$, the *subgraph complementation* operation (acting on G with respect to G') replaces every edge present in G' by a non-edge, and vice versa. Similarly, for two disjoint vertex subsets X and Y in G, the *bipartite complementation* operation with respect to X and Y acts on G by replacing every edge with one end-vertex in X and the other one in Y by a non-edge and vice versa.

We now state some useful facts for dealing with clique-width. We will use these facts throughout the paper. Let $k \geq 0$ be a constant and let γ be some graph operation. We say that a graph class $\mathcal{G}'$ is (k, γ)-*obtained* from a graph class $\mathcal{G}$ if the following two conditions hold:

(i) every graph in $\mathcal{G}'$ is obtained from a graph in $\mathcal{G}$ by performing γ at most k times, and
(ii) for every $G \in \mathcal{G}$ there exists at least one graph in $\mathcal{G}'$ obtained from G by performing γ at most k times.

We say that γ *preserves* boundedness of clique-width if for any finite constant k and any graph class $\mathcal{G}$, any graph class $\mathcal{G}'$ that is (k, γ)-obtained from $\mathcal{G}$ has bounded clique-width if and only if $\mathcal{G}$ has bounded clique-width.

Fact 1. Vertex deletion preserves boundedness of clique-width [24].

Fact 2. Subgraph complementation preserves boundedness of clique-width [22].

Fact 3. Bipartite complementation preserves boundedness of clique-width [22].

The following lemma is well-known and straightforward to check.

Lemma 3. *The clique-width of a forest is at most* 3.

Let G be a graph. The size of a largest independent set and a largest clique in G are denoted by $\alpha(G)$ and $\omega(G)$, respectively. The chromatic number of G is denoted by $\chi(G)$. We say that G is *perfect* if $\chi(H) = \omega(H)$ for every induced subgraph H of G.

We need the following well-known result, due to Chudnovsky, Robertson, Seymour and Thomas.

Theorem 4 (The Strong Perfect Graph Theorem [10]). *A graph is perfect if and only if it is C_r-free and $\overline{C_r}$-free for every odd $r \geq 5$.*

The *clique covering number* $\overline{\chi}(G)$ of a graph G is the smallest number of (mutually vertex-disjoint) cliques such that every vertex of G belongs to exactly one clique. If G is perfect, then $\overline{G}$ is also perfect (by Theorem 4). By definition, $\overline{G}$ can be partitioned into $\omega(\overline{G}) = \alpha(G)$ independent sets. This leads to the following well-known lemma.

Lemma 5. *Let G be any perfect graph. Then $\overline{\chi}(G) = \alpha(G)$.*

We say that a graph G is *chordal* if G contain no induced cycle on four or more vertices. Bipartite graphs and chordal graphs are perfect (by Theorem 4).

The following three lemmas give us a number of subclasses of perfect graphs with bounded clique-width. We will make use of these lemmas later on in the proofs as part of our technique.

Lemma 6 ([16]). *Let H be a graph. The class of H-free bipartite graphs has bounded clique-width if and only if $H \subseteq_i K_{1,3} + 3P_1, K_{1,3} + P_2, P_1 + S_{1,1,3}$ or $S_{1,2,3}$ or $H = sP_1$ for some $s \geq 1$.*

Lemma 7 ([19]). *The class of chordal $(\overline{2P_1 + P_2})$-free graphs has clique-width at most* 3.

Lemma 8 ([15]). *The class of $(K_3, K_{1,3} + P_2)$-free graphs has bounded clique-width.*

Finally, we also need the following lemma, which corresponds to the first lemma of [14] by complementing the graphs under consideration.

Lemma 9 ([14]). *Let $s \geq 0$ and $t \geq 0$. Then every $(\overline{sP_1 + P_2}, tP_1 + P_2)$-free graph is $(K_{s+1}, tP_1 + P_2)$-free or $(\overline{sP_1 + P_2}, (s^2(t-1)+2)P_1)$-free.*

3   The Clique Covering Lemma

In Section 2 we stated several lemmas that can be used to bound the clique-width if we can manage to reduce to some specific graph class. As we shall see, such a reduction is not always sufficient and the following lemma forms a crucial part of our technique (we use it in the proofs of each of our main results). We omit the proof due to space restrictions.

Lemma 10. *Let $k \geq 1$ be a constant and let G be a $(\overline{2P_1 + P_2}, 2P_2 + P_4)$-free graph. If $\overline{\chi}(G) \leq k$ then $\mathrm{cw}(G) \leq f(k)$ for some function f that only depends on k.*

It is easy to see that for any fixed constant $s \geq 2$ we can generalize Lemma 10 to be valid for $(\overline{2P_1 + P_2}, 2K_s + P_4)$-free graphs. By more complicated arguments it is also possible to generalize it to other graph classes, such as $(\overline{2P_1 + P_2}, K_s + P_6)$-free graphs for any fixed $s \geq 0$. However, this is not necessary for the main results of this paper.

4   The Proof of Theorem 1

Theorem 1 (i). *The class of $(\overline{2P_1 + P_2}, 3P_1 + P_2)$-free graphs has bounded clique-width.*

To prove this theorem, suppose G is a $(\overline{2P_1 + P_2}, 3P_1 + P_2)$-free graph. Applying Lemma 9 we find that G is $(K_3, 3P_1 + P_2)$-free or $(\overline{2P_1 + P_2}, 10P_1)$-free. If G is $(K_3, 3P_1 + P_2)$-free then it has bounded clique-width by Lemma 8, so we may assume it is $(\overline{2P_1 + P_2}, 10P_1, 3P_1 + P_2)$-free. We can then show that the vertex set of the graph can be partitioned into a bounded number of cliques, so the clique-width is bounded by Lemma 10. We omit the proof details.

We also omit the proof of our second main result.

Theorem 1 (ii). *The class of $(\overline{2P_1 + P_2}, 2P_1 + P_3)$-free graphs has bounded clique-width.*

We now prove the last of our three main results, namely that the class of $(\overline{2P_1 + P_2}, P_2 + P_3)$-free graphs has bounded clique-width. We first establish, via a series of lemmas, that we may restrict ourselves to graphs in this class that are also (C_4, C_5, C_6, K_5)-free. We omit the proofs for the first two of these lemmas.

Lemma 11. *The class of those $(\overline{2P_1 + P_2}, P_2 + P_3)$-free graphs that contain a K_5 has bounded clique-width.*

Lemma 12. *The class of those $(\overline{2P_1 + P_2}, P_2 + P_3, K_5)$-free graphs that contain an induced C_5 has bounded clique-width.*

Lemma 13. *The class of those $(\overline{2P_1 + P_2}, P_2 + P_3, K_5, C_5)$-free graphs that contain an induced C_4 has bounded clique-width.*

Proof. Suppose that G is a $(\overline{2P_1 + P_2}, P_2 + P_3, K_5, C_5)$-free graph containing a C_4, say on vertices v_1, v_2, v_3, v_4 in order. Let Y be the set of vertices adjacent to v_1 and v_2 (and possibly other vertices on the cycle). If $y_1, y_2 \in Y$ are non-adjacent then $G[v_1, v_2, y_1, y_2]$ would be a $\overline{2P_1 + P_2}$. Therefore Y is a clique. Since G is K_5-free, there are at most four such vertices. Therefore by Fact 1 we may assume that no vertex in G has two consecutive neighbours on the cycle. For $i \in \{1, 2\}$ let V_i be the set of vertices outside the cycle adjacent to v_{i+1} and v_{i+3} (where $v_5 = v_1$). For $i \in \{1, 2, 3, 4\}$ let W_i be the set of vertices whose unique neighbour on the cycle is v_i. Let X be the set of vertices with no neighbours on the cycle.

We first prove the following properties:

(i) V_i are independent sets for $i = 1, 2$.
(ii) W_i are independent sets for $i = 1, 2, 3, 4$.
(iii) X is an independent set.
(iv) X is anti-complete to W_i for $i = 1, 2, 3, 4$.
(v)   Without loss of generality $W_3 = \emptyset$ and $W_4 = \emptyset$.
(vi) Without loss of generality W_1 is anti-complete to W_2.

To prove Property (i), if $x, y \in V_i$ are adjacent then $G[x, y, v_{i+1}, v_{i+3}]$ is a $\overline{2P_1 + P_2}$. For $i = 1, \ldots, 4$, the set $W_i \cup X$ must also be independent, since if $x, y \in W_1 \cup X$ were adjacent then $G[x, y, v_2, v_3, v_4]$ would be a $P_2 + P_3$. This proves Properties (ii)–(iv).

To prove Property (v), suppose that $x \in W_1$ and $y \in W_3$ are adjacent. In that case $G[v_1, v_2, v_3, y, x]$ would be a C_5. This contradiction means that no vertex of W_1 is adjacent to a vertex of W_3. Now suppose that $x, x' \in W_1$ and $y \in W_3$. Then $G[y, v_3, x, v_1, x']$ would be a $P_2 + P_3$ by Property (ii). Therefore, if both W_1 and W_3 are non-empty, then they each contain at most one vertex and we can delete these vertices by Fact 1. Without loss of generality we may therefore assume that W_3 is empty. Similarly, we may assume W_4 is empty. Hence we have shown Property (v).

We are left to prove Property (vi). Suppose that $x \in W_1$ is adjacent to $y \in W_2$. Then x cannot have a neighbour in V_2. Indeed, suppose for contradiction that x has a neighbour $z \in V_2$. Then $G[x, z, y, v_1]$ is a $\overline{2P_1 + P_2}$ if y and z are adjacent, and $G[x, y, v_2, v_3, z]$ is a C_5 if y and z are not adjacent. By symmetry, y cannot have a neighbour in V_1. Now y must be complete to V_2. Indeed, if y has a non-neighbour $z \in V_2$ then $G[x, y, z, v_3, v_4]$ is a $P_2 + P_3$. By symmetry, x is complete to V_1. Recall that $W_1 \cup X$ is an independent set by Properties (ii)–(iv). We conclude that any vertex in W_1 with a neighbour in W_2 is complete to V_1 and anti-complete to $V_2 \cup X$. Similarly, any vertex in W_2 with a neighbour in W_1 is complete to V_2 and anti-complete to $V_1 \cup X$.

Let W_1^* (respectively W_2^*) be the set of vertices in W_1 (respectively W_2) that have a neighbour in W_2 (respectively W_1). Then, by Fact 3, we may apply two bipartite complementations, one between W_1^* and $V_1 \cup \{v_1\}$ and the other between W_2^* and $V_2 \cup \{v_2\}$. After these operations, G will be split into two disjoint parts, $G[W_1^* \cup W_2^*]$ and $G \setminus (W_1^* \cup W_2^*)$, both of which are induced subgraphs of G. The first of these is a bipartite $(P_2 + P_3)$-free graph and therefore has bounded clique-width by Lemma 6. We therefore only need to consider the second graph $G\backslash(W_1^*\cup W_2^*)$. In other words, we may assume without loss of generality that W_1 is anti-complete to W_2. This proves Property (vi).

If a vertex in X has no neighbours in $V_1 \cup V_2$ then it is an isolated vertex by Property (iv) and the definition of the set X. In this case we may delete it without affecting the clique-width. Hence, we may assume without loss of generality that every vertex in X has at least one neighbour in $V_1\cup V_2$. We partition X into three sets X_0, X_1, X_2 as follows. Let X_1 (respectively X_2) denote the set of vertices in X with at least one neighbour in V_1 (respectively V_2), but no neighbours in V_2 (respectively V_1). Let X_0 denote the set of vertices in X adjacent to at least one vertex of V_1 and at least one vertex of V_2.

Let $G^* = G[V_1 \cup V_2 \cup W_1 \cup W_2 \cup X_1 \cup X_2]$. We prove the following additional properties:

(vii) G^* is bipartite.
(viii) Without loss of generality $X_0 \neq \emptyset$.
(ix) Every vertex in V_1 that has a neighbour in X is complete to V_2.
(x) Every vertex in V_2 that has a neighbour in X is complete to V_1.
(xi) Every vertex in X_0 has exactly one neighbour in V_1 and exactly one neighbour in V_2.
(xii) Without loss of generality, every vertex in $V_1 \cup V_2$ has at most one neighbour in X_0.
(xiii) Without loss of generality, V_1 is anti-complete to W_2.
(xiv) Without loss of generality, V_2 is anti-complete to W_1.

Property (vii) can be seen has follows. Because G is $(P_2 + P_3, C_5)$-free, G^* has no induced odd cycles of length at least 5. Suppose, for contradiction, that G^* is not bipartite. Then it must contain an induced C_3. Now V_1, V_2, W_1, W_2, X_1 and X_2 are independent sets, so at most one vertex of the C_3 can be in any one of these sets. The set X_1 is anti-complete to V_2, W_1, W_2 and X_2 (by definition of V_2 and Properties (iii) and (iv)). Hence no vertex of the C_3 can be in X_1. Similarly, no vertex of the C_3 be be in X_2. The sets W_1 and W_2 are anti-complete to each other by Property (vi), so the C_3 must therefore consist of one vertex from each of V_1 and V_2, along with one vertex from either W_1 or W_2. However, in this case, these three vertices, along with either v_1 or v_2, respectively would induce a $\overline{2P_1 + P_2}$ in G, which would be a contradiction. Hence we have proven Property (vii).

We now prove Property (viii). Suppose X_0 is empty. Then, since G^* is $(P_2 + P_3)$-free and bipartite (by Property (vii)), it has bounded clique-width by Lemma 6. Hence, G has bounded clique-width by Fact 1, since we may delete v_1, v_2, v_3 and v_4 to obtain G^*. This proves Property (viii).

We now prove Property (ix). Let $y_1 \in V_1$ have a neighbour $x \in X$. Suppose, for contradiction, that y_1 has a non-neighbour $y_2 \in V_2$. Then $G[x, y_2, v_1, v_2, y_1]$ is a C_5 if x is adjacent to y_2 and $G[x, y_1, v_1, y_2, v_3]$ is a $P_2 + P_3$ if x is non-adjacent to y_2, a contradiction. This proves Property (ix). By symmetry, Property (x) holds.

We now prove Property (xi). By definition, every vertex in X_0 has at least one neighbour in V_1 and at least one neighbour in V_2. Suppose, for contradiction, that a vertex $x \in X_0$ has two neighbours $y, y' \in V_1$. By definition, x must also have a neighbour $z \in V_2$. Then z must be adjacent to both y and y' by Property (x). However, then $G[x, z, y, y']$ is a $\overline{2P_1 + P_2}$ by Property (i), a contradiction. This proves Property (xi).

We now prove Property (xii). Suppose a vertex $y \in V_1$ has two neighbours $x, x' \in X_0$. If there is another vertex $z \in X_0$ then z must have a unique neighbour z' in V_1. If z' is a different vertex from y then $G[z, z', x, y, x']$ would be a $P_2 + P_3$ by Properties (i) and (iii). Thus $z' = y$, that is, every vertex in X_0 must be adjacent to y and to no other vertex of V_1. By Fact 1, we may delete y. In the resulting graph no vertex of X would have neighbours in both V_1 and V_2. So X_0 would become empty, in which case we can argue as in the proof of Property (viii). This proves Property (xii).

We now prove Property (xiii). First, for $i \in \{1, 2\}$, suppose that a vertex $y \in V_i$ is adjacent to a vertex $x \in X$. Then y can have at most one non-neighbour in W_i. Indeed, suppose for contradiction that $z, z' \in W_i$ are non-neighbours of y. Then $G[x, y, z, v_i, z']$ is a $P_2 + P_3$ by Properties (ii) and (vi), a contradiction. We claim that at most one vertex of W_2 has a neighbour in V_1. Suppose, for contradiction, that W_2 contains two vertices w and w' adjacent to (not necessarily distinct) vertices z and z' in V_1, respectively. Since $X_0 \neq \emptyset$ by Property (viii), there must be a vertex $y \in V_2$ with a neighbour in X_0. As we just showed that such a vertex y can have at most one non-neighbour in W_2, we may assume without loss of generality that y is adjacent to w. Since y has a neighbour in X, it must also be adjacent to z by Property (x). Now $G[w, z, y, v_2]$ is a $\overline{2P_1 + P_2}$, which is a contradiction. Therefore at most one vertex of W_2 has a neighbour in V_1 and similarly, at most one vertex of W_1 has a neighbour in V_2. By Fact 1, we may delete these vertices if they exist. This proves Properties (xiii) and (xiv).

For $i = 1, 2$ let V_i' be the set of vertices in V_i that have a neighbour in X_0. We show two more properties:

(xv) Every vertex in $W_1 \cup X_1$ is adjacent to either none, precisely one or all vertices of V_1'.

(xvi) Every vertex of $W_2 \cup X_2$ is adjacent to either none, precisely one or all vertices of V_2'.

We prove Property (xv) as follows. Suppose a vertex $x \in X_1 \cup W_1$ has at least two neighbours in $z, z' \in V_1$. We claim that x must be complete to V_1'. Suppose, for contradiction, that x is not adjacent to $y \in V_1'$. By definition, y has a neighbour $y' \in X_0$. Then $G[y, y', z, x, z']$ is a $P_2 + P_3$ by Properties (i), (iii) and (iv), a contradiction. This proves Property (xv). Property (xvi) follows by symmetry.

Let W_i' and X_i' be the sets of vertices in W_i and X_i respectively that are adjacent to precisely one vertex of V_i'. We delete v_1, v_2, v_3 and v_4, which we may do by Fact 1. We do a bipartite complementation between V_1' and those vertices in $W_1 \cup X_1$ that are complete to V_1'. We also do this between V_2' and those vertices in $W_2 \cup X_2$ that are complete to V_2'. Finally, we perform a bipartite complementation between V_1' and $V_2 \setminus V_2'$ and also between V_2' and $V_1 \setminus V_1'$. We may do all of this by Fact 3. Afterwards, Properties (i)–(vi), (ix), (x), (xiii)–(xvi) and the definitions of V_1', V_2', W_1', W_2', X_1, X_2 imply that there are no edges between the following two vertex-disjoint graphs:

1. $G[W_1' \cup W_2' \cup X_1' \cup X_2' \cup V_1' \cup V_2' \cup X_0]$ and
2. $G \setminus (W_1' \cup W_2' \cup X_1' \cup X_2' \cup V_1' \cup V_2' \cup X_0 \cup \{v_1, v_2, v_3, v_4\})$

Both of these graphs are induced subgraphs of G. The second of these graphs does not contain any vertices of X_0. So it is bipartite by Property (vii) and therefore has bounded clique-width, as argued before (in the proof of Property (viii)).

Now consider the first graph, which is $G[W_1' \cup W_2' \cup X_1' \cup X_2' \cup V_1' \cup V_2' \cup X_0]$. By Fact 3, we may complement the edges between V_1' and V_2'. This yields a new graph G'. By definition of V_1', V_2' and Properties (ix) and (x), we find that V_1' is anti-complete to V_2' in G'. Hence, by definition of V_1', V_2' and Properties (i), (iii), (xi) and (xii), we find that $G'[V_1' \cup V_2' \cup X_0]$ is a disjoint union of P_3's. For $i \in \{1, 2\}$, every vertex in $W_i' \cup X_i'$ is adjacent to precisely one vertex in V_i' by definition. As the last bipartite complementation operation did not affect these sets, this is still the case in G'. By Properties (ii)–(iv) and (vi), we find that $W_1' \cup W_2' \cup X_0 \cup X_1' \cup X_2'$ is an independent set. Then, by also using Properties (xiii) and (xiv) together with the definitions of X_1 and X_2, we find that no vertex in $W_i' \cup X_i'$ has any other neighbour in G' besides its neighbour in V_i'. Therefore G' is a disjoint union of trees and thus has bounded clique-width by Lemma 3. We conclude that G has bounded clique-width. This completes the proof of Lemma 13. $\square$

We omit the proof of the next lemma.

Lemma 14. *The class of those $(\overline{2P_1 + P_2}, P_2 + P_3, K_5, C_5, C_4)$-free graphs that contain an induced C_6 has bounded clique-width.*

We now use Lemmas 11–14 and the fact that $(\overline{2P_1 + P_2}, P_2 + P_3, C_4, C_5, C_6)$-free graphs are chordal graphs, and so have bounded clique-width by Lemma 7, to obtain:

Theorem 1 (iii). *The class of $(\overline{2P_1 + P_2}, P_2 + P_3)$-free graphs has bounded clique-width.*

References

1. Boliac, R., Lozin, V.V.: On the clique-width of graphs in hereditary classes. In: Bose, P., Morin, P. (eds.) ISAAC 2002. LNCS, vol. 2518, pp. 44–54. Springer, Heidelberg (2002)
2. Bonomo, F., Grippo, L.N., Milanič, M., Safe, M.D.: Graphs of power-bounded clique-width. arXiv abs/1402.2135 (2014)

3. Brandstädt, A., Engelfriet, J., Le, H.O., Lozin, V.V.: Clique-width for 4-vertex forbidden subgraphs. Theory of Computing Systems **39**(4), 561–590 (2006)
4. Brandstädt, A., Klembt, T., Mahfud, S.: P_6- and triangle-free graphs revisited: structure and bounded clique-width. Discrete Mathematics and Theoretical Computer Science **8**(1), 173–188 (2006)
5. Brandstädt, A., Kratsch, D.: On the structure of (P_5, gem)-free graphs. Discrete Applied Mathematics **145**(2), 155–166 (2005)
6. Brandstädt, A., Le, H.O., Mosca, R.: Gem- and co-gem-free graphs have bounded clique-width. International Journal of Foundations of Computer Science **15**(1), 163–185 (2004)
7. Brandstädt, A., Le, H.O., Mosca, R.: Chordal co-gem-free and (P_5, gem)-free graphs have bounded clique-width. Discrete Applied Mathematics **145**(2), 232–241 (2005)
8. Brandstädt, A., Mahfud, S.: Maximum weight stable set on graphs without claw and co-claw (and similar graph classes) can be solved in linear time. Information Processing Letters **84**(5), 251–259 (2002)
9. Brandstädt, A., Mosca, R.: On variations of P_4-sparse graphs. Discrete Applied Mathematics **129**(2–3), 521–532 (2003)
10. Chudnovsky, M., Robertson, N., Seymour, P., Thomas, R.: The strong perfect graph theorem. Annals of Mathematics **164**, 51–229 (2006)
11. Corneil, D.G., Rotics, U.: On the relationship between clique-width and treewidth. SIAM Journal on Computing **34**, 825–847 (2005)
12. Courcelle, B.: Clique-width and edge contraction. Information Processing Letters **114**(1–2), 42–44 (2014)
13. Courcelle, B., Makowsky, J.A., Rotics, U.: Linear time solvable optimization problems on graphs of bounded clique-width. Theory of Computing Systems **33**(2), 125–150 (2000)
14. Dabrowski, K.K., Golovach, P.A., Paulusma, D.: Colouring of graphs with Ramsey-type forbidden subgraphs. Theoretical Computer Science **522**, 34–43 (2014)
15. Dabrowski, K.K., Lozin, V.V., Raman, R., Ries, B.: Colouring vertices of triangle-free graphs without forests. Discrete Mathematics **312**(7), 1372–1385 (2012)
16. Dabrowski, K.K., Paulusma, D.: Classifying the clique-width of H-free bipartite graphs. In: Cai, Z., Zelikovsky, A., Bourgeois, A. (eds.) COCOON 2014. LNCS, vol. 8591, pp. 489–500. Springer, Heidelberg (2014)
17. Dabrowski, K.K., Paulusma, D.: Clique-width of graph classes defined by two forbidden induced subgraphs. CoRR abs/1405.7092 (2014)
18. Fellows, M.R., Rosamond, F.A., Rotics, U., Szeider, S.: Clique-width is NP-Complete. SIAM Journal on Discrete Mathematics **23**(2), 909–939 (2009)
19. Golumbic, M.C., Rotics, U.: On the clique-width of some perfect graph classes. International Journal of Foundations of Computer Science **11**(03), 423–443 (2000)
20. Gurski, F.: Graph operations on clique-width bounded graphs. CoRR abs/cs/0701185 (2007)
21. Johansson, Ö.: Clique-decomposition, NLC-decomposition, and modular decomposition - relationships and results for random graphs. Congressus Numerantium **132**, 39–60 (1998)
22. Kamiński, M., Lozin, V.V., Milanič, M.: Recent developments on graphs of bounded clique-width. Discrete Applied Mathematics **157**(12), 2747–2761 (2009)
23. Kobler, D., Rotics, U.: Edge dominating set and colorings on graphs with fixed clique-width. Discrete Applied Mathematics **126**(2–3), 197–221 (2003)

24. Lozin, V.V., Rautenbach, D.: On the band-, tree-, and clique-width of graphs with bounded vertex degree. SIAM Journal on Discrete Mathematics **18**(1), 195–206 (2004)
25. Lozin, V.V., Rautenbach, D.: The tree- and clique-width of bipartite graphs in special classes. Australasian Journal of Combinatorics **34**, 57–67 (2006)
26. Lozin, V.V., Volz, J.: The clique-width of bipartite graphs in monogenic classes. International Journal of Foundations of Computer Science **19**(02), 477–494 (2008)
27. Makowsky, J.A., Rotics, U.: On the clique-width of graphs with few P_4's. International Journal of Foundations of Computer Science **10**(03), 329–348 (1999)
28. Oum, S.I.: Approximating rank-width and clique-width quickly. ACM Transactions on Algorithms **5**(1), 10 (2008)
29. Oum, S.I., Seymour, P.D.: Approximating clique-width and branch-width. Journal of Combinatorial Theory, Series B **96**(4), 514–528 (2006)
30. Rao, M.: MSOL partitioning problems on graphs of bounded treewidth and clique-width. Theoretical Computer Science **377**(1–3), 260–267 (2007)

Order Structures for Subclasses
of Generalised Traces

Ryszard Janicki[1], Jetty Kleijn[2], Maciej Koutny[3]($\boxtimes$), and Łukasz Mikulski[3,4]

[1] Department of Computing and Software, McMaster University,
Hamilton, ON L8S 4K1, Canada
janicki@mcmaster.ca
[2] LIACS, Leiden University, P.O.Box 9512, 2300 RA Leiden, The Netherlands
h.c.m.kleijn@liacs.leidenuniv.nll
[3] School of Computing Science, Newcastle University,
Newcastle upon Tyne NE1 7RU, UK
maciej.koutny@ncl.ac.uk
[4] Faculty of Mathematics and Computer Science,
Nicolaus Copernicus University, Chopina 12/18, Toruń, Poland
lukasz.mikulski@mat.umk.pl

Abstract. Traces are equivalence classes of action sequences which can be represented by partial orders capturing the causality in the behaviour of a concurrent system. Generalised traces, on the other hand, are equivalence classes of step sequences. They are represented by order structures that can describe non-simultaneity and weak causality, phenomena which cannot be expressed by partial orders alone. In this paper, we provide a systematic classification of different subclasses of generalised traces in terms of the order structures representing them. We also show how the original trace model fits into the overall framework.

Keywords: Trace · Independence · Dependence graph · Partial order · Simultaneity · Serialisability · Interleaving · Generalised causal order structure

1 Introduction

Mazurkiewicz traces [14,15] are a well-established, classical, and basic model for representing and structuring sequential observations of concurrent behaviour; see, e.g., [1,10]. The fundamental assumption underlying trace theory is that independent events (occurrences of actions) may be observed in any order. Sequences that differ only w.r.t. the ordering of independent events are identified as belonging to the same concurrent run of the system under consideration. Thus a trace is an equivalence class of sequences comprising all (sequential) observations of a single concurrent run. The dependencies between the events of a trace are invariant among (common to) all elements of the trace. They define an acyclic dependence graph which — through its transitive closure — determines the underlying causality structure of the trace as a (labelled) partial order [16].

© Springer International Publishing Switzerland 2015
A.-H. Dediu et al. (Eds.): LATA 2015, LNCS 8977, pp. 689–700, 2015.
DOI: 10.1007/978-3-319-15579-1_54

In fact, this partial order can also be obtained as the intersection of the labelled total orders corresponding to the sequences forming the trace. Moreover, the sequences belonging to the trace correspond exactly to the linearisations (saturations) of this partial order. In [17] the necessary connection between the causal structures (partial orders) and observations (total orders) is provided by showing that each partial order is the intersection of all its linearisations (Szpilrajn's property). Consequently, each trace can also be viewed as a labelled partial order which is unique up to isomorphism; see, e.g., [1,3,10]. Thus, to capture the essence of equivalence between different observations of the same run of a concurrent system, Mazurkiewicz traces bring together two mathematical ideas both based on a notion of independence between actions expressed as a binary independence relation ind. On the one hand, there are equations $ab = ba$ generating the equivalence by expressing the commutativity of occurrences of certain actions as determined by the independence relation. As a result, sequences $wabu$ and $wbau$ of action occurrences are considered equivalent whenever $\langle a, b \rangle \in$ ind. On the other hand, there is the idea of a common partial order structure that underlies equivalent observations defined by the ordering of the occurrences of dependent actions. However, being based on equating independence and lack of ordering, the model of Mazurkiewicz traces with the corresponding partial order interpretation of concurrency is rather restricted [6].

In [5], a full generalisation of the theory of Mazurkiewicz traces is presented for the case that actions could occur and may be observed as occurring simultaneously. Thus observations consist of sequences of *steps*, i.e., sets of one or more actions that occur simultaneously. In order to retain the philosophy underlying Mazurkiewicz traces, the extended set-up is based on a few explicit and simple design choices. Instead of the single independence relation ind, now three basic relations between pairs of different actions are distinguished: *simultaneity* indicating that actions may occur together in a step; *serialisability* indicating a possible execution order for potentially simultaneous actions; and *interleaving* indicating that actions can *not* occur simultaneously though no specific ordering is required. These three relations are used to define *fundamental concurrency alphabets* and then applied to identify step sequences as observations of the same concurrent run. In this more general case, the equations are of the form $A_1 A_2 = B_1 B_2$ where the A_i and B_j are steps, and defined in terms of simultaneity, serialisability, and interleaving. The resulting equivalence classes of step sequences are called *generalised traces*. Actually, in this paper we will work with the definition of generalised traces provided by *generalised concurrency alphabets* also introduced in [5]. These alphabets have only two relations: simultaneity as before and *sequentialisability* combining serialisability and interleaving.

It is the main aim of this paper to characterise and discuss generalised traces in more detail. As demonstrated in [5], the clear semantical meaning of the three relations — simultaneity, serialisability, interleaving — allows for an intuitive classification of some natural subclasses of fundamental concurrency alphabets. A hierarchy of interesting families of generalised traces is presented in [5], including new non-trivial classes of traces as well as the original Mazurkiewicz traces, comtraces [7,13], and g-comtraces [8]. Comtraces are equivalence classes of step sequences derived from equations of the form $AB = A \uplus B$ using the two relations simultaneity and serialisability. Likewise, g-comtraces are equivalence classes of step sequences derived from equations of the form $AB = A \uplus B$ and $AB = BA$ — using simultaneity, serialisability as well as interleaving. Actually, as shown in [11], the equations used in [8] do not model the relevant aspects of concurrent behaviours in a fully adequate way. This has been corrected in the general set-up of [5] with generalised traces and fundamental concurrency alphabet providing the full generalisation of Mazurkiewicz traces to step sequences. There a complete picture is presented including extended dependence graphs and a characterisation of the causal order structures underlying generalised traces as the most general *order structures* from [4].

Modelling concurrency with order structures stems from the results of [2,6] and [12]. The basic idea is that general concurrent causal behaviour is represented by *a pair* of relations, instead of just one, as in the standard (partial order) approach (see, e.g., [16]). Depending on the assumptions for the chosen model of concurrency, details vary, but basically there are two versions: one in which the two relations are interpreted as standard *causality* (dependence or precedence) and *weak causality* (not later than), respectively (see, e.g., [2,6,7]); and an extended, general, version (suggested in [6,11] but eventually defined in [4]) with the two relations *mutual exclusion* and *weak causality* (causality is now a derived notion). The first version has a relatively well developed theory and substantial applications (see, e.g., [2,6,7,9]). The second one, however, is relatively new and as such the starting point for this paper where we identify the order structures that characterise the subfamilies of generalised traces from the classification in [5].

The paper is organised as follows. In the next section, we recall the definitions of generalised concurrency alphabets and the corresponding generalised traces. We also discuss two ways of partitioning the causal dependencies between actions which leads to the identification of five interesting subclasses of generalised concurrency alphabets and the induced generalised traces. After that, we recall the definition of ordered structures corresponding to the generalised traces. In the following section, we present the main results of the paper, providing a full characterisation of the relationships between the various subclasses of generalised traces and the corresponding subclasses of order structures.

2   Generalised Traces

For a binary relation R, the notations R^{-1}, R^+ and R^* are standard. Moreover, $R^{sym} = R \cup R^{-1}$ is the symmetric closure, $R = R^+ \setminus id_X = R^* \setminus id_X$ is the irreflexive transitive closure of R; and $R^\circledast = R^* \cap (R^*)^{-1}$ is the largest equivalence relation contained in R^*. R is a partial order relation if it is irreflexive and transitive, and a total order relation if, in addition, $R^{sym} = (X \times X) \setminus id_X$.

Throughout the paper, $\Sigma \neq \varnothing$ is a finite *alphabet* of actions, $\mathbb{S} = 2^\Sigma \setminus \{\varnothing\}$ is the set of all *steps*, and $\mathbb{S}^*$ is the set of *step sequences*. For every $a \in \Sigma$

and $u = A_1 \ldots A_k \in \mathbb{S}^*$, $\#_u(a)$ is the number of occurrences of a within u; $occ(u) = \{\langle a, i \rangle \mid a \in \Sigma \wedge 1 \le i \le \#_u(a)\}$ is the set of *action occurrences* of u; and the *position* $pos_u(\alpha)$ within u of $\alpha = \langle a, i \rangle \in occ(u)$ is the smallest index $j \le k$ such that the number of occurrences of a within $A_1 \ldots A_j$ is exactly i.

Let EQ be a finite set of equations of the form $u = v$, where u and v are nonempty step sequences. EQ induces a relation $\approx$ on step sequences comprising all pairs $\langle tuw, tvw \rangle$ such that $t, w \in \mathbb{S}^*$, and $u = v$ or $v = u$ is an equation. Furthermore, $\equiv$ is the equivalence relation on step sequences defined as $\approx^*$.

The report [5] presents a full generalisation of the theory of Mazurkiewicz traces to the case that the smallest unit of observation is a set of actions (a step) rather than a single action. Thus observation sequences consist of sequences of *steps*, i.e., sets of actions that occur simultaneously. In order to extend the Mazurkiewicz trace approach to this more general situation, [5] proposes *generalised concurrency alphabets* Θ employing two relations defined for a set of atomic actions Σ, namely *simultaneity* sim defining legal steps, and *sequentialisation* seq specifying actions which can be swapped, or actions whose simultaneous occurrence means that they can also occur one after another. Together sim and seq define a set of equations and an equivalence relation for step sequences over Σ.

A *generalised concurrency alphabet* is a triple $\theta = \langle \Sigma, \text{sim}, \text{seq} \rangle \in \Theta$, where Σ is a finite nonempty set, and sim and seq are two irreflexive relations over Σ such that sim and seq $\setminus$ sim are symmetric. The sets of *steps* and *step sequences* defined by θ are given by $\mathbb{S}_\theta = \{A \subseteq \Sigma \mid A \ne \varnothing \wedge (A \times A) \setminus id_\Sigma \subseteq \text{sim}\}$ and $\text{SSEQ}_\theta = \mathbb{S}_\theta^*$; and the induced *equations* are as follows, where $A, B \in \mathbb{S}_\theta$:

$$
\begin{aligned}
AB = BA \quad &\text{if } A \times B \subseteq \text{seq} \cap \text{seq}^{-1} \quad &(interleaving) \\
AB = A \cup B \quad &\text{if } A \times B \subseteq \text{seq} \cap \text{sim} \quad &(serialisability)
\end{aligned} \tag{1}
$$

Note that if $A, B \in \mathbb{S}_\theta$ and $A \times B \subseteq \text{seq} \cap \text{sim}$ then $A \cap B = \varnothing$ and $A \cup B \in \mathbb{S}_\theta$, and so the above equations (1) can never transform a step sequence in $\mathbb{S}_\theta^*$ into a sequence of sets outside $\mathbb{S}_\theta^*$.

Similarly as in the case of Mazurkiewicz traces, the equations (1) induce an equivalence relation $\equiv$ on the step sequences SSEQ_θ defined by θ. The equivalence classes TSSEQ_θ of the relation $\equiv$ are called *(generalised) traces*, and the generalised trace containing a step sequence $u \in \text{SSEQ}_\theta$ is denoted by $[\![u]\!]_\theta$.

There are six meaningful relationships between pairs of actions which together form a partition of $\Sigma \times \Sigma$: (i) con $=$ seq $\cap$ seq^{-1} $\cap$ sim is *concurrency* identifying actions which can be executed simultaneously as well as in any order; (ii) inl $=$ (seq $\cap$ seq^{-1}) $\setminus$ sim is *interleaving* allowing a pair of actions to be swapped, but disallowing simultaneity; (iii) ssi $=$ sim $\setminus$ (seq $\cup$ seq^{-1}) is *strong simultaneity* allowing a pair of actions to be executed simultaneously, but disallowing serialisation and interleaving; (iv) sse $=$ (seq $\setminus$ seq^{-1}) $\cap$ sim is *semi-serialisability* allowing a pair of simultaneously executed actions to be executed in the order given, but not in the reverse order; (v) wdp $=$ (seq^{-1} $\setminus$ seq) $\cap$ sim is *weak dependence*, the inverse of semi-serialisability; and (vi) rig $=$ ($\Sigma \times \Sigma$) $\setminus$ (sim $\cup$ (seq $\cap$ seq^{-1})) is *rigid order* allowing neither simultaneity nor changing of the order of actions.

The Venn diagram of sim and seq consists of three components: sim $\setminus$ seq, seq $\setminus$ sim, and sim $\cap$ seq. Hence, one can distinguish in a natural way eight classes of generalised concurrency alphabets, as shown in the diagram below, where the subscripts indicate which relations are empty. Out of the seven proper subclasses of Θ, there is little to be gained from studying $\Theta_{\text{sim} \cup \text{seq}}$ and Θ_{seq} as for these each trace is a singleton. Hence we will concentrate on the remaining types of generalised concurrency alphabets, viz. $\Theta_{\text{seq} \setminus \text{sim}}, \Theta_{\text{sim}}, \Theta_{\text{sim} \setminus \text{seq}}, \Theta_{\text{seq} \cap \text{sim}},$ and $\Theta_{\text{sim} \triangle \text{seq}},$ where $\text{sim} \triangle \text{seq} = (\text{sim} \setminus \text{seq}) \cup (\text{seq} \setminus \text{sim}).$

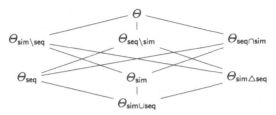

3 Order Structures for Generalised Traces

The order theoretic treatment of generalised traces is based on *relational structures* $\langle \Delta, \rightleftharpoons, \sqsubset, \ell \rangle$ comprising a finite *domain* Δ, two binary relations $\rightleftharpoons$ and $\sqsubset$ on Δ, and a domain labelling $\Delta \xrightarrow{\ell} \Sigma$. To represent observational and causal relationships in the behaviours of concurrent systems we use OS, the *order structures* from [4] which are an extension of an idea first proposed in [2,6,12]. Individual observations (step sequences) are represented by *saturated* order structures, or SO-structures for short, and causal relationships are represented by *invariant* order structures (IO-structures). Formal definitions follow below.

An *order structure* is a relational structure $os = \langle \Delta, \rightleftharpoons, \sqsubset, \ell \rangle$ with a symmetric and irreflexive *mutex* relation $\rightleftharpoons$ and an irreflexive *weak causality* relation $\sqsubset$. Intuitively, Δ is the set of events that have happened during some execution of a concurrent system; $x \rightleftharpoons y$ means that x occurred *not simultaneously* with y, and $x \sqsubset y$ that x occurred *not later* than y, i.e., *before or simultaneously* with y. Hence if $x \sqsubset y$ and $x \rightleftharpoons y$, then x must have occurred *before* y. We will therefore refer to the intersection $\sqsubset \cap \rightleftharpoons$ as *causality* (or *precedence*), denoting it by $\prec$. Note that $x \sqsubset y \sqsubset x$ intuitively means that x and y were observed as *simultaneous*. It is assumed that os is *separable* meaning that $\rightleftharpoons \cap \sqsubset^\circledast = \varnothing$. Separability excludes situations where events forming a weak causality cycle in $\sqsubset^\circledast$ are also involved in the mutex relationship. Furthermore, it is assumed that os is *label-linear* meaning that $\rightleftharpoons \cap \sqsubset$ is a total order relation when restricted to the domain elements labelled by the same action. Referring to the set-up of Mazurkiewicz traces, order structures correspond to (labelled) acyclic relations.

An *extension* of the order structure os is any order structure $\langle \Delta, \rightleftharpoons', \sqsubset', \ell \rangle$ such that $\rightleftharpoons \subseteq \rightleftharpoons'$ and $\sqsubset \subseteq \sqsubset'$. An SO-*structure* is a relational structure $sos = \langle \Delta, \rightleftharpoons, \sqsubset, \ell \rangle$ satisfying

$$x \neq y \wedge \quad x \neq y \iff x \sqsubset y \sqsubset x \qquad\qquad x \rightleftharpoons y \implies x \sqsubset^{sym} y$$
$$x \neq y \wedge x \sqsubset z \sqsubset y \implies x \sqsubset y \qquad x \neq y \wedge \ell(x) = \ell(y) \implies x \rightleftharpoons y$$

One can see that saturated order structures are the only order structures without proper extensions. Referring to the set-up of Mazurkiewicz traces, SO-structures correspond to total order relations, i.e., the only acyclic relations which cannot be extended without violating their acyclicity. We denote by $\mathsf{satext}(os)$ the set of all saturated extensions of $os \in \mathsf{OS}$.

An IO-*structure* is a relational structure $ios = \langle \Delta, \rightleftharpoons, \sqsubset, \ell \rangle$ satisfying

$$
\begin{array}{ccc}
x \not\sqsubset x & x \neq y \wedge x \sqsubset z \sqsubset y \implies x \sqsubset y \\
y \rightleftharpoons x \neq y \impliedby x \rightleftharpoons y & x \prec z \sqsubset y \vee x \sqsubset z \prec y \implies x \rightleftharpoons y \\
x \neq y \wedge \ell(x) = \ell(y) \implies x \prec^{sym} y & z \rightleftharpoons y \wedge z \sqsubset x \sqsubset z \implies x \rightleftharpoons y \\
& z \rightleftharpoons z' \wedge x \sqsubset z \sqsubset y \wedge x \sqsubset z' \sqsubset y \implies x \rightleftharpoons y
\end{array}
$$

Invariant order structures are the only order structures which cannot be extended without making the set of their saturated extensions smaller (follows from the results of [5]). Referring to the set-up of Mazurkiewicz traces, IO-structures correspond to partial order relations, the only acyclic relations which cannot be extended without making the set of their total order extensions smaller. Crucially, IOS are exactly those order structures os for which $\mathsf{satext}(os) \neq \varnothing$ and $os = \bigcap \mathsf{satext}(os)$. In other words, IO-structures are exactly those order structures which can be represented by their saturated extensions. This fundamental property is a counterpart of Szpilrajn's Theorem [17] which implies that partial orders are exactly those acyclic relations which can be represented by their total order extensions.

The *order structure closure* $\mathsf{OS} \xrightarrow{\text{os2ios}} \mathsf{IOS}$ is given by $\langle \Delta, \rightleftharpoons, \sqsubset, \ell \rangle \xmapsto{\text{os2ios}} \langle \Delta, \sqsubset^{\circledast} \circ \rightleftharpoons \circ \sqsubset^{\circledast} \ \cup \ \sqsubset^{\circledast} \circ \nabla^{sym} \circ \sqsubset^{\circledast}, \sqsubset \ , \ell \rangle$ where $\nabla = \{\langle x, y \rangle \mid \exists z, w : z \rightleftharpoons w \wedge x \sqsubset^{*} z \sqsubset^{*} y \wedge x \sqsubset^{*} w \sqsubset^{*} y\}$. Order structure closure corresponds to the transitive closure for acyclic relations. It is also the unique mapping $\mathsf{OS} \xrightarrow{\mathsf{f}} \mathsf{IOS}$ such that $\mathsf{f}(ios) = ios$, for every $ios \in \mathsf{IOS}$, and $\mathsf{satext}(os) = \mathsf{satext} \circ \mathsf{f}(os)$, for every $os \in \mathsf{OS}$ (see [5]). This corresponds to the fact that transitive closure is the unique mapping from acyclic relations to partial orders which preserves the total order extensions.

4   Relating Generalised Traces and Order Structures

In this section we will identify the order structures corresponding to the five subclasses of generalised concurrency alphabets identified in Section 2, but first we recall from [5] the main results established for the general case.

Let $\theta = \langle \Sigma, \mathsf{sim}, \mathsf{seq} \rangle$ be a generalised concurrency alphabet. An *event domain* (for θ) is a set $\Delta \subseteq \Sigma \times \mathbb{N}$ for which there is a mapping $\Sigma \xrightarrow{\epsilon} \mathbb{N}$ such that $\Delta = \{\langle a, i \rangle \mid a \in \Sigma \wedge 1 \leq i \leq \epsilon(a)\}$.

An SO-structure $sos = \langle \Delta, \rightleftharpoons, \sqsubset, \ell \rangle$ is *consistent* with θ if Δ is an event domain for θ, $\langle a, i \rangle \xmapsto{\ell} a$ is the default labelling of Δ, and, for all distinct $\langle a, i \rangle, \langle a, j \rangle, \langle b, k \rangle \in \Delta$, we have: $\langle a, i \rangle \prec \langle a, j \rangle \iff i < j$ and $\langle a, i \rangle \sqsubset^{\circledast} \langle b, k \rangle \implies \langle a, b \rangle \in \mathsf{sim}$.

We let SOS_θ denote the set of all SO-structures *consistent* with θ. Step sequences defined by θ correspond to SO-structures in SOS_θ via the bijection

$\mathsf{SSEQ}_\theta \xrightarrow{\mathsf{sseq2sos}} \mathsf{SOS}_\theta$ such that $\mathsf{sseq2sos}(u) = \langle occ(u), \rightleftharpoons, \sqsubset, \ell \rangle$, where, for all $\alpha, \beta \in occ(u)$ with $pos_u(\alpha) = k$ and $pos_u(\beta) = m$ we have: $k \neq m \Longrightarrow \alpha \rightleftharpoons \beta$ and $k \leq m \wedge \alpha \neq \beta \Longrightarrow \alpha \sqsubset \beta$.

Dependencies between events are captured by the map $\mathsf{SSEQ}_\theta \xrightarrow{\mathsf{sseq2os}_\theta} \mathsf{OS}$ such that $\mathsf{sseq2os}_\theta(u) = \langle occ(u), \rightleftharpoons, \sqsubset, \ell \rangle$, where, for all $\alpha, \beta \in occ(u)$ with $pos_u(\alpha) = k$ and $pos_u(\beta) = m$:

$$
\begin{aligned}
\alpha \rightleftharpoons \beta \quad &\text{if} \quad \langle \ell(\alpha), \ell(\beta) \rangle \in \mathsf{ssi} \cup \mathsf{wdp} \cup \mathsf{rig} \cup \mathsf{inl} \quad \wedge \quad k < m \\
&\text{or} \quad \langle \ell(\alpha), \ell(\beta) \rangle \in \mathsf{ssi} \cup \mathsf{sse} \cup \mathsf{rig} \cup \mathsf{inl} \quad \wedge \quad k > m \\
\alpha \sqsubset \beta \quad &\text{if} \quad \langle \ell(\alpha), \ell(\beta) \rangle \in \mathsf{ssi} \cup \mathsf{sse} \cup \mathsf{wdp} \cup \mathsf{rig} \quad \wedge \quad k < m \\
&\text{or} \quad \langle \ell(\alpha), \ell(\beta) \rangle \in \mathsf{ssi} \cup \mathsf{sse} \quad\quad\quad\quad\quad \wedge \quad k = m
\end{aligned}
\tag{2}
$$

We refer to $\mathsf{sseq2os}_\theta(u)$ as the *dependence graph* of u. Crucially, if $u \equiv w$, then $\mathsf{sseq2os}_\theta(u) = \mathsf{sseq2os}_\theta(w)$, and so dependence graphs can be lifted to the level of generalised traces via $\mathsf{sseq2os}_\theta([\![u]\!]_\theta) = \mathsf{sseq2os}_\theta(u)$. Hence there are two kinds of order structures capturing causal dependencies in the step sequences of SSEQ_θ and traces in TSSEQ_θ, namely dependence graphs and their closures, i.e., $\mathsf{OS}_\theta = \mathsf{sseq2os}_\theta(\mathsf{SSEQ}_\theta)$ and $\mathsf{IOS}_\theta = \mathsf{os2ios}(\mathsf{OS}_\theta)$. In what follows, for every $\Phi \subseteq \Theta$, we will denote $\mathsf{OS}_\Phi = \bigcup_{\theta \in \Phi} \mathsf{OS}_\theta$ and $\mathsf{IOS}_\Phi = \bigcup_{\theta \in \Phi} \mathsf{IOS}_\theta$.

Generalised traces in TSSEQ_θ can be identified with the IO-structures in IOS_θ and a suitable correspondence is established by the pair of inverse bijections $\mathsf{TSSEQ}_\theta \xrightarrow{\mathsf{os2ios} \,\circ\, \mathsf{sseq2os}_\theta} \mathsf{IOS}_\theta \xrightarrow{\mathsf{sseq2sos}^{-1} \,\circ\, \mathsf{satext}} \mathsf{TSSEQ}_\theta$. Moreover, if an order structure os has injective labelling, then there is a generalised concurrency alphabet θ and a step sequence $u \in \mathsf{SSEQ}_\theta$ such that os is isomorphic to $\mathsf{sseq2os}_\theta(u)$. Thus generalised concurrency alphabets can generate all the complex patterns involving causal relationships captured by IO-structures.

An example system model for which generalised traces and IO-structures provide a suitable semantical treatment are the elementary net systems with inhibitor and mutex arcs [11].

The restriction to subclasses of generalised concurrency alphabets can lead to striking simplifications in the order structures involved and the corresponding order structure closure. Such simplifications enable, e.g., a more efficient treatment of the computational aspects involving generalised traces and their corresponding order structures. In what follows, we will consider the five non-trivial subclasses of generalised concurrency alphabets, aiming at as simple as possible descriptions of the order structures capturing the corresponding IO-structures.

Order Structures for Θ_{sim}. An alphabet $\mu \in \Theta_{\mathsf{sim}}$ has $\mathsf{sim} = \varnothing$ and so does not allow for true step sequences and there are no serialisability equations as in (1). Moreover, $\mathsf{con} = \mathsf{ssi} = \mathsf{sse} = \mathsf{wdp} = \varnothing$, $\mathsf{seq} = \mathsf{seq}^{-1} = \mathsf{inl}$ and $\mathsf{rig} = (\Sigma \times \Sigma) \setminus \mathsf{inl}$. As a result, one can simplify the definition of the dependence graph of a step sequence $u \in \mathsf{SSEQ}_\mu$, by replacing (2) with:

$$
\alpha \rightleftharpoons \beta \quad \text{if} \quad k \neq m \qquad \alpha \sqsubset \beta \quad \text{if} \quad \langle \ell(\alpha), \ell(\beta) \rangle \in \mathsf{rig} \,\wedge\, k < m .
$$

It is possible to treat μ as a Mazurkiewicz concurrency alphabet $\langle \Sigma, \mathsf{seq} \rangle$ with seq and rig playing the roles of the standard independence and dependence relations, respectively. As all step sequences in SSEQ_μ consist of singleton steps, they correspond one-to-one to the sequences in Σ^*. Moreover, the saturated order structures in SOS_μ correspond one-to-one to the sequences in Σ^*. Indeed, since $\mathsf{sim} = \varnothing$, we have that for every $sos = \langle \Delta, \rightleftharpoons, \sqsubset, \ell \rangle \in \mathsf{SOS}_\mu$ it is the case that $\sqsubset^{\circledast} = id_\Delta$, and so $\prec$ is a total order relation.

The order structures $\mathsf{OS}_{\mathsf{sim}}$ reflecting the causal dependencies in the generalised traces over the alphabets of Θ_{sim} are those $os \in \mathsf{OS}$ for which $\rightleftharpoons = (\Delta \times \Delta) \setminus id_\Delta$. The resulting simplified definitions are then as follows. A relational structure $\langle \Delta, \rightleftharpoons, \sqsubset, \ell \rangle$ belongs to $\mathsf{IOS}_{\mathsf{sim}}$ if

$$x \not\sqsubset x \qquad\qquad x \sqsubset z \sqsubset y \implies x \sqsubset y$$
$$x \neq y \Longleftrightarrow x \rightleftharpoons y \qquad\qquad x \neq y \wedge \ell(x) = \ell(y) \implies x \sqsubset^{sym} y$$

and the simplified order closure $\mathsf{OS}_{\mathsf{sim}} \xrightarrow{\mathsf{os2ios}_{\mathsf{sim}}} \mathsf{IOS}_{\mathsf{sim}}$ corresponds to the transitive closure of an acyclic relation through $\mathsf{os2ios}_{\mathsf{sim}}(os) = \langle \Delta, \rightleftharpoons, \sqsubset^+, \ell \rangle$.

Theorem 1. $\mathsf{IOS}_{\Theta_{\mathsf{sim}}} \subset \mathsf{OS}_{\Theta_{\mathsf{sim}}} \subset \mathsf{OS}_{\mathsf{sim}} \subset \mathsf{OS}$ *and* $\mathsf{IOS}_{\Theta_{\mathsf{sim}}} \subset \mathsf{IOS}_{\mathsf{sim}} \subset \mathsf{IOS} \subset \mathsf{OS}$ *and* $\mathsf{IOS}_{\mathsf{sim}} \subset \mathsf{OS}_{\mathsf{sim}}$.

Proposition 2. $\mathsf{os2ios}_{\mathsf{sim}}$ *is a surjection with* $\mathsf{os2ios}_{\mathsf{sim}} = \mathsf{os2ios}|_{\mathsf{OS}_{\mathsf{sim}}}$. *Moreover, if* $os \in \mathsf{OS}_{\mathsf{sim}}$ *has an injective labelling, then there are* $\mu \in \Theta_{\mathsf{sim}}$ *and* $u \in \mathsf{SSEQ}_\mu$ *such that* os *is isomorphic to* $\mathsf{sseq2os}_\mu(u)$.

Following Mazurkiewicz [15], the classical example of a system model for which the generalised concurrency alphabets in Θ_{sim} and IO-structures $\mathsf{IOS}_{\mathsf{sim}}$ provide a suitable semantical treatment are the elementary net systems with sequential execution semantics.

Order Structures for $\Theta_{\mathsf{seq}\setminus\mathsf{sim}}$. An alphabet $\sigma \in \Theta_{\mathsf{seq}\setminus\mathsf{sim}}$ has $\mathsf{seq}\setminus\mathsf{sim} = \varnothing$, and so $\mathsf{seq} \subseteq \mathsf{sim}$, $\mathsf{rig} = (\Sigma \times \Sigma) \setminus \mathsf{sim}$, and $\mathsf{inl} = \varnothing$. As a result, one can simplify the definition of the dependence graph of a step sequence $u \in \mathsf{SSEQ}_\sigma$, by replacing (2) with:

$$
\begin{aligned}
\alpha \rightleftharpoons \beta \quad & if \quad \langle \ell(\alpha), \ell(\beta) \rangle \in \mathsf{ssi} \cup \mathsf{rig} \cup \mathsf{sse} \quad \wedge \quad k < m \\
& or \quad \langle \ell(\alpha), \ell(\beta) \rangle \in \mathsf{ssi} \cup \mathsf{rig} \cup \mathsf{wdp} \quad \wedge \quad k > m
\end{aligned}
$$

$$
\begin{aligned}
\alpha \sqsubset \beta \quad & if \quad \langle \ell(\alpha), \ell(\beta) \rangle \in \mathsf{rig} \cup \mathsf{wdp} \qquad\quad \wedge \quad k < m \\
& or \quad \langle \ell(\alpha), \ell(\beta) \rangle \in \mathsf{ssi} \cup \mathsf{sse} \qquad\quad \wedge \quad k \leq m
\end{aligned}
$$

Alphabets in $\Theta_{\mathsf{seq}\setminus\mathsf{sim}}$ do not allow true interleaving, and swapping of steps can be achieved by splitting and combining. In [6], such alphabets are referred to as *comtrace alphabets*.

The order structures $\mathsf{OS}_{\mathsf{seq}\setminus\mathsf{sim}}$ needed to reflect causal dependencies in the generalised traces over the concurrent alphabets of $\Theta_{\mathsf{seq}\setminus\mathsf{sim}}$ are all those order structures $os \in \mathsf{OS}$ for which $x \rightleftharpoons y \implies x \sqsubset^{sym} y$. The resulting simplified definitions are then as follows. A relational structure $\langle \Delta, \rightleftharpoons, \sqsubset, \ell \rangle$ belongs to $\mathsf{IOS}_{\mathsf{seq}\setminus\mathsf{sim}}$ if $x \not\sqsubset x$ and

$$x \neq y \wedge x \sqsubset z \sqsubset y \implies x \sqsubset y \qquad x \sqsubset^{sym} y \wedge y \rightleftharpoons x \Longleftarrow x \rightleftharpoons y$$
$$x \prec z \sqsubset y \vee x \sqsubset z \prec y \implies x \rightleftharpoons y \qquad x \neq y \wedge \ell(x) = \ell(y) \implies x \rightleftharpoons y$$

and the simplified order closure $OS_{seq\setminus sim} \xrightarrow{os2ios_{seq\setminus sim}} IOS_{seq\setminus sim}$ is such that
$os2ios_{seq\setminus sim}(os) = \langle \Delta, (\sqsubset^* \circ \prec \circ \sqsubset^*)^{sym}, \sqsubset, \ell \rangle$.

Theorem 3. $IOS_{\Theta_{seq\setminus sim}} \subset OS_{\Theta_{seq\setminus sim}} \subset OS_{seq\setminus sim} \subset OS$ *and* $IOS_{seq\setminus sim} \subset OS_{seq\setminus sim}$ *and* $IOS_{\Theta_{seq\setminus sim}} \subset IOS_{seq\setminus sim} \subset IOS$.

Proposition 4. $os2ios_{seq\setminus sim}$ *is a surjection with* $os2ios_{seq\setminus sim} = os2ios|_{OS_{seq\setminus sim}}$. *Moreover, if* $os \in OS_{seq\setminus sim}$ *has an injective labelling* $\ell : \Delta \to \Sigma$, *then there are* $\sigma \in \Theta_{seq\setminus sim}$ *and* $u \in SSEQ_\sigma$ *such that* os *is isomorphic to* $sseq2os_\sigma(u)$.

A system model for which $\Theta_{seq\setminus sim}$ and $IOS_{seq\setminus sim}$ provide a suitable semantical treatment are the elementary net systems with inhibitor arcs [7].

Finally, traces generated by the alphabets in $\Theta_{seq\setminus sim}$ are histories satisfying the concurrency paradigm π_3 of [6] by which actions that can be executed in any order can also be executed simultaneously (but not necessarily vice versa).

Proposition 5. *Let* $\alpha \neq \beta$ *be two action occurrences of a generalised trace* τ *generated by* $\sigma \in \Theta_{seq\setminus sim}$. *If there are* $u, w \in \tau$ *with* $pos_u(\alpha) < pos_u(\beta)$ *and* $pos_w(\alpha) > pos_w(\beta)$ *then there is* $v \in \tau$ *with* $pos_v(\alpha) = pos_v(\beta)$.

Order Structures for $\Theta_{sim\setminus seq}$. An alphabet $\kappa \in \Theta_{sim\setminus seq}$ has $sim \setminus seq = \varnothing$, and so $ssi = sse = wdp = \varnothing$ and $rig = (\Sigma \times \Sigma) \setminus seq$. As a result, one can simplify the definition of the dependence graph of a step sequence $u \in SSEQ_\mu$, by replacing (2) with:

$$\alpha \rightleftharpoons \beta \quad if \quad \langle \ell(\alpha), \ell(\beta) \rangle \notin sim \qquad \alpha \sqsubset \beta \quad if \quad \langle \ell(\alpha), \ell(\beta) \rangle \notin seq \wedge k < m$$

For the alphabets in $\Theta_{sim\setminus seq}$ the serialisability equations are rich enough to split any step in every possible way.

The order structures $OS_{sim\setminus seq}$ are all those $os \in OS$ for which $x \sqsubset^{sym} y \Longrightarrow x \rightleftharpoons y$. The resulting simplified definitions are then as follows. A relational structure $\langle \Delta, \rightleftharpoons, \sqsubset, \ell \rangle$ belongs to $IOS_{sim\setminus seq}$ if:

$$x \sqsubset z \sqsubset y \Longrightarrow x \sqsubset y \qquad\qquad x \rightleftharpoons y \Longrightarrow y \rightleftharpoons x \neq y$$
$$x \sqsubset^{sym} y \Longrightarrow x \rightleftharpoons y \qquad x \neq y \wedge \ell(x) = \ell(y) \Longrightarrow x \sqsubset^{sym} y$$

and the simplified order closure $OS_{sim\setminus seq} \xrightarrow{os2ios_{sim\setminus seq}} IOS_{sim\setminus seq}$ is such that
$os2ios_{sim\setminus seq}(os) = \langle \Delta, \rightleftharpoons \cup (\sqsubset^+)^{sym}, \sqsubset^+, \ell \rangle$.

Theorem 6. $IOS_{\Theta_{sim\setminus seq}} \subset OS_{\Theta_{sim\setminus seq}} \subset OS_{sim\setminus seq} \subset OS$ *and* $IOS_{sim\setminus seq} \subset OS_{sim\setminus seq}$ *and* $IOS_{\Theta_{sim\setminus seq}} \subset IOS_{sim\setminus seq} \subset IOS$.

Proposition 7. $os2ios_{sim\setminus seq}$ *is a surjection with* $os2ios_{sim\setminus seq} = os2ios|_{OS_{sim\setminus seq}}$. *Moreover, if* $os \in OS_{sim\setminus seq}$ *has an injective labelling* $\ell : \Delta \to \Sigma$, *then there are* $\kappa \in \Theta_{sim\setminus seq}$ *and* $u \in SSEQ_\kappa$ *such that* os *is isomorphic to* $sseq2os_\kappa(u)$.

Finally, traces generated by the alphabets in $\Theta_{sim\setminus seq}$ are histories satisfying the concurrency paradigm π_2 of [6].

Proposition 8. *Let* $\alpha \neq \beta$ *be action occurrences of a generalised trace* τ *generated by* $\kappa \in \Theta_{sim\setminus seq}$. *If there is* $v \in \tau$ *with* $pos_v(\alpha) = pos_v(\beta)$ *then there are* $u, w \in \tau$ *with* $pos_u(\alpha) < pos_u(\beta)$ *and* $pos_w(\alpha) > pos_w(\beta)$.

Order Structures for $\Theta_{\mathsf{seq}\cap\mathsf{sim}}$. An alphabet $\nu \in \Theta_{\mathsf{sim}\cap\mathsf{seq}}$ has $\mathsf{sim} \cap \mathsf{seq} = \varnothing$, and so we have $\mathsf{ssi} = \mathsf{sim}$, $\mathsf{sse} = \mathsf{wdp} = \mathsf{con} = \varnothing$, and $\mathsf{rig} = (\Sigma \times \Sigma) \setminus (\mathsf{sim} \uplus \mathsf{seq})$. As a result, one can simplify the definition of the dependence graph of a step sequence $u \in \mathsf{SSEQ}_\mu$, by replacing (2) with:

$$\alpha \rightleftharpoons \beta \ \ \text{if} \ k \neq m \qquad \alpha \sqsubset \beta \ \ \text{if} \ \langle \ell(\alpha), \ell(\beta) \rangle \notin \mathsf{seq} \wedge k \leq m \wedge \alpha \neq \beta$$

For the alphabets in $\Theta_{\mathsf{sim}\cap\mathsf{seq}}$ steps can be only manipulated through the interleaving equations.

The order structures $\mathsf{OS}_{\mathsf{sim}\cap\mathsf{seq}}$ are all those $os \in \mathsf{OS}$ for which $x \neq y \Longrightarrow x \rightleftharpoons y \vee x \sqsubset y \sqsubset x$. The resulting simplified definitions are then as follows. A relational structure $\langle \Delta, \rightleftharpoons, \sqsubset, \ell \rangle$ belongs to $\mathsf{IOS}_{\mathsf{sim}\cap\mathsf{seq}}$ if:

$$x \neq y \wedge x \sqsubset z \sqsubset y \implies x \sqsubset y \qquad\qquad x \neq y \wedge x \neq y \Longleftrightarrow x \sqsubset y \sqsubset x$$
$$x \neq x \qquad\qquad\qquad\qquad x \neq y \wedge \ell(x) = \ell(y) \implies x \prec^{sym} y$$

and the simplified order closure $\mathsf{OS}_{\mathsf{sim}\cap\mathsf{seq}} \xrightarrow{\mathsf{os2ios}_{\mathsf{sim}\cap\mathsf{seq}}} \mathsf{IOS}_{\mathsf{sim}\cap\mathsf{seq}}$ is such that $\mathsf{os2ios}_{\mathsf{sim}\cap\mathsf{seq}}(os) = \langle \Delta, \rightleftharpoons, \sqsubset, \ell \rangle$.

Theorem 9. $\mathsf{OS}_{\Theta_{\mathsf{seq}\cap\mathsf{sim}}} \subset \mathsf{OS}_{\mathsf{seq}\cap\mathsf{sim}} \subset \mathsf{OS}$ *and* $\mathsf{IOS}_{\Theta_{\mathsf{seq}\cap\mathsf{sim}}} \subset \mathsf{IOS}_{\mathsf{seq}\cap\mathsf{sim}} \subset \mathsf{IOS}$ *and* $\mathsf{IOS}_{\Theta_{\mathsf{seq}\cap\mathsf{sim}}} \subset \mathsf{OS}_{\mathsf{seq}\cap\mathsf{sim}}$ *and* $\mathsf{IOS}_{\mathsf{seq}\cap\mathsf{sim}} \subset \mathsf{OS}_{\mathsf{seq}\cap\mathsf{sim}}$.

Proposition 10. $\mathsf{os2ios}_{\mathsf{seq}\cap\mathsf{sim}}$ *is a surjection with* $\mathsf{os2ios}_{\mathsf{seq}\cap\mathsf{sim}} = \mathsf{os2ios}|_{\mathsf{OS}_{\mathsf{seq}\cap\mathsf{sim}}}$. *Moreover, if* $os \in \mathsf{OS}_{\mathsf{seq}\cap\mathsf{sim}}$ *has an injective labelling* $\ell : \Delta \to \Sigma$, *then there are* $\nu \in \Theta_{\mathsf{seq}\cap\mathsf{sim}}$ *and* $u \in \mathsf{SSEQ}_\nu$ *such that* os *is isomorphic to* $\mathsf{sseq2os}_\nu(u)$.

Order Structures for $\Theta_{\mathsf{sim}\triangle\mathsf{seq}}$. An alphabet $\omega \in \Theta_{\mathsf{sim}\triangle\mathsf{seq}}$ has $\mathsf{sim}\triangle\mathsf{seq} = \varnothing$, and so $\mathsf{sim} = \mathsf{seq} = \mathsf{con}$, $\mathsf{ssi} = \mathsf{sse} = \mathsf{wdp} = \mathsf{inl} = \varnothing$ and $\mathsf{rig} = (\Sigma \times \Sigma) \setminus \mathsf{con}$. As a result, one can simplify the definition of the dependence graph of a step sequence $u \in \mathsf{SSEQ}_\mu$, by replacing (2) with:

$$\alpha \rightleftharpoons \beta \ \ \text{if} \ \langle \ell(\alpha), \ell(\beta) \rangle \in \mathsf{rig} \qquad \alpha \sqsubset \beta \ \ \text{if} \ \langle \ell(\alpha), \ell(\beta) \rangle \in \mathsf{rig} \wedge k < m$$

For the alphabets in $\Theta_{\mathsf{sim}\triangle\mathsf{seq}}$ the interleaving equations are not really needed, and the serialisability equations are rich enough to split and reorder steps in every possible way. As a result, all steps can be completely sequentialised.

The order structures $\mathsf{OS}_{\mathsf{sim}\triangle\mathsf{seq}}$ are all those $os \in \mathsf{OS}$ for which $x \rightleftharpoons y \Longleftrightarrow x \sqsubset^{sym} y$. The resulting simplified definitions are then as follows. A relational structure $\langle \Delta, \rightleftharpoons, \sqsubset, \ell \rangle$ belongs to $\mathsf{IOS}_{\mathsf{sim}\triangle\mathsf{seq}}$ if

$$x \not\sqsubset x \qquad\qquad\qquad x \sqsubset z \sqsubset y \implies x \sqsubset y$$
$$x \rightleftharpoons y \Longleftrightarrow x \sqsubset^{sym} y \qquad x \neq y \wedge \ell(x) = \ell(y) \implies x \sqsubset^{sym} y$$

and the simplified order closure $\mathsf{OS}_{\mathsf{sim}\triangle\mathsf{seq}} \xrightarrow{\mathsf{os2ios}_{\mathsf{sim}\triangle\mathsf{seq}}} \mathsf{IOS}_{\mathsf{sim}\triangle\mathsf{seq}}$ is such that $\mathsf{os2ios}_{\mathsf{sim}\triangle\mathsf{seq}}(os) = \langle \Delta, (\sqsubset^+)^{sym}, \sqsubset^+, \ell \rangle$.

Theorem 11. $\mathsf{IOS}_{\mathsf{sim}\triangle\mathsf{seq}} \subset \mathsf{OS}_{\mathsf{sim}\triangle\mathsf{seq}}$ *and* $\mathsf{IOS}_{\Theta_{\mathsf{sim}\triangle\mathsf{seq}}} \subset \mathsf{IOS}_{\mathsf{sim}\triangle\mathsf{seq}} \subset \mathsf{IOS}$ *and* $\mathsf{IOS}_{\Theta_{\mathsf{sim}\triangle\mathsf{seq}}} \subset \mathsf{OS}_{\Theta_{\mathsf{sim}\triangle\mathsf{seq}}} \subset \mathsf{OS}_{\mathsf{sim}\triangle\mathsf{seq}} \subset \mathsf{OS}$.

Proposition 12. $\mathsf{os2ios}_{\mathsf{seq}\triangle\mathsf{sim}}$ *is a surjection with* $\mathsf{os2ios}_{\mathsf{seq}\triangle\mathsf{sim}}=\mathsf{os2ios}|_{\mathsf{OS}_{\mathsf{seq}\triangle\mathsf{sim}}}$. *Moreover, if* $os \in \mathsf{OS}_{\mathsf{sim}\triangle\mathsf{seq}}$ *has an injective labelling* $\ell : \Delta \to \Sigma$, *then there are* $\omega \in \Theta_{\mathsf{sim}\triangle\mathsf{seq}}$ *and* $u \in \mathsf{SSEQ}_\omega$ *such that* os *is isomorphic to* $\mathsf{sseq2os}_\omega(u)$.

It may come as a surprise that although the structures $\mathsf{IOS}_{\mathsf{sim}\triangle\mathsf{seq}}$ are in a one-to-one correspondence with partial orders, similarly as for $\mathsf{IOS}_{\mathsf{sim}}$, the actual definition of the two classes of order structures is different.

Finally, the generalised traces generated by the alphabets in $\Theta_{\mathsf{sim}\triangle\mathsf{seq}}$ are histories satisfying the true concurrency paradigm π_8 of [6] and a system model for which this subclass provides a suitable semantical treatment are the elementary net systems with step sequence semantics.

Proposition 13. *Let* $\alpha \neq \beta$ *be action occurrences of a generalised trace* τ *generated by* $\omega \in \Theta_{\mathsf{sim}\triangle\mathsf{seq}}$. *Then there is* $v \in \tau$ *with* $pos_v(\alpha) = pos_v(\beta)$ *if and only if there are* $u, w \in \tau$ *with* $pos_u(\alpha) < pos_u(\beta)$ *and* $pos_w(\alpha) > pos_w(\beta)$.

5 Conclusions

In [5] we introduced and investigated how to extend Mazurkiewicz trace theory to the case of step sequences and we established that the general traces defined through general concurrency alphabets are indeed the most general in terms of their underlying order structures. In this paper we have continued our investigations and identified for the five natural subclasses of generalised traces their corresponding – simplified – IO-order structures. We have also established connections between some of these subclasses and the concurrency paradigms of [6].

As observed in [5], there are IO-structures that cannot be generated by any generalised concurrency alphabet. The intuitive reason is that the latter can only capture *static* dependencies between actions, whereas in the former different occurrences of the same pair of actions may exhibit different causality dependencies. In our future work we will aim at a precise characterisation of the labellings of IO-structures which correspond to statically defined causality relationships between actions, for each subclass of IO-structures considered in this paper.

Acknowledgments. We would like to thank the anonymous referees for their comments and useful suggestions. This research was supported by Polish National Science Center under the grant No.2013/09/D/ST6/03928, the EPSRC Uncover project, and the NSERC grant of Canada.

References

1. Diekert, V., Rozenberg, G. (eds.): The Book of Traces. World Scientific (1995)
2. Gaifman, H., Pratt, V.R.: Partial order models of concurrency and the computation of functions. In: LICS, pp. 72–85. IEEE Computer Society (1987)

3. Hoogeboom, H.J., Rozenberg, G.: Dependence graphs. In: Diekert, V., Rozenberg, G. (eds.) The Book of Traces, pp. 43–67. World Scientific (1995)
4. Janicki, R., Kleijn, J., Koutny, M., Mikulski, Ł.: Causal structures for general concurrent behaviours. In: Szczuka, M.S., Czaja, L., Kacprzak, M. (eds.) CS&P. CEUR Workshop Proceedings, vol. 1032, pp. 193–205. CEUR-WS.org (2013)
5. Janicki, R., Kleijn, J., Koutny, M., Mikulski, Ł.: Generalising traces. TR-CS 1436. Newcastle University (2014)
6. Janicki, R., Koutny, M.: Structure of concurrency. Theor. Comput. Sci. 112(1), 5–52 (1993)
7. Janicki, R., Koutny, M.: Semantics of inhibitor nets. Inf. Comput. 123(1), 1–16 (1995)
8. Janicki, R., Le, D.T.M.: Modelling concurrency with comtraces and generalized comtraces. Inf. Comput. 209(11), 1355–1389 (2011)
9. Juhás, G., Lorenz, R., Mauser, S.: Complete process semantics of Petri nets. Fundam. Inform. 87(3–4), 331–365 (2008)
10. Kleijn, J., Koutny, M.: Formal languages and concurrent behaviours. In: Enguix, G.B., Jiménez-López, M.D., Martín-Vide, C. (eds.) New Developments in Formal Languages and Applications. SCI, vol. 113, pp. 125–182. Springer, Heidelberg (2008)
11. Kleijn, J., Koutny, M.: Mutex causality in processes and traces of general elementary nets. Fundam. Inform. 122(1–2), 119–146 (2013)
12. Lamport, L.: The mutual exclusion problem: part I - a theory of interprocess communication. J. ACM 33(2), 313–326 (1986). http://doi.acm.org/10.1145/5383.5384
13. Le, D.T.M.: On three alternative characterizations of combined traces. Fundam. Inform. 113(3–4), 265–293 (2011)
14. Mazurkiewicz, A.: Concurrent program schemes and their interpretations. DAIMI Rep. PB 78. Aarhus University (1977)
15. Mazurkiewicz, A.W.: Basic notions of trace theory. In: de Bakker, J.W., de Roever, W.P., Rozenberg, G. (eds.) REX Workshop. LNCS, vol. 354, pp. 285–363. Springer, Heidelberg (1988)
16. Pratt, V.: Modeling concurrency with partial orders. International Journal of Parallel Programming 15(1), 33–71 (1986)
17. Szpilrajn, E.: Sur l'extension de l'ordre partiel. Fundam. Math. 16, 386–389 (1930)

Transducers, Tree Automata, and Weighted Automata

A Nivat Theorem for Weighted Picture Automata and Weighted MSO Logic

Parvaneh Babari[(✉)] and Manfred Droste

Institut für Informatik, Universität Leipzig, 04109 Leipzig, Germany
{babari,droste}@informatik.uni-leipzig.de

Abstract. Picture languages have been intensively investigated by several research groups. In this paper, we define weighted two-dimensional on-line tessellation automata (W2OTA) taking weights from a new weight structure called picture valuation monoid. The behavior of this automaton model is a picture series mapping pictures over an alphabet to a picture valuation monoid. As one of our main results, we prove a Nivat theorem for W2OTA. It shows that recognizable picture series can be obtained precisely as projections of particularly simple unambiguously recognizable series restricted to unambiguous recognizable picture languages. In addition, we introduce a weighted MSO logic which can model average density of pictures. As the other main result of this paper, we show that W2OTA and a suitable fragment of our weighted MSO logics are expressively equivalent.

Keywords: Picture valuation monoids · Weighted two-dimensional on-line tessellation automata · Picture series · Nivat's theorem · Weighted logic · Average behavior

1 Introduction

The theory of picture languages as a generalization of formal string languages was motivated by problems arising from image processing and pattern recognition [19,31], and also plays a role in the theory of cellular automata and other devices of parallel computing [28,34]. In the nineties, the family of recognizable picture languages was defined and characterized by many different devices [20,22]. Several research groups obtained a description of recognizable picture languages in terms of automata, sets of tiles, rational operations, and existential monadic second-order logic [21,23,24,27]. Bozapalidis and Grammatikopoulou introduced the interesting model of weighted (quadrapolic) picture automata whose transitions carry weights taken as elements from a given commutative semiring [3]. The behavior of such a picture automaton is a picture series which maps pictures over an arbitrary alphabet to elements of the semiring. In 2006, Fichtner provided a notion of a weighted MSO logic over pictures [16–18]. She proved that for commutative semirings, the class of picture series defined

P. Babari: Supported by DFG Graduiertenkolleg 1763 (QuantLA).

A.-H. Dediu et al. (Eds.): LATA 2015, LNCS 8977, pp. 703–715, 2015.
DOI: 10.1007/978-3-319-15579-1_55

by sentences of the weighted logics coincides with those computed by weighted picture automata [16].

In this paper we define picture valuation monoids as the abstract model for the weight structures and we introduce weighted two-dimensional on-line tessellation automata (W2OTA) taking weights from picture valuation monoids. By this, we can model several application examples, e.g., the average light of picture (interpreting the alphabet as different levels of light) which can not be modelled with commutative semirings. Weighted automata over words computing objectives like the average cost were introduced recently by Chatterjee, Doyen, and Henzinger [4–7].

As our first main result, we prove a Nivat-like theorem for recognizable picture series, i.e., for the behaviors of W2OTA. Nivat's Theorem is a fundamental characterization of rational transductions and provides a connection between rational transductions and rational languages; see [9] for a version of this result for semiring-weighted automata on words. Recently, Droste and Perevoshchikov [11] proved a Nivat-like theorem for recognizable quantitative timed languages. Here, we will derive such a result for recognizable picture series. We show that recognizable picture series can be obtained precisely as projections of particularly simple unambiguously recognizable series restricted to unambiguously recognizable picture languages. In addition, we show that if the underlying picture valuation monoid is idempotent, then we do not need unambiguity of the underlying picture language.

In the second part of this paper we define a new weighted MSO logic which can model average density of pictures. The weighted MSO logic used here is a combination of the ideas from [2], [10], [11] and [16]. In [16], disjunction and existential quantification were interpreted by the sum, and the semantics of both conjunction and universal quantification were defined by the product operation of the semiring. In this paper, using picture valuation monoids as the abstract model, the semantics of universal quantification will be interpreted by a picture valuation function, which for example provides the average value of light of a picture.

Our second main result states that the weighted automata device of W2OTA and a fragment of weighted MSO logic are expressively equivalent. To reach this result, we define a suitable fragment of our logic in which the application of universal first order (FO) quantification is restricted to almost boolean FO formulas, and the application of conjunction is restricted to either almost boolean FO formulas or boolean FO formulas. In addition, we restrict the use of constants in the formula by allowing their occurrence only in the scope of an FO universal quantifier. This enables us to derive our second main result for arbitrary product picture valuation monoids, not requiring regularity as in [10]. Also, our results differ from the ones in [16] which required commutative semirings as weight structure.

We would like to mention that our results do not need distributivity of multiplication over addition or commutativity or even associativity of multiplication, while considering a commutative semiring as the weight structure was previously an essential assumption in the weighted picture automata theory.

2 Valuation Monoids and Weighted Picture Automata

We assume that the reader is familiar with notions and results of two-dimensional languages. For more details see [19, 20, 36]. We just give the notion of pictures and some basic notations that we use here.

Let $\mathbb{N} = \{0, 1, 2, ...\}$ be the set of natural numbers. Let Σ be a non-empty set. A *picture* over Σ is a non-empty rectangular array of elements in Σ [1]. A *picture language* is a set of pictures. The set of all pictures over Σ is denoted by Σ^{++}. Let $p \in \Sigma^{++}$. We write $p(i, j)$ or $p_{i,j}$ for the component of p at position (i, j) and let $\ell_v(p)$ and $\ell_h(p)$ be the number of rows and columns of p, respectively. The pair $(\ell_v(p), \ell_h(p))$ is the size of p. The set $\Sigma^{m \times n}$ comprises all pictures of size (m, n).

Definition 1. *A picture valuation monoid, or for short pv-monoid, is a tuple $\mathcal{D} = (D, +, val, \mathbb{0})$ consisting of a commutative monoid $(D, +, \mathbb{0})$ and a valuation function $val : D^{++} \to D$ with $val(d) = d$ for all $d \in D$ and*

$$val\begin{pmatrix} d_{1,1} & d_{1,2} & \cdots & d_{1,n} \\ d_{2,1} & d_{2,2} & \cdots & d_{2,n} \\ \vdots & \vdots & \ddots & \vdots \\ d_{m,1} & d_{m,2} & \cdots & d_{m,n} \end{pmatrix} = \mathbb{0} \text{ whenever } d_{ij} = \mathbb{0} \text{ for some } i \text{ and } j, \text{ for}$$

$d_{11}, ..., d_{mn} \in D$. We say that $\mathcal{D}$ is idempotent if $+$ is idempotent, i.e., $d + d = d$ for all $d \in D$.

Example 2. Consider $(\mathbb{R} \cup \{-\infty\}, sup, avg, -\infty)$ where $avg((d_{ij})_{\substack{1 \leq i \leq m \\ 1 \leq j \leq n}}) = \frac{1}{m \times n} \sum_{\substack{1 \leq i \leq m \\ 1 \leq j \leq n}} d_{ij}$. Let $B \subseteq [0, 1]$ be a finite set of values and let $L \subseteq B^{++}$ be any picture language over B. Consider the function $S : B^{++} \to \mathbb{R} \cup \{-\infty\}$ defined for $p \in B^{++}$ by:

$$S(p) = \begin{cases} \frac{1}{m \times n} \sum_{\substack{1 \leq i \leq m \\ 1 \leq j \leq n}} p_{i,j} & \text{if } p \in L, \\ -\infty & \text{otherwise.} \end{cases}$$

We could interpret the values in B as different levels of light. Then for each picture p in the language L, the function S provides the average value $S(p)$ of light of p.

Example 3. Consider $(\mathbb{N} \cup \{\infty\}, min, maj, \infty)$ where the valuation function is a majority function, i.e., $maj(d_1, ..., d_n)$ is the greatest value among all values that occur most frequently in $d_1, ..., d_n$, e.g. $maj(4, 6, 6, 6, 8, 8, 8, 12) = 8$. Now let B

---

[1] We assume a picture to be non-empty for technical simplicity reason, as in [18, 21].

be a finite set, interpreted for instance as a set of different colors, and consider the function $S : B^{++} \to \mathbb{N}$, defined for $p \in B^{++}$ by:

$$S(p) = maj\{m \times n \mid q^{m \times n} \text{ is a monochrome subpicture of } p\}.$$

Then for each picture p, S provides the largest area of a monochrome rectangle, enclosed as a subpicture within p, which can be found most frequently among all monochrome subpictures of p.

Definition 4. *A weighted 2-dimensional on-line tessellation automaton (W2OTA) $\mathcal{A} = (Q, T, I, F, \gamma)$ over the alphabet Σ and a picture valuation monoid $\mathcal{D} = (D, +, val, \mathbb{0})$ consists of a finite set Q of states, a set of transitions $T \subseteq Q \times Q \times \Sigma \times Q$, sets of initial and final states $I, F \subseteq Q$, respectively, and a weight function $\gamma : T \to D$.*

For a transition $t = (q_h, q_v, a, q) \in T$, we set $\sigma_h(t) = q_h$, $\sigma_v(t) = q_v$ and $\sigma(t) = q$. We denote by label(t) its *label* a and by $\gamma(t) = d$ its *weight*. We extend both functions to pictures by setting for a picture $c = (c_{i,j})_{\substack{1 \le i \le m \\ 1 \le j \le n}} \in T^{m \times n}$ over the set of transitions:

$$label(c)_{i,j} := label(c_{i,j}) \text{ and } \gamma(c) = val(\gamma(c_{i,j}))_{i,j}.$$

This defines functions $label : T^{++} \to \Sigma^{++}$ and $\gamma : T^{++} \to D$. We call label(c) the *label* of c and $\gamma(c)$ the *weight* of c.

A *run* c in $\mathcal{A}$ is an element in $T^{m \times n}$, for some m, n, satisfying

$$\forall\, 1 \le i \le m, 2 \le j \le n : \sigma_h(c_{i,j}) = \sigma(c_{i,j-1}),$$

$$\forall\, 2 \le i \le m, 1 \le j \le n : \sigma_v(c_{i,j}) = \sigma(c_{i-1,j}).$$

We call the run $c \in T^{m \times n}$ *successful* if for all $1 \le i \le m, 1 \le j \le n$ we have $\sigma_v(c_{1,j}), \sigma_h(c_{i,1}) \in I$ and $\sigma(c_{m,n}) \in F$. The set of all successful runs labeled with a picture p in $\mathcal{A}$ is denoted by $I \xrightarrow{p}_{\mathcal{A}} F$.

We define the *behavior* of $\mathcal{A}$ as the function $\|\mathcal{A}\| : \Sigma^{++} \to D$ given by

$$\|\mathcal{A}\|(p) = \sum (val(\gamma(c)) \mid c \in I \xrightarrow{p}_{\mathcal{A}} F).$$

If $p \in \Sigma^{++}$ has no successful run in $\mathcal{A}$, then $\|\mathcal{A}\|(p) = \mathbb{0}$. Intuitively, the weight of a picture p is the sum of the weights of all successful runs in $\mathcal{A}$ that read p.

Example 5. For instance, a run in a W2OTA over $(\mathbb{R} \cup \{-\infty\}, sup, avg, -\infty)$ and the alphabet $\{a, b, c\}$ where p, q, s are states, can be shown by the following image:

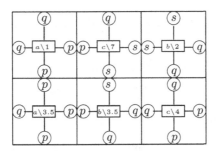

The label of this run is the picture $\begin{pmatrix} a & c & b \\ a & b & c \end{pmatrix}$ and its weight is

$$avg\left(\begin{pmatrix} 1 & 7 & 2 \\ 3.5 & 3.5 & 4 \end{pmatrix}\right) = 4.$$

In Definition 4, when $\mathcal{D}$ is considered as two-valued boolean algebra, we get precisely the definition of a *2-dimensional on-line tessellation automaton* (2OTA). For an alphabet Σ, 2OTA over Σ define picture languages and compute the family of recognizable picture languages [20].

A weighted or unweighted 2OTA $\mathcal{A}$ is called *unambiguous* if for any picture there exists at most one successful run in $\mathcal{A}$. A picture language L is called *unambiguously 2OTA recognizable* if it can be recognized by an unambiguous 2OTA.

Any function $S : \Sigma^{++} \to D$ is called a *picture series* or a *quantitative picture language* over Σ and $\mathcal{D}$. If $S = \|\mathcal{A}\|$ for a weighted 2-dimensional on-line tessellation automaton $\mathcal{A}$ taking weights from D, then S is called *W2OTA-recognizable*.

Example 6. Let $(\mathbb{R} \cup \{-\infty\}, sup, avg, -\infty)$ be a pv-monoid with $avg((d_{ij})_{\substack{1 \leq i \leq m \\ 1 \leq j \leq n}}) = \frac{1}{m \times n} \sum_{\substack{1 \leq i \leq m \\ 1 \leq j \leq n}} d_{ij}$. Consider $\mathcal{A} = (\{q\}, T, \{q\}, \{q\}, \gamma)$ as a W2OTA over the alphabet $\Sigma = \{b, w\}$ with $T = \{(q, q, b, q), (q, q, w, q)\}$, $\gamma(q, q, b, q) = -1$, and $\gamma(q, q, w, q) = 1$. If we let the letter b interpret black color and the letter w interpret white color, then $\mathcal{A}$ computes the average difference of darkness and brightness for every monochrome picture $p \in \Sigma^{++}$. For example

$$\|\mathcal{A}\|\left(\begin{pmatrix} b & b & w \\ w & b & w \\ w & w & b \end{pmatrix}\right) = \tfrac{1}{9}, \quad \|\mathcal{A}\|\left(\begin{pmatrix} w & w & w \\ w & w & w \\ w & w & w \end{pmatrix}\right) = 1, \text{ and } \|\mathcal{A}\|\left(\begin{pmatrix} b & b \\ w & w \end{pmatrix}\right) = 0.$$

Now we give some closure properties of W2OTA-recognizable picture series which we will use for the proof of our main results in the following two sections and which could be also of independent interest.

Let Σ and Γ be two sets. We can extend any mapping $\pi : \Sigma \to \Gamma$ to a function $\pi : \Sigma^{++} \to \Gamma^{++}$ in the natural way, mapping a picture $p \in \Sigma^{m \times n}$ to $p' \in \Gamma^{m \times n}$

such that $\pi(p(i,j)) = p'(i,j)$, for all $1 \le i \le m$, $1 \le j \le n$. Now let Σ and Γ be finite. We define for every $S : \Sigma^{++} \to D$ the projection $\pi(S) : \Gamma^{++} \to D$ by

$$\pi(S)(p) = \sum (S(p') \mid p' \in \Sigma^{++}, \pi(p') = p)$$

for every $p \in \Gamma^{++}$.

Proposition 7. *Let $\mathcal{D}$ be a picture valuation monoid, and let $\pi : \Sigma \to \Gamma$ be a mapping. If $S : \Sigma^{++} \to D$ is W2OTA-recognizable, then the projection $\pi(S) : \Gamma^{++} \to D$ is also W2OTA-recognizable.*

To prove this lemma, we apply a similar idea as used in [10,13].

Let Σ be an alphabet, $\mathcal{D}$ a picture valuation monoid and $r : \Sigma \to D$ be a mapping. We denote by $val \circ r : \Sigma^{++} \to D$ the picture series over $\mathcal{D}$ defined for all $p \in \Sigma^{++}$ by $(val \circ r)(p) = val(r(p))$.

Lemma 8. *Let Σ be an alphabet, $\mathcal{D}$ a picture valuation monoid and $r : \Sigma \to D$ a mapping. Then, $val \circ r$ is unambiguously W2OTA-recognizable.*

Let $\mathcal{D} = (D, +, val, \mathbb{0})$ be a pv-monoid and $S, S' : \Sigma^{++} \to D$. Then $S + S'$ is defined pointwise by $(S+S')(p) = S(p)+S'(p)$ for all $p \in \Sigma^{++}$. Let $L \subseteq \Sigma^{++}$ be a picture language. Then the intersection $(S \cap L) : \Sigma^{++} \to D$ is a picture series over $\mathcal{D}$ defined by $(S \cap L)(p) = S(p)$ if $p \in L$ and $(S \cap L)(p) = \mathbb{0}$ if $p \in \Sigma^{++} \backslash L$.

Proposition 9. *Let $(D, +, val, \mathbb{0})$ be a pv-monoid and let $S : \Sigma^{++} \to D$ and $S' : \Sigma^{++} \to D$ be W2OTA-recognizable picture series. Then $S + S'$ is W2OTA-recognizable.*

Lemma 10. *Let $\mathcal{D} = (D, +, val, \mathbb{0})$ be a pv-monoid, S a W2OTA-recognizable picture series and L a 2OTA-recognizable picture language.*

1. *If L is unambiguously 2OTA-recognizable, then $S \cap L$ is W2OTA-recognizable.*
2. *If L is unambiguously 2OTA-recognizable and S is unambiguously W2OTA-recognizable, then $S \cap L$ is unambiguously W2OTA-recognizable.*
3. *If $\mathcal{D}$ is idempotent, then $S \cap L$ is W2OTA-recognizable.*

Proof. (Sketch) We use the standard product construction of two automata $\mathcal{A}$ and $\mathcal{A}'$ where $\mathcal{A}$ is a 2OTA over Σ such that $L(\mathcal{A}) = L$, and $\mathcal{A}'$ is a W2OTA over Σ and $\mathcal{D}$ such that $\|\mathcal{A}'\| = S$. We call the automaton obtained by this construction $\tilde{\mathcal{A}}$. Now in (1), since $\mathcal{A}$ is unambiguous, then for every $p \in \Sigma^{++}$ there is either no successful run on p in $\tilde{\mathcal{A}}$ or there are as many successful runs on p in $\tilde{\mathcal{A}}$ as there are in $\mathcal{A}'$. Hence, for $p \in L$ we have $\|\tilde{\mathcal{A}}\|(p) = \|\mathcal{A}'\|(p) = (S \cap L)(p)$. In (2), we can choose $\mathcal{A}$ and $\mathcal{A}'$ both unambiguous, then obviously $\tilde{\mathcal{A}}$ is unambiguous, and the result is obtained similarly to (1). To prove case (3), if $p \in \Sigma^{++} \backslash L$, then $\|\tilde{\mathcal{A}}\|(p) = (S \cap L)(p) = 0$. Now let $N > 0$ be the number of successful runs on $p \in \Sigma^{++}$ in $\mathcal{A}$. Then, every successful run on p in $\mathcal{A}'$ can be simulated by N runs in $\tilde{\mathcal{A}}$ having the same weight. If c is a successful run in $\mathcal{A}'$ over p, then by idempotency of $\mathcal{D}$ we have $\sum_{i=1}^{N} val(\gamma(c)) = val(\gamma(c))$. It follows that $\|\tilde{A}\|(p) = (S \cap L)(p)$.

We just note that the assumptions on L being unambiguous are necessary, cf. [11].

3   A Nivat Theorem for Weighted Picture Automata

Nivat's theorem is a fundamental characterization of rational transductions and provides a connection between rational transductions and rational languages. Recently, Droste and Perevoshchikov [11] proved a Nivat-like theorem for recognizable quantitative timed languages. Here we want to prove such a result for recognizable picture series.

Let Σ be an alphabet and $\mathcal{D} = (D, +, val, 0)$ a pv-monoid. Let $\mathcal{D}^{rec}(\Sigma^{++}, W2OTA)$ denote the family of picture series recognized by W2OTA over Σ and $\mathcal{D}$. The abbreviation $\mathcal{D}^{\mathcal{N}}(\Sigma^{++})$ (with $\mathcal{N}$ meaning Nivat) stands for the set of all picture series $S : \Sigma^{++} \rightarrow D$ over $\mathcal{D}$ such that there exist an alphabet Γ, mappings $h : \Gamma \rightarrow \Sigma$ and $r : \Gamma \rightarrow D$ and a recognizable picture language $L \subseteq \Gamma^{++}$ such that $S = h((val \circ r) \cap L)$. Finally, the abbreviation $\mathcal{D}^{\mathcal{N}}(\Sigma^{++}, UNAMB)$ is defined like $\mathcal{D}^{\mathcal{N}}(\Sigma^{++})$ with the difference that L is an unambiguous picture language.

Theorem 11. *Let Σ be an alphabet and $\mathcal{D} = (D, +, val, 0)$ a pv-monoid. Then we have the following,*

- $\mathcal{D}^{rec}(\Sigma^{++}, W2OTA) = \mathcal{D}^{\mathcal{N}}(\Sigma^{++}, UNAMB);$
- *Moreover, if $\mathcal{D}$ is idempotent, then $\mathcal{D}^{rec}(\Sigma^{++}, W2OTA) = \mathcal{D}^{\mathcal{N}}(\Sigma^{++}).$*

Proof. First we show that $\mathcal{D}^{rec}(\Sigma^{++}, W2OTA) \subseteq \mathcal{D}^{\mathcal{N}}(\Sigma^{++}, UNAMB)$. Let $\mathcal{A} = (Q, T, I, F, \gamma)$ be a W2OTA over Σ and $\mathcal{D}$ such that $\|\mathcal{A}\| = S$. Let $\Gamma = T$. We define the mapping $h : \Gamma \rightarrow \Sigma$ for all $\lambda = (q_h, q_v, a, q) \in \Gamma$ by $h(\lambda) = a$ and we let $r = \gamma : \Gamma \rightarrow D$. From $\mathcal{A}$ we construct the 2OTA $\mathcal{A}' = (Q, T', I, F)$ over the enlarged alphabet Γ such that

$$T' = \{(q_h, q_v, (q_h, q_v, a, q), q) \mid \exists a \in \Sigma : (q_h, q_v, a, q) \in T\}.$$

With this construction, clearly for every input label $(q_h, q_v, a, q) \in \Gamma$ there is at most one transition in T' with label (q_h, q_v, a, q). So $\mathcal{A}'$ is unambiguous and we put $L(\mathcal{A}') = L$. It remains to show that $S = h((val \circ r) \cap L)$. For this let $p \in \Sigma^{++}$. Note that $I \xrightarrow{p}_{\mathcal{A}} F = \{p' \in L(\mathcal{A}') \mid h(p') = p\}$. Moreover, if $c \in I \xrightarrow{p}_{\mathcal{A}} F$, we have $(val \circ r)(c) = val(r(c)) = val(\gamma(c))$. Therefore,

$$h((val \circ r) \cap L)(p) = \sum ((val \circ r)(p') \mid p' \in L \text{ and } h(p') = p)$$
$$= \sum (val(\gamma(c)) \mid c \in I \xrightarrow{p}_{\mathcal{A}} F)$$
$$= \|\mathcal{A}\|(p).$$

The converse inclusions are immediate by Lemmas 8 and 10 and Proposition 7.

Let Σ be an alphabet and $\mathcal{D} = (D, +, val, 0)$ a pv-monoid. Let $\mathcal{D}^{\mathcal{H}}(\Sigma^{++}, UNAMB)$ denote the family of picture series $S : \Sigma^{++} \to D$ over $\mathcal{D}$ such that there exist an alphabet Γ, a mapping $\pi : \Gamma \to \Sigma$ and an unambiguously recognizable picture series $T : \Gamma^{++} \to D$ over $\mathcal{D}$ such that $S = \pi(T)$. Now as a corollary of Theorem 11, we have the following result:

Corollary 12. *Let Σ be an alphabet and $\mathcal{D} = (D, +, val, 0)$ a pv-monoid. Then $\mathcal{D}^{\mathcal{H}}(\Sigma^{++}, UNAMB) = \mathcal{D}^{rec}(\Sigma^{++}, W2OTA)$.*

4 Weighted MSO Logic

In this section we introduce the syntax and semantics of the weighted MSO logic on pictures. The syntax is a combination of the idea introduced in [10], [11] and [16]. We define the syntax as a combination of almost boolean first-order formulas and formulas of weighted monadic second-order logic. The idea of boolean formulas was introduced first by Bollig and Gastin [2]. Here we fix an alphabet Σ. For $a \in \Sigma$, P_a denotes a unary predicate symbol. We provide a countable set $\mathcal{V}$ of first-order and second-order variables. Lower-case letters x, y denote first-order variables, and capital letters like X, Y denote second-order variables. We also need to equip our picture valuation monoid with a product operation and a unit element.

Definition 13. *A product picture valuation monoid, or a ppv-monoid for short, is a tuple $\mathcal{D} = (D, +, val, \diamond, 0, 1)$ consisting of a picture valuation monoid $(D, +, val, 0)$, a binary operation $\diamond : D^2 \to D$, and $1 \in D$ with $val((1)_{i,j})_{\substack{1 \le i \le m \\ 1 \le j \le n}} = 1$ for all $m, n \ge 1$ and $0 \diamond d = d \diamond 0 = 0$, $1 \diamond d = d \diamond 1 = d$ for all $d \in D$.*

Definition 14. *The syntax of weighted MSO logics over a ppv-monoid $(D, +, val, \diamond, 0, 1)$ is defined as follows:*

$$\beta ::= P_a(x) \mid xS_v y \mid xS_h y \mid x \in X \mid x = y \mid \neg\beta \mid \beta \wedge \beta \mid \forall x \beta$$
$$\varphi ::= d \mid \beta \mid \varphi \vee \varphi \mid \varphi \wedge \varphi \mid \exists x \varphi \mid \forall x \varphi \mid \exists X \varphi$$

where $d \in D$, $a \in \Sigma$, $x, y, X \in \mathcal{V}$. The formulas β are called boolean first-order, for short FO, formulas and the formulas φ are called weighted MSO-formulas, for short wMSO-formulas.

Note that negation is only applied in boolean FO formulas.

The set $Free(\varphi)$ of free variables in φ is defined as usual. Let $p \in \Sigma^{m \times n}$. We put $Dom(p) = \{1, 2, ..., m\} \times \{1, 2, ..., n\}$ and denote the component of p at position (i, j) by $p(i, j)$. Intuitively, $P_a(x)$ means that the position x of some considered picture carries the letter a. S_v and S_h represent the two successor relations of both directions, defined by $(i, j)S_v(i + 1, j)$ and $(i, j)S_h(i, j + 1)$. Formulas containing no set quantification but possibly including atomic formulas of the form $(x \in X)$ are called first-order formulas.

Table 1. The semantics of weighted MSO formulas over a ppv-monoid

$$\llbracket d \rrbracket_{\mathcal{V}}(p,\sigma)=d$$

$$\llbracket P_a(x) \rrbracket_{\mathcal{V}}(p,\sigma)=\begin{cases}1, & \text{if } p(\sigma(x))=a,\\ 0, & \text{otherwise}\end{cases} \qquad \llbracket \neg\beta \rrbracket_{\mathcal{V}}(p,\sigma)=\begin{cases}1, & \text{if } \llbracket\beta\rrbracket_{\mathcal{V}}(p,\sigma)=0,\\ 0, & \text{if } \llbracket\beta\rrbracket_{\mathcal{V}}(p,\sigma)=1\end{cases}$$

$$\llbracket xS_vy \rrbracket_{\mathcal{V}}(p,\sigma)=\begin{cases}1, & \text{if } \sigma(x)S_v\sigma(y),\\ 0, & \text{otherwise}\end{cases} \qquad \begin{aligned}&\llbracket\varphi_1\vee\varphi_2\rrbracket_{\mathcal{V}}(p,\sigma)=\llbracket\varphi_1\rrbracket_{\mathcal{V}}(p,\sigma)+\llbracket\varphi_2\rrbracket_{\mathcal{V}}(p,\sigma)\\ &\llbracket\varphi_1\wedge\varphi_2\rrbracket_{\mathcal{V}}(p,\sigma)=\llbracket\varphi_1\rrbracket_{\mathcal{V}}(p,\sigma)\diamond\llbracket\varphi_2\rrbracket_{\mathcal{V}}(p,\sigma)\end{aligned}$$

$$\llbracket xS_hy \rrbracket_{\mathcal{V}}(p,\sigma)=\begin{cases}1, & \text{if } \sigma(x)S_h\sigma(y),\\ 0, & \text{otherwise}\end{cases} \qquad \llbracket\exists x\varphi\rrbracket_{\mathcal{V}}(p,\sigma)=\sum_{(i,j)\in\mathrm{Dom}(p)}\llbracket\varphi\rrbracket_{\mathcal{V}\cup\{x\}}(p,\sigma[x/(i,j)])$$

$$\llbracket x=y \rrbracket_{\mathcal{V}}(p,\sigma)=\begin{cases}1, & \text{if } \sigma(x)=\sigma(y),\\ 0, & \text{otherwise}\end{cases} \qquad \llbracket\exists X\varphi\rrbracket_{\mathcal{V}}(p,\sigma)=\sum_{I\subseteq\mathrm{Dom}(p)}\llbracket\varphi\rrbracket_{\mathcal{V}\cup\{X\}}(p,\sigma[X/I])$$

$$\llbracket x\in X \rrbracket_{\mathcal{V}}(p,\sigma)=\begin{cases}1, & \text{if } \sigma(x)\in\sigma(X),\\ 0, & \text{otherwise}\end{cases} \qquad \llbracket\forall x\varphi\rrbracket_{\mathcal{V}}(p,\sigma)=val\big(\llbracket\varphi\rrbracket_{\mathcal{V}\cup\{x\}}(p,\sigma[x/(i,j)])\big)_{(i,j)\in\mathrm{Dom}(p)}$$

The picture language $L(\varphi)$ defined by the sentence φ is the set of all pictures $p\in\Sigma^{++}$ satisfying φ. We say a picture language L is first-order definable ($L\in FO$) if there is a first-order sentence φ such that $L=L(\varphi)$. We refer the reader for more details on classical results to [23].

Let $\mathcal{V}$ be a finite set of variables with $Free(\varphi)\subseteq\mathcal{V}$. We define a $(\mathcal{V},p)$-assignment as a function:

$$\sigma:\mathcal{V}\to Dom(p)\cup 2^{Dom(p)}$$

mapping FO variables in $\mathcal{V}$ to elements of $Dom(p)$ and second-order variables in $\mathcal{V}$ to subsets of $Dom(p)$. If x is an FO variable and $(i,j)\in Dom(p)$, then the update $\sigma[x/(i,j)]$ for $(i,j)\in\mathbb{N}\times\mathbb{N}$ is defined as $\sigma[x/(i,j)](x)=(i,j)$ and $\sigma[x/(i,j)]\restriction_{\mathcal{V}\setminus\{x\}}=\sigma\restriction_{\mathcal{V}\setminus\{x\}}$. Similarly, the update $\sigma[X/I]$ is defined for $I\subseteq\mathbb{N}\times\mathbb{N}$. We encode a pair (p,σ), where σ is a $(\mathcal{V},p)$-assignment, as a picture over the enriched alphabet $\Sigma_{\mathcal{V}}=\Sigma\times\{0,1\}^{\mathcal{V}}$. Conversely, an element $r\in\Sigma_{\mathcal{V}}^{++}$ can be viewed as a pair (p,σ) where $p\in\Sigma^{++}$ is the projection over Σ and $\sigma\in(\{0,1\}^{\mathcal{V}})^{++}$ is the projection of r over $\{0,1\}^{\mathcal{V}}$. Now, if the latter projection $\sigma(\{0,1\}^{\mathcal{V}})^{++}$ is such that for each FO variable $x\in\mathcal{V}$, the projection of σ to the x-coordinate contains exactly one pixel carrying a 1, then σ will be called a valid assignment.

Now, similar to [16] we give the semantics of wMSO formulas. However, here the semantics of universal quantification will be interpreted by a valuation function, which for example provides the average value of light of a picture and the semantics of conjunction is interpreted by a product operation.

Let φ be a weighted MSO-formula and $\mathcal{V}$ be a finite set of variables containing $Free(\varphi)$. The semantics of φ will be a series $\llbracket\varphi\rrbracket_{\mathcal{V}}:\Sigma_{\mathcal{V}}^{++}\to D$ such that if σ is not a valid $(p,\mathcal{V})$-assignment, then $\llbracket\varphi\rrbracket_{\mathcal{V}}(p,\sigma)=0$. Otherwise, we define $\llbracket\varphi\rrbracket_{\mathcal{V}}(p,\sigma)\in D$ inductively as in Table 1. We write $\Sigma_\varphi=\Sigma_{Free(\varphi)}$ and $\llbracket\varphi\rrbracket=\llbracket\varphi\rrbracket_{Free(\varphi)}$. In case φ is a sentence, then the semantics is a picture series over Σ.

Example 15. Consider $(\mathbb{R}\cup\{-\infty\}, sup, avg, +, -\infty, 0)$ where $avg((d_{ij})_{\substack{1\le i\le m\\1\le j\le n}})=$ $\frac{1}{m\times n}\sum_{\substack{1\le i\le m\\1\le j\le n}}d_{ij}$. Let $B\subseteq[0,1]$ be the finite set of values considered in

Example 2. Let B^{++} be the set of all pictures over B. Consider the formula $\psi = \forall x \, (\bigvee_{b \in B}(P_b(x) \wedge b))$. Since universal quantification is interpreted by average, so, the semantics of the formula ψ provides the average value $[\![\psi]\!](p)$ of light of p. Therefore, we have $[\![\psi]\!] = S$ for the picture series S from Example 2 (where $L = B^{++}$).

Lemma 16. [16] *Let Σ be any alphabet and let φ be a first-order formula. Then $\mathcal{L}(\varphi)$ is an unambiguously recognizable picture language.*

For a language $L \subseteq \Sigma^{++}$, the characteristic series $\mathbb{1}_L : \Sigma^{++} \to D$ is defined for $p \in \Sigma^{++}$ by $\mathbb{1}_L(p) = \mathbb{1}$ if $p \in L$, and $\mathbb{1}_L(p) = \mathbb{0}$ otherwise. Hence, for any picture series $S : \Sigma^{++} \to D$ and a picture language $L \subseteq \Sigma^{++}$ we have $S \diamond \mathbb{1}_L = S \cap L$.

Lemma 17. *Let $\mathcal{D}$ be a ppv-monoid. If L is unambiguously 2OTA-recognizable, then $\mathbb{1}_L$ is W2OTA-recognizable.*

Definition 18. *The class of almost boolean FO formulas of weighted-MSO is the smallest class containing all constant $d \in D$ and all boolean FO formulas and which is closed under disjunction and conjunction.*

The following result can be shown by an easy induction on the structure of φ which we will use it in Lemma 21.

Lemma 19. *Let D be a ppv-monoid. If φ is an almost boolean FO formula, then $[\![\varphi]\!]$ can be written in the form $[\![\varphi]\!] = \sum_{k=1}^{n} d_k \mathbb{1}_{L_k}$ with $d_1, ..., d_n \in D$ for some $n \in \mathbb{N}$, and $(L_k)_{k=1,...,n}$ forming a partition of Σ_φ^{++} of FO definable picture languages over Σ_φ.*

We already know that for semiring weighted automata on words, the full weighted MSO logic without any restriction on universal quantification is expressively stronger than weighted automata [8]. In [16], the author considers commutative semirings as the abstract model for weight structure. Here we do not require commutativity of the product operation $\diamond$, but the application of conjunction is restricted to either almost boolean FO formulas or boolean FO formulas. Moreover, using an idea of Perevoshchikov [11], we restrict the use of constants in the formula by allowing their occurrence only in the scope of FO universal quantifiers. By this, we can avoid the assumption of regularity of [10] for the valuation monoids.

Definition 20. *A weighted MSO formula φ is called*

- *$\forall$-restricted if whenever it contains a sub-formula of the form $\forall x \psi$, then ψ is an almost boolean FO formula.*
- *$\wedge$-restricted if for every sub-formula $\psi_1 \wedge \psi_2$ of φ either both ψ_1 and ψ_2 are almost boolean FO formulas, or ψ_1 or ψ_2 is a boolean FO formula.*
- *D-restricted if every constant $d \in D$ occurring in φ is in the scope of a first-order universal quantifier.*

Lemma 21. *Let $\mathcal{D}$ be a ppv-monoid and let φ be an almost boolean FO formula. Then $[\![\forall x \; \varphi]\!]$ is W2OTA-recognizable over $im([\![\varphi]\!])$.*

Proof. (Sketch). Let φ be an almost boolean FO formula. By Lemma 19, we have $[\![\varphi]\!] = \sum_{k=1}^{n} d_k \mathbb{1}_{L_k}$ where $(L_k)_{k=1,\ldots,n}$ is a partition of Σ_φ^{++} of FO definable picture languages over Σ_φ. Let $\widetilde{\Sigma} = \Sigma \times \{1, 2, \ldots, n\}$, and let $\widetilde{L}$ be the picture language of all $(p, \nu, \sigma) \in (\widetilde{\Sigma}_\mathcal{V})^{++}$ such that (p, σ) is valid and for all $(i, j) \in Dom(p)$ and $k \in \{1, 2, \ldots, n\}$ we have $\nu(i, j) = k \Leftrightarrow (p, \sigma[x/(i, j)]) \in L_k$. In [27], it was shown that $\widetilde{L}$ is FO definable and as a result, there exists an unambiguous 2OTA $\widetilde{\mathcal{A}}$ computing $\widetilde{L}$. Now from $\widetilde{\mathcal{A}}$ we can construct a W2OTA $\mathcal{A}$, by letting the transitions of $\mathcal{A}$ have a suitable value d_k as weight, and it follows that $[\![\forall x \; \varphi]\!] = \pi(\|\mathcal{A}\|)$ for a projection π. Then $[\![\forall x \; \varphi]\!]$ is W2OTA-recognizable by Proposition 7.

Our main result of this section is the following.

Theorem 22. *Let D be a ppv-monoid and let $S : \Sigma^{++} \to D$ be a picture series. Then S is W2OTA-recognizable if and only if $S = [\![\varphi]\!]$ for a $\forall$-restricted, $\wedge$-restricted and D-restricted wMSO-sentence φ.*

Proof. (Sketch). ($\Longrightarrow$) Given a W2OTA $\mathcal{A}$, we can use its structure to construct a $\forall$- and $\wedge$- restricted sentence α such that $[\![\alpha]\!] = \|\mathcal{A}\|$ and all the constants of α can occur in the scope of the FO universal quantifier. Hence, α is also D-restricted.

($\Longleftarrow$) Let φ be a $\forall$-restricted, $\wedge$-restricted and D-restricted wMSO-sentence. By induction over the structure of a suitable subformula ψ of φ, we construct a W2OTA $\mathcal{A}_\psi$ such that $\|\mathcal{A}_\psi\| = [\![\psi]\!]$. If ψ is a boolean FO formula, then it can be regarded as a classical boolean FO formula defining the picture language $L(\psi)$, which is unambiguous due to Lemma 16. We know that $[\![\psi]\!] = \mathbb{1}_{L(\psi)}$. Now by Lemma 17, $[\![\psi]\!]$ is W2OTA-recognizable. If $\psi = \psi_1 \vee \psi_2$, $\varphi = \exists x \; \psi'$, and $\psi = \exists X \; \psi'$ we apply Propositions 9 and 7. If $\psi = \forall x \; \psi'$, we use Lemma 21. Now, consider $\psi = \psi_1 \wedge \psi_2$. Since ψ is $\wedge$-restricted, either both ψ_1 and ψ_2 are almost boolean FO formulas, or ψ_1 or ψ_2 is a boolean FO formula. In the first case, ψ can occur as a subformula of $\forall x \varphi'$ and then the result follows by Lemma 21. In the second case, we apply Lemma 10.

5   Conclusion

We defined weighted two-dimensional on-line tessellation automata taking weights from picture valuation monoids which are more general than semirings. These automata can model average density of pictures. We proved a Nivat's theorem for W2OTA. This result makes a connection between the behaviors of W2OTA and 2OTA. In fact, we showed that recognizable picture series can be obtained precisely as projections of particularly simple unambiguously recognizable series restricted to unambiguous recognizable picture languages. In addition, we showed that if we consider idempotent picture valuation monoids, then we get this result without any unambiguity condition.

We also defined a new weighted MSO logic. We considered a suitable fragment of this logic and as the second result of this paper we proved that our W2OTA and this fragment of weighted MSO logic are expressively equivalent.

We would like to mention that, because of space constraints here we did not include the following results: Under additional assumptions on the underlying picture valuation monoid, there are several extended fragments of our weighted MSO logic which are expressively equivalent to our W2OTA model.

References

1. Anselmo, M., Giammarresi, D., Madonia, M., Restivo, A.: Unambiguous recognizable two-dimensional languages. Theoretical Information and Application **40**(2), 277–293 (2006)
2. Bollig, B., Gastin, P.: Weighted versus probabilistic logics. In: Diekert, V., Nowotka, D. (eds.) DLT 2009. LNCS, vol. 5583, pp. 18–38. Springer, Heidelberg (2009)
3. Bozapalidis, S., Grammatikopoulou, A.: Recognizable picture series. Journal of Automata, Languages and Combinatorics **10**, 159–183 (2005)
4. Chatterjee, K., Doyen, L., Henzinger, T.A.: Quantitative languages. In: Kaminski, M., Martini, S. (eds.) CSL 2008. LNCS, vol. 5213, pp. 385–400. Springer, Heidelberg (2008)
5. Chatterjee, K., Doyen, L., Henzinger, T.A.: Alternating weighted automata. In: Kutyłowski, M., Charatonik, W., Ębala, M. (eds.) FCT 2009. LNCS, vol. 5699, pp. 3–13. Springer, Heidelberg (2009)
6. Chatterjee, K., Doyen, L., Henzinger, T.A.: Expressiveness and closure properties for quantitative languages. Logical Methods in Computer Science **6**(3–10), 1–23 (2010)
7. Chatterjee, K., Doyen, L., Henzinger, T.A.: Probabilistic weighted automata. In: Bravetti, M., Zavattaro, G. (eds.) CONCUR 2009. LNCS, vol. 5710, pp. 244–258. Springer, Heidelberg (2009)
8. Droste, M., Gastin, P.: Weighted automata and weighted logics. Theoretical Computer Science **380**(1–2), 69–86 (2007)
9. Droste, M., Kuske, D.: Weighted automata. In: Pin, J.-E. (ed.) Handbook: "Automata: from Mathematics to Applications". European Mathematical Society (to appear)
10. Droste, M., Meinecke, I.: Weighted automata and weighted MSO logics for average- and longtime-behaviors. Information and Computation **220–221**, 44–59 (2012)
11. Droste, M., Perevoshchikov, V.: A Nivat theorem for weighted timed automata and weighted relative distance logic. In: Esparza, J., Fraigniaud, P., Husfeldt, T., Koutsoupias, E. (eds.) ICALP 2014, Part II. LNCS, vol. 8573, pp. 171–182. Springer, Heidelberg (2014)
12. Droste, M., Rahonis, G.: Weighted automata and weighted logics on infinite words. In: Ibarra, O.H., Dang, Z. (eds.) DLT 2006. LNCS, vol. 4036, pp. 49–58. Springer, Heidelberg (2006)
13. Droste, M., Vogler, H.: Weighted automata and multi-valued logics over arbitrary bounded lattices. Theoretical Computer Science **418**, 14–36 (2012)
14. Droste, M., Vogler, H.: Weighted tree automata and weighted logics. Theoretical Computer Science **366**, 228–247 (2006)

15. Eilenberg, S.: Automata, Languages, and Machines, vol. A. Academic Press (1974)
16. Fichtner, I.: Weighted picture automata and weighted logics. Theory of Computing Systems **48**(1), 48–78 (2011)
17. Fichtner, I.: Characterizations of recognizable picture series. Theoretical Computer Science **374**, 214–228 (2007)
18. Mäurer, I.: Weighted picture automata and weighted logics. In: Durand, B., Thomas, W. (eds.) STACS 2006. LNCS, vol. 3884, pp. 313–324. Springer, Heidelberg (2006)
19. Fu, K.S.: Syntactic Methods in Pattern Recognition. Academic Press, New York (1974)
20. Giammarresi, D., Restivo, A.: Recognizable picture languages. International Journal of Pattern Recognition and Artificial Intelligence **6**(2, 3), 256 (1992)
21. Giammarresi, D., Restivo, A.: Two-dimensional finite state recognizability. Fundamental Informaticae **25**(3), 399–422 (1996)
22. Giammarresi, D., Restivo, A.: Two-dimensional languages. In: Rozenberg, G., Salomaa, A., (eds.) Handbook of Formal Languages, vol. 3, pp. 215–267. Springer (1997)
23. Giammarresi, D., Restivo, A., Seibert, S., Thomas, W.: Monadic second-order logic over rectangular pictures and recognizability by tiling systems. Information and Computation **125**(1), 32–45 (1996)
24. Inoue, K., Nakamura, A.: Some properties of two-dimensional on-line tessellation acceptors. Information Sciences **13**, 95–121 (1977)
25. Inoue, K., Takanami, I.: A survey of two-dimentional automata theory. Information Sciences **55**, 99–121 (1991)
26. Kuich, W., Salomaa, A.: Semirings, Automata, Languages, vol. 6. EATCS Monographs, Theoretical Computer Science. Springer (1986)
27. Latteux, M., Simplot, D.: Recognizable picture languages and domino tiling. Theoretical Computer Science **178**, 275–283 (1997)
28. Lindgren, K., Moore, C., Nordahl, M.: Complexity of two-dimensional patterns. Journal of Statistical Physics **91**(5–6), 909–951 (1998)
29. Matz, O.: On piecewise testable, starfree, and recognizable picture languages. In: Nivat, M. (ed.) FOSSACS 1998. LNCS, vol. 1378, pp. 203–210. Springer, Heidelberg (1998)
30. Meinecke, I.: Weighted logics for traces. In: Grigoriev, D., Harrison, J., Hirsch, E.A. (eds.) CSR 2006. LNCS, vol. 3967, pp. 235–246. Springer, Heidelberg (2006)
31. Minski, M., Papert, S.: Perceptron. M.I.T Press, Cambridge (1969)
32. Salomaa, A., Soittola, M.: Automata-Theoretic Aspects of Formal Power Series. Texts and Monographs on Computer Science. Springer (1978)
33. Simplot, D.: A characterization of recognizable picture languages by tilings by finite sets. Theoretical Computer Science **218**(2), 297–323 (1999)
34. Smith, R.A.: Two-dimensional formal languages and pattern recognition by cellular automata. In: 12th IEEE FOCS Conference Record, pp. 144–152 (1971)
35. Thomas, W.: On logics, tilings, and automata. In: Leach Albert, J., Monien, B., Rodríguez-Artalejo, M. (eds.) ICALP 1991. LNCS, vol. 510, pp. 441–454. Springer, Heidelberg (1991)
36. Wilke, T.: Star-free picture expressions are strictly weaker than first-order logic. In: Degano, P., Gorrieri, R., Marchetti-Spaccamela, A. (eds.) ICALP 1997. LNCS, vol. 1256, pp. 347–357. Springer, Heidelberg (1997)

Rational Selecting Relations and Selectors

Luc Boasson and Olivier Carton[✉]

Université Paris Diderot – LIAFA, Paris, France
olivier.carton@liafa.univ-paris-diderot.fr

Abstract. We consider rational relations made of pairs (u, v) of finite words such that v is a subword of u. We show that such a selecting relation can be realized by a transducer such that the output label of each transition is a subword of its input label. We also show that it is decidable whether a given relation has this property.

Keywords: Transducers · Synchronization

1 Introduction

Selecting elements from a sequence to build a new sequence is an ubiquitous process in mathematics and computer science. Its use ranges from topology to the theory of quasi-orderings. In this paper, we study selecting processes realized by transducers which are finite automata equipped with inputs and outputs. The subword relation is one of the most natural relations on words. It gives raise to Higman's theorem [7] and can be realized by a transducer. These selecting processes also occur in [1] where it is shown that selecting symbols by oblivious sequential transducers preserves normality. Let us recall that an infinite word x over some alphabet is said to be normal if, for each length, all finite words of that length occur with the same frequency in x. The term *oblivious* means here that the selection of an input symbol is only based upon the prefix of the input read before that symbol, not including the symbol itself. The work presented here was actually motivated by this use of transducers but our results only hold for finite words and cannot be extended to infinite words as it was initially expected (see Sect. 6).

Transducers considered in [1] have a very special form. For each input symbol, the transducer either outputs nothing (the empty word) or copies the input to the output. The natural transducer realizing the subword relation has also this special form. This property of the transitions insures that the output is always a subword of the input. The global relation between the input and the output is guaranteed by local properties of the transitions. It is not true however that each transducer realizing a selecting process has this local property (see Example 1 for instance). As usual, such an observation raises the two related following questions: first, is every selecting process realized by a transducer having the local property and second, is it decidable whether some process realized by a given transducer is a selecting one. In this paper, we answer the two questions

© Springer International Publishing Switzerland 2015
A.-H. Dediu et al. (Eds.): LATA 2015, LNCS 8977, pp. 716–726, 2015.
DOI: 10.1007/978-3-319-15579-1_56

positively. We show that each relation such that the output is always a subword of the input is realized by a transducer where the transitions already satisfy this property locally. We also show how to decide whether a given rational relation satisfies that for every of its pairs (u, v), v is a subword of u.

The first question follows a line of research started by Elgot and Mezei in [6] where it is shown that a rational length preserving relation, that is input and output always have the same length, can be realized by a transducer such that for each transition input and output have the same length (see [5, Thm IX.6.1] and [11, Thm IV.6.1]). This result has been extended by Leguy in [8] where it is shown that each length-decreasing relation can be realized by a transducer in which each transition has an output which is shorter than the input. Notice that this latter result is really an extension of Elgot and Mezei's one which can easily be recovered from it. This result was further extended by Sakarovitch using the notion of a stable monoid of $\mathbb{N}^2$ [11, Thm IV.6.3].

The first question can also be rephrased as a Fatou property of a submonoid M of $A^* \times B^*$ [4]. Let R be a rational relation of $A^* \times B^*$ contained in a sub-monoid M of $A^* \times B^*$. A very natural question is whether the relation R can be realized by a transducer T in which all transitions are labelled by elements of M. If the family of rational subsets of a given monoid M is denoted by $\mathrm{Rat}(M)$, this question is whether the following implication holds for all relations R.

$$R \in \mathrm{Rat}(A^* \times B^*) \wedge R \subseteq M \Longrightarrow R \in \mathrm{Rat}(M)$$

The questions answered in [6] and [8] correspond to the two sub-monoids $M = \{(u, v) : |u| = |v|\}$ and $M = \{(u, v) : |u| \geqslant |v|\}$. The case $M = \{(u, v) : |u|/|v| = \alpha\}$ for some fixed rational number α has also been answered positively in [3]. It turns out that the answer to this question is negative in general (see example 13). In this paper, we answer positively this question for the sub-monoid $M = \{(u, v) : v|u\}$ made of pairs (u, v) such that v is a subword of u. Notice that in all previous results, the monoid M is defined by constraints regarding lengths of the words and making thus all symbols equivalent. Our result is the first positive one where the monoid M is defined by other constraints.

The second question is a very classical one. For each subclass of rational relations, it is natural to check whether a given relation is a member of that subclass. It turns out that most of these questions are undecidable [2, Thm 8.4] but it is decidable in the two cases considered above. It is always surprising when this membership problem is decidable.

The paper is organized as follows. Basic definitions and transducers are recalled in Sect. 2. Sect. 3 introduces the crucial notion of delays and shifts of a transducer. It is proved in Sect. 4 that each selecting relation is realized by a transducer with the local property. Sect. 5 is devoted to the decidability result. In Sect. 6 are discussed some positive and negative extensions of our result. Due to lack of space, all proofs have been omitted.

2 Preliminaries

Let A be finite alphabet. We let A^* and A^ω respectively denote the sets of all finite and infinite words. The empty word is denoted by ε and the length of a word w is denoted $|w|$. The number of occurrences of a symbol a in a word w is denoted by $|w|_a$. The cardinality of a finite set E is also denoted $|E|$.

A word $v = b_1 \cdots b_n$ is a *subword* of a word $u = a_1 \cdots a_m$ if there exists a strictly increasing function ι from $\{1, \ldots, n\}$ to $\{1, \ldots, m\}$ such that $b_j = a_{\iota(j)}$ for each $1 \leqslant j \leqslant n$, or equivalently if there exist $n + 1$ words $u_0, \ldots, u_n$ such that $u = u_0 b_1 u_1 \cdots b_n u_n$ [9, Chap. 6]. The increasing function ι is called an *embedding of v in u* and the relation is denoted by $v|u$. Note that there may exist several embeddings of a word v in a word u. There are, for instance, 5 embeddings of $v = aabb$ in $u = aababb$. Three of them are depicted in Fig. 1. Among all possible embeddings, there is an embedding ι_0 such that for any other embedding ι, $\iota_0(j) \leqslant \iota(j)$ holds for each $1 \leqslant j \leqslant n$. This embedding ι_0 is called the *leftmost* embedding of v in u. Note that if $\iota_0(j) = k$, then $b_1 \cdots b_j$ is the longest prefix of v which is a subword of $a_1 \cdots a_k$. The *rightmost* embedding can be defined similarly. The first embedding shown in Fig. 1 is the leftmost one of $aabb$ in $aababb$. If ι is an embedding of v in u and w is a factor of v starting at position k, the restriction of ι to w is the function ι' from $\{1, \cdots, |w|\}$ to $\{1, \cdots, m\}$ defined by $\iota'(j) = \iota(j + k - 1)$. It is an embedding of w in u.

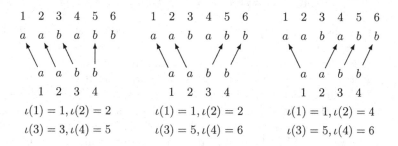

Fig. 1. Three embeddings of $aabb$ in $aababb$

A relation $R \subseteq A^* \times B^*$ is called *selecting* if, for any words u and v, $(u, v) \in R$ implies $v|u$.

A transducer is a finite automaton, the transitions of which are labelled by pairs of words. More formally, a *transducer* is a tuple $\langle Q, A, B, \Delta, I, F \rangle$, where Q is a finite set of states, A and B are the input and output alphabets, $\Delta \subseteq Q \times A^* \times B^* \times Q$ is a finite set of transitions, $I \subseteq Q$ is the subset of initial states, and $F \subseteq S$ is the subset of final states. A transition (p, u, v, q) is written $p \xrightarrow{u:v} q$. A *run* from a state p to a state q is a sequence of consecutive transitions $p \xrightarrow{u_1:v_1} q_1 \xrightarrow{u_2:u_2} q_2 \cdots q_{n-1} \xrightarrow{u_n:v_n} q$. Its *label* is the pair (u, v) where $u = u_1 \cdots u_n$ and $v = v_1 \cdots v_n$. The run is also written $p \xrightarrow{u:v} q$. The run is *successful* if $p \in I$

and $q \in F$. The relation realized by $\mathcal{T}$, denoted by $|\mathcal{T}|$, is the subset of $A^* \times B^*$ consisting of the labels of successful runs. A state is *useful* if it occurs in some successful run. Such states can be computed in linear time. Since the realized relation remains unchanged when useless states are removed, we always assume in the sequel that the set Q only contains useful states. A relation $R \subseteq A^* \times B^*$ is said to be *rational* if it equal to $|\mathcal{T}|$ for some transducer $\mathcal{T}$. A transducer is *normalized* if each of its transitions is either labelled by a pair (a, ε) where $a \in A$ or by a pair (ε, b) where $b \in B$. It is well known [11, Thm IV.1.1] that any transducer can be transformed into an equivalent normalized one. All transitions of the form $p \xrightarrow{\varepsilon:\varepsilon} q$ are first removed in the same way as ε-transitions are removed from automata. Then each transition $p \xrightarrow{u:v} q$ where $u = a_1 \cdots a_m$ and $v = b_1 \cdots b_n$ is replaced by transitions $p \xrightarrow{a_1:\varepsilon} q_1 \cdots q_{m-1} \xrightarrow{a_m:\varepsilon} q_m \xrightarrow{\varepsilon:b_1} q_{m+1} \cdots q_{m+n-1} \xrightarrow{\varepsilon:b_n} q$ where $q_1, \ldots, q_{m+n-1}$ are newly introduced states. In this paper, we are interested in transducers realizing selecting relations.

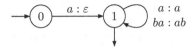

Fig. 2. Transducer realizing a selecting relation

Example 1. It can be checked that the transducer depicted in Fig. 2 realizes a selecting relation.

3   Shifts and Delays

The aim of the section is to introduce the notion of delay of runs in a transducer realizing a selecting relation. We begin by defining the shift and delay of a pair of factorized words with respect to a given embedding. This allows us to define, first the delay of a run with respect to a given embedding and second, the delay of this run as the minimal delay over all possible embeddings.

3.1   Definitions

So, we start by defining the *shift* and *delay* of a pair of factorizations $u = xx'$ and $v = yy'$ of words such that $v|u$. Let $v = b_1 \cdots b_n$ be a subword of $u = a_1 \cdots a_m$. Let us denote by r and s the lengths of x and y. We now define the *shift* and the *delay* of an embedding ι of v in u. The set $\{j : \iota(j) \leqslant r\}$ is an interval of the form $\{1, \ldots, k\}$ for some integer $0 \leqslant k \leqslant n$. The position k is the rightmost position of v which is mapped to a position of x by the embedding. If no letter of v is mapped to a letter of x, then $k = 0$. If $k = s$, the shift of the embedding is the empty word and the delay is zero. If $k > s$, the shift is said to be *negative*, it is the prefix $b_{s+1} \cdots b_k$ of y' and the delay is the length of this prefix. If $k < s$, the shift is said to be *positive*, it is the suffix $b_{k+1} \ldots b_s$ of y and the delay is the

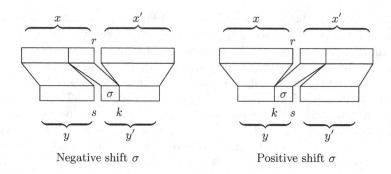

Negative shift σ Positive shift σ

Fig. 3. Negative and positive shifts

length of this suffix. An empty shift is considered as either positive or negative. The delay is always the length of the shift.

We now come to the delay of runs in a transducer. Let $T = \langle Q, A, B, \Delta, I, F \rangle$ be a transducer realizing a selecting relation. Consider a successful run $\rho = q_0 \xrightarrow{u_1:v_1} q_1 \cdots q_{n-1} \xrightarrow{u_n:v_n} q_n$ of T. Consider also an embedding ι of $v = v_1 \cdots v_n$ in $u = u_1 \cdots u_n$. For each integer $1 \leqslant j \leqslant n - 1$, we consider the factorization $u = x_j x_j'$ and $v = y_j y_j'$ where $x_j = u_1 \cdots u_j$, $x_j' = u_{j+1} \cdots u_n$, $y_j = v_1 \cdots v_j$ and $v_j' = v_{j+1} \cdots v_n$. With these factorizations and the embedding ι, is associated a negative or positive shift $\sigma(\rho, \iota, j)$ and a delay $d(\rho, \iota, j)$. Then, by definition, the *delay* of ρ with respect to ι is $d(\rho, \iota) = \max\{d(\rho, \iota, j) : 1 \leqslant j \leqslant n - 1\}$. Finally, the delay $d(\rho)$ of ρ is the minimum over all possible embeddings ι of v in u of $d(\rho, \iota)$. A transducer is said to be of *bounded delay* if there is an integer N such that the delay of any of its successful run is bounded by N.

3.2 Reformulation in the Free Group

The free group provides a easy way to denote shifts in a run that we now describe. It allows us to have uniform formulations for negative and positive shifts. We start by briefly recalling the definition of the free group F_A over some alphabet A.

Let A be alphabet. We denote by $\bar{A}$ a disjoint copy of the alphabet A, that is $\bar{A} = \{\bar{a} : a \in A\}$. The function which maps each letter a to $\bar{a}$ is first extended to an involution of $A + \bar{A}$ by setting $\bar{\bar{a}} = a$. It is then extended further to an involution of $(A + \bar{A})^*$ by setting $\bar{w} = \bar{a}_n \cdots \bar{a}_1$ for each word $w = a_1 \cdots a_n$. Note that this involution in an anti-morphism: $\overline{uv} = \bar{v}\bar{u}$. The free group F_A generated by A is the group obtained by quotienting the monoid $(A + \bar{A})^*$ by the congruence generated by all the pairs $\{(a\bar{a}, \varepsilon) : a \in A + \bar{A}\}$. Its element can be identified with the reduced words of $(A + \bar{A})^*$, that is words with no factor of the form $a\bar{a}$ or $\bar{a}a$. Suppressing occurrences of factors of that form from a word w until there is no occurrence yields a unique reduced word. The inverse of an element w is $\bar{w}$. We call an element of F_A *positive* (respectively *negative*) if its reduced word belongs to A^* (respectively $\bar{A}^*$).

We now come to the use of the free group to denote shifts. Let u and v be two words such that $v|u$. Given two factorizations $u = u_1 u_2$ and $v = v_1 v_2$, and an embedding of v in u, we denote by σ the associated shift which is a word over A. We now associate with σ an element w of the free group as follows. If the shift is positive, we set $w = \sigma$ and if it is negative, we set $w = \bar{\sigma}$. In both cases, $v_1 \bar{w}$ and $w v_2$ are positive and satisfy $v_1 \bar{w}|u_1$ and $w v_2|u_2$. This notion simplifies notations in constructions and proofs given later.

Let $\mathcal{T} = \langle Q, A, B, \Delta, I, F \rangle$ be a transducer realizing a selecting relation. Consider a successful run $\rho = q_0 \xrightarrow{u_1 : v_1} q_1 \cdots q_{n-1} \xrightarrow{u_n : v_n} q_n$ of $\mathcal{T}$ and an embedding ι of $v = v_1 \cdots v_n$ in $u = u_1 \cdots u_n$. As described before, there is, for each integer $1 \leqslant j \leqslant n-1$, a shift $\sigma(\rho, \iota, j)$. Let w_j be defined by $w_j = \sigma(\rho, \iota, j)$ if the shift is positive and $w_j = \overline{\sigma(\rho, \iota, j)}$ if it is negative. Setting $w_0 = w_n = \varepsilon$, it is easily seen that $w_{j-1} v_j \bar{w}_j$ is positive and that $w_{j-1} v_j \bar{w}_j | u_j$ for each $1 \leqslant j \leqslant n$.

4   Selectors

A transducer is a *selector* if each of its transitions is labelled by a pair (u, v) of words such that $v|u$. The output is then obtained by selecting some symbols of the input and deleting the others. This justifies the terminology. Clearly, the relation realized by a selector is a selecting one. Our main result is a converse, namely that any selecting relation can be realized by a selector.

It can be easily assumed that each transition of a selector is labelled by either a pair (a, a) or a pair (a, ε) for some symbol a. Indeed, if a transition has an input word of length greater than 1, it can be replaced by a bunch of transitions using newly introduced states, one transition for each symbol of the input word.

As shown by Example 1, it is not true that any transducer realizing a selecting relation is a selector. The transducer of Fig. 2 has indeed a transition $1 \xrightarrow{ba:ab} 1$ although it realizes a selecting relation. However, the following theorem states that any selecting relation is realized by a selector.

Theorem 2. *Any rational selecting relation can be realized by a selector.*

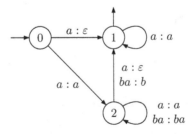

Fig. 4. A selector realizing the same relation as in Example 1

Example 3. The transducer depicted in Fig. 4 is a selector realizing the relation considered in Example 1.

The proof of the theorem is carried out as follows. For a given transducer T, we define from T a sequence of selectors T_n. The construction is in the same spirit as the resynchronization lemma given in [11, Lemma IV.6.2] but it is based on transducers rather than on rational expressions. Each selector T_n realizes a sub-approximation of the relation realized by T and each of these approximations improves on the previous one, that is $|T_n| \subseteq |T_{n+1}| \subseteq |T|$. We finally show that if T realizes a selecting relation, then T_n realizes the same relation as T for n large enough. The transducer depicted in Fig. 4 is actually the selector T_1 obtained from the transducer T depicted in Fig. 2.

We now come to the definition of the selector T_n. Let $T = \langle Q, A, B, \Delta, I, F \rangle$ be a transducer and let n be a non-negative integer. The transducer T_n is defined as follows. Its state set is $Q_n = Q \times (A^{\leqslant n} + \bar{A}^{\leqslant n})$. Its initial and final states are respectively $I \times \{\varepsilon\}$ and $F \times \{\varepsilon\}$. Its set of transitions Δ_n is given by

$$\Delta_n = \{(q, w) \xrightarrow{u:v} (q', w') : v | u \text{ and } q \xrightarrow{u:\bar{w}vw'} q' \text{ in } T\}.$$

In the definition above, using the label $\bar{w}vw'$ in the transition $q \xrightarrow{u:\bar{w}vw'} q'$ assumes implicitly that $\bar{w}vw'$ computed in the free group F_A is a positive word. In the constructed transducer, some states might be useless as they cannot occur in a successful run. These useless states and all transitions involving them are implicitly removed from T_n. The transducer T_0 has the same state set as T but its transition set Δ_0 only contains transitions $p \xrightarrow{u:v} q$ of T where $v | u$.

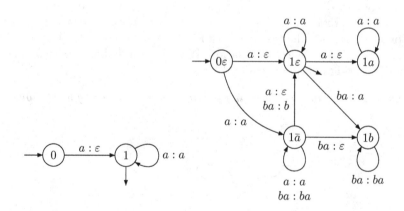

Fig. 5. Transducers T_0 and T_1

Example 4. Consider the transducer T given in Example 1. The first transducer depicted in Fig. 5 is indeed T_0. The second one is the transducer T_1 with some extra useless states. If useless states $1a$ and $1b$ are removed, one gets T_1 which is the selector given in Example 3.

The following lemma allows us to show that the relation realized by T_n is contained in the relation realized by T.

Lemma 5. *If there is run* $(q, w) \xrightarrow{u:v} (q', w')$ *in* T_n, *then* $v|u$ *and there is a run* $q \xrightarrow{u:\bar{w}vw'} q'$ *in* T.

Corollary 6. *One has* $|T_n| \subseteq |T|$.

Example 7. The relation realized by the selector T_0 of Example 4 is strictly contained in the relation realized by T. The selector T_1 realizes the same relation as T.

The following remark is useful in constructing runs in T_n. Let $q \xrightarrow{u:v} q'$ be a transition of T and let w and w' be negative or positive elements of the free group of length at most n such that $wv\bar{w}'$ is positive and $wv\bar{w}'|u$. Then $(q, w) \xrightarrow{u:wv\bar{w}'} (q', w')$ is a transition of T_n. The following lemma is a partial converse of Lemma 5.

Lemma 8. *Let* T *be a transducer realizing a selecting relation. Let* $q \xrightarrow{u:v} q'$ *be a successful run in* T *of delay* d. *If* $d \leqslant n$, *there is a successful run* $(q, \varepsilon) \xrightarrow{u:v} (q', \varepsilon)$ *in* T_n.

Lemma 9. *Let* T *be a normalized transducer realizing a selecting relation. For any successful run* $\rho = q_0 \xrightarrow{u:v} q_n$, *the leftmost embedding of* v *in* u *does not yield any positive shift of length greater than the number of state of* T.

Obviously, a similar result with rightmost embeddings and negative shifts also holds. However, this does not imply that there always exists an embedding with bounded delay. Nevertheless, this result holds and it is proved in the following lemma.

Lemma 10 (Fundamental). *Let* T *be a normalized transducer realizing a selecting relation. Then, the delay of any successful run of* T *is bounded by the number of states of* T.

5   Decidability

In this section, we show that this is decidable whether a given rational relation is selecting or not.

Let us recall that, for a transducer T, the transition set Δ_n of the transducer T_n is given by the following formula.

$$\Delta_n = \{(q, w) \xrightarrow{u:v} (q', w') : v|u \text{ and } q \xrightarrow{u:\bar{w}vw'} q' \text{ in } T, |w|, |w'| \leqslant n\}.$$

This allows us to define a function π from Δ_n to the set Δ transitions of the transducer T which maps each transition $(q, w) \xrightarrow{u:v} (q', w')$ to the transition $q \xrightarrow{u:\bar{w}vw'} q'$. This function π is then extended to a morphism from Δ_n^* to the set Δ^*. Since π maps consecutive transitions of T_n to consecutive transitions of T, it also maps runs in T_n to runs in T. Furthermore, it maps each successful run $(q, \varepsilon) \xrightarrow{u:v} (q', \varepsilon)$ in T_n to the successful run $q \xrightarrow{u:v} q'$ in T with the same label (u, v).

We consider each run in T (respectively in T_n) as a word over the alphabet Δ (respectively Δ_n). We respectively denote by E and E_n the set of successful runs in T and T_n. As sets of words over Δ and Δ_n, the two sets E and E_n are rational. Indeed, E is accepted by the transducer T considered as an automaton over Δ in the following way. Each transition $\tau = p \xrightarrow{u:v} q$ of T is replaced by the transition $p \xrightarrow{\tau} q$ in the automaton. Similarly the transducer T_n is considered as an automaton accepting E_n.

Lemma 11. *Let T be a normalized transducer with N states. Let E and E_N be set of successful runs in T and T_N. The relation realized by T is selecting if and only if $\pi(E_N) = E$.*

Theorem 12. *It can be decided whether the relation realized by a given transducer is selecting.*

Regarding the complexity of the decision procedure, the following remarks can be done. The size of a transition $p \xrightarrow{u:v} q$ is defined as $1 + |uv|$. The size of a transducer is then the sum of the sizes of all its transitions. The first step of the decision procedure is the normalization of the transducer T which may increase its size quadratically. The size of T_N is then exponential in the size of T. Checking equality $\pi(E_N) = E$ can be done in exponential time in the size of T and T_N. The complete decision procedure is then doubly exponential in the size of the given transducer.

6   Extensions and Open Problems

As a conclusion, we consider some natural questions about our result. These questions deal with variants as well as extensions. In particular, the natural extension to infinite words fails unexpectedly. We begin by some variants.

6.1   Variants

We present here two variants of our result. The first one gives raise easily to the same conclusion whence the second one leaves the problem open.

Let us consider the set $G = \{(a, a) : a \in A\} \cup \{(a, \varepsilon) : a \in A\}$. Our main theorem states that if a relation R is contained in G^*, then it can be realized by a transducer with transitions labelled by elements of G^*. This result can be extended to any given subset $H \subseteq G$ as follows. Let R be a relation such that $R \subseteq H^*$. By Theorem 2, there is a selector T realizing R. It can be easily checked that for each transition $p \xrightarrow{u:v} q$ of T, the pair (u, v) belongs to H^*.

We now turn to the second variant. Let us fix a subset I of $(A \times A) \cup (A \times \{\varepsilon\})$. A word $v = b_1 \cdots b_n$ is a *subword with respect to* I of a word $u = a_1 \cdots a_m$ if there exists a strictly increasing function ι from $\{1, \ldots, n\}$ to $\{1, \ldots, m\}$ such that, for each $1 \leqslant k \leqslant m$, if $\iota(j) = k$ for some $1 \leqslant j \leqslant n$ then $(a_k, b_j) \in I$ and if $\iota^{-1}(k) = \varnothing$ then $(a_k, \varepsilon) \in I$. This is denoted by $v|_I u$. If $I = G$, the relation $|_I$ is the classical subword relation $|$. If the set I contains $A \times \{\varepsilon\}$, our proof can be adapted to get the result. It is an open problem whether our result can be extended to the general setting. In the same vein, many variants of the subword relation could be considered.

6.2   Finitely Generated Monoid

We show here that a very general extension of our result as a Fatou property is not possible. It would be tempting to state that if a rational relation R is contained in a finitely generated sub-monoid M of $A^* \times B^*$, then R can be realized by a transducer transitions of which are labelled by elements of M. Unfortunately, this fairly general statement does not hold as the following example shows.

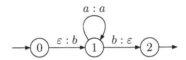

Fig. 6. Transducer realizing $(\varepsilon, b)(a, a)^*(b, \varepsilon)$

Example 13. Let A be the alphabet $\{a, b\}$ and let R be the relation $\{(a^n b, b a^n) : n \geqslant 0\} = (\varepsilon, b)(a, a)^*(b, \varepsilon)$. It is contained in the sub-monoid $M = \{(u, v) : |u|_b = |v|_b\}$ of $A^* \times A^*$ which is generated by the its finite subset $\{(\varepsilon, a), (a, \varepsilon), (b, b)\}$. However, R cannot be realized by a transducer labelled by elements of M.

A first open problem is to find a characterization of the finitely generated sub-monoids M of $A^* \times B^*$ for which the Fatou property holds. So far, for each sub-monoid M of $A^* \times B^*$ that has been proved to have the Fatou property, it is also decidable whether a given rational relation R is contained in M. This naturally raises the second problem whether the Fatou property implies the decidability of the inclusion $R \subseteq M$.

6.3   Infinite Words

We consider finally the extension of our result to infinite words. Surprisingly, the result cannot be extended to infinite words. As the following example shows, there exists a rational selecting relation on infinite words that cannot be realized by a selector.

Example 14. Let A be the alphabet $\{a, b\}$ and let R be the relation on A^ω defined by

$$R = \{(a^n (ba)^\omega, ba^n b^\omega) : n \geqslant 0\} = (\varepsilon, b)(a, a)^*(ba, b)^\omega.$$

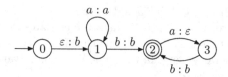

Fig. 7. Büchi transducer realizing $(\varepsilon, b)(a, a)^*(ba, b)^\omega$

This relation is realized by the transducer depicted in Fig. 7 and equipped with a
Büchi acceptance condition [10]. For each integer n, the output ω-word $ba^n b^\omega$ is
a subword of the input ω-word $a^n(ba)^\omega$. However, this relation cannot be realized
by a transducer labelled by elements from $\{(a, \varepsilon) : a \in A\} \cup \{(a, a) : a \in A\}$.

References

1. Agafonov, V.N.: Normal sequences and finite automata. Soviet Mathematics Doklady **9**, 324–325 (1968)
2. Berstel, J.: Transductions and Context-Free Languages. B.G. Teubner (1979)
3. Carton, O.: The growth ratio of synchronous rational relations is unique. Theoret. Comput. Sci. **376**, 52–59 (2007)
4. Choffrut, C., Karhumäki, J.: On Fatou properties of rational languages. In: Where Mathematics, Computer Science, Linguistics and Biology Meet: Essays in Honour of Gheorghe Paun. Kluwer Academic Publishers (2001)
5. Eilenberg, S.: Automata, Languages and Machines, vol. A. Academic Press, New York (1972)
6. Elgot, C.C., Mezei, J.E.: On relations defined by generalized finite automata. IBM Journal Res. and Dev. **9**, 47–68 (1965)
7. Higman, G.: Ordering by divisibility in abstract algebra. Proc. London Math. Soc. **2**, 326–336 (1952)
8. Leguy, J.: Transductions rationnelles décroissantes. R.A.I.R.O.-Informatique Théorique et Applications **5**, 141–148 (1981)
9. Lothaire, M.: Combinatorics on Words, Encyclopedia of Mathematics and its Applications, vol. 17. Addison-Wesley, Reading (1983)
10. Perrin, D., Pin, J.É.: Infinite Words. Elsevier (2004)
11. Sakarovitch, J.: Elements of Automata Theory. Cambridge University Press (2009)

A Hierarchy of Transducing Observer Systems

Peter Leupold[1]([⊠]) and Norbert Hundeshagen[2]

[1] Institut für Informatik, Universität Leipzig, Leipzig, Germany
Peter.Leupold@web.de
[2] Fachbereich Elektrotechnik/Informatik, Universität Kassel, Kassel, Germany
hundeshagen@theory.informatik.uni-kassel.de

Abstract. We mainly investigate the power of weight-reducing string-rewriting systems in the context of transducing observer systems. First we relate them to a special type of restarting transducer. Then we situate them between painter and length-reducing systems. Further we show that for every weight-reducing system there is an equivalent one that uses only weight-reducing painter rules. This result enables us to prove that the class of relations that is computed by transducing observer systems with weight-reducing rules is closed under intersection.

1 Transducing by Observing

The paradigm of *Computing by Observing* was originally introduced for generating and accepting formal languages [3,4]. The basic architecture is depicted in Figure 1. At the basis there is some system that evolves in discrete steps. Every configuration of this system is mapped to one single letter by the so-called observer. In this way a kind of protocol of the computation is built. This idea of observing and writing a protocol translates maybe even more naturally into transductions if we consider the input and the observation as a pair.

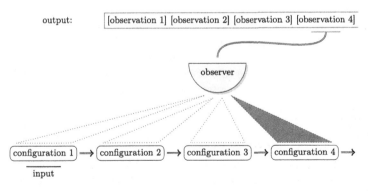

Fig. 1. Schematic representation of a transducing observer system

This approach was first investigated with painter string-rewriting systems as underlying systems [5]. They turned out to be quite powerful and compute

© Springer International Publishing Switzerland 2015
A.-H. Dediu et al. (Eds.): LATA 2015, LNCS 8977, pp. 727–738, 2015.
DOI: 10.1007/978-3-319-15579-1_57

classes of transductions that are beyond the best-known classes like rational and pushdown transductions. This corresponds to the fact that they are equivalent to linear bounded automata, when we use them for accepting languages. In the sequel, length-reducing string-rewriting systems are shown to compute a smaller class of transductions that is actually similar to the one computed by RRWW-transducers [6], another recently introduced model for computing word transductions based on string-rewriting systems [7]. The main problem in showing the equivalence between these two classes was that a length-reducing system can make at most n reduction steps on a string of length n. In this respect weight-reducing systems are slightly more permissive, because they can allow a number of steps that is linear in the length of the input string. Therefore, they are a good candidate for further investigations on transductions realized by string-rewriting systems that are linearly bounded in the length of their derivations. For restarting automata, the relation between length-reducing and weight-reducing rewriting is a longstanding open question [9]. Thus, investigating these systems in the context of transductions might open up new insights into the latter question.

Recently, the Computing by Observing architecture has also received interest from an entirely different perspective: it formalizes the concept of observer that was postulated by Searle as an indispensable constitutent of any computation [14]. Here the architecture offers ways to reason about such observers not only philosophically but in a formal approach [10,11].

2   Transducing Observer Systems

We use standard terminology and notations from Formal Language Theory as they are exposed for example by Salomaa [13]. Concerning transductions we mainly follow Aho and Ullman [1]; in contrast to this standard approach, we exempt pairs with the empty word on the left hand side and some nonempty right hand side. Since the input is the space on which we computate, such pairs might make classes different that are equal for all other pairs.

The observed systems in our architecture will be string-rewriting systems. Concerning these we follow notations and terminology as exposed by Book and Otto [2]. A *string-rewriting system* W on an alphabet Σ is a subset of $\Sigma^* \times \Sigma^*$. Its elements are called rewrite rules, and are written either as ordered pairs (ℓ, r) or as $\ell \to r$ for $\ell, r \in \Sigma^*$. In the role of observers we use a variant of the devices that have become standard in this function: monadic transducers.

Definition 1. A *generalized monadic transducer* is a tuple $\mathcal{O} = (Q, \Sigma, \Delta, \delta, q_0, \phi)$, where the set of states Q, the input alphabet Σ, the transition function δ, and the start state q_0 are the same as for deterministic finite automata. Δ is the output alphabet, and ϕ is the output function, a mapping $Q \mapsto \Delta^*$ which assigns an output word or the empty word to each state. The class of all generalized monadic transducers is denoted by $g\mathcal{MT}$.

The mode of operation is as follows: the monadic transducer reads the input word; then the output is the image under ϕ of the state it stops in. For a sequence of strings $(w_1, w_2, \ldots, w_k)$ we write $\mathcal{O}(w_1, w_2, \ldots, w_k)$ for $\mathcal{O}(w_1) \cdot \mathcal{O}(w_2) \cdots \mathcal{O}(w_k)$.

Now we combine the two components, a string-rewriting system and a generalized monadic transducer in the way described in the introduction.

Definition 2. A *transducing observer system*, short T/O system is a triple $\Omega = (\Sigma, W, \mathcal{O})$, where Σ is the input alphabet, W is a string-rewriting system over an alphabet Γ such that $\Sigma \subseteq \Gamma$ which consists of all the symbols that occur in the rule set W, and the *observer* $\mathcal{O}$ is a generalized monadic transducer, whose input alphabet is Γ.

The mode of operation of a transducing observer system $\Omega = (\Sigma, W, \mathcal{O})$ is as follows: the string-rewriting system starts to work on an input word u. After every reduction step the observer reads the new string and produces an output. The concatenation of all observations of a terminating derivation forms the output word v. The relation that Ω computes consists of all possible pairs (u, v). Note that already the input string is the first observation; thus there can be an output even if no rewriting rule can be applied to the first string.

Further, the observer is equipped with an important feature: By outputting the special symbol $\perp$ it can abort a computation. In that case no output is produced. The other way in which no output might be produced is, if the string-rewriting system does not terminate. Formally, the relation computed is

$$\mathrm{Rel}(\Omega) = \{(u, v) \mid \exists w : w \in W(u) \text{ and } v = \mathcal{O}(w) \text{ and } |v|_\perp = 0\},$$

where $W(u)$ denotes all sequences of words $(u, u_2, \ldots, u_k)$ that form terminating derivations $u \Rightarrow_W u_2 \Rightarrow_W \cdots \Rightarrow_W u_k$ of W; $|v|_\perp$ is the number of occurrences of $\perp$ in v. Thus $\mathrm{Rel}(\Omega)$ consists of all pairs of input words combined with the observations of possible terminating derivations on the given input word.

Different types of string-rewriting rules result in different types of transducing observer systems. Here we will use three types of string-rewriting systems. A string-rewriting system is called a *painter* system if for all its rules (ℓ, r), we have $|\ell| = |r| = 1$, that is, every rule just replaces one letter by another one. A string-rewriting system over an alphabet Σ is called *weight-reducing* if there exists a weight function $\omega : \Sigma \mapsto \mathbb{N}_+$ from the alphabet to the set of all positive integers such that for all the rewrite rules (ℓ, r) we have $\omega(\ell) > \omega(r)$. Thus every rule reduces the string's weight by at least one. Finally, a string-rewriting system is called *length-reducing* if for all its rules (ℓ, r) we have $|\ell| > |r|$, that is, every rule shortens the string by at least one symbol.

The class of all string-rewriting systems that have painter, weight-reducing, and length-reducing rules are denoted by pnt-*SRS*, wr-*SRS*, and lr-*SRS*, respectively. The corresponding transducing observer systems are denoted by pnt-T/O, wr-T/O, and lr-T/O. Finally and in what follows the class of relations defined by a system of a special type is denoted by $\mathcal{R}\mathrm{el}([\text{system}])$.

In order to see how a transducing observer system works, we first look at an example that uses painter rules.

Example 3. We construct a transducing observer system that computes the relation $\{(a^n, (a^n b^n)^n) \mid n \geq 2\}$. Note that here the right-hand sides of the pairs are much longer than the left-hand sides; more precisely, a left-hand side of length n has a right-hand side of length $2n^2$.

The tactics our system follows to generate the right-hand side is the following: if the input consists only of letters a, then these are changed to A from left to right one by one in n steps. During each of these steps, the entire string is marked with underline from left to right; every time a letter is marked the output a is produced. Then the marks are removed from left to right and in every step b is output. When the entire input string has been converted to A this will have produced exactly one factor $a^n b^n$ for every input letter.

The string-rewriting rules that we employ are $a \to \alpha$ and $\underline{\alpha} \to \underline{A}$ for converting the input letters a in two steps to A; the rules $a \to \underline{a}$, $\alpha \to \underline{\alpha}$, and $A \to \underline{A}$ are there for marking the letters; finally, $\underline{a} \to a$ and $\underline{A} \to A$ are used for unmarking the letters. Of course, these rules could be applied in arbitrary orders by the string-rewriting system. Thus the observer must be constructed in such a way that it rejects any derivation that deviates from the sequence described above.

We introduce two mappings that help us specify the observer clauses in a more concise manner. For a string w we denote by $\overset{\hookrightarrow}{\mapsto}(w)$ the set of all strings that have the same letter sequence as w and have exactly one non-empty continuously underlined factor that starts in w's first letter. So for a string of length n we obtain $n-1$ different strings. $\overset{\hookleftarrow}{\mapsto}(w)$ is the symmetric version denoting underlines that start somewhere in the string and are continuous till its end. Both apply also to sets of strings in the obvious way. Finally we must introduce a special symbol f by a rule $A \to f$, which cannot be rewritten anymore. The application of this rule results in irreducible strings of the form f^* that are needed to stop the system.

The observer realizes the following mapping:

$$
\mathcal{O}(w) = \begin{cases}
\varepsilon & \text{if } w \in a^+ \cup A^+ a^* \cup A^+ \alpha a^* \cup (A \cup f)^+, \\
a & \text{if } w \in A^* \alpha a^* \cup \overset{\hookrightarrow}{\mapsto}(A^* \alpha a^*), \\
b & \text{if } w \in \overset{\hookleftarrow}{\mapsto}(A^*) \underline{\alpha a}^* \cup A^+ \overset{\hookleftarrow}{\mapsto}(a^*), \\
\bot & \text{else}
\end{cases}
$$

In this way a derivation sequence

$$AAAaaa \Rightarrow AAA\alpha aa \Rightarrow^* \underline{AAA\alpha}aa \Rightarrow^* AAA\underline{\alpha a}a \Rightarrow AAA\underline{Aa}a \Rightarrow^* AAAAaa$$

produces exactly the output $a^6 b^6$. Further it is important to notice that the application of any rule out of this order leads to a string that is not treated in the first three clauses of $\mathcal{O}$ and thus results in the abortion of the computation. For example, if in the first string $AAAaaaa$ any symbol is underlined or an a different from the left-most one is converted to α, this leads to a rejection. Other rule applications are not possible, and for the other configurations the situation is similar. □

3   The Relation to Restarting Transducers

One aim of the work on transducing observer systems has been the search for new characterizations of known classes of transductions. We have been especially interested in relations to *restarting transducers* which were derived from restarting automata that were introduced by Jancar et al. [8]. A *restarting transducer* (RRWW-Td for short) is a 9-tuple $T = (Q, \Sigma, \Delta, \Gamma, \math065, \$, q_0, k, \delta)$ where Q is the finite set of states, Σ and Γ are the finite input and tape alphabet, Δ is the finite output alphabet, $\math065, \$ \notin \Gamma$ are the markers for the left and right border of the tape, $q_0 \in Q$ is the initial state and $k \geq 1$ is the size of the read/write window. Additionally the transition function δ is defined by:

$$\delta : Q \times \mathcal{PC}^{(k)} \rightarrow \mathcal{P}(Q \times (\{\mathsf{MVR}\} \cup \mathcal{PC}^{\leq(k-1)}) \cup \{\mathsf{Restart}, \mathsf{Accept}\} \times \Delta^*),$$

where $\mathcal{PC}^{(k)}$ denotes the set of possible contents (over Γ) of the read/write window of T. The transducer works in cycles, where each cycle is a combination of a number of move-right-steps, one rewrite-step and a restart- or accept-step. In case of a tail computation the rewrite-step is optional.

Every rewrite step of the form $(q, v) \in \delta(p, u)$ shortens the tape, i. e. $|u| > |v|$. After a rewrite step is applied, the read/write window is placed immediately to the right of the string v. Further, the output of the transducer, that is a word in Δ^*, is produced during a restart-step at every end of a cycle or during an accept-step in a tail computation.

Such a transducer T defines a transduction as every input word $w \in \Sigma^*$ is mapped onto the set of words $z \in \Delta^*$ for which there exists an accepting computation of T during which the output z is produced.

Equivalences of classes computed by restarting transducers and transducing observer systems would strongly connect the two models. So far, however, only close similarities have been found; most noteworthy the following relation between $\mathcal{Rel}(\mathsf{lr\text{-}T}/\mathsf{O})$ and $\mathcal{Rel}(\mathsf{RRWW\text{-}Td})$, the class of transductions computed by RRWW-transducers, via a morphism:

Theorem 4 ([6]) . *For every relation $R \subseteq \Sigma^* \times \Delta^*$ and $R \in \mathcal{Rel}(\mathsf{RRWW\text{-}Td})$, there is a uniform morphism φ and a relation $S \in \mathcal{Rel}(\mathsf{lr\text{-}T}/\mathsf{O})$ such that $R = \{(u, v) \mid (\varphi(u), v) \in S\}$.*

Looking at weight-reducing string-rewriting systems instead of length-reducing ones, we see that the resulting class of relations contains $\mathcal{Rel}(\mathsf{RRWW\text{-}Td})$, but it is not clear whether this inclusion is proper. Before we can prove this we need to recall the formal definition of a restarting transducer.

To increase the readability of the behavior of restarting automata we use meta-instructions (see [12]). We recall that a tuple of the form $(E_1, u \mapsto u', E_2)$ mirrors the cycle of an RRWW-automaton that reads across the tape content E_1, rewrites a subword u by a shorter subword u' and finally the automaton checks if the part of the tape unseen until then corresponds to E_2. As these meta-instruction describe the rewriting behavior of an automaton they can easily be extended to restarting transducers. Now

$$(E_1, u \mapsto u', E_2; v)$$

is a restarting transducer's meta-instruction, where E_1, E_2, u, u' are defined as for the corresponding automaton and v is the output word produced at the end of this cycle. To describe accepting tail computations we use meta-instructions of the form $(E, \mathsf{Accept}; v)$.

Theorem 5 . $\mathcal{R}\mathrm{el}(\mathsf{RRWW\text{-}Td}) \subseteq \mathcal{R}\mathrm{el}(\mathsf{wr\text{-}T/O})$.

Proof. As mentioned above, for every relation $R \in \mathcal{R}\mathrm{el}(\mathsf{RRWW\text{-}Td})$ there is a relation S realized by a length reducing system and a morphism φ such that $R = \{(u, v) \mid (\varphi(u), v) \in S\}$. Actually this morphism just transforms each input word into a redundant representation with a copy of each letter, and thus it provides some additional space to simulate one cycle of the restarting transducer in two steps of the $\mathsf{lr\text{-}T/O}$-system. The latter was needed to "clean up" after applying a rule. Here we extend this idea to weight reducing systems.

Let $T = (Q, \Sigma, \Delta, \Gamma, \mathfrak{c}, \$, q_0, k, \delta)$ be an RRWW-transducer. The weight function ω is defined for all $x \in \Gamma$ as $\omega(x) = 2$. The string-rewriting rules that the $\mathsf{wr\text{-}T/O}$-system Ω uses are derived from the meta-instructions of T. Let

$$t : (E_1, u \cdot x \to u', E_2; v)$$

be such a transition for $u, u' \in \Gamma^*$ and $x \in \Gamma$. We now associate with each transition a unique label, here t. From this description we build a $\mathsf{wr\text{-}T/O}$-system $\Omega = (\Sigma, W, \mathcal{O})$ such that for each of the meta-instructions above two rules are added to $W \subseteq \Gamma'^* \times \Gamma'^*$: $u \cdot x \to u' \cdot t$ and $t \to \varepsilon$. Note that Γ is a subset of Γ', where each label $t \in \Gamma' \backslash \Gamma$. Furthermore, the weight assigned to each t equals 1. Finally for every meta-instruction of T the observer's mapping includes the clause

$$\mathcal{O}(w) = v; \text{ if } w \in E_1 \cdot u' \cdot t \cdot E_2,$$

and $\mathcal{O}(w) = \varepsilon$ if no such label t is present in w. Observe that after the application of any rule from W the weight of a string is at least decreased by 1. Further note, that in this way the proof of correctness is a direct consequence of the simulation of restarting transducers by length reducing systems, cited above. Consequently, accepting meta-instructions of T are simulated by introducing a special symbol t_a, which can not be rewritten anymore.

Finally, T rejects an input word simply by getting stuck, that is, no transition is applicable in the current configuration. As the clauses of the observer mirror directly the move-right steps of T, it will also get stuck. In this situation we have to output $\bot$ to abort the computation of Ω. This can be done by making the observer "complete", that is, the transition function of $\mathcal{O}$ is extended such that every input word w, which is not described by the regular expressions above leads to the following output:

$$\mathcal{O}(w) = \bot.$$

This completes the proof.   □

So, for the moment we have the following chain of inclusions: $\mathcal{R}el(\mathsf{lr}\text{-}\mathsf{T}/\mathsf{O}) \subseteq$ $\mathcal{R}el(\mathsf{RRWW}\text{-}\mathsf{Td}) \subseteq \mathcal{R}el(\mathsf{wr}\text{-}\mathsf{T}/\mathsf{O})$. But we do not know if either of them is proper. We suspect that this really is a chain and the three classes are different.

One indicator that $\mathcal{R}el(\mathsf{wr}\text{-}\mathsf{T}/\mathsf{O})$ is similar to $\mathcal{R}el(\mathsf{RRWW}\text{-}\mathsf{Td})$ is the fact that the class $\mathcal{R}el(\mathsf{wr}\text{-}\mathsf{T}/\mathsf{O})$ is subject to the same *length-bounded property* as the class $\mathcal{R}el(\mathsf{RRWW}\text{-}\mathsf{Td})$. This is a property that holds for every pair from relations from these classes; it bounds the length of the right-hand side in terms of the length of the left-hand side. In general, we call a relation R *length-bounded*, if there is a constant c, such that for each pair $(u,v) \in R$ with $u \neq \varepsilon$, $|v| \leq c \cdot |u|$ holds. All relations in $\mathcal{R}el(\mathsf{RRWW}\text{-}\mathsf{Td})$ are length-bounded; this can be seen in a way similar to the following proof that this property also holds for $\mathcal{R}el(\mathsf{wr}\text{-}\mathsf{T}/\mathsf{O})$.

Lemma 6. *For every relation R in $\mathcal{R}el(\mathsf{wr}\text{-}\mathsf{T}/\mathsf{O})$, there is a constant c such that for each pair $(u,v) \in R$ with $u \neq \varepsilon$, $|v| \leq c \cdot |u|$ holds.*

Proof. Let k be the sum of the weights of all the symbols in a given input string w. Every rule application in a wr-T/O system must reduce the weight by at least one. Therefore there can be at most k computation steps. In every step the observer outputs one of its output strings. If n is the length of the longest string that can be output in one step, then the total length of the output for the string w cannot exceed $k \cdot n$. If m is the highest possible weight of an input symbol, then k is at most $|w| \cdot m$. Thus setting $c = |w| \cdot m \cdot n$ makes the statement true. $\square$

4   A Hierarchy of Transductions Realized by Observer Systems

Now we compare the computational power that transducing observer systems achieve with the three types of string-rewriting systems introduced above. From earlier work we know that the relations computed by length-reducing T/O-systems are included in those that are computed by painter T/O-systems [6]. The same is true for weight-reducing systems. In order to show this we first establish a kind of normal form for weight-reducing systems.

Lemma 7. *Every transducing observer system with weight-reducing rules can be simulated by one with rules that do not increase the string's length.*

Proof. The key observation is the following: a symbol of a given weight s can only be rewritten to a maximum of $s-1$ symbols. If the maximum weight of an input symbol is $m+1$, then a string w cannot increase in length beyond $m \cdot |w|$.

So we code m symbols into one with space symbols filling the unused spots. In the beginning, every letter is written into the first one of the m slots in its position. The main problems are then how to implement rule applications and when to delete empty spaces. If we just leave empty spaces there, the distance between different parts of the left-hand side of a rule could become arbitrarily large. If we delete them too early, later expansions might become impossible.

For the latter problem we delete as late as possible. This means that a rewriting rule that has k compound symbols on its left-hand side will always write k

new symbols unless the total of coded original symbols on its right-hand side is less than k. In that case, each symbol is written by itself into one compound symbol, the remaining positions are deleted. If there are more than k original symbols, say ℓ, then we distribute them in order from left to right in such a way that the first $k - \lfloor \frac{\ell}{k} \rfloor k$ original symbols positions contain $\lfloor \frac{\ell}{k} \rfloor + 1$ original symbols; the remaining positions contain $\lfloor \frac{\ell}{k} \rfloor$ original symbols. In this way the total weight of the original symbols can never exceed the number of available slots.

For a rule in the original system we need to introduce several new rules, because we do not know what the left-hand side will look like. The original symbols can be distributed in various ways over compound symbols with space slots in between, and the factor might even start and end inside compound symbols. So for an original rule $u \to v$ we need to implement all the possible left-hand sides starting with at least one original symbol in the left-most position and ending with at least one symbol in the right-most position. The symbols between the two ends might occur in just one compound symbol each or with several in one compound symbol. All possibilities must be implemented. Because we do not have completely empty compound symbols, there are only finitely many possibilities.

Note that at the right-most and left-most positions the distribution of symbols on the right-hand side described above might theoretically not work. If the rule's left-most symbol occupies the right-most slot in a compound symbol and more than one symbol should be written into that position, then this is not possible. However, if a symbol occupies the right-most slot, then this means that some rule before (or a sequence of rules) expanded a string roughly by the factor m; this in turn means that the weight of nearly all the original symbols in this string must be one. Thus we can adapt the distribution of original letters slightly to putting only one into the left-most compound symbol. We can proceed in a similar fashion for other small numbers greater than one and on the right end of the rewritten factor.

It is obvious how the observer of the new systems must work: it reads the compound symbols and after each one it changes its state as if it had read all the original symbols contained in the compound one.   □

Theorem 8. $\mathcal{R}el(\text{wr-T}/\text{O}) \subsetneq \mathcal{R}el(\text{pnt-T}/\text{O})$.

Proof. Transducing observer systems with weight-reducing rules fulfill the length-bounded property from Lemma 6. The relation $\{(a^n, (a^n b^n)^n) \mid n \geq 2\}$ from Example 3 shows that with painter systems relations without such a linear bound can be computed. Thus the inclusion is proper.

The key for showing that the inclusion $\mathcal{R}el(\text{wr-T}/\text{O}) \subseteq \mathcal{R}el(\text{pnt-T}/\text{O})$ holds is Lemma 7, which shows us that for every wr-T/O-system there is an equivalent one that uses only non-increasing rules. On the other hand, it has been shown several times how general non-increasing (context-sensitive) rules can be simulated by painter systems. The technique was used for example in the proof of Theorem 4.1 and Corollary 4.1 on accepting observer systems [4]. Further, the proof that all relations of deterministic pushdown transducers can be computed via painter systems shows how to accommodate output in this context [5]. For technical details we refer to the referenced descriptions of these constructions.   □

So this results follows in a straight-forward manner from earlier proofs. Now we take a look at what happens, if we put a further restriction on the painter systems in Theorem 8. Namely, we require the painter rules to be weight-reducing, too. Systems with this type of rules can still simulate all systems, whose rules are weight-reducing but not necessarily of size one. Thus the length bound of one on the size of the rules does not affect the computational power in this case.

Theorem 9. $\mathcal{R}el(\mathsf{wr\text{-}T/O}) = \mathcal{R}el(\mathsf{wr\text{-}pnt\text{-}T/O})$.

Proof (Sketch). Again, following Lemma 7 we can suppose that we are dealing with a weight-reducing system without length-increasing rules. The main problem in splitting a rule that rewrites more than one symbol into several painter rules is the following: a rule like $728 \to 555$ increases the weight in the second position, although the total weight is decreased. Here and in what follows the digits are symbols with the corresponding weight. So the painter rule resulting from the rewriting in the central position is $2 \to 5$ and would be weight-increasing. The basic idea to solve this is the introduction of new symbols. In this case, between the weights 2 and 1 we would use a symbol like 2_5. The index indicates the symbol that is in that position after application of the original rule.

The main difficulty lies in an increase in weight that might return to the same symbol in a loop. Such a loop can run several, actually an unbounded number of times in the same position. Therefore no static solution like the fixed index above can work. Instead, we make use of the fact that an increase in the weight in one position must be compensated by a decrease in some other position in the same rule. In the example above there are two options: $7 \to 5$ in the first position and $8 \to 5$ in the last. So we just mark the 2 with the name of the loop and let an index run for example in the position of the 8.

Two consecutive runs of this same loop could need the same index in the same position. To avoid this, we actually introduce some additional space by a rule $8 \to \quad 5$ instead of $8 \to 5$ every time the loop is started. The resulting system is not a painter system anymore. But from another application of Lemma 7 we can obtain an equivalent painter system.

One crucial point is the definition of what we call a loop. Since we reserve a special index and use special painter rules (that mark the 2 and introduce the   in our example), we can only deal with a finite number of loops. With a sufficiently restrictive definition of loops we can ensure that there are only finitely many distinct ones.   □

As we have explained, we use the reduction of weight at the start of a loop for the (temporary) creation of some additional space. This is a technique similar to one that Jurdzinski and Otto used, when they showed the equivalence between finite-change automata and shrinking restarting-automata [9].

From the last part of the proof or simply from Lemma 7 we can see that the use of context-free rules (left-hand side of length one, right-hand side arbitrarily long) does not increase the computational power compared to the use of painter rules as long as we use only weight-reducing rules. Denoting by wr-cf- the class of all weight-reducing context-free string-rewriting systems we can slightly generalize Theorem 9.

Corollary 10. $\mathcal{R}el(\text{wr-T}/\text{O}) = \mathcal{R}el(\text{wr-pnt-T}/\text{O}) = \mathcal{R}el(\text{wr-cf-T}/\text{O})$.

This is in contrast to the situation for general painter and context-free systems. There the former allow linearly bounded computations, while the additional use of just one context-free rule leads to computational completeness [4].

The next question concerns the relation between T/O systems with length-reducing and those with weight-reducing rules. The simple observation that every length-reducing string-rewriting system is weight-reducing for the weight function that assigns weight one to every symbol places weight-reducing T/O-systems above the latter class.

Theorem 11. $\mathcal{R}el(\text{lr-T}/\text{O}) \subseteq \mathcal{R}el(\text{wr-T}/\text{O})$.

Actually, this result also follows directly from Theorem 5 and the fact that length-reducing T/O-systems can be simulated by RRWW-transducers. However, we are unable to determine whether this inclusion is proper. Intuitively, we suspect that it is. A weight-reducing system can in some sense use the input several times, while length-reducing rules necessarily consume it during the computation. To show what we mean by using the input several times, we sketch the proof of the following result.

Theorem 12. *The class $\mathcal{R}el(\text{wr-T}/\text{O})$ is closed under intersection.*

Proof. After Theorem 9 we can assume that all relations in wr-T/O are computed by systems that only use painter rules. We only sketch how a wr-pnt-T/O can compute the intersection between two relations that are computed by systems from this class, let us call them Ω_1 and Ω_2. Further, let their observers be $\mathcal{O}_1$ and $\mathcal{O}_2$, respectively. For both relations the length-bounded property holds, and let k be a constant that fulfills this property for both relations.

The system that computes the intersection works on four tapes that are simulated on the symbols of the input string. First, two copies of the input string w are created. Then the computation of Ω_1 on w is simulated and the output is written on the third tape. Here we need to write k symbols into one. Then the output of Ω_2 on w is computed and written on the fourth tape.

It remains to compare the two outputs. Only if they are equal we now output exactly this string. During the steps before no output has been produced. Since Ω_1 and Ω_2 are both weight-reducing, appropriate weights can be chosen such that the two successive computations are simulated by weight-reducing rules. Also storing the outputs can be done in a weight-reducing way, since symbols that have been written are not changed any more; thus the empty output should produce the highest weight, every further symbols from left to right should decrease the weight. □

The same tactics does not work in a direct way for length-reducing rules, just as it would not work for general weight-reducing rules. Already the computation of the first output could reduce the string to just one symbol, and no copy can be stored on a second track. It also does not seem to be possible to simulate the two computations in parallel, since the rewritings might take place at different positions. However, it is not clear how to formalize this. Therefore the properness of the inclusion remains an open problem.

This parallels the situation for RRWW-automata. These are like RRWW-transducers, but do not produce output; rather they accept or reject a word. There, automata with length-reducing rules can be simulated by those with weight-reducing rules (which are called shrinking restarting automata) [9]. But it remains unknown whether the weight-reducing rules lead to bigger computational power.

5   Conclusions

The results of the preceding sections are summarized in Figure 2. Lines indicate inclusions from bottom to top, arrows indicate proper inclusions. Dotted arrows show the relation via morphisms in the style of Theorem 4.

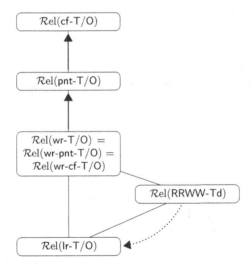

Fig. 2. The hierarchy of transducing observer systems with the relation to RRWW-tranducers

What we have not treated explicitly is the relation of the two classes on top, but it follows from results on accepting observer systems. Painters are equally powerful as linear bounded automata, context-free rules allow the simulation of arbitrary Turing Machines [4]. Thus $\mathcal{R}el(\text{pnt-T/O})$ is related to the class of transduction that can be computed by Turing Machines with a linear space bound, while $\mathcal{R}el(\text{cf-T}/O)$ are all computable relations; these two classes are obviously not equal.

There are mainly three questions left open; these regard the properness of the inclusions $\mathcal{R}el(\text{lr-T}/O) \subseteq \mathcal{R}el(\text{wr-T}/O)$, $\mathcal{R}el(\text{lr-T}/O) \subseteq \mathcal{R}el(\text{RRWW-Td})$, and $\mathcal{R}el(\text{RRWW-Td}) \subseteq \mathcal{R}el(\text{wr-T}/O)$. As argued above and in earlier work [6], we expect all three of these inclusions to be proper. Theorem 12 might be a hint that $\mathcal{R}el(\text{RRWW-Td})$ and $\mathcal{R}el(\text{wr-T}/O)$ are different, because the former might not be able to compute intersections. We also doubt that the morphism in Theorem 4 can be eliminated, which would mean that also $\mathcal{R}el(\text{lr-T}/O)$ and $\mathcal{R}el(\text{RRWW-Td})$ are different.

References

1. Aho, A., Ullman, J.: The Theory of Parsing, Translation, and Compiling. Prentice-Hall, Upper Saddle River (1972)
2. Book, R., Otto, F.: String-Rewriting Systems. Springer, Berlin (1993)
3. Cavaliere, M., Leupold, P.: Evolution and Observation – A Non-Standard Way to Generate Formal Languages. Theoretical Computer Science **321**, 233–248 (2004)
4. Cavaliere, M., Leupold, P.: Observation of String-Rewriting Systems. Fundamenta Informaticae **74**(4), 447–462 (2006)
5. Hundeshagen, N., Leupold, P.: Transducing by observing. In: Bordihn, H., Freund, R., Hinze, T., Holzer, M., Kutrib, M., Otto, F. (eds.) NCMA. books@ocg.at, vol. 263, pp. 85–98. Österreichische Computergesellschaft (2010)
6. Hundeshagen, N., Leupold, P.: Transducing by Observing Length-Reducing and Painter Rules. RAIRO - Theor. Inf. and Applic. **48**(1), 85–105 (2014)
7. Hundeshagen, N., Otto, F.: Characterizing the rational functions by restarting transducers. In: Dediu, A.-H., Martín-Vide, C. (eds.) LATA 2012. LNCS, vol. 7183, pp. 325–336. Springer, Heidelberg (2012)
8. Jančar, P., Mráz, F., Plátek, M., Vogel, J.: Restarting automata. In: Reichel, H. (ed.) FCT 1995. LNCS, vol. 965, pp. 283–292. Springer, Heidelberg (1995)
9. Jurdzinski, T., Otto, F.: Shrinking Restarting Automata. Int. J. Found. Comput. Sci. **18**(2), 361–385 (2007)
10. Leupold, P.: Is computation observer-relative? In: Bensch, S., Freund, R., Otto, F. (eds.) Proceedings of Sixth Workshop on Non-Classical Models for Automata and Applications - NCMA 2014, Kassel, Germany, July 28–29, 2014. books@ocg.at, vol. 304, pp. 13–27. Österreichische Computer Gesellschaft (2014)
11. Leupold, P.: What is the role of the observer in a computation? In: Proceedings of 7th AISB Symposium on Computing and Philosophy: Is computation observer-relative? London (2014)
12. Otto, F.: Restarting automata. In: Ésik, Z., Martín-Vide, C., Mitrana, V. (eds.) Recent Advances in Formal Languages and Applications. SCI, vol. 25, pp. 269–303. Springer, Heidelberg (2006)
13. Salomaa, A.: Formal Languages. Academic Press, New York (1973)
14. Searle, J.R.: The Rediscovery of the Mind. MIT Press (1992)

Sublinear DTD Validity

Antoine Ndione[1,3]([✉]), Aurélien Lemay[2,3], and Joachim Niehren[1,3]

[1] INRIA, Paris, France
ndionea@gmail.com
[2] University of Lille 3, Paris, France
[3] Links Project of INRIA Lille and CRISTAL (UMR 9189 of CNRS), Paris, France

Abstract. We present an efficient algorithm for testing approximate DTD validity modulo the strong tree edit distance. Our algorithm inspects XML documents in a probabilistic manner. It detects with high probability the nonvalidity of XML documents with a large fraction of errors, measured in terms of the strong tree edit distance from the DTD. The run time depends polynomially on the depth of the XML document tree but not on its size, so that it is sublinear in most cases (because in practice XML documents tend to be shallow). Therefore, our algorithm can be used to speed up exact DTD validators that run in linear time.

Keywords: Property testing · Regular tree automata · DTD · XML

1 Introduction

Validity checking for collections of large XML documents may quickly become time consuming. With today's technology, more than 10 minutes are needed to validate a single document of more than 20 giga bytes, so that the treatment of hundreds such documents may take days or weeks. This difficulty could be overcome by sublinear algorithms that can quickly detect invalid documents without reading them entirely.

Whether sublinear algorithms for XML schema validation exist is a principle question. One approach to obtain sublinear algorithms for schema validation is to use algorithms that evaluate schemas on XML streams in an online manner [9,11]. In this manner, errors can be in sublinear time when they are localized in a prefix of sublinear size of the XML streams, but not otherwise. In contrast, our objective is to develop probabilistic approximation algorithms inspired by property testing [2,7,8] which access a random fragment of constant size only, in order to detect invalidity with high probability, if the input structure contains many errors wheresoever located.

For approximate membership testing for unranked ordered trees as with XML, we need a storage model that permits to randomly draw descendants of any node from a uniform distribution, while giving deterministic access to its first child, next sibling and parent. Such a storage model is easy to implement with the techniques from XML databases. The number of errors is measured by the

© Springer International Publishing Switzerland 2015
A.-H. Dediu et al. (Eds.): LATA 2015, LNCS 8977, pp. 739–751, 2015.
DOI: 10.1007/978-3-319-15579-1_58

minimal number of edit operations needed to repair the XML document so that it satisfies the schema, but normalized with respect to the documents size. The more edit operations are permitted, the smaller is the error measured, and the easier becomes approximate membership testing. The only positive result so far applies to testing DTD validity modulo the tree edit distance with subtree moves [7]. But this edit distance is weaker than the usual edit distance, in that it permits subtree moves beside of all usual operations. Thus the edit distance with move can be very small compared to the usual edit distance and therefore it detects less errors. Approximate membership for tree automata modulo the usual edit distance would be nice to have, but its existence was stated as an open question in [5]. It should also be noticed that property testers for graphs are usually limited to local properties [13,15].

The first contribution of this paper is an approximate membership tester for unranked tree automata modulo the usual tree edit distance [14,18], closing the open problem from [5]. Indeed such a test can be obtained by linearization of unranked trees into words. In order to show this, we use the nontrivial observation that the usual edit distance between two trees is bounded in function of the edit distance of their linearizations [1]. Thereby, can we apply the recent approximate membership tester [12] for non-deterministic finite automata (NFAs) modulo the edit distance on words. This tester improves on a previous tester by Alon, Krivelevich, and Newman [2] for the Hamming distance, so that it runs in polynomial time in the size of the automaton and the inverse error precision, and still independently of the size of the input word. However, the time complexity of the so obtained tester for appoximate membership for unranked tree automata depends exponentially on depth of the input tree. This is not a problem for shallow trees, that are frequent in the case of XML, but leaves open whether this depth dependence can be removed, or whether a polynomial depth dependence can be obtained.

The second and main contribution of this paper is an efficient probabilistic algorithm testing approximate DTD validity modulo the strong tree edit distance from [17], which restricts the usual tree edit operations to leaf insertion, leaf deletion, and node relabeling (while ruling out node inserting and deletion). Its run time depends only on the depth of the XML document but not of its size. Trivially, the same tester is also correct for all weaker distances such as the usual tree edit distance. With inputs: an error precision $\epsilon > 0$, a DTD D, and an XML documents t that is ϵ-far (normalized by the size of t) from satisfying the DTD D modulo the strong edit distance; the algorithm returns NO with high probability. It answers CLOSE for valid trees, and either CLOSE or NO for all others. The running time is polynomially bounded in the depth of t, $1/\epsilon$, and the mintree size m_D of the DTD, which is the maximum over element names $a \in \Sigma$ of the minimal sizes of a-labeled subtrees of D-valid trees. Even though m_D may grow exponentially with D, it seems to be close to the size of D for all practically relevant DTDs. Furthermore, m_D can be computed in quadratic time in the size of D, so unusual cases can be recognized efficiently and passed directly to exact DTD validity checking.

The next difficulty is that we cannot use the linearization approach for approximate membership testing modulo the *strong* tree edit distance. To remedy the situation we study weighted words, i.e., words in which all positions are assigned a weight. The edit distance on words is also lifted to weighted words, such that the costs of edit operations are given by these weights. Then, we extend the algorithm from [12] to a polynomial time NFA membership tester for weighted words modulo the stong edit distance. We next contribute a direct reduction from approximate DTD validity to approximate NFA membership of weighted words.

Outline. In Section 2 we recall preliminaries on XML data models and schemas. In Section 3, we recall edit distances for trees and words. In Section 4, we present our main result. In Section 5, we introduce weighted words, lift the edit distance, and present our tester for membership of weighted words to regular languages modulo the edit distance. In Section 6, we prove the main result. A long version with full proofs is available at https://hal.inria.fr/hal-00803696.

2   Data Models and Schemas

We recall preliminaries on the XML data model and on XML schemas.

Words. An alphabet Σ is a finite set. We denote the set of words over alphabet Σ by Σ^*. The length of a word $w \in \Sigma^n$ is $|w| = n$ and the set of its positions is $pos(w) = \{1, \ldots, n\}$. The empty word is denoted ε and $w \cdot w'$ is the concatenation of w and w'.

A *nondeterministic finite automaton with ε-transitions* (NFA) is a tuple $A = (\Sigma, Q, init, fin, \Delta)$, where alphabet Σ is a finite set, Q is a finite set of states with subsets *init* and *fin* of initial and final states, and $\Delta \subseteq Q \times (\Sigma \uplus \{\varepsilon\}) \times Q$ a transition relation.

For states q, $q' \in Q$, $\xrightarrow{\varepsilon}$ is the relation such that $q \xrightarrow{\varepsilon} q'$ if and only if $(q, \varepsilon, q') \in \Delta$. In analogy, for any $a \in \Sigma$, $q \xrightarrow{a} q'$ if and only if $(q, a, q') \in \Delta$. The relation $\xrightarrow{\varepsilon}^*$ is the reflexive transitive closure of $\xrightarrow{\varepsilon}$. The relation $\xRightarrow{a}$ includes multiple ε-transitions and a single a-transition, i.e, $\xRightarrow{a}$ is the composition of relations $\xrightarrow{\varepsilon}^* \circ \xrightarrow{a} \circ \xrightarrow{\varepsilon}^*$.

A *quasi-run* of an NFA A on a word $w = a_1 \ldots a_n$ over Σ is a function $r : pos(w) \to Q$ such that $r(i-1) \xRightarrow{a_i} r(i)$. A *run* is a quasi-run such that : $\exists q \in init, q \xrightarrow{\varepsilon}^* r(0)$. A run is called *successful* if $r(n) \in fin$. The language $\mathcal{L}(A)$ recognized by A is the set of all words w that permit a successful run.

An NFA $A = (\Sigma, Q, init, fin, \Delta)$ is *productive* if every state in Q is reachable from *init* and co-reachable from *fin*. Without loss of generality we might assume all automata input by our algorithms as productive.

A *fragment* of a word w is a subset of its positions. A fragment of consecutive positions (or without holes) is called an *interval* and denoted by $I =]i, j]$ as usual. A *factor* of w is the word located at an interval, and a *subword* is the word located at a fragment. The subword of w at fragment F is denoted wF.

An interval I of w is called *blocking* for an NFA A if starting from any state of A, every possible quasi-run of A on wI gets stuck, that is, none of those possible quasi-runs occurs in some successful run on a word in $\mathcal{L}(A)$. As an example, if A is an automaton recognizing $L = ab^*$, then the interval $]0,2]$ of aab is blocking, since after reading the first a, A cannot proceed with any second "a". Whether a fragment is *blocking* is defined similarly, except that the automaton is allowed to jump over holes to arbitrary accessible states. We can decide whether a fragment F is blocking for A in time $O(|F|\,|A|)$ without reading the entire word [12].

XML Data Model. The XML data model essentially boils down to finite unranked data trees when ignoring details of attributes, processing instructions and comments. Since we only consider structural aspects of XML documents described by DTDs, we can safely ignore data values and thus simplify the XML data model further to finite unranked trees over a finite alphabet (fixed by the DTD).

The set of unranked trees over an alphabet Σ is the least set $\mathcal{T}_\Sigma^*$ containing all tuples $a(t_1, \cdots, t_i)$ where $a \in \Sigma$ and $t_1 \cdots t_i$ in $\mathcal{T}_\Sigma^*$. The set $nod(t)$ of nodes of an unranked tree t is a prefix closed subset of $\mathbb{N}^*$ (words with labels in $\mathbb{N}$). The size $|t|$ is the number of nodes of t. As usual, we have the binary relations $parent^t$, fc^t (firstchild) and ns^t (nextsibling) on $nod(t)$. The root of t is denoted by $root^t = root = \varepsilon$ and is the unique node without parent. The i-th child of a node v is $v \cdot i$. A leaf is a node without children. The depth $d(t)$ of a tree is the maximal number of edges on paths from some leaf to the root with parent edges only. For any $v \in nod(t)$ we denote by $t_{|v}$ the subtree of t rooted at node v, and by $t[v] \in \Sigma$ the label of v. Furthermore, we define $word(t) \in \Sigma^*$ to be the sequence of labels of the children of the root of t. For instance, if $t = c(b(a,a), b(a,a,a))$ then $word(t) = bb$ and $word(t_{|root \cdot 2}) = aaa$.

Schemas. Various languages for defining schemas of XML documents were proposed in the literature. Document type descriptors (DTDs) are most basic, while XML SCHEMAS are more expressive. Our choice of DTDs is motivated by the fact that equally efficient membership testers for more expressive formalisms such as tree automata are difficult to find or may even not exist.

Standard DTDs define regular languages of unranked trees by using regular expressions. These can be compiled into NFAs in linear time, but only when permitting ε-transitions as we do [10,16]. It should also be noticed that all regular expressions in DTDs are deterministic (see the W3C recommendation). Therefore they can be converted into deterministic finite automata in polynomial time. However, this conversion might require quadratic time if not fixing the alphabet [4]. For our purpose, it is therefore advantageous to define DTDs based on NFAs with ε-transitions. In some examples we will use regular expressions for illustration nevertheless.

Definition 1. *A* DTD *D over an alphabet Σ is a tuple $(\Sigma, init, (A_a)_{a \in \Sigma})$ where init is an element of Σ, and all A_a are* NFAs *with alphabet Σ.*

An unranked tree t over Σ is valid for a DTD D iff $t[root] = init$ and $word(t_{|v}) \in \mathcal{L}(A_{t[v]})$ for all $v \in nod(t)$. We denote the set of all D-valid trees by $\mathcal{L}(D)$. For all labels $a \in \Sigma$ and DTD $D = (\Sigma, init, (A_a)_{a \in \Sigma})$, we denote by D_a the DTD $(\Sigma, a, (A_a)_{a \in \Sigma})$. The mintree size m_D is the maximum for all $a \in \Sigma$ of all minimal sizes of trees belonging to $\mathcal{L}(D_a)$: $m_D = \max_{a \in \Sigma, \mathcal{L}(D_a) \neq \emptyset} \min\{|t| \mid t \in$

$\mathcal{L}(D_a)\}$. Note that one can compute m_D in quadratic time from D even though this number might be exponentially bigger than the size of D.

3   Edit Distances

We recall the the edit distance for words and trees.

Edit Operations. The (usual) edit operations on words permit to relabel, insert, and delete a letter at a given position. The edit distance between two words w and w' is the least number of usual edit operations needed to transform w into w'. It is denoted by $e(w, w')$. The usual edit operations on trees allow for node relabelling, node inserting, and node deletion [18]. The (usual) edit distance on unranked trees t and t', that we will denote by $e_{stand}(t, t')$ is the least number of usual edit operations required to transform t into t'. The strong edit operations [17] restricts the usual edit operations to node relabelling, leaf insertion and leaf deletion. We consider a tree $t = C(a(t_1, \ldots, t_n))$ where C is a context with hole marker at node v, so that $t_{|v} = a(t_1, \ldots, t_n)$. The relabelling of v to b in t is the tree: $rel_{v,b}(t) = C(b(t_1, \ldots, t_n))$. The insertion of a b-leaf at a position $0 \leq i \leq n$ below node v yields the tree: $ins_{v,i,b}(t) = C(a(t_1, \ldots, t_i, b, t_{i+1}, \ldots, t_n))$. The deletion of a leaf $v \cdot i$ with $1 \leq i \leq n$ yields: $del_{v \cdot i}(t) = C(a(t_1, \ldots t_{i-1}, t_{i+1}, \ldots, t_n))$. The strong edit distance between two trees t and t' is the least number of strong edit operation to transform t into t'. It is denoted by $e(t, t')$.

It always holds that $e_{stand}(t, t') \leq e(t, t')$. Furthermore, $e(t, t') \leq |t| + |t'| - 1$ since we can first delete all nodes of t except the root, then relabel the root, and finally add all non-root nodes of t' one by one.

Farness. Let S be a set of structures and $e : S \times S \to \mathbb{N}_0$ a function called the distance for S. We assume that any structure $s \in S$ has a finite size $|s| \in \mathbb{N}_0$. We define the distance of a structure s to a language $L \subseteq S$ as the least number of edit operations needed to transform s into a member of L, i.e., $E(s, L) = \min_{s' \in L} E(s, s')$.

Definition 2. *Let $\epsilon > 0$ and $\mathcal{L}(A) \subseteq S$ for some language definition A. A structure s is called ϵ-far from A modulo distance E if the normalized distance of s to $\mathcal{L}(A)$ is greater than ϵ, that is if $E(s, \mathcal{L}(A))/|s| \geq \epsilon$, and ϵ-close otherwise.*

Note that ϵ-farness from a DTD D with respect to the usual edit distance implies ϵ-farness from D with respect to the strong edit distance. Furthermore, since $e(t, t') \leq |t| + |t'| - 1$ for any two unranked trees, it follows that $e(t, D_a) \leq |t| + m_D - 1 \leq m_D |t|$ for all labels a in the alphabet of a non-empty DTD D (a tree always has a root, so $|t|, m_D \geq 1$). Since emptiness of DTDs is linearly decidable we only consider non empty DTDs in the rest of the paper.

Linearization. The relationship from [1] between the usual edit distance on trees t and t' and the edit distance of their respective XML linearizations w and w' depends on the minimal depth of the two trees d:

$$\frac{e(w, w')}{2} \leq e_{stand}(t, t') \leq (2d + 1) \, e(w, w')$$

These estimations are thight up to a constant factor. For any tree t of depth d, if t is ϵ-far from a DTD D modulo the usual tree edit distance, then its XML linearization w is $\frac{\epsilon}{2d+1}$-far from the linearizations of any D-valid tree.

Note however, that the same upper bound does not hold for the strong tree edit distance $e(t, t')$. This indicates already, that we will need a more general method for testing DTD membership modulo the strong tree edit distance, than for testing membership for finite word automata modulo the edit distance.

4   Main Results

Sublinear membership testers are not allowed to read the whole input structure. Instead they only access some elements of the structure randomly and navigate from there on. Which access operations are permitted can be defined by a randomized data model. In this section, we introduce appropriate randomized data models for words and trees, and then formulate our positive results.

Randomized Data Models. As usual, any word w with alphabet Σ defines a unique relational structure S_w with domain $dom(w) = pos(w) \cup \{0\}$, that is $S_w = (dom(w), start^w, succ^w, (lab_a^w)_{a \in \Sigma})$ where $start^w = \{0\}$, $succ^w = \{(i, i+1) \mid 0 \leq i < |w|\}$, and lab_a^w is the set of positions of w labeled by a. The randomized data model is similar except that it gives random access to elements of some structure isomorphic to S_w. More formally, the randomized data model of a word w and a bijection $\theta : dom(w) \to V$ is the tuple $rdm_w^\theta = (ele, start, succ, lab)$, which contains a random generator ele that draws an arbitrary element of V from a uniform distribution, a start position $start \in V$, a successor function $succ : V \to V \cup \{\bot\}$, and a labeling function $lab : V \to \Sigma$, such that θ is an isomorphism between S_w and the relational structure $(V, \{start\}, \{(v, succ(v)) \mid succ(v) \neq \bot\}, (\theta(lab_a^w))_{a \in \Sigma})$.

Any XML document t, as an unranked tree over some alphabet Σ, defines a unique relational structure $S_t = (nod(t), root^t, parent^t, fc^t, ns^t, (lab_a^t)_{a \in \Sigma})$ with domain $nod(t)$. The randomized data model is similar except that it gives random access to elements of some structure isomorphic to S_t. More formally, the randomized data model of a word w and a bijection $\theta : nod(t) \to V$ is the tuple $rdm_t^\theta = (desc, root, parent, fc, ns, lab, depth)$. For any node v of t, $desc(v)$ is a random generator that draws descendants of v from a uniform distribution or returns $\bot$ if v is a leaf, a root element $root \in V$, a parent function $parent : V \backslash \{root\} \to V$, the firstchild function $fc : V \to V \cup \{\bot\}$, the nextsibling function $ns : V \to V \cup \{\bot\}$, and labeling functions $lab : V \to \Sigma$. We require that θ is an isomorphism from S_t to the relational structure $(V, \{root\}, \{(v, parent(v)) \mid v \in V\}, \{(v, fc(v)) \mid fc(v) \neq \bot\}, \{(v, ns(v)) \mid ns(v) \neq \bot\}, (\theta(lab_a^t))_{a \in \Sigma}))$. Finally, $depth = d(t)$ is the depth of t.

Approximate Membership. Approximate membership is a special case of property testing, aiming for probabilistic algorithms that read a sublinear part of the input structure based on a randomized data model. For the formal definition, we fix a class S of structures such that each structure s in S has randomized data models denoted by rdm_s, and a class A of language definitions such that each definition A in $\mathcal{A}$ defines a language $\mathcal{L}(A)$ included in $\mathcal{S}$ and has a size $|A|$ in $\mathbb{N}$.

Definition 3. *An approximate membership tester for $\mathcal{S}$ and $\mathcal{A}$ is an algorithm (possibly randomized) that receives as inputs a randomized data model rdm_s for some structure $s \in \mathcal{S}$, an error precision $\epsilon > 0$ and a language definition $A \in \mathcal{A}$, and answers with probability $\frac{2}{3}$: CLOSE if $s \in \mathcal{L}(A)$ and NO if s is ϵ-far from A.*

The *query complexity* of a tester is the number of times it uses rdm_s during the computation in dependence of the input (size). Its *time complexity* accounts for all other operation performed by the algorithm in addition to the query complexity.

Tree Edit Distance. We next sketch how to test approximate membership for tree automata on unranked trees [6] modulo the usual tree edit distance, based on tree linearization. Note that such tree automata subsume our DTDs.

The idea is to use the upper bound $e_{stand}(t, t') \le (2d + 1)e(w, w')$ from [1], where w is the XML linearization of t and w' the XML linearization of t'. We want to test approximately whether an unranked tree t of depth d is recognized by a tree automaton B. If t is ϵ-far from B modulo the usual tree edit distance, then its linearization is $\frac{\epsilon}{2d+1}$-far from the language of linearizations of trees recognized by B of depth at most d. The tree automaton B can then be compiled into finite automata A of exponential size $|B|^d$ that accepts all these linearizations. This can be done by first compiling B into a nested word automaton [3] in linear time, which in turn is compiled to a finite automaton by moving stacks up to depth d into states. One can then apply the polynomial time membership tester for finite automata modulo the edit distance on words from [12]. In order to do so, one has to verify that the randomized data model of words can be simulated by a randomized data model of the corresponding tree, which is straigthforward. Since the tester in [12] never errs for correct words, and there is no requirement on close trees, this method gives indeed a valid tester. The query complexity of this test is in $O(p(|A|, 1/\epsilon, d))$ where p is the polynomially bounded function that satisfies for all positive real numbers $\mathbf{a}, \mathbf{e}, \mathbf{d}$: $p(\mathbf{a}, \mathbf{e}, \mathbf{d}) = \mathbf{a}^3 \, \mathbf{e} \, \mathbf{d} \, \log^3(\mathbf{a}^2 \, \mathbf{e} \, \mathbf{d})$ The time complexity is in $O(|A| \, p(|A|, 1/\epsilon, d))$. In combination we obtain:

Theorem 4. *Whether an unranked tree t is approximatively recognized by a tree automaton B can be tested with query complexity and time complexity in $O(p(|B|^{d(t)}, 1/\epsilon, d(t)))$ and $O(|B|^{d(t)} \, p(|B|^{d(t)}, 1/\epsilon, d(t)))$ respectively; modulo the tree edit distance with error precision ϵ.*

Even though nontrivial, this theorem has three weaknesses. First of all, the finite automaton A constructed from the tree automaton B and the depth d may be of exponential size $O(|B|^d)$. Second, the tester does not apply to the strong tree edit distance. And third, the query complexity of the tester depends exponentially on the depth of the tree.

Strong Tree Edit Distance. Our main result is that all three problems can be solved for DTD membership modulo the strong tree edit distance, as stated in the following theorem.

Theorem 5. *Whether an unranked tree t is valid for a* DTD $D = (\Sigma, init, (A_a)_{a \in \Sigma})$ *modulo the strong tree edit distance with error precision ϵ can be tested with query complexity in $O(d^2 \, p(\boldsymbol{a}, d/\epsilon, m_D))$ and time complexity in $O(\boldsymbol{a} \, d^2 \, p(\boldsymbol{a}, d/\epsilon, m_D) + |D|)$, where $d = d(t)$ and $\boldsymbol{a} = \max_{A \in \Sigma} |A_a|$ is smaller than $|D|$.*

The dependency on the depth is reduced from exponential to polynomial. In contrast, approximate membership of NFAs can be done with constant query complexity [2,7,12]. Nevertheless, as DTDs are naturally connected to NFAs, one might want to reduce approximate membership of the former to the one of the latter. We believe that this cannot be archieved. Instead, we will present a reduction to a more general property tester for so called weighted words that we will develop for this purpose.

5   Weighted Words

We present an approximate membership tester for finite automata on weighted words modulo a weighted edit distance.

From Trees to Weighted Words. A weighted word over an alphabet Σ is a word over the alphabet $\Sigma \times \mathbb{N}$. The idea for the introduction of weigthed words is as follows. To any node v of a tree t we assign the weight $|t_{|v}|$. The weighted word associated to a node v is then the word $word(t_{|v})$, in which each position is weighted by the weight of the corresponding child of v.

We next illustrate the close link from trees to weighted words by example. We consider the DTD D with rules $r \to ab^*, a \to a^*, b \to b^*$. For any $i \geq 0$, let a_i be the tree $a(a, \ldots, a)$ with i a-leaves and b_i the tree $b(b, \ldots, b)$ with i b-leaves. The tree $t = r(a_1, b_2, b_3, a_4)$ of depth 2 is clearly invalid for D. Its distance is $e(t, D) = 5$ since one must delete the whole last subtree to become valid and this subtree has size 5. However, if we consider the regular language below the root $L = ab^*$ and pick the word at the root $w = word(t_{|\varepsilon}) = abba$, then we have $e(w, L) = 1$ for the edit distance for words. One way to understand the problem is that we cannot simply ignore the sizes of the subtrees as we did. Instead, we should associate a weight to each position, and consider the weighted word $\omega = (a, 2)(b, 3)(b, 4)(a, 5)$ for the above example. We also need to adapt the costs of deleting a weighted letter such as (c, i) to its weight i. In this way, the weighted distance of ω to L becomes 5 which is equal to the distance of t to D.

Any weighted word has the form $w * p$ for some $w \in \Sigma^*$ and $p \in \mathbb{N}^*$, where w and p have the same length. We call w the word part and p the weight part of $w * p$. We will also say that ω has at position i the weight $k \in \mathbb{N}$ and the label $a \in \Sigma$ if $\omega[i] = (a, k)$. The weight $|\omega|_*$ is the sum of the weights at all positions of ω. The word part of a weighted word is used to define its membership to word regular languages while the weight part is used to define weighted words edit distance. We say that a weighted word $w * p$ is recognized by an NFA A if and only if $w \in \mathcal{L}(A)$. The set of weighted words recognized by A is denoted by $\mathcal{L}_*(A)$. The notions of blocking fragment and interval are lifted to weighted words by deletion of the weights. For example, if A is a productive automaton recognizing

$L = ab^*$, then the interval $I =]0,2]$ of $\omega = (a,1)(a,3)(b,4)$ is blocking for A, since aa is the word part of the weighted word ωI located at I, and after reading the first a, A cannot proceed with any second "a".

The edit operations for weighted words are essentially the same as for words, i.e, insertions, relabeling, and deletions. The only difference is that the costs of these operations depend on the weights of the letters that are edited. For a weighted word $\omega = \sigma_1 \ldots \sigma_n$ and a natural number $i \in [0,n]$, the insertion of a weighted letter $\sigma \in \Sigma \times \mathbb{N}$ following position i in ω yields the weighted word: $ins_{i,\sigma}(\omega) = \sigma_1 \ldots \sigma_i \sigma \sigma_{i+1} \ldots \sigma_n$. The cost of this insertion is the weight of σ. The deletion of position $1 \leq i \leq n$ of w yields the following weighted word: $del_i(\omega) = \sigma_1 \ldots \sigma_{i-1}\sigma_{i+1} \ldots \sigma_n$. The cost of such a deletion operation is the weight of the deleted letter σ_i. The relabeling at position $1 \leq i \leq n$ of ω into a letter b changes only the letter at this position but not its weight. Let $\sigma_i = (a,k)$ then the relabeling operation at position i costs k and yields: $rel_{i,b}(\omega) = \sigma_1 \ldots \sigma_{i-1}(b,k)\sigma_{i+1} \ldots \sigma_n$.

Testing Weighted Words. We next show that approximate NFA membership modulo the edit distance can be tested efficiently for weighted words. We will prove the following result for the randomized data model of weighted words defined below.

Theorem 6. *Let A be an NFAs that has k strongly connected components. Whether a weighted word ω is approximately a member of $\mathcal{L}_*(A)$ modulo the weighted edit distance with error precision ϵ can be tested with query complexity $O(\frac{k^2|A|}{\epsilon}\log^3(\frac{k|A|}{\epsilon}))$ and time complexity $O(\frac{k^2|A|^2}{\epsilon}\log^3(\frac{k|A|}{\epsilon}))$ independently of the weight or size of ω.*

So far, the ideas to Theorem 6 are essentially the same as for usual words [12]. What changes for weighted words is that many errors can be concentrated at some position of high weight. This can be accounted by adapting the random drawing of fragments. Instead of using a generator that draws positions uniformly in the word, we use a random generator that draws positions depending on weights. The probability to draw the i-th position of a weighted word $\omega = w * p$ should be $p(i)/|\omega|_*$. We call such a random generator a drawing from a weighted distribution. We can now define the random data model of a weighted word ω and a bijection $\theta : pos(\omega) \to V$ in analogy to the case of words: $rdm_\omega^\theta = (ele, \ start, \ next, \ lab)$. Here, ele is a random generator of positions for the weighted distribution. It might be disturbing that such random generator cannot be obtained from a weighted word without reading it entirely. However, as we will see, we can obtain it from the randomized data model of a tree.

With respect to such random data models for weighted words, Theorem 6 become true. We prove this in two steps. We first consider the case where the NFA is strongly connected, and second study the general case of automaton with several strongly connected components.

Strongly Connected Automata. Let A be an NFA that is strongly connected. An approximate membership testing for weighted words can proceed as follows. The input is a randomized data model rdm_ω^θ for some weighted word ω. The tester then generates randomly sufficiently many positions of the word according to their weights, reads sufficiently long factors starting there, and returns NO if one of them is blocking. What "sufficient" here means can be deduced from the following Lemma.

Lemma 7. *Let* $A = (\Sigma, Q, init, fin, \Delta)$ *be an* NFA *that is strongly connected,* ω *a weighted word,* m *a natural number bigger than* $|\omega|_*$, $\epsilon > 0$ *an error precision, and* $\gamma = \frac{8|Q|}{\epsilon}$. *If* $e(\omega, A) > \epsilon m$ *and* $m \geq 8\gamma \lceil \log(\gamma) \rceil$, *then there is a length* $l \in [2, \gamma]$ *which is a power of 2, and a set of disjoint intervals* $\mathcal{I}_l$ *with weight* $m\beta_l$ *such that: all intervals of* $\mathcal{I}_l$ *are of lentgh* $2l$ *and blocking for* A. *Where* $\beta_l = l/(2\gamma \lceil \log(\gamma) \rceil)$.

General Automata. We generalised the previous result to automata with multiple strongly connected components. We refine the results from [12], which in turn adapts the schema from [2], and prove that drawing positions from the weight distribution of ϵ-far weighted words, yields a blocking fragment with high probability.

For integers l, α, weighted word ω and a sequence $S = (i_1, \cdots, i_\alpha)$ of α positions in ω, we denote by F_S the fragment $\cup_{1 \leq j \leq \alpha} [i_j, \min(i_j + l, |\omega|)]$. And we define $\mathcal{S}(\omega, l, \alpha)$ as the set of all fragments F_S.

Lemma 8. *Let* A *be a productive* NFA *with state set* Q *and* k *strongly connected components. Let* $\epsilon > 0$, $\gamma' = \frac{16k|Q|}{\epsilon}$ *and* ω *be a weighted word of weight greater than* $8\gamma' \lceil \log(\gamma') \rceil$. *If* ω *is* ϵ-far *from* A, *then there exists a power of two* $l \in [2, \gamma']$ *such that: with probability* $\frac{5}{6}$, *drawing* $\alpha_l = 30k\gamma' \lceil \log(\gamma') \rceil^2 /l$ *positions with the weight distribution of* ω *yields some blocking fragment in* $\mathcal{S}(\omega, 2l, \alpha_l)$.

Finally, Theorem 6 is a consequence of Lemma 8. Indeed, the algorithm in Figure 1 is a one sided membership tester for weighted words. In fact by the previous lemma, drawing enough positions according to the weight distribution gives a blocking fragment with probability at least $\frac{5}{6} \geq \frac{2}{3}$. The case of weighted words with small lengths is easily detected using $start, succ, lab$ and the exact membership is checked; this case includes all light weighted words. Furthermore weighted words in $\mathcal{L}(A)$ have no blocking fragments.

6   Testing Unranked Trees

We reduce DTD membership of trees to NFA membership of weighted words. For a tree t, the reduction is based on the weighted words $\omega_v = word(t_{|v}) * p_v$ for nodes v of t, where p_v is the sequence of sizes $|t_{|v'}|$ of subtrees rooted at the children v' of v in document order. Note that we do not need to compute the values $|t_{|v'}|$. However, we can draw a child v' of some node v in a tree t with probability $|t_{|v'}|/|t_{|v}|$, and this is the only thing needed by our weighted word tester (Figure 1), as this drawing corresponds to the drawing of positions for the weighted distribution of ω_v. Therefore, by Theorem 6, for all nodes v of t, we can test membership of ω_v to NFAs efficiently. However, the query complexity measured in terms of accesses to tree t belongs only to $O(d(t) \cdot \frac{k^2 |A|}{\epsilon} \log^3(\frac{k|A|}{\epsilon}))$,

```
fun  member_A(r, ε)
    //  r = rdm_ω^θ  for some  weighted  word  ω
    //  and  ε  an  error  precision
    let  (ele, start, succ, lab) = r
    let  k  be the  number  of  strongly  connected  components  of  A
    let  γ' = (16k|Q|)/ε
    if  |ω| < 8γ'⌈log(γ')⌉  then
        if  ω ∈ L(A)//run  A  via  start, succ, lab
        then return CLOSE else return NO
    else
        for  i = 1  to  ⌈log(γ')⌉  do
            let  l = min(2^i, γ')
            let  α_l = 30kγ'⌈log(γ')⌉²/l
            let  S  be  sequence  of  α_l  positions  of  ω  randomly  drawn  by  ele
            let  F  be  the  union  of  all  intervals  of  ω  of  length  2l  starting
                at  positions  in  S
            if  F  is  blocking  wrt.  A
                //  run  A  via  succ  and  lab
            then return NO; exit else skip
        end
        return CLOSE
```

Fig. 1. An approximate membership tester for weighted words

since we need to draw children of v as explained above in order to draw positions in ω_v with the correct weight.

We now link the strong tree edit distance to DTDs to the edit distance of weighted words to NFAs.

Lemma 9. *Let $D = (\Sigma, init, (A_a)_{a \in \Sigma})$ be a DTD, $\epsilon > 0$ a precision, and t a tree. If all nodes $v \in nod(t)$ satisfy $e(\omega_v, A_{t[v]}) \le \frac{\epsilon}{m_D}|\omega_v|_*$ and $lab(root^t) = init$ then $e(t, D) \le d(t)\epsilon|t|$.*

Lemma 9 shows that trees ϵ-farness is witnessed by nodes with weighted words far from their appropriate regular language. We next explain how to detect such nodes. Indeed the next lemma states that the overall subtree sizes of nodes whose corresponding weighted words are far from their regular word language is important. A node $v \in nod(t)$ is ϵ-bad if $e(\omega_v, A_{t_{|v}}) > \frac{\epsilon}{2m_D \cdot d(t)}|\omega_v|_*$. Let B_t be the set of bad nodes whose ancestors aren't bad. $|B_t| = \sum_{v \in B_t} |t_{|v}|$ is the size of B_t.

Lemma 10. *For DTD D, precision $\epsilon > 0$, and t a tree ϵ-far from D, one has $|B_t| > \frac{\epsilon}{2}|t|$.*

We describe now how this lemma translates to a membership tester. Let μ be the random process that uniformly selects a node of t and returns its path to the root π. The size of π is at most $d(t)$ and for all nodes $v \in nod(t)$, the probability that v is in π is $Pr_{\pi \sim \mu}[v \in \pi] = |t_{|v}|/|t|$. Therefore, by Lemma 10, if t is ϵ-far from D, then the probability that π contains an element of B_t is at least $\frac{\epsilon}{2}$. Hence for $\lceil \log(5)/\epsilon \rceil$ drawn paths, with probability $\frac{4}{5}$ one drawn node is bad. However we do not know which selected node is bad, so we need to verify them all to detect ϵ-farness. We next use the membership tester of weighted words in section 5, which answers correctly with probability at least $\frac{5}{6}$. It follows that with probability at least $\frac{4}{5} \cdot \frac{5}{6} = \frac{2}{3}$ we would find error in t.

7   Conclusion and Future Work

We have presented the first approximate membership tester for DTDs modulo the strong tree edit distance. The most difficult part was to extend previous results for regular words languages to regular tree languages that are restricted to locality in vertical direction (but not horizontally). Some questions remain open. First of all, it might be possible that approximate membership modulo the edit distance can be tested efficiently for XML SCHEMAs by extending the methods presented here. In such a setting one would preserve top-down determinism but give up vertical locality. A second more difficult question is whether approximate membership can be tested efficiently for bottom-up tree automata for ranked trees, while depending only on their depth. The third yet more difficult question is whether efficient algorithms exist for testing RELAXNG validity. Fourth, it might be interesting to study property testing for schemas with key constraints.

References

1. Akutsu, T.: A relation between edit distance for ordered trees and edit distance for Euler strings. Inf. Process. Lett., 105–109 (2006)
2. Alon, N., Krivelevich, M., Newman, I., Szegedy, M.: Regular Languages are Testable with a Constant Number of Queries. SIAM J. Comput., 1842–1862 (2000)
3. Alur, R., Madhusudan, P.: Adding nesting structure to words. Journal of the ACM, 1–43 (2009)
4. Brüggemann-Klein, A.: Regular Expressions to Finite Automata. Theoretical Computer Science, 197–213 (1993)
5. Chockler, H., Kupferman, O.: w-Regular languages are testable with a constant number of queries. Theor. Comput. Sci., 71–92 (2004)
6. Comon, H., Dauchet, M., Gilleron, R., Löding, C., Jacquemard, F., Lugiez, D., Tison, S., Tommasi, M.: Tree Automata Techniques and Applications (2007)
7. Fischer, E., Magniez, F., de Rougemont, M.: Approximate satisfiability and equivalence. In: LICS, pp. 421–430 (2006)
8. Goldreich, O.: Combinatorial property testing (a survey). In: Randomization Methods in Algorithm Design, pp. 45–60 (1998)
9. Green, T.J., Gupta, A., Miklau, G., Onizuka, M., Suciu, D.: Processing XML streams with deterministic automata and stream indexes. ACM Trans. Database Syst., 752–788 (2004)
10. Hagenah, C., Muscholl, A.: Computing epsilon-free nfa from regular expressions in $O(nlog^2(n))$ time. ITA, 257–278 (2000)
11. Martens, W., Neven, F., Schwentick, T., Bex, G.J.: Expressiveness and complexity of XML schema. ACM Transactions of Database Systems, 770–813 (2006)
12. Ndione, A., Lemay, A., Niehren, J.: Approximate membership for regular languages modulo the edit distance. Theor. Comput. Sci., 37–49 (2013)

13. Newman, I., Sohler, C.: Every property of hyperfinite graphs is testable. In: STOC, pp. 675–684 (2011)
14. Pawlik, M., Augsten, N.: RTED: A Robust Algorithm for the Tree Edit Distance. PVLDB, 334–345 (2011)
15. Ron, D.: Property Testing: A Learning Theory Perspective. Foundations and Trends in Machine Learning, 307–402 (2008)
16. Schnitger, G.: Regular expressions and NFAs without ε-transitions. In: Durand, B., Thomas, W. (eds.) STACS 2006. LNCS, vol. 3884, pp. 432–443. Springer, Heidelberg (2006)
17. Selkow, S.M.: The Tree-to-Tree Editing Problem. Inf. Process. Lett., 184–186 (1977)
18. Zhang, K., Shasha, D.: Simple Fast Algorithms for the Editing Distance Between Trees and Related Problems. SIAM J. Comput., 1245–1262 (1989)

Author Index